Pathology and Epidemiology of Cancer

Massimo Loda · Lorelei A. Mucci
Megan L. Mittelstadt
Mieke Van Hemelrijck
Maura Bríd Cotter
Editors

# Pathology and Epidemiology of Cancer

Springer

*Editors*
Massimo Loda
Department of Pathology
Dana-Farber Cancer Institute
and Brigham & Women's Hospital,
Harvard Medical School
Boston, MA
USA

Lorelei A. Mucci
Department of Epidemiology
Harvard T.H. Chan School of Public Health
Boston, MA
USA

Megan L. Mittelstadt
Department of Genetics
Harvard Medical School
Boston, MA
USA

Mieke Van Hemelrijck
Guy's Hospital
King's College London
London
UK

Maura Bríd Cotter
Department of Pathology
Dana-Farber Cancer Institute
and Brigham & Women's Hospital,
Harvard Medical School
Boston, MA
USA

ISBN 978-3-319-35151-3 ISBN 978-3-319-35153-7 (eBook)
DOI 10.1007/978-3-319-35153-7

Library of Congress Control Number: 2016940001



Printed on acid-free paper

This Springer imprint is published by Springer Nature
The registered company is Springer International Publishing AG Switzerland

# Preface

The conception of this book arose from an ongoing course created in 2010 for Cancer Biology Ph.D. students at Harvard Medical School, entitled "The Epidemiology and Molecular Pathology of Cancer." A series of general lectures begins the course, upon which the introductory chapters of this text are based. Each subsequent day of the course is dedicated to a single cancer type, with lecturers providing an overview of the cancer from the points-of-view of both an epidemiologist and a pathologist. The course, and thus this text, is unique in that it integrates the disciplines of cancer pathology and epidemiology, ranging from population-based understandings to detailed molecular mechanisms to provide a synergistic and comprehensive view of cancer pathogenesis.

Pathology is at the cornerstone of cancer pathogenesis, diagnosis, prognosis, and treatment. While other books may focus on either the morphological aspects or the mechanisms of cancer etiology and pathogenesis, this text will provide relevant information on the diagnostic, prognostic, and predictive molecular pathology of cancer. Epidemiological studies, both descriptive and analytical, provide insights into the burden of cancer, its causes and opportunities for prevention, and contribute to the understanding of the molecular mechanisms of disease from a population-based prospective. The two disciplines pathology and epidemiology are symbiotic in their objectives and tackle the understanding of disease from complementary paths. By integrating these two disciplines with basic and medical science, as well as with population-based studies, we provide a unique and comprehensive overview that helps the reader to understand neoplastic disease processes.

This book will concentrate on several of the major cancers that are prevalent and for which substantial molecular, pathological, and epidemiological data are currently available. Part I of this book introduces readers to some basic concepts in pathology, epidemiology, screening, genetics, and biostatistical approaches. Following this, Part II consists of paired chapters presenting basic biology, current epidemiological data, and common practices and challenges related to molecular pathology of a given cancer type—one chapter written from the point of view of a pathologist and one from the point of view of an epidemiologist. That said, we have collectively taken great care to ensure that the chapters provide complementary and nonoverlapping information for the cancer types discussed within.

# Contents

# Contributors

**Ayal Aizer** Harvard Medical School, Dana-Farber/Brigham and Women's Cancer Center, Boston, MA, USA

**Brian Alexander** Harvard Medical School, Dana-Farber/Brigham and Women's Cancer Center, Boston, MA, USA

**Christopher B. Allard** Department of Urology, Massachusetts General Hospital, Boston, MA, USA

**Justine A. Barletta** Brigham and Women's Hospital, Harvard Medical School, Boston, MA 02115, USA

**Julie L. Batista** Channing Division of Network Medicine, Department of Medicine, Brigham and Women's Hospital and Harvard Medical School, Boston, MA 02115, USA

**Brenda M. Birmann** Channing Division of Network Medicine, Department of Medicine, Brigham and Women's Hospital and Harvard Medical School, Boston, MA 02115, USA

**Luis G. Carvajal-Carmona** Genome Center and School of Medicine, University of California at Davis, Davis, CA, USA

**Sue Chang** David Geffen School of Medicine, UCLA, CHS Los Angeles, CA, USA

**Lucian R. Chirieac** Harvard Medical School, Brigham and Women's Hospital, Boston, MA, USA

**Maura Bríd Cotter** Department of Pathology, Dana-Farber Cancer Institute and Brigham & Women's Hospital, Harvard Medical School, Boston, MA, USA

**Danielle Crawley** Cancer Epidemiology Group, Research Oncology, Bermondsey Wing Guy's Hospital, King's College London, London SE1 9RT, UK

**Lyn McDivitt Duncan** WRN825 Dermatopathology Unit, Massachusetts General Hospital, 55 Fruit Street, Boston, MA 02114, USA

**Mara Meyer Epstein** Department of Medicine, University of Massachusetts Medical School, Worcester, MA 01605, USA

**Michelangelo Fiorentino** S.Orsola-Malpighi Teaching Hospital, Addarii Institute of Oncology, 40138 Bologna, Italy

**Wai Chin Foo** MD Anderson Cancer Center, University of Texas, Houston, TX 77030, USA

**Felipe C. Geyer** Hospital Israelita Albert Einstein, Sao Paolo, Brazil; Memorial Sloan Kettering Cancer Center, New York, NY, USA

**Jessica J. Gierut** Beth Israel Deaconess Medical Center, Cambridge, MA, USA

**Edward L. Giovannucci** Harvard T.H. Chan School of Public Health, Boston, MA, USA

**Francesca Giunchi** S.Orsola-Malpighi Teaching Hospital, Addarii Institute of Oncology, 40138 Bologna, Italy

**Amrita Goyal** University of Minnesota, Minneapolis, MN 55455, USA

**Sameer Gupta** Harvard Medical School, Boston, MA 02114, USA

**Kevin M. Haigis** Harvard Medical School, Cancer Research Institute, Beth Israel Deaconess Medical Center, Boston, MA, USA

**Mieke Van Hemelrijck** Division of Cancer Studies, Cancer Epidemiology Group, Guy's Hospital, King's College London, London, UK

**Lars Holmberg** Division of Cancer Studies, Cancer Epidemiology Group, Guy's Hospital, King's College London, London, UK

**Eric C. Huang** School of Medicine, University of California at Davis, Sacramento, CA, USA

**NaNa Keum** Harvard T.H. Chan School of Public Health, Boston, MA, USA

**Elizabeth L. Kehr** Brigham and Women's Hospital, Harvard Medical School, Boston, MA 02115, USA

**David W. Kindelberger** School of Medicine, Boston University, Boston, MA, USA

**Lester Kobzik** Harvard Medical School, Brigham and Women's Hospital, Boston, MA, USA

**Keith L. Ligon** Harvard Medical School, Boston, MA, USA

**Massimo Loda** Department of Pathology, Dana-Farber Cancer Institute and Brigham & Women's Hospital, Harvard Medical School, Boston, MA, USA

**Caterina Marchiò** University of Turin, Turin, Italy

**Dominique S. Michaud** Tufts University School of Medicine, Boston, MA 02111, USA

**Stacey A. Missmer** Brigham and Women's Hospital and Harvard Medical School, Harvard T.H. Chan School of Public Health, Boston, MA, USA

**Megan L. Mittelstadt** Harvard Medical School, Boston, MA, USA

**Lorelei A. Mucci** Harvard T.H. Chan School of Public Health, Boston, MA, USA

**Giovanni Parmigiani** Department of Biostatistics and Computational Biology, Dana-Farber Cancer Institute, Boston, MA, USA; Department of Biostatistics, Harvard T.H. Chan School of Public Health, Boston, MA, USA

**Kathryn L. Penney** Harvard Medical School, Brigham and Women's Hospital, Boston, MA, USA

**Emily Jean Poulin** Beth Israel Deaconess Medical Center, Boston, MA, USA

**Dinesh Rao** David Geffen School of Medicine, UCLA Medical Center, Los Angeles, CA 90095, USA

**Jorge S. Reis-Filho** Memorial Sloan Kettering Cancer Center, New York, NY, USA

**Dimitra Repana** Guy's and St. Thomas' NHS Foundation Trust, London, UK

**Jennifer R. Rider** Harvard T.H. Chan School of Public Health, Boston, MA, USA; Department of Epidemiology, Boston University School of Public Health, 715 Albany Street Talbot T317E, Boston, MA 02118, USA

**Sarah M. Rudman** Guy's and St. Thomas' NHS Foundation Trust, Bermondsey Wing Guy's Hospital, London SE1 9RT, UK

**Jonathan Said** David Geffen School of Medicine, UCLA Medical Center, Le Conte Ave., 13-222 CHS, Los Angeles, CA 90095, USA

**Alejandro Sanchez** Department of Urology, Massachusetts General Hospital, Boston, MA, USA

**Jeanne Shen** Southwestern Medical Center, University of Texas, Dallas, TX, USA

**James Spicer** Guy's Hospital, King's College London, London, UK

**Matthew D. Stachler** Harvard Medical School, Brigham and Women's Hospital, Boston, MA, USA

**Charles D. Stiles** Department of Neurobiology, Harvard Medical School, Boston, MA, USA; Department of Cancer Biology, Dana-Farber Cancer Institute, Boston, MA, USA

**Rulla M. Tamimi** Harvard Medical School, Channing Division of Network Medicine, Brigham and Women's Hospital, Boston, MA, USA

**Nairi Tchrakian** Department of Pathology, St. James's Hospital, Dublin 8, Ireland

**Kathryn L. Terry** Brigham and Women's Hospital and Harvard Medical School, Harvard T.H. Chan School of Public Health, Boston, MA, USA

**Hensin Tsao** Harvard Medical School, Boston, MA 02114, USA

**Svitlana Tyekucheva** Department of Biostatistics and Computational Biology, Dana-Farber Cancer Institute, Boston, MA, USA; Department of Biostatistics, Harvard T.H. Chan School of Public Health, Boston, MA, USA

**Francesco Vasuri** S.Orsola-Malpighi Teaching Hospital, Addarii Institute of Oncology, 40138 Bologna, Italy

**Huamin Wang** MD Anderson Cancer Center, University of Texas, Houston, TX 77030, USA

**Kim Wilkinson** Dana-Farber Cancer Institute, Boston, MA, USA

**Kathryn M. Wilson** Harvard T.H. Chan School of Public Health, Boston, MA, USA

# Part I

# Basic Principles of Patho-epidemiology

# The Intersection of Epidemiology and Pathology

# 1

Megan L. Mittelstadt, Edward L. Giovannucci, Lorelei A. Mucci and Massimo Loda

## 1.1 The Intersection of Pathology and Epidemiology

As is the trend in many disciplines, research groups are approaching cancer using interdisciplinary approaches to gain a more complete understanding of the disease as well as to identify predictive biomarkers at molecular, individual, and population levels [1]. Pathology and epidemiology are central to these ambitions, and work together toward the common goal of elucidating disease etiology and progression. Modern cancer biology research is increasingly focused on identification of genetic heterogeneity via molecular pathology, but it remains important to contextualize individual molecular profiles within the individual's lifetime environmental exposures ("exposome"), resulting gene: environment interactions, and disease trajectory [2]. Furthermore, population-level trends gathered via epidemiological models can be integrated with molecular pathologic data to elucidate etiology simultaneously at molecular, individual, and population levels. Identification of trends and associations driving cancers on a population level provide opportunity for predictive markers, screening opportunities, and therapeutic advances with a greater impact. It is our hope that in making pathology accessible to epidemiologists, and epidemiology accessible to pathologists, trends that are important in populations of cells and of people might be identified and acted upon with greater frequency.

Beyond this, the two fields work more symbiotically. It is increasingly clear that cancers, such as prostate, colon, or breast cancer, are not single diseases but rather are comprised of many subtypes defined by molecular pathology and histology. Patho-epidemiology incorporates pathological and tumor biomarker data for individuals diagnosed with cancer or other conditions that are participants in well-defined epidemiological studies. A more detailed classification of tumors can be achieved by adding molecular annotation based on biomarker assessment in pathology specimens from patients in existing epidemiologic cohorts to existing clinical data available in these databases. On the other side, pathology studies are enriched by the principles of epidemiological methods to define study

M.L. Mittelstadt (✉)
Harvard Medical School, 77 Avenue Louis Pasteur, NRB 241A, Boston, MA 02115, USA
e-mail: megan_mittelstadt@hms.harvard.edu

E.L. Giovannucci
Harvard T.H. Chan School of Public Health, 655 Huntington Avenue, Boston, MA 02115, USA

L.A. Mucci
Harvard T.H. Chan School of Public Health, 677 Huntington Avenue, Kresge 920a, Boston, MA 02115, USA

M. Loda
Department of Pathology, Dana Farber Cancer Institute and Brigham & Women's Hospital, Harvard Medical School, 450 Brookline Avenue, Boston, MA 02115, USA
e-mail: massimo_loda@dfci.harvard.edu


M. Loda et al. (eds.), *Pathology and Epidemiology of Cancer*,
DOI 10.1007/978-3-319-35153-7_1

populations and design. Patho-epidemiology is uniquely derived from the interaction between investigators in these two disciplines.

## 1.2 Examples from the Intersection of Epidemiology and Pathology

One of the earliest cancer studies integrating the disciplines of epidemiology and pathology was undertaken to examine histological changes in lung tissue associated with exposure to passive smoking. This work was initiated following the publication of two seminal studies in 1983 showing nonsmoking women exposed to environmental tobacco from their husbands had an increased risk of lung cancer [3]. To provide follow-up evidence for this epidemiological finding, a team of epidemiologists and pathologists retrieved lung tissue at autopsy from a cohort of 283 adults not known to have died of cancer [4]. The team interviewed next of kin to collect data on the smoking habits of both the deceased individual and his/her family. A single pathologist reviewed the histological specimens without knowledge of the smoking status and characterized a variety of preneoplastic epithelial lesions. The study found nonsmoking women married to smoking husbands had a much higher prevalence of these precursor lesions in lung tissue than women married to nonsmoking husbands, and at a level similar to that of smoking women. The results from this study were pivotal in establishing the causal link between passive smoking and lung cancer risk.

### 1.2.1 Breast Cancer

Some of the clearest examples of the importance of integrating epidemiologists and pathologists working together to identify unique risk factors based on molecular subtypes are in breast cancer. Gene expression profiling studies identified unique molecular subtypes of breast cancer [5, 6], and these subtypes are not only prognostic and predictive, but also appear to be etiologically unique. Eliassen et al. [7] undertook a study to examine prediagnostic circulating levels of α-carotene, β-carotene, lycopene, and total carotenoids as related to breast cancer risk. Higher levels of each carotenoid were associated with a significant 18–28 % lower breast cancer risk during 20-year follow-up. Moreover, high plasma carotenoids were specifically associated with reduced risk for ER- cancers as well as cancers that were ultimately fatal.

### 1.2.2 Prostate Cancer

Prostate specific antigen (PSA), widely hailed as a victory in cancer screening, became a common screening test in the 1990s and eventually led to overdiagnosis and overtreatment of prostate cancer, causing some to advocate that routine screening be abandoned [8–10]. In the PSA screening era, detected prostate cancer is mainly indolent disease and there is a need for diagnostic measures that will narrow the focus on cancers with lethal potential, requiring that better surrogates be identified and validated [11]. Approaching this issue from the field of epidemiology, it appears that the risk factor patterns for potentially lethal prostate cancer differ greatly from that of indolent disease, suggesting different etiologies and distinct subtypes [12]. It was only through the integration of large epidemiologic databases and careful molecular annotation in prostate cancer specimens from these cohorts that novel molecular biomarkers such as gene expression profile signatures have been identified. For example, researchers identified a molecular signature of Gleason grade that strongly predicts lethal disease [13]. Gene expression profile signatures can also provide novel predictive tools to clinicians to decide on important therapeutic options.

The somatic gene fusion *TMPRSS2:ERG,* involving the androgen-regulated *TMPRSS2* and *ERG* (a member of the ETS family of oncogenes), has been proposed as a common prostate cancer molecular subtype that can readily be identified with molecular pathology techniques. A polymorphic CAG repeat sequence in the androgen receptor gene influences activity of the gene product. Men with shorter CAG repeats have a significantly higher risk of *TMPRSS2:ERG* prostate cancer, while there is no association of repeat length with TMPRSS2:ERG negative cancer [14]. There also exists a link between obesity and *TMPRSS2:ERG*, which stands out given the association of obesity and insulin signaling with prostate cancer mortality. *TMPRSS2:ERG* tumors have increased insulin and IGF1 expression. Likewise, obesity was strongly associated with mortality among patients whose tumors harbored *TMPRSS2:ERG* while no association was shown for men whose tumors lacked the gene fusion [15].

### 1.2.3 Glioblastoma

Great progress has been made in the field of glioblastoma (GBM), a cancer associated with very poor survival and high resistance to radiotherapy and chemotherapy. The identification of GBM patients also harboring *IDH1* mutations, and the subsequent pathologic separation of these tumors from type I glioblastomas, has translated to effective stratification of prognostic groups and potential therapeutic opportunities. However, the identification of many risk factors associated with this disease (exposure to high doses of ionizing radiation, inherited mutations of highly penetrant genes associated with rare syndromes, etc.) have not yet translated to the identification of valuable therapeutic opportunities. Collaborative studies should continue to examine the interaction between exposures to therapeutic doses of ionizing radiation and identified signaling pathways disrupted in GBM (increased activation of receptor tyrosine kinase/RAS/PI3K signaling, loss of function in p53 signaling, and reduced signaling of the RB pathway) to identify links, potentially surrounding DNA repair genes [16].

### 1.2.4 Hematologic Malignancies

In the field of hematologic malignancies, epidemiologists and pathologists work together under the purview of the International Lymphoma Epidemiology Consortium (InterLymph) Pathology Working Group to facilitate uniformity in the investigation of lymphoma subtypes in epidemiologic research [17]. This initiative has completed significant work on non-genetic risk factors to determine end points for epidemiologic studies based on pathologic classification. InterLymph has also sought to delineate major risk factors associated with specific hematologic malignancies as well as across all hematologic malignancies. As an example, certain autoimmune diseases (e.g. systemic lupus erythematosus) are significantly associated with many hematologic malignancies [18]. However, there is a lack of data on risk factors for diffuse large B-cell lymphoma molecular subtypes, an area that would benefit from further collaborative work from this or other working groups.

The examples included thus far only scratch the surface of their respective fields. In order to provide a deep view of how the interactions between epidemiology and pathology can drive progress in a field, we will use colorectal cancer as a representative example.

### 1.2.5 Colorectal Cancer

Fearon and Vogelstein first described the classic model of colorectal carcinogenesis in 1990. Their model outlined the sequence of steps involved in the progression from normal epithelium, to benign precursors and into invasive adenocarcinoma [19]. Subsequently, additional genetic and epigenetic changes have been identified for colorectal carcinogenesis. Only some of these alterations are considered to be drivers of the development and progression of these cancers [20, 21]. Currently, colorectal cancer (CRC) has been characterized by three key molecular subtypes: chromosomal instability (CIN), microsatellite instability (MSI), and the CpG island methylator (CIMP) pathway [22–24]. CIN is the most common subtype, and is observed in approximately 80 % of the sporadic cases of CRC. The second major subtype, which accounts for approximately 10–15 % of CRC, is the CIMP subtype. Patho-epidemiology studies have helped to elucidate the effect of various exposures on CRC risk and survival.

An association between obesity and CRC risk overall is well established. In recent years, some studies have examined the association between measures of obesity and molecular subtypes or specific molecular alterations in CRC. Interestingly, based on studies conducted to date, obesity does not appear to act preferentially on many of the common molecular pathways identified for CRC. For example, obesity is not associated with risk of developing CRCs with high MSI that contains MLH1 methylation [25–28], BRAF mutation [27], CTNNB1 overexpression [29], or loss of expression in TP53 [30], in CDKN1B (or p27) [31], and in CDKN1A (or p21) [32]. Overall, it appears that obesity is not differentially related to tumors with MSI, but rather, may be even more strongly associated with those with a MS-stable phenotype. The timing of excess or deficiency of energy over the life course may be important. For example, in the Netherlands Cohort Study on Diet and Cancer, individuals who experienced severe war-related energy restriction during adolescence and early adulthood had a 35 % lower risk for CIMP-positive CRC [33]. It is unclear if severe energy restriction during the growth period would operate on similar mechanisms as excessive energy intake in adulthood.

Beyond the major subtypes of CRC, insights might also be provided by molecular markers specifically related to energy balance. Fatty acid synthase (FASN) is a potential candidate. FASN has been suggested to act as a "metabolic oncogene," conferring a selective growth advantage to cells upon nutritional deprivation [34]. Experimentally, FASN activity appears to promote tumor cell proliferation and survival and allow neoplastic cells to attain autonomy from the regulation of host metabolic status. In one analysis based on the Nurses' Health Study, women who were overweight or obese were at higher risk of FASN-negative CRC, but not at increased risk of FASN-positive CRC relative to normal weight women [35]. The authors hypothesized that FASN-inactive cells are dependent on excess energy balance for malignant progression, whereas FASN-active cells may progress to cancer independent of energy balance status.

Tobacco has had a moderate association with risk of CRC, though the mechanism has not been established. Interestingly, although effects of tobacco-related carcinogens have been emphasized in discussing potential mechanisms, in experimental studies, tobacco triggers multistep epigenetic alterations at several sites within an exposed tissue field; these altered cells precede multifocal lesions and some foci may progress to cancer overtime [36]. In parallel, epidemiologic studies fairly consistently show smokers have an increased risk of CRC showing epigenetic alterations in genes associated with CIMP- and MSI-related CRC. Although the increased risk of total CRC associated with smoking has been modest, for example about 1.2–1.3-fold, current smoking has been associated with an approximately two-fold increased risk of developing CRCs with features of high MSI or CIMP [37–39].

Aspirin and nonaspirin NSAIDs have been shown to reduce risk of CRC. The mechanism of action of these agents has been typically attributable to their anti-inflammatory effects. In particular, these compounds inhibit the actions of PTGS2 or COX2. PTGS2 converts arachidonic acid to prostaglandins, which exert numerous pro-inflammatory effects. Supportive of a role in cancer, PTGS2 is overexpressed in the majority of CRC [40]. One study showed that regular aspirin use was associated with an approximately 40 % lower risk for CRCs that overexpressed PTGS2, but was not associated with CRCs that did not overexpress PTGS2 [41]. In addition, post-diagnostic use of aspirin also appeared to improve overall survival among CRC patients with PTGS2-overexpression but not those without PTGS2 overexpression [42]. These findings, if confirmed, suggest the interesting possibility that post-diagnostic aspirin may be more beneficial among those who did not use aspirin before diagnosis because the CRCs in these patients will tend to be enriched in those with PTGS2 overexpression.

The incorporation of molecular pathologic concepts into the epidemiology of CRC has been increasing in the past decade. Taking into account the heterogeneous nature of CRC should enhance our understanding of factors involving etiology and prognosis.

### 1.2.6 Genome-Wide Association Studies

One key example of the synergy of epidemiology and pathology comes from recent genome-wide association studies (GWAS) of cancer that have identified hundreds of genetic risk variants, but the functional effects of these variants remain largely unexplained. Epidemiologic analysis together with molecular pathology annotation allows an in-depth understanding of the biological function of candidate genes or regions identified in GWAS studies. The Genotype-Tissue Expression (GTEx) project, a human tissue bank that includes tissues from the Cancer Human Biobank program at the National Institutes of Health (NIH), will aid in the identification of genetic variation and gene expression causally related to cancer initiation and progression in a variety of tissues [43]. This project, at the intersection of molecular pathology and epidemiology, includes tissue sampling upon death to distinguish whether SNPs previously associated with particular cancer contexts through GWAS studies, are influencing gene expression in a tissue-specific or tissue-wide manner.

## 1.3 Summary

The fields of epidemiology and pathology together investigate various lifetime exposures, molecular alterations associated with disease, and progression of disease overtime. With epidemiology providing insights into the burden and molecular mechanisms of cancer and pathology providing the complimentary molecular and histological diagnoses of cancer, patho-epidemiological investigators are better able to stratify and characterize cancer subtypes to promote diagnosis and prevention. Analysis of cancer incidence and mortality stratified by tumor subtypes may provide etiologic clues, provide evidence basis for precision prevention of cancer, and improve outcomes for patients. Future goals at the intersection of these fields include using pathologic and epidemiologic data to determine the type of tumor likely to develop, the molecular mechanisms that mediate its development, and the threat that tumor poses to an individual's health. New knowledge will provide the information necessary to better establish causal associations, determine dose-response and timing issues, and inform on recommendations regarding optimal disease prevention, early detection, and treatment. In summary, cancer etiology, pathogenesis, maintenance, and progression are all influenced by context. To this end, epidemiology provides context at a population level while pathology provides context at the cellular, tissue, organ, and systemic levels. The integration of these two disciplines is therefore essential to cancer research and to the ultimate defeat of this disease.

## References


1. Ogino S, King EE, Beck AH, Sherman ME, Milner DA, Giovannucci E. Interdisciplinary education to integrate pathology and epidemiology: towards molecular and population-level health science. Am J Epidemiol. 2012;176(8):659–67.
2. Wild C. Complementing the genome with an "exposome": The outstanding challenge of environmental exposure measurement in molecular epidemiology. Cancer Epidemiol Biomarkers Prev. 2005;14(8):1847–50.
3. Trichopoulos D, Kalandidi A, Sparros L. Lung cancer and passive smoking: conclusion of Greek study. Lancet. 1983;2(8351):677–8.
4. Trichopoulos D, Mollo F, Tomatis L, Agapitos E, Delsedime L, Zavitsanos X, et al. Active and passive smoking and pathological indicators of lung cancer risk in an autopsy study. JAMA. 1992;268(13):1697–701.

5. Waldron L, Haibe-Kains B, Culhane AC, Riester M, Ding J, Wang XV, et al. Comparative meta-analysis of prognostic gene signatures for late-stage ovarian cancer. J Natl Cancer Inst. 2014;106(5):1–12.
6. Arason A, Gunnarsson H, Johannesdottir G, Jonasson K, Bendahl PO, Gillanders EM, et al. Genome-wide search for breast cancer linkage in large Icelandic non-BRCA1/2 families. Breast Cancer Res. 2010;12(4):R50.
7. Eliassen AH, Liao X, Rosner B, Tamimi RM, Tworoger SS, Hankinson SE. Plasma carotenoids and risk of breast cancer over 20 y of follow-up. Am J Clin Nut. 2015;101(6):1197–205.
8. Loeb S, Bjurlin MA, Nicholson J, Tammela TL, Penson DF, Carter HB, et al. Overdiagnosis and overtreatment of prostate cancer. Eur Urol. 2014;65: 1046–55.
9. Carter HB. American urological association (AUA) guideline on prostate cancer detection: process and rationale. BJU Int. 2013;112:543–7.
10. Qaseem A, Barry MJ, Denberg TD, Owens DK, Shekelle P, Clinical Guidelines Committee of the American College of Physicians. Screening for prostate cancer: a guidance statement from the Clinical Guidelines Committee of the American College of Physicians. Ann Intern Med. 2013;158: 761–9.
11. Jahn JL, Giovannucci EL, Stampfer MJ. The high prevalence of undiagnosed prostate cancer at autopsy: implications for epidemiology and treatment of prostate cancer in the Prostate-specific Antigen-era. Int J Cancer. 2015;137(12):2795–802.
12. Giovannucci E, Liu Y, Platx EA, Stampfer MJ, Willett WC. Risk Factors for prostate cancer incidence and progression in the health professionals follow-up study. Int J Cancer. 2007;121:1571–8.
13. Penney KL, Sinnott JA, Fall K, Pawitan Y, Hoshida Y, Kraft P, et al. mRNA expression signature of gleason grade predicts lethal prostate cancer. J Clin Oncol. 2011;29(17):2391–6.
14. Yoo S, Pettersson A, Jordahl KM, Lis RT, Lindstrom S, Meisner A, et al. Androgen receptor CAG repeat polymorphism and risk of TMPRSS2: ERG-positive prostate cancer. Cancer Epidemiol Biomarkers Prev. 2014;23(10):2027–31.
15. Pettersson A, Lis RT, Meisner A, Flavin R, Stack EC, Fiorentino M, et al. Modification of the association between obesity and lethal prostate cancer by TMPRSS2:ERG. J Natl Cancer Inst. 2013;105 (24):1881–90.
16. Schwartzbaum JA, Fisher JL, Aldape KD, Wrensch M. Epidemiology and molecular pathology of glioma. Nature Clin Prac Neurology. 2006;2(9): 494–503.
17. Morton LM, Turner JJ, Cerhan JR, Linet MS, Treseler PA, Clarke CA, et al. Proposed classification of lymphoid neoplasms for epidemiologic research from the pathology working group of the International Lymphoma Epidemiology Consortium (InterLymph). Blood. 2007;110(2):695–708.
18. Morton LM, Slager SL, Cerhan JR, Wang SS, Vaidic CM, Skibola CF, et al. Etiologic heterogeneity among non-Hodgkin lymphoma subtypes: the InterLymph non-hodgkin lymphoma subtypes project. J Natl Cancer Inst Monogr. 2014;48:130–44.
19. Fearon ER, Vogelstein B. A genetic model for colorectal tumorigenesis. Cell. 1990;61:759–67.
20. Wood LD, Parsons DW, Jones S, Lin J, Sjoblom T, Leary RJ, Shen D, Boca SM, Barber T, Ptak J, Silliman N, Szabo S, et al. The genomic landscapes of human breast and colorectal cancers. Science. 2007;318:1108–13.
21. Comprehensive molecular characterization of human colon and rectal cancer. Nature. 2012;487:330–7.
22. Samowitz WS. Genetic and epigenetic changes in colon cancer. Exp Mol Pathol. 2008;85:64–7.
23. Markowitz SD, Bertagnolli MM. Molecular origins of cancer: molecular basis of colorectal cancer. N Engl J Med. 2009;361:2449–60.
24. Kang GH. Four molecular subtypes of colorectal cancer and their precursor lesions. Arch Pathol Lab Med. 2011;135:698–703.
25. Slattery ML, Curtin K, Anderson K, Ma KN, Ballard L, Edwards S, Schaffer D, Potter J, Leppert M, Samowitz WS. Associations between cigarette smoking, lifestyle factors, and microsatellite instability in colon tumors. J. 2000;92:1831–6.
26. Campbell PT, Jacobs ET, Ulrich CM, Figueiredo JC, Poynter JN, McLaughlin JR, Haile RW, Jacobs EJ, Newcomb PA, Potter JD, Le Marchand L, Green RC, et al. Case-control study of overweight, obesity, and colorectal cancer risk, overall and by tumor microsatellite instability status. J Natl Cancer Inst. 2010;102:391–400.
27. Hughes LA, Williamson EJ, van Engeland M, Jenkins MA, Giles GG, Hopper JL, Southey MC, Young JP, Buchanan DD, Walsh MD, van den Brandt PA, Alexandra Goldbohm R, et al. Body size and risk for colorectal cancers showing BRAF mutations or microsatellite instability: a pooled analysis. Int J Epidemiol. 2012;41:1060–72.
28. Hoffmeister M, Blaker H, Kloor M, Roth W, Toth C, Herpel E, Frank B, Schirmacher P, Chang-Claude J, Brenner H. Body mass index and microsatellite instability in colorectal cancer: A population-based study. Cancer Epidemiol Biomarkers Prev. 2013;22:2303–11.
29. Morikawa T, Kuchiba A, Lochhead P, Nishihara R, Yamauchi M, Imamura Y, Liao X, Qian ZR, Ng K, Chan AT, Meyerhardt JA, Giovannucci E, et al. Prospective analysis of body mass index, physical activity, and colorectal cancer risk associated with beta-catenin (CTNNB1) status. Cancer Res. 2013;73:1600–10.
30. Morikawa T, Kuchiba A, Liao X, Imamura Y, Yamauchi M, Qian ZR, Nishihara R, Sato K, Meyerhardt JA, Fuchs CS, Ogino S. Tumor TP53 expression status, body mass index and prognosis in colorectal cancer. International Journal of Cancer (Journal international du cancer). 2012;131:1169–78.

31. Ogino S, Shima K, Nosho K, Irahara N, Baba Y, Wolpin BM, Giovannucci EL, Meyerhardt JA, Fuchs CS. A cohort study of p27 localization in colon cancer, body mass index, and patient survival. Cancer Epidemiol Biomarkers Prev. 2009;18:1849–58.
32. Ogino S, Nosho K, Shima K, Baba Y, Irahara N, Kirkner GJ, Hazra A, De Vivo I, Giovannucci EL, Meyerhardt JA, Fuchs CS. p21 expression in colon cancer and modifying effects of patient age and body mass index on prognosis. Cancer Epidemiol Biomarkers Prev. 2009;18:2513–21.
33. Hughes LA, van den Brandt PA, de Bruine AP, Wouters KA, Hulsmans S, Spiertz A, Goldbohm RA, de Goeij AF, Herman JG, Weijenberg MP, van Engeland M. Early life exposure to famine and colorectal cancer risk: a role for epigenetic mechanisms. PLoS ONE. 2009;4:e7951.
34. Menendez JA, Lupu R. Fatty acid synthase and the lipogenic phenotype in cancer pathogenesis. Nat Rev Cancer. 2007;7:763–77.
35. Kuchiba A, Morikawa T, Yamauchi M, Imamura Y, Liao X, Chan AT, Meyerhardt JA, Giovannucci E, Fuchs CS, Ogino S. Body mass index and risk of colorectal cancer according to fatty acid synthase expression in the nurses' health study. J Natl Cancer Inst. 2012;104:415–20.
36. DeMarini DM. Genotoxicity of tobacco smoke and tobacco smoke condensate: a review. Mutat Res. 2004;567:447–74.
37. Limsui D, Vierkant RA, Tillmans LS, Wang AH, Weisenberger DJ, Laird PW, Lynch CF, Anderson KE, French AJ, Haile RW, Harnack LJ, Potter JD, et al. Cigarette smoking and colorectal cancer risk by molecularly defined subtypes. J Natl Cancer Inst. 2010;102(14):1012–22.
38. Curtin K, Samowitz WS, Wolff RK, Herrick J, Caan BJ, Slattery ML. Somatic alterations, metabolizing genes and smoking in rectal cancer. Int J Cancer. 2009;125:158–64.
39. Poynter JN, Haile RW, Siegmund KD, Campbell PT, Figueiredo JC, Limburg P, Young J, Le Marchand L, Potter JD, Cotterchio M, Casey G, Hopper JL, et al. Associations between smoking, alcohol consumption, and colorectal cancer, overall and by tumor microsatellite instability status. Cancer Epidemiol Biomarkers Prev. 2009;18:2745–50.
40. Soumaoro LT, Uetake H, Higuchi T, Takagi Y, Enomoto M, Sugihara K. Cyclooxygenase-2 expression: a significant prognostic indicator for patients with colorectal cancer. Clin Cancer Res. 2004;10:8465–71.
41. Chan AT, Ogino S, Fuchs CS. Aspirin and the risk of colorectal cancer in relation to the expression of COX-2. N Engl J Med. 2007;356:2131–42.
42. Chan AT, Ogino S, Fuchs CS. Aspirin use and survival after diagnosis of colorectal cancer. JAMA. 2009;302:649–59.
43. GTEx Consortium. The genotype-tissue expression (GTEx) pilot analysis: multitissue gene regulation in human. Science. 2015;348(6235):648–60.

# Introduction to Histology

# 2

Maura Bríd Cotter and Massimo Loda

## 2.1 Introduction to Histology

"Histology" is the examination of normal cells and tissue and is performed with the aid of a microscope. In contrast, "histopathology," a subdiscipline within pathology, refers to the study of diseased tissues, and will be discussed separately in Chap. 3. Histologists have the specialized skills necessary to process and stain various tissue samples, while histopathologists are physicians with the skills necessary to interpret the histological slides. The routine specimen types received in a histology laboratory, and their preparation using routine histological techniques, are first introduced below, followed by an introduction to the important components of a normal human cell and the histology of various normal tissue types.

## 2.2 Specimen Types

Specimens received for histological examination include both cytology specimens and histopathology specimens, examples of which are listed in Table 2.1. Cytology specimens are taken with the aim of examining tissue at a cellular level [1]. These specimens, therefore, include samples of free cells or tissue fragments. The most common sample type is a fine needle aspiration (FNA), where a very thin needle and a syringe are used to acquire a small amount of cells or fluids from a lesion (e.g. thyroid cyst [2]). Bodily fluids, such as urine [3, 4], or cerebrospinal fluid [5], etc., can also be processed. Another special sampling technique is when cells are gently scraped or brushed from an organ (e.g. cervical smear [6]). In contrast, histopathology specimens include whole organs or small samples of larger tissues. A needle core biopsy is the most common type of sample where, in comparison to a FNA, a large needle is used to remove a greater quantity of tissue. Other sampling techniques available to clinicians include excisional biopsies where an entire lesion is surgically excised, or incisional biopsies where part of a larger lesion is removed.

M.B. Cotter · M. Loda (✉)
Department of Pathology, Dana-Farber Cancer Institute and Brigham & Women's Hospital, Harvard Medical School, Boston, MA 02115, USA
e-mail: massimo_loda@dfci.harvard.edu


M. Loda et al. (eds.), *Pathology and Epidemiology of Cancer*,
DOI 10.1007/978-3-319-35153-7_2

**Table 2.1** Examples of cytopathological and histopathological specimens

| Cytopathology | Histopathology |
|---|---|
| Fine needle aspiration (e.g. thyroid cyst) | Biopsies (e.g. biopsy of a breast mass) |
| Smears (e.g. cervical) | Surgical specimens (e.g. prostatectomy) |
| Bodily fluids (e.g. urine) | Autopsy specimens (e.g. kidney) |

## 2.3 Specimen Examination and Sampling

Grossing of histopathology specimens [7] involves careful examination by the pathologist including a specimen description, weight, and measurement of dimensions [8]. Photographs can be taken and relevant surgical margins inked. Thorough dissection is performed in order to locate representative areas suitable for sampling. Biobanking [9] can also be completed at this stage, which involves taking small samples of fresh tissue and is used, for instance, to create cell lines, isolate stem cells, generate organoids or ex vivo organotypic cultures [10, 11] for storage in a tissue biobank. Touch preparations (or "touch preps") can also be made using fresh tissue [12, 13], where the specimen is gently touched against a clean glass slide, an imprint made, and later examined.

## 2.4 Preparation of Histological Slides

The sampled tissue is placed into a plastic cassette and undergoes a series of steps in order to prepare the tissue for histological examination. Table 2.2 highlights the various stages of tissue preparation and summarizes the purpose of each stage mentioned below.

### 2.4.1 Fixation of Tissue

Fixation involves submerging the sampled tissue in chemical substances (i.e. fixatives) in order to prevent tissue digestion by enzymes or bacteria and to preserve as much as possible of its morphologic and chemical characteristics. Fixatives promote cross-links between proteins and form a gel that maintains the in vivo relations of tissue components to each other [14]. There are a number of reagents that can be used for fixation, each of which has differing penetration rates. Formaldehyde [15] is the most commonly used agent for histopathology and when dissolved in water, it is referred to as "formalin." One part formalin is typically diluted with nine parts water to produce a 10 % formalin solution, a concentration that is optimal for tissue fixation [16]. This solution penetrates tissue at about 1 mm an hour [17], therefore, biopsies are generally submitted for processing the same day as received, while larger specimens (e.g. a mastectomy) are not processed the same day as received as they require a longer fixation period.

**Table 2.2** The stages of tissue preparation and the purpose of each stage

| Stage | Purpose |
|---|---|
| Sampling | To choose the most representative areas of the specimen |
| Fixation | To preserve tissue morphology and chemical composition |
| Dehydration | To remove fixative and cell water and replace with dehydrating fluid |
| Clearing | To remove dehydrating fluid and replace with clearing fluid |
| Embedding | To impregnate with liquid paraffin and make the tissue resistant to sectioning |
| Sectioning | To make tissue sections available for histological analysis |

### 2.4.2 Processing of Tissue

The processing of tissue includes dehydration, clearing, and embedding steps. Dehydration involves the removal of fixative and water from the tissue and their replacement with dehydrating agents by placement in increasing concentrations of ethanol. Next, the clearing step involves replacing the dehydrating fluid with a lipid solvent (e.g. xylene). The tissue is subsequently removed from the cassette and placed in a molten wax-filled mold for embedding. At this point, orientation of the tissue within the mold is critical, as it will determine the plane through which the section will be cut. Incorrect placement of tissues may result in diagnostically important areas being missed or damaged later [18]. Elongate tissues should be placed diagonally across the block (e.g. core biopsy), while tubular structures (e.g. vas deferens) are embedded so as to provide transverse sections showing all tissue layers. Specimens that have an epithelial surface (e.g. skin surface) are embedded in such a way as to provide sections in a plane at right angles to the surface.

### 2.4.3 Sectioning of Tissue

The waxed cassette is next placed in an instrument with fine blades called a microtome. Rotation of the drive wheel moves the block holder a controlled distance forwards, the blade's edge strikes the tissue block, and thin sections are cut and affixed to a glass slide. Sections are usually four microns thick so that a single layer of cells can later be seen under the microscope. Following thorough drying of the tissue sections are ready for staining.

## 2.5 Staining of Tissue

As most tissues are colorless, methods of staining tissues have been developed to make them visible, while also allowing distinctions to be made between tissue components. This is done by using mixtures of acidic or basic dyes that selectively stain various tissue elements. The constituents that react with basic dyes do so because of acid in their composition (e.g. nucleoproteins), while acidic dyes stain basic tissue components (e.g. cytoplasmic proteins). Of all routine stains, the combination of hematoxylin and eosin, or the "H&E stain" [19], is the most commonly used and dates as far back as the 1870s. This stain is considered the gold standard in histology and in a typical tissue section, nuclei are stained blue/purple, whereas the cytoplasm and surrounding matrix have varying degrees of pink staining [20]. Therefore, the H&E stain has the ability to reveal structural information with specific functional implications. Special stains [21] use a slightly different technique to stain particular structures (e.g. Masson's trichrome for muscle and collagen fibers [22]) or pathogens (e.g. Ziehl–Neelsen for acid-fast bacteria [23]). When routine or special staining cannot provide all the diagnostic answers required, histopathologists can use advanced staining techniques including immunohistochemistry (IHC) or in situ hybridization (ISH).

### 2.5.1 Immunohistochemistry

IHC is a multistep technique involving the interaction of a target antigen (i.e. the protein of interest) with a specific antibody tagged with a visible label [24]. The aim is to detect the presence of elevated levels or the absence of a particular target antigen. Antibodies can be coupled with an enzyme, such as horseradish peroxidase (HRP), requiring the use of a light microscope for visualization, or with fluorescent chemical compounds requiring the use of a fluorescent microscope. Different labeling techniques are available including the direct or indirect labeling method. In a direct assay, a fluorophore-labeled antibody, such as fluorescein isothiocyanate, can react directly with the target antigen. This tag allows immediate visualization of the antigen. In contrast, in an indirect assay, an unlabeled primary antibody is used. This binds to the target antigen and an enzyme-labeled secondary antibody binds to the primary antibody. In general,

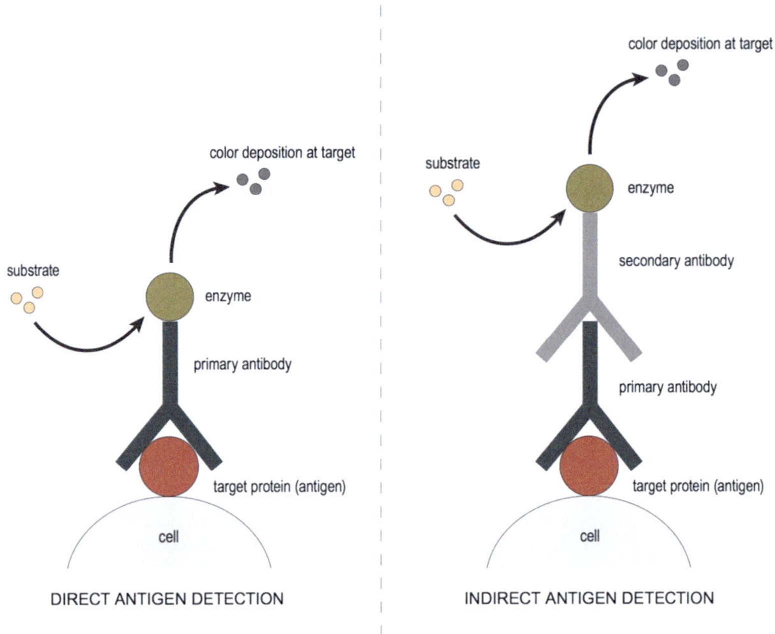


**Fig. 2.1** Illustration of immunohistochemistry

primary antibodies are raised against the antigen of interest and are unlabeled, while secondary antibodies are raised against IgG of the primary antibody. Finally, a chromogenic substrate must be added for visualization (e.g. diaminobenzidine), which reacts to produce a brown precipitate in the presence of the HRP enzyme. A simplistic illustration of these two assays is shown in Fig. 2.1. Recent advances in this area include multiplexing of antigens [25, 26], whereby multiple stains can be performed on the same tissue section, followed by analysis using digital imaging software.

### 2.5.2 In Situ Hybridization

While IHC involves the detection of marker proteins in tissue, ISH can detect target ribonucleic acid (RNA) or deoxyribonucleic acid (DNA) sequences [27]. Different detection systems are then used to visualize the presence of the target sequence. Fluorescence ISH (FISH) uses fluorescent dyes and fluorescent microscopy while chromogenic ISH (CISH) uses chromogenic dyes and brightfield microscopy [28]. In this process, a complementary DNA strand (or probe), or RNA strand (or riboprobe) is used to localize a specific DNA or RNA sequence. The probe hybridizes to the target sequence with elevated temperature (i.e. denaturation) and excess probe is washed away. If the probe is already fluorescent, it will detect the site of hybridization directly. If the probe is chromogenic, an additional step is needed to visualize the probe [29]. A simplistic illustration of ISH is shown in Fig. 2.2. Multiplexing using ISH can also be performed [30], followed by spectral imaging for the detection and subsequent deconvolution of multiple signals.

## 2.6 The Frozen Section

The "frozen section" is an alternative tissue preparation technique where, in contrast to routine processing, is a rapid histological examination done on fresh tissue. The sample is quickly placed into cryoprotective embedding medium and cut in a refrigerated microtome (i.e. cryostat). The sections are then stained with H&E. This technique is used, for example, where the surgeon needs a tumor margin to be examined to ensure that it has been adequately removed, or to

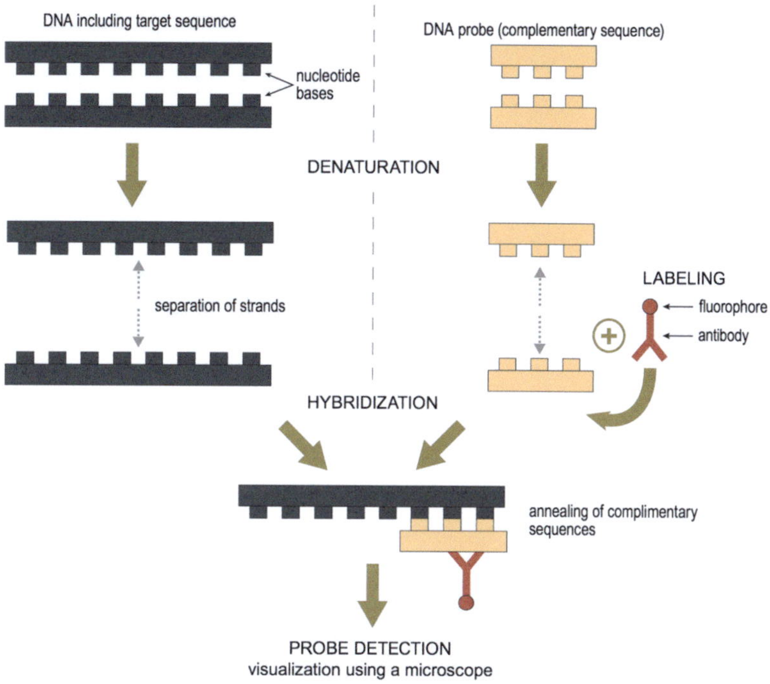


**Fig. 2.2** Illustration of in situ hybridization

confirm a diagnosis of cancer intraoperatively [31, 32]. The diagnosis should ideally be conveyed to the clinician within 20 min after receipt of the tissue within the pathology laboratory, termed "the turnaround time" [33]. Despite the speed of the procedure, one major disadvantage of the frozen section technique is that "freezing artifacts" are frequently seen which can obscure tissue morphology and cellular detail. Examples of freezing artifacts seen, include nuclear ice crystals, bubbles, vacuolated cytoplasm, nuclear chromatin changes, tissue cracking, etc. Therefore, this method is only used when an urgent intraoperative diagnosis is needed.

## 2.7 Tissue Microarray Construction and Evaluation

The tissue microarray (TMA) construction process [34] involves a region of interest first being identified and marked on a histology slide. This same area is then marked on the corresponding paraffin tissue block (i.e. the donor block) and core biopsies are taken [35]. The cores are then inserted into a separate paraffin block (i.e. the recipient block) [36] in a precisely spaced, array pattern. This process is repeated multiple times where finally, one paraffin block is made up of hundreds of tissue core biopsies from many donor blocks (Fig. 2.3). Simultaneous analysis of molecular targets at the DNA, mRNA, and protein levels can then be performed under identical conditions. Epidemiologists, histopathologists, and researchers can subsequently analyze data following quantitative digital image analysis [37]. Therefore, the development of TMA technology has allowed the efficient study hundreds of different tissue samples concurrently [38] and is an invaluable research tool.

## 2.8 Microscopes, Automated Imaging, and Digital Software

Many different microscopes are available today, each of which have varied applications and modifications that contribute to their usefulness

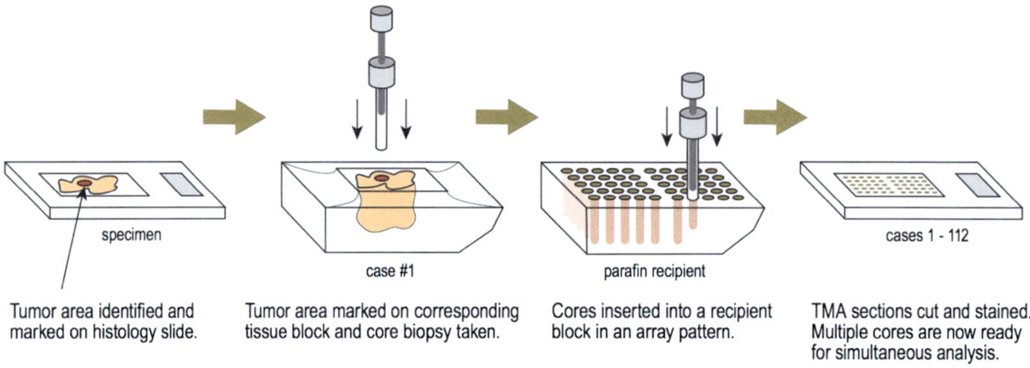


**Fig. 2.3** Tissue microarray construction

**Table 2.3** Classification, type, light source, and function of various microscopes

| Classification | Microscope type | Light source | Description |
|---|---|---|---|
| Optical | Light microscope (or "compound") | Visible light | This is the most commonly used microscope with strong magnifying power, used for the study of cells, chromosomes, and DNA |
| | Dissecting microscope (or "stereoscope") | Visible light | This contains lenses in different angles that provides 3D viewing, used for forensics, fine repair, microsurgery |
| | Fluorescence microscope | UV light | This is a special type of light microscope where, instead of light reflection and absorption it uses UV light to view cells |
| | Digital microscope | Visible light | This makes use of the optical lens and charge-coupled device (CCD) sensors to magnify objects and includes a camera for high quality recording |
| Electron | Transmission electron microscope (TEM) | Electron beam | This microscope is used for studying cells and microorganisms and can produce images as small as 1 nm in size |
| | Scanning electron microscope (SEM) | Electron beam | This is less powerful than the TEM but can provide 3D viewing of objects and is used for studying cells and small particles of matter |

(Table 2.3). The upright microscope is the most common configuration and has binocular eyepieces, high power compound objective lenses, and a precision sample stage [39]. In contrast, an inverted microscope is essentially an upside-down, upright microscope. As brightfield and fluorescence microscopy are the most frequent types performed in histology labs, only these will be introduced in this chapter.

Brightfield microscopy is performed using a light microscope with a light source commonly projecting from the back of the microscope. This light travels upwards from beneath, projects through the field diaphragm and condenser lens beneath the stage, up through the histology slide containing tissue, into the objective lens, and finally up to the camera and/or eyepiece. A simplistic illustration of this light path is shown in Fig. 2.4. The fluorescence microscope is similar to the conventional brightfield microscope with added features to enhance its capabilities [39]. While the conventional microscope uses visible light (~400–700 nm), the fluorescence microscope uses much higher intensity light source that causes excitation of fluorophores (i.e. the excitation light). The light is absorbed by the

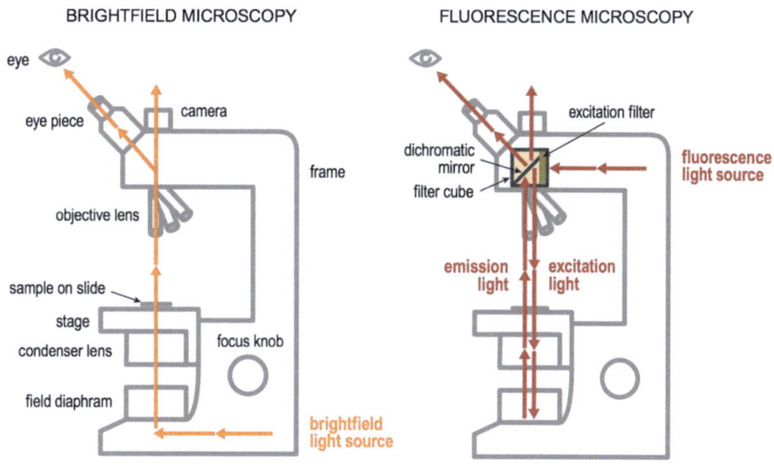


**Fig. 2.4** Light paths of brightfield and fluorescence microscopes

fluorophores, which causes them to emit a longer, lower energy wavelength light. This fluorescent light (i.e. the emission light) can be separated with filters designed for that specific wavelength. In the fluorescence microscope (Fig. 2.4), the light initially travels to a filter cube. Inside the filter cube it passes through an excitation filter to a dichroic mirror (or beam-splitter), which sends light down through the objective lens and onto the histology slide containing tissue. The emitted fluorescence from the specimen then travels back up through the objective lens, into the filter cube, through the dichroic mirror and through an emission filter. From here it continues traveling upwards toward the camera and/or eyepiece and can be recorded or viewed [39].

Modern digital pathology combines the power of the microscope with electronic detection and advanced computerized analysis and is now progressively replacing previously subjective, semiquantitative manual scoring with precise quantification of protein expression [40]. A sensor is used to obtain an image, which is then displayed on a computer monitor using charge-coupled device technology. Automated scanning can also be performed using a robotic loader. Associated imaging software packages for both brightfield and fluorescence purposes provide complex algorithms for quantitation of immunostaining. Tissue can be automatically segmented into gland or stromal targets and cells can also be segmented into nuclei and cytoplasm using various algorithms. This allows the translation of extent and intensity of immunostaining into a continuous variable, more amenable to large-scale bioinformatics analyses.

## 2.9 The Normal Human Cell and its Components

The human cell is the basic structural and functional unit of the body and its main compartments are the nucleus and the cytoplasm, which are completely separate entities that work together to keep the cell functioning [41]. Histologically, the nucleus, nucleolus, and cytoplasm can easily be distinguished on routine H&E. Antibodies can also be used to selectively stain the nucleus, cytoplasm, or cytoplasmic membrane also, depending on the target of interest. While the structures within the nucleus and cytoplasm cannot be seen using routine microscopy, their functions are introduced briefly, as knowledge of this basic information is imperative for an understanding of the topics discussed in forthcoming chapters.

### 2.9.1 The Cytoplasm and its Organelles

The cytoplasm surrounds the cell nucleus and is surrounded by a plasma membrane, which separates the interior of the cell from the outside environment. This membrane is selectively permeable and controls the movement of substances in and out of the cell. The cytoplasm is made largely of cytosol fluid containing multiple organelles ("little organs") suspended within it that carry out specific directions of the nucleus. These organelles include mitochondria, endoplasmic reticula, the Golgi apparatus, vacuoles, and lysosomes (Fig. 2.5) among others. Mitochondria function to produce most of the cell's energy in the form of adenosine triphosphate (ATP) [42] and are also involved in processes such as cell signaling [43], cellular differentiation [44], cell growth [45], cell cycle control [46], and cell death [47]. There are two types of endoplasmic reticula, the rough endoplasmic reticulum (RER) and the smooth endoplasmic reticulum (SER) [48]. The SER is involved in lipid metabolism [49], carbohydrate metabolism, and detoxification, while the RER contains ribosomes on the surface where active protein synthesis occurs [50, 51]. The Golgi apparatus concentrates and packages proteins from the RER inside the cytoplasm prior to being transferred to their appropriate destinations [52]. Vacuoles are involved in the storage and intracellular digestion of molecules and lysosomes contain enzymes responsible for the breakdown of proteins, nucleic acid, carbohydrates, lipids, cellular debris, and foreign organisms [53].

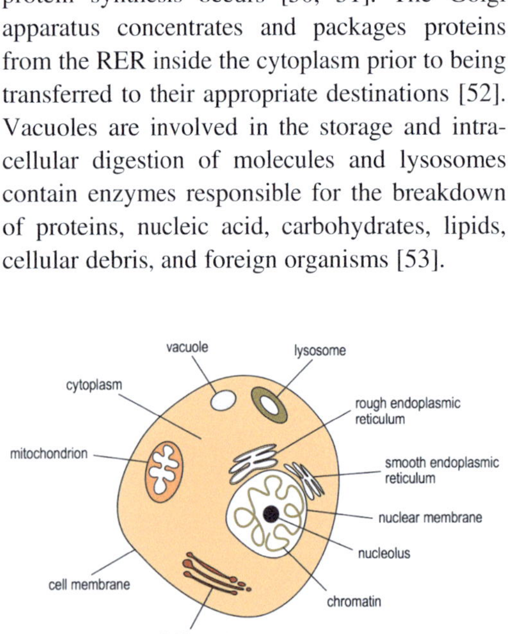


**Fig. 2.5** The human cell and its components

### 2.9.2 The Nucleus and Gene Expression

The nucleus is the largest organelle found in human cells and is surrounded by a nuclear envelope with nuclear pore complexes that allow material to move in and out [54]. It contains genetic information in the form of DNA. DNA is a complex molecule consisting of two antiparallel strands of nucleotide bases, each with a backbone of sugar (deoxyribose) molecules linked together by phosphate groups [55]. Each sugar molecule is linked to a base, which is attached by hydrogen bonds to a base on the other strand in a complementary fashion so that adenine (A) bonds with thymine (T) and guanine (G) bonds with cytosine (C) [56]. DNA is organized into highly compact, regular units called chromosomes. Within the nucleus is a structure called the nucleolus, which contains RNA, ribosomal proteins, and functions as the site of ribosome synthesis. RNA differs from DNA in that RNA molecules are single-stranded, the backbone sugar is ribose, and it contains uracil (U) in place of thymine [57].

Human genetic testing has made significant advances in the past decade [58]. The identification of certain sequences in order to diagnose genetic diseases can be done by performing sequencing on a blood sample, fresh tissue, or paraffin embedded tissue. DNA sequencing [59] is the process of determining the nucleotide order of DNA fragment. As mRNA is generated by transcription from DNA, reverse transcription must be performed (using a reverse transcriptase enzyme) for RNA sequencing [60]. Following this, complementary DNA (cDNA) fragments are generated, PCR amplification executed, a library created, and sequencing performed comparing results to a reference genome.

## 2.10 Basic Histology of Normal Tissues

"Tissues" refer to groups of similar cells performing similar functions (e.g. cardiac myocytes). "Organs" refer to groups of tissues (e.g. the heart) and "organ systems" include groups of organs that function together (e.g. the cardiovascular system). Tissue is composed of various cells together with a surrounding extracellular matrix (ECM). There are four fundamental tissue types including epithelia, connective tissue, muscle and nervous tissue, each of which will be introduced separately below.

### 2.10.1 Epithelia

Epithelial cells are generally classified as "covering and lining epithelia" which are found lining the cavities of the body and surfaces of structures, or as "glandular (or secretory) epithelia" [41]. Glands refer to single cells or groups of cells that secrete protein, mucus, or lipid. This includes "endocrine glands" which secrete into extracellular spaces (i.e. secrete internally) and "exocrine glands" which secrete into ducts (i.e. secrete into the external environment). Epithelial cells can also be classified based on the number of cell layers present and the shape of the cells in the top layer (Fig. 2.6). Epithelial tissue can, therefore, be one cell thick (i.e. simple epithelium), or two or more cells thick (i.e. stratified epithelium) [61]. There are three basic cell shapes based on microscopic appearance, including squamous (flat and wide), columnar (tall), and cuboidal (cube shaped) cells. Consequently, by describing the number of cell layers and the surface cell shape the different forms of epithelia can easily be classified. In some cases a third feature, namely specialization of the cell surface, e.g., keratinization [62] or the presence of cilia [63], is included. Stratified squamous epithelium that is exposed directly to the environment (e.g. the skin), can show keratinization (i.e. a layer of dead cells is present on the surface), while those that are not directly exposed (e.g. the oral cavity) are only partially keratinized or nonkeratinized. The presence of cilia (i.e. hair-like motile processes) is another specialization seen in simple columnar epithelium present on the surface of these cells. This surface adaptation helps propel substances along, e.g., the airways contain cilia to propel mucus.

Cells of epithelial tissues are usually tightly packed together and form a continuous sheet or a solid aggregation of cells. They lack intracellular spaces and are united by several types of junctional specializations [64] (i.e. tight junctions [65], desmosomes [66], hemidesmosomes [67] etc.). Therefore, epithelia have only one free surface (i.e. apical surface), which is exposed at the body surface or at the lumen of a duct, tube, or vessel. The lower surface of an epithelium (i.e. basal surface) [68] rests on an underlying basement membrane, which is a thin sheet of collagen and glycoproteins, which acts as both a scaffold and a selectively permeable membrane allowing water and small molecules through [69].

Two special categories of epithelium are pseudostratified and transitional epithelium [41]. Pseudostratified columnar epithelium is so called, because of an apparent stratification, however, all of the cells are attached to the basement membrane (Fig. 2.6). Therefore, it is really simple epithelium, despite giving the impression of stratification. Transitional epithelium (or urothelium) is stratified epithelium lining the walls of the urinary tract. The term refers to the fact that it may appear as stratified cuboidal to squamous in appearance depending on the extent of bladder distention. Specific names are also given to epithelium in certain locations. Endothelium is a term given to simple epithelium lining blood vessels (vascular endothelial cells) [70] or lymphatics (lymphatic endothelial cells) [71]. Mesothelium is a name given simple squamous epithelium lining the major body cavities [72], for example, the peritoneal epithelium lining the abdominal organs [73].

### 2.10.2 Connective Tissues

Connective tissues form a scaffold that epithelial tissues lie on, and nerve and muscle tissues are embedded. They are classified as connective tissue proper and specialized connective tissues, including adipose tissue, cartilage, bone, and blood (Fig. 2.7). All connective tissues are characterized by various individual cells scattered within an extracellular space filled with an ECM

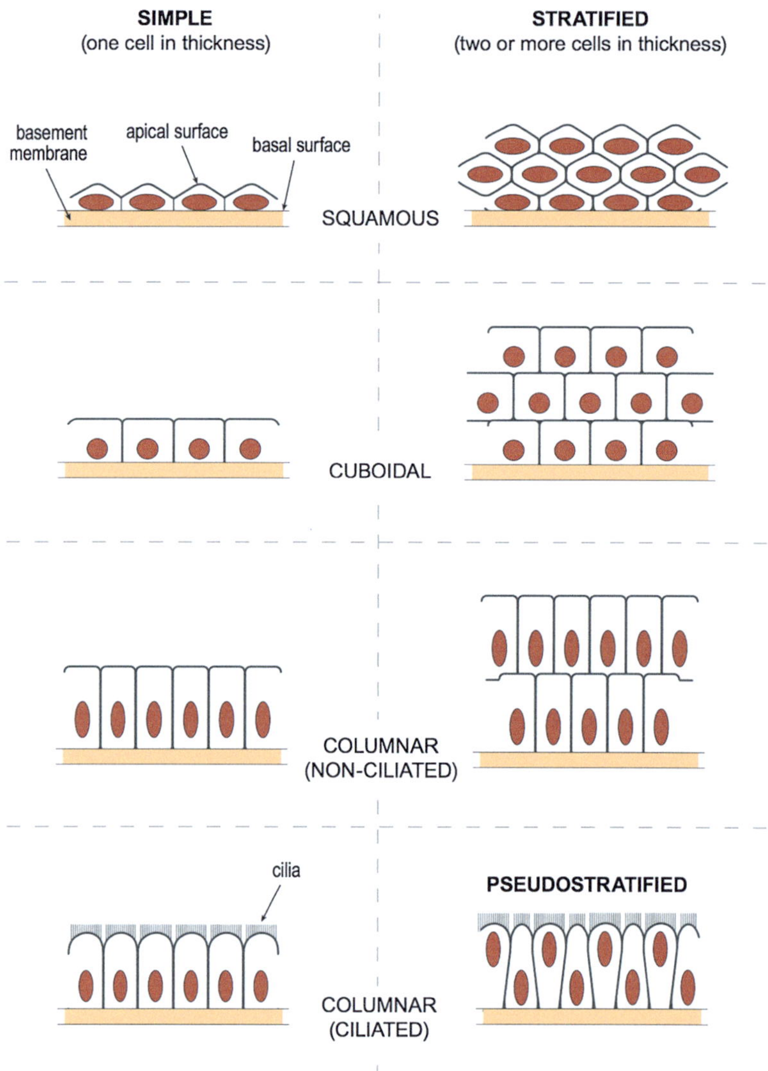


**Fig. 2.6** Illustration of epithelial cell types

[41]. Variations in the composition of the ECM determine the properties of the connective tissue. In general, the ECM comprises ground substance and various fibers (i.e. collagen, reticular or elastic fibers) woven into a network. Ground substance supports the connective tissue cells, binds them together, and permits the diffusion of nutrients and other dissolved substances between capillaries and cells. There are many known types of collagens, with type I being the most abundant. Histologically, collagen appears as irregular, wavy fibers arranged singly or in small groups. Reticular fibers are very fine fibrils consisting of another type of collagen (type III). They are usually not visible using routine H&E, but can be demonstrated using special stains (e.g. reticulin) [22]. Like reticular fibers, elastic fibers [74] require special stains to be visualized also (e.g. elastin). Once stained, elastic fibers appear as fine, dark, undulating fibers within the tissue.

The principle cells in connective tissue proper include fibroblasts that secrete collagen fibers and ground substance [75], together with macrophages, mast cells, and adipocytes among others. Fibroblasts have elongated nuclei with a moderate amount of cytoplasm that tapers at the

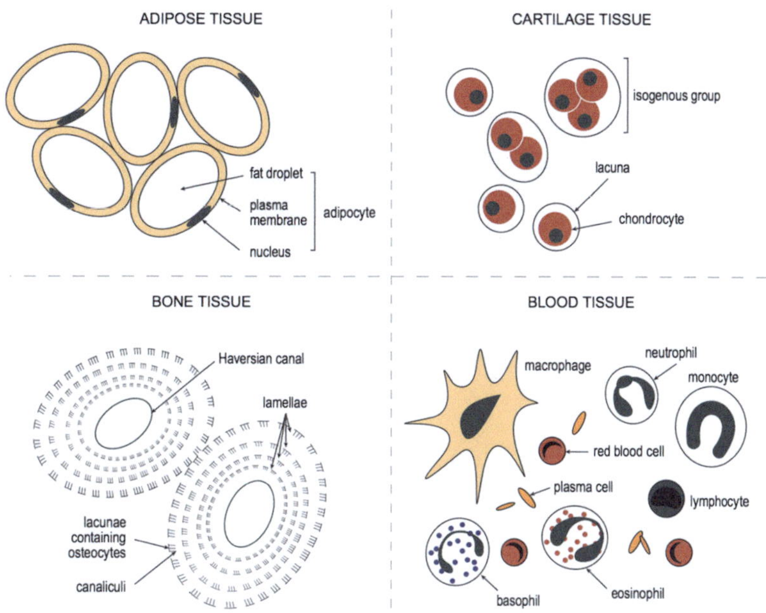


**Fig. 2.7** Illustration of connective tissue cell types

ends under the microscope. They are usually found alone and contain an elliptical nucleus showing finely stippled (i.e. "dot-like") chromatin and one or two nucleoli. Macrophages are large, round cells, with vesicular nuclei [76]. In some cases, a brown pigment is seen within them, which is the result of lysosomal action on ingested red blood cells. Mast cells are small, ovoid cells with spherical, eccentric nuclei, and basophilic granules [77]. In adipocyte cells, the nucleus appears flattened with the cytoplasm forming a very narrow rim around a large central lipid droplet. During routine preparation of histological slides fat is dissolved, therefore, the adipocytes actually appear empty. When adipocytes are seen in large numbers, the tissue is referred to as adipose tissue.

There are three types of cartilage tissues that differ in the type of fibers they contain within the ECM. These include hyaline, fibrocartilage, and elastic cartilage, with hyaline being the most abundant type. Histologically, hyaline cartilage has a basophilic appearance on H&E. Chondrocytes (cartilage cells) produce the matrix of cartilage and are seen within lacunae (i.e. matrix cavities), singly or in clusters of 2–8 cells. These groups are called isogenous groups and are derived by mitosis from a single chondrocyte. In contrast, the principle cells in bone tissue are osteoblasts, osteocytes, and osteoclasts. The osteoblast is involved in bone deposition at the bone surface and secretes the matrix of bone (i.e. osteoid), which becomes calcified following deposition [78]. As they are trapped within the ECM, they become osteocytes [79]. The osteocytes are located in lacunae and fine channels (canaliculi) containing osteocyte cell processes, connecting lacunae to each other. The third cell type, the osteoclast, is associated with bone resorption [80] and does this by secreting enzymes that acidify the matrix. These are large, multinucleated cells with a ruffled border histologically. In order to be able to visualize bone tissue, the specimen is usually placed in a decalcifying solution (e.g. formic acid) [81], which removes calcified material so that good quality paraffin sections can be prepared that will preserve the microscopic elements.

Blood is traditionally classified as a specialized form of connective tissue, even though it has a different function in comparison to other connective tissue types. It has a highly fluid ground substance (i.e. plasma), which comprises mainly water together with salts, proteins, nutrients, hormones, and waste material. The cellular component of blood is produced by the bone

**Table 2.4** Histologic description and function of hematopoietic cells

| Cell type | Histologic description | Function of hematopoietic cell |
|---|---|---|
| Erythrocyte | Flat or oval-shaped cell with no nucleus | Responsible for the transportation of oxygen in the body |
| Neutrophil | Multilobated nucleus, often with 3–5 lobes | Involved in the acute inflammatory response |
| Eosinophil | Eosinophilic granular cytoplasm and a bilobed nucleus | Associated with the allergic response and parasitic infections |
| Basophil | Basophilic granular cytoplasm and a bilobed nucleus | Responsible for allergic response by releasing histamine |
| Monocyte | Large cell, with a large, indented ("kidney-bean" shape) nucleus and abundant cytoplasm | Precursors for tissue macrophages, which engulf and digest foreign microorganisms, dead cells, or debris (i.e. phagocytosis) |
| Lymphocyte | Small, round cell with a deeply staining spherical nucleus surrounded by a thin rim of basophilic cytoplasm | Precursors of natural killer cells, B and T lymphocytes involved in the acute inflammatory response. B cells further differentiate into plasma cells |
| Megakaryocyte | Large cell, with a large pale multilobated nucleus and abundant cytoplasm | Precursors of thrombocytes (or platelets) formed by budding from megakaryocytes and contribute to hemostasis |

marrow in a process termed hematopoiesis [82]. These hematopoietic cells are derived from multipotent hematopoietic stem cells. Following division, the resulting daughter cells (myeloid or lymphoid progenitor cells) can commit to alternative differentiation pathways depending on growth factors involved. Finally, blood cells are divided into three lineages including the erythroid lineage (reticulocytes and erythrocytes), the lymphoid lineage (T cells, B cells [83] and natural killer cells [84]) and the myeloid lineage (granulocytes [85, 86], megakaryocytes [87], and macrophages [88]). Table 2.4 lists the histologic description and function of various hematopoietic cells, and their basic structures are also illustrated in Fig. 2.7.

### 2.10.3 Muscles

Muscles are responsible for maintaining posture, locomotion, and movement of the internal organs (e.g. contraction of the heart). Three kinds of muscle tissues are found in different organs of the body, including skeletal muscle, cardiac muscle, and smooth muscle (Fig. 2.8). Skeletal muscle generally forms the muscles attached to bones and by contracting, these muscles move joints. Cardiac muscle (myocardium) forms the mass of the heart. Smooth muscle is a component of the walls of many hollow organs within the body, such as the digestive tract. By contracting, smooth muscle propels the contents along the tube it surrounds (e.g. the intestine), or regulates

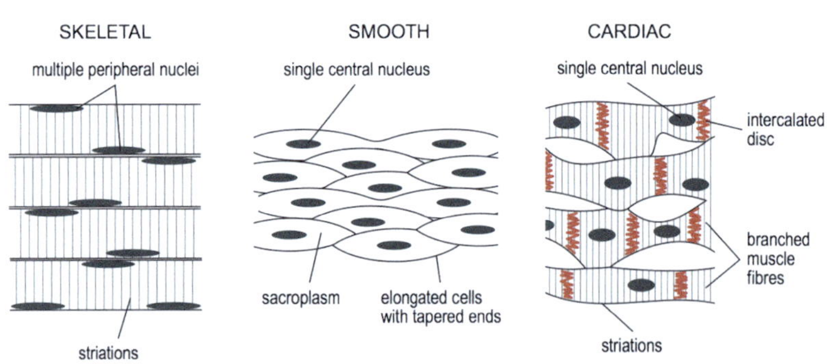


**Fig. 2.8** Illustration of three kinds of muscle tissue cell types

the amount of fluid flowing through it (e.g. the blood vessels). Skeletal and cardiac muscle is referred to as "striated" muscle as they show light and dark bands when viewed under the microscope. Smooth muscle cells do not have visible striations, however, they do contain the same contractile proteins arranged in a different pattern. While muscles are conventionally classified based on morphology (i.e. striated or smooth muscle), they can also be classified based on function (i.e. voluntary or involuntary muscle). Histologically, muscle tissue cells (myocytes) are elongated and spindle-shaped with little intervening extracellular material. Smooth muscle and cardiac muscle myocytes contain one nucleus, while skeletal muscle is multinucleate. The unusual microstructure of myocytes has lead to the use of specialized terminology, including the sarcolemma (plasma membrane), the sarcoplasm (cytoplasm), the sarcoplasmic reticulum (endoplasmic reticulum), and sarcosomes (mitochondria).

### 2.10.4 Neural Tissues

The nervous system is divided anatomically into the central nervous system (CNS) and the peripheral nervous system (PNS). The CNS consists structurally of the brain (within the skull) and the spinal cord (within the vertebral canal), while the PNS is composed of nerves (i.e. cranial nerves from the brain and spinal nerves from the spinal cord) and ganglia (i.e. nerve cell clusters). Neurons (nerve cells) are specialized cells that respond to stimuli and conduct electrical impulses to and from all body organs. All neurons have the same basic structure consisting of the cell body (soma) containing the nucleus, cytoplasm and organelles, and nerve processes that conduct the signals [89]. Nerve processes include the axon, which carries signals away from the cell body, and dendrites, which carry signals toward the cell body. There are three basic shapes to neurons, bipolar (i.e. consisting of a single axon and single dendrite), pseudounipolar (i.e. consisting of a single axon with a central and a peripheral branch), and multipolar (i.e. consisting of a single axon and numerous dendrites), illustrated in Fig. 2.9. There are four types of supporting cells (or central neuroglia) in the CNS, including oligodendrocytes [90], microglia [91], astrocytes [92], and ependymal cells [93]. In contrast, the Schwann cell is the principle supporting cell in the PNS (or peripheral neuroglia) [94]. The oligodendrocyte or Schwann cell wraps around axons of neurons to form the myelin sheath that ensures rapid conduction of nerve impulses [95]. This sheath is not continuous and gaps between neighboring cells

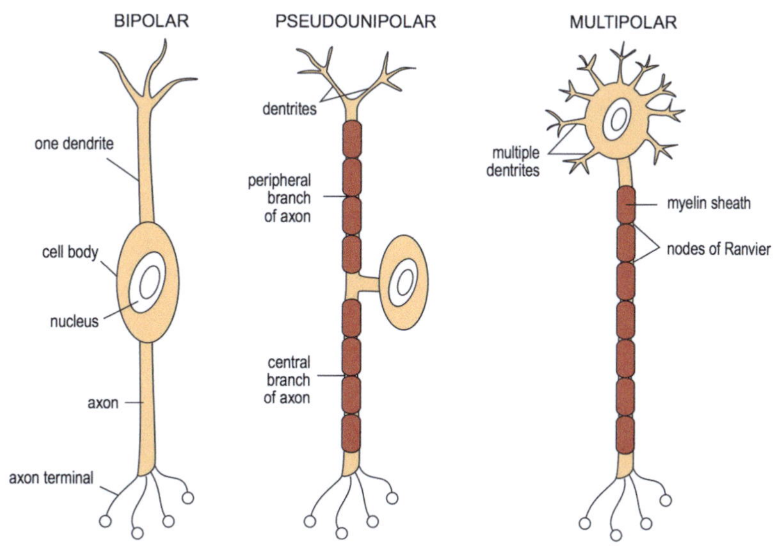


**Fig. 2.9** Illustration of basic neuron types

are called nodes of Ranvier [96]. Neurons are seen histologically as irregular or stellate in shape and are multipolar. They have a large cell body with a large, round, and pale (euchromatic) nucleus and a single prominent nucleolus. The supporting cells are quite difficult to distinguish using routine H&E, however, and immunocytochemical methods are therefore necessary to demonstrate them adequately, e.g., glial fibrillary acidic protein (GFAP) highlights astrocytes [97].

## 2.11 Summary

In summary, this chapter has introduced basic terminology and important techniques and instruments used in histology. This information should be of practical use to the reader and will help to develop and refine the body of knowledge necessary for understanding the forthcoming chapters.

## References


1. Hall-Craggs MA, Lees WR. Fine needle biopsy: cytology, histology or both? Gut. 1987;28(3):233–6.
2. Cibas ES, Ali SZ. NCI, thyroid FNA, state of the science conference. The Bethesda system for reporting thyroid cytopathology. Am J Clin Pathol. 2009;132(5):658–65.
3. Brimo F, Vollmer RT, Case B, Aprikian A, Kassouf W, Auger M. Accuracy of urine cytology and the significance of an atypical category. Am J Clin Pathol. 2009;132(5):785–93.
4. Raitanen MP, Leppilahti M, Tuhkanen K, Forssel T, Nylund P, Tammela T. The dilemma of suspicious urine cytology in patients being followed for bladder cancer. Ann Chir Gynaecol. 2001;90(4):256–9.
5. Bigner SH. Cerebrospinal fluid (CSF) cytology: current status and diagnostic applications. J Neuropathol Exp Neurol. 1992;51(3):235–45.
6. Massad LS, Einstein MH, Huh WK, Katki HA, Kinney WK, Schiffman M, Solomon D, Wentzensen N. 2012 updated consensus guidelines for the management of abnormal cervical cancer screening tests and cancer precursors. Obstet Gynecol. 2013;121(4):829–46.
7. Bell WC, Young ES, Billings PE, Grizzle WE. The efficient operation of the surgical pathology gross room. Biotech Histochem. 2008;83(2):71–82.
8. Huo L. A practical approach to grossing breast specimens. Ann Diagn Pathol. 2011;15(4):291–301.
9. Elger BS, Caplan AL. Consent and anonymization in research involving biobanks: differing terms and norms present serious barriers to an international framework. EMBO Rep. 2006;7(7):661–6.
10. Helgesson G, Helgesson G, Dillner J, Carlson J, Bartram CR, Hansson MG. Ethical framework for previously collected biobank samples. Nat Biotechnol. 2007;25(9):973–6.
11. Vaira V, Fedele G, Pyne S, Fasoli E, Zadra G, Bailey D, Snyder E. Preclinical model of organotypic culture for pharmacodynamic profiling of human tumors. Proc Natl Acad Sci USA. 2010;107(18):8352–6.
12. Valdes EK, Boolbol SK, Cohen JM, Feldman SM. Intra-operative touch preparation cytology; does it have a role in re-excision lumpectomy? Ann Surg Oncol. 2007;14(3):1045–50.
13. Yildiz-Aktas IZ, Monaco SE, Khalbuss WE, Parwani AV, Jaffe TM, Pantanowitz L. Testicular touch preparation cytology in the evaluation of male infertility. Cytojournal. 2011;8:24.
14. Hopwood D. Fixation and fixatives: In: Bancroft JD, Gamble M, editors. Theory and practice of histological techniques. 5th ed. Philadelphia. Churchill Livingstone Publishers; 2002. p. 63–84.
15. Thavarajah R, Mudimbaimannar VK, Elizabeth J, Rao UK, Ranganathan K. Chemical and physical basics of routine formaldehyde fixation. J Oral Maxillofac Pathol. 2012;16(3):400–5.
16. Chesnick IE, Mason JT, Leary T, Fowler C. Elevated pressure improves the rate of formalin penetration while preserving tissue morphology. J Cancer. 2010;1:178–83.
17. Grizzle WE. Special symposium: fixation and tissue processing models. Biotech Histochem. 2009;84(5):185–93.
18. Anderson G. Tissue Processing and Microtomy including Frozen. In: Bancroft JD, Gamble M, editors. Theory and practice of histological techniques. 5th ed. Philadelphia: Churchill Livingstone Publishers; 2002. p. 85–108.
19. Chan JK. The wonderful colors of the hematoxylin-eosin stain in diagnostic surgical pathology. Int J Surg Pathol. 2014;22(1):12–32.
20. Wilson I. The Hematoxylins and Eosin. In: Bancroft JD, Gamble M, editors. Theory and practice of histological techniques. 5th ed. Philadelphia: Churchill Livingstone Publishers; 2002. p. 125–138.
21. Collier FC, Bretschneider AM, Dowling EA. Histochemical and special stains in the routine laboratory. J Natl Med Assoc. 1962;54:76–8.
22. Jones L. Connective tissues and stains. In: Bancroft JD, Gamble M, editors. Theory and practice of histological techniques. 5th ed. Philadelphia: Churchill Livingstone Publishers; 2002. p. 139–162.
23. Wu RI, Mark EJ, Hunt JL. Staining for acid-fast bacilli in surgical pathology: practice patterns and variations. Hum Pathol. 2012;43(11):1845–51.
24. Miller K. Immunocytochemical techniques. In: Bancroft JD, Gamble M, editors. Theory and practice of

histological techniques. 5th ed. Philadelphia: Churchill Livingstone Publishers; 2002. p. 421–464.
25. Fiore C, Bailey D, Conlon N, Wu X, Martin N, Fiorentino M, Finn S, Fall K, Andersson SO, Andren O, Loda M, Flavin R. Utility of multispectral imaging in automated quantitative scoring of immunohistochemistry. J Clin Pathol. 2012;65(6):496–502.
26. Snyder EL, Bailey D, Shipitsin M, Polyak K, Loda M. Identification of CD44v6(+)/CD24- breast carcinoma cells in primary human tumors by quantum dot-conjugated antibodies. Lab Invest. 2009;89(8):857–66.
27. McNicol AM, Farquharson MA. In situ hybridization and its diagnostic applications in pathology. J Pathol. 1997;182(3):250–61.
28. Hicks DG, Longoria G, Pettay J, Grogan ST, Tubbs R. In situ hybridization in the pathology laboratory: general principles, automation, and emerging research applications for tissue-based studies of gene expression. J Mol Histol. 2004;35(6): 595–601.
29. Jones M. Molecular Pathology and in-situ Hybridization. In: Bancroft JD, Gamble M, editors. Theory and practice of histological techniques. 5th ed. Philadelphia: Churchill Livingstone Publishers; 2002. p. 553–78.
30. Byers RJ, Di Vizio D. O'connell F, Tholouli E, Levenson RM, Gossage K. Semiautomated multiplexed quantum dot-based in situ hybridization and spectral deconvolution. J Mol Diagn. 2007;9(1):20–9.
31. Jaafar H. Intra-operative frozen section consultation: concepts, applications and limitations. Malays J Med Sci. 2006;13(1):4–12.
32. Ilvan S, Ramazanoglu R, Ulker AE, Calay Z, Bese T, Oruc N. The accuracy of frozen section (intraoperative consultation) in the diagnosis of ovarian masses. Gynecol Oncol. 2005;97(2):395–9.
33. Novis DA, Zarbo RJ. Interinstitutional comparison of frozen section turnaround time. A College of American Pathologists Q-Probes study of 32868 frozen sections in 700 hospitals. Arch Pathol Lab Med. 1997;121(6):559–67.
34. Fowler CB, Man YG, Zhang S, O'Leary TJ, Mason JT, Cunningham RE. Tissue microarrays: construction and uses. Methods Mol Biol. 2011;724:23–35.
35. Packeisen J, Korsching E, Herbst H, Boecker W, Boecker H. Demystified … tissue microarray technology. Mol Pathol. 2003;56(4):198–204.
36. Mirlacher M, Simon R. Recipient block TMA technique. Methods Mol Biol. 2010;664:37–44.
37. Rizzardi AE, Johnson AT, Vogel RI, Pambuccian SE, Henriksen J, Skubitz AP, Metzger GJ, Schmechel SC. Quantitative comparison of immunohistochemical staining measured by digital image analysis versus pathologist visual scoring. Diagn Pathol. 2012;7:42.
38. Voduc D, Kenney C, Nielsen TO. Tissue microarrays in clinical oncology. Semin Radiat Oncol. 2008;18(2):89–97.
39. Bancroft J. Light Microscopy. In: Bancroft JD, Gamble M, editors. Theory and practice of histological techniques. 5th ed. Philadelphia: Churchill Livingstone Publishers. 2002; 43–62.
40. Chen X, Zheng B, Liu H. Optical and digital microscopic imaging techniques and applications in pathology. Anal Cell Pathol (Amst). 2011;34(1–2):5–18.
41. Mills S. Histology for pathologists. 4th ed. Philadelphia: Lippincott, Williams & Wilkins; 2012. p. 1328.
42. Schapira AH. Mitochondrial disease. Lancet. 2006;368(9529):70–82.
43. Tait SW, Green DR. Mitochondria and cell signalling. J Cell Sci. 2012;125(4):807–15.
44. Mandal S, Lindgren AG, Srivastava AS, Clark AT, Banerjee U. Mitochondrial function controls proliferation and early differentiation potential of embryonic stem cells. Stem Cells. 2011;29(3):486–95.
45. Duchen MR. Roles of mitochondria in health and disease. Diabetes. 2004;53(Suppl 1):96–102.
46. Antico Arciuch VG, Elguero ME, Poderoso JJ, Carreras MC. Mitochondrial regulation of cell cycle and proliferation. Antioxid Redox Signal. 2012;16(10):1150–80.
47. Parsons MJ, Green DR. Mitochondria in cell death. Essays Biochem. 2010;47:99–114.
48. Xu C, Bailly-Maitre B, Reed JC. Endoplasmic reticulum stress: cell life and death decisions. J Clin Invest. 2005;115(10):2656–64.
49. Fu S, Yang L, Li P, Hofmann O, Dicker L, Hide W, Lin X, Watkins SM, Ivanov AR, Hotamisligil GS. Aberrant lipid metabolism disrupts calcium homeostasis causing liver endoplasmic reticulum stress in obesity. Nature. 2011;473(7348):528–31.
50. Ellgaard L, Helenius A. Quality control in the endoplasmic reticulum. Nat Rev Mol Cell Biol. 2003;4(3):181–91.
51. Boelens J, Lust S, Offner F, Bracke ME, Vanhoecke B. Review. The endoplasmic reticulum: a target for new anticancer drugs. In Vivo. 2007;21(2):215–26.
52. Morre DM. Role of the golgi apparatus in cellular pathology. J Electron Microsc Tech. 1991;17(2): 200–11.
53. Cox TM, Cachon-Gonzalez MB. The cellular pathology of lysosomal diseases. J Pathol. 2012;226(2):241–54.
54. Chow KH, Factor RE, Ullman KS. The nuclear envelope environment and its cancer connections. Nat Rev Cancer. 2012;12(3):196–209.
55. Watson JD, Crick FH. Molecular structure of nucleic acids; a structure for deoxyribose nucleic acid. Nature. 1953;171(4356):737–8.
56. Watson JD, Crick FH. Genetical implications of the structure of deoxyribonucleic acid. Nature. 1953;171(4361):964–7.
57. Mortimer SA, Kidwell MA, Doudna JA. Insights into RNA structure and function from genome-wide studies. Nat Rev Genet. 2014;15(7):469–79.

58. Crotwell PL, Hoyme HE. Advances in whole-genome genetic testing: from chromosomes to microarrays. Curr Probl Pediatr Adolesc Health Care. 2012;42 (3):47–73.
59. Boyd SD. Diagnostic applications of high-throughput DNA sequencing. Annu Rev Pathol. 2013;8:381–410.
60. Wang Z, Gerstein M, Snyder M. RNA-Seq: a revolutionary tool for transcriptomics. Nat Rev Genet. 2009;10(1):57–63.
61. Koster MI, Roop DR. Mechanisms regulating epithelial stratification. Annu Rev Cell Dev Biol. 2007;23: 93–113.
62. Smack DP, Korge BP, James WD. Keratin and keratinization. J Am Acad Dermatol. 1994;30(1): 85–102.
63. Klysik M. Ciliary syndromes and treatment. Pathol Res Pract. 2008;204(2):77–88.
64. Lai-Cheong JE, Arita K, McGrath JA. Genetic diseases of junctions. J Invest Dermatol. 2007;127 (12):2713–25.
65. Shen L, Turner JR. Role of epithelial cells in initiation and propagation of intestinal inflammation. Eliminating the static: tight junction dynamics exposed. Am J Physiol Gastrointest Liver Physiol. 2006;290(4): 577–82.
66. Runswick SK, Runswick SK, O'Hare MJ, Jones L, Streuli CH, Garrod DR. Desmosomal adhesion regulates epithelial morphogenesis and cell positioning. Nat Cell Biol. 2001;3(9):823–30.
67. Borradori L, Sonnenberg A. Structure and function of hemidesmosomes: more than simple adhesion complexes. J Invest Dermatol. 1999;112(4):411–8.
68. Martin-Belmonte F, Perez-Moreno M. Epithelial cell polarity, stem cells and cancer. Nat Rev Cancer. 2012;12(1):23–38.
69. Martin GR, Rohrbach DH, Terranova VP, Liotta LA. Structure, function, and pathology of basement membranes. Monogr Pathol. 1983;24:16–30.
70. Sumpio BE, Riley JT, Dardik A. Cells in focus: endothelial cell. Int J Biochem Cell Biol. 2002;34 (12):1508–12.
71. Ji RC. Characteristics of lymphatic endothelial cells in physiological and pathological conditions. Histol Histopathol. 2005;20(1):155–75.
72. Mutsaers SE. The mesothelial cell. Int J Biochem Cell Biol. 2004;36(1):9–16.
73. Daya D, McCaughey WT. Pathology of the peritoneum: a review of selected topics. Semin Diagn Pathol. 1991;8(4):277–89.
74. Christiano AM, Uitto J. Molecular pathology of the elastic fibers. J Invest Dermatol. 1994;103(5 Suppl): 53–7.
75. Quan TE, Cowper S, Wu SP, Bockenstedt LK, Bucala R. Circulating fibrocytes: collagen-secreting cells of the peripheral blood. Int J Biochem Cell Biol. 2004;36(4):598–606.
76. Wynn TA, Chawla A, Pollard JW. Macrophage biology in development, homeostasis and disease. Nature. 2013;496(7446):445–55.
77. Kalesnikoff J, Galli SJ. New developments in mast cell biology. Nat Immunol. 2008;9(11):1215–23.
78. Neve A, Corrado A, Cantatore FP. Osteoblast physiology in normal and pathological conditions. Cell Tissue Res. 2011;343(2):289–302.
79. Neve A, Corrado A, Cantatore FP. Osteocytes: central conductors of bone biology in normal and pathological conditions. Acta Physiol (Oxf). 2012;204(3): 317–30.
80. Blair HC, Athanasou NA. Recent advances in osteoclast biology and pathological bone resorption. Histol Histopathol. 2004;19(1):189–99.
81. Lillie RD. Studies on the decalcification of Bone. Am J Pathol. 1944;20(2):291–6.
82. Smith JN, Calvi LM. Concise review: Current concepts in bone marrow microenvironmental regulation of hematopoietic stem and progenitor cells. Stem Cells. 2013;31(6):1044–50.
83. Pieper K, Grimbacher B, Eibel H. B-cell biology and development. J Allergy Clin Immunol. 2013;131 (4):959–71.
84. Kwong YL, Chan AC, Liang RH. Natural killer cell lymphoma/leukemia: pathology and treatment. Hematol Oncol. 1997;15(2):71–9.
85. Blanchard C, Rothenberg ME. Biology of the eosinophil. Adv Immunol. 2009;101:81–121.
86. Amulic B, Cazalet C, Hayes GL, Metzler KD, Zychlinsky A. Neutrophil function: from mechanisms to disease. Annu Rev Immunol. 2012;30:459–89.
87. Wickrema A, Crispino JD. Erythroid and megakaryocytic transformation. Oncogene. 2007;26(47): 6803–15.
88. Ginhoux F, Jung S. Monocytes and macrophages: developmental pathways and tissue homeostasis. Nat Rev Immunol. 2014;14(6):392–404.
89. Brini M, Calì T, Ottolini D, Carafoli E. Neuronal calcium signaling: function and dysfunction. Cell Mol Life Sci. 2014;71(15):2787–814.
90. Morrison BM, Lee Y, Rothstein JD. Oligodendroglia: metabolic supporters of axons. Trends Cell Biol. 2013;23(12):644–51.
91. Garden GA, Moller T. Microglia biology in health and disease. J Neuroimmune Pharmacol. 2006; 1(2):127–37.
92. Sofroniew MV, Vinters HV. Astrocytes: biology and pathology. Acta Neuropathol. 2010;119(1):7–35.
93. Del Bigio MR. Ependymal cells: biology and pathology. Acta Neuropathol. 2010;119(1):55–73.
94. Kidd GJ, Ohno N, Trapp BD. Biology of Schwann cells. Handb Clin Neurol. 2013;115:55–79.
95. Aggarwal S, Yurlova L, Simons M. Central nervous system myelin: structure, synthesis and assembly. Trends Cell Biol. 2011;21(10):585–93.
96. Arancibia-Carcamo IL, Attwell D. The node of Ranvier in CNS pathology. Acta Neuropathol. 2014;128(2):161–75.
97. Yung WK, Luna M, Borit A. Vimentin and glial fibrillary acidic protein in human brain tumors. J Neurooncol. 1985;3(1):35–8.

# 3 Introduction to Pathology

Maura Bríd Cotter and Massimo Loda

## 3.1 Introduction to Pathology

The term "pathology" comes from the Greek words pathos (suffering/disease) and—logia (science/study) and refers to the scientific study of disease. This includes everything from the causes (etiology) of disease, the underlying mechanisms (pathogenesis), the cellular, molecular, and genetic changes, and the clinical manifestations (signs and symptoms) of disease. Pathology is performed by examination of human tissues, bodily fluids, organs, and in some cases, whole bodies (i.e., autopsy examination). Pathologists are medical specialists that have the knowledge necessary to understand the changes seen in cells, in order to come to a diagnosis to guide patient therapy. We begin this chapter by introducing the normal formation of tissues, organ systems, and germ layer derivatives, before moving on to the pathological changes seen in cells and tissues as a response to cellular adaptation, and the characteristics of neoplasia.

## 3.2 Formation of Tissues and Organ Systems

The basic structure, components, and function of a normal cell and the four major cell groups (tissues) have been described in Chap. 2. Here we describe the formation of these tissues and the development of organ systems.

The entire body evolves following human fertilization of one single cell (or fertilized ovum) [1]. This cell undergoes multiple divisions to eventually produce a hollow cluster of cells called a blastocyst approximately one week following fertilization. The outer layer of the blastocyst (i.e., trophoblast) implants into the endometrium and eventually produces the placenta and other structures that support the developing embryo [1]. The inner cell mass gives rise to the embryonic body. The region of embryonic development within the inner cell mass (i.e., embryonic shield) consists of two epithelial layers, an ectoderm (or outer layer), and an endoderm (or inner layer) [1, 2]. At a

M.B. Cotter · M. Loda (✉)
Department of Pathology, Dana-Farber Cancer Institute and Brigham & Women's Hospital, Harvard Medical School, 450 Brookline Ave., Boston, MA 02115, USA
e-mail: massimo_loda@dfci.harvard.edu


M. Loda et al. (eds.), *Pathology and Epidemiology of Cancer*,
DOI 10.1007/978-3-319-35153-7_3

region called the primitive streak, superficial cells migrate between these two layers forming an intermediate layer called mesoderm [1, 2]. All three germ layers participate in formation of the organs and organ systems.

## 3.3 Germ Layer Derivatives

The three germ layers are roughly equivalent to the position of structures in the fully formed human body, with ectoderm forming external structures, mesoderm forming central connective structures, and endoderm forming internal organs [1]. Each germ layer differentiates to give rise to specific tissues and while many organs are described as developing from one germ layer (e.g., the large bowel from endoderm), their accompanying structures develop from the other two layers (e.g., the colonic muscular wall from mesoderm and nervous innervation from ectoderm). Therefore the three germ layers actually interact with and complement each other. Examples of specific tissues derived from each germ layer are described below and illustrated in Fig. 3.1.

### 3.3.1 Ectoderm

Ectoderm produces the epidermis of the skin and its appendages (i.e., hair follicles and sebaceous glands), the cornea and lens of the eye and the epithelial linings of the mouth, nasal cavity, salivary glands, and anal canal. Portions of the skull, teeth, and the adrenal medulla are also derived from ectoderm together with the nervous system (including the hypothalamus, pituitary, and pineal gland) [1, 2].

### 3.3.2 Mesoderm

The mesoderm layer forms the lining of the pericardial, pleural, and peritoneal cavities and

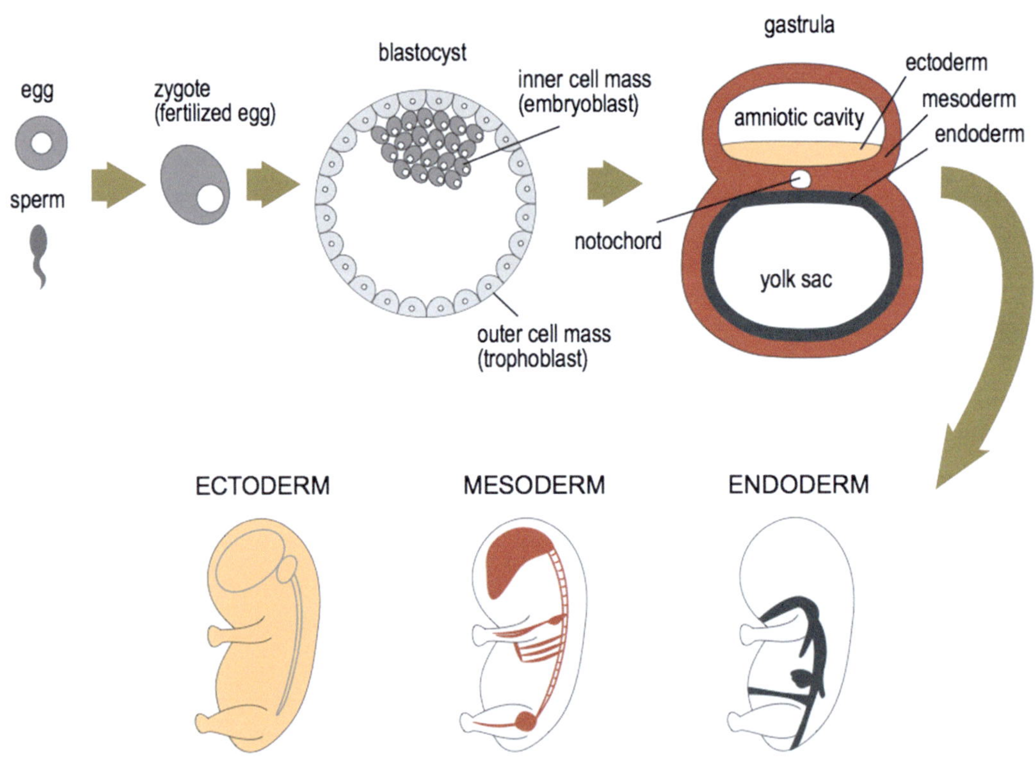


**Fig. 3.1** Formation of tissues and germ layer derivatives

produces the musculoskeletal, cardiovascular, and lymphatic systems. The adrenal cortex, spleen, kidneys, reproductive organs and dermis of the skin also develop from this layer [1, 2].

### 3.3.3 Endoderm

This inner layer produces most of the epithelium of the gastrointestinal tract and its organs (i.e., liver, gallbladder, and pancreas), the lining of the respiratory system (i.e., trachea, bronchi, and alveoli) the bladder and portions of the urethra. The thymus gland, thyroid gland, and parathyroid glands also develop from the endoderm layer [1, 2].

## 3.4 Cellular Adaptation and Death

Cells constantly adapt within a narrow range of physiological parameters so that internal conditions remain relatively constant. This is referred to as cellular "homeostasis" [3]. Cells are capable of making changes in response to unfavorable environmental changes (i.e., injury) in an attempt to maintain this internal stability, which may be physiological (or normal) or pathological (or abnormal) changes [4]. The most common types of cellular adaptation include cell atrophy, hypertrophy, hyperplasia, and metaplasia and are illustrated in Fig. 3.2.

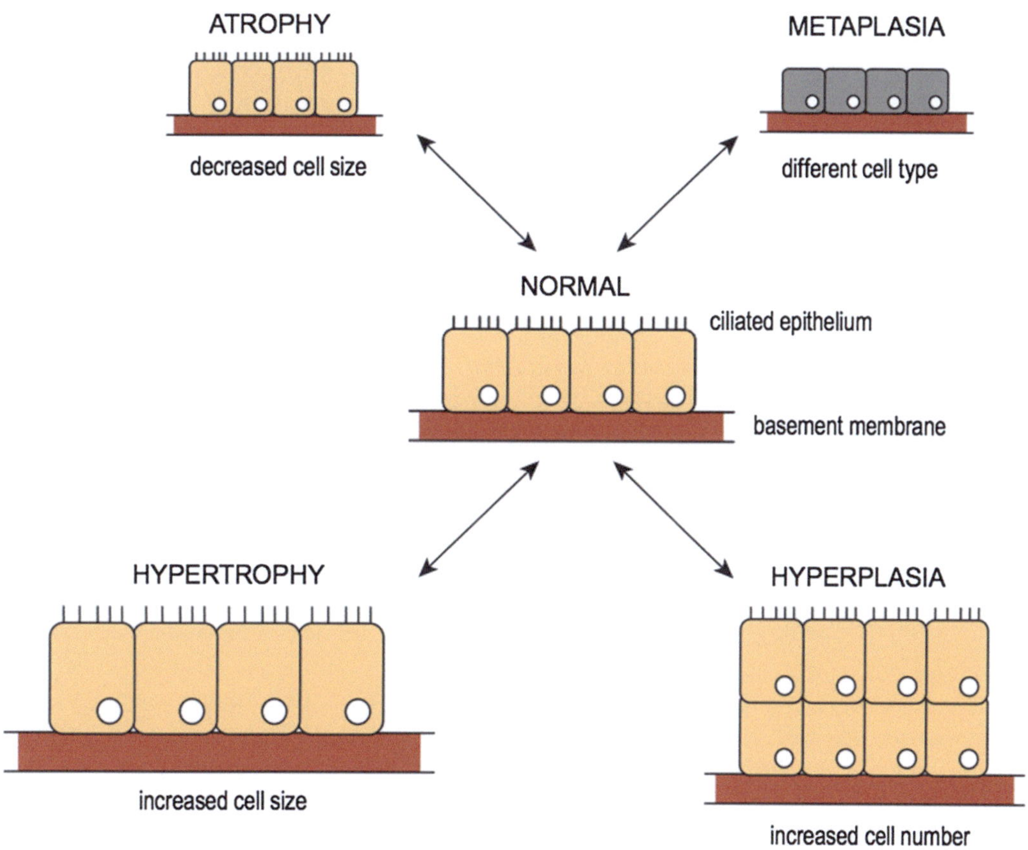


**Fig. 3.2** Reversible cell injury and cellular adaptation

### 3.4.1 Atrophy

Atrophy is the reduction in size of cells in response to cell injury [5]. It results from decreased protein synthesis and increased protein degradation within the cells. Atrophy of various cells and organs is normal at certain points in the human life cycle. An example of atrophy as part of normal development includes the involution of the thymus gland in early childhood. Examples of pathologic causes of atrophy include atrophying muscle diseases (e.g., muscular dystrophy [6] or myotonic dystrophy [7]) characterized by progressive muscle weakness over time resulting in functional disability [6, 7].

### 3.4.2 Hypertrophy and Hyperplasia

Hypertrophy refers to an increase in volume of an organ or tissue due to the enlargement of cell size. There is no increase in cell number. Hypertrophy should be distinguished from hyperplasia where cells remain approximately the same size but increase in number, although both can occur simultaneously. In hypertrophy, cells are enlarged by an increased amount of structural proteins and organelles. This can be both physiologic or pathologic due to increased functional requirements. A physiological example of hypertrophy is increased muscle size in response to normal exercise [8, 9] (e.g., lifting weights), whereas a pathologic example would include cardiac enlargement as a result of hypertension [10, 11]. An example of a normal hyperplastic response on the other hand, is the normal proliferation of glandular epithelium in the breast as a response to pregnancy [12]. A pathologic example of hyperplasia would include endometrial hyperplasia as a result of unopposed estrogen [13], a risk factor for endometrial carcinoma [14].

### 3.4.3 Metaplasia

The definition of metaplasia is the change of a cell from one cell type to another cell type. This occurs when the original cell type is unable to withstand new environmental stress and changes into another phenotype more suited to the new environment. Metaplasia is usually reversible once the cause of environmental stress is removed. An example of metaplasia is intestinal metaplasia, whereby the change in cell type resembles that found in the intestine (i.e., intestinalized columnar epithelium with goblet cells). When intestinal metaplasia occurs in the esophagus [15] (normally lined by squamous mucosa), it is referred to as "Barrett's esophagus" [16]. This diagnosis is clinically important as these patients are at risk of developing esophageal adenocarcinoma [17], and therefore it is considered to be a premalignant condition.

### 3.4.4 Cell Death: Necrosis and Apoptosis

If the injury to the cell is too severe (i.e., irreversible injury), the affected cells are no longer able to adapt to stress and subsequently die [18]. There are two main types of cell death, necrosis [19], and apoptosis [20]. These differ in their roles in disease and in their mechanisms [18] and are compared in Fig. 3.3. Cellular necrosis is caused by factors outside the cell (e.g., infection, toxins, or trauma) that result in the unregulated breakdown of the cell's internal components. When the damage to a cell's plasma membrane is severe, enzymes leak from lysosomes into the cytoplasm and cause autolysis [21]. The leakage of cellular contents through the damaged cell membrane prompts a host reaction (i.e., an inflammatory response). In contrast, apoptosis (or programmed cell death) is a targeted cause of cellular demise [22] where activated enzymes degrade the internal contents of the cell. Although the plasma membrane remains intact, it becomes altered so that it fragments into apoptotic bodies which ultimately become phagocytosed by host immune cells [23]. In this form of cellular death an inflammatory response is not elicited [20].

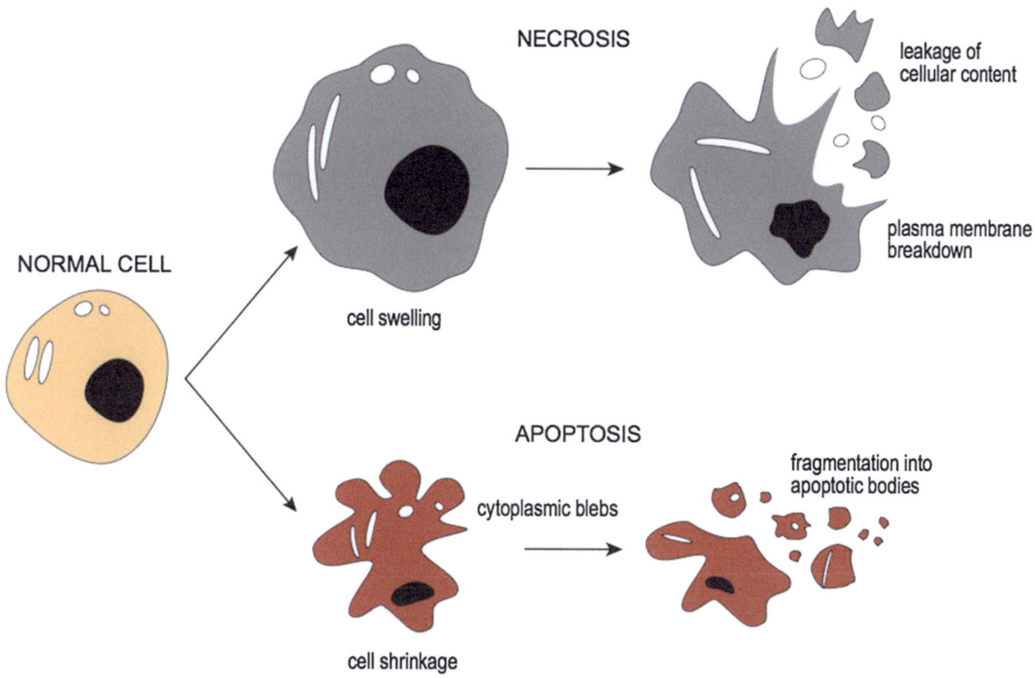


**Fig. 3.3** Cellular death: necrosis and apoptosis

## 3.5 Dysplasia and Carcinoma In Situ

Dysplasia is the abnormal growth or development of cells where the cells are structurally changed in shape, size, and appearance from the original cell type. The tissue becomes disordered in appearance under the microscope, often with an increase in the number of immature cells. Dysplasia is often a precursor to tumor formation (or neoplasia) [24–26] and the chances of development into a carcinoma rise with worsening degrees of dysplasia. Low grade cervical dysplasia usually does not usually evolve into cervical carcinoma for example, whereas high grade dysplasia usually does [27, 28]. High grade, or full thickness dysplasia is often referred to as "carcinoma in situ" and refers to neoplastic cells within the normal boundaries of the tissue of origin (i.e., neoplastic cells do not migrate beyond the basement membrane). This is often abbreviated as CIS and some refer it as "pre-cancer" or "non-invasive cancer" [29]. In contrast, invasive carcinoma refers to when neoplastic cells have moved beyond the basement membrane layer [30] and therefore have the potential to spread to other areas within the body. The progression from normal epithelium to invasive carcinoma is illustrated in Fig. 3.4.

## 3.6 Neoplasia

Neoplasia is the process of tumor formation resulting in a neoplasm (or tumor). Tumors are classified as either benign or malignant based on their pathology and known clinical behaviors [5]. It is very important for a histopathologist to be able to distinguish between a benign and a malignant tumor, as the treatment required is usually very different. Certain long-established histological features indicate innocence while other features indicate malignancy, therefore the distinction between the two tumor classifications is made with remarkable accuracy. Histopathologic features assessed include evaluation of architectural (i.e., the arrangement of cells) and

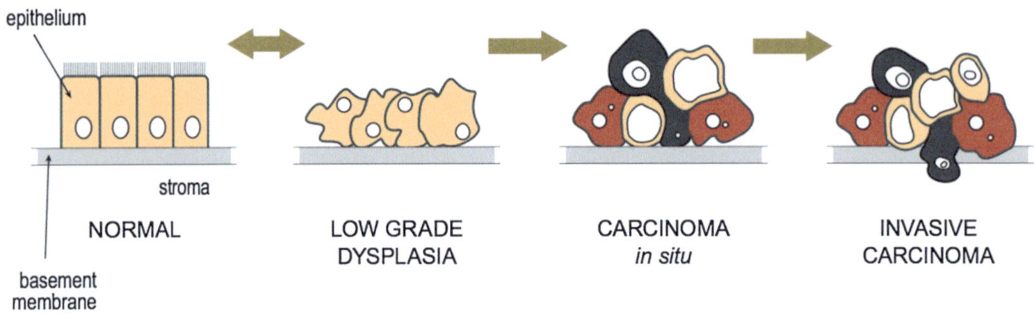


**Fig. 3.4** Progression from normal epithelium to invasive carcinoma

cytological characteristics (i.e., the nuclear and cytoplasmic features) of the tissue under examination. The normal architecture of an organ involves standard relationships between specific groups of cells. In neoplastic growth, the relationships between cells are significantly distorted. The architecture of neoplastic cells in comparison to normal cells is not orderly and many layers of cells are arranged in irregular patterns. Various malignancies show particular altered architectural growth patterns also, for example, a "papillary" (or finger-like) growth pattern in thyroid carcinoma [31, 32] in comparison to normal thyroid follicles lined by a single layer of cuboidal epithelium [33]. Other histological patterns of growth are listed in Table 3.1.

Neoplastic cells are also distinguished from normal cells by a loss of "cellular differentiation" [4, 5]. Differentiation refers to how different the tumor cells are from the cells from which they originated. If neoplastic cells are almost like normal cells they are said to be "well differentiated." In contrast, if they are very unlike normal cells they are referred to as "poorly differentiated." In pathology, tumor grade is a measure of cellular differentiation [34] where low grade tumors are well differentiated, high grade tumors are poorly differentiated and intermediate grade tumors are intermediate between these two. Each type of cancer is graded using a different system, e.g., the Gleason grading system [35, 36] for prostate carcinoma or the Nottingham grading system [37, 38] for breast cancer.

The cytological changes seen in tumors include increased variations in cellular and nuclear size and shape (or pleomorphism) compared to corresponding normal cells, large hyperchromatic (i.e., stain more deeply basophilic) nuclei containing coarse, irregular

**Table 3.1** Histological patterns of growth

| Histological pattern | Description |
|---|---|
| Papillary | "Finger-like" projections lined by neoplastic cells with central fibrovascular cores |
| Cribriform | Clusters of neoplastic cells with sharply punched out round spaces, described as "swiss cheese-like" |
| Solid | Sheets of multiple neoplastic cells together with little intervening stroma |
| Nested | Small solid groups or "nests" of neoplastic cells which cluster together |
| Infiltrative | Neoplastic cells growing in single-file and intersecting into surrounding structures |
| Rosettes | Neoplastic cells growing circumferentially around a lumen forming a "halo-like" or "spoke-wheel" shape |
| Whorls | Neoplastic cells sweeping and swirling in different directions |
| Fascicular | Bundles of neoplastic cells intersecting at right angles to each other |

chromatin, increased nuclear to cytoplasmic ratios, large nucleoli, and a high rate of cell division according to the number of cells with the nucleus showing the characteristic pattern of separating chromosomes (or mitotic figures [39]).

### 3.6.1 Benign Tumors

Benign lesions refer to non-cancerous, localized tumors, which do not have the capacity to spread to other sites and are generally amenable to surgical removal. These tumors are named by the cell type from which they originate, followed by the suffix "-oma." An example is a "lipoma" [40], a benign tumor of fat cells (or lipocytes) [41]. Other examples of benign tumors are listed in Table 3.2. Benign tumors tend to be histologically and cytologically similar to their tissues of origin (i.e., they are well differentiated). However, the gross structure of a benign tumor may stray from the normal and assume papillary or polypoid configurations, for example, in squamous papillomas of the skin [42] in comparison to the flat epidermal layer of normal skin [33]. Once excised, benign tumors do not tend to recur or spread to distant sites.

### 3.6.2 Malignant Tumors

Malignant lesions are cancers which show aggressive behavior including invasion into and destruction of adjacent tissues. These tumors can also recur if incompletely excised and also have the capacity to spread to other sites. There are two general pathologic categories including "carcinomas" derived from epithelial cells (i.e., from endodermal or ectodermal layers) or "sarcomas" of mesenchymal origin (i.e., from the mesodermal layer). Lymphomas and leukemias are a separate category of malignancies that arise from hematopoietic cells. Malignant lesions are further defined by their tissue of origin (i.e., prostatic [43]). Some examples of malignant tumors are listed in Table 3.2. Malignant neoplasms exhibit malignant cytologic features, disorganized growth patterns, abundant mitotic

**Table 3.2** Examples of benign and malignant neoplasms and relationship to cell of origin

| | | Neoplasm | |
|---|---|---|---|
| Cell origin | Cell type | Benign | Malignant |
| Ectoderm | Skin cells | Nevus | Malignant melanoma |
| | Nerve cells | Neurofibroma | Neurofibrosarcoma |
| Mesoderm | Fat cells | Lipoma | Liposarcoma |
| | Cartilage cells | Chondroma | Chondrosarcoma |
| | Bone | Osteoma | Osteogenic sarcoma |
| | Blood vessels | Hemangioma | Angiosarcoma |
| | Lymph vessels | Lymphangioma | Lymphangiosarcoma |
| | Smooth muscle | Leiomyoma | Leiomyosarcoma |
| | Striated muscle | Rhabdomyoma | Rhabdomyosarcoma |
| | Hematopoietic cells | NA | Lymphoma/leukemia |
| Endoderm | Colonic cells | Adenoma | Adenocarcinoma |
| | Liver cells | Liver cell adenoma | Hepatocellular carcinoma |

figures, and arrangements around blood vessels. Malignant tumors often outgrow their blood supply and show ischemic necrosis. The rate of growth of malignant lesions usually correlates inversely with their level of differentiation.

## 3.7 Important Biological Characteristics of Solid Tumors

Tumors are believed to be acquired through a multistep process involving genetic alterations which result in the loss of function of genes which normally inhibit cell growth (i.e., tumor suppressor genes) [44], increased activation of genes that stimulate proliferation (i.e., oncogenes) [45] or alteration of the function of genes involved in DNA repair [46]. Genes can be mutated in several different ways including altered arrangement of chromosomes (or rearrangement) [47] resulting in translocations [48], insertions [49], deletions or duplications [47], which may lead to subsequent gene activation or inactivation. Mutations are seen in many important regulatory pathways including genes that activate and deactivate carcinogens and those that govern the cell cycle, cell senescence (or aging), apoptosis, angiogenesis (or new blood vessel formation), cell signaling, and cellular differentiation to name a few.

### 3.7.1 Proliferative Activity

The normal human cell cycle involves a complex system of signaling pathways that a cell undergoes in order to copy itself precisely [50]. As illustrated in Fig. 3.5, the normal human cell cycle is divided into four phases: the first growth phase (G1), the DNA synthesis phase (S), the second growth phase (G2), and mitosis (i.e., the cellular division phase) (M). The cycle is checked at three main checkpoints: the G1/S checkpoint, the G2/M checkpoint, and the M checkpoint [50]. If a cell reaches a checkpoint

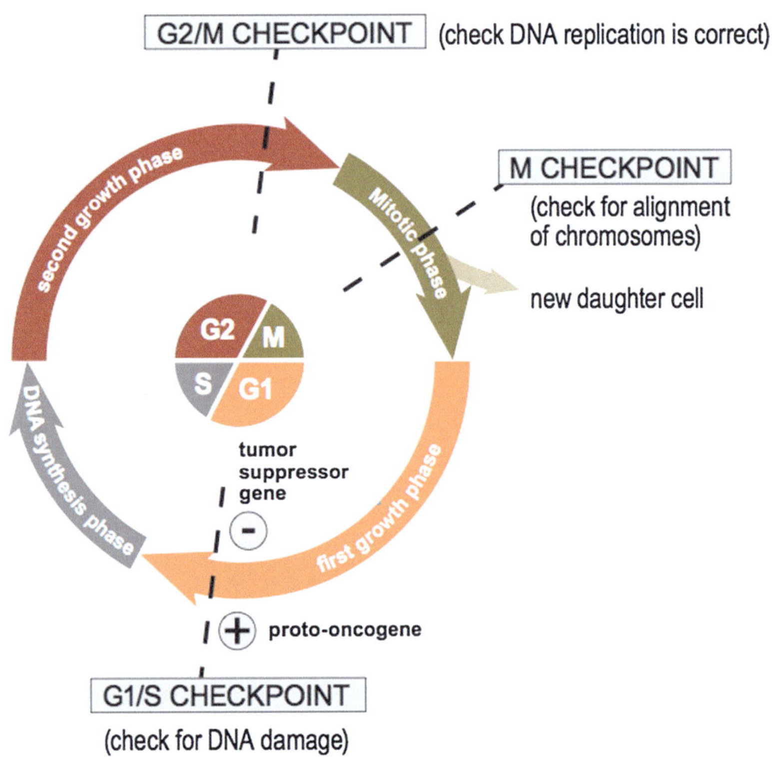


**Fig. 3.5** The normal human cell cycle

and damage is detected, the cell cycle ceases for a time. During this time, the cell has the opportunity to repair the DNA damage using various DNA repair genes and resume cycling [51]. If this repair is not successful, the cell is then triggered into apoptosis. In contrast, in the tumor cell cycle [50, 52], neoplastic cells may undergo multiple consecutive cycles of mitosis [53] or they may leave the cell cycle, remain dormant for a period and reenter the cycle with the appropriate stimulus.

Proto-oncogenes and tumor suppressor genes are important components of the cell cycle involved in checkpoint control. Firstly, proto-oncogenes (e.g., RAS [54], MYC [55], CDK [56], etc.) encode components of cellular proliferation within the cell cycle including growth factors, receptors, signaling enzymes, and transcription factors. Oncogenes arise from the mutation or increased expression of proto-oncogenes [57] and disrupt the cell's normal signaling pathway allowing for uncontrolled cell proliferation. In contrast, tumor suppressor genes (e.g., TP53 [58], PTEN [59], etc.) are a family of genes that instruct cells to produce proteins that inhibit cell growth within the cell cycle. The loss of these proteins allows a cell to grow and divide in an uncontrolled manner [60]. Finally, DNA repair genes are also essential components of the cell cycle and code for proteins whose normal function is to correct errors that arise during DNA duplication. There are multiple DNA repair genes and various pathways including base excision repair (BER) [61], nucleotide excision repair (NER) [62], homologous recombination (HR) repair [63], and mismatch repair (MMR) [64]. Mutations in DNA repair genes involved in these pathways can lead to a failure in DNA repair, which in turn allows subsequent mutations to accumulate. An example is a hereditary mutation of MMR genes (i.e., Lynch syndrome) [65] leading to a high lifetime risk of colonic carcinoma [66] and other cancers.

### 3.7.2 Tumor Stroma

In general, neoplasms are composed of two main components, the tumor parenchyma (i.e., the main tumor mass composed of proliferating tumor cells) and the tumor stroma (i.e., the microenvironment around the tumor). In addition to all the molecular changes that occur within a cancer cell, the tumor stroma is thought to also undergo alterations [67] and contribute both to tumorigenesis and resistance to chemotherapeutic agents. Stroma consists of a supportive framework, which includes the basement membrane, fibroblasts, inflammatory cells, vascular cells, and the extracellular matrix (ECM) [68]. Fibroblasts represent the principal cellular stromal component [69]. Normal fibroblasts are usually inactive within the ECM. Once they are recruited and activated around the tumor cells, they are referred to as cancer-associated fibroblasts (CAFs) [70] or tumor-associated fibroblasts and have different activity (e.g., higher proliferative activity) in comparison to normal fibroblasts. Some authors report that tumor cells are thought to cause the proliferation of CAFs and secretion of collagen (i.e., the reactive stroma hypothesis). Others believe that tumor cells dedifferentiate into CAFs and then secrete collagens (i.e., the tumor-induced change hypothesis). Under the microscope, this distinctive fibrous response (or desmoplastic reaction) can vary from a very dense hyalinised stroma with a minimal cellular infiltrate to a predominantly cellular stroma with little collagenous tissue. Certain malignancies are known to have a more pronounced desmoplastic reaction (e.g., breast cancer [71] or pancreatic cancer [72]) than others (e.g., colon cancer).

Angiogenesis is another important feature of cancer progression [73]. It begins when neoplastic cells secrete molecules (e.g., VEGF) [74] into the surrounding stroma activating vascular endothelial cells, which in turn produce enzymes (matrix metalloproteinases) to breakdown the ECM. This permits migration and division into networks of blood vessels. Multiple proteins have been identified as angiogenic activators [73] including VEGF, FGF, TGF beta, EGF, PDGF, etc. For many years, researchers have concentrated on the malignant tumor cells as the main target of cancer therapy with the majority of chemotherapeutics either selectively killing tumor cells (cytotoxic) or restricting tumor growth. In recent years, however, the tumor stroma is now being pursued as an anti-cancer target itself [75]. Consequently, stromal molecules are being targeted by inhibitors, e.g., CAF inhibitors [76] (e.g., PDGF receptor inhibitors [77] like Dasatinib [78]) or angiogenesis inhibitors [79] (e.g., VEGF receptor inhibitors [80, 81] like Bevacizumab [82]).

### 3.7.3 Tumor Tissue Heterogeneity

Intratumoral heterogeneity (i.e., coexistence of distinct clones within a tumor) has been the subject of much debate. Authors have shown that tumor subclones may differ from the original clone in various characteristics [83], for example in invasiveness [84], in metastatic potential [85], and in response to chemotherapy [86]. Also it is thought that primary and metastatic lesions can also be genetically distinct [87, 88], which again may affect patient therapeutic options. Two concepts have been proposed as causes for tumor phenotypic heterogeneity, "the clonal evolution theory" (or "the conventional cancer theory") [83] and "the cancer stem cell theory" [89]. Simplistic illustrations comparing each of these theories are outlined in Fig. 3.6. The

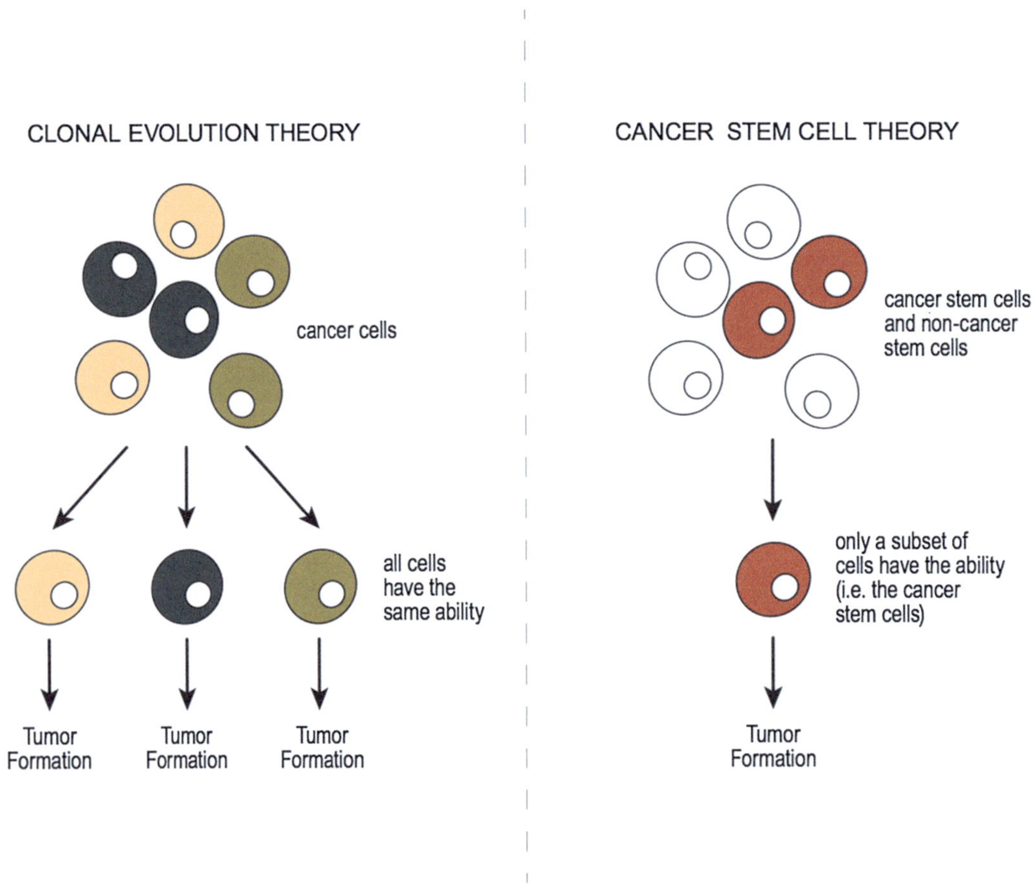


**Fig. 3.6** The two main theories behind tumor phenotypic heterogeneity

conventional cancer theory proposes that any cancer cell within a tumor can accumulate mutations and initiate tumor growth. In this multistep theory, a series of clonal expansions occurs, each of which is triggered by the acquisition of further mutations allowing mutant clone to expand [83]. In contrast, the stem cell theory suggests that among all cancerous cells within a tumor, only a subset of cells act as stem cells [89]. Normal stem cells are cells that can reproduce themselves and give rise to other kinds of cells. Cancer stem cells (CSC) are therefore cancerous cells that possess the same characteristics as are associated with normal stem cells. Research has shown that cancerous cells within a tumor may, therefore, not all be the same, in that some may be CSCs while others may be non-CSCs [90, 91].

Intratumor heterogeneity can also involve other cell types (i.e., inflammatory cells, fibroblasts, vascular endothelial cells, smooth muscle cells, and the ECM) together with the tumor cells. This can have profound clinical implications for disease progression, diagnosis, and therapeutic responses. Tumor core biopsies sample only small regions of tissue and therefore, as a researcher or histopathologist, adequate sampling of different tumor regions to sufficiently show intratumor clonal heterogeneity is a challenge.

### 3.7.4 Epithelial-Mesenchymal Transition

Epithelial cells and mesenchymal cells differ greatly in phenotype as well as in function. Epithelial cells are closely connected to each other by tight junctions, have polarity, and are bound by a basal lamina at their basal surface. Mesenchymal cells, on the other hand, lack polarization, have a spindle-shaped morphology and interact with each other through focal points. Epithelial cells express high levels of transmembrane proteins involved in cell–cell adhesion (E-cadherin [92, 93]) whereas mesenchymal cells express different proteins (e.g., fibronectin or vimentin). Epithelial–mesenchymal transition (EMT) is the process by which epithelial cells are believed to lose their normal epithelial "traits" (e.g., polarity [94] and cell–cell adhesion [95]) and gain the migratory and invasive properties of mesenchymal cells [96]. The EMT process is known to be essential for numerous developmental processes including the formation of the mesodermal germ layer (i.e., Type I EMT). EMT has also been shown to occur in relation to normal processes such as fibrosis and wound healing (i.e., Type II EMT) and in the initiation of metastasis for cancer progression (i.e., Type III EMT).

In regard to metastases, EMT is thought to occur where carcinoma cells in primary tumors lose their cell–cell adhesion (mediated by loss of E-cadherin), break through the basement membrane and enter the bloodstream. Later, when these circulating tumor cells exit the bloodstream to form metastases, they undergo mesenchymal–epithelial transition (MET) for clonal outgrowth at these metastatic sites [97]. It is, therefore, believed by some authors that EMT and MET form the initiation and completion of the invasion–metastasis process of tumors [97, 98]. Loss of E-cadherin expression has been documented in many malignancies [99, 100] for example, invasive lobular breast carcinoma is negative for E-cadherin [101–103] by immunohistochemical testing. The role of EMT in metastases remains controversial, however, as while a large body of the literature is available in studies involving cell lines and animal models, little is available from studies involving human tumor samples. EMT in tumor cells is a transient process [104], therefore, once invasion by the metastatic cell has occurred, its mesenchymal features disappear. The heterogeneity of tumors, as mentioned previously, also makes it difficult to distinguish tumor cells that have undergone EMT, from stromal cells that show a mesenchymal phenotype (e.g., using standard EMT markers such as vimentin) [105, 106]. Some authors also argue that distant metastases originating from a variety of primary carcinomas, show an epithelial phenotype histologically, raising the possibility that tumor cells may disseminate without switching to a mesenchymal phenotype [107]. In addition, the finding of circulating tumor cells (i.e., tumor cells found within the bloodstream) in clusters/glandular forms (rather than singly) [108] raises the possibility that cells within these groups may not undergo loss of cell–cell adhesion and may be protected from a loss of basement membrane attachment. Finally, histological features of some tumors, such as paradoxical differentiation seen in invasive cervical carcinoma (i.e., where invasive tongues of tumor show less pleomorphism than the surface tumor) [109], loss of E-cadherin expression in lobular carcinoma in situ [110] or the expression of vimentin in localized renal cell carcinoma [111], are all examples challenging EMT as a theory for cancer metastases.

## 3.8 Stage of Disease

Once cancer has been diagnosed, the extent to which the cancer has spread must be assessed. This process is termed cancer staging and features such as tumor size, depth of invasion, lymph node status, and the presence or absence of metastases are examined. The tumor being examined is assigned a stage based on the results of these features. A number from I to IV is generally assigned, with I corresponding to an isolated cancer and IV corresponding to a tumor that has metastasized. Stage has prognostic significance where lower stage tumors are associated with longer patient survival and vice versa. Many staging systems are available, including the Ann Arbor staging classification [112, 113] commonly used to stage lymphomas, or the FIGO (International Federation of Gynecology and Obstetrics) staging system [114, 115] used to stage gynecological cancers. Additionally, most childhood cancers are staged using the staging criteria of the Children's Oncology Group (COG) [116], which conducts pediatric clinical trials worldwide.

The "TNM" system is one of the most widely used cancer staging systems and was developed by the American Joint Committee on Cancer (AJCC) and also used by the Union for International Cancer Control (UICC). It is based on three key features: the size of the primary tumor (T), whether the tumor has spread to nearby lymph nodes (N), and whether metastases (M) are present. Each tumor type has their own staging system based on multiple research results and clinical trial results. Clinical stage (indicated by "c") is based on all of the available information obtained prior to surgical excision (e.g., physical examination, radiological results, etc.). Pathologic stage (represented by "p") includes information gathered following histological examination (e.g., margin status, tumor type, etc.). Because they use different criteria, clinical stage and pathologic stage often differ, however, at the same time complement each other.

## 3.9 Summary

In summary, pathology is an important medical discipline that requires an in-depth knowledge not only of normal tissues but also of diseased tissues. In this chapter, everything from the beginning of cell fertilization to the final process of tumor staging has been discussed. The reader should now be better able to understand basic pathological terms mentioned in the forthcoming chapters or can return to this chapter and use it as a reference guide.

## References


1. Sadler TW. Langman's medical embryology, vol. 1. Lippincott Williams & Wilkins; 2006.
2. Moore KL. Before we are born: essentials of embryology and birth defects. 8th ed. Elsevier; 2012.
3. Lockshin RA, Zakeri Z. Cell death in health and disease. J Cell Mol Med. 2007;11(6):1214–24.
4. Rubin R, Strayer DS, Rubin E. Rubin's pathology: clinicopathologic foundations of medicine. 6th ed. Philadelphia: Lippincott, Williams & Wilkins; 2012.
5. Kumar V, Abbas AK, Fausto N, Aster J. Robbins and Cotran pathologic basis of disease. 8th ed. Elsevier Saunders; 2012.
6. Nowak KJ, Davies KE. Duchenne muscular dystrophy and dystrophin: pathogenesis and opportunities for treatment. EMBO Rep. 2004;5(9):872–6.
7. Nadaj-Pakleza A, Lusakowska A, Sułek-Piątkowska A, Krysa W, Rajkiewicz M, Kwieciński H,

Kamińska A. Muscle pathology in myotonic dystrophy: light and electron microscopic investigation in eighteen patients. Folia Morphol (Warsz). 2011;70(2):121–9.
8. Harber MP, Konopka AR, Douglass MD, Minchev K, Kaminsky. LA, Trappe TA, Trappe S. Aerobic exercise training improves whole muscle and single myofiber size and function in older women. Am J Physiol Regul Integr Comp Physiol. 2009;297(5):1452–9.
9. Russell B, Motlagh D, Ashley WW. Form follows function: how muscle shape is regulated by work. J Appl Physiol (1985). 2000;8(3):1127–32.
10. Devereux RB, Pickering TG, Harshfield GA, Kleinert HD, Denby L, Clark L. Left ventricular hypertrophy in patients with hypertension: importance of blood pressure response to regularly recurring stress. Circulation. 1983;68(3):470–6.
11. Anversa P, Palackal T, Sonnenblick EH, Olivetti G, Capasso JM. Hypertensive cardiomyopathy. Myocyte nuclei hyperplasia in the mammalian rat heart. J Clin Invest. 1990;85(4):994–7.
12. Battersby S, Anderson TJ. Proliferative and secretory activity in the pregnant and lactating human breast. Virchows Arch A Pathol Anat Histopathol. 1988;413(3):189–96.
13. Furness S, Roberts H, Marjoribanks J, Lethaby A, Hickey M, Farquhar C. Hormone therapy in postmenopausal women and risk of endometrial hyperplasia. Cochrane Database Syst Rev. 2009;2: CD000402.
14. Lacey JV, Chia VM. Endometrial hyperplasia and the risk of progression to carcinoma. Maturitas. 2009;63(1):39–44.
15. Flejou JF. Barrett's oesophagus: from metaplasia to dysplasia and cancer. Gut. 2005;54(Suppl 1):i6–12.
16. Spechler SJ. Barrett's oesophagus: diagnosis and management. Baillieres Best Pract Res Clin Gastroenterol. 2000;14(5):857–79.
17. Mueller J, Werner M, Siewert JR. Malignant progression in Barrett's esophagus: pathology and molecular biology. Recent Results Cancer Res. 2000;155:29–41.
18. Fulda S, Gorman AM, Hori O, Samali A. Cellular stress responses: cell survival and cell death. Int J Cell Biol. 2010;214074.
19. Golstein P, Kroemer G. Cell death by necrosis: towards a molecular definition. Trends Biochem Sci. 2007;32(1):37–43.
20. Elmore S. Apoptosis: a review of programmed cell death. Toxicol Pathol. 2007;35(4):495–516.
21. Vanlangenakker N, Vanden Berghe T, Krysko DV, Festjens N, Vandenabeele P. Molecular mechanisms and pathophysiology of necrotic cell death. Curr Mol Med. 2008;8(3):207–20.
22. Samali A, Zhivotovsky B, Jones D, Nagata S, Orrenius S. Apoptosis: cell death defined by caspase activation. Cell Death Differ. 1999;6(6):495–6.
23. Eskelinen EL. New insights into the mechanisms of macroautophagy in mammalian cells. Int Rev Cell Mol Biol. 2008;266:207–47.
24. Christopherson WM. Dysplasia, carcinoma in situ, and microinvasive carcinoma of the uterine cervix. Hum Pathol. 1977;8(5):489–501.
25. Nelson JH, Averette HE, Richart RM. Dysplasia, carcinoma in situ, and early invasive cervical carcinoma. CA Cancer J Clin. 1984;34(6):306–27.
26. Haggitt RC. Barrett's esophagus, dysplasia, and adenocarcinoma. Hum Pathol. 1994;25(10):982–93.
27. Maniar KP, Nayar R. HPV-related squamous neoplasia of the lower anogenital tract: an update and review of recent guidelines. Adv Anat Pathol. 2014;21(5):341–58.
28. International Agency for Research on Cancer. WHO classification of tumours of female reproductive organs. 4th ed. vol. 6. WHO; 2014.
29. Frick HC. Management of noninvasive carcinoma of the cervix. Surg Clin North Am. 1978;58(1):55–60.
30. Barsky H, Siegal GP, Jannotta F, Liotta LA. Loss of basement membrane components by invasive tumors but not by their benign counterparts. Lab Invest. 1983;49(2):140–7.
31. Brito JP, Hay ID, Morris JC. Low risk papillary thyroid cancer. BMJ. 2014;348:3045.
32. International Agency for Research on Cancer. Pathology and genetics of tumours of endocrine organs. 3rd ed. vol. 8. WHO; 2004.
33. Mills SE. Histology for pathologists. 4th ed. Philadelphia: Lippincott, Williams & Wilkins; 2012.
34. Jogi A, Vaapil M, Johansson M, Påhlman S. Cancer cell differentiation heterogeneity and aggressive behavior in solid tumors. Ups J Med Sci. 2012;117(2):217–24.
35. Epstein JI, Allsbrook WC Jr, Amin MB, Egevad LL, ISUP Grading Committee. The 2005 international society of urological pathology (ISUP) consensus conference on gleason grading of prostatic carcinoma. Am J Surg Pathol. 2005;29(9):1228–42.
36. Humphrey PA. Gleason grading and prognostic factors in carcinoma of the prostate. Mod Pathol. 2004;17(3):292–306.
37. Elston CW, Ellis IO. Pathological prognostic factors in breast cancer. I. The value of histological grade in breast cancer: experience from a large study with long-term follow-up. Histopathology. 1991;19 (5):403–10.
38. Frkovic-Grazio S, Bracko M. Long term prognostic value of Nottingham histological grade and its components in early (pT1N0M0) breast carcinoma. J Clin Pathol. 2002;55(2):88–92.
39. Medri L, Volpi A, Nanni O, Vecci AM, Mangia A, Schittulli F, Padovani F, Giunchi DC, Vito A, Amadori D, Paradiso A, Silvestrini R. Prognostic relevance of mitotic activity in patients with node-negative breast cancer. Mod Pathol. 2003;16 (11):1067–75.

40. International Agency for Research on Cancer. WHO classification of tumours of soft tissue and bone. 4th ed. vol. 5. WHO; 2013.
41. Lanng C, Eriksen BO, Hoffmann J. Lipoma of the breast: a diagnostic dilemma. Breast. 2004;13 (5):408–11.
42. Weedon D. Weedon's skin pathology. 3rd ed. London: Churchill Livingstone; 2009.
43. International Agency for Research on Cancer. WHO classification of tumours of the urinary system and male genital organs. 3rd ed. vol. 7. WHO; 2004.
44. Yeo CJ. Tumor suppressor genes: a short review. Surgery. 1999;125(4):363–6.
45. Vicente-Duenas C, Romero-Camarero I, Cobaleda C, Sanchez-Garcia I. Function of oncogenes in cancer development: a changing paradigm. EMBO J. 2013;32(11):1502–13.
46. Helleday T, Petermann E, Lundin C, Hodgson B, Sharma RA. DNA repair pathways as targets for cancer therapy. Nat Rev Cancer. 2008;8 (3):193–204.
47. Tsai AG, Lieber MR. Mechanisms of chromosomal rearrangement in the human genome. BMC Genom. 2010;11(Suppl 1):S1.
48. Nambiar M, Kari V, Raghavan SC. Chromosomal translocations in cancer. Biochim Biophys Acta. 2008;1786(2):139–52.
49. Van Hemel JO, Eussen HJ. Interchromosomal insertions. Identification of five cases and a review. Hum Genet. 2000;107(5):415–32.
50. Vermeulen K, Van Bockstaele DR, Berneman ZN. The cell cycle: a review of regulation, deregulation and therapeutic targets in cancer. Cell Prolif. 2003;36(3):131–49.
51. Branzei D, Foiani M. Regulation of DNA repair throughout the cell cycle. Nat Rev Mol Cell Biol. 2008;9(4):297–308.
52. Collins K, Jacks T, Pavletich NP. The cell cycle and cancer. Proc Natl Acad Sci USA. 1997;94 (7):2776–8.
53. Kastan MB, Bartek J. Cell-cycle checkpoints and cancer. Nature. 2004;432(7015):316–23.
54. Pylayeva-Gupta Y, Grabocka E, Bar-Sagi D. RAS oncogenes: weaving a tumorigenic web. Nat Rev Cancer. 2011;11(11):761–74.
55. Bouchard C, Staller P, Eilers M. Control of cell proliferation by Myc. Trends Cell Biol. 1998;8 (5):202–6.
56. Lim S, Kaldis P. Cdks, cyclins and CKIs: roles beyond cell cycle regulation. Development. 2013;140(15):3079–93.
57. Chial H. Proto-oncogenes to cancer. Nat Educ. 2008;1(1).
58. Vousden KH, Prives C. Blinded by the light: the growing complexity of p53. Cell. 2009;137(3):413–31.
59. Song MS, Salmena L, Pandolfi P. The functions and regulation of the PTEN tumour suppressor. Nat Rev Mol Cell Biol. 2012;13(5):283–96.
60. Payne SR, Kemp CJ. Tumor suppressor genetics. Carcinogenesis. 2005;26(12):2031–45.
61. Zharkov DO. Base excision DNA repair. Cell Mol Life Sci. 2008;65(10):1544–65.
62. Marteijn JA, Lans H, Vermeulen W, Hoeijmakers JH. Understanding nucleotide excision repair and its roles in cancer and ageing. Nat Rev Mol Cell Biol. 2014;15(7):465–81.
63. Li X, Heyer WD. Homologous recombination in DNA repair and DNA damage tolerance. Cell Res. 2008;18(1):99–113.
64. Jiricny J. The multifaceted mismatch-repair system. Nat Rev Mol Cell Biol. 2006;7(5):335–46.
65. Silva FC, Valentin MD, Ferreira Fde O, Carraro DM, Rossi BM. Mismatch repair genes in lynch syndrome: a review. Sao Paulo Med J. 2009;127 (1):46–51.
66. Bellcross CA, Bedrosian SR, Daniels E, Duquette D, Hampel H, Jasperson K, Joseph DA, Kaye C, Lubin I, Meyer LJ, Reyes M, Scheuner MT, Schully SD, Senter L, Stewart SL, St Pierre J, Westman J, Wise P, Yang VW, Khoury MJ. Implementing screening for lynch syndrome among patients with newly diagnosed colorectal cancer: summary of a public health/clinical collaborative meeting. Genet Med. 2012;14(1):152–62.
67. Weinberg RA. Coevolution in the tumor microenvironment. Nat Genet. 2008;40(5):494–5.
68. Pietras K, Ostman A. Hallmarks of cancer: interactions with the tumor stroma. Exp Cell Res. 2010;316 (8):1324–31.
69. Kalluri R, Zeisberg M. Fibroblasts in cancer. Nat Rev Cancer. 2006;6(5):392–401.
70. Ostman A, Augsten M. Cancer-associated fibroblasts and tumor growth—bystanders turning into key players. Curr Opin Genet Dev. 2009;19(1):67–73.
71. Iacobuzio-Donahue CA, Argani P, Hempen PM, Jones J, Kern SE. The desmoplastic response to infiltrating breast carcinoma: gene expression at the site of primary invasion and implications for comparisons between tumor types. Cancer Res. 2002;62 (18):5351–7.
72. Apte MV, Park S, Phillips PA, Santucci N, Goldstein D, Kumar R. Desmoplastic reaction in pancreatic cancer: role of pancreatic stellate cells. Pancreas. 2004;29(3):179–87.
73. Weis SM, Cheresh DA. Tumor angiogenesis: molecular pathways and therapeutic targets. Nat Med. 2011;17(11):1359–70.
74. Goel HL, Mercurio AM. VEGF targets the tumour cell. Nat Rev Cancer. 2013;13(12):871–82.
75. Ahmed F, Steele JC, Herbert JM, Steven NM, Bicknell R. Tumor stroma as a target in cancer. Curr Cancer Drug Targets. 2008;8(6):447–53.
76. Loeffler M, Kruger JA, Niethammer AG, Reisfeld RA. Targeting tumor-associated fibroblasts improves cancer chemotherapy by increasing intratumoral drug uptake. J Clin Invest. 2006;116 (7):1955–62.

77. Ball SG, Shuttleworth A, Kielty CM. Inhibition of platelet-derived growth factor receptor signalling regulates Oct 4 and nanog expression, cell shape, and mesenchymal stem cell potency. Stem Cells. 2012;30(3):548–60.
78. Haubeiss S, Schmid JO, Muerdter TE, Sonnenberg M, Friedel G, van der Kuip H, Aulitzky WE. Dasatinib reverses cancer-associated fibroblasts (CAFs) from primary lung carcinomas to a phenotype comparable to that of normal fibroblasts. Mol Cancer. 2010;9:168.
79. El-Kenawi AE, El-Remessy AB. Angiogenesis inhibitors in cancer therapy: mechanistic perspective on classification and treatment rationales. Br J Pharmacol. 2013;170(4):712–29.
80. Ellis LM, Hicklin DJ. VEGF-targeted therapy: mechanisms of anti-tumour activity. Nat Rev Cancer. 2008;8(8):579–91.
81. Sitohy B, Nagy JA, Dvorak HF. Anti-VEGF/VEGFR therapy for cancer: reassessing the target. Cancer Res. 2012;72(8):1909–14.
82. McCormack PL, Keam SJ. Bevacizumab: a review of its use in metastatic colorectal cancer. Drugs. 2008;68(4):487–506.
83. Nowell PC. The clonal evolution of tumor cell populations. Science. 1976;194(4260):23–8.
84. Fidler IJ. Tumor heterogeneity and the biology of cancer invasion and metastasis. Cancer Res. 1978;38 (9):2651–60.
85. Spreafico F, Mantovani A, Giavazzi R, Conti G, Anaclerio A. Metastatic potential of metastases, tumor cell heterogeneity, and therapeutic implications. Recent Results Cancer Res. 1982;80:1–8.
86. Calderwood SK. Tumor heterogeneity, clonal evolution, and therapy resistance: an opportunity for multitargeting therapy. Discov Med. 2013;15 (82):188–94.
87. Minn AJ, Kang Y, Serganova I, Gupta GP, Giri DD, Doubrovin M, Ponomarev V, Gerald WL, Massague J. Distinct organ-specific metastatic potential of individual breast cancer cells and primary tumors. J Clin Invest. 2005;115(1):44–55.
88. Almendro V, Kim HJ, Cheng YK, Gönen M, Itzkovitz S, Argani P, van Oudenaarden A, Sukumar S, Michor F, Polyak K. Genetic and phenotypic diversity in breast tumor metastases. Cancer Res. 2014;74(5):1338–48.
89. Marotta LL, Polyak K. Cancer stem cells: a model in the making. Curr Opin Genet Dev. 2009;19 (1):44–50.
90. Cho RW, Clarke MF. Recent advances in cancer stem cells. Curr Opin Genet Dev. 2008;18(1):48–53.
91. Quintana E, Shackleton M, Sabel MS, Fullen DR, Johnson TM, Morrison SJ. Efficient tumour formation by single human melanoma cells. Nature. 2008;456(7222):593–8.
92. Takeichi M. Cadherins: a molecular family important in selective cell-cell adhesion. Annu Rev Biochem. 1990;59:237–52.
93. Berx G, Staes K, van Hengel J, Molemans F, Bussemakers MJG, van Bokhoven A. Cloning and characterization of the human invasion suppressor gene E-cadherin (CDH1). Genomics. 1995;26 (2):281–9.
94. Royer C, Lu X. Epithelial cell polarity: a major gatekeeper against cancer? Cell Death Differ. 2011;18(9):1470–7.
95. Frixen UH, Behrens J, Sachs M, Eberle G, Voss B, Warda A, Lochner D, Birchmeier W. E-cadherin-mediated cell-cell adhesion prevents invasiveness of human carcinoma cells. J Cell Biol. 1991;113(1):173–85.
96. Kalluri R, Weinberg RA. The basics of epithelial-mesenchymal transition. J Clin Invest. 2009;119(6):1420–8.
97. Brabletz T. EMT and MET in metastasis: where are the cancer stem cells? Cancer Cell. 2012;22 (6):699–701.
98. Alderton GK. Metastasis: epithelial to mesenchymal and back again. Nat Rev Cancer. 2013;13(1):3.
99. Tang A, Amagai M, Granger LG, Stanley JR, Udey MC. Adhesion of epidermal Langerhans cells to keratinocytes mediated by E-cadherin. Nature. 1993;361(6407):82–5.
100. Handschuh G, Candidus S, Luber B, Reich U, Schott C, Oswald S. Tumour-associated E-cadherin mutations alter cellular morphology, decrease cellular adhesion and increase cellular motility. Oncogene. 1999;18(30):4301–12.
101. Singhai R, Patil VW, Jaiswal SR, Patil SD, Tayade MB, Patil AV. E-Cadherin as a diagnostic biomarker in breast cancer. N Am J Med Sci. 2011;3 (5):227–33.
102. Wahed A, Connelly J, Reese T. E-cadherin expression in pleomorphic lobular carcinoma: an aid to differentiation from ductal carcinoma. Ann Diagn Pathol. 2002;6(6):349–51.
103. Moll R, Mitze M, Frixen UH, Birchmeier W. Differential loss of E-cadherin expression in infiltrating ductal and lobular breast carcinomas. Am J Pathol. 1993;143(6):1731–42.
104. Krebs MG, Metcalf RL, Carter L, Brady G, Blackhall FH, Dive C. Molecular analysis of circulating tumour cells-biology and biomarkers. Nat Rev Clin Oncol. 2014;11(3):129–44.
105. Armstrong AJ, Marengo MS, Oltean S, Kemeny G, Bitting RL, Turnbull JD, Herold CI, Marcom PK, George DJ, Garcia-Blanco MA. Circulating tumor cells from patients with advanced prostate and breast cancer display both epithelial and mesenchymal markers. Mol Cancer Res. 2011;9(8):997–1007.
106. Balasubramanian P, Lang JC, Jatana KR, Miller B, Ozer E, Old M, Schuller DE, Agrawal A, Teknos TN, Summers TA, Lustberg MB, Zborowski M, Chalmers JJ. Multiparameter analysis, including EMT markers, on negatively enriched blood samples from patients with squamous cell carcinoma of the head and neck. PLoS ONE. 2012;7 (7):42048.

107. Brabletz T, Jung A, Reu S, Porzner M, Hlubek F, Kunz-Schughart LA, Knuechel R, Kirchner T. Variable beta-catenin expression in colorectal cancers indicates tumor progression driven by the tumor environment. Proc Natl Acad Sci USA. 2001;98(18):10356–61.
108. Aceto N, Bardia A, Miyamoto DT, Donaldson MC, Wittner BS, Spencer JA, Yu M, Pely A, Engstrom A, Zhu H, Brannigan BW, Kapur R, Stott SL, Shioda T, Ramaswamy S, Ting DT, Lin CP, Toner M, Haber DA, Maheswaran S. Circulating tumor cell clusters are oligoclonal precursors of breast cancer metastasis. Cell. 2014;158(5):1110–22.
109. Burghardt E, Holzer E. Diagnosis and treatment of microinvasive carcinoma of the cervix uteri. Obstet Gynecol. 1977;49(6):641–53.
110. Mastracci TL, Tjan S, Bane AL, O'Malley FP, Andrulis IL. E-cadherin alterations in atypical lobular hyperplasia and lobular carcinoma in situ of the breast. Mod Pathol. 2005;18(6):741–51.
111. Sabo E, Miselevich I, Bejar J, Segenreich M, Wald M, Moskovitz B, Nativ O. The role of vimentin expression in predicting the long-term outcome of patients with localized renal cell carcinoma. Br J Urol. 1997;80(6):864–8.
112. Armitage JO. Staging non-Hodgkin lymphoma. CA Cancer J Clin. 2005;55(6):368–76.
113. Carbone PP, Kaplan HS, Musshoff K, Smithers DW, Tubiana M. Report of the committee on Hodgkin's disease staging classification. Cancer Res. 1971;31(11):1860–1.
114. Odicino F, Pecorelli S, Zigliani L, Creasman WT. History of the FIGO cancer staging system. Int J Gynaecol Obstet. 2008;101(2):205–10.
115. Benedet JL, Pecorelli S. Why cancer staging? FIGO 26th annual report on the results of treatment in gynecological cancer. Int J Gynaecol Obstet. 2006;95(Suppl 1):S3.
116. Castleberry RP, Shuster JJ, Smith EI. The pediatric oncology group experience with the international staging system criteria for neuroblastoma. Member institutions of the pediatric oncology group. J Clin Oncol. 1994;12(11):2378–81.

# Basic Techniques in Molecular Pathology

4

Matthew D. Stachler

## 4.1 Introduction

Molecular pathology is a rapidly changing field. What was cutting edge four or five years ago is now considered outdated. Many new molecular techniques are focused around the general technology of massively parallel sequencing (MPS) (or so-called "next generation sequencing"). While this chapter will go over several more traditional techniques, it will pay particular attention to MPS as it presents several unique challenges. With modern highly multiplex technologies, whether one is planning on analyzing DNA, RNA, or proteins; a huge amount of data is often generated. While this can lead to many new discoveries, the sheer amount of data can be overwhelming for someone not used to dealing with these kinds of datasets. Due to this, it is often necessary to perform this kind of research in a collaborative setting. Working in a group where one is a computational expert and someone else is more knowledgeable on the actual biology of what is being studied can be a very successful and rewarding strategy. This chapter is intended to be an introduction to several commonly used molecular techniques (focusing on MPS) and will go into significant detail covering the requirements and tissue selection criteria for these tests.

## 4.2 Material (Tissue) Selection

Molecular analysis requires the input of nucleic acid (or protein), which must come from some kind of source. While this may seem like the simplest part of molecular pathology, in reality it is by far the most important aspect. Without a precise knowledge of what exactly is being tested, it is impossible to interpret and know the biological significance of any results. In general, molecular pathology usually deals with human tissue samples, however, this is not always the case. Available sources of nucleic acid can usually be broken down into several categories. These include, but are not necessarily limited to: cell culture cells, tissue, or body fluids. When dealing with human tissue it is generally acquired either as fresh tissue or formalin-fixed and paraffin-embedded (FFPE) tissue. As explained later in the chapter, the process of purifying nucleic acid from FFPE material introduces a couple of complicating factors.

A major emphasis of molecular pathology focuses on cancer. Molecular analysis of tumor samples introduces another complicating factor, tumor cell percentage. Any solid tumor is composed of a varying degree of neoplastic cells, inflammatory cells, and stromal cells (Fig. 4.1). When analyzing a tumor, one is usually only interested in a specific cellular component. For

M.D. Stachler (✉)
Harvard Medical School, Brigham and Women's Hospital, 75 Francis St., Boston, MA 02115, USA
e-mail: MStachler@partners.org


M. Loda et al. (eds.), *Pathology and Epidemiology of Cancer*,
DOI 10.1007/978-3-319-35153-7_4

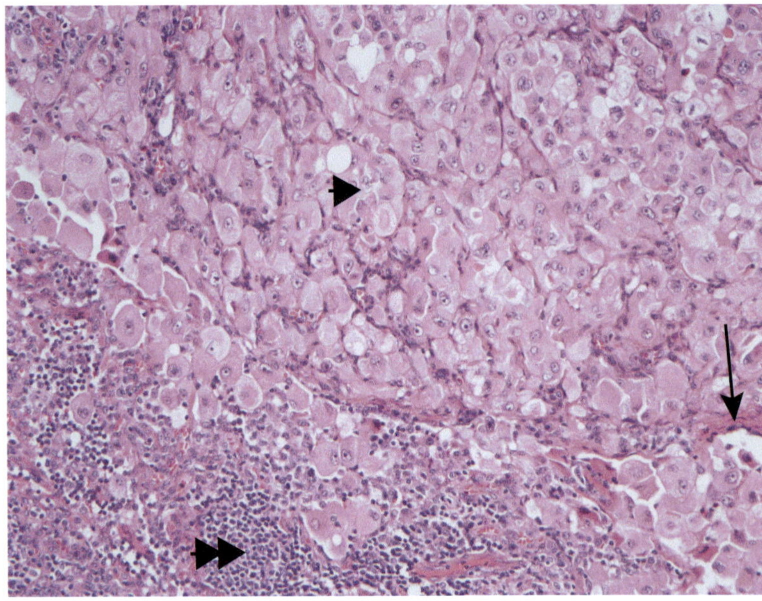

**Fig. 4.1** Hematoxylin and eosin (H&E) picture of a tumor showing neoplastic, inflammatory, and stromal cells. H&E stained section of a tumor showing neoplastic cells (*arrow head*), inflammatory cells (*double arrow head*), and stromal cells (*arrow*) that compromise the typical tumor environment

example, if trying to determine if a colon cancer sample contains a mutation in the KRAS gene, one is only interested in looking at the neoplastic cells and not the associated inflammatory cells or stromal fibroblasts. It is important to have a general idea of what percentage of the tumor sample is actually comprised of these neoplastic cells. If the percentage of neoplastic cells is too small, it will be increasingly difficult to find a somatic mutation as it will be drowned out by the DNA of the non-neoplastic cells. The lower the tumor cell percentage the lower the percentage of a variant allele will be. For example, if a heterozygous mutation is present in a tumor that contains 30 % neoplastic cells only 15 % of the total DNA will contain that mutation (assuming the tumor has a normal diploid content). Different assays have a different level of minimum tumor percentage required for accurate detection of the mutant allele. While newer techniques like MPS and digital droplet pcr generally have a better (able to detect a lower allele fractions) detection than traditional methods such as Sanger sequencing, when initially performing an assay it is important to perform validation testing to determine the limit of detection as this will be highly dependent on how the assay is set up and ran.

### 4.2.1 Cell Culture

Purifying nucleic acid from cells grown in tissue culture is straightforward and will provide high quality, unfragmented DNA or RNA (assuming cells are not fixed in formalin). In addition, since cells are usually grown as a pure cell line, the percentage of DNA actually coming from the cells of interest should be 100 % assuming no others cells contaminating the culture. The difficulty in using nucleic acid from cell culture revolves more around having a thorough knowledge of what cells are actually growing in the culture. As older cell lines have undergone many passages and have been shared from laboratory to laboratory, it is not always certain that the cell line being used is actually the right one. A study in 2008 reported that 23 % of the cell lines tested had discrepant TP53 mutations [1]. Therefore, it is suggested to confirm the molecular traits of cell lines before they are used in crucial experiments [2, 3]. In addition to cell mix up, every time a cell replicates there is a chance for a mutation to occur. With some cell lines having been passaged for decades, there is a high likelihood that the cell lines of today contain many more alterations than they originally did.

### 4.2.2 Direct Human Samples

#### 4.2.2.1 Fluids

Body fluids can often be a good source for human nucleic acid as they are often easier to access than solid tissues and require less invasive techniques to sample. Blood, more specifically the white blood cells within blood, can serve as a high quality source for germline DNA. Drawing a small tube of blood is easily performed in the clinic or outpatient setting and requires little equipment. It should be noted, however, that certain blood tubes may contain anti-clotting agents that can inhibit molecular analysis. It is generally recommended to use tubes containing ethylenediamine tetra acetic acid (EDTA) as an anticoagulant and to avoid tubes containing heparin as it is an inhibitor of many enzymes used in molecular analysis [4]. In addition to germline DNA, in the setting of leukemias, myelodysplastic syndrome, or now even circulating tumor cells from solid tumors, blood can also serve as a good source for somatic studies. Saliva, commercially available mouth rinses, and cheek swabs can also serve as a source of germline DNA. While these are often even easier to obtain than fresh blood, they can sometimes suffer from degradation and bacterial contamination [5–7]. The acquisition of nucleic acid from tumors can often be obtained from ascites or pleural effusions, often taken for therapeutic reasons from patients with cancer. However, a disadvantage in relation to these sample types is that the cells can suffer from varying degrees of degradation. Regardless of the source of fluid, if the intent is to obtain nucleic acid from tumor cells it is important for the fluid to also undergo cytologic analysis so it is known what cells are actually being captured.

#### 4.2.2.2 Tissue

##### Fresh/Frozen Tissue

Fresh tissue will usually provide high quality DNA since it is not subjected to the crosslinking formalin causes, however, nucleic acid must be isolated from it quickly, or the tissue needs to be preserved. Typically, to store fresh tissue it is frozen in either a cutting media such as optimum cutting temperature (OTC) compound or snap frozen. OTC allows for the cutting of frozen section slides, which can be used for histological analysis and for nucleic acid purification. Freezing fresh tissue allows for the long-term storage of high quality nucleic acid. Fresh tissue is typically taken at the time of surgery, requiring protocols in place to quickly select and freeze the tissue. Due to the need for clinical pathologic analysis, the amount of tissue that can be taken for molecular analysis is often limited, especially if not for clinical testing. This tissue is usually selected by gross examination only, which can lead to missing the desired tissue. Therefore, it is recommended to always confirm the type of tissue through histologic analysis.

##### Formalin-Fixed Paraffin-Embedded Tissue

The typical workflow for any pathology department involves receiving specimens, performing a gross examination/description, and then preparing the specimen for histologic analysis., as described in Chap. 2. As formalin fixation and paraffin embedding of tissue is a standard procedure, there is a wealth of archived material in almost all pathology departments. FFPE tissue has several advantages over fresh frozen tissue that make it so widely used. These include being able to store blocks at room temperature and proving better histology. Formalin fixes tissue by forming cross-links between proteins, DNA, and RNA. Unfortunately, the purification of nucleic acid from FFPE material causes shearing of the nucleic acid. This is a major drawback and can limit the ability to use FFPE material in molecular analysis, depending on how long of nucleic acid fragments are needed for testing. Figure 4.2a shows an agarose gel of DNA purified from both fresh frozen and FFPE tissue. The large smears in the FFPE columns show sheared DNA of varying lengths, when in comparison, the fresh frozen column has a large band of DNA at the top of the gel indicating large-sized DNA. Figure 4.2b shows an RNA electropherogram showing the size of RNA fragments from fresh and FFPE tissue. The FFPE RNA sample is predominately composed of small fragments, when in contrast, in the fresh sample, one can see large 18S and 28S ribosomal RNA

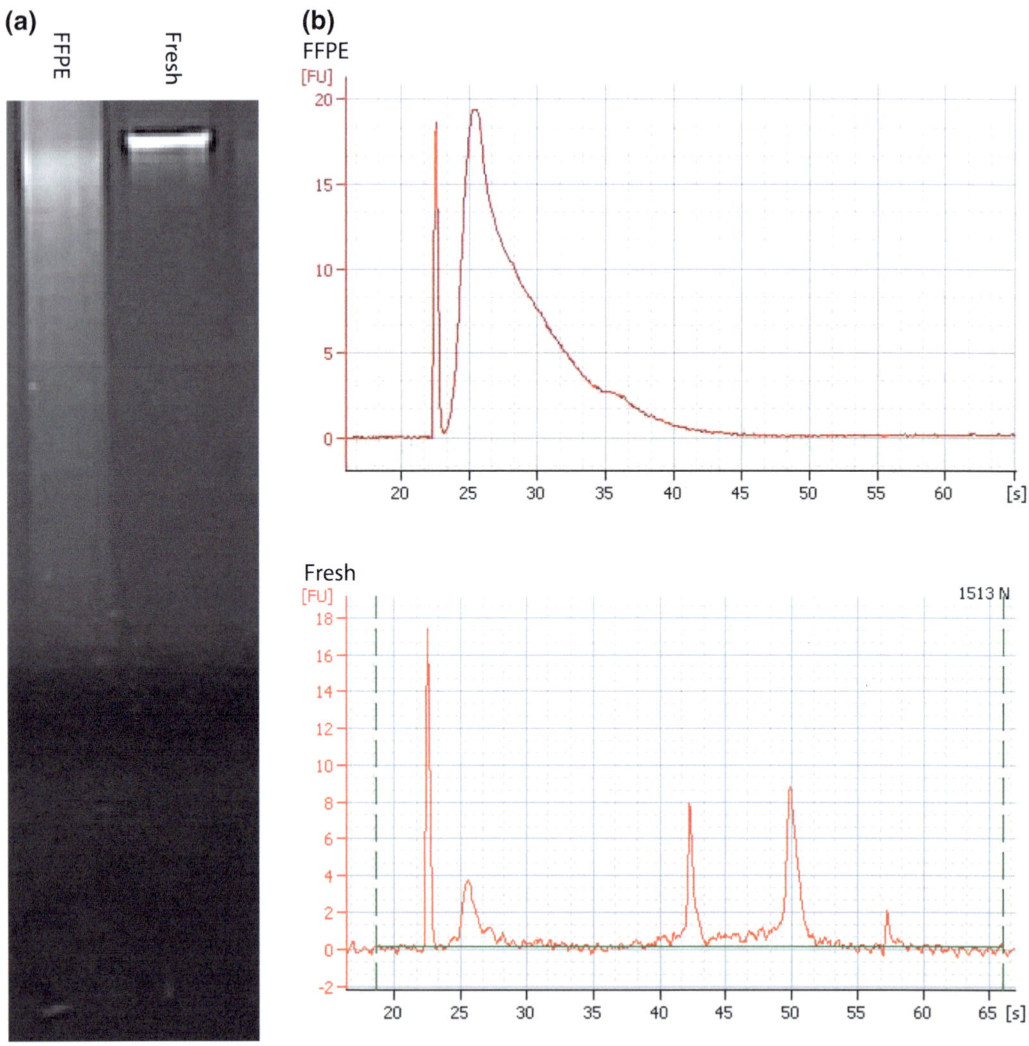


**Fig. 4.2** Comparison of fresh and formalin-fixed paraffin-embedded (FFPE) nucleic acid quality. **a** Agorose gel of FFPE and fresh DNA showing a degraded smear for FFPE and a large band for fresh DNA. **b** Agilent bioanalyzer electropherograms showing the size of FFPE and fresh RNA showing degraded low-sized fragments of FFPE RNA compared to RNA from fresh tissue with large 18S and 28S rRNA peaks

(rRNA) peaks. Despite these issues, FFPE tissue can often be used for a variety of testing [8].

### Other Fixatives Used in Pathology

There are several other fixatives apart from formalin that can be sometimes used in pathology departments. Alcohol fixation (with either ethanol or methanol) will provide higher quality DNA and RNA than formalin. Bouin's fixative or solution is a combination of formalin, picric acid, and acetic acid. While it provides excellent histology, the acid component degrades nucleic acid making most Bouin's-fixed tissue unusable for molecular analysis. The same is true for many of the acid-based decalcification solutions used in pathology [9].

## 4.3 Nucleic Acid Purification

Following specimen collection and storage, whether fresh frozen or FFPE, the material to use for nucleic acid purification must be selected.

Depending on how much material is available and how that material was stored, this process may need the assistance of a pathologist to review slides and select the appropriate area. Knowing that tumors can be very heterogeneous with widely varying amounts of actual tumor cells, it is important to think about proper material selection and isolation.

### 4.3.1 Type of Tissue Isolation

The least accurate, but usually the easiest way to get nucleic acid is to isolate it from a bulk piece of tissue. This is usually performed on frozen tissue, which can be pulverized and digested. The major drawback of using this technique is the lack of histological correlation and therefore, one does not know the amount of tumor in the sample.

Isolating nucleic acid by coring tissue blocks allows for fairly accurate selection of material when there are paired slides for histological analysis. It is also fairly fast and easy, as extra slides do not need to be made and scraped. When using a paired slide to select the areas to core, one is only visualizing the tissue in two dimensions (*x* and *y*-axis), while coring takes a lot of material vertically from the tissue block (*z*-axis). Therefore, it is a bit uncertain how much the tissue will change as one goes deeper into the block.

Macrodissection is possibly a more accurate method of obtaining the desired tissue. In this method, a series of sections are cut from a block and the first and last slide are mounted and stained for histological analysis. Using these two slides one can circle the area of interest to be dissected. Then, using a razor blade the region corresponding to the circled area can be scraped off the unstained slides for nucleic acid isolation. This technique allows for good localization and requires no special equipment. However, having to cut and scrape slides is time consuming and requires more resources than the methods mentioned above.

By far the most accurate method for tissue selection and isolation is laser capture microdissection. This technique is similar to macrodissection, except instead of scraping the unstained slides, the slides are stained, visualized under a microscope, and a special instrument is used to cut out the desired tissue [10]. The laser capture microdissection instrument consists of a microscope, slide scanner, and laser dissector. The operator marks the area to be dissected by direct visualization of a digital histological image. A laser then is used to cut out and grab the selected tissue. As a result, this technique allows for much finer areas to be dissected, even down to a single cell in some instances [11]. The major drawbacks to this method, however, are that it is time consuming and requires an expensive piece of specialized equipment.

No matter which technique is used to isolate nucleic acid, it is imperative to have some form of histological analysis. As new molecular techniques generate massive amounts of data, one must know what type of tissue the data is being generated on.

## 4.4 Analysis

There are an ever-growing number of molecular techniques used to analyze DNA and RNA. It is impossible to cover all of them in a single chapter. Therefore, this chapter will focus on the general technique of MPS or so-called "next-generation sequencing" (NGS) and a few other techniques that are often used in a molecular pathology laboratory.

### 4.4.1 Focused Analytical Techniques

#### 4.4.1.1 Quantitative Real-Time Polymerase Chain Reaction

Quantitative real-time polymerase chain reaction (qPCR) utilizes a traditional PCR reaction with the addition of a reporter probe (Fig. 4.3) [12]. The reporter probe has both a fluorescent reporter and a quencher with a sequence of complimentary nucleic acid for binding. Due to the close proximity of the reporter and quencher, no

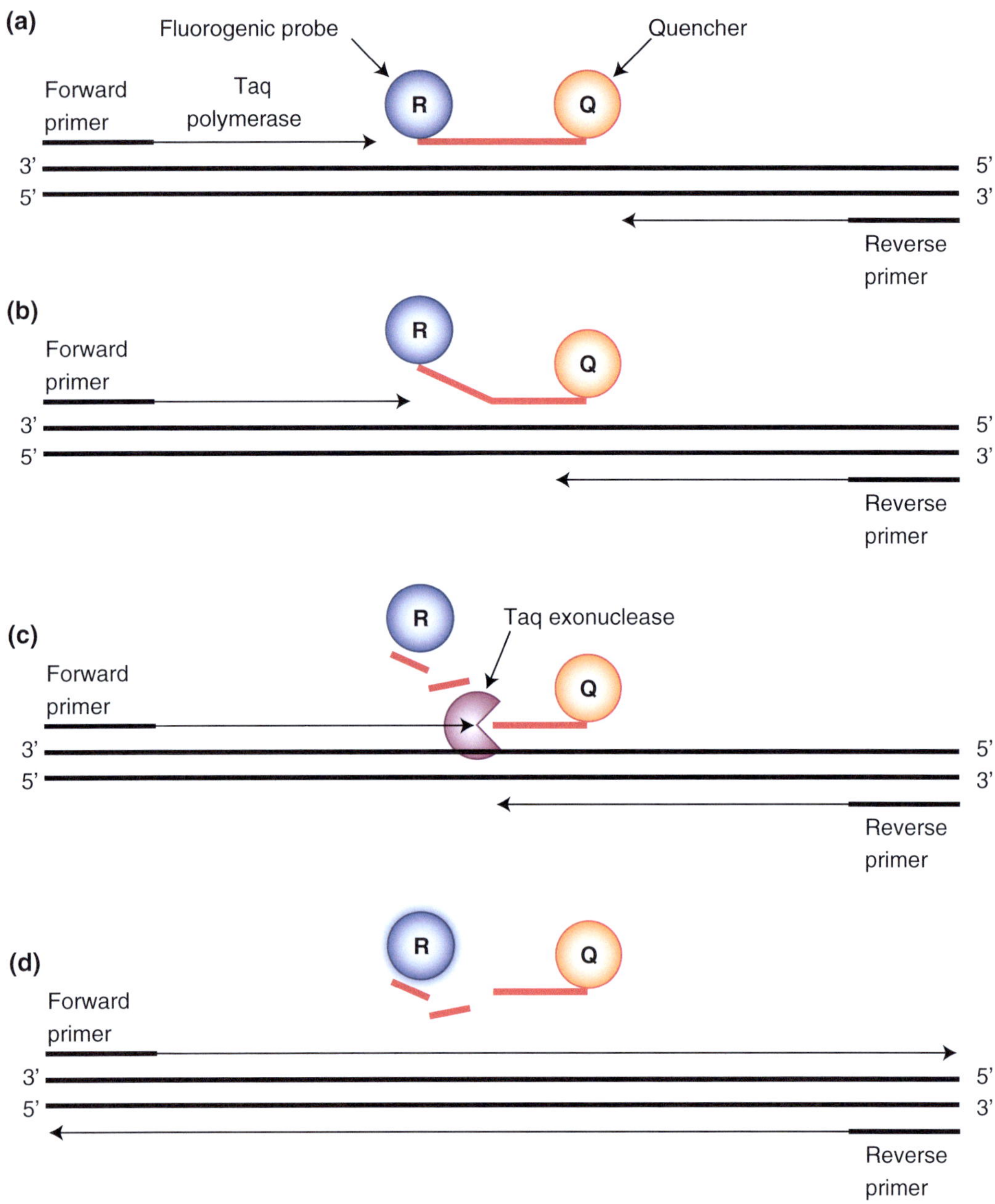


**Fig. 4.3** Real-time PCR. **a** A nucleic acid probe that contains a fluorescent marker and quencher is bound to template DNA along with PCR primers. **b** Taq polymerase extends the PCR primer until it reaches the bound probe. **c** Taq exonuclease activity then cleaves the bound probe releasing the fluorescent marker. **d** Since the fluorescent marker is no longer in close proximity to the quencher, its fluorescence can be detected allowing detection and quantification

fluorescence is seen. As the PCR primers get extended and approach the reporter probe, the 5′–3′ exonuclease activity of the Taq polymerase degrades the bound probe, releasing the fluorescent marker. The fluorescent marker can then be detected since it is no longer in close proximity to the quencher. Like the name of the technique suggests, qPCR can be used to quantify the amount of DNA. The more DNA is present, the more reporter probe is bound and thus more fluorescence is detected once the marker is released. A variation of this technique is reverse transcription (RT) qPCR. This method can be used to quantify the amount of an RNA transcript. For this, messenger RNA is purified and undergoes a RT step to form complementary DNA (cDNA) before the qPCR is performed. One example of a typical use of this technique is quantifying the amount of cDNA to determine titers of nonhuman RNA/DNA (e.g., viral) within a human sample [13]. The results can either be compared to a set of standards to determine the absolute amount of the pathogenic DNA or more commonly when analyzing human DNA and with RTqPCR, the DNA or mRNA amount is compared to an internal standard. In this way, the results can be reported as a ratio of the specific nucleic acid to the standard. These results can then easily be compared with other samples. For expression analysis, this internal standard is usually a common (or housekeeper) gene with a known, relatively constant expression. In addition to the above-mentioned uses, qPCR can also be used for the detection of a PCR product similar to standard PCR except the analyte is detected in "real time" so a detection method such as running the DNA out on a gel is not needed.

#### 4.4.1.2 Mutation-Specific PCR

Mutation-specific PCR uses primers in the PCR reaction that are specific for a certain single nucleotide pleomorphism (SNP), or variant. In this reaction, the primer is designed so that it will only extend if the nucleotide being interrogated matches the primer (Fig. 4.4). There are multiple ways in which the primer can be designed to achieve this, however, one of the most common is to make the 3′ most nucleotide the SNP. Even if all of the other nucleotides match and the primer binds, the primer cannot get extended by the polymerase. A variant of this procedure intentionally designs a mismatch next to the nucleotide being interrogated. Therefore, if the interrogated nucleotide also does not match it will create two mismatches next to each other,

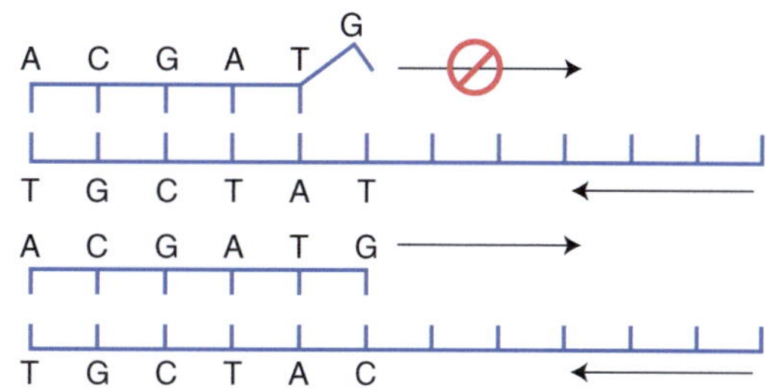


**Fig. 4.4** Mutation-specific PCR. A PCR primer is designed so the 3′ most position is the nucleotide to be interrogated. **a** If that nucleotide does not match the template, a DNA polymerase will not extend the primer. **b** When the 3′ position matches, PCR will continue as normal and the product can be detected

the probe cannot stay bound, and amplification will not occur. In this technique, when designing the primer, the base of interrogation is placed in position-1 from the 3′ base and the intentional mismatch is placed at-2 from the 3′ base. Mutation-specific PCR is a fast, sensitive way to quickly determine if a specific mutation is present. The downside is that the alteration being interrogated must be known in advance. Due to the amplification of the target, mutation-specific PCR can be a very sensitive means of detecting rare sequence variations. However, as with any PCR reaction the utmost care needs to be taken to avoid contamination.

#### 4.4.1.3 In Situ Hybridization

In situ hybridization (ISH), introduced in Chap. 2, is a technique that can be used to detect RNA expression or DNA copy number. There are many variations of ISH depending on what is being detected, the type of probe, and the detection method. In general, it uses a nucleic acid probe to detect DNA or RNA in situ. This is often performed on touch preps, frozen section slides, or FFPE material. The probes are designed to be complimentary to the target sequence and are often labeled with a fluorescent (i.e., FISH) or chromogenic (i.e., CISH) reporter. A signal amplification step may be utilized for increased detection. DNA rearrangements and amplifications are typically detected with FISH (Fig. 4.5a) while RNA is often detected by CISH (Fig. 4.5b). For detection and quantification of the amount of

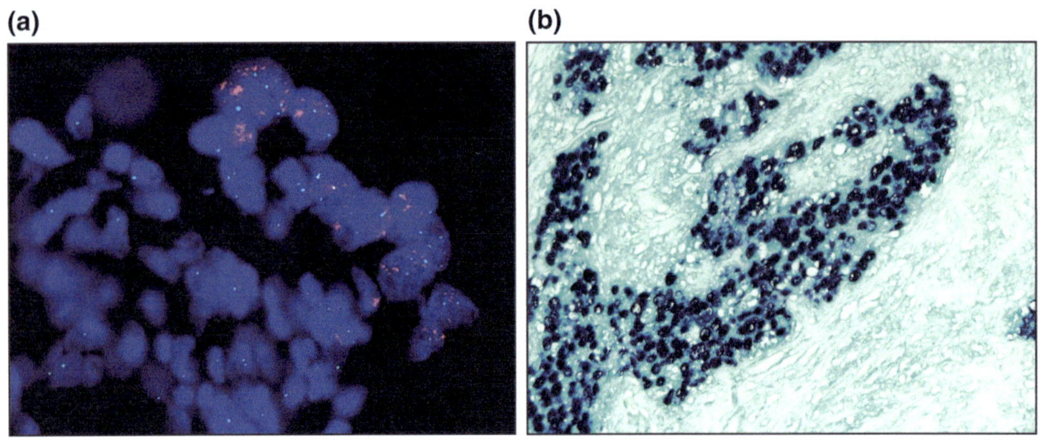


**Fig. 4.5** In situ hybridization. **a** Fluorescent in situ hybridization with ERBB2 (*red*) and CEP 17 (*green*) showing amplification of ERBB2. **b** Chromogenic in situ hybridization for epstein-barr encoding RNA. With the assay used, positive cells are *dark blue*, which is all of the cells in the picture

a specific DNA or RNA, the probe is bound and the number of markers are counted. In the clinical setting, CISH is often used to detect Epstein-Barr virus-encoded RNA (EBER) inside human tissue [14]. One of the major advantages of ISH is the fact that it can be performed on a normal tissue slide allowing correlation with histology and confirmation that the proper cells are being interrogated. One of the drawbacks of ISH, similar to PCR, is that sequences near the target must be known. In addition, ISH is somewhat less sensitive as compared to PCR-based tests.

## 4.4.2 Broad Analytical Techniques

### 4.4.2.1 Microarrays

Microarrays are somewhat similar to ISH in that they utilize a nucleic acid probe designed to be complimentary to a specific sequence, however, microarrays do it on a massive scale. Arrays have been designed to analyze both DNA and RNA. For DNA they are predominately used to analyze copy number changes, SNP sites, and loss of heterozygosity (LOH) [15]. RNA-based arrays are used to look at mRNA expression levels [15]. Microarrays are a fairly mature technology that have been improved greatly over time. They work well with not only fresh tissue/nucleic acid but many also perform well with FFPE tissue. The basic principle involves immobilizing nucleic acid probes onto a known spot on a solid surface, fragmenting the nucleic acid to be analyzed, and labeling it with a fluorescent marker. The labeled nucleic acid is bound to the probes and a sensitive camera detects where there is signal. A computational algorithm then matches up the signal to the specific probes and converts the data into a viewable format. Originally the probes were derived from a variety of sources, however, now almost all are composed of synthetically made oligonucleotides. Modern arrays can contain over a million different probes. The leading manufacturers of arrays include Affymetrix (Santa Clara, CA) and Agilent Technologies (Santa Clara, CA) among others.

#### DNA Arrays

There are two types of standard DNA microarrays. Array comparative genomic hybridization (aCGH) is used to determine copy number changes while SNP chips are used to determine the genotype at specified loci as well as copy number and LOH. In aCGH, the test DNA to be analyzed is labeled with one color fluorescent marker and a "normal" or standard DNA is labeled with a separate fluorescent marker. Both labeled DNAs are mixed together and bound to the probes. The array instrument then reads the relative amount of the different colors. If a greater amount of fluorescence from the test DNA is present as compared to the standard DNA it is read as copy number gain, whereas if less test color fluorescence is present it is read as a DNA copy number loss. Modern aCGH platforms can contain around a million different probes that allow for much finer detail than previous arrays. For example, the Agilent SurePrint 3G array contains 60-mer oligonucleotide probes covering over 900,000 unique biological features [16]. Depending on the manufacturer, SNP arrays use varying techniques to bind and detect the sample DNA in an allele-specific manner. They therefore, as an output, provide genotyping data for each SNP that they capture. As SNP arrays have progressed they now also include copy number probes for areas of the genome that do not contain common SNPs. Like aCGH, the number of probes on SNP arrays has greatly increased with some platforms containing over 900,000 SNP and 900,000 copy number probes [17]. In addition to SNP and copy number data, SNP arrays can also give information about copy neutral LOH. One important area of molecular pathology where arrays are frequently used is in pediatric developmental delay and other disorders. In fact, it is now suggested as the first line standard of care in working up developmental delay [18–20]. Another common use of SNP arrays is in genome-wide association studies (GWAS) where they are used to identify allele variants that may be associated with a certain disorder.

#### RNA Expression Arrays

RNA-based expression arrays are a common format that is frequently used in a research setting to compare the level of RNA expression in different samples. Modern expression arrays typically use millions of oligonucleotide probes directly synthesized or "printed" onto the solid support. Depending on the array, the oligonucleotide probes are generated to be complimentary to known mRNAs or other RNA species such as microRNAs or long noncoding RNAs. The number and targeting location of probes for each RNA species can also vary depending on the array type (e.g., gene array, exon array, transcriptome array, etc.). The general principles of expression arrays do not differ much from DNA-based arrays except that RNA is isolated from the source and is used to construct a cDNA library. With the low quality of RNA that is extracted from FFPE material, expression arrays were traditionally primarily used on fresh frozen material. However, with the recent advances in arrays specifically designed for FFPE material and the algorithms used to analyze the data, the use of FFPE material is now standard practice.

### 4.4.2.2 Sequencing

Since the original joint publication of the draft human genome in 2001 [21, 22], there has been an ever-increasing interest and use of sequencing in molecular pathology. The majority of this interest now focuses on MPS, however, traditional Sanger remains a technique that one must be familiar with and still has its uses. The advantage that sequencing has over many other molecular techniques is the fact that it provides precise information at the nucleotide level across the region analyzed, not just at a specific point. Another factor that has increased the use of sequencing in research is the rapid reduction in cost of sequencing. It is estimated that the initial human genome project cost approximately $2.7 billion, and now an entire genome can easily be sequenced for under $5000 [23].

#### Sanger Sequencing

Sanger or traditional sequencing is a good technique to sequence one segment of a couple of hundred base pairs and can work well with either fresh or FFPE material. It relies on the use of a DNA polymerase and a low concentration of chain terminating nucleotides. A complimentary oligonucleotide primer is designed to bind in close proximity to the area that is to be sequenced and a DNA polymerase reaction is set up similar to a PCR reaction including the primer, template DNA to be sequenced, and the standard four deoxynucleotide triphosphates (dNTP) (e.g., dTTP, dATP, dCTP, dGTP). However, in addition to the standard deoxynucleotides, a low concentration (typically 1 %) of individually labeled dideoxynucleotide

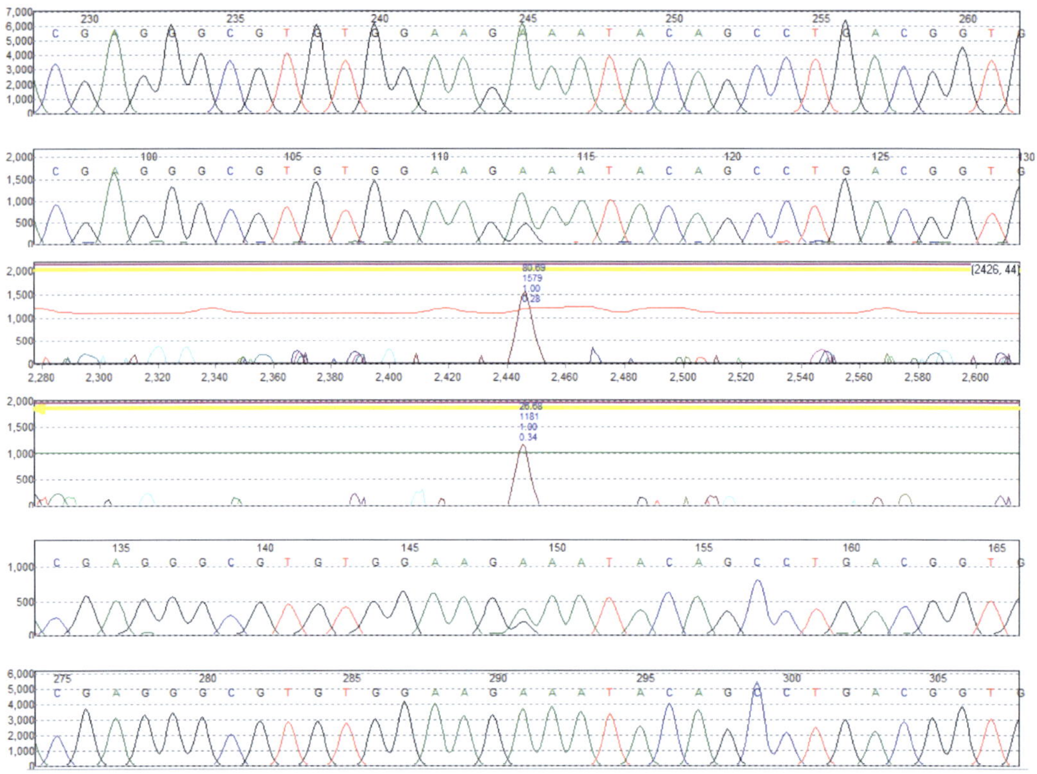

**Fig. 4.6** Sanger sequencing analysis. Sanger trace showing sequencing from forward (*top*) and reverse (*bottom*) sequencing. Control sequence shown on *very top* and *bottom*, with actual sequence in *middle top* and *middle bottom*. Heterozygous A > G mutation shown at position 245 (*top*) and 291 (*bottom*)

triphosphates (ddNTP) are also added. The ddNTPs can be incorporated into the extended primer, however, another dNTP cannot be added as the ddNTP lacks a hydroxyl group at the 3′ position. Therefore the DNA polymerase reaction is terminated which gives Sanger sequencing its other name, chain terminator sequencing. As the ddNTPs are in a much lower concentration than the dNTPs, they will be randomly incorporated and terminate the DNA polymerase reaction at varying lengths. If each type of ddNTP (e.g., ddTTP, ddATP, ddCTP, and ddGTP) is differently labeled, the reaction can be run out on capillary electrophoresis and the output can be read (Fig. 4.6). Since Sanger sequencing will display all base signals at a given nucleotide on top of each other, it is important to know the percentage of the cell population you wish to sequence in the overall sample if one is dealing with nonhomogeneous samples (such a human tissue). For example, for sequencing of a heterozygous mutation in a tumor with only 20 % tumor nuclei (not an uncommon occurrence), the mutated allele would only compromise 10 % of the total bases/alleles at that location. This would produce a relatively small peak compared to the wild type peak and become increasingly difficult to distinguish from background as the tumor percentage gets smaller. As with other assays, if one wishes to sequence RNA instead of DNA, RT can be used to generate DNA from the RNA and sequencing can proceed as above.

### Massively Parallel Sequencing

Massively parallel sequencing (MPS) is better known as "next-generation sequencing" (NGS), however, this is a misnomer because what was originally called NGS is now outdated technology. Currently, most systems that are used would

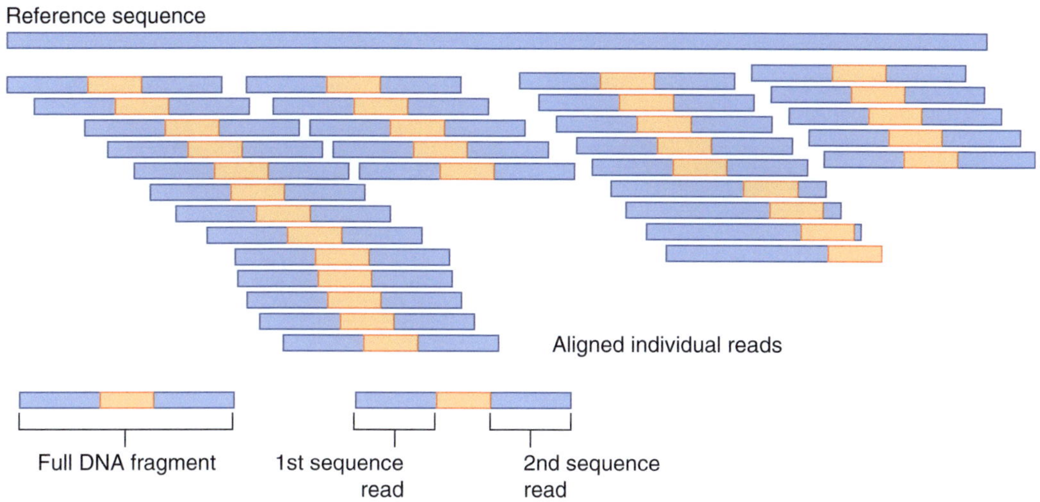


**Fig. 4.7** MPS read alignment. After all of the sequencing reads are created, computational algorithms are used to align the individual reads to a reference sequence

be considered "next, next" or "next, next, next" generation sequencing. As one can see this could get a bit long and confusing. Therefore, it is much more appropriate to consider all of the sequencing systems that align sequencing reads together on a large scale, using the term MPS. Whether one refers to MPS or NGS, it is a relatively generic term and each company that produces MPS equipment has their own somewhat unique system. MPS is a little different than some other technologies as it is very dependent on the specific manufacturer of the equipment. Therefore, it is almost impossible to learn about MPS without learning about several of the leading companies that produce MPS equipment. While constantly changing, the field is currently dominated by two companies/technologies, Illumina (San Diego, CA) and Ion Torrent/Life Technologies (Grand Island, NY). Regardless of the technology used, all of the systems rely on sequencing millions of short reads that are then computationally aligned to each other and to a reference sequence (Fig. 4.7). By performing all of the reads in parallel it is now possible to sequence large amounts of DNA much faster and cheaper than sequencing each fragment of DNA separately. In addition, by the addition of barcodes to each DNA sample, multiple samples can be sequenced at once with the downstream computational algorithms used to separate each one. Using this technology it is now possible to routinely detect mutations (e.g., point and insertion/deletions), large and small copy number variants, and even large chromosomal rearrangements (Fig. 4.7).

### Presequencing

There are several steps that need to be performed in order to get DNA ready for sequencing. First, as discussed earlier in the chapter, the DNA must be isolated and purified. While originally the amount and quality of DNA needed for MPS required fresh/frozen material, it is now common practice to perform MPS on FFPE material [24]. Once the DNA is isolated and purified, it must be fragmented into the appropriate length fragments for further processing. There are several methods for this including sonication, enzymatic digestion, and PCR among others. The length of DNA fragments needed will also vary depending on the technology used and the desired read length. While somewhat variable, typical average fragment lengths are usually somewhere around 200–1000 base pairs. The ideal fragment size will be determined by the planned read length, technology used, and desired results (what you are looking for). For example, when identifying translocations are particularly important, larger fragment sizes are sometimes chosen. Once the DNA is fragmented, adapters for future binding, amplification, and identification are ligated to each end of the fragment. Within these adapters can be a unique sequence of DNA used to identify the sample in a group of pooled samples. This is referred to as the "barcode." After the adapters are bound to the fragmented DNA, the next step involves amplification of the DNA. This step is where there is significant deviation depending on the technology used. For Illumina-based sequencing, "bridge amplification" is utilized (Fig. 4.8). This process, also called "clustering," utilizes a flow cell, which is covered with a lawn of two oligonucleotide probes. One of the probes is complimentary and thus binds to one of the adapters ligated to the end of the DNA fragment. After binding the processed, fragmented DNA, using the probe as a primer, a DNA polymerase then makes a DNA strand complimentary to the bound DNA fragment. The original DNA template is then denatured and washed away. The adaptor on the non-bound end of the intact DNA fragment folds over and binds to the second type of oligonucleotide probe on the flow cell as it is complementary to it. This creates a single-stranded DNA bridge with a bound probe/primer, which is then used to generate the complimentary DNA strand (i.e., identical to the original DNA but now bound to the flow cell). The double-stranded DNA is denatured and the process is repeated over and over so that each fragment of DNA forms its own cluster on the flow cell. The last step involves cleaving off and washing away the complimentary strands so only the original DNA strand remains. The Ion Torrent technology uses a bead-based amplification (emulsion PCR) where each fragment of DNA is individually bound to a metal bead (one fragment of DNA per bead) [25]. That DNA fragment is then amplified to cover the bead in copies of the single fragment of DNA. This is done on a large scale so that millions of beads are covered with millions of DNA fragments. The beads are then flowed across a chip that contains small wells so that each bead can individual deposit in one well. At the end of the presequencing steps, regardless of technology, you have fragments of DNA individually amplified and separated from other DNA fragments in either clusters on a chip or isolated on individual beads that rest within a well.

### Sequencing

On the Illumina platform, sequencing is initiated by binding a sequencing primer to the adapter on the free end of the DNA strand. Fluorescently labeled nucleotides (with each type of nucleotide differently labeled) that have been reversibly modified so that the polymerase reaction is terminated after it is incorporated, are added. The complimentary nucleotide binds and is incorporated by a polymerase. After washing, the fluorescence is read. Afterwards, the reversible terminator is cleaved off and the process is repeated until the desired read length is achieved. Once finished, the entire read product is denatured off and washed away. The 3′ end of the template DNA (i.e., the non-bound end) is then allowed to bind to the second oligo probe on the flow cell in a similar manner to what occurred during bridge amplification. A DNA polymerase is then used to generate the complimentary strand to form double-stranded DNA using the second oligo as a primer. The double-stranded DNA is denatured to form two single-stranded complimentary strands and the original forward strand is cleaved and washed away. The second read primer is then bound and sequencing commences as it did during the first round (Fig. 4.8). Through this process, both ends of the DNA fragment are sequenced giving valuable additional information besides just the sequencing read. If only single end reads are desired only the forward strand is sequenced. The Ion Torrent technology uses a different approach. Once the beads with the amplified fragments are dispersed in the wells, the wells are flooded with a solution containing a single type of nucleotide. If the nucleotide is complimentary to the next base, it is incorporated by a DNA polymerase. During this chemical reaction, a single hydrogen ion is released which slightly shifts the pH of the solution in the well. A sensitive semiconductor at the bottom of each well then detects this pH shift. This process is then repeated base after base to sequence the desired read length (Fig. 4.9). While the maximum read length possible for each technology is constantly changing, currently the maximum read lengths for Illumia is around 150–250 base pairs depending on platforms and up to 400 base pairs with the Ion Torrent.

#### Post-Sequencing

Once sequencing is complete, the real difficulty of interpreting the data begins. Post-sequencing analysis can get extremely complicated, requires advanced computational and statistical analysis, and is constantly changing. Therefore, an in-depth look into these processes is beyond the scope of this introductory book chapter. However, as someone who is responsible for the interpretation of the biological significance of the molecular assays, it is important to have an understanding of the processes involved as well as the advantages and disadvantages of each process used. The same is true for the type of sequencing used or any other technique. This is not to say one has to have expertise in computational biology and programming, just an understanding of what is being used and an ability to communicate with those that do. The opposite also holds true, while most computationalists should not be expected to know all the intricacies of the clinical significance of the findings, a basic understanding is needed in order to properly communicate with collaborators that do. The sequence data initially needs to go through a series of quality control steps so that only high quality reads are used for further analysis. After the quality control steps, the sequence reads can be aligned to each other and to a standard known sequence, as was shown in Fig. 4.7. There are multiple aligning and calling programs that can be used for this process including ones provided by the companies that manufacture the sequencers and many developed for specific tasks by researchers throughout the world.

#### Uses of Massively Parallel Sequencing

Generally speaking, MPS tests fall into three categories. First, whole genome sequencing (WGS), where all of the DNA is sequenced, leads to information on all sequenceable areas of the genome including exons, introns, and other noncoding areas. While this type of sequencing will provide the most information, it is a huge amount of data, is still relatively expensive, and currently we do not know the significance of many of the discovered alterations in the noncoding areas of the genome. Whole exome sequencing (WES) attempts to take the area of the genome we know the most about (i.e., exons) and sequence them only. With WES, a large portion of the usable data obtained from WGS is retained with an overall reduction in the amount of DNA needed to be sequenced, which reduces the cost of both sequencing and computation. Targeted sequencing further reduces the amount of data collected, and only a selected number of genes are sequenced. Targeted sequencing gene panels can be anywhere from a small handful to many hundreds of genes. For any of the targeted sequencing panels (including WES), a DNA selection step is needed to first isolate only the DNA desired. These selection steps can be probe (binding) based or PCR based. Targeted sequencing provides the advantage of being the cheapest to perform while providing data on the genes of interest (i.e., the ones selected for the panel).

#### Other MPS Techniques

MPS is also used for several other techniques including RNA-sequencing (RNA-seq) and chromatin immunoprecipitation sequencing (CHIP-seq). For RNA-seq, sequencing is done in a similar manor as DNA sequencing with the exception that RNA is isolated and either RNA fragments or generated cDNA is sequenced. Since the number of reads will correlate with the number of original copies, RNA expression can be quantified. In addition, since mature mRNA will have removed introns, translocations that place exons from multiple proteins can be recognized. Currently, RNA-seq works well for fresh, high quality RNA but this technology is in its infancy with the use of FFPE. This is largely due to how fragmented and degraded FFPE RNA can become. Another technique that utilizes MPS is CHIP-seq. CHIP-seq combines chromatin

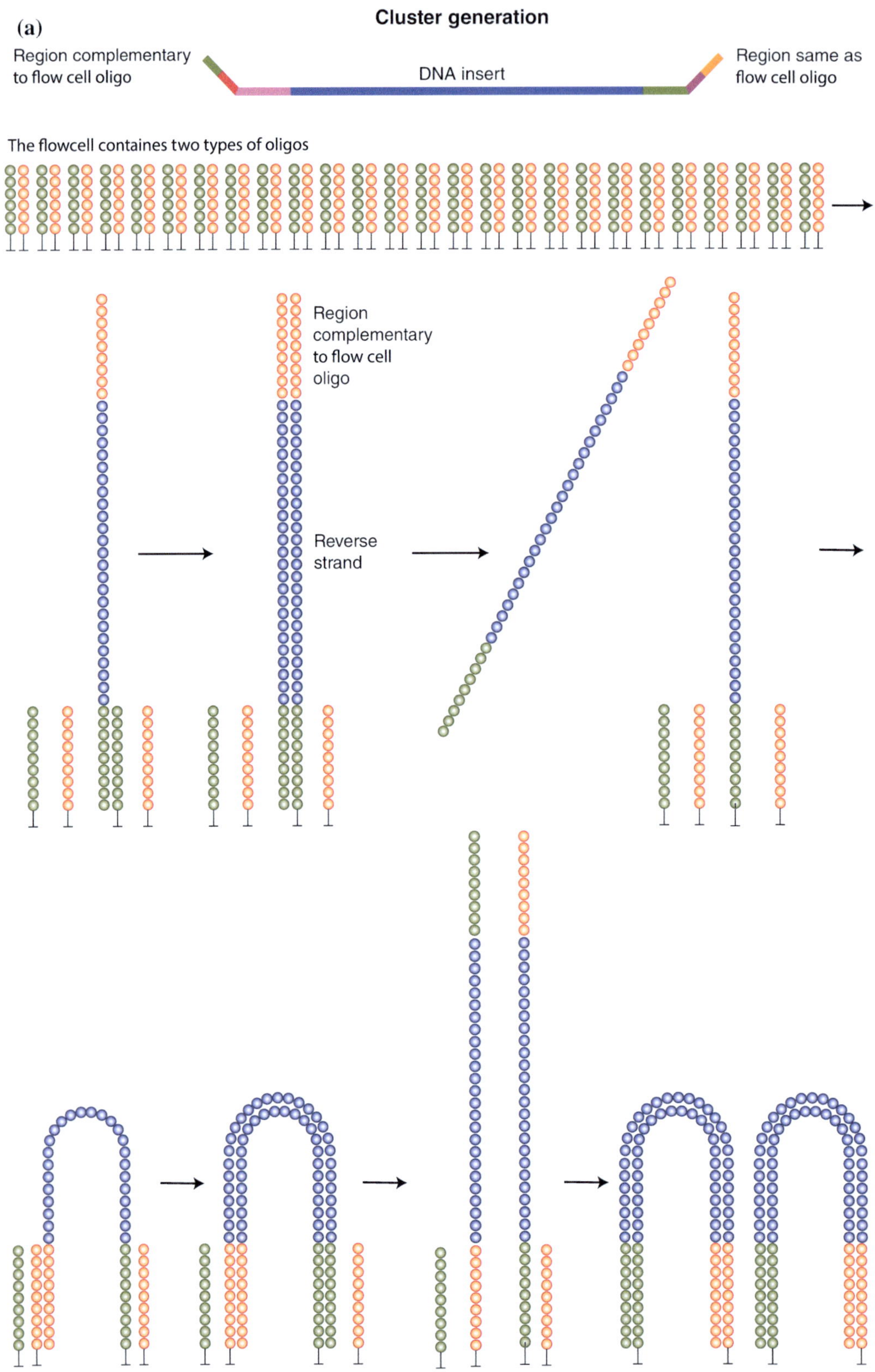
(a)
Cluster generation
Region complementary to flow cell oligo
DNA insert
Region same as flow cell oligo
The flowcell containes two types of oligos
Region complementary to flow cell oligo
Reverse strand

◀ **Fig. 4.8** Illumina seq. **a** Bridge amplification: two primers are annealed onto the ends of the fragmented DNA. The primers are designed to be complementary to the oligos on the sequencing flow cell as well as containing other areas such as the sample barcode. These DNA fragments are then bound to the flow cell oligos, where the oligo is used to generate the complementary DNA strand. After which, the original DNA is washed away. The distal single-stranded DNA is then free to bend over and bind to the second flow cell oligo, allowing for priming for complementary strand synthesis. Through this process individual clusters of DNA are created. **b** Sequencing: after the reverse strands are cleaved and washed off, a sequencing primer is bound and fluorescently labeled nucleotides with a cleavable chain terminator are incorporated one at a time and fluorescence is read. Once the desired read length is completed, the read sequence is washed off and the free end of the DNA fragment can bind to the complementary flow cell oligo. The complementary DNA fragment is created and sequenced similar to the first sequencing read

**(b)**

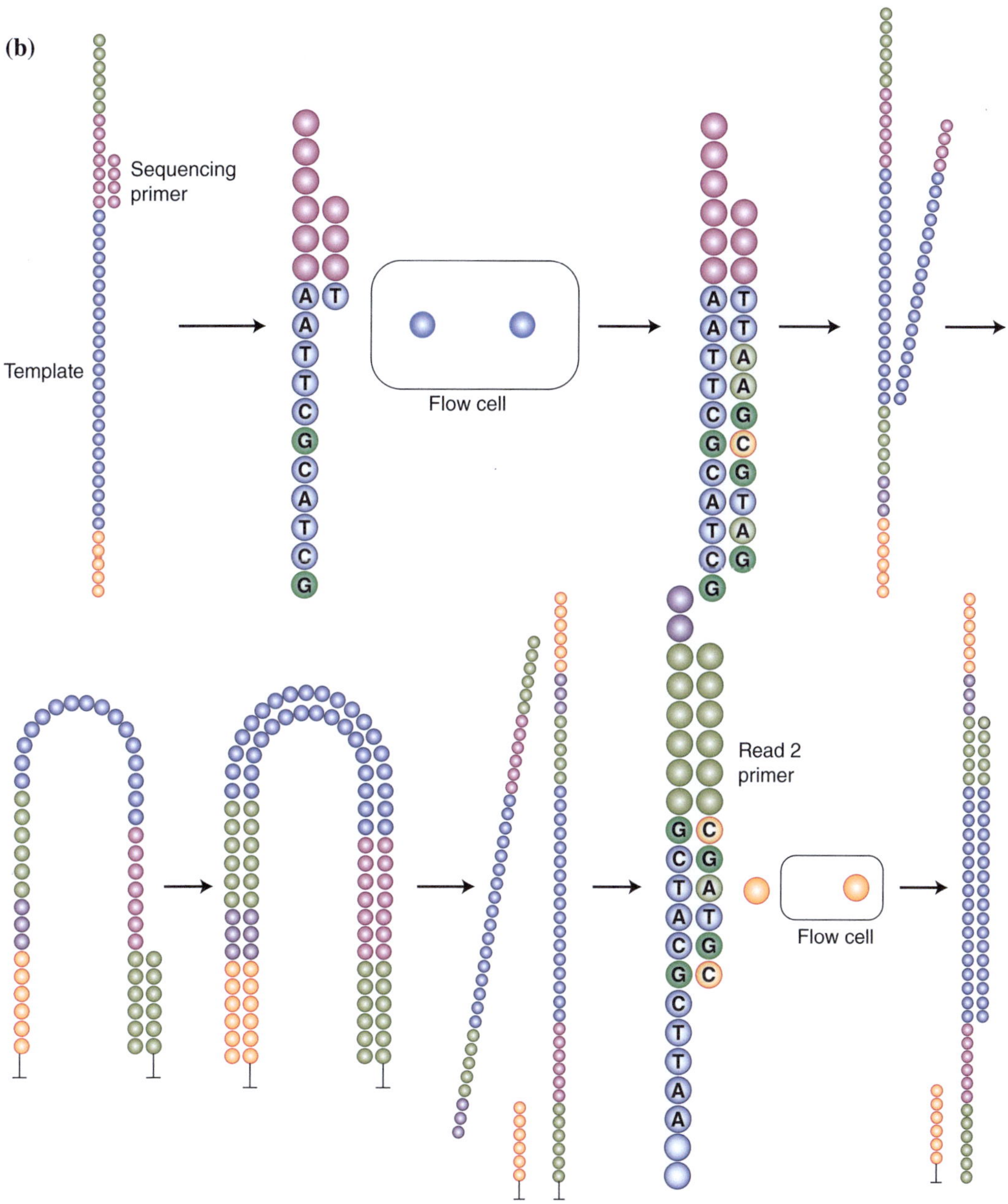


**Fig. 4.8** (continued)

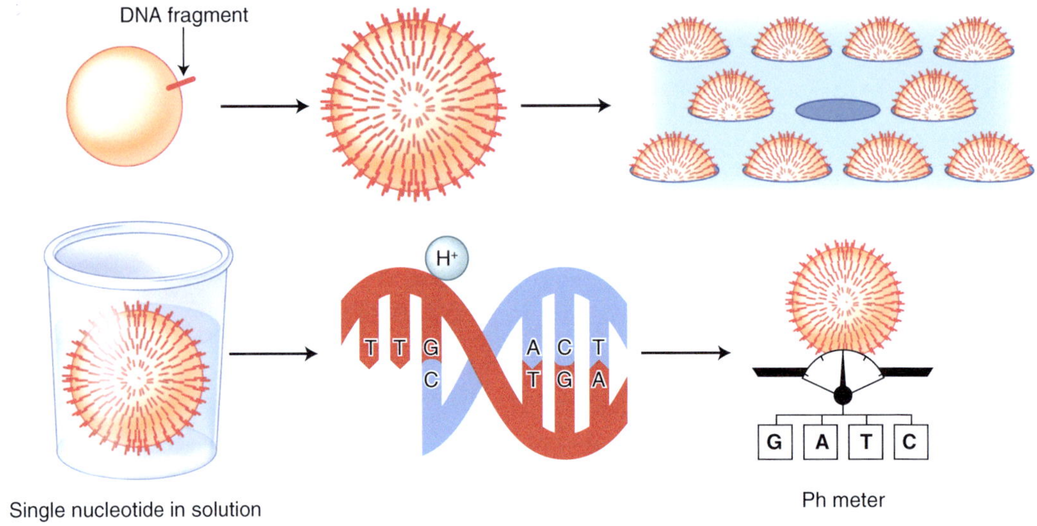


**Fig. 4.9** Ion Torrent sequencing. A single DNA fragment is bound to a bead where it is amplified to cover the bead. The beads are then flowed onto a chip containing small wells large enough to hold an individual bead. The plate is then flooded with a solution containing a polymerase and a single nucleotide. If the nucleotide is complementary to the base on the DNA fragment it gets incorporated, releasing a single hydrogen ion, slightly altering Ph. Each well contains a small Ph meter that can detect this which is registered and the base is recorded. This process is then repeated over and over for all of the bases

immunoprecipitation with DNA MPS to identify the binding sites of DNA-associated proteins, and is often used to determine where transcription factors interact with the genome. Like RNA-seq, this is a technology that is focused on using DNA from fresh cells/tissue. As CHIP-seq depends on binding between protein and DNA, any fixative that artificially binds/cross-links products together (for example formalin) will complicate detection.

Conclusions
With the recent advances in both the sequencing technology and the computational algorithms used to analyze data, MPS is now the standard assay used for many research and clinical tests. Unlike many other techniques that specialize in detecting either small local changes (i.e., Sanger sequencing) or large chromosomal level changes (i.e., FISH), standard DNA MPS can detect the full range of alterations that occur (Fig. 4.10). Getting this much information from a single assay can be extremely powerful, however, one must have a plan in place to not only accurately detect these changes, but also to organize and make sense of them all. In addition, as MPS is indifferent to the DNA that is used for sequencing (quality not withstanding), results always have to be carefully considered in the context of the experiment. For example, if a tumor is supposed to be sequenced but the biopsy missed the neoplastic cells, the sequencer will still output a DNA sequence. As stated earlier in the chapter, it is imperative to have a good understanding of the source of the DNA. At times, however, using the fact that whatever DNA is present will be sequenced can be used as an advantage. For example, a site of infection can be sequenced to try to identify unknown infectious organisms by comparing the obtained sequence with known sequences of infectious agents.

## 4.5 Summary

There are many molecular techniques available to today's researchers and clinicians. These range from simple, fast assays that interrogate a single area to large, complex assays that can sequence the entire genome. It is always important to

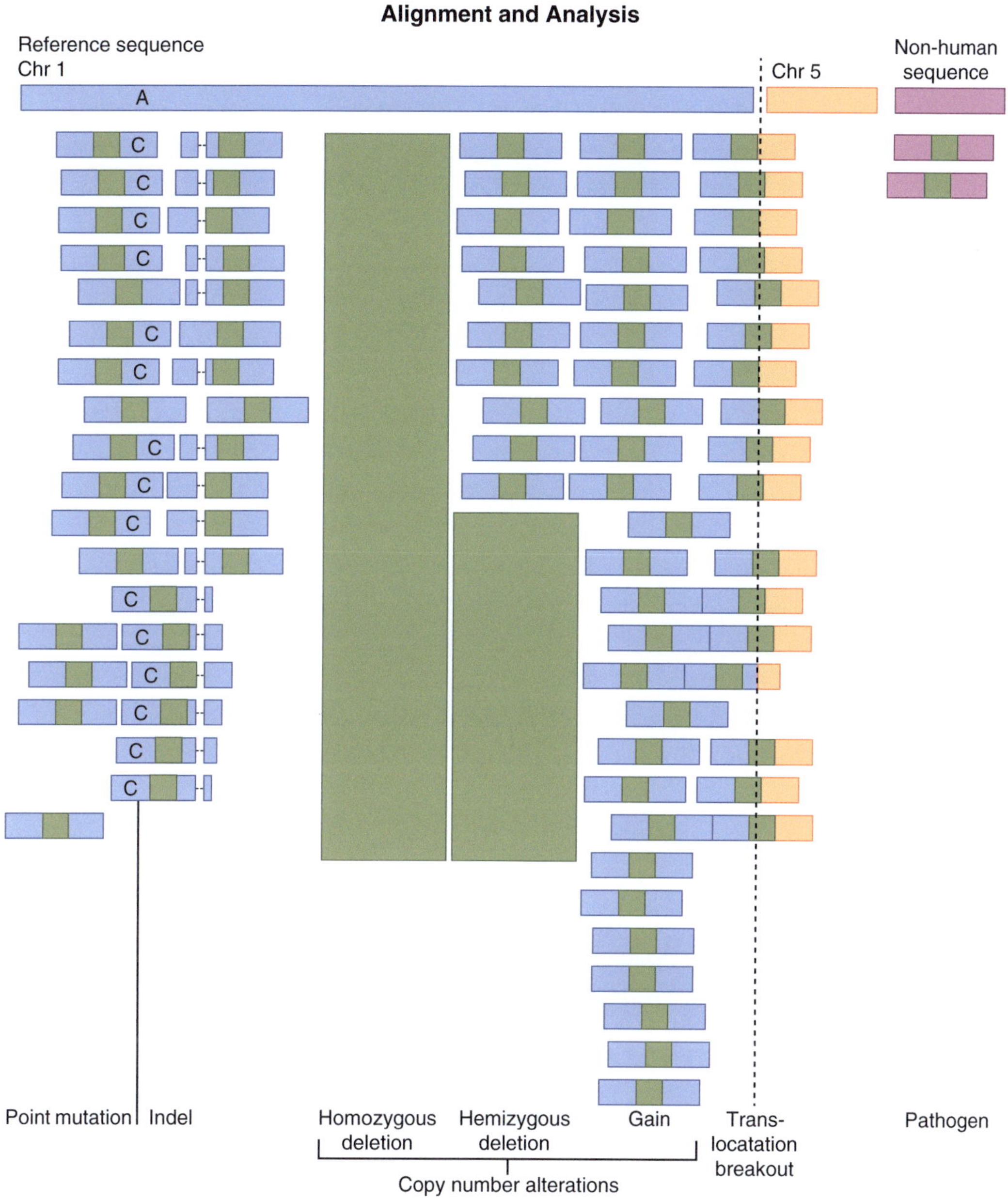


**Fig. 4.10** Uses for Massively parallel sequencing (MPS). DNA sequencing through MPS allows the detection of single point mutations, small insertions and deletion mutations, copy number alterations (including hemi/homozygous deletions and gains), translocations, and even nonhuman DNA

carefully consider the desired results with what can be realistically achieved with the given technologies. Once a technique is selected, it is important to set up the assay with proper validation and controls as well as careful consideration of the input (including type, location, and quality of the nucleic acid). Finally, many of the more complex techniques (especially MPS) require significant resources and generate tremendous amounts of data. Often a single person does not have all of the desired expertise to effectively handle all of the requirements necessary to properly generate and interpret this amount of data. Therefore, collaborations between biologists, bioinformaticians, computational biologists, and others with different expertise often can lead to very rewarding outcomes.

## References


1. Berglind H, Pawitan Y, Kato S, Ishioka C, Soussi T. Analysis of p53 mutation status in human cancer cell lines. Cancer Biol Ther. 2008;5(May):699–708.
2. Yu M, Selvaraj SK, Liang-Chu MMY, Aghajani S, Busse M, Yuan J, et al. A resource for cell line authentication, annotation and quality control. Nature. 2015;520(7547):307–11. Available from http://www.nature.com/doifinder/10.1038/nature14397. Accessed 16 Apr 2015.
3. Geraghty RJ, Capes-Davis A, Davis JM, Downward J, Freshney RI, Knezevic I, et al. Guidelines for the use of cell lines in biomedical research. Br J Cancer. 2014;111(6):1021–46. Available from http://www.nature.com/doifinder/10.1038/bjc.2014.166\n, http://www.ncbi.nlm.nih.gov/pubmed/25117809. Accessed 9 Sept 2014 (Nature Publishing Group).
4. Bowen RAR, Remaley AT. Interferences from blood collection tube components on clinical chemistry assays. Biochem Medica. 2014;24(1):31–44. Available from http://www.pubmedcentral.nih.gov/articlerender.fcgi?artid=3936985&tool=pmcentrez&rendertype=abstract.
5. Philibert RA, Zadorozhnyaya O, Beach SRH, Brody GH. A comparison of the genotyping results using dna obtained from blood and saliva. Psychiatr Genet. 2008;18(6):275–81.
6. Hansen TVO, Simonsen MK, Nielsen FC, Hundrup YA. Collection of blood, saliva, and buccal cell samples in a pilot study on the Danish nurse cohort: comparison of the response rate and quality of genomic DNA. Cancer Epidemiol Biomarkers Prev. 2007;16(10):2072–6. Available from http://www.ncbi.nlm.nih.gov/pubmed/17932355. Accessed 22 Nov 2014.
7. Kidd JM, Sharpton TJ, Bobo D, Norman PJ, Martin AR, Carpenter ML, et al. Exome capture from saliva produces high quality genomic and metagenomic data. BMC Genomics. 2014;15(1):262. Available from http://www.pubmedcentral.nih.gov/articlerender.fcgi?artid=4051168&tool=pmcentrez&rendertype=abstract. Accessed 19 Nov 2014.
8. Spencer DH, Sehn JK, Abel HJ, Watson MA, Pfeifer JD, Duncavage EJ. Comparison of clinical targeted next-generation sequence data from formalin-fixed and fresh-frozen tissue specimens. J Mol Diagn. 2013;15(5):623–33. Available from http://linkinghub.elsevier.com/retrieve/pii/S1525157813000913 (American Society for Investigative Pathology).
9. Howat WJ, Wilson BA. Tissue fixation and the effect of molecular fixatives on downstream staining procedures. Methods. 2014;70(1):12–9. Available from: http://linkinghub.elsevier.com/retrieve/pii/S1046202314000607 (Elsevier Inc.).
10. Espina V, Wulfkuhle JD, Calvert VS, VanMeter A, Zhou W, Coukos G, et al. Laser-capture microdissection. Nat Protoc. 2006;1(2):586–603. Available from http://www.nature.com/doifinder/10.1038/nprot.2006.85.
11. Keaysa KM, Owensa GP, Ritchiea AM, Gildena DH, Burgoon MP. Laser capture microdissection and single-cell RT-PCR without RNA purification. J Immunol Methods. 2005;302(1–2):90–8.
12. Heid CA, Stevens J, Livak KJ, Williams PM. Real time quantitative PCR. Genome Res. 1996;6(10):986–94. Available from http://www.ncbi.nlm.nih.gov/pubmed/8908518.
13. Keyaerts E, Vijgen L, Maes P, Duson G, Neyts J, Van Ranst M. Viral load quantitation of SARS-coronavirus RNA using a one-step real-time RT-PCR. Int J Infect Dis. 2006;10(1):32–7. Available from http://www.ncbi.nlm.nih.gov/pubmed/16023880.
14. Weiss LM, Chen Y-Y. EBER in situ hybridization for epstein-barr virus. Methods Mol Biol. 2013;999:223–30. Available from http://www.ncbi.nlm.nih.gov/pubmed/23666702.
15. Sealfon SC, Chu TT. RNA and DNA microarrays. Methods Mol Biol. 2011;671:3–34. Available from http://www.ncbi.nlm.nih.gov/pubmed/20967621.
16. Genomics.agilent.com. Sureprint G3 gene expression microarrays; 2015. Available from http://www.genomics.agilent.com/article.jsp?pageId=2197&_requestid=340872. Accessed 1 Jan 2015.
17. Affymetrix. Genome-wide human SNP array 6.0; 2015. Available from http://www.afymetrix.com/catalog/131533/AFFY/Genome-Wide+Human+SNP+Array+6.0#1_1. Accessed 1 Jan 2015.
18. Manning M, Hudgins L. Array-based technology and recommendations for utilization in medical genetics practice for detection of chromosomal abnormalities. Genet Med. 2010;12:742–5.
19. Michelson D, Shevell M, Sherr E, Moeschler J, Gropman A, Ashwal S. Evidence report: genetic and metabolic testing on children with global developmental delay: report of the quality standards subcommittee of the American Academy of Neurology and the Practice Committee of the Child Neurology Society. Neurology. 2011;77:1629–35.
20. American Academy of Pediatrics. genetic and metabolic testing on children with global developmental delay. Pediatrics. 2012;129:129e825.
21. Venter JC, Adams MD, Myers EW, Li PW, Mural RJ, Sutton GG, et al. The sequence of the human genome. Science. 2001;291(5507):1304–51.

22. Lander ES, Linton LM, Birren B, Nusbaum C, Zody MC, Baldwin J, et al. Initial sequencing and analysis of the human genome. Nature. 2001;409 (6822):860–921. Available from http://www.ncbi.nlm.nih.gov/pubmed/11237011. Accessed 15 Feb 2001.
23. Genome.gov (National Intitute of Health US. The human genome project completion: frequently asked questions; 2015. Available from http://www.genome.gov/11006943. Accessed 1 Jan 2015.
24. Stachler MD, Taylor-weiner A, Peng S, Mckenna A, Agoston AT, Odze RD, et al. Paired exome analysis of Barrett's esophagus and adenocarcinoma. Nat Genet. 2015;1–12. Available from http://dx.doi.org/10.1038/ng.3343 (Nature Publishing Group).
25. Scientific T. Ion Torrent next-genration sequencing. Available from https://www.thermofisher.com/us/en/home/brands/ion-torrent.html.

# Introduction to Cancer Epidemiology 5

Mieke Van Hemelrijck and Lars Holmberg

## 5.1 What Is Epidemiology?

According to the World Health Organisation (WHO), epidemiology is "the study of the distribution and determinants of health-related states or events, and the application of this study to the control of diseases and other health problems" [1]. One of the pioneers of epidemiological methods was John Snow (1813–1858), who was a physician who lived and worked in London [2]. His landmark discovery was that the spread of cholera in London was due to transmission via water pipes connected to an infected public water well, which he discovered through careful interview and annotation of affected and unaffected individuals during an outbreak in 1854 (Fig. 5.1). His findings had important public health implications by impacting the waste water systems.

The scope of epidemiology is much broader than the study of epidemics; it also involves the study of causal factors for all kinds of diseases, including both acute and chronic disease states. In cancer, epidemiology has had major public health impacts in defining causal associations between smoking and second hand smoke with lung cancer, hepatitis B infections and hepatocellular carcinoma and exposure to aromatic amines and bladder cancer.

To get a better understanding of epidemiology it is important to clarify some of the terminology and methods commonly used.

## 5.2 Measures of Disease Occurrence

To study the occurrence of a disease, the frequency and distribution of the events of interest must be measured. Prevalence describes the number of people living with a defined disease state at a certain point in time, whereas incidence describes how many people were diagnosed with a defined disease state during a specified time. For example, it is estimated that worldwide in 2012 there were 952,000 people living with stomach cancer [2]. One can also define how many people are still living with a defined outcome five years post the development of this outcome. Worldwide, the five-year prevalence of stomach cancer is estimated at 1,548,000 in 2012. In addition, the incidence of stomach cancer was estimated to be about 18 per 100,000 men per year in 2012 worldwide, whereas this was about 7 per 100,000 per year for women [2]. These incidences are reported as age-standardised rates, which allow comparisons of the incidences between different countries or geographical areas independent of the different age distribution in the areas.

M. Van Hemelrijck (✉) · L. Holmberg
Division of Cancer Studies, King's College London, Cancer Epidemiology Group, Guy's Hospital, 3rd Floor Bermondsey Wing, London SE1 9RT, UK
e-mail: mieke.vanhemelrijck@kcl.ac.uk


M. Loda et al. (eds.), *Pathology and Epidemiology of Cancer*,
DOI 10.1007/978-3-319-35153-7_5

**Fig. 5.1** John Snow Pub, Broadwick Road, London, UK

Mortality rates are estimated as the number of people who died of a defined outcome during a specified time in a population. For instance, worldwide 13 per 100,000 men and 7 per 100,000 women died of stomach cancer in 2012 [2]. One key attribute of mortality rates is that the denominator is the number of individuals in the population, rather than the number of individuals diagnosed with the disease.

Another central measure in epidemiology is risk, or a person's probability of developing a health state during a certain time period. Risk is defined as a probability between zero and one (often expressed as a percentage).

## 5.3 Exposure and Outcome

In addition to information about the occurrence of disease, information about measurement of exposures is needed to study different disease patterns. An exposure denotes a factor that one wants to study in relation to the incidence or the risk of an outcome. For instance, in the case of the cholera epidemic, Snow mapped cholera deaths to the distribution of water from different pumps and concluded that infected water, the exposure, was causing cholera, the outcome. Smoking as an exposure or risk factor for developing bladder cancer is another example and would refer to the etiology of bladder cancer.

The distinction between causation and association is imperative in epidemiology. Using different study designs, measurements and biostatistical methods, epidemiology can strictly only study the associations between exposures and outcomes. As can be seen from the examples above, however, we often aim conceptually at a study of cause and effect. David Savitz, in his book "Interpreting Epidemiologic Evidence", explains this as follows: "In our epidemiological studies we can in principle only look at statistical associations between exposure and outcome, but by interpreting evidence given the pros and cons in a given study, we can make an interpretation of whether what we have studied is causal" [3]. Detailed descriptions of epidemiological statistical methods can be found elsewhere [4].

Austin B. Hill proposed in 1965 a set of criteria to help assess causation in biomedical research [5]. Apart from strength and consistency of the statistical association, he suggested that specificity in terms of outcome, temporality (exposure before outcome), biological gradient, theoretical plausibility, coherence between observational and experimental evidence and analogy also play a role in judging whether an exposure is causing an outcome. All these criteria may help in interpreting evidence for causality, but none of them (except that the exposure must precede the effect) are a formal test of causation.

Kenneth Rothman has provided another useful framework for reasoning about causation by introducing the concept of sufficient and component causes [6]. This concept aims to take into account the complexity of a chain of events and their potential interactions in study design as well as interpretation. For instance, to understand why some chain-smokers never get lung cancer and others are at high risk even after only a few years of cigarette smoking, we can move away from measuring just the average risk in smokers by looking at multiple putative causative factors of those exposed to get a better understanding of the different causative components in developing lung cancer.

## 5.4 Measures of Association

Measuring the association between an exposure and an outcome requires an estimate of the difference of disease occurrence between exposed and non-exposed study subjects. Ratios, i.e. an incidence rate ratio or a risk ratio, belong to the family of relative risks (RR), while the differences between incidences or risks are referred to as absolute differences. The relative risk is a measure of the strength of the association between the exposure and outcome, and thus important when trying to understand the aetiology of a disease. However, when the baseline risk is low, even a seemingly important relative risk for instance 2 (doubling of risk) or 0.5 (halving of risk) may have limited clinical or public health impacts, when we look at the absolute difference. To get the full picture of an association between an exposure and an outcome, both absolute and relative measures are needed. For instance, in a study of radical prostatectomy compared to watchful waiting [7], radical prostatectomy reduced the relative risk of dying from prostate cancer by 38 % (RR = 0.62). However, the absolute risk difference as measured at 15 years was reduced from 21–15 %, an absolute difference of 6 %. In this example, one can see that the interpretation of the strength of the association may differ when looking at the relative versus absolute differences. Another key epidemiological concept is the Number Needed to Treat (NNT) to avoid one event. This can be calculated by taking 1/absolute difference. In the above example, this would be estimated as 1/0.06 = 15 treated to avert 1 prostate cancer death.

## 5.5 Study Population

A research question about cause and effect can thus be addressed with an investigation of an association between an exposure and an outcome. In doing so it is essential to ascertain the group of persons in which we can quantify this association. In the typical epidemiological study we need some follow-up time for the outcomes of interest to develop. Hence, the quantification of the above-mentioned incidences and risks relies on using the observed person-time in our studied individuals. It is common to refer to this defined group of people as the study population and the total person-year experience as the study base. To enumerate the total person-time in the study base it is thus necessary to clearly define the time at risk that each person contributes to a study. In a study of prognosis in breast cancer, the study population could be women diagnosed with invasive breast cancer in a certain institution, person-time could be time from breast cancer diagnosis until breast cancer death, death from other causes or end of study, whichever comes first.

The term target population refers to the target for our findings to be generalised to. Generalisability relates to the validity of a study by relating the people in the study population to people outside the study population. For instance, a study based on men with prostate cancer diagnosed between 2008 and 2013 at Guy's Hospital in London (study population) to identify the effect of ethnicity (exposure) on prostate cancer-specific death (outcome) aims to be generalisable for men with prostate cancer in the UK (target population).

## 5.6 Study Design

There is a wide variety of study designs available in epidemiology. For a detailed description of each study design we refer to Modern Epidemiology by Rothman et al. [8]. A brief description of the most common study designs is provided below.

Study designs can be classified as prospective and retrospective. If the researcher has collected information about the exposure before the outcome event, a study is considered to be prospective. In contrast, if information about the exposure is collected after the outcome, the study is retrospective. Thus, this classification refers to time of information collection.

### 5.6.1 Experimental Studies

In an experimental study investigators manipulate the conditions under study, i.e. they assign the study subjects to the exposure, most often a medical intervention. However, it is often very expensive to conduct a large experimental study and sometimes it is ethically not possible to allow for the investigator to assign people to different exposures (i.e. smoking or histopathological tumour type), so that observational or non-experimental studies are required.

A randomised clinical trial (RCT) is the gold standard of experimental studies. It studies the effect of an intervention applied by the investigator to two or more groups that are built up by random assignment. The major advantage of a trial over an observational study is its ability to circumvent selection bias, which is a major threat to the validity of observational studies (see below). By randomising participants to the intervention or the control arm, it is assumed that all baseline characteristics of the study population (i.e. demographics, medical history) are distributed similarly between both arms.

Another advantage with experimental studies is the possibility to design and follow a strict protocol, not only for the assignment and conduct of the intervention, but also for the information collection. Randomisation helps preventing selection bias, but there is no guarantee against information bias (see below). To ensure that information about treatment effects is obtained as objective as possible, studies are often designed as double-blinded, i.e. neither the investigators nor the participants know who was given the treatment and who was not. Thus, blinding refers to whether the participant, treating physician or researchers have knowledge about randomisation to the intervention arm. However, sometimes blinding is only possible in one-way (single-blinded) or even impossible due to the exposure being studied (i.e., new surgical techniques).

In the context of testing new therapies, experimental studies are often subdivided into phase I, II, III and IV studies. This categorisation depends on the stage of testing. Once preclinical or experimental studies have been conducted, a phase I study is performed to test the safety of the therapy in a few 'unblinded' volunteers. When the phase I study has shown safe results, a phase II study is conducted to test tolerability and define intensity of dose of the intervention. A phase III study is a much larger RCT to test the effect of the new therapy on the specified clinical outcome. Following a successful outcome of a phase III study, a new therapy may get approved by the relevant agencies (i.e. the Food and Drug Administration in the United States). However, the approval still requires follow-up with a phase IV study to identify the rate of serious adverse effects over a longer follow-up time. This study is either performed as an RCT or as an observational study.

BOLERO-3 is an example of a recently published randomised, double-blinded, placebo controlled phase III trial. It was designed to investigate a new drug regime for women with HER2-positive advanced breast cancer [9]. Eligible patients were randomly assigned to two different cytotoxic regimens. The primary outcome was progression-free survival. The statistical analyses showed that the addition of one drug to the standard treatment prolonged progression-free survival.

### 5.6.2 Observational Studies

In an observational study, the exposure is not manipulated by the investigators, but observed in a way that would allow some degree of causal interpretation—using what is sometimes referred to as the "natural experiment". The above-described example of the cholera epidemic is a prime example.

The most common observational study design in epidemiology is a cohort study, which refers to a cohort of people who are followed overtime to identify whether they develop an event of interest. Usually a cohort is considered to be a fixed roster of study participants who stop contributing person-time either when they develop the outcome of interest, die, are lost to follow-up or at the end of the study. The cohort members can be defined at one point in time or recruited over some inclusion time. An RCT is a special case of a cohort study comparing two (or more) cohorts based on their exposure status.

The European Prospective Investigation into Cancer and Nutrition (EPIC) is a cohort study with more than half a million participants recruited across ten European countries and followed for almost 15 years. It was designed to investigate the associations between different exposures to diet, nutritional status, lifestyle and environmental factors and the incidence of cancer and other chronic diseases [10]. Another well-known cohort study is the American Nurses' Health Study (NHS) [11], which is one of the largest and longest running investigations of factors that influence women's health. It started in 1976 and the information provided by the 238,000 dedicated nurse-participants has led to many new insights into health and disease.

Case-control studies are the other typical design for non-experimental or observational studies. The design starts with identifying those who have the outcome (cases) and a set of controls. A case-control study can be seen as the sampling of an existing or tentative cohort. Cases and controls should come from the same person-time experience, i.e. study base. The term "nested case control study" is often used when the cases and controls are drawn from an existing cohort study. Cases and controls are compared on the basis of an exposure variable for which information was collected either retrospectively or prospectively. Most case-control studies are retrospective because information about the exposure is collected after the cases and controls are defined. The Northern California Childhood Leukaemia Study is an example of a population-based case-control study of risk factors for childhood leukaemia [12]. Cases were identified from paediatric oncology centres in the Northern California region and controls were identified from birth certificates. Information on maternal diet was collected using a food frequency questionnaire to assess the diet 12 months prior to pregnancy. The statistical analyses showed that women with a high intake of fruits and vegetables had children with a lower risk of leukaemia.

Case control studies can be very effective in several situations. When the outcome under study is rare, it would be extremely expensive and time-consuming to collect all relevant information on exposures in a large cohort study as it would only result in a few cases. Moreover, when one wants to study an exposure that is very expensive or time-consuming to measure (i.e. staining for a specific biomarker in cancer patients to predict recurrence), it may be more suitable to conduct a case-control study.

Cross-sectional studies, also called prevalence studies, involve all persons (or a subset) in the population at a specific time of ascertainment of their health status. These studies do not include a longitudinal component and are hence often used to describe the prevalence of a disease since both newly and previously diagnosed disease will be enumerated. The American National Health and Nutrition Examination Survey (NHANES) is a programme of studies designed to assess the health and nutritional status of adults and children. It is run by the National Centre for Health Statistics with the aim to produce vital and health statistics for the nation [13]. NHANES includes information from an interview on demographic, socioeconomic, dietary and health-related issues as well as an examination involving medical, dental and physiological assessments. It is thus used to determine the prevalence of major diseases and risk factors for diseases.

Ecological studies use data from a population level, defined based on geographical or temporal characteristics. Individual information on the potential relation between exposure and outcome is therefore missing. In the context of cancer, there are many examples investigating dietary and lifestyle risk factors using geographical or temporal ecological study designs. A main threat to the validity of ecological studies is the lack of individual information. We usually have little or no control over the distribution of important exposures and/or confounders in the populations and a specific sub-group of the population may be over or underexposed. Therefore, biological interpretation from ecological data is limited. Even if one population is at large much more exposed to a certain factor, it may not be the exposed ones but rather the un-exposed minority in that population is the source of most of the outcomes we are looking for.

## 5.7 Bias and Confounding

Assessment and attempts to control for systematic errors (bias) in a study is a prime concern in epidemiology to ensure that the results of a study reflect a real-world situation as much as possible. Errors will always be present to some extent, but an understanding of these errors makes the reader and the researcher able to judge whether the problems can be tackled and minimised. It would allow assessing whether these errors have influenced the results in an important way and whether a study can contribute to the existing evidence in a specific research area. There is no external validity without the study being internally valid.

Internal validity of a study can be affected by biases due to errors in the selection of the study population, the collection of information on the exposure or the outcome and failure to account for other external factors or individual characteristics associated with exposure status that may influence outcome. These systematic errors in the study design may thus cause bias by showing an association between the exposure and the outcome that is different from what is the causal association. The following section describes in more detail what is meant in epidemiology by selection bias, information bias and confounding [14].

Selection bias refers to errors in the selection of the study population. For instance, when a cohort study aims to compare the outcome for two cohorts with different exposures, selection bias would be present if the cohorts are assembled so that they would have a different outcome regardless of exposure. Selection bias is particularly common in observational studies of medical interventions and may appear in many different ways. For example, when studying patient outcomes in a group of patients who underwent a small surgical procedure compared to a group of patients undergoing a large surgical procedure, it is generally known that those selected for the latter have a better general health status as they are considered to be suitable to undergo such a procedure. In this case the attending physician would have made the selection. In addition, self-selection is often an issue when performing cancer screening studies. For instance, a study analysed the epidemiology of irritable bowel syndrome (IBS) symptoms in a random sample of the general population and a subsample consenting to a colonoscopy to assess how IBS may introduce selection bias [15]. All 3347 randomly selected adults were mailed an abdominal symptom questionnaire and all responders were contacted to consent for a colonoscopy. All non-responders were also contacted to ask seven key questions of this validated questionnaire. It was found that the prevalence of IBS-like symptoms was highest among the responders who consented for a colonoscopy and lowest for the non-responders, suggesting that when studying the effect of colonoscopy on cancer outcomes self-selection would introduce bias as symptomatic subjects—and some of the symptoms might be due to cancer—are more likely to go for screening than the general population.

Information bias refers to errors in the collection of information on the exposure or outcome. We already mentioned above that not even a RCT is completely free from information bias and hence double-blinding is often required when studying new medications. In retrospective data collection of exposure status (i.e. through use of medical records), it is an advantage to blind the assessor to outcome or case status of a study participant. In a case-control study, recall bias is a well-known example of differential misclassification of the exposure. For example, a nested case-control study in the NHS examined whether diet during preschool age affects a woman's risk of breast cancer later in life [16]. Information concerning the nurses' childhood diet at ages 3–5 was obtained from their mothers with a food frequency questionnaire. Recall bias is an issue not only because of genuine forgetfulness, but also due to potential under or over-reporting of specific dietary habits by mothers of women who have developed breast cancer especially if some dietary habits have been publicly discussed as associated with the risk under study. Misclassifications of the kind discussed above are called differential, i.e. different by exposure or outcome status. Misclassification can also be non-differential, meaning that it is similar for exposed/non-exposed or cases/non-cases. In general, this would lead to less statistical precision to study the observed associations.

Confounding is the problem of confusing or mixing of the exposure effects under study with other "extraneous" effects [17]. Factors responsible for confounding are called confounders and need to fulfil the following three criteria (Fig. 5.2):

- associated with the outcome irrespective of exposure;
- associated with exposure in the study base irrespective of outcome and
- not an intermediate in the pathway between exposure and outcome.

For instance, when studying the association between obesity (exposure) and risk of bladder cancer (outcome), smoking may be a confounder [18]:

- smoking is associated with risk of bladder cancer in both obese and non-obese populations;
- if smoking is non-causally associated with obesity in the particular population giving rise to the study base and
- smoking is not an intermediate in the pathway between obesity and risk of bladder cancer.

Therefore, confounding by smoking needs to be addressed in this study.

Ideally, bias and confounding should be dealt with as much as possible in the study design and the study protocol. This would include definition of the study population, a detailed plan for data

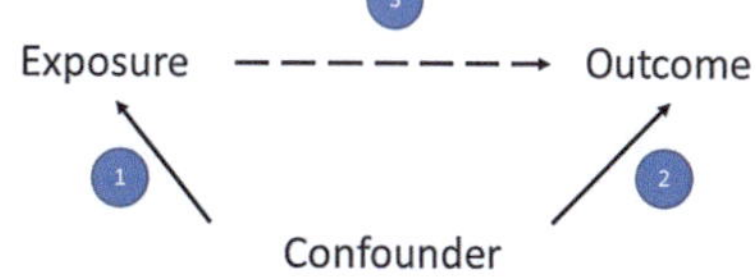


**Fig. 5.2** Do the criteria for confounding hold?

collection on exposure, outcome and confounding variables, as well as a statistical analysis plan. Even when an appropriate strategy and statistical plan are in place, if the study design is seriously flawed and important pieces of information are lacking, no statistical method can vouch for the internal validity. A more detailed description of multivariate statistical methods in epidemiology can be found elsewhere [4].

## 5.8 Summary: Epidemiological Methods as a Toolbox

Epidemiology is thus much more than the study of epidemics. It is a toolbox of research methods that can be used whenever studying some event or characteristic of a population (i.e., humans, animals, etc.). An epidemiological study is always defined by a research question that wants to describe the frequency distribution of an event or address the association between an exposure and an outcome. Once you have defined the ideal study base for the study you have to select a study population in which you then according to a study plan (protocol) obtain information on exposure and outcome. Statistical methods are then applied to quantify this association. The researchers have to be very careful throughout this entire process to minimise bias and maximise the internal and external validity of the study.

Epidemiology very often thus addresses hypotheses about the association between an exposure and an outcome, with a final aim to understand the causation of a disease or health status. The testing of new medical interventions in RCTs has developed into a field of its own. However, the many situations where an RCT cannot be undertaken, has spurred the recent development of epidemiological research methods trying to utilise the "natural experiment" as an RCT—these new methods are often referred to as "causal inference methods" [19]. Hopefully, this toolbox of epidemiological methods will assist researchers in becoming the John Snows of modern cancer studies.

## References


1. World Health Organisation. Epidemiology 2014. Available from http://www.who.int/topics/epidemiology/en/.
2. International Agency for Research on Cancer. GLOBOCAN 2012: estimated cancer incidence, mortality and prevalence worldwide in 2012. World Health Organisation;2014. Cited 16 July 2014. Available from http://globocan.iarc.fr/Pages/fact_sheets_cancer.aspx.
3. Savitz D. Interpreting epidemiologic evidence: strategies for study design and analysis. New York: Oxford University Press; 2003.
4. Holford T. Multivariate methods in epidemiology. New York: Oxford University Press Inc.; 2002.
5. Hill AB. The environment and disease: association or causation? Proc R Soc Med. 1965;58:295–300.
6. Rothman KJ. Causes. Am J Epidemiol. 1976;104(6):587–92.
7. Bill-Axelson A, Holmberg L, Ruutu M, Garmo H, Stark JR, Busch C, et al. Radical prostatectomy versus watchful waiting in early prostate cancer. N Engl J Med. 2011;364(18):1708–17.
8. Rothman K, Greenland S, Lash T. Modern epidemiology. 3rd ed. Philadelphia, PA: Lippincott, Williams & Wilkins; 2008.
9. Andre F, O'Regan R, Ozguroglu M, Toi M, Xu B, Jerusalem G, et al. Everolimus for women with trastuzumab-resistant, HER2-positive, advanced breast cancer (BOLERO-3): a randomised, double-blind, placebo-controlled phase 3 trial. Lancet Oncol. 2014;15(6):580–91.
10. International Agency for Research on Cancer. EPIC study 2014. Cited 16 July 2014. Available from http://epic.iarc.fr/.
11. Channing Laboratory. Nurses' health study 2014. Cited 16 July 2014. Available from http://www.channing.harvard.edu/nhs/.
12. Jensen CD, Block G, Buffler P, Ma X, Selvin S, Month S. Maternal dietary risk factors in childhood acute lymphoblastic leukemia (United States). Cancer Causes Control (CCC). 2004;15(6):559–70.
13. Centers for Disease Control and Prevention. National health and nutrition examination survey 2014. Cited 16 July 2014. Available from http://www.cdc.gov/nchs/nhanes/about_nhanes.htm.
14. Rothman K, Greenland S. Precision and validity in epidemiologic studies. In: Rothman K, Greenland S, editors. Modern epidemiology. 2nd ed. Philadelphia, PA: Lippincott Williams & Wilkins;1998. p. 115–34.
15. Kjellstrom L, Molinder H, Agreus L, Nyhlin H, Talley NJ, Andreasson A. A randomly selected population sample undergoing colonoscopy: prevalence of the irritable bowel syndrome and the impact of selection factors. Eur J Gastroenterol Hepatol. 2014;26(3):268–75.

16. Michels KB, Rosner BA, Chumlea WC, Colditz GA, Willett WC. Preschool diet and adult risk of breast cancer. Int J Cancer (Journal international du cancer). 2006;118(3):749–54.
17. Greenland S, Robins JM. Identifiability, exchangeability and confounding revisited. Epidemiol Perspect Innovations. 2009;6:4.
18. Koebnick C, Michaud D, Moore SC, Park Y, Hollenbeck A, Ballard-Barbash R, et al. Body mass index, physical activity, and bladder cancer in a large prospective study. Cancer Epidemiol Biomarkers Prev. 2008;17 (5):1214–21 (A publication of the American Association for Cancer Research, cosponsored by the American Society of Preventive Oncology).
19. Hernan MA, Robins JM. Estimating causal effects from epidemiological data. J Epidemiol Community Health. 2006;60(7):578–86.

# Cancer Screening

# 6

Lorelei A. Mucci, Kathryn M. Wilson and Jennifer R. Rider

## 6.1 Overview

Screening involves the examination of asymptomatic individuals in order to classify them as likely or unlikely to have the disease of interest. In 1968, the World Health Organization (WHO) first described the following additional criteria for screening programs [1]:

- The disease must be a serious health problem that causes significant morbidity and/or mortality in the general or a particular population.
- Effective treatments are available for earlier stages of disease, these potentially decreasing morbidity and mortality.
- Screening tests used to diagnose various forms of disease must be safe, relatively easy to apply, rapid, inexpensive, reproducible, and must have acceptable performance (sensitivity, specificity, accurate predictive values).

Screening is common for several different cancers using a variety of modalities, including breast (mammography), colon (colonoscopy), prostate (prostate-specific antigen, PSA), cervical (Pap Smear), and lung (CT scan) cancers. The goal of cancer screening is to reduce the morbidity and mortality associated with cancer by diagnosing the disease at a stage when early treatment is presumed to be less intrusive, less prone to complications, and more effective in reducing mortality than delaying treatment until when symptoms occur [2]. There is considerable debate around the effectiveness of screening for several of the above cancers. Understanding the benefits and risks of screening relies on a number of key attributes of the disease itself, the screening test, and ultimately the screening program in the population selected for testing.

## 6.2 Characteristics of a Cancer Suitable for Screening

For cancer screening to be effective, the disease must pass through a preclinical phase during which the cancer does not cause symptoms but is detectable by clinical examination or through measure of a biological marker [3]. In addition, the duration of this preclinical phase must be sufficiently long and provide a window of time for screening to occur. Screening is not useful if it is unable to detect the cancer much before the

L.A. Mucci (✉)
Harvard T.H. Chan School of Public Health, 677 Huntington Ave., Kresge 920a, Boston, MA 02115, USA
e-mail: lmucci@hsph.harvard.edu

K.M. Wilson
Harvard T.H. Chan School of Public Health, 677 Huntington Ave., Kresge 920b, Boston, MA 02115, USA

J.R. Rider
Department of Epidemiology, Boston University School of Public Health, 715 Albany Street Talbot T317E, Boston, MA 02118, USA

J.R. Rider
Harvard T.H. Chan School of Public Health, 677 Huntington Ave., Kresge 920d, Boston, MA 02115, USA


M. Loda et al. (eds.), *Pathology and Epidemiology of Cancer*,
DOI 10.1007/978-3-319-35153-7_6

onset of symptoms, since symptoms would likely eventually bring an individual to clinical attention. In addition, the prevalence of the disease must be high enough to warrant screening in a population. This point is key both with respect to health care costs associated with screening as well as the potential for false positives in a population where most individuals do not have the cancer of interest. Finally, if available treatment of the cancer at the preclinical phase is no more effective than treatment at the symptomatic stage, then the cancer is not suitable to screening.

### 6.2.1 The Preclinical Phase

Figure 6.1 presents an overview of the preclinical phase in the context of cancer's natural history. The preclinical phase theoretically begins at a point after the initial tumor forms and lasts until symptomatic disease occurs [4]. This window of time, which is critical for assessing the value of screening, is not always easily determined. Indeed, the window of the preclinical phase for screening actually depends on the screening tools available. For example, if one is using a clinical examination to screen, the preclinical phase will not begin until the tumor is of sufficient size for the clinical assessment to identify the growth. The use of blood or urine-based biomarkers assays can move the start of the preclinical phase to an earlier time when the tumor produces sufficient amounts of the biomarker for detection. It is noteworthy that there may be considerable between-person variability in the length of the preclinical phase for a specific cancer based on biological characteristics of each individual's disease. Some tumors may grow more slowly, while others proceed rapidly to symptomatic or metastatic disease.

### 6.2.2 Cancer Prevalence

Another key attribute for screening is a cancer's prevalence, particularly in its preclinical phase [4]. Prevalence is a function of the incidence of the disease and its duration in its preclinical phase. Thus, if the incidence of a cancer is low or the duration between biological onset of a tumor and symptomatic disease is short, the cancer's prevalence will be too low to make population-based screening effective. Recent screening in a population can also decrease the frequency of preclinical disease in a population. Although the prevalence of a cancer may be low in the general population, it may be possible to define a subgroup of individuals with an increased likelihood of the cancer based on risk factor profiles. For example, the prevalence may be higher for lung cancer among long-term cigarette smokers, hepatocellular carcinoma among carriers of the hepatitis B virus, or mammography screening among women who carry mutations in *BRCA1*. If the relation between these factors and cancer incidence is of sufficient strength, prevalence in these subgroups may be high enough for screening to be useful.

### 6.2.3 Cancer Metastases and Mortality

The development of metastatic disease in cancer is the major cause of cancer death in patients. A key goal of cancer screening is to identify and

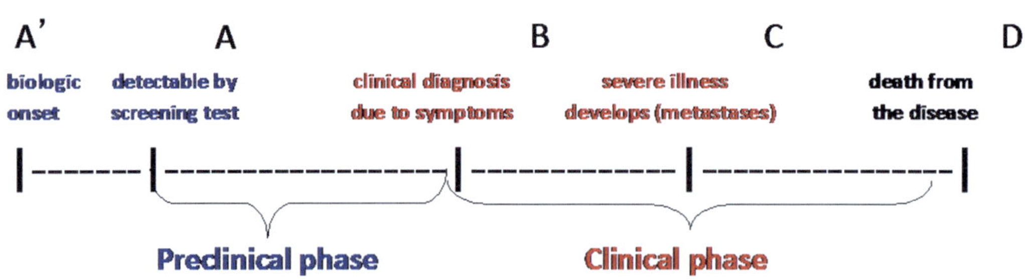


**Fig. 6.1** The preclinical phase in cancer screening

treat cancer at a window of time before the metastasis has occurred. For a cancer to be suitable for screening, the rate of metastasis or death should be sufficiently high. Moreover, earlier intervention and treatment before symptoms occurs should lead to substantial reductions both in morbidity from the cancer and its treatment and in its mortality. As a corollary, screening requires a clear understanding of the cancer's natural history, including the rate at which the disease in its various forms progresses, an understanding of the signs and symptoms of the different forms of disease, and the cancer's amenability at each phase to treatment. Most cancer sites do have sufficiently high risks of metastasis and cancer death to warrant screening.

## 6.3 Characteristics of a Suitable Screening Test

There are several prerequisite criteria for the suitability of a screening test. The test should be relatively simple to administer and perform. For example, if the screening tool requires extensive training or its methodology is complex, there can be considerable variation in how well the test is administered across centers. The test should be rapid, both in its conduct and in its turnaround time to obtain results. The test should have a relatively low cost to benefit ratio. The screening tool and its follow-up exams should be safe, and should cause as little discomfort or potential harm as possible. This is of particular importance since the majority of individuals screened will not have the cancer of interest. Colonoscopy for colorectal cancers notably challenges some of these key tenets of suitability. For the patient, there is some discomfort associated with the colonoscopy preparation, and the procedure itself often requires sedation. The diagnosis of most cancers requires a follow-up biopsy after a positive screening result, and the biopsy can carry risk of discomfort, pain, and infection.

The immediate goal of screening is to correctly identify as positive those individuals who have a cancer in the preclinical phase and as negative those without cancer. Thus, the reliability and validity of the screening tool are essential to effective cancer screening. Reliability refers to the ability of the test to give the same result, whether correct or incorrect, on repeated applications of the test in the same person with a given level of disease [5]. Reliability depends on the intrinsic variability of the factor being measured, the variability of the method used, the skill of the individual performing the measurement and the accuracy of interpretation of the test results. For example, if the screening tool uses a biomarker level, it is critical to understand whether there are diurnal or other variations in these levels independent of the disease itself. Moreover, it is important that the test is reliably measured, whether in a small clinic or large teaching hospital.

### 6.3.1 Sensitivity and Specificity

Sensitivity and specificity are the most commonly used measures of a screening tool's validity to estimate how well the tool correctly classifies someone as having or not having the disease [5]. The screening tool is compared to a gold standard that is accepted as the means by which the presence or absence of a particular disease is established. Figure 6.2 provides a contingency table, which cross-classifies individuals based on the results of a screening test compared to information on an individual's disease status as determined by the gold standard. In this table, "*a*" represents true positives, or individuals who truly have the disease of interest and are correctly classified by the screening test, "*d*" represents true negatives, or individuals who

| | | Gold Standard | | |
|---|---|---|---|---|
| | | Disease Present | Disease Absent | Total |
| Test Result | Positive | a | b | a + b |
| | Negative | c | d | c + d |
| | Total | a + c | b + d | |

**Fig. 6.2** A 2 × 2 contingency table for determining test characteristics of a screening test: sensitivity, specificity, positive, and negative predictive values

truly do not have the disease, "*b*" represents false positives, or individuals who are incorrectly labeled by the screening test as having the disease, and "*c*" represents false negatives, or individuals incorrectly labeled by the test as being negative when in fact they have the disease.

Sensitivity indicates the proportion of individuals with cancer correctly classified by the screening test as having the disease. It is calculated as: True positives (*a*)/[True positives (*a*) + False negatives (*c*)]. The number of false negatives is determined largely by the sensitivity of a test, and also depends on the distribution of preclinical disease stages in the population selected for testing [3, 5]. When cancer is early in the preclinical phase, it is more difficult to detect by a screening test, and the detectability of the disease tends to increase as the preclinical phase progresses [4]. Thus, change in detectability of each case as the disease progresses suggests a sensitivity function, whereby sensitivity increases according to the point in the preclinical phases at which cases are tested [4].

There can therefore be real challenges in determining sensitivity, since the number of individuals who truly have the disease must be determined by another (diagnostic) test. Depending on the cancer, it may be challenging to apply a definitive diagnostic test to asymptomatic individuals. In addition, a screening test may not be as sensitive when applied to a different population [5]. Finally, sensitivity will appear relatively high in the first screening of a population since there is a pool of prevalent preclinical cases.

Specificity is the probability that individuals without cancer will be correctly classified by the screening test as being disease-free. This is calculated as the number of patients without disease who test negative divided by the total number without cancer [True negatives (*d*)/True negatives (*d*) + False positives (*b*)]. The same conceptual difficulties emerge in estimating specificity as in determining sensitivity, since knowledge of numbers in the denominator must be determined on the basis of a given diagnostic test that may or may not in actuality be a true gold standard. For many cancers, the diagnostic tests have inherent risks, and therefore would not be done on asymptomatic people with negative screening test results.

Biomarkers often provide information on a continuous scale. Screening tests that use a biomarker will make decisions about positivity and negativity at a dichotomous cutoff point, such that values above a certain level will indicate a decision for further testing. *Receiver operator characteristic* (*ROC*) *curves* provide a graphic display for studying the effects of using different cutoff points on the performance of a screening test or as a method for comparing competing screening tests. The ROC plots the true-positive rate (Sensitivity) against the false positive rate (1—Specificity) and illustrates the trade-off that exists between them. Defining a specific cutoff point requires tradeoffs. Changing a cutoff point to improve sensitivity will lead to concomitant decreases in specificity, and conversely changing it to improve specificity will decrease sensitivity. The area under the curve is a measure of the test's accuracy. An area that equals 1 indicates a perfect test, while an area that equals 0.5 indicates that the test did no better in predicting disease than chance alone.

For serious diseases it is often best to optimize sensitivity (and thus reduce the number of false negatives), particularly when the subsequent diagnostic test is of low risk to the individual and of relatively low cost. For cancer, this will avoid a missed early diagnosis, which could allow the disease to progress to a more advanced phase before the emergence of symptoms prompt further diagnostic tests. However, in this setting the specificity will be decreased, prompting more falsely positive diagnoses, more retrospectively unnecessary tests and patient anxiety over a disease that ultimately is proven not to be present.

In contrast, it may be preferable to optimize specificity (and thus reduce false positives) when the disease is of low risk for progression and therefore of low consequence, particularly when the subsequent diagnostic tests are of high attendant risk and cost. In these instances, not making a diagnosis earlier may be of benefit in avoiding intensive and protracted treatments. The exception to this might involve an opportunity of facilitating the efficacy of treatments for an earlier phase of disease or even of reversing its full manifestation so that a later expression, even if indolent, can be avoided.

The primary screening tool for prostate cancer is the blood biomarker prostate-specific antigen, PSA. Clinically, a man would come to his physician for a usual-care medical visit without any symptoms, and have a blood drawn. The blood level of PSA provides the patient and his physician information on how likely or less likely he is to have prostate cancer. PSA blood levels are continuous, and a cut point of 4.0 ng/ml or higher generally is used to determine if a follow-up prostate biopsy is needed. Within the Prostate Cancer Prevention Trial (PCPT) study, a PSA of 4.0 ng/ml or higher has a sensitivity of 20 % and a specificity of 98 %. Given this estimate of sensitivity, 20 % of men with prostate cancer will be correctly classified as having prostate cancer with a cutpoint of 4.0 or higher, and the false negative rate is 80 %. Given this estimate of specificity, 98 % of men without prostate cancer will be correctly classified as *not* having prostate cancer, and the false positive rate is 2 %.

### 6.3.2 Positive and Negative Predictive Value

Two additional measures of the validity of a test are its positive predictive value (PPV) and negative predictive value (NPV). PPV is the proportion of individuals classified by the screening test as having the disease among the total number of individuals who have tested positive [True positives ($a$)/True positives ($a$) + False positives ($b$)]. The number of false positives is determined primarily by the specificity of the test because the number of non-diseased individuals in most settings greatly exceeds the number of diseased individuals. Thus, a small decrease in specificity may lead to a very large increase in the number of false positives and thus a large decrease in the PPV. If a positive test result is followed by a repeat screen test or other noninvasive procedure, then a low PPV may be acceptable to the population. However, if a positive screening test is followed by an expensive or potentially harmful diagnostic evaluation, then it is important to use a test that has a high PPV (i.e., a low number of false positive results, indicated by a good specificity). The PPV may influence acceptance of a screening program by the target population, since when the PPV is low a positive screening test represents a false alarm more often, consequently leading to unnecessary testing and anxiety. For instance, the PPV of mammography for breast cancer in screened populations such as the United States is estimated to be between 15 and 30 %.

The NPV is the proportion of individuals classified by the screening test as being disease-free who do not have the disease. This is represented in the contingency table as true negatives ($d$)/False negatives ($c$) + True negatives ($d$). A test with an NPV that approximates a value of 1 indicates that testing negative on the test will be reassuring. However, if the NPV is $<1$ by a value comparable to that of the prevalence of preclinical disease, then most of the preclinical disease will be missed, and the screening test will have a large number of false negative results. A low NPV is more likely to be the result of poor sensitivity than poor specificity.

NPV and PPV are determined by both the test characteristics (i.e., sensitivity and specificity) *and* the prevalence of the disease in the population being tested by the screening test. When the prevalence of a disease increases, the PPV increases, and the NPV decreases. Therefore, the same screening test, with a given sensitivity and specificity, will have different predictive values in populations with different underlying disease prevalence.

## 6.4 Characteristics of a Suitable Screening Program

Beyond its potential impact on reducing mortality, a screening program should meet several additional criteria related to its suitability to the population being tested [4]. The test should be relatively free of discomfort and attendant risks, and it should be convenient and attractive to the target population. The test should be economical, both to the individual and to society. The test should have a high PPV and NPV. Indeed, if the frequency of case detection is low there is little or no return to justify the costs of screening. Moreover, if the frequency of false positives is high, the additional costs and adverse effects from further diagnostic tests will result in the population deriving no benefit and possibly even experiencing a deleterious outcome.

There are methods to improve the suitability of a screening program even if the screening test characteristics are fixed. For example, as described above, it is possible to improve the PPV of a screening program by restricting the program to high-risk individuals or by screening at a lower frequency to maintain a high prevalence of the disease in its preclinical phases in the target population [4]. The latter is more feasible and acceptable in low-risk patients. Targeting high-risk groups of individuals can greatly improve the feasibility of a screening program if the cancer is rare. If it is possible to easily characterize a population as being high risk, such as by the presence of particular risk factors such as smoking or genetics, screening only in this population may capture the majority of cases that would conceivably be detected by screening the entire population, and the overall cost of screening could be reduced. One could also decrease the number of false positives by raising the criteria of positivity, such as screening with two sequential tests and only considering an individual as "screen-positive" if they tested positive on both tests [6]. If there is concern that such a program will miss too many cases, one can lower the criteria for positivity, screen at more frequent intervals, or use two different screening tests in combination and consider a positive result if an individual tests positive on at least one.

It is critical for any screening program that there be appropriate follow-up for individuals who test positive. Thus, there should be a clear procedure for follow-up diagnostic testing that can be instituted quickly for screen-positive cases, and then subsequent therapeutic intervention if the individual is indeed found to have the disease. For any screening program to be successful at reducing mortality and morbidity, a substantial proportion of cases must be detected during the preclinical phase, with enough time for treatment to be more effective than it would have been if given later. Thus, the availability and timeliness of effective treatment are critical components of an appropriate and suitable screening program.

Determining an optimal interval for screening is a further important consideration [7]. With implementation of a one-time screening program, the incidence rate of cancer for the period of screening will be quite high as the prevalent pool of preclinical, screen-detectable cases will be diagnosed. After the one-time screening, the incidence rate will drop to a level lower than expected (without screening) since the cases already diagnosed have been removed from this population [4]. The cumulative incidence (and prevalence) of diagnosed disease, however, will be increased. After the initial jump, the cumulative incidence will then increase more slowly than expected because of the reduced incidence rate.

After a second screening, cancer incidence will again rise. However, the rate of increase will depend on the interval since the initial screen. If the timing between screening tests is too short, then the prevalence of cancer cases in the preclinical phase will be too low. If screening, once begun, is continued indefinitely, the average incidence rate of the disease may increase above the baseline rate indefinitely because of the detection of nonprogressive cancer cases that would not have otherwise come to light clinically and only arise as a result of additional screening. Such cases are called "pseudodisease," and can include both cases that never would have progressed to a symptomatic state and cases that would have progressed except that the individual died of a competing cause before symptoms occurred. There are several examples of cancers with a high prevalence of pseudodisease, including breast and prostate cancer.

## 6.5 Evaluating the Effectiveness of a Cancer Screening Program

The benefit of a screening program in a population can be measured in several ways. Improvements in cancer-specific or overall survival are perhaps the most commonly used measures of effectiveness [8]. In addition, improvements in quality of life and changes in the proportion of individuals diagnosed with advanced disease can be used to show effectiveness of a program, and may be worthwhile measures even if survival is not changed.

The number needed to screen (NNS) is defined as the number of people who need to be screened for a given duration to prevent one death or adverse event. For example, the NNS for Hemoccult testing (i.e., fecal occult blood testing) to prevent a death from colon cancer is 1374, and for mammography to prevent a death from breast cancer is 2451 among women aged 50–59 [9]. The number of years gained by screening can be calculated by multiplying the number of years lost without screening by (1—Relative risk), where the relative risk refers to the estimate of screening effectiveness in the population.

Certain biases may complicate the interpretation of some measures of screening effectiveness [8]. For example, comparing the number of patients with newly diagnosed disease in a screened versus unscreened population may not be useful, since the overall goal of screening is to diagnose disease at an earlier (or otherwise preclinical) stage, and thus the expected incidence rate should be higher. Lead time bias and length bias are two other important issues to consider in estimating screening effectiveness.

Lead time is the interval from disease detection with screening to the time at which diagnosis would have been made without screening, and is thus the amount of time by which the diagnosis was advanced owing to screening. The length of the lead time interval is a function of the timing of screening in relation to the preclinical window, and may be different for different individuals because of the heterogeneity of cancers. Although the lead time interval is not directly observable, a distribution of lead times can be estimated from randomized controlled trials by comparing time of diagnosis between the screened group and the unscreened group from the same baseline start point [10].

The lead time gained by screening and the degree that these lead times improve the effectiveness of early intervention are primary determinants of the effectiveness of a screening program. However, lead time may allow an earlier diagnosis but not necessarily a later death, for instance, if earlier treatment is not more effective, or if screening is not associated with appropriate follow-up care. In this case, the length of time from diagnosis to death will still be longer among screen-detected cases compared to routinely diagnosed cases. If 5-year survival rates are compared between these groups, screening will appear to be beneficial, even though there may be no real gain in survival time. This is known as lead time bias [4].

If a screening program is beneficial, there should be reductions in mortality after the lead time interval [4, 8, 10]. Therefore, the primary outcome of interest in the evaluation of screening should be the late manifestation of disease or death from disease. Moreover, in this situation, the lead time forces an individual to live longer with the knowledge of their cancer diagnosis. In addition, their exposure to more treatments without benefit in terms of actual survival introduces the risk of additional morbidity which these patients might not ordinarily have experienced until later in the course of their disease.

Another potential bias is length bias, which is the phenomenon whereby screen-detected cancer may not be representative of all cases [11]. Cancer can be a disease of considerable biological heterogeneity with different rates of progression. Screen-detected cases tend to have longer preclinical phases, biologically slower progression and somewhat better prognosis than cases detected through diagnostic evaluation of symptoms [10, 11]. Unless the screening test is conducted frequently in a population, screening will preferentially detect cases with a more slowly developing disease, and therefore an intrinsically better prognosis. Thus, because of length bias, screen-detected cases will appear to have improved survival, which may be erroneously attributed to the screening.

The selection of study design, study population, outcome and exposure assessment is critical to evaluate screening effectiveness and avoid the potential introduction of these biases. The randomized controlled trial, introduced in Chap. 5, is considered the gold standard for evaluating the effectiveness of cancer screening. In this design, asymptomatic individuals without cancer are randomly assigned to be part of the screening or unscreened arm and followed prospectively for the outcome of interest, primarily cancer mortality [12]. The length of observation period must account for the natural history of the cancer, and should at least encompass the time by which most screen-detected cases would die of cancer if they were not treated early. Random allocation can take place at the individual or community level. The main strength of randomization is that the distribution of measured and unmeasured factors that could impact cancer survival is randomly distributed among the screened and unscreened groups.

In an intention-to-screen analysis, the disease-specific mortality experience is compared among those randomized to screening and those randomized to no screening, regardless of compliance of the individuals. Indeed, a limitation of randomized trials is that noncompliance or crossover between the groups can result in misclassification so that an effect of screening on mortality is obscured [12]. Other limitations include the large number of participants required, particularly for cancers with few expected deaths, the long-term follow-up generally needed, that the screening technology may have changed by the time the study results are available, that the screening protocol may not be acceptable to physician and/or patient to participate in random allocation, and finally that the unscreened and screened arms do not have equipoise [13].

Observational studies, also introduced in Chap. 5, can provide a cost-effective and valid design to evaluate screening tests or programs when a randomized trial is not feasible [14, 15]. In particular, a case-control study design can be used in which cases are defined as individuals who died of the cancer of interest or have advanced stage or metastatic cancer. Controls should represent a source population that gave rise to the cases, and would include all living members, including people with and without cancer. This is critical, since systematic exclusion of individuals with cancer would tend to remove screened individuals preferentially and reduce the apparent size of the true beneficial effect of screening. Cases and controls should be selected independently of whether they have been screened [14, 15]. In defining screening in a case-control study, the screen period for cases and matched controls ends at the time the cancer diagnosis is made, since the person is then no longer eligible for screening. An exposure window can be defined to approximate the detectable preclinical period in order to find the relevant period based on the natural history of the cancer.

The strength of the case-control design is that it provides an opportunity to examine the efficacy of screening in the absence of randomized controlled trial data and when a screening test has already become widespread in a test population. Moreover, it is a faster, more efficient and lower cost design than a randomized trial. However, case-control studies are prone to biases, including selection bias, the inability to distinguish screening from diagnostic tests, and the lack of randomization, which can result in confounding [14, 15].

## 6.6 Cancers for Which Screening Is Recommended

Recommendations for cancer screening are made by a number of organizations and panels. There can be a lack of consensus across these organizations, which can create challenges for patients and clinicians in interpreting the appropriate cancer screening strategies. In the United States, the Preventive Services Task Force (USPSTF) is one of the leading bodies making consensus statements around screening. USPSTF is an independent group of physicians in primary care and preventive care who review peer-reviewed literature and provide an evaluation on the weight of evidence.

Table 6.1 presents an overview of the recommendations by the Task Force for several of the major cancers. There is good consensus of recommendations for screening for cancers of the lung, cervix and colorectum. However, for other cancer sites there is a lack of consensus agreement. For example, while the USPSTF suggests there is no evidence to begin mammography screening before age 50 [16], the American Cancer Society (ACS) recommends annual mammography screening starting at age 40 without any upper age restriction [17]. For prostate cancer screening, the ACS and American Urological Association recommend informed decision-making starting at age 50 or 55 and older (however, younger for African American men and those with a positive family history) [18]. A consensus statement by the European Urologic Association recommends a baseline prostate-specific antigen (PSA) screening among men age 40–45 and then offering early detection for men with a life expectancy of greater than 10 years [19].

The development of new screening tools is an active area of research, including both for cancers with existing screening tools (e.g., breast and prostate cancer) as well as for cancers without any established screening methods (pancreatic cancer). These new tools include the development of novel biological markers that could be measured in blood, urine or other biospecimens, as well as noninvasive imaging technologies. Collaborations such as the National Cancer Institute's Early Detection Research Network are working to accelerate the development and validation of these screening tools.

**Table 6.1** Summary of recommendations for cancer screening by the United States preventive services task force

| Cancer site | Date | Sufficient evidence | Insufficient evidence |
|---|---|---|---|
| Breast | 2015 | Mammography every two years for women age 50–74<br>Individual choice for women age 40–49 | Screening women age 75 and older |
| Cervical | 2012 | Pap smear for women age 21–65, or<br>Pap smear and HPV test for women age 30–65 | HPV testing before age 30<br>Any screening before age 21<br>Women age 65 and older who have had adequate screening<br>women who have had a hysterectomy |
| Colorectal | 2008 | FOBT, sigmoidoscopy or colonoscopy for men and women age 50–74 | Screening in adults age 75 and older<br>CT colonography or fecal DNA testing |
| Lung | 2013 | Annual chest CT for men and women age 55–79 with 30-year history of smoking | Screening individuals after smoking cessation of 15 years |
| Ovarian | | | Ovarian cancer screening for women of any age (asymptomatic) |
| Prostate | 2012 | | PSA screening for men of any age |
| Skin | | Counseling of children and young adults age 10–24 about reducing UV exposure | Whole body skin exam by physician or self-exam among adults |

*CT* Computed tomography
*FOBT* Fecal occult blood test
*HPV* Human papilloma virus
*Pap smear* Papanicolaou smear
*UV* Ultraviolet

## 6.7 Summary

Screening can be a critical component of cancer prevention, and in the right setting can lead to reductions in cancer morbidity and mortality. A detailed understanding of the principles of screening is needed in order to assess the value of cancer screening in the context of the disease itself, the screening tools, and program. While there is sufficient evidence for a benefit of screening for some cancers, there is an active debate on the benefits of screenings for others, and indeed for some cancers there is an urgent need for the identification of screening tools that could improve mortality outcomes for cancer patients.

## References


1. Wilson JMG, Jungner G. Principles and practice of screening for disease. Geneva: World Health Organization; 1968. p. 164.
2. Hulka BS. Cancer screening. Degrees of proof and practical application. Cancer. 1988;62(8 Suppl):1776–80.
3. Cole P, Morrison AS. Basic issues in population screening for cancer. J Natl Cancer Inst. 1980;64 (5):1263–72.
4. Morrison AS. Screening. In: Rothman KJ, editor. Modern epidemiology. 2nd ed. Philadelphia, PA: Lippincott-Raven; 1998. p. xiii, 737.
5. Mahe C, Gaffikin L. Screening test accuracy studies: how valid are our conclusions? Application to visual inspection methods for cervical screening. Cancer Causes Control CCC. 2005;16(6):657–66.
6. Seed CR, Margaritis AR, Bolton WV, Kiely P, Parker S, Piscitelli L. Improved efficiency of national HIV, HCV, and HTLV antibody testing algorithms based on sequential screening immunoassays. Transfusion. 2003;43(2):226–34.
7. Tabar L, Faberberg G, Day NE, Holmberg L. What is the optimum interval between mammographic screening examinations? An analysis based on the latest results of the Swedish two-county breast cancer screening trial. Br J Cancer. 1987;55(5):547–51.
8. Alibhai SM. Cancer screening: the importance of outcome measures. Critical Rev Oncol/Hematol. 2006;57(3):215–24.
9. Rembold CM. Number needed to screen: development of a statistic for disease screening. Br Med J. 1998;317(7154):307–12.
10. Walter SD, Day NE. Estimation of the duration of a pre-clinical disease state using screening data. Am J Epidemiol. 1983;118(6):865–86.
11. Morrison AS. The effects of early treatment, lead time and length bias on the mortality experienced by cases detected by screening. Int J Epidemiol. 1982;11 (3):261–7.
12. Paci E, Alexander FE. Study design of randomized controlled clinical trials of breast cancer screening. J Natl Cancer Inst Monogr. 1997;22:21–5.
13. Taylor KL, Shelby R, Gelmann E, McGuire C. Quality of life and trial adherence among participants in the prostate, lung, colorectal, and ovarian cancer screening trial. J Natl Cancer Inst. 2004;96(14): 1083–94.
14. Cronin KA, Weed DL, Connor RJ, Prorok PC. Case-control studies of cancer screening: theory and practice. J Natl Cancer Inst. 1998;90(7):498–504.
15. Weiss NS, McKnight B, Stevens NG. Approaches to the analysis of case-control studies of the efficacy of screening for cancer. Am J Epidemiol. 1992;135 (7):817–23.
16. Force USPST. Screening for breast cancer: U.S. preventive services task force recommendation statement. Ann Int Med. 2009;151(10):716–26, W-236.
17. Society AC. Breast cancer screening guidelines 2015.
18. Gomella LG, Liu XS, Trabulsi EJ, Kelly WK, Myers R, Showalter T, et al. Screening for prostate cancer: the current evidence and guidelines controversy. Can J Urol. 2011;18(5):5875–83.
19. Heidenreich A, Bastian PJ, Bellmunt J, Bolla M, Joniau S, van der Kwast T, et al. EAU guidelines on prostate cancer. Part II: treatment of advanced, relapsing, and castration-resistant prostate cancer. Eur Urol. 2014;65(2):467–79.

# 7 Molecular and Genetic Epidemiology of Cancer

Kathryn L. Penney

## 7.1 Twin Studies and Heritability

Cancer is a genetic disease, with inherited variants contributing to its risk and somatic mutations directly driving carcinogenesis and tumor progression. Studies from decades ago determined that cancer is an inherited disease from studies of family aggregation. The family history of individuals with and without a particular cancer can be compared to generate a relative risk measurement for the impact of family history. For example, a woman with one first-degree female relative with breast cancer has a twofold higher risk of developing breast cancer compared to a woman without a family history [1–3]. Men whose father had prostate cancer at any age are more than twice as likely to develop prostate cancer (RR = 2.35, 95 % CI 2.02–2.72) [4]. Individuals with a family history of colon cancer in any first-degree relatives have a 37 % increased risk of developing the disease (OR = 1.37, 95 % CI 1.02–1.85) [5]. However, since behaviors and environmental exposures can also aggregate in families, these types of studies do not allow the familial component to be decomposed into an inherited susceptibility and shared environmental factors.

Heritability is the proportion of observed differences among individuals due to genetics. Therefore, determining heritability can help assess the impact of genetics on the risk of a particular cancer. Twin studies provide important information to disentangle genetic from environmental effects. The components contributing to disease development in a pair of twins are genetic (heritable component), shared environment, and individual environment. Monozygotic (MZ), or "identical", twins share ~100 % of their genome while dizygotic (DZ), or "fraternal", twins share ~50 %, like any siblings. The correlation of disease status ($R$) in DZ and MZ twins can be compared with the following equations:

- $R_{MZ}$ = genetics + shared environment
- $R_{DZ}$ = ½ genetics + shared environment
- Genetics = $2 \times (R_{MZ} - R_{DZ})$

The genetic contribution can range dramatically for different cancers. A classic paper by Lichtenstein et al. [6] analyzed a dataset of 44,788 twin pairs from the Swedish, Danish, and Finnish twin registries to calculate the heritability for many cancers. The authors determined a heritability of 27 % for breast cancer (95 % CI 4–41 %), 35 % for colorectal cancer (95 % CI 10–48 %), and 45 % for prostate cancer (95 % CI 29–50 %). In a recent paper from Hjelmborg et al. [7] within the Nordic Twin Study of Cancer (NorTwinCan), the authors utilized time-to-event analyses to estimate risk concordance and heritability while accounting

K.L. Penney (✉)
Harvard Medical School, Brigham and Women's Hospital, 181 Longwood Ave., Boston, MA 02115, USA
e-mail: kpenney@hsph.harvard.edu


M. Loda et al. (eds.), *Pathology and Epidemiology of Cancer*,
DOI 10.1007/978-3-319-35153-7_7

for censoring and competing risks of death, which had not previously been considered (see reference for methods). After adjusting for possible biases, their analyses demonstrated that the genetic contribution may be even higher than previously thought—the heritability of prostate cancer was estimated to be 58 %. This finding highlights the importance of accounting for additional factors to improve the accuracy of analyses that examine the heritability of disease.

## 7.2 Types of Samples and Genetic Variants

Deoxyribonucleic acid (DNA) is the molecule that provides a blueprint for all the basic functions of life: survival, development, and reproduction. DNA is inherited from our parents—one chromosome from mother, one from father, for each of the 23 chromosomes. DNA is transcribed into RNA, then RNA is translated into proteins. No two individuals have the same DNA; even monozygotic twins acquire some genetic differences during fetal development.

Single nucleotide polymorphisms (SNPs) are the most widely studied inherited variant. SNPs are single nucleotide differences across a population, generally present at a frequency >1 %. Each possible nucleotide is called an allele; almost all SNPs are biallelic, meaning that only two possible alleles exist at that location (locus) across the population. These variants can be coding—located within gene exons that are transcribed—or noncoding. Coding variants include synonymous variants, which do not change the amino acid encoded for the protein, or nonsynonymous, which result in an amino acid substitution. Noncoding variants include those located in the transcriptional regulatory regions of genes, introns of genes, or intergenic regions of the genome. In addition to SNPs, other common types of genetic variation include indels, the insertion or deletion of one or more nucleotides; tandem repeats (such as microsatellites), the adjacent repetition of short sequences; and copy number variants, the deletion or duplication of large sections of DNA.

The existence of particular variants and the frequency of variation differ across ethnic populations, an important consideration when designing genetic epidemiology studies. Linkage disequilibrium (LD)—the nonrandom association of alleles at two or more loci that descend from a single ancestral chromosome, where no recombination has taken place between loci—is common across regions of the genome. The patterns of LD are also different across populations, which is important when assessing haplotypes (discussed in Sect. 7.5) [8].

Germline DNA (inherited DNA) is often studied to assess the risk of cancer and can be obtained from cells in blood or saliva. Additionally, an important source of DNA for many cancer studies is normal tissue that is removed during diagnostic or therapeutic procedures. While the vast majority of the genome will be identical to the true germline, it is possible that somatic mutations (mutations that occur after conception) have occurred in these "normal" tissues. However, the relatively constant nature of the germline genome throughout life makes it an appealing exposure to study.

As cancer is a disease caused by the accumulation of mutations, DNA can also be obtained from tumor tissue in order to identify the many somatic mutations present in cancer. A comparison of DNA from the tumor to the germline DNA will identify these somatic alterations. Determining which of these mutations are cancer "drivers" of the disease, and therefore drug targets, or merely "passengers" is currently a major focus of cancer genetic research.

## 7.3 Family-Based and Association Studies

Two types of association study designs are commonly used in genetic epidemiology. The first is family-based studies. These are often trios, comprised of parents and an affected offspring, or other combinations of affected and unaffected family members. The second is case-control studies, where unrelated individuals with the disease and individuals without the disease are compared. Case-control studies are more common in the study of cancer as most cancers have a later age at onset and diagnosis, and it is therefore more difficult to study a patient's parents or other family members. A family history of cancer is often an important covariate to include in genetic studies.

In a case-control study, individuals with cancer (cases) are compared to those without cancer (controls). These study participants can be matched on different factors, such as age or race; however, further matching is less crucial in genetic studies than in other epidemiologic studies due to the limited possibility of confounding (discussed in Sect. 7.5). If cases and controls are selected from a larger cohort study, this can be called a "nested" case-control study. An advantage to including participants from a cohort is that often other types of exposure or outcome data have been collected, and can therefore be interrogated using the same genetic data and set of participants.

## 7.4 Types of Genetic Studies and Resulting Data

In the years since the sequencing of the human genome [9], several different types of genetic studies have been conducted, each generating increasingly more data. This is largely in part to advances in technology with ever decreasing costs to generate more data, coupled with the improved understanding of the characteristics of the genome. Candidate gene or candidate SNP studies, based on an investigator's hypotheses about cancer, were considered cutting-edge in the 1990s and early 2000s. Researchers deposited SNPs they discovered into the National Center for Biotechnology Information (NCBI) SNP database (dbSNP) [10]. For candidate studies, SNPs were chosen from this database within a region or gene of interest, without preexisting knowledge of the relationship between those SNPs within a population.

The International HapMap Project, initiated in 2002, changed this process [11]. The HapMap cataloged genetic variants, their frequencies in different global ethnic populations, and the correlation between variants. This catalog made it possible for researchers to choose a subset of SNPs that "tagged" for entire haplotypes. This tagging is possible because LD is common across regions of the genome. This haplotype tagging allowed for fewer SNPs to be genotyped while still capturing a larger amount of information, as shown in Fig. 7.1.

One of the issues that plagued early genetic epidemiologic research was the lack of replication of findings across studies. Few of the SNPs identified during this time have actually been confirmed as bona fide cancer risk SNPs. During the mid-2000s, investigators began performing genome-wide association studies (GWAS). The information from the HapMap and innovations in array-based genotyping technology allowed for the interrogation of hundreds of thousands of SNPs simultaneously. Given the very large number of tests performed, a SNP must reach a genome-wide p value threshold to be considered significant (generally $p < 5 \times 10^{-8}$—which is 0.05/1,000,000 tests). To encourage the

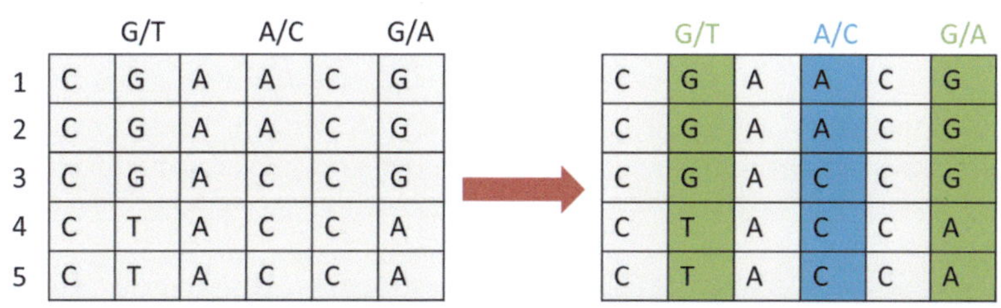


**Fig. 7.1** The *left table* shows six nucleotides from five individuals; three of these nucleotides vary and are therefore SNPs. From the *right table*, we can see that the two SNPs in green (G/T and G/A) are strongly correlated. If we genotype only the G/T SNP, we will also know the allele present at the G/A location

publication of valid findings, very specific study design and publication criteria were developed for GWAS. To publish a manuscript in Nature Genetics, associations from GWAS studies must be observed in at least two independent cohorts [12]. In 2012, the editorial board additionally stipulated that authors report the co-location of disease-associated variants with gene regulatory elements identified by epigenetic, functional, and conservation criteria. Authors are also asked to publish or include in a public database genotype frequencies or p values for associations for all SNPs investigated, regardless of whether they reached genome-wide significance [13].

Hundreds of GWAS have now been published, identifying thousands of SNPs associated with various diseases, including cancers. A good resource for updated results on confirmed genetic variants of disease can be found at the National Human Genome Research Institute's website (https://www.genome.gov/26525384) [14]. One early story of the success of GWAS in cancer was the discovery of a locus on chromosome 8q24 that was associated with prostate cancer [15]. Many more SNPs in this region have since been identified, and several of these are associated with the risk of multiple cancer types, including breast, colorectal, and ovarian cancer [16]. The majority of the genetic variants associated with cancer risk has very small effect sizes (odds ratios commonly $\sim$1.1–1.2 per each risk allele). The risk estimates can also differ by ethnicity, highlighting the importance of conducting these large-scale studies in multiple ethnic populations.

Technology has continued to advance making it possible to obtain the sequence of the entire exome or genome. The 1000 Genomes Project, which is the first project to sequence the genomes of a large amount of people, allows for imputation of more variants from GWAS-level data [17]. The goal of the project is to find genetic variants that are present in at least 1 % of study populations, which can be done through light sequencing of individuals. Currently, finding the complete genomic sequence of one person requires sequencing an individuals' DNA 28 times (28×), however, due to expense, data is usually combined across many samples to detect most of the variants in a particular region. Currently, the 1000 Genomes Project plans to sequence each sample 4× to detect variants with frequencies as low as 1 %. As the price continues to decline, more studies will include whole genome sequencing to detect variants associated with cancer. This leads to the identification of many rare (<1 %) and private (<0.01 %) variants. Given that power will be limited to study these variants individually, methods have been developed that study variation across a gene or region in aggregate.

## 7.5 Bias in Genetic Studies

As explained in Chap. 5, confounding is a bias that results when there is a third factor that influences (is causally related to) the exposure and the outcome. The presence of this bias and methods to correct for it, either through stratification or adjustment in statistical models, is typically a large component of epidemiologic studies However, given that there are not many "causes" of germline genetics, confounding is much less of a concern in genetic epidemiology. One major exception to this is confounding by ethnicity or race, often referred to as "population stratification" or "population structure." The frequency of SNPs differs in different ethnic groups, as described above; race or ethnicity is additionally often associated through other mechanisms with disease outcome. Prostate cancer risk is substantially higher in African Americans than European Americans, and the frequency of many SNPs varies greatly between African Americans and European Americans. If ethnicity were ignored when designing a study, a larger percentage of the cases than of the controls would be African Americans—any SNPs discovered to be associated with case/control status could potentially only be markers for ethnicity.

Most studies are more carefully conducted and avoid blatant bias by restricting to one self-reported ethnicity. However, even in this situation more "cryptic" population stratification can exist. For instance, African Americans have varying percentage of African ancestry and there are genetic differences between Northern and Southern Europeans. To statistically correct for this, a principal components analysis can be run on hundreds of SNPs or even the entire GWAS dataset, using software such as Eigenstrat [18]. The principal components are modeling the ancestral differences in the cases and controls. By correcting for the significant principal components, any underlying bias by ethnicity will be removed.

This type of correction with principal components can also remove misclassification which could result from technical errors. While new genotyping and sequencing technologies are extremely accurate, missing data or incorrect genotype calls can occur. Again, serious bias can be avoided by careful study design—interspersing cases and controls across an array, for example—but a small amount of bias may still be present. Investigators tend to exclude SNPs with >5 % missing data as well as individuals who are missing >5 % of data from an analysis to avoid the possible inclusion of flawed data.

## 7.6 Gene x Environment Interactions

The impact of many traditional epidemiologic risk factors on disease may be modified by an individual's genetic background (the reciprocal is therefore also true, that the impact of the genetic background may be modified by lifestyle, diet, or environmental exposures). This type of effect modification is referred to as "gene x environment interaction." A classic example is Phenylketonuria (PKU) and phenylalanine—individuals with this rare inherited disease caused by a variant in the PAH gene cannot metabolize phenylalanine, which accumulates and leads to mental retardation. However, these downstream phenotypic effects can be avoided by adopting a phenylalanine-free diet high in fruits and vegetables and low-protein breads and pastas [19]. PKU is actually quite common, affecting 1 in 15,000 infants in the United States, but due to widespread screening programs and awareness most grow up unaffected [20].

Gene x environment interactions are appealing in the field of cancer prevention, as an individual's genetic background cannot be altered, but one's lifestyle can. It has been thought that individuals at greater risk due to their genetics may be more likely to change their behavior, though this has not always transpired in practice. Results from a recent randomized controlled trial of 783 patients at average risk of colorectal cancer within four medical school affiliated primary care practices found that individuals who were informed they were at an increased risk for colorectal cancer based on gene environmental risk assessment (GERA) were no more likely to be screened than individuals who received usual care [21]. Gene x environment studies are often performed using candidate genes/variants and environmental factors thought to be involved independently with the disease process. These types of studies often have limited power, especially when carried out at the genome-wide level, and are subject to all the usual biases present in epidemiologic studies (see references for methodology) [22, 23].

Epistasis is when the effect of one gene (or variant) is modified by another gene (or variant), leading to a nonadditive effect. However, testing all pairwise sets of genetic variants in a GWAS leads to so many tests that power is always a limitation, so few examples of epistasis exist in the literature.

## 7.7 GWAS Follow-Up and Linking SNPs to Function

One of the original hopes for GWAS was that common variants identified for common diseases could be utilized for risk prediction. Now that dozens of variants have been identified for many cancers, research is beginning to demonstrate that this may be possible. For example, using 25 of the known prostate cancer risk SNPs, a genetic risk score was developed and applied to 40,414 individuals. The men in the top 1 % of the risk distribution were 4.2 times more likely to develop prostate cancer than men with the median risk [24]. This type of information may prove incredibly useful in the future to make decisions about who should receive cancer screening.

A major contribution of GWAS has been the improvement in the understanding of the biology of many diseases. In retrospect, many of our "candidate" genes were poor hypotheses and there have been surprising findings about the underlying genetic causes of disease. Discovering the mechanisms by which these risk variants impact disease has furthered knowledge about specific diseases, and also about genetic architecture in general. While many assumed that these risk variants would be protein coding, many of the identified risk loci are actually located in introns or even intergenic regions. Therefore, while most variants are not directly changing protein structure or function, many of them are thought to alter gene expression through transcription. For example, using publicly available data from The Cancer Genome Atlas (TCGA), researchers tested the association of 149 known cancer risk loci with gene expression from five different tumor types and found 28.2 % were significantly associated with at least one gene [25]. Identifying the mechanism by which the variants impact cancer risk and the genes involved may help lead to improved prevention and treatment strategies.

## 7.8 The Personal Genome and Personalized Medicine

Identification of genetic risk loci can eventually lead to personalized prevention and risk prediction. Already companies, such as 23&Me, have provided a service to the public that genotypes many risk SNPs for numerous diseases and provides an interpretation of the results [26]. However, the widespread use of genetic analysis tools like this is currently in question, after the FDA in 2013 issue a warning that 23&Me had not yet received marketing clearance or approval to be a medical test [27]. In compliance with this warning, 23&Me briefly stopped offering comprehensive genetic testing related directly to health, but after discussions with the FDA has reimplemented their test [28]. These types of tests are likely to be commonplace in the future and soon everyone may know the sequence of their own personal genome. Helping individuals understand that carrying an allele that is associated with an increase in risk of a specific disease is not an absolute, but only relative to the average risk (which is often relatively low for each specific cancer) can be challenging. Careful interpretation, explanation, and implementation will be necessary for this type of information to be useful to individuals. Additionally, the SNPs need to lead to improved prediction (as described in Sect. 7.7) so that physicians can modify policies and apply different follow-up actions to those with differing genetic risk. Personalized medicine in oncology today often focuses on the sequencing of mutations within a tumor to identify drug targets. But germline variants may prove to be equally useful determining risk, improving cancer screening practices, and evaluating drug efficacy and the development of side effects. Pharmacogenetics, the study of genetics and response to therapeutics, has had great success identifying germline variants that predict cancer treatment toxicity (reviewed in [29]). This field will most likely expand and identify many more germline and somatic variants that influence treatment response.

## 7.9 Conclusion

Epidemiologic methods are well suited to the study of genetic variants and molecular data on a population level. Many different types of genetic data can be analyzed, with many study designs to consider. Genetic epidemiologic studies have identified hundreds of genetic risk loci for cancer, and molecular studies are underway to identify the biological mechanisms. The discovery of the variants that impact cancer risk may improve screening practices and risk prediction, and the identification of variants, genes, and pathways underlying carcinogenesis may lead to new drug development and personalized medicine.

## References


1. Collaborative Group on Hormonal Factors in Breast Cancer. Familial breast cancer: collaborative reanalysis of individual data from 52 epidemiological studies including 58,209 women with breast cancer and 101,986 women without the disease. Lancet. 2001;358(9291):1389–99.
2. Pharoah PD, Day NE, Duffy S, Easton DF, Ponder BA. Family history and the risk of breast cancer: a systematic review and meta-analysis. Int J Cancer (Journal international du Cancer). 1997;71(5):800–9.
3. Kharazmi E, Chen T, Narod S, Sundquist K, Hemminki K. Effect of multiplicity, laterality, and age at onset of breast cancer on familial risk of breast cancer: a nationwide prospective cohort study. Breast Cancer Res Treat. 2014;144(1):185–92.
4. Kicinski M, Vangronsveld J, Nawrot TS. An epidemiological reappraisal of the familial aggregation of prostate cancer: a meta-analysis. PLoS ONE. 2011;6(10):e27130.
5. Slattery ML, Levin TR, Ma K, Goldgar D, Holubkov R, Edwards S. Family history and colorectal cancer: predictors of risk. Cancer Causes Control CCC. 2003;14(9):879–87.
6. Lichtenstein P, Holm NV, Verkasalo PK, Iliadou A, Kaprio J, Koskenvuo M, et al. Environmental and heritable factors in the causation of cancer—analyses of cohorts of twins from Sweden, Denmark, and Finland. N Engl J Med. 2000;343(2):78–85.
7. Hjelmborg JB, Scheike T, Holst K, Skytthe A, Penney KL, Graff RE, et al. The heritability of prostate cancer in the Nordic Twin Study of Cancer. Cancer Epidemiol Biomarkers Prev (A Publication of the American Association for Cancer Research, Cosponsored by the American Society of Preventive Oncology). 2014;23(11):2303–10.
8. Gabriel SB, Schaffner SF, Nguyen H, Moore JM, Roy J, Blumenstiel B, et al. The structure of haplotype blocks in the human genome. Science. 2002;296(5576):2225–9.
9. Lander ES, Linton LM, Birren B, Nusbaum C, Zody MC, Baldwin J, et al. Initial sequencing and analysis of the human genome. Nature. 2001;409(6822):860–921.
10. Database of Genomic Structural Variation. http://www.ncbi.nlm.nih.gov/dbvar. Accessed 29 May.
11. The International HapMap Project. http://hapmap.ncbi.nlm.nih.gov/. Accessed 29 May.
12. Nature Genetics Editorial Staff. Framework for a fully powered risk engine. Nat Genet. 2005;37(11):1153.
13. Nature Genetics Editorial Staff. Asking for more. Nat Genet. 2012;44(7):733.
14. National Institutes of Health. A catalog of published genome-wide association studies. 2015.
15. Amundadottir LT, Sulem P, Gudmundsson J, Helgason A, Baker A, Agnarsson BA, et al. A common variant associated with prostate cancer in European and African populations. Nat Genet. 2006;38(6):652–8.
16. Ghoussaini M, Song H, Koessler T, Al Olama AA, Kote-Jarai Z, Driver KE, et al. Multiple loci with different cancer specificities within the 8q24 gene desert. J Natl Cancer Inst. 2008;100(13):962–6.
17. The 1000 Genomes Project. 1000 Genomes: a deep catalog of human genetic variation. http://www.1000genomes.org/about.
18. Price AL, Patterson NJ, Plenge RM, Weinblatt ME, Shadick NA, Reich D. Principal components analysis corrects for stratification in genome-wide association studies. Nat Genet. 2006;38(8):904–9.
19. Start K. Treating phenylketonuria by a phenylalanine-free diet. Prof Care Mother Child. 1998;8(4):109–10.
20. National Institutes of Health. Phenylketonuria: screening and management. NIH Consens Statement. 2000;17(3):1–27.
21. Weinberg DS, Myers RE, Keenan E, Ruth K, Sifri R, Ziring B, et al. Genetic and environmental risk assessment and colorectal cancer screening in an average-risk population: a randomized trial. Ann Int Med. 2014;161(8):537–45.
22. Liu CY, Maity A, Lin X, Wright RO, Christiani DC. Design and analysis issues in gene and environment studies. Environ Health (A Global Access Science Source). 2012;11:93.
23. Thomas D. Methods for investigating gene-environment interactions in candidate pathway and genome-wide association studies. Annu Rev Public Health. 2010;31:21–36.
24. Al Olama AA, Benlloch S, Antoniou AC, Giles GG, Severi G, Neal D, et al. Risk analysis of prostate cancer in practical, a multinational consortium, using 25 known prostate cancer susceptibility loci. Cancer Epidemiol Biomarkers Prev (A Publication of the American Association for Cancer Research, Cosponsored by the American Society of Preventive Oncology). 2015.
25. Li Q, Stram A, Chen C, Kar S, Gayther S, Pharoah P, et al. Expression QTL-based analyses reveal candidate causal genes and loci across five tumor types. Hum Mol Genet. 2014;23(19):5294–302.
26. 23andMe, Inc. https://www.23andme.com/. Accessed 1 June 2015.
27. Office of in vitro Diagnostics and Radiological Health, Center for Devices and Radiological Health. "23andMe, Inc. 11/22/13: Warning Letter. Document Number: GEN1300666, Re: Personal Genome Service (PGS)".
28. Food and Drug Administration. FDA permits marketing of first direct-to-consumer genetic carrier test for bloom syndrome. 2015.
29. Hertz DL, Rae J. Pharmacogenetics of cancer drugs. Annu Rev Med. 2015;66:65–81.

# Bioinformatic Analysis of Epidemiological and Pathological Data

# 8

Svitlana Tyekucheva and Giovanni Parmigiani

## 8.1 Introduction

Analysis of high-throughput 'omics' data is an important part of contemporary cancer research. Well-designed and properly analyzed high-throughput genomic studies greatly aid our understanding of cancer biology and can be used for prognostic and predictive studies. A demonstration of the power of genomic data analysis is comprehensive characterization of the molecular features of over twenty-five different cancers produced by The Cancer Genome Atlas (TCGA) project [1]. Extensively validated and carefully developed prognostic tests help to guide personalized treatment decisions. One prognostic gene expression-based test, Oncotype Dx®, was developed to predict the risk of recurrence for node-negative, estrogen receptor-positive breast cancer [2]. Initial candidate genes for the test were identified by analysis of microarray-based gene expression data, and patients with high risk of recurrence as predicted by Oncotype Dx® benefited from neoadjuvant chemotherapy [3]. The test is now being evaluated in two ongoing prospective clinical trials, Trial Assigning IndividuaLized Options for Treatment (Rx) (TAILORx: to determine benefits of chemotherapy for patients with mid range risk [4]) and Rx for Positive Node, Endocrine Responsive Breast Cancer (RxPONDER: to determine benefits of chemotherapy for node positive breast cancer for patients who have low to intermediate risk [5]).

On the other hand, lack of attention to proper statistical analysis of gene expression data as well as lack of proper validation have already had catastrophic consequences [6–12]. An extreme example of this were the clinical trials that used unvalidated genomic tests to predict response to chemotherapy and that were terminated to prevent potential harm to the patients. This incident warranted establishing a committee on the Review of Omics-Based Tests for Predicting Patient Outcomes in Clinical Trials to recommend ways to strengthen omics-based test development and evaluation. In 2012 the work of the committee resulted in a comprehensive report entitled "Evolution of Translational Omics: Lessons Learned and the Path Forward" [13]. The main recommendations from this report are reflected throughout this chapter, but the original report is recommended for supplementary reading.

Because of the complexity of high-throughput data, we often employ sophisticated statistical and bioinformatical methodology and the results of the analyses might depend on the chosen analysis strategy. Different statistical methods applied to the same dataset are likely to yield

S. Tyekucheva (✉) · G. Parmigiani
Department of Biostatistics and Computational Biology, Dana-Farber Cancer Institute, Boston, MA, USA
e-mail: svitlana@jimmy.harvard.edu

S. Tyekucheva · G. Parmigiani
Department of Biostatistics, Harvard T.H. Chan School of Public Health, Boston, MA, USA


M. Loda et al. (eds.), *Pathology and Epidemiology of Cancer*,
DOI 10.1007/978-3-319-35153-7_8

different predictive models, and even different versions of the same algorithm or the same software could produce different results. Therefore, careful annotation of all analysis steps, including versions of the used software, complete and unequivocal description of the computational models and meticulous validation of the findings are essential in high-throughput "omics." The raw data from the genomic assays, as well as the values of clinical covariates used in the study should be made publicly available, so that other researches can check and reproduce conclusions of any given "omics"-based study.

Many of the bioinformatic algorithms for analyzing genomic data are available as stand-alone software and as freely available Bioconductor packages. Bioconductor is an open-source and open development software project that uses R programming language [14]. It provides a wide variety of statistical and bioinformatical tools for analysis of high-dimensional genomic data. Developing bioinformatical tools for analyzing high-throughput data is a quickly evolving field, with new algorithms and methods appearing every day, but the basic principles of analyses remain the same. Further in the chapter, we will be mentioning examples of both Bioconductor packages and stand-alone software that are commonly used to work with genomic data mainly for illustration purposes, and this list of software is not comprehensive.

The principles of experimental design that we will discuss in terms of gene expression analysis also apply to biomarker discovery and validation from any other molecular data type, such as copy number variation, methylation, metabolomics, proteomics, and mutational profiles. Rigorous preprocessing and quality control is also essential for any molecular data type, but different data types, and even different technologies within the same data type might require different algorithms. Moreover, emerging new technologies for generating genomic data often require updated and sometimes entirely new, analysis tools. As an example, we will discuss how it is wrong to employ inference tools developed for microarrays to analyze RNA-Seq data, even though both technologies measure gene expression.

## 8.2 Statistical Considerations in Designing Genomic Experiments

### 8.2.1 Discovery Studies

Proper design of a high-throughput genomic experiment is imperative to a successful and reproducible study. Measurements obtained from any molecular assay have multiple sources of variability. Some variability is of scientific interest, and will come from biological differences between samples, such as molecular differences between healthy and malignant samples, cancer subtypes. Other sources of variability will be unwanted, such as technical variability in sample preparation. Importantly, the same sources of biological variability could be either of interest or unwanted, depending on the goal of the study. For example, we might be interested in studying differences between high and low grades of a certain cancer type. In this case the variability between high and low grade cases is the focus of our study. On the other hand, it might be better to stratify by grade, or to perform separate analysis for high and low grade cases when we are interested in studying survival, because the grade is not the main focus of the study, and variability associated with it might interfere with our ability to identify biomarkers that are related to survival.

All principles of designing epidemiology and pathology studies should be followed when conducting genomic experiments. The goal is to minimize technical variability, and to maximize information about important biological variability such as the difference between phenotypes of interest. Therefore, it is important to identify all potential sources of variability in the samples before collection, and to design the study so that research questions can be properly addressed by statistical methods. If the study is not properly designed, even sophisticated computation will be unable to salvage the data and produce meaningful conclusions from the expensive experiment. Extra caution should be used to avoid potential confounding, and information about relevant covariates should be collected and used to the greatest possible extent.

One of the often-overlooked sources of confounding in genomic studies of cancer is tumor cellularity. Unsupervised analysis methods, such as clustering, could be greatly affected by cellularity, because some of the variation in gene expression values in the set of samples with varying amounts of tumor cells will be explained by that difference in cell composition. Therefore, some of the clusters could be driven by differences between tumor and normal tissue (or between tumor and adjacent stroma) and not by underlying molecular subtypes of tumor tissue, leading to erroneous conclusions. If it is not possible to obtain tissue samples with approximately equal, comparable cell composition the pathologist involved in the study should record the cellularity of each sample in order to control for the effects of cellularity computationally.

Genomic studies must be designed with appropriate controls. For example, if you are interested in describing gene expression patterns in tumors you may want to include measurements from normal tissue from the same individuals in order to assess if the gene expression patterns are tumor-specific.

### 8.2.2 Batch Effects

The data generated using high-throughput assays is sensitive to a large variety of technical variability, i.e., not related to the underlying biological differences. This variability can usually be attributed to differences in sample handling and preparation, differences in sensitivity of microarray scanners, calibration of the instruments and instrument drift over time. It is not uncommon to see that the major variability in the data arises from handling the samples in different facilities, preparation done by different technicians within the same facility, using different batches of reagents, or simply by extracting, labeling DNA/RNA, running hybridization or sequencing of the prepared libraries on different days. This systematic variability between groups of samples is known as batch effects [15]. If the experiment is not planned carefully, this variability can cause biases and false discoveries [16]. For example, if all or almost all tumor samples were run on one day and all normal samples several days later, among the genes found to be differentially expressed between tumor and normal tissues only some will be related to true biological differences and others will simply reflect technical variability not related to the disease status. It would be impossible to distinguish between underlying biological mechanisms responsible for the differences between the phenotypes and genes differentially expressed merely due to technical artifacts. This will compromise an experiment and render the collected data useless [17].

For large studies, it will be impossible to process all samples on the same day and they will be processed in batches. Therefore, to avoid confounding and minimize encumbering batch effects, the samples should be randomized to the processing batches. It is desirable to keep the design balanced. Randomization might be especially difficult for studies where the complete set of all samples is not available simultaneously at the beginning of the study. This can happen, for example, in studies that involve patient samples from ongoing clinical trials. In such cases it is best to consult a statistician, who will help design a randomization protocol given the particular properties of the trial. It is also a good idea to obtain a large quantity of RNA by pooling several samples and to include an aliquot of pooled RNA in every batch of library preparations and hybridizations/sequencing to track and to correct for potential batch effects.

It is advisable to plan ahead to ensure that all samples are prepared using the same reagent lot. This might be impossible when wishing to profile additional samples to augment previously collected data and increase sample size. In this case, a good strategy is to use the same gene expression profiling platform. Probe design for different microarray platforms might be different even for the arrays made by the same manufacturer, and probes representing the same transcript might have different hybridization efficiencies. Moreover, for different platforms each transcript is likely to be represented by different sets of probes, so the measurement errors for the same gene or transcript might differ between the platforms. This could alter conclusions about expression levels, and statistical models developed based on one platform might not properly calibrate to the data from another platform, making combining of the data challenging. Similar arguments hold for different library preparation methods for RNA sequencing libraries—different methods introduce different biases at each step of their protocols [18] leading to different accuracies of the gene expression quantification.

In practice it is difficult to control for all nonbiological technical variation, because in many cases the detailed information about the sample preparation is not available. It is often difficult to remove this kind of technical variability and it might persist even after normalization (discussed later in the chapter) of the raw data that is aimed at making data comparable across the samples.

Several methods have been proposed to mitigate batch effects, but these methods should be used in addition to, not instead of, a well-designed experimental design. One of the most popular methods for batch effect correction is called ComBat [19], it implements a robust empirical Bayes procedure to model and estimate gene-specific additive and multiplicative batch effects, and produces adjusted gene expression values. There exist several modifications of the procedure using a slower nonparametric version and faster parametric adjustments. While nonparametric adjustments should theoretically allow for more flexibility and relax distributional assumptions on the data, in practice both versions perform similarly. ComBat method requires upfront knowledge of the samples' assignment to the batches.

Another popular method is called Surrogate Variables Analysis (SVA) [20, 21]. An advantage of SVA is that it does not require knowledge of the batches and identifies and estimates surrogate variables for unknown sources of variation from the observed data. The SVA method uses singular value decomposition on the residual gene expression matrix after regressing out primary design variables to identify signatures of the unmodeled patterns of variation in the data and build surrogate variables. The inferred surrogate variables can then be used to adjust the original gene expression data for the identified sources of variability. A shortcoming of SVA is that it cannot distinguish unobserved cancer subtypes from potential batch effects and might remove biological signal relevant for the subtypes. For some research questions this drawback is not critical, but if unsupervised analysis is necessary, such as clustering to identify and describe expression-based disease subtypes, it might not be ideal to preprocess the data using original SVA. A recent update to the method has been launched to improve its performance to preserve biological heterogeneity [22].

### 8.2.3 Validation Studies

When a computational model is built to predict outcomes using high-throughput genomic data, one of the major concerns is overfitting. In a genomic experiment the number of observations (samples) is usually much smaller than the number of measured molecular features (for e.g., gene expression measurements). It is thus easy to choose a relatively large number of features that together will provide perfect or almost perfect predictions. However, when the same model is applied to additional datasets, it is likely to perform poorly, and this phenomenon is known as overfitting. Overfitting can occur when the algorithms optimizing performance of the model may unintentionally fit the noise specific to that dataset in addition to a biological signal. High-dimensional data and large overly flexible models derived using a small number of observations are prone to overfitting, therefore performance of predictive models obtained using high-throughput genomic data might be overoptimistic and the models will not generalize to other datasets—even if the samples for these new datasets are drawn from the same population.

To avoid overfitting the model should ideally be fit and then evaluated using independent datasets. To increase the chances of successful validation, it is recommended to avoid models with very large numbers of parameters and prefer parsimonious models. The data used to define a model and to estimate model parameters is called a training set, and the data used to evaluate the performance of the model is called a test set. When a validation study is planned, it is best to use a test set of completely independent samples (individuals) chosen from a different cohort if possible. It is also best to prepare the datasets independently from the original samples to help ensure that results are not specific to one cohort and are generalizable to different patient populations. Replication using a different genomics platform is another approach to filter out false positives due to unaccounted batch effects or platform-specific measurement errors. However, model parameters might need to be recalibrated for the new measurement scales.

Sometimes, performing a validation study is not feasible right away, either because of budgetary constraints or lack of independent samples for genomic analysis. In this case, a computational approach called cross-validation should be employed to evaluate model performance. In cross-validation, the data is split into nonintersecting training and test sets of predefined sizes. The model is fit using the training set and evaluated on the test set. This process is repeated several times and the values used to benchmark model performance are averaged across the test sets. K-fold cross-validation stands for cross-validation with K repetitions, where the data is partitioned into K sets, and on each repetition (K-1) subsets are used for training with the remaining one used for validation. Each of the K sets is used as a test set once. Another popular version of cross-validation is leave-one-out. This is a special case of K-fold cross-validation with K equal to sample size, a useful approach for smaller sample sizes, where withholding too many samples might compromise variable selection and model fitting. In this approach, at each fold all but one sample are used for training, and prediction is performed for the single withheld sample.

It is important to design cross-validation to include all steps of the model building procedure. For example, if data-driven variable selection is performed, or dimension reduction techniques are used such as principal components directions, the explanatory variables should be reselected or dimension reduction directions re-estimated at every cross-validation fold. Otherwise, using a complete dataset at certain steps will lead to overoptimistic estimates of the model performance. It is useful to save selected sets of variables at each cross-validation fold and assess how stable these sets are. Then, the final model might use the variables that are consistently selected to be in the model at each cross-validation fold.

### 8.2.4 Using Publicly Available Data

Over the past years, thousands of genomic experiments have accumulated in publicly available databases such as Gene Expression Omnibus (GEO; [23]), Array Express [24], and Sequence Reads Archive (SRA; [25]). These databases are a great source for additional analysis and validation. When using publicly available data one should follow all principles of design, preprocessing, quality control, and analysis as when working on an original experiment. Special attention should be paid to the quality assessment of the data, including checking for batch effects and confounding, as some published studies have been reported to be compromised by batch effects [17]. As discussed in the previous section, a validation study should not include samples used for model training for similar reasons, as using duplicated samples for validation might result in incorrect conclusions. Sometimes samples from the same patient are included in multiple studies (especially in series of studies performed by the same research group). These samples might be difficult to trace, because they are included in the databases with different accession numbers or run on different platforms. Therefore, matching records should be carefully inspected, and unusually high correlations of the gene expression profiles (higher than expected between profiles of different patients) are examined, especially when several datasets are used. A recently developed Bioconductor compatible package doppelgangR [26] implements computational tools to identify duplicated samples within and across datasets.

## 8.3 Basic Principles of the Analysis of 'Omic-Data'

### 8.3.1 Preprocessing and Quality Control

As mentioned in the previous section, gene expression profiling data is subject to nonbiological variability. Therefore, the raw expression values either from microarrays or RNA-Seq experiments are not comparable across samples. In order to analyze the data and draw biological conclusions, one needs to first preprocess the data.

The raw data from microarray experiments are the fluorescence intensity values from the oligonucleotide probes to which labeled cDNA from the prepared samples was hybridized. The preprocessing procedures for the microarray data usually consist of three steps: (i) background correction, (ii) normalization, and (iii) summarization. The observed intensity values of the probes consist of two sources: actual fluorescence from the labeled cDNA that hybridized to the probe and background noise fluorescence, such as laser-induced auto fluorescence of the chip surface, nonspecific binding and other related spatial effects. The goal of the background correction is to separate signal measuring the mRNA abundances in samples from the noise. The measurements from each individual chip might be on a different scale due to varying uncontrolled conditions for the hybridization reactions. The aim of the normalization step is to bring the values to the same scale making the values from different chips comparable. Depending on the particular array design, each gene or transcript could be represented by multiple probes, called the probeset. To obtain an expression value for the gene the signal from all probes that represent that gene should be summarized across those probes.

Over the years researchers have suggested many different methods for normalizing microarray data. Two of the most popular methods are Robust Multichip Average (RMA; [27]) and dChip [28].

For background correction RMA models the observed intensities as a convolution of an exponentially distributed noise and normally distributed true signal. Note that RMA uses only the fluorescence intensities of the perfect match probes to estimate and subtract background effects. Assuming that the majority of the genes will not be differentially expressed between any experimental conditions, the distributions of background corrected intensities of all the probes for each array should be the same. The quantile normalization implemented in the RMA algorithm ranks the probe intensity values for each array, and then substitutes the values for each rank by the average of the values with the same rank across arrays. This technique produces identical distributions of the probe intensities for each array. Next, a linear model is fit to estimate an expression of each probeset corresponding to a gene. In this model log-transformed observed intensities of perfect match probes are modeled as a linear function of probe effects, log-transformed underlying true expression values plus an error term. A robust fitting approach, such as median polish, is used to minimize the effects of outlying probes (not all probes in a probeset have same properties, and some might be noisier than others potentially producing outlying values not representative of the true gene expression). As the summarization model works on a log scale, the summarized values (i.e. estimated from the model expression values) are reported as log-transformed values. RMA method is implemented in several Bioconductor packages.

The dChip method uses an image gradient correction algorithm to correct for local image artifacts by adjusting the background brightness of the irregular region to a similar level as the background of the surrounding region. For normalization, dChip uses the Invariant Set method. In this method a reference chip is chosen, for example an array closest to the median chip, and for each of the remaining arrays a robust normalization curve is fit using a set of rank-invariant probes, i.e., probes with only small within-subset rank differences. The set of invariant probes are either inferred from the data, or a list of housekeeping genes of the user's choice may be used. All probes on the array are then corrected based on this curve. As a summary value for a gene/probeset, dChip uses a model-based expression index estimated from a linear fit as a weighted average of the observed normalized perfect match and mismatch probe differences, with weights being probe-sensitivity indices, also estimated from the data. The dChip summarization model works on a linear scale, and it is advised to log-transform resulting values for further analysis. The dChip algorithm is implemented as a stand-alone Windows software package and can be performed using Bioconductor as well.

The raw data from an RNA-Seq experiment is a set of short reads generated from a prepared RNA-Seq library. The first step in processing RNA-Seq data is aligning the reads to the reference genome or transcriptome. Several efficient algorithms that allow mapping of splice junctions have been proposed to align RNA-Seq reads. Among the most popular software packages for this task are TopHat [29, 30], STAR [31], SOAPsplice [32]. The next step is to quantify the gene/transcript expression. A number of algorithms exist to complete this task: i.e., HTSeq [33], SAMTools [34], RSEM [35], Cufflinks [36, 37]. Each of these tools implement different models for assigning reads to transcripts, such as choice of gene models, ways of handling the reads that map to multiple locations of the genome, accounting for read quality, and detangling expression for different isoforms. Importantly, major preprocessing steps of the RNA-Seq data analysis, such as alignments and counting aligned reads that map to each gene, are very computationally intensive, and running these tasks on a personal computer is rarely feasible even for a moderately sized study.

Developing methods for normalization of RNA-Seq data is an active area of research in bioinformatics. A key issue is correction for the different library size or sequencing depth of the RNA-Seq. The number of reads for each gene should be proportional to the mRNA abundances. If two samples are sequenced with a twofold difference in the total number of reads, twice as many reads mapping to each gene of the sample with higher coverage are expected, making direct comparisons between these two samples impossible. The first and the simplest method of normalization was proposed by Mortazavi et al. [38], who suggested to use reads per kilobase per million sequenced reads (RPKM). The reads are scaled per kilobase (of the gene length). Such scaling takes into account gene length, because more reads are expected for longer genes. It has been found that scaling to the total number of sequenced reads might cause biases for lower expressed genes if some small proportion of the reads belongs to a small number of very highly expressed genes. Several modifications to the scaling procedure have been proposed, such as scaling to the upper quartile of the gene counts [39]. Other sources of biases in the read counts exist, i.e., the GC-content. Interestingly, GC-content effects tend to be sample specific. Some normalization methods have been developed to account for this bias [40, 41]. RNA-Seq data is prone to batch effect as much as the data from microarray experiments [15].

Both raw and normalized data should be inspected to ensure that normalization produced data with desirable distributional properties. Graphical exploratory analysis tools include density plots for expression values across each sample, boxplots and so-called MA-plots. MA-plots usually show a rotated scatterplot of the expression values for a chosen sample against the median across all samples (each point is a gene). The scatterplot is rotated so that the vertical axis $M$ represents the difference between these values and horizontal axis $A$ represents average of the observed and median expression values. When the expression values are log-transformed, the vertical axis shows a log-fold-change between the expressions. Example of MA-plots is shown in Fig. 8.1. After normalization we expect to observe similar distributions of the expression values for all samples. The MA-plots should not exhibit much curvature, and only a moderate spread on the vertical axis, because we expect the majority of genes not to be different across samples. Another useful quality assessment plot is an RNA degradation plot. Using information about mapping coordinates to the genome of the probes on the microarray or mapping locations of the reads

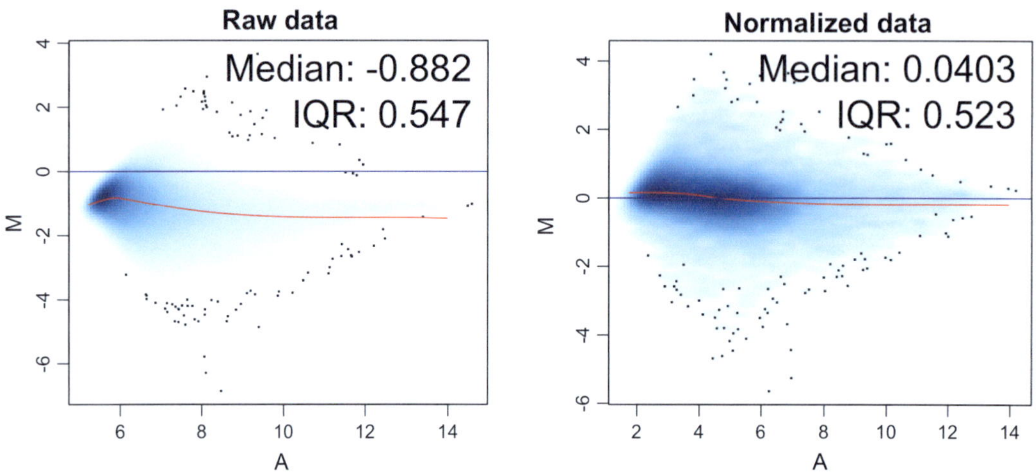


**Fig. 8.1** An example of MA-plots for one microarray sample with the raw (*left*) and normalized (*right*) expression values. *Red lines* show LOWESS smoothing curves, *blue lines* correspond to zero log-fold change in expression. The log-fold changes between gene expressions in the sample and median array (vertical axis *M*) are not centered around zero for the raw data. Also, raw data show more pronounced curvature (an undesired feature in an MA-plot) than normalized data

from RNA-Seq and comparing observed expression of the gene at 3′ and 5′ ends, one can estimate and plot the slopes of the RNA degradation for the samples. RNA is often better preserved at 3′ end. As long as all samples show similar degradation patterns, RNA degradation should not influence the analysis, because all samples will be influenced consistently.

Samples with outlier expression may be a concern, because outlying samples might influence (bias) a normalization procedure in an undesirable way. Thus, it might be advisable to remove these outlying samples and re-normalize the data without them. However, caution should be exercised when removing outliers, as they might represent interesting biological outliers and not technical artifacts.

Particular attention should be paid to the gene expression data generated using archival formalin fixed paraffin embedded (FFPE) samples. Nucleic acids extracted from archived FFPE samples are typically degraded, their overall quality is significantly poorer than that of the frozen tissues and corresponding gene expression data are much noisier [42]. However, microarray techniques have developed that can partially mitigate this degradation and provide valid mRNA profiling data within FFPE materials [43]. Samples that fail to produce data of a reasonable quality must be identified and excluded from the analysis (prior to normalization, to avoid biases). It might be useful to inspect distributional properties of the samples as a function of the FFPE block age.

After preprocessing, proper summarization at the gene or transcript level and, if necessary, removing outlying observations, a data matrix is obtained. Typically, rows represent genes and columns represent samples. This matrix is ready for downstream analysis and hypothesis testing.

### 8.3.2 Differential Expression Analysis

One of the most basic questions that can be addressed using microarray and RNA-Seq data is 'what genes are differentially expressed between phenotypes or conditions of the experiment.' When comparing mean expression values between two experimental conditions, such as comparing tumor and normal tissue, or short and long term survival, statistical procedures include simple two-sample or paired $t$-tests of each individual gene or transcript, provided that distributional assumptions are met. Log-transformed normalized gene expression values from microarray experiments usually do not display critical departures from normality, but if in a particular experiment they do, nonparametric tests, such as Mann–Whitney or Wilcoxon tests could replace *t*-tests. For study designs with multiple experimental conditions and/or additional categorical and continuous covariates a linear models approach is taken. Most familiar are simple and multiple linear regressions, Analysis of variance (ANOVA), and Analysis of covariance (ANCOVA), which belong to the family of linear models. Other commonly used models are logistic regression to analyze binary outcomes or Cox proportional hazards regression used for analysis of time-to-event data.

The most commonly used tool for fitting linear models for gene expression analysis is a Bioconductor package called limma [44]. Functions of the package use an empirical Bayes approach and compute moderated *t*-statistics. In moderated *t*-statistics, sample variances for each gene are shrunk toward the pooled estimate of variance. This approach provides robust inference even for the smaller sample sizes [45].

Distributional properties of the microarray and RNA-Seq data are fundamentally different. As mentioned above, microarray-based gene expression values, that are summarized from the intensities of multiple probes that comprise a probeset, and typically log-transformed, usually approximately meet normality conditions. RNA-Seq data is different, in that the expression of the transcript is represented by read counts, and is described best by a negative binomial distribution, therefore it is inappropriate to apply a linear models pipeline developed for microarrays directly.

Recent additions to the limma package allow for analysis of the RNA-Seq data. Using a transformation algorithm called voom [46] the counts are transformed using a robust LOWESS (locally weighted scatterplot smoothing) regression that estimates the mean-variance relation and transforms the log-counts scaled to the library size for linear modeling using the limma pipeline.

The Bioconductor package DeSeq2 takes a direct approach and models observed raw read counts as a negative binomial distribution with mean equal to a value proportional to the concentration of the cDNA fragments from that gene, in a sample scaled by a normalization factor to account for different sequencing depth, GC-content, gene length, etc. [47]. That value proportional to the concentration of the cDNA fragments is further modeled using a generalized linear models approach with a design matrix describing experimental conditions. An empirical Bayes approach is used for shrinkage of the count's variances.

### 8.3.3 Correcting for Multiple Testing

When comparing groups for differentially expressed genes several thousands of statistical tests are performed at a time, some of which are found to be statistically significant just by chance due a to a type I error—incorrect rejection of the true null hypothesis. Multiple testing correction is applied to avoid high numbers of false positive results. One approach is to control for the family-wise error rate (FWER; probability of making one or more type I errors) by multiplying the *p*-values by the total number of tests, known as the Bonferroni correction. In practice, a Bonferroni correction is usually too conservative, as it drastically reduces the size of the test, and therefore leads to an unaffordable loss of power, leading to type II errors—failing to reject the null hypothesis when the alternative is true. Less conservative modifications of the Bonferroni correction to control for FWER are also available, for example Holm's step-wise procedure. Another adopted alternative to controlling FWER is controlling False Discovery Rate (FDR). FDR is an expected proportion of errors among the rejected hypothesis [48], and is less conservative than controlling FWER while providing a good control for multiple testing. Several procedures have been proposed to calculate FDR, such a Benjamini and Hochberg [48], Benjamini and Yekutieli [49], *q*-value [50], empirical Bayes [51], etc. Most bioinformatics software reports both *p*-values and FDR.

### 8.3.4 Principal Components Analysis

Principal components analysis (PCA) is an unsupervised statistical technique that identifies uncorrelated (orthogonal) directions of the largest variability in the data by finding an appropriate rotation. It is usually used to obtain a lower dimensional representation of the high-dimensional data. Modifications of PCA for sparse data and supervised versions have also been proposed. While there is no guarantee that the component representing the largest variability found by PCA will be associated with the variables of biological interest, in practice for well-designed experiments with a strong signal that defines the phenotypes in the gene expression data, PCA works well and can be useful for data visualization.

However, sometimes the largest variability in the data will be associated with known technical variables, such as batches. In this case these unwanted effects could be removed by regressing out principal components associated with non-biological variability. This approach is widely used in genome-wide association studies to adjust genotypes for individual's ancestry [52].

Another useful PCA application is summarizing expression activity of a pathway with a reduced number of observations (as compared to representing a pathway by a vector of expression values of all genes that belong to the pathway) or even by a single number [53]. The principal components directions are calculated using the expression values of the genes that belong to the pathway of interest, and projection of the original data onto first several principal components directions is then used for further analysis. The number of directions to use is usually decided based on the percent of total variation explained by each component.

### 8.3.5 Gene Sets (Pathway) Analysis

Gene sets, or pathways analysis (GSA) is one of the easiest and most popular ways to interpret a list of genes resulting from differential gene expression or similar type of analysis in terms of biological concepts. GSA allows to determine whether a list of genes share features like participating in the same biological processes or metabolic pathway, are a target of a common transcription factor, or belong to the same functional module. It is not uncommon that similarly designed studies report nonintersecting lists of 'top' genes. This could be due to differences in the assays that were used to obtain expression data, differences in power of the studies, and other sources of variability that are beyond control. Results obtained at the gene sets level tend to be more stable across the studies. Additionally, each gene might not contribute enough to the difference between the phenotypes to be detected at the individual level, but moderate changes across many genes that belong to the same pathway might.

There are several computational methods to perform gene sets analysis, each answering a slightly different scientific question [54]. For the purposes of this section, by gene sets and pathways we will mean collections of the genes assigned to non-exclusive groups according to some annotation. While there exist sophisticated methods that take into account directional relationships between the genes, here we describe basic methods that do not take those relationships into account. The choice of gene set collections for a particular analysis is entirely dictated by the scientific question one is trying to answer, and it is common to consider several collections. The Broad Institute, for example, maintains a database of curated gene sets that are organized into several collections that is called the Molecular Signatures Database (MSigDb; [55]).

GSA methods can be divided into two classes: "cut-off" (or overrepresentation) and "non cut-off" based. Cut-off methods take a list of genes, usually resulting from filtering of a ranked list of differentially expressed genes, and tests whether the genes that belong to a gene set are overrepresented in that list. Hypergeometric or Fisher's exact tests are usually used to test the null hypothesis that observing the genes from the gene set in a predefined list of a given size happened by chance, i.e., if the list was selected from a universe of all genes at random. The cut-off methods are subjective, in a sense that if one chooses a different threshold for selecting genes the resulting significantly overrepresented genes sets or pathways could be different. However, these methods are still useful when the ranking for all genes is not available, or when the gene list of interest was not obtained directly from the differential gene expression analysis, but was selected in some other way, for example by a variable selection procedure that optimizes discrimination between two phenotypes in a multivariate model.

The non cut-off based methods, as evident from the name, use an entire list of genes measured in an experiment. The inferential procedure typically consists of two steps. In the first step, a gene-to-phenotype association score is calculated. Almost any gene-level statistic that measures the strength of the association between the expression of a gene and phenotype can be used. For example, one can use a *t*-statistic, a likelihood ratio statistic, their corresponding *p*-values, a standardized coefficient from a regression model, where the expression of a gene was used as predictor and phenotype as a response variable, etc. In the second step, a summary statistic for each gene set is calculated and a statistical test is performed to find gene sets for which such a value is unlikely to be observed by chance. For example, we can compare the distribution of ranks of genes that belong to a pathway of interest to the genes that do not belong to that pathway using the Wilcoxon test. From the standpoint of the null hypothesis, this is a 'competitive' test, because a given gene set is compared to the rest of the observed genes and phenotype labels are fixed. Another type of null hypothesis that can be tested is called 'self-contained.' In this case, one tests if the distribution of the ranks of the fixed set of genes constituting the set is different from the distribution when the gene set is not associated with the phenotype. We can obtain the distribution of the gene set statistic under no association with phenotype by permuting the phenotype labels and repeating the procedure multiple times.

One of the frequently used non cut-off GSA methods is called Gene Sets Enrichment Analysis (GSEA; [56]). This method uses the modified Kolmogorov–Smirnov statistic to summarize the distribution of the gene-to-phenotype association scores within each gene set and tests its significance by permuting phenotype labels. Another group of GSA methods are called 'global' tests. These methods directly test for an association between a group of genes constituting a gene set or pathway with a phenotype. For example, after fitting a generalized linear model with phenotype as the outcome and genes that belong to a set of interest as predictors, one tests the null hypothesis that none of the genes in the model are predictive of the phenotype (i.e., all model coefficients are zero).

There is a wide variety of methods for gene sets analysis, and depending on the procedure, the methods will be effective for specific patterns. Some methods will be tailored to discover gene sets where a small number of genes have a very large difference between the phenotypes, while others tend to prefer sets where the majority of the genes have at least a modest difference in the comparisons of interest.

Another useful modification to traditional GSA is the single sample Gene Sets Enrichment Analysis (ssGSEA) that extends the GSEA method [57]. This method uses a similar statistic as GSEA, but instead of gene-to-phenotype association scores, it considers ranking of the absolute genes expression values within each sample, and evaluates the tendency of coordinated over-or under-expression of the genes in a set. The ssGSEA can be used as a dimension reduction technique to obtain pathway summary scores for further analysis, for example, as variables for building classifiers, unsupervised clustering, etc.

Since GSA involves performing multiple statistical tests, *p*-values should be corrected to control for multiple comparisons.

## 8.4 Summary

In this chapter, we introduced the basic principles of designing and analyzing studies of high-throughput molecular data. We discussed the importance of proper planning and design of the initial discovery studies, the dangers of batch effects and how to minimize their damage. We explained why validation studies are necessary and how to use computational method cross-validation when additional samples to conduct separate validation studies cannot be obtained. We mentioned some hidden obstacles in the analysis of publicly available data and got acquainted with several common algorithms and methods for gene expression data preprocessing and analysis. This information should help the readers to understand the powers and limitations of genomic data in order to lead successful, reproducible genomic studies.

## References


1. The Cancer Genome Atlas. 2015. Available from http://cancergenome.nih.gov/abouttcga/overview. Cited July 2015.
2. Paik S, Shak S, Tang G, Kim C, Baker J, Cronin M, et al. A multigene assay to predict recurrence of tamoxifen-treated, node-negative breast cancer. N Engl J Med. 2004 Dec 30;351(27):2817–26. PubMed PMID: 15591335.
3. Paik S, Tang G, Shak S, Kim C, Baker J, Kim W, et al. Gene expression and benefit of chemotherapy in women with node-negative, estrogen receptor-positive breast cancer. J Clin Oncol. 2006 Aug 10;24(23):3726–34. PubMed PMID: 16720680.
4. The TAILORx Breast Cancer Trial. 2015. Available from http://www.cancer.gov/types/breast/research/tailorx. Cited Dec 2015.
5. The RxPONDER Breast Cancer Trial. 2015. Available from http://www.cancer.gov/about-cancer/treatment/clinical-trials/search/view?cdrid=692475. Cited Dec 2015.
6. Retraction. Pharmacogenomic strategies provide a rational approach to the treatment of cisplatin-resistant patients with advanced cancer. J Clin Oncol. 2007;25:4350–7. J Clin Oncol. 2010 Dec 10;28(35):5229. PubMed PMID: 21148129.
7. Retraction. An integrated genomic-based approach to individualized treatment of patients with advanced-stage ovarian cancer. J Clin Oncol. 2012 Feb 20;30(6):678. PubMed PMID: 22451975.
8. Baggerly K, Coombes K. Deriving chemosensitivity from cell lines: forensic bioinformatics and reproducible research in high-throughput biology. Ann Appl Stat. 2009;26(7):1186–7.
9. Baron AE, Bandeen-Roche K, Berry DA, Bryan J, Carey VJ, Chaloner K, et al. Letter to Harold Varmus: concerns about prediction models used in duke clinical trials. MD: Bethesda; 2010.
10. Bonnefoi H, Potti A, Delorenzi M, Mauriac L, Campone M, Tubiana-Hulin M, et al. Retraction-validation of gene signatures that predict the response of breast cancer to neoadjuvant chemotherapy: a substudy of the EORTC 10994/BIG 00–01 clinical trial. Lancet Oncol. 2011 Feb;12(2):116. PubMed PMID: 21277543.
11. Potti A, Dressman HK, Bild A, Chan G, Sayer R, Cragun J, et al. Retraction: genomic signatures to guide the use of chemotherapeutics. Nature Med. 2011;17(1):135.
12. Potti A, Mukherjee S, Petersen R, Dressman HK, Bild A, Koontz J, et al. Retraction: a genomic strategy to refine prognosis in early-stage non-small-cell lung cancer. N Engl J Med. 2011;364(12):1176.
13. Committee on the review of omics-based tests for predicting patient outcomes in clinical trials. In: Micheel CM, Nass SJ, Omenn GS, editors. Washington, DC; 2012.
14. Gentleman RC, Carey VJ, Bates DM, Bolstad B, Dettling M, Dudoit S, et al. Bioconductor: open software development for computational biology and bioinformatics. Genome Biol. 2004;5(10):R80. PubMed PMID: 15461798. Pubmed Central PMCID: 545600.
15. Leek JT, Scharpf RB, Bravo HC, Simcha D, Langmead B, Johnson WE, et al. Tackling the widespread and critical impact of batch effects in high-throughput data. Nat Rev Genet. 2010 Oct;11(10):733–9. PubMed PMID: 20838408. Pubmed Central PMCID: 3880143.
16. Parker HS, Leek JT. The practical effect of batch on genomic prediction. Stat Appl Genet Mol Biol. 2012;11(3):Article 10. PubMed PMID: 22611599. Pubmed Central PMCID: 3760371.
17. Baggerly KA, Coombes KR, Neeley ES. Run batch effects potentially compromise the usefulness of genomic signatures for ovarian cancer. J Clin Oncol. 2008 Mar 1;26(7):1186–7; Author reply 7–8. PubMed PMID: 18309960.
18. van Dijk EL, Jaszczyszyn Y, Thermes C. Library preparation methods for next-generation sequencing: tone down the bias. Exp Cell Res. 2014 Mar 10;322(1):12–20. PubMed PMID: 24440557.
19. Johnson EW, Li C, Rabinovic A. Adjusting batch effects in microarray expression data using empirical Bayes methods. Biostatistics. 2007;8(1):118–27.
20. Leek JT, Johnson WE, Parker HS, Jaffe AE, Storey JD. The sva package for removing batch effects and other unwanted variation in high-throughput experiments. Bioinformatics. 2012 Mar 15;28(6):882–3. PubMed PMID: 22257669. Pubmed Central PMCID: 3307112.
21. Leek JT, Storey JD. Capturing heterogeneity in gene expression studies by surrogate variable analysis. PLoS Genet. 2007 Sept;3(9):1724–35. PubMed PMID: 17907809. Pubmed Central PMCID: 1994707.

22. Parker HS, Leek JT, Favorov AV, Considine M, Xia X, Chavan S, et al. Preserving biological heterogeneity with a permuted surrogate variable analysis for genomics batch correction. Bioinformatics. 2014 Oct;30(19):2757–63. PubMed PMID: 24907368. Pubmed Central PMCID: 4173013.
23. Gene Expression Omnibus. 2015. Available from http://www.ncbi.nlm.nih.gov/geo/. Cited July 2015.
24. Array Express. 2015. Available from http://www.ebi.ac.uk/arrayexpress/. Cited July 2015.
25. Sequence Reads Archive. 2015. Available from http://www.ncbi.nlm.nih.gov/sra/. Cited July 2015.
26. Waldron L, Riester M. doppelgangR: identify possibly duplicate samples in a list of ExpressionSets (July 2015). Available from https://github.com/lwaldron/doppelgangR.
27. Irizarry RA, Hobbs B, Collin F, Beazer-Barclay YD, Antonellis KJ, Scherf U, et al. Exploration, normalization, and summaries of high density oligonucleotide array probe level data. Biostatistics. 2003 Apr;4(2):249–64. PubMed PMID: 12925520.
28. Li C, Wong WH. Model-based analysis of oligonucleotide arrays: expression index computation and outlier detection. Proc Natl Acad Sci USA. 2001 Jan 2;98(1):31–6. PubMed PMID: 11134512. Pubmed Central PMCID: 14539.
29. Trapnell C, Pachter L, Salzberg SL. TopHat: discovering splice junctions with RNA-Seq. Bioinformatics. 2009 May 1;25(9):1105–11. PubMed PMID: 19289445. Pubmed Central PMCID: 2672628.
30. Kim D, Pertea G, Trapnell C, Pimentel H, Kelley R, Salzberg SL. TopHat2: accurate alignment of transcriptomes in the presence of insertions, deletions and gene fusions. Genome Biol. 2013;14(4):R36. PubMed PMID: 23618408. Pubmed Central PMCID: 4053844.
31. Dobin A, Davis CA, Schlesinger F, Drenkow J, Zaleski C, Jha S, et al. STAR: ultrafast universal RNA-Seq aligner. Bioinformatics. 2013 Jan 1;29 (1):15–21. PubMed PMID: 23104886. Pubmed Central PMCID: 3530905.
32. Huang S, Zhang J, Li R, Zhang W, He Z, Lam TW, et al. SOAPsplice: genome-wide ab initio detection of splice junctions from RNA-Seq data. Front Genet. 2011;2:46. PubMed PMID: 22303342. Pubmed Central PMCID: 3268599.
33. Anders S, Pyl PT, Huber W. HTSeq—a Python framework to work with high-throughput sequencing data. Bioinformatics. 2015 Jan 15;31(2):166–9. PubMed PMID: 25260700. Pubmed Central PMCID: 4287950.
34. Li H, Handsaker B, Wysoker A, Fennell T, Ruan J, Homer N, et al. The sequence alignment/map format and SAMtools. Bioinformatics. 2009 Aug 15;25 (16):2078–9. PubMed PMID: 19505943. Pubmed Central PMCID: 2723002.
35. Li B, Dewey CN. RSEM: accurate transcript quantification from RNA-Seq data with or without a reference genome. BMC Bioinformatics. 2011;12:323. PubMed PMID: 21816040. Pubmed Central PMCID: 3163565.
36. Trapnell C, Williams BA, Pertea G, Mortazavi A, Kwan G, van Baren MJ, et al. Transcript assembly and quantification by RNA-Seq reveals unannotated transcripts and isoform switching during cell differentiation. Nat Biotechnol. 2010 May;28(5):511–5. PubMed PMID: 20436464. Pubmed Central PMCID: 3146043.
37. Roberts A, Pimentel H, Trapnell C, Pachter L. Identification of novel transcripts in annotated genomes using RNA-Seq. Bioinformatics. 2011 Sep 1;27(17):2325–9. PubMed PMID: 21697122.
38. Mortazavi A, Williams BA, McCue K, Schaeffer L, Wold B. Mapping and quantifying mammalian transcriptomes by RNA-Seq. Nat Methods. 2008 July;5 (7):621–8. PubMed PMID: 18516045.
39. Bullard JH, Purdom E, Hansen KD, Dudoit S. Evaluation of statistical methods for normalization and differential expression in mRNA-Seq experiments. BMC Bioinformatics. 2010;11:94. PubMed PMID: 20167110. Pubmed Central PMCID: 2838869.
40. Risso D, Schwartz K, Sherlock G, Dudoit S. GC-content normalization for RNA-Seq data. BMC Bioinformatics. 2011;12:480. PubMed PMID: 22177264. Pubmed Central PMCID: 3315510.
41. Hansen KD, Irizarry RA, Wu Z. Removing technical variability in RNA-Seq data using conditional quantile normalization. Biostatistics. 2012 Apr;13(2):204–16. PubMed PMID: 22285995. Pubmed Central PMCID: 3297825.
42. Waldron L, Ogino S, Hoshida Y, Shima K, McCart Reed AE, Simpson PT, et al. Expression profiling of archival tumors for long-term health studies. Clin Cancer Res. 2012 Nov 15;18(22):6136–46. PubMed PMID: 23136189. Pubmed Central PMCID: 3500412.
43. Tyekucheva S, Martin NE, Stack EC, Wei W, Vathipadiekal V, Waldron L, et al. Comparing platforms for messenger RNA expression profiling of archival formalin-fixed, Paraffin-embedded tissues. J Mol Diagn. 2015 July;17(4):374–81. PubMed PMID: 25937617. Pubmed Central PMCID: 4483460.
44. Ritchie ME, Phipson B, Wu D, Hu Y, Law CW, Shi W, et al. limma powers differential expression analyses for RNA-Sequencing and microarray studies. Nucleic Acids Res. 2015 Apr 20;43(7):e47. PubMed PMID: 25605792. Pubmed Central PMCID: 4402510.
45. Smyth GK. Linear models and empirical Bayes methods for assessing differential expression in microarray experiments. Stat Appl Genet Mol Biol. 2004;3:Article3. PubMed PMID: 16646809.
46. Law CW, Chen Y, Shi W, Smyth GK. voom: precision weights unlock linear model analysis tools for RNA-Seq read counts. Genome Biol. 2014;15(2): R29. PubMed PMID: 24485249. Pubmed Central PMCID: 4053721.
47. Love MI, Huber W, Anders S. Moderated estimation of fold change and dispersion for RNA-Seq data with

DESeq2. Genome Biol. 2014;15(12):550. PubMed PMID: 25516281. Pubmed Central PMCID: 4302049.
48. Benjamini Y, Hochberg Y. Controlling the false discovery rate: a practical and powerful approach to multiple testing. J Roy Stat Soc B. 1995;57(1):289–300.
49. Benjamini Y, Yekutieli D. The control of the false discovery rate in multiple testing under dependency. Ann Stat. 2001;29(4):1165–88.
50. Storey JD. A direct approach to false discovery rates. J Roy Stat Soc B. 2002;64(3):479–98.
51. Efron B, Tibshirani R. Empirical Bayes methods and false discovery rates for microarrays. Genet Epidemiol. 2002 Jun;23(1):70–86. PubMed PMID: 12112249.
52. Price AL, Patterson NJ, Plenge RM, Weinblatt ME, Shadick NA, Reich D. Principal components analysis corrects for stratification in genome-wide association studies. Nat Genet. 2006 Aug;38(8):904–9. PubMed PMID: 16862161.
53. Braun R, Cope L, Parmigiani G. Identifying differential correlation in gene/pathway combinations. BMC Bioinformatics. 2008;9:488. PubMed PMID: 19017408. Pubmed Central PMCID: 2613418.
54. Ackermann M, Strimmer K. A general modular framework for gene set enrichment analysis. BMC Bioinformatics. 2009;10:47. PubMed PMID: 19192285. Pubmed Central PMCID: 2661051.
55. Broad Institute. Molecular signatures database. Available from http://www.broadinstitute.org/gsea/msigdb/index.jsp. July 2015.
56. Subramanian A, Tamayo P, Mootha VK, Mukherjee S, Ebert BL, Gillette MA, et al. Gene set enrichment analysis: a knowledge-based approach for interpreting genome-wide expression profiles. Proc Natl Acad Sci USA. 2005 Oct 25;102(43):15545–50. PubMed PMID: 16199517. Pubmed Central PMCID: 1239896.
57. Barbie DA, Tamayo P, Boehm JS, Kim SY, Moody SE, Dunn IF, et al. Systematic RNA interference reveals that oncogenic KRAS-driven cancers require TBK1. Nature. 2009 Nov 5;462(7269):108–12. PubMed PMID: 19847166. Pubmed Central PMCID: 2783335.

# Part II

# Cancer Types

# Epidemiology of Prostate Cancer

# 9

Lorelei A. Mucci, Kathryn M. Wilson and Edward L. Giovannucci

## 9.1 Clinical Presentation of Prostate Cancer

### 9.1.1 Introduction

Prostate cancer is among the commonly diagnosed cancers among men around the world. The clinical presentation and diagnosis of prostate cancer has changed substantially over time, with the proportion of advanced stage tumors decreasing considerably over time. This decrease is partially explained by a shift in detecting cancers earlier at a more treatable stage, as well as an increase in the proportion of latent cancers that are overdiagnosed since the introduction of screening by prostate specific antigen (PSA) in the 1990s.

### 9.1.2 Symptoms

The anatomical location of the prostate sometimes results in symptomatic disease among men with prostate cancer. These symptoms include increased urinary frequency, nocturia, and urgency which results from the cancer obstructing the urethra. In addition, erectile dysfunction can be an early symptom of cancer. The urinary and sexual symptoms are not specific to prostate cancer, and can arise from benign conditions including benign prostatic hyperplasia. The primary site of metastasis in prostate cancer is to bone, and thus some men with metastatic disease can present with pain in the hips, spine or ribs. In the current era of PSA screening, most men with prostate cancer are asymptomatic at the time of diagnosis.

### 9.1.3 Diagnosis

Screening with the blood marker PSA became widespread in the United States (US) in the early 1990s, and late 1990s in other western countries. In the era of screening, most prostate cancers are initially identified with a PSA blood test and/or a digital rectal exam [1]. PSA is used as a screening tool among asymptomatic men or as a follow-up test for men whose symptoms may raise the suspicion of prostate cancer. Most guidelines suggest using PSA levels of above 4.0 ng/ml to warrant an ultrasound guided prostate biopsy, which is ultimately used to diagnose the cancer. After a prostate cancer diagnosis, a patient may undergo a series of imaging tests to assess whether the cancer has spread beyond the prostate including bone scans to detect presence of bone metastases, computed tomography (CT) to assess whether the cancer has spread to local lymph nodes, or a magnetic resonance imaging (MRI) scan to assess potential spread to seminal vesicles or adjacent organs [1]. For most men whose cancer is suspected to still be confined to the prostate, no imaging is performed.

L.A. Mucci (✉) · K.M. Wilson · E.L. Giovannucci
Harvard T.H. Chan School of Public Health, 677 Huntington Ave., Kresge 920a, Boston, MA 02115, USA
e-mail: lmucci@hsph.harvard.edu


M. Loda et al. (eds.), *Pathology and Epidemiology of Cancer*,
DOI 10.1007/978-3-319-35153-7_9

### 9.1.4 Treatment

Initial treatment for men with prostate cancer is based on the stage and grade of the cancer, as well as other clinical factors that predict the likelihood that a man's cancer will progress to metastatic disease. For men with a cancer that appears to still be localized to the prostate, there are three main options at diagnosis: active surveillance, prostatectomy (surgery) or radiation therapy [1, 2]. During active surveillance, a man receives no initial treatment but is monitored closely by his physician through regular PSA tests, biopsies and clinical assessments. Active surveillance is appropriate for men whose cancer has the lowest risk of progression. Treatment with a curative intent through surgery or radiation has been shown to reduce the risk of cancer-specific mortality [3], although many men with low risk prostate cancer are "over-treated" as their risk of progression during their lifetime is low [4]. For men with a localized but high risk prostate cancer, adjuvant androgen deprivation therapy (ADT) has been shown to improve cancer outcomes in conjunction with surgery or radiation [5]. For men with advanced prostate cancer, ADT is the first line of therapy, and recent data suggests chemotherapy and ADT together may significantly prolong life among men with newly diagnosed metastatic disease [6]. In the past few years, several new therapies have been approved for the treatment of men whose cancer is no longer responsive to ADT [7].

## 9.2 Descriptive Epidemiology of Prostate Cancer

### 9.2.1 Introduction

Prostate cancer contributes substantially to cancer incidence and mortality rates among men internationally, particularly among men in westernized countries [8]. A comparison of incidence and mortality across populations as well as trends over time can provide hints about the role of lifestyle factors or screening patterns.

### 9.2.2 Incidence

More than 1.1 million men are diagnosed with prostate cancer globally, making it the second most commonly diagnosed cancer after lung [9]. In the US, prostate cancer is the most common cancer among men and 180,890 are expected to be diagnosed in 2016. The lifetime risk of being diagnosed with prostate cancer in the US is 1 in 7 [10].

Prostate cancer incidence shows some of the greatest variation globally of any cancer (Fig. 9.1), with a 40-fold difference in age-adjusted incidence rates between men with the highest (African–American men in the United States) and lowest (Japanese and Chinese men living in their native countries) incidence. Part of the variation in incidence rates across populations can be explained by differences in diagnostic intensity, primarily due to PSA screening. However, geographic differences in prostate cancer incidence were evident even before the introduction of PSA screening in the early 1990s, highlighting a potential role of lifestyle factors in disease risk. Migrant studies further support a role of lifestyle factors. Prostate cancer incidence and mortality rates increase among men who migrate from low-risk (e.g. Asia) to high-risk (e.g. US) countries compared to those in their native countries, although rates remain below the host countries rates [11, 12].

Age-adjusted incidence rates have increased in notable patterns over time across the world (Fig. 9.2), particularly in the United States, Europe, and Australia, paralleling the uptake of PSA screening. However, incidence rates have also increased in Japan and some other Asian and Eastern European countries where PSA testing has to date not been widely used [13]. PSA screening has also led to a shift in stage presentation, with an

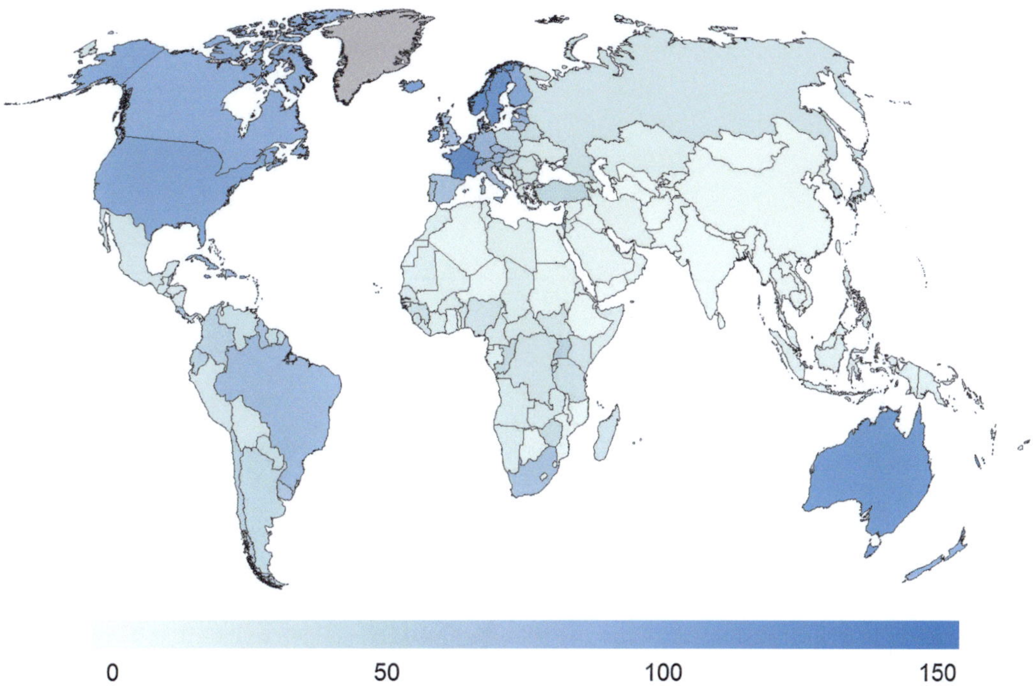


**Fig. 9.1** Comparison of prostate cancer incidence rates globally. Rates are age-adjusted for comparisons across countries and are presented per 100,000 in the population. Globocan 2012

increased proportion of localized prostate cancer disease as well as an earlier age at diagnosis [14]. Another consequence has been the substantial overdiagnosis of prostate cancer, i.e. the detection of a significant number of cancers that may never have come to light clinically nor harmed a man during his lifetime [15, 16].

Additionally, PSA screening has likely been changes in the observed associations between specific lifestyle factors and risk of total prostate cancer in epidemiological studies. First, lifestyle factors may impact prostate cancer at various stages from initiation to progression to metastases. As such, the associations may differ according to disease clinical characteristics, such as those defined by cancer stage or tumor grade [17]. Indeed, it seems unlikely that the factors associated with development of indolent cancers would be similar to those associated with cancers demonstrating malignant potential. Second, PSA screening is a strong potential confounder, as screening behaviors tend to be associated with other healthy behaviors as well as strongly associated with prostate cancer incidence. Thus, an assessment of the quality of epidemiological studies in prostate cancer should include an evaluation of the ability of the study to integrate information on PSA screening.

### 9.2.3 Mortality

An estimated 307,000 men died of prostate cancer worldwide [9], with a 10-fold variation in mortality rates among countries (Fig. 9.3). The highest prostate cancer mortality rates are among men in Caribbean countries as well as parts of Africa. Prostate cancer is the second most

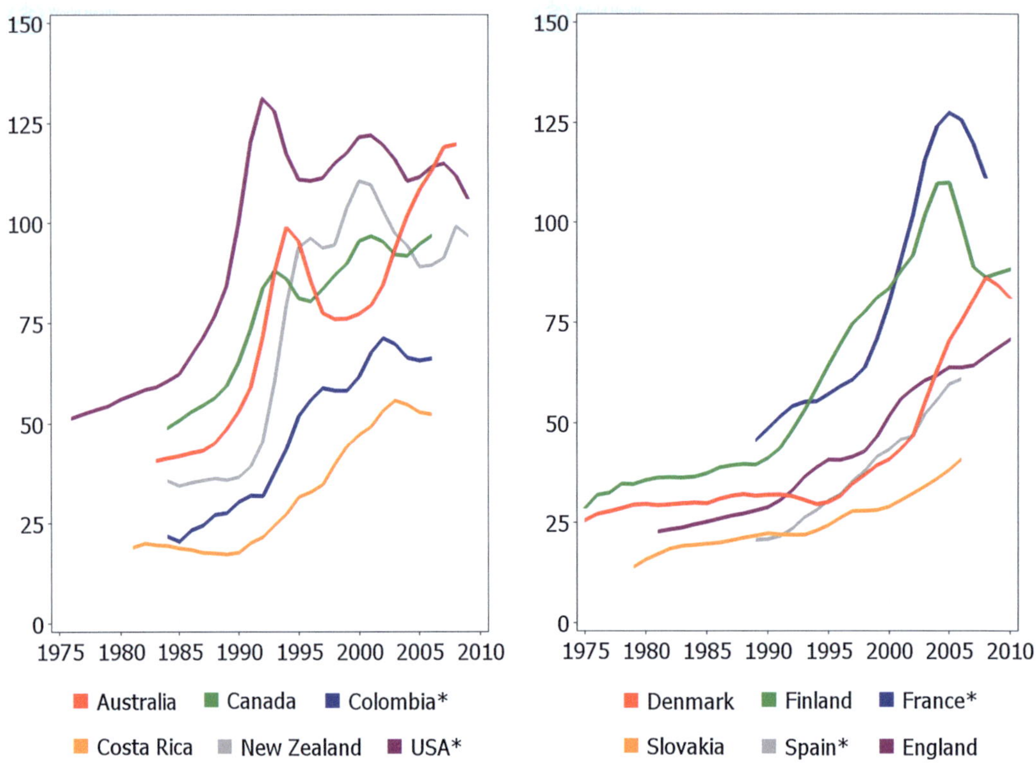


**Fig. 9.2** Trends in age-adjusted prostate cancer incidence rates over time in selected populations. Rates are presented per 100,000 in the population. Globocan 2012

common cause of cancer death among men in the United States, with 26,120 cancer deaths expected in 2016 [10]. Over the past decade, prostate cancer mortality rates have shown declines in many westernized countries. The reasons for this decline remain controversial, but may be attributable in part to earlier detection of prostate cancer through PSA screening and subsequent earlier treatment [18]. In contrast, mortality rates from prostate cancer are rising in countries throughout Africa.

Mortality rates are estimated as the number of cancer deaths per 100,000 in the population, and these rates are influenced as a function of both the incidence of the disease and survival among prostate cancer patients. The ratio of incidence to mortality rates range from 10:1 in North America and Australia to 2:1 in Central America and Caribbean to 1.2:1 in parts of Africa. Part of these differences can be attributed on the one hand to the slow growing cancers diagnosed as a result of PSA screening [19, 20] and on the other due to later presentation of disease in countries with little diagnostic intensity.

More than 4 million men are prostate cancer survivors living with a cancer diagnosis around the world, of whom 2.7 million are in the United States [10].

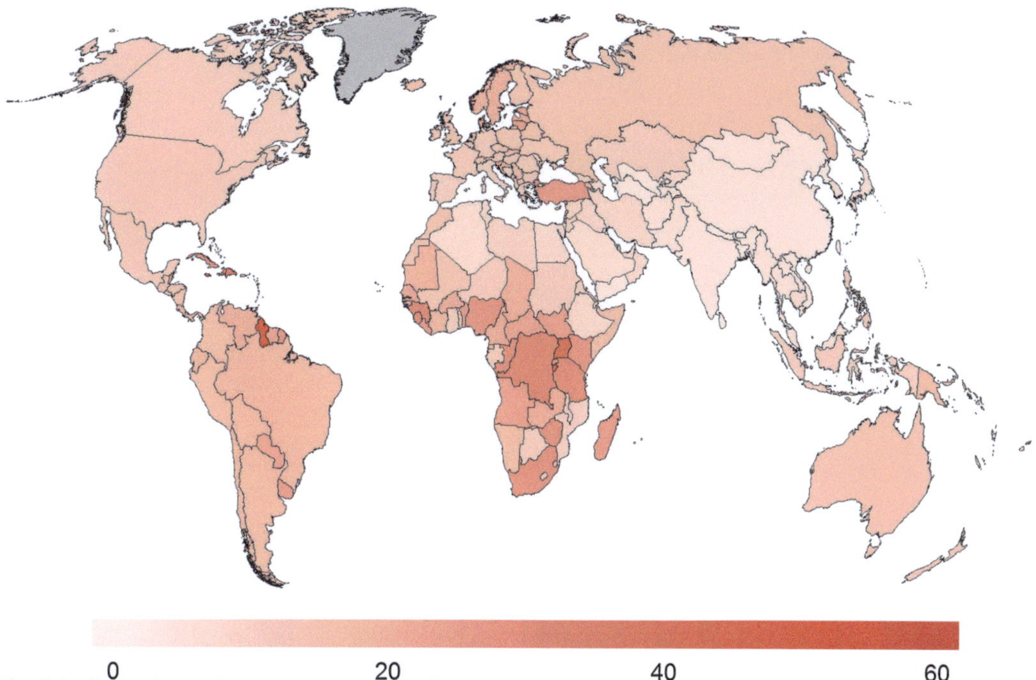


**Fig. 9.3** Comparison of prostate cancer mortality rates globally. Rates are age-adjusted for comparisons across countries and are presented per 100,000 in the population. Globocan 2012

## 9.3 Risk Factors

### 9.3.1 Introduction

The prevention of prostate cancer has the potential to improve health and reduce suffering from this common disease. The disease is heterogeneous in its biological potential, and this heterogeneity is an important feature of the disease. While some men have an aggressive form of prostate cancer, most others have a slow growing or indolent form of disease, and risk factors for total versus aggressive prostate cancer may differ. Below, we discuss the evidence surrounding specific lifestyle and dietary factors as potential risk factors for prostate cancer overall as well as for cancers with a lethal potential.

### 9.3.2 Risk Factors for Total Prostate Cancer

There are few established risk factors for the incidence of total prostate cancer: older age, African–American race, and positive family history. Moreover, there are now more than 105 genetic risk loci that have been identified and confirmed in genome wide association studies [21, 22] in ethnically diverse populations. Taller height is also a probable risk factor for total prostate cancer [23]. It is important to note that none of these factors are modifiable.

Older age is one of the strongest risk factors for prostate cancer. Prostate cancer rarely is diagnosed among men before the age of 40 years. As with other epithelial cancers, the incidence rates of prostate cancer increase exponentially from around age 55 years, a pattern observed across multiple populations. PSA screening diagnoses cancers 10-years earlier than through symptomatic disease (lead-time), and thus widespread screening has led to a shift to an earlier average age of cancer diagnosis. The median age at diagnosis among US men is 66 years.

Prostate cancer incidence and mortality rates differ substantially by race/ethnicity. In the US, incidence and mortality rates are highest among black men (Fig. 9.4), with mortality rates that are 2.4 times greater than among white men. The

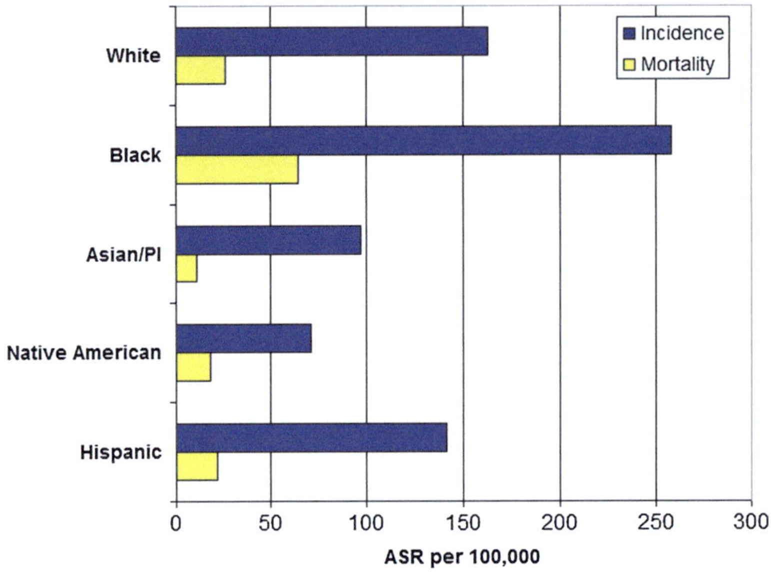


**Fig. 9.4** Incidence and mortality of prostate cancer by race/ethnicity in the United States. Rates are age-standardized for comparisons and presented per 100,000 in the population. SEER Registry data

reasons for the disparity in prostate cancer rates among black men are unknown, although there is some data to suggest differences in access to care and stage at diagnosis may in part explain the differences in prostate cancer mortality [24]. Both incidence and mortality rates are lower among Asian/Pacific-Islanders, Native Americans, and Hispanic men than among non-Hispanic whites [10].

Data from family and twin studies provide strong evidence of a role of family history in overall prostate cancer risk. Men whose father or brother is diagnosed with prostate cancer have a 2–3-fold higher risk than men without a family history. For men with a positive family history in both the father and a brother, the risk increases almost 9-fold [25]. Family history has also been associated with an increased risk of lethal prostate cancer. The risk of death from prostate cancer is approximately two-fold higher for men with a father or a brother who died of prostate cancer compared to men with prostate cancer who do not have a positive family history [26].

The familial aggregation of prostate cancer incidence is in large part due to genetic factors [27] with an estimated heritability from twin studies of 56 % [28]. Multiple genome-wide association studies have been conducted to identify common single nucleotide polymorphisms (SNPs) associated with prostate cancer incidence [29]. To date, more than 105 prostate cancer risk loci have been confirmed across multiple studies [21, 30], and these loci explain about one-third of the heritability. Most of the identified germline risk loci do not appear to be more strongly associated with lethal or nonlethal prostate cancer [31, 32], suggesting that inherited factors may play a role quite early in the pathogenesis of the disease. There are notable differences in the prevalence of several of the genetic risk loci by race/ethnicity men [33], which could account for at least part of the difference in incidence rates.

### 9.3.3 Risk Factors for Advanced Prostate Cancer

#### 9.3.3.1 Obesity and Weight Change

The obesity epidemic looms large globally, with 1.5 billion adults estimated in 2008 to be overweight or obese [34]. In the US, one-third of adults were obese defined as having a body mass index (BMI) ≥30.0 kg/m$^2$ [35]. Obesity dysregulates multiple hormonal pathways, including higher levels of insulin, lower levels of adiponectin, lower levels of testosterone and sex hormone binding globulin, higher estradiol and higher levels of inflammatory cytokines [36–39].

The relation between body size and incidence of prostate cancer is complex [17, 23, 36, 40–42]. Obese men are at higher risk of developing advanced stage prostate cancer and have higher rates of recurrence and cancer-specific mortality after diagnosis [33, 42]. A meta-analysis of 6 cohort studies found that among men with prostate cancer, a 5 kg/m$^2$ increase in BMI was associated with a 20 % (95 % CI: 0.99–1.46) increased risk of prostate cancer-specific mortality [42]. The association between obesity and poor prostate cancer outcomes do not appear to reflect solely differences in screening, as similar associations are seen after adjusting for stage and grade at diagnosis.

Higher pre-diagnosis levels of C-peptide, a circulating marker of insulin secretion, were associated with increased cancer-specific mortality, independent of BMI [36]. Men who were both overweight and who had high insulin levels had a 4-fold greater risk of death. However, two prospective studies found no association between pre-diagnosis C-peptide and risk of aggressive or advanced disease [36, 43]. Understanding drivers of the association with obesity are critical to understand mechanisms and guide prevention.

Abdominal obesity, as measured by waist circumference, may indicate a more metabolically active obesity. In the European Prospective Investigation into Cancer and Nutrition (EPIC) of 150,000 European men, waist circumference was positively associated with risk of advanced prostate cancer with a 1.06 times greater risk with a 5 cm increase in circumference [44]. Waist circumference was also significantly associated with more aggressive disease in the Melbourne Collaborative Cohort Study [45], but not in the Health Professionals Follow-up Study (HPFS) [46].

Several cohort studies have examined adult weight change and the risk of prostate cancer. Overall, weight gain from early adulthood (age 18 or 21) to mid-life was not associated with prostate cancer incidence in all [46–54] but one study [55]. Only one study has examined weight change in the period shortly before and after prostate cancer diagnosis and the risk of recurrence, measured by post-treatment PSA increase [56]. This retrospective cohort study found that weight gain from five years before treatment by prostatectomy to one year after treatment was associated with statistically significant increase in recurrence, while weight loss was non-statistically significantly associated with lower risk of recurrence.

#### 9.3.3.2 Physical Activity

Physical activity has not been associated with overall prostate cancer risk. However, several studies report an inverse association between recreational physical activity and the risk of advanced prostate cancer. The HPFS [57] and the Cancer Prevention Study (CPS) II [58] studies both reported lower risks of more advanced disease with increasing physical activity. In the CPS II, men reporting the greatest physical activity per week had a relative risk for aggressive cancer (high stage or grade) of 0.69 (95 % CI: 0.52–0.92). The EPIC cohort found no association between recreational physical activity and advanced or high-stage disease; however, activity levels were substantially higher in this cohort, and the reference group included men with up to 25 MET-hours per week.

Among 2705 men with prostate cancer, those who exercised vigorously for 3 or more hours per week had a 61 % lower risk of prostate cancer-specific mortality than those with less than one hour per week of vigorous activity (RR 0.4, 95 % CI: 0.2–0.8) [59]. Both vigorous and non-vigorous activities were associated with lower risk of all-cause mortality among these men with prostate cancer. Similarly, brisk walking was associated with a lower risk of recurrence (RR 0.4, 95 % CI: 0.2–0.9) for those walking 3 or more hours per week versus easy walking for less than 3 h per week [60].

#### 9.3.3.3 Smoking

As with other factors, smoking is not associated with total prostate cancer incidence. However, the latest review of evidence by the United States

Surgeon General concluded that smoking is a "probable" risk factor for prostate cancer mortality [61]. In HPFS, greater pack-years of smoking in the 10 years prior to prostate cancer diagnosis were associated with an increased risk of lethal disease, whereas total lifetime smoking was not associated with risk [62]. However, current smokers report less PSA testing than non-smokers [63], and the positive associations between smoking and prostate cancer mortality may be due in part to later diagnosis and treatment of these cancers among smokers.

Smoking may also influence cancer-specific outcomes by influencing response to treatment. Studies in specific treatment populations have consistently reported worse outcomes for smokers than non-smokers among prostate cancer patients treated with radiation, ADT, and radical prostatectomy [64–68].

To date, one prospective study of smoking and cancer-specific mortality among men with prostate cancer has been published [69]. Among 5366 men diagnosed with prostate cancer in the HPFS, there were 524 prostate cancer deaths. The relative risk of prostate cancer-specific mortality was 60 % higher (95 % CI: 1.1–2.3) among current versus never smokers after adjusting for potential confounders. The relative risk was attenuated, although still elevated, when models were further adjusted for stage and grade, which may suggest that part of the relationship between smoking and prostate cancer mortality is through its influence on these clinical parameters. The increased risk of prostate cancer-specific mortality was restricted to men diagnosed with localized or locally advanced cancer (stage T1–T3). Former smokers who quit 10 or more years before diagnosis or who had smoked less than 20 pack-years had the same risk as never smokers.

The possible biological basis for an association between smoking and risk of advanced prostate cancer or survival among men with prostate cancer is not clear, but several mechanisms have been proposed [69]. Tumor promotion through carcinogens from tobacco smoke is a possibility, with several studies finding prostate-cancer specific mechanisms in animal and in vitro studies.

#### 9.3.3.4 Antioxidants

Several dietary antioxidants, including selenium, Vitamin E, and lycopene/tomato sauce have been investigated with respect to prostate cancer incidence. Antioxidants are compounds that inhibit the oxidation of other species, thereby limiting the damaging effects of oxidation in animal tissues. Oxidative stress may damage molecules including proteins and DNA, and has been implicated in carcinogenesis.

*Vitamin E and Selenium.* Vitamin E generally refers to a group of fat-soluble compounds that include tocopherols and tocotrienols. Alpha-tocopherol is the biologically most active form, and current US dietary recommendations are based on alpha-tocopherol alone. Possible anti-carcinogenic actions of vitamin E include its ability to reduce DNA damage and inhibit malignant cellular transformation [70, 71]. In experimental models, derivatives of vitamin E inhibit growth, induce apoptosis [72] and enhance therapeutic effects in human prostate cancer cells [73].

Secondary results of the Alpha-Tocopherol Beta-Carotene Cancer Prevention (ATBC) Study [74] showed a 32 % reduction in prostate cancer risk among men assigned to alpha-tocopherol supplementation compared to placebo [75]. Another trial of a variety of nutrients found that vitamin E (in combination with selenium and beta-carotene) reduced overall cancer mortality [76]. These results, along with laboratory evidence and some epidemiologic support, motivated two trials of vitamin E supplementation on the risk of prostate cancer.

The Selenium and Vitamin E Cancer Prevention Trial (SELECT) primary prevention study of 50,000 men, planned for 7–12 years, was stopped early because of lack of efficacy for risk reduction. The initial report based on an average of 5.5 years of treatment, found a non-significant suggestion of *increased* prostate cancer risk among men receiving 400 IU/day of alpha-tocopherol [77]. With additional follow-up the vitamin E group was found to have a statistically significant increase in prostate cancer risk (RR 1.17, 99 % CI: 1.00–1.36, $P = 0.008$, among 1149 cases) [78]. Interestingly, there was no statistically significant increased risk of prostate cancer in the vitamin E and selenium combination group, suggesting the two may interact. The Physicians Health Study II (PHS II), conducted contemporaneously with SELECT, found no effect on the incidence of prostate cancer (RR 0.97, 95 % CI: 0.85–1.09), with a dose of 400 IU/day for a median of 8 years of follow-up [79].

All men in the ATBC trial were smokers, and the cancers were diagnosed outside the context of PSA screening, and were generally aggressive. In the VITamins And Lifestyle (VITAL) study, a cohort study specifically designed to examine supplement use and future cancer risk, a 10-year average intake of supplemental vitamin E was not associated with a reduced prostate cancer risk overall but it was associated with a reduced risk for advanced prostate cancer (regionally invasive or distant metastatic, $n = 123$) (HR 0.43, 95 % CI: 0.19–1.0 for 10-year average intake ≥400 IU/day vs. non-use) [80]. In a prospective study of plasma vitamin E and prostate cancer mortality, there was a reduced risk associated with higher circulating levels limited to smokers, although the number of cases was small (<30) [81]. Other epidemiological studies have similarly found a protective effect limited to smokers [82–84].

The SELECT and PHS II trials were done in the PSA screening era, and had small numbers of current smokers. Thus neither trial could address the effect of alpha-tocopherol specifically on advanced or fatal prostate cancers, or among current smokers. However, the results overall do not support the use of supplemental vitamin E for prostate cancer prevention.

The trace element selenium is not an anti-oxidant per se, but plays an important role as an essential element for the antioxidant enzyme glutathione peroxidase [85] as well as other selenoproteins involved in exerting anti-tumor effects [86, 87]. Dietary intake of selenium depends on the selenium content of soil in which foods are grown, which varies greatly by geographic area. Ecologic studies have suggested an inverse association between selenium soil content and prostate cancer incidence [88]. Because selenium content in specific foods vary as a function of the selenium content of the soil, epidemiological studies of selenium require biomarkers, primarily measuring levels in blood or toenails. Since the activity of some selenoenzymes plateau with higher selenium level [89], the chemopreventive effect of selenium may be greatest in populations with low selenium exposure [90].

The Nutritional Prevention of Cancer Trial found a 63 % reduction in prostate cancer risk among men taking selenium supplements [91]; with additional follow-up time, the protective effect was limited to those with low baseline levels of PSA or selenium [90]. Another trial of selenium (with vitamin E and beta-carotene) found a reduction in total cancer mortality in China [76]. The SELECT trial found no association between selenium and prostate cancer risk (RR 1.09, 99 % CI: 0.93–1.27; $P = 0.18$). Moreover, baseline selenium levels were not associated with total prostate cancer risk, nor did levels modify the association between selenium supplementation and risk [92].

Six prospective biomarker studies have reported significant associations between higher levels of selenium and reduced prostate cancer risk [92–98], particularly for advanced disease [93, 94, 97], however, not all epidemiological studies have reported a protective association of selenium [99–101]. Furthermore, two randomized studies found no effect of selenium supplementation, alone or in combination, in reducing progression of high-grade prostatic intraepithelial neoplasia (PIN) to invasive cancer [102, 103].

In conclusion, there is some evidence that selenium may play a role in prostate cancer biology; however, there is no evidence to support the use of selenium supplements to prevent prostate cancer.

*Lycopene and Tomato-Based Products* The carotenoid lycopene is found in high quantities in tomato and tomato-based products, as well as pink grapefruit, and watermelon [104]. Lycopene accumulates in high levels in prostate tissue, and given its role as a potent antioxidant, is plausible as a potential protective factor for prostate cancer. This hypothesis has been tested in multiple studies that have investigated lycopene, or lycopene-rich food, such as tomato and tomato-based products, in relation to prostate cancer risk [105–125]. In a meta-analysis of studies published up to 2003 [126], high intakes of tomato or tomato-based products was associated with a 10–20 % reduction in prostate cancer risk. For the serum-or plasma-based studies, high concentrations of lycopene conferred a 25 % reduction in prostate cancer risk. More recent epidemiological studies of lycopene and prostate cancer showed mixed results, with some supporting an inverse association [118, 119, 127, 128], and others null [108, 109, 117, 120, 122, 129].

The association between tomatoes and prostate cancer has been studied extensively in the epidemiological literature, with evidence suggesting a significant benefit associated with a higher intake of tomatoes, particularly cooked tomatoes, or lycopene, the major antioxidant in tomatoes. In a meta-analysis of 10 prospective cohort or nested case-control studies, the relative risk of prostate cancer among consumers of higher amounts of raw tomato (5th quantile of intake) was 0.89 (95 % CI 0.80–1.00) [126]. For cooked tomato products, which are more bioavailable sources of lycopene than fresh tomatoes [130], the summary RR was 0.81 (95 % CI 0.71–0.92) comparing extreme categories of intake. The results from cohort studies generally indicate a 25–30 % reduction in risk of prostate cancer, whereas dietary-based case-control studies are not supportive of an association. For example, the summary RR for intake of one serving/day of raw tomato was 0.97 (95 % CI: 0.85–1.10) for the case-control studies and 0.78 (95 % CI: 0.66–0.92) for cohort studies [126].

The 2004 meta-analysis found an inverse association in studies of plasma lycopene and prostate cancer risk, with corresponding summary relative risks of 0.55 (95 % CI: 0.32–0.94) for case-control studies and 0.78 (95 % CI: 0.61–1.00) for cohort studies [126]. An additional nested case-control study not included in the meta-analysis found a modest, not statistically significant, inverse association overall, and a significantly reduced risk with higher levels among men over 65 years old and among those without a family history of prostate cancer [127]. However, more recent studies have found no associations with serum lycopene [118, 120, 122, 123, 129]. It is possible that these conflicting results are due, in part, to the changing mix of prostate cancer cases diagnosed with PSA screening [131], which has increased the pool of biologically indolent cancers.

Indeed, epidemiological studies generally point to a stronger reduction in risk of advanced stage or lethal prostate cancer, suggesting that tomato products and lycopene may play a role in prostate cancer progression. For example, in the HPFS, the associations comparing high and low quintiles of lycopene intake were 0.91 (0.84–1.00) for total prostate cancer and 0.72 (0.56–0.94) for fatal or metastatic disease [92]. This study also found that higher lycopene intake was associated with biomarkers indicating lower angiogenic potential in tumor specimens. In the EPIC study based on 966 total cases and 205 advanced stage cases of prostate cancer, there was no association between plasma lycopene and overall risk, but men in the top quintile of plasma lycopene had a significantly reduced risk of advanced stage prostate cancer (RR 0.40, 95 % CI: 0.19–0.88) [118].

Although not definitive, the available data suggest that increased consumption of tomato and tomato-based products is associated with lower prostate cancer risk and progression. Whether the effect is driven through lycopene or other aspects of tomatoes remains undetermined. The relationship appears to be stronger for advanced prostate cancer than for indolent disease.

#### 9.3.3.5 Calcium, Dairy Products, and Vitamin D

Calcium intake has been associated with an increased risk of prostate cancer in many but not all epidemiological studies. A meta-analysis in 2005 found an increased risk of 1.39 (95 % CI = 1.09–1.77), for extreme categories of calcium intake [132]. Since the meta-analysis, four new prospective cohorts studies found some suggestion of an increased risk of prostate cancer with higher calcium [133–136] while five studies found no associations [109, 135, 137–140]. Total calcium intake varied widely across study populations: the highest category of intake was less than 1000 mg/day in three studies, whereas the highest category was greater than 2000 mg/day in three other studies [134, 135, 137–140]. Some, but not all, studies have reported stronger associations between high intake of calcium and risks of aggressive forms of prostate cancer, defined by high grade, or advanced or lethal prostate cancer [17, 141, 142].

The association between serum calcium and risk of prostate cancer also has been examined in several prospective studies. Serum calcium was associated with an increased risk of fatal disease in National Health and Nutrition Examination Survey (NHANES) I and NHANES III [143, 144] with a RR of fatal prostate cancer of 2.68 (95 % CI: 1.02–6.99). Similar increases in risk were seen for both total serum calcium and ionized serum calcium, the biologically active component. Two nested studies of Swedish men found no association between serum calcium and overall risk of prostate cancer [145, 146]; in fact, there was a weak inverse association with overall risk in one study [146]. This study also found no indication of an association between serum calcium and risk of fatal prostate cancer. Circulating calcium levels are tightly regulated and are related to diet only at very high levels of intake, so it is unclear how this finding relates to dietary calcium intake, if at all. However, it suggests a role for calcium, vitamin D, and perhaps related factors, such as parathyroid hormone, in the etiology of lethal prostate cancer.

Dairy foods, a major dietary source of calcium, have also been associated with risk, with a meta-analysis reporting a RR of 1.11 (95 % CI: 1.03–1.19) for total dairy, 1.06 (95 % CI: 0.91–1.23) for milk; and 1.11 (95 % CI: 0.99–1.25) per serving for cheese [132]. Most [133, 139, 147], but not all [135, 148] studies published since this meta-analysis have tended to support an association between higher milk or dairy consumption and total prostate cancer risk. However, findings specifically for advanced or lethal cancer are mixed [107]. The correlation between dairy foods and calcium and other nutrients creates challenges in trying to disentangle the independent effects of these compounds; however, studies that have tried to separate effects generally suggest calcium may be the predominant player in explaining positive associations with prostate cancer. As a result, the World Cancer Research Fund 2007 Expert Report on Diet and Cancer concluded that calcium is a "probable" risk factor for prostate cancer, while the evidence for dairy was weak/inconclusive [149]. Since then, the EPIC study found that dairy calcium, but not non-dairy calcium, was associated with total and high grade prostate cancer risk [133].

One proposed mechanism is that calcium acts by suppressing circulating levels of dihydroxyvitamin D (1,25$(OH)_2$D), the bioactive metabolite of vitamin D. The main source of vitamin D is endogenous production in the skin resulting from sun exposure, and diet is a secondary source. 1,25 $(OH)_2$D is the most biologically active form, whereas 25(OH)D is found in much higher concentrations and better reflects sun and dietary exposure [150]. Dairy protein also increases levels of insulin-like growth factor (IGF) [151], which may thus influence risk of advanced or lethal prostate cancer [152].

No study of dietary or supplemental vitamin D have reported protective effects for prostate cancer incidence [153–156]. Results of studies using prediagnostic circulating vitamin D metabolites have reported mainly null results [157–169], in addition to significant positive [170] inverse [171], and U-shaped [172] associations. There is, however, a suggestion that vitamin D plays a role in prostate cancer progression. Genetic variants in the vitamin D pathway are associated with risk of recurrence or progression and prostate cancer-specific mortality [173]. In addition, high expression of the vitamin D receptor protein in prostate cancer tissue has been associated with lower risk of lethal cancer among men with prostate cancer in the HPFS and PHS [174]. Prostate cancer patients with the lowest levels of pre-diagnostic 25(OH)D had significantly greater risk of prostate cancer-specific mortality, with a RR of 1.59

(95 % CI: 1.06, 2.39) for the highest versus lowest quartiles [175]. Pre-diagnostic vitamin D levels were significantly associated with both stage and grade in this study. Thus, while vitamin D exposure does not seem to be associated with lower risk of incident prostate cancer, multiple lines of evidence suggest that the vitamin D pathway may play a role in prostate cancer progression.

#### 9.3.3.6 Coffee

Most prior epidemiological studies of coffee and prostate cancer have focused on total incidence of disease, with generally null results. However, recent meta-analyses support an inverse association of coffee intake and risk of fatal or advanced disease [176–179]. Discacciati et al. [179] found a summary RR of 0.89 (95 % CI: 0.82–0.97) per 3 cups/day increment in coffee intake, as well as an inverse association with high grade (Gleason 8–10) disease. These intriguing data, while biologically plausible, need to be confirmed in additional study populations with large numbers of fatal or advanced disease.

Coffee is rich in several biologically active compounds including caffeine, minerals, and phytochemicals. In observational and animal studies, long-term coffee drinking has been associated with improved glucose metabolism and insulin secretion in observational and animal studies, and coffee is a potent antioxidant.

#### 9.3.3.7 Statins

The class of lipid lowering medications known as statins have been proposed to have anti-tumor effects in prostate by influencing cell proliferation, inflammation, and steroidogenesis. The first study to look at the association was by Platz et al. [180] who found a RR of 0.39 (95 % CI: 0.19–0.77) of advanced prostate cancer comparing men who were statin users to nonusers. In a 2012 meta-analysis of 27 observational studies, the pooled RR of statins of 0.93 (95 % CI: 0.87–0.99) for total prostate cancer and 0.80 (95 % CI: 0.70–0.90) for advanced prostate cancer based on seven studies [181]. Since publication of the meta-analysis, six additional epidemiological studies have reported on associations between statin use and lethal prostate cancer, all suggesting inverse associations [182–185]. The largest to date included more than 11,000 prostate cancer patients in the United Kingdom and studied both prediagnostic and postdiagnostic statin use [186]. Post-diagnostic statins were associated with 34 % (95 % CI: 0.66–0.88) lower risk of prostate cancer death, and the effect was stronger among men who were using statins before diagnosis. Additional studies are needed to disentangle the relevant etiological window as well as identify mechanisms of association.

## 9.4 Summary

Prostate cancer epidemiology is complex, in part due to the biological heterogeneity of the disease as well as PSA screening. The established risk factors for prostate cancer incidence—age, race/ethnicity, family history, and genetic variants—are not modifiable, and thus primary prevention of prostate cancer is challenging. However, there are a number of promising lifestyle and dietary factors that may lower risk of developing a more aggressive cancer or hold promise in secondary prevention among prostate cancer patients.

## References

1. Nelson WG, De Marzo AM, Isaacs WB. Prostate cancer. N Engl J Med. 2003;349(4):366–81.
2. Heidenreich A, Bastian PJ, Bellmunt J, Bolla M, Joniau S, van der Kwast T, et al. EAU guidelines on prostate cancer. Part II: treatment of advanced, relapsing, and castration-resistant prostate cancer. Eur Urol. 2014;65(2):467–79.
3. Bill-Axelson A, Holmberg L, Garmo H, Rider JR, Taari K, Busch C, et al. Radical prostatectomy or watchful waiting in early prostate cancer. N Engl J Med. 2014;370(10):932–42.
4. Loeb S, Bjurlin MA, Nicholson J, Tammela TL, Penson DF, Carter HB, et al. Overdiagnosis and overtreatment of prostate cancer. Eur Urol. 2014;65(6):1046–55.
5. Lin CC, Gray PJ, Jemal A, Efstathiou JA. Androgen deprivation with or without radiation therapy for clinically node-positive prostate cancer. J Natl Cancer Inst. 2015;107(7).
6. Sweeney CJ, Chen YH, Carducci M, Liu G, Jarrard DF, Eisenberger M, et al. Chemohormonal

therapy in metastatic hormone-sensitive prostate cancer. N Engl J Med. 2015;373(8):737–46.
7. Gillessen S, Omlin A, Attard G, de Bono JS, Efstathiou E, Fizazi K, et al. Management of patients with advanced prostate cancer: recommendations of the St. Gallen advanced prostate cancer consensus conference (APCCC) 2015. Ann Oncol. 2015;26(8):1589–604.
8. Ferlay J, Bray F, Pisani P, Parkin MX. GLOBOCAN 2002: cancer incidence, mortality and prevalence worldwide. Lyon: IARC Press. Report No. 2004.
9. GLOBOCAN 2008 cancer fact sheet: prostate cancer incidence and mortality worldwide in 2008 summary [Internet]. 2008. Available from http://globocan.iarc.fr/factsheets/cancers/prostate.asp. Cited 16 Nov 2011.
10. SEER cancer statistics review 1975–2008. Bethesda, MD: National Cancer Institute. http://seer.cancer.gov/csr/1975_2008/, based on November 2010 SEER data submission, posted to the SEER web site [Internet]. 2011.
11. Shimizu H, Ross RK, Bernstein L, Yatani R, Henderson BE, Mack TM. Cancers of the prostate and breast among Japanese and white immigrants in Los Angeles County. Br J Cancer. 1991;63(6):963–6.
12. Yu H, Harris RE, Gao YT, Gao R, Wynder EL. Comparative epidemiology of cancers of the colon, rectum, prostate and breast in Shanghai, China versus the United States. Int J Epidemiol. 1991;20(1):76–81.
13. Jemal A, Center MM, DeSantis C, Ward EM. Global patterns of cancer incidence and mortality rates and trends. Cancer Epidemiol Biomarkers Prev. 2010;19(8):1893–907.
14. Etzioni R, Gulati R, Falcon S, Penson DF. Impact of PSA screening on the incidence of advanced stage prostate cancer in the United States: a surveillance modeling approach. Med Decis Making. 2008;28(3):323–31.
15. Ciatto S, Gervasi G, Bonardi R, Frullini P, Zendron P, Lombardi C, et al. Determining overdiagnosis by screening with DRE/TRUS or PSA (Florence pilot studies, 1991–1994). Eur J Cancer. 2005;41(3):411–5.
16. Etzioni R, Penson DF, Legler JM, di Tommaso D, Boer R, Gann PH, et al. Overdiagnosis due to prostate-specific antigen screening: lessons from U.S. prostate cancer incidence trends. J Natl Cancer Inst. 2002;94(13):981–90.
17. Giovannucci E, Liu Y, Platz EA, Stampfer MJ, Willett WC. Risk factors for prostate cancer incidence and progression in the health professionals follow-up study. Int J Cancer. 2007;121(7):1571–8.
18. Chu KC, Tarone RE, Freeman HP. Trends in prostate cancer mortality among black men and white men in the United States. Cancer. 2003;97(6):1507–16.
19. Albertsen PC, Hanley JA, Fine J. 20-year outcomes following conservative management of clinically localized prostate cancer. JAMA. 2005;293(17):2095–101.
20. Johansson JE, Andren O, Andersson SO, Dickman PW, Holmberg L, Magnuson A, et al. Natural history of early, localized prostate cancer. JAMA. 2004;291(22):2713–9.
21. Eeles RA, Olama AA, Benlloch S, Saunders EJ, Leongamornlert DA, Tymrakiewicz M, et al. Identification of 23 new prostate cancer susceptibility loci using the iCOGS custom genotyping array. Nat Genet. 2013;45(4):385–91, 91e1–2.
22. Al Olama AA, Kote-Jarai Z, Berndt SI, Conti DV, Schumacher F, Han Y, et al. A meta-analysis of 87,040 individuals identifies 23 new susceptibility loci for prostate cancer. Nat Genet. 2014;46(10):1103–9.
23. MacInnis RJ, English DR. Body size and composition and prostate cancer risk: systematic review and meta-regression analysis. Cancer Causes Control. 2006;17(8):989–1003.
24. Taksler GB, Keating NL, Cutler DM. Explaining racial differences in prostate cancer mortality. Cancer. 2012;118(17):4280–9.
25. Hemminki K, Czene K. Attributable risks of familial cancer from the family-cancer database. Cancer Epidemiol Biomarkers Prev. 2002;11(12):1638–44.
26. Brandt A, Sundquist J, Hemminki K. Risk for incident and fatal prostate cancer in men with a family history of any incident and fatal cancer. Ann Oncol. 2011;23:251–6.
27. Lichtenstein P, Holm NV, Verkasalo PK, Iliadou A, Kaprio J, Koskenvuo M, et al. Environmental and heritable factors in the causation of cancer—analyses of cohorts of twins from Sweden, Denmark, and Finland. N Engl J Med. 2000;343(2):78–85.
28. Hjelmborg JB, Scheike T, Holst K, Skytthe A, Penney KL, Graff RE, et al. The heritability of prostate cancer in the nordic twin study of cancer. Cancer Epidemiol Biomarkers Prev. 2014;23(11):2303–10.
29. Thomas G, Jacobs KB, Yeager M, Kraft P, Wacholder S, Orr N, et al. Multiple loci identified in a genome-wide association study of prostate cancer. Nat Genet. 2008;40(3):310–5.
30. Hoffmann TJ, Van Den Eeden SK, Sakoda LC, Jorgenson E, Habel LA, Graff RE, et al. A large multiethnic genome-wide association study of prostate cancer identifies novel risk variants and substantial ethnic differences. Cancer Discov. 2015;5(8):878–91.
31. Pomerantz MM, Freedman ML. Genetics of prostate cancer risk. Mt Sinai J Med. 2010;77(6):643–54.
32. Shui IM, Lindstrom S, Kibel AS, Berndt SI, Campa D, Gerke T, et al. Prostate cancer (PCa) risk variants and risk of fatal PCa in the National Cancer Institute Breast and Prostate Cancer Cohort Consortium. Eur Urol. 2014;65(6):1069–75.
33. Haiman CA, Chen GK, Blot WJ, Strom SS, Berndt SI, Kittles RA, et al. Characterizing genetic risk at known prostate cancer susceptibility loci in

African Americans. PLoS Genet. 2011;7(5):e1001387.
34. Finucane MM, Stevens GA, Cowan MJ, Danaei G, Lin JK, Paciorek CJ, et al. National, regional, and global trends in body-mass index since 1980: systematic analysis of health examination surveys and epidemiological studies with 960 country-years and 9.1 million participants. Lancet. 2011;377 (9765):557–67.
35. Flegal KM, Carroll MD, Ogden CL, Curtin LR. Prevalence and trends in obesity among US adults, 1999–2008. JAMA. 2010;303(3):235–41.
36. Ma J, Li H, Giovannucci E, Mucci L, Qiu W, Nguyen PL, et al. Prediagnostic body-mass index, plasma C-peptide concentration, and prostate cancer-specific mortality in men with prostate cancer: a long-term survival analysis. Lancet Oncol. 2008;9(11):1039–47.
37. Li H, Stampfer MJ, Mucci L, Rifai N, Qiu W, Kurth T, et al. A 25-year prospective study of plasma adiponectin and leptin concentrations and prostate cancer risk and survival. Clin Chem. 2010;56(1):34–43.
38. Platz EA, Giovannucci E. The epidemiology of sex steroid hormones and their signaling and metabolic pathways in the etiology of prostate cancer. J Steroid Biochem Mol Biol. 2004;92(4):237–53.
39. De Marzo AM, Platz EA, Sutcliffe S, Xu J, Gronberg H, Drake CG, et al. Inflammation in prostate carcinogenesis. Nat Rev Cancer. 2007;7(4):256–69.
40. Discacciati A, Orsini N, Andersson SO, Andren O, Johansson JE, Wolk A. Body mass index in early and middle-late adulthood and risk of localised, advanced and fatal prostate cancer: a population-based prospective study. Br J Cancer. 2011;105 (7):1061–8.
41. Robinson WR, Poole C, Godley PA. Systematic review of prostate cancer's association with body size in childhood and young adulthood. Cancer Causes Control. 2008;19(8):793–803.
42. Cao Y, Ma J. Body mass index, prostate cancer-specific mortality, and biochemical recurrence: a systematic review and meta-analysis. Cancer Prev Res (Phila). 2011;4(4):486–501.
43. Stevens VL, Jacobs EJ, Sun J, Gapstur SM. No association of plasma levels of adiponectin and c-peptide with risk of aggressive prostate cancer in the cancer prevention study II nutrition cohort. Cancer Epidemiol Biomarkers Prev. 2014;23 (5):890–2.
44. Pischon T, Boeing H, Weikert S, Allen N, Key T, Johnsen NF, et al. Body size and risk of prostate cancer in the European prospective investigation into cancer and nutrition. Cancer Epidemiol Biomarkers Prev. 2008;17(11):3252–61.
45. MacInnis RJ, English DR, Gertig DM, Hopper JL, Giles GG. Body size and composition and prostate cancer risk. Cancer Epidemiol Biomarkers Prev. 2003;12(12):1417–21.
46. Giovannucci E, Rimm EB, Stampfer MJ, Colditz GA, Willett WC. Height, body weight, and risk of prostate cancer. Cancer Epidemiol Biomarkers Prev. 1997;6(8):557–63.
47. Nomura A, Heilbrun LK, Stemmermann GN. Body mass index as a predictor of cancer in men. J Natl Cancer Inst. 1985;74(2):319–23.
48. Cerhan JR, Torner JC, Lynch CF, Rubenstein LM, Lemke JH, Cohen MB, et al. Association of smoking, body mass, and physical activity with risk of prostate cancer in the Iowa 65+ rural health study (United States). Cancer Causes Control. 1997;8 (2):229–38.
49. Putnam SD, Cerhan JR, Parker AS, Bianchi GD, Wallace RB, Cantor KP, et al. Lifestyle and anthropometric risk factors for prostate cancer in a cohort of Iowa men. Ann Epidemiol. 2000;10 (6):361–9.
50. Schuurman AG, Goldbohm RA, Dorant E, van den Brandt PA. Anthropometry in relation to prostate cancer risk in the Netherlands cohort study. Am J Epidemiol. 2000;151(6):541–9.
51. Spitz MR, Strom SS, Yamamura Y, Troncoso P, Babaian RJ, Scardino PT, et al. Epidemiologic determinants of clinically relevant prostate cancer. Int J Cancer. 2000;89(3):259–64.
52. Jonsson F, Wolk A, Pedersen NL, Lichtenstein P, Terry P, Ahlbom A, et al. Obesity and hormone-dependent tumors: cohort and co-twin control studies based on the Swedish twin registry. Int J Cancer. 2003;106(4):594–9.
53. Friedenreich CM, McGregor SE, Courneya KS, Angyalfi SJ, Elliott FG. Case-control study of anthropometric measures and prostate cancer risk. Int J Cancer. 2004;110(2):278–83.
54. Littman AJ, White E, Kristal AR. Anthropometrics and prostate cancer risk. Am J Epidemiol. 2007;165 (11):1271–9.
55. Hernandez BY, Park SY, Wilkens LR, Henderson BE, Kolonel LN. Relationship of body mass, height, and weight gain to prostate cancer risk in the multiethnic cohort. Cancer Epidemiol Biomarkers Prev. 2009;18(9):2413–21.
56. Joshu CE, Mondul AM, Menke A, Meinhold C, Han M, Humphreys EB, et al. Weight gain is associated with an increased risk of prostate cancer recurrence after prostatectomy in the PSA era. Cancer Prev Res (Phila). 2011;4(4):544–51.
57. Giovannucci EL, Liu Y, Leitzmann MF, Stampfer MJ, Willett WC. A prospective study of physical activity and incident and fatal prostate cancer. Arch Intern Med. 2005;165(9):1005–10.
58. Patel AV, Rodriguez C, Jacobs EJ, Solomon L, Thun MJ, Calle EE. Recreational physical activity and risk of prostate cancer in a large cohort of U.S. men. Cancer Epidemiol Biomarkers Prev. 2005;14 (1):275–9.
59. Kenfield SA, Stampfer MJ, Giovannucci E, Chan JM. Physical activity and survival after

prostate cancer diagnosis in the health professionals follow-up study. J Clin Oncol. 2011;29(6):726–32.
60. Richman EL, Kenfield SA, Stampfer MJ, Paciorek A, Carroll PR, Chan JM. Physical activity after diagnosis and risk of prostate cancer progression: data from the cancer of the prostate strategic urologic research endeavor. Cancer Res. 2011;71 (11):3889–95.
61. US Department of Health and Human Services. The health consequences of smoking. A report of the surgeon general. Washington, DC: US Dept of Health and Human Services. 2004.
62. Giovannucci E, Rimm EB, Ascherio A, Colditz GA, Spiegelman D, Stampfer MJ, et al. Smoking and risk of total and fatal prostate cancer in United States health professionals. Cancer Epidemiol Biomarkers Prev. 1999;8(4 Pt 1):277–82.
63. Byrne MM, Davila EP, Zhao W, Parker D, Hooper MW, Caban-Martinez A, et al. Cancer screening behaviors among smokers and non-smokers. Cancer Epidemiol. 2010;34(5):611–7.
64. Pickles T, Liu M, Berthelet E, Kim-Sing C, Kwan W, Tyldesley S. The effect of smoking on outcome following external radiation for localized prostate cancer. J Urol. 2004;171(4):1543–6.
65. Moreira DM, Antonelli JA, Presti JC Jr, Aronson WJ, Terris MK, Kane CJ, et al. Association of cigarette smoking with interval to biochemical recurrence after radical prostatectomy: results from the SEARCH database. Urology. 2010;76(5):1218–23.
66. Joshu CE, Mondul AM, Meinhold CL, Humphreys EB, Han M, Walsh PC, et al. Cigarette smoking and prostate cancer recurrence after prostatectomy. J Natl Cancer Inst. 2011;103(10):835–8.
67. Pantarotto J, Malone S, Dahrouge S, Gallant V, Eapen L. Smoking is associated with worse outcomes in patients with prostate cancer treated by radical radiotherapy. BJU Int. 2007;99(3):564–9.
68. Oefelein MG, Resnick MI. Association of tobacco use with hormone refractory disease and survival of patients with prostate cancer. J Urol. 2004;171(6 Pt 1):2281–4.
69. Kenfield SA, Stampfer MJ, Chan JM, Giovannucci E. Smoking and prostate cancer survival and recurrence. JAMA. 2011;305(24):2548–55.
70. Meydani M, Vitamin E. Lancet. 1995;345 (8943):170–5.
71. Meydani SN, Hayek MG. Vitamin E and aging immune response. Clin Geriatr Med. 1995;11 (4):567–76.
72. Gunawardena K, Murray DK, Meikle AW. Vitamin E and other antioxidants inhibit human prostate cancer cells through apoptosis. Prostate. 2000;44 (4):287–95.
73. Ripoll EA, Rama BN, Webber MM. Vitamin E enhances the chemotherapeutic effects of adriamycin on human prostatic carcinoma cells in vitro. J Urol. 1986;136(2):529–31.
74. The effect of vitamin E and beta carotene on the incidence of lung cancer and other cancers in male smokers. The alpha-tocopherol, beta carotene cancer prevention study group. N Engl J Med. 1994;330 (15):1029–35.
75. Heinonen OP, Albanes D, Virtamo J, Taylor PR, Huttunen JK, Hartman AM, et al. Prostate cancer and supplementation with alpha-tocopherol and beta-carotene: incidence and mortality in a controlled trial. J Natl Cancer Inst. 1998;90(6):440–6.
76. Blot WJ, Li JY, Taylor PR, Guo W, Dawsey S, Wang GQ, et al. Nutrition intervention trials in Linxian, China: supplementation with specific vitamin/mineral combinations, cancer incidence, and disease-specific mortality in the general population. J Natl Cancer Inst. 1993;85(18): 1483–92.
77. Lippman SM, Klein EA, Goodman PJ, Lucia MS, Thompson IM, Ford LG, et al. Effect of selenium and vitamin E on risk of prostate cancer and other cancers: the selenium and Vitamin E cancer prevention trial (SELECT). JAMA. 2009;301(1):39–51.
78. Klein EA, Thompson IM Jr, Tangen CM, Crowley JJ, Lucia MS, Goodman PJ, et al. Vitamin E and the risk of prostate cancer: the selenium and Vitamin E cancer prevention trial (SELECT). JAMA. 2011;306(14):1549–56.
79. Gaziano JM, Glynn RJ, Christen WG, Kurth T, Belanger C, MacFadyen J, et al. Vitamins E and C in the prevention of prostate and total cancer in men: the physicians' health study II randomized controlled trial. JAMA. 2009;301(1):52–62.
80. Peters U, Littman AJ, Kristal AR, Patterson RE, Potter JD, White E. Vitamin E and selenium supplementation and risk of prostate cancer in the Vitamins and lifestyle (VITAL) study cohort. Cancer Causes Control. 2008;19(1):75–87.
81. Eichholzer M, Stahelin HB, Gey KF, Ludin E, Bernasconi F. Prediction of male cancer mortality by plasma levels of interacting vitamins: 17-year follow-up of the prospective Basel study. Int J Cancer. 1996;66(2):145–50.
82. Kirsh VA, Hayes RB, Mayne ST, Chatterjee N, Subar AF, Dixon LB, et al. Supplemental and dietary vitamin E, beta-carotene, and vitamin C intakes and prostate cancer risk. J Natl Cancer Inst. 2006;98(4):245–54.
83. Chan JM, Stampfer MJ, Ma J, Rimm EB, Willett WC, Giovannucci EL. Supplemental vitamin E intake and prostate cancer risk in a large cohort of men in the United States. Cancer Epidemiol Biomarkers Prev. 1999;8(10):893–9.
84. Gann PH, Ma J, Giovannucci E, Willett W, Sacks FM, Hennekens CH, et al. Lower prostate cancer risk in men with elevated plasma lycopene levels: results of a prospective analysis. Cancer Res. 1999;59(6):1225–30.
85. Combs GF Jr, Combs SB. The nutritional biochemistry of selenium. Annu Rev Nutr. 1984;4:257–80.
86. Menter DG, Sabichi AL, Lippman SM. Selenium effects on prostate cell growth. Cancer Epidemiol Biomarkers Prev. 2000;9(11):1171–82.

87. Redman C, Scott JA, Baines AT, Basye JL, Clark LC, Calley C, et al. Inhibitory effect of selenomethionine on the growth of three selected human tumor cell lines. Cancer Lett. 1998;125(1–2):103–10.
88. Rayman MP. The importance of selenium to human health. Lancet. 2000;356(9225):233–41.
89. Neve J. Human selenium supplementation as assessed by changes in blood selenium concentration and glutathione peroxidase activity. J Trace Elem Med Biol. 1995;9(2):65–73.
90. Duffield-Lillico AJ, Dalkin BL, Reid ME, Turnbull BW, Slate EH, Jacobs ET, et al. Selenium supplementation, baseline plasma selenium status and incidence of prostate cancer: an analysis of the complete treatment period of the nutritional prevention of cancer trial. BJU Int. 2003;91(7):608–12.
91. Clark LC, Combs GF Jr, Turnbull BW, Slate EH, Chalker DK, Chow J, et al. Effects of selenium supplementation for cancer prevention in patients with carcinoma of the skin. A randomized controlled trial. Nutritional prevention of cancer study group. JAMA. 1996;276(24):1957–63.
92. Kristal AR, Darke AK, Morris JS, Tangen CM, Goodman PJ, Thompson IM, et al. Baseline selenium status and effects of selenium and vitamin e supplementation on prostate cancer risk. J Natl Cancer Inst. 2014;106(3):djt456.
93. Li H, Stampfer MJ, Giovannucci EL, Morris JS, Willett WC, Gaziano JM, et al. A prospective study of plasma selenium levels and prostate cancer risk. J Natl Cancer Inst. 2004;96(9):696–703.
94. Yoshizawa K, Willett WC, Morris SJ, Stampfer MJ, Spiegelman D, Rimm EB, et al. Study of prediagnostic selenium level in toenails and the risk of advanced prostate cancer. J Natl Cancer Inst. 1998;90(16):1219–24.
95. Brooks JD, Metter EJ, Chan DW, Sokoll LJ, Landis P, Nelson WG, et al. Plasma selenium level before diagnosis and the risk of prostate cancer development. J Urol. 2001;166(6):2034–8.
96. Helzlsouer KJ, Huang HY, Alberg AJ, Hoffman S, Burke A, Norkus EP, et al. Association between alpha-tocopherol, gamma-tocopherol, selenium, and subsequent prostate cancer. J Natl Cancer Inst. 2000;92(24):2018–23.
97. Nomura AM, Lee J, Stemmermann GN, Combs GF Jr. Serum selenium and subsequent risk of prostate cancer. Cancer Epidemiol Biomarkers Prev. 2000; 9(9):883–7.
98. van den Brandt PA, Zeegers MP, Bode P, Goldbohm RA. Toenail selenium levels and the subsequent risk of prostate cancer: a prospective cohort study. Cancer Epidemiol Biomarkers Prev. 2003; 12(9):866–71.
99. Goodman GE, Schaffer S, Bankson DD, Hughes MP, Omenn GS. Predictors of serum selenium in cigarette smokers and the lack of association with lung and prostate cancer risk. Cancer Epidemiol Biomarkers Prev. 2001;10(10):1069–76.
100. Hartman TJ, Albanes D, Pietinen P, Hartman AM, Rautalahti M, Tangrea JA, et al. The association between baseline vitamin E, selenium, and prostate cancer in the alpha-tocopherol, beta-carotene cancer prevention study. Cancer Epidemiol Biomarkers Prev. 1998;7(4):335–40.
101. Peters U, Takata Y. Selenium and the prevention of prostate and colorectal cancer. Mol Nutr Food Res. 2008;52(11):1261–72.
102. Fleshner NE, Kapusta L, Donnelly B, Tanguay S, Chin J, Hersey K, et al. Progression from high-grade prostatic intraepithelial neoplasia to cancer: a randomized trial of combination vitamin-E, soy, and selenium. J Clin Oncol. 2011;29(17):2386–90.
103. Marshall JR, Tangen CM, Sakr WA, Wood DP Jr, Berry DL, Klein EA, et al. Phase III trial of selenium to prevent prostate cancer in men with high-grade prostatic intraepithelial neoplasia: SWOG S9917. Cancer Prev Res (Phila). 2011;4(11):1761–9.
104. Ilic D, Forbes KM, Hassed C. Lycopene for the prevention of prostate cancer. Cochrane Database System Rev. 2011;11:CD008007.
105. Giovannucci E, Ascherio A, Rimm EB, Stampfer MJ, Colditz GA, Willett WC. Intake of carotenoids and retinol in relation to risk of prostate cancer. J Natl Cancer Inst. 1995;87(23):1767–76.
106. Cerhan JR, Cantor KP, Williamson K, Lynch CF, Torner JC, Burmeister LF. Cancer mortality among Iowa farmers: recent results, time trends, and lifestyle factors (United States). Cancer Causes Control. 1998;9(3):311–9.
107. Mills PK, Beeson WL, Phillips RL, Fraser GE. Cohort study of diet, lifestyle, and prostate cancer in adventist men. Cancer. 1989;64(3):598–604.
108. Kirsh VA, Mayne ST, Peters U, Chatterjee N, Leitzmann MF, Dixon LB, et al. A prospective study of lycopene and tomato product intake and risk of prostate cancer. Cancer Epidemiol Biomarkers Prev. 2006;15(1):92–8.
109. Kristal AR, Arnold KB, Neuhouser ML, Goodman P, Platz EA, Albanes D, et al. Diet, supplement use, and prostate cancer risk: results from the prostate cancer prevention trial. Am J Epidemiol. 2010;172(5):566–77.
110. Giovannucci E, Rimm EB, Liu Y, Stampfer MJ, Willett WC. A prospective study of tomato products, lycopene, and prostate cancer risk. J Natl Cancer Inst. 2002;94(5):391–8.
111. Le Marchand L, Hankin JH, Kolonel LN, Wilkens LR. Vegetable and fruit consumption in relation to prostate cancer risk in Hawaii: a reevaluation of the effect of dietary beta-carotene. Am J Epidemiol. 1991;133(3):215–9.
112. Key TJ, Silcocks PB, Davey GK, Appleby PN, Bishop DT. A case-control study of diet and prostate cancer. Br J Cancer. 1997;76(5):678–87.
113. Meyer F, Bairati I, Fradet Y, Moore L. Dietary energy and nutrients in relation to preclinical prostate cancer. Nutr Cancer. 1997;29(2):120–6.

114. Deneo-Pellegrini H, De Stefani E, Ronco A, Mendilaharsu M. Foods, nutrients and prostate cancer: a case-control study in Uruguay. Br J Cancer. 1999;80 (3–4):591–7.
115. Jain MG, Hislop GT, Howe GR, Ghadirian P. Plant foods, antioxidants, and prostate cancer risk: findings from case-control studies in Canada. Nutr Cancer. 1999;34(2):173–84.
116. Vogt TM, Mayne ST, Graubard BI, Swanson CA, Sowell AL, Schoenberg JB, et al. Serum lycopene, other serum carotenoids, and risk of prostate cancer in US blacks and whites. Am J Epidemiol. 2002;155 (11):1023–32.
117. Bosetti C, Talamini R, Montella M, Negri E, Conti E, Franceschi S, et al. Retinol, carotenoids and the risk of prostate cancer: a case-control study from Italy. Int J Cancer. 2004;112(4):689–92.
118. Key TJ, Appleby PN, Allen NE, Travis RC, Roddam AW, Jenab M, et al. Plasma carotenoids, retinol, and tocopherols and the risk of prostate cancer in the European prospective investigation into cancer and nutrition study. Am J Clin Nutr. 2007;86(3):672–81.
119. Pourmand G, Salem S, Mehrsai A, Lotfi M, Amirzargar MA, Mazdak H, et al. The risk factors of prostate cancer: a multicentric case-control study in Iran. Asian Pacific J Cancer Prev (APJCP). 2007;8(3):422–8.
120. Kristal AR, Till C, Platz EA, Song X, King IB, Neuhouser ML, et al. Serum lycopene concentration and prostate cancer risk: results from the prostate cancer prevention trial. Cancer Epidemiol Biomarkers Prev. 2011;20(4):638–46.
121. Schuurman AG, Goldbohm RA, Dorant E, van den Brandt PA. Vegetable and fruit consumption and prostate cancer risk: a cohort study in The Netherlands. Cancer Epidemiol Biomarkers Prev. 1998;7 (8):673–80.
122. Peters U, Leitzmann MF, Chatterjee N, Wang Y, Albanes D, Gelmann EP, et al. Serum lycopene, other carotenoids, and prostate cancer risk: a nested case-control study in the prostate, lung, colorectal, and ovarian cancer screening trial. Cancer Epidemiol Biomarkers Prev. 2007;16(5):962–8.
123. Beilby J, Ambrosini GL, Rossi E, de Klerk NH, Musk AW. Serum levels of folate, lycopene, beta-carotene, retinol and vitamin E and prostate cancer risk. Eur J Clin Nutr. 2010;64(10):1235–8.
124. Huang HY, Alberg AJ, Norkus EP, Hoffman SC, Comstock GW, Helzlsouer KJ. Prospective study of antioxidant micronutrients in the blood and the risk of developing prostate cancer. Am J Epidemiol. 2003;157(4):335–44.
125. Jian L, Lee AH, Binns CW. Tea and lycopene protect against prostate cancer. Asia Pacific J Clin Nutr. 2007;16(Suppl 1):453–7.
126. Etminan M, Takkouche B, Caamano-Isorna F. The role of tomato products and lycopene in the prevention of prostate cancer: a meta-analysis of observational studies. Cancer Epidemiol Biomarkers Prev. 2004;13(3):340–5.
127. Wu K, Erdman JW Jr, Schwartz SJ, Platz EA, Leitzmann M, Clinton SK, et al. Plasma and dietary carotenoids, and the risk of prostate cancer: a nested case-control study. Cancer Epidemiol Biomarkers Prev. 2004;13(2):260–9.
128. Jian L, Du CJ, Lee AH, Binns CW. Do dietary lycopene and other carotenoids protect against prostate cancer? Int J Cancer. 2005;113(6):1010–4.
129. Karppi J, Kurl S, Nurmi T, Rissanen TH, Pukkala E, Nyyssonen K. Serum lycopene and the risk of cancer: the Kuopio ischaemic heart disease risk factor (KIHD) study. Ann Epidemiol. 2009;19 (7):512–8.
130. Maiani G, Caston MJ, Catasta G, Toti E, Cambrodon IG, Bysted A, et al. Carotenoids: actual knowledge on food sources, intakes, stability and bioavailability and their protective role in humans. Mol Nutr Food Res. 2009;53(Suppl 2):S194–218.
131. Giovannucci E. Commentary: serum lycopene and prostate cancer progression: a re-consideration of findings from the prostate cancer prevention trial. Cancer Causes Control. 2011;22(7):1055–9.
132. Gao X, LaValley MP, Tucker KL. Prospective studies of dairy product and calcium intakes and prostate cancer risk: a meta-analysis. J Natl Cancer Inst. 2005;97(23):1768–77.
133. Allen NE, Key TJ, Appleby PN, Travis RC, Roddam AW, Tjonneland A, et al. Animal foods, protein, calcium and prostate cancer risk: the European prospective investigation into cancer and nutrition. Br J Cancer. 2008;98(9):1574–81.
134. Mitrou PN, Albanes D, Weinstein SJ, Pietinen P, Taylor PR, Virtamo J, et al. A prospective study of dietary calcium, dairy products and prostate cancer risk (Finland). Int J Cancer. 2007;120(11):2466–73.
135. Park Y, Mitrou PN, Kipnis V, Hollenbeck A, Schatzkin A, Leitzmann MF. Calcium, dairy foods, and risk of incident and fatal prostate cancer: the NIH-AARP diet and health study. Am J Epidemiol. 2007;166(11):1270–9.
136. Kesse E, Bertrais S, Astorg P, Jaouen A, Arnault N, Galan P, et al. Dairy products, calcium and phosphorus intake, and the risk of prostate cancer: results of the French prospective SU.VI.MAX (Supplementation en Vitamines et Mineraux Antioxydants) study. Br J Nutr. 2006;95(3):539–45.
137. Kurahashi N, Inoue M, Iwasaki M, Sasazuki S, Tsugane AS. Dairy product, saturated fatty acid, and calcium intake and prostate cancer in a prospective cohort of Japanese men. Cancer Epidemiol Biomarkers Prev. 2008;17(4):930–7.
138. Ahn J, Albanes D, Peters U, Schatzkin A, Lim U, Freedman M, et al. Dairy products, calcium intake, and risk of prostate cancer in the prostate, lung, colorectal, and ovarian cancer screening trial. Cancer Epidemiol Biomarkers Prev. 2007;16 (12):2623–30.

139. Rohrmann S, Platz EA, Kavanaugh CJ, Thuita L, Hoffman SC, Helzlsouer KJ. Meat and dairy consumption and subsequent risk of prostate cancer in a US cohort study. Cancer Causes Control. 2007;18(1):41–50.
140. Butler LM, Wong AS, Koh WP, Wang R, Yuan JM, Yu MC. Calcium intake increases risk of prostate cancer among Singapore Chinese. Cancer Res. 2010;70(12):4941–8.
141. Giovannucci E, Liu Y, Stampfer MJ, Willett WC. A prospective study of calcium intake and incident and fatal prostate cancer. Cancer Epidemiol Biomarkers Prev. 2006;15(2):203–10.
142. Tseng M, Breslow RA, Graubard BI, Ziegler RG. Dairy, calcium, and vitamin D intakes and prostate cancer risk in the National health and nutrition examination epidemiologic follow-up study cohort. Am J Clin Nutr. 2005;81(5):1147–54.
143. Skinner HG, Schwartz GG. Serum calcium and incident and fatal prostate cancer in the National health and nutrition examination survey. Cancer Epidemiol Biomarkers Prev. 2008;17(9):2302–5.
144. Skinner HG, Schwartz GG. A prospective study of total and ionized serum calcium and fatal prostate cancer. Cancer Epidemiol Biomarkers Prev. 2009;18(2):575–8.
145. Halthur C, Johansson AL, Almquist M, Malm J, Gronberg H, Manjer J, et al. Serum calcium and the risk of prostate cancer. Cancer Causes Control. 2009;20(7):1205–14.
146. Van Hemelrijck M, Hermans R, Michaelsson K, Melvin J, Garmo H, Hammar N, et al. Serum calcium and incident and fatal prostate cancer in the Swedish AMORIS study. Cancer Causes Control. 2012;23(8):1349–58.
147. Park Y, Leitzmann MF, Subar AF, Hollenbeck A, Schatzkin A. Dairy food, calcium, and risk of cancer in the NIH-AARP diet and health study. Arch Intern Med. 2009;169(4):391–401.
148. Koh KA, Sesso HD, Paffenbarger RS Jr, Lee IM. Dairy products, calcium and prostate cancer risk. Br J Cancer. 2006;95(11):1582–5.
149. Wiseman M. The second World Cancer Research Fund/American Institute for Cancer Research expert report. Food, nutrition, physical activity, and the prevention of cancer: a global perspective. Proc Nutr Soc. 2008;67(3):253–6.
150. Ali MM, Vaidya V. Vitamin D and cancer. J Cancer Res Ther. 2007;3(4):225–30.
151. Giovannucci E, Pollak M, Liu Y, Platz EA, Majeed N, Rimm EB, et al. Nutritional predictors of insulin-like growth factor I and their relationships to cancer in men. Cancer Epidemiol Biomarkers Prev. 2003;12(2):84–9.
152. Chan JM, Stampfer MJ, Ma J, Gann P, Gaziano JM, Pollak M, et al. Insulin-like growth factor-I (IGF-I) and IGF binding protein-3 as predictors of advanced-stage prostate cancer. J Natl Cancer Inst. 2002;94(14):1099–106.
153. Giovannucci E. Dietary influences of 1,25(OH)2 vitamin D in relation to prostate cancer: a hypothesis. Cancer Causes Control. 1998;9(6):567–82.
154. Chan JM, Giovannucci E, Andersson SO, Yuen J, Adami HO, Wolk A. Dairy products, calcium, phosphorous, vitamin D, and risk of prostate cancer (Sweden). Cancer Causes Control. 1998;9(6):559–66.
155. Chan JM, Pietinen P, Virtanen M, Malila N, Tangrea J, Albanes D, et al. Diet and prostate cancer risk in a cohort of smokers, with a specific focus on calcium and phosphorus (Finland). Cancer Causes Control. 2000;11(9):859–67.
156. Kristal AR, Cohen JH, Qu P, Stanford JL. Associations of energy, fat, calcium, and vitamin D with prostate cancer risk. Cancer Epidemiol Biomarkers Prev. 2002;11(8):719–25.
157. Corder EH, Guess HA, Hulka BS, Friedman GD, Sadler M, Vollmer RT, et al. Vitamin D and prostate cancer: a prediagnostic study with stored sera. Cancer Epidemiol Biomarkers Prev. 1993;2(5):467–72.
158. Braun MM, Helzlsouer KJ, Hollis BW, Comstock GW. Prostate cancer and prediagnostic levels of serum vitamin D metabolites (Maryland, United States). Cancer Causes Control. 1995;6(3):235–9.
159. Gann PH, Ma J, Hennekens CH, Hollis BW, Haddad JG, Stampfer MJ. Circulating vitamin D metabolites in relation to subsequent development of prostate cancer. Cancer Epidemiol Biomarkers Prev. 1996;5(2):121–6.
160. Nomura AM, Stemmermann GN, Lee J, Kolonel LN, Chen TC, Turner A, et al. Serum vitamin D metabolite levels and the subsequent development of prostate cancer (Hawaii, United States). Cancer Causes Control. 1998;9(4):425–32.
161. Platz EA, Leitzmann MF, Hollis BW, Willett WC, Giovannucci E. Plasma 1,25-dihydroxy-and 25-hydroxyvitamin D and subsequent risk of prostate cancer. Cancer Causes Control. 2004;15(3):255–65.
162. Jacobs ET, Giuliano AR, Martinez ME, Hollis BW, Reid ME, Marshall JR. Plasma levels of 25-hydroxyvitamin D, 1,25-dihydroxyvitamin D and the risk of prostate cancer. J Steroid Biochem Mol Biol. 2004;89–90(1–5):533–7.
163. Travis RC, Crowe FL, Allen NE, Appleby PN, Roddam AW, Tjonneland A, et al. Serum vitamin D and risk of prostate cancer in a case-control analysis nested within the European prospective investigation into cancer and nutrition (EPIC). Am J Epidemiol. 2009;169(10):1223–32.
164. Li H, Stampfer MJ, Hollis JB, Mucci LA, Gaziano JM, Hunter D, et al. A prospective study of plasma vitamin D metabolites, vitamin D receptor polymorphisms, and prostate cancer. PLoS Med. 2007;4(3):e103.
165. Barnett CM, Nielson CM, Shannon J, Chan JM, Shikany JM, Bauer DC, et al. Serum 25-OH vitamin D levels and risk of developing prostate cancer in

older men. Cancer Causes Control. 2010;21 (8):1297–303.
166. Faupel-Badger JM, Diaw L, Albanes D, Virtamo J, Woodson K, Tangrea JA. Lack of association between serum levels of 25-hydroxyvitamin D and the subsequent risk of prostate cancer in Finnish men. Cancer Epidemiol Biomarkers Prev. 2007;16 (12):2784–6.
167. Park SY, Cooney RV, Wilkens LR, Murphy SP, Henderson BE, Kolonel LN. Plasma 25-hydroxyvitamin D and prostate cancer risk: the multiethnic cohort. Eur J Cancer. 2010;46(5):932–6.
168. Gilbert R, Metcalfe C, Fraser WD, Donovan J, Hamdy F, Neal DE, et al. Associations of circulating 25-hydroxyvitamin D with prostate cancer diagnosis, stage and grade. Int J Cancer. 2012;131 (5):1187–96.
169. Ahn J, Peters U, Albanes D, Purdue MP, Abnet CC, Chatterjee N, et al. Serum vitamin D concentration and prostate cancer risk: a nested case-control study. J Natl Cancer Inst. 2008;100(11):796–804.
170. Albanes D, Mondul AM, Yu K, Parisi D, Horst RL, Virtamo J, et al. Serum 25-hydroxy vitamin D and prostate cancer risk in a large nested case-control study. Cancer Epidemiol Biomarkers Prev. 2011;20 (9):1850–60.
171. Meyer HE, Robsahm TE, Bjorge T, Brustad M, Blomhoff R. Vitamin D, season, and risk of prostate cancer: a nested case-control study within Norwegian health studies. Am J Clin Nutr. 2013;97 (1):147–54.
172. Kristal AR, Till C, Song X, Tangen CM, Goodman PJ, Neuhauser ML, et al. Plasma vitamin D and prostate cancer risk: results from the selenium and Vitamin E cancer prevention trial. Cancer Epidemiol Biomarkers Prev. 2014;23(8):1494–504.
173. Holt SK, Kwon EM, Koopmeiners JS, Lin DW, Feng Z, Ostrander EA, et al. Vitamin D pathway gene variants and prostate cancer prognosis. Prostate. 2010;70(13):1448–60.
174. Hendrickson WK, Flavin R, Kasperzyk JL, Fiorentino M, Fang F, Lis R, et al. Vitamin D receptor protein expression in tumor tissue and prostate cancer progression. J Clin Oncol. 2011;29 (17):2378–85.
175. Fang F, Kasperzyk JL, Shui I, Hendrickson W, Hollis BW, Fall K, et al. Prediagnostic plasma vitamin D metabolites and mortality among patients with prostate cancer. PLoS ONE. 2011;6(4):e18625.
176. Lu Y, Zhai L, Zeng J, Peng Q, Wang J, Deng Y, et al. Coffee consumption and prostate cancer risk: an updated meta-analysis. Cancer Causes Control. 2014;25(5):591–604.
177. Cao S, Liu L, Yin X, Wang Y, Liu J, Lu Z. Coffee consumption and risk of prostate cancer: a meta-analysis of prospective cohort studies. Carcinogenesis. 2014;35(2):256–61.
178. Zhong S, Chen W, Yu X, Chen Z, Hu Q, Zhao J. Coffee consumption and risk of prostate cancer: an up-to-date meta-analysis. Eur J Clin Nutr. 2014;68(3):330–7.
179. Discacciati A, Orsini N, Wolk A. Coffee consumption and risk of nonaggressive, aggressive and fatal prostate cancer—a dose-response meta-analysis. Ann Oncol. 2014;25(3):584–91.
180. Platz EA, Leitzmann MF, Visvanathan K, Rimm EB, Stampfer MJ, Willett WC, et al. Statin drugs and risk of advanced prostate cancer. J Natl Cancer Inst. 2006;98(24):1819–25.
181. Bansal D, Undela K, D'Cruz S, Schifano F. Statin use and risk of prostate cancer: a meta-analysis of observational studies. PLoS ONE. 2012;7(10): e46691.
182. Geybels MS, Wright JL, Holt SK, Kolb S, Feng Z, Stanford JL. Statin use in relation to prostate cancer outcomes in a population-based patient cohort study. Prostate. 2013;73(11):1214–22.
183. Nielsen SF, Nordestgaard BG, Bojesen SE. Statin use and reduced cancer-related mortality. The New Engl J Med. 2012;367(19):1792–802.
184. Margel D, Urbach DR, Lipscombe LL, Bell CM, Kulkarni G, Austin PC, et al. Metformin use and all-cause and prostate cancer-specific mortality among men with diabetes. J Clin Oncol Official J Am Soc Clin Oncol. 2013;31(25):3069–75.
185. Grytli HH, Fagerland MW, Fossa SD, Tasken KA. Association between use of beta-blockers and prostate cancer-specific survival: a cohort study of 3561 prostate cancer patients with high-risk or metastatic disease. Eur Urol. 2013.
186. Yu O EM, Benayoun S, Aprikian A, Baptist G, Suissa S, Azoulay L. Use of statins and the risk of death in patients with prostate cancer. J Clin Oncol: Official J Am Soc Clin Oncol 2013;31(27).

# 10 Pathology and Molecular Pathology of Prostate Cancer

Nairi Tchrakian, Maura Bríd Cotter and Massimo Loda

## 10.1 Introduction

In this chapter, we begin by discussing the embryological development of the prostate gland, in addition to its normal anatomy and histology in the adult male. The various pathological processes that affect the prostate will then be outlined, and we will devote the remainder of the discussion to premalignant and malignant disease. In particular, we will focus on the molecular pathogenesis of prostate cancer (PCa) as this is an evolving area that is continually challenging our current diagnostic and therapeutic practices.

N. Tchrakian
Department of Pathology, St. James's Hospital, James's Street, Dublin 8, Ireland

M.B. Cotter · M. Loda (✉)
Department of Pathology, Dana-Farber Cancer Institute and Brigham & Women's Hospital, Harvard Medical School, 450 Brookline Ave., Boston, MA 02115, USA
e-mail: massimo_loda@dfci.harvard.edu

## 10.2 Embryology and Development of the Prostate Gland

The prostate is the largest accessory gland of the male reproductive system, and its function is to contribute to the production of seminal fluid and the ejaculation of semen [1]. Growth and development of the prostate is a continuous, hormonally regulated process, which commences in foetal life, and is completed at sexual maturity [2]. The prostate develops as part of the embryological urogenital system which consists of both the urinary and genital systems. The urogenital system arises from the intermediate mesodermal layer, which forms a urogenital ridge on either side of the aorta [3]. At the end of the third month of development, the urogenital sinus develops, which later forms the urethra and the prostate. The prostate initially develops as outpouchings from the proximal urethra, which become tubular and subsequently fuse, finally forming the glandular substance of the two lobes [3]. Figure 10.1 illustrates the embryological development of the prostate and its associated structures.

The differentiation and growth of the prostate gland is androgen dependent [4]. The primary androgen involved is testosterone, the majority of which is produced by Leydig cells of the testes [5] and a smaller proportion is produced in the adrenal cortex. With continued development, the epithelium is organised into two distinct populations of cells; namely, basal cells and luminal


M. Loda et al. (eds.), *Pathology and Epidemiology of Cancer*,
DOI 10.1007/978-3-319-35153-7_10

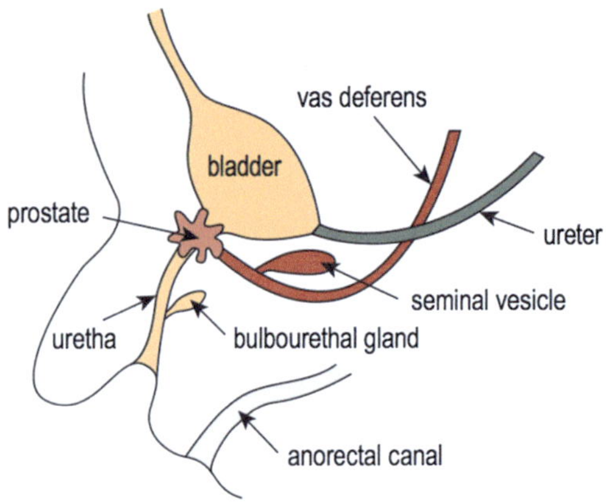


**Fig. 10.1** Embryological development of the prostate

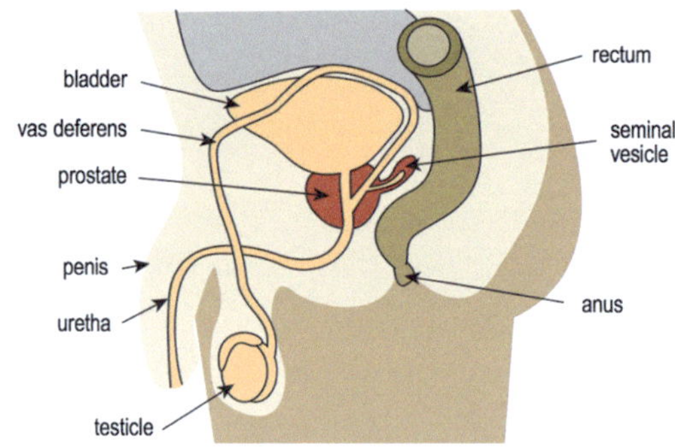


**Fig. 10.2** Sagittal view of the prostate and surrounding structures

cells [2]. Each cell type expresses its own characteristic subset of cytokeratins. Simultaneous to this process of epithelial differentiation, the prostatic stroma differentiates into smooth muscle, which is periductal in distribution [6]. At puberty, as serum testosterone levels rise, the prostate increases rapidly in size and undergoes morphologic changes to bring it to the adult phenotype, reaching an approximate weight of 20 g by 25 years of age [7].

## 10.3 Anatomy of the Prostate Gland

In the adult male, the prostate is a walnut-shaped gland, approximately 4 cm in maximum dimension [8]. It is situated between the bladder and the penis, and the prostatic base surrounds the bladder neck and the pre-prostatic portion of the urethra [9] (Fig. 10.2). The apex of the gland is in contact with the superior aspect of the urethral sphincter and the deep perineal muscle. The rectum lies posterior to the prostate [8]. The posterolateral portion of the gland is bounded by a dense fibrous capsule which contains neurovascular plexuses. Anterior and apical surfaces are covered by the anterior fibromuscular stroma, which is thinner than the fibrous capsule covering the posterior gland [10]. Bilateral seminal vesicles are located on the superoposterior aspect of the prostate. The seminal vesicle ducts join the vasa deferentia to form the ejaculatory ducts. The ejaculatory ducts then pass anteroinferiorly through the prostate to open into the posterior aspect of the prostatic urethra. The smaller prostatic ducts (20–30 in number) open into the prostatic sinuses, which are located on the posterior wall of the prostatic urethra on either side of the seminal colliculus [8].

The prostate comprises several lobes. The anterior lobe is the portion of the gland in front of the urethra and is mainly fibromuscular. The median lobe is situated between the ejaculatory ducts and the urethra. The lateral lobes (right and left lobes) form the main mass of the gland and are separated by the prostatic urethra. The posterior lobe is the area most easily palpable during digital rectal examination [8]. The zones, of the prostate include the central, peripheral and transitional zones. The transitional zone surrounds the prostatic urethra and the central zone envelops the ejaculatory ducts, while the peripheral zone provides the bulk of the gland [10] (Fig. 10.3).

Arterial supply of the prostate is provided by branches of the internal iliac artery; chiefly the inferior vesical artery, but also the internal pudendal and middle rectal arteries [8]. Prostatic veins join to form the prostatic venous plexus, which is located around the basal and lateral aspects of the prostate. The prostatic venous plexus in turn drains into the internal iliac veins. Lymphatic vessels of the prostate terminate predominantly in the internal iliac lymph nodes, and a minor component drains to the sacral nodes. Similar to the other internal male genital organs,

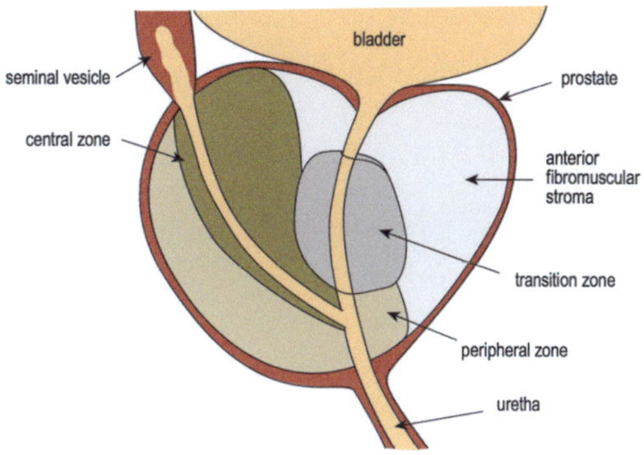


**Fig. 10.3** Zones of the prostate

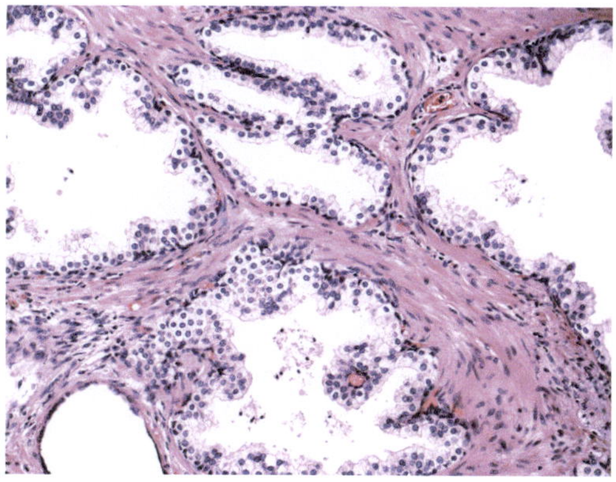

**Fig. 10.4** Benign prostatic glands

the prostate is richly innervated by the autonomic nervous system [11]. Sympathetic fibres arise from T12-L2/3 spinal cord segments, and parasympathetic fibres originate from S2 and S3 segments. Both sympathetic and parasympathetic components join the inferior hypogastric and pelvic plexuses, which innervate the gland. Knowledge of this anatomy is crucial, as nerves can be injured following radical prostatectomy, resulting in erectile dysfunction [12].

## 10.4 Histology of the Benign Prostate

Microscopically, the prostate parenchyma is best visualised in young males under 50 years of age, when significant distortion by inflammation [13], atrophy [14] and hyperplasia [15] is not yet present. Histologically, the prostate parenchyma consists of two distinct compartments, including the glandular epithelium and the surrounding stromal compartment, which will be discussed separately below.

### 10.4.1 Benign Glandular Component

The normal glandular epithelium of the prostate (Fig. 10.4) is classically defined as having two cell layers, namely a luminal (or secretory) layer and a basal layer. A third cell type is also present, the neuroendocrine cell, which is usually infrequently seen [16]. The secretory epithelial cells make up the majority of the epithelial volume. These cells are tall and columnar, with clear to pale cytoplasm and basally located nuclei. They are situated on the luminal aspect of the gland, as part of the double layer of epithelium. The nuclei are small and round with fine, evenly dispersed chromatin and nucleoli are usually not prominent. Between the secretory cell layer and the basement membrane lies the basal cell layer. The nuclei of the basal cells are small, and hyperchromatic with scant surrounding cytoplasm. These nuclei are flattened and oriented parallel to the basement membrane. The basal layer is often inconspicuous and incomplete, and it may be difficult to distinguish individual basal cells from underlying stromal fibroblasts in benign prostatic tissue. Recognition of the basal cell layer is, however, important, as these cells are absent in prostatic adenocarcinoma, and should be present in benign conditions that mimic cancer [17]. The final epithelial group, the neuroendocrine cells of the prostate, are irregularly distributed throughout the epithelial compartment [18]. They are inconspicuous, but can occasionally be recognised on haematoxylin and eosin (H&E) staining by deeply eosinophilic fine cytoplasmic granules [16]. The intraglandular contents of normal glands must also be examined and include corpora amylacea [19], (inspissated secretions that appear as pink, concentrically

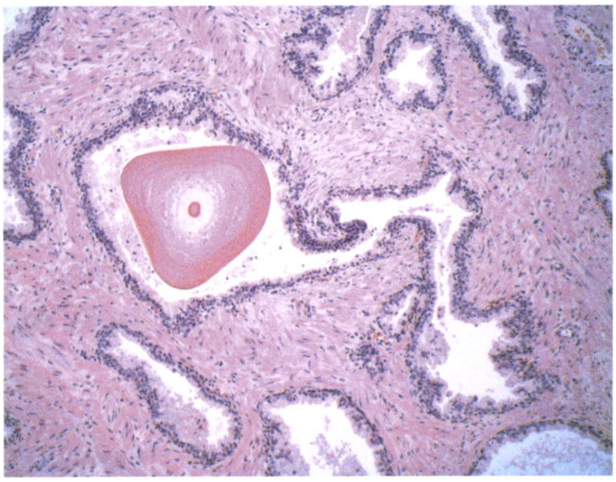

**Fig. 10.5** Corpora amylacea

lamellated rings) (Fig. 10.5), calculi and degenerating epithelial cells.

### 10.4.2 Benign Stromal Component

Histologically, the non-glandular components of the prostate gland include the pseudocapsule, a fibromuscular layer most prominent along the base and posterior portion of the lateral lobes, the fibrous septa, the periprostatic adipose tissue and the stroma. Carcinoma invading adipose tissue is considered to represent extraprostatic extension [20]. Skeletal muscle fibres can also be seen in sections of prostatic tissue, admixed with the stroma at the distal apex of the gland. The prostatic stroma accounts for about half the volume of the prostate gland. The most abundant normal stromal cell type is the smooth muscle cell, derived from the embryonic urogenital sinus. In addition to being the most abundant cell type, the smooth muscle cell appears to be the most important cell type, with respect to homeostasis. Other stromal cells commonly present include fibroblasts, endothelial cells, nerve fibres with associated ganglia, and various immune cells. The extracellular matrix is rich in collagen fibres that intervene between the glandular epithelium. The benign prostatic stroma is affected by ageing, with inflammatory cells becoming more abundant and stromal fibroblasts becoming senescent with age.

## 10.5 Inflammatory Conditions of the Prostate

Even though in a protected anatomical site, the prostate can be subject to various infections and inflammatory processes, whether of infectious origin or not. Inflammation of the prostate, or prostatitis, is seen commonly in clinical practice, and is usually treated empirically with antibiotics [21]. Histologic examination of specimens removed specifically for prostatitis is therefore uncommon. The inflammatory processes involving the prostate encompass acute and chronic bacterial prostatitis, chronic abacterial prostatitis and granulomatous prostatitis. Acute bacterial prostatitis usually results from similar bacteria to those causing urinary tract infections, namely: *Escherichia coli* and other gram-negative bacilli, enterococci and staphylococci. Generally, the organisms infect the prostate via reflux of urine from the urethra or urinary bladder. Chronic bacterial prostatitis is caused by similar organisms, and there is often an accompanying history of recurrent urinary tract infections. Mixed acute and chronic prostatitis is depicted in Fig. 10.6. Granulomatous prostatitis is another subtype [22], which may be due to fungal or mycobacterial infection. An attenuated form of the latter may be seen following intravesical therapy with Bacillus Calmette–Guerin (BCG) for bladder cancer. 'Nonspecific granulomatous prostatitis' is the most commonly diagnosed granulomatous

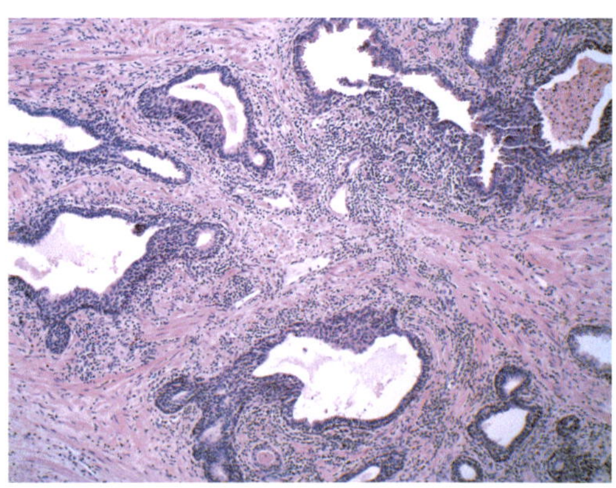

**Fig. 10.6** Benign prostatic glands showing mixed acute and chronic prostatitis

process within the prostate, and results from an immune-mediated reaction to prostatic secretions from obstructed or ruptured ducts and acini [23]. Lastly, prostatic granulomas are frequent post-surgical sequelae of transrectal biopsy and transurethral resection.

## 10.6 Benign Prostatic Hyperplasia

Benign Prostatic Hyperplasia (BPH) is a common disorder in men over the age of 50 years [21]. It typically affects the central zone of the gland, and is characterised by nodular hyperplasia of either or both stromal and epithelial compartments. The enzyme 5α[alpha]-reductase, located predominantly in stromal cells, converts testosterone to dihydrotestosterone (DHT) [2]. DHT drives the hyperplastic process by increasing stromal proliferation and decreasing epithelial apoptosis. 5α[alpha]-reductase inhibitors (5ARIs) are widely used in the pharmacologic treatment of BPH. Inhibiting DHT production leads to a decrease in prostatic volume, thereby relieving the symptoms and averting the consequences of the disorder. Indeed, the widespread use of 5ARIs (and other medical therapies) in the management of BPH has resulted in a marked shift from surgery to drug therapy. 5ARIs have also been proposed as chemopreventive agents for PCa. Two randomised controlled trials examining the efficacy of 5ARIs for PCa prevention showed a reduction in incidence of up to 25 % when compared with placebo. This decrease was attributed to a reduced incidence of low-grade cancers. However, there was an unexpected concomitant increase of high-grade cancer in the groups treated with 5ARIs; namely, absolute increases of 0.5 and 0.7 % with the use of the inhibitors dutasteride and finasteride, respectively. The Food and Drug Administration (FDA) evaluated the potential risks and benefits of 5ARIs in PCa chemoprevention, and concluded that finasteride and dutasteride do not have favourable risk-benefit profiles. As a result, the use of 5ARIs for PCa chemoprevention has not been endorsed. However, the effect of these drugs on PCa incidence remains controversial [24].

## 10.7 Premalignant Prostatic Disease

Prostatic intraepithelial neoplasia (PIN) and proliferative inflammatory atrophy (PIA) are the two candidate precursor lesions for PCa [25]. Much of the evidence supporting PIN as precursor to adenocarcinoma is also true for PIA [25]. Compellingly, histologic transitions have been noted between foci of PIA and high-grade PIN (HGPIN), and also between PIA and invasive carcinoma [26]. Both PIN and PIA show particular molecular alterations characteristic of malignancy, which occur in pre-existing epithelium and are confined by the basement membrane. For example, chromosome 8 gain has been observed in PIN, PIA and invasive cancer, lending further support to the precursor status of the former two entities [27, 28]. Although both are precursors of invasive carcinoma, it is unclear whether PIN and PIA represent diseases on the same spectrum or distinct pathways, and they are discussed separately in greater detail below.

### 10.7.1 Prostatic Intraepithelial Neoplasia

PIN consists of architecturally benign acini lined by cells showing cytologic atypia; namely, nuclear crowding, enlargement, hyperchromasia and prominent nucleolation [29]. There is a surrounding basal cell layer and an intact basement membrane (Fig. 10.7). Most agree that PIN is a precursor of invasive adenocarcinoma, a theory which is supported by the fact that prostates with invasive carcinoma have an increased incidence of PIN, an increase in the number and size of PIN foci, and, on occasion, zones of HGPIN where there appears to be a 'budding off' of invasive carcinoma glands. PIN is subdivided into two grades, high and low grades, and the distinction between the two is based on the presence of

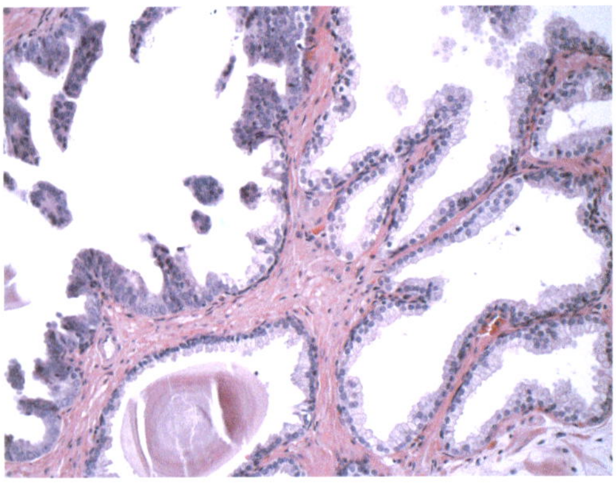

**Fig. 10.7** High-grade prostatic intraepithelial neoplasia and adjacent benign glands

prominent nucleoli, a diagnostic feature of HGPIN [29]. Most cases of low-grade PIN do not progress, and therefore, many pathologists no longer comment on its presence in biopsy specimens. The presence of multiple foci of HGPIN has a high predictive value for PCa, which develops within 1–2 years in an estimated 30 % of men with multiple cores containing HGPIN [30]. Great variation exists in the literature regarding the incidence of HGPIN on needle biopsy, which is reported to range from 1.5 % to about 16 % [31]. When HGPIN is identified on needle biopsy, a careful search is typically made by the pathologist for invasive cancer in the remainder of the tissue. If no cancer is found, follow-up is unnecessary if only a small amount of HGPIN is present in 1–2 cores. However, if there is a more significant amount of HGPIN, a follow-up protocol is adopted, consisting of serum prostate-specific antigen (PSA) testing, physical examination, and possibly repeat biopsies. However, recommendations for following men with HGPIN have varied widely [32].

### 10.7.2 Proliferative Inflammatory Atrophy

PIA refers to foci of epithelial atrophy, usually in the periphery of the gland, which show a high proliferative rate, and the presence of mononuclear and polymorphonuclear inflammatory cells within both epithelial and stromal compartments [26]. PIA is associated with chronic prostatitis [26], and it has been hypothesised that repeated tissue injury and cell loss results in proliferative regeneration of damaged epithelium which may subsequently progress to in situ and/or frank invasive carcinoma [33]. The glutathione-S-transferase Pi 1 (GSTP1) gene is located on chromosome 11q13 [21], and plays an important part in cell detoxification [34]. Hypermethylation of the GSTP1 gene (which results in downregulation of GSTP1 protein expression) [21] is seen in PIA, as well as in >90 % of PCa [34]. Importantly, GSTP1 protein detected in free plasma circulating DNA has proved to be a promising biomarker for detection of primary and recurrent PCa [34].

## 10.8 Histopathology of Prostate Cancer

Grossly, neoplastic prostate tissue can be very difficult to recognise. With widespread PSA screening, a shift towards the detection of smaller tumours and tumour volumes has occurred, making tumours even more subtle. When visible, tumour tissue is nodular and white, and may blend into surrounding benign, tan-coloured parenchyma. Most adenocarcinomas arise in the peripheral lateral or posterior zones, with anterior peripheral localization infrequently occurring. Transitional (20 %) and central zone (5 %) lesions are also less common in comparison to peripheral lesions (70 %). Prostatic adenocarciomas are multifocal in 50–97 % of cases and recent studies now suggest that these lesions are different, in that they are thought to arise from multiple, independent clonal expansions.

### 10.8.1 Malignant Glandular Component

Histologically, when prostatic adenocarcinoma is referred to without further qualification, it can be assumed to be the common, or acinar, variant [21]. The glands are smaller and more crowded than their benign counterparts under the

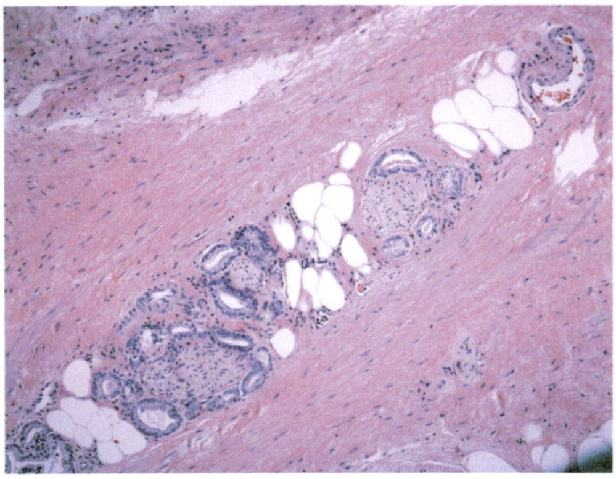

**Fig. 10.8** Perineural invasion

microscope, and have straight luminal borders rather than papillary infoldings. Having lost the outer basal cell layer, the neoplastic glands are lined by a single layer of cuboidal or low columnar epithelium. The cytoplasm of the malignant cells ranges from amphophilic to the pale/clear cytoplasm seen in benign glands. Nuclei are enlarged and often contain one or more prominent nucleoli. Typically, nuclear pleomorphism and mitotic activity are not prominent. Intraluminal crystalloids or mucinous secretions, when present, are observed mostly in carcinomas, but only rarely identified in benign glands [35]. Although there are a few specific features of malignancy on biopsy (e.g. perineural invasion (Fig. 10.8), which is seen in approximately 20 % of prostate biopsies), the diagnosis is typically made on a combination of architectural and cytologic features on routine H&E. The diagnosis of PCa may be difficult because the clues to malignancy are often subtle, and also because there are several benign mimickers of PCa, such as adenosis, atrophy, and basal cell hyperplasia.

### 10.8.2 Malignant Stromal Component

The majority of prostate tumours are adenocarcinomas arising from epithelial cells lining the prostatic glands. As a consequence, prostatic epithelial cells have been the major focus of histopathological examination and research studies to date. While the development of an altered stromal microenvironment in response to carcinoma is a common feature of many other tumours [36, 37], emerging evidence suggests that changes occur in the surrounding prostatic stroma also, that may serve to enhance the malignant potential of the nearby epithelium [38]. Tumour-associated stroma is referred to as 'reactive' in the prostate, as the phenotypic and genotypic alterations seen are similar to those seen in wound repair, including matrix remodelling and altered expression of growth factors and cytokines [39]. The stroma normally consists predominantly of smooth muscle cells, however, during carcinogenesis, these cells are gradually replaced by the activated form of fibroblasts, termed carcinoma-associated fibroblasts (CAFs). There is some debate as to whether CAFs and myofibroblasts represent different cell types, or whether they are the same cell type with different gene expression profiles [40, 41]. It is also unclear as to whether these CAFs are generated by activation of fibroblasts and perivascular cells already present in the prostatic stroma, or whether they are recruited from other locations, i.e. from the bone marrow. CAFs have been shown to affect cancer cells by secreting growth factors [42], extracellular matrix (ECM) components [43] and proteases [44].

Tumour development is also associated with an influx of macrophages [45], lymphocytes and mast cells into the tumour stroma [34]. These inflammatory cells secrete cytokines, which have stimulatory or inhibitory effects on adjacent CAFs, blood vessels and epithelial cells. While inflammatory cells are generally viewed as beneficial, e.g. in wound repair and in preventing the spread of infection, tumour-associated inflammation appears to be associated with an increased rate of growth and the spread of cancer [46, 47]. This large and varied list of stromal cell types may explain the heterogeneity observed in PCa and further research into stromal–epithelial interactions [48] must be performed.

### 10.8.3 The Use of Immunohistochemistry in Prostate Pathology

The most valuable adjunctive study for the diagnosis of adenocarcinoma of the prostate is immunohistochemistry (IHC), which is used currently in up to 20 % of all prostate needle core biopsies. Any IHC marker, however, carries a risk of false-positive and false-negative results, and therefore must be used in conjunction with the H&E morphology. The immunophenotype of prostatic basal cells is distinctive and can be of diagnostic value, as positive staining with antibodies directed against basal epithelial cells (34β[beta]E12 and p63) rules out the presence of invasive carcinoma [21, 35]. Nuclear basal markers can also be used when the differential diagnosis is between atypia, atypical adenomatous hyperplasia, PIN and atrophy. In some cases, a basal cell marker cocktail can be used to further improve detection of basal cells [49]. In contrast, the normal secretory cells of the prostate stain with pan-cytokeratins, cytokeratins 8 and 18, androgen receptor (AR), PSA and prostate- specific acid phosphatase (PSAP). This immunoprofile distinguishes them from the underlying basal cells. Normal neuroendocrine cells are characterised by their immunoreactivity for synaptophysin, chromogranin or CD56. They variably express AR, an important feature in relation to castration-resistant disease in prostatic carcinoma with neuroendocrine differentiation [50, 51].

Alpha-methylacyl-CoA racemase (AMACR, also known as 'P504S' or 'racemase') [35] is a commonly used cytoplasmic marker which is selective and very sensitive for carcinoma. While some benign glands (e.g. atrophic glands) and glands showing PIN can also stain positively for AMACR [52], the staining is usually more focal and weaker in comparison to neoplastic glands. Some laboratories combine antibodies in a cocktail also, using a basal marker together with AMACR for definitive diagnosis [53]. A 'triple stain' of p63, 34β[beta]E12, and AMACR in prostatic adenocarcinoma can be seen in Fig. 10.9. Stromal tumour markers are not used as routinely in the diagnosis of PCa, however, it is known that decreased expression of smooth muscle cell markers (desmin, smooth muscle actin) and increased fibroblast markers (vimentin) are seen [54]. As no therapeutic targets have currently been identified in PCa, IHC for proteins such as AR, HER2, CD117 (c-Kit), EGFR, etc., are not routinely employed as they are not predictive of response to treatment. Therefore, there is an urgent need to identify molecular signatures to discriminate lethal and indolent PCa and to identify predictive and prognostic biomarkers. While most studies in the past have concentrated on epithelial-based signatures, the stromal microenvironment may hold the answer to this question.

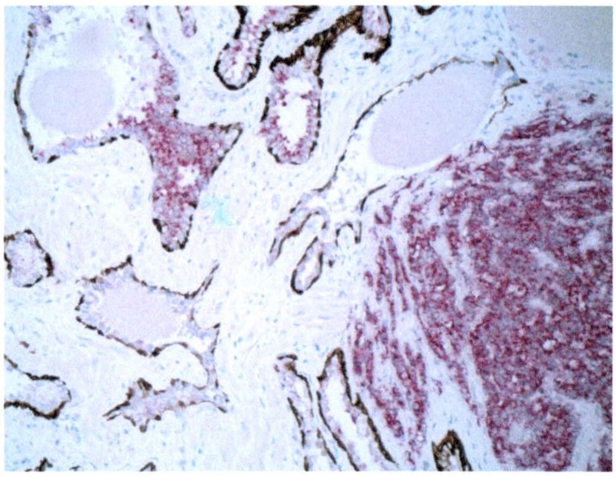

**Fig. 10.9** Photomicrograph of a 'triple stain' highlighting benign glands, high-grade prostatic intraepithelial neoplasia and prostate adenocarcinoma

### 10.8.4 Histologic Grading

Histologic grading is the most useful tissue-based prognostic predictor in PCa [55]. The quintessential grading scheme for PCa, the Gleason grading system, is based solely on the architectural pattern of the tumour. Both the predominant ('primary') and the second most prevalent ('secondary') patterns are identified [56]. Each is assigned a numeric value, and these are added together to yield the 'Gleason Score'. The grading system originally devised in 1966 by Donald Gleason categorised prostate cancer into five separate grades ranging from 1–5, representing increasingly more aggressive and less differentiated morphologic patterns. In practice, however, grades 1 and 2 are no longer in use, and the primary and secondary patterns of the tumour are generally assigned a value from 3 (well differentiated, with individual, discrete, well-formed glands) to 5 (poorly differentiated/undifferentiated tumour, or tumour with comedonecrosis). A Gleason grade of 4 indicates that the tumour displays an intermediate level of differentiation; poorly formed glands, fused glands, or glands with cribriform architecture [56]. In practice, therefore, the Gleason score on prostatic specimens usually falls within the range of 6–10 [56]. Examples of Gleason patterns (i.e. grades 3+3, 3+4, 4+4 and 5+4) are shown in Figs. 10.10, 10.11, 10.12 and 10.13. In addition, it is advisable to report the percentage of Gleason pattern 4 tumour in a biopsy specimen with an overall Gleason Score of 7; this is in order that clinicians are aware of the approximate extent of the less differentiated (and therefore more aggressive) tumour present in the biopsy material. Increasingly, pathologists also comment on the presence of a 'tertiary' component—this situation arises where a third, typically more aggressive pattern is identified. Studies have shown that the presence of a tertiary higher grade component is associated with an increased risk of biochemical recurrence, and it is currently recommended that tertiary grade patterns be recorded in the pathology report in order to accurately reflect the overall grade [57]. The approach to tertiary grading differs between biopsy and prostatectomy specimens. For biopsies, the primary pattern and the pattern of highest grade should be added together to formulate the Gleason score. In the case of a higher grade, tertiary pattern, this would result in the exclusion of the secondary pattern from the overall Gleason score. Conversely, in radical prostatectomy specimens, the Gleason score should be based on primary and secondary patterns, with a comment added on the tertiary pattern [58]. Some special rules apply to Gleason grading of uncommon histologic variants of PCa [55], but these are beyond the scope of this discussion.

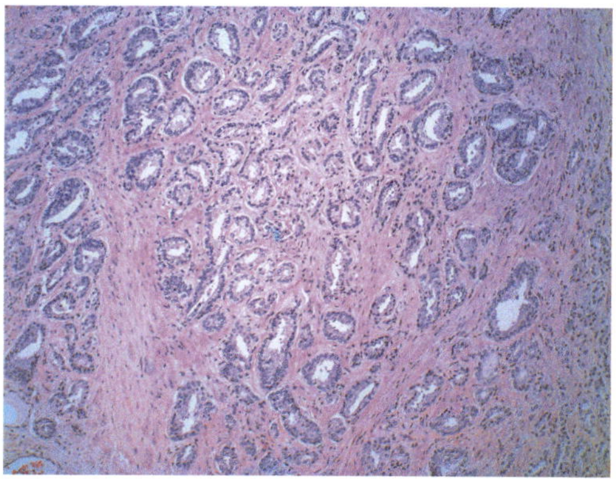

**Fig. 10.10** Prostatic adenocarcinoma: Gleason grade 3+3

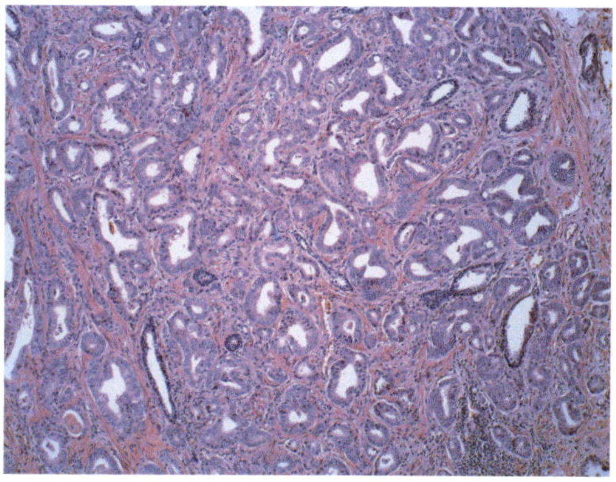

**Fig. 10.11** Prostatic adenocarcinoma: gleason grade 3+4

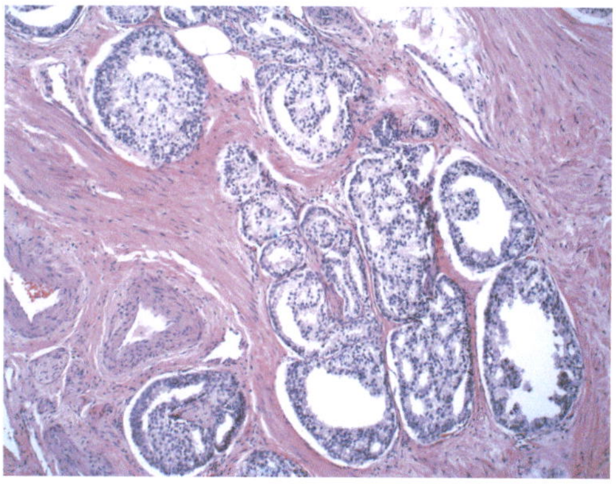

**Fig. 10.12** Prostatic adenocarcinoma: gleason grade 4+4

The Gleason grade remains the most reliable prognostic factor in PCa [59], however, there has been a gradual shift in how the Gleason grade is

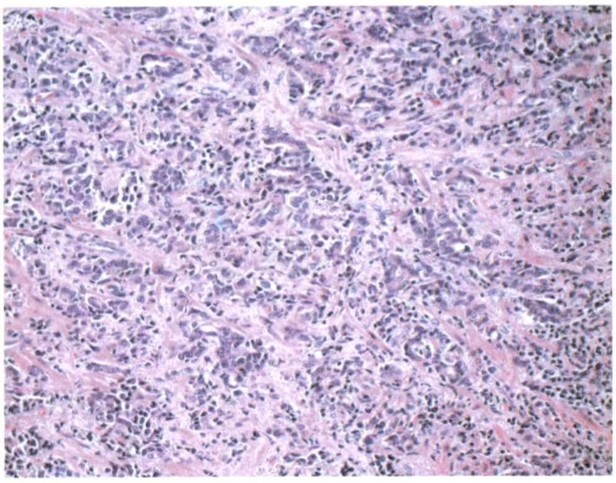

**Fig. 10.13** Prostatic adenocarcinoma: gleason grade 5+4

applied in practice, with a general trend towards upgrading. In addition, the Gleason grading system has been adapted over time in order to accommodate changing clinical practices [55]. In particular, a consensus conference which took place in 2005 [60] was organised in order to standardise the perception of the various patterns, as well as the manner in which the information relating to tumour grading was reported. The resultant 'Modified Gleason System' reflected this effort at standardisation, and one of its effects was a shift towards upgrading [55]. Studies have shown that needle biopsies using the Modified Gleason System correlated with progression after radical prostatectomy [61] and predicted biochemical recurrence after radical prostatectomy [62]. More recently (November 2014), a second major consensus conference was held with the aim of further updating the Gleason System. Among the parameters examined, an effort was made to standardise the grading of particular morphologic patterns and variants of prostate cancer, including those commonly encountered (e.g. cribriform glands), in addition to relatively uncommon patterns/variants (e.g. glomeruloid glands, mucinous adenocarcinoma, intraductal carcinoma). With regard to the grading of cribriform glands in PCa, according to the updated (2014) guidelines, all cribriform glands are now assigned a Gleason grade of 4, in contrast to the

**Fig. 10.14** Tissue microarray of prostatic adenocarcinoma

previous iteration of the Gleason System, which designated a Gleason Grade of 3 to small and relatively round/regular cribriform glands. [63] Interestingly, further prognostic information can be gleaned from examining the relative contributions of the individual patterns that ultimately make up the Gleason score. The most striking example of this is the dramatic difference seen between the clinical stages of tumours with a Gleason score of 7, depending on whether there was a primary pattern of 3 and a secondary pattern of 4, and vice versa (3 + 4 = 7, and 4 + 3 = 7, respectively). In one study, 95 % of the 3 + 4 = 7 tumours were clinically staged as pT2, while 79 % of 4 + 3 = 7 cancers were pT3 or pT4 [64]. The prognostic implications of both the overall Gleason score and the relative proportions of the individual component patterns were also discussed at the 2014 consensus meeting, and a novel grading system was proposed. The practice of combining the most common and second most common Gleason patterns to obtain an overall Gleason Score was retained, but the new system stratifies tumours with Gleason Scores of 2–10 into prognostically distinct 'Grade Groups' numbered 1–5. Grade Group 1 tumours are composed of only discrete, well-formed glands (therefore with a Gleason Score of 6 or lower), while Grade Groups 2 and 3 comprise tumours of Gleason Score 7 whose component patterns are 3 + 4 = 7 and 4 + 3 = 7, respectively. Grade Group 4 contains tumours of Gleason Score 8, while the most aggressive tumours (Gleason Score of 9 or 10) make up Grade Group 5. These 'Grade Groups' are felt to carry greater prognostic value and to more accurately reflect the biology of prostate cancer than the previous Gleason system and various algorithms which had been used for prognostic 'grouping' based on Gleason Score. Furthermore, a perceived flaw of the older system was that although 6 was the lowest score assigned to a tumour in practice, a score of 6 out of 10 nevertheless might contribute to patient fear by implying an intermediate rather than an excellent prognosis. It is thought that the new Grade Groups would—for the foreseeable future—be used in conjunction with the Gleason System. The new system and its terminology 'Grade Groups 1–5' have been accepted by the World Health Organisation for the 2016 edition of Pathology and Genetics: Tumours of the Urinary System and Male Genital Organs [63].

Lastly, because of heterogeneity and sampling error, some studies report that up to 28 % of cases are upgraded in radical prostatectomies compared to biopsies [65, 66]. This may have a significant impact on clinical decision-making, and can constitute a confusing and problematic factor with regard to the choice of therapeutic approach for a given patient (e.g. active surveillance versus surgery or radiotherapy) [59, 60, 67].

### 10.8.5 Spread and Staging

Local extension of prostatic carcinoma is usually into periprostatic tissue, seminal vesicles, and bladder base [21]. Cancer staging takes into account whether the cancer is unilateral or bilateral and the presence or absence of invasion beyond the prostate or into surrounding structures. [24] Metastases initially spread via the lymphatics to the obturator nodes, and eventually to the paraaortic nodes. Haematogenous spread is chiefly to the bones, in particular the axial skeleton, proximal femurs and ribs. However, rare cancers can spread widely to the viscera [21]. The reader should refer back to Chap. 3 for an overview of the principles of cancer staging. Organ-specific TNM staging information can be obtained from the current American Joint Committee on Cancer (AJCC) staging manual [68].

### 10.8.6 Uncommon Histologic Variant: Carcinoma with Neuroendocrine Features

Aside from the common, acinar variant, several distinct morphologic subtypes of prostatic carcinoma exist. These rare subtypes are beyond the scope of this book chapter, but they are described in detail in the current World Health Organisation/ International Agency for Research on Cancer's

(WHO/IARC) Classification of Tumours publication relating to the urinary system and male genital organs [69]. Of the uncommon variants of PCa, carcinoma with neuroendocrine (NE) features merits a brief discussion, as it is increasingly recognised as an aggressive variant that is associated with castration-resistant disease [51]. The NE cells of the prostate lack ARs, and NE differentiation increases in the late stages of castration-resistant PCa and in response to androgen deprivation therapy. Transformation to a predominantly AR-negative PCa with increasing NE differentiation may be an important resistance mechanism in castration-resistant disease, and indeed may be more common than was previously recognised. NE tumours of the prostate comprise a heterogeneous group, and the current WHO classification of these cancers includes the following subtypes: Focal NE differentiation in otherwise conventional prostatic adenocarcinoma, carcinoid tumour (well-differentiated NE tumour) and small cell carcinoma (poorly differentiated NE tumour). It has been suggested, however, that this basic classification should be broadened to include three further variants; namely, adenocarcinoma with paneth cell-like NE differentiation, large cell NE carcinoma and mixed NE-acinar carcinoma. Diagnosis of carcinoma with NE features is important, as it is considered an aggressive phenotype of advanced PCa with treatment strategies that differ from those of conventional adenocarcinoma (notably, the use of platinum-based therapies as opposed to AR-targeted treatment). The diagnosis generally depends on histology, as no reliable serum markers are currently available for identification of patients who are transforming to an NE phenotype. However, the detection and molecular characterisation of circulating tumour cells (CTCs) has been suggested for this purpose.

## 10.9 The Molecular Pathology of Prostate Cancer

PCa is a disease that exhibits great variation in its clinical behaviour. For this reason, prognosis, treatment approach and outcome can vary widely between individuals. The goal of molecular staging in PCa is to identify the genes involved in the pathways that lead to malignancy, and to exploit them as prognostic and/or predictive markers and, as outlined above, to distinguish indolent from aggressive disease. In PCa, as in other malignancies, this field has evolved rapidly in the past few years. Technologies that enable high-throughput analysis have led to a more comprehensive understanding of the progression of the disease, and there has been a dramatic increase in the number of candidate biomarkers for PCa. The identification of these genes and biomarkers uses both serum and tissue-based assays. In the following sections, we examine the resources and techniques presently employed in the molecular pathology of PCa, in addition to the important biomarkers, both current and emerging.

## 10.10 Resources

Multiple resources are currently utilised to further our understanding of the molecular pathogenesis of PCa, including tissue culture models and mouse models. Examples of tissue culture models include cell lines, organ cultures and organotypic cultures. Cell lines refer to single cell suspensions of cancerous cells, which can be grown for prolonged periods in vitro due to their ability to evade senescence and to continually proliferate [70]. In PCa, the LNCaP cell line, isolated in 1977 from metastatic prostatic carcinoma, has been intensively exploited in research applications [71]. However, cell lines grown as a monolayer in tissue culture dishes lack the same features present in intact tissue. Therefore, three-dimensional representations of tumour microenvironment are now available, using organ cultures, consisting of tissue slices and organotypic cultures consisting of cells grown in an extracellular matrix to recapitulate the biological behaviour of the prostate [72, 73].

Mouse models are another important resource used in PCa research due to the genomic similarities between mice and humans, the relative ease with which genetic modification can be performed, and the fact that mice are comparatively easy to keep and breed [74]. They can be divided into two broad categories, including genetically engineered mouse (GEM) models [75], in which a gene is either knocked out or overexpressed, and mouse xenografts, which involve the injection of human tumour cells into the mouse using various techniques. The main advantage of GEM models is that they reflect tumour progression over time from the initiation of preinvasive lesions, to invasive and metastatic disease. An additional benefit is that this progression is observed within the prostatic microenvironment. The chief disadvantage associated with GEM models is the fact that important biologic differences may exist between murine and human prostates, which may affect model phenotypes [76]. The TRAMP model, one of the original PCa GEM models, was created by inducing the prostate-specific rat probasin promoter (PB) to regulate the SV40 tumour antigens [74]. The TRAMP mice developed epithelial hyperplasia by the time of sexual maturity (approximately 8 weeks of age), followed by PIN at ~18 weeks. By 28 weeks, 100 % had developed invasive tumours with lymphatic metastases, while two-thirds had pulmonary metastases [74]. In addition, this model exhibited hormone-resistant, NE disease after androgen withdrawal. Subsequently, a large number of other GEM models have been developed [77], including *Phosphate and Tensin Homolog* (*PTEN*) knockout, *Myc* overexpression [78], *TMPRSS2-ERG* fusion, AR manipulation, *Rb* inactivation [79], *APC* gene deletion [80], RAS/RAF/MAPK pathway mutations, and TGF-β[beta] signalling loss models, to name a few [76]. The second broad mouse model category are the pure xenograft models, in which fresh human primary PCa tissues are implanted subcutaneously or orthotopically into the prostate of host immunocompromised mice [81]. The latter technique carries the advantage of growth within the prostatic microenvironment, and a number of such models show high rates of metastasis. The LuCaP series is important to note, as it includes a large number of PCa xenograft lines, mostly derived from metastatic disease [82]. Finally, tissue recombination models are another model type, where benign or malignant epithelial cells combined with mesenchymal cells, are implanted subcutaneously or under the renal capsule in immunocompromised host mice [76]. These models represent a powerful method in the study of tumour–stromal interactions in particular.

Tissue microarrays (TMAs), CTCs, disseminated tumour cells and free plasma DNA are also used as resources in PCa research. TMAs, introduced in Chap. 2, are used extensively in research applications relating to PCa (Fig. 10.14). Epidemiologists and pathologists use cancer TMAs to validate multiple biomarkers concurrently in large numbers of patients, and correlate biomarker expression with long-term follow-up utilising sophisticated statistical tools. A significant limitation of TMA use in PCa in particular, is that due to tissue heterogeneity [83], the small cores may not always be representative of the whole tumour, with areas of variable protein expression thus missed on sampling [84]. Several different assays are used to identify CTCs in the peripheral blood, and disseminated tumour cells in the bone marrow, including immunological techniques [85, 86] or size-trapping methods [87] that identify cell surface markers. High CTC counts in the peripheral blood have been associated with higher Gleason score and increased disease stage [88, 89]. Finally, free plasma DNA levels can be analysed and high levels have been correlated with both tumour stage and the presence of CTCs [90, 91]. This suggests that free plasma DNA may constitute a candidate biomarker for monitoring disease progression and metastasis [92–94] and further research is ongoing.

## 10.11 Molecular Pathology Techniques

Available molecular pathology techniques today include laser capture microdissection (LCM), fluorescence in situ hybridization (FISH), quantum dot analysis and DNA methylation analysis. First, LCM is a specialised technique used to isolate specific groups of cells with high precision using laser technology. In this way, pure samples of the cells of interest, e.g. tumour only, normal only, or dysplastic only, may be obtained from heterogeneous solid tumours. DNA, RNA and proteins may subsequently be extracted from the tissue and analysed [95]. FISH techniques, introduced in Chap. 2, are commonly used in the molecular staging of PCa, notably for detection of the TMPRSS2-ERG fusion gene [96], loss of NKX3.1 [97], gain of Myc, and PTEN inactivation [98]. Quantum dots are fluorescent semiconductor nanocrystals, which are gaining increasing attention in the field of bioimaging. Using quantum dot-labelling of DNA and RNA probes followed by spectral imaging, this technique can be used to apply gene expression signatures to formalin-fixed paraffin-embedded (FFPE) tissue biopsies at diagnosis [99]. Finally, DNA methylation is the most intensively studied epigenetic alteration in human malignancies and aberrant hypermethylation of promoter-associated CpG islands (i.e. regions of DNA where a cytosine nucleotide occurs adjacent to a guanine nucleotide in a sequence of bases) is the most common known epigenetic abnormality in human cancers. In PCa, abnormal DNA methylation has been identified in precursor lesions and in early cancers, suggesting that this change may be a driver of malignant transformation [100]. Alterations in DNA methylation also appear to be present in progressive and metastatic disease, meaning that they may play an additional role in disease expansion and dissemination. DNA methylation markers, therefore, hold great promise as prognostic biomarkers, both for early detection of PCa, as well as for prediction of clinical outcome [101].

## 10.12 The Genetics of Hereditary Prostate Cancer

There is compelling evidence for a strong genetic component in PCa. It is widely accepted that the phenotypic heterogeneity observed in PCa is either genetically determined, or modulated by a complex interaction of environmental and genetic factors [102]. It follows that the clinical behaviour of the disease can be predicted using genetic stratification. Genetic variation in PCa can result from germline mutations, which are heritable, or somatic mutations, which accumulate over the course of a person's lifetime. Unfortunately, unlike other malignancies such as breast cancer, where familial predisposition to disease has been strongly linked to specific genes, the identification of genes clearly responsible for the development of PCa has proven extremely challenging. There are two main reasons why this has been the case. First, because PCa arises at an advanced age, it is difficult to identify more than two generations affected by the disease in order to perform molecular studies. Second, given the high prevalence of the disease, where a family with a high rate of PCa is identified, it is difficult to distinguish sporadic from hereditary disease. This is because hereditary PCa does not have any specific pathologic or clinical characteristics other than an earlier age of onset in some cases. In addition, hereditary PCa does not occur in any of the known cancer syndromes [103].

Over the last few years, it has been possible to conduct genome-wide association studies (GWAS) in order to identify single nucleotide polymorphisms (SNPs) (i.e. a DNA sequence variation) associated with PCa. The long arm of chromosome 8 (8q24) was found to contain multiple SNPs that showed significant associations with PCa across multiple ethnic groups. This was the first region to be identified as related to PCa development. Though the mechanisms by which this particular region of chromosome 8 confers an increased risk of PCa are still unclear, it is known that the coding region closest to 8q24 is the oncogene Myc. It is therefore hypothesised that polymorphisms in this region may be modulating the expression of Myc, which is discussed in greater detail in the section below. Genetic studies have also identified germline mutations in the HOXB13 gene (present in the 17q21–22 region) in many hereditary PCa families. In particular, the HOXB13 G84E variant is associated with a significantly increased risk of hereditary PCa [104].

The catalogue of loci associated with PCa development has now expanded beyond 8q24 and 17q21–22, with over one hundred additional SNPs [105] having been identified throughout the genome. Many SNPs are only moderately associated with PCa risk; however, there is a cumulative, dose-dependent effect when multiple SNPs are present in combination [106]. These risk-associated variants are currently undergoing further exploration, and data from recent years has indicated that there may be several new potential pathways implicated in the development of PCa. [27] Notwithstanding this fact, the majority of SNPs that are disease associated are present in non-coding regions and thus have unexplained functions. One means to ascribe functionality to certain SNPs is to link them with expression quantitative trait loci [105], which are genomic loci that partly account for the genetic variance of a gene expression phenotype [107].

### 10.12.1 Genetic Sequencing

Techniques such as gene expression profiling, comparative genomic hybridisation (CGH) and SNP arrays analyse tumours at a genome-wide level. New insights into the molecular profiles of various cancers have been gained from large-scale projects characterising and classifying cancer genomes such as The Cancer Genome Atlas (www.cancergenome.nih.gov) and the International Cancer Genome Consortium (www.icgc.org). The datasets for such classifications, including those for PCa, are now available as public resources, and are invaluable tools in the cancer research community. The integration of traditional pathology with genomic-based techniques offers the potential of a more accurate and personalised classification with more informed decision-making on the part of clinicians, in addition to the identification of novel therapeutic targets [108].

Regarding PCa in particular, analyses of transcriptomes (i.e. collections of all the RNA transcripts within a cell) and copy number alterations (CNAs) (i.e. somatic changes to chromosome structure that result in the gain or loss in copies of sections of DNA [109] have been reported by several groups. Consistent findings include TMPRSS2-ERG fusion ($\sim$50 %), 8p loss ($\sim$30–50 %), and 8q gain ($\sim$20–40 %). In addition, numerous general PCa signatures have been identified through transcriptome studies; however, unlike breast cancer, defined subtypes with distinct outcomes have not yet been characterised [108]. Associations have been identified between relapse and CNA pattern in primary PCa. Recent work has found that CNA burden across the genome, defined as the percentage of the genome affected by CNA, was associated with biochemical recurrence and metastatic disease over a broad range of clinical presentations, independent of PSA status or Gleason Grade. These results indicate that CNA burden may be of use for risk prediction in the pretreatment setting [110].

## 10.13 Altered Pathways in Prostate Cancer

The use of tools such as gene expression signatures derived from prostate tumours has led the field to the discovery of activated pathways that drive tumour biology. In PCa, a variety of cellular pathways show significant dysregulation. An understanding of them furnishes us with diagnostic, prognostic and predictive information, as well as opportunities to target certain steps therapeutically. These pathways include AR signalling, tyrosine kinase receptor pathways, cell cycle regulation, and lipogenesis, among others. Considerable overlap and cross-interaction exists between the different pathways, however, rendering the molecular pathology of PCa rather complex. In addition, gene expression profiling has lead to the identification of both prognostic markers [111] and metabolic programs that support tumorigenic genomic alterations [112]. These metabolic programs can be analysed by other high-throughput methodologies such as metabolic profiling, which can be applied to serum as well as tissues. These novel insights into tumour biology that are derived from advances in technology and high-dimensional data analysis, pave the way to the possibility of inferring the genomic make up of a tumour in serum profiling [113]. Furthermore, the integration of different technologies in the analysis of pathway activation, can lead to the discovery and utilisation of these biomarkers as in vivo radiotracer and novel therapeutic targets. Taken together, these approaches are likely to improve outcomes while reducing the overtreatment of indolent disease.

Selected important molecular pathways, candidate genes and biomarkers pertaining to Pca are outlined in this section.

### 10.13.1 The Androgen Receptor Signalling Pathway

The AR gene is located at Xq12 and codes for the AR protein, which mediates the development and differentiation of prostatic cells under normal physiologic circumstances. IHC studies have demonstrated persistent AR expression in many subsets of PCa, including hormone-refractory disease. This suggests that not only is PCa progression almost never associated with loss of AR expression, it may even require AR signalling, regardless of the upstream activator. AR expression shows considerable heterogeneity within specimens, a feature which is more marked with increasing Gleason grade [114].

Hormone ablation therapy causes a rapid decrease in tumour volume. This decrease, unfortunately, is often followed by PCa recurrence which is hormone refractory (castration-resistant) [115]. AR mutations, generation of ligand-independent splice variants, AR copy number increase and AR cistrome reprogramming contribute to the development of hormone-refractory PCa. The AR cistrome, or the genome-wide set of AR binding sites, undergoes substantial reprogramming during prostate tumorigenesis. In this context, HOXB13 and FOXA1 appear to function as co-factors, enabling cancer cell survival by co-occupying tumour-only AR binding sites. Amplifications or mutations of the AR gene could constitute the mechanism of progression of PCa, by resulting in a selective growth advantage for the malignant cells [108]. Indeed, AR gene amplification has been observed in the androgen-independent state.

### 10.13.2 Phosphate and Tensin Homolog and the Phosphoinositol-3 Kinase Pathway

The PTEN gene, a tumour suppressor gene, is located at 10q23 [116], and modulates epithelial proliferation and normal cellular homeostasis. The gene acts as a negative regulator of the Phosphoinositol-3 Kinase (PI3K) signalling cascade, thereby inhibiting several cellular functions such as cell growth, proliferation, differentiation and survival. It follows that loss of function of PTEN leads to dysregulation of PI3K signalling, and subsequent activation of downstream targets which promote cell proliferation and survival. These downstream targets include the β[beta]-alanyl-α [alpha]-ketoglutarate transaminase/protein kinase B family (mercifully abbreviated to AKT). AKT is a protein kinase which acts within the PI3K pathway to mediate multiple signals upstream [117].

Studies on transgenic mice with PTEN deletion or AKT overexpression have demonstrated the critical role of this signalling pathway in the progression of PCa [118–120]. Through the loss of PTEN function, PI3K signalling upregulation (which is central to a number of solid tumours) is thought to be present in 30–50 % of PCas. In addition, aberrant signalling of this pathway is strongly associated with metastasis and reduced survival, and there is some evidence that it is associated with treatment resistance in addition [121]. Genetic inactivation of PTEN and consequent activation of the PI3K pathway can be demonstrated by FISH or IHC studies.

Interestingly, it has been found that molecular characterisation of PTEN loss in biopsy specimens of low-grade (Gleason score 6) PCa can help identify which cases are more likely to be upgraded at subsequent radical prostatectomy.

Tumours with PTEN loss are more likely to undergo upgrading in this setting than those without loss of PTEN, even after adjustment for age, preoperative PSA, clinical stage, and race. Use of PTEN as a biomarker in this fashion thus has the potential to augment the traditional preoperative pathologic grading by Gleason score [122].

### 10.13.3 TMPRSS2-ERG and TMPRSS2-ETV1 Translocations

A relatively recent discovery in PCa has been the translocation resulting in the fusion of TMPRSS2 (21q22.2), which codes for an androgen-regulated transmembrane protein, with members of the ETS transcription factors family. The latter family includes ERG (21q22.3) and ETV1 (7p21.2). The consequence of this translocation is that the oncogenes ERG and ETV1 are then regulated by the androgen-responsive promoter of TMPRSS2. This may constitute a mechanism for early carcinogenesis [123]. Both translocations (i.e. TMPRSS2-ERG and TMPRSS2- ETV1) appear to be mutually exclusive. TMPRSS2-ERG is the more common of the two, and is present in 85 % of all ETS fusion-positive samples, and approximately 50 % of localised PSA-screened PCa. It can be detected using IHC, FISH or reverse transcription polymerase chain reaction (RT-PCR) [76]. The prognostic significance of carrying a fusion protein remains unclear [124–126], one of the splice variants has been associated with aggressive disease [123]. In addition, TMPRSS2-ERG has been shown in mouse models to act in synergy with PTEN loss in the development of PCa. ERG and ETV1 control a common transcriptional network but largely in an opposing fashion. In particular, while ERG negatively regulates the AR transcriptional program, ETV1 cooperates with AR signalling by favouring activation of the AR transcriptional program and only ETV1 appeared to support development of invasive adenocarcinoma under the background of full PTEN loss [127].

### 10.13.4 Myc

CGH studies have shown that one of the most frequently altered regions in the setting of PCa progression is 8q24 where the Myc proto-oncogene is located. More than a fifth of recurrent and metastatic PCas carry high-level amplification of Myc [128], which has been associated with advanced histological grade and worse prognosis. Furthermore, Myc amplification is often concurrent with other chromosome 8 aberrations, and the cumulative effects can lead to more aggressive biologic behaviour. One such gene is prostate stem cell antigen (PSCA), which is located nearby (8q24.2). Increased expression of PSCA may be linked to PCa progression [129]. The Myc gene codes for a transcription factor which is integral to the regulation of various cellular processes ranging from cellular proliferation and differentiation to metabolism and apoptosis. Transgenic mice expressing human Myc in the prostate showed that these mice developed PIN and subsequent invasive carcinoma with 100 % penetrance.

USP2a, a deubiquinating enzyme which is overexpressed in around 40 % of PCas, has recently been shown to regulate micro RNA (miRNA) expression which subsequently results in Myc upregulation. It has also been suggested that USP2a can confer chemoresistance through upregulation of Myc [130, 131]. In a recent study of ours, (Pettersson, unpublished) no significant associations between Myc protein expression and clinicopathologic factors, biochemical recurrence, or lethal PCa were seen. Upregulation of nuclear Myc protein expression is a highly prevalent and early change in PCa and suggest that increased nuclear Myc may be a critical oncogenic event driving human PCa initiation and progression [132].

### 10.13.5 Lipogenic Prostate Cancer

Associations between high-fat diet and PCa risk have been reported in numerous studies. In addition, there is growing epidemiological evidence to support a relationship between obesity and PCa progression. Though the mechanisms that underlie these associations are unclear, the insulin-mediated release of insulin-like growth factor 1 receptor (IGF-1R) has been implicated in the relationship between PCa and obesity [133]. 5′AMP-activated protein kinase (AMPK) is an energy-sensing serine/threonine kinase which is activated by metabolic stressors that deplete intracellular ATP while increasing AMP, such as glucose deprivation and hypoxia [134]. Once activated, AMPK reduces plasma insulin, suppresses ATP-consuming metabolic functions, and increases ATP-producing activities in an effort to restore energy homeostasis. In this way, it functions as a metabolic switch that mediates glucose and lipid metabolism. Decreased AMPK activation has been implicated in obesity and the metabolic syndrome (MS), both of which are associated with increased cancer risk. Importantly, drugs that improve MS conditions through AMPK activation may also be beneficial for the prevention and treatment of PCa. In this context, there is increasing interest in targeting the metabolic pathways that may be altered during the development and progression of PCa.

PCa cells synthesise large quantities of fatty acids and cholesterol de novo, regardless of circulating lipid levels, thus conferring a selective growth advantage, as well greater self-survival and drug resistance. Numerous studies have shown that inactivation of most lipogenic enzymes, including fatty acid synthase (FASN), acetyl-CoA-carboxylase (ACC) and 3-hydroxy-3-methylglutaryl CoA (HMG-CoA) reductase, leads to either growth suppression or death of tumour cells. FASN is a metabolic enzyme that catalyses the synthesis of palmitate from the condensation of malonyl-CoA and acetyl-CoA de novo [134]. By converting excess carbon intake into fatty acids for storage it plays a central role in energy homeostasis. The activity and expression of FASN are closely regulated by growth factors, hormones and diet. In most normal cells the enzyme is expressed at low levels. However, greatly increased expression is seen in many cancers, including almost all PCas, as well as in some benign and preinvasive lesions of the prostate. Indeed, increased FASN expression has been associated with worse prognosis and a reduction in disease-free survival in PCa. FASN thus represents an important biomarker of a subset of PCas and is an important therapeutic target. Its inhibition can be achieved by direct means (e.g. with FASN inhibitors) or indirect means (e.g. with USP2a inhibitors or AMPK activators). In addition to FASN and AMPK, other key lipid enzymes have been identified as potential therapeutic targets in the treatment of PCa [134]. To this end, HMG-CoA reductase inhibitors (i.e. statins) and ACC inhibitors (e.g. soraphen A) have shown promising preclinical results, both in vitro and in vivo. The use of anti-IGF-1R antibodies and IGF-1R tyrosine kinase inhibitors are currently under investigation for the treatment of androgen-independent PCa.

Despite significant efforts, conventional imaging in PCa does not contribute to patient management to the same degree as imaging performed for most other malignancies. In particular, positron emission tomography (PET) and PET-CT with the glucose analogue 18F-FDG have become routine staging tests in clinical oncology. However, since PCa is less glycolytic than most other malignancies [135], these particular techniques have no role in early, localised disease, and are of very limited benefit in advanced cancer. Using lipid precursor tracers in order to monitor the lipogenic profile of the cancer, these methods exploit the fact that increased fatty acid synthesis occurs as a relatively early event in prostate tumorigenesis [136] and correlates with disease progression [135, 137]. Imaging with lipid precursor tracers such as 11C acetate, 11C choline, 18F fluoroacetate or 18F choline has been evaluated to monitor lipogenic phenotype in PCa. In particular, both 11C acetate and 11C choline have shown increased sensitivity in the detection of primary, metastatic and recurrent PCa.

## 10.14 Summary

We have set out to emphasise the phenotypic heterogeneity of PCa in this chapter, which manifests itself in the wide range of biologic behaviours it exhibits. We have concluded with the molecular underpinnings of the disease, as this expanding area of research has furnished us with promising diagnostic and therapeutic applications, and it continues to evolve.

## References


1. Kumar VL, Majumder PK. Prostate gland: structure, functions and regulation. Int Urol Nephrol. 1995;27 (3):231–43.
2. Marker PC, Donjacour AA, Dahiya R, Cunha GR. Hormonal, cellular, and molecular control of prostatic development. Dev biol. 2003;253(2):165–74.
3. Sadler T. Urogenital system. Langman's medical embryology. 11th ed. Lippincott Williams & Wilkins; 2010. p. 235–64.
4. Waltregny D, Leav I, Signoretti S, Soung P, Lin D, Merk F, et al. Androgen-driven prostate epithelial cell proliferation and differentiation in vivo involve the regulation of p27. Mol Endocrinol. 2001;15 (5):765–82.
5. Shima Y, Miyabayashi K, Haraguchi S, Arakawa T, Otake H, Baba T, et al. Contribution of Leydig and Sertoli cells to testosterone production in mouse fetal testes. Mol Endocrinol. 2013;27(1):63 73.
6. Hayward SW, Baskin LS, Haughney PC, Foster BA, Cunha AR, Dahiya R, et al. Stromal development in the ventral prostate, anterior prostate and seminal vesicle of the rat. Acta Anat. 1996;155(2):94–103.
7. Hricak H. Anatomy of the prostate gland and surgical pathology of prostate cancer. Prostate cancer: contemporary issues in cancer imaging. Cambridge University Press; 2008. p. 1–5.
8. Moore K. Pelvis and perineum. Clinically oriented anatomy. 5th ed. Lippincott Williams & Wilkins; 2006. p. 402–9.
9. Villers A, Steg A, Boccon-Gibod L. Anatomy of the prostate: review of the different models. Eur Urol. 1991;20(4):261–8.
10. Young B. Male reproductive system. Wheater's functional histology. 4th ed. Churchill Livingstone; 2000. p. 337–9.
11. McVary KT, McKenna KE, Lee C. Prostate innervation. Prostate Suppl. 1998; 8:2–13.
12. McCullough AR. Prevention and management of erectile dysfunction following radical prostatectomy. Urol Clin N Am. 2001;28(3):613–27.
13. Di Silverio F, Gentile V, De Matteis A, Mariotti G, Giuseppe V, Luigi PA, et al. Distribution of inflammation, pre-malignant lesions, incidental carcinoma in histologically confirmed benign prostatic hyperplasia: a retrospective analysis. Eur Urol. 2003;43(2):164–75.
14. Billis A. Prostatic atrophy. Clinicopathological significance. Int Braz J Urol (Official Journal of the Brazilian Society of Urology). 2010;36(4):401–9.
15. Untergasser G, Madersbacher S, Berger P. Benign prostatic hyperplasia: age-related tissue-remodeling. Exp Gerontol. 2005;40(3):121–8.
16. Epstein J. Gross anatomy and normal histology. Biopsy interpretation of the prostate. 4th ed. Lippincott Williams & Wilkins; 2008. p. 13–21.
17. Srigley JR. Benign mimickers of prostatic adenocarcinoma. Mod Pathol (An Official Journal of the United States and Canadian Academy of Pathology, Inc.). 2004;17(3):328–48.
18. Nelson EC, Cambio AJ, Yang JC, Ok JH, Lara PN Jr, Evans CP. Clinical implications of neuroendocrine differentiation in prostate cancer. Prostate Cancer Prostatic Dis. 2007;10(1):6–14.
19. Christian JD, Lamm TC, Morrow JF, Bostwick DG. Corpora amylacea in adenocarcinoma of the prostate: incidence and histology within needle core biopsies. Mod Pathol (An Official Journal of the United States and Canadian Academy of Pathology, Inc.). 2005;18(1):36–9.
20. Magi-Galluzzi C, Evans AJ, Delahunt B, Epstein JI, Griffiths DF, van der Kwast TH, et al. International Society of Urological Pathology (ISUP) consensus conference on handling and staging of radical prostatectomy specimens. Working group 3: extraprostatic extension, lymphovascular invasion and locally advanced disease. Mod Pathol (An Official Journal of the United States and Canadian Academy of Pathology, Inc.). 2011;24(1):26–38.
21. Kumar V. The lower urinary tract and male genital system. Robbins and cotran pathologic basis of disease. 8th ed. Elsevier Health Sciences; 2009. p. 993–1002.
22. Pavlica P, Barozzi L, Bartolone A, Gaudiano C, Menchi M, Veneziano S. Nonspecific granulomatous prostatitis. Ultraschall Med. 2005;26(3):203–8.
23. Mohan H, Bal A, Punia RP, Bawa AS. Granulomatous prostatitis—an infrequent diagnosis. Int J Urol (Official Journal of the Japanese Urological Association). 2005;12(5):474–8.
24. Preston MA, Wilson KM, Markt SC, Ge R, Morash C, Stampfer MJ, et al. 5alpha-Reductase inhibitors and risk of high-grade or lethal prostate cancer. JAMA Intern Med. 2014;174(8):1301–7.
25. De Marzo AM, Marchi VL, Epstein JI, Nelson WG. Proliferative inflammatory atrophy of the prostate: implications for prostatic carcinogenesis. Am J Pathol. 1999;155(6):1985–92.
26. Vykhovanets EV, Maclennan GT, Vykhovanets OV, Gupta S. IL-17 Expression by macrophages is associated with proliferative inflammatory atrophy lesions in prostate cancer patients. Int J Clin Exp Pathol. 2011;4(6):552–65.

27. Shah R, Mucci NR, Amin A, Macoska JA, Rubin MA. Postatrophic hyperplasia of the prostate gland: neoplastic precursor or innocent bystander? Am J Pathol. 2001;158(5):1767–73.
28. Macoska JA, Trybus TM, Wojno KJ. 8p22 loss concurrent with 8c gain is associated with poor outcome in prostate cancer. Urology. 2000;55 (5):776–82.
29. Epstein J. Prostatic intraepithelial neoplasia and its mimickers. Biopsy interpretation of the prostate. 4th ed. Lippincott Williams & Wilkins; 2008. p. 35–43.
30. Alsikafi NF, Brendler CB, Gerber GS, Yang XJ. High-grade prostatic intraepithelial neoplasia with adjacent atypia is associated with a higher incidence of cancer on subsequent needle biopsy than high-grade prostatic intraepithelial neoplasia alone. Urology. 2001;57(2):296–300.
31. Adamczyk P, Wolski Z, Butkiewicz R, Nussbeutel J, Drewa T. Significance of atypical small acinar proliferation and extensive high-grade prostatic intraepithelial neoplasm in clinical practice. Cent Eur J Urol. 2014;67(2):136–41.
32. Bostwick DG, Qian J. High-grade prostatic intraepithelial neoplasia. Mod Pathol (An Official Journal of the United States and Canadian Academy of Pathology, Inc.). 2004;17(3):360–79.
33. Vral A, Magri V, Montanari E, Gazzano G, Gourvas V, Marras E, et al. Topographic and quantitative relationship between prostate inflammation, proliferative inflammatory atrophy and low-grade prostate intraepithelial neoplasia: a biopsy study in chronic prostatitis patients. Int J Oncol. 2012;41(6):1950–8.
34. Sfanos KS, De Marzo AM. Prostate cancer and inflammation: the evidence. Histopathology. 2012;60 (1):199–215.
35. Epstein J. Diagnosis of limted adenocarcinoma of the prostate. Biopsy interpretation of the prostate. 4th ed. Lippincott, Williams & Wilkins; 2008. p. 81–3.
36. Martinez-Outschoorn U, Sotgia F, Lisanti MP. Tumor microenvironment and metabolic synergy in breast cancers: critical importance of mitochondrial fuels and function. Semin Oncol. 2014;41 (2):195–216.
37. Cammarota R, Bertolini V, Pennesi G, Bucci EO, Gottardi O, Garlanda C, et al. The tumor microenvironment of colorectal cancer: stromal TLR-4 expression as a potential prognostic marker. J Transl Med. 2010;8:112.
38. Hagglof C, Bergh A. The stroma-a key regulator in prostate function and malignancy. Cancers. 2012;4 (2):531–48.
39. Barron DA, Rowley DR. The reactive stroma microenvironment and prostate cancer progression. Endocr Relat Cancer. 2012;19(6):R187–204.
40. Bianchini F, Giannoni E, Serni S, Chiarugi P, Calorini L. 22: 6n-3 DHA inhibits differentiation of prostate fibroblasts into myofibroblasts and tumorigenesis. Br J Nutr. 2012;108(12):2129–37.
41. Kharaishvili G, Simkova D, Bouchalova K, Gachechiladze M, Narsia N, Bouchal J. The role of cancer-associated fibroblasts, solid stress and other microenvironmental factors in tumor progression and therapy resistance. Cancer Cell Int. 2014;14:41.
42. Liao CP, Adisetiyo H, Liang M, Roy-Burman P. Cancer-associated fibroblasts enhance the gland-forming capability of prostate cancer stem cells. Cancer Res. 2010;70(18):7294–303.
43. Stewart DA, Cooper CR, Sikes RA. Changes in extracellular matrix (ECM) and ECM-associated proteins in the metastatic progression of prostate cancer. Reprod Biol Endocrinol (RB&E). 2004;2:2.
44. Kalluri R, Zeisberg M. Fibroblasts in cancer. Nat Rev Cancer. 2006;6(5):392–401.
45. Comito G, Giannoni E, Segura CP, Barcellos-de-Souza P, Raspollini MR, Baroni G, et al. Cancer-associated fibroblasts and M2-polarized macrophages synergize during prostate carcinoma progression. Oncogene. 2014;33(19):2423–31.
46. Nakai Y, Nonomura N. Inflammation and prostate carcinogenesis. Int J Urol (Official Journal of the Japanese Urological Association). 2013;20(2): 150–60.
47. Omabe M, Ezeani M. Infection, inflammation and prostate carcinogenesis. Infect Genet Evol (Journal of Molecular Epidemiology and Evolutionary Genetics in Infectious Diseases). 2011;11(6):1195–8.
48. Wang X, Lee SO, Xia S, Jiang Q, Luo J, Li L, et al. Endothelial cells enhance prostate cancer metastasis via IL-6→androgen receptor→TGF-beta→MMP-9 signals. Mol Cancer Ther. 2013;12(6):1026–37.
49. Zhou M, Shah R, Shen R, Rubin MA. Basal cell cocktail (34betaE12 + p63) improves the detection of prostate basal cells. Am J Surg Pathol. 2003;27 (3):365–71.
50. Sun Y, Niu J, Huang J. Neuroendocrine differentiation in prostate cancer. Am J Transl Res. 2009; 1(2):148–62.
51. Epstein JI, Amin MB, Beltran H, Lotan TL, Mosquera JM, Reuter VE, et al. Proposed morphologic classification of prostate cancer with neuroendocrine differentiation. Am J Surg Pathol. 2014;38(6): 756–67.
52. Ananthanarayanan V, Deaton RJ, Yang XJ, Pins MR, Gann PH. Alpha-methylacyl-CoA racemase (AMACR) expression in normal prostatic glands and high-grade prostatic intraepithelial neoplasia (HGPIN): association with diagnosis of prostate cancer. Prostate. 2005;63(4):341–6.
53. Herawi M, Epstein JI. Immunohistochemical antibody cocktail staining (p63/HMWCK/AMACR) of ductal adenocarcinoma and Gleason pattern 4 cribriform and noncribriform acinar adenocarcinomas of the prostate. Am J Surg Pathol. 2007;31(6):889–94.
54. Wei J, Xu G, Wu M, Zhang Y, Li Q, Liu P, et al. Overexpression of vimentin contributes to prostate cancer invasion and metastasis via src regulation. Anticancer Res. 2008;28(1A):327–34.
55. Egevad L, Mazzucchelli R, Montironi R. Implications of the International Society of Urological

Pathology modified Gleason grading system. Arch Pathol Lab Med. 2012;136(4):426–34.
56. Epstein J. Grading of prostatic adenocarcinomas. Biopsy interpretation of the prostate. 4th ed. Lippincott, Williams & Wilkins; 2008. p. 175–87.
57. Pierorazio PM, Walsh PC, Partin AW, Epstein JI. Prognostic Gleason grade grouping: data based on the modified Gleason scoring system. BJU Int. 2013;111(5):753–60.
58. Epstein JI. An update of the Gleason grading system. J Urol. 2010;183(2):433–40.
59. Lotan TL, Epstein JI. Clinical implications of changing definitions within the Gleason grading system. Nat Rev Urol. 2010;7(3):136–42.
60. Epstein JI, Allsbrook WC, Jr., Amin MB, Egevad LL, Committee IG. The 2005 International Society of Urological Pathology (ISUP) consensus conference on Gleason grading of prostatic carcinoma. Am J Surg Pathol. 2005;29(9):1228–42.
61. Billis A, Guimaraes MS, Freitas LL, Meirelles L, Magna LA, Ferreira U. The impact of the 2005 International Society of Urological Pathology consensus conference on standard Gleason grading of prostatic carcinoma in needle biopsies. J Urol. 2008;180(2):548–52, discussion 52–3.
62. Tsivian M, Sun L, Mouraviev V, Madden JF, Mayes JM, Moul JW, et al. Changes in Gleason score grading and their effect in predicting outcome after radical prostatectomy. Urology. 2009;74 (5):1090–3.
63. Epstein JI, Evegad, L, Amin MB, Delahunt, B, Srigley JR, Humphrey PA. The 2014 International Society of Urological Pathology (ISUP) consensus conference on Gleason grading of prostatic carcinoma. Definition of grading patterns and proposal for a new grading system. Am J Surg Pathol. 2015 [epub ahead of print].
64. Helpap B, Egevad L. Correlation of modified Gleason grading with pT stage of prostatic carcinoma after radical prostatectomy. Anal Quant Cytol Histol (The International Academy of Cytology [and] American Society of Cytology). 2008;30(1):1–7.
65. Corcoran NM, Hong MK, Casey RG, Hurtado-Coll A, Peters J, Harewood L, et al. Upgrade in Gleason score between prostate biopsies and pathology following radical prostatectomy significantly impacts upon the risk of biochemical recurrence. BJU Int. 2011;108(8):E202–10.
66. Suer E, Gokce MI, Gulpinar O, Guclu AG, Haciyev P, Gogus C, et al. How significant is upgrade in Gleason score between prostate biopsy and radical prostatectomy pathology while discussing less invasive treatment options? Scand J Urol. 2014;48(2):177–82.
67. D'Elia C, Cerruto MA, Cioffi A, Novella G, Cavalleri S, Artibani W. Upgrading and upstaging in prostate cancer: from prostate biopsy to radical prostatectomy. Mol Clin Oncol. 2014;2(6):1145–9.
68. Edge S. Genitourinary sites. AJCC Cancer staging manual. 7th ed. Springer; 2010. p. 457–68.
69. Epstein JI, et al. Tumours of the prostate. In: Eble JN, Sauter G, Epstein JI, Sesterhenn IA, editors. IARC WHO classification of tumours tumours of the urinary system and male genital organs. 2004. p. 159–216.
70. Sobel RE, Sadar MD. Cell lines used in prostate cancer research: a compendium of old and new lines —part 2. J Urol. 2005;173(2):360–72.
71. Johnson IR, Parkinson-Lawrence EJ, Butler LM, Brooks DA. Prostate cell lines as models for biomarker discovery: performance of current markers and the search for new biomarkers. Prostate. 2014;74(5):547–60.
72. Gao D, Vela I, Sboner A, Iaquinta PJ, Karthaus WR, Gopalan A, et al. Organoid cultures derived from patients with advanced prostate cancer. Cell. 2014;159(1):176–87.
73. Karthaus WR, Iaquinta PJ, Drost J, Gracanin A, van Boxtel R, Wongvipat J, et al. Identification of multipotent luminal progenitor cells in human prostate organoid cultures. Cell. 2014;159(1): 163–75.
74. Valkenburg KC, Williams BO. Mouse models of prostate cancer. Prostate Cancer. 2011;2011:895238.
75. Parisotto M, Metzger D. Genetically engineered mouse models of prostate cancer. Mol Oncol. 2013;7(2):190–205.
76. Ittmann M, Huang J, Radaelli E, Martin P, Signoretti S, Sullivan R, et al. Animal models of human prostate cancer: the consensus report of the New York meeting of the Mouse Models of Human Cancers Consortium Prostate Pathology Committee. Cancer Res. 2013;73(9):2718–36.
77. Irshad S, Abate-Shen C. Modeling prostate cancer in mice: something old, something new, something premalignant, something metastatic. Cancer Metastasis Rev. 2013;32(1–2):109–22.
78. Cho H, Herzka T, Zheng W, Qi J, Wilkinson JE, Bradner JE, et al. RapidCaP, a novel GEM model for metastatic prostate cancer analysis and therapy, reveals myc as a driver of Pten-mutant metastasis. Cancer Discov. 2014;4(3):318–33.
79. Song Y, Gilbert D, O'Sullivan TN, Yang C, Pan W, Fathalizadeh A, et al. Carcinoma initiation via RB tumor suppressor inactivation: a versatile approach to epithelial subtype-dependent cancer initiation in diverse tissues. PLoS ONE. 2013;8(12):e80459.
80. Bjerke GA, Pietrzak K, Melhuish TA, Frierson HF Jr, Paschal BM, Wotton D. Prostate cancer induced by loss of Apc is restrained by TGFbeta signaling. PLoS ONE. 2014;9(3):e92800.
81. Lawrence MG, Taylor RA, Toivanen R, Pedersen J, Norden S, Pook DW, et al. A preclinical xenograft model of prostate cancer using human tumors. Nat Protoc. 2013;8(5):836–48.
82. Priolo C, Agostini M, Vena N, Ligon AH, Fiorentino M, Shin E, et al. Establishment and genomic characterization of mouse xenografts of human primary prostate tumors. Am J Pathol. 2010;176(4):1901–13.

83. Rubin MA, Dunn R, Strawderman M, Pienta KJ. Tissue microarray sampling strategy for prostate cancer biomarker analysis. Am J Surg Pathol. 2002;26(3):312–9.
84. Jawhar NM. Tissue microarray: a rapidly evolving diagnostic and research tool. Ann Saudi Med. 2009;29(2):123–7.
85. Esmaeilsabzali H, Beischlag TV, Cox ME, Parameswaran AM, Park EJ. Detection and isolation of circulating tumor cells: principles and methods. Biotechnol Adv. 2013;31(7):1063–84.
86. Pantel K, Alix-Panabieres C. Detection methods of circulating tumor cells. J Thorac Dis. 2012;4(5):446–7.
87. Sollier E, Go DE, Che J, Gossett DR, O'Byrne S, Weaver WM, et al. Size-selective collection of circulating tumor cells using Vortex technology. Lab Chip. 2014;14(1):63–77.
88. Lohr JG, Adalsteinsson VA, Cibulskis K, Choudhury AD, Rosenberg M, Cruz-Gordillo P, et al. Whole-exome sequencing of circulating tumor cells provides a window into metastatic prostate cancer. Nat Biotechnol. 2014;32(5):479–84.
89. Amato RJ, Melnikova V, Zhang Y, Liu W, Saxena S, Shah PK, et al. Epithelial cell adhesion molecule-positive circulating tumor cells as predictive biomarker in patients with prostate cancer. Urology. 2013;81(6):1303–7.
90. Papadopoulou E, Davilas E, Sotiriou V, Koliopanos A, Aggelakis F, Dardoufas K, et al. Cell-free DNA and RNA in plasma as a new molecular marker for prostate cancer. Oncol Res. 2004;14(9):439–45.
91. Boddy JL, Gal S, Malone PR, Harris AL, Wainscoat JS. Prospective study of quantitation of plasma DNA levels in the diagnosis of malignant versus benign prostate disease. Clin Cancer Res (An Official Journal of the American Association for Cancer Research). 2005;11(4):1394–9.
92. Chun FK, Muller I, Lange I, Friedrich MG, Erbersdobler A, Karakiewicz PI, et al. Circulating tumour-associated plasma DNA represents an independent and informative predictor of prostate cancer. BJU Int. 2006;98(3):544–8.
93. Altimari A, Grigioni AD, Benedettini E, Gabusi E, Schiavina R, Martinelli A, et al. Diagnostic role of circulating free plasma DNA detection in patients with localized prostate cancer. Am J Clin Pathol. 2008;129(5):756–62.
94. Schwarzenbach H, Alix-Panabieres C, Muller I, Letang N, Vendrell JP, Rebillard X, et al. Cell-free tumor DNA in blood plasma as a marker for circulating tumor cells in prostate cancer. Clin Cancer Res (An Official Journal of the American Association for Cancer Research). 2009;15(3):1032–8.
95. Joseph A, Gnanapragasam VJ. Laser-capture microdissection and transcriptional profiling in archival FFPE tissue in prostate cancer. Methods Mol Biol. 2011;755:291–300.
96. Chaux A, Albadine R, Toubaji A, Hicks J, Meeker A, Platz EA, et al. Immunohistochemistry for ERG expression as a surrogate for TMPRSS2-ERG fusion detection in prostatic adenocarcinomas. Am J Surg Pathol. 2011;35(7):1014–20.
97. Ornstein DK, Cinquanta M, Weiler S, Duray PH, Emmert-Buck MR, Vocke CD, et al. Expression studies and mutational analysis of the androgen regulated homeobox gene NKX3.1 in benign and malignant prostate epithelium. J Urol. 2001;165(4):1329–34.
98. Zhang Y, Perez T, Blondin B, Du J, Liu P, Escarzaga D, et al. Identification of FISH biomarkers to detect chromosome abnormalities associated with prostate adenocarcinoma in tumour and field effect environment. BMC Cancer. 2014;14:129.
99. Byers RJ, Di Vizio D, O'Connell F, Tholouli E, Levenson RM, Gossage K, et al. Semiautomated multiplexed quantum dot-based in situ hybridization and spectral deconvolution. J Mol Diagn (JMD). 2007;9(1):20–9.
100. Goering W, Kloth M, Schulz WA. DNA methylation changes in prostate cancer. Methods Mol Biol. 2012;863:47–66.
101. Strand SH, Orntoft TF, Sorensen KD. Prognostic DNA methylation markers for prostate cancer. Int J Mol Sci. 2014;15(9):16544–76.
102. Loeb S, Peskoe SB, Joshu CE, Huang WY, Hayes RB, Carter HB, et al. Do environmental factors modify the genetic risk of prostate cancer? Cancer epidemiology, biomarkers & prevention. A publication of the American Association for Cancer Research, cosponsored by the American Society of Preventive Oncology. 2014.
103. Bova GS, Partin AW, Isaacs SD, Carter BS, Beaty TL, Isaacs WB, et al. Biological aggressiveness of hereditary prostate cancer: long-term evaluation following radical prostatectomy. J Urol. 1998;160(3):660–3.
104. Ewing CM, Ray AM, Lange EM, Zuhlke KA, Robbins CM, Tembe WD, et al. Germline mutations in HOXB13 and prostate-cancer risk. N Engl J Med. 2012;366(2):141–9.
105. Jiang J, Jia P, Shen B, Zhao Z. Top associated SNPs in prostate cancer are significantly enriched in cis-expression quantitative trait loci and at transcription factor binding sites. Oncotarget. 2014;5(15):6168–77.
106. Van den Broeck T, Joniau S, Clinckemalie L, Helsen C, Prekovic S, Spans L, et al. The role of single nucleotide polymorphisms in predicting prostate cancer risk and therapeutic decision making. BioMed Res Int. 2014;2014:627510.
107. Nica AC, Dermitzakis ET. Expression quantitative trait loci: present and future. Philos Trans R Soc Lond B Biol Sci. 2013;368(1620):20120362.
108. Taylor BS, Schultz N, Hieronymus H, Gopalan A, Xiao Y, Carver BS, et al. Integrative genomic profiling of human prostate cancer. Cancer Cell. 2010;18(1):11–22.

109. Beroukhim R, Mermel CH, Porter D, Wei G, Raychaudhuri S, Donovan J, et al. The landscape of somatic copy-number alteration across human cancers. Nature. 2010;463(7283):899–905.
110. Hieronymus H, Schultz N, Gopalan A, Carver BS, Chang MT, Xiao Y, et al. Copy number alteration burden predicts prostate cancer relapse. Proc Natl Acad Sci USA. 2014;111(30):11139–44.
111. Ding Z, Wu CJ, Chu GC, Xiao Y, Ho D, Zhang J, et al. SMAD4-dependent barrier constrains prostate cancer growth and metastatic progression. Nature. 2011;470(7333):269–73.
112. Penney KL, Sinnott JA, Fall K, Pawitan Y, Hoshida Y, Kraft P, et al. mRNA expression signature of Gleason grade predicts lethal prostate cancer. J Clin Oncol (Official Journal of the American Society of Clinical Oncology). 2011;29(17):2391–6.
113. Priolo C, Pyne S, Rose J, Regan ER, Zadra G, Photopoulos C, et al. AKT1 and MYC induce distinctive metabolic fingerprints in human prostate cancer. Cancer Res. 2014.
114. Magi-Galluzzi C, Xu X, Hlatky L, Hahnfeldt P, Kaplan I, Hsiao P, et al. Heterogeneity of androgen receptor content in advanced prostate cancer. Mod Pathol (An Official Journal of the United States and Canadian Academy of Pathology, Inc.). 1997;10 (8):839–45.
115. Hodgson MC, Bowden WA, Agoulnik IU. Androgen receptor footprint on the way to prostate cancer progression. World J Urol. 2012;30(3):279–85.
116. Kumar V. Neoplasia. Robbins and Cotran pathologic basis of disease. Elsevier Health Sciences; 2009. p. 294.
117. Martin NE, et al. Measuring PI3K activation in prostate cancer using immunohistochemistry and RNA expression. Mol Can Res. 2014;(in revision).
118. Lee SH, Poulogiannis G, Pyne S, Jia S, Zou L, Signoretti S, et al. A constitutively activated form of the p110beta isoform of PI3-kinase induces prostatic intraepithelial neoplasia in mice. Proc Natl Acad Sci USA. 2010;107(24):11002–7.
119. Majumder PK, Grisanzio C, O'Connell F, Barry M, Brito JM, Xu Q, et al. A prostatic intraepithelial neoplasia-dependent p27 Kip1 checkpoint induces senescence and inhibits cell proliferation and cancer progression. Cancer Cell. 2008;14(2):146–55.
120. Chen ML, Xu PZ, Peng XD, Chen WS, Guzman G, Yang X, et al. The deficiency of Akt1 is sufficient to suppress tumor development in Pten$^{+/-}$ mice. Genes Dev. 2006;20(12):1569–74.
121. Fiorentino M, Capizzi E, Loda M. Blood and tissue biomarkers in prostate cancer: state of the art. Urol Clin N Am. 2010;37(1):131–41 (Table of Contents).
122. Lotan TL, Carvalho FL, Peskoe SB, Hicks JL, Good J, Fedor HL, et al. PTEN loss is associated with upgrading of prostate cancer from biopsy to radical prostatectomy. Mod Pathol (An Official Journal of the United States and Canadian Academy of Pathology, Inc.). 2014.
123. Benedettini E, Nguyen P, Loda M. The pathogenesis of prostate cancer: from molecular to metabolic alterations. Diagn Histopathol. 2008;14(5):195–201.
124. St John J, Powell K, Conley-Lacomb MK, Chinni SR. TMPRSS2-ERG fusion gene expression in prostate tumor cells and its clinical and biological significance in prostate cancer progression. J Cancer Sci Ther. 2012;4(4):94–101.
125. Clark JP, Cooper CS. ETS gene fusions in prostate cancer. Nat Rev Urol. 2009;6(8):429–39.
126. Narod SA, Seth A, Nam R. Fusion in the ETS gene family and prostate cancer. Br J Cancer. 2008;99 (6):847–51.
127. Baena E, Shao Z, Linn DE, Glass K, Hamblen MJ, Fujiwara Y, et al. ETV1 directs androgen metabolism and confers aggressive prostate cancer in targeted mice and patients. Genes Dev. 2013;27(6):683–98.
128. Clegg NJ, Couto SS, Wongvipat J, Hieronymus H, Carver BS, Taylor BS, et al. MYC cooperates with AKT in prostate tumorigenesis and alters sensitivity to mTOR inhibitors. PLoS ONE. 2011;6(3):e17449.
129. Dong L, Zhang X, Yu C, Yu T, Liu S, Hou L, et al. Monitoring luciferase-labeled human prostate stem cell antigen-expressing tumor growth in a mouse model. Exp Ther Med. 2013;6(5):1208–12.
130. Benassi B, Marani M, Loda M, Blandino G. USP2a alters chemotherapeutic response by modulating redox. Cell Death Dis. 2013;4:e812.
131. Benassi B, Flavin R, Marchionni L, Zanata S, Pan Y, Chowdhury D, et al. MYC is activated by USP2a-mediated modulation of microRNAs in prostate cancer. Cancer Discov. 2012;2(3):236–47.
132. Gurel B, Iwata T, Koh CM, Jenkins RB, Lan F, Van Dang C, et al. Nuclear MYC protein overexpression is an early alteration in human prostate carcinogenesis. Mod Pathol (an official journal of the United States and Canadian Academy of Pathology, Inc.). 2008;21(9):1156–67.
133. Allott EH, Masko EM, Freedland SJ. Obesity and prostate cancer: weighing the evidence. Eur Urol. 2013;63(5):800–9.
134. Flavin R, Zadra G, Loda M. Metabolic alterations and targeted therapies in prostate cancer. J Pathol. 2011;223(2):283–94.
135. Zadra G, Photopoulos C, Loda M. The fat side of prostate cancer. Biochim Biophys Acta. 2013;1831 (10):1518–32.
136. Migita T, Ruiz S, Fornari A, Fiorentino M, Priolo C, Zadra G, et al. Fatty acid synthase: a metabolic enzyme and candidate oncogene in prostate cancer. J Natl Cancer Inst. 2009;101(7):519–32.
137. Hamada S, Horiguchi A, Kuroda K, Ito K, Asano T, Miyai K, et al. Increased fatty acid synthase expression in prostate biopsy cores predicts higher Gleason score in radical prostatectomy specimen. BMC Clin Pathol. 2014;14(1):3.

# 11 Epidemiology of Breast Cancer

Rulla M. Tamimi

## 11.1 Clinical Picture of Disease

### 11.1.1 Symptoms

Prior to mammographic screening, the most common initial symptom for breast cancer was a palpable tumor. Occasionally, however, changes in the nipple or spread to axillary lymph nodes or distant metastases, often in the lung or bone, prompted medical consultation. Mammographic screening programs began in many developed countries in the 1980s and uptake of such programs has increased globally. Today, many breast cancers are diagnosed before symptoms occur due to mammographic screening.

### 11.1.2 Diagnosis

When breast cancer is suspected on the basis of clinical examination or mammography, pathologic confirmation is necessary before definitive primary treatment. A core biopsy, fine needle aspiration, or surgical biopsy can be used to establish the diagnosis.

R.M. Tamimi (✉)
Harvard Medical School, Channing Division of Network Medicine, Brigham and Women's Hospital, 181 Longwood Avenue, Boston, MA 02115, USA
e-mail: rulla.tamimi@channing.harvard.edu

### 11.1.3 Histopathologic and Molecular Subtypes

Almost all breast cancers are adenocarcinomas. Cancer in situ of the breast is now detected more often due to widespread use of mammography. The etiology and natural history of in situ cancers and their relation to invasive cancer are largely unknown. Therefore, this chapter focuses on invasive cancer only. Invasive breast carcinomas are often described as either ductal, the most common type, or lobular. Current methods of breast cancer classification group tumors into genetically and molecularly defined intrinsic subtypes including, but not limited to: luminal A, luminal B, HER2-overexpressing, and triple negative cancers [1]. Characteristics such as estrogen receptor (ER), progesterone receptor (PR), and HER2 status of the tumor are important for prognosis and treatment decisions.

### 11.1.4 Treatment

Today, breast-conserving surgery is used increasingly in combination with postoperative radiation therapy, which limits local recurrences. Moreover, adjuvant treatment with chemotherapy or tamoxifen, an antiestrogen, has become part of routine treatment for ER-positive tumors. The most recent clinical guidelines are now recommending the inclusion of aromatase inhibitors as an initial therapy, or after tamoxifen therapy, for


M. Loda et al. (eds.), *Pathology and Epidemiology of Cancer*,
DOI 10.1007/978-3-319-35153-7_11

postmenopausal women with estrogen receptor positive (ER+) breast cancer. Women with HER2-positive cancers are likely to benefit from treatment with HER2 targeted therapies such as trastuzumab (Herceptin) or lapatinib (Tykerb).

### 11.1.5 Prognosis

In the United States (U.S.), overall 5-year breast cancer relative survival rates have been increasing over time. Among women diagnosed in 1986–1993, 5-year relative survival was 84.2 %, and this increased to 91 % in 2007. This reflects both earlier detection through mammography and improved treatment. It should be noted that the 5-year relative survival rates are substantially higher among white women compared with black women.

## 11.2 Descriptive Epidemiology of Breast Cancer

### 11.2.1 Burden of Disease

Breast cancer is one of the most common cancers among women in the U.S., accounting for 32 % of all incident cancers among women [2, 3] and affecting approximately 232,670 women per year [4]. Breast cancer is rare among men of all ages and women who are younger than 30 years[2]. Incidence rates increase over a lifetime, slowing down around menopause (Fig. 11.1); the incidence rate for women age 30–34 is 25 per 100,000, for women age 45–49 is 190 per 100,000, and for women age 70–74 is 455 per 100,000. Additionally, the lifetime risk of being diagnosed through age 85 for U.S. women is 1 in 8 (12.3 %) [4].

Breast cancer is the second leading cause of cancer death among U.S. women, after lung cancer, which is the leading cause of cancer death among women [2]. Approximately 40,000 women die per year from breast cancer [4]. Breast cancer is the leading cause of death among women age 40–55 [2]. However, the lifetime risk of death from breast cancer is low (3.4%) [5]. As of 2012, there are approximately 2.9 million breast cancer survivors (i.e., women who have been diagnosed with breast cancer who are alive) in the U.S. [4].

Breast cancer incidence and mortality rates vary significantly by race and ethnicity. Hispanic, Asian, and American Indian women have the lowest incidence rates of breast cancer, while

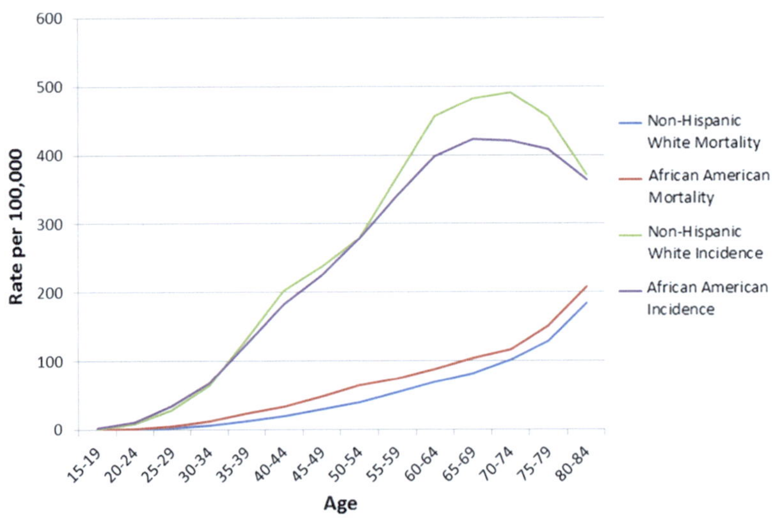


**Fig. 11.1** Age incidence and mortality curves for breast cancer in the U.S. for African American women and non-Hispanic white women (SEER Research Data 1973–2012; Fast Stats: an interactive tool for access to SEER cancer statistics. Surveillance Research Program, National Cancer Institute. http://seer.cancer.gov/faststats.) (Accessed on 11-21-2015)

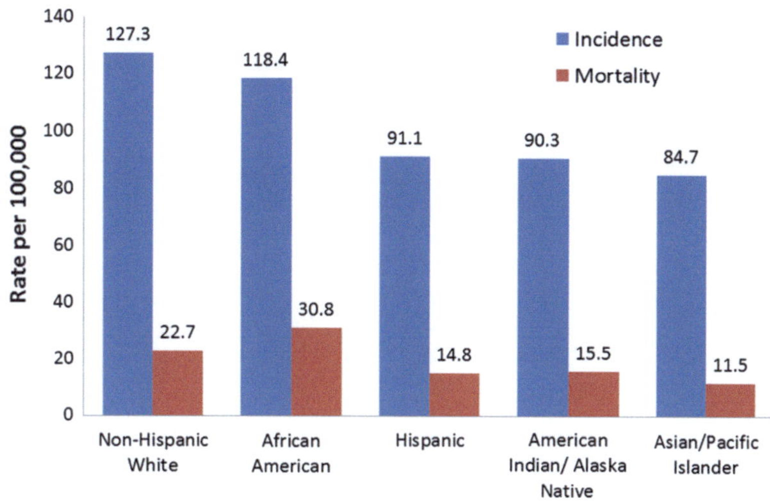


**Fig. 11.2** Breast Cancer Incidence and Mortality Rates (per 100,000) by Race and Ethnicity (SEER Research Data 1973-2012; Fast Stats: An interactive tool for access to SEER cancer statistics. Surveillance Research Program, National Cancer Institute. http://seer.cancer.gov/faststats.) (Accessed on 11-21-2015)

white and African American women have the highest incidence rates. Overall, white U.S. women have the highest lifetime risk (12.8 %) of breast cancer, while African American women have a slightly lower life time risk of 10.1 % [6]. However, African American women have higher incidence rates before age 40 and also have higher rates of more aggressive breast cancer subtypes such as estrogen receptor negative (ER −) breast cancers compared with white women [7]. White women also have the highest age-adjusted incidence rate (137 per 100,000 vs. 118 per 100,000) compared to all other race/ethnicities. Although white women have higher lifetime risk and age-adjusted incidence rates, they have lower mortality rates than African American women (Fig. 11.2) [6, 7]. African American women have a 3.4 % (30.8 per 100,000) risk of dying from breast cancer compared to a 3.1 % (22.7 per 100,000) for white women [5, 6].

### 11.2.2 Incidence and Mortality Trends

Breast cancer incidence (Fig. 11.3) and mortality trends (Fig. 11.4) have varied over time in the U. S. Breast cancer incidence has increased in all age groups since 1930, averaging 1.4 % increase in age-adjusted incidence per year from 1950–2000 [8]. This increase has increased more sharply in older women and young African American women (22 %) [6, 7]. Increases since the 1930s may be attributable to the changing prevalence of breast cancer risk factors including changes in reproductive patterns, increasing exogenous hormone use, and increasing postmenopausal body mass index (BMI). Moreover, increases in incidence rates seen since the 1980s can be attributed in part to widespread uptake of screening mammography [9–14]. Additionally, from 2001 to 2004, a decrease in incidence rates (3.5 %) occurred possibly due to the 2002 Women's Health Initiative (WHI) finding that estrogen plus progestin menopausal hormone therapy increased breast cancer risk for postmenopausal women (See Sect. 11.4.6) [8, 15]. From 2006 to 2010, overall breast cancer incidence rates increased slightly among African Americans (0.2 % per year) and decreased among Hispanic women, with no change in other racial groups [7].

In 2008, breast cancer incidence rates varied by more than 13-fold [16], with the highest rates observed in Europe and North America and the lowest rates in Asia. This likely reflects a

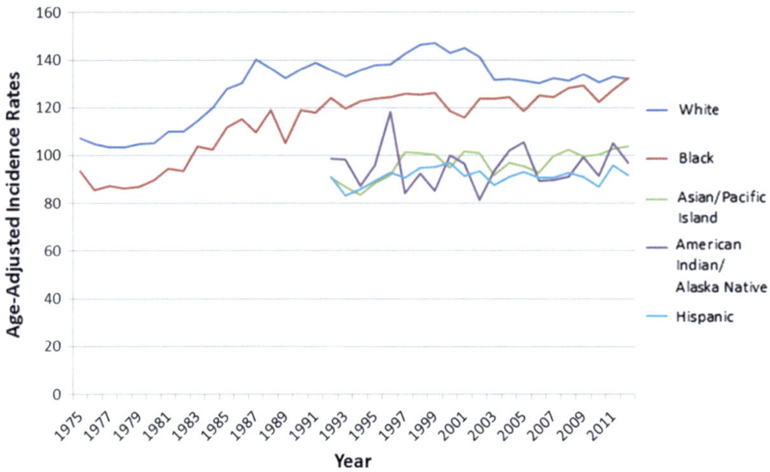


**Fig. 11.3** Age-Adjusted Incidence Rates (per 100,000) of Female Breast Cancer from 1975–2012 (SEER Research Data 1973-2012; Fast Stats: An interactive tool for access to SEER cancer statistics. Surveillance Research Program, National Cancer Institute. http://seer.cancer.gov/faststats.) (Accessed on 11-21-2015)

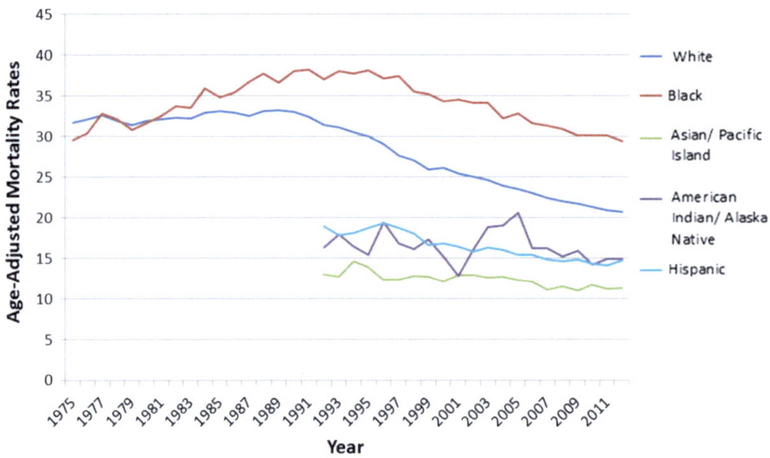


**Fig. 11.4** Age-Adjusted Mortality Rates (per 100,000) of Female Breast Cancer from 1975–2012 (SEER Research Data 1973-2012; Fast Stats: An interactive tool for access to SEER cancer statistics. Surveillance Research Program, National Cancer Institute. http://seer.cancer.gov/faststats.) (Accessed on 11-21-2015)

combination of differences in breast cancer screening, reporting, and risk factor exposure rates. Migrant studies, in which changes in breast cancer rates are evaluated in women who move from low- to high-risk countries—or vice versa—have shown that the rates of the host country are assumed over time, frequently one or two generations later [17, 18]. These data indicate that international differences in breast cancer rates may be due, at least in part, to environmental and lifestyle differences.

Age-adjusted mortality rates were stable from 1950s to 1980s [19], with a slight decrease in mortality in the 1980s–1990s likely due to advances in treatment and screening [3]. Mortality rates from the 1970s to 1990s continued to decrease for young and old (>60) white women, while increasing for African American women in all age groups [20]. Recently, both white and African American women have seen declines in breast cancer mortality rates, although the decline was smaller for African American women (1.6 % per year) when compared to white women (2.3 % per year) [6]. However, there are still disparities in 5-year survival rates between African American and white women (79 % vs. 92 %, respectively) [6].

## 11.3 Breast Cancer Subtypes

### 11.3.1 Introduction

Breast cancer is a heterogeneous disease with respect to its etiology, prognosis, and response to therapy. ER status and tumor grade were among the first important prognostic and predictive factors. In the early 1970s, the first reliable assay for testing ER status in tumors was available [21]. Presence of ER is important for response to specific treatments (e.g., tamoxifen, aromatase inhibitors) and prognosis, and more recently may define etiologic subtypes. Additionally, breast cancer can also be characterized by menopausal status at time of diagnosis and by histologic subtype.

Recently, there have been large-scale efforts to further characterize the distinct genetic and genomic variation of breast tumors [22–24]. It was not until the early 2000s that four major breast cancer molecular subtypes were identified: luminal A-like, luminal B-like, HER2+ type, and triple negative (or basal-like) identified through gene expression profiling and hierarchical clustering analyses. As with ER status, these breast cancer subtypes are associated with different etiologies, risk factors, clinical outcomes, and treatment options (Table 11.1) [25]. Large-scale epidemiologic studies have utilized immunohistochemical staining for markers including ER, progesterone receptor (PR), and human epidermal receptor 2 (HER2) (can also be assessed through fluorescent in situ hybridization), epidermal growth factor receptor (EGFR) and cytokeratin (CK) 5/6 as a proxy for the gene expression [22]. Below is an overview of these breast cancer subtypes as summarized by the 2013 St. Gallen Consensus report [22].

### 11.3.2 Luminal A-like

Luminal A-like breast cancer is the most common breast cancer subtype accounting for 42–59 % of all breast cancer cases and includes ER+/PR+/HER2-cancers [26]. These cancers are low grade 1 or 2 and/or have low Ki-67 expression (proliferative marker). Luminal A-like breast cancer tends to be less aggressive, slow growing, and more endocrine sensitive, thus associated with better prognosis [7]. Endocrine therapy is the

**Table 11.1** Summary of breast cancer subtypes, molecular markers, prevalence, and clinical characteristics

| Subtype | Markers | Prevalence (%) | Clinical characteristics |
|---|---|---|---|
| Luminal A-like | ER+ and/or PR +/HER2−/low grade | 42–59 | Less aggressive, slow growing, endocrine sensitive. Low recurrence rates |
| Luminal B-like | ER+ and/or PR +/HER+<br>Or<br>ER+ and/or PR +/HER2-/high grade | 10–20 | Aggressive, poor-prognosis, less estrogen sensitive. High recurrence rates. High recurrence rates |
| HER2+ type | ER-/PR-/HER2 + | 10–20 | Aggressive, poor short-term prognosis, more common in younger women. High recurrence rates |
| Triple negative/Basal-like | ER−/PR−/ HER2− | 10–20 | Fast growing, aggressive, often has a higher grade, and tends to metastasize. High recurrence rates. More common in African Americans, premenopausal women, and those with the BRCA1 mutations |

primary treatment method and recurrence rates for women diagnosed with luminal A-like tumors are low [22].

### 11.3.3 Luminal B-like

Luminal B-like cancer account for 10–20 % of breast cancer [7]. Similar to luminal A-like breast cancer, most luminal B tumors are ER+/PR+. However, luminal B-like breast cancer usually has high expression of Ki-67 and includes both HER2+ and HER2−. For luminal B-like HER2− cancer, Ki-67 is high and PR is negative or low, while luminal B-like HER+ exhibit a wide range of Ki-67 and PR levels. Treatment for luminal B-like tumors incorporates estrogen therapy with chemotherapy and anti-HER2, as these have a worse prognosis, are more aggressive, and less endocrine sensitive. Furthermore, luminal B-like HER2-breast cancer has a high recurrence rate [22].

### 11.3.4 HER2+ Type (Nonluminal) or Erb-B2 Overexpressing

HER2+ type of breast cancer is defined by ER−/PR−/HER2+, where HER2 is over expressed and ER and PR are absent. HER2+ type of breast cancer accounts for approximately 10 % of all breast cancers [7]. These cancers tend to be more aggressive and are associated with a poorer short-term prognosis. Chemotherapy and more recently anti-HER2 therapy are used in treatment. The anti-HER2 targeted treatment regimens have substantially improved prognosis associated with this subtype [7].

### 11.3.5 Triple Negative/Basal-like

Triple negative breast cancer is defined by ER−/PR−/HER2−, and represents a diverse group of tumors. Basal-like cancers occur in about 10–20 % of breast cancer, are more common in African American women, premenopausal women, and those with BRCA1 mutations, and are associated with poorer short-term prognosis [7]. In epidemiologic studies, triple negative tumors can be further classified as basal-like if they express either CK5/6 and/or EGFR. The only standard treatment option available for triple negative cancer is chemotherapy [22].

## 11.4 Breast Cancer Risk Factors

### 11.4.1 Introduction

Epidemiological studies have convincingly established a number of risk factors for breast cancer. Many of these are reproductive factors during the course of a woman's life. A unifying concept of these risk factors is that ovarian hormones initiate breast development and that monthly menstrual cycles induce regular breast cell proliferation. Puberty is an important period during breast development and is marked by a surge of hormones that induce regular breast cell proliferation. Pregnancy is also associated with higher circulating hormone levels, and is associated with a transient short-term increased risk of breast cancer. However, pregnancy and lactation are also associated with terminal differentiation of breast tissue and are associated with a reduced risk of breast cancer long term. The monthly pattern of cell division associated with regular menstrual cycles terminates with menopause, as indicated by cessation of ovulation and menstrual periods. The integration of pathology data into epidemiological studies of breast cancer has been the key, as the etiology of breast cancer based on molecular subtypes varies. The section below summarizes established breast cancer risk factors, as well as how these risk factors relate to specific breast cancer subtypes. Table 11.2 provides a summary of the established risk factors as well as the strength of the association overall and by ER/PR status [27–67]. Additionally, although the data are more limited, we have also summarized the current state of knowledge as it relates to risk factors and intrinsic subtypes in Table 11.3.

**Table 11.2** Summary of risk factors for breast cancer, strength of association, and association with hormone receptor status

| | Comparisons | Overall risk ratio | Association by hormone receptor (HR) status |
|---|---|---|---|
| Age at menarche | 1 year delay | 0.95 [28] | Association evident for both HR+ and HR− |
| Parity | Nulliparous versus parous | 1.2–1.7 [27] | Association evident for HR+, and maybe HR− |
| Breastfeeding | Each year woman breastfeeds | 0.93–0.96 [29] | Association evident for both HR+ and HR− |
| Oral contraceptives | Longer duration | 1.24–1.54 [30–32] | Association evident for both HR+ and HR− |
| Menopausal hormone therapy | Longer duration (5+ years vs. never) | 1.30–1.47 [33–40] | Association evident only for HR+ |
| Early-life adiposity | per 1-unit increase | 0.88–0.91 [41] | Association evident for both HR+ and HR- |
| BMI (only premenopausal) | 2 unit increment in BMI | 0.9 [42] | Association evident for both HR+ and HR− |
| Weight gain since age 18 | Gained 25 kg after age 18 versus those that remained withing 2 kg of weight since they were 18 | 2.0 [43] | Association evident only for HR+ |
| Physical activity | Recent total physical activity (≥27 MET vs. <3 MET) and total physical activity (active or moderately active vs. not active) | 0.81–0.92 [44, 45] | Association evident for both HR+ and HR− |
| Family history | Increasing number of affected relatives (1, 2, or 3+ vs. no family history) | 1.5–3.90 [46–48] | Association evident for both HR+ and HR- |
| Alcohol | Increases with 1 drink per day | 1.8–2.2 [49–59] | Association evident only for HR+ |
| Age at menopause | Increase in risk per increase in year | 1.03 [30] | Association evident only for HR+ |
| Mammographic density | >75 % vesus <5 % | 4.64 [60] | Association evident for both HR+ and HR− |
| Circulating estrogen (postmenopausal) | Increasing quartiles (all levels vs. lowest) | 1.2–2.4 [61–64] | Association evident only for HR+ |
| Circulation androgens (postmenopausal) | Increasing quintiles (highest *Q* vs. lowest *Q*) | 1.3–2.2 [61, 65–67] | Association evident only for HR+ |

### 11.4.2 Age at Menarche

Later age at menarche has been consistently associated with a lower risk of breast cancer [27]. Risk of cancer decreases by 5 % for every 1-year delay in the start of menarche [28]. Brinton et al [68] observed a 23 % lower breast cancer risk in women who started menstruating after the age of 15 when compared to women who started menstruation at 12 years or younger. More recent results from the large pooling efforts of the Collaborative Group on Hormonal Factors in Breast Cancer [69] are consistent with previous findings of a dose–response relationship between age at menarche and risk of breast cancer.

Earlier menarche may be associated with earlier onset of regular ovulation menstrual cycles, which leads to a greater lifetime exposure of endogenous hormones [70]. The relation between age at menarche and breast cancer may be modified by

**Table 11.3** Summary of association between breast cancer risk factors by subtype and direction of association

| | Risk factors | | | | | | | |
|---|---|---|---|---|---|---|---|---|
| | Older age at menarche | Higher parity | Breastfeeding | Oral contraceptives | Higher BMI in premenopausal women | Weight gain since age 18 | Family History | Alcohol |
| Luminal A-like | – | – | – | – | – | + | + | + |
| Luminal B-like | – | ‡ | – | ‡ | + * | + | + | ‡ |
| HER2+ type | ‡ | ‡ | ‡ | ‡ | ‡ | ‡ | + | + |
| Triple Negative/ Basal-like | – | + | – | + | + | ‡ | + | ‡ |

‡ Lack of evidence/no association
– Inverse association
+ Positive association
* Postmenopausal women only

menopausal status. A pooled analysis across studies found that each additional year in delay of menarche was associated with a stronger 9 % decrease in premenopausal breast cancer compared to a 4 % decrease in postmenopausal women [71]. The relationship between age at menarche and breast cancer also varies by molecular subtype. Although earlier age at menarche is inversely associated with both ER+/PR+ and ER−/PR− breast cancers, the magnitude of effect is greater for hormone receptor positive cancers [72]. In a systematic review of 39 studies, older age at menarche was consistently associated with moderately decreased risk of triple negative breast cancer [73]. These results were further confirmed by a population-based study that found that increases in age at menarche (per 2 years) are inversely associated with risk of basal-like breast cancer (Odds Ratio (OR) 0.8, 95 % Confidence Interval (95 % CI) 0.7–0.9) [74]. Further supporting these results, Millikan et al. [75] found that earlier-onset menarche (age < 13 years old) increased the risk of basal-like breast cancer (OR 1.4, 95 % CI 1.1–1.9). Older age at menarche appears to also decrease risk for luminal A-like and luminal B-like breast cancer [73]. There is no clear relationship between age at menarche and between HER2+ breast cancer.

### 11.4.3 Parity and Age at First Birth

Parity is often defined as the number of times a women has had a full-term pregnancy or pregnancy lasting >20 weeks [76]. In general, nulliparous women have increased risk of breast cancer compared with parous women [27]. However, the strength of association depends on the age of a woman's first birth; a younger age at first term pregnancy is associated with a lower lifetime risk of breast cancer [27]. Research suggests that the protective effect of pregnancy takes 10–15 years to manifest [77]. In fact, there is a transient increase in risk of breast cancer for the first ten years after pregnancy [78, 79]. The dual effects of pregnancy on breast cancer are attributed to the proliferation of breast cells during pregnancy which may lead to growth of mutated cells (increasing risk) as well as the differentiation of mature breast cells making them less susceptible to carcinogens (decreasing risk). A higher number of births have also been consistently associated with a reduced risk of breast cancer [80]. Some studies also suggest that more closely spaced births are more protective for breast cancer.

The relationship between parity varies by breast cancer subtype. Greater parity is associated with a reduced risk for luminal A-like cancer [73], while it may increase risk for triple negative/basal-like cancer [75, 81–85]. There is no clear relationship between parity and HER2+ or luminal B-like breast cancer [73]. Younger age at first birth was also associated with decreased risk of luminal A-like and luminal B-like cancer, with stronger protective effect for luminal-A cancer in women with greater parity [83–85].

### 11.4.4 Breastfeeding

Longer breast feeding duration (≥6 months) has consistently been associated with a reduced risk of breast cancer. There is also strong evidence that the relationship is dose-dependent and is independent of parity. In a pooled analysis of 47 studies from 30 countries ($N = 50{,}302$ cases and $N = 96{,}973$ controls), the relative risk of breast cancer is reduced by 4.3 % (95 % CI 2.9–5.8 %) for each year that a women breastfeeds and by 7.0 % (95 % CI 5.0–9.0 %) for each birth [29]. After adjusting for parity, the relative risk for ever versus never having breastfed was 0.96 ($p = 0.04$) [29]. The two main mechanisms by which lactation may reduce risk of breast cancer are: (1) delaying regular ovulatory cycles, and (2) further terminal differentiation of breast tissue.

Longer breastfeeding duration is protective for luminal A-like, luminal B-like, and triple negative (or basal-like) breast cancer [75], with inconclusive results for HER2+ type [73]. Additionally, lactation may mitigate the increased risk of ER−/PR− breast cancer subtypes associated with parity [86]. The relationship between breast feeding and basal-like tumors is one of the most consistent protective factors for basal-like tumors and represents an opportunity for preventing this aggressive breast cancer subtype.

### 11.4.5 Oral Contraceptives

The hypothesis that oral contraceptive use increases the risk of breast cancer was first proposed several decades, and more than 50 studies to date have investigated the association.

Most epidemiological studies find no significant increase in breast cancer risk associated with ever use or duration of use and risk of breast cancer [30]. However, current and recent users of oral contraceptives have a small increased risk of breast cancer compared with never users (Relative Risk (RR) 1.2, 95 % CI 1.2–1.3) [30], and the increased risk appears to be no longer evident within 10 years after stopping use of oral contraceptives. Longer durations of use in young women <35 years old appear to increase breast cancer risk [31, 32], however it is unclear if this relationship is confounded by the recency of use. Additionally, age at first use of oral contraceptives may play a role in breast cancer risk [30].

It is important to note that majority of data in epidemiologic studies contributing to the literature and large pooled analyses are based on early formulations of oral contraceptives that had higher doses of ethinyl estradiol and different types of progestins than are currently available [87, 88]. It is noteworthy that the patterns, dose, and combination of use of oral contraceptives have varied over time. Since the 1960s, age at initiation of oral contraceptive has decreased, duration of use has increased while the doses have decreased. Most oral contraceptives contain a combination of ethinyl estradiol and a progestin. In the 1960s, the ethinyl estradiol dose in oral contraceptives was ≥100 mg. Today, the ethinyl estradiol dose is around 20–30 mg [87, 88]. Additionally, the formulations have changed over time with at least nine different progestins. There is limited long-term data evaluating the currently available formulations and breast cancer risk. Additionally, progestin-only contraceptives, long-acting contraceptives, and implantable levonorgestrel (Norplant) have not been thoroughly investigated and more research is needed, as these are continuing to grow in popularity.

Relatively few studies have evaluated the association between oral contraceptives and breast cancer subtype. There is some evidence that oral contraceptives may be more strongly associated with risk of triple negative cancers than luminal A-like cancers [73, 84, 89]. In a recent systematic review [73], there was insufficient data on HER2+ and Luminal B-like cancers to understand their relationship with oral contractive use.

### 11.4.6 Menopausal Hormone Therapy

Menopausal hormone therapies containing estrogens have been used for over half a century, and over three dozen epidemiological studies, six meta-analyses, and one larger pooled analysis published over the past 30 years that investigate associations with breast cancer risk. Most studies have found that menopausal hormone therapy use increases breast cancer risk; however, the magnitude of risk depends on the formulations used and duration of use [33–39]. Meta-analyses have reported a 30–45 % increased risk of breast cancer with greater than five years of use compared with never use [33–38]. A large, prospective analysis in the Nurses' Health Study [40] found that the increased risk of breast cancer was limited to women with current or recent use of menopausal hormone therapy. Risk increased with longer duration of current use ($RR_{\text{Current 5+ years versus never users}}$ 1.5, 95 % CI 1.2–1.8). In a large, pooled analysis of epidemiological studies, the association between current use of menopausal hormone therapy and risk of breast cancer was also strongest for those with the longest duration of use; RR for <1 years was 1.1, 1–4 years was 1.1, 1.2 for 5–9 years, 1.1 for 10–14 years, and 1.6 for 15+ years [39].

A woman's body weight may play a role in the association of menopausal hormone therapy and breast cancer risk. The magnitude of association between menopausal hormone use and risk of breast cancer appears to be higher among women who are leaner compared with heavier women [39]. This different effect of hormone therapy by BMI is consistent over many studies including the Women's Health Initiative, a randomized control trial [90]. Further, for women who quit using menopausal hormones, the increased risk of breast cancer decreases and is similar to never users after quitting for 5 or more years, regardless of their duration of use [39].

Estrogen only and estrogen plus progestin (E&P) menopausal hormone therapies are both associated with increased risk in breast cancer, with a slightly higher risk seen in women who use E&P compared to estrogen only. Increases in breast cancer risk were seen in the Breast Cancer Detection and Demonstration Project [91] with recent use of estrogen only (RR 1.2, 95 % CI 1.0–1.4) and E&P (RR 1.4, 95 % CI 1.1–1.8). Similarly, the Million Women's Study [92] observed increased breast cancer risk for current users of preparations containing estrogens only (RR 1.3, 95 % CI 1.2–1.4), and a higher risk for E&P (RR 2.0, 95 % CI 1.9–2.1). However, results from the Million Women's Study suggested little differences in the associations between specific estrogens and progestins, doses or regimen types (e.g., sequential vs. continuous) [92]. Menopausal hormone therapy use appears to be associated with an increased risk of ER+ breast cancers, but not ER-cancers [93]. Given the large body of consistent evidence, the International Agency for Research on Cancer (IARC) has classified estrogen plus progestin therapy as a human carcinogen [94].

A number of studies in the U.S. and globally have reported declines in breast cancer incidence rates after 2002 [8, 95–99], the year that the Women's Health Initiative trial published their results on the positive association between E&P therapy and increased breast cancer risk [100, 101]. Following this publication, the prescribing pattern for menopausal hormone therapy dramatically declined in the U.S. Although based on ecologic data, the relatively rapid declines in breast cancer incidence, specifically hormone receptor positive cancers, mirrored declines in menopausal hormone therapy prescriptions suggesting that the decline in incidence is attributable to reduced exposures to combined hormone therapies [8, 99, 102].

### 11.4.7 Body Size Throughout the Life Course

#### 11.4.7.1 Introduction

The relationship between adiposity and breast cancer risk varies over the life course. In general, there is consistent evidence that adult premenopausal BMI is inversely related to risk of premenopausal breast cancer, while postmenopausal BMI is positively associated with postmenopausal breast cancer risk. More recent studies have provided consistent evidence that adiposity early in life (e.g., childhood and adolescence) is inversely associated with breast cancer risk. Interestingly, this protective effect of early-life body size is associated with both premenopausal and postmenopausal breast cancer. Below we discuss early-life body adiposity premenopausal BMI, postmenopausal BMI, and weight gain in relation to breast cancer risk.

#### 11.4.7.2 Early-Life Body Size

There is consistent evidence that body fatness during childhood and adolescence is associated with a reduced risk of breast cancer, with the effect lasting throughout lifetime [43, 103]. In the Nurses' Health Study II cohort [104], women who were heavier at ages 5 and 10 had half the risk of premenopausal breast cancer compared to those who were leanest at these ages [41]. Similarly, there is an inverse association between BMI at age 18 (or 20) and breast cancer observed in a number of other countries and racial/ethnic groups [105–109]. Additionally, the strong inverse association is observed for both ER+ and ER− breast cancers [41, 106]. Few studies have investigated early-life body size and breast cancer based on molecular subtype. Results from the Carolina Breast Cancer Study show that women who reported being heavier than their peers in 5th grade had a nonsignificant reduced risk of basal-like breast cancer (OR 0.5, 95 % CI 0.2–1.4) [75]. The mechanisms by which adiposity early in life may reduce breast cancer risk are not well understood. A few potential mechanisms have been suggested including that girls who are overweight may have slower sexual maturation, slower pubertal growth [110], and more anovulatory cycles [111].

#### 11.4.7.3 Premenopausal and Postmenopausal BMI

Prospective studies [43, 112] and meta-analyses [42] have found an inverse relationship between adult body weight and incidence of premenopausal breast cancer. In a recent meta-analysis [42], the relative risk was 0.94 (95 % CI 0.92–0.95) for a two unit (kg/m$^2$) increase in BMI in premenopausal women. One hypothesis is that heavier premenopausal women have more irregular menstrual cycles and increased rates of anovulatory infertility, thus decreasing risk due to fewer ovulatory cycles and less exposure to ovarian hormones [113].

Although there is an inverse association between BMI and premenopausal breast cancer, there is only a weakly positive relationship or no association observed in postmenopausal women [112, 114, 115]. The lack of a strong association appears to be due to the influence of the protective effects of early pregnancy and lasting protective effects of overweight early in life [43, 103].

#### 11.4.7.4 Weight Gain

Weight gain in adulthood has been positively associated with increasing breast cancer risk in postmenopausal women. After menopause, the main source of circulating estrogens is the adipose tissue. Therefore, a higher body fat percentage after menopause translates to a higher woman's exposure to estrogen. After menopause, obese women have both higher levels of endogenous estrogen and higher risk of breast cancer. An increased breast cancer risk is also seen in individuals who gain 25 kg after age 18; women who gained 25 kg had two times the risk of breast cancer compared to women who maintained their weight within two kg [43]. Weight gain since age 18 has consistently been associated with ER+ breast cancer subtypes (luminal A and luminal B-like) [73]. Although the data are more limited, there is a suggestion that weight gain since age 18 may also be associated with basal-like cancer [75].

### 11.4.8 Physical Activity

Epidemiological evidence suggests a possible relationship between physical activity and breast cancer risk, specifically in postmenopausal women. Results from one of the first pooled studies [116] (two case-control studies) on physical activity and breast cancer observed a significant inverse association between total physical activity and breast cancer risk (OR 0.9, 95 % CI 0.8–1.0) as well as specifically leisure time activity (OR 0.8, 95 % CI 0.7–0.9) for the highest quintile versus the lowest quintile [116]. Most recent studies from large, prospective studies also observe inverse associations for postmenopausal breast cancer. The European Prospective Investigation into Cancer (EPIC) cohort observed inverse associations with household and recreational physical activity [44]. Total physical activity was also associated with a decreased risk of breast cancer (Hazard ratio $(HR)_{active\ vs.\ inactive}$ 0.87, 95 % CI 0.79–0.97; $HR_{moderately\ active\ vs.\ inactive}$ 0.92, 95 % CI 0.86–0.99) [44]. Additionally, there was a suggestion that the inverse association was strongest for ER+/PR+ tumors. In the Nurses' Health Study [45], postmenopausal women who engaged in higher amounts of recent total physical activity also had a lower breast cancer risk (HR 0.9, 95 % CI 0.8–0.9; ≥27 MET [approximately 1 h/day of brisk walking] versus <3 MET (<1 h/week walking). In this study, there was no evidence that the association varied by hormone receptor status.

The modest inverse association between physical activity and breast cancer appears to be limited to hormone receptor positive cancers in most but not all studies. In the pooled case-control study described above [116], leisure time activity since age 50 was inversely associated with ER+/ PR+ cancers but not other subtypes. Recent results from EPIC [44] also observed the strongest association between recreational and household activity and ER+/PR+ tumors (HR 0.84, 95 % CI 0.74–0.96, active vs. inactive) and total physical activity (HR 0.88, 95 % CI 0.78–0.99, moderate active vs. inactive), in line with results from the NIH-AARP Diet and Health Study [117]. Total weekly energy expenditure from recreational physical activity (modest-low intensity) was inversely associated with ER+ breast cancer.

### 11.4.9 Family History

Family history is a well-established and strong risk factor for breast cancer. Results from the Nurses' Health Study [46] suggest that the age-adjusted risk ratio was 1.8 (95 % CI 1.5–2.0) for women with a maternal history of breast cancer versus women who did not have a family history of breast cancer. The strength of the association with family history is stronger for women whose mothers were diagnosed with breast cancer at younger ages. For example, women whose mothers were diagnosed before the age of 40 had a more than twofold greater risk (RR 2.1, 95 % CI 1.6–2.8). In contrast, for women whose mothers were diagnosed at age 70 years or older, the RR was 1.5 (95 % CI 1.1–2.2) [46]. In a pooled study by Pharoah et al. [47], the relative risk estimates compared to women with no family history were: for any relative (RR 1.9, 95 % CI 1.7–2.0), first degree relative (RR 2.1, 95 % CI 2.0–2.2), mother (RR 2.0, 95 % CI 1.8–2.1), sister (RR 2.3, 95 % CI 2.1–2.4), daughter (RR 1.8, 95 % CI 1.6–2.0), mother and sister (RR 3.6, 95 % CI 2.5–5.0), and second-degree relative(RR 1.5, 95 % CI 1.4–1.6). The Collaborative Group on Hormonal Factors in Breast Cancer [48] found that the risk ratio increased with increasing number of affected relatives, with RR for one, two, and three or more affected relatives of 1.8 (99 % CI 1.7–1.9), 2.9 (99 % CI 2.4–3.6), and 3.9 (99 % CI 2.0–7.5), respectively [48]. Results from a systematic review observed that a positive family history was associated with a 1.5- to two-fold increased risk of breast cancer for all subtypes compared with women without a family history of breast cancer [73].

Genetic epidemiology studies have sought to uncover the extent to which inherited genetic factors underlie the strong family history. The proportion of breast cancer estimated to be due to rare highly penetrant genetic mutations such as *BRCA1* and *BRCA2* is small, approximately 3 [118]–10 % [119]. Hereditary syndromes such as Li-Fraumeni syndrome and Cowden syndrome are associated with increased risk of breast cancer. Li-Fraumeni syndrome is due to germline mutations in the *p53* gene [120], while Cowden syndrome is due to germline mutations of the *PTEN* gene [121]. The cumulative lifetime risk of breast cancer in *BRCA1* and *BRCA2* carriers is estimated to range from 50 to 85 % [119]. These highly penetrant genetic mutations are rare in the population and account for less than 25 % of familial breast cancer risk [122, 123].

There are also rare moderate penetrance mutations in genes such as *CHEK2, ATM,* and *PALB2* that also increase breast cancer risk [123]. These moderate penetrance gene mutations are associated with an approximately two-fold increased risk of breast cancer and only explain 2–3 % of familial cases [124]. Most recently, efforts have been focused on identifying low-penetrance common genetic variants associated with breast cancer. To date, more than 90 single nucleotide polymorphisms (SNPs) have been identified through genome-wide association studies [125–128]. Together, these SNPs are estimated to account for 16 % of the familial risk of breast cancer [128].

### 11.4.10 Alcohol

In a pooled analysis of epidemiological studies, the risk of breast cancer increased monotonically with increasing intake of alcohol [49]. A 10 g per day (approximately 1 drink/day) increase in alcohol increased breast cancer by 9 %. There was also a modest effect observed for just one alcoholic drink per day, with a 7 % increased risk compared to never drinkers [49–51]. There does not appear to be any difference in the association by type of alcohol consumed, with similar effects on breast cancer risk associated with beer, wine, and liquor [51–53]. The positive association between alcohol consumption and breast cancer risk is possibly due to influences on circulating hormone levels. In short-term feeding studies, alcohol has been associated with increased total and bioavailable estrogen in premenopausal women [54], increased elevated plasma levels of estrone [55], and increased plasma estradiol levels in postmenopausal women [56].

Both recent adult alcohol consumption and consumption early in life appear to influence risk of breast cancer [57]. Women who drank regularly before age 30 and later stopped had an elevated risk of breast cancer similar to those who kept drinking [58]. The Nurses' Health Study [59] reported that alcohol consumption between 18–40 years of age and recent alcohol consumption after age 40 were both independently associated with an increased risk of breast cancer.

Further supporting this finding, Liu et al. [129] observed that cumulative drinking before first pregnancy was strongly associated with risk of ER+/PR+ breast tumors (RR = 1.8 per 10 g per day; 95 % CI = 1.03–1.34). Moreover, alcohol consumption has been positively associated with both HER2+ and luminal A-like breast cancer [82]. Together, there is compelling evidence for a causal relationship between alcohol consumption and breast cancer risk. However, the public health implications of this are complex given that low to moderate alcohol consumption is beneficial against cardiovascular disease risk for both women and men [130]. However, limiting alcohol consumption is one of the most modifiable risk factors to reduce risk of breast cancer.

### 11.4.11 Age at Menopause

Initial studies of age at menopause and breast cancer risk were based on women who had undergone bilateral oophorectomy at a young age [131, 132]. Results from these early studies suggested that women with a bilateral oophorectomy before age 45 years have approximately half the risk of breast cancer compared to women who underwent natural menopause at age 55 years or older [131, 132]. Among women experiencing natural menopause, for each year delay in age at menopause breast cancer risk increased by 3 % [30]. Earlier age at menopause causes a reduction in endogenous hormone levels with the termination of the menstrual cycle as well as the number of breast cell divisions. There is some evidence that age at menopause has different associations with different breast cancer subtypes. Four studies [74, 82, 85, 133] observed an increased risk between older age at menopause and luminal A-like cancer. Interestingly, a similar positive relationship has also been observed for triple negative cancers, although more research is needed. No obvious pattern has been documented for age at menopause and luminal B-like or HER2 positive breast cancer [73].

## 11.4.12 Mammographic Density

Mammographic density can be defined as percentage of breast area on a mammogram composed of dense breast tissue [60]. Radiologically dense tissue can comprise connective and epithelial tissue and appears light on mammogram, while fat is radiologically lucent and appears dark on mammogram. Mammographic density is one of the strongest risk factors for breast cancer; only age and BRCA carrier status are associated with larger relative risks for breast cancer. A meta-analysis by McCormack and dos Santos Silva [60] found a strong dose–response relationship between percent mammographic density and risk of breast cancer. Relative to women with <5 % mammographic density, women with ≥75 % mammographic density had a 4.6-fold (95 % CI 3.6–5.9) increased risk of breast cancer. Importantly, dense tissue can mask tumors on a mammogram [134–136]. Although bias due to masking does exist, it cannot explain the strong effects of breast density on breast cancer risk, as noted by the associations evident even 10 years after a mammogram [137–139].

The mechanism by which mammographic density increases breast cancer risk is unclear, although a number of hypotheses have been put forward. The primary mechanisms suggested are that mammographic density reflects: the number of mammary stem cells 'at risk' of developing breast cancer [140, 141]; the combined effects of cell proliferation and genetic damage to proliferating cells by mutagens [142]; and local estrogen production in the breast [143].

Importantly, mammographic density has been strongly associated with all breast cancer subtypes. Although the magnitude of this association may vary by subtype [144, 145] Bertrand et al. [144] observed a positive association between mammographic density and both ER+ and ER− breast cancer, however the association was stronger for ER− ($OR_{51\ \%\ \text{vs.}\ 11\text{–}25\ \%}$ =2.8, 95 % CI 1.8–4.4) cancers than ER+ disease ($OR_{51\ \%\ \text{vs.}\ 11\text{–}25\ \%}$ 2.0, 95% CI 1. 6–2. 5) in women younger than 55.

## 11.4.13 Circulating Hormones

### 11.4.13.1 Postmenopausal Hormone Levels

Estrogens, progesterone, and prolactin all promote mammary tumors in animal models. In clinical practice, antiestrogens are used to treat breast cancer and as chemoprevention in high-risk women. Estradiol circulates in the blood either unbound ("free estradiol") or bound to sex hormone binding globulin (SHBG). Free estradiol is believed to be readily available in breast tissue, and therefore may be more strongly related to breast cancer than total estradiol. In postmenopausal women, estrone is the major source of most circulating estradiol, while estrone sulfate is the most abundant circulating estrogen [146]. Results from a pooled analysis on 663 breast cancer cases and 1765 healthy controls suggest that increasing prediagnostic levels of postmenopausal estrogen are associated with increased risk of breast cancer [61]. The relative risk of breast cancer comparing extreme quintiles of estradiol was 2.0 (95 % CI 1.5–2.7). Circulating estrone, estrone sulfate, and free estradiol were similarly related to an increased risk.

Androgens have been hypothesized to increase breast cancer risk either directly, by increasing the growth and proliferation of breast cancer cells, or indirectly, by their conversion to estrogen [70]. In the pooled analysis described above [61], circulating androgens were positively associated with breast cancer risk with a relative risk for extreme quintiles of 2.2 (95 % CI 1.6–3.1).

As expected, the association between postmenopausal circulating estrogens and breast cancer are limited to hormone receptor positive cancers [147]. The relative risk of estradiol, highest versus lowest category, was 3.3 (95 % CI 2.0–5.4) for ER+/PR+ tumors and 1.0 (95 % CI 0.4–2.4) for ER−/PR− tumors. Similarly, the association between postmenopausal androgens and breast cancer risk also appears to be limited to hormone receptor positive cancers [65–67].

#### 11.4.13.2 Premenopausal Hormone Levels

Data on premenopausal estrogen levels and breast cancer risk are limited due to the complexities of biospecimen sampling during the menstrual cycle. Results from studies on premenopausal estrogen are varied [62, 148, 149]. A case-control study nested within the EPIC cohort [148] found no association between premenopausal estradiol or estrone levels and breast cancer risk. In the most recent analysis in the Nurses' Health Study II, Fortner et al. [149], found no association between follicular levels of estradiol, estrone, and free estradiol and risk of total or invasive breast cancer. However, luteal estradiol levels were positively associated with ER+/ PR+ cancer (OR 1.7, 95 % CI 1.0–2.9). The Endogenous Hormones and Breast Cancer Collaborative Group [150] conducted a pooled analysis to evaluate premenopausal hormone levels and breast cancer risk in 767 breast cancer cases and 1699 controls. Premenopausal estradiol, free estradiol, estrone, androstenedione, DHEAS, and testosterone were all positively associated with breast cancer risk, with ORs in the highest quintile relative to the lowest equal to 1.4 (95 % CI 1.0–2.0), 1.2 (95 % CI 0.9–1.6), 1.5 (95 % CI 1.0–2.2), 1.7 (95 % CI 1.2–2.4), 1.5 (95 % CI 1.1–2.0), and 1.3 (95 % CI 1.0–1.8), respectively.

## 11.5 Possible Breast Cancer Risk Factors

### 11.5.1 Introduction

A number of proposed risk factors are currently under investigation. The majority of risk factors described above are associated with hormone receptor positive breast cancers, which make up the majority of breast cancers. It is plausible that other risk factors that operate independent of sex hormones may also influence breast cancer risk. If these associations are limited to or are stronger for hormone receptor negative breast cancer subtypes, early studies that did not evaluate subtype may have missed an association or have been underpowered to evaluate hormone receptor negative subtypes.

Recent literature has focused on identifying modifiable risk factors to decrease breast cancer risk, as modifying behavior would be the easiest and most effective way for a person to decrease their risk of breast cancer. Below, we summarize the literature for two modifiable risk factors for which there are accumulating data.

### 11.5.2 Tobacco

The relationship between cigarette smoking and breast cancer risk has been evaluated in many studies, however the data have been quite inconsistent. Initial reports, overall, did not support any important association [151, 152]. However, more recent studies suggest that there may be an association particularly among women with early exposure prior to first pregnancy, long durations of smoking and among women with specific genetic mutations.

Large cohort studies have reported increased risks for breast cancer among women with the longest durations of smoking; those with significant findings had RRs ranging from 1.2 to 1.5 comparing long-term smokers with never smokers [153–156]. A meta-analysis of 15 cohort studies, found that current (HR 1.1, 95 % CI 1.1–1.2) and former smoking (HR 1.1, 95 % CI 1.0–1.2) were weakly associated with breast cancer risk [157]. A stronger association (HR 1.2, 95 % CI 1.1–1.3) was reported in women who initiated smoking before first birth. The Canadian Expert Panel on Tobacco Smoke and Breast Cancer Risk (2009) [158] concluded that the relationship between active smoking and breast cancer was consistent with causality. They also considered the association between second-hand smoke and premenopausal breast cancer as consistent with causality, however there was insufficient evidence for postmenopausal breast cancer.

### 11.5.3 Vegetables and Carotenoids

Vegetables contain a number of micronutrients including carotenoids, which are potent antioxidants that may provide a defense against reactive oxygen species which damage DNA and regulate cell differentiation. Fruit and vegetable consumption may decrease breast cancer risk, by providing protection against oxidative stress [159]. In early case-control studies, fruit and vegetable [52] consumption was inversely associated with breast cancer. However, in a more recent pooled analysis of prospective studies [160], there was no overall association between fruit and vegetable consumption and breast cancer risk. However, there was a significant inverse association between vegetable consumption and ER− breast cancer ($HR_{\text{highest vs. lowest quintile of total vegetable consumption}}$ 0.8, 95 % CI 0.7–0.9). Dietary intake of fruits and vegetables may not be the best measure of carotenoid intake. Blood levels may better reflect the more biologically relevant exposures of interest. A recent pooled analysis of circulating carotenoids [161] found that total carotenoids were associated with a significant 19 % reduced breast cancer risk. For many of the carotenoids, the inverse association was stronger for ER− cancer. For example, circulating β[beta]-carotene had a 48 % reduced risk of ER− cancer ($RR_{\text{highest versus lowest quintile}}$ 0.5, 95 % CI 0.4–0.8); in contrast, only a 17 % reduced risk of ER+ breast cancer (RR 0.8, 95 % CI 0.7–1.0; p-heterogeneity = 0.01). In sum, the data is suggestive but not conclusive of a protective effect, specifically for ER-negative breast cancer.

### 11.5.4 Vitamin D

Vitamin D is acquired through diet and sun exposure. Vitamin D and its metabolites can reduce cell proliferation, enhance apoptosis, and inhibit tumor progression [162]. Higher dietary intakes of vitamin D were associated with reduced risk of breast cancer in the French E3N cohort (HR 0.68, 95 % CI 0.54–0.85). Circulating levels of vitamin D are likely a more integrated and better measure of vitamin D exposure. A meta-analysis of nine prospective studies found a nonlinear inverse association between prediagnostic levels of 25-hydroxyvitamin D (25(OH)D) and breast cancer among postmenopausal women [163]. In contrast, no association was found among premenopausal women. Because there is strong mechanistic evidence and 25(OH)D levels can easily be raised through supplementation, there still remains a great deal of interest in resolving the association between vitamin D and breast cancer. A large, on-going, randomized control of vitamin D supplementation may help to better understand this association [164, 165].

## 11.6 Summary

In summary, there are a number of well-established risk factors for breast cancer. Many of these are related to reproductive factors and circulating hormones. As such, they are associated with hormone receptor positive breast cancers. The understanding of hormones and breast cancer etiology has been a cornerstone of breast cancer chemoprevention and treatment. Although many of the risk factors are non-modifiable, there is evidence around dietary factors and obesity that suggests opportunities for breast cancer prevention. Future work to identify risk factors and mechanisms for less common subtypes of breast cancer such as HER2 type and basal-like will be important for risk prediction and prevention.

## References

1. Goldhirsch A, Winer EP, Coates AS, Gelber RD, Piccart-Gebhart M, Thurlimann B, et al. Personalizing the treatment of women with early breast cancer: highlights of the St Gallen International Expert Consensus on the Primary Therapy of Early Breast Cancer 2013. Ann Oncol. 2013;24(9):2206–23.
2. Jemal A, Siegel R, Ward E, Murray T, Xu J, Thun MJ. Cancer statistics, 2007. CA Cancer J Clin. 2007;57(1):43–66.
3. Weir HK, Thun MJ, Hankey BF, Ries LA, Howe HL, Wingo PA, et al. Annual report to the

nation on the status of cancer, 1975–2000, featuring the uses of surveillance data for cancer prevention and control. J Natl Cancer Inst. 2003;95 (17):1276–99.
4. SEER. Stat Fact Sheets: Breast Cancer. 2014.
5. Carey LA, Perou CM, Livasy CA, Dressler LG, Cowan D, Conway K, et al. Race, breast cancer subtypes, and survival in the Carolina Breast Cancer Study. JAMA. 2006;295(21):2492–502.
6. SEER. Cancer Statistics Review, 1975–2011. 2014.
7. Breast Cancer Facts and Figures 2013–2014. Atlanta, GA: American Cancer Society, Inc.; 2013.
8. Ravdin PM, Cronin KA, Howlader N, Berg CD, Chlebowski RT, Feuer EJ, et al. The decrease in breast-cancer incidence in 2003 in the United States. N Engl J Med. 2007;356(16):1670–4.
9. Feuer EJ, Wun LM. How much of the recent rise in breast cancer incidence can be explained by increases in mammography utilization? A dynamic population model approach. Am J Epidemiol. 1992;136(12):1423–36.
10. Lantz PM, Remington PL, Newcomb PA. Mammography screening and increased incidence of breast cancer in Wisconsin. J Natl Cancer Inst. 1991;83(21):1540–6.
11. Liff JM, Sung JF, Chow WH, Greenberg RS, Flanders WD. Does increased detection account for the rising incidence of breast cancer? Am J Public Health. 1991;81(4):462–5.
12. Miller BA, Feuer EJ, Hankey BF. The increasing incidence of breast cancer since 1982: relevance of early detection. Cancer Causes Control. 1991;2 (2):67–74.
13. White E, Lee CY, Kristal AR. Evaluation of the increase in breast cancer incidence in relation to mammography use. J Natl Cancer Inst. 1990;82 (19):1546–52.
14. Miller BA, Feuer EJ, Hankey BF. Recent incidence trends for breast cancer in women and the relevance of early detection: an update. CA Cancer J Clin. 1993;43(1):27–41.
15. Rossouw JE, Anderson GL, Prentice RL, LaCroix AZ, Kooperberg C, Stefanick ML, et al. Risks and benefits of estrogen plus progestin in healthy postmenopausal women: principal results From the Women's Health Initiative randomized controlled trial. JAMA. 2002;288(3):321–33.
16. Society. AC. Global Cancer Facts and Figures. 2nd edn. Atlanta: American Cancer Society, 2011.
17. Tominaga S. Cancer incidence in Japanese in Japan, Hawaii, and western United States. Natl Cancer Inst Monogr. 1985;69:83–92.
18. Ziegler RG, Hoover RN, Pike MC, Hildesheim A, Nomura AM, West DW, et al. Migration patterns and breast cancer risk in Asian-American women. J Natl Cancer Inst. 1993;85(22):1819–27.
19. Chu KC, Tarone RE, Kessler LG, Ries LA, Hankey BF, Miller BA, et al. Recent trends in U. S. breast cancer incidence, survival, and mortality rates. J Natl Cancer Inst. 1996;88(21):1571–9.
20. Chevarley F, White E. Recent trends in breast cancer mortality among white and black US women. Am J Public Health. 1997;87(5):775–81.
21. Jensen EV, Block GE, Smith S, Kyser K, DeSombre ER. Estrogen receptors and breast cancer response to adrenalectomy. Natl Cancer Inst Monogr. 1971;34:55–70.
22. Goldhirsch A, Winer EP, Coates AS, Gelber RD, Piccart-Gebhart M, Thurlimann B, et al. Personalizing the treatment of women with early breast cancer: highlights of the St Gallen International Expert Consensus on the Primary Therapy of Early Breast Cancer 2013. Ann Oncol. 2013;24(9):2206–23.
23. Vargo-Gogola T, Rosen JM. Modelling breast cancer: one size does not fit all. Nat Rev Cancer. 2007;7(9):659–72.
24. Cancer Genome Atlas. N. Comprehensive molecular portraits of human breast tumours. Nature. 2012;490 (7418):61–70.
25. Curigliano G, Criscitiello C, Andre F, Colleoni M, Di Leo A. Highlights from the 13th St Gallen International Breast Cancer Conference. Access to innovation for patients with breast cancer: how to speed it up? Ecancermedicalscience. 2013;(7):299.
26. Haque R, Ahmed SA, Inzhakova G, Shi J, Avila C, Polikoff J, et al. Impact of breast cancer subtypes and treatment on survival: an analysis spanning two decades. Cancer Epidemiol Biomarkers Prev. 2012;21(10):1848–55.
27. Kelsey JL, Gammon MD, John EM. Reproductive factors and breast cancer. Epidemiol Rev. 1993;15 (1):36–47.
28. Hunter DJ, Spiegelman D, Adami HO, van den Brandt PA, Folsom AR, Goldbohm RA, et al. Non-dietary factors as risk factors for breast cancer, and as effect modifiers of the association of fat intake and risk of breast cancer. Cancer Causes Control. 1997;8(1):49–56.
29. Collaborative Group on Hormonal Factors in Breast. Breast cancer and breastfeeding: collaborative reanalysis of individual data from 47 epidemiological studies in 30 countries, including 50302 women with breast cancer and 96973 women without the disease. Lancet. 2002;360(9328):187–95.
30. Collaborative Group on Hormonal Factors in Breast. Breast cancer and hormonal contraceptives: collaborative reanalysis of individual data on 53 297 women with breast cancer and 100 239 women without breast cancer from 54 epidemiological studies. Lancet. 1996;347(9017):1713–27.
31. Romieu I, Berlin JA, Colditz G. Oral contraceptives and breast cancer. Review and meta-analysis. Cancer. 1990;66(11):2253–63.
32. Thomas DB. Oral contraceptives and breast cancer: review of the epidemiologic literature. Contraception. 1991;43(6):597–642.
33. Colditz GA, Egan KM, Stampfer MJ. Hormone replacement therapy and risk of breast cancer: results from epidemiologic studies. Am J Obstet Gynecol. 1993;168(5):1473–80.

34. Dupont WD, Page DL. Menopausal estrogen replacement therapy and breast cancer. Arch Intern Med. 1991;151(1):67–72.
35. Grady D, Rubin SM, Petitti DB, Fox CS, Black D, Ettinger B, et al. Hormone therapy to prevent disease and prolong life in postmenopausal women. Ann Intern Med. 1992;117(12):1016–37.
36. Sillero-Arenas M, Delgado-Rodriguez M, Rodigues-Canteras R, Bueno-Cavanillas A, Galvez-Vargas R. Menopausal hormone replacement therapy and breast cancer: a meta-analysis. Obstet Gynecol. 1992;79(2):286–94.
37. Steinberg KK, Thacker SB, Smith SJ, Stroup DF, Zack MM, Flanders WD, et al. A meta-analysis of the effect of estrogen replacement therapy on the risk of breast cancer. JAMA. 1991;265(15):1985–90.
38. Steinberg KK, Smith SJ, Thacker SB, Stroup DF. Breast cancer risk and duration of estrogen use: the role of study design in meta-analysis. Epidemiology. 1994;5(4):415–21.
39. Breast cancer and hormone replacement therapy: collaborative reanalysis of data from 51 epidemiological studies of 52,705 women with breast cancer and 108,411 women without breast cancer. Collaborative Group on Hormonal Factors in Breast Cancer. Lancet. 1997;350(9084):1047–59.
40. Colditz GA, Hankinson SE, Hunter DJ, Willett WC, Manson JE, Stampfer MJ, et al. The use of estrogens and progestins and the risk of breast cancer in postmenopausal women. N Engl J Med. 1995;332(24):1589–93.
41. Baer HJ, Tworoger SS, Hankinson SE, Willett WC. Body fatness at young ages and risk of breast cancer throughout life. Am J Epidemiol. 2010;171(11):1183–94.
42. Physical activity, and the prevention of cancer: a global perspective. Washington, D.C.: American Institute for Cancer Research, 2007.
43. Huang Z, Hankinson SE, Colditz GA, Stampfer MJ, Hunter DJ, Manson JE, et al. Dual effects of weight and weight gain on breast cancer risk. JAMA. 1997;278(17):1407–11.
44. Steindorf K, Ritte R, Eomois PP, Lukanova A, Tjonneland A, Johnsen NF, et al. Physical activity and risk of breast cancer overall and by hormone receptor status: the European prospective investigation into cancer and nutrition. Int J Cancer. 2013;132(7):1667–78.
45. Eliassen AH, Hankinson SE, Rosner B, Holmes MD, Willett WC. Physical activity and risk of breast cancer among postmenopausal women. Arch Intern Med. 2010;170(19):1758–64.
46. Colditz GA, Willett WC, Hunter DJ, Stampfer MJ, Manson JE, Hennekens CH, et al. Family history, age, and risk of breast cancer. Prospective data from the Nurses' Health Study. JAMA. 1993;270(3): 338–43.
47. Pharoah PD, Day NE, Duffy S, Easton DF, Ponder BA. Family history and the risk of breast cancer: a systematic review and meta-analysis. Int J Cancer. 1997;71(5):800–9.
48. Collaborative Group on Hormonal Factors in Breast. C. Familial breast cancer: collaborative reanalysis of individual data from 52 epidemiological studies including 58,209 women with breast cancer and 101,986 women without the disease. Lancet. 2001;358(9291):1389–99.
49. Smith-Warner SA, Spiegelman D, Yaun SS, van den Brandt PA, Folsom AR, Goldbohm RA, et al. Alcohol and breast cancer in women: a pooled analysis of cohort studies. JAMA. 1998;279(7): 535–40.
50. Missmer SA, Smith-Warner SA, Spiegelman D, Yaun SS, Adami HO, Beeson WL, et al. Meat and dairy food consumption and breast cancer: a pooled analysis of cohort studies. Int J Epidemiol. 2002;31(1):78–85.
51. Alcohol beverage consumption. Lyon, France: IARC, 2007.
52. Food, nutrition and the prevention of cancer: a global perspective. Washington, D.C.: American Institute for Cancer Research, 1997.
53. Tjonneland A, Christensen J, Olsen A, Stripp C, Thomsen BL, Overvad K, et al. Alcohol intake and breast cancer risk: the European Prospective Investigation into Cancer and Nutrition (EPIC). Cancer Causes Control. 2007;18(4):361–73.
54. Reichman ME, Judd JT, Longcope C, Schatzkin A, Clevidence BA, Nair PP, et al. Effects of alcohol consumption on plasma and urinary hormone concentrations in premenopausal women. J Natl Cancer Inst. 1993;85(9):722–7.
55. Hankinson SE, Willett WC, Manson JE, Hunter DJ, Colditz GA, Stampfer MJ, et al. Alcohol, height, and adiposity in relation to estrogen and prolactin levels in postmenopausal women. J Natl Cancer Inst. 1995;87(17):1297–302.
56. Ginsburg ES, Walsh BW, Gao X, Gleason RE, Feltmate C, Barbieri RL. The effect of acute ethanol ingestion on estrogen levels in postmenopausal women using transdermal estradiol. J Soc Gynecol Investig. 1995;2(1):26–9.
57. Kinney AY, Millikan RC, Lin YH, Moorman PG, Newman B. Alcohol consumption and breast cancer among black and white women in North Carolina (United States). Cancer Causes Control. 2000;11(4):345–57.
58. Harvey EB, Schairer C, Brinton LA, Hoover RN, Fraumeni JF Jr. Alcohol consumption and breast cancer. J Natl Cancer Inst. 1987;78(4):657–61.
59. Chen WY, Rosner B, Hankinson SE, Colditz GA, Willett WC. Moderate alcohol consumption during adult life, drinking patterns, and breast cancer risk. JAMA. 2011;306(17):1884–90.
60. McCormack VA, dos Santos Silva I. Breast density and parenchymal patterns as markers of breast cancer risk: a meta-analysis. Cancer Epidemiol Biomarkers Prev. 2006;15(6):1159–69.

61. Key T, Appleby P, Barnes I, Reeves G, Endogenous H. Breast Cancer Collaborative G. Endogenous sex hormones and breast cancer in postmenopausal women: reanalysis of nine prospective studies. J Natl Cancer Inst. 2002;94(8):606–16.
62. Eliassen AH, Missmer SA, Tworoger SS, Spiegelman D, Barbieri RL, Dowsett M, et al. Endogenous steroid hormone concentrations and risk of breast cancer among premenopausal women. J Natl Cancer Inst. 2006;98(19):1406–15.
63. Eliassen AH, Colditz GA, Rosner B, Willett WC, Hankinson SE. Adult weight change and risk of postmenopausal breast cancer. JAMA. 2006;296 (2):193–201.
64. Tworoger SS, Missmer SA, Barbieri RL, Willett WC, Colditz GA, Hankinson SE. Plasma sex hormone concentrations and subsequent risk of breast cancer among women using postmenopausal hormones. J Natl Cancer Inst. 2005;97(8):595–602.
65. Zhang X, Tworoger SS, Eliassen AH, Hankinson SE. Postmenopausal plasma sex hormone levels and breast cancer risk over 20 years of follow-up. Breast Cancer Res Treat. 2013;137(3):883–92.
66. Farhat GN, Cummings SR, Chlebowski RT, Parimi N, Cauley JA, Rohan TE, et al. Sex hormone levels and risks of estrogen receptor-negative and estrogen receptor-positive breast cancers. J Natl Cancer Inst. 2011;103(7):562–70.
67. James RE, Lukanova A, Dossus L, Becker S, Rinaldi S, Tjonneland A, et al. Postmenopausal serum sex steroids and risk of hormone receptor-positive and-negative breast cancer: a nested case-control study. Cancer Prev Res (Phila). 2011;4(10):1626–35.
68. Brinton LA, Schairer C, Hoover RN, Fraumeni JF Jr. Menstrual factors and risk of breast cancer. Cancer Invest. 1988;6(3):245–54.
69. Collaborative Group on Hormonal Factors in Breast. C. Menarche, menopause, and breast cancer risk: individual participant meta-analysis, including 118 964 women with breast cancer from 117 epidemiological studies. Lancet Oncol. 2012;13(11): 1141–51.
70. Bernstein L, Ross RK. Endogenous hormones and breast cancer risk. Epidemiol Rev. 1993;15(1):48–65.
71. Clavel-Chapelon F, Gerber M. Reproductive factors and breast cancer risk. Do they differ according to age at diagnosis? Breast Cancer Res Treat. 2002;72 (2):107–15.
72. Ma H, Bernstein L, Pike MC, Ursin G. Reproductive factors and breast cancer risk according to joint estrogen and progesterone receptor status: a meta-analysis of epidemiological studies. Breast Cancer Res. 2006;8(4):R43.
73. Barnard ME, Boeke CE, Tamimi RM. Established breast cancer risk factors and risk of intrinsic tumor subtypes. Biochim Biophys Acta. 2015;1856(1): 73–85.
74. Yang XR, Sherman ME, Rimm DL, Lissowska J, Brinton LA, Peplonska B, et al. Differences in risk factors for breast cancer molecular subtypes in a population-based study. Cancer Epidemiol Biomarkers Prev. 2007;16(3):439–43.
75. Millikan RC, Newman B, Tse CK, Moorman PG, Conway K, Dressler LG, et al. Epidemiology of basal-like breast cancer. Breast Cancer Res Treat. 2008;109(1):123–39.
76. Cunningham F, Leveno K, Bloom S, Hauth J, Rouse D, Spong C. Williams obstetrics. 23rd ed. NY: McGraw-Hill Companies; 2009.
77. Bruzzi P, Negri E, La Vecchia C, Decarli A, Palli D, Parazzini F, et al. Short term increase in risk of breast cancer after full term pregnancy. BMJ. 1988;297(6656):1096–8.
78. Rosner B, Colditz GA. Nurses' health study: log-incidence mathematical model of breast cancer incidence. J Natl Cancer Inst. 1996;88(6):359–64.
79. Pike MC, Krailo MD, Henderson BE, Casagrande JT, Hoel DG. 'Hormonal' risk factors, 'breast tissue age' and the age-incidence of breast cancer. Nature. 1983;303(5920):767–70.
80. La Vecchia C, Negri E, Boyle P. Reproductive factors and breast cancer: an overview. Soz Praventivmed. 1989;34(3):101–7.
81. Phipps AI, Chlebowski RT, Prentice R, McTiernan A, Wactawski-Wende J, Kuller LH, et al. Reproductive history and oral contraceptive use in relation to risk of triple-negative breast cancer. J Natl Cancer Inst. 2011;103(6):470–7.
82. Tamimi RM, Colditz GA, Hazra A, Baer HJ, Hankinson SE, Rosner B, et al. Traditional breast cancer risk factors in relation to molecular subtypes of breast cancer. Breast Cancer Res Treat. 2012;131 (1):159–67.
83. Ma H, Wang Y, Sullivan-Halley J, Weiss L, Marchbanks PA, Spirtas R, et al. Use of four biomarkers to evaluate the risk of breast cancer subtypes in the women's contraceptive and reproductive experiences study. Cancer Res. 2010;70 (2):575–87.
84. Gaudet MM, Press MF, Haile RW, Lynch CF, Glaser SL, Schildkraut J, et al. Risk factors by molecular subtypes of breast cancer across a population-based study of women 56 years or younger. Breast Cancer Res Treat. 2011;130 (2):587–97.
85. Xing P, Li J, Jin F. A case-control study of reproductive factors associated with subtypes of breast cancer in Northeast China. Med Oncol. 2010;27(3):926–31.
86. Palmer JR, Boggs DA, Wise LA, Ambrosone CB, Adams-Campbell LL, Rosenberg L. Parity and lactation in relation to estrogen receptor negative breast cancer in African American women. Cancer Epidemiol Biomarkers Prev. 2011;20(9):1883–91.
87. Annegers JF. Patterns of oral contraceptive use in the United States. Br J Rheumatol. 1989;28(Suppl 1):48–50.
88. Burkman RT. Oral contraceptives: current status. Clin Obstet Gynecol. 2001;44(1):62–72.

89. Razzaghi H, Troester MA, Gierach GL, Olshan AF, Yankaskas BC, Millikan RC. Association between mammographic density and basal-like and luminal A breast cancer subtypes. Breast Cancer Res. 2013;15(5):R76.
90. Anderson GL, Limacher M, Assaf AR, Bassford T, Beresford SA, Black H, et al. Effects of conjugated equine estrogen in postmenopausal women with hysterectomy: the Women's Health Initiative randomized controlled trial. JAMA. 2004;291(14):1701–12.
91. Schairer C, Lubin J, Troisi R, Sturgeon S, Brinton L, Hoover R. Menopausal estrogen and estrogen-progestin replacement therapy and breast cancer risk. JAMA. 2000;283(4):485–91.
92. Beral V. Million Women Study C. Breast cancer and hormone-replacement therapy in the Million Women Study. Lancet. 2003;362(9382):419–27.
93. Colditz GA, Rosner BA, Chen WY, Holmes MD, Hankinson SE. Risk factors for breast cancer according to estrogen and progesterone receptor status. J Natl Cancer Inst. 2004;96(3):218–28.
94. Combined estrogen-progestogen postmenopausal therapy. Combined Estrogen-progestogen contraceptives and combined estrogen-progestogen menopausal therapy. Lyon, France: International Agency for Research on Cancer; 2007.
95. Jemal A, Ward E, Thun MJ. Recent trends in breast cancer incidence rates by age and tumor characteristics among U.S. women. Breast Cancer Res. 2007;9(3):R28.
96. Kerlikowske K, Miglioretti DL, Buist DS, Walker R, Carney PA. National Cancer Institute-Sponsored Breast Cancer Surveillance. Declines in invasive breast cancer and use of postmenopausal hormone therapy in a screening mammography population. J Natl Cancer Inst. 2007;99(17):1335–9.
97. Canfell K, Banks E, Moa AM, Beral V. Decrease in breast cancer incidence following a rapid fall in use of hormone replacement therapy in Australia. Med J Aust. 2008;188(11):641–4.
98. Katalinic A, Rawal R. Decline in breast cancer incidence after decrease in utilisation of hormone replacement therapy. Breast Cancer Res Treat. 2008;107(3):427–30.
99. Coombs NJ, Cronin KA, Taylor RJ, Freedman AN, Boyages J. The impact of changes in hormone therapy on breast cancer incidence in the US population. Cancer Causes Control. 2010;21(1):83–90.
100. Clarke CA, Glaser SL, Uratsu CS, Selby JV, Kushi LH, Herrinton LJ. Recent declines in hormone therapy utilization and breast cancer incidence: clinical and population-based evidence. J Clin Oncol. 2006;24(33):e49–50.
101. Robbins AS, Clarke CA. Regional changes in hormone therapy use and breast cancer incidence in California from 2001 to 2004. J Clin Oncol. 2007;25(23):3437–9.
102. Colditz GA. Decline in breast cancer incidence due to removal of promoter: combination estrogen plus progestin. Breast Cancer Res. 2007;9(4):108.
103. Barnes-Josiah D, Potter JD, Sellers TA, Himes JH. Early body size and subsequent weight gain as predictors of breast cancer incidence (Iowa, United States). Cancer Causes Control. 1995;6(2):112–8.
104. Baer HJ, Colditz GA, Rosner B, Michels KB, Rich-Edwards JW, Hunter DJ, et al. Body fatness during childhood and adolescence and incidence of breast cancer in premenopausal women: a prospective cohort study. Breast Cancer Res. 2005;7(3):R314–25.
105. Ahlgren M, Melbye M, Wohlfahrt J, Sorensen TI. Growth patterns and the risk of breast cancer in women. N Engl J Med. 2004;351(16):1619–26.
106. Suzuki R, Iwasaki M, Inoue M, Sasazuki S, Sawada N, Yamaji T, et al. Body weight at age 20 years, subsequent weight change and breast cancer risk defined by estrogen and progesterone receptor status—the Japan public health center-based prospective study. Int J Cancer. 2011;129(5):1214–24.
107. Palmer JR, Adams-Campbell LL, Boggs DA, Wise LA, Rosenberg L. A prospective study of body size and breast cancer in black women. Cancer Epidemiol Biomarkers Prev. 2007;16(9):1795–802.
108. Weiderpass E, Braaten T, Magnusson C, Kumle M, Vainio H, Lund E, et al. A prospective study of body size in different periods of life and risk of premenopausal breast cancer. Cancer Epidemiol Biomarkers Prev. 2004;13(7):1121–7.
109. Robinson WR, Tse CK, Olshan AF, Troester MA. Body size across the life course and risk of premenopausal and postmenopausal breast cancer in Black women, the Carolina Breast Cancer Study, 1993–2001. Cancer Causes Control. 2014;25(9):1101–17.
110. Berkey CS, Frazier AL, Gardner JD, Colditz GA. Adolescence and breast carcinoma risk. Cancer. 1999;85(11):2400–9.
111. Stoll BA. Teenage obesity in relation to breast cancer risk. Int J Obes Relat Metab Disord. 1998;22(11):1035–40.
112. van den Brandt PA, Spiegelman D, Yaun SS, Adami HO, Beeson L, Folsom AR, et al. Pooled analysis of prospective cohort studies on height, weight, and breast cancer risk. Am J Epidemiol. 2000;152(6):514–27.
113. Rich-Edwards JW, Goldman MB, Willett WC, Hunter DJ, Stampfer MJ, Colditz GA, et al. Adolescent body mass index and infertility caused by ovulatory disorder. Am J Obstet Gynecol. 1994;171(1):171–7.
114. Howe GR, Hirohata T, Hislop TG, Iscovich JM, Yuan JM, Katsouyanni K, et al. Dietary factors and risk of breast cancer: combined analysis of 12 case-control studies. J Natl Cancer Inst. 1990;82(7):561–9.

115. Hunter DJ, Willett WC. Diet, body size, and breast cancer. Epidemiol Rev. 1993;15(1):110–32.
116. Enger SM, Ross RK, Paganini-Hill A, Carpenter CL, Bernstein L. Body size, physical activity, and breast cancer hormone receptor status: results from two case-control studies. Cancer Epidemiol Biomarkers Prev. 2000;9(7):681–7.
117. Peters TM, Schatzkin A, Gierach GL, Moore SC, Lacey JV Jr, Wareham NJ, et al. Physical activity and postmenopausal breast cancer risk in the NIH-AARP diet and health study. Cancer Epidemiol Biomarkers Prev. 2009;18(1):289–96.
118. Lux MP, Fasching PA, Beckmann MW. Hereditary breast and ovarian cancer: review and future perspectives. J Mol Med (Berl). 2006;84(1):16–28.
119. Whittemore AS. The Eighth AACR American Cancer Society Award lecture on cancer epidemiology and prevention. Genetically tailored preventive strategies: an effective plan for the twenty-first century? American Association for Cancer Research. Cancer Epidemiol Biomarkers Prev. 1999;8(8):649–58.
120. Malkin D, Li FP, Strong LC, Fraumeni JF Jr, Nelson CE, Kim DH, et al. Germ line p53 mutations in a familial syndrome of breast cancer, sarcomas, and other neoplasms. Science. 1990;250(4985):1233–8.
121. Liaw D, Marsh DJ, Li J, Dahia PL, Wang SI, Zheng Z, et al. Germline mutations of the PTEN gene in Cowden disease, an inherited breast and thyroid cancer syndrome. Nat Genet. 1997;16 (1):64–7.
122. Antoniou AC, Easton DF. Models of genetic susceptibility to breast cancer. Oncogene. 2006;25 (43):5898–905.
123. Walsh T, Casadei S, Coats KH, Swisher E, Stray SM, Higgins J, et al. Spectrum of mutations in BRCA1, BRCA2, CHEK2, and TP53 in families at high risk of breast cancer. JAMA. 2006;295 (12):1379–88.
124. Shiovitz S, Korde LA. Genetics of breast cancer: a topic in evolution. Ann Oncol. 2015.
125. Hunter DJ, Kraft P, Jacobs KB, Cox DG, Yeager M, Hankinson SE, et al. A genome-wide association study identifies alleles in FGFR2 associated with risk of sporadic postmenopausal breast cancer. Nat Genet. 2007;39(7):870–4.
126. Easton DF, Pooley KA, Dunning AM, Pharoah PD, Thompson D, Ballinger DG, et al. Genome-wide association study identifies novel breast cancer susceptibility loci. Nature. 2007;447(7148):1087–93.
127. Stacey SN, Manolescu A, Sulem P, Rafnar T, Gudmundsson J, Gudjonsson SA, et al. Common variants on chromosomes 2q35 and 16q12 confer susceptibility to estrogen receptor-positive breast cancer. Nat Genet. 2007;39(7):865–9.
128. Michailidou K, Beesley J, Lindstrom S, Canisius S, Dennis J, Lush MJ, et al. Genome-wide association analysis of more than 120,000 individuals identifies 15 new susceptibility loci for breast cancer. Nat Genet. 2015;47(4):373–80.
129. Liu Y, Colditz GA, Rosner B, Berkey CS, Collins LC, Schnitt SJ, et al. Alcohol intake between menarche and first pregnancy: a prospective study of breast cancer risk. J Natl Cancer Inst. 2013;105 (20):1571–8.
130. Pownall HJ, Rosales C, Gillard BK, Gotto AM Jr. Alcohol: a nutrient with multiple salutary effects. Nutrients. 2015;7(3):1992–2000.
131. Lilienfeld AM. The relationship of cancer of the female breast to artificial menopause and martial status. Cancer. 1956;9(5):927–34.
132. Trichopoulos D, MacMahon B, Cole P. Menopause and breast cancer risk. J Natl Cancer Inst. 1972;48 (3):605–13.
133. Pollan M, Ascunce N, Ederra M, Murillo A, Erdozain N, Ales-Martinez J, et al. Mammographic density and risk of breast cancer according to tumor characteristics and mode of detection: a Spanish population-based case-control study. Breast Cancer Res. 2013;15(1):R9.
134. Ma L, Fishell E, Wright B, Hanna W, Allan S, Boyd NF. Case-control study of factors associated with failure to detect breast cancer by mammography. J Natl Cancer Inst. 1992;84(10):781–5.
135. Mandelson MT, Oestreicher N, Porter PL, White D, Finder CA, Taplin SH, et al. Breast density as a predictor of mammographic detection: comparison of interval-and screen-detected cancers. J Natl Cancer Inst. 2000;92(13):1081–7.
136. Bailey SL, Sigal BM, Plevritis SK. A simulation model investigating the impact of tumor volume doubling time and mammographic tumor detectability on screening outcomes in women aged 40–49 years. J Natl Cancer Inst. 2010;102(16):1263–71.
137. Byrne C, Schairer C, Wolfe J, Parekh N, Salane M, Brinton LA, et al. Mammographic features and breast cancer risk: effects with time, age, and menopause status. J Natl Cancer Inst. 1995;87 (21):1622–9.
138. McCormack VA, dos Santos Silva I. Breast density and parenchymal patterns as markers of breast cancer risk: a meta-analysis. Cancer epidemiology, biomarkers and prevention: a publication of the American Association for Cancer Research, cosponsored by the American Society of Preventive. Oncology. 2006;15(6):1159–69.
139. Boyd NF, Guo H, Martin LJ, Sun L, Stone J, Fishell E, et al. Mammographic density and the risk and detection of breast cancer. N Engl J Med. 2007;356(3):227–36.
140. Trichopoulos D, Lipman RD. Mammary gland mass and breast cancer risk. Epidemiology. 1992;3 (6):523–6.
141. Trichopoulos D, Adami HO, Ekbom A, Hsieh CC, Lagiou P. Early life events and conditions and breast cancer risk: from epidemiology to etiology. Int J Cancer. 2008;122(3):481–5.
142. Martin LJ, Boyd NF. Mammographic density. Potential mechanisms of breast cancer risk associated with mammographic density: hypotheses based

on epidemiological evidence. Breast Cancer Res. 2008;10(1):201.
143. Vachon CM, Sasano H, Ghosh K, Brandt KR, Watson DA, Reynolds C, et al. Aromatase immunoreactivity is increased in mammographically dense regions of the breast. Breast Cancer Res Treat. 2011;125(1):243–52.
144. Bertrand KA, Tamimi RM, Scott CG, Jensen MR, Pankratz V, Visscher D, et al. Mammographic density and risk of breast cancer by age and tumor characteristics. Breast Cancer Res. 2013;15(6): R104.
145. Yaghjyan L, Colditz GA, Collins LC, Schnitt SJ, Rosner B, Vachon C, et al. Mammographic breast density and subsequent risk of breast cancer in postmenopausal women according to tumor characteristics. J Natl Cancer Inst. 2011;103(15):1179–89.
146. Roberts KD, Rochefort JG, Bleau G, Chapdelaine A. Plasma estrone sulfate levels in postmenopausal women. Steroids. 1980;35(2):179–87.
147. Missmer SA, Eliassen AH, Barbieri RL, Hankinson SE. Endogenous estrogen, androgen, and progesterone concentrations and breast cancer risk among postmenopausal women. J Natl Cancer Inst. 2004;96(24):1856–65.
148. Kaaks R, Berrino F, Key T, Rinaldi S, Dossus L, Biessy C, et al. Serum sex steroids in premenopausal women and breast cancer risk within the European Prospective Investigation into Cancer and Nutrition (EPIC). J Natl Cancer Inst. 2005;97(10):755–65.
149. Fortner RT, Eliassen AH, Spiegelman D, Willett WC, Barbieri RL, Hankinson SE. Premenopausal endogenous steroid hormones and breast cancer risk: results from the Nurses' Health Study II. Breast Cancer Res. 2013;15(2):R19.
150. Endogenous H, Breast Cancer Collaborative G, Key TJ, Appleby PN, Reeves GK, Travis RC, et al. Sex hormones and risk of breast cancer in premenopausal women: a collaborative reanalysis of individual participant data from seven prospective studies. Lancet Oncol. 2013;14(10):1009–19.
151. Palmer JR, Rosenberg L. Cigarette smoking and risk of breast cancer. Kelsey JL, editor. Baltimore: John Hopkins University of Hygiene and Public Health; 1993.
152. Laden F, Hunter DJ. Environmental risk factors and female breast cancer. Annu Rev Public Health. 1998;19:101–23.
153. Al-Delaimy WK, Cho E, Chen WY, Colditz G, Willet WC. A prospective study of smoking and risk of breast cancer in young adult women. Cancer Epidemiol Biomarkers Prev. 2004;13(3):398–404.
154. Calle EE, Miracle-McMahill HL, Thun MJ, Heath CW Jr. Cigarette smoking and risk of fatal breast cancer. Am J Epidemiol. 1994;139(10):1001–7.
155. Cui Y, Miller AB, Rohan TE. Cigarette smoking and breast cancer risk: update of a prospective cohort study. Breast Cancer Res Treat. 2006;100(3):293–9.
156. Gram IT, Braaten T, Terry PD, Sasco AJ, Adami HO, Lund E, et al. Breast cancer risk among women who start smoking as teenagers. Cancer Epidemiol Biomarkers Prev. 2005;14(1):61–6.
157. Gaudet MM, Gapstur SM, Sun J, Diver WR, Hannan LM, Thun MJ. Active smoking and breast cancer risk: original cohort data and meta-analysis. J Natl Cancer Inst. 2013;105(8):515–25.
158. Johnson KC, Miller AB, Collishaw NE, Palmer JR, Hammond SK, Salmon AG, et al. Active smoking and secondhand smoke increase breast cancer risk: the report of the Canadian Expert Panel on Tobacco Smoke and Breast Cancer Risk (2009). Tob Control. 2011;20(1):e2.
159. Thomson CA, Thompson PA. Fruit and vegetable intake and breast cancer risk: a case for subtype-specific risk? J Natl Cancer Inst. 2013;105 (3):164–5.
160. Jung S, Spiegelman D, Baglietto L, Bernstein L, Boggs DA, van den Brandt PA, et al. Fruit and vegetable intake and risk of breast cancer by hormone receptor status. J Natl Cancer Inst. 2013;105(3):219–36.
161. Eliassen AH, Hendrickson SJ, Brinton LA, Buring JE, Campos H, Dai Q, et al. Circulating carotenoids and risk of breast cancer: pooled analysis of eight prospective studies. J Natl Cancer Inst. 2012;104(24):1905–16.
162. Colston KW, Berger U, Coombes RC. Possible role for vitamin D in controlling breast cancer cell proliferation. Lancet. 1989;1(8631):188–91.
163. Bauer SR, Hankinson SE, Bertone-Johnson ER, Ding EL. Plasma vitamin D levels, menopause, and risk of breast cancer: dose-response meta-analysis of prospective studies. Medicine (Baltimore). 2013;92 (3):123–31.
164. Manson JE, Bassuk SS, Lee IM, Cook NR, Albert MA, Gordon D, et al. The VITamin D and OmegA-3 TriaL (VITAL): rationale and design of a large randomized controlled trial of vitamin D and marine omega-3 fatty acid supplements for the primary prevention of cancer and cardiovascular disease. Contemp Clin Trials. 2012;33(1):159–71.
165. Pradhan AD, Manson JE. Update on the Vitamin D and OmegA-3 trial (VITAL). J Steroid Biochem Mol Biol. 2015.

# Pathology and Molecular Pathology of Breast Cancer 12

Caterina Marchiò, Felipe C. Geyer and Jorge S. Reis-Filho

## 12.1 Introduction

In this chapter, we begin by giving an overview of embryology and development of the mammary gland, together with a summary of the features of breast anatomy, histology and physiology that are germane to the understanding and recognition of pathologic alterations. The distinct pathological processes affecting the mammary gland, encompassing hyperplastic conditions to malignant lesions, are presented. The molecular pathology features are discussed in each section, reporting the main findings of studies based on high-throughput techniques that have reshaped the way breast cancer pathology is currently practiced and breast cancer research is conducted. Rather than being a reference text for morphology, this chapter aims to provide the readers with combined information stemming from both morphological features and molecular data.

C. Marchiò
University of Turin, via Santena 7,
10126 Turin, Italy

F.C. Geyer
Hospital Israelita Albert Einstein,
Avenida Albert Einstein, Sao Paolo, Brazil

F.C. Geyer · J.S. Reis-Filho (✉)
Memorial Sloan Kettering Cancer Center,
1275 York Ave., New York, NY 10065, USA
e-mail: reisfilj@mskcc.org

## 12.2 Embryology and Development of the Mammary Gland

Development of the mammary glands commences from the milk lines that appear on the ventral surface of the fetus at the 5th week of gestation [1]. Milk lines, also known as mammary ridges, represent thickenings of the epidermis and extend from the axilla to the upper medial region of the thigh [1]. In humans, most of the milk lines disappear during fetal development; persisting segments of the milk line may give rise to ectopic mammary glandular tissue, most commonly at the extreme ends of the mammary ridge, i.e., in the axilla or vulva [1].

In the 15th week of gestation, the breast bud, which is an epithelial stalk on the chest wall at the site of mammary development, undergoes mesenchymal condensation. Mammary gland lobes are then formed by outgrowth of solid epithelial cords into the mesenchyme [1]. At this stage, differential expression of cytokeratins (CK) 14, 18, and 19, and of actin in the ducts and lobular buds characterizes the development of fetal breast [2]. Between weeks 23 and 28 of gestation, the basal cells differentiate toward myoepithelial cells [3], which play an important role in the branching morphogenesis of the mammary gland through the synthesis of basement membrane constituents such as laminin, type IV collagen, and fibronectin, as well as metalloproteinases and growth factors [4]. While


M. Loda et al. (eds.), *Pathology and Epidemiology of Cancer*,
DOI 10.1007/978-3-319-35153-7_12

the papillary layer of the fetal dermis encloses these growing epithelial cords and develops into the vascularized fibrous tissue surrounding individual ducts and their branches of ducts which form lobules [1], the less cellular, more collagenized stroma from the reticular dermis extends into the breast to encompass lobes and their subdivisions, forming the suspensory ligaments of Cooper, which attach the breast parenchyma to the skin [3].

Coincidentally, differentiation of the mesenchyme into fat within the collagenous stroma occurs between weeks 20 and 32. In the last 2 months of gestation, canalization of the epithelial cords takes place, followed by development of branching lobuloalveolar glandular structures. The mammary pit is a depression in the epidermis where the lactiferous ducts converge, that, near birth, forms the nipple by evagination [1].

The development of the breast structure is dependent on steroid hormones only after the 15th week, when it seems to be influenced largely by testosterone [1]. In the last weeks of gestation, maternal and placental steroid hormones and prolactin induce secretory activity in the mammary gland, which is manifested after birth by the secretion of colostrum and palpable enlargement of the breast bud. Due to the disappearance of maternal hormones from the infant's bloodstream, the gland shrinks during the first trimester after birth and returns to an inactive state (branching of lactiferous ducts without progressive alveolar differentiation, lobular structures may persist) [1].

At puberty, the onset of cyclical estrogen and progesterone secretion reinitiates the normal breast development. Estrogen stimulates epithelial proliferation, causing the ducts to elongate and to acquire a thickened epithelium [5], while also orchestrating the differentiation of hormonally responsive periductal stroma. Growth hormone and glucocorticoids also contribute to ductal growth, whereas insulin, progesterone, and growth hormone are responsible for lobuloalveolar differentiation and growth during this period. The lobules derive from solid masses of cells that form at the ends of terminal ducts. Although the greatest amount of breast glandular differentiation occurs during puberty, this process continues for at least a decade and is actually enhanced by pregnancy [6]. In fact, full maturation of the female mammary gland occurs only during pregnancy: at this stage, the lobuloalveolar differentiation is completed due to hormonal stimulus, enabling lactation.

In contrast, due to the lack of estrogen stimulation, the male breast does not undergo lobuloalveolar differentiation and remains composed predominantly of ducts, with no/little specialized stroma.

## 12.3 Anatomy, Histology, and Physiology of the Mammary Gland

### 12.3.1 Anatomy

The mammary gland is an even and symmetric gland localized in the subcutaneous tissue of the thorax over the pectoral muscle between the second and the sixth rib in the vertical axis and from the sternal edge to the mid-axillary line in the horizontal axis [1, 7]. The peripheral anatomic boundaries are ill defined except for the deep surface where the gland overlies the pectoralis fascia. Given that this gland is a repository for fat, the size may be highly variable among individuals, depending on the body habitus (from 30 to >1000 g) [7].

Anatomically, the breast resides in a space within the superficial fascia, which superiorly is in continuity with the cervical fascia and inferiorly with the superficial abdominal fascia of Cooper. It is important to note that microscopic extensions of the mammary gland beyond these boundaries can be present [1, 7]: clinically this means that even total mastectomy does not result in complete removal of all glandular breast tissue [7]. Fibrous strands extend from the dermis into the breast thus forming the suspensory ligament of Cooper, which is responsible for attaching the skin and nipple to the breast. The nipple is centered (and elevated) in the so-called nipple/areolar complex situated on the skin overlying the mammary gland. The nipple represents the site where major/principal ducts end (the tip of the nipple contains 15–20 orifices) [7]. In a way akin to the microscopic extensions of glandular tissue beyond the breast boundaries, it should be noted that there is no definite anatomical plan between the skin and the breast and that terminal duct-lobular unit may be found close/within the nipple/areolar complex; therefore, skin-sparing mastectomy may leave a non-negligible amount of glandular tissue underneath the skin.

The arterial supply of the breast is derived from the internal mammary (60 % of blood supply), lateral thoracic (30 %), and intercostal arteries (minor contribution) [1, 7]. Venous drainage is more variable but tends to mirror the distribution of the arterial supply [1]. For lymphatic drainage, three dominant routes have been identified. The most important is that of the axilla, which drains >75 % of lymphatic flow into the axillary nodes. Lymph nodes located in the interpectoral fascia constitute Rotter's nodes, which sometimes are excised during intervention. Drainage through the internal lymphatics accounts for less than 25 % of the whole lymphatic flow. These vessels drain to the internal thoracic mammary nodes located along the sternal borders of the internal thoracic trunks. Finally, the lymphatic vessels drain into the supraclavicular and infraclavicular as well as to intramammary nodes [1].

Operationally speaking, the breast can be schematically subdivided into quadrants (Q), conventionally numbered from 1 to 6, the external ones being Q1 and Q3 (upper and lower, respectively), and the internal Q2–Q4 (upper and lower, respectively), the central region, i.e., that of the retroareolar parenchyma, is Q5 and the axillary extension is known as Q6.

### 12.3.2 Histology and Physiology

Histologically, the mammary gland is defined as a modified sweat gland composed of acini and ducts within stromal tissue, which represents the major portion of the nonlactating adult breast and is composed of a variable amount of fibrous and adipose tissue. The structure of the gland is like a "flowering tree," with major ducts budding in smaller ductules up to the smallest structures which are the ductules of the terminal duct-lobular unit (TDLU) that end in acini, i.e., blind-ending tubules that compose the lobule (Fig. 12.1). Ducts, ductules, and acini are tubular structures with an empty lumen and a wall. The wall is made up of a bilayered epithelium composed of an inner epithelial layer of so-called "luminal cells" (cuboidal or low columnar in

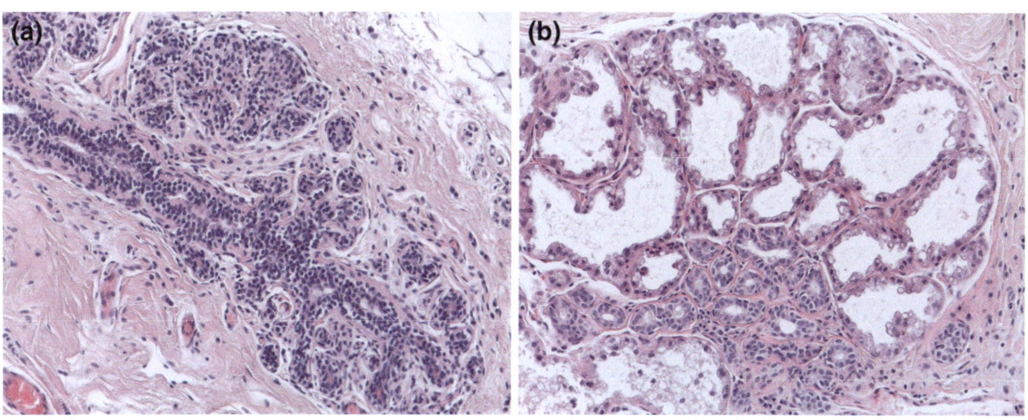


**Fig. 12.1** Normal breast (**a**), and lactational changes (**b**)

shape) and an external layer of myoepithelial cells (see Table 12.1 for specific markers). The TDLU is the functional unit of the mammary gland and represents the site where most carcinomas take origin (both *in situ* and invasive). Occasionally, the epithelial cells of the TDLU display prominent clear cell changes in the cytoplasm (so called "lamprocytes"), that may be appreciated in both premenopausal and postmenopausal women and seem unrelated to hormone stimulation [8].

The stromal counterpart recognizes the intervening stroma between lobules (interlobular stroma) and the stroma within lobules (intralobular stroma). The first is composed of highly collagenized paucicellular fibroadipose tissue that occasionally may be populated by multinucleated giant cells of unknown significance (that should not be mistaken for malignant or inflammatory cells); the latter is made of loose fibrovascular stroma populated also by lymphocytes, plasma cells, macrophages, and mast cells [7]. The intralobular stroma responds to hormonal changes and upon estradiol stimulation becomes more edematous.

Hormones elicit their action also on epithelial cells, where estrogen, progesterone, and androgen promote differentiation and proliferation of luminal cells and oxytocin is responsible for contraction of myoepithelial cells.

Hormone receptors (estrogen and progesterone receptors, ER and PR, respectively) are expressed in the normal breast but at low levels (and lower in the ducts than in the lobules): typically, single positive cells are scattered in the epithelial population, never reaching 100 %. In addition, there is heterogeneity of expression among different lobules [7].

The gland produces milk and is active only during pregnancy, when the full development of the breast occurs in humans. In this period of the adult life, epithelial cell proliferation resumes under the influence of ER, PR, prolactin, and growth hormones, leading to an increase in the number of lobules at the expenses of both intralobular and interlobular stroma [7]. In addition, the mammary gland undergoes secretory changes featuring luminal cells acquiring a foamy cytoplasm (they contain numerous lipid vacuoles) and a characteristic hobnail appearance with nuclei and prominent nucleoli. Secretions accumulate in the expanded lobules (Fig. 12.1).

So-called "lactational changes" can also occur outside of pregnancy and may be alarming due to features of cytologic atypia; therefore, a differential diagnosis including atypical lesions and carcinoma, in particular ductal carcinoma *in situ* (DCIS) of clinging pattern, must be considered. In the context of fine-needle aspirates, extreme caution is needed to avoid false-positive results.

Over time, with the decrease of ER and PR stimulation in the postmenopausal period, the gland undergoes regressive changes with fibroadipose involution. More specifically, there is atrophy of the TDLUs (reduction in size and complexity of the acini), loss of specialized intralobular stroma, and ducts may become ectatic [7].

The nipple and the areola are lined by keratinizing squamous epithelium with minimal extension into the terminal portion of the lactiferous ducts. Clear cells can occasionally be present in the epidermis of the nipple/areola

**Table 12.1** Markers of luminal epithelial cells and myoepithelial cells in the terminal ductular-lobular unit (TDLU)

| Luminal cells | Myoepithelial cells |
|---|---|
| CK18 | CK14 |
| CK19 | CK5/6 |
| CK8 | P63 |
| CK7 | Calponin |
| GFCD15 | Caveolin |
| EMA | D2-40 |
| KIT | Smooth muscle actin (SMA) |

complex. These are benign epithelial cells and should not be confused with Paget's disease. Some are simply clear keratinocytes, others are known as "Toker cells" [9].

## 12.4 Benign Epithelial Proliferations

### 12.4.1 Adenosis

Adenosis is a lobulocentric proliferative lesion largely derived from the TDLU that occurs most often as part of a spectrum of proliferative abnormalities commonly referred to as fibrocystic changes [10]. Both epithelial and myoepithelial cells participate in adenosis [10]. When by itself it may either be isolated to single lobules and represent a microscopic lesion that comes to attention clinically only if it contains mammographically detected calcifications, or it can rarely form a palpable or a radiographically detectable mass. This is the case when a confluence or fusion of the affected lobules creates an "adenosis tumor" [11]. Presentation with a mass is more frequently found when adenosis is seen in fibrocystic disease [10].

Microscopically, adenosis tends to have a more prominent glandular pattern in premenopausal women, whereas sclerosis and diminished gland formation are conspicuous after menopause. The most cellular type of adenosis is florid adenosis, characterized by hyperplasia of epithelial and myoepithelial cells. Proliferation of ductules and lobular glands severely distorts the architecture of the underlying lobules. The hyperplastic structures appear to elongate, becoming tortuous and entwined in a fashion that results in many more ductular cross sections than are present in an anatomically normal lobule [10].

Epithelial cells lining the tubules and glands of adenosis are most often flattened, cuboidal, or slightly columnar, and are arranged in one or two orderly layers surrounded by myoepithelial cells. Mitoses are vanishingly rare, and are more numerous during pregnancy. Apocrine metaplasia is uncommon in florid adenosis. Luminal secretions may undergo calcification, but this is less common and less extensive in florid than in sclerosing adenosis [10].

#### 12.4.1.1 Sclerosing Adenosis

In sclerosing adenosis there is preferential preservation of myoepithelial cells with variable atrophy of epithelial cells, accompanied by lobular fibrosis. The swirling lobulocentric pattern encountered in florid adenosis is retained, but epithelial cells are less conspicuous, the ductular structures and their lumens are largely attenuated, and the myoepithelial cells predominate [10]. The identification of sclerosing adenosis is important given that it may display an infiltrative pattern in the stroma and fat especially when sclerosing adenosis is not limited to a lobulocentric pattern [10]. The differentiation between invasive carcinoma must be made and such a scenario may be challenging especially in needle core biopsy samples, which lack the orientation provided by surrounding tissue of a surgical biopsy [10]. Low-power microscopic assessment reveals the remaining lobulocentric pattern and immunohistochemistry (IHC), including cytoplasmic markers of smooth muscle differentiation and the nuclear marker P63, both label myoepithelial cells (Table 12.1). Microcalcifications are more frequently formed in the sclerosing type of adenosis and become progressively more numerous with increasing sclerosis [10].

Several epidemiological studies have demonstrated that a diagnosis of benign entities, such as sclerosing adenosis, as well as fibrosis and cysts, confers a low relative risk of development of cancer [12–14]. These lesions, however, do not display genetic aberrations in common with those of true precursors and invasive breast cancer [14, 15].

### 12.4.2 Usual Ductal Hyperplasia

Usual ductal hyperplasia (UDH) is an intraductal proliferation of a mixed population of epithelial cells, leading to the formation of secondary lumens. The latter are often distributed at the periphery of the ducts and associated with streaming of the central proliferating cells [16,

17], producing the so-called "slit-like" spaces. This mixed population of cells displays features of both luminal and myoepithelial differentiation, with heterogeneous positivity for ER and PR, and consistent expression of high molecular weight cytokeratins, such as CK5/6. IHC for hormone receptors and CK5/6 can indeed be very helpful in the context of lesions with equivocal morphological features (Fig. 12.2; Table 12.2) to help differentiate UDH from atypical ductal hyperplasia (ADH)/low-grade DCIS. Another feature that helps in differentiating between these two entities is the presence of intranuclear pseudoinclusions, which are typically found in florid UDH and are rare in ADH/low-grade DCIS [18, 19].

The role of UDH in the evolution of breast cancer has largely been debated [20, 21]. UDH has been shown to be associated with a low risk of breast cancer development and was initially considered to be a nonobligate precursor of both ADH and DCIS [22, 23].

Genetic data suggest that the majority of UDH constitutes only a risk indicator rather than a true breast cancer precursor [24]; however, loss of heterozygosity (LOH) analyses of UDHs have revealed that at least a subgroup of these lesions are clonal and that the prevalence of LOH in UDHs (4.5–13 %) is lower than that found in ADH and DCIS [25]. Chromosomal comparative genomic hybridization (CGH) studies have produced conflicting results on the presence of numerical chromosomal aberrations in UDHs [22, 23]. While some have demonstrated the presence of unbalanced chromosomal aberrations in UDHs [22, 23], others have failed to identify recurrent genetic aberrations [26] or any aberration at all [27]. It should be noted, however, that the studies demonstrating unbalanced chromosomal changes in UDHs employed whole genome amplification methods [22, 23], which have been shown to introduce artifacts in CGH and microarray-based CGH (aCGH) [28], or failed to correct the results for some of the known artifacts of CGH analysis performed with DNA extracted from formalin-fixed paraffin-embedded (FFPE) samples subjected to whole genome amplification [14].

In a similar way to copy number analyses, phosphatidylinositol-4,5-bisphosphate 3-kinase, catalytic subunit alpha (PIK3CA) mutational studies on UDH have provided conflicting results. While, Li et al. [29] found no *PIK3CA* mutations in 16 cases analyzed by Sanger sequencing, Ang et al. [30] used a PCR-mass spectroscopy-based technique and found *PIK3CA* mutations in around 50 % of cases. Interestingly, *PIK3CA* mutations in UDH tended to be in exon 20 while in carcinomas the majority of mutations were in exon 9. Additionally, in the vast majority of cases, concurrent invasive carcinomas had discordant genotype, while concordance was high for matched *in situ* and invasive carcinomas. It is likely that differences

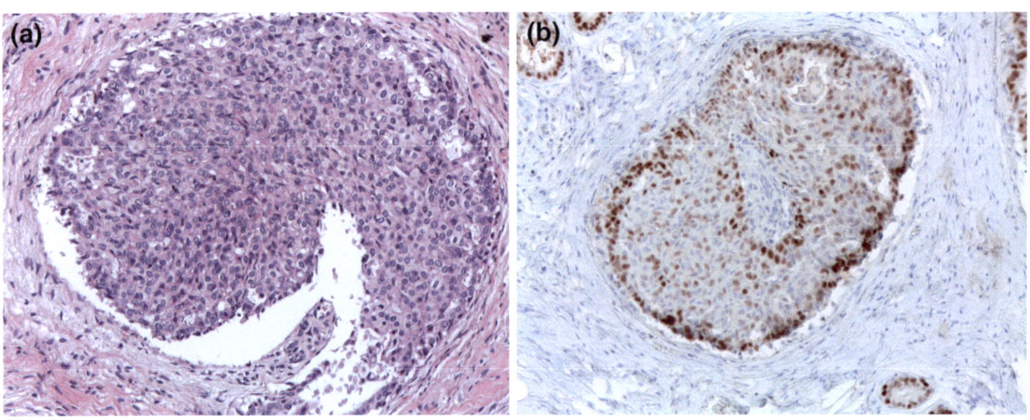


**Fig. 12.2** Usual ductal hyperplasia (**a** H&E; **b** ER)

from these two studies are related to the different techniques and their distinct sensitivities. In addition, due to the highly sensitive methodology used by Ang et al., it is unclear whether *PIK3CA* mutations found in UDH samples were clonal (i.e., present in all/nearly all lesional cells) or were present in only a small fraction of cells.

Taken together the available evidence suggests that the vast majority of UDHs lacks clonal genetic changes; when these are present, they are usually randomly distributed and do not affect regions usually involved in invasive breast cancers. Therefore, it is reasonable to conclude that the majority of UDHs are risk indicators rather than nonobligate precursors of breast cancer development; however, a small minority may be associated with the development of breast cancer [14].

### 12.4.3 Sclerosing Lesions

#### 12.4.3.1 Radial Scar, Complex Sclerosing Lesion

A radial scar (RS) is defined as a benign lesion with a typical stellate profile, which is microscopically characterized by obliterated ducts and pseudoinfiltrative tubules, immersed in an elastotic stroma (Fig. 12.3). These tubules are lined by epithelial and myoepithelial cells, surrounded by contracted ducts and lobules exhibiting a variety of features including epithelial hyperplasia, duct ectasia, adenosis, papillomatosis, and apocrine cysts (Fig. 12.3; Table 12.2) [31].

The term "complex sclerosing lesion" (CSL) is applied to those lesions larger in size and more complex in features. Most pathologists

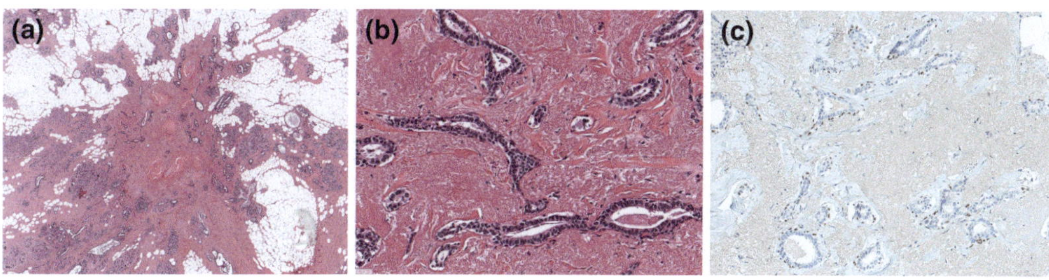


**Fig. 12.3** Radial scar/complex sclerosing lesion (**a**, **b** H&E; **c** P63)

**Table 12.2** Utility of most commonly used IHC antibodies in breast pathology, other than evaluation of prognostic and predictive factors for primary carcinomas

| Marker | Diagnostic scenario |
|---|---|
| High molecular weight cytokeratins | Heterogeneous in UDH, negative in ADH/low-grade DCIS |
| ER | Heterogeneous in UDH, homogeneous (≅100 %) in ADH/low-grade DCIS |
| P63 | Suspicion for invasion: sclerosing lesions *versus* invasive carcinoma, microinvasion in DCIS;<br>Diagnosis of myoepithelial lesions and spindle cell MBC |
| HER2 | Microinvasion in (HER2+) high-grade DCIS |
| E-CAD | IC-NST *versus* ILC and LCIS *versus* DCIS (mind possible aberrant E-CAD expression in lobular lesions) |

diagnose "complex sclerosing lesion" in lesions larger than 1 cm.

Contradictory results have been reported on the role of RS/CSL in the development of breast cancer, namely whether RSs/CSLs are risk indicators of breast cancer development or if they constitute true nonobligate precursors of DCIS and invasive breast cancer [32–34]. Long-term follow-up of women with RS/CSL indicates a 1.5–2 fold increase in subsequent breast cancer risk [35–37], which persists after adjusting for concurrent proliferative disease [38].

Some have observed clonal differences between hyperplasic lesions and other areas within RSs. 16q and 8p allelic imbalance were detected, as well as different genetic losses. There is also some evidence to suggest that RSs may be associated with the development of breast cancer. Manfrin et al. [39] reported on a high incidence (32 %) of carcinomas in a series of 117 asymptomatic patients with mammography detected RSs. Given the observation that RSs harbor 16q LOH, it is not surprising that the majority of invasive breast cancers developing in the context of RSs are of low histologic grade [39], such as tubular and classical lobular carcinomas. Nevertheless, a rare association between RSs/CSLs and low-grade metaplastic carcinomas has also been described [40, 41]. Overall, the presence of a coexisting carcinoma in RSs/CSLs ranges from 3.6–32 % [39, 42, 43].

From a genetic standpoint, RSs/CSLs have been shown to harbor *PIK3CA*-activating mutations in 63.6 % [44], and these were more prevalent when epithelial atypia was superimposed (56.3 % vs. 83.3 %) [44]. Interestingly, activating mutations affecting *PIK3CA*, the second most frequently mutated gene in breast cancer [45], have also been documented in papillomas [46], columnar cell lesions (CCLs) [47], and UDH [30], and it has been hypothesized that *PIK3CA* mutations may be more relevant for proliferation than for malignant transformation in breast epithelium [44].

In conclusion, despite some studies finding molecular changes in RSs/CSLs, it is not clear whether those lesions should be broadly categorized as clonal neoplastic lesions and considered true nonobligate precursors. It is possible that those detected molecular changes reflect the epithelial proliferations that may be or may not be present, rather than the lesion as a whole.

"Infiltrating epitheliosis" (IE), is sometimes perceived as synonym of RS and CSL, however, some authors have suggested that these terms should not be used interchangeably [48]. IE may be associated with a higher risk of carcinoma, and defined morphological criteria for this differentiation have been published [48]. Historically, the term IE was first used by John Azzopardi in 1979 [49].

The overall appearance of IE is more infiltrative than usual sclerosing lesions, as the involved ducts often have jagged or irregular edges and the proliferating epithelium often appears to "flow-out" into adjacent stroma [50]. As a matter of fact, invasive carcinoma is often considered in the differential diagnosis. In addition, florid UDH-like epithelial proliferation is an integral component of IE. As in UDH, IE displays a heterogeneous epithelial phenotype (admixture of CK5/6-positive basal/intermediate-type cells and both ER-positive and ER-negative/C-KIT-positive luminal-type cells) [50]. However, at a variance to UDH and most sclerosing lesions, myoepithelial cells, if not absent, are usually immunophenotypically altered and not detected by routinely used antibodies [50]. Despite these distinctive histologic features, most pathologists currently classify IE in the RS/CSL spectrum and therefore its true incidence and association with carcinoma are unknown [50].

A recent study has analyzed the mutational repertoire of eight IE cases by targeted capture massive parallel sequencing and consistently found mutations in components of the PI3K pathway: seven samples harbored *PIK3CA* hotspot mutations, while the remaining case displayed a *PIK3R1* somatic mutation. Of note, analysis of one case composed of IE, DCIS, and low-grade adenosquamous carcinoma revealed the three components were clonally related [50]. In conclusion, these data supports the notion that IE is part of the RS/CSL spectrum, with a high prevalence of *PIK3CA* mutations, but a subset of them may constitute the substrate for carcinoma development.

### 12.4.4 Benign Tumors of the Nipple

#### 12.4.4.1 Nipple Adenoma

A nipple adenoma is defined as a compact proliferation of small tubules showing both epithelial and myoepithelial layers arising around the collecting ducts of the nipple [16]. Epithelial hyperplasia may be florid within the tubules or in the collecting ducts. It should be noted that necrosis of comedo-type may occasionally be present in cases with florid UDH and should not be interpreted as a sign of malignancy. Furthermore, a nipple adenoma may mimick an invasive carcinoma as marked sclerosis may impart a pseudoinfiltrative pattern.

Clinically, it usually presents with nipple discharge and erosion of the nipple itself or with an underlying nodule. An association with carcinoma is on record but it represents a rare event [16]. Therefore, simple excision of those lesions, and whenever possible, conservation of the nipple, is the treatment of choice.

#### 12.4.4.2 Syringomatous Adenoma

Syringomatous adenoma is a rare lesion, defined as a nonmetastasizing, locally recurrent, and locally invasive tumor of the nipple/areolar region displaying sweat duct differentiation [16]. Clinically it presents as a firm, discrete mass that on gross observation shows ill-defined margins. Microscopically, it is composed of nests and branching cords of cells, glandular structures, and small keratinous cysts that permeate the stroma of the nipple (bundles of the muscle and perineural spaces) [16]. This infiltrative pattern is coupled with bland cytology with regular nuclei and rare mitoses. Frequently, the glandular structures show an inner luminal layer and an outer layer of basal cells occasionally positive for smooth muscle actin.

The optimal treatment is excision with wide free margins. Of note, extension of the tumor can be appreciated also at a great distance from the main mass. Recurrence has been reported [16].

Syringomatous adenoma has to be differentiated from tubular carcinoma (which rarely involves the nipple) and low-grade adenosquamous carcinoma (which occurs in the breast parenchyma) [16]. The latter has a similar tendency for local recurrence and minimal metastatic potential, and may be morphologically indistinguishable from syringomatous adenoma, being therefore the anatomical site the main determinant for differential diagnosis.

## 12.5 Breast Cancer Precursors

The introduction of mammographic breast cancer screening programs has dramatically increased the detection of risk indicators, premalignant/preinvasive lesions, thus posing important issues related to patient management. A multitude of proliferative hyperplastic and premalignant alterations have been documented not uncommonly occurring synchronously with invasive breast cancer.

Observational and correlative studies have identified some of these lesions as risk indicators or breast cancer precursors [51–55]. A risk indicator can be defined as lesions that have been reported to be associated with increased risk of breast cancer development, whereas breast cancer precursors are those preinvasive lesions with the potential to progress to an overtly malignant phenotype [14]. Typically, the latter group includes risk indicators that have been shown to be clonal, neoplastic proliferations and to have histologic, IHC, and molecular features identical to those of matched invasive breast cancers, either synchronous or metachronous [14]. Based on the observation that the chances of one of these precursors progressing to invasive breast cancer rarely, if ever equates to 100 %, these lesions are best named nonobligate precursors [14].

Observational and molecular studies have recently demonstrated that a family of *in situ* lesions of the breast coexist at frequencies that could not be justified by chance alone [56, 57].

This family encompasses CCLs, flat epithelial atypia (FEA), ADH, atypical lobular hyperplasia (ALH), lobular carcinoma *in situ* (LCIS) and low-grade DCIS, and invasive lesions reported to be found in association with these nonobligatory precursors (i.e., invasive tubular carcinoma, invasive cribriform carcinoma, classic invasive lobular carcinoma, and low-grade invasive ductal carcinoma) [14]. These lesions are characterized by low histologic grade, expression of hormone receptors, lack of HER2 expression or HER2 gene amplification, lack of high molecular weight cytokeratin expression, and by the presence of genetic aberrations usually found in low-grade breast cancers (i.e., deletions of 16q and gains of 1q) [14]. In fact, frequent co-existence of tubular carcinoma, lobular carcinoma *in situ*, and CLLs (the so-called "Rosen triad" [58]) is on record since late 1990s., the Nottingham group coined the term "low-grade breast neoplasia family" to refer to these lesions [56, 57, 59].

Although different stages of progression have been identified for low-grade lesions, until recently only high-grade DCIS was recognized as a precursor of high-grade breast cancer [24, 60]. It has been demonstrated, however, that microglandular adenosis, a lesion considered by many only to be hyperplastic and an incidental finding, is often associated with high-grade breast cancer and harbors genomic aberrations identical to those found in adjacent synchronous invasive cancers [61–63].

Based on the observations above, we can subdivide low-grade precursors from high-grade precursors [14]. Low-grade precursors are almost exclusively of luminal phenotype (ER-positive/HER2-negative) and give rise to luminal invasive breast cancers. By contrast, high-grade precursors are more heterogeneous, including luminal, HER2-positive, and triple negative (TN) lesions. Microarray-based gene expression analyses have confirmed that breast cancer evolution follows two main pathways according to histologic grade, as preinvasive and invasive lesions when subjected to hierarchical clustering analysis cluster together according to grade rather than stage [64]. Nevertheless, it has been later documented that progression from low to high grade is not an uncommon event within the luminal subtype [65]. Therefore, given that ER defines two fundamentally different large subgroups of breast cancer, a modified hypothetical model of breast cancer evolution has been proposed, which follows two main pathways according to ER pathway activation [14]. In this model, the ER-positive branch includes all well-established low-grade ER-positive preinvasive lesions and their low-grade ER-positive invasive counterparts, which may progress to high-grade ER-positive lesions. The ER-negative branch includes ER-negative DCIS and microglandular adenosis as preinvasive lesions, and invasive carcinomas, which are mostly high-grade, frequently HER2-positive and display high levels of genetic instability. It should be mentioned that a subset of low-grade ER- and HER2-negative carcinomas do exist and display low levels of genetic instability, such as adenoid cystic and secretory carcinomas. Of note, these special types of ER-negative carcinomas are often driven by specific fusion genes.

### 12.5.1 Low-Grade ER-Positive Precursors

#### 12.5.1.1 Columnar Cell Lesions

CCLs of the breast encompass a spectrum of lesions that feature distended acini lined by tightly packed columnar epithelial cells with apical snouts [66–68]. They can display varying degrees of cytological and/or architectural atypia, ranging from columnar cell change and hyperplasia to FEA and are frequently associated with microcalcifications (Table 12.3), thus justifying their increasing incidence as a result of mammographic screening [67]. Importantly, their behavior and significance are still poorly understood [69, 70] and this impacts on therapeutic decisions about interventions. Of note, complex architectural atypia such as well-defined micropapillae, rigid arches, and secondary lumen formation cannot be present in FEA; whenever these features are present, a diagnosis of ADH or low-grade DCIS according to the size of the lesion should be rendered.

**Table 12.3** Morphological details of columnar cell lesions

| | CCC | CCC-a | CCH | CCH-a |
|---|---|---|---|---|
| Architecture | TDLUs with variably dilated acini lined by one or two layers of epithelial cells<br>Flocculent secretion and luminal calcifications may be present | | TDLUs with variably dilated acini with stratification of with more than two cell layers. Small mounds, tufts, or abortive micropapillations may be formed. Abundant flocculent secretion is frequent, as well as luminal calcifications that may have the configuration of psammoma bodies | |
| Cytology | 1. Columnar epithelial cells with uniform, ovoid to elongated nuclei oriented in a regular fashion perpendicular to the basement membrane<br>2. Evenly dispersed chromatin, no conspicuous nucleoli<br>3. Mitotic figures rarely encountered | 1. Increase in the nuclear/ cytoplasmic ratio (akin to that seen in DCIS)<br>2. Nuclei may show stratification<br>3. Nuclear chromatin may be evenly dispersed or slightly marginated<br>4. Nucleoli variably prominent<br>5. Mitotic figures may be seen but uncommon | 1. Nuclei are ovoid to elongated and for the most part oriented perpendicular to the basement membrane<br>2. Some cells may have hobnail appearance<br>Crowding or overlapping of nuclei may give appearance of nuclear hyperplasia | 1. Increase in the nuclear/ cytoplasmic ratio (akin to that seen in DCIS)<br>2. Nuclei may show stratification<br>3. Nuclear chromatin may be evenly dispersed or slightly marginated<br>4. Nucleoli variably prominent<br>5. Mitotic figures may be seen but uncommon |

Legend: *a* atypia; *CCC* columnar cell changes; *CCH* columnar cell hyperplasia

A striking feature of CCLs is the constant expression of ER and PR and low Ki67 labeling indices. They also lack expression of HER2 and "basal" keratins [26, 66, 69, 70]. In accordance with this homogeneous immunoprofile, molecular studies have provided evidence that the majority of CCLs are clonal and neoplastic rather than hyperplastic [24, 26, 69, 71, 72]. Interestingly, the degree of genetic changes appears to mirror the degree of architectural and cytological atypia found in different types of CCLs [26, 71]. Allelic imbalances are most commonly located at 3p, 9q, 10q, 11q, 16q, 17p, and 17q [71, 72], while recurrent copy number alterations include losses of 16q and chromosome X and gains of 15q, 16p, 17q, and 19q [26]. These molecular analyses, combined with morphological and IHC observations have provided strong circumstantial evidence to suggest that CCLs are part of the above described "low-grade breast neoplasia family" [56, 57], and they constitute the first morphologically identifiable precursor of low-grade ER-positive breast cancers [69]. Indeed, CCLs frequently coexist with ADH, DCIS, and lobular neoplasia (LN) in the same breast or even in the same TDLU [56, 73] with whom they also share similar IHC profiles [57] and identical genomic aberrations [26, 71, 72].

Although there is scientific evidence that CCLs, in particular those with atypia (aka FEA), are nonobligate precursors of breast cancer, their rate of progression and their relative risk of subsequent invasive cancer is still a matter of debate. There is some evidence that the risk of developing invasive cancer conferred by a diagnosis of FEA as the most advanced lesion in a breast biopsy is low and perhaps comparable to that of UDH [74, 75]. Therefore, the best therapeutic option in this setting is yet to be demonstrated. It is unknown if surgical excision and/or hormonal prophylaxis should be offered. Adequate radiological–pathological correlation plays an important role in this setting, however, FEA on its own should not be routinely interpreted in the same way as ADH or LN [16].

#### 12.5.1.2 Lobular Neoplasia

The initial classifications for preinvasive lesions assumed that some breast cancers would arise from the ducts and others would arise from the lobules and therefore *in situ* proliferations were named "ductal" carcinoma *in situ* and "lobular" carcinoma *in situ* [14]. It is important to note that the terms "ductal" and "lobular" carcinoma have no specific implications with regard to the site of origin within the mammary ductal-lobular system. The seminal works of Wellings et al. [76, 77] have documented that the vast majority of preinvasive lesions of the breast would arise in the TDLUs. Despite this, which constituted the beginning of the end of the histogenetic implications of the ductal and lobular terminology [14, 78], the terms ductal and lobular have been perpetuated, as they do identify lesions with distinct morphological and molecular features, and, more importantly, different therapeutic implications.

Lobular carcinoma *in situ* (LCIS) is composed of a monomorphic population of generally small and loosely discohesive cells that expand the TDLUs with or without pagetoid involvement of terminal ducts [79–81]. ALH was coined to refer to a morphologically similar but less well-developed lesion, i.e., a partial involvement of acini by LN cells [82, 83]. Morphological distinction of these lesions, however, is somewhat arbitrary and partly subjective [14]. In 1978, Haagensen et al. [80] retrospectively analyzed 211 examples of *in situ* lobular proliferations and introduced the term LN to encompass both ALH and LCIS. The rationale for combining these lesions into a single category was based on the observation that in that study the microscopic distinction of ALH from LCIS based on morphological parameters was not found to have any value in predicting subsequent carcinoma [80]. Importantly, the levels of genetic instability found in ALH and LCIS appear to be similar and both ALH and LCIS display similar patterns of recurrent unbalanced chromosomal aberrations, including losses of 16q, 16p, and 17p, and gains of 6q [84]. In fact, the molecular data available to date provide support for the LN concept and for the notion that distinction between ALH and LCIS is largely quantitative (i.e., extent of disease) rather than qualitative. A recent study [85], however, demonstrated a surprisingly greater amount of copy number alterations in ALH than in LCIS. It is important to specify that even though the term LN helps to avoid somewhat arbitrary diagnostic decisions, there is strong evidence to suggest that LCIS carries a higher risk of breast cancer development than ALH (8–10 vs. 4–5, respectively [86]).

In a way akin to other lesions of the "low-grade breast neoplasia family," LN is characterized by ER and PR expression and is typically HER2-negative [81, 87]. From a genetic standpoint, LN harbors recurrent deletions of 16q and gains of 1q [84, 85].

ALH and LCIS have been accepted as risk indicators of breast cancer development, either in the ipsi-or contralateral breast; however, the risk is higher in the ipsilateral breast [88]. Conversely, their role as nonobligate precursors of ILC has been a matter of contention [89, 90]. The identification of CDH1 (E-cadherin gene) as the target gene of 16q deletions in lobular carcinomas [51, 91, 92] was crucial in elucidating the role of LN in breast cancer progression. Vos et al. [93] demonstrated the presence of identical *CDH1* truncating mutations in matched LCIS and adjacent ILCs, providing strong circumstantial evidence to suggest that at least some LNs may evolve to ILCs [87]. This hypothesis is further corroborated by the similarity between LN and matched ILCs at the genetic level

[94–96]. A transcriptomic analysis focused on normal epithelium, LCIS, and ILC has recently identified differentially expressed genes between the three groups and identified 169 candidate precursor genes, which likely play a role in LCIS progression, including PIK3R1, GOLM1, and GPR137B. Interestingly, these potential precursor genes map significantly more frequently to 1q and 16q, regions frequently targeted by gene copy number alterations in LCIS and ILC [97].

Taken together, the clinical and molecular evidence available suggests that ALH and LCIS are clonal and neoplastic, and that these lesions are both risk indicators and nonobligate precursors of breast cancer. It should be noted, however, that as a group the proclivity of LN to progress to invasive breast cancer is low. Conservative management of these lesions remains the mainstay of treatment [98].

LCIS can be subclassified as classic and variants, of which the most important are the florid and pleomorphic. Classic LCIS refers to the lesion above described, and should not display marked pleomorphism, comedonecrosis, or extreme distension of the involved TDLUs, and is best managed as an indolent disease. The pleomorphic variant of LCIS is defined by the presence of marked nuclear pleomorphism and is described below. More recently, a florid variant has been cataloged, which is cytologically similar to classic LCIS, but induces extreme distension of the TDLUs, with frequent comedonecrosis. Observational analyses have stressed that florid LCIS is often associated with invasive lobular carcinoma, while molecular analyses have documented that this lesion is molecularly as advanced as the pleomorphic variant [99]. Therefore, it has been suggested that florid LCIS should be treated more aggressively and completely resected. It should be noted, however, that histologic subtyping of LCIS has not been fully integrated in the last World Health Organization (WHO) classification of breast tumors, in part due to its subjective nature and the lack of prospective validation of its clinical relevance.

#### 12.5.1.3 Atypical Ductal Hyperplasia (ADH) and Low-Grade DCIS

Low-grade DCIS is defined by a proliferation of monomorphic cells with uniform-sized nuclei and rare mitotic figures growing in arcades, micropapillae, cribriform, or solid patterns (Fig. 12.4; Table 12.4) [16]. Lesions classified as ADH display some but not all morphological features of low-grade DCIS. Similarly to the distinction between ALH and LCIS, the distinction between low-grade DCIS and ADH is mainly quantitative and somewhat subjective [16]; however, the risk of invasive breast cancer development reported for low-grade DCIS is higher than that for ADH [14, 37, 52, 83, 100, 101]. Whenever asked, breast pathology experts do acknowledge that they use both qualitative and quantitative criteria for this distinction in their practice. Despite being arbitrarily set, quantitative criteria are important to avoid overtreatment of minute lesions in the era of mammographic screening.

ADH has long been recognized as both a risk indicator and a nonobligate precursor of low-grade DCIS and invasive breast cancer [52, 86]. In fact, the similarities between ADH and low-grade DCIS are striking. Immunophenotypically, both lesions are consistently positive for ER and PR, and lack HER2 overexpression and gene amplification, and expression of high molecular weight cytokeratins (Table 12.2), thus are part of the "low-grade breast neoplasia family" [52, 56, 57, 86]. Close similarities have also been repeatedly reported at the genetic level and these data have corroborated the idea that ADH is neoplastic and a nonobligate precursor with an unequivocal role in the development of low-grade DCIS and invasive lesions [20, 102, 103]. ADH and low-grade DCIS harbor allelic imbalances at similar frequencies [104, 105]. In addition, the study of matched samples of ADH, DCIS and invasive lesions identified concordant allelic imbalances. Recurrent regions displaying

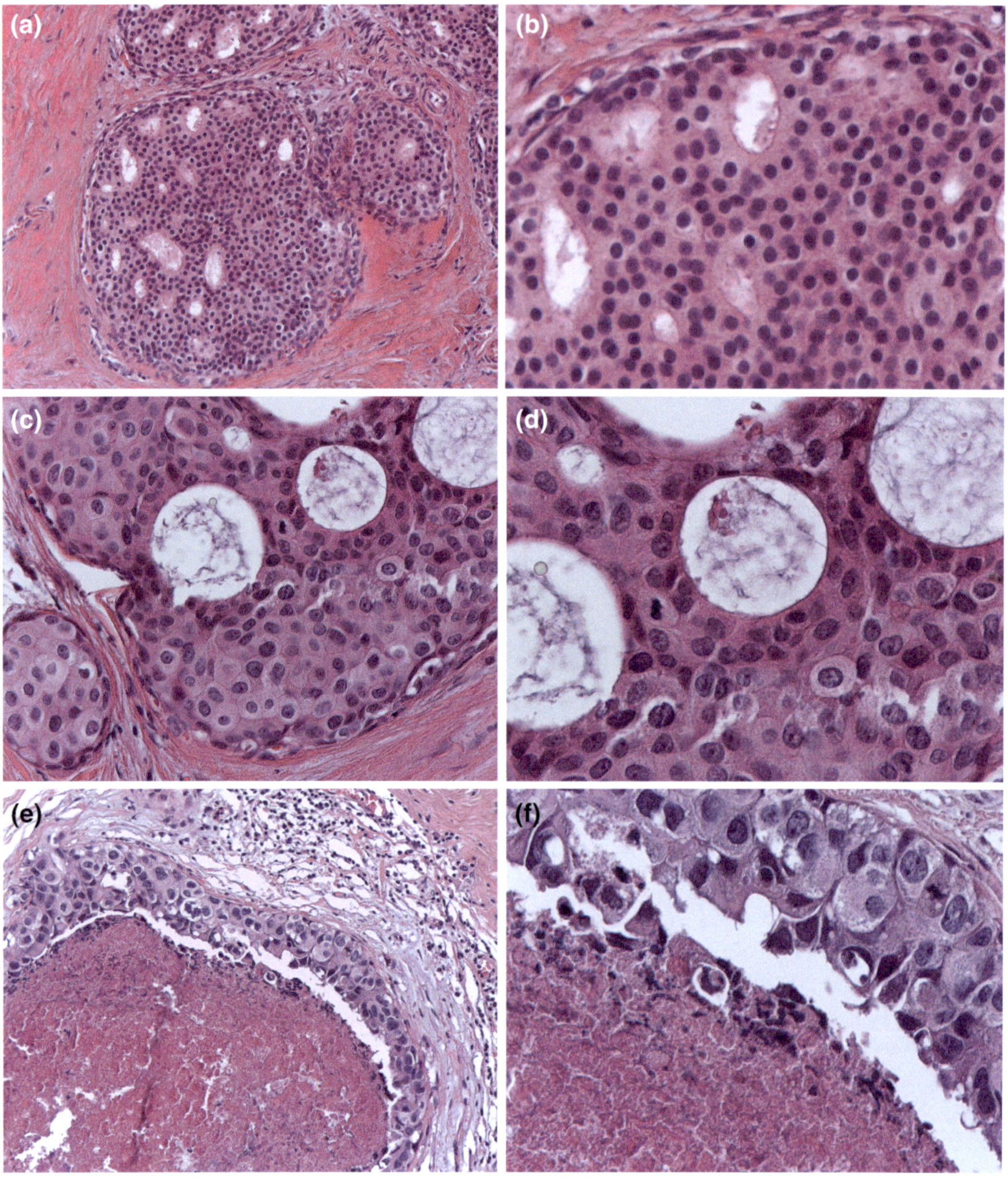


**Fig. 12.4** DCIS (**a**, **b** low nuclear grade; **c**, **d** intermediate nuclear grade; **e**, **f** high nuclear grade)

LOH include loci on 1q, 16q, and 17p [20, 102, 103]. CGH studies have confirmed these observations and demonstrated that ADH and low-grade DCIS are clonal, neoplastic, and have similar number, type, and complexity of unbalanced chromosomal aberrations [22, 24, 106, 107]. Not surprisingly for lesions of low histologic grade, ADH and low-grade DCIS are characterized by recurrent losses of 16q and 17p and gains of 1q. In conclusion, at the molecular level, ADH and low-grade DCIS seem to be nearly, if not completely, identical.

**Table 12.4** Morphological features for diagnosis of ductal carcinoma *in situ* (DCIS)

| | Low-grade DCIS | DCIS of intermediate grade | High-grade DCIS |
|---|---|---|---|
| Architecture | Any<br>(very frequently cribriform pattern; solid and/or micropapillary pattern may also be present) | Any<br>(solid, cribriform or micropapillary pattern) | Any<br>(solid, cribriform or micropapillary pattern) |
| Cytologic features | Monotonous and uniform rounded cell with round nuclei<br>Minimal increase in nuclear-cytoplasmic ratio<br>Regular chromatin pattern, inconspicuous nuclei | Mild to moderate variability in size, shape and placement<br>Cell polarization not well developed as in low nuclear grade<br>Nuclei with variably coarse chromatin and variably prominent nucleoli | Highly atypical cells<br>Markedly pleomorphic nuclei, poorly polarized with irregular contour<br>Coarse, clumped chromatin with prominent nucleoli |
| Necrosis | Uncommon but possible* | May be present | Frequently present* |
| Mitosis | Absent or rare | May be present | Usually common* |
| Typical pattern of calcifications | Calcification of psammomatous type common | Either similar to low grade or amorphous calcifications like in high grade | Amorphous calcification often associated with necrotic intraluminal debris |

*Not obligate criterion for diagnosis or exclusion of diagnosis

### 12.5.2 High-Grade Precursors

#### 12.5.2.1 High-Grade DCIS

High-grade DCIS is composed of a population of atypical cells displaying marked nuclear pleomorphism, arranged in multiple architectural patterns, including solid, cribriform, and micropapillary [16] (Fig. 12.4; Table 12.4). Comedonecrosis is often present, but its detection does not necessarily equate to a diagnosis of high-grade DCIS [16]. There is one particular scenario, called Paget disease of the nipple, in which malignant and mostly HER2-positive glandular epithelial cells populate the squamous epithelium of the nipple and this is almost always associated with an underlying high-grade DCIS, which typically involves more than one collecting ducts and also more distant ducts deep in the breast gland (Fig. 12.5). An associated infiltrating carcinoma can also be seen in up to 90 % of cases [16].

High-grade DCIS is a paradigmatic lesion for the concept of nonobligate precursor of invasive breast cancer, given that it has a proclivity to progress to invasive cancer but does not always [14]. A diagnosis of high-grade DCIS is associated with a significantly higher risk of invasive breast cancer development and with earlier recurrences than low-grade DCIS [37, 52, 86, 108]. In fact, the risk of an ipsilateral breast recurrence either in the form of DCIS or invasive breast cancer after local excision alone of high-grade DCIS is estimated at around 15 % at 5 years [108].

As mentioned above, the immunoprofile and patterns of genetic aberrations of high-grade DCIS are more heterogeneous than those observed in low-grade DCIS [14]. Importantly, despite the greater complexity of the pattern of genetic aberrations found in high-grade than in low-grade DCIS, deletions of the whole of 16q are found in a minority of the former suggesting that the majority of high-grade DCIS arise either *de novo* or from a precursor other than ADH/low-grade DCIS [14]. However, most likely, most ER-positive DCIS may still display 16q whole arm loss and possibly originate from low-grade lesions. In contrast, ER-negative high-grade DCIS probably arise

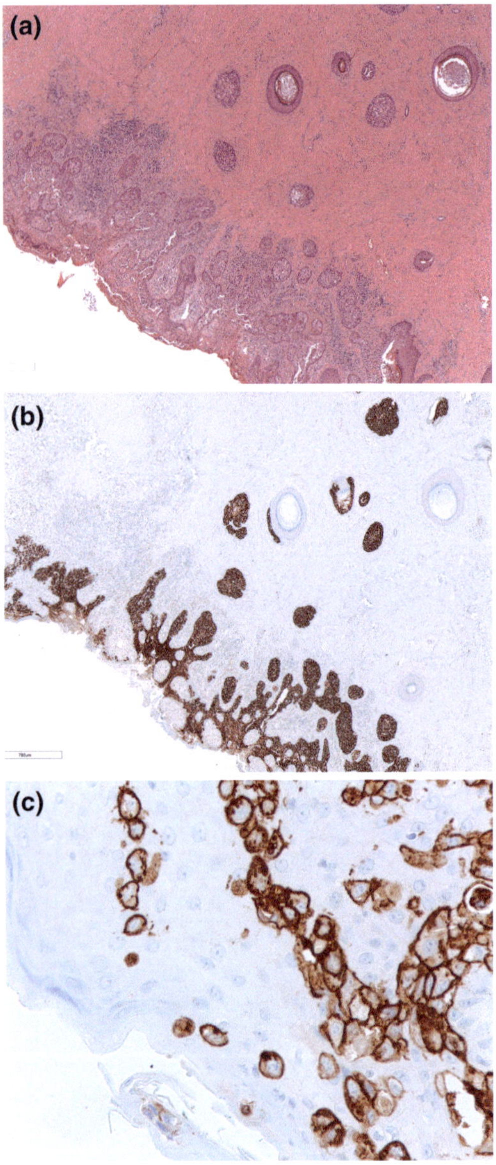


**Fig. 12.5** Paget disease of the nipple (**a**, **b** H&E; **c** HER2)

either *de novo* or from a precursor other than ADH/low-grade DCIS.

As a last remark, intratumor heterogeneity has been definitely documented in DCIS [109, 110]. For instance, it is not uncommon to find different nuclear grades in the same lesion [109], a fact that has been associated with our poor ability to predict clinical behavior of DCIS based on morphological parameters [111, 112]. aCGH and fluorescence *in situ* hybridization (FISH) analyses of matched DCIS and invasive carcinomas have also demonstrated the existence of subclones in DCIS and that stromal invasion may occur due to a clonal selection process [110].

#### 12.5.2.2 Pleomorphic Lobular Carcinoma *In Situ*

Pleomorphic lobular carcinoma *in situ* (PLCIS) typically displays cytological and architectural features of both classic LCIS and high-grade DCIS [113–115]. The growth pattern exhibits large, discohesive cells, often with finely granular apocrine cytoplasm and intracytoplasmic vacuoles with marked nuclear atypia and pleomorphism. Foci of classic LN can be found in association with PLCIS [113, 116–118]. PLCIS is characterized by moderate proliferation index and may show comedonecrosis [113–115]. These lesions often show low levels of ER and PR expression, lack of E-cadherin expression, and occasionally display HER2 overexpression and amplification [115–119]. Although traditionally described in association with its invasive counterpart (pleomorphic invasive lobular carcinoma, ILC) [113, 114, 120], PLCIS can also be found as an isolated lesion [115].

Molecular studies have demonstrated that PLCIS and pleomorphic ILC harbor remarkably similar genetic profiles, and that both have the hallmark features of lobular carcinomas, namely 16q loss, 17p loss, 1q gain, and E-cadherin loss of expression [117, 121]. Of note, PLCIS and pleomorphic ILC display additional genetic aberrations, including amplification of key oncogenes, deletion of 8p and 13q, and gains of 8q. These alterations may account for the higher nuclear grade and reported more aggressive clinical behavior of PLCIS and pleomorphic ILCs [117, 118, 121]. Chen et al. [116] analyzed a large series of PLCIS and classic LCIS confirming the similarities between these lesions at the genetic level. More specifically, PLCIS samples were divided into two groups, non-apocrine and apocrine: the former had similar levels of genetic instability as those observed in classic LCIS, whereas the latter displayed more and specific genomic changes as amplification at

17p11.2–17q12 and 11q.13.3, gain of 16p and losses of 11q and 13q [116]. These alterations were very much similar to those previously described [117].

The available evidence suggests that PLCIS is a genetically advanced lesion and is likely to be a nonobligate precursor of pleomorphic ILC [116–118, 121].

#### 12.5.2.3 Microglandular Adenosis

Microglandular adenosis (MGA) is an uncommon entity characterized by a haphazard proliferation of homogeneous small and rounded glands lined by a single layer of epithelial cells around lumen containing secretions and/or calcifications (Fig. 12.6) [122–124]. It has long been described to occur in association with invasive carcinomas (Fig. 12.6) [123, 125, 126], however, the precursor nature of this lesion was called into question by some experts in the field [124]. This lesion displays a typical immunophenotype, being strongly positive for S100 protein and negative for ER, PR, and HER2 expression (TN phenotype), an immunophenotype that is also shared by concurrent invasive carcinomas, which are often of high histologic grade and frequently express basal markers [61, 127]. aCGH-based analyses have identified concordant copy number changes in this spectrum (MGA, atypical MGA, associated DCIS, and invasive carcinomas), providing molecular evidence of the neoplastic and nonobligate precursor nature of a subset of MGA [61].

## 12.6 Papillary Lesions

Papillary breast lesions encompass a group of lesions that share a typical architectural pattern, being defined as epithelial proliferations supported by fibrovascular stalks with or without a layer of myoepithelial cells occurring anywhere

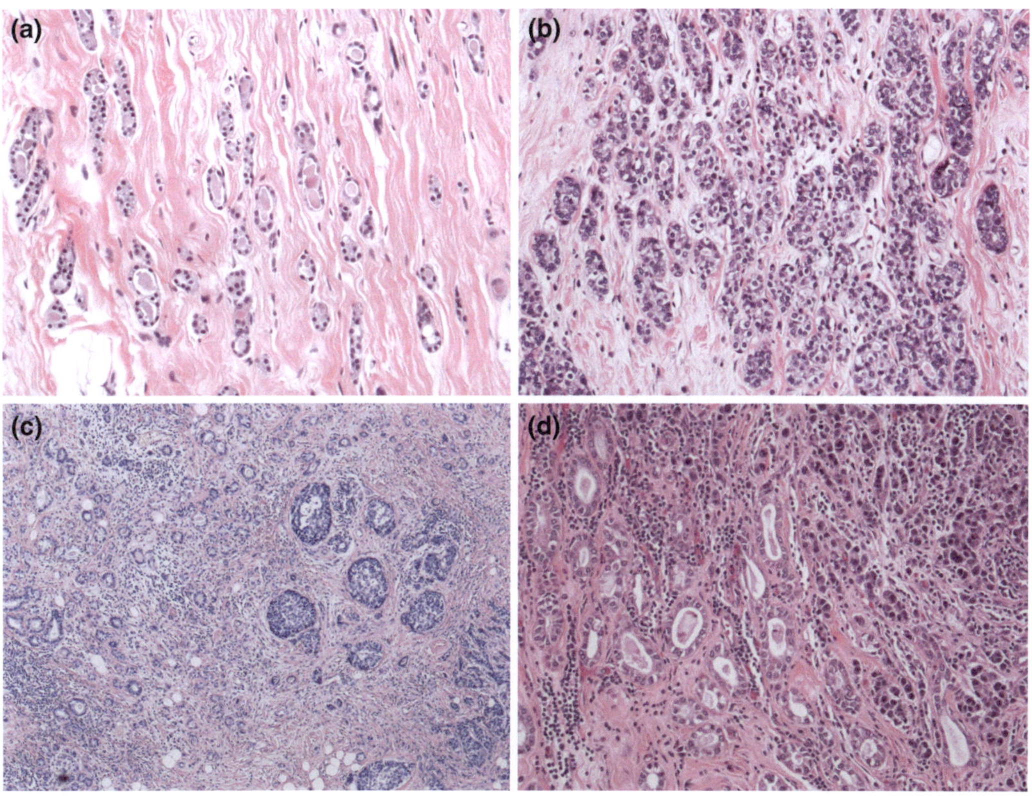


**Fig. 12.6** MGA (**a** typical; **b** atypical; **c, d** two examples with concurrent invasive carcinoma)

in the ductal system (from the large retroareolar ducts to the TDLU) [16]. They can be benign (intraductal papilloma), atypical (atypical papilloma), or carcinomas (encapsulated papillary carcinoma, solid papillary carcinoma, and invasive papillary carcinoma).

### 12.6.1 Intraductal Papilloma

An intraductal papilloma is defined as a proliferation of epithelial and myoepithelial cells overlying fibrovascular stalks, thus creating an arborescent structure within the lumina of a duct [16]. They can be central (large ducts involved) or peripheral (arising in the TDLU). Central papillomas usually present with unilateral sanguineous nipple discharge, while palpable masses are less frequent. Mammography can detect a circumscribed retroareolar mass with dilated duct, whereas small lesions can be occult. Calcifications are rare. Galactography, by detecting the irregular filling defect and obstruction may be of help to the surgeon to localize the discharging duct before intervention [16].

Histopathologically, the arborescent structure may present a coexistence of papillary and ductal patterns; whenever the latter predominates and is also associated with marked sclerosis, the term sclerosing papilloma is used.

Papillomas can undergo a series of changes, such as inflammation, necrosis, and metaplasia (of different types: apocrine, squamous, chondroid, osseous, mucinous). In addition, the whole range of atypical/neoplastic proliferations may arise in a papilloma or secondarily involve it.

Peripheral papillomas are often clinically occult and multiple, involve terminal ducts and TDLUs rather than large ducts, and nipple discharge is much less frequent. Morphologically, they share the same architectural pattern and histologic features; the main distinction resides on the more frequent association with concomitant sclerosing adenosis, radial scars, UDH, ADH, and *in situ* or invasive carcinomas [16].

### 12.6.2 Encapsulated Papillary Carcinoma and Solid Papillary Carcinoma

Breast papillary carcinomas are histologically defined as malignant neoplasms with a papillary growth pattern, where the epithelial cells reside directly in fine and delicate papillary fronds, with no underlying myoepithelial cell layer. A papillary carcinoma that is clearly intraductal (i.e., surrounded by a ductal wall with a myoepithelial cell layer) is considered as a papillary pattern of DCIS. The spectrum of breast papillary carcinomas encompasses, however, three other histologic entities, namely encapsulated papillary carcinoma (EPC), solid papillary carcinoma (SPC), and invasive papillary carcinoma (IPC) [128, 129]. IPCs will be discussed in the section of special histologic types of invasive carcinoma. Here we discuss the two controversial papillary lesions (EPC and SPC), whose status as intraductal or invasive lesions has been a matter of debate.

EPC is a well-circumscribed nodule of papillary carcinoma surrounded by a thick fibrous capsule. In EPCs, the neoplastic cells are of low to intermediate nuclear pleomorphism and arranged in papillary fronds in the majority of cases; however, areas with cribriform and/or solid patterns can be found [16, 130]. Although initially perceived as a variant of *in situ* papillary carcinoma, recent studies have demonstrated that EPCs consistently lack a myoepithelial cell layer not only in the papillary fronds, but also at the periphery of the tumor nodules [129, 131–133]. Therefore, what was thought to be a malignant papillary intraductal/intracystic neoplasm, is now best considered as an invasive carcinoma with circumscribed borders, slow growth, and surrounded by a reactive fibrous capsule. In agreement with this notion, recurrence in the chest wall and rare lymph node metastasis with the same papillary architecture have been documented [129, 134]. Despite this invasive nature, its long-term prognosis is excellent and comparable to DCIS; thus, most experts still recommend EPC to be staged and clinically managed as an *in situ* carcinoma.

SPC is also a well-circumscribed lesion that is densely cellular and composed of one or multiple expansile nodules, sheets, or coalescent papillae of ovoid-to-spindle-shaped cells growing in a solid pattern. Frequently, the fine and delicate fibrovascular cores are not evident at first glance. Neuroendocrine differentiation is a frequent feature, which may be of diagnostic utility [135]. In addition, when SPC is associated with a clearly invasive carcinoma, the latter often displays neuroendocrine features [136]. In a way akin to EPC, the classification of SPCs as invasive or *in situ* disease remains a matter of controversy. For instance, a myoepithelial cell layer may also be absent in solid-papillary nodules, but if these nodules are histologically well circumscribed, the lesion can still be considered *in situ*. Nevertheless, an invasive counterpart of solid-papillary carcinoma is accepted by many experts and it has been coded in the last WHO classification of breast tumors [16]. Solid nodules in a geographic, jigsaw-like pattern within a fibrous, focally desmoplastic background, coupled with absence of myoepithelial cells, may suggest an invasive disease. As EPCs, these tumors may also have the potential to disseminate to axillary lymph nodes; distant metastases, although rare, are on record [129, 130]. However, it has been suggested that, if there is uncertainty about invasion, SPCs should be regarded for staging purposes as *in situ* carcinoma [16].

Finally, EPCs and SPCs are not uncommonly associated with a clearly invasive carcinoma component. This invasive component may be of no special type, usually of low to intermediate histologic grade, or of some special types, in particular carcinomas of mucinous type or with neuroendocrine differentiation. Whenever a clearly invasive carcinoma component is present, its size and features should be used for staging, histologic grading, and IHC profiling purposes.

From a molecular standpoint, a recent exploratory analysis of the transcriptomic profiles of all three subtypes of PCs has revealed that despite displaying similar patterns of gene copy number alterations, transcriptomic profiles showed significant differences among EPCs, SPCs, and IPCs: EPCs expressed a subset of genes involved in cell migration at significantly lower levels than SPCs and IPCs, and SPCs displayed transcriptomic and IHC features consistent with those associated with neuroendocrine differentiation. One may hypothesize that such differences may account for their different histologic features [137].

A recent study described a high-grade variant of EPC, which was frequently ER-negative and, in one case, had a fatal outcome despite the lack of usual stromal invasion in the primary lesion [138]. We argue that perhaps those cases with papillary architecture, marked nuclear pleomorphism, and ER-negative phenotype would be best categorized as invasive carcinomas of no special type (IC-NSTs) with papillary features to avoid under-treatment.

## 12.7 Invasive Carcinoma

An invasive breast carcinoma is defined by a morphological infiltrative pattern, often coupled with a desmoplastic stromal reaction, and/or, with very few exceptions, the lack of a myoepithelial cell layer.

The introduction of annual screening programs has dramatically changed the presentation and natural history of breast cancer over the past decades. In screening populations smaller lesions are usually diagnosed at lower stages, allowing more conservative surgical interventions. Invasive carcinomas are detected at mammography as opacities, distortions, or stellate lesions with or without calcifications. This pattern largely mirrors what pathologists observe at gross analysis where invasive carcinomas are usually described as opaque whitish nodules, distortions, or lesions. Occasionally, well-circumscribed lesions of soft texture and/or gelatinous appearance can be encountered, typically for mucinous, papillary, and some high-grade carcinomas that at ultrasound examination may be misinterpreted as dense cysts or benign nodules.

The WHO classification [16] is currently based on morphological features and roughly separates invasive carcinoma of no special type (IC-NST), which is the commonest form of breast cancer, from a large group of so-called "special histologic types," which together account for approximately 25 % of all newly diagnosed breast carcinomas. It is important to note that histologic subtyping holds prognostic significance, accordingly to earlier studies conducted by Elston and Ellis [139], where it was shown that patients affected by some special types of breast cancer have a better or worse outcome than those with an IC-NST (formerly known as invasive ductal carcinoma of no special type, IDC-NST). The St. Gallen International Expert Consensus stressed that attention should be posed on the recognition of histologic types, as some of them, like tubular and cribriform carcinomas, have an excellent prognosis and may be treated by surgery alone [140]. On the other hand, interobserver agreement rates of histologic subtyping are modest and the existence of some entities is controversial [16].

Since the last decade of the twentieth century, the application of high-throughput molecular techniques, in particular microarray-based gene expression profiling, has definitely reshaped our understanding of breast cancer. Studies using this set of techniques have changed some aspects of breast cancer classification, and provided a working model for the taxonomy for breast cancer [141]. Undoubtedly, the most important contribution of microarrays to breast cancer research has been the realization that breast cancer is not a single disease. The seminal expression microarray-based class discovery studies by the Stanford group devised the so-called molecular classification of breast cancer, based on the identification of the "intrinsic gene list" and subsequent hierarchical clustering of cases based on the expression of genes pertaining to this list [142]. This approach divides breast cancers into two main groups: ER-positive and ER-negative disease. The ER-positive group comprises the luminal tumors: these typically show the expression of ER and of genes pertaining to ER pathway, and other transcripts usually found in luminal epithelial cells [141, 142]. Luminal cancers can be further subdivided into two distinct subgroups: luminal A (characteristically showing high levels of ER and ER pathway activation) and luminal B cancers (which show lower levels of ER and high levels of genes pertaining to the proliferation cluster) [141, 142]. The ER-negative cluster encompasses HER2, and basal-like cancers. The microarray-defined group of HER2-expressing tumors is characterized by high levels of expression of genes pertaining to the HER2 amplicon, including *HER2*, *GRB7*, *GATA4*, and a high level of NF-κB activation [141, 142]. ER-positive HER2-amplified cancers often cluster together with luminal B tumors [141]. The last group comprises cancers that lack ER and HER2 expression and have been named basal-like carcinomas, as they are characterized by the expression of genes known to be preferentially expressed by basal/myoepithelial cells, such as CK 5 and 17, integrin 4, laminin, c-KIT, α6 integrin, metallothionein IX, fatty acid binding protein 7, P-cadherin, EGFR, and NF-κB [141, 142].

This molecular classification is not a mere academic exercise, given that it holds prognostic significance, with tumors of luminal A subtype being associated with the best outcome, and tumors of basal-like or HER2 subtype with the worst outcome [143–145].

In addition, subsequent studies have refined this classification and recognized further subgroups. For instance, the ER-negative group beyond the originally described basal-like and HER2-enriched subtypes [142, 146] has been shown to encompass other subgroups. These include claudin-low tumors, of which 60–70 % are of TN phenotype (i.e., lacking ER, PR, and HER2 expression) and are potentially enriched for the so-called cancer stem cells [147] and the molecular apocrine subtype, characterized by the expression of androgen receptor, transcriptomic features consistent with the activation of the androgen receptor pathway and poor clinical outcome [148–150].

It has long been debated about carcinomas of basal-like intrinsic subtype being equal to IHC-defined TN breast cancers (TNBCs). Although there is a wide overlap between the two entities (about 80 %), it has been demonstrated that one is not the surrogate of the other [151–153]. The clinically-defined umbrella "TN subgroup" is more heterogeneous than the basal-like intrinsic subtype [151, 154]. While perhaps the major feature of basal-like subtype is a notable association with *BRCA1*-mutant patients ("BRCAness") [155–158], recent studies have shown the existence within TNBCs of either six (basal-like I, basal-like II, mesenchymal, mesenchymal stem-like, immunomodulatory, and luminal androgen receptor) [159] or four subtypes (luminal androgen receptor, mesenchymal, basal-like immune-suppressed, and basal-like immune-activated) [160]. These TNBC classification systems also provide an interesting framework to match subtypes of the disease with specific targeted therapies [160, 161], given that the six-subtype classification is associated with distinct responses to neoadjuvant chemotherapy [162].

A further step about molecular classification of breast cancer has been added by the Molecular Taxonomy of Breast Cancer International Consortium (METABRIC), which has reported on a breast cancer classification based on a genomic analysis that integrates gene expression and genome-wide copy number alterations (CNAs) [163]. From the first study, which included the analysis of approximately 2000 tumors, it has become apparent that the most parsimonious number of molecular subtypes of breast cancer is ten [163]. ER-positive and ER-negative tumors differ also in the pattern and type of gene CNAs: while the majority of ER-positive breast cancers (grade 1, 80 %; grade 3, 50 %) are characterized by concurrent deletions of 16q and gains of 1q, these concurrent alterations appear to be remarkably rare in ER-negative tumors [65]. On the other hand, TNBCs harbor complex patterns of copy number gains and losses throughout the genome [164, 165]. The proponents of this classification system have further developed a gene expression-based approach to classify breast carcinomas into the ten integrative clusters [165]. This methodology, in a way akin to PAM50 for the "intrinsic" subtypes [146] could enable a simpler translation of such integrative analysis that otherwise would require both gene expression and CNA information. The analysis of 7544 BCs with the new classifier has revealed that the METABRIC classification may be more informative in the contextualization of the genomic drivers identified by massively parallel sequencing studies of breast cancer [165] than the "intrinsic" subtypes [166].

Although the molecular classification of breast cancer has been largely adopted by the scientific and clinical community, the histologic classification should not be neglected. The St. Gallen International Expert Consensus has indeed suggested that early breast cancer therapy should be defined according to the intrinsic subtypes [140]. Nevertheless, as mentioned above, St. Gallen experts advised that some special histologic types may be particularly endocrine sensitive and therefore may not need chemotherapy. In addition, in the context of TN disease, proper histologic classification is of extreme importance to avoid unnecessary systemic treatment for patients with some low-grade forms of TN carcinoma, such as acinic cell, adenoid cystic, and secretory carcinoma. Therefore, despite the great value of the molecular classification for treatment decision, histologic classification still plays a role and should not be relegated to an academic exercise.

### 12.7.1 Invasive Carcinoma of no Special Type (IC-NST)

Invasive carcinoma of no special type (IC-NST), despite accounting for the vast majority (up to 75 %) of all invasive breast carcinomas, represents a diagnosis of exclusion, as it is best defined as any epithelial invasive neoplasm that does not fulfill criteria for any of the special histologic subtypes [16]. This entity was formerly known as invasive ductal carcinoma of no special type. As explained above, the ductal/lobular terminology has no histogenetic implications, thus the WHO committee decided to change the nomenclature.

Due to this negative definition, morphologic heterogeneity is to be expected. An IC-NST can be composed of several growth patterns, such as duct-forming structures of variable size or large solid sheets of cells. To circumvent this histologic heterogeneity and provide some prognostic information, histologic grading according to the Elston and Ellis grading system, also known as Nottingham grading system [167] plays a major role in this subtype of breast carcinoma (see below).

The stromal compartment is usually desmoplastic and may be populated by inflammatory cells, such as lympho-mononuclear cells that infiltrate the tumor (tumor infiltrating lymphocytes, TILs). Recently, due to the increasing knowledge on the immune response elicited by tumors and the availability of immune therapies, attention has been given to the TILs in breast cancer. Presence of TILs has been reported in different breast cancer subtypes: brisk lymphocytic infiltrates represent, for instance, one of the typical features of basal-like and BRCA1-associated breast carcinomas [168], and increased lymphocytic infiltration into tumors has been associated with ductal histology, high grade, absence of expression of hormone receptors or high expression of the proliferation antigen Ki67 [169, 170]. Lymphocyte-rich tumors have been associated with effective therapeutic responses and favorable clinical outcomes [171–174]. Multiple retrospective analyses of prospective clinical trial samples have provided level I evidence for the assessment of TILs as a prognostic factor in breast cancer, and international guidelines for their assessment have been published [175]. It should be noted, however, that TILs are a prognostic factor in patients with TN or HER2-positive tumors treated with chemotherapy. Therefore, for the time being, TIL assessment may not have a clear clinical utility, as the management of these patients is still the same independent of the presence or lack of TILs. Perhaps, the introduction of immune therapies or distinct chemotherapies for those patients that lack an immune response may change this scenario.

In a way akin to the morphological heterogeneity, IC-NSTs is also heterogeneous at the immunophenotypical and molecular level, as highlighted by the molecular classification of breast cancer, which was originally described in a cohort of IC-NSTs and a few lobular carcinomas [142]. The vast majority is ER-positive (approximately 75 %), 15 % of cases may be HER2-positive and the remaining cases fall into the category of TNBC. As for histologic special types, definition of prognosis is based on evaluation of morphologic features (size, histologic grade, mitotic index, lymph node status), immunophenotypical profiling (assessment of ER, PR, proliferation index by Ki67), and genomics (*HER2* gene status, use of gene expression signatures), which will be discussed more in detail in the next sections.

IC-NSTs can be admixed to any histologic special type and the WHO classification has defined some arbitrary cut-offs. A 50 % cut-off is adopted (i.e., the special type component has to represent at least 50 % of the lesion to allow for a diagnosis of invasive mixed carcinoma [16]). When the special type makes up for less than 50 % of the tumor, a diagnosis of IC-NST with a focal special type component should be rendered.

### 12.7.2 Special Types of Invasive Carcinoma

Breast cancer special types account for up to 25 % of all breast cancers. The WHO classification recognizes the existence of at least 20 entities, which are mainly defined by unique histologic features [16]. It is important to note that histologic subtyping holds some prognostic significance: patients affected by some special histologic types do differently than those affected by IC-NST [139].

It should be noted that the molecular classification was originally devised for IC-NSTs and a small subset of ILCs, however, a subsequent study [176] analyzed a series of 113 tumors from 11 special histologic types of breast cancer and provided the first transcriptomic description of histologic special types. Importantly, gene expression [176–181] and aCGH studies [177, 180, 182–184] of special histologic types of breast cancer have revealed that at the genomic and transcriptomic level, tumors from each of the special histologic types of breast cancer are more homogeneous amongst themselves than IC-NSTs. In addition, some special histologic types appear to be almost exclusively ER-positive (micropapillary, mucinous, tubular/cribriform, lobular and papillary carcinomas), whereas others (adenoid cystic, secretory and metaplastic breast cancers) are uniformly ER-negative [176, 185–187].

Genotypic–phenotypic correlations between specific genetic aberrations and special histologic types of breast cancer have emerged [185]. For instance, adenoid cystic carcinomas and secretory carcinomas are underpinned by the recurrent fusion genes *MYB–NFIB* [188] and *ETV6–NTRK3* [189], respectively, and lobular carcinomas are underpinned by loss of function of E-cadherin [187, 190]. Furthermore, micropapillary [183, 184, 191], mucinous [182], and adenoid cystic [192] carcinomas display different patterns of genomic alterations when compared to grade- and ER-matched IC-NSTs. Finally, the genes whose expression determines the clinical behavior of special types of breast cancer may differ from those implicated in the prognosis of IC-NSTs, as exemplified by the relatively poor discriminatory power of prognostic gene signatures observed in special types of breast cancer [182, 193, 194].

Below we provide an overview of some special histologic types where morphologic features and genomic make-ups are described.

#### 12.7.2.1 Lobular Carcinoma

Invasive lobular carcinomas (ILC) is the most frequently diagnosed special histologic type and the second most frequently diagnosed form of invasive breast cancer, accounting for about 10–15 % of all newly diagnosed breast cancers [16]. It is characterized by small discohesive neoplastic cells invading the stroma in a single cell-file (aka "indian file") pattern (classic type) (Figs. 12.7 and 12.8) [79]. ILC variants have also been described including the solid, alveolar, trabecular, and pleomorphic patterns. The solid pattern features sheets of cells, whereas in the alveolar variant we encounter globular aggregated of up to 20 cells. In both cases cells show the typical discohesiveness of the classic variant, but in the solid variant, more significant pleomorphism and higher mitotic activity may be present [16]. The pleomorphic variant displays the distinctive growth pattern of lobular carcinoma but with a marked degree of cellular atypia and nuclear pleomorphism [16] and may show apocrine or histiocytoid differentiation [16]. Signet ring cells are typically encountered in all ILC variants, if making up most of the tumor volume a signet ring cell variant may be diagnosed.

Lack of E-Cadherin protein expression is the hallmark feature of ILC (Fig. 12.7), also encountered in the *in situ* lesion (see section on LN) and represents the underlying reason for the discohesive growth pattern [16, 78]. This phenotype is determined by inactivating *CDH1* mutations in the large majority of cases. The *CDH1* gene is located on 16q, thus truncating mutations, loss of 16q, gene promoter methylation and/or transcriptional alterations co-occur leading to biallelic inactivation of the gene and loss of protein expression [78]. Interestingly, CDH1 inactivation may not only be responsible for the histologic pattern, but also for the metastatic behavior of ILCs, which more commonly metastasize to anatomical sites such peritoneum, meninges, and gynecological and gastrointestinal organs. In a mouse model where *CDH1* and

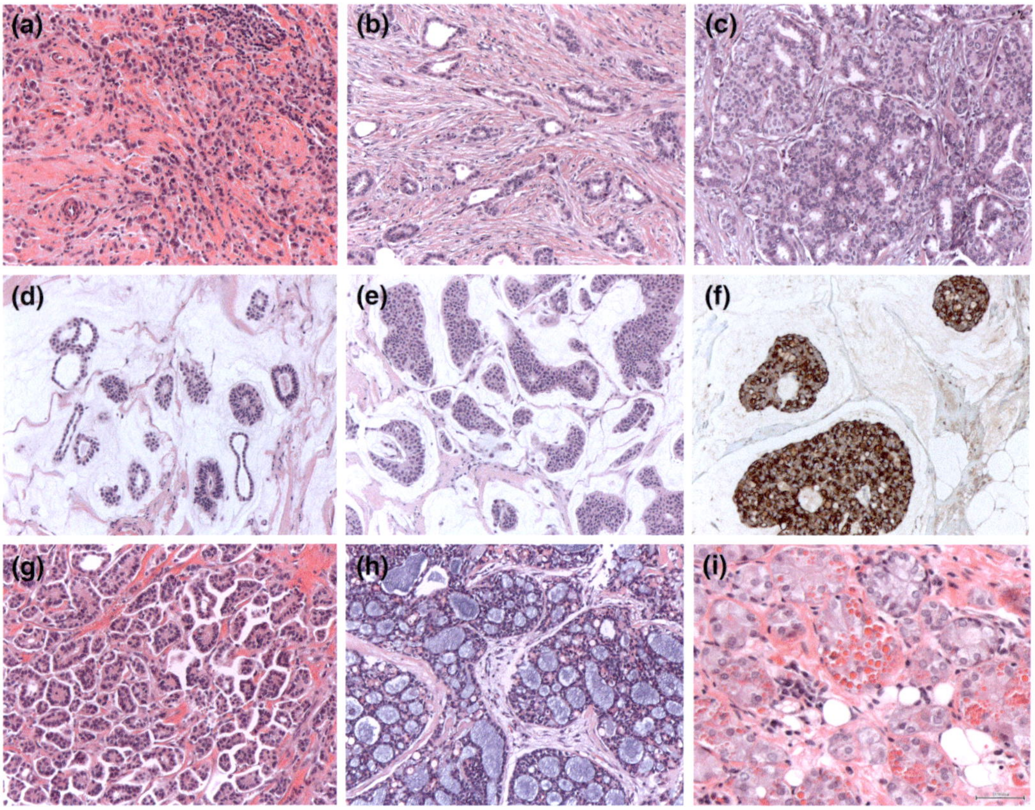


**Fig. 12.7** Panel of special histologic types of invasive breast cancer (**a** lobular carcinoma; **b** tubular carcinoma; **c** cribriform carcinoma; **d** mucinous carcinoma, type A; **e** mucinous carcinoma, type B; **b**; **f** synaptophysin staining in a mucinous carcinoma, type B; **g** inverted micropapillary carcinoma; **h** adenoid cystic carcinoma; **i** acinic cell carcinoma)

*TP53* were inactivated, the animals developed tumors that recapitulated both morphological and metastatic patterns of ILCs [91].

IHC for E-cadherin can be used to confirm the histologic type in lesions with borderline ductal (no special type) *versus* lobular histologic features, even though it has to be pointed out that aberrant E-cadherin expression in ILCs may be misleading [78]. Another characteristic immunophenotypic feature of ILC is the presence of cytoplasmic expression of p120 catenin [78]. p120 catenin, normally bound to the intracytoplasmic domain of E-cadherin, becomes upregulated when E-cadherin is lost [195]. Cytoplasmic p120 staining can be of help to resolve discordance between morphology and the E-cadherin pattern of staining.

A diagnosis of ILC on presurgical biopsy material should be made whenever possible; indeed, this detail may be of help to clinicians since: (i) ILC is frequently multicentric and bilateral, and careful study of both mammary glands with magnetic resonance may be considered (lobular carcinoma represents one of the indications in the guidelines of the EUSOMA working group for magnetic resonance imaging of the breast [196]); (ii) it has been reported that lobular carcinomas show a low rate of pathologic complete response (pCR) when subjected to primary systemic treatment, therefore, this histologic type does not represent a good candidate for neoadjuvant treatment [197, 198].

These carcinomas are typically ER-positive (only 5 % of ILCs are actually ER-negative)

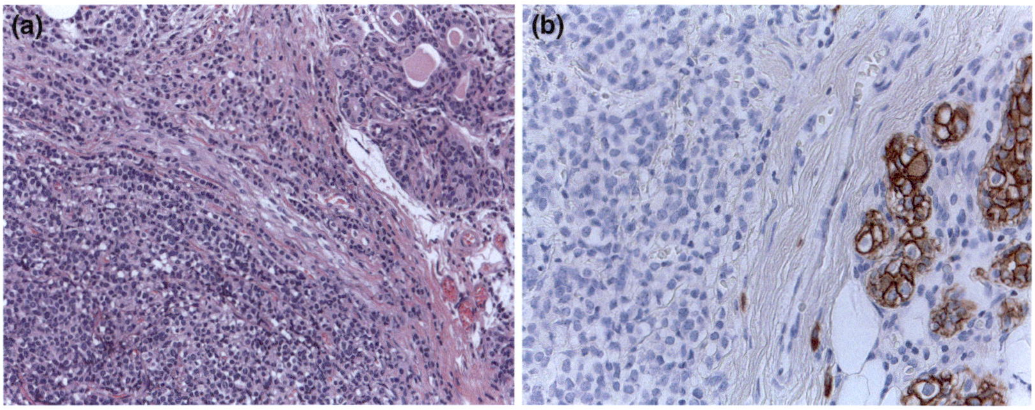


**Fig. 12.8** Invasive lobular carcinoma (**a** H&E; **b** E-CAD staining with positive internal control, i.e. a breast lobule)

[16] and most of them are of luminal A subtype [176]. However, some degree of molecular heterogeneity is still observed in ILCs, in particular in the pleomorphic form, which is not uncommonly ER-negative, HER2-positive by IHC or ISH, and/or harbor *HER2* gene mutation [199]. Notably, gene expression analyses have confirmed that pleomorphic ILCs are closer to classic ILCs than grade- and ER-matched IC-NSTs [190].

Very recently, The Cancer Genome Atlas (TCGA) consortium has carried out the most comprehensive genomic comparison between ILCs and IC-NSTs and identified, in addition to the best-known ILC genetic hallmark E-cadherin loss, some ILC-enriched features, including mutations targeting *PTEN*, *TBX3*, and *FOXA1* [200]. *HER2* mutations were also detected, at a lower frequency (4 %) than reported before (23 % in [201]). A striking difference between ILCs and IC-NST was a higher PI3K/Akt activation in ILCs as compared with IC-NSTs within the luminal A subtype, a finding that may have therapeutic implications as anti-PI3K therapy (i.e., Everolimus) is a therapeutic option for ER-positive/HER2-negative metastatic breast cancer [202]. Interestingly, contrary to the previous notion that *CDH1* gene promoter methylation could be causative of loss of E-cadherin expression in ILCs, the results of the ILC TCGA study failed to observe an impact of methylation on E-cadherin protein expression [202]. Further studies to determine the mechanism by which E-cadherin is lost in ILCs lacking biallelic inactivation of *CDH1* are warranted.

#### 12.7.2.2 Tubular Carcinoma

Tubular carcinoma (TC) accounts for approximately 2–4 % of all invasive breast cancers [16]. Histologically, TCs feature well-defined round, ovoid or angulated tubules with open lumina dispersed in a cellular fibrous or fibroelastotic cellular stroma (Fig. 12.7), which is characteristically reproduced also in rare lymph node metastasis. The tubules are lined by a single layer of relatively uniform epithelial cells with little nuclear pleomorphism and low proliferation [16]. Given the high percentage of tubule formation, low levels of nuclear pleomorphism, and low mitotic rates, TCs always fall within histologic grade 1 carcinomas [16]. It is important to adhere to stringent criteria when diagnosing TC to ascertain its excellent prognosis.

TC is typically hormone receptor positive, HER2-negative, and low proliferative (Ki67 < 10 %), and the great majority of these tumors are classified as luminal A molecular subtype [176], which has been consistently shown to display a better outcome than the remaining molecular subtypes [143–146]. We have observed, however, morphological *bona fide* examples of TCs that are PR-negative, despite a very low Ki67 index. This may have some clinical impact as PR levels below 20 % have been suggested to correlate with luminal B subtype [203–205], whereas the identification of TCs denotes a particularly good prognosis [16, 206]. A large retrospective study carried out by Rakha et al. [206] has demonstrated that the outcome of patients with pure TCs is significantly better than that of patients with grade 1 IC-NSTs. Pure TCs, low-grade IC-NSTs, and low-grade ILCs have been demonstrated to share immunophenotypic and genetic similarities, and pertain to the so-called "low-grade breast neoplasia family" [56, 57]. At the transcriptomic level, TCs are indeed very similar to histologic grade- and molecular subtype-matched IC-NSTs, supporting the concept that these two entities may evolve through common molecular pathways and have similar precursor lesions. However, subtle transcriptomic differences between these two entities were detected. Pure TCs were shown to be characterized by an upregulation of several components of the ER canonical pathway, including *ESR1*, *CREBBP1*, and *NCOR1*. Furthermore, TCs were shown to display higher expression levels of *INPP4B*, a tumor suppressor gene with inhibitory effect on PI3K pathway. It should be noted, however, that differences were small and it is very unlikely that markers or a gene signature for TC could be developed. Therefore, the correct histologic identification of TCs remains an important prognostic factor within the luminal A subtype. For instance, patients with TCs may not need a good-prognosis gene expression signature to forgo chemotherapy.

#### 12.7.2.3 Cribriform Carcinoma

Similar to TCs, this lesion represents an invasive carcinoma with an excellent prognosis that shows a cribriform growth pattern similar to that of the *in situ* carcinoma, with which it is frequently associated (Fig. 12.7). A mixture of cribriform and tubular components may be appreciated in some cases. As in TCs, tumor cells are small and nuclei show a low or moderate degree of pleomorphism. Mitotic figures are rare. Cribriform carcinoma is invariably (100 %) ER positive and frequently (about 70 %) PR positive. The main and most important differential diagnosis is with adenoid cystic carcinoma, which mimics the growth pattern but is consistently of TN phenotype [16].

#### 12.7.2.4 Mucinous Carcinoma

Also called mucin-producing or colloid carcinoma, this histologic type includes a variety of carcinomas accounting for about 2 % of all breast cancers characterized by production of abundant extracellular and/or intracellular mucin [16]. Mucinous carcinomas preferentially affect older women and are usually associated with a good clinical outcome [207–209].

Histologically, mucinous carcinomas feature small cell clusters floating in large amounts of extracellular mucin (Fig. 12.7). Nuclear atypia and mitotic figures are uncommon. There is some controversy on how to diagnose breast carcinomas with mucin production but with marked nuclear pleomorphism. Some advocate that these cases should be called "IC-NSTs with mucin production," while the term mucinous carcinoma should be used only for those with low-grade histology. The WHO classification, however, accepts the existence of some high-grade mucinous carcinomas. Capella et al. [210] have described two variants, based on cellularity, size of the clusters and cellular patterns: (i) type A, or hypocellular; (ii) type B, or hypercellular. Type B is more frequently associated with neuroendocrine differentiation as a cellular/nuclear pattern and at IHC [181, 211].

Mucinous tumors are typically ER-positive and classified as "luminal" tumors at the transcriptomic level. The histologic variants, type A and type B, have been shown to harbor significantly different gene expression profiles. Mucinous B tumors often display features of neuroendocrine differentiation and transcriptomic profiles remarkably similar to those of neuroendocrine breast cancers [181].

Mucinous carcinomas have been shown to be genomically distinct from IC-NSTs. In a comparative study with grade- and ER-matched IC-NSTs, these tumors significantly less frequently harbored gains of 1q and 16p and losses of 16q and 22q. Notably, no pure mucinous carcinoma displayed concurrent 1q gain and 16q loss, a hallmark genetic feature of low-grade

ER-positive IC-NSTs [182]. In addition, mucinous carcinomas have been shown to completely lack *PIK3CA* mutations, another feature of low-grade ER-positive IC-NSTs [212]. Taken together, these findings suggest that mucinous carcinomas may evolve from a different molecular pathway as compared to usual ER-positive breast cancers.

#### 12.7.2.5 Micropapillary Carcinoma

Pure micropapillary carcinomas (MPCs) account for 0.7–3 % of all breast carcinomas [16]. MPCs display a distinctive growth pattern, featuring clusters of cells with "inverted polarity" [213] surrounded by empty spaces in a spongy stroma [16] (Figs. 12.7 and 12.9). A more objective identification of this subtype can be achieved with the help of IHC analysis with antibodies against epithelial membrane antigen (EMA, aka MUC-1) [214–216], which typically decorate the stroma-facing border of the cell clusters (Fig. 12.9). Importantly, pure MPCs of the breast are associated with a peculiar proclivity for lymphovascular invasion (Fig. 12.9), a high incidence of lymph node metastases and arguably a poorer prognosis than unselected IC-NSTs [214, 217, 218]. Importantly, when compared with IC-NSTs matched for age, tumor size, and grade, peritumoral vascular invasion, IHC-defined molecular subtype, number of positive lymph nodes and year of surgery, micropapillary histologic type did not add any independent information to the risk of locoregional or distant relapse, or overall survival [219]; however, MPCs more frequently presented as locally advanced disease than IC-NSTs [219].

MPCs are usually ER-positive/HER2-negative with a minority being ER-positive/HER2-positive and accordingly classified at the transcriptomic level as luminal carcinomas [176]. We have observed, however, some rare cases of

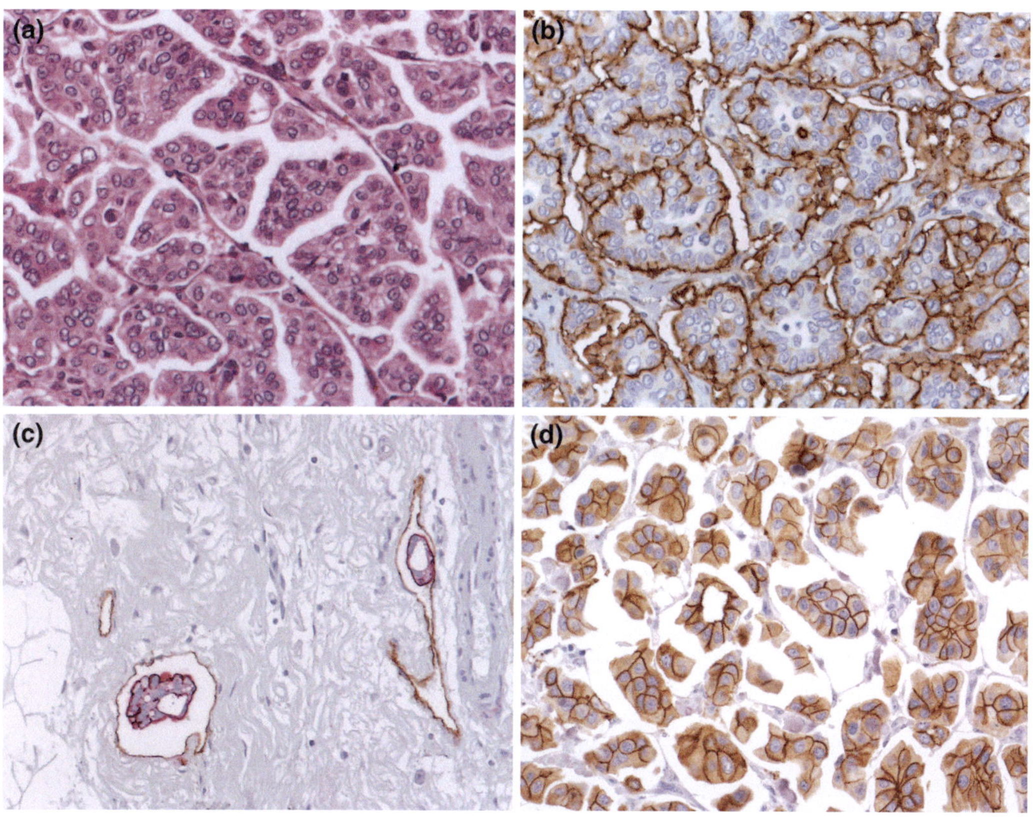


**Fig. 12.9** MPC (**a** H&E; **b** EMA; **c** tumor emboli, double staining for EMA (*red*) and for D2-40 (*brown*); **d** HER2)

MPCs with apocrine differentiation, which are ER-negative and frequently HER2-positive. Interestingly, positivity for HER2, even when strong, is typically not complete and is typically mutually exclusive with EMA (Fig. 12.9). This pattern has been taken into account in the latest ASCO/CAP guidelines as a particular category that needs ISH testing to confirm the HER2 status, in case membrane positivity is moderate/intense but incomplete in more than 10 % of cells [220]. Proliferation indices are usually medium to high, leading to the concept MPCs may be closer to luminal B than luminal A carcinomas [184]. This has been corroborated by results obtained with aCGH analysis, which have revealed that: (i) pure and mixed MPCs are remarkably similar at the genetic level [183]; (ii) they harbor a constellation of genetic aberrations that is distinct from that of grade- and ER-matched IC-NSTs [183, 184], and (iii) they typically display genomic features of luminal B breast carcinomas, such as gains of 1q, 8q, 17q, and 20q and losses of 1p 8p, 13q, 16q, and 22q [183, 184, 191].

In a small series of pure MPCs subjected to targeted capture sequencing for genes recurrently mutated in breast cancer and DNA repair-related genes, single cases harbored mutations described in luminal B carcinomas corroborating the concept that MPCs have a constellation of genetic aberrations similar to those of luminal B breast cancers [191]. No pathognomonic mutation or expressed fusion gene underpinning the micropapillary morphology has so far been identified.

#### 12.7.2.6 Invasive Papillary Carcinoma

Invasive papillary carcinoma (IPC) of the breast is a histologic special type that accounts for approximately 1 % of all invasive breast cancers. This is a poorly-characterized histologic special type of breast cancer; thus, clinical features and outcome features are unknown. As already described in general for papillary lesions, histologically they feature arborescent fibrovascular stalks lined by a layer of neoplastic epithelial cells devoid of an intervening myoepithelial cell layer, frankly invading surrounding tissue [128]. Whenever facing a clearly invasive papillary carcinoma in the breast parenchyma, a metastatic carcinoma from another organ to the breast should be ruled out [16].

At the genomic level, one study revealed papillary carcinomas constitute a relatively homogeneous entity displaying a luminal phenotype and lacking *HER2* gene amplification. They display relatively simple genomes [185], consistent with those of low-grade ER-positive breast cancers (i.e., 16q losses, 16p gains, and 1q gains) [14], with a significantly lower number of gene copy number aberrations than grade- and ER-matched IC-NSTs. It should be noted that this study included EPCs and SPCs. IPCs have been reported to harbor *PIK3CA* mutations in about 40 % of cases, a feature of ER-positive IC-NSTs of good prognosis [221]. A recent transcriptomic analysis has revealed that IPCs display lower levels of expression of proliferation-related, cell assembly and organization and cellular movement and migration genes when compared to IC-NSTs of the same histologic grade and ER status [137].

#### 12.7.2.7 Carcinoma with Neuroendocrine Features

Primary neuroendocrine breast carcinoma (NBC) was originally described in 1977 by Cubilla and Woodruff [222], based on the recognition of carcinoid-like structures in breast cancer showing a solid/alveolar pattern. In 2003, the WHO included a category of neuroendocrine tumors that in the latest edition has been rephrased as carcinoma with neuroendocrine features [16]. These lesions, which account for 2–5 % of all BCs, are defined as carcinomas exhibiting morphological features similar to those of neuroendocrine neoplasms of other organs, including the gut [16]. Typically, tumor cells are arranged in solid nests and/or trabeculae separated by delicate fibrovascular stroma; rosettes, peripheral palisading, and solid papillary formations are also considered features of neuroendocrine cancers [16]. Some variants have been described, i.e., solid/carcinoid-like type, usually of low to intermediate grade, the small/oat cell variant and the large cell type, which are both poorly differentiated NBCs [211]. In addition to morphological features, the diagnosis of NBCs relies on the expression of neuroendocrine markers, which ought to be present in >50 % of tumor cells. These apparently easy-to-apply criteria are equivocal and controversial (except for the extremely rare classical small/oat cell carcinomas), since NBC morphology may be highly heterogeneous displaying features shown by non-NBCs. Of note, other histologic types can show neuroendocrine differentiation, such as the cellular mucinous type B (Fig. 12.7) and SPC [16].

These lesions usually affect elderly women and have an indolent clinical behavior; they are typically ER-positive/HER2-negative and at the transcriptomic level have been shown to pertain to luminal A subtype more frequently than to the luminal B subgroup [181]. Interestingly, in a transcriptomic study of mucinous and NBCs, where it was demonstrated that both are transcriptionally distinct from IC-NST, mucinous carcinomas type B and NBCs shared similar transcriptomic profiles [181].

#### 12.7.2.8 Carcinoma with Apocrine Differentiation

Apocrine carcinoma of the breast is a rare subtype of breast carcinoma constituting approximately 1 % of all mammary carcinomas [16]. In routine work, these tumors are often diagnosed using standard hematoxylin and eosin (H&E) staining because of their characteristic cytological features, i.e., cells with abundant eosinophilic and granular cytoplasm, large nuclei with prominent nucleoli, and visible cell membrane (Fig. 12.7). Some authors, however, have suggested that (pure) apocrine carcinoma should be diagnosed only when morphology correlates with the expected immunoprofile, that is, ER-, PR- and HER2-negative and androgen receptor (AR)-positive [223, 224]. The cytomorphological appearance of apocrine differentiation in a breast tumor does not always correspond to a classic apocrine immunophenotype reported in the literature as specific for apocrine tumors [225, 226]. In fact, only 73 % of these carcinomas fulfill the IHC criteria of true apocrine breast tumors (i.e., ER-/PR-/AR+) [226]. Diffuse expression of the 15-kDa gross cystic disease glycoprotein (GCDFP-15) is also a common feature [226], whereas in usual breast carcinomas GCDFP-15 expression is usually focal.

Transcriptomic analyses [176] have suggested that apocrine carcinomas hold heterogeneous gene expression profiles and pertain to multiple molecular subtypes, consistent with the notion that these tumors are unlikely to constitute a distinct entity. As breast carcinomas of any type and grade may display features of apocrine differentiation, these data suggest that it might be more clinically and biologically relevant to identify the group of "molecular apocrine" tumors, which show not only features of apocrine differentiation at the histologic level, but also increased androgen signaling [176]. Recently, Lehmann-Che et al. [227] have described the morphological and IHC features of a series of "molecular apocrine"-defined tumors. From the histologic point of view, the retrospective analysis described them all as IC-NSTs, but only 7 % presented with morphological apocrine features. From the IHC point of view, the signature "HER2(3+) OR GCDFP-15(+)" had a sensitivity and a specificity for molecular apocrine tumors of 94 and 100 %, respectively, and displayed the best possible marker combination to discriminate molecular apocrine carcinomas from basal-like carcinomas in the context ER negative breast cancer [227].

Targeted sequencing analysis on a small subset of apocrine carcinomas (7 cases) has revealed mutations affecting mTOR signaling pathway genes including *PIK3CA* (2/7) and *PTEN* genes (3/7) (coupled to PTEN loss of protein expression), and *TP53* (2/7) [225]. *PIK3CA* mutations have been previously described in a small series of benign and malignant apocrine lesions, whereas germline *PTEN* mutations have been described in patients with Cowden syndrome that are prone to develop breast cancer with apocrine differentiation. Single cases also harbored *KRAS* and *BRAF* gene mutations [225].

Given that apocrine differentiation can occur in any histologic type and is associated with heterogeneous molecular profiles, and that, at this moment, the clinical significance of apocrine differentiation is not clear, apart from a definite association with androgen signaling and possible benefit from antiandrogen therapy, the latest WHO classification does not recognize pure apocrine carcinoma. It has been recommended to type an invasive carcinoma according to its structural type and mention the apocrine differentiation. Thus, for instance, according to the WHO classification, a case with non special morphology, but with diffuse apocrine cytology, should be diagnosed as IC-NST with apocrine differentiation [16].

#### 12.7.2.9 Metaplastic Carcinoma

Metaplastic breast carcinoma (MBC) accounts for 0.2–5 % of all invasive breast cancers and is an aggressive histologic type of breast cancer. The adjective metaplastic is actually an "umbrella term" for a heterogeneous group of cancers that are characterized by the presence of neoplastic cells displaying differentiation toward squamous epithelium and/or mesenchymal components, such as spindle, chondroid, osseous, or rhabdoid cells [16, 228]. They are perceived clinically as a single entity, however, there is evidence to demonstrate that these tumors are heterogeneous in regards to their biological characteristics and, potentially, clinical behaviors [16, 228, 229]. For the time being, the WHO adopted a descriptive histologic classification of MBC. Thus, when diagnosing a MBC, one should clearly describe the distinct morphological components present. In fact, this may have some clinical implications, as, for instance, MBCs with complete (>95 % of the tumor) sarcomatoid differentiation display a significantly worse prognosis and seem not to benefit from conventional chemotherapy offered for TNBC patients [230].

As a group, MBCs (>90 %) display a TN phenotype, and frequent expression of basal-markers [228]. Like TNBCs, MBCs are generally characterized by high levels of genetic instability and these two entities also share a similar pattern of gene copy number alterations [231]. At the transcriptomic level, MBCs preferentially have a basal-like or claudin-low molecular subtype and frequently harbor *TP53* mutations [231–233]. MBCs with spindle cell morphology are the ones more likely to be classified as claudin-low [228, 234–236], which is not surprising given that these cases display overt mesenchymal differentiation akin to epithelial-to-mesenchymal transition phenotype, a defining-feature of the claudin-low subtype. In addition, it has been shown that MBCs were preferentially of mesenchymal-like and mesenchymal stem-like using the six molecular subtype classification of TNBC [159]. When applying the integrative clustering approach, MBCs were of IntClust 4, IntClust 1, IntClust 8, and IntClust 9 [229].

As a further level of complexity, it has recently been shown that different histologic components of MBCs are associated with specific molecular features [229]. Samples exclusively or predominantly composed of areas of spindle cell metaplasia or chondroid metaplasia were of claudin-low intrinsic molecular subtype and of mesenchymal-like TNBC subtype, respectively, whereas those samples exclusively or predominantly composed of squamous metaplasia were more heterogeneous. These findings imply that the molecular subtype of a metaplastic breast carcinoma may be different depending on the morphologic component sampled and subjected to molecular analysis. It is currently unknown, however, whether the different histologic components of metaplastic breast carcinomas are underpinned by distinct constellations of point mutations, structural genetic rearrangements, and/or epigenetic alterations; however, a previous study has demonstrated that histologically distinct components of a given metaplastic breast carcinoma may harbor different somatic genetic alterations [237].

The information on the mutational landscape of MBCs is still scarce. No pathognomonic mutation or recurrent fusion gene that accounts for the histologic characteristics of these cancers has been identified, and it is unknown whether different repertoires of somatic mutations drive the distinct subtypes of MBCs. The morphologically and phenotypically distinct components of individual MBCs have been shown to be clonal on the basis of the presence of identical *TP53* mutations [237]. However, intratumor genetic heterogeneity (i.e., differences in the somatic genetic alterations between phenotypically distinct components of individual cases) is also evident [237]. Mutations in *TP53* and *PIK3CA*, and losses of *PTEN* and p16 cyclin-dependent kinase inhibitor 2A (*CDKN2A*) have been recurrently found in MBCs; however, they do not seem to be restricted to tumors of specific phenotypes [231, 237] and are not unique to MBCs. Overexpression and amplification of the epidermal growth factor receptor (*EGFR*) gene have been reported in a subset of MBCs and seem to be more prevalent in tumors with squamous and/or spindle cell metaplasia [238].

Although MBCs are mostly high-grade and display an aggressive behavior, which appears to be worse than usual TNBCs, there are two low-grade MBC subtypes, namely, fibromatosis-like spindle cell carcinoma and low-grade adenosquamous carcinoma. Both are also of TN phenotype, but carry distinct clinical implications. First, from a histologic point of view, fibromatosis-like spindle cell carcinoma has to be distinguished from desmoid-type fibromatosis and other benign spindle cell lesions of the breast, as, though less than high-grade MBCs, they still have a long-term metastatic potential and must be surgically treated as carcinomas [41, 239]. Low-grade adenosquamous carcinoma display infiltrative borders and a tendency for local recurrence, thus extreme attention to surgical margins are needed. In contrast, the metastatic potential is minimal and therefore chemotherapy should be avoided.

#### 12.7.2.10 Adenoid Cystic Carcinoma

Adenoid cystic carcinomas (AdCCs) are malignant tumors most commonly affecting the salivary glands, but they can also be found in other anatomical sites, including the breast, lungs, and prostate [240]. AdCCs provide a clear example of genotypic–phenotypic correlation, as they display similar histologic characteristics, irrespective of the site of origin, and often harbor the recurrent t(6, 9)(q22–23;p23–24) translocation that results in the formation of the *MYB–NFIB* fusion gene [188, 240].

The existence of AdCC in the breast has been questioned, given its morphological similarity with cribriform carcinoma [241]. In fact, misdiagnosis between these two entities is not an uncommon event. IHC and ultrastructural studies, however, have proven the existence of true breast AdCC [16, 240]. AdCCs account for 0.1–1 % of all breast cancers [16, 241–243] and their most striking feature is the excellent long-term prognosis, which is in stark contrast with that of salivary gland AdCCs or common-type TNBCs (i.e., IC-NSTs) [240, 241, 243].

At the morphological level AdCCs can feature a variety of growth patterns, including cribriform, trabecular, and solid variants (Fig. 12.7) [243]. Not uncommonly, single cases display all growth patterns. A sebaceous differentiation is found in up to 14 % of cases, and foci of adenosquamous differentiation may also be encountered [243, 244]. The proportion of solid growth has been reported to be clinically meaningful, as Ro et al. [245] have reported tumors with solid components were more likely to develop recurrences; however, independent validation of the prognostic impact of this classification has yet to be reported. In addition, in the same cohort, the only patient who experienced metastases was affected by a high-grade tumor.

Phenotypically, AdCCs display a TN phenotype and express basal CKs (CK5/6). Not surprisingly, at the transcriptomic level AdCCs share similar gene expression patterns with metaplastic and medullary carcinomas within the ER-negative subgroups. From a molecular standpoint, AdCCs often display the recurrent chromosomal translocation t(6, 9)(q22–23;p23–24), which generates a fusion transcript involving the genes *MYB* and *NFIB* [188]. The prevalence of the *MYB–NFIB* fusion gene ranges from 23 to 100 % [188, 192, 246–248]. Differences may be partly explained to sample type (frozen vs. FFPE) and methodology (FISH vs. PCR), and the potential inclusion of other mimics of AdCC (e.g., breast cylindroma) [249]; however, the clinical significance of this observation remains unclear. Although a substantial proportion of AdCCs lack the *MYB–NFIB* fusion gene, the majority of MYB–NFIB fusion gene-negative salivary gland AdCCs likely displays activation of *MYB* due to mechanisms other than the t(6, 9) chromosomal translocation [250].

AdCCs display lower levels of genetic instability, as defined by gene copy number alterations, than basal-like IC-NSTs [192, 251], as well as a low mutation rate with a heterogeneous repertoire of somatic genetic alterations. Mutations affecting *TP53* and *PIK3CA*, and 5q losses and 8q gains [252, 253], which are frequent in TNBCs, are not found in breast AdCCs. On the other hand, AdCCs of the breast were found to harbor mutations in genes rarely mutated in basal-like breast cancers, including *RASA1*, *PTPN11*, or *BRAF*, and recurrent 12q losses. Importantly, recurrent losses of 12q and somatic mutations affecting *SF3B1*, *MYB*, *PRKD1*, and *FGFR2* have been documented in salivary gland AdCCs [250, 254–256]. Taken together, these results demonstrate that breast AdCCs are more similar to salivary gland AdCCs than to common forms of TNBCs. In fact, contrary to common forms of TNBC, patients with breast AdCCs have a favorable outcome [16, 187, 240]. Based on these observations, as also stressed by the St. Gallen recommendations, although AdCCs are of TN and basal-like phenotype, a diagnosis of breast AdCC should not prompt the use of systemic chemotherapy, given the excellent prognosis patients with this tumor type have [140]. In addition, it has been reported that chemotherapy response rates in patients with breast AdCCs are low, akin to those observed in patients with salivary gland AdCCs [257, 258].

#### 12.7.2.11 Exceptionally Rare Types

##### Acinic Cell Carcinoma

This vanishingly rare special histologic type of breast cancer (only 39 cases of ACCs have been described to date) is included in the spectrum of salivary gland-type tumors of the breast and is reported to have a favorable outcome [16, 259]. Distant recurrences have been described, but these are rare and usually related to the presence of a poorly differentiated invasive component [260–262]. Metastases from low-grade breast ACCs are still to be documented. Morphologically, they show infiltrative microglandular or solid-nest structures composed of cells with round-to-ovoid nuclei, discrete nucleoli, and abundant eosinophilic-to-amphophilic cytoplasm containing small or coarse Paneth cell-like granules (Fig. 12.7). Areas of clear cells with hypernephroid appearance may be present [263]. Phenotypically ACCs are of TN phenotype and express S100 protein, EMA, and serous differentiation markers, including amylase, lysozyme, and alpha 1-antichymotrypsin [264].

A recent genomic characterization has demonstrated that despite their low-grade and indolent clinical behavior, ACCs display the cardinal genomic features documented in high-grade forms of TNBC (i.e., high mutational burden, recurrent *TP53* mutations, *BRCA1* germline, and somatic pathogenic mutations, and complex patterns of gene copy number alterations). Furthermore, it has been demonstrated that, at a variance of AdCC, breast ACC does not resemble salivary gland ACC at the molecular level [265]. Based on these observations, breast ACCs are best considered as part of a spectrum of TNBCs, with low/minimal metastatic capacity, but with a potential to progress to high-grade aggressive

TNBC [266]. In agreement with that, a case report has described an ACC of the breast in a patient with a germline *BRCA1* mutation [267].

**Secretory Carcinoma**

This is a rare (<0.15 %) lesion most commonly found in young patients (median ages: 25). It is defined as a low-grade invasive carcinoma with a solid, microcystic, and tubular architecture composed of cells that produce intracellular and extracellular secretory material [16]. Cells are polygonal with granular eosinophilic to foamy cytoplasm. Nuclei are regular with inconspicuous nucleoli and mitoses are rare. It typically shows a TN phenotype. EMA, S100 are frequently expressed [16].

The vast majority of secretory carcinomas (>95 %) harbor a recurrent balanced chromosomal translocation t(12, 15)(p13;q25), which leads to a fusion gene between the *ETV6* gene from chromosome 12 and the *NTRK3* gene from chromosome 15 [189]. This alteration is useful for differential diagnosis with ACCs.

**Sebaceous Carcinoma**

Sebaceous carcinoma is defined as an invasive breast carcinoma with prominent sebaceous differentiation in at least 50 % of cells. Lack of evidence of origin from cutaneous adnexal sebaceous glands is required [16].

Histologically, it shows a nested structure, where sebaceous cells with finely vacuolated cytoplasm intermingle with smaller ovoid to spindle cells showing eosinophilic cytoplasm without vacuolization. The nuclei (of both cell types) are globoid with up to two nucleoli and mitoses can be numerous. ER, PR, AR, and HER2 can be expressed [16].

Genetic information on these lesions is currently unavailable.

**Lipid-Rich Carcinoma**

Lipid-rich carcinomas account for less than 1 % of cases, and are defined as an invasive carcinoma that contains no less than 90 % of cells with abundant cytoplasmic neutral lipids. Due to histologic difficulties to prove lipids' cytoplasmic accumulation, these special types are classified as a G3 carcinoma with clear cell features [16]. Tumor cells are positive for EMA, CEA, alpha-lactoglobulin, lactoferrin, and typically negative for ER and PR and HER2.

There is no genetic information reported to date on these lesions.

**Glycogen-Rich Carcinoma**

Glycogen-rich carcinomas account for about 1–3 % of all breast carcinomas. A diagnosis of glycogen-rich carcinoma is reached when >90 % of neoplastic cells have abundant clear cytoplasm containing glycogen [16].

Tumor cells display sharply demarcated borders and polygonal contours. The cytoplasm contains PAS-positive diastase-labile glycogen. Hyperchromatic nuclei with clumped chromatin are typically found. It is an ER-positive carcinoma in 50 % of cases, but PR is usually negative and ER-negative/HER2-positive cases have also been described [16].

There are not genetic data reported on these carcinomas to date.

**Polymorphous Low-Grade Adenocarcinoma**

Polymorphous low-grade adenocarcinoma (PLGA) is a very rare entity, whose true incidence is not known as only three cases have been reported so far [16]. In a way akin to polymorphous low-grade adenocarcinoma of the salivary glands, this lesion features solid nests surrounded by alveolar, cribriform, trabecular, and single file patterns. It is reported to be negative for hormone receptors and HER2 and to show strong positivity for Bc12 and faint positivity for E-CAD. Molecular data on this special type are yet to be provided; in particular, it remains to be determined whether PLGAs of the breast would harbor the *PRKD1* activating hotspot mutation encoding p.Glu710Asp that has been recently described in the majority (72.9 %) of PLGAs of the salivary glands [268].

### 12.7.2.12 Inflammatory Carcinoma

The definition of this form of breast cancer is based on a distinct clinical presentation, i.e., edema, redness, warmth, and tenderness of the skin of the affected breast (the so-called "peau d'orange"), which is believed to be due to lymphatic obstruction from an underlying carcinoma [16]. Dermal lymphatic invasion without the typical clinical presentation does not fulfill the definition of inflammatory carcinoma (Fig. 12.10); on the other hand, dermal lymphatic invasion is not needed for defining and staging (T4d) a case as inflammatory carcinoma. The adjective "inflammatory" to define this lesion may be misleading, as no particular degree of inflammatory cell infiltration is observed in these carcinomas. The cutaneous signs are the epiphenomenon of a massive neoplastic embolization of the lymphatic vessels in the derma. The underlying carcinoma is not reported to have particular features, but usually is a G3 IC-NSTs.

Little is known about the molecular pathology of these lesions, which are classified as T4d and more frequently ER and HER2 negative than other IC-NSTs and definitely represent a challenge for clinical management.

In the last two years, there have been reports about detection of *ALK* gene alterations (mainly amplification) in inflammatory carcinomas, leading to possible new therapeutic interventions for this challenging entity. However, it should be noted that the range of copy number threshold for amplification was relatively low with the large majority showing 3–4 copies [269]. Subsequent studies have provided indirect evidence that these results may stem from polysomy of chromosome 2 rather than a true *ALK* gene amplification [270, 271]. Finally, a recent study has identified clinically relevant genomic alterations, where those were defined as genomic alterations associated with label-targeted therapies and targeted therapies in mechanism-driven clinical trials. The most frequently altered genes were *TP53* (62 %), *MYC* (32 %), *PIK3CA* (28 %), *ERBB2* (26 %), *FGFR1* (17 %), *BRCA2* (15 %), and *PTEN* (15 %). In the TNBC subset of inflammatory breast carcinoma, 8/19 (42 %) showed *MYC* amplification as compared to 9/32 (28 %) in non-TNBC inflammatory breast carcinoma [272].

## 12.8 Histologic Grading and Staging

Histologic grading, a measure of the differentiation of the lesion, is a central evaluation for each newly diagnosed breast cancer, since it is an independent prognostic factor and correlates with response to chemotherapy [273]. According to the system proposed by Elston and Ellis in 1991 [167] the percentage of tubule formation, the number of mitosis, and the degree of nuclear pleomorphism are routinely examined (Table 12.5): a score from 1 to 3 is assigned to each feature, thus leading to a final value that according to the three-tiered system equates to grade 1 (well differentiated), grade 2 (intermediately differentiated), and grade 3 (poorly differentiated) (Table 12.5). Of all histologic features,

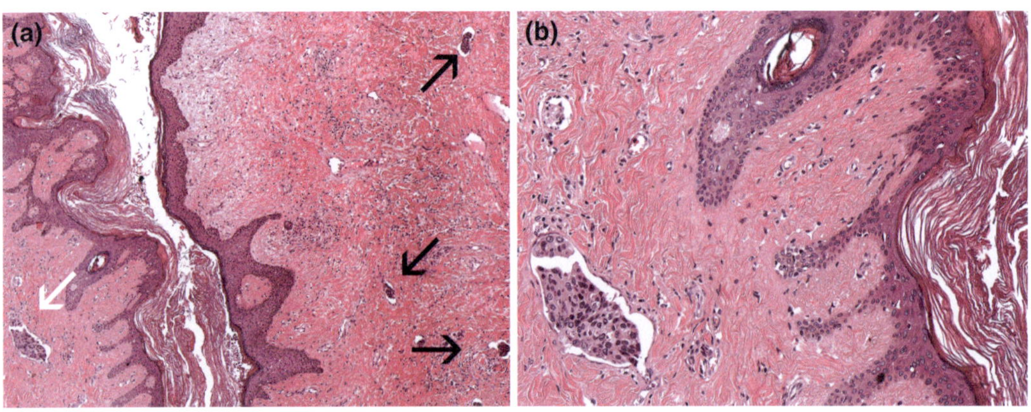


**Fig. 12.10** Inflammatory breast carcinoma

**Table 12.5** Histologic grading system for invasive breast cancer

| | | Score 1 | Score 2 | Score 3 | G |
|---|---|---|---|---|---|
| *Tubule formation*: % of luminal structures with polarized epithelial cells | | >75 % | 10–75 % | <10 % | G1[3–5] G2[6, 7] *G*3[8, 9] |
| *Nuclear pleomorphism*: evaluated in the area where it is worse | | Nuclei non significantly different from those of normal breast, uniform and regular in size and shape; uniform chromatin | Variability in shape and size; nucleoli present | Marked variability in shape and size, possible bizarre nuclei; prominent nucleoli | |
| *Mitotic count*: depending on the diameter of the microscopic field | 0.46 mm | ≤5 | 6–11 | ≥12 | |
| | 0.50 mm | ≤6 | 7–13 | ≥14 | |
| | 0.55 mm | ≤8 | 9–16 | ≥17 | |

Score for mitotic count depends on the number of mitoses and on the diameter of the microscopic field; here reported three different microscopic fields as an example

grade best reflects the complexity, pattern and type of molecular genetic changes found in breast cancer [24, 274, 275]. Several groups have independently shown that grade 1 and grade 3 invasive breast cancers have different transcriptomes and patterns of genetic aberrations [64, 274–277]. Grade 1 tumors are characterized at the transcriptomic level by high levels of ER expression and ER pathway activation and low levels of proliferation [278, 279]. CGH and aCGH studies have demonstrated that these cancers harbor recurrent losses of 16q and gains of 1q and 16p [24, 65, 274, 280], which often stem from an unbalanced chromosomal translocation involving chromosomes 1 and 16 [281, 282]. On the other hand, grade 3 lesions display varying levels of ER expression and ER pathway activation, high levels of proliferation, and complex karyotypes. Concurrent deletions of 16q and gains of 1q and 16p, however, are only seen in approximately half of ER-positive G3 cancers [65, 276, 278, 279], indicating that progression from G1 to G3 cancers occur within ER-positive tumors. Conversely, progression from ER-positive G1 to ER-negative G3 tumors is an unlikely biological phenomenon [65, 276].

Additional morphological features holding an important prognostic value are represented by two elements of the pTNM staging, namely size and lymph node status, which stresses the importance of proper staging in everyday practice. There is some evidence that tumor size may not be that relevant in the context of TNBCs [283, 284], however, within the ER-positive/HER2-negative tumors, tumor burden is a significant prognostic factor which remains independent of the proliferation-based gene expression signatures [285, 286]. Staging can be particularly troublesome in specimens following neoadjuvant chemotherapy, as assessment of residual disease is not trivial especially when the tumor lesion is no longer macroscopically evident. Description of specimen handling is beyond the scope of this chapter; however, readers are encouraged to refer to recent reviews that have defined operational procedures to best handle these specimens [287, 288].

## 12.9 Immunohistochemistry and Molecular Techniques in Breast Pathology

### 12.9.1 Immunohistochemistry

In breast cancer diagnostic pathology, IHC is an important ancillary tool. In the previous paragraphs, we have touched on the use of E-cadherin to distinguish ductal from lobular lesions, on the employment of ER and CK5/6 to differentiate UDH from ADH and on the use of myoepithelial markers to ascertain the lack of the myoepithelial cell layer in challenging lesions, such as scleroelastotic lesions, in particular on presurgical diagnostic setting (core biopsy material). Another scenario in which IHC is frequently applied is the diagnosis of microinvasion in specimens showing large DCIS lesions, typically featuring also an intense inflammatory infiltrate around the ducts (suspicious for invasion) but without clear morphological signs of infiltration of the surrounding stromal tissue. In this scenario, a simple pan-cytokeratin such as AE1/AE3 can be employed that may highlight single cell permeation of the surrounding stroma. Likewise, based on the observation that HER2 is positive in up to 40 % of DCIS lesions and at a much higher frequency in high-grade than low-grade DCIS [289], HER2 immunostaining may be of some help in this scenario.

The major role played by IHC, however, is related to the definition of prognosis (prognostication beyond H&E) and prediction of response to therapy. It is indeed a companion test for targets of drugs currently validated for clinical use [290]. At present four markers represent the cornerstone of such an assessment. ER and PR are evaluated by using clinically validated antibodies and positivity predicts response to endocrine treatments (tamoxifen and aromatase inhibitors) [291]. Guidelines recommend reporting the percentage of positive cells as well as the intensity [291]. Percentage of positive cells has been shown to correlate with likely degree of response to treatment [292]. ASCO/CAP recommendations have set a 1 % cut-off to define positivity for ER and PR. These guidelines notwithstanding, how breast cancers displaying 1–9 % of positive cells should be classified remains a matter of debate. Gene expression data shows that breast cancers with 1–9 % ER-positive cells constitute a heterogeneous group, and >50 % of them may be classified as nonluminal by gene expression profiling [293, 294].

Since the first description of its overexpression in breast carcinomas in the 90's, HER2 has dramatically changed the natural history of a fraction of breast cancer, which is esteemed to be around 10–15 % [290, 295]. Overexpression is measured based on intensity and completeness of membrane staining as well as on percentage of positive cells. The 4-tier scoring system (from score 0 to 3+) categorizes samples into negative (score 0 and 1+), positive (score 3+), and equivocal (score 2+) (Fig. 12.11) [220]. If the clinical decision is crystal clear in the first two scenarios (anti-HER2 therapy for the positive

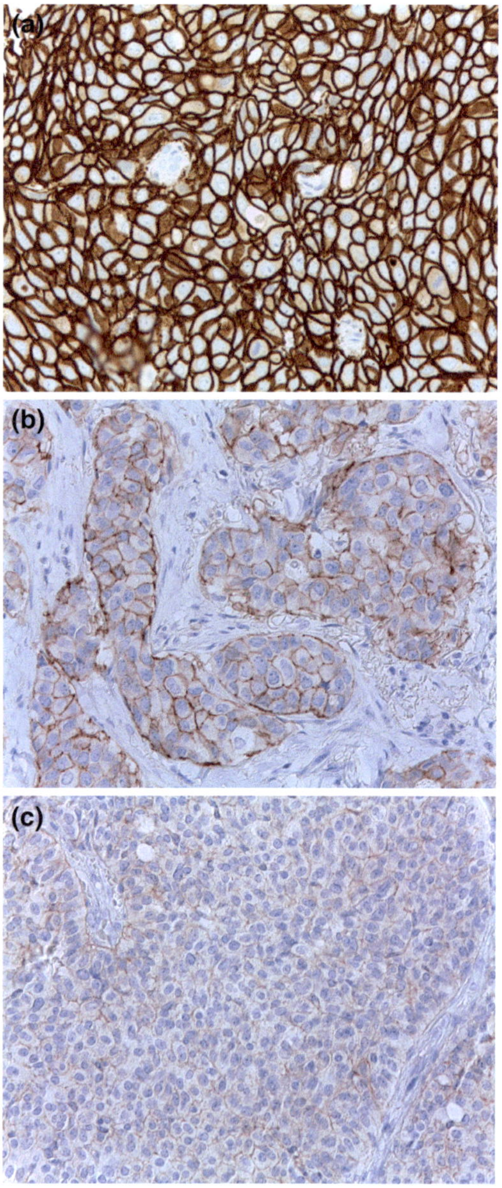


**Fig. 12.11** Evaluation of HER2 overexpression by immunohistochemistry (**a** score 3+; **b** score 2+; **c** score 1+)

cases), the latter should undergo ISH testing for evaluation of the *HER2* gene status (see below) [220, 296].

At present several therapeutic options are available, spanning from monoclonal antibodies (trastuzumab and pertuzumab) to tyrosine kynase inhibitors (lapatinib) and new compounds that couple a monoclonal antibody to chemotherapeutic agents (e.g., TDM1). In this respect, the role of pathologists in the definition of the right candidate to targeted treatment is emphasized, given that the administration of chemotherapy would utterly depend on proper evaluation of HER2 positivity [290, 295].

The last prognostic and predictive marker that is routinely tested at most institutions in IHC is Ki67, an antigen expressed by cells in any phase of the cycle but G0. Ki67, however, is much less validated than ER, PR, and HER2, its analytical validity is still under consideration and it is not mandatory for all invasive breast cancers. For instance, some institutions have decided not to include it in their minimal dataset, neither to score outside Ki67 slides. Ki67 is a proliferation marker and has been proven prognostic and predictive of pathological complete response after neoadjuvant chemotherapy, in particular within ER-positive cancers [297–299]. Scoring recommendations are on record [300] though of difficult and time-consuming implementation [301]. Ki67 index intratumor heterogeneity occurs, posing additional difficulties for its interpretation and use in core biopsies as a predictive factor of benefit from neoadjuvant chemotherapy. Some international studies have demonstrated poor interlaboratory concordance, in particular for locally-stained slides [299]. Concordance was improved after a web-based methodological training, but the study was performed using tissue microarrays [301]. Thus, it is very likely that the use of full sections as in clinical practice would add another variable and decrease the interpathologist agreement.

A final remark about proper IHC assessment should be made about the so-called "pre-analytical phase." Since biomarker analysis dictates whether or not the patients will be deemed suitable to a potentially life-saving treatment, it is not surprising that ASCO/CAP guidelines for both hormone receptors and HER2 assays stress the importance of preanalytical conditions to guarantee reliable results for best treatment planning [220, 291]. A great source of variability is represented by fixation, in terms of type of fixative, time to fixation, and duration of the whole process from start to end. Over the past few years time to fixation, aka "cold ischemia time," has received a great deal of attention and it has been show how deleterious can be the effect on breast cancer patient management [295, 302]. Guidelines recommend ensuring that time to fixation and time in fixative are recorded and considered in defining the test result: the time from removal from the patient to incision of the specimen should be as short as possible (ideally no longer than one hour) and time in fixative should be ranging between to 6 and 72 h [220, 291]. Excessive delay from tissue collection to the initiation of formalin fixation (60–120 min) has been reported to impact adversely on the analysis of both hormone receptors and HER2 IHC and ISH testing. As a consequence, due to loss of staining or hybridization signal intensity may intervene and tumors with excessive cold ischemic times may give false negative results. In terms of duration of tissue fixation, the immune-reactivity for ER, PR, and HER2 seems to be reduced by very long, extended formalin overfixation. Participation to quality controls may definitely help improve both the preanalytical phase as well as the analytical phase, which may suffer from subjectivity and partly overlaps with a matter of interpretation.

### 12.9.2 *In Situ* Hybridization

One of the molecular techniques that have become part of our diagnostic armamentarium is ISH. It has gained a lot of success since the very beginning because it allows visualization of genes on a glass slide [141], thus allowing for semi-quantitative assessment of gains, losses and amplifications directly on tissue sections. Three methods of ISH have been introduced in breast pathology: FISH, chromogenic ISH (CISH), and silver ISH (SISH). ISH tests, either in dark or in bright field can be performed with a single color probe (gene probe) or with a dual-color probe (probes for the gene and the centromere of the corresponding chromosome).

In diagnostic breast pathology, ISH is mainly used for HER2 testing; however, ISH can also be employed to look for amplification of other oncogenes that have emerged as prognostic or predictive markers (e.g., *MYC*, *TOP2A*, and *CCND1*) or potential therapeutic targets (such as *FGFR1*) [303]. In addition, as seen in diagnostic hematopathology and soft tissue pathology, FISH has also found its use as a diagnostic tool in breast pathology: as discussed above, secretory carcinoma harbors a recurrent balanced chromosomal translocation t(12;15)(p13;q25), which leads to a fusion gene between the *ETV6* gene from chromosome 12 and the *NTRK3* gene from chromosome 15 [189] and adenoid cystic carcinomas harbor the recurrent t(6;9)(q22–23; p23–24) translocation that results in the formation of the *MYB–NFIB* fusion gene [188].

For *HER2* gene assessment, FISH (Fig. 12.12), CISH, and SISH are all FDA-approved and can be employed in routine practice. ISH can be used as a first line test or as a second level approach. In the latter scenario, it is applied to score 2+ carcinomas (carcinomas with equivocal HER2 overexpression) [220, 296].

Over the past few years assessment of *HER2* gene amplification by ISH testing has undergone several changes. Following the demonstration of the phenomenon of chromosome 17 centromere (CEP17) amplification [304], independently validated by several groups [295, 305–307] we have come to terms that true chromosome 17 (Chr17) polysomy is quite a rare event and whenever an abnormal number of CEP17 is encountered this is most likely to be due to high level gains or amplification of the centromeric region of Chr17. This has lead to a paradigm shift in the way we score ISH results when adopting a dual-color probe. The recently updated ASCO/CAP guidelines have taken into account this issue and changed the scoring system accordingly by introducing the so-called "ISH algorithm," which first calculates the *HER2*/CEP17 ratio followed by analysis of the *HER2* copy number [220] (Table 12.6). This has allowed *HER2* gene amplification not to be underestimated in a subset of tumors showing high *HER2* copy numbers together with CEP17 gain or amplification. On the other hand, a big debate is currently ongoing about the possibility to score as "ISH positive" cases showing a HER2/CEP17 ratio <2 and mean HER2 copy number <4 [296, 308]. This rare scenario typically occurs in tumors harboring monosomy of Chr17 that easily leads to ratio values >2 despite a not high absolute *HER2* copy number (mean of 2 or 3) [296, 309].

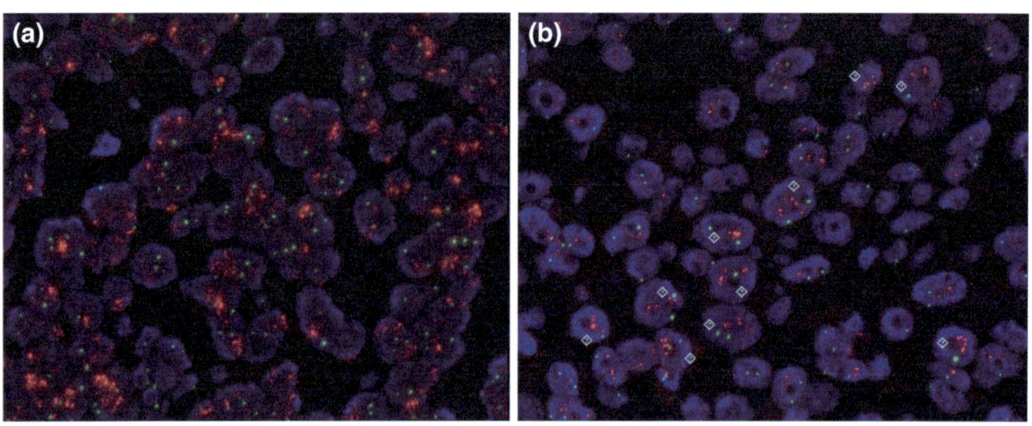


**Fig. 12.12** *HER2* evaluation by FISH (**a** homogeneous population of *HER2* amplified cells; **b** heterogeneous *HER2* amplification, rhomboidal symbols identify amplified cells)

**Table 12.6** Summary of *HER2* gene scoring by *in situ* hybridization (ISH) according to ASCO/CAP 2013

| ASCO/CAP 2013 | | Results | | |
|---|---|---|---|---|
| Probe | Scoring method | Positive | Equivocal | Negative |
| Dual-color | ISH algorithm | (a) *HER2*/CEP17 ratio ≥2, regardless of *HER2* copy number<br>(b) *HER2*/CEP17 <2 but *HER2* copy number ≥6 | *HER2*/CEP17 <2 with *HER2* copy number ≥4 and <6 | *HER2*/CEP17 <2 with *HER2* copy number <4 |
| Single-color | *HER2* copy number | ≥6 | ≥4 and <6 | <4 |

The rationale for the panelist to include these cases as positive resides in the first generation trials of adjuvant trastuzumab, where patients with a *HER2*/CEP17 >2 (i.e., regardless of *HER2* copy number) were deemed suitable to anti-HER2 treatment and these included also a minority of cases with low (<4) *HER2* copy number. It should be noted, however, that a head to head comparison between patients with *HER2* copy number >4 and <4, respectively, was not performed [290]. Of note, cases with copy number less than 4 would be considered negative if a single ISH assay were employed [220].

Another potential pitfall in ISH testing is encountered during retesting after neoadjuvant treatment. Occasionally, some carcinomas may show giant syncytial multinucleated-looking cells that can have a very high number of *HER2* signals scattered within a background of negative cells. These ISH findings are most likely to be due to a polyploidy status induced by chemotherapy rather than to a focal amplification of the *HER2* locus, as confirmed by the presence of additional copies of CEP17 as well as many regions in other chromosomes [310]. These findings should not be misinterpreted as heterogeneous *HER2* amplification (Fig. 12.13).

Finally, *HER2* intratumor heterogeneity (Fig. 12.12), a well-documented phenomenon in breast cancer [311–314] poses considerable interpretational challenges [295, 296, 312]. ASCO/CAP guidelines [220] define as significant only spatially clustered heterogeneity; however, scattered *HER2*-amplified cells, although of unknown clinical relevance, are more frequent [296] and usually present a source of interobserver disagreement.

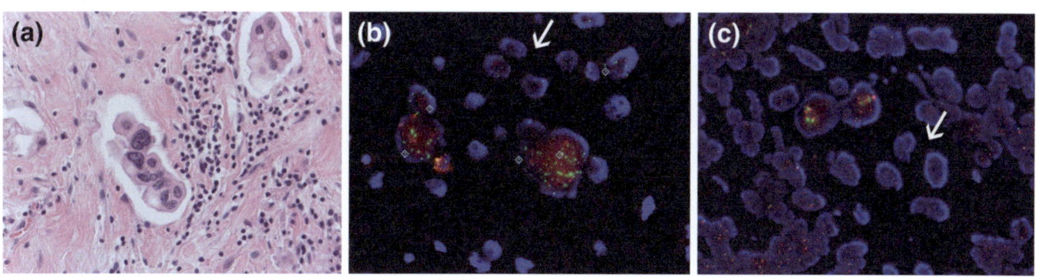


**Fig. 12.13** FISH for *HER2*/CEP17 and *EGFR*/CEP7 in a surgical sample post neoadjuvant treatment (**a** H&E; **b** *HER2*/CEP17, rhomboidal symbols identify syncytial cells with a high number of signals, *arrows* indicate neoplastic cells with 2–3 signals; **c** *EGFR*/CEP7, syncytial cells with a high number of signals are evident in the *middle*, *arrows* indicate neoplastic cells with 2–3 signals)

### 12.9.3 Gene Expression Signatures

In addition to the unraveling of breast cancer complexity due to heterogeneity, microarray-based technologies have contributed the development of gene expression prognostic signatures and at present many commercial molecular tests are available (Table 12.7). Indeed, Oncotype DX [315], Mammaprint [316, 317], Breast Cancer Index (BCI) [318], PAM50 ROR/Prosigna [146], and EndoPredict [319], have been implemented in the clinical setting, at least in the subset of ER-positive disease where they provide independent prognostic information [286, 320]. The prognostic value offered by these proliferation-based prognostic signatures is complementary to that of classical clinicopathologic parameters [321] and these tests help physicians decide which patients could be spared from chemotherapy [320, 322]. It should be noted, however, that pair-wise agreement for individual patients between different commercial tests is suboptimal [323]. The UK OPTIMA-Prelim trial has showed that although there is a modest agreement between tests when dichotomizing results between high

**Table 12.7** Some of the most used commercially available genomic prognostic signatures

| Test and company | Type of specimen required | Type of methodology | Output | Level I evidence |
|---|---|---|---|---|
| MammaPrint® by Agendia BV, Amsterdam, Netherlands | Originally fresh-frozen, possible also on FFPE | Microarray-based gene expression profiling (70 genes) | *Two risk categories*<br>Risk of developing metastasis at 10 years<br>– Low<br>– High | N |
| OncotypeDX™ by Genomic Health Inc., Redwood City, CA, USA | Either fresh-frozen or FFPE tissue | qRT-PCR (21 genes) | *Recurrence score (RS)* Risk of 10-year distant recurrence<br>– Low<br>– Intermediate<br>– High | Y |
| Breast Cancer Index℠ by Biotheranostics, San Diego, USA | Either fresh-frozen or FFPE tissue | qRT-PCR (measures the expression of H/I, MGI and four normalization genes) | *BCI score*<br>Risk of early and late distant recurrence<br>– Low<br>– Intermediate<br>– High | N |
| *Endopredict by Sividon Diagnostics GmbH, Cologne, Germany | FFPE tissue | qRT-PCR<br>(8 genes of interest related to proliferation or *ESR1* pathway + 3 normalization genes) | *EP risk score*<br>Risk of recurrence at 10 years (low risk: <9 % or recurrence at 10 years)<br>– Low<br>– High | Y |
| *Prosigna™ Breast Cancer Prognostic Gene Signature Assay by NanoString Technologies Inc., Seattle, WA, USA | FFPE tissue | NanoString (multiplexed measurement of expression of 50 genes, based on the PAM50 gene signature) | *ROR score*<br>Risk of recurrence<br>– Low<br>– Intermediate<br>– High | Y |

Legend: *Indicates those that can be performed in house; *FFPE* formalin-fixed paraffin-embedded; *H/I*: HOXB13: IL17R ratio (Ma XJ et al., Cancer Cell 2004); *MGI* molecular grade index; *RT-PCR* reverse-transcriptase polymerase chain reaction *versus* low/intermediate grade risk, disagreement between different tests in assigning individual tumors to risk categories is not uncommon: up to 52 % of tumors were indeed assigned to different risk categories by different tests [324].

The prognosis of ER-negative disease has been recently associated with the expression of immune-related genes [325] and prognostic signatures linked to genes involved in immune, inflammatory, and/or chemokine pathways have been developed for ER–negative/TNBCs (STAT1 cluster [285], the IFN cluster [143], the IR-7 [325, 326], the Buck-14 [327], TN-45 [261], and a B-cell/IL-8 metagene ratio [328]). An immune-based signature has also been described for HER2-positive disease to predict benefit from trastuzumab [329]. Nevertheless, until appropriate clinical validation takes place and a better alternative for systemic treatment of patients with HER2-positive and TN tumors is developed, the clinical utility of those second-generation signatures (i.e., unrelated to proliferation) will remain negligible.

### 12.9.4 Lessons from Massively Parallel Sequencing in Brief

The advent of massively parallel sequencing (MPS) has allowed characterization of breast cancer genomes at base-pair resolution increasing dramatically the complexity of intertumor heterogeneity. The TCGA project [166] has analyzed by several high-throughput platforms a large series of breast cancer patients (825 in total) and an integrative semi-supervised clustering analysis including data of MPS, copy number and gene expression analyses (507 patients) revealed four major subgroups which to some extent correlate with luminal A and B, HER2-enriched and basal-like PAM50-defined molecular subtypes. Importantly, however, although the PAM50-defined subtypes display distinct repertoires of somatic mutations, there is no highly recurrent gene or highly recurrent mutation that defines each subtype [45]. The average of somatic mutations in breast cancer is comprised between 1.02 and 1.66 per Mb in coding regions [166, 252, 330–332], thus resulting in a mean of 56.9 (range 5–374) somatic mutations per cancer [333]. Such frequencies are fairly similar to those of ovarian or renal clear cell carcinomas, but lower than those of bladder urothelial (8.03 somatic mutations/Mb) or lung squamous cell carcinoma (9.92 somatic mutations/Mb) [333].

The genes significantly affected by mutations included genes and pathways known to be aberrant in breast cancers, such as *TP53* or the PI3K pathway (i.e., *PIK3CA*, *PTEN*, and *AKT1*) but also genes of functional or cellular processes previously not considered to be major determinants of breast cancer biology, including the MAPK/JNK signaling (i.e., *MAP3K1*, *MAP2K4*, and *NF1*), transcription factors and regulators (i.e., *GATA3*, *RUNX1*, and *CBFB*), splicing factors (i.e., *SF3B1*), and chromatin remodelers (i.e., *MLL3*, *ARID1A*) [166, 333, 334].

These analyses have also showed that the first hopes to find a short list of highly recurrently mutated genes have been largely unattended and indeed only *TP53*, *PIK3CA*, and *GATA3* were found to be mutated in >10 % of unselected breast carcinomas. The remaining genes are actually mutated in less than 7.7 % of cases, with a very long list of genes mutated in less than 1 % of cases [166, 330–334]. Finally, although the list of significantly mutated genes may vary in different studies mostly due to case selection and methods employed, a set of genes composed of *PIK3CA*, *TP53*, *GATA3*, *MAP3K1*, *AKT1*, and *CBFB* genes is quite constant [166, 330, 333, 334] and these are likely to be the drivers of the disease. In the landscape of genes that appear to be mutated at low frequencies, there are some that may be clinically relevant, such as for instance *AKT1*, *NF1*, *ESR1*, and *HER2*, as these are potential therapeutic targets and/or drivers of resistance to known therapies. Somatic *ESR1* mutations have been detected at a much higher frequency in the metastases of breast cancer patients previously treated with aromatase inhibitors than in primary tumors [335–338]. *HER2* mutations have been initially described in ER-positive carcinomas with low levels of HER2 (score 0, score 1+, score 2+ without gene amplification) [339] with an enrichment (up to 23 %) in ILCs [201] and pleomorphic ILCs [199]. The TCGA study of ILCs has found a 4 % of *HER2* mutations, which is in line with that of PAM50-defined luminal carcinomas (cbioportal, queried Nov 7th 2015). In addition, data on *HER2* mutations in trastuzumab-treated (i.e., HER2-positive) patients [340, 341] and in breast cancers with *HER2* heterogeneous gene amplification [313] have been recently reported.

## 12.10 The Genetics of Hereditary Breast Cancer

Hereditary breast cancer accounts for ≤10 % of breast cancers, half of which are due to germline mutations in two high-penetrance genes: *BRCA1* and *BRCA2* [342]. Pathogenic germline mutations in *BRCA1* or *BRCA2* genes are associated with high risk of development of breast and ovarian cancer [343]. Inactivation of the wild type allele is found in most breast cancers developing in *BRCA* germline mutation carriers, and most likely this is needed for their complete functional deficiency, which leads to a DNA repair defect and oncogenic properties [141].

Selection of individuals with familial breast cancer for genetic testing is difficult and based on strong family history, young age, and other clinical and familial characteristics depending on the model adopted [141]; however, it should be noted that these models have been shown to have suboptimal specificity (25–30 %) [344].

BRCA1 tumors display characteristic morphological features [343], i.e., they are more frequently of nonspecial type with medullary features, of grade 3, and more often display pushing borders, brisk lymphocytic infiltrate and necrosis when compared to grade-matched controls and tumors arising in *BRCA2* mutation carriers [343]. Most BRCA1 cancers are of TN phenotype and express basal makers [157]. Conversely, differences between BRCA2 cancers and grade-matched controls are not as conspicuous. BRCA2 cancers seem to be more often of high histologic grade and have pushing borders than matched controls [343], but are more frequently ER-positive despite the higher frequency of higher histologic grade [345]. In addition, these tumors have been found to be more often invasive lobular, pleomorphic lobular, tubular, and cribriform histologic types than sporadic controls in one study [343].

Based on these data, it is not surprising that if morphological features of BRCA2 cancers are of limited help in identifying patients to be screened for mutations, histopathologic models to predict *BRCA1* germline mutations have been developed. Farshid et al. have proposed a system based on ER, PR, and the above morphological features of BRCA1 tumors. It has similar sensitivity when compared to clinical models, but much higher specificity (86 %) and positive and negative predictive values (61 and 98 %, respectively) [344]. Of note, BRCA1 tumors share also similarities with sporadic basal-like breast cancers [157] and it has been demonstrated that the BRCA1 pathway is dysfunctional in most sporadic basal-like cancers [158]. An IHC predictor of *BRCA1* germline mutation using ER and CK5/6 has been shown to have a sensitivity of 56 %, a specificity of 87 %, and positive and negative predictive values of 28 and 99 %, respectively [346]. It should be noted though that the association between tumors of TN phenotype and negative for basal CKs still holds true.

The knowledge of the phenotype and pathological characteristics of BRCA cancers is also important with respect to therapeutic options. Targeted therapies addressing one of the defining features of BRCA cancers have been developed: inactivation of *BRCA1* or *BRCA2* leads to deficiency in homologous recombination repair (HR) of DNA double-strand breaks and interstrand crosslinks. It is not surprising that these tumors are often sensitive to crosslinking agents (e.g., platinum salts) [347], given that platinum salts generate interstrand cross-links that can only be adequately repaired by HR-based DNA repair. In addition, BRCA cancers have an exquisite sensitivity to PARP enzyme inhibitors. These agents block one of the alternative mechanisms of DNA repair (i.e., base excision repair). In cells deficient in HR repair, the inhibition of base excision repair has been shown to result in dramatic levels of chromosomal instability, cell cycle arrest, and subsequent apoptosis [149]. Objective clinical responses for PARP inhibitors in clinical trials have been seen in patients with *BRCA1* and *BRCA2* mutated cancers [347].

*PALB2* (Partner And Localizer of BRCA2) has been recognized as a breast cancer predisposition gene, given that *PALB2* loss of function germline mutations confer a high risk of breast cancer development [348], probably similar risk to that conferred by *BRCA2* mutations. Of note, the loss of function caused by *PALB2* mutations is associated with sensitivity to platinum salts/PARP inhibitors in preclinical models. These mutations account for approximately 2.4 % of the breast cancer familial aggregation and *PALB2* should be added to genetic testing for *BRCA1*/*BRCA2*. Some issues remain still to be addressed, such as the proper evaluation of risk of male breast cancer, ovarian and pancreatic cancer, and the prevalence of *PALB2* mutation in different populations.

## 12.11 Fibroepithelial Lesions

Fibroepithelial lesions of the breast are a heterogeneous group of biphasic neoplasms composed of epithelial and stromal components in a typical architecture. Fibroadenoma and phyllodes tumors (PTs) represent the major entities. The stromal component has been shown to be the neoplastic component of these lesions and is thought to originate from the specialized breast stromal cells of TDLUs. This is partly supported by the expression of CD34 in these cells and the mesenchymal component of these tumors. Notably, hotspot mutations in exon 2 of *MED12* gene are likely to be the founder genetic event both in fibroadenomas and PTs. Due to histologic overlap, differentiation between fibroadenomas and PTs, in particular benign PTs in core biopsies, can represent a diagnostic challenge. As discussed below, novel molecular findings may assist in this setting.

### 12.11.1 Fibroadenoma

Fibroadenoma is a common benign biphasic tumor, well-circumscribed grossly and at imaging, occurring most frequently in women of childbearing age, especially those of <30 years [16]. It may occur, however, at any age; indeed, one study described a second peak incidence in the late 40s to early 50s [349]. They present typically as a solitary, firm, and slow-growing nodules, but synchronous or metachronous multifocal lesions are not uncommon.

Histologically, fibroadenomas are composed of bland stromal cells with no/rare mitotic figures admixed with benign epithelium in pericanalicular (stromal cells proliferate around rounded ducts) and/or intracanalicular patterns (proliferating stromal cells compress the ducts into clefts). Focal or diffuse hypercellularity may occur (cellular fibroadenomas), in particular in young patients (so-called "juvenile fibrodenomas"). Importantly, the proliferation rate of stromal cells should be low; some authors have set a cut-off of two mitoses in 10 high power field [350]. Higher mitotic activity may be found in young and pregnant patients [16]. The stroma may undergo morphological changes such as mixoid degeneration and hyalinization with dystrophic calcification. The epithelial component may be involved by atypical hyperplasia and *in situ* carcinomas; when the latter does not extend to surrounding breast parenchyma, the relative risk of subsequent carcinoma is apparently not increased [351].

Whole exome sequencing of fibroadenomas followed by a validation in a larger series (total $n = 98$) revealed hotspot exon 2 *MED12* mutations in 59 % of cases [352], with a mutational spectrum similar to that of uterine leiomyomas [353]. Therefore, it has been hypothesized that *MED12* mutations may promote tumorigenesis in association with estrogen stimulation.

Fibroadenomas were not found to display additional recurrent mutations [352].

Fibroadenomas are considered lesions of limited growth potential and are adequately treated by local resection or radiologic follow-up. Progression from fibroadenomas to PTs is a controversial topic; there is molecular evidence, however, that this phenomenon may happen [354, 355].

### 12.11.2 Phyllodes Tumor

Phyllodes tumors (PTs) are rare fibroepithelial lesions characterized by a proliferation of stromal cells with varying degree of cellularity, atypia and proliferation, resulting in the formation of leaf-like projections protruding into cleft-like or cystically-dilated spaces lined by benign epithelium [16]. PTs bear resemblance to intracanalicular fibroadenomas, in particular those with hypercellular stroma. According to a variety of histologic features, PTs are classified as benign, borderline or malignant. Local recurrences may occur regardless of grade, and are strongly associated with positive surgical margins in lumpectomy specimens [356]. Overall, approximately 29 % of malignant PTs display metastatic behavior; distant metastases have been documented in 2–3 % and up to 11 % of benign PT and borderline PTs, respectively [357–359].

Clinically, PTs usually present as single and firm breast mass, which may be rapidly growing and large (>10 cm). Multifocal lesions are rare. Grossly, PTs are well circumscribed, but satellite nodules may be present, while microscopically borders may be infiltrative.

Grading is based on stromal cellularity, nuclear pleomorphism and mitotic rate, borders, and presence of stromal overgrowth and/or heterologous components. Briefly, benign PTs are usually of moderate cellularity, low nuclear pleomorphism and display mitotic rate of <5/10 high power fields and well-defined borders. Malignant PTs are usually highly cellular and pleomorphic, have a high mitotic rate (>10/10 high power fields), infiltrative borders, stromal expansion, and may show heterologous stromal components. Borderline PTs display some, but do not fulfill all criteria for malignancy.

As fibroadenomas, PTs are characterized by highly recurrent exon 2 *MED12* mutations [360–363], which seem to be more prevalent in benign than in malignant cases [361, 362, 364]. In contrast, alterations in *bona fide* cancer genes, such as *TP53* and *RB1*, are found only in borderline and malignant samples [363, 365]. Of note, *TERT* hotspot promoter mutations and/or *TERT* gene amplification have been found in up to 65 % of cases [363, 366], and, at a variance with *MED12* mutations, seem to be more prevalent in malignant (68 %) than in benign (18 %) PTs [363]. In addition, *TERT* genetic alterations were extremely rare [366] or not found in fibroadenomas [363]. Therefore, it has been suggested that *TERT* hotspot promoter mutations and/or *TERT* gene amplification may drive the progression of PTs [363, 366].

Additionally, testing for *TERT* genetic changes may help in the differential diagnosis between benign PTs and cellular fibroadenomas, which remains a diagnostic issue with important surgical implications. Due to their tendency for local recurrence, PTs (including those classified as benign) are treated with surgical excision with clear margins, whereas fibroadenomas may be followed-up. These molecular tests display an excellent positive predictive factor for PT ($\sim$100 %), but it should be pointed out that the negative predictive factor is suboptimal.

Treatment of metastatic lesions from PTs has limited options, as PTs tend to be resistant to available systemic therapies. Nevertheless, malignant PTs were found to harbor mutations in actionable genes, such as *PIK3CA*, *ERBB2*, and *ERBB3* [363, 365]. Therefore, genetic profiling of malignant PTs may assist in the identification of targeted therapies for patients with this locally aggressive and potentially metastatic tumor type.

## 12.12 Summary

In this chapter, we have discussed the basics about development of the mammary gland, its anatomy and physiology, which are instrumental to understand the breast pathological features. The chapter mainly focused on epithelial lesions, including risk indicators, preinvasive lesions, and invasive carcinomas. The sections provided the basics about morphology and the most recent findings about molecular pathology of each entity. At the end, an overview of fibroepithelial lesion is provided, given the recently described molecular findings associated to these lesions.

## References


1. Rosen PP. Anatomy and physiological morphology. In: Rosen's breast pathology. Philadelphia: Lippincott Williams and Wilkins; 2009. p. 1–25.
2. Anbazhagan R, Osin PP, Bartkova J, Nathan B, Lane EB, Gusterson BA. The development of epithelial phenotypes in the human fetal and infant breast. J Pathol. 1998;184(2):197–206.
3. Ham AW, Comrack DH. The breast. In: Histology. Philadelphia: Lipponcott; 1979. p. 866–74.
4. Adriance MC, Inman JL, Petersen OW, Bissell MJ. Myoepithelial cells: good fences make good neighbors. Breast Cancer Res. 2005;7(5):190–7.
5. Topper YJ, Freeman CS. Multiple hormone interactions in the developmental biology of the mammary gland. Physiol Rev. 1980;60(4):1049–106.
6. Sternlicht MD. Key stages in mammary gland development: the cues that regulate ductal branching morphogenesis. Breast Cancer Res. 2006;8(1):201.
7. Collins LC, Schnitt S. Breast. In: Histology for pathologists. Philadelphia: Lippincott Williams and Wilkins; 2007. p. 55–72.
8. Bussolati G. Clear cell metaplasia of the breast. Histopathology. 1990;16(3):316.
9. Toker C. Clear cells of the nipple epidermis. Cancer. 1970;25(3):601–10.
10. Rosen PP. Adenosis and microglandular adenosis. In: Rosen's breast pathology. Philadelphia: Lippincott Williams and Wilkins; 2009. p. 161–86.
11. Haagensen CD. Adenosis tumor. In: Diseases of the breast. Philadelphia: WB Saunders; 1971.
12. Bodian CA, Perzin KH, Lattes R, Hoffmann P, Abernathy TG. Prognostic significance of benign proliferative breast disease. Cancer. 1993;71(12):3896–907.
13. Kamel OW, Kempson RL, Hendrickson MR. In situ proliferative epithelial lesions of the breast. Pathology (Phila). 1992;1(1):65–102.
14. Lopez-Garcia MA, Geyer FC, Lacroix-Triki M, Marchio C, Reis-Filho JS. Breast cancer precursors revisited: molecular features and progression pathways. Histopathology. 2010;57(2):171–92.
15. Washington C, Dalbegue F, Abreo F, Taubenberger JK, Lichy JH. Loss of heterozygosity in fibrocystic change of the breast: genetic relationship between benign proliferative lesions and associated carcinomas. Am J Pathol. 2000;157(1):323–9.
16. Lakhani SR, Ellis IO, Schnitt SJ, Hoon Tan P, Van de Vijver M. WHO classification of tumors of the breast. Lyon: IARC Press; 2012.
17. Rosen PP. Ductal hyperplasia: usual and atypical. In: Rosen's breast pathology. Philadelphia: Lippincott Williams and Wilkins; 2009. p. 230–63.
18. Harada O, Hoe R, Lin J, Thike AA, Jara-Lazaro AR, Petersson F, et al. Intranuclear inclusions in epithelial cells of benign proliferative breast lesions. J Clin Pathol. 2011;64(9):776–80.
19. Lauer S, Oprea-Ilies G, Cohen C, Adsay V, Adams AL. Acidophilic nuclear inclusions are specific for florid ductal hyperplasia among proliferative breast lesions. Arch Pathol Lab Med. 2011;135(6):766–9.
20. O'Connell P, Pekkel V, Fuqua SA, Osborne CK, Clark GM, Allred DC. Analysis of loss of heterozygosity in 399 premalignant breast lesions at 15 genetic loci. J Natl Cancer Inst. 1998;90(9):697–703.
21. Reis-Filho JS, Lakhani SR. The diagnosis and management of pre-invasive breast disease: genetic alterations in pre-invasive lesions. Breast Cancer Res. 2003;5(6):313–9.
22. Aubele MM, Cummings MC, Mattis AE, Zitzelsberger HF, Walch AK, Kremer M, et al. Accumulation of chromosomal imbalances from intraductal proliferative lesions to adjacent in situ and invasive ductal breast cancer. Diagn Mol Pathol. 2000;9(1):14–9.
23. Jones C, Merrett S, Thomas VA, Barker TH, Lakhani SR. Comparative genomic hybridization analysis of bilateral hyperplasia of usual type of the breast. J Pathol. 2003;199(2):152–6.
24. Simpson PT, Reis-Filho JS, Gale T, Lakhani SR. Molecular evolution of breast cancer. J Pathol. 2005;205(2):248–54.
25. Lakhani SR, Slack DN, Hamoudi RA, Collins N, Stratton MR, Sloane JP. Detection of allelic imbalance indicates that a proportion of mammary hyperplasia of usual type are clonal, neoplastic proliferations. Lab Invest. 1996;74(1):129–35.
26. Simpson PT, Gale T, Reis-Filho JS, Jones C, Parry S, Sloane JP, et al. Columnar cell lesions of the breast: the missing link in breast cancer progression? A morphological and molecular analysis. Am J Surg Pathol. 2005;29(6):734–46.
27. Boecker W, Moll R, Dervan P, Buerger H, Poremba C, Diallo RI, et al. Usual ductal hyperplasia of the breast is a committed stem (progenitor) cell lesion distinct from atypical ductal hyperplasia and ductal carcinoma in situ. J Pathol. 2002;198(4):458–67.
28. Arriola E, Lambros MB, Jones C, Dexter T, Mackay A, Tan DS, et al. Evaluation of Phi29-based whole-genome amplification for microarray-based

comparative genomic hybridisation. Lab Invest. 2007;87(1):75–83.
29. Li H, Zhu R, Wang L, Zhu T, Li Q, Chen Q, et al. PIK3CA mutations mostly begin to develop in ductal carcinoma of the breast. Exp Mol Pathol. 2010;88(1):150–5.
30. Ang DC, Warrick AL, Shilling A, Beadling C, Corless CL, Troxell ML. Frequent phosphatidylinositol-3-kinase mutations in proliferative breast lesions. Mod Pathol. 2014;27(5):740–50.
31. Rosen PP. Papilloma and related benign tumors. In: Rosen's breast pathology. Philadelphia: Lippincott Williams and Wilkins; 2009. p. 85–136.
32. Andersen JA, Gram JB. Radial scar in the female breast. A long-term follow-up study of 32 cases. Cancer. 1984;53(11):2557–60.
33. Anderson TJ, Battersby S. Radial scars of benign and malignant breasts: comparative features and significance. J Pathol. 1985;147(1):23–32.
34. Fisher ER, Palekar AS, Kotwal N, Lipana N. A nonencapsulated sclerosing lesion of the breast. Am J Clin Pathol. 1979;71(3):240–6.
35. Berg JC, Visscher DW, Vierkant RA, Pankratz VS, Maloney SD, Lewis JT, et al. Breast cancer risk in women with radial scars in benign breast biopsies. Breast Cancer Res Treat. 2008;108(2):167–74.
36. Jacobs TW, Byrne C, Colditz G, Connolly JL, Schnitt SJ. Radial scars in benign breast-biopsy specimens and the risk of breast cancer. N Engl J Med. 1999;340(6):430–6.
37. Sanders ME, Schuyler PA, Dupont WD, Page DL. The natural history of low-grade ductal carcinoma in situ of the breast in women treated by biopsy only revealed over 30 years of long-term follow-up. Cancer. 2005;103(12):2481–4.
38. Aroner SA, Collins LC, Connolly JL, Colditz GA, Schnitt SJ, Rosner BA, et al. Radial scars and subsequent breast cancer risk: results from the Nurses' Health Studies. Breast Cancer Res Treat. 2013;139(1):277–85.
39. Manfrin E, Remo A, Falsirollo F, Reghellin D, Bonetti F. Risk of neoplastic transformation in asymptomatic radial scar. Analysis of 117 cases. Breast Cancer Res Treat. 2008;107(3):371–7.
40. Denley H, Pinder SE, Tan PH, Sim CS, Brown R, Barker T, et al. Metaplastic carcinoma of the breast arising within complex sclerosing lesion: a report of five cases. Histopathology. 2000;36(3):203–9.
41. Gobbi H, Simpson JF, Jensen RA, Olson SJ, Page DL. Metaplastic spindle cell breast tumors arising within papillomas, complex sclerosing lesions, and nipple adenomas. Mod Pathol. 2003;16(9):893–901.
42. Sloane JP, Mayers MM. Carcinoma and atypical hyperplasia in radial scars and complex sclerosing lesions: importance of lesion size and patient age. Histopathology. 1993;23(3):225–31.
43. Wellings SR, Alpers CE. Subgross pathologic features and incidence of radial scars in the breast. Hum Pathol. 1984;15(5):475–9.
44. Wolters KL, Ang D, Warrick A, Beadling C, Corless CL, Troxell ML. Frequent PIK3CA mutations in radial scars. Diagn Mol Pathol. 2013;22(4):210–4.
45. Ng CK, Schultheis AM, Bidard FC, Weigelt B, Reis-Filho JS. Breast cancer genomics from microarrays to massively parallel sequencing: paradigms and new insights. J Natl Cancer Inst. 2015;107(5).
46. Troxell ML, Levine J, Beadling C, Warrick A, Dunlap J, Presnell A, et al. High prevalence of PIK3CA/AKT pathway mutations in papillary neoplasms of the breast. Mod Pathol. 2010;23(1):27–37.
47. Troxell ML, Brunner AL, Neff T, Warrick A, Beadling C, Montgomery K, et al. Phosphatidylinositol-3-kinase pathway mutations are common in breast columnar cell lesions. Mod Pathol. 2012;25(7):930–7.
48. Eusebi V, Millis RR. Epitheliosis, infiltrating epitheliosis, and radial scar. Semin Diagn Pathol. 2010;27(1):5–12.
49. Azzopardi JG. Overdiagnosis of malignancy. London, UK: W.B. Saunders Company, Ltd; 1979.
50. Eberle CA, Piscuoglio S, Rakha EA, Ng CK, Geyer FC, Edelweiss M, et al. Infiltrating epitheliosis of the breast: characterization of histologic features, immunophenotype and genomic profile. Histopathology. 2015.
51. Droufakou S, Deshmane V, Roylance R, Hanby A, Tomlinson I, Hart IR. Multiple ways of silencing E-cadherin gene expression in lobular carcinoma of the breast. Int J Cancer. 2001;92(3):404–8.
52. Dupont WD, Page DL. Risk factors for breast cancer in women with proliferative breast disease. N Engl J Med. 1985;312(3):146–51.
53. Page DL, Dupont WD. Anatomic markers of human premalignancy and risk of breast cancer. Cancer. 1990;66(6 Suppl):1326–35.
54. Page DL, Dupont WD. Proliferative breast disease: diagnosis and implications. Science. 1991;253(5022):915–6.
55. Page DL, Dupont WD. Indicators of increased breast cancer risk in humans. J Cell Biochem Suppl. 1992;16G:175–82.
56. Abdel-Fatah TM, Powe DG, Hodi Z, Lee AH, Reis-Filho JS, Ellis IO. High frequency of coexistence of columnar cell lesions, lobular neoplasia, and low grade ductal carcinoma in situ with invasive tubular carcinoma and invasive lobular carcinoma. Am J Surg Pathol. 2007;31(3):417–26.
57. Abdel-Fatah TM, Powe DG, Hodi Z, Reis-Filho JS, Lee AH, Ellis IO. Morphologic and molecular evolutionary pathways of low nuclear grade invasive breast cancers and their putative precursor lesions: further evidence to support the concept of low nuclear grade breast neoplasia family. Am J Surg Pathol. 2008;32(4):513–23.
58. Brandt SM, Young GQ, Hoda SA. The "Rosen Triad": tubular carcinoma, lobular carcinoma in situ,

and columnar cell lesions. Adv Anat Pathol. 2008;15(3):140–6.
59. Gewinner C, Wang ZC, Richardson A, Teruya-Feldstein J, Etemadmoghadam D, Bowtell D, et al. Evidence that inositol polyphosphate 4-phosphatase type II is a tumor suppressor that inhibits PI3K signaling. Cancer Cell. 2009;16(2):115–25.
60. Reis-Filho JS, Simpson PT, Gale T, Lakhani SR. The molecular genetics of breast cancer: the contribution of comparative genomic hybridization. Pathol Res Pract. 2005;201(11):713–25.
61. Geyer FC, Kushner YB, Lambros MB, Natrajan R, Mackay A, Tamber N, et al. Microglandular adenosis or microglandular adenoma? A molecular genetic analysis of a case associated with atypia and invasive carcinoma. Histopathology. 2009;55(6):732–43.
62. Geyer FC, Lacroix-Triki M, Colombo PE, Patani N, Gauthier A, Natrajan R, et al. Molecular evidence in support of the neoplastic and precursor nature of microglandular adenosis. Histopathology. 2012;60(6B):E115–30.
63. Shin SJ, Simpson PT, Da Silva L, Jayanthan J, Reid L, Lakhani SR, et al. Molecular evidence for progression of microglandular adenosis (MGA) to invasive carcinoma. Am J Surg Pathol. 2009;33(4):496–504.
64. Ma XJ, Salunga R, Tuggle JT, Gaudet J, Enright E, McQuary P, et al. Gene expression profiles of human breast cancer progression. Proc Natl Acad Sci USA. 2003;100(10):5974–9.
65. Natrajan R, Lambros MB, Geyer FC, Marchio C, Tan DS, Vatcheva R, et al. Loss of 16q in high grade breast cancer is associated with estrogen receptor status: evidence for progression in tumors with a luminal phenotype? Genes Chromosom Cancer. 2009;48(4):351–65.
66. Feeley L, Quinn CM. Columnar cell lesions of the breast. Histopathology. 2008;52(1):11–9.
67. Fraser JL, Raza S, Chorny K, Connolly JL, Schnitt SJ. Columnar alteration with prominent apical snouts and secretions: a spectrum of changes frequently present in breast biopsies performed for microcalcifications. Am J Surg Pathol. 1998;22(12):1521–7.
68. Goldstein NS, O'Malley BA. Cancerization of small ectatic ducts of the breast by ductal carcinoma in situ cells with apocrine snouts: a lesion associated with tubular carcinoma. Am J Clin Pathol. 1997;107(5):561–6.
69. Pinder SE, Reis-Filho JS. Non-operative breast pathology. J Clin Pathol. 2007;60(12):1297–9.
70. Schnitt SJ. The diagnosis and management of pre-invasive breast disease: flat epithelial atypia—classification, pathologic features and clinical significance. Breast Cancer Res. 2003;5(5):263–8.
71. Dabbs DJ, Carter G, Fudge M, Peng Y, Swalsky P, Finkelstein S. Molecular alterations in columnar cell lesions of the breast. Mod Pathol. 2006;19(3):344–9.
72. Moinfar F, Man YG, Bratthauer GL, Ratschek M, Tavassoli FA. Genetic abnormalities in mammary ductal intraepithelial neoplasia-flat type ("clinging ductal carcinoma in situ"): a simulator of normal mammary epithelium. Cancer. 2000;88(9):2072–81.
73. Rosen PP. Columnar cell hyperplasia is associated with lobular carcinoma in situ and tubular carcinoma. Am J Surg Pathol. 1999;23(12):1561.
74. Boulos FI, Dupont WD, Schuyler PA, Sanders ME, Page DL, Fedda FA, et al. Clinicopathologic characteristics of carcinomas that develop after a biopsy containing columnar cell lesions: evidence against a precursor role. Cancer. 2012;118(9):2372–7.
75. Boulos FI, Dupont WD, Simpson JF, Schuyler PA, Sanders ME, Freudenthal ME, et al. Histologic associations and long-term cancer risk in columnar cell lesions of the breast: a retrospective cohort and a nested case-control study. Cancer. 2008;113(9):2415–21.
76. Wellings SR, Jensen HM. On the origin and progression of ductal carcinoma in the human breast. J Natl Cancer Inst. 1973;50(5):1111–8.
77. Wellings SR, Jensen HM, Marcum RG. An atlas of subgross pathology of the human breast with special reference to possible precancerous lesions. J Natl Cancer Inst. 1975;55(2):231–73.
78. Dabbs DJ, Schnitt SJ, Geyer FC, Weigelt B, Baehner FL, Decker T, et al. Lobular neoplasia of the breast revisited with emphasis on the role of E-cadherin immunohistochemistry. Am J Surg Pathol. 2013;37(7):e1–11.
79. Foote FW Jr, Stewart FW. A histologic classification of carcinoma of the breast. Surgery. 1946;19:74–99.
80. Haagensen CD, Lane N, Lattes R, Bodian C. Lobular neoplasia (so-called lobular carcinoma in situ) of the breast. Cancer. 1978;42(2):737–69.
81. Hanby AM, Hughes TA. In situ and invasive lobular neoplasia of the breast. Histopathology. 2008;52(1):58–66.
82. Page DL, Dupont WD, Rogers LW. Ductal involvement by cells of atypical lobular hyperplasia in the breast: a long-term follow-up study of cancer risk. Hum Pathol. 1988;19(2):201–7.
83. Page DL, Dupont WD, Rogers LW, Rados MS. Atypical hyperplastic lesions of the female breast. A long-term follow-up study. Cancer. 1985;55(11):2698–708.
84. Lu YJ, Osin P, Lakhani SR, Di Palma S, Gusterson BA, Shipley JM. Comparative genomic hybridization analysis of lobular carcinoma in situ and atypical lobular hyperplasia and potential roles for gains and losses of genetic material in breast neoplasia. Cancer Res. 1998;58(20):4721–7.
85. Mastracci TL, Shadeo A, Colby SM, Tuck AB, O'Malley FP, Bull SB, et al. Genomic alterations in lobular neoplasia: a microarray comparative genomic hybridization signature for early neoplastic proliferationin the breast. Genes Chromosom Cancer. 2006;45(11):1007–17.

86. Fitzgibbons PL, Henson DE, Hutter RV. Benign breast changes and the risk for subsequent breast cancer: an update of the 1985 consensus statement. Cancer Committee of the College of American Pathologists. Arch Pathol Lab Med. 1998;122 (12):1053–5.
87. Reis-Filho JS, Pinder SE. Non-operative breast pathology: lobular neoplasia. J Clin Pathol. 2007;60(12):1321–7.
88. Page DL, Schuyler PA, Dupont WD, Jensen RA, Plummer WD Jr, Simpson JF. Atypical lobular hyperplasia as a unilateral predictor of breast cancer risk: a retrospective cohort study. Lancet. 2003;361 (9352):125–9.
89. Lakhani SR, Audretsch W, Cleton-Jensen AM, Cutuli B, Ellis I, Eusebi V, et al. The management of lobular carcinoma in situ (LCIS). Is LCIS the same as ductal carcinoma in situ (DCIS)? Eur J Cancer. 2006;42(14):2205–11.
90. Schnitt SJ, Morrow M. Lobular carcinoma in situ: current concepts and controversies. Semin Diagn Pathol. 1999;16(3):209–23.
91. Derksen PW, Liu X, Saridin F, van der Gulden H, Zevenhoven J, Evers B, et al. Somatic inactivation of E-cadherin and p53 in mice leads to metastatic lobular mammary carcinoma through induction of anoikis resistance and angiogenesis. Cancer Cell. 2006;10(5):437–49.
92. Sarrio D, Moreno-Bueno G, Hardisson D, Sanchez-Estevez C, Guo M, Herman JG, et al. Epigenetic and genetic alterations of APC and CDH1 genes in lobular breast cancer: relationships with abnormal E-cadherin and catenin expression and microsatellite instability. Int J Cancer. 2003;106 (2):208–15.
93. Vos CB, Cleton-Jansen AM, Berx G, de Leeuw WJ, ter Haar NT, van Roy F, et al. E-cadherin inactivation in lobular carcinoma in situ of the breast: an early event in tumorigenesis. Br J Cancer. 1997;76 (9):1131–3.
94. Hwang ES, Nyante SJ, Yi Chen Y, Moore D, DeVries S, Korkola JE, et al. Clonality of lobular carcinoma in situ and synchronous invasive lobular carcinoma. Cancer. 2004;100(12):2562–72.
95. Morandi L, Marucci G, Foschini MP, Cattani MG, Pession A, Riva C, et al. Genetic similarities and differences between lobular in situ neoplasia (LN) and invasive lobular carcinoma of the breast. Virchows Arch. 2006;449(1):14–23.
96. Nyante SJ, Devries S, Chen YY, Hwang ES. Array-based comparative genomic hybridization of ductal carcinoma in situ and synchronous invasive lobular cancer. Hum Pathol. 2004;35(6):759–63.
97. Andrade VP, Morrogh M, Qin LX, Olvera N, Giri D, Muhsen S, et al. Gene expression profiling of lobular carcinoma in situ reveals candidate precursor genes for invasion. Mol Oncol. 2015;9(4):772–82.
98. King TA, Pilewskie M, Muhsen S, Patil S, Mautner SK, Park A, et al. Lobular carcinoma in situ: a 29-year longitudinal experience evaluating clinicopathologic features and breast cancer risk. J Clin Oncol. 2015.
99. Shin SJ, Lal A, De Vries S, Suzuki J, Roy R, Hwang ES, et al. Florid lobular carcinoma in situ: molecular profiling and comparison to classic lobular carcinoma in situ and pleomorphic lobular carcinoma in situ. Hum Pathol. 2013;44(10):1998–2009.
100. Page DL, Dupont WD, Rogers LW, Landenberger M. Intraductal carcinoma of the breast: follow-up after biopsy only. Cancer. 1982;49 (4):751–8.
101. Tavassoli FA, Norris HJ. A comparison of the results of long-term follow-up for atypical intraductal hyperplasia and intraductal hyperplasia of the breast. Cancer. 1990;65(3):518–29.
102. Amari M, Suzuki A, Moriya T, Yoshinaga K, Amano G, Sasano H, et al. LOH analyses of premalignant and malignant lesions of human breast: frequent LOH in 8p, 16q, and 17q in atypical ductal hyperplasia. Oncol Rep. 1999;6(6):1277–80.
103. Larson PS, de las Morenas A, Cerda SR, Bennett SR, Cupples LA, Rosenberg CL. Quantitative analysis of allele imbalance supports atypical ductal hyperplasia lesions as direct breast cancer precursors. J Pathol. 2006;209(3):307–16.
104. Lakhani SR, Collins N, Stratton MR, Sloane JP. Atypical ductal hyperplasia of the breast: clonal proliferation with loss of heterozygosity on chromosomes 16q and 17p. J Clin Pathol. 1995;48 (7):611–5.
105. Stratton MR, Collins N, Lakhani SR, Sloane JP. Loss of heterozygosity in ductal carcinoma in situ of the breast. J Pathol. 1995;175(2):195–201.
106. Buerger H, Otterbach F, Simon R, Poremba C, Diallo R, Decker T, et al. Comparative genomic hybridization of ductal carcinoma in situ of the breast-evidence of multiple genetic pathways. J Pathol. 1999;187(4):396–402.
107. Gong G, DeVries S, Chew KL, Cha I, Ljung BM, Waldman FM. Genetic changes in paired atypical and usual ductal hyperplasia of the breast by comparative genomic hybridization. Clin Cancer Res. 2001;7(8):2410–4.
108. Hughes LL, Wang M, Page DL, Gray R, Solin LJ, Davidson NE, et al. Local excision alone without irradiation for ductal carcinoma in situ of the breast: a trial of the Eastern Cooperative Oncology Group. J Clin Oncol. 2009;27(32):5319–24.
109. Allred DC, Wu Y, Mao S, Nagtegaal ID, Lee S, Perou CM, et al. Ductal carcinoma in situ and the emergence of diversity during breast cancer evolution. Clin Cancer Res. 2008;14(2):370–8.
110. Hernandez L, Wilkerson PM, Lambros MB, Campion-Flora A, Rodrigues DN, Gauthier A, et al. Genomic and mutational profiling of ductal carcinomas in situ and matched adjacent invasive breast cancers reveals intra-tumour genetic heterogeneity and clonal selection. J Pathol. 2012;227 (1):42–52.

111. Solin LJ, Gray R, Baehner FL, Butler SM, Hughes LL, Yoshizawa C, et al. A multigene expression assay to predict local recurrence risk for ductal carcinoma in situ of the breast. J Natl Cancer Inst. 2013;105(10):701–10.
112. Solin LJ, Gray R, Hughes LL, Wood WC, Lowen MA, Badve SS, et al. Surgical excision without radiation for ductal carcinoma in situ of the breast: 12-year results from the ECOG-ACRIN E5194 study. J Clin Oncol. 2015.
113. Eusebi V, Magalhaes F, Azzopardi JG. Pleomorphic lobular carcinoma of the breast: an aggressive tumor showing apocrine differentiation. Hum Pathol. 1992;23(6):655–62.
114. Middleton LP, Palacios DM, Bryant BR, Krebs P, Otis CN, Merino MJ. Pleomorphic lobular carcinoma: morphology, immunohistochemistry, and molecular analysis. Am J Surg Pathol. 2000;24 (12):1650–6.
115. Sneige N, Wang J, Baker BA, Krishnamurthy S, Middleton LP. Clinical, histopathologic, and biologic features of pleomorphic lobular (ductal-lobular) carcinoma in situ of the breast: a report of 24 cases. Mod Pathol. 2002;15(10):1044–50.
116. Chen YY, Hwang ES, Roy R, DeVries S, Anderson J, Wa C, et al. Genetic and phenotypic characteristics of pleomorphic lobular carcinoma in situ of the breast. Am J Surg Pathol. 2009;33 (11):1683–94.
117. Reis-Filho JS, Simpson PT, Jones C, Steele D, Mackay A, Iravani M, et al. Pleomorphic lobular carcinoma of the breast: role of comprehensive molecular pathology in characterization of an entity. J Pathol. 2005;207(1):1–13.
118. Vargas AC, Lakhani SR, Simpson PT. Pleomorphic lobular carcinoma of the breast: molecular pathology and clinical impact. Future Oncol. 2009;5(2):233–43.
119. Chivukula M, Haynik DM, Brufsky A, Carter G, Dabbs DJ. Pleomorphic lobular carcinoma in situ (PLCIS) on breast core needle biopsies: clinical significance and immunoprofile. Am J Surg Pathol. 2008;32(11):1721–6.
120. Weidner N, Semple JP. Pleomorphic variant of invasive lobular carcinoma of the breast. Hum Pathol. 1992;23(10):1167–71.
121. Simpson PT, Reis-Filho JS, Lambros MB, Jones C, Steele D, Mackay A, et al. Molecular profiling pleomorphic lobular carcinomas of the breast: evidence for a common molecular genetic pathway with classic lobular carcinomas. J Pathol. 2008;215 (3):231–44.
122. Clement PB, Azzopardi JG. Microglandular adenosis of the breast—a lesion simulating tubular carcinoma. Histopathology. 1983;7(2):169–80.
123. Rosen PP. Microglandular adenosis. A benign lesion simulating invasive mammary carcinoma. Am J Surg Pathol. 1983;7(2):137–44.
124. Tavassoli FA, Norris HJ. Microglandular adenosis of the breast. A clinicopathologic study of 11 cases with ultrastructural observations. Am J Surg Pathol. 1983;7(8):731–7.
125. Acs G, Simpson JF, Bleiweiss IJ, Hugh J, Reynolds C, Olson S, et al. Microglandular adenosis with transition into adenoid cystic carcinoma of the breast. Am J Surg Pathol. 2003;27(8):1052–60.
126. Koenig C, Dadmanesh F, Bratthauer GL, Tavassoli FA. Carcinoma arising in microglandular adenosis: an immunohistochemical analysis of 20 intraepithelial and invasive neoplasms. Int J Surg Pathol. 2000;8(4):303–15.
127. Khalifeh IM, Albarracin C, Diaz LK, Symmans FW, Edgerton ME, Hwang RF, et al. Clinical, histopathologic, and immunohistochemical features of microglandular adenosis and transition into in situ and invasive carcinoma. Am J Surg Pathol. 2008;32 (4):544–52.
128. Pal SK, Lau SK, Kruper L, Nwoye U, Garberoglio C, Gupta RK, et al. Papillary carcinoma of the breast: an overview. Breast Cancer Res Treat. 2010;122 (3):637–45.
129. Rakha EA, Gandhi N, Climent F, van Deurzen CH, Haider SA, Dunk L, et al. Encapsulated papillary carcinoma of the breast: an invasive tumor with excellent prognosis. Am J Surg Pathol. 2011;35 (8):1093–103.
130. Wynveen CA, Nehhozina T, Akram M, Hassan M, Norton L, Van Zee KJ, et al. Intracystic papillary carcinoma of the breast: an in situ or invasive tumor? Results of immunohistochemical analysis and clinical follow-up. Am J Surg Pathol. 2011;35 (1):1–14.
131. Collins LC, Carlo VP, Hwang H, Barry TS, Gown AM, Schnitt SJ. Intracystic papillary carcinomas of the breast: a reevaluation using a panel of myoepithelial cell markers. Am J Surg Pathol. 2006;30(8):1002–7.
132. Collins LC, Schnitt SJ. Papillary lesions of the breast: selected diagnostic and management issues. Histopathology. 2008;52(1):20–9.
133. Hill CB, Yeh IT. Myoepithelial cell staining patterns of papillary breast lesions: from intraductal papillomas to invasive papillary carcinomas. Am J Clin Pathol. 2005;123(1):36–44.
134. Mulligan AM, O'Malley FP. Metastatic potential of encapsulated (intracystic) papillary carcinoma of the breast: a report of 2 cases with axillary lymph node micrometastases. Int J Surg Pathol. 2007;15(2):143–7.
135. Tan PH, Schnitt SJ, van de Vijver MJ, Ellis IO, Lakhani SR. Papillary and neuroendocrine breast lesions: the WHO stance. Histopathology. 2015;66 (6):761–70.
136. Nassar H, Qureshi H, Adsay NV, Visscher D. Clinicopathologic analysis of solid papillary carcinoma of the breast and associated invasive carcinomas. Am J Surg Pathol. 2006;30(4):501–7.
137. Piscuoglio S, Ng CK, Martelotto LG, Eberle CA, Cowell CF, Natrajan R, et al. Integrative genomic and transcriptomic characterization of papillary

carcinomas of the breast. Mol Oncol. 2014;8 (8):1588–602.
138. Rakha EA, Badve S, Eusebi V, Reis-Filho JS, Fox SB, Dabbs DJ, et al. Breast lesions of uncertain malignant nature and limited metastatic potential: proposals to improve their recognition and clinical management. Histopathology. 2015.
139. Ellis IO, Galea M, Broughton N, Locker A, Blamey RW, Elston CW. Pathological prognostic factors in breast cancer. II. Histological type. Relationship with survival in a large study with long-term follow-up. Histopathology. 1992;20 (6):479–89.
140. Goldhirsch A, Wood WC, Coates AS, Gelber RD, Thurlimann B, Senn HJ. Strategies for subtypes—dealing with the diversity of breast cancer: highlights of the St. Gallen International Expert Consensus on the Primary Therapy of Early Breast Cancer 2011. Ann Oncol. 2011;22(8):1736–47.
141. Marchiò C, Reis-Filho JS. Molecular diagnosis of breast cancer. Diagn Mol Pathol. 2008.
142. Perou CM, Sorlie T, Eisen MB, van de Rijn M, Jeffrey SS, Rees CA, et al. Molecular portraits of human breast tumours. Nature. 2000;406 (6797):747–52.
143. Hu Z, Fan C, Oh DS, Marron JS, He X, Qaqish BF, et al. The molecular portraits of breast tumors are conserved across microarray platforms. BMC Genomics. 2006;7:96.
144. Sorlie T, Perou CM, Tibshirani R, Aas T, Geisler S, Johnsen H, et al. Gene expression patterns of breast carcinomas distinguish tumor subclasses with clinical implications. Proc Natl Acad Sci USA. 2001;98 (19):10869–74.
145. Sorlie T, Tibshirani R, Parker J, Hastie T, Marron JS, Nobel A, et al. Repeated observation of breast tumor subtypes in independent gene expression data sets. Proc Natl Acad Sci USA. 2003;100(14):8418–23.
146. Parker JS, Mullins M, Cheang MC, Leung S, Voduc D, Vickery T, et al. Supervised risk predictor of breast cancer based on intrinsic subtypes. J Clin Oncol. 2009;27(8):1160–7.
147. Prat A, Parker JS, Karginova O, Fan C, Livasy C, Herschkowitz JI, et al. Phenotypic and molecular characterization of the claudin-low intrinsic subtype of breast cancer. Breast Cancer Res. 2010;12(5): R68.
148. Doane AS, Danso M, Lal P, Donaton M, Zhang L, Hudis C, et al. An estrogen receptor-negative breast cancer subset characterized by a hormonally regulated transcriptional program and response to androgen. Oncogene. 2006;25(28):3994–4008.
149. Farmer P, Bonnefoi H, Becette V, Tubiana-Hulin M, Fumoleau P, Larsimont D, et al. Identification of molecular apocrine breast tumours by microarray analysis. Oncogene. 2005;24(29):4660–71.
150. Guedj M, Marisa L, de Reynies A, Orsetti B, Schiappa R, Bibeau F, et al. A refined molecular taxonomy of breast cancer. Oncogene. 2012;31 (9):1196–206.
151. Rakha EA, Elsheikh SE, Aleskandarany MA, Habashi HO, Green AR, Powe DG, et al. Triple-negative breast cancer: distinguishing between basal and nonbasal subtypes. Clin Cancer Res. 2009;15(7):2302–10.
152. Rakha EA, Reis-Filho JS, Ellis IO. Basal-like breast cancer: a critical review. J Clin Oncol. 2008;26 (15):2568–81.
153. Rakha EA, Tan DS, Foulkes WD, Ellis IO, Tutt A, Nielsen TO, et al. Are triple-negative tumours and basal-like breast cancer synonymous? Breast Cancer Res. 2007;9(6):404; author reply 5.
154. Prat A, Adamo B, Cheang MC, Anders CK, Carey LA, Perou CM. Molecular characterization of basal-like and non-basal-like triple-negative breast cancer. Oncologist. 2013;18(2):123–33.
155. Holstege H, Horlings HM, Velds A, Langerod A, Borresen-Dale AL, van de Vijver MJ, et al. BRCA1-mutated and basal-like breast cancers have similar aCGH profiles and a high incidence of protein truncating TP53 mutations. BMC Cancer. 2010;10:654.
156. Turner N, Tutt A, Ashworth A. Hallmarks of 'BRCAness' in sporadic cancers. Nat Rev Cancer. 2004;4(10):814–9.
157. Turner NC, Reis-Filho JS. Basal-like breast cancer and the BRCA1 phenotype. Oncogene. 2006;25 (43):5846–53.
158. Turner NC, Reis-Filho JS, Russell AM, Springall RJ, Ryder K, Steele D, et al. BRCA1 dysfunction in sporadic basal-like breast cancer. Oncogene. 2007;26(14):2126–32.
159. Lehmann BD, Bauer JA, Chen X, Sanders ME, Chakravarthy AB, Shyr Y, et al. Identification of human triple-negative breast cancer subtypes and preclinical models for selection of targeted therapies. J Clin Invest. 2011;121(7):2750–67.
160. Burstein MD, Tsimelzon A, Poage GM, Covington KR, Contreras A, Fuqua SA, et al. Comprehensive genomic analysis identifies novel subtypes and targets of triple-negative breast cancer. Clin Cancer Res. 2014;21(7):1688–98.
161. Lehmann BD, Pietenpol JA. Identification and use of biomarkers in treatment strategies for triple-negative breast cancer subtypes. J Pathol. 2014;232(2):142–50.
162. Masuda H, Baggerly KA, Wang Y, Zhang Y, Gonzalez-Angulo AM, Meric-Bernstam F, et al. Differential response to neoadjuvant chemotherapy among 7 triple-negative breast cancer molecular subtypes. Clin Cancer Res. 2013;19(19):5533–40.
163. Curtis C, Shah SP, Chin SF, Turashvili G, Rueda OM, Dunning MJ, et al. The genomic and transcriptomic architecture of 2,000 breast tumours reveals novel subgroups. Nature. 2012;486 (7403):346–52.
164. Hicks J, Krasnitz A, Lakshmi B, Navin NE, Riggs M, Leibu E, et al. Novel patterns of genome rearrangement and their association with survival in breast cancer. Genome Res. 2006;16(12):1465–79.

165. Ali H, Rueda OM, Chin SF, Curtis C, Dunning MJ, Aparicio S, et al. Genome-driven integrated classification of breast cancer validated in over 7,500 samples. Genome Biol. 2014;15(8):431.
166. Cancer Genome Atlas Network. Comprehensive molecular portraits of human breast tumours. Nature. 2012;490(7418):61–70.
167. Elston CW, Ellis IO. Pathological prognostic factors in breast cancer. I. The value of histological grade in breast cancer: experience from a large study with long-term follow-up. Histopathology. 1991;19 (5):403–10.
168. Fulford LG, Easton DF, Reis-Filho JS, Sofronis A, Gillett CE, Lakhani SR, et al. Specific morphological features predictive for the basal phenotype in grade 3 invasive ductal carcinoma of breast. Histopathology. 2006;49(1):22–34.
169. Mahmoud SM, Paish EC, Powe DG, Macmillan RD, Grainge MJ, Lee AH, et al. Tumor-infiltrating CD8+ lymphocytes predict clinical outcome in breast cancer. J Clin Oncol. 2011;29(15):1949–55.
170. Mohammed ZM, Going JJ, Edwards J, Elsberger B, Doughty JC, McMillan DC. The relationship between components of tumour inflammatory cell infiltrate and clinicopathological factors and survival in patients with primary operable invasive ductal breast cancer. Br J Cancer. 2012;107(5):864–73.
171. Denkert C, Loibl S, Noske A, Roller M, Muller BM, Komor M, et al. Tumor-associated lymphocytes as an independent predictor of response to neoadjuvant chemotherapy in breast cancer. J Clin Oncol. 2010;28(1):105–13.
172. Gu-Trantien C, Loi S, Garaud S, Equeter C, Libin M, de Wind A, et al. CD4(+) follicular helper T cell infiltration predicts breast cancer survival. J Clin Invest. 2013;123(7):2873–92.
173. Loi S, Michiels S, Salgado R, Sirtaine N, Jose V, Fumagalli D, et al. Tumor infiltrating lymphocytes are prognostic in triple negative breast cancer and predictive for trastuzumab benefit in early breast cancer: results from the FinHER trial. Ann Oncol. 2014;25(8):1544–50.
174. Loi S, Sirtaine N, Piette F, Salgado R, Viale G, Van Eenoo F, et al. Prognostic and predictive value of tumor-infiltrating lymphocytes in a phase III randomized adjuvant breast cancer trial in node-positive breast cancer comparing the addition of docetaxel to doxorubicin with doxorubicin-based chemotherapy: BIG 02–98. J Clin Oncol. 2013;31(7):860–7.
175. Salgado R, Denkert C, Demaria S, Sirtaine N, Klauschen F, Pruneri G, et al. The evaluation of tumor-infiltrating lymphocytes (TILs) in breast cancer: recommendations by an International TILs Working Group 2014. Ann Oncol. 2014;26 (2):259–71.
176. Weigelt B, Horlings HM, Kreike B, Hayes MM, Hauptmann M, Wessels LF, et al. Refinement of breast cancer classification by molecular characterization of histological special types. J Pathol. 2008;216(2):141–50.
177. Bertucci F, Orsetti B, Negre V, Finetti P, Rouge C, Ahomadegbe JC, et al. Lobular and ductal carcinomas of the breast have distinct genomic and expression profiles. Oncogene. 2008;27 (40):5359–72.
178. Gruel N, Lucchesi C, Raynal V, Rodrigues MJ, Pierron G, Goudefroye R, et al. Lobular invasive carcinoma of the breast is a molecular entity distinct from luminal invasive ductal carcinoma. Eur J Cancer. 2010;46(13):2399–407.
179. Lopez-Garcia MA, Geyer FC, Natrajan R, Kreike B, Mackay A, Grigoriadis A, et al. Transcriptomic analysis of tubular carcinomas of the breast reveals similarities and differences with molecular subtype-matched ductal and lobular carcinomas. J Pathol. 2010;222(1):64–75.
180. Vincent-Salomon A, Gruel N, Lucchesi C, MacGrogan G, Dendale R, Sigal-Zafrani B, et al. Identification of typical medullary breast carcinoma as a genomic sub-group of basal-like carcinomas, a heterogeneous new molecular entity. Breast Cancer Res. 2007;9(2):R24.
181. Weigelt B, Geyer FC, Horlings HM, Kreike B, Halfwerk H, Reis-Filho JS. Mucinous and neuroendocrine breast carcinomas are transcriptionally distinct from invasive ductal carcinomas of no special type. Mod Pathol. 2009;22(11):1401–14.
182. Lacroix-Triki M, Suarez PH, MacKay A, Lambros MB, Natrajan R, Savage K, et al. Mucinous carcinoma of the breast is genomically distinct from invasive ductal carcinomas of no special type. J Pathol. 2010;222(3):282–98.
183. Marchio C, Iravani M, Natrajan R, Lambros MB, Geyer FC, Savage K, et al. Mixed micropapillary-ductal carcinomas of the breast: a genomic and immunohistochemical analysis of morphologically distinct components. J Pathol. 2009;218(3):301–15.
184. Marchio C, Iravani M, Natrajan R, Lambros MB, Savage K, Tamber N, et al. Genomic and immunophenotypical characterization of pure micropapillary carcinomas of the breast. J Pathol. 2008;215(4):398–410.
185. Duprez R, Wilkerson PM, Lacroix-Triki M, Lambros MB, Mackay A, Hern RA, et al. Immunophenotypic and genomic characterization of papillary carcinomas of the breast. J Pathol. 2011;226 (3):427–41.
186. Weigelt B, Geyer FC, Reis-Filho JS. Histological types of breast cancer: how special are they? Mol Oncol. 2010;4(3):192–208.
187. Weigelt B, Reis-Filho JS. Histological and molecular types of breast cancer: is there a unifying taxonomy? Nat Rev Clin Oncol. 2009;6 (12):718–30.
188. Persson M, Andren Y, Mark J, Horlings HM, Persson F, Stenman G. Recurrent fusion of MYB and NFIB transcription factor genes in carcinomas of the breast and head and neck. Proc Natl Acad Sci USA. 2009;106(44):18740–4.

189. Tognon C, Knezevich SR, Huntsman D, Roskelley CD, Melnyk N, Mathers JA, et al. Expression of the ETV6-NTRK3 gene fusion as a primary event in human secretory breast carcinoma. Cancer Cell. 2002;2(5):367–76.
190. Weigelt B, Geyer FC, Natrajan R, Lopez-Garcia MA, Ahmad AS, Savage K, et al. The molecular underpinning of lobular histological growth pattern: a genome-wide transcriptomic analysis of invasive lobular carcinomas and grade-and molecular subtype-matched invasive ductal carcinomas of no special type. J Pathol. 2010;220(1):45–57.
191. Natrajan R, Wilkerson PM, Marchio C, Piscuoglio S, Ng CK, Wai P, et al. Characterization of the genomic features and expressed fusion genes in micropapillary carcinomas of the breast. J Pathol. 2014;232(5):553–65.
192. Wetterskog D, Lopez-Garcia MA, Lambros MB, A'Hern R, Geyer FC, Milanezi F, et al. Adenoid cystic carcinomas constitute a genomically distinct subgroup of triple-negative and basal-like breast cancers. J Pathol. 2011;226(1):84–96.
193. Reis-Filho JS, Lakhani SR. Breast cancer special types: why bother? J Pathol. 2008;216(4):394–8.
194. Weigelt B, Baehner FL, Reis-Filho JS. The contribution of gene expression profiling to breast cancer classification, prognostication and prediction: a retrospective of the last decade. J Pathol. 2010;220 (2):263–80.
195. Sarrio D, Perez-Mies B, Hardisson D, Moreno-Bueno G, Suarez A, Cano A, et al. Cytoplasmic localization of p120ctn and E-cadherin loss characterize lobular breast carcinoma from preinvasive to metastatic lesions. Oncogene. 2004;23 (19):3272–83.
196. Sardanelli F, Boetes C, Borisch B, Decker T, Federico M, Gilbert FJ, et al. Magnetic resonance imaging of the breast: recommendations from the EUSOMA working group. Eur J Cancer. 2010;46 (8):1296–316.
197. Balmativola D, Marchio C, Maule M, Chiusa L, Annaratone L, Maletta F, et al. Pathological non-response to chemotherapy in a neoadjuvant setting of breast cancer: an inter-institutional study. Breast Cancer Res Treat. 2014;148(3):511–23.
198. Loibl S, Volz C, Mau C, Blohmer JU, Costa SD, Eidtmann H, et al. Response and prognosis after neoadjuvant chemotherapy in 1,051 patients with infiltrating lobular breast carcinoma. Breast Cancer Res Treat. 2014;144(1):153–62.
199. Lien HC, Chen YL, Juang YL, Jeng YM. Frequent alterations of HER2 through mutation, amplification, or overexpression in pleomorphic lobular carcinoma of the breast. Breast Cancer Res Treat. 2015;150 (2):447–55.
200. Ciriello G, Sinha R, Hoadley KA, Jacobsen AS, Reva B, Perou CM, et al. The molecular diversity of luminal A breast tumors. Breast Cancer Res Treat. 2013;141(3):409–20.
201. Ross JS, Wang K, Sheehan CE, Boguniewicz AB, Otto G, Downing SR, et al. Relapsed classic E-cadherin (CDH1)-mutated invasive lobular breast cancer shows a high frequency of HER2 (ERBB2) gene mutations. Clin Cancer Res. 2013;19 (10):2668–76.
202. Baselga J, Campone M, Piccart M, Burris HA 3rd, Rugo HS, Sahmoud T, et al. Everolimus in postmenopausal hormone-receptor-positive advanced breast cancer. N Engl J Med. 2012;366 (6):520–9.
203. Nishimukai A, Yagi T, Yanai A, Miyagawa Y, Enomoto Y, Murase K, et al. High Ki-67 expression and low progesterone receptor expression could independently lead to a worse prognosis for postmenopausal patients with estrogen receptor-positive and HER2-negative breast cancer. Clin Breast Cancer. 2015;15(3):204–11.
204. Prat A, Cheang MC, Martin M, Parker JS, Carrasco E, Caballero R, et al. Prognostic significance of progesterone receptor-positive tumor cells within immunohistochemically defined luminal A breast cancer. J Clin Oncol. 2013;31(2):203–9.
205. Sato K, Miyashita M, Ishida T, Suzuki A, Tada H, Watanabe G, et al. Prognostic significance of the progesterone receptor status in Ki67-high and-low luminal B-like HER2-negative breast cancers. Breast Cancer. 2014.
206. Rakha EA, Lee AH, Evans AJ, Menon S, Assad NY, Hodi Z, et al. Tubular carcinoma of the breast: further evidence to support its excellent prognosis. J Clin Oncol. 2010;28(1):99–104.
207. Barkley CR, Ligibel JA, Wong JS, Lipsitz S, Smith BL, Golshan M. Mucinous breast carcinoma: a large contemporary series. Am J Surg. 2008;196 (4):549–51.
208. Li CI, Uribe DJ, Daling JR. Clinical characteristics of different histologic types of breast cancer. Br J Cancer. 2005;93(9):1046–52.
209. Vo T, Xing Y, Meric-Bernstam F, Mirza N, Vlastos G, Symmans WF, et al. Long-term outcomes in patients with mucinous, medullary, tubular, and invasive ductal carcinomas after lumpectomy. Am J Surg. 2007;194(4):527–31.
210. Capella C, Eusebi V, Mann B, Azzopardi JG. Endocrine differentiation in mucoid carcinoma of the breast. Histopathology. 1980;4(6):613–30.
211. Righi L, Sapino A, Marchio C, Papotti M, Bussolati G. Neuroendocrine differentiation in breast cancer: established facts and unresolved problems. Semin Diagn Pathol. 2010;27(1):69–76.
212. Kehr EL, Jorns JM, Ang D, Warrick A, Neff T, Degnin M, et al. Mucinous breast carcinomas lack PIK3CA and AKT1 mutations. Hum Pathol. 2012;43(12):2207–12.
213. Peterse JL. Breast carcinomas with an unexpected inside out growth pattern. Rotation of polarisation associated with angioinvasion. Pathol Res Pract. 1993;189:780.

214. Li YS, Kaneko M, Sakamoto DG, Takeshima Y, Inai K. The reversed apical pattern of MUC1 expression is characteristics of invasive micropapillary carcinoma of the breast. Breast Cancer. 2006;13 (1):58–63.
215. Luna-More S, Gonzalez B, Acedo C, Rodrigo I, Luna C. Invasive micropapillary carcinoma of the breast. A new special type of invasive mammary carcinoma. Pathol Res Pract. 1994;190(7):668–74.
216. Nassar H, Pansare V, Zhang H, Che M, Sakr W, Ali-Fehmi R, et al. Pathogenesis of invasive micropapillary carcinoma: role of MUC1 glycoprotein. Mod Pathol. 2004;17(9):1045–50.
217. Kim MJ, Gong G, Joo HJ, Ahn SH, Ro JY. Immunohistochemical and clinicopathologic characteristics of invasive ductal carcinoma of breast with micropapillary carcinoma component. Arch Pathol Lab Med. 2005;129(10):1277–82.
218. Nassar H, Wallis T, Andea A, Dey J, Adsay V, Visscher D. Clinicopathologic analysis of invasive micropapillary differentiation in breast carcinoma. Mod Pathol. 2001;14(9):836–41.
219. Vingiani A, Maisonneuve P, Dell'orto P, Farante G, Rotmensz N, Lissidini G, et al. The clinical relevance of micropapillary carcinoma of the breast: a case-control study. Histopathology. 2013;63 (2):217–24.
220. Wolff AC, Hammond ME, Hicks DG, Dowsett M, McShane LM, Allison KH, et al. Recommendations for human epidermal growth factor receptor 2 testing in breast cancer: American Society of Clinical Oncology/College of American Pathologists clinical practice guideline update. J Clin Oncol. 2013;31 (31):3997–4013.
221. Di Cosimo S, Baselga J. Phosphoinositide 3-kinase mutations in breast cancer: a "good" activating mutation? Clin Cancer Res. 2009;15(16):5017–9.
222. Cubilla AL, Woodruff JM. Primary carcinoid tumor of the breast. A report of eight patients. Am J Surg Pathol. 1977;1:283–92.
223. Dellapasqua S, Maisonneuve P, Viale G, Pruneri G, Mazzarol G, Ghisini R, et al. Immunohistochemically defined subtypes and outcome of apocrine breast cancer. Clin Breast Cancer. 2013;13(2):95–102.
224. Tsutsumi Y. Apocrine carcinoma as triple-negative breast cancer: novel definition of apocrine-type carcinoma as estrogen/progesterone receptor-negative and androgen receptor-positive invasive ductal carcinoma. Jpn J Clin Oncol. 2012;42(5):375–86.
225. Vranic S, Marchio C, Castellano I, Botta C, Scalzo MS, Bender RP, et al. Immunohistochemical and molecular profiling of histologically defined apocrine carcinomas of the breast. Hum Pathol. 2015;46(9):1350–9.
226. Vranic S, Schmitt F, Sapino A, Costa JL, Reddy S, Castro M, et al. Apocrine carcinoma of the breast: a comprehensive review. Histol Histopathol. 2013;28 (11):1393–409.
227. Lehmann-Che J, Hamy AS, Porcher R, Barritault M, Bouhidel F, Habuellelah H, et al. Molecular apocrine breast cancers are aggressive estrogen receptor negative tumors overexpressing either HER2 or GCDFP15. Breast Cancer Res. 2013;15(3):R37.
228. Weigelt B, Eberle C, Cowell CF, Ng CK, Reis-Filho JS. Metaplastic breast carcinoma: more than a special type. Nat Rev Cancer. 2015;14(3):147–8.
229. Weigelt B, Ng CK, Shen R, Popova T, Schizas M, Natrajan R, et al. Metaplastic breast carcinomas display genomic and transcriptomic heterogeneity [corrected]. Mod Pathol. 2015;28(3):340–51.
230. Lester TR, Hunt KK, Nayeemuddin KM, Bassett RL Jr, Gonzalez-Angulo AM, Feig BW, et al. Metaplastic sarcomatoid carcinoma of the breast appears more aggressive than other triple receptor-negative breast cancers. Breast Cancer Res Treat. 2012;131 (1):41–8.
231. Hennessy BT, Gonzalez-Angulo AM, Stemke-Hale K, Gilcrease MZ, Krishnamurthy S, Lee JS, et al. Characterization of a naturally occurring breast cancer subset enriched in epithelial-to-mesenchymal transition and stem cell characteristics. Cancer Res. 2009;69(10):4116–24.
232. Lien HC, Lin CW, Mao TL, Kuo SH, Hsiao CH, Huang CS. p53 overexpression and mutation in metaplastic carcinoma of the breast: genetic evidence for a monoclonal origin of both the carcinomatous and the heterogeneous sarcomatous components. J Pathol. 2004;204(2):131–9.
233. Weigelt B, Kreike B, Reis-Filho JS. Metaplastic breast carcinomas are basal-like breast cancers: a genomic profiling analysis. Breast Cancer Res Treat. 2009;117(2):273–80.
234. Gerhard R, Ricardo S, Albergaria A, Gomes M, Silva AR, Logullo AF, et al. Immunohistochemical features of claudin-low intrinsic subtype in metaplastic breast carcinomas. Breast. 2012;21 (3):354–60.
235. Kim HM, Kim do H, Jung WH, Koo JS. Molecular classification of metaplastic carcinoma using surrogate immunohistochemical staining. Pathobiology. 2013;81(2):69–77.
236. Rito M, Schmitt F, Pinto AE, Andre S. Fibromatosis-like metaplastic carcinoma of the breast has a claudin-low immunohistochemical phenotype. Virchows Arch. 2014;465(2):185–91.
237. Geyer FC, Weigelt B, Natrajan R, Lambros MB, de Biase D, Vatcheva R, et al. Molecular analysis reveals a genetic basis for the phenotypic diversity of metaplastic breast carcinomas. J Pathol. 2010;220 (5):562–73.
238. Reis-Filho JS, Pinheiro C, Lambros MB, Milanezi F, Carvalho S, Savage K, et al. EGFR amplification and lack of activating mutations in metaplastic breast carcinomas. J Pathol. 2006;209(4):445–53.
239. Sneige N, Yaziji H, Mandavilli SR, Perez ER, Ordonez NG, Gown AM, et al. Low-grade (fibromatosis-like) spindle cell carcinoma of the breast. Am J Surg Pathol. 2001;25(8):1009–16.

240. Marchio C, Weigelt B, Reis-Filho JS. Adenoid cystic carcinomas of the breast and salivary glands (or 'The strange case of Dr Jekyll and Mr Hyde' of exocrine gland carcinomas). J Clin Pathol. 2010;63 (3):220–8.
241. Ellis IO, Pinder S, Lee AHS. Tumors of the breast. Aadenoid cystic carcinoma. In: Fletcher C, editor. Diagnostic histopathology of tumors. London: Churchill Livingstone Elsevier; 2007. p. 953–4.
242. Bennett AK, Mills SE, Wick MR. Salivary-type neoplasms of the breast and lung. Semin Diagn Pathol. 2003;20(4):279–304.
243. Rosen PP. Adenoid cystic carcinoma. In: Rosen's breast pathology. Philadelphia: Lippincott Williams and Wilkins; 2009. p. 590–604.
244. Tavassoli FA, Norris HJ. Mammary adenoid cystic carcinoma with sebaceous differentiation. A morphologic study of the cell types. Arch Pathol Lab Med. 1986;110(11):1045–53.
245. Ro JY, Silva EG, Gallager HS. Adenoid cystic carcinoma of the breast. Hum Pathol. 1987;18 (12):1276–81.
246. Brill LB 2nd, Kanner WA, Fehr A, Andren Y, Moskaluk CA, Loning T, et al. Analysis of MYB expression and MYB-NFIB gene fusions in adenoid cystic carcinoma and other salivary neoplasms. Mod Pathol. 2011;24(9):1169–76.
247. D'Alfonso TM, Mosquera JM, MacDonald TY, Padilla J, Liu YF, Rubin MA, et al. MYB-NFIB gene fusion in adenoid cystic carcinoma of the breast with special focus paid to the solid variant with basaloid features. Hum Pathol. 2014;45 (11):2270–80.
248. Wetterskog D, Wilkerson PM, Rodrigues DN, Lambros MB, Fritchie K, Andersson MK, et al. Mutation profiling of adenoid cystic carcinomas from multiple anatomical sites identifies mutations in the RAS pathway, but no KIT mutations. Histopathology. 2013;62(4):543–50.
249. Fusco N, Colombo PE, Martelotto LG, De Filippo MR, Piscuoglio S, Ng CK, et al. Resolving quandaries: basaloid adenoid cystic carcinoma or breast cylindroma? The role of massively parallel sequencing. Histopathology. 2015.
250. Persson M, Andren Y, Moskaluk CA, Frierson HF Jr, Cooke SL, Futreal PA, et al. Clinically significant copy number alterations and complex rearrangements of MYB and NFIB in head and neck adenoid cystic carcinoma. Genes Chromosom Cancer. 2012;51(8):805–17.
251. Horlings HM, Weigelt B, Anderson EM, Lambros MB, Mackay A, Natrajan R, et al. Genomic profiling of histological special types of breast cancer. Breast Cancer Res Treat. 2013;142(2):257–69.
252. Shah SP, Roth A, Goya R, Oloumi A, Ha G, Zhao Y, et al. The clonal and mutational evolution spectrum of primary triple-negative breast cancers. Nature. 2012;486(7403):395–9.
253. Turner N, Lambros MB, Horlings HM, Pearson A, Sharpe R, Natrajan R, et al. Integrative molecular profiling of triple negative breast cancers identifies amplicon drivers and potential therapeutic targets. Oncogene. 2010;29(14):2013–23.
254. El-Rifai W, Rutherford S, Knuutila S, Frierson HF Jr, Moskaluk CA. Novel DNA copy number losses in chromosome 12q12-q13 in adenoid cystic carcinoma. Neoplasia. 2001;3(3):173–8.
255. Ho AS, Kannan K, Roy DM, Morris LG, Ganly I, Katabi N, et al. The mutational landscape of adenoid cystic carcinoma. Nat Genet. 2013;45(7):791–8.
256. Stephens PJ, Davies HR, Mitani Y, Van Loo P, Shlien A, Tarpey PS, et al. Whole exome sequencing of adenoid cystic carcinoma. J Clin Invest. 2013;123(7):2965–8.
257. Papaspyrou G, Hoch S, Rinaldo A, Rodrigo JP, Takes RP, van Herpen C, et al. Chemotherapy and targeted therapy in adenoid cystic carcinoma of the head and neck: a review. Head Neck. 2010;33 (6):905–11.
258. Yerushalmi R, Hayes MM, Gelmon KA. Breast carcinoma—rare types: review of the literature. Ann Oncol. 2009;20(11):1763–70.
259. Foschini MP, Krausz T. Salivary gland-type tumors of the breast: a spectrum of benign and malignant tumors including "triple negative carcinomas" of low malignant potential. Semin Diagn Pathol. 2010;27(1):77–90.
260. Coyne JD, Dervan PA. Primary acinic cell carcinoma of the breast. J Clin Pathol. 2002;55(7):545–7.
261. Huo L, Bell D, Qiu H, Sahin A, Wu Y, Sneige N. Paneth cell-like eosinophilic cytoplasmic granules in breast carcinoma. Ann Diagn Pathol. 2010;15 (2):84–92.
262. Peintinger F, Leibl S, Reitsamer R, Moinfar F. Primary acinic cell carcinoma of the breast: a case report with long-term follow-up and review of the literature. Histopathology. 2004;45(6):645–8.
263. Damiani S, Pasquinelli G, Lamovec J, Peterse JL, Eusebi V. Acinic cell carcinoma of the breast: an immunohistochemical and ultrastructural study. Virchows Arch. 2000;437(1):74–81.
264. Limite G, Di Micco R, Esposito E, Sollazzo V, Cervotti M, Pettinato G, et al. Acinic cell carcinoma of the breast: review of the literature. Int J Surg. 2014;12(Suppl 1):S35–9.
265. Piscuoglio S, Hodi Z, Katabi N, Guerini-Rocco E, Macedo GS, Ng CK, et al. Are acinic cell carcinomas of the breast and salivary glands distinct diseases? Histopathology. 2014;67(4):529–37.
266. Guerini-Rocco E, Hodi Z, Piscuoglio S, Ng CK, Rakha EA, Schultheis AM, et al. The repertoire of somatic genetic alterations of acinic cell carcinomas of the breast: an exploratory, hypothesis-generating study. J Pathol. 2015.
267. Ripamonti CB, Colombo M, Mondini P, Siranoush M, Peissel B, Bernard L, et al. First description of an acinic cell carcinoma of the breast in a

BRCA1 mutation carrier: a case report. BMC Cancer. 2013;13:46.
268. Weinreb I, Piscuoglio S, Martelotto LG, Waggott D, Ng CK, Perez-Ordonez B, et al. Hotspot activating PRKD1 somatic mutations in polymorphous low-grade adenocarcinomas of the salivary glands. Nat Genet. 2014.
269. Robertson FM, Petricoin Iii EF, Van Laere SJ, Bertucci F, Chu K, Fernandez SV, et al. Presence of anaplastic lymphoma kinase in inflammatory breast cancer. Springerplus. 2013;2:497.
270. Hanna MG, Najfeld V, Irie HY, Tripodi J, Nayak A. Analysis of ALK gene in 133 patients with breast cancer revealed polysomy of chromosome 2 and no ALK amplification. Springerplus. 2015;4:439.
271. Krishnamurthy S, Woodward W, Yang W, Reuben JM, Tepperberg J, Ogura D, et al. Status of the anaplastic lymphoma kinase (ALK) gene in inflammatory breast carcinoma. Springerplus. 2013;2:409.
272. Ross JS, Ali SM, Wang K, Khaira D, Palma NA, Chmielecki J, et al. Comprehensive genomic profiling of inflammatory breast cancer cases reveals a high frequency of clinically relevant genomic alterations. Breast Cancer Res Treat. 2015;154(1):155–62.
273. Rakha EA, Reis-Filho JS, Baehner F, Dabbs DJ, Decker T, Eusebi V, et al. Breast cancer prognostic classification in the molecular era: the role of histological grade. Breast Cancer Res. 2010;12(4):207.
274. Buerger H, Mommers EC, Littmann R, Simon R, Diallo R, Poremba C, et al. Ductal invasive G2 and G3 carcinomas of the breast are the end stages of at least two different lines of genetic evolution. J Pathol. 2001;194(2):165–70.
275. Roylance R, Gorman P, Harris W, Liebmann R, Barnes D, Hanby A, et al. Comparative genomic hybridization of breast tumors stratified by histological grade reveals new insights into the biological progression of breast cancer. Cancer Res. 1999;59(7):1433–6.
276. Natrajan R, Lambros MB, Rodriguez-Pinilla SM, Moreno-Bueno G, Tan DS, Marchio C, et al. Tiling path genomic profiling of grade 3 invasive ductal breast cancers. Clin Cancer Res. 2009;15(8):2711–22.
277. Vincent-Salomon A, Lucchesi C, Gruel N, Raynal V, Pierron G, Goudefroye R, et al. Integrated genomic and transcriptomic analysis of ductal carcinoma in situ of the breast. Clin Cancer Res. 2008;14(7):1956–65.
278. Sotiriou C, Piccart MJ. Taking gene-expression profiling to the clinic: when will molecular signatures become relevant to patient care? Nat Rev Cancer. 2007;7(7):545–53.
279. Sotiriou C, Wirapati P, Loi S, Harris A, Fox S, Smeds J, et al. Gene expression profiling in breast cancer: understanding the molecular basis of histologic grade to improve prognosis. J Natl Cancer Inst. 2006;98(4):262–72.
280. Cleton-Jansen AM, Buerger H, Haar N, Philippo K, van de Vijver MJ, Boecker W, et al. Different mechanisms of chromosome 16 loss of heterozygosity in well-versus poorly differentiated ductal breast cancer. Genes Chromosom Cancer. 2004;41(2):109–16.
281. Flagiello D, Gerbault-Seureau M, Sastre-Garau X, Padoy E, Vielh P, Dutrillaux B. Highly recurrent der(1; 16)(q10; p10) and other 16q arm alterations in lobular breast cancer. Genes Chromosom Cancer. 1998;23(4):300–6.
282. Tsuda H, Takarabe T, Fukutomi T, Hirohashi S. der(16)t(1;16)/der(1;16) in breast cancer detected by fluorescence in situ hybridization is an indicator of better patient prognosis. Genes Chromosom Cancer. 1999;24(1):72–7.
283. Dent R, Hanna WM, Trudeau M, Rawlinson E, Sun P, Narod SA. Time to disease recurrence in basal-type breast cancers: effects of tumor size and lymph node status. Cancer. 2009;115(21):4917–23.
284. Foulkes WD, Reis-Filho JS, Narod SA. Tumor size and survival in breast cancer—a reappraisal. Nat Rev Clin Oncol. 2010;7(6):348–53.
285. Desmedt C, Haibe-Kains B, Wirapati P, Buyse M, Larsimont D, Bontempi G, et al. Biological processes associated with breast cancer clinical outcome depend on the molecular subtypes. Clin Cancer Res. 2008;14(16):5158–65.
286. Wirapati P, Sotiriou C, Kunkel S, Farmer P, Pradervand S, Haibe-Kains B, et al. Meta-analysis of gene expression profiles in breast cancer: toward a unified understanding of breast cancer subtyping and prognosis signatures. Breast Cancer Res. 2008;10(4):R65.
287. Bossuyt V, Provenzano E, Symmans WF, Boughey JC, Coles C, Curigliano G, et al. Recommendations for standardized pathological characterization of residual disease for neoadjuvant clinical trials of breast cancer by the BIG-NABCG collaboration. Ann Oncol. 2015;26(7):1280–91.
288. Provenzano E, Bossuyt V, Viale G, Cameron D, Badve S, Denkert C, et al. Standardization of pathologic evaluation and reporting of postneoadjuvant specimens in clinical trials of breast cancer: recommendations from an international working group. Mod Pathol. 2015;28(9):1185–201.
289. Leonard GD, Swain SM. Ductal carcinoma in situ, complexities and challenges. J Natl Cancer Inst. 2004;96(12):906–20.
290. Marchio C, Balmativola D, Castiglione R, Annaratone L, Sapino A. Predictive diagnostic pathology in the target therapy era in breast cancer. Curr Drug Targets. 2015.
291. Hammond ME, Hayes DF, Dowsett M, Allred DC, Hagerty KL, Badve S, et al. American Society of

Clinical Oncology/College Of American Pathologists guideline recommendations for immunohistochemical testing of estrogen and progesterone receptors in breast cancer. J Clin Oncol. 2010;28(16):2784–95.
292. Viale G, Regan MM, Maiorano E, Mastropasqua MG, Dell'Orto P, Rasmussen BB, et al. Prognostic and predictive value of centrally reviewed expression of estrogen and progesterone receptors in a randomized trial comparing letrozole and tamoxifen adjuvant therapy for postmenopausal early breast cancer: BIG 1–98. J Clin Oncol. 2007;25(25):3846–52.
293. Iwamoto T, Booser D, Valero V, Murray JL, Koenig K, Esteva FJ, et al. Estrogen receptor (ER) mRNA and ER-related gene expression in breast cancers that are 1% to 10% ER-positive by immunohistochemistry. J Clin Oncol. 2012;30(7):729–34.
294. Yi M, Huo L, Koenig KB, Mittendorf EA, Meric-Bernstam F, Kuerer HM, et al. Which threshold for ER positivity? a retrospective study based on 9639 patients. Ann Oncol. 2014;25(5):1004–11.
295. Sapino A, Goia M, Recupero D, Marchio C. Current challenges for HER2 testing in diagnostic pathology: state of the art and controversial issues. Front Oncol. 2013;3:129.
296. Sapino A, Maletta F, Verdun di Cantogno L, Macri L, Botta C, Gugliotta P, et al. Gene status in HER2 equivocal breast carcinomas: impact of distinct recommendations and contribution of a polymerase chain reaction-based method. Oncologist. 2014;19(11):1118–26.
297. Bertucci F, Finetti P, Roche H, Le Doussal JM, Marisa L, Martin AL, et al. Comparison of the prognostic value of genomic grade index, Ki67 expression and mitotic activity index in early node-positive breast cancer patients. Ann Oncol. 2012;24(3):625–32.
298. Luporsi E, Andre F, Spyratos F, Martin PM, Jacquemier J, Penault-Llorca F, et al. Ki-67: level of evidence and methodological considerations for its role in the clinical management of breast cancer: analytical and critical review. Breast Cancer Res Treat. 2011;132(3):895–915.
299. Polley MY, Leung SC, McShane LM, Gao D, Hugh JC, Mastropasqua MG, et al. An international Ki67 reproducibility study. J Natl Cancer Inst. 2013;105(24):1897–906.
300. Dowsett M, Nielsen TO, A'Hern R, Bartlett J, Coombes RC, Cuzick J, et al. Assessment of Ki67 in breast cancer: recommendations from the International Ki67 in Breast Cancer working group. J Natl Cancer Inst. 2011;103(22):1656–64.
301. Polley MY, Leung SC, Gao D, Mastropasqua MG, Zabaglo LA, Bartlett JM, et al. An international study to increase concordance in Ki67 scoring. Mod Pathol. 2015;28(6):778–86.
302. Sapino A, Annaratone L, Marchio C. Current projects in Pre-analytics: where to go? Recent Results Cancer Res. 2015;199:65–70.
303. Turner N, Pearson A, Sharpe R, Lambros M, Geyer F, Lopez-Garcia MA, et al. FGFR1 amplification drives endocrine therapy resistance and is a therapeutic target in breast cancer. Cancer Res. 2010;70(5):2085–94.
304. Marchio C, Lambros MB, Gugliotta P, Di Cantogno LV, Botta C, Pasini B, et al. Does chromosome 17 centromere copy number predict polysomy in breast cancer? A fluorescence in situ hybridization and microarray-based CGH analysis. J Pathol. 2009;219(1):16–24.
305. Moelans CB, de Weger RA, van Diest PJ. Absence of chromosome 17 polysomy in breast cancer: analysis by CEP17 chromogenic in situ hybridization and multiplex ligation-dependent probe amplification. Breast Cancer Res Treat. 2010;120(1):1–7.
306. Varga Z, Tubbs RR, Wang Z, Sun Y, Noske A, Kradolfer D, et al. Co-amplification of the HER2 gene and chromosome 17 centromere: a potential diagnostic pitfall in HER2 testing in breast cancer. Breast Cancer Res Treat. 2012;132(3):925–35.
307. Yeh IT, Martin MA, Robetorye RS, Bolla AR, McCaskill C, Shah RK, et al. Clinical validation of an array CGH test for HER2 status in breast cancer reveals that polysomy 17 is a rare event. Mod Pathol. 2009;22(9):1169–75.
308. Bhargava R, Dabbs DJ. Interpretation of human epidermal growth factor receptor 2 (HER2) in situ hybridization assays using 2013 update of American Society of Clinical Oncology/College of American Pathologists HER2 Guidelines. J Clin Oncol. 2014;32(17):1855.
309. Brunelli M, Nottegar A, Bogina G, Calio A, Cima L, Eccher A, et al. Monosomy of chromosome 17 in breast cancer during interpretation of HER2 gene amplification. Am J Cancer Res. 2015;5(7):2212–21.
310. Valent A, Penault-Llorca F, Cayre A, Kroemer G. Change in HER2 (ERBB2) gene status after taxane-based chemotherapy for breast cancer: polyploidization can lead to diagnostic pitfalls with potential impact for clinical management. Cancer Genet. 2013;206(1–2):37–41.
311. Cottu PH, Asselah J, Lae M, Pierga JY, Dieras V, Mignot L, et al. Intratumoral heterogeneity of HER2/neu expression and its consequences for the management of advanced breast cancer. Ann Oncol. 2008;19(3):595–7.
312. Hanna WM, Ruschoff J, Bilous M, Coudry RA, Dowsett M, Osamura RY, et al. HER2 in situ hybridization in breast cancer: clinical implications of polysomy 17 and genetic heterogeneity. Mod Pathol. 2013;27(1):4–18.
313. Ng CK, Martelotto LG, Gauthier A, Wen HC, Piscuoglio S, Lim RS, et al. Intra-tumor genetic heterogeneity and alternative driver genetic

alterations in breast cancers with heterogeneous HER2 gene amplification. Genome Biol. 2015;16:107.
314. Seol H, Lee HJ, Choi Y, Lee HE, Kim YJ, Kim JH, et al. Intratumoral heterogeneity of HER2 gene amplification in breast cancer: its clinicopathological significance. Mod Pathol. 2012;25(7):938–48.
315. Paik S, Shak S, Tang G, Kim C, Baker J, Cronin M, et al. A multigene assay to predict recurrence of tamoxifen-treated, node-negative breast cancer. N Engl J Med. 2004;351(27):2817–26.
316. van de Vijver MJ, He YD, van't Veer LJ, Dai H, Hart AA, Voskuil DW, et al. A gene-expression signature as a predictor of survival in breast cancer. N Engl J Med. 2002;347(25):1999–2009.
317. van't Veer LJ, Dai H, van de Vijver MJ, He YD, Hart AA, Mao M, et al. Gene expression profiling predicts clinical outcome of breast cancer. Nature. 2002;415(6871):530–6.
318. Zhang Y, Schnabel CA, Schroeder BE, Jerevall PL, Jankowitz RC, Fornander T, et al. Breast cancer index identifies early-stage estrogen receptor-positive breast cancer patients at risk for early-and late-distant recurrence. Clin Cancer Res. 2013;19 (15):4196–205.
319. Filipits M, Rudas M, Jakesz R, Dubsky P, Fitzal F, Singer CF, et al. A new molecular predictor of distant recurrence in ER-positive, HER2-negative breast cancer adds independent information to conventional clinical risk factors. Clin Cancer Res. 2011;17(18):6012–20.
320. Reis-Filho JS, Pusztai L. Gene expression profiling in breast cancer: classification, prognostication, and prediction. Lancet. 2011;378(9805):1812–23.
321. Azim HA Jr, Michiels S, Zagouri F, Delaloge S, Filipits M, Namer M, et al. Utility of prognostic genomic tests in breast cancer practice: the IMPAKT 2012 Working Group Consensus Statement. Ann Oncol. 2013;24(3):647–54.
322. Goldhirsch A, Winer EP, Coates AS, Gelber RD, Piccart-Gebhart M, Thurlimann B, et al. Personalizing the treatment of women with early breast cancer: highlights of the St. Gallen International Expert Consensus on the Primary Therapy of Early Breast Cancer 2013. Ann Oncol. 2013;24 (9):2206–23.
323. Varga Z, Sinn P, Fritzsche F, von Hochstetter A, Noske A, Schraml P, et al. Comparison of EndoPredict and Oncotype DX test results in hormone receptor positive invasive breast cancer. PLoS ONE. 2013;8(3):e58483.
324. Bartlett JMS, Stein RC, Bayani J, Marshall A, Dunn JA, Campbell AF, et al. Comparison of multiparameter tests in the UK OPTIMA-Prelim trial. Cancer Res. 2014;75(P4–11–07).
325. Teschendorff AE, Miremadi A, Pinder SE, Ellis IO, Caldas C. An immune response gene expression module identifies a good prognosis subtype in estrogen receptor negative breast cancer. Genome Biol. 2007;8(8):R157.
326. Teschendorff AE, Caldas C. A robust classifier of high predictive value to identify good prognosis patients in ER-negative breast cancer. Breast Cancer Res. 2008;10(4):R73.
327. Yau C, Esserman L, Moore DH, Waldman F, Sninsky J, Benz CC. A multigene predictor of metastatic outcome in early stage hormone receptor-negative and triple-negative breast cancer. Breast Cancer Res. 2010;12(5):R85.
328. Rody A, Karn T, Liedtke C, Pusztai L, Ruckhaeberle E, Hanker L, et al. A clinically relevant gene signature in triple negative and basal-like breast cancer. Breast Cancer Res. 2011;13(5):R97.
329. Perez EA, Thompson EA, Ballman KV, Anderson SK, Asmann YW, Kalari KR, et al. Genomic analysis reveals that immune function genes are strongly linked to clinical outcome in the North Central Cancer Treatment Group n9831 Adjuvant Trastuzumab Trial. J Clin Oncol. 2015;33 (7):701–8.
330. Banerji S, Cibulskis K, Rangel-Escareno C, Brown KK, Carter SL, Frederick AM, et al. Sequence analysis of mutations and translocations across breast cancer subtypes. Nature. 2012;486 (7403):405–9.
331. Ellis MJ, Ding L, Shen D, Luo J, Suman VJ, Wallis JW, et al. Whole-genome analysis informs breast cancer response to aromatase inhibition. Nature. 2012;486(7403):353–60.
332. Stephens PJ, Tarpey PS, Davies H, Van Loo P, Greenman C, Wedge DC, et al. The landscape of cancer genes and mutational processes in breast cancer. Nature. 2012;486(7403):400–4.
333. Kandoth C, McLellan MD, Vandin F, Ye K, Niu B, Lu C, et al. Mutational landscape and significance across 12 major cancer types. Nature. 2013;502 (7471):333–9.
334. Lawrence MS, Stojanov P, Mermel CH, Robinson JT, Garraway LA, Golub TR, et al. Discovery and saturation analysis of cancer genes across 21 tumour types. Nature. 2014;505(7484):495–501.
335. Jeselsohn R, Yelensky R, Buchwalter G, Frampton G, Meric-Bernstam F, Gonzalez-Angulo AM, et al. Emergence of constitutively active estrogen receptor-alpha mutations in pretreated advanced estrogen receptor-positive breast cancer. Clin Cancer Res. 2014;20(7):1757–67.
336. Li S, Shen D, Shao J, Crowder R, Liu W, Prat A, et al. Endocrine-therapy-resistant ESR1 variants revealed by genomic characterization of breast-cancer-derived xenografts. Cell Rep. 2013;4 (6):1116–30.
337. Robinson DR, Wu YM, Vats P, Su F, Lonigro RJ, Cao X, et al. Activating ESR1 mutations in hormone-resistant metastatic breast cancer. Nat Genet. 2013;45(12):1446–51.
338. Toy W, Shen Y, Won H, Green B, Sakr RA, Will M, et al. ESR1 ligand-binding domain mutations in hormone-resistant breast cancer. Nat Genet. 2013;45 (12):1439–45.

339. Bose R, Kavuri SM, Searleman AC, Shen W, Shen D, Koboldt DC, et al. Activating HER2 mutations in HER2 gene amplification negative breast cancer. Cancer Discov. 2012;3(2):224–37.
340. Fang Y, Jiang Y, Wang X, Yang X, Gao Y, Wang J. Somatic mutations of the HER2 in metastatic breast cancer. Tumour Biol. 2014;35 (12):11851–4.
341. Park YH, Shin HT, Jung HH, Choi YL, Ahn T, Park K, et al. Role of HER2 mutations in refractory metastatic breast cancers: targeted sequencing results in patients with refractory breast cancer. Oncotarget. 2015;6(31):32027–38.
342. Eccles DM, Pichert G. Familial non-BRCA1/BRCA2-associated breast cancer. Lancet Oncol. 2005;6(9):705–11.
343. Honrado E, Benitez J, Palacios J. The molecular pathology of hereditary breast cancer: genetic testing and therapeutic implications. Mod Pathol. 2005;18 (10):1305–20.
344. Farshid G, Balleine RL, Cummings M, Waring P. Morphology of breast cancer as a means of triage of patients for BRCA1 genetic testing. Am J Surg Pathol. 2006;30(11):1357–66.
345. Bane AL, Beck JC, Bleiweiss I, Buys SS, Catalano E, Daly MB, et al. BRCA2 mutation-associated breast cancers exhibit a distinguishing phenotype based on morphology and molecular profiles from tissue microarrays. Am J Surg Pathol. 2007;31(1):121–8.
346. Lakhani SR, Reis-Filho JS, Fulford L, Penault-Llorca F, van der Vijver M, Parry S, et al. Prediction of BRCA1 status in patients with breast cancer using estrogen receptor and basal phenotype. Clin Cancer Res. 2005;11(14):5175–80.
347. Lord CJ, Tutt AN, Ashworth A. Synthetic lethality and cancer therapy: lessons learned from the development of PARP inhibitors. Annu Rev Med. 2015;66:455–70.
348. Antoniou AC, Casadei S, Heikkinen T, Barrowdale D, Pylkas K, Roberts J, et al. Breast-cancer risk in families with mutations in PALB2. N Engl J Med. 2014;371(6):497–506.
349. Nigro DM, Organ CH Jr. Fibroadenoma of the female breast. Some epidemiologic surprises. Postgrad Med. 1976;59(5):113–7.
350. Jacobs TW, Chen YY, Guinee DG Jr, Holden JA, Cha I, Bauermeister DE, et al. Fibroepithelial lesions with cellular stroma on breast core needle biopsy: are there predictors of outcome on surgical excision? Am J Clin Pathol. 2005;124(3):342–54.
351. Carter BA, Page DL, Schuyler P, Parl FF, Simpson JF, Jensen RA, et al. No elevation in long-term breast carcinoma risk for women with fibroadenomas that contain atypical hyperplasia. Cancer. 2001;92(1):30–6.
352. Lim WK, Ong CK, Tan J, Thike AA, Ng CC, Rajasegaran V, et al. Exome sequencing identifies highly recurrent MED12 somatic mutations in breast fibroadenoma. Nat Genet. 2014;46(8):877–80.
353. Makinen N, Mehine M, Tolvanen J, Kaasinen E, Li Y, Lehtonen HJ, et al. MED12, the mediator complex subunit 12 gene, is mutated at high frequency in uterine leiomyomas. Science. 2011;334(6053):252–5.
354. Noguchi S, Yokouchi H, Aihara T, Motomura K, Inaji H, Imaoka S, et al. Progression of fibroadenoma to phyllodes tumor demonstrated by clonal analysis. Cancer. 1995;76(10):1779–85.
355. Hodges KB, Abdul-Karim FW, Wang M, Lopez-Beltran A, Montironi R, Easley S, et al. Evidence for transformation of fibroadenoma of the breast to malignant phyllodes tumor. Appl Immunohistochem Mol Morphol. 2009;17(4):345–50.
356. Tan PH, Thike AA, Tan WJ, Thu MM, Busmanis I, Li H, et al. Predicting clinical behaviour of breast phyllodes tumours: a nomogram based on histological criteria and surgical margins. J Clin Pathol. 2011;65(1):69–76.
357. Belkacemi Y, Bousquet G, Marsiglia H, Ray-Coquard I, Magne N, Malard Y, et al. Phyllodes tumor of the breast. Int J Radiat Oncol Biol Phys. 2008;70(2):492–500.
358. Ben Hassouna J, Damak T, Gamoudi A, Chargui R, Khomsi F, Mahjoub S, et al. Phyllodes tumors of the breast: a case series of 106 patients. Am J Surg. 2006;192(2):141–7.
359. Tan PH, Thike AA, Tan WJ, Thu MM, Busmanis I, Li H, et al. Predicting clinical behaviour of breast phyllodes tumours: a nomogram based on histological criteria and surgical margins. J Clin Pathol. 2012;65(1):69–76.
360. Nagasawa S, Maeda I, Fukuda T, Wu W, Hayami R, Kojima Y, et al. MED12 exon 2 mutations in phyllodes tumors of the breast. Cancer Med. 2015;4 (7):1117–21.
361. Ng CC, Tan J, Ong CK, Lim WK, Rajasegaran V, Nasir ND, et al. MED12 is frequently mutated in breast phyllodes tumours: a study of 112 cases. J Clin Pathol. 2015;68(9):685–91.
362. Piscuoglio S, Murray M, Fusco N, Marchio C, Loo FL, Martelotto LG, et al. MED12 somatic mutations in fibroadenomas and phyllodes tumours of the breast. Histopathology. 2015;67(5):719–29.
363. Piscuoglio S, Ng CKY, Murray M, Burke KA, Edelweiss M, Geyer FC, et al. Massively parallel sequencing of Phyllodes tumors of the breast reveals actionable mutations, and TERT promoter hotspot mutations and TERT gene amplification as likely drivers of progression. J Pathol. 2015;in press.

364. Pfarr N, Kriegsmann M, Sinn P, Klauschen F, Endris V, Herpel E, et al. Distribution of MED12 mutations in fibroadenomas and phyllodes tumors of the breast — implications for tumor biology and pathological diagnosis. Genes Chromosom Cancer. 2015;54(7):444–52.
365. Tan J, Ong CK, Lim WK, Ng CC, Thike AA, Ng LM, et al. Genomic landscapes of breast fibroepithelial tumors. Nat Genet. 2015;47 (11):1341–5.
366. Yoshida M, Ogawa R, Yoshida H, Maeshima A, Kanai Y, Kinoshita T, et al. TERT promoter mutations are frequent and show association with MED12 mutations in phyllodes tumors of the breast. Br J Cancer. 2015;113(8):1244–8.

# Epidemiology of Ovarian and Endometrial Cancers

# 13

Kathryn L. Terry and Stacey A. Missmer

## 13.1 Clinical Picture of the Disease

### 13.1.1 Overview

Approximately 20,000 women in the United States (U.S.) and 125,000 worldwide die annually of ovarian cancer [1, 2]. While ovarian cancer is relatively rare, compared to other types of cancer, it is deadly with less than half of women diagnosed surviving for 5 years [1]. In contrast, endometrial cancer is the most common gynecologic malignancy in the U.S. [3], but is frequently detected at early stage due to post-menopausal bleeding. Consequently, endometrial cancer survival is much better with 82 % surviving 5 years [1].

Since about 90 % of ovarian tumors are epithelial in origin, the focus of this chapter will be epithelial tumors. However, even within epithelial ovarian cancer there are a wide range of ovarian cancer subtypes that likely arise from distinct etiologic pathways; therefore, the epidemiology of ovarian cancer needs to be viewed with an eye toward differences that may exist by ovarian cancer subtypes which are best defined by histology (serous, endometrioid, clear cell, mucinous) and grade (low grade, high grade). Histologic subtypes of ovarian cancer differ in their pathologic appearance and generally do not resemble the ovarian surface epithelium, which had been the presumed source of all epithelial ovarian tumors [4]. Recent findings, most notably evidence of frequent fallopian tube cancers in women with ovarian cancer or at high risk for ovarian cancer [5–7], have led to a paradigm shift in ovarian cancer research in which some ovarian tumors are thought to arise from non-ovarian origins like the fallopian tube for serous tumors, endometrial tissue for endometrioid and clear cell tumors, and possibly gastrointestinal origins for mucinous tumors (reviewed in [8, 9]).

Similarly, endometrial cancer has different subtypes that have different etiologic pathways. Traditionally, endometrial cancer was classified into two major subtypes—Type I, which is serous and is related to excess estrogen, and accounts for 80 % of the cases and Type II, which includes a much broader range of histologic subtypes. However, recent analyses by The Cancer Genome Atlas (TCGA) Research Network, including genomic, transcriptomic, and proteomic characterization of 373 endometrial cancers, identified four subtypes (POLE ultramutated, microsatellite instability hypermutated, copy number low, and copy number high) [10].

K.L. Terry · S.A. Missmer (✉)
Brigham and Women's Hospital and Harvard Medical School, Harvard T.H. Chan School of Public Health, 221 Longwood Ave., Boston, MA 02115, USA
e-mail: stacey.missmer@channing.harvard.edu


M. Loda et al. (eds.), *Pathology and Epidemiology of Cancer*,
DOI 10.1007/978-3-319-35153-7_13

## 13.1.2 Ovarian Cancer

### 13.1.2.1 Symptoms

Ovarian cancer is known as the "silent killer" due to its ambiguous symptoms until disease is at an advanced stage [11–13]. However, most women with ovarian cancer experience symptoms before diagnosis. In fact, a meta-analysis of existing studies shows that less than 10 % of women with ovarian cancer report having no symptoms at the time of diagnosis [14]. The proportion of asymptomatic women with ovarian cancer is slightly higher when documentation in medical records were used to determine the presence of symptoms at diagnosis [14], suggesting that symptoms were not always communicated to and/or recorded by the clinician or suggesting errors in recall for self-reported symptoms. These symptoms include back pain, fatigue, bloating, constipation, difficulty eating or feeling full quickly, abdominal pain and urinary symptoms, which are difficult to distinguish from normal variation or benign conditions. However, symptoms tend to be more common, frequent, and severe in women with ovarian cancer [15]. Importantly, many symptoms are reported in women with early stage disease, suggesting that awareness of these symptoms could aid in early detection of ovarian cancer and consequently improve survival [15–20].

### 13.1.2.2 Diagnosis

Definitive diagnosis of ovarian cancer requires surgery and pathologic examination of the mass to distinguish it from pelvic inflammatory disease, endometriosis, pelvic kidney, or a colonic mass (either inflammatory or neoplastic) [21]. To date, no non-invasive marker or screening test has been identified. Therefore, algorithms have been developed to aid in the differential diagnosis of ovarian tumors. Notably, the risk of malignancy index (RMI) which uses a combination of menopausal status, ultrasound, and serum levels of the tumor marker CA125, has been shown to correctly identify 77 % of benign masses, 59 % borderline ovarian tumors, and 91 % of invasive ovarian cancers [22, 23].

### 13.1.2.3 Treatment

Ovarian cancer treatment includes surgical removal/excision (oophorectomy) with staging and debulking followed by chemotherapy (usually carboplatin and paclitaxel) for most cases [24]. Recent studies suggest that intraperitoneal chemotherapy, which is sometimes offered to women with less than one-centimeter postoperative residual disease, provides a survival benefit [25, 26]. However, intraperitoneal chemotherapy is frequently discontinued prematurely due its increased toxicity [27]. The marker CA-125 is used to assess whether a complete response has been achieved. However, a recent study suggests that initiation of chemotherapy based on a rise in CA125 in the absence of symptoms may not improve mortality and may result in poorer quality of life [28, 29].

### 13.1.2.4 Prognosis

Overall, only 40–45 % of women diagnosed with ovarian cancer survive with the disease for 5 years or more. Early stage disease (confined to ovary) is generally responsive to treatment and therefore has a good prognosis (92 % 5-year survival). Unfortunately, only 15 % of women are diagnosed with ovarian cancer at an early stage [1].

Ovarian cancer survival varies by histologic subtype, with the shortest survival for serous tumors. In a study of nearly 2000 women with ovarian cancer, those with serous borderline (Hazard Ratio (HR) 0.35, 95 % Confidence Interval (CI) 0.18–0.67), endometrioid (HR 0.28, 95 % CI 0.22–0.37), clear cell (HR 0.30, 95 % CI 0.21–0.44) and mucinous (HR 0.38, 95 % CI 0.25–0.58) had lower mortality than women with serous invasive tumors [30]. However, differences in survival by histologic type were attenuated after accounting for clinical factors including debulking status, stage, and grade.

Women with a younger age at diagnosis are more likely to have tumors with a lower stage and less aggressive histology; therefore, they tend to have a better prognosis. Similarly, cases with minimal residual disease (optimal debulking) have a good prognosis [31]. Interestingly, women with germline BRCA1/2 mutations have a good prognosis because the homologous recombination repair deficiencies in these tumors make them responsive to platinum-based therapy and PARP inhibitors [32].

### 13.1.3 Endometrial Cancer

#### 13.1.3.1 Symptoms

Endometrial cancer is characterized by bleeding, particularly in postmenopausal women. In fact, 75–90 % of women with endometrial cancer present clinically with abnormal bleeding [33, 34], making population screening of the disease unnecessary.

#### 13.1.3.2 Diagnosis

Evaluation of women with a suspected endometrial tumor, due to abnormal bleeding or incidental findings, includes physical examination, a pregnancy test, and pelvic ultrasonography to rule out benign conditions [35]. Definitive diagnosis requires evaluation of an endometrial biopsy, endometrial curettage, or hysterectomy specimen [36].

#### 13.1.3.3 Treatment

Hysterectomy with removal of the tumor is the standard treatment for endometrial cancer and generally curative for early stage of recurrence. Other tumors may benefit from adjuvant therapy [36].

#### 13.1.3.4 Prognosis

Unlike ovarian cancer, endometrial cancer is often diagnosed at early stage (68 % are confined to the uterus at diagnosis) [36] and prognosis is good since it has a largely curative treatment (hysterectomy). Overall, an estimated 82 % of women with endometrial cancer survive 5 years after diagnosis [1]. When examined by stage, prognosis is best for early stage with 80–90 % survival for stage I, 70–80 % for stage II, 20–60 % for stages III and IV.

## 13.2 Descriptive Epidemiology

### 13.2.1 Ovarian Cancer

#### 13.2.1.1 Trends Over Time

Ovarian cancer incidence has exhibited a modest decline in the U.S. over the last 30 years with a slight drop in rates since the mid-1990s (Fig. 13.1). However, the Surveillance, Epidemiology, and End Result (SEER) agency began excluding borderline tumors from their assessment of ovarian cancer incidence in 2004, which may explain part of the decline [37]. For every 100,000 women in the US, the age-adjusted number of ovarian cancer cases in 1980 was 15.5, 15.4 in 1990, 14.3 in 2000, and 12.7 in 2010 [38]. However, the drop in incidence between 2004 and 2006 can be largely explained by SEER's exclusion of low malignant potential (LMP) or borderline tumors, which account for approximately 17 % of ovarian cancers. LMP tumors are ovarian tumors that can extend beyond the ovary, but have an indolent course and usually do not require surgery or radiation [37]. Ovarian cancer deaths have been stable with a slight drop in rates since the mid-1990s, which coincides with the introduction of taxol treatment [39–43].

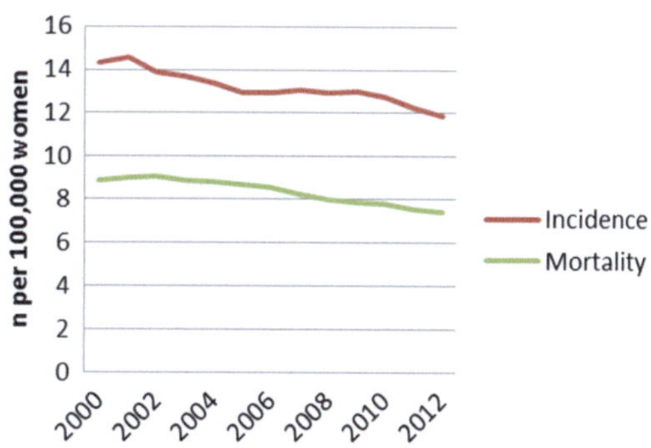


**Fig. 13.1** Ovarian cancer incidence and mortality in the United States, 2000–2012 (data from surveillance, epidemiology, and end results program)

#### 13.2.1.2 Geographic Variation

Ovarian cancer incidence is higher in developed than developing countries with the highest rates in Europe, North America, and Australia and lowest rates in Africa, after accounting for age (Fig. 13.2). Interestingly, while ovarian cancer is relatively rare in Asia, most cases from this region are clear cell tumors [44, 45], which often have a poor prognosis since they generally do not respond well to standard chemotherapy at advanced stages [46].

#### 13.2.1.3 Variation by Age

Epithelial ovarian cancer is rare at younger ages, and increases steadily starting in a woman's reproductive years (Fig. 13.3). Most ovarian cancer cases are diagnosed after menopause with a median age at diagnosis of 63 years [38].

### 13.2.2 Endometrial Cancer

#### 13.2.2.1 Trends Over Time

Endometrial cancer incidence in the U.S. has remained relatively steady since 2000 (Fig. 13.4), with an age-adjusted estimate of 24.8 endometrial cancer cases per 100,000 women in 2000, and 27.5 in 2012. The slight increase in recent years may be attributable to increasing

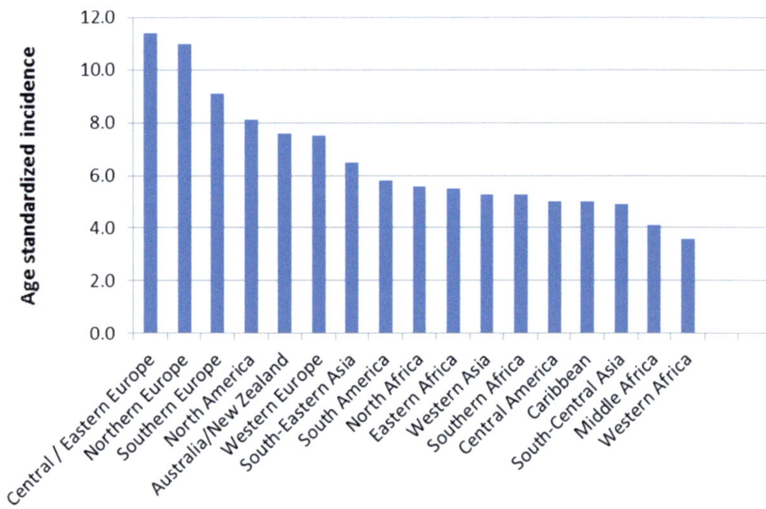


**Fig. 13.2** Worldwide variation in ovarian cancer incidence (data from Globocan 2012, available at http://globocan.iarc.fr)

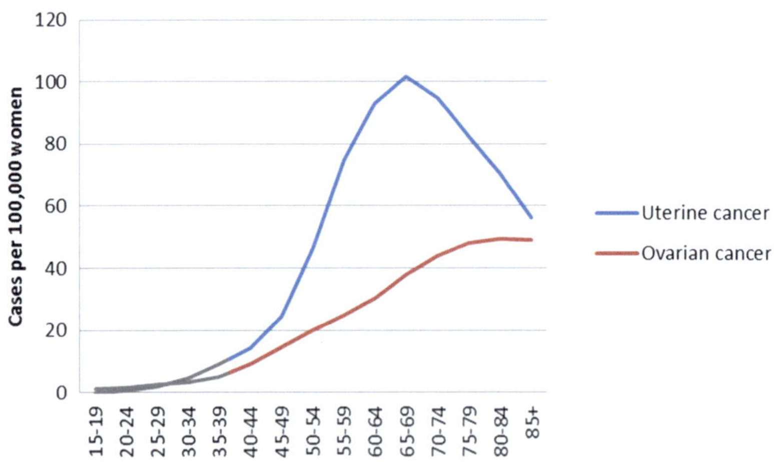


**Fig. 13.3** Age-specific incidence rates for ovarian and endometrial cancer, United States, 2008–2012 (data from surveillance, epidemiology, and end results program)

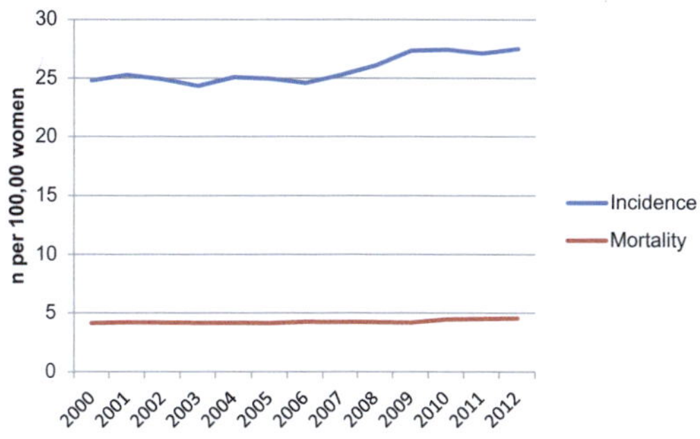


**Fig. 13.4** Endometrial cancer incidence and mortality in the United States, 2000–2012. (data from surveillance, epidemiology, and end results program)

average body mass index (BMI), which is expected to continue rising in the future.

#### 13.2.2.2 Geographic Variation

Endometrial cancer is the most common gynecologic cancer in developed areas like North American and Europe [47], and second in developing areas, like Africa, where endometrial cancer incidences follow behind cervical cancer (Fig. 13.5).

#### 13.2.2.3 Variation by Age

Similar to ovarian cancer, endometrial cancer incidence accelerates around menopause. However, endometrial cancer incidence peaks about 10 years later with a median age of 61 years [36] and then tapers off at older ages (Fig. 13.3).

## 13.3 Risk Factors

### 13.3.1 Ovarian Cancer

#### 13.3.1.1 Reproductive History

The strongest and most consistent "risk" factors for ovarian cancer are oral contraceptive (OC) use and parity—both of which reduce ovarian cancer risk. These two factors also reduce the lifetime number of ovulatory cycles, which inflict repeated damage and repair to the ovarian surface that may initiate or promote carcinogenesis (Fig. 13.6). In a meta-analysis of 55 epidemiological studies, women who had ever used OCs had more than a 25 % reduction in risk (Odds Ratio (OR) 0.73, 95 % CI 0.66–0.81). Longer duration was correlated with greater reduction in risk in a dose–response fashion with the greatest reduction in risk for women who had used OCs more than 10 years (OR 0.43, 95 % CI 0.37–0.51) [48]. Since OC use is common in reproductive aged women, the population impact of this significant reduction in risk is notable for this rare disease with an estimated two ovarian cancer diagnoses and one ovarian cancer death avoided before age 75 for every 5000 woman-years of OC use [49]. To date, the reduction in ovarian cancer risk with OC use has persisted across calendar time despite the changes in formulations [49] and generally across ovarian cancer subtypes though there may be a

**Fig. 13.5** Worldwide variation in endometrial cancer incidence (data from Globocan 2012, available at http://globocan.iarc.fr)

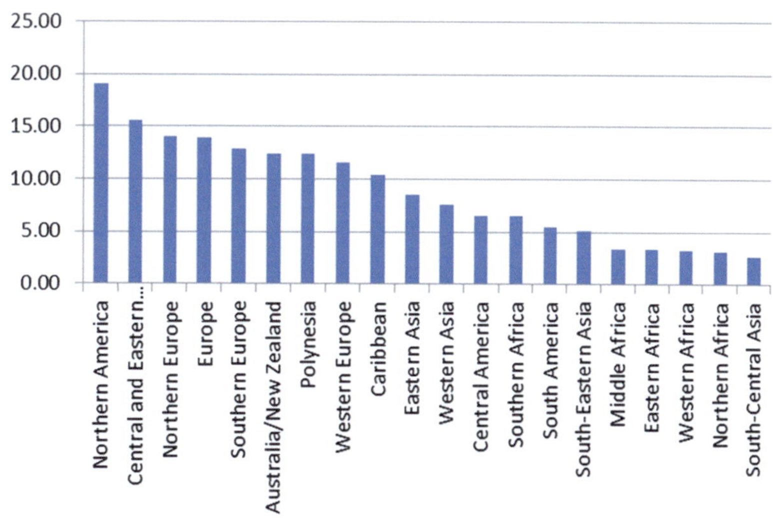

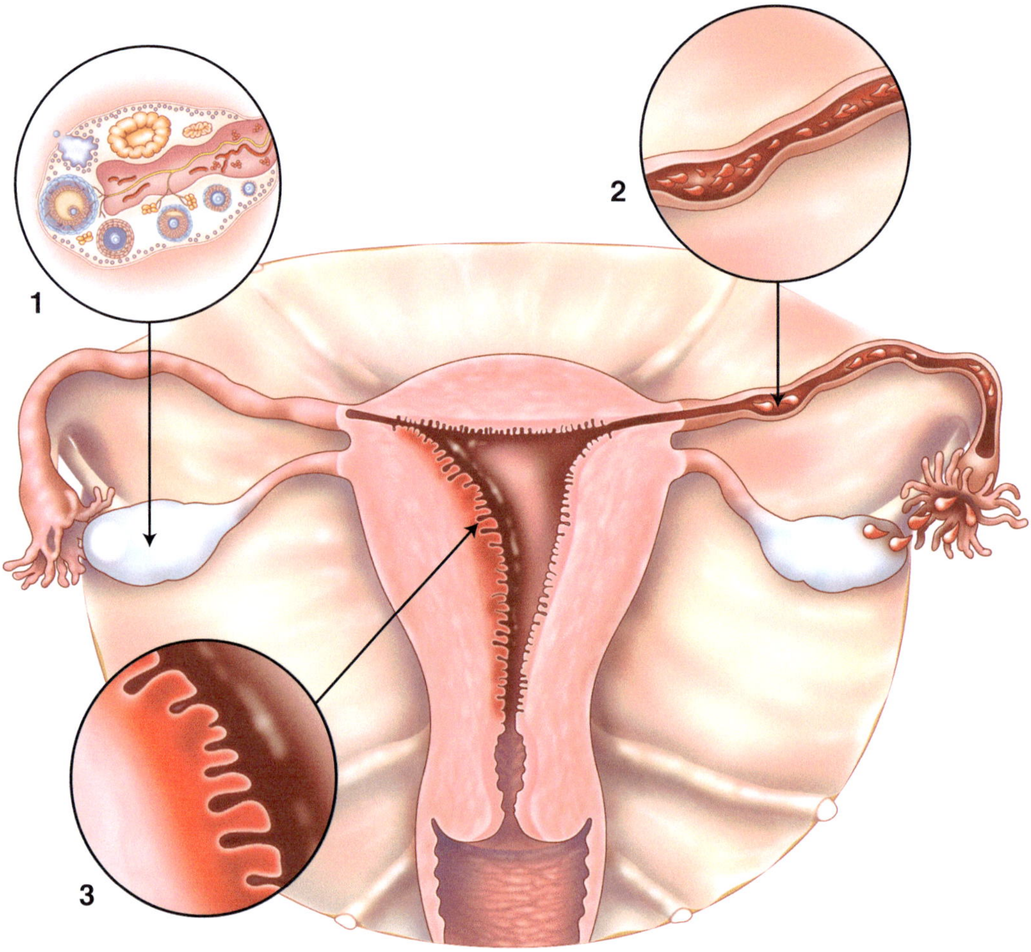


**Fig. 13.6** Epidemiologic evidence that supports etiologic hypotheses for ovarian and endometrial carcinogenesis. *1* Increased ovarian cancer risk due to a greater number of ovulatory cycles is supported by decreased risk with pregnancy and oral contraceptive use, *2* Increased risk of endometrioid and clear cell ovarian cancer due to retrograde menstruation is supported by increased risk of those subtypes for women with endometriosis and decreased risk with tubal ligation, *3* Increased risk of endometrial cancer developing from estrogen induced endometrial hyperplasia is supported by an increased risk with estrogen-only hormone therapy and higher body mass index

greater reduction in risk for more aggressive tumors [50, 51].

While parous women have a lower ovarian cancer risk than nulliparous women, all pregnancies are not equally protective. A single pregnancy lowers ovarian cancer risk by approximately 40 % while each subsequent pregnancy lowers risk by an additional 10–15 % per pregnancy [52, 53]. Interestingly, parity is more protective for endometrioid, clear cell, low-grade serous, and mucinous tumors than high-grade serous, though studies differ on whether it is the first or subsequent pregnancies that differ [50, 54, 55]. Later age at first and last births reduce ovarian cancer risk [56–58]. However, later age at first birth is likely only protective due to its correlation with age at last birth since only age at last birth remains significantly associated with ovarian cancer risk when both are included in a multivariate model [58].

#### 13.3.1.2 Hormone Replacement Therapy

The association between hormone replacement therapy (HRT) and ovarian cancer has been inconsistent and may vary by HRT formulation and ovarian cancer histology. Although prior pooled analyses and meta-analyses reported no association between HRT use and ovarian cancer risk [59–64], some studies suggest that long duration of HRT, particularly estrogen-only formulations, can increase risk [65–68]. For instance, in the Nurses' Health Study, women who used HRT for more than 5 years had a significant 40–50 % increase in ovarian cancer risk compared to women who had never used HRT [62]. In the United Kingdom (U.K.) Million Women's Study, increased risk with HRT use was restricted to women who were current users [63].

#### 13.3.1.3 Endometriosis

Endometriosis, which affects approximately 10% of reproductive aged women, is the occurrence of endometrial-like tissue outside the uterus that is often associated with pain and infertility [69–71]. Women with endometriosis have three times the risk of developing clear cell ovarian cancer and twice the risk of developing endometrioid or low-grade serous ovarian cancer as women without endometriosis [72]. On pathologic review, ovarian tumors are sometimes found adjacent to endometriosis and these tissues share genetic mutations and mRNA expression patterns [73–75].

A pooled analysis from the Ovarian Cancer Association Consortium showed increased risk of clear cell (OR 3.1, 95 % CI 2.4–3.8), endometrioid (OR 2.0, 95 % CI 1.7–2.5) and low-grade serous (OR 2.1, 95 % CI 1.4–3.2) ovarian cancer for women with self-reported endometriosis, compared to the general population but not of high-grade serous or mucinous ovarian cancer [72].

#### 13.3.1.4 Tubal Ligation

Tubal ligation, surgical sterilization by closure of the fallopian tube, reduces a woman's overall ovarian cancer risk by 30–34 %, according to recent meta-analyses [76, 77]. Interestingly, this association has held up across time, despite changes in the how tubal ligations have been performed (cutting, burning, banding) [76, 78], though a recent study suggests that tubal sterilization by excision may be more protective than other methods [79]. Tubal ligation is thought to reduce ovarian cancer risk by blocking retrograde menstruation or inflammatory contaminants from ascending the reproductive tract (Fig. 13.6) and thereby preventing exposure of the ovaries to these potential carcinogens, such as talc and endometrial tissue [80]. Recently, many studies and pooled analyses have demonstrated that the reduction in risk afforded by tubal ligation is restricted to endometrioid and clear cell types [81, 82]. In a pooled analysis including 10,157 cases from the Ovarian Cancer Association Consortium, Sieh et al. observed that women with a tubal ligation had a 29 % reduction in serous ovarian cancer risk, but a 52 and 48 % reduction in risk for endometrioid and clear cell subtypes, respectively [82].

#### 13.3.1.5 Genital Powder Use

Powder use, which generally includes talc, is associated with a 25–35 % increase in ovarian cancer risk when applied to the genital area [83, 84]. While the association between ever use of genital powder and ovarian cancer is fairly consistent over 20 years of research, evidence of a dose response has been inconsistent [85, 86], which has brought into doubt the biologic plausibility of the association.

#### 13.3.1.6 Smoking

The association between smoking and ovarian cancer differs by histologic subtype. In an analysis within the Ovarian Cancer Association Consortium, including more than 14,000 cases, current smoking was associated with a 30 % increase in mucinous invasive, but not other histologic subtypes [87]. Former versus never smoking was associated with a 30 % increase in risk of serous borderline ovarian cancer. Interestingly, current smoking was associated with a reduced risk of endometrioid (OR 0.84, 95 % CI 0.69–1.02) and clear cell ovarian cancer (OR

0.74, 95 % CI 0.56–0.98), which is not entirely surprising since smoking also decreases the risk of endometriosis and endometrial cancer.

#### 13.3.1.7 Body Mass Index

Observations regarding BMI and ovarian cancer risk are inconsistent. A pooled analysis of 12 cohort studies, including 2036 cases, showed no significant increase in ovarian cancer risk with each 4 kg/m$^2$ increase in prediagnostic adult life BMI (OR 1.01, 95 % CI 0.95–1.07) [88]. In contrast, a pooled analysis of 11 case-control studies in OCAC, including 13,548 cases and 17,913 controls reported a small, but statistically significant increase in risk with each 5 kg/m$^2$ increase in BMI for both invasive (OR 1.04, 95 % CI 1.00–1.08) and borderline tumors (OR 1.18, 95 % CI 1.14–1.23) [89]. Furthermore, the association with BMI was restricted to mucinous and endometrioid tumors. These results are consistent with a meta-analysis of 47 studies showing a statistically significant increase in ovarian cancer risk with each 5 kg/m$^2$ increase in BMI in both prospective (Relative Risk (RR) 1.03) and population-based case-control studies (RR 1.10). However, the association between BMI and ovarian cancer risk per 5 kg/m$^2$ differed by use of HRT use with a positive association between BMI and risk in never users (RR 1.10, 95 % 1.07–1.13) and an inverse association in ever HRT users (RR 0.95, 95 % CI 0.92–0.99) [90].

#### 13.3.1.8 Polycystic Ovarian Syndrome

Recent meta-analyses disagree on whether ovarian cancer risk is higher in women with polycystic ovarian syndrome (PCOS), a condition characterized by irregular menstrual cycles, infertility, hyperandrogenism, and frequently a higher BMI. One epidemiological study reported a two- to threefold increased risk associated with PCOS [91], while another study found no significant association [92]. Differences might be explained by variability in the classification of PCOS or considerations of potential confounders like OCs, which are often a first line therapy to treat irregular menstrual cycles.

#### 13.3.1.9 Family History

There are a number of inherited genetic risk factors emerging for ovarian cancer. Having a mother or sister with the disease increases a woman's risk for ovarian cancer by approximately two- to threefold [93]. Most hereditary ovarian cancers are due to BRCA1 and BRCA2 mutations, accounting for 10–15 % of ovarian cancers [94–98]. Less commonly, hereditary ovarian cancer occurs due to the Lynch syndrome or hereditary non-polyposis colorectal cancer syndrome (HNPCC), which is characterized by mutations in mismatch repair genes (hMLH1, hMSH2, hPMS1, hPMS2, and hMSH6). These syndromes account for about 2 % of ovarian cancers [94]. While these higher penetrant genetic factors are more likely to be found in families in which a number of relatives have been affected with breast or ovarian cancer, an estimated 10 % of "sporadic" ovarian cancer [99] and up to 40 % among women with a Jewish ethnic background [100] also carry these genetic variants.

### 13.3.2 Endometrial Cancer

Unlike ovarian cancer, endometrial cancer risk factors can be largely explained by a single mechanism—excess estrogen exposure that promotes endometrial hyperplasia that can transform into endometrial cancer (Fig. 13.6). For instance, estrogen therapy unopposed by progestin is one of the strongest risk factors for endometrial cancer, which is why this type of hormone therapy is only recommended for women who have had a hysterectomy.

#### 13.3.2.1 Reproductive History

Most reproductive factors that influence endometrial cancer risk can be attributed to increased estrogen exposure. For example, earlier age at menarche and later age at menopause both extend the duration of exposure to ovarian hormones and increase endometrial cancer risk [101]. Interestingly, a later age at last birth lowers endometrial cancer risk and largely explains the inverse association between number of children and endometrial cancer risk [101]. In an international pooled analysis including 8671 cases and 16,562 controls, Setiawan et al. reported an 18 % decrease in endometrial cancer risk for each 5-year increase in age at last birth [102].

#### 13.3.2.2 Hormone Replacement Therapy

The strong and consistent elevation in endometrial cancer risk with HRT has lead to recommendations that women with an intact uterus avoid estrogen-only therapy. In a meta-analysis of 30 studies, estrogen-only users had a twofold increase in endometrial cancer risk and nearly a tenfold increase in risk for women who had used estrogen-only therapy for more than 10 years [103].

#### 13.3.2.3 Body Mass Index

Given the growing obesity epidemic, the strong and consistent association between increasing BMI and endometrial cancer risk is concerning and may explain a large proportion of the recent rise in endometrial cancer incidence in the U.S. [104]. A meta-analysis of 24 prospective studies, including more than 17,000 cases, reported a 60 % increase in risk for each 5 kg/m$^2$ increase in BMI. Compared to normal weight women (BMI 20–25 kg/m$^2$), the heaviest women (BMI > 42 kg/m$^2$) had more than a ninefold increase in risk. The association was strongest among women with a BMI > 42 kg/m$^2$ who had never used HRT (RR 20, 95 % CI 8–52). Estrogenic proliferation is thought to be the primary mechanism underlying the BMI endometrial cancer association since fat cells convert androgens to estrogens and are the leading source of estrogen in postmenopausal women. A stronger association in non-HRT users lends support to this mechanism.

However, associations between diabetes and/or metabolic syndrome, which consists of a combination of cardiovascular risk factors including dysglycemia, elevated blood pressure, low high-density lipoprotein cholesterol levels, with endometrial cancer suggest that BMI may increase endometrial cancer mechanisms through non-estrogenic pathways as well. Recent meta-analyses reported a nearly twofold increase in endometrial cancer risk for women with diabetes or metabolic syndrome compared to those without [105]. While higher BMI largely accounts for the association between these diabetes and endometrial cancer, a significant association remains after adjusting for BMI [106], which suggests that insulin dysregulation or some other aspect of diabetes encourages endometrial carcinogenesis above and beyond increased body weight alone.

#### 13.3.2.4 Smoking

Compared to women who have never smoked, former and current smokers have a 20–60 % reduction in endometrial cancer risk [107, 108]. Risk reduction is greater for current smokers (RR 0.65, 95 % CI 0.55, 0.78) than for former smokers (RR 0.89, 95 % CI 0.80, 1.00), but there is no dose response association with greater number of cigarettes per day [109]. Potential mechanisms for this reduction in risk relates to reduced or modified estrogen, including a younger age of menopause due to destruction of oocytes, lower BMI, a shift in estrogen metabolism towards the anti-carcinogenic 2-hydroxyestrone form, and increased progesterone [107, 109].

#### 13.3.2.5 Polycystic Ovarian Syndrome

An elevated risk of endometrial cancer for women with PCOS, a condition in which women are often overweight and have insulin dysregulation, is not surprising given these are both established endometrial cancer risk factors. Women with PCOS have approximately a fourfold increase in risk of endometrial cancer [92, 110]. Although the association between PCOS and endometrial cancer is attenuated after adjustment for BMI, suggesting that some of the association is explained by higher BMI in women with PCOS, a significant association remains indicating that PCOS increases risk independent of increased body weight [110].

#### 13.3.2.6 Family History

Having a mother or sister with endometrial cancer doubles a woman's own risk [111], and there are several familial syndromes known to include endometrial cancer. Most notably, Lynch syndrome is associated with a 27–71 % lifetime risk of endometrial cancer compared to a 3 % risk in the general population [47]. Cowden syndrome is a rare familial syndrome, characterized by a mutation in Phosphatase and tensin homolog (PTEN), and is associated with 13–28 % lifetime risk of endometrial cancer [47, 112].

## 13.4 Conclusion

The last several years have been a time of great epidemiologic and basic science discovery for ovarian and endometrial cancer. These cancers of the reproductive tract have both concordant and discordant risk factor profiles. For the former, identification of distinct subtypes has resulted in paradigm-shifting focus for etiologic hypotheses as well as diagnostic and treatment discovery. For the latter, etiologic hypotheses which expand beyond estrogenic proliferation as risk factors that do not fit the classic etiologic framework, like age at last birth and PCOS, have been appreciated. As distinct subtypes of these cancers are identified, the need for multicenter international collaborations that contribute "big data" to foster disease subtype discovery will be critical to moving the field forward.

## References

1. American Cancer Society. Cancer facts and figures. Atlanta, GA: American Cancer Society; 2015.
2. Ferlay J, Soerjomataram I, Dikshit R, et al. Cancer incidence and mortality worldwide: sources, methods and major patterns in GLOBOCAN 2012. Int J Cancer. 2015;136(5):E359–86.
3. Group SGOCPECW, Burke WM, Orr J, et al. Endometrial cancer: a review and current management strategies: part II. Gynecol Oncol. 2014;134(2):393–402.
4. Dubeau L. The cell of origin of ovarian epithelial tumors and the ovarian surface epithelium dogma: does the emperor have no clothes? Gynecol Oncol. 1999;72(3):437–42.
5. Lee Y, Miron A, Drapkin R, et al. A candidate precursor to serous carcinoma that originates in the distal fallopian tube. J Pathol. 2007;211(1):26–35.
6. Levanon K, Ng V, Piao HY, et al. Primary ex vivo cultures of human fallopian tube epithelium as a model for serous ovarian carcinogenesis. Oncogene. 2010;29(8):1103–13.
7. Piek JM, van Diest PJ, Zweemer RP, et al. Dysplastic changes in prophylactically removed fallopian tubes of women predisposed to developing ovarian cancer. J Pathol. 2001;195(4):451–6.
8. Kurman RJ, Shih IM. The origin and pathogenesis of epithelial ovarian cancer: a proposed unifying theory. Am J Surg Pathol. 2010;34(3):433–43.
9. Vaughan S, Coward JI, Bast RC Jr, et al. Rethinking ovarian cancer: recommendations for improving outcomes. Nat Rev Cancer. 2011;11(10):719–25.
10. Cancer Genome Atlas Research Network, Kandoth C, Schultz N, et al. Integrated genomic characterization of endometrial carcinoma. Nature. 2013;497(7447):67–73.
11. Ihde DC, Longo DL. Presentations of the patient with cancer: solid tumors in adults. 14th ed. New York: McGraw-Hill; 1998.
12. Clement KD, Connor PD. Tumors of the female reproductive organs. In: Taylor RB, David AK, Johnson TAJ, Phillips DM, Scherger JE, editors. Family medicine: principles and practice. 5th ed. New York: Spinger-Verlag Inc; 1998. p. 916–24.
13. Young RC. Gynecologic malignancies. In: Fauci AS, Braunwald E, Isselbacher KJ, Wilson JD, Martin JB, Kasper DL, editors. Harrison's principles of internal medicine. 14th ed. New York: McGraw-Hill; 1998. p. 605–11.
14. Bankhead CR, Kehoe ST, Austoker J. Symptoms associated with diagnosis of ovarian cancer: a systematic review. BJOG. 2005;112(7):857–65.
15. Goff BA, Mandel LS, Melancon CH, Muntz HG. Frequency of symptoms of ovarian cancer in women presenting to primary care clinics. JAMA. 2004;291(22):2705–12.
16. Yawn BP, Barrette BA, Wollan PC. Ovarian cancer: the neglected diagnosis. Mayo Clin Proc. 2004;79(10):1277.
17. Goff BA, Mandel L, Muntz HG, Melancon CH. Ovarian carcinoma diagnosis. Cancer. 2000;89(10):2068–75.
18. Goff BA, Mandel LS, Drescher CW, et al. Development of an ovarian cancer symptom index: possibilities for earlier detection. Cancer. 2007;109(2):221–7.
19. Olson SH, Mignone L, Nakraseive C, Caputo TA, Barakat RR, Harlap S. Symptoms of ovarian cancer. Obstet Gynecol. 2001;98(2):212–7.
20. Vine MF, Ness RB, Calingaert B, Schildkraut JM, Berchuck A. Types and duration of symptoms prior to diagnosis of invasive or borderline ovarian tumor. Gynecol Oncol. 2001;83(3):466–71.
21. Berek JS, Friedlander M, Hacker N. Epithelial ovarian, fallopian tube, and peritoneal cancer. In:

Berek and Hacker's gynecologic oncology 2015;475.
22. Chia YN, Marsden DE, Robertson G, Hacker NF. Triage of ovarian masses. Aust N Z J Obstet Gynaecol. 2008;48(3):322–8.
23. Menon U, Talaat A, Jeyarajah AR, et al. Ultrasound assessment of ovarian cancer risk in postmenopausal women with CA125 elevation. Br J Cancer. 1999;80(10):1644–7.
24. Chen L, Berek JS. Overview of epithelial carcinoma of the ovary, fallopian tube, and peritoneum. The Netherland: Wolters Kluwer; 2015 (UpToDate).
25. Hess LM, Benham-Hutchins M, Herzog TJ, et al. A meta-analysis of the efficacy of intraperitoneal cisplatin for the front-line treatment of ovarian cancer. Int J Gynecol Cancer. 2007;17(3):561–70.
26. Jaaback K, Johnson N. Intraperitoneal chemotherapy for the initial management of primary epithelial ovarian cancer. Cochrane Database Syst Rev. 2006;(1):CD005340.
27. Armstrong DK, Bundy B, Wenzel L, et al. Intraperitoneal cisplatin and paclitaxel in ovarian cancer. N Engl J Med. 2006;354(1):34–43.
28. Rustin G, van der Burg M, Griffin C, Qian W, Swart AM. Early versus delayed treatment of relapsed ovarian cancer. Lancet. 2011;377(9763):380–1.
29. Rustin GJ. Follow-up with CA125 after primary therapy of advanced ovarian cancer has major implications for treatment outcome and trial performances and should not be routinely performed. Ann Oncol. 2011;22(Suppl 8):viii45–48.
30. Poole EM, Cramer DW, Vitonis AF, Terry KL. Ovarian cancer survival by tumor subtype. Cancer Res. 2014;74: 4124.
31. Aletti GD, Gostout BS, Podratz KC, Cliby WA. Ovarian cancer surgical resectability: relative impact of disease, patient status, and surgeon. Gynecol Oncol. 2006;100(1):33–7.
32. Konstantinopoulos PA, Spentzos D, Karlan BY, et al. Gene expression profile of BRCAness that correlates with responsiveness to chemotherapy and with outcome in patients with epithelial ovarian cancer. J Clin Oncol. 2010;28(22):3555–61.
33. Kimura T, Kamiura S, Yamamoto T, Seino-Noda H, Ohira H, Saji F. Abnormal uterine bleeding and prognosis of endometrial cancer. Int J Gynaecol Obstet. 2004;85(2):145–50.
34. Seebacher V, Schmid M, Polterauer S, et al. The presence of postmenopausal bleeding as prognostic parameter in patients with endometrial cancer: a retrospective multi-center study. BMC Cancer. 2009;9:460.
35. Chen L-M, Berek JS. Endometrial carcinoma: clinical features and diagnosis. In: Barbara Goff M, Falk SJ, Facog D, editors. The Netherlands: Wolters Kluwer Health; 2015 (UpToDate).
36. Plaxe SC, Mundt AJ. Overview of endometrial carcinoma. In: Barbara Goff M, Don S, Dizon MD F, Sandy J, Falk M, Facog D, editors. The Netherlands: Wolters Kluwer Health; 2015 (UpToDate).
37. Cramer DW. The epidemiology of endometrial and ovarian cancer. Hematol Oncol Clin North Am. 2012;26(1):1–12.
38. Howlader N, Noone AM, Krapcho M, et al. SEER cancer statistics review, 1975–2011. Bethesda, MD: National Cancer Institute. http://seer.cancer.gov/csr/1975_2011/, based on November 2013 SEER data submission, posted to the SEER web site, April 2014.
39. du Bois A, Luck HJ, Meier W, et al. A randomized clinical trial of cisplatin/paclitaxel versus carboplatin/paclitaxel as first-line treatment of ovarian cancer. J Natl Cancer Inst. 2003;95(17):1320–9.
40. McGuire WP, Hoskins WJ, Brady MF, et al. Cyclophosphamide and cisplatin compared with paclitaxel and cisplatin in patients with stage III and stage IV ovarian cancer. N Engl J Med. 1996;334(1):1–6.
41. Neijt JP, Engelholm SA, Tuxen MK, et al. Exploratory phase III study of paclitaxel and cisplatin versus paclitaxel and carboplatin in advanced ovarian cancer. J Clin Oncol. 2000;18(17):3084–92.
42. Ozols RF, Bundy BN, Greer BE, et al. Phase III trial of carboplatin and paclitaxel compared with cisplatin and paclitaxel in patients with optimally resected stage III ovarian cancer: a gynecologic oncology group study. J Clin Oncol. 2003;21(17):3194–200.
43. Piccart MJ, Bertelsen K, James K, et al. Randomized intergroup trial of cisplatin-paclitaxel versus cisplatin-cyclophosphamide in women with advanced epithelial ovarian cancer: three-year results. J Natl Cancer Inst. 2000;92(9):699–708.
44. Tay SK, Cheong MA. Evidence for ethnic and environmental contributions to frequency of ovarian clear cell carcinoma. Aust N Z J Obstet Gynaecol. 2014;54(3):225–30.
45. Tung KH, Goodman MT, Wu AH, et al. Reproductive factors and epithelial ovarian cancer risk by histologic type: a multiethnic case-control study. Am J Epidemiol. 2003;158(7):629–38.
46. Goff BA, Sainz de la Cuesta R, Muntz HG, et al. Clear cell carcinoma of the ovary: a distinct histologic type with poor prognosis and resistance to platinum-based chemotherapy in stage III disease. Gynecol Oncol. 1996;60(3):412–7.
47. Chen L-M, Berek JS. In: Barbara Goff M, Sandy J, Falk M, Facog D, editors. Endometrial carcinoma: epidemiology and risk factors. 2015 (UpToDate).
48. International Agency for Research on Cancer (IARC). IARC monographs on the evaluation of carcinogenic risks to man. Pharmaceuticals: combined estrogen-progestogen contraceptives. Lyon: IARC. Vol 100A-192012:p. 283e311.
49. Collaborative Group on Epidemiological Studies of Ovarian Cancer, Beral V, Doll R, Hermon C, Peto R, Reeves G. Ovarian cancer and oral contraceptives: collaborative reanalysis of data from 45 epidemiological studies including 23,257 women with ovarian cancer and 87,303 controls. Lancet. 2008;371(9609):303–14.

50. Merritt MA, De Pari M, Vitonis AF, Titus LJ, Cramer DW, Terry KL. Reproductive characteristics in relation to ovarian cancer risk by histologic pathways. Hum Reprod. 2013;28(5):1406–17.
51. Poole EM, Merritt MA, Jordan SJ, et al. Hormonal and reproductive risk factors for epithelial ovarian cancer by tumor aggressiveness. Cancer Epidemiol Biomarkers Prev. 2013;22(3):429–37.
52. Adami HO, Hsieh CC, Lambe M, et al. Parity, age at first childbirth, and risk of ovarian cancer. Lancet. 1994;344(8932):1250–4.
53. Permuth-Wey J, Sellers TA. Epidemiology of ovarian cancer. Methods Mol Biol. 2009;472:413–37.
54. Kotsopoulos J, Terry KL, Poole EM, et al. Ovarian cancer risk factors by tumor dominance, a surrogate for cell of origin. Int J Cancer. 2013;133(3):730–9.
55. Fortner RT, Ose J, Merritt MA, et al. Reproductive and hormone-related risk factors for epithelial ovarian cancer by histologic pathways, invasiveness, and histologic subtypes: results from the EPIC cohort. Int J Cancer. 2015.
56. Titus-Ernstoff L, Perez K, Cramer DW, Harlow BL, Baron JA, Greenberg ER. Menstrual and reproductive factors in relation to ovarian cancer risk. Br J Cancer. 2001;84(5):714–21.
57. Bevier M, Sundquist J, Hemminki K. Does the time interval between first and last birth influence the risk of endometrial and ovarian cancer? Eur J Cancer. 2011;47(4):586–91.
58. Whiteman DC, Siskind V, Purdie DM, Green AC. Timing of pregnancy and the risk of epithelial ovarian cancer. Cancer Epidemiol Biomarkers Prev. 2003;12(1):42–6.
59. Whittemore AS, Harris R, Itnyre J. Characteristics relating to ovarian cancer risk: collaborative analysis of 12 US case-control studies. II. Invasive epithelial ovarian cancers in white women. Collaborative Ovarian Cancer Group. Am J Epidemiol. 1992;136(10):1184–203.
60. Coughlin SS, Giustozzi A, Smith SJ, Lee NC. A meta-analysis of estrogen replacement therapy and risk of epithelial ovarian cancer. J Clin Epidemiol. 2000;53(4):367–75.
61. Garg PP, Kerlikowske K, Subak L, Grady D. Hormone replacement therapy and the risk of epithelial ovarian carcinoma: a meta-analysis. Obstet Gynecol. 1998;92(3):472–9.
62. Danforth KN, Tworoger SS, Hecht JL, Rosner BA, Colditz GA, Hankinson SE. A prospective study of postmenopausal hormone use and ovarian cancer risk. Br J Cancer. 2007;96(1):151–6.
63. Million Women Study Collaborators, Beral V, Bull D, Green J, Reeves G. Ovarian cancer and hormone replacement therapy in the million women study. Lancet. 2007;369(9574):1703–10.
64. Risch HA. Estrogen replacement therapy and risk of epithelial ovarian cancer. Gynecol Oncol. 1996;63(2):254–7.
65. Folsom AR, Anderson JP, Ross JA. Estrogen replacement therapy and ovarian cancer. Epidemiology. 2004;15(1):100–4.
66. Lacey JV Jr, Brinton LA, Leitzmann MF, et al. Menopausal hormone therapy and ovarian cancer risk in the National Institutes of Health-AARP Diet and Health Study Cohort. J Natl Cancer Inst. 2006;98(19):1397–405.
67. Lacey JV Jr, Mink PJ, Lubin JH, et al. Menopausal hormone replacement therapy and risk of ovarian cancer. JAMA. 2002;288(3):334–41.
68. Rodriguez C, Patel AV, Calle EE, Jacob EJ, Thun MJ. Estrogen replacement therapy and ovarian cancer mortality in a large prospective study of US women. JAMA. 2001;285(11):1460–5.
69. Giudice LC, Kao LC. Endometr Lancet. 2004;364(9447):1789–99.
70. Eskenazi B, Warner ML. Epidemiology of endometriosis. Obstet Gynecol Clin North Am. 1997;24(2):235–58.
71. Missmer SA, Hankinson SE, Spiegelman D, Barbieri RL, Marshall LM, Hunter DJ. Incidence of laparoscopically confirmed endometriosis by demographic, anthropometric, and lifestyle factors. Am J Epidemiol. 2004;160(8):784–96.
72. Pearce CL, Templeman C, Rossing MA, et al. Association between endometriosis and risk of histological subtypes of ovarian cancer: a pooled analysis of case-control studies. Lancet Oncol. 2012;13(4):385–94.
73. Wiegand KC, Shah SP, Al-Agha OM, et al. ARID1A mutations in endometriosis-associated ovarian carcinomas. N Engl J Med. 2010;363(16):1532–43.
74. Jones S, Wang TL, Shih Ie M, et al. Frequent mutations of chromatin remodeling gene ARID1A in ovarian clear cell carcinoma. Science. 2010;330(6001):228–31.
75. Munksgaard PS, Blaakaer J. The association between endometriosis and ovarian cancer: a review of histological, genetic and molecular alterations. Gynecol Oncol. 2012;124(1):164–9.
76. Rice MS, Murphy MA, Tworoger SS. Tubal ligation, hysterectomy and ovarian cancer: a meta-analysis. J Ovarian Res. 2012;5(1):13.
77. Cibula D, Widschwendter M, Majek O, Dusek L. Tubal ligation and the risk of ovarian cancer: review and meta-analysis. Hum Reprod Update. 2011;17(1):55–67.
78. Rice MS, Murphy MA, Vitonis AF, et al. Tubal ligation, hysterectomy and epithelial ovarian cancer in the New England case-control study. Int J Cancer (Journal international du cancer). 2013;133(10):2415–21.
79. Lessard-Anderson CR, Handlogten KS, Molitor RJ, et al. Effect of tubal sterilization technique on risk of serous epithelial ovarian and primary peritoneal carcinoma. Gynecol Oncol. 2014;135(3):423–7.

80. Ness RB, Cottreau C. Possible role of ovarian epithelial inflammation in ovarian cancer. J Natl Cancer Inst. 1999;91(17):1459–67.
81. Rice MS, Hankinson SE, Tworoger SS. Tubal ligation, hysterectomy, unilateral oophorectomy, and risk of ovarian cancer in the nurses' health studies. Fertil Steril. 2014;102(1):192–198, e193.
82. Sieh W, Salvador S, McGuire V, et al. Tubal ligation and risk of ovarian cancer subtypes: a pooled analysis of case-control studies. Int J Epidemiol. 2013;42(2):579–89.
83. Langseth H, Hankinson SE, Siemiatycki J, Weiderpass E. Perineal use of talc and risk of ovarian cancer. J Epidemiol Commun Health. 2008;62(4):358–60.
84. Terry KL, Karageorgi S, Shvetsov YB, et al. Genital powder use and risk of ovarian cancer: a pooled analysis of 8,525 cases and 9,859 controls. Cancer Prev Res. 2013;6(8):811–21.
85. Booth M, Beral V, Smith P. Risk factors for ovarian cancer: a case-control study. Br J Cancer. 1989;60(4):592–8.
86. Whittemore AS, Wu ML, Paffenbarger RS Jr, et al. Personal and environmental characteristics related to epithelial ovarian cancer. II: Exposures to talcum powder, tobacco, alcohol, and coffee. Am J Epidemiol. 1988;128(6):1228–40.
87. Faber MT, Kjaer SK, Dehlendorff C, et al. Cigarette smoking and risk of ovarian cancer: a pooled analysis of 21 case-control studies. Cancer Causes Control. 2013;24(5):989–1004.
88. Schouten LJ, Rivera C, Hunter DJ, et al. Height, body mass index, and ovarian cancer: a pooled analysis of 12 cohort studies. Cancer Epidemiol Biomarkers Prev. 2008;17(4):902–12.
89. Olsen CM, Nagle CM, Whiteman DC, et al. Obesity and risk of ovarian cancer subtypes: evidence from the Ovarian Cancer Association Consortium. Endocr Relat Cancer. 2013;20(2):251–62.
90. Collaborative Group on Epidemiological Studies of Ovarian Cancer. Ovarian cancer and body size: individual participant meta-analysis including 25,157 women with ovarian cancer from 47 epidemiological studies. PLoS Med. 2012;9(4):e1001200.
91. Barry JA, Azizia MM, Hardiman PJ. Risk of endometrial, ovarian and breast cancer in women with polycystic ovary syndrome: a systematic review and meta-analysis. Hum Reprod Update. 2014;20(5):748–58.
92. Gottschau M, Kjaer SK, Jensen A, Munk C, Mellemkjaer L. Risk of cancer among women with polycystic ovary syndrome: a Danish cohort study. Gynecol Oncol. 2015;136(1):99–103.
93. Kerlikowske K, Brown JS, Grady DG. Should women with familial ovarian cancer undergo prophylactic oophorectomy? Obstet Gynecol. 1992;80(4):700–7.
94. Malander S, Rambech E, Kristoffersson U, et al. The contribution of the hereditary nonpolyposis colorectal cancer syndrome to the development of ovarian cancer. Gynecol Oncol. 2006;101(2):238–43.
95. Pal T, Permuth-Wey J, Betts JA, et al. BRCA1 and BRCA2 mutations account for a large proportion of ovarian carcinoma cases. Cancer. 2005;104(12):2807–16.
96. Risch HA, McLaughlin JR, Cole DE, et al. Prevalence and penetrance of germline BRCA1 and BRCA2 mutations in a population series of 649 women with ovarian cancer. Am J Hum Genet. 2001;68(3):700–10.
97. Stratton JF, Gayther SA, Russell P, et al. Contribution of BRCA1 mutations to ovarian cancer. N Engl J Med. 1997;336(16):1125–30.
98. Van Der Looij M, Szabo C, Besznyak I, et al. Prevalence of founder BRCA1 and BRCA2 mutations among breast and ovarian cancer patients in Hungary. Int J Cancer. 2000;86(5):737–40.
99. Rubin SC, Blackwood MA, Bandera C, et al. BRCA1, BRCA2, and hereditary nonpolyposis colorectal cancer gene mutations in an unselected ovarian cancer population: relationship to family history and implications for genetic testing. Am J Obstet Gynecol. 1998;178(4):670–7.
100. Lu KH, Cramer DW, Muto MG, Li EY, Niloff J, Mok SC. A population-based study of BRCA1 and BRCA2 mutations in Jewish women with epithelial ovarian cancer. Obstet Gynecol. 1999;93(1):34–7.
101. Karageorgi S, Hankinson SE, Kraft P, De Vivo I. Reproductive factors and postmenopausal hormone use in relation to endometrial cancer risk in the nurses' health study cohort 1976–2004. Int J Cancer. 2010;126(1):208–16.
102. Setiawan VW, Pike MC, Karageorgi S, et al. Age at last birth in relation to risk of endometrial cancer: pooled analysis in the epidemiology of endometrial cancer consortium. Am J Epidemiol. 2012;176(4):269–78.
103. Grady D, Gebretsadik T, Kerlikowske K, Ernster V, Petitti D. Hormone replacement therapy and endometrial cancer risk: a meta-analysis. Obstet Gynecol. 1995;85(2):304–13.
104. Renehan AG, Soerjomataram I, Leitzmann MF. Interpreting the epidemiological evidence linking obesity and cancer: A framework for population-attributable risk estimations in Europe. Eur J Cancer. 2010;46(14):2581–92.
105. Esposito K, Chiodini P, Capuano A, Bellastella G, Maiorino MI, Giugliano D. Metabolic syndrome and endometrial cancer: a meta-analysis. Endocrine. 2014;45(1):28–36.
106. Luo J, Beresford S, Chen C, et al. Association between diabetes, diabetes treatment and risk of developing endometrial cancer. Br J Cancer. 2014;111(7):1432–9.

107. Terry PD, Rohan TE, Franceschi S, Weiderpass E. Cigarette smoking and the risk of endometrial cancer. Lancet Oncol. 2002;3(8):470–80.
108. Zhou B, Yang L, Sun Q, et al. Cigarette smoking and the risk of endometrial cancer: a meta-analysis. Am J Med. 2008;121(6):501–8, e503.
109. Felix AS, Yang HP, Gierach GL, Park Y, Brinton LA. Cigarette smoking and endometrial carcinoma risk: the role of effect modification and tumor heterogeneity. Cancer Causes Control. 2014;25(4):479–89.
110. Fearnley EJ, Marquart L, Spurdle AB, et al. Polycystic ovary syndrome increases the risk of endometrial cancer in women aged less than 50 years: an Australian case-control study. Cancer Causes Control. 2010;21(12):2303–8.
111. Kelsey JL, LiVolsi VA, Holford TR, et al. A case-control study of cancer of the endometrium. Am J Epidemiol. 1982;116(2):333–42.
112. Tan MH, Mester JL, Ngeow J, Rybicki LA, Orloff MS, Eng C. Lifetime cancer risks in individuals with germline PTEN mutations. Clin Cancer Res. 2012;18(2):400–7.

# 14 Pathology and Molecular Pathology of Uterine and Ovarian Cancers

Eric C. Huang, David W. Kindelberger and Luis G. Carvajal-Carmona

## 14.1 Introduction

In this chapter, we will first discuss the embryology of the Müllerian tract. The chapter is then further separated into two major sections: uterus and ovary. The normal histology, pathologic abnormalities, and neoplastic processes for each organ will be described, followed by genetics and genomic information.

## 14.2 Embryology and Development of the Müllerian Tract

A basic understanding of the development of female genital tract can provide insights into a variety of developmental and other disorders. This developmental program may be imprecisely divided into early and late stages, with the production of anti-Müllerian hormone (AMH) by the embryo being the critical step in this division. Prior to this point, the primitive coelomic epithelium (in both males and females) begins to invaginate at multiple foci to form bilateral tubular structures known as the Müllerian ducts. These tubes follow the already established course demarcated by the mesonephric ducts. Because of this spatial relationship, the Müllerian ducts are also termed the paramesonephric ducts. As the Müllerian ducts expand caudally, they cross the mesonephric ducts and ultimately fuse in the midline to form the precursor of the uterovaginal canal [1, 2]. This entire process occurs from approximately the 5th week of embryonic development through the 8th week. At this time, production of AMH signals the beginning of development of the "male" program, which culminates in the regression of the Müllerian duct system. AMH has multiple functions, working to prevent development of the early fallopian tubes and ultimately, of both uterus and vagina. Of note, production of AMH by each primitive (ovo)testis only affects ipsilateral fallopian tube development, such that bilateral production is needed to fully switch off bilateral development of female structures. Ultimately, female structures become insensitive to effects of AMH. In females, very little AMH is produced which leads to the default development of the Müllerian system into the fallopian tubes, uterus, and vaginal wall [3].

E.C. Huang (✉)
School of Medicine, University of California at Davis, 4400V. St., PATH Building, Sacramento CA 95817, USA
e-mail: ecchuang@ucdavis.edu

D.W. Kindelberger
School of Medicine, Boston University, 670 Albany St., Boston, MA 02118, USA

L.G. Carvajal-Carmona
Genome Center and School of Medicine, University of California at Davis, GBSF, 451 Health Science Dr., Davis, CA 95616, USA


M. Loda et al. (eds.), *Pathology and Epidemiology of Cancer*,
DOI 10.1007/978-3-319-35153-7_14

As fetal development continues past 8 weeks, if the embryo is female, there will be bilateral fallopian tubes which enter the uterine corpus and fuse. Simultaneously, epithelium from the urogenital sinus proliferates. Caudally, this differentiates into squamo-transitional epithelium which will come to form the vagina and ectocervix. More cranially, endocervical glands begin to develop at approximately 15 weeks. Differentiation continues cranially as primitive endometrial glands may be seen as early as week 19–20 [4]. Also by 20 weeks, smooth muscle cells begin to appear in the walls of the tubular genital tract. This leads to essentially complete exclusion of the mesonephric duct remnants, as the muscular wall becomes increasingly well-developed around the Müllerian duct system. After approximately 20 weeks, maternal estrogen levels are high enough to trigger maturation of the vaginal squamous epithelium, which shows evidence of intracellular glycogen accumulation [5].

In summary, hormonal production early in genital tract development is crucial to the determination of male versus female internal structures. Thus, only if there are elevated levels of AMH before week 10, does the embryos develop significant male internal structures through persistence of the mesonephric system and inhibition of the Müllerian system.

Critical to the development of functioning gonads and, by extension, to the determination of biologic sex of the embryo is the presence of a functional SRY gene, located on the Y chromosome. In general, testes are formed following *SRY* expression very early in development (before the urogenital ridge differentiates). Likewise, the *RSP01* gene has been linked to development of the ovary through activation of β-catenin signaling pathways [6]. Very early in development (week 3), primordial germ cells migrate from the yolk sac to the regions of the urogenital ridges, bilaterally. Over the next several weeks, the ridges develop into physically well-defined primitive gonads with early sex cords composed of primordial germ cells and supporting mesonephric cells. At this point, under the influence of SRY gene products, the sex cords begin forming tubular structures and immature Sertoli-like cells are present (i.e., the male phenotype becomes dominant). In the absence of a functional SRY gene, germ cells continue to increase in number until genes associated with ovarian development become active [7]. These germ cells become admixed with surrounding ovarian stromal cells. Each germ cell will become encapsulated within its own primordial follicle. These cells arrest during meiosis and do not proliferate further. The ovary develops structurally beginning at approximately the 15th week. Under inductive influence of mesonephric cells that migrated into the developing ovary early in development, the germ cells progress from primordial follicles and into primary follicles, with development occurring from the ovarian center toward the periphery. Germ cells that do not develop into follicles undergo apoptosis; thus, the primary sex cords will ultimately regress so that only primary follicles are present at birth [8, 9].

## 14.3 The Uterus

### 14.3.1 Cycling Endometrium

The endometrium undergoes morphologic changes continuously, secondary to hormonal changes. Endometrial biopsies from cycling patients (typically in the third or fourth decades) are among the most common specimens received. The principal questions being addressed in the vast majority of these cases are whether the patient has ovulated and the corollary: whether the luteal phase is progressing normally. The criteria most commonly used to answer these questions by assigning a "date" (or date range) to endometrial biopsies, based on morphologic grounds, were put forth by Noyes, Hertig, and Rock in 1950 [10]. Several modifications have been made to their initial system, but the basic principles remain largely unchanged. To create a reproducible conceptual framework, cycling endometrium may be divided into six categories, based on morphologic features: proliferative endometrium, sixteen day endometrium, early (vacuolar) secretory endometrium, mid (exhausted) secretory endometrium, late (predecidual) secretory endometrium, and menstrual endometrium [11]. An idealized setting of a 28-day cycle is assumed with day one representing the first day of menses. In reality, many women have cycles that differ from that 28-day ideal. These differences are most often accounted for within the pre-ovulatory interval of the cycle (the proliferative phase), while the post-ovulatory period (secretory phase to menses) remains remarkably constant at 14 days.

#### 14.3.1.1 Proliferative Endometrium

In response to estrogen, endometrial glands are stimulated to proliferate and show the characteristic morphologic pattern of proliferative endometrium: tubular glands with cells having pseudostratified, deeply basophilic nuclei with coarse chromatin and numerous mitotic figures (Fig. 14.1).

#### 14.3.1.2 Sixteen Day Endometrium

Sixteen day endometrium is characterized by glands with cells displaying subnuclear vacuoles combined with abundant mitotic figures (Fig. 14.2). The overall gland architecture is the simple tubular glands of proliferative endometrium and, although mitoses are prominent, they are not as numerous as during the proliferative phase. In sixteen day endometrium, the presence of numerous mitoses prevents one from confirming that ovulation has occurred [11].

#### 14.3.1.3 Early (Vacuole Phase) Secretory Endometrium

Entry into the vacuole phase of the cycle suggests that ovulation has occurred. Glands with cells showing uniform subnuclear vacuoles and rare mitotic figures warrant the diagnosis of day seventeen secretory endometrium (Fig. 14.3a). In day eighteen secretory endometrium, the glandular epithelial cells have both sub- and supranuclear vacuoles where the nuclei have migrated toward the gland lumen and now occupy the center of the cell (Fig. 14.3b). By day nineteen, the nuclei have nearly all returned to the base of the cell. There are abundant intraluminal secretions and the glands begin to show secretory exhaustion with low columnar to cuboidal cells, loss of vacuoles, and a lack of mitotic activity [12].

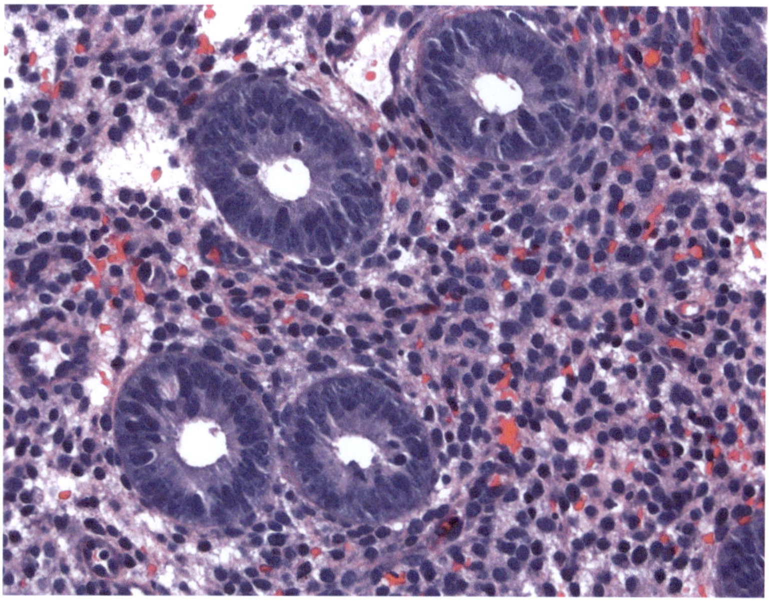

**Fig. 14.1** Proliferative endometrium. The glands are round to tubular with pseudostratified nuclei and mitotic figures

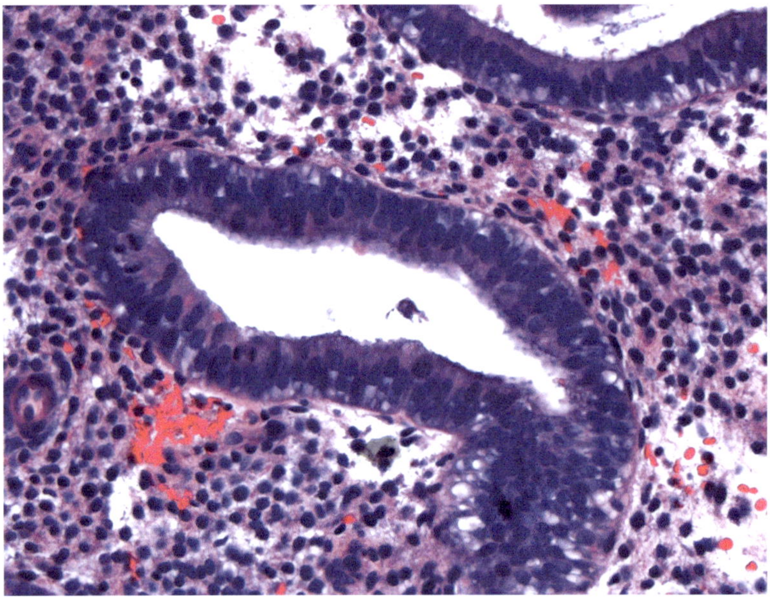

**Fig. 14.2** Sixteen day endometrium. The glands are similar to proliferative endometrium and contain scattered subnuclear vacuoles

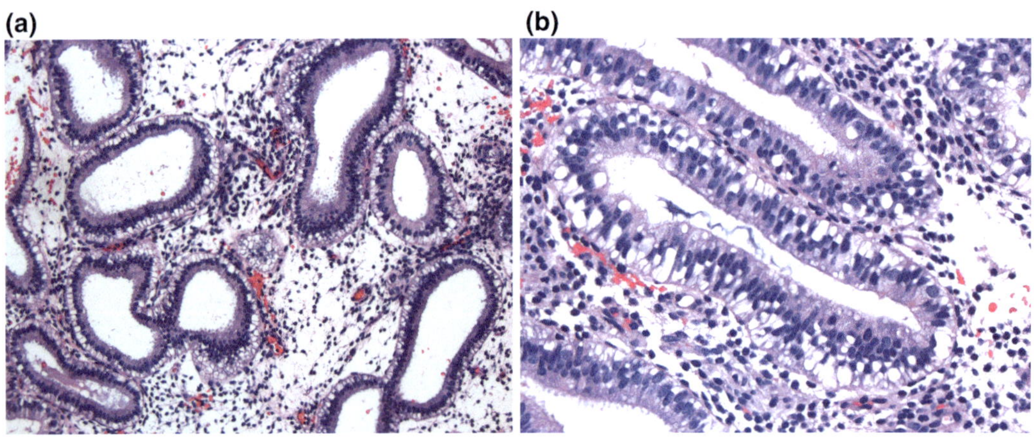

**Fig. 14.3** Early (vacuole phase) secretory endometrium. **a** Day 17 has uniform prominent subnuclear vacuoles. **b** Day 18 has both sub-and supranuclear vacuoles

#### 14.3.1.4 Mid (Secretory Exhaustion Phase) Secretory Endometrium

The loss of secretory vacuoles signals the end of the early secretory phase and the beginning of the mid phase. From this point forward, the focus in assigning dates to endometrial samples will be on stromal changes. Day twenty secretory endometrium exhibits maximal intraluminal secretions with only very rare residual subnuclear vacuoles (Fig. 14.4a). The stroma beneath the endometrial surface becomes more compact and can mimic predecidualized stroma. However, the stromal cells still have dark nuclei and a high nuclear to cytoplasmic ratio, unlike predecidualized cells. Day twenty-one secretory endometrium shows increased stromal edema, which peaks at day twenty-two. The stromal cells may appear widely spaced and their high nuclear to cytoplasmic ratios make them appear to have "naked nuclei." There are no conspicuous predecidual changes (Fig. 14.4b).

#### 14.3.1.5 Late (Predecidual Phase) Secretory Endometrium

There are three well-defined components of predecidual change. Initially, there are aggregates of stromal cells surrounding spiral arterioles. As the changes develop, the stromal cells acquire distinct cell borders and slightly basophilic cytoplasm. Finally, the chromatin texture becomes much finer and the nuclei take on much less basophilia. For day twenty-three secretory endometrium, the predecidual cells are found surrounding individual spiral arterioles (Fig. 14.5a). By day twenty-four, the aggregates of predecidua bridge multiple vessels. Day twenty-five secretory endometrium is characterized by predecidua in thin aggregates below the endometrial surface (Fig. 14.5b). A thick subsurface band of predecidua is present by day twenty-six, accompanied by the presence of endometrial stromal granulocytes. Day twenty-seven endometrium shows abundant predecidua expanding downward from the endometrial surface, with increased numbers of granulocytes [11] (Fig. 14.5c).

#### 14.3.1.6 Menstrual (Breakdown) Endometrium

The menstrual phase is defined by the presence of discrete aggregates of condensed predecidual cells (stromal breakdown), admixed with inflammatory cells and blood [13]. Interspersed between these fragments of condensed stroma are glands showing secretory exhaustion (Fig. 14.6). The presence of both components (stromal breakdown and secretory exhausted glands) is evidence that breakdown has occurred.

### 14.3.2 Dysfunctional Uterine Bleeding

A number of conditions may lead to dysfunctional uterine bleeding, which can be thought of as bleeding resulting from alterations in the normal cycling pattern of the endometrium. The most common causes include anovulation

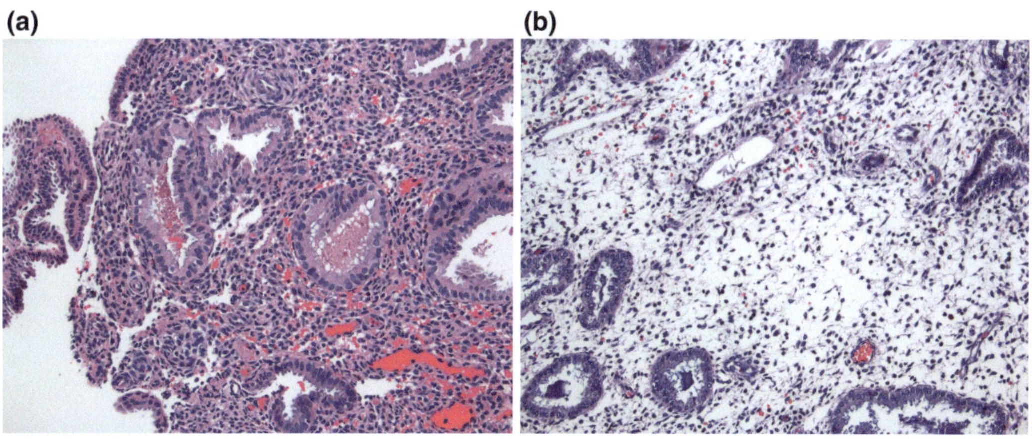


**Fig. 14.4** Mid (secretory exhaustion phase) secretory endometrium. **a** Day 20 has rare subnuclear vacuoles and exhibits maximal intraluminal secretion. **b** Day 22 shows prominent endometrial stromal edema that has an appearance of "naked nuclei"

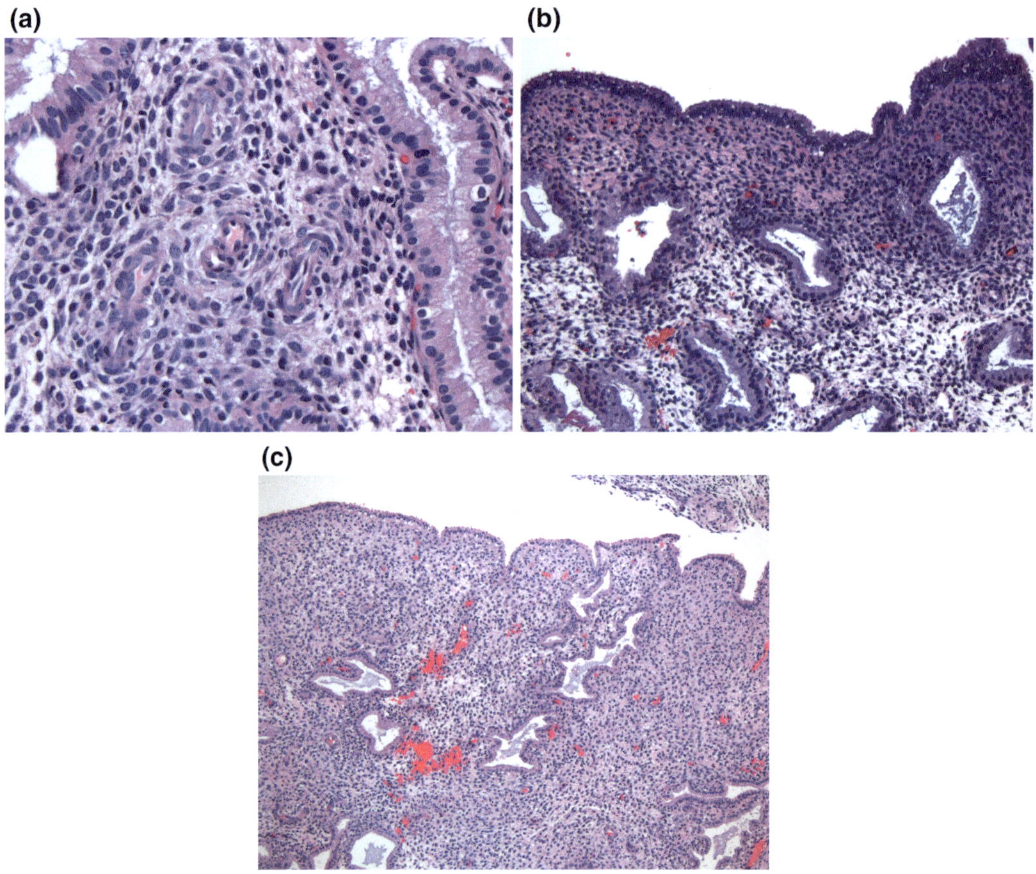


**Fig. 14.5** Late (predecidual phase) secretory endometrium. **a** Day 23 has spiral arterioles surrounded by predecidual cells. **b** Day 25 contains a thin layer of predecidua underneath the surface. **c** Day 27 shows abundant predecidua expanding downward from the endometrial surface, with increased numbers of granulocytes

(typically seen in patients in their fifth decade), chronic endometritis, endometrial polyps, and submucosal leiomyomata [11].

### 14.3.3 Anovulation

Anovulatory cycles are very common and these changes can lead to multiple histologic patterns depending on the presence or absence of elevated estrogen levels. Persistent estrogen stimulation leads to the formation of cystically dilated endometrial glands with an irregular distribution (Fig. 14.7a). Stromal breakdown may be patchy due to ischemia from focal fibrin thrombi in spiral arterioles. In the absence of elevated estrogen, the cystic, dilated glands are not prominent, but other features of anovulation including tubal metaplasia (Fig. 14.7b), patchy breakdown due to fibrin thrombi, and surface repair are present. Another pattern seen when the proliferative phase is followed by abrupt loss of estrogen (absence of a persistent ovarian follicle) is that of uniform tubal glands with diffuse stromal breakdown. This can be distinguished from normal ovulatory breakdown by the absence of predecidual changes in stroma and the lack of secretory changes in the glands in the case of anovulation [14].

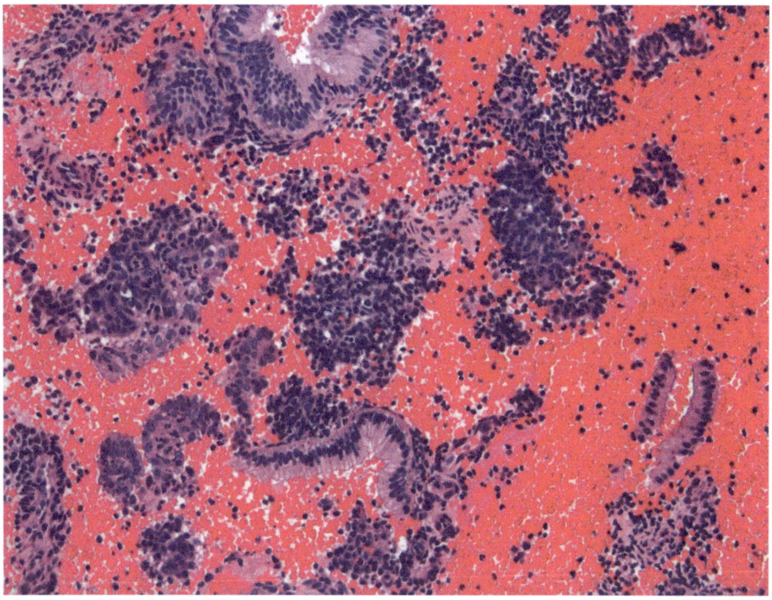

**Fig. 14.6** Menstral (breakdown) endometrium. Diffuse stromal breakdown admixed with inflammatory cells, blood and associated exhausted secretory glands

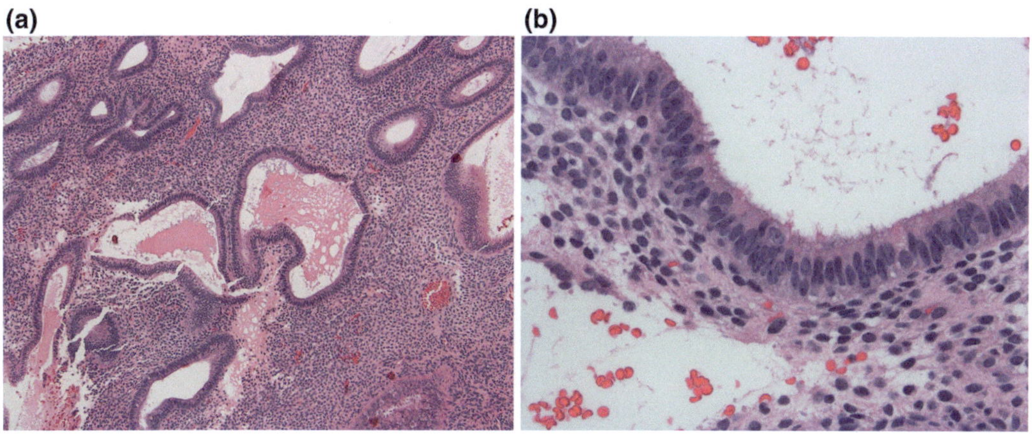


**Fig. 14.7** Anovulatory endometrium. **a** Cystically dilated glands. **b** Tubal metaplasia

### 14.3.4 Benign Endometrial Hyperplasia

Benign endometrial hyperplasia (or hyperplasia without atypia) typically occurs in perimenopausal women with a presenting symptom of abnormal uterine bleeding. It is a condition that results from unopposed estrogen that may be caused by multiple factors including chronic anovulation, absence of progesterone, androgen to estrogen conversion within adipose tissue secondary to obesity, polycystic ovarian syndrome (PCOS), estrogen-secreting ovarian tumors (e.g., granulosa cell tumor), or exogenous estrogen (e.g., Tamoxifen therapy). Women with excess estrogen carry a 3–4 fold increased risk of developing endometrial carcinoma and 10-fold increased risk after a decade of exposure [15]. Women with abnormal uterine bleeding usually will undergo endometrial biopsy or curettage to rule out the possibility of uterine cancer [16]. Histologically, the samples contain endometrial glands that vary in size and shape with an associated increase in gland to stromal ratio; however, there is no significant cytologic demarcation from the background endometrium [17, 18].

### 14.3.5 Endometrial Intraepithelial Neoplasia

Endometrial intraepithelial neoplasia (EIN) (or atypical hyperplasia) commonly affects peri-menopausal to postmenopausal women. Similar to benign endometrial hyperplasia, this condition is developed secondary to excess endogenous or exogenous estrogen, causing abnormal uterine bleeding as the presenting clinical symptom. Much evidence supports EIN as the precursor lesion to endometrioid type of endometrial cancer [18–22]. In addition, ~25–35 % of patients with EIN will also develop concurrent or subsequent endometrial carcinoma [19, 21]. The diagnostic criteria for EIN include: (1) an area where glandular content exceeds that of stroma (glands/stroma > 1), (2) nuclear and/or cytoplasmic features of epithelial cells differ between architecturally abnormal glands and normal background glands, (3) maximum linear dimension exceeds 1 mm, and (4) excluding benign mimics (e.g., endometrial polyps or anovulatory endometrium) and adenocarcinoma (Fig. 14.8). All of these criteria must be satisfied in order to make the diagnosis of EIN.

### 14.3.6 Endometrial Carcinoma

Endometrial carcinoma is the most common malignant tumor in the female genital tract with an age-adjusted incident rate of 24.6 per 100,000 women per year in the United States (US) [23]. Patients often present with abnormal uterine bleeding that initiates endometrial sampling [16]. Based on clinical and pathologic information, endometrial carcinomas have been divided into type 1 and type 2 [24, 25]. Type 1 cancers are usually low-grade tumors associated with estrogen overexposure, whereas type 2 cancers are aggressive high-grade tumors which occur independently of estrogen stimulation.

#### 14.3.6.1 Endometrioid Adenocarcinoma

This tumor is the prototypic type 1 cancer. It is the most commonly encountered cancer in the

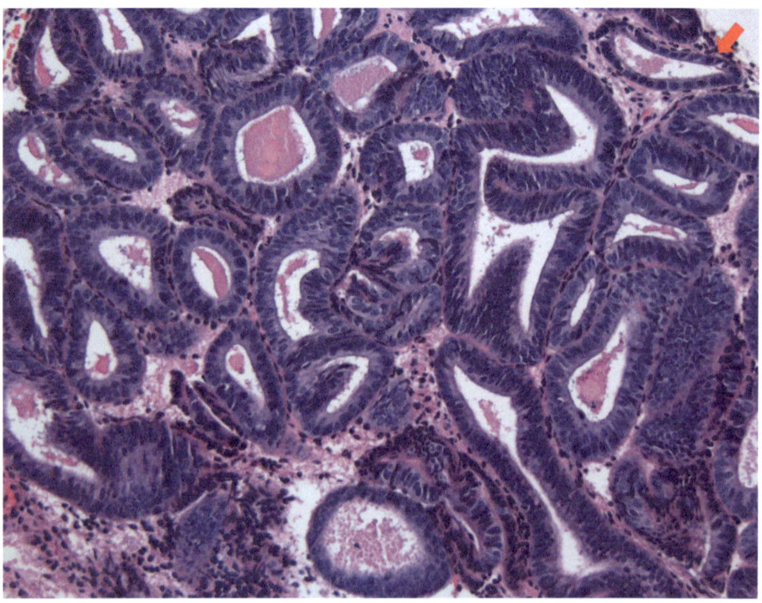

**Fig. 14.8** Endometrial intraepithelial neoplasia. There is increased gland to stromal ratio that exceeds 1 mm in linear dimension. In addition, the nuclear cytology differs from the normal background glands (highlighted by *red arrow*)

endometrium, affecting predominantly perimenopausal to postmenopausal women with a mean age around 60 years [26]. The most common presenting symptom is abnormal uterine bleeding where endometrial thickening can sometimes be visualized by transvaginal ultrasound imaging. As discussed earlier, EIN is the precursor lesion for the endometrioid type of endometrial adenocarcinoma [18–22]; therefore, they share similar risk factors. Genetically, individuals with Lynch syndrome or Cowden syndrome also have higher risk of developing endometrioid adenocarcinoma [27–29].

Histologically, endometrioid adenocarcinoma has glands that are similar to benign endometrium. The columnar lining cells have a stratified or pseudostratified appearance and the glands are crowded with complex maze-like or cribriform architecture. Gland fusion and solid areas develop as the tumor becomes less differentiated. The grading of endometrioid adenocarcinoma is based on the amount of solid growth pattern (non-squamoid) in these cancers. Grade 1 tumors have <5 % solid component (Fig. 14.9a), grade 2 tumors have solid areas between 5 and 50 % (Fig. 14.9b), and grade 3 tumors have >50 % solid growth pattern (Fig. 14.9c). The tumor nuclei typically demonstrate a low to moderate amount of cytologic atypia in grade 1–2 tumors. The presence of severe cytologic atypia in >50 % of the tumor is associated with more aggressive clinical behavior and justifies increasing the tumor by one grade in architecturally grade 1 or 2 tumors [30]. Squamous, villoglandular and

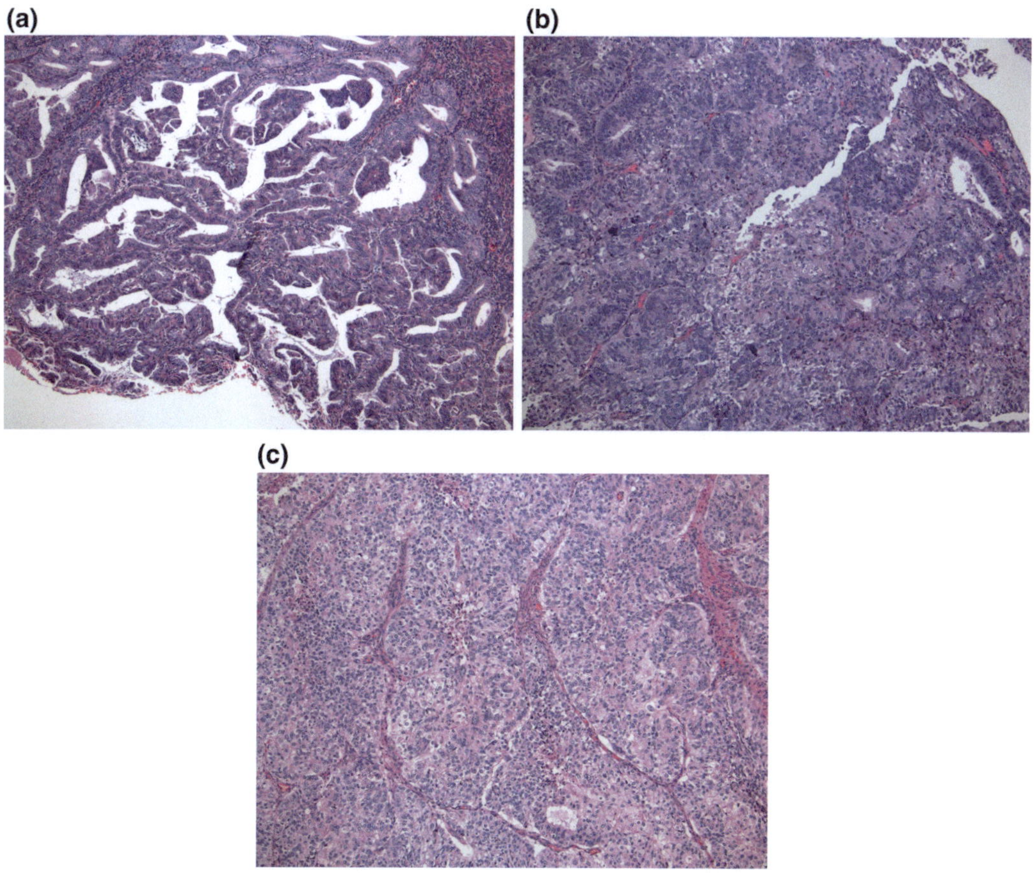


**Fig. 14.9** Endometrial adenocarcinoma. **a** Grade 1 tumor showing glands with complex architecture. **b** Grade 2 tumor showing glands similar to grade 1; however, >5 % of solid areas can be seen. **c** Grade 3 tumor with mostly solid pattern (>50 %)

secretory differentiations can also be seen in endometrioid adenocarcinomas.

In biopsy/curettage material, when the lesion is present in both endometrial and endocervical samples, it can be difficult to decipher the primary site of origin based on histomorphology alone. However, it is important to make this distinction as endometrial and endocervical cancers have different clinical and surgical managements. Identification of precursor lesions (EIN or endocervical adenocarcinoma in situ) may assist in this scenario. However, when absent, immunohistochemistry with a panel of antibodies including ER, vimentin, monoclonal CEA and p16 can be helpful in making this distinction. Endometrial cancers typically are positive for ER and vimentin while endocervical lesions are positive for p16 and monoclonal CEA. On the other hand, it should be noted that high-grade endometrial cancers can occasionally stain diffusely positive for p16, therefore, the staining should be interpreted with caution and in conjunction with histomorphology.

#### 14.3.6.2 Mucinous Adenocarcinoma

This tumor is defined by >50 % of neoplastic cells containing cytoplasmic mucin. It is an uncommon lesion, found in <10 % of all endometrial cancers [31]. Clinically, these tumors are usually associated with low-stage disease, although 50 % of cases can develop deep myometrial invasion [31]. Some experts consider mucinous adenocarcinoma as a variant of endometrioid adenocarcinoma with mucinous differentiation, as these tumors behave similarly and the distinction poses no clinical significance. However, this entity is listed separately under the 2014 World Health Organization classification of endometrial tumors [32]. Histologically, the neoplastic glands are lined by mucin-rich cells with minimal stratification (Fig. 14.10). The mucin is producing cells that are positive for mucicarmine and CEA. There is a mild to moderate cytologic atypia.

#### 14.3.6.3 Serous Adenocarcinoma

Serous carcinoma is one of the most aggressive endometrial carcinomas and is considered a type 2 tumor. Women with this disease are generally postmenopausal and older than patients with type 1 tumors. Estrogen excess is less commonly associated with this tumor. Histologically, the tumor consists of complex papillary and/or glandular architecture where neoplastic cells contain pleomorphic high-grade nuclei and

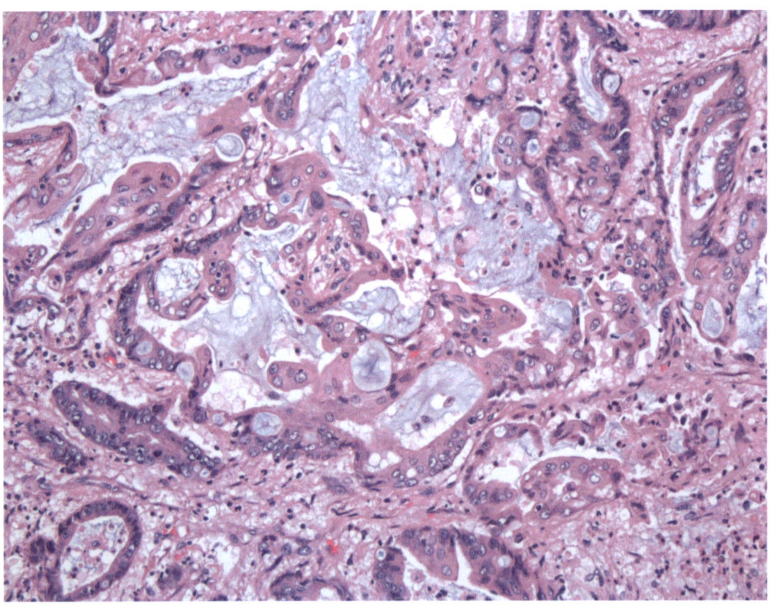

**Fig. 14.10** Mucinous adenocarcinoma. There are abundant mucin pools seen within the tumor and the tumor cells contain cytoplasmic mucin

macronucleoli (Fig. 14.11a). Mitoses are readily identified. Papillary tuftings and irregular slit-like spaces are other morphologic features seen in serous carcinoma (Fig. 14.11a). By definition, serous carcinoma is a high-grade tumor; therefore, grading is not required. Serous carcinoma has an aggressive clinical behavior and is often associated with deep myometrial and lymphovascular invasion. Even in the absence of myometrial invasion, it is not uncommon to find extrauterine metastasis.

Serous carcinoma and grade 3 endometrioid adenocarcinoma can have overlapping morphologic features. Immunohistochemistry may help in differentiating these two different entities. Aberrant p53 expression (either diffuse positivity or complete absence) is typically seen in serous carcinoma (Fig. 14.11b). However, it is important to note that some grade 3 endometrioid adenocarcinomas can harbor p53 mutations, making this immunostain unhelpful in this situation. Fortunately, this distinction is not critically important for clinicians as both of these tumors often present with high stage disease that requires similar aggressive clinical management.

#### 14.3.6.4 Clear Cell Adenocarcinoma

Clear cell carcinoma is a rare type 2 tumor that is encountered in <5 % of all endometrial cancers. Histologically, these tumors have papillary, tubulocystic and solid growth patterns (Fig. 14.12a). Hobnail cells with clear cytoplasm are the hallmark characteristic of this tumor (Fig. 14.12b). Occasionally, the cytoplasm may appear eosinophilic rather than clear. Severe nuclear atypia is always present. Similar to serous carcinoma, this tumor is by definition a high-grade neoplasm; therefore, grading is unnecessary.

#### 14.3.6.5 Carcinosarcoma

Carcinosarcoma (or malignant mixed Müllerian tumor) is a highly aggressive malignant neoplasm that consists of both malignant epithelial and mesenchymal components. It affects postmenopausal women and represents <5 % of malignant uterine tumors. Carcinosarcomas have been associated with tamoxifen or exogenous estrogen therapies [33, 34]. The majority of women with carcinosarcoma present as high-stage disease with tumor extending beyond the uterus. Clinically, the tumor may present as a polypoid mass that protrudes through the cervical os. Histologically, the epithelial component more commonly consists of endometrioid or serous adenocarcinomas (Fig. 14.13); however, other types of epithelial malignancy, such as clear cell carcinoma, can also be identified. The malignant

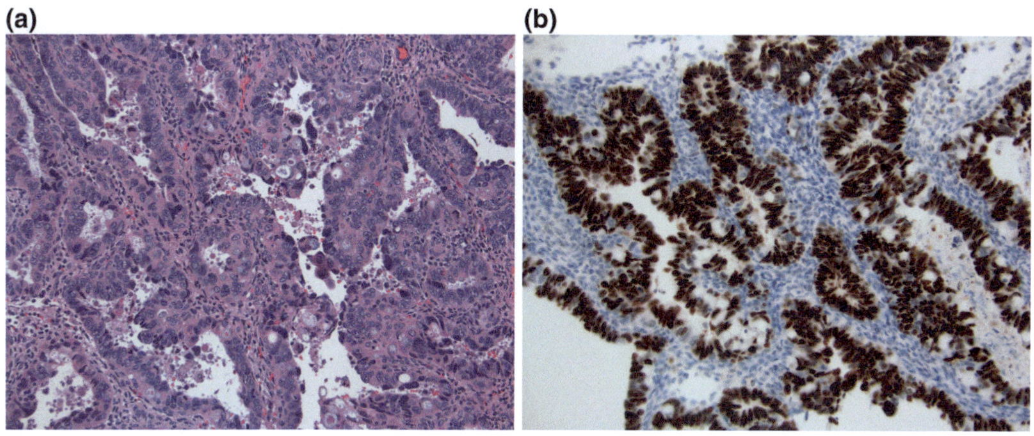

**Fig. 14.11** Serous adenocarcinoma. **a** The tumor contains papillary/glandular architecture with high-grade nuclei and mitoses. **b** This tumor harbors p53 mutation, which is demonstrated by p53 immunostain

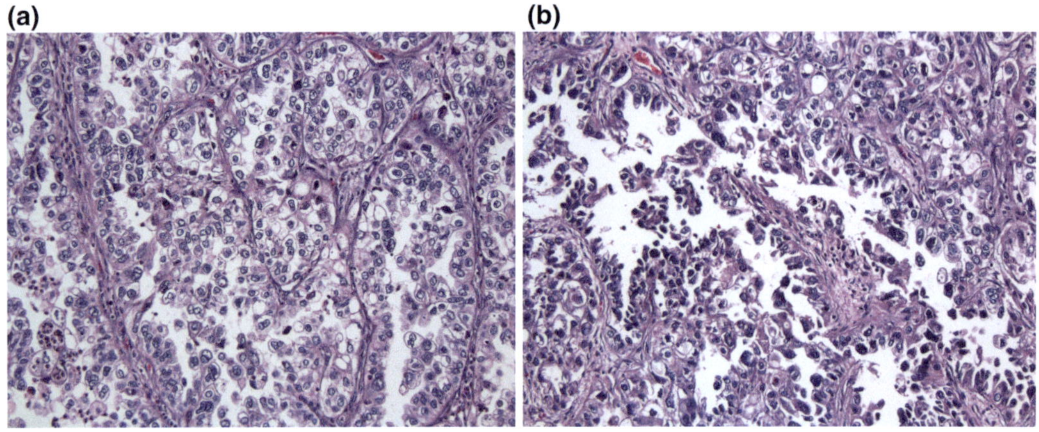


**Fig. 14.12** Clear cell adenocarcinoma. **a** As its name implies, the tumor cell contains clear cytoplasm. **b** Hobnail cells with high nuclear grade can be appreciated

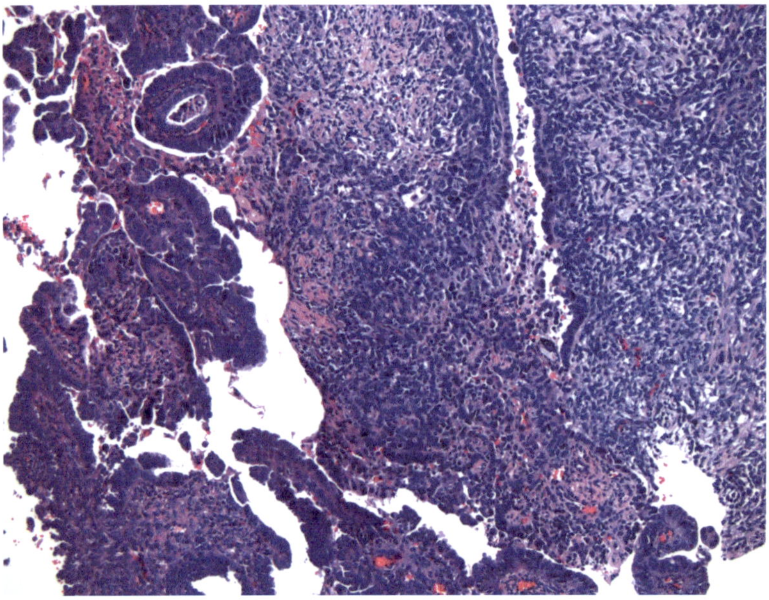

**Fig. 14.13** Carcinosarcoma. This tumor has a mixture of high-grade malignant epithelial (*left*) and mesenchymal (*right*) components

mesenchymal component is typically composed of non-specific high-grade sarcoma; however, heterologous elements such as benign cartilage or rhabdomyosarcoma can be present. Recent studies have suggested that carcinosarcomas are of epithelial in origin with transitions to mesenchymal differentiation [35–37].

#### 14.3.6.6 Less Common Tumors

There are other less commonly seen uterine neoplasms, such as neuroendocrine tumors, undifferentiated carcinoma, leiomyosarcoma, adenosarcoma and endometrial stromal sarcoma. However, for the scope of this chapter, these rare entities will not be discussed.

### 14.3.7 Spread and Staging

Staging of endometrial cancers is based on the International Federation of Gynecology and Obstetrics (FIGO) and the American Joint Committee on Cancer (AJCC) staging systems, which are similar. These systems take into account the depth of the myometrial invasion and the involvement of cervical stroma, uterine serosa, adnexa, vagina, parametrium, bladder, or bowel mucosa [38]. Lymphatic spread initially involves the pelvic lymph nodes, eventually reaching the para-aortic lymph nodes. However, a subset of endometrial cancers do metastasize directly to the para-aortic lymph nodes.

### 14.3.8 Genetics and Genomics

As mentioned previously, endometrial carcinoma have been divided into type 1 and 2 tumors. Type 1 tumors are typically associated with microsatellite instability (MSI), *PTEN* and *KRAS* mutations, and β-catenin nuclear accumulation, while type 2 tumors are associated with p53 abnormalities and loss of heterozygosity at different loci.

#### 14.3.8.1 Familial Syndromes

It has been estimated that about 5 % of all newly diagnosed endometrial cancers can be attributed to inherited familial cancer syndromes [39]. Of such familial syndromes, Lynch syndrome accounts for the majority of inherited endometrial cancers and is also characterized by a high risk of colorectal tumors [40–43]. This familial cancer syndrome results from germline mutations in the mismatch repair (MMR) genes *MLH1*, *MSH2*, *MSH6* and *PMS2* [40–43]. The cumulative lifetime risk (typically calculated by age 75) of developing endometrial cancer in MMR gene mutation carriers ranges from 20 to 70 % and depends on the mutated gene, with patients carrying MSH6 mutations having the highest risk (up to 71 % by age 75) [44]. On average, the age of diagnosis of endometrial cancer in Lynch syndrome is ~2 decades younger than in sporadic cases [44]. Histologically, most Lynch syndrome cancers are endometrioid tumors. In addition to a personal and/or familial history of colorectal and endometrial tumors, or early age of diagnosis and endometrioid histology, another hallmark of Lynch syndrome cancers is the presence of MSI, which can be diagnosed using microsatellite testing or MMR protein immunohistochemistry. Although MSI and MMR testing can aid in the identification of Lynch Syndrome-associated cases, they are not completely diagnostic since ~10–15 % of all endometrial cancers have a sporadic/non- germline-related MSI phenotype that is caused by methylation of the *MLH1* gene promoter [45].

A small fraction of inherited endometrial cancer cases can also be seen in Cowden syndrome and in the polymerase proofreading- associated polyposis (PPAP) syndrome [46]. Cowden syndrome is caused by very rare inherited *PTEN* gene mutations and has a cumulative lifetime risk of developing endometrial cancer that ranges from 19 to 28 % [47–50]. Women with Cowden syndrome also have a high risk of developing breast (up to 50 % by age 70) and thyroid (up to 10 % by age 70) cancers [47, 49–52]. Therefore, a family or personal history of breast and thyroid tumors could help in establishing a diagnosis of Cowden syndrome [48]. We recently described the PPAP syndrome, a condition that is associated with tumors of the colon and endometrium (a syndromic association also seen in Lynch syndrome, see above), which results from rare mutations in the polymerase epsilon (*POLE*) and delta (*POLD1*) genes [53]. Endometrial cancers in PPAP have been described in only a handful of patients, who typically develop early onset disease (age 35–55) and synchronic or metachronic colorectal tumors and cancers [53–55].

In summary, endometrial cancer cases can be found in three familial syndromes (Lynch, Cowden and PPAP) that are characterized by an early onset and a higher risk of other malignant conditions such as colorectal, breast or thyroid cancers. The use of clinical, pathological, and molecular information can be informative and identify such patients who will benefit from genetic counseling and testing, which can help them minimize the risk of developing additional malignancies through early detection.

#### 14.3.8.2 Low-Penetrance Risk Variants

The genetic basis of endometrial cancer has been recently investigated at the population level using genome-wide associations (GWA) and candidate gene studies. Two genomic regions, near the HIF1B and TERT genes, have been identified by these population-based studies and harbor several independent endometrial cancer risk variants [56–58]. Unlike very rare causal mutations that result in familial syndromes, these low-penetrance HIF1B and TERT variants are very common in the population and are not necessarily causal of endometrial cancer on their own. However, the identification of low-penetrance variants is important because they highlight potentially important pathways, such as hypoxia and telomere biology in endometrial carcinogenesis. Furthermore, these variants can be used in the future, in combination with other clinical, genetic, and environmental information, to improve risk prediction models for early cancer detection [58, 59].

#### 14.3.8.3 Somatic Genetics

Like most other tumors, endometrial cancers are characterized by unique patterns of genomic instability as well as by the presence of mutations in oncogenes and tumor suppressor genes. Several studies have found that endometrial tumors harbor somatic mutations in well-established cancer genes such as *PTEN*, *FGFR2*, *ARID1A*, *CTNNB1*, *PIK3CA*, *PIK3R1*, and *KRAS* [60, 61]. Also, as mentioned above, the MSI tumor phenotype, a pattern of genomic instability that involves mutations in di-, tri-and tetra-nucleotide repeats, is found ~15–30 % of all endometrial cancers [44, 45]. Our knowledge of endometrial cancer genomics was greatly enhanced by the recent molecular analysis carried out by The Cancer Genome Atlas (TCGA) study [62], which performed a multi-omic characterization of 373 endometrial tumors that included 307 endometrioid, 66 serous and 13 mixed histology cancers. These studies led to the identification of four major endometrial cancer subtypes including: (1) near-diploid and ultra-mutated tumors with somatic *POLE* gene mutations, (2) near-diploid tumors with MSI, (3) genomically stable/microsatellite stable tumors and (4) genomically instable/serous-like tumors. The patterns of mutations and genomic instability in these four tumor subtypes appeared to be unique and provided an alternative classification to that based on histological features [62].

Near-diploid/*POLE* mutant tumors represented about 8 % of all endometrial cancer analyzed by the TCGA. All of these tumors had endometrioid histology and were characterized by the presence of mutations in the proofreading domain (exons 9 and 13) of the POLE gene, a region that was also found to be mutated in the germline DNA of patients with PPAP syndrome (see above) [46]. The lack of proofreading capacity of POLE (a DNA polymerase) is likely to explain why these tumors are characterized by an extremely high mutation rate (~232 mutations per megabase, the highest of all human tumors) and why they are characterized by a mutation signature that involves mostly G:C > T:A transversions [46]. Interestingly, the TCGA study, as well as more recent studies, have provided evidence suggesting that endometrial tumors with *POLE* mutations have a relatively good prognosis, independent of other clinical and histological variables, particularly for high-grade tumors [53]. A recent study by Church et al. [53], which assessed the prognostic effect of *POLE* mutations in 1416 patients, including 788 from two clinical trials, suggested that *POLE* mutations could be used as a novel biomarker to identify endometrial cancer patients with a good prognosis that could benefit from less intensive treatments.

The second group defined by the TCGA study, which represented ~28 % of all tumors, included endometrial cancers with the MSI phenotype. Consistent with previous studies, MSI endometrial tumors have a predominant endometrioid histology and *MLH1* silencing [62]. The third group of TCGA tumors represented about 40 % of all samples and included low-grade endometrioid tumors that had a low mutation rate, microsatellite stability, a low proportion of copy number changes and a high frequency of β-catenin (*CTNNB1*) mutations. Lastly, a genomically instable/serous-like tumor group was also identified by the TCGA study and was characterized by extensive copy number variation and a very high rate of *TP53* mutations (which were rare in the other three subtypes). Interestingly, this was a histologically heterogeneous group which included all serous cancers as well as about 25 % of all high-grade endometrioid tumors analyzed in the study. The latter finding was significant as it suggested that such a molecular subtype could benefit from the therapies used for serous tumors, which typically involves chemotherapy rather than adjuvant radiotherapy [62].

In summary, recent advances in the genomic characterization of endometrial tumors have identified four major subtypes that offer an alternative to histological classification. The most clinically relevant findings in this regard include the identification of the *POLE* mutation group, representing ~8 % of all tumors, that have a particularly good prognosis and that could benefit from less aggressive treatment and the discovery of a particularly aggressive type of endometrioid tumors that will benefit from the chemotherapeutic approaches used in serous tumors [62]. Furthermore, a plethora of therapeutic targets is now available and will offer unique opportunities for the future development of molecularly guided treatments. While these observations need to be tested under rigorous clinical trials, they represent a unique opportunity for the establishment of personalized medicine in endometrial cancer in the coming years.

## 14.4 The Ovary

### 14.4.1 Histology of the Normal Ovary

The ovary is lined by a single layer of low columnar to cuboidal cells which have been come to be known as "ovarian surface epithelium" although in reality it is a specialized mesothelium (Fig. 14.14). This surface layer commonly invaginates to form cortical (or epithelial) inclusion cysts which have lost their surface connections. Occasionally, invaginations of transitional-type epithelium may be encountered. These are known as Walthard nests (or rests).

As discussed above, the primary oocytes with their surrounding specialized stromal (granulosa) cells are known as primordial follicles (Fig. 14.15). As the oocyte and its surrounding

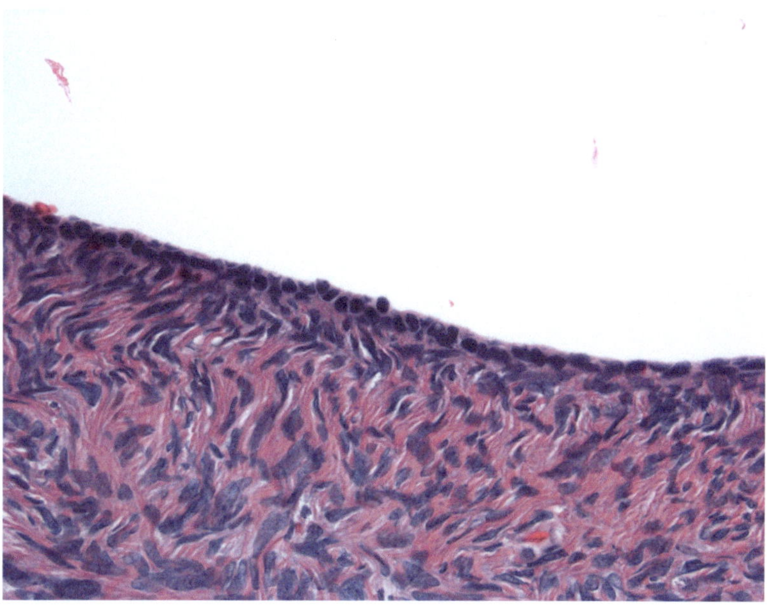

**Fig. 14.14** Ovarian surface epithelium. The ovarian surface is lined by a single layer of low columnar to cuboidal cells

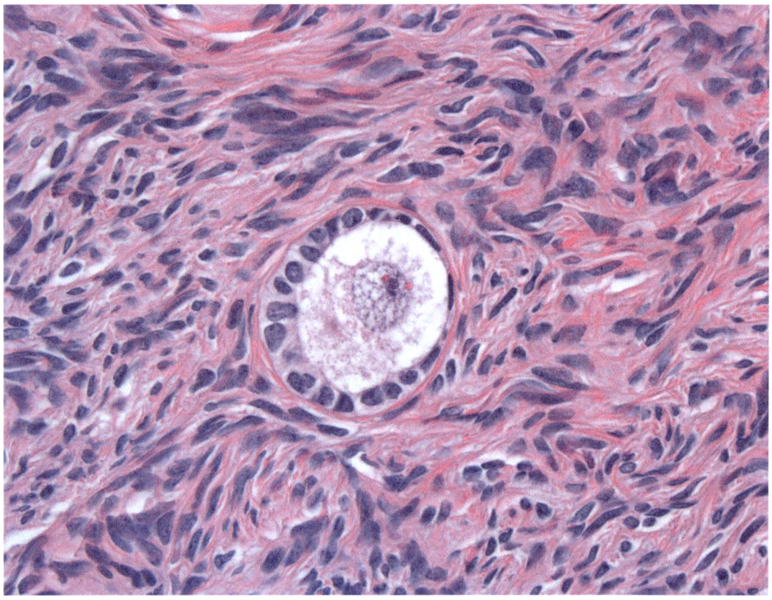

**Fig. 14.15** Primordial follicles. They are composed of a primary oocyte surrounded by a single layer of flattened follicular cells

cells become larger, the primordial follicle transitions into a primary follicle. As the surrounding granulosa cells proliferate, a strongly eosinophilic band forms around the oocyte—the zona pellucida. Simultaneously, through inductive effects, the cortical stromal cells immediately surrounding the granulosa cells develop into the theca interna and externa. The oocyte takes on an eccentric position within the follicle structure, giving rise to an antral (or Graafian) follicle. During reproductive years, one follicle will become dominant during each ovulatory cycle and will continue to grow into a dominant follicle. Immediately following the luteal surge (peak level of luteinizing hormone) ovulation occurs, and the oocyte is liberated into the peritoneal cavity. Following ovulation, the granulosa cells—in response to progestin stimulation—become enlarged with pale eosinophilic cytoplasm. If pregnancy occurs, the granulosa cells continue to enlarge and accumulate eosinophilic material in the cytoplasm, forming a corpus luteum of pregnancy (Fig. 14.16). At the conclusion of the pregnancy, the corpus luteum undergoes regression and hyalinization to form a persistent structure: the corpus albicans [63].

### 14.4.2 Non-neoplastic Disorders of the Ovary

#### 14.4.2.1 Follicle Cysts

If the luteal surge does not occur properly (which commonly occurs around menarche and menopause), follicle cysts lined by luteinized granulosa cells or theca cells may result [64]. They typically are small (less than 4–5 cm), thin-walled cysts with smooth surfaces, containing either blood or serous fluid.

#### 14.4.2.2 Corpus Luteum Cysts

If a corpus luteum attains a diameter over 3 cm, it may be termed a corpus luteum cyst. These structures may rupture, leading to irritation of the peritoneum.

#### 14.4.2.3 Polycystic Ovarian Syndrome (PCOS)

The hallmark finding is bilateral ovarian enlargement with numerous small cystic follicles surrounded by an eosinophilic collagenized stroma. The etiology of the disorder is not known and disease presentation may be variable with infertility (due to persistent anovulation), obesity,

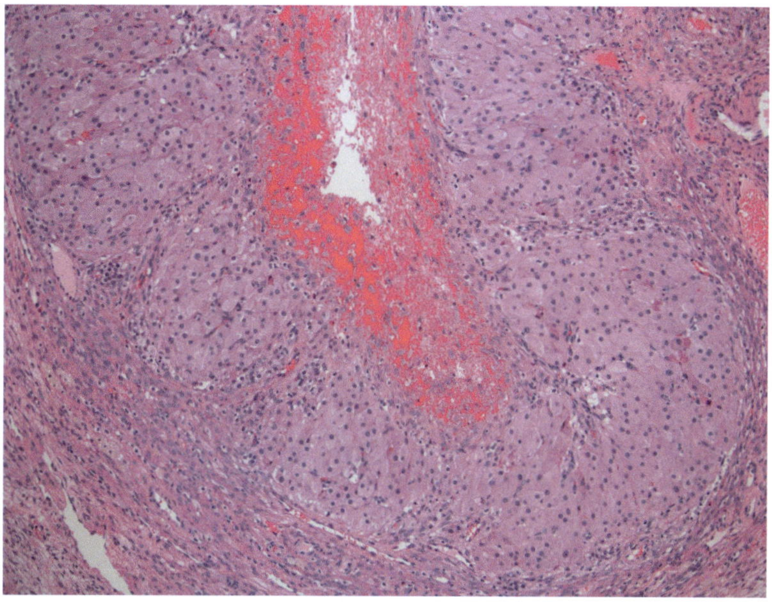

**Fig. 14.16** Corpus luteum of pregnancy. Note the eosinophilic cytoplasm of the granulosa cells in response to progestin stimulation

and hirsutism. Elevated levels of luteinizing hormone are seen which are thought to stimulate theca cells to synthesize excess androgens, some of which are converted to estrogens; thus, these patients may be susceptible to estrogen-driven endometrial carcinomas [65, 66].

##### 14.4.2.4 Stromal Hyperthecosis (Cortical Stromal Hyperplasia)

There is significant overlap in clinical presentation between PCOS and stromal hyperthecosis, although the latter is more commonly seen in postmenopausal women. In this case, there is bilateral ovarian enlargement (up to 7–8 cm), but instead of numerous cysts, there is proliferation of luteinized stromal cells which are often arranged in nests [67]. Clinically, women with stromal hyperthecosis often display striking virilizing symptoms, along with obesity. Similar to PCOS, patients with stromal hyperthecosis are at increased risk of developing estrogen associated endometrial carcinomas.

### 14.4.3 Ovarian Neoplasms

Ovarian cancer is one of the deadliest diseases in the female genital tract with an incidence rate of 12.3 per 100,000 women per year in the US [23]. Sadly, the 5 year survival rate is only at 44.6 % [23]. This high mortality rate is contributed mainly by our inability to detect ovarian cancer at an early stage. More than 70 % of the patients with ovarian cancer present with late stage disease, where the tumor has already disseminated throughout the abdominal and peritoneal cavities. Surgical debulking and chemotherapy can produce complete response in some patients; however, many will relapse after a brief period of treatment.

Clinically, the initial presentation of women with ovarian neoplasms includes abdominal enlargement, pain, swelling or vaginal bleeding. However, many are discovered incidentally during radiologic work-up for other diseases. Ascites is more suggestive of a malignant tumor as it is rarely seen in benign tumors.

As the majority of malignant ovarian neoplasms are of epithelial origin, for the scope of this chapter, rare cancers originating from germ cells and sex-cord stromal cells will not be discussed. Epithelial tumors are categorized as benign, borderline (or atypical proliferative tumor), or malignant. A malignant diagnosis is based on the presence of invasion. In the sections below, common epithelial ovarian tumors will be discussed.

#### 14.4.3.1 Serous Neoplasms

"Serous" is a term used to describe linings that resemble fallopian tubal epithelium. Traditionally, serous tumors were thought to arise from the ovarian surface epithelium. However, recent data has suggested that there are two distinct types of serous carcinomas: low-grade and high-grade [68]. Both low-grade serous carcinoma and serous borderline tumor harbor *KRAS* and *BRAF* mutations; therefore, a serous borderline tumor is considered the precursor lesion for low-grade serous carcinoma [68, 69]. In contrast, high-grade serous carcinoma harbors p53 mutation and often presents at advanced stage. In addition, recent developments have identified tubal intraepithelial carcinoma in the fallopian tube as the precursor lesion to high-grade serous carcinoma [70, 71].

**Benign Serous Tumors**

This is the most commonly encountered benign epithelial tumor of the ovary. Patients often are asymptomatic and the tumors are incidentally found. Symptomatic patients typically present with large ovarian masses. Histologically, the tumor is lined by cuboidal to columnar, tubal-like epithelium with minimal stratification (Fig. 14.17). Oophorectomy is curative.

**Serous Borderline Tumors**

Serous borderline tumors (or atypical proliferative serous tumors) often present bilaterally and occur a decade younger than high-grade serous carcinomas. Histologically, this tumor type is non-invasive and composed of hierarchically branching papillae with cellular stratification and tufting arranged in a complex architecture (Fig. 14.18). The epithelial proliferation and cytologic atypia are greater than those of benign serous tumors. The nuclei show mild to moderate atypia and mitoses are infrequent. As discussed earlier, these are thought to represent the precursor lesion for low-grade serous carcinoma [68, 69]. While most of these tumors have an indolent course, extraovarian spread, recurrence, and death can occur.

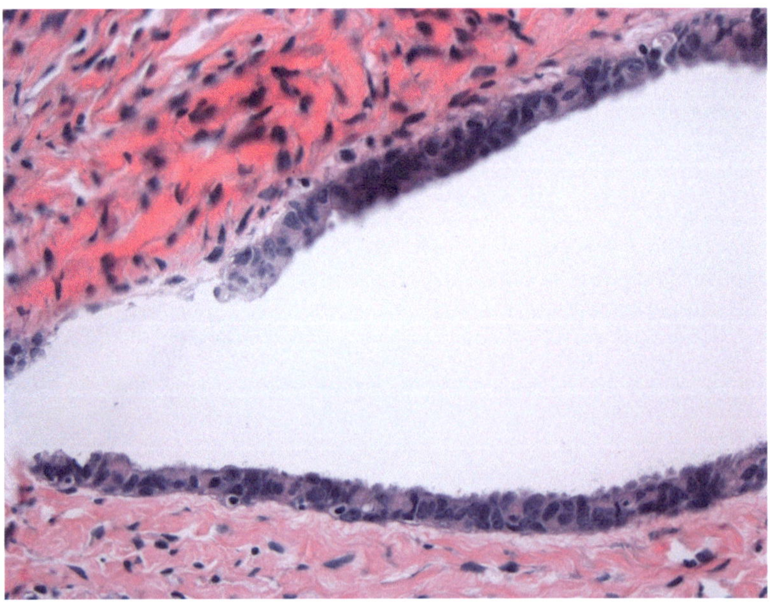

**Fig. 14.17** Serous cystadenoma. This benign tumor is lined by ciliated tubal-like epithelium without nuclear atypia

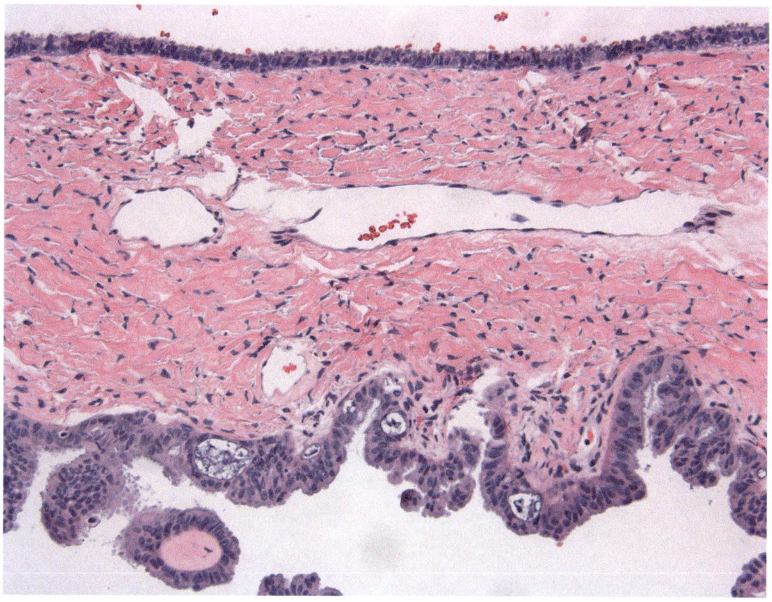

**Fig. 14.18** Serous borderline tumor. There is architectural complexity and pseudostratification of the nuclei (*bottom*). Serous cystadenoma can be seen on *top* of the figure

## Low-Grade Serous Adenocarcinoma

This tumor is an invasive carcinoma with low-grade cytology that often affects both ovaries. This tumor is uncommon and represents ~5 % of all serous carcinomas. Patients tend to be a decade younger when compared to high-grade serous carcinoma. Histologically, this tumor has greater architectural complexity and cytologic atypia when compared to serous borderline tumor. Invasion of the ovarian stroma is the defining diagnostic feature for low-grade serous carcinoma. Invasive tumor cells form small papillae surrounded by stromal clefts (Fig. 14.19). It is not uncommon to find a serous borderline tumor

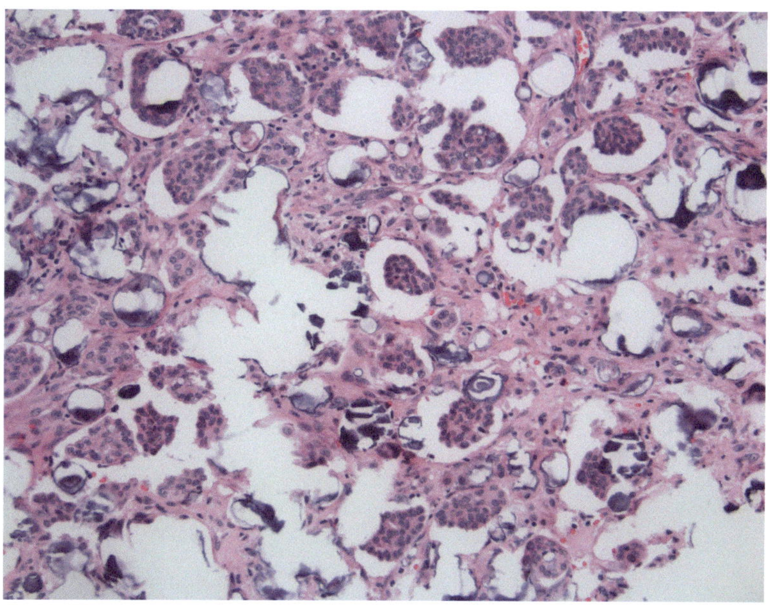

**Fig. 14.19** Low-grade serous adenocarcinoma. Invasive small nests of low-grade cells surrounded by stromal clefts. Psammomatous calcifications can be appreciated

adjacent to a low-grade serous carcinoma, supporting the notion that a serous borderline tumor is the precursor lesion for low-grade serous carcinoma. Clinically, this tumor is difficult to treat as the tumor tends to recur and the available chemotherapeutic agents are ineffective [72].

#### High-Grade Serous Adenocarcinoma

This is the most common malignant neoplasm of the ovary and occurs most frequently during the sixth and seventh decade. The majority of patients with high-grade serous carcinoma present with advanced stage disease where the tumor involves bilateral ovaries and has disseminated throughout the abdominal and pelvic regions. Histologically, this tumor demonstrates complex papillary, glandular or solid architecture with slit-like spaces (Fig. 14.20). Nuclei are large and pleomorphic with prominent nucleoli and numerous mitoses, including atypical forms. Most of the time, the tumor involvement is so extensive that invasion can be easily recognized by its confluent growth pattern. Clinically, many of these tumors respond to platinum-based chemotherapeutic regimens; however, the relapse rate is high with poor overall survival.

### 14.4.3.2 Mucinous Neoplasms

Mucinous tumors of the ovary are the second most commonly encountered epithelial ovarian tumors and account for ~15 % of the surface epithelial ovarian neoplasms in the western world. These tumors encompass ovarian masses lined by mucinous epithelium, resembling those of gastrointestinal tract (intestinal-type) or endocervix (endocervical-type or seromucinous- type). Similar to serous tumors, mucinous neoplasms are divided into three general categories: benign tumors, borderline tumors and malignant carcinomas. *KRAS* mutation can be found in all three categories, suggesting the continuum of this disease.

#### Benign Mucinous Tumors

Benign mucinous neoplasms account for ~80 % of all mucinous ovarian tumors. The age range is wide, however, they are more commonly seen in reproductive age women. The majority of these tumors are unilateral and cystic with an average size of around 10 cm. However, large tumors greater than 30 cm have been reported [73, 74]. Microscopically, the cyst is lined by columnar cells with basally located bland nuclei and abundant mucinous cytoplasm (Fig. 14.21). There is minimal nuclear stratification or atypia.

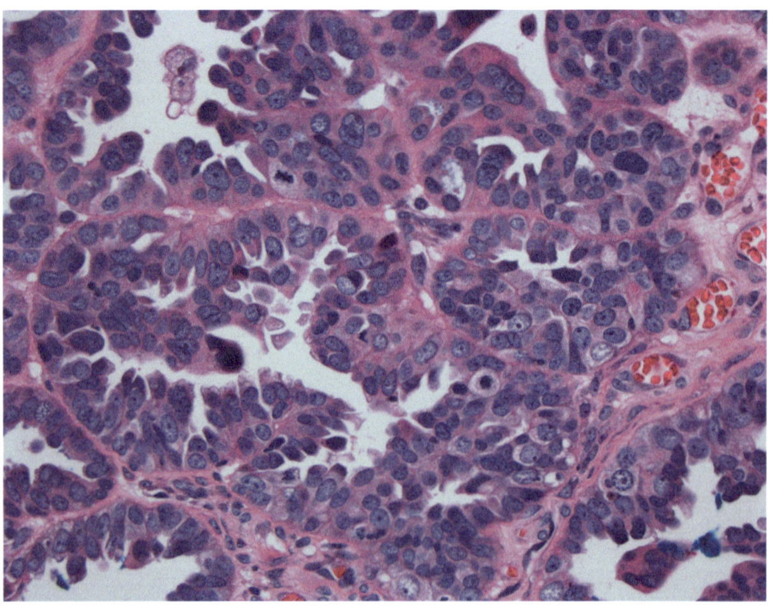

**Fig. 14.20** High-grade serous adenocarcinoma. The tumor forms pseudoglandular spaces and papillae. The cells have high-grade nuclear cytology and mitoses are readily identified

## Mucinous Borderline Tumors

Mucinous borderline tumors consist of ~15 % of all mucinous neoplasms. This tumor is often unilateral and seen in a wide age range with a mean age of ~45 years. When bilateral ovaries are involved, a metastasis should be considered. The majority of mucinous borderline tumors are stage I disease, and tend to have a benign course with less than 1 % of recurrence rate [75]. Histologically, the mucinous epithelium shows nuclear proliferation and stratification that forms various papillary folds and architectural complexity (Fig. 14.22). Mild to moderate nuclear cytologic atypia can be seen,

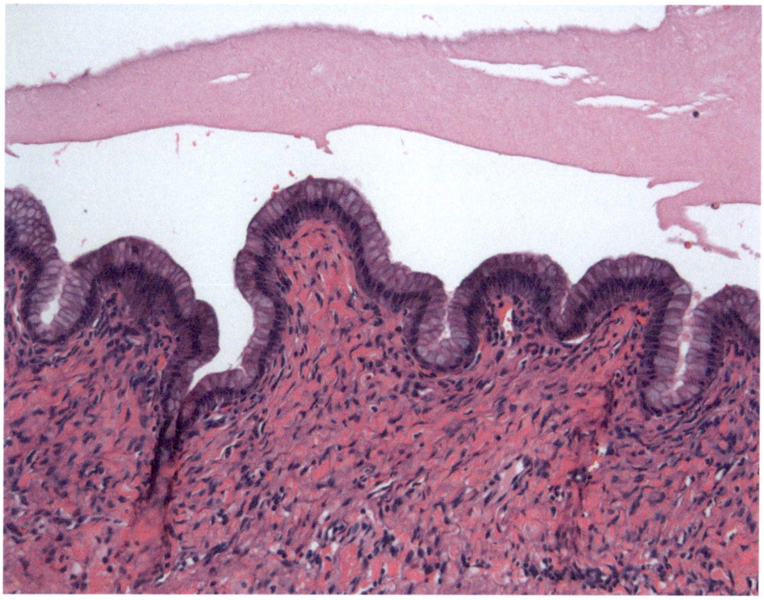

**Fig. 14.21** Mucinous cystadenoma. The cells lining the cyst contain mucinous cytoplasm with basally oriented nuclei. Stratification and nuclear atypia are absent

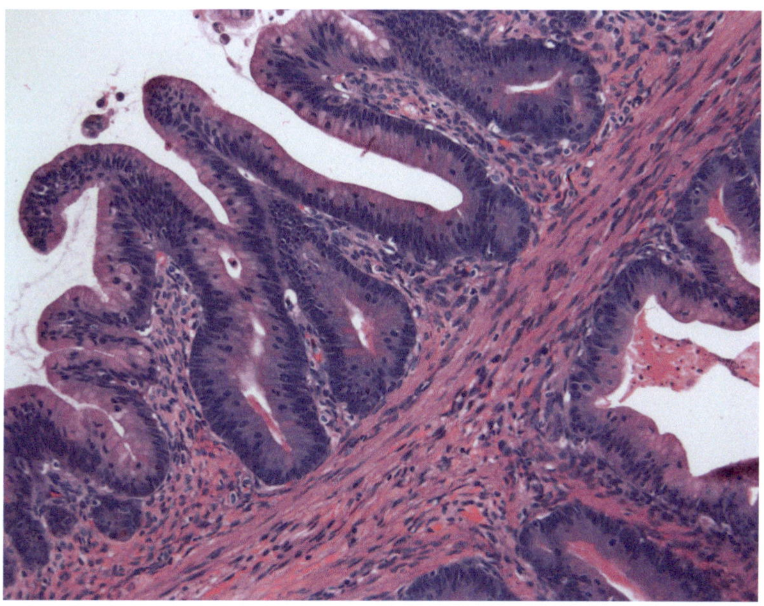

**Fig. 14.22** Mucinous borderline tumor. The tumor has complex villous architecture and is lined by cells with psedostratified nuclei resembling an adenomatous polyp of the colon

but there should be no evidence of stromal invasion.

**Mucinous Adenocarcinoma**

Mucinous adenocarcinoma is uncommon and represents only 3–4 % of all primary ovarian carcinomas. The age at presentation is comparable to mucinous borderline tumors (a mean of 45 years). Similar to mucinous borderline tumors, adenocarcinomas also tend to be unilateral; therefore, when bilateral ovaries are involved, metastatic disease has to be excluded. Most of these tumors present as stage I disease where the tumor is limited to the ovary. Histologically, a diagnosis of malignancy is based on the presence of stromal invasion. Two types of invasion have been described: expansile and infiltrative [75]. The expansile type is more common with tumor forming confluent back-to-back glands with minimal to no intervening stroma, creating a maze-like pattern (Fig. 14.23). A cribriform pattern can also be present. The infiltrative type is less common, and demonstrates irregular nests of malignant tumor infiltrating the stroma. It is common to see a continuum of benign, borderline and carcinomatous components in the same lesion where *KRAS* mutations have been identified in each component, suggesting *KRAS* mutation may be an early event in mucinous carcinogenesis [76, 77].

### 14.4.3.3 Endometrioid Neoplasms

Endometrioid tumors resemble glands and stroma from the endometrium. The majority of these tumors are malignant and unilateral. The diagnostic criteria for borderline tumor and malignancy are similar to EIN and endometrioid type of endometrial carcinoma, respectively.

**Benign Endometrioid Tumors**

Benign endometrioid tumors are cystic ovarian masses lined by benign endometrial epithelium without endometrial stroma. These are uncommon and typically unilateral. When the cyst is lined by endometrial epithelium and underlying endometrial stroma, the tumor is called an endometrioma (or endometriotic cyst), which is a more common lesion and frequently bilateral. Hemorrhage is usually associated with this lesion, which gives it a dark brown appearance

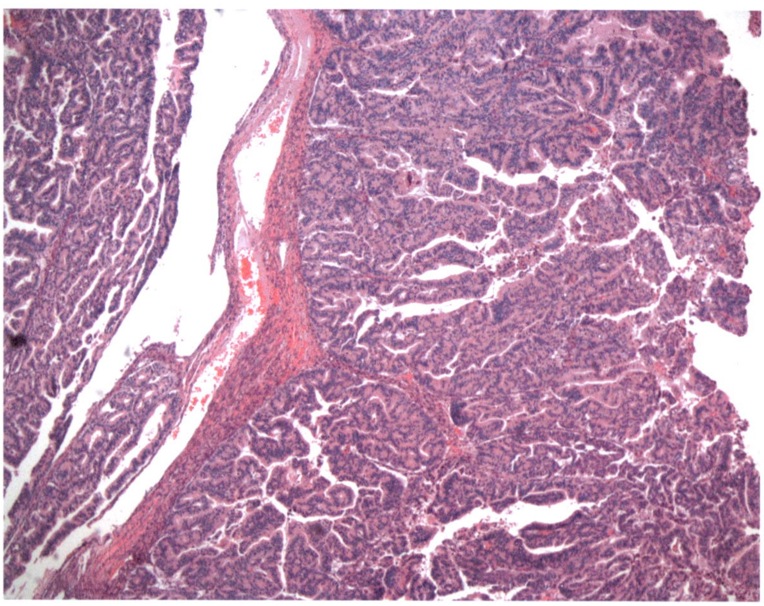

**Fig. 14.23** Mucinous adenocarcinoma. The expansile type invasion contains back-to-back glands with minimal intervening stroma

(chocolate cyst). Some patients have concurrent endometriosis elsewhere in the pelvis, causing pelvic pain and infertility.

#### Endometrioid Borderline Tumors

An endometrioid borderline tumor is an uncommon lesion and morphologically similar to EIN of the endometrium with closely packed endometrial glands lacking stromal invasion. Cribriform architecture can occasionally be present and squamous morules are frequently observed in this tumor.

#### Endometrioid Adenocarcinoma

This is by far the most common ovarian neoplasm in the endometrioid group. The histologic criteria for malignancy are similar to endometrioid adenocarcinoma of the endometrium showing back-to-back glands with fusion of glands or confluent cribriform architecture (Fig. 14.24). The grading schema is the same as the endometrial counterpart: <5 % solid pattern—grade 1; 5–50 % solid pattern—grade 2; >50 % solid pattern—grade 3. It is not uncommon to find endometriosis associated with an adenocarcinoma.

In ~15–20 % of endometrioid adenocarcinomas, a concurrent endometrial endometrioid adenocarcinoma is present [78–80]. Most of these patients show early FIGO stage I or II disease in both the ovary and endometrium [81, 82]. In this scenario, it is difficult to ascertain whether these are synchronous primaries or metastasis from either a primary ovarian or uterine malignancy. When bilateral ovaries are involved or lymphovascular invasion is identified, a metastasis should be considered. In the presence of precursor lesions, a synchronous primary would be favored. However, clinical and pathologic correlation is needed to make this distinction. When the tumors are confined within the uterus and ovary, the overall prognosis is good [83].

### 14.4.3.4 Clear Cell Neoplasms

Clear cell neoplasms consist of epithelium lined by cells with glycogen-rich clear cytoplasm.

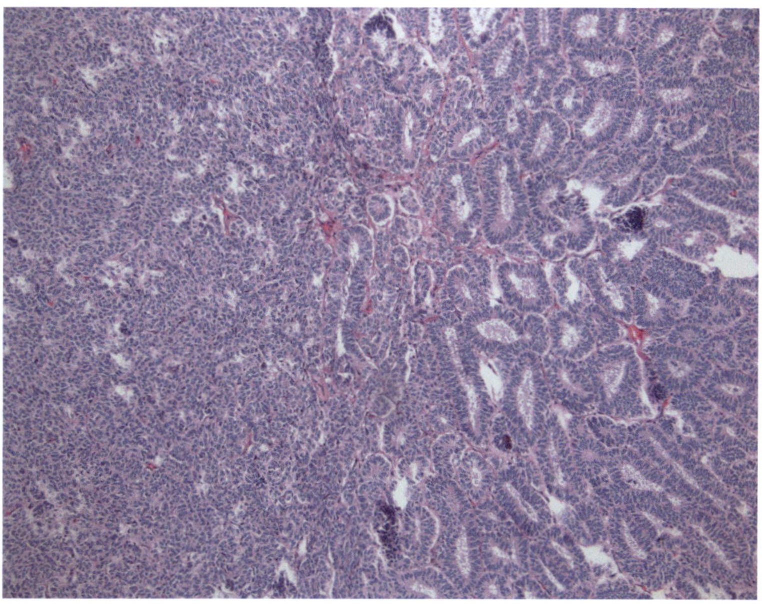

**Fig. 14.24** Endometrioid adenocarcinoma. This tumor shows back-to-back glandular (*right*) and solid (*left*) patterns

Benign and borderline clear cell tumors are exceedingly rare; therefore, they will not be discussed in this chapter.

#### Clear Cell Carcinoma

Approximately 99 % of clear cell neoplasms are clear cell carcinomas. This group consists of ~3 % of all epithelial neoplasms and is more commonly seen in Japan in comparison to western societies [84]. Patients usually present in the fifth to seventh decade with a mean age of 55 years [84]. This disease is commonly unilateral and associated with endometriosis [85]. The microscopic appearance of these tumors displays tubulocystic, papillary and solid architectures. The tumor cells have a hobnail appearance and contain high-grade nuclei with abundant clear, or less commonly, eosinophilic cytoplasm (Fig. 14.25). Recent studies have shown that napsin A is highly expressed in ovarian clear cell carcinomas and can help to differentiate them from ovarian serous or endometrioid carcinomas in difficult cases [86].

### 14.4.3.5 Transitional Cell Neoplasms

#### Brenner Tumor

A Brenner tumor is a benign neoplasm lined by transitional-type epithelium resembling the urothelial lining of the urinary tract. Most of these tumors affect patients in fifth to seventh decades. As most Brenner tumors are <2 cm in size, the majority of the patients are asymptomatic and the lesions are incidentally found during investigation for other unrelated pelvic conditions. Histologically, the tumor forms nests of transitional-type epithelial cells with bland nuclei that can have occasional grooves (what some have called "coffee bean" nuclei) in a background of fibromatous stroma (Fig. 14.26). As these lesions are benign, complete oophorectomy is curative.

#### Malignant Brenner Tumor

Malignant Brenner tumors are uncommon, consisting of <5 % of all transitional differentiated ovarian tumors. The tumor typically affects patients older than 50 years. A malignant Brenner tumor is an entity defined as a tumor which morphologically resembles invasive urothelial carcinoma of the urinary tract and is associated with a benign or borderline Brenner tumor (Fig. 14.27). Historically, tumors that consist exclusively of invasive urothelial cells were classified as transitional cell carcinoma. However, recent immunohistochemical and molecular data demonstrate that "transitional cell carcinoma" is not a distinct entity, but rather a poorly differentiated form of serous carcinoma, or less commonly, endometrioid carcinoma [87].

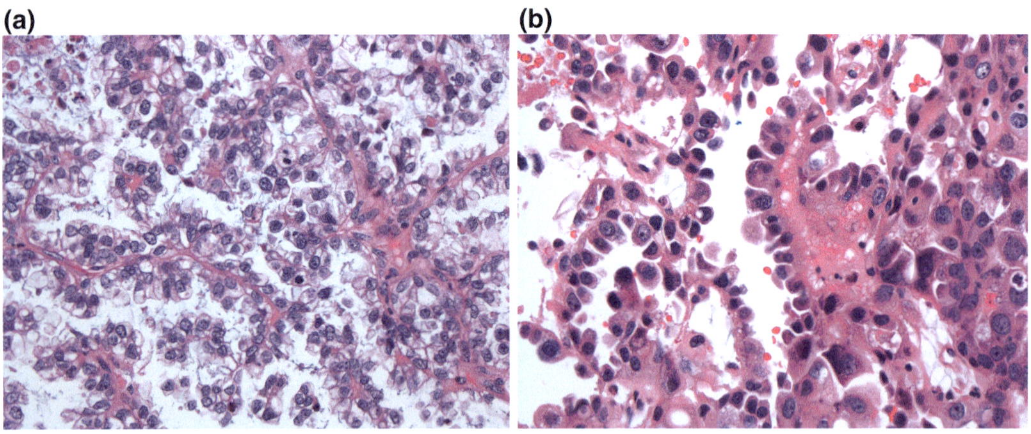


**Fig. 14.25** Clear cell adenocarcinoma. **a** The tumor has papillary architecture with high-grade nuclei and clear cytoplasm. **b** Hobnail cells and eosinophilic cytoplasm can also be seen

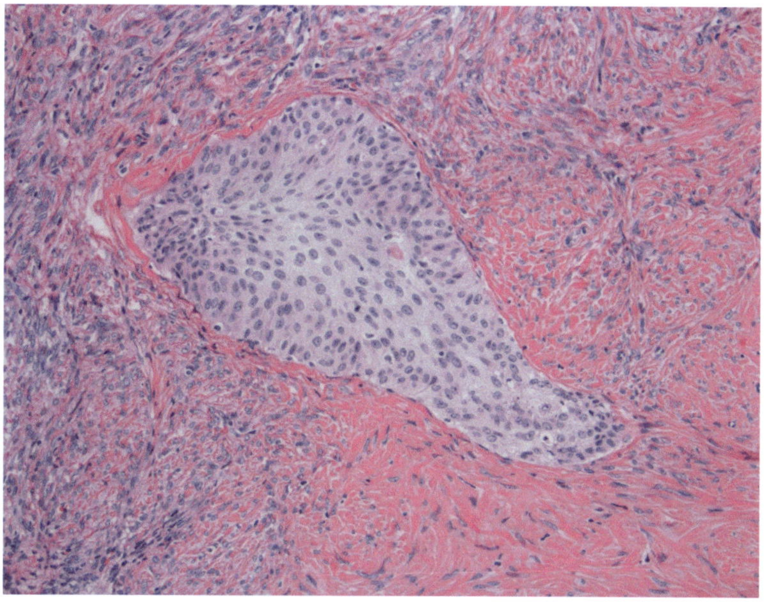

**Fig. 14.26** Brenner tumor. Bland appearing transitional cell nest within fibromatous stroma

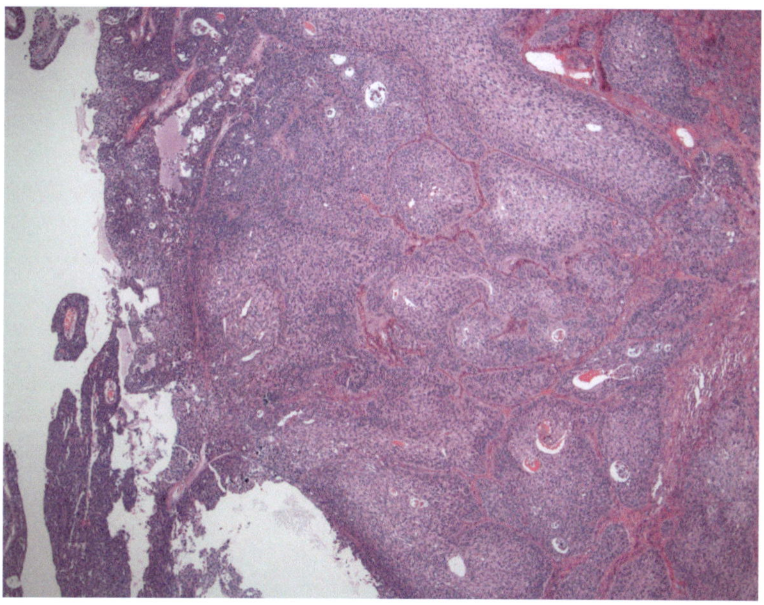

**Fig. 14.27** Malignant Brenner tumor. The tumor resembles high-grade urothelial carcinoma of the bladder with nests of transitional epithelium invading the stroma

### 14.4.4 Spread and Staging

Staging of ovarian cancers is based on the FIGO and AJCC staging systems, which are similar. These systems take into account the status of the ovarian capsule (intact or ruptured) and the involvement of ovarian surface, peritoneal fluid, pelvic extension, and peritoneal metastasis outside the pelvis (including liver capsule metastasis and/or regional lymph node metastasis) [88].

### 14.4.5 Genetics and Genomics

Ovarian cancers are heterogeneous groups of tumors with various morphology and molecular biology. As aforementioned, high-grade serous carcinomas are typically associated with p53 alteration and low-grade serous carcinomas frequently show *BRAF* and *KRAS* mutations. Mucinous adenocarcinomas often carry *KRAS* mutations, and less commonly, amplification of c-erbB2 gene. Ovarian endometrioid adenocarcinomas have MSI, mutations in *PTEN* and *KRAS* genes, and accumulation of β-catenin, similar to the uterine counterpart. Clear cell adenocarcinomas have *PIK3CA* and *ARID1A* mutations, *PTEN* inactivation and up-regulation of HNF-1β.

#### 14.4.5.1 Familial Syndromes

Ovarian cancer is diagnosed in ~22,000 women every year in the US and represents the fifth most common cancer in American women [89]. About 20 % of all ovarian cancer cases develop due to inherited mutations [90]. Having a family history of the disease is the most important and well-established risk factor for ovarian cancer. It has been estimated that having one first-degree relative diagnosed with ovarian cancer increases the risk of developing this gynecological malignancy three-fold, compared to the risk of the general population [91].

#### 14.4.5.2 Breast and Ovarian Cancer (BRCA) Syndrome

Among the 20 % of cases who develop ovarian cancer due to inherited gene mutations of high penetrance, about half harbor mutations in the breast cancer susceptibility genes *BRCA1* and *BRCA2*. The lifetime risk of developing ovarian cancer, by age 70, is 39 % in *BRCA1* mutation carriers and 11 % in *BRCA2* carriers [92]. This compares to a lifetime risk of ~1.3 % of ovarian cancer risk in the general population [89]. Most cases with *BRCA1* and *BRCA2* mutations are diagnosed with premenopausal ovarian cancer while the average age of onset of this malignancy in the general population is 63 years [23]. Therefore, both early onset and family history suggest the presence of mutations in these genes. The importance of mutations in *BRCA1* and *BRCA2* in ovarian cancer is therefore very high. In a recent evaluation, the US Preventive Services Task Force (USPSTF) recommended that "primary care providers screen women who have family members with breast, ovarian, tubal, or peritoneal cancer with one of several screening tools designed to identify a family history that may be associated with an increased risk for potentially harmful mutations in breast cancer susceptibility genes (*BRCA1* or *BRCA2*). Women with positive screening results should receive genetic counseling and, if indicated after counseling, BRCA testing" [93].

The importance of *BRCA1* and *BRCA2* mutations in ovarian cancer is two-fold. Besides the important role in cancer susceptibility and risk assessment, these genes are critically involved in repairing DNA double-strand breaks, which represents actionable information for treatment by Poly (ADP-ribose) polymerase (PARP) inhibitors. Indeed, in December 2014, the FDA approved one PARP inhibitor, Olaparib, as monotherapy for the treatment of BRCA-related ovarian cancer.

#### 14.4.5.3 Lynch Syndrome

Lynch syndrome, a common cause of colorectal cancer caused by mutations in MMR genes, is also a condition where ovarian cancer can be diagnosed. Indeed, after colorectal and endometrial cancer, ovarian cancer is the third most commonly diagnosed in Lynch syndrome patients [40–43]. As in cancers associated with *BRCA1* and *BRCA2* mutations, most of ovarian cancer cases in Lynch syndrome are diagnosed in young and premenopausal women. The lifetime risk of ovarian cancer in carriers of MMR gene mutations has been estimated to be ~7 % or 7-fold higher than that of the general population [40–42, 94]. Ovarian cancers in Lynch syndrome tend to have endometrioid, serous or clear cell histology [43].

#### 14.4.5.4 Other Syndromes

Ovarian cancer is also rarely diagnosed in familial cancer syndromes such as Peutz-Jeghers,

Li-Fraumeni and ataxia telangiectasia [41], although the rarity of these syndromes in the population has not allowed reliable ovarian cancer lifetime risk estimates. More recently, *RAD51C* and *RAD51D*, two closely related genes that act in the homologous recombination DNA repair pathway, have been associated with ovarian cancer risk. Meindl and collaborators [95] suggested that ~1.3 % of BRCA1/2 mutation negative BRCA families had mutations in *RAD51C*. In another study, Loveday et al. [96] found that mutations in *RAD51B* were associated with a relative risk for ovarian cancer that was 6.3-fold higher than of the general population. Because *RAD51C* and *RAD51D* act in the same pathway as *BRCA1* and *BRCA2*, it is possible that patients with mutations in these two genes may also benefit from PARP inhibitor treatment [96].

#### 14.4.5.5 Low-Penetrance Risk Variants

GWA studies have been also used to investigate the genetic susceptibility to ovarian cancer in the general population. So far, 18 genomic regions have been associated with increased risk of ovarian cancer [97–103]. These low-penetrance ovarian cancer risk alleles typically increase the risk of developing ovarian cancer by ~10–30 % when compared to the risk in the general population. Despite the large scale of the ovarian cancer association studies (the latest study used DNA samples from ~70,000 individuals), the 18 known ovarian cancer genes only explain ~4 % of the heritability of the disease and therefore are still unlikely to be useful on their own to predict risk [103]. It is, therefore, possible that there exist many more common risk alleles with even weaker effect sizes or that most of ovarian cancer cases in the population may be explained by very rare variants of intermediate penetrance [104]. The latter scenario, which follows the "common-disease rare-variant hypothesis" is likely to be further investigated using genome-wide sequencing approaches [104].

#### 14.4.5.6 Common SNPs and the Risk of Ovarian Cancer in BRCA1 and BRCA2 Mutation Carriers

GWA studies have been also used to identify common low-penetrance variants that modify the phenotype of ovarian cancer patients who have germline mutations in *BRCA1* and *BRCA2*. These efforts have been mainly led by the Consortium of Investigators of Modifiers of BRCA1/2 (CIMBA) [105]. CIMBA studies have identified several SNPs that modify the risk of ovarian cancer in *BRCA1* and *BRCA2* and have shown that about half of the risk variants that affect ovarian cancer risk in the sporadic ovarian cancer (see above) also modify the risk to this malignancy in BRCA1/2 mutation carriers [106–108]. CIMBA studies have also shown that all known modifier SNPs account for ~7.5 % of the ovarian cancer polygenic modifying variance in *BRCA1* mutation carriers [103] and suggest that incorporating the information of polygenic scores could improve risk prediction of ovarian cancer among BRCA1/2 mutation carriers.

#### 14.4.5.7 Somatic Genetics

Ovarian tumors are characterized by complex mutation patterns, where the predominant mutation driver is *TP53*, which is found to be mutated in >90 % of all high-grade serous adenocarcinomas [109]. The second most commonly reported somatic mutations in ovarian cancers are in the tumor suppressor genes BRCA1 and BRCA2, which are mutated in ~15 % of all high-grade serous adenocarcinomas. In the largest report of somatic mutations in ovarian cancer published to date, the TCGA study [110] identified six additional cancer driver genes in a sample of 489 high-grade serous ovarian adenocarcinomas which included *RB1*, *NF1*, *FAT3*, *CSMD3*, *GABRA6*, and *CDK12*. The frequency of mutations in the latter genes ranged from 1.8 % for *RB1* and *GABRA6* to 19 % for *CSMD3*. Collectively, these six newly identified genes were mutated in 24% of all ovarian tumors sequenced by the TCGA study. Interestingly, about half of all tumors analyzed by the TCGA study showed either mutations or other somatic defects in genes involved in DNA recombination. For example, about 24 % of the tumors showed inactivation of either *BRCA1* or *BRCA2*. These findings suggest that most ovarian cancers with homologous recombination deficiency could benefit from targeted therapies that use PARP inhibitors.

## 14.5 Summary

Traditionally, cancer therapeutics has been based on the morphology, grade, and stage of the tumors. However, with advances in molecular techniques, our understanding of cancer biology continues to evolve. The newly acquired information from TCGA has allowed scientists to identify important mutations that may play crucial roles in cancer development that cannot be readily visualized under the microscope. The goal of this chapter is to introduce the readers to cancer morphology and conclude with our current understanding of molecular biology for uterine and ovarian cancers. We hope the readers will gain a deeper knowledge of cancer biology through the understanding and appreciation of both histology and molecular biology.

## References


1. Crosby WM, Hill EC. Embryology of the mullerian duct system. Obstet Gynecol. 1962;20:507–15.
2. T.W.S. Langman's Medical Embryology. 5 ed. London: Williams and Wilkins; 1985.
3. Josso N, Racine C, di Clemente N, Rey R, Xavier F. The role of anti-Mullerian hormone in gonadal development. Mol Cell Endocrinol. 1998;145:3–7.
4. Hunter RHF. Differentiation of the genital duct system. In: Hunter RHF, editor. Sex determination, differentiation, and intersexuality in placental mammals. Cambridge: Cambridge University Press; 1995. p. 107–38.
5. Moore KL. Before we are born: basic enbryology and birth defects. Philadelphia: Saunders; 1974.
6. Brennan J, Capel B. One tissue, two fates: molecular genetic events that underlie testis versus ovary development. Nat Rev Genet. 2004;5:509–21.
7. Mackay S. Gonadal development in mammals at the cellular and molecular level. Int Rev Cytol. 2000;200.
8. Hunter RHF. Differentiation of the gonads. In: Hunter RHF, editor. Sex determination, differentiation, and intersexuality in placental mammals. Cambridge: Cambridge University Press; 1995. p. 69–106.
9. Nicosia SV. Morphological changes of the human ovary throughout life. In: Serra GB, editor. The ovary. New York: Raven Press; 1983. p. 57–81.
10. Noyes RW, Hertig AT, Rock J. Dating the endometrial biopsy. Fertil Steril. 1950;1(1):3–25.
11. Crum CP, Hornstein MD, Nucci MR, Mutter GL. Hertig and beyond: a systematic and practical approach to the endometrial biopsy. Advances in anatomic pathology. 2003;10(6):301–18.
12. Crum CP, Nucci MR, Lee KR. Diagnostic gynecologic and obstetric pathology. 2nd ed. Philadelphia, PA: Saunders/Elsevier; 2011. xiv, 1202 p. p.
13. Ferenczy A. Pathophysiology of endometrial bleeding. Maturitas. 2003;45(1):1–14.
14. Vakiani M, Vavilis D, Agorastos T, Stamatopoulos P, Assimaki A, Bontis J. Histopathological findings of the endometrium in patients with dysfunctional uterine bleeding. Clin Exp Obstet Gynecol. 1996;23(4):236–9.
15. Parazzini F, La Vecchia C, Bocciolone L, Franceschi S. The epidemiology of endometrial cancer. Gynecol Oncol. 1991;41(1):1–16.
16. Montgomery BE, Daum GS, Dunton CJ. Endometrial hyperplasia: a review. Obstet Gynecol Surv. 2004;59(5):368–78.
17. Kurman RJ, Kaminski PF, Norris HJ. The behavior of endometrial hyperplasia. A long-term study of "untreated" hyperplasia in 170 patients. Cancer. 1985;56(2):403–12.
18. Mutter GL, Zaino RJ, Baak JP, Bentley RC, Robboy SJ. Benign endometrial hyperplasia sequence and endometrial intraepithelial neoplasia. Int J Gynecol Pathol. 2007;26(2):103–14.
19. Baak JP, Mutter GL, Robboy S, van Diest PJ, Uyterlinde AM, Orbo A, et al. The molecular genetics and morphometry-based endometrial intraepithelial neoplasia classification system predicts disease progression in endometrial hyperplasia more accurately than the 1994 World Health Organization classification system. Cancer. 2005;103 (11):2304–12.
20. Mutter GL, Baak JP, Crum CP, Richart RM, Ferenczy A, Faquin WC. Endometrial precancer diagnosis by histopathology, clonal analysis, and computerized morphometry. J Pathol. 2000;190 (4):462–9.
21. Mutter GL, Kauderer J, Baak JP, Alberts D. Biopsy histomorphometry predicts uterine myoinvasion by endometrial carcinoma: a Gynecologic Oncology Group study. Hum Pathol. 2008;39(6):866–74.

22. Mutter GL. Histopathology of genetically defined endometrial precancers. Int J Gynecol Pathol. 2000;19(4):301–9.
23. National Cancer Institute. Surveillance epidemiology and end results 2014 (Internet). (cited 11 Feb 2015). Available from http://seer.cancer.gov/statfacts/html/ovary.html.
24. Bokhman JV. Two pathogenetic types of endometrial carcinoma. Gynecol Oncol. 1983;15(1):10–7.
25. Sherman ME. Theories of endometrial carcinogenesis: a multidisciplinary approach. Mod Pathol. 2000;13(3):295–308.
26. Creasman WT, Odicino F, Maisonneuve P, Quinn MA, Beller U, Benedet JL, et al. Carcinoma of the corpus uteri. FIGO 26th annual report on the results of treatment in gynecological cancer. Int J Gynaecol Obstet Official Organ Int Fed Gynaecol Obstet. 2006;95 Suppl 1:S105–43.
27. Barrow E, Robinson L, Alduaij W, Shenton A, Clancy T, Lalloo F, et al. Cumulative lifetime incidence of extracolonic cancers in Lynch syndrome: a report of 121 families with proven mutations. Clin Genet. 2009;75(2):141–9.
28. Bonadona V, Bonaiti B, Olschwang S, Grandjouan S, Huiart L, Longy M, et al. Cancer risks associated with germline mutations in MLH1, MSH2, and MSH6 genes in Lynch syndrome. JAMA. 2011;305(22):2304–10.
29. Tan MH, Mester JL, Ngeow J, Rybicki LA, Orloff MS, Eng C. Lifetime cancer risks in individuals with germline PTEN mutations. Clin Cancer Res. 2012;18(2):400–7.
30. Zaino RJ, Kurman RJ, Diana KL, Morrow CP. The utility of the revised International Federation of Gynecology and Obstetrics histologic grading of endometrial adenocarcinoma using a defined nuclear grading system. A Gynecologic Oncology Group study. Cancer. 1995;75(1):81–6.
31. Ross JC, Eifel PJ, Cox RS, Kempson RL, Hendrickson MR. Primary mucinous adenocarcinoma of the endometrium. A clinicopathologic and histochemical study. Am J Surg Pathol. 1983;7(8):715–29.
32. Kurman RJ. International Agency for Research on Cancer, World Health Organization. WHO classification of tumours of female reproductive organs. 4th ed. Lyon: International Agency for Research on Cancer; 2014. 307 pp.
33. Evans MJ, Langlois NE, Kitchener HC, Miller ID. Is there an association between long-term tamoxifen treatment and the development of carcinosarcoma (malignant mixed Mullerian tumor) of the uterus? Int J Gynecol Cancer. 1995;5(4):310–3.
34. Hubalek M, Ramoni A, Mueller-Holzner E, Marth C. Malignant mixed mesodermal tumor after tamoxifen therapy for breast cancer. Gynecol Oncol. 2004;95(1):264–6.
35. Costa MJ, Guinee D Jr. CD34 immunohistochemistry in female genital tract carcinosarcoma (malignant mixed mullerian tumors) supports a dominant role of the carcinomatous component. Appl Immunohistochem Mol Morphol. 2000;8(4):293–9.
36. Seidman JD, Chauhan S. Evaluation of the relationship between adenosarcoma and carcinosarcoma and a hypothesis of the histogenesis of uterine sarcomas. Int J Gynecol Pathol. 2003;22(1):75–82.
37. Sreenan JJ, Hart WR. Carcinosarcomas of the female genital tract. A pathologic study of 29 metastatic tumors: further evidence for the dominant role of the epithelial component and the conversion theory of histogenesis. Am J Surg Pathol. 1995;19(6):666–74.
38. Pecorelli S. Revised FIGO staging for carcinoma of the vulva, cervix, and endometrium. Int J Gynaecol Obstet Official Organ Int Fed Gynaecol Obstet. 2009;105(2):103–4.
39. Lancaster JM, Powell CB, Chen LM, Richardson DL. Committee SGOCP. Society of Gynecologic Oncology statement on risk assessment for inherited gynecologic cancer predispositions. Gynecol Oncol. 2015;136(1):3–7.
40. Bonadona V, Bonaiti B, Olschwang S, Grandjouan S, Huiart L, Longy M, et al. Cancer risks associated with germline mutations in MLH1, MSH2, and MSH6 genes in Lynch syndrome. JAMA. 2011;305(22):2304–10.
41. Engel C, Loeffler M, Steinke V, Rahner N, Holinski-Feder E, Dietmaier W, et al. Risks of less common cancers in proven mutation carriers with lynch syndrome. J Clin Oncol. 2012;30(35):4409–15.
42. Watson P, Vasen HF, Mecklin JP, Bernstein I, Aarnio M, Jarvinen HJ, et al. The risk of extra-colonic, extra-endometrial cancer in the Lynch syndrome. Int J Cancer. 2008;123(2):444–9.
43. Ketabi Z, Bartuma K, Bernstein I, Malander S, Gronberg H, Bjorck E, et al. Ovarian cancer linked to Lynch syndrome typically presents as early-onset, non-serous epithelial tumors. Gynecol Oncol. 2011;121(3):462–5.
44. Buchanan DD, Tan YY, Walsh MD, Clendenning M, Metcalf AM, Ferguson K, et al. Tumor mismatch repair immunohistochemistry and DNA MLH1 methylation testing of patients with endometrial cancer diagnosed at age younger than 60 years optimizes triage for population-level germline mismatch repair gene mutation testing. J Clin Oncol. 2014;32(2):90–100.
45. Simpkins SB, Bocker T, Swisher EM, Mutch DG, Gersell DJ, Kovatich AJ, et al. MLH1 promoter methylation and gene silencing is the primary cause of microsatellite instability in sporadic endometrial cancers. Hum Mol Genet. 1999;8(4):661–6.
46. Palles C, Cazier JB, Howarth KM, Domingo E, Jones AM, Broderick P, et al. Germline mutations affecting the proofreading domains of POLE and POLD1 predispose to colorectal adenomas and carcinomas. Nat Genet. 2013;45(2):136–44.
47. Riegert-Johnson DL, Gleeson FC, Roberts M, Tholen K, Youngborg L, Bullock M, et al. Cancer

and Lhermitte-Duclos disease are common in Cowden syndrome patients. Hered Cancer Clin Pract. 2010;8(1):6.
48. Pilarski R, Burt R, Kohlman W, Pho L, Shannon KM, Swisher E. Cowden syndrome and the PTEN hamartoma tumor syndrome: systematic review and revised diagnostic criteria. J Natl Cancer Inst. 2013;105(21):1607–16.
49. Pilarski R, Stephens JA, Noss R, Fisher JL, Prior TW. Predicting PTEN mutations: an evaluation of Cowden syndrome and Bannayan-Riley-Ruvalcaba syndrome clinical features. J Med Genet. 2011;48(8):505–12.
50. Tan MH, Mester JL, Ngeow J, Rybicki LA, Orloff MS, Eng C. Lifetime cancer risks in individuals with germline PTEN mutations. Clin Cancer Res. 2012;18(2):400–7.
51. Bubien V, Bonnet F, Brouste V, Hoppe S, Barouk-Simonet E, David A, et al. High cumulative risks of cancer in patients with PTEN hamartoma tumour syndrome. J Med Genet. 2013;50(4):255–63.
52. Tan MH, Mester J, Peterson C, Yang Y, Chen JL, Rybicki LA, et al. A clinical scoring system for selection of patients for PTEN mutation testing is proposed on the basis of a prospective study of 3042 probands. Am J Hum Genet. 2011;88(1):42–56.
53. Church DN, Briggs SE, Palles C, Domingo E, Kearsey SJ, Grimes JM, et al. DNA polymerase epsilon and delta exonuclease domain mutations in endometrial cancer. Hum Mol Genet. 2013;22(14):2820–8.
54. Briggs S, Tomlinson I. Germline and somatic polymerase epsilon and delta mutations define a new class of hypermutated colorectal and endometrial cancers. J Pathol. 2013;230(2):148–53.
55. Valle L, Hernandez-Illan E, Bellido F, Aiza G, Castillejo A, Castillejo MI, et al. New insights into POLE and POLD1 germline mutations in familial colorectal cancer and polyposis. Hum Mol Genet. 2014;23(13):3506–12.
56. Carvajal-Carmona LG, O'Mara TA, Painter JN, Lose FA, Dennis J, Michailidou K, et al. Candidate locus analysis of the TERT-CLPTM1L cancer risk region on chromosome 5p15 identifies multiple independent variants associated with endometrial cancer risk. Hum Genet. 2015;134(2):231–45.
57. Painter JN, O'Mara TA, Batra J, Cheng T, Lose FA, Dennis J, et al. Fine-mapping of the HNF1B multicancer locus identifies candidate variants that mediate endometrial cancer risk. Hum Mol Genet. 2015;24(5):1478–92.
58. Spurdle AB, Thompson DJ, Ahmed S, Ferguson K, Healey CS, O'Mara T, et al. Genome-wide association study identifies a common variant associated with risk of endometrial cancer. Nat Genet. 2011;43(5):451–4.
59. Dunlop MG, Tenesa A, Farrington SM, Ballereau S, Brewster DH, Koessler T, et al. Cumulative impact of common genetic variants and other risk factors on colorectal cancer risk in 42,103 individuals. Gut. 2013;62(6):871–81.
60. O'Hara AJ, Bell DW. The genomics and genetics of endometrial cancer. Adv Genomics Genet. 2012;2012(2):33–47.
61. Le Gallo M, Bell DW. The emerging genomic landscape of endometrial cancer. Clin Chem. 2014;60(1):98–110.
62. Cancer Genome Atlas Research N, Kandoth C, Schultz N, Cherniack AD, Akbani R, Liu Y, et al. Integrated genomic characterization of endometrial carcinoma. Nature. 2013;497(7447):67–73.
63. Satoh M. Histogenesis and organogenesis of the gonad in human embryos. J Anat. 1991;77:85–107.
64. Strickler RC, Kelly RW, Askin FB. Postmenopausal ovarian follicle cyst: an unusual cause of estrogen excess. Int J Gynecol Pathol. 1984;3:318–22.
65. Dunaif A, Thomas A. Current concepts in polycystic ovary syndrome. Annu Rev Med. 2001;52:401–19.
66. Hardiman P, Pillary OS, Atiomo W. Polycystic ovary syndrome and endometrial carcinoma. Lancet. 2003;361:1810–2.
67. Sasamo H, Fukunaga M, Rojas M, Silverberg SG. Hyperthecosis of the ovary: clinicopathologic study of 19 caseswith immunohistochemical analysis of steroidogenic enzymes. Int J Gynecol Pathol. 1989;8:311–20.
68. Shih IM, Kurman RJ. Ovarian tumorigenesis: a proposed model based on morphological and molecular genetic analysis. Am J Pathol. 2004;164(5):1511–8.
69. Kurman RJ, Shih IM. The origin and pathogenesis of epithelial ovarian cancer: a proposed unifying theory. Am J Surg Pathol. 2010;34(3):433–43.
70. Kindelberger DW, Lee Y, Miron A, Hirsch MS, Feltmate C, Medeiros F, et al. Intraepithelial carcinoma of the fimbria and pelvic serous carcinoma: evidence for a causal relationship. Am J Surg Pathol. 2007;31(2):161–9.
71. Medeiros F, Muto MG, Lee Y, Elvin JA, Callahan MJ, Feltmate C, et al. The tubal fimbria is a preferred site for early adenocarcinoma in women with familial ovarian cancer syndrome. Am J Surg Pathol. 2006;30(2):230–6.
72. Della Pepa C, Tonini G, Santini D, Losito S, Pisano C, Di Napoli M, et al. Low grade serous ovarian carcinoma: from the molecular characterization to the best therapeutic strategy. Cancer Treat Rev. 2015;41(2):136–43.
73. Chao A, Chao A, Yen YS, Huang CH. Abdominal compartment syndrome secondary to ovarian mucinous cystadenoma. Obstet Gynecol. 2004;104(5 Pt 2):1180–2.
74. Bicer S, Erkul Z, Demiryilmaz I, Peker N. A 9-kg ovarian mucinous cystadenoma in a 14-year-old premenarchal girl. Am J Case Rep. 2014;15:326–9.
75. Lee KR, Scully RE. Mucinous tumors of the ovary: a clinicopathologic study of 196 borderline tumors (of intestinal type) and carcinomas, including an

evaluation of 11 cases with 'pseudomyxoma peritonei'. Am J Surg Pathol. 2000;24(11):1447–64.
76. Cuatrecasas M, Villanueva A, Matias-Guiu X, Prat J. K-ras mutations in mucinous ovarian tumors: a clinicopathologic and molecular study of 95 cases. Cancer. 1997;79(8):1581–6.
77. Garrett AP, Lee KR, Colitti CR, Muto MG, Berkowitz RS, Mok SC. K-ras mutation may be an early event in mucinous ovarian tumorigenesis. Int J Gynecol Pathol. 2001;20(3):244–51.
78. Falkenberry SS, Steinhoff MM, Gordinier M, Rappoport S, Gajewski W, Granai CO. Synchronous endometrioid tumors of the ovary and endometrium. A clinicopathologic study of 22 cases. J Reprod Med. 1996;41(10):713–8.
79. Sheu BC, Lin HH, Chen CK, Chao KH, Shun CT, Huang SC. Synchronous primary carcinomas of the endometrium and ovary. Int J Gynaecol Obstet Official Organ Int Fed Gynaecol Obstet. 1995;51(2):141–6.
80. Zaino RJ, Unger ER, Whitney C. Synchronous carcinomas of the uterine corpus and ovary. Gynecol Oncol. 1984;19(3):329–35.
81. Heitz F, Amant F, Fotopoulou C, Battista MJ, Wimberger P, Traut A, et al. Synchronous ovarian and endometrial cancer—an international multicenter case-control study. Int J Gynecol Cancer. 2014;24(1):54–60.
82. Liu Y, Li J, Jin H, Lu Y, Lu X. Clinicopathological characteristics of patients with synchronous primary endometrial and ovarian cancers: a review of 43 cases. Oncol Lett. 2013;5(1):267–70.
83. Zaino R, Whitney C, Brady MF, DeGeest K, Burger RA, Buller RE. Simultaneously detected endometrial and ovarian carcinomas—a prospective clinicopathologic study of 74 cases: a gynecologic oncology group study. Gynecol Oncol. 2001;83(2):355–62.
84. Chan JK, Teoh D, Hu JM, Shin JY, Osann K, Kapp DS. Do clear cell ovarian carcinomas have poorer prognosis compared to other epithelial cell types? A study of 1411 clear cell ovarian cancers. Gynecol Oncol. 2008;109(3):370–6.
85. Ogawa S, Kaku T, Amada S, Kobayashi H, Hirakawa T, Ariyoshi K, et al. Ovarian endometriosis associated with ovarian carcinoma: a clinicopathological and immunohistochemical study. Gynecol Oncol. 2000;77(2):298–304.
86. Kandalaft PL, Gown AM, Isacson C. The lung-restricted marker napsin A is highly expressed in clear cell carcinomas of the ovary. Am J Clin Pathol. 2014;142(6):830–6.
87. Takeuchi T, Ohishi Y, Imamura H, Aman M, Shida K, Kobayashi H, et al. Ovarian transitional cell carcinoma represents a poorly differentiated form of high-grade serous or endometrioid adenocarcinoma. Am J Surg Pathol. 2013;37(7):1091–9.
88. Prat J, Oncology FCoG. Staging classification for cancer of the ovary, fallopian tube, and peritoneum. Int J Gynaecol Obstet Official Organ Int Fed Gynaecol Obstet. 2014;124(1):1–5.
89. Siegel RL, Miller KD, Jemal A. Cancer statistics, 2015. CA Cancer J Clin. 2015;65(1):5–29.
90. Weissman SM, Weiss SM, Newlin AC. Genetic testing by cancer site: ovary. Cancer J. 2012;18(4):320–7.
91. Stratton JF, Pharoah P, Smith SK, Easton D, Ponder BA. A systematic review and meta-analysis of family history and risk of ovarian cancer. Br J Obstet Gynaecol. 1998;105(5):493–9.
92. Antoniou A, Pharoah PD, Narod S, Risch HA, Eyfjord JE, Hopper JL, et al. Average risks of breast and ovarian cancer associated with BRCA1 or BRCA2 mutations detected in case Series unselected for family history: a combined analysis of 22 studies. Am J Hum Genet. 2003;72(5):1117–30.
93. Moyer VA, Force USPST. Risk assessment, genetic counseling, and genetic testing for BRCA-related cancer in women: U.S. Preventive Services Task Force recommendation statement. Ann Intern Med. 2014;160(4):271–81.
94. Lu KH, Daniels M. Endometrial and ovarian cancer in women with Lynch syndrome: update in screening and prevention. Fam Cancer. 2013;12(2):273–7.
95. Meindl A, Hellebrand H, Wiek C, Erven V, Wappenschmidt B, Niederacher D, et al. Germline mutations in breast and ovarian cancer pedigrees establish RAD51C as a human cancer susceptibility gene. Nat Genet. 2010;42(5):410–4.
96. Loveday C, Turnbull C, Ramsay E, Hughes D, Ruark E, Frankum JR, et al. Germline mutations in RAD51D confer susceptibility to ovarian cancer. Nat Genet. 2011;43(9):879–82.
97. Bolton KL, Tyrer J, Song H, Ramus SJ, Notaridou M, Jones C, et al. Common variants at 19p13 are associated with susceptibility to ovarian cancer. Nat Genet. 2010;42(10):880–4.
98. Goode EL, Chenevix-Trench G, Song H, Ramus SJ, Notaridou M, Lawrenson K, et al. A genome-wide association study identifies susceptibility loci for ovarian cancer at 2q31 and 8q24. Nat Genet. 2010;42(10):874–9.
99. Song H, Ramus SJ, Tyrer J, Bolton KL, Gentry-Maharaj A, Wozniak E, et al. A genome-wide association study identifies a new ovarian cancer susceptibility locus on 9p22.2. Nat Genet. 2009;41(9):996–1000.
100. Pharoah PD, Tsai YY, Ramus SJ, Phelan CM, Goode EL, Lawrenson K, et al. GWAS meta-analysis and replication identifies three new susceptibility loci for ovarian cancer. Nat Genet. 2013;45(4):362–70, 70e1–2.
101. Permuth-Wey J, Lawrenson K, Shen HC, Velkova A, Tyrer JP, Chen Z, et al. Identification and molecular characterization of a new ovarian cancer susceptibility locus at 17q21.31. Nat Commun. 2013;4:1627.
102. Bojesen SE, Pooley KA, Johnatty SE, Beesley J, Michailidou K, Tyrer JP, et al. Multiple independent

variants at the TERT locus are associated with telomere length and risks of breast and ovarian cancer. Nat Genet. 2013;45(4):371–84, 84e1–2.
103. Kuchenbaecker KB, Ramus SJ, Tyrer J, Lee A, Shen HC, Beesley J, et al. Identification of six new susceptibility loci for invasive epithelial ovarian cancer. Nat Genet. 2015;47(2):164–71.
104. Carvajal-Carmona LG. Challenges in the identification and use of rare disease-associated predisposition variants. Curr Opin Genet Dev. 2010; 20:277–281.
105. Chenevix-Trench G, Milne RL, Antoniou AC, Couch FJ, Easton DF, Goldgar DE, et al. An international initiative to identify genetic modifiers of cancer risk in BRCA1 and BRCA2 mutation carriers: the Consortium of Investigators of Modifiers of BRCA1 and BRCA2 (CIMBA). Breast Cancer Res. 2007;9(2):104.
106. Couch FJ, Gaudet MM, Antoniou AC, Ramus SJ, Kuchenbaecker KB, Soucy P, et al. Common variants at the 19p13.1 and ZNF365 loci are associated with ER subtypes of breast cancer and ovarian cancer risk in BRCA1 and BRCA2 mutation carriers. Cancer Epidemiol Biomarkers Prev. 2012;21(4):645–57.
107. Couch FJ, Wang X, McGuffog L, Lee A, Olswold C, Kuchenbaecker KB, et al. Genome-wide association study in BRCA1 mutation carriers identifies novel loci associated with breast and ovarian cancer risk. PLoS Genet. 2013;9(3): e1003212.
108. Ramus SJ, Kartsonaki C, Gayther SA, Pharoah PD, Sinilnikova OM, Beesley J, et al. Genetic variation at 9p22.2 and ovarian cancer risk for BRCA1 and BRCA2 mutation carriers. J Natl Cancer Inst. 2011;103(2):105–16.
109. Ahmed AA, Etemadmoghadam D, Temple J, Lynch AG, Riad M, Sharma R, et al. Driver mutations in TP53 are ubiquitous in high grade serous carcinoma of the ovary. J Pathol. 2010;221 (1):49–56.
110. Cancer Genome Atlas Research N. Integrated genomic analyses of ovarian carcinoma. Nature. 2011;474(7353):609–15.

# 15 Brain Tumors—Epidemiology

Ayal Aizer and Brian Alexander

## 15.1 Introduction

While metastases to the central nervous system (CNS) account for a large proportion of tumors discovered clinically, the epidemiology associated with CNS metastases is closely related to the tumor of origin so this chapter will focus exclusively on primary tumors of the CNS. Primary tumors of the CNS are both common and heterogeneous in terms of clinical behavior. A nontrivial percentage of the general population likely harbor CNS tumors, but many are benign tumors of questionable clinical significance. For example, the Rotterdam Scan Study showed that 1.6 % of the general population harbors a benign brain tumor, with meningioma (0.9 %), pituitary adenoma (0.3 %), and vestibular schwannoma (0.2 %) being the most common [1]. Other studies have shown higher rates of such tumors [2]. Many people with incidentally discovered brain tumors are asymptomatic. Conversely, malignant brain tumors such as glioblastoma are associated with a very poor prognosis [3–5]. Prevalence and basic demographic information, based on data from the United States (U.S.) Surveillance, Epidemiology, and End Results (SEER) program [6], for patients diagnosed with common brain tumors is presented in Table 15.1.

Approximately one-third of primary CNS tumors are located in the meninges, while the supratentorium, ventricle, cerebellum, brainstem, cranial nerves/cauda equina, pituitary gland, and pineal gland account for 23, 1, 3, 2, 7, 3, and 14 % of tumors, respectively [7]. Tumors in other parts of the CNS comprise the remainder [7]. Patient age significantly impacts the distribution of tumors of the CNS. Among patients 0–4 years old, embryonal tumors/medulloblastoma is the most common tumor of the CNS, while between the ages of 5–14 years, 15–34 years, and 35 years and above pilocytic astrocytoma, pituitary adenoma, and meningioma are the most common, respectively [7]. Histopathologic classification of CNS tumors show a diverse range of diagnoses [8], most of which are relatively rare, making epidemiologic associations difficult. This chapter will therefore focus on two of the more common tumors of the CNS, meningioma and glioblastoma.

## 15.2 Meningioma

As noted above, meningioma represents the most common intracranial brain tumor [1, 2]. Meningioma originate from the dura that surround the brain and spinal cord, often within the sites of reflection (falx cerebri and tentorium cerebelli).

A. Aizer · B. Alexander (✉)
Harvard Medical School, Dana-Farber/Brigham and Women's Cancer Center, 450 Brookline Avenue, Boston, MA 02115, USA
e-mail: Brian_Alexander@dfci.harvard.edu


M. Loda et al. (eds.), *Pathology and Epidemiology of Cancer*,
DOI 10.1007/978-3-319-35153-7_15

**Table 15.1** Estimated prevalence and demographic characteristics of primary brain tumors in year 2012, from extrapolation of SEER data

| Tumor | Number | Female (%) | White (%) | Black (%) | Hispanic (%) | Asian (%) | Other (%) |
|---|---|---|---|---|---|---|---|
| Meningioma | 25,946 | 73 | 69 | 12 | 11 | 7 | 0 |
| Pituitary | 11,921 | 55 | 57 | 17 | 18 | 7 | 0 |
| Glioblastoma | 10,854 | 42 | 80 | 6 | 10 | 5 | 0 |
| Nerve sheath | 5932 | 52 | 74 | 6 | 10 | 10 | 1 |
| Astrocytoma | 3089 | 45 | 73 | 7 | 14 | 6 | 0 |
| Other | 2996 | 51 | 67 | 11 | 16 | 6 | 0 |
| Mixed/unclassified glioma | 1936 | 42 | 67 | 10 | 17 | 6 | 0 |
| Hemangioma/hemangioblastoma | 1571 | 53 | 69 | 11 | 12 | 8 | 0 |
| Lymphoma | 1464 | 43 | 63 | 10 | 17 | 10 | 0 |
| Oligodendroglioma | 1064 | 49 | 72 | 3 | 14 | 9 | 1 |
| Ependymoma | 1054 | 50 | 66 | 6 | 22 | 7 | 0 |
| Pilocytic astrocytoma | 1018 | 46 | 63 | 13 | 19 | 4 | 1 |
| Glioneuronal | 814 | 46 | 67 | 12 | 15 | 5 | 0 |
| Medulloblastoma | 664 | 38 | 54 | 9 | 30 | 5 | 1 |
| Craniopharyngioma | 557 | 51 | 49 | 19 | 18 | 13 | 1 |
| Choroid plexus | 168 | 47 | 62 | 9 | 19 | 11 | 0 |
| Pineal | 150 | 57 | 66 | 24 | 7 | 2 | 0 |
| Chordoma/chondrosarcoma | 50 | 57 | 71 | 0 | 29 | 0 | 0 |

The most important prognostic factor in patients with meningioma is grade (see Fig. 15.1). World Health Organization (WHO) grade I meningioma are the most common type (Fig. 15.2) and are associated with very high cure rates following surgical resection. Many patients with presumed grade I tumors discovered incidentally on imaging may need no treatment at all. Historical series

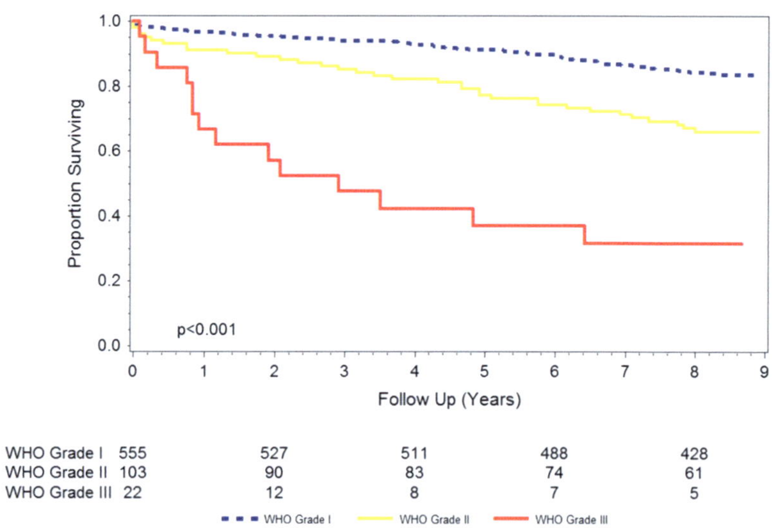


**Fig. 15.1** Overall survival by grade in patients diagnosed with meningioma in 2003. *Source* SEER

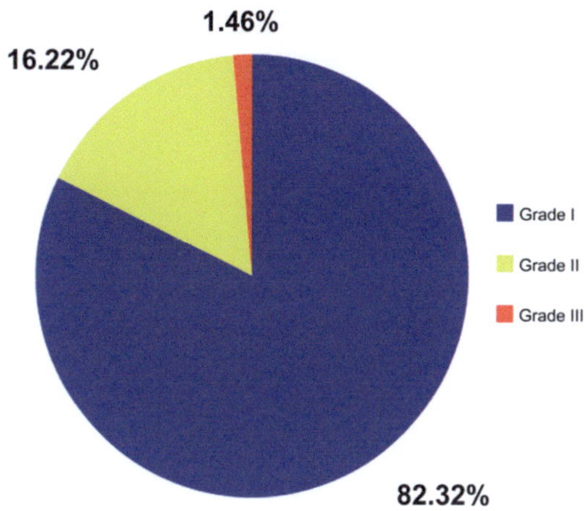


**Fig. 15.2** Distribution of WHO grade among patients with newly diagnosed meningioma in 2012. *Source* SEER

suggest that approximately 7 % of patients presenting with meningioma have WHO grade II (i.e., atypical) lesions [9], while more modern series, which incorporate the 2000 and 2007 WHO pathologic reclassification systems for meningioma, suggest that an even higher proportion may be atypical [10, 11]. WHO grade III (malignant) meningioma represents the rarest of the three entities, accounting for only a few percent of cases. The prognosis for patients with malignant meningioma is relatively poor.

### 15.2.1 Symptoms

Meningioma are often asymptomatic. Many are diagnosed incidentally, typically for imaging obtained for other reasons [1]. When symptomatic, presenting symptoms may be nonspecific (headache, nausea, seizures) or may correlate with the specific region of involvement of the underlying brain. Frontal lobe lesions are associated with mental status changes and weakness, while dominant temporal lesions can cause aphasia. Cerebellar tumors often cause ataxia and gait disturbances. Occipital lesions can impact vision while parietal lesions can cause problems with higher order sensorimotor function. Lesions in the skull base can be associated with cranial nerve palsies, and spinal lesions can cause back pain, numbness, weakness, gait disturbances, and rarely bowel/bladder incontinence. Larger tumors near narrow portions of the ventricular system can cause obstructive hydrocephalus.

### 15.2.2 Diagnosis

Meningioma can be diagnosed via characteristic findings on diagnostic imaging, such as contrast-enhanced computed tomography (CT) or magnetic resonance imaging (MRI). On CT some meningiomas can calcify, and meningioma typically take up contrast. On MRI meningioma are located extra axially, are typically isointense on T1 precontrast imaging, and become hyperintense in a typically homogeneous manner with contrast administration. Contours are typically smooth and well-defined. A dural tail may be present, but dural tails can be seen in other entities as well, such as lymphoma, sarcoidosis, and chloroma.

### 15.2.3 Management

Small, incidentally discovered meningioma can typically be monitored with surveillance imaging, as the growth rate is often 1 mm/year or less, and some lesions may never grow [12, 13]. Treatment for symptomatic lesions, large lesions, or observed lesions which grow significantly in patients with otherwise favorable prognoses entails maximal safe resection as a first step. The impact of extent of resection on survival and recurrence in patients with meningioma has been most notably demonstrated by Simpson for benign meningioma, with emphasis on the negative prognostic influence of trace residual disease in the dural tail or venous sinuses [14]. Apart from grade, extent of resection represents the most important prognostic factor for patients with meningioma [15–17].

Adjuvant management typically depends on the grade and extent of resection. WHO grade I lesions that are gross totally resected should be observed, and observation is reasonable for many subtotally resected lesions as well. Whether to offer radiation therapy to patients with gross totally resected WHO grade II meningioma remains controversial, but several series support a local control benefit with radiation in such settings [18–20]. Generally, patients with subtotally resected WHO grade II meningioma and all patients with WHO grade III meningioma should receive radiation, given the very high rates of recurrence in such populations. Systemic therapy has marginal efficacy and is usually reserved for multiply recurrent lesions. Hormonal agents, chemotherapy, platelet-derived growth factor inhibitors, angiogenesis inhibitors, epidermal growth factor inhibitors, somatostatin analogs, interferon, and other agents have been tested in patients with advanced/recurrent meningiomas, often with disappointing results [21–23].

## 15.3 Epidemiology of Meningioma

### 15.3.1 Prevalence and Incidence

As discussed above, meningioma are very common among the general population, with MRI studies indicating that 1–2 % of the population harbors a meningioma [1, 2]. Vernooij et al. [1] as part of the Rotterdam Scan Study, prospectively performed thin-slice (1.6 mm), 1.5 T MRIs of the brain in approximately 2000 individuals in the general population with a mean age of 63 years. The study found a prevalence of meningioma of 0.9 %; ranging in size from 0.5 to 6.0 cm [1]. Notably, contrast was not given, and scans were initially read by radiology or neurology residents and abnormal findings were reviewed by neuroradiologists. Both schema may have led to underestimation of the true prevalence of meningioma. The incidence of meningioma has been estimated to be 7.8/100,000 per year, although the rate in men (2.9/100,000) is significantly lower than in women (13.0/1000,000) [24]. Many are incidentally found.

### 15.3.2 Mortality

Most meningioma will have minimal clinical impact. Detailed population-based estimates for mortality rates are lacking. WHO grade I meningioma are typically cured with surgery alone, with radiation typically reserved for recurrent disease. Death due to a WHO grade I meningioma is rare. WHO grade II tumors, and especially WHO grade III tumors, carry a more significant mortality risk (Fig. 15.1), warranting aggressive treatment with maximal surgical resection and often adjuvant radiation therapy.

### 15.3.3 Risk Factors

#### 15.3.3.1 Sex

Most meningioma occur in women (Table 15.2). In the Rotterdam Scan Study, the prevalence of meningioma varied by gender (1.1 % in women; 0.7 % in men). Autopsy studies have also shown that meningioma more commonly occur in women, by a factor of up to 3:1 [25]. Among patients diagnosed with a meningioma in 2012, per the SEER database, 73 % were female [6]. Notably, the female preponderance was primarily for WHO I tumors; patients with WHO III tumors were 50 % female. Spinal meningioma may be especially more likely to occur in females as well.

#### 15.3.3.2 Race

Meningioma may be slightly more common in patients of African American race, relative to white race. Rates in Asian Americans may be

**Table 15.2** Gender distribution among patients with WHO grade I, II, and III meningioma, SEER data, 2004–2012

| Meningioma grade | Male (%) | Female (Number, %) |
|---|---|---|
| WHO I | 28 | 72 |
| WHO II | 44 | 56 |
| WHO III | 50 | 50 |

lower. Aggregated data from Central Brain Tumor Registry of the U.S. (CBTRUS)/SEER from 2006 to 2010 reveal an age-adjusted incidence (as stratified by race) of 7.2/100,000 for whites, 8.8/100,000 for African Americans, and 5.1/100,000 for Asian Americans. Rates in Hispanic Americans appear to be similar to whites, with an incidence of 7.3/100,000 [7].

#### 15.3.3.3 Age

The risk of meningioma appears to increase with age. In the Rotterdam Scan Study, the prevalence of meningiomas increased from 0.5 % in those age 45–59 years to 1.6 % in persons 75 years of age or older. Autopsy studies have shown that the risk of meningioma indeed increases with age [25]. Among patients diagnosed with a meningioma in 2012 per the SEER database, 1, 6, 30, 43, and 20 % were age <20, 20–39, 40–59, 60–79, and ≥80 years of age, respectively [6]. There does not appear to be an association between age and grade of meningioma, per the SEER data.

#### 15.3.3.4 Neurofibromatosis Type 2

Neurofibromatosis type 2 is an autosomal dominant condition that predisposes patients to the development of multiple intracranial neoplasms such as vestibular schwannomas (often bilateral), ependymal tumors, and intracranial/spinal meningioma. Patients can often present with cataracts, retinal hamartomas, and cutaneous/subcutaneous tumors. The condition is caused by a mutation in the NF2 gene, which codes the protein merlin, a tumor-suppressor gene. Estimates of prevalence have ranged from 1:25,000 to closer to 1:100,000 [26–28]. When meningioma are present in a patient with neurofibromatosis type 2, they are often multiple. Meningioma in such patients often develop by the patient's early 20s, with increasing prevalence as age increases. Most patients will develop meningioma should they live to late adulthood, and meningioma in this population are more likely to be atypical or malignant [29, 30].

#### 15.3.3.5 Ionizing Radiation

Ionizing radiation represents one of the strongest risk factors for the development of meningioma, particularly when given in childhood. The latency period between receipt of radiation and development of meningioma is often long. Taylor et al. [31] followed approximately 13,000 patients in Great Britain diagnosed with cancer when they were less than 15 years of age, between 1940 and 1991. The most common intracranial tumor that developed was a meningioma. The risk of meningioma increased strongly, linearly, and independently with dose of radiation to meningeal tissue. Those whose meningeal tissue received 0.01–9.99, 10.00–19.99, 20.00–29.99, 30.00–39.99, and ≥40 Gy had risks (of development of meningioma) that were twofold, eightfold, 52-fold, 568-fold, and 479-fold, respectively, as compared to those whose meningeal tissue was unexposed to radiation. Of note, receipt of intrathecal methotrexate was also linked to the development of meningioma in this study [31]. The Childhood Cancer Survivor Study, which followed patients treated for pediatric malignancies, concluded that radiation exposure was significantly associated with the development meningioma (Odds Ratio (OR) 9.94, 95 % Confidence Interval (CI) 2.17–45.6) [32]. Other studies of patients treated with therapeutic radiation involving the CNS have also identified higher than usual rates of development on meningioma in patients, often with a significant latency of years—decades [33].

Patients exposed to radiation for reasons other than cancer also appear to be predisposed to development of meningioma. Follow up of patients treated with radiation as part of the therapeutic regimen for tinea capitis display higher rates of meningioma, with a long latency from the delivery of radiation to the development of meningioma [34]. Atomic bomb survivors have similarly displayed higher rates of development of meningioma, also in a dose dependent manner [35]. Whether the lower doses associated with medical/dental imaging can contribute to the development of meningioma is more controversial. Several studies have examined this issue and there is some evidence for a correlation [36–38]. However, whether more modern imaging techniques, which often use lower doses, are associated with meningioma is more debatable.

Unfortunately, radiation-induced meningioma tend to be more aggressive than sporadic meningioma [39, 40]. The higher incidence of meningioma after radiation to the CNS as well as the potentially poorer prognosis associated with such tumors should promote clinicians to minimize diagnostic and therapeutic radiation exposure to patients.

#### 15.3.3.6 Hormones and Body Mass Index

Meningioma tumors may contain progesterone, estrogen, and androgen receptors. Benson et al. [41] recently conducted a meta-analysis of published studies evaluating the relationship between menopausal hormonal therapy and the development of meningioma. The authors found that estrogen increased the risk of meningioma (Relative Risk (RR) 1.31, 95 % CI 1.20–1.43) relative to patients who did not use estrogen, but combined estrogen–progestin hormonal treatment did not (RR 1.05, 95 % CI 0.95–1.16). Additional evidence linking hormones and development of meningioma relates to higher rates of meningioma in obese patients, given the link between adiposity and hormonal levels [42, 43].

## 15.4 Glioblastoma

Glioblastoma represents one of the most common intraparenchymal tumors of the CNS. The prognosis for patients with glioblastoma is extremely poor [4, 5] and long-term survival is unfortunately uncommon. With aggressive treatment in favorable prognosis patients, 2–10 % of patients may survive five years or longer. Although treatment advances have been made, durable treatment options for patients with glioblastoma remain lacking.

### 15.4.1 Symptoms

Unlike patients with meningioma, most patients with glioblastoma present to medical attention as a result of symptoms. When symptomatic, presenting symptoms may be nonspecific (headache, nausea, mental status changes, and seizures) or may correlate with the specific region of involvement of the underlying brain. In addition, the invasive and sometimes diffuse nature of glioblastoma often results in compromise of performance status and reduced ability to complete activities of daily living.

### 15.4.2 Management

The current management of glioblastoma in non-elderly patients with a good performance status, based on published randomized controlled trials, includes maximal safe resection with adjuvant temozolomide-based chemoradiation followed by additional chemotherapy [44, 45]. In elderly patients or patients with poor performance status, an abbreviated course of radiation without concurrent chemotherapy [46] or chemotherapy alone [47, 48] may be viable options. $O^6$-methylguanine–DNA methyltransferase (MGMT) promoter methylation status may be an important predictor of the response to chemotherapy [49].

## 15.5 Epidemiology of Glioblastoma

### 15.5.1 Incidence

Glioblastoma represents the most common intraparenchymal brain tumor. Age-adjusted incidence of glioblastoma, the most common type of glioma in adults, ranges from 0.59 to 3.69 per 100,000 persons [50]. The SEER data suggests that 11,000 cases were diagnosed in 2012, although this figure may be an underestimate given that some cases of glioblastoma are categorized under other types of glioma (Table 15.1). Approximately 20,000 cases of glioma are diagnosed per year in the U.S. (Table 15.1), accounting for 28 % of all intracranial tumors and 80 % of malignant brain tumors [7].

### 15.5.2 Mortality

The prognosis for patients with glioblastoma is unfortunately very poor [3–5]. A landmark clinical trial of surgery followed by temozolomide-based chemoradiation and then adjuvant temozolomide alone yielded a median survival of 14.6 months, even though all patients in this study were ≤70 years of age and most had a favorable performance status [45]. The prognosis in elderly patients and patients with a poor prognosis is significantly worse [46, 48]. SEER data depicting survival over time among patients diagnosed with a glioblastoma in the year 2007 are presented in Fig. 15.3. The median survival in this cohort was only 6 months. The presence of MGMT promoter methylation appears to be one of the most favorable prognostic factors, and also predicts response to temozolomide-based therapy [49].

### 15.5.3 Risk Factors

#### 15.5.3.1 Sex

Male sex represents a slight risk factor for development of glioblastoma, as indicated by the SEER data (see Table 15.1). As a result, most clinical trials evaluating patients with glioblastoma have also reflected this male predominance [44, 45, 48].

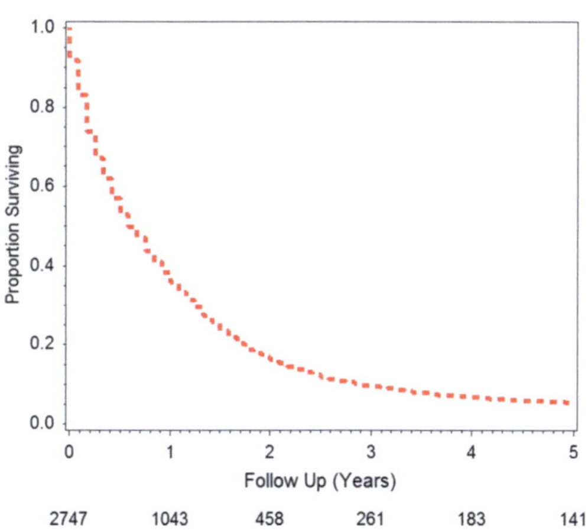


**Fig. 15.3** Overall survival of patients diagnosed with glioblastoma in 2007. *Source* SEER

#### 15.5.3.2 Race

The SEER data suggests that, more than any other type of brain tumor, patients diagnosed with glioblastoma are more likely to be white (see Table 15.1) [51].

#### 15.5.3.3 Age

The risk of glioblastoma increases with age [51]. Most patients are diagnosed after the age of 60 years [48, 52]. Moreover, the incidence of glioblastoma in the elderly appears to be increasing at a faster rate [53, 54], although how much of this increase is accounted for by the increasing use of MRI remains unclear [54].

#### 15.5.3.4 Ionizing Radiation

Ionizing radiation represents one of the most clearly defined risk factors for the development of glioblastoma. The Childhood Cancer Survivor Study, which followed patients treated for pediatric malignancies, concluded that radiation exposure was significantly associated with the development of glioma (OR 6.78, 95 % CI 1.54–29.7), as compared to no radiation exposure [32]. Medical diagnostic imaging may also increase the risk for glioma, although studies have yielded conflicting results [55, 56]. In one study, the relative risk of developing a brain malignancy in patients receiving a mean dose of only 6 cGy of exposure was 2.82, as compared to those receiving no radiation [56].

#### 15.5.3.5 Occupational Exposures

Certain occupations have been linked to the development of gliomas. A Swedish study linking cancer incidence with occupation identified an association with certain occupations and malignant gliomas [57]. These occupations were not statistically significantly linked in any readily apparent way (e.g., public prosecutors and brick and tile workers). Patients whose primary occupation exposes them to pesticides may display higher rates of development of brain tumors [58, 59], although studies have yielded conflicting results [60]. Mixed data exist linking electric workers with the development of brain tumors [61–63].

#### 15.5.3.6 Allergies

Allergic conditions such as asthma, hay fever, eczema, and food allergies seem to decrease the risk of development of malignant gliomas and meningioma. Linos et al. [64] conducted a meta-analysis of eight observational studies, encompassing 3450 patients with glioma. They found that a history of atopy was inversely correlated with the development of glioma, with pooled relative risks for the development of glioma in patients with allergies, asthma, and eczema of 0.61, 0.68, and 0.69, respectively (all highly significant), as compared to those without any history of allergy [64].

#### 15.5.3.7 Genetic Factors

Patients with certain genetic conditions are at increased risk for developing malignant gliomas. These conditions include neurofibromatosis type 1, Turcot Syndrome, Li-Fraumeni syndrome, melanoma-neural system tumor syndrome, and Ollier disease/Maffucci syndrome [50]. However, the above syndromes cannot explain all familial cases of glioma and research to identify novel mutations is ongoing [65].

## 15.6 Other Potential Global Risk Factors for Brain Tumors

### 15.6.1 Cellular Phones and Radiofrequency Exposure

Cellular phones expose users to radiofrequency electromagnetic fields (RF-EMF) and the possible association with brain tumors has been studied extensively. In 2011, the International Agency for Research on Cancer (IARC) summarized their review of the available literature and concluded that RF-EMF was "possibly carcinogenic to humans" (Group 2B) based on limited evidence of associations with glioma and vestibular schwannoma in human and animal studies [66]. IARC acknowledged that there was weak evidence to hypothesize a causative mechanism for RF-EMF. RF-EMF is distinct from ionizing radiation in that there is not free radical and DNA damage resulting from exposure. IARC considered evidence from time-trend studies, one cohort study [67], and five case-control studies [68–72]. The time-trend studies and three of the case-control studies [69, 71, 72], while showing no association between brain tumor risk and RF-EMF, were felt to be generated at a time when exposure was low and latency periods were not long enough to see meaningful evidence. The largest and most well-designed of the remaining two case-control studies was the INTERPHONE study, an interview-based study with 2708 glioma and 2409 meningioma cases and matched controls that was conducted in 13 countries using a common protocol that was coordinated by IARC [68]. Overall, the authors concluded that there was no increased risk of glioma or meningioma observed with the use of mobile phones. There was a suggestion of an association of elevated glioma risk with heavy cell phone users, but the authors suggested this may be related to biases in generating the data. For example, there was also a lower overall risk of glioma among regular users of cell phones in comparison to non-regular users making the data difficult to interpret. Nevertheless, IARC considered the high-exposure data in forming their recommendations along with that of another pooled analysis of two case-controlled studies. Hardell et al. [70] reported a pooled analysis of two Swedish studies and showed that use of a cellular phone was associated with development of malignant brain tumors, though the methods and results have been questioned by some [50]. Since the publication of the IARC report, an update of the large Danish cohort study of 358,403 people continued to show no association of cell phone use and the development of brain tumors [73]. Additionally, the United Kingdom (U.K.) Million Women Study analyzed a prospective cohort of 791,710 women and found no association between glioma and >10 years of cell phone use [74].

Ultimately, given the limited follow up in the above studies and the long latency between radiation exposure and tumorigenesis, longer follow up may be needed in order to more definitely determine whether the radiofrequency exposure attributable to cellular phone use has a causal relationship with the development of brain tumors. For the present, however, there is no conclusive evidence to suggest a link between brain tumor development and RF-EMF.

### 15.6.2 Electromagnetic Radiation

Electromagnetic radiation has not been definitively shown to predispose patients to the development of brain tumors. An early epidemiologic study linked electric wiring configurations to brain tumors in pediatric patients [75]. Another early study of patients with jobs in the electrical and electronics industry also linked exposure with the development of brain tumors [62]. However, other studies have not suggested such an association. Preston-Martin et al. [76] conducted a case-control study involving approximately 600 children with and without primary brain tumors. They quantified the intensity of the magnetic fields by mapping and coding the wiring configurations outside the home and by taking a series of exterior spot and profile measurements, and ultimately found no association linking pediatric brain tumors with the magnitude of these fields. Kheifets et al. [77] evaluated nine studies examining the relationship between electromagnetic fields and brain tumor incidence among pediatric patients and concluded that the data do not support such an association.

### 15.6.3 Diet

*N*-nitroso compounds have been linked to the development of benign and malignant brain tumors and are found in tobacco smoke, cured meats, rubber products, soaps, shampoos, and other products as well. They can also be made endogenously. Cured meat ingestion has been linked with development of malignant glioma and meningioma in several studies [78, 79]. Conversely, fruits, vegetables, and vitamins appear to be associated with lower risk of development of such tumors [80–82]. Tobacco and alcohol have not been as definitively associated with the development of brain tumors, although studies have yielded conflicting results [83–85].

## 15.7 Summary

Tumors of the CNS are both common and heterogeneous in terms of their nature, aggressiveness, clinical impact, and risk factors. Each histology likely possesses a somewhat unique set of predisposing features.

## References

1. Vernooij MW, Ikram MA, Tanghe HL, et al. Incidental findings on brain MRI in the general population. N Engl J Med. 2007;357(18):1821–8.
2. Koppelmans V, Schagen SB, Poels MM, et al. Incidental findings on brain magnetic resonance imaging in long-term survivors of breast cancer treated with adjuvant chemotherapy. Eur J Cancer. 2011;47(17):2531–6.
3. Curran WJ Jr, Scott CB, Horton J, et al. Recursive partitioning analysis of prognostic factors in three Radiation Therapy Oncology Group malignant glioma trials. J Natl Cancer Inst. 1993;85(9):704–10.
4. Gorlia T, van den Bent MJ, Hegi ME, et al. Nomograms for predicting survival of patients with newly diagnosed glioblastoma: prognostic factor analysis of EORTC and NCIC trial 26981–22981/CE.3. Lancet Oncol. 2008;9(1):29–38.
5. Mirimanoff RO, Gorlia T, Mason W, et al. Radiotherapy and temozolomide for newly diagnosed glioblastoma: recursive partitioning analysis of the EORTC 26981/22981-NCIC CE3 phase III randomized trial. J Clin Oncol. 2006;24(16):2563–9.
6. Hollander M, Wolfe DA. Nonparametric statistical methods. 2nd ed. New York, NY: Wiley; 1999. pp 189–269.
7. Ostrom QT, Gittleman H, Liao P, et al. CBTRUS statistical report: primary brain and central nervous system tumors diagnosed in the United States in 2007–2011. Neuro Oncol. 2014;16 Suppl 4:iv1–63.
8. Louis DN, Ohgaki H, Wiestler OD, et al. The 2007 WHO classification of tumours of the central nervous system. Acta Neuropathol. 2007;114(2):97–109.
9. Goyal LK, Suh JH, Mohan DS, Prayson RA, Lee J, Barnett GH. Local control and overall survival in atypical meningioma: a retrospective study. Int J Radiat Oncol Biol Phys. 2000;46(1):57–61.

10. Willis J, Smith C, Ironside JW, Erridge S, Whittle IR, Everington D. The accuracy of meningioma grading: a 10-year retrospective audit. Neuropathol Appl Neurobiol. 2005;31(2):141–9.
11. Combs SE, Schulz-Ertner D, Debus J, von Deimling A, Hartmann C. Improved correlation of the neuropathologic classification according to adapted world health organization classification and outcome after radiotherapy in patients with atypical and anaplastic meningiomas. Int J Radiat Oncol Biol Phys. 2011;81(5):1415–21.
12. Nakamura M, Roser F, Michel J, Jacobs C, Samii M. The natural history of incidental meningiomas. Neurosurgery. 2003;53(1):62–70, discussion 70–61.
13. Yano S, Kuratsu J. Indications for surgery in patients with asymptomatic meningiomas based on an extensive experience. J Neurosurg. 2006;105(4):538–43.
14. Simpson D. The recurrence of intracranial meningiomas after surgical treatment. J Neurol Neurosurg Psychiatry. 1957;20(1):22–39.
15. Durand A, Labrousse F, Jouvet A, et al. WHO grade II and III meningiomas: a study of prognostic factors. J Neurooncol. 2009;95(3):367–75.
16. Palma L, Celli P, Franco C, Cervoni L, Cantore G. Long-term prognosis for atypical and malignant meningiomas: a study of 71 surgical cases. J Neurosurg. 1997;86(5):793–800.
17. Zaher A, Abdelbari Mattar M, Zayed DH, Ellatif RA, Ashamallah SA. Atypical meningioma: a study of prognostic factors. World Neurosurg. 2013;80(5):549–553.
18. Aizer AA, Arvold ND, Catalano P, et al. Adjuvant radiation therapy, local recurrence, and the need for salvage therapy in atypical meningioma. Neuro Oncol. 2014.
19. Aghi MK, Carter BS, Cosgrove GR, et al. Long-term recurrence rates of atypical meningiomas after gross total resection with or without postoperative adjuvant radiation. Neurosurgery. 2009;64(1):56–60; discussion 60.
20. Komotar RJ, Iorgulescu JB, Raper DM, et al. The role of radiotherapy following gross-total resection of atypical meningiomas. J Neurosurg. 2012;117(4):679–86.
21. Wen PY, Quant E, Drappatz J, Beroukhim R, Norden AD. Medical therapies for meningiomas. J Neurooncol. 2010;99(3):365–78.
22. Goodwin JW, Crowley J, Eyre HJ, Stafford B, Jaeckle KA, Townsend JJ. A phase II evaluation of tamoxifen in unresectable or refractory meningiomas: a Southwest Oncology Group study. J Neurooncol. 1993;15(1):75–7.
23. Simo M, Argyriou AA, Macia M, et al. Recurrent high-grade meningioma: a phase II trial with somatostatin analogue therapy. Cancer Chemother Pharmacol. 2014;73(5):919–23.
24. Larjavaara S, Haapasalo H, Sankila R, Helen P, Auvinen A. Is the incidence of meningiomas underestimated? A regional survey. Br J Cancer. 2008;99(1):182–4.
25. Nakasu S, Hirano A, Shimura T, Llena JF. Incidental meningiomas in autopsy study. Surg Neurol. 1987;27(4):319–22.
26. Antinheimo J, Sankila R, Carpen O, Pukkala E, Sainio M, Jaaskelainen J. Population-based analysis of sporadic and type 2 neurofibromatosis-associated meningiomas and schwannomas. Neurology. 2000;54(1):71–6.
27. Evans DG, Moran A, King A, Saeed S, Gurusinghe N, Ramsden R. Incidence of vestibular schwannoma and neurofibromatosis 2 in the North West of England over a 10-year period: higher incidence than previously thought. Otol Neurotol. 2005;26(1):93–7.
28. Evans DG, Howard E, Giblin C, et al. Birth incidence and prevalence of tumor-prone syndromes: estimates from a UK family genetic register service. Am J Med Genet A. 2010;152A(2):327–32.
29. Evans DG, Huson SM, Donnai D, et al. A clinical study of type 2 neurofibromatosis. Q J Med. 1992;84(304):603–18.
30. Perry A, Giannini C, Raghavan R, et al. Aggressive phenotypic and genotypic features in pediatric and NF2-associated meningiomas: a clinicopathologic study of 53 cases. J Neuropathol Exp Neurol. 2001;60(10):994–1003.
31. Taylor AJ, Little MP, Winter DL, et al. Population-based risks of CNS tumors in survivors of childhood cancer: the British Childhood Cancer Survivor Study. J Clin Oncol. 2010;28(36):5287–93.
32. Neglia JP, Robison LL, Stovall M, et al. New primary neoplasms of the central nervous system in survivors of childhood cancer: a report from the Childhood Cancer Survivor Study. J Natl Cancer Inst. 2006;98(21):1528–37.
33. Friedman DL, Whitton J, Leisenring W, et al. Subsequent neoplasms in 5-year survivors of childhood cancer: the Childhood Cancer Survivor Study. J Natl Cancer Inst. 2010;102(14):1083–95.
34. Ron E, Modan B, Boice JD Jr, et al. Tumors of the brain and nervous system after radiotherapy in childhood. N Engl J Med. 1988;319(16):1033–9.
35. Preston DL, Ron E, Yonehara S, et al. Tumors of the nervous system and pituitary gland associated with atomic bomb radiation exposure. J Natl Cancer Inst. 2002;94(20):1555–63.
36. Lin MC, Lee CF, Lin CL, et al. Dental diagnostic X-ray exposure and risk of benign and malignant brain tumors. Ann Oncol. 2013;24(6):1675–9.
37. Claus EB, Calvocoressi L, Bondy ML, Schildkraut JM, Wiemels JL, Wrensch M. Dental X-rays and risk of meningioma. Cancer. 2012;118(18):4530–7.
38. Longstreth WT Jr, Phillips LE, Drangsholt M, et al. Dental X-rays and the risk of intracranial meningioma: a population-based case-control study. Cancer. 2004;100(5):1026–34.
39. Harrison MJ, Wolfe DE, Lau TS, Mitnick RJ, Sachdev VP. Radiation-induced meningiomas: experience at the Mount Sinai Hospital and review of the literature. J Neurosurg. 1991;75(4):564–74.

40. De Tommasi A, Occhiogrosso M, De Tommasi C, Cimmino A, Sanguedolce F, Vailati G. Radiation-induced intracranial meningiomas: review of six operated cases. Neurosurg Rev. 2005;28 (2):104–14.
41. Benson VS, Kirichek O, Beral V, Green J. Menopausal hormone therapy and central nervous system tumor risk: large UK prospective study and meta-analysis. Int J Cancer. 2015;136 (10):2369–77.
42. Johnson DR, Olson JE, Vierkant RA, et al. Risk factors for meningioma in postmenopausal women: results from the Iowa Women's Health Study. Neuro Oncol. 2011;13(9):1011–9.
43. Wiedmann M, Brunborg C, Lindemann K, et al. Body mass index and the risk of meningioma, glioma and schwannoma in a large prospective cohort study (The HUNT Study). Br J Cancer. 2013;109(1):289–94.
44. Keime-Guibert F, Chinot O, Taillandier L, et al. Radiotherapy for glioblastoma in the elderly. N Engl J Med. 2007;356(15):1527–35.
45. Stupp R, Mason WP, van den Bent MJ, et al. Radiotherapy plus concomitant and adjuvant temozolomide for glioblastoma. N Engl J Med. 2005;352 (10):987–96.
46. Roa W, Brasher PM, Bauman G, et al. Abbreviated course of radiation therapy in older patients with glioblastoma multiforme: a prospective randomized clinical trial. J Clin Oncol. 2004;22(9):1583–8.
47. Wick W, Platten M, Meisner C, et al. Temozolomide chemotherapy alone versus radiotherapy alone for malignant astrocytoma in the elderly: the NOA-08 randomised, phase 3 trial. Lancet Oncol. 2012;13 (7):707–15.
48. Malmstrom A, Gronberg BH, Marosi C, et al. Temozolomide versus standard 6-week radiotherapy versus hypofractionated radiotherapy in patients older than 60 years with glioblastoma: the Nordic randomised, phase 3 trial. Lancet Oncol. 2012;13 (9):916–26.
49. Hegi ME, Diserens AC, Gorlia T, et al. MGMT gene silencing and benefit from temozolomide in glioblastoma. N Engl J Med. 2005;352(10):997–1003.
50. Ostrom QT, Bauchet L, Davis FG, et al. The epidemiology of glioma in adults: a "state of the science" review. Neuro Oncol. 2014;16(7):896–913.
51. Dubrow R, Darefsky AS. Demographic variation in incidence of adult glioma by subtype, United States, 1992–2007. BMC Cancer. 2011;11:325.
52. Elia-Pasquet S, Provost D, Jaffre A, et al. Incidence of central nervous system tumors in Gironde, France. Neuroepidemiology. 2004;23(3):110–7.
53. Fleury A, Menegoz F, Grosclaude P, et al. Descriptive epidemiology of cerebral gliomas in France. Cancer. 1997;79(6):1195–202.
54. Chakrabarti I, Cockburn M, Cozen W, Wang YP, Preston-Martin S. A population-based description of glioblastoma multiforme in Los Angeles County, 1974–1999. Cancer. 2005;104(12):2798–806.
55. Davis F, Il'yasova D, Rankin K, McCarthy B, Bigner DD. Medical diagnostic radiation exposures and risk of gliomas. Radiat Res. 2011;175(6):790–6.
56. Pearce MS, Salotti JA, Little MP, et al. Radiation exposure from CT scans in childhood and subsequent risk of leukaemia and brain tumours: a retrospective cohort study. Lancet. 2012;380(9840):499–505.
57. McLaughlin JK, Malker HS, Blot WJ, et al. Occupational risks for intracranial gliomas in Sweden. J Natl Cancer Inst. 1987;78(2):253–7.
58. Musicco M, Sant M, Molinari S, Filippini G, Gatta G, Berrino F. A case-control study of brain gliomas and occupational exposure to chemical carcinogens: the risk to farmers. Am J Epidemiol. 1988;128(4):778–85.
59. Heineman EF, Gao YT, Dosemeci M, McLaughlin JK. Occupational risk factors for brain tumors among women in Shanghai, China. J Occup Environ Med. 1995;37(3):288–93.
60. Bohnen NI, Kurland LT. Brain tumor and exposure to pesticides in humans: a review of the epidemiologic data. J Neurol Sci. 1995;132(2):110–21.
61. Sahl JD, Kelsh MA, Greenland S. Cohort and nested case-control studies of hematopoietic cancers and brain cancer among electric utility workers. Epidemiology. 1993;4(2):104–14.
62. Thomas TL, Stolley PD, Stemhagen A, et al. Brain tumor mortality risk among men with electrical and electronics jobs: a case-control study. J Natl Cancer Inst. 1987;79(2):233–8.
63. Lin RS, Dischinger PC, Conde J, Farrell KP. Occupational exposure to electromagnetic fields and the occurrence of brain tumors. An analysis of possible associations. J Occup Med. 1985;27(6):413–9.
64. Linos E, Raine T, Alonso A, Michaud D. Atopy and risk of brain tumors: a meta-analysis. J Natl Cancer Inst. 2007;99(20):1544–50.
65. Blumenthal DT, Cannon-Albright LA. Familiality in brain tumors. Neurology. 2008;71(13):1015–20.
66. IARC Working Group on the Evaluation of Carcinogenic Risks to Humans, World Health Organization, International Agency for Research on Cancer. Non-ionizing radiation. Part 2, Radiofrequency electromagnetic fields.
67. Schuz J, Jacobsen R, Olsen JH, Boice JD Jr, McLaughlin JK, Johansen C. Cellular telephone use and cancer risk: update of a nationwide Danish cohort. J Natl Cancer Inst. 2006;98(23):1707–13.
68. Brain tumour risk in relation to mobile telephone use: results of the INTERPHONE international case-control study. Int J Epidemiol. 2010;39(3):675–694.
69. Auvinen A, Hietanen M, Luukkonen R, Koskela RS. Brain tumors and salivary gland cancers among cellular telephone users. Epidemiology. 2002;13 (3):356–9.
70. Hardell L, Carlberg M, Soderqvist F, Mild KH. Case-control study of the association between malignant brain tumours diagnosed between 2007 and 2009

and mobile and cordless phone use. Int J Oncol. 2013;43(6):1833–45.
71. Inskip PD, Tarone RE, Hatch EE, et al. Cellular-telephone use and brain tumors. N Engl J Med. 2001;344(2):79–86.
72. Muscat JE, Malkin MG, Thompson S, et al. Handheld cellular telephone use and risk of brain cancer. JAMA. 2000;284(23):3001–7.
73. Frei P, Poulsen AH, Johansen C, Olsen JH, Steding-Jessen M, Schuz J. Use of mobile phones and risk of brain tumours: update of Danish cohort study. BMJ. 2011;343:d6387.
74. Benson VS, Pirie K, Schuz J, Reeves GK, Beral V, Green J. Mobile phone use and risk of brain neoplasms and other cancers: prospective study. Int J Epidemiol. 2013;42(3):792–802.
75. Wertheimer N, Leeper E. Electrical wiring configurations and childhood cancer. Am J Epidemiol. 1979;109(3):273–84.
76. Preston-Martin S, Navidi W, Thomas D, Lee PJ, Bowman J, Pogoda J. Los Angeles study of residential magnetic fields and childhood brain tumors. Am J Epidemiol. 1996;143(2):105–19.
77. Kheifets LI, Sussman SS, Preston-Martin S. Childhood brain tumors and residential electromagnetic fields (EMF). Rev Environ Contam Toxicol. 1999;159:111–29.
78. Chen H, Ward MH, Tucker KL, et al. Diet and risk of adult glioma in Eastern Nebraska, United States. Cancer Causes Control. 2002;13(7):647–55.
79. Preston-Martin S, Mack W. Gliomas and meningiomas in men in Los Angeles county: investigation of exposures to *N*-nitroso compounds. IARC Sci Publ. 1991;105:197–203.
80. Preston-Martin S, Pogoda JM, Mueller BA, et al. Prenatal vitamin supplementation and risk of childhood brain tumors. Int J Cancer Suppl. 1998;11:17–22.
81. Bunin GR, Kuijten RR, Buckley JD, Rorke LB, Meadows AT. Relation between maternal diet and subsequent primitive neuroectodermal brain tumors in young children. N Engl J Med. 1993;329(8):536–41.
82. Lee M, Wrensch M, Miike R. Dietary and tobacco risk factors for adult onset glioma in the San Francisco Bay Area (California, USA). Cancer Causes Control. 1997;8(1):13–24.
83. Silvera SA, Miller AB, Rohan TE. Cigarette smoking and risk of glioma: a prospective cohort study. Int J Cancer. 2006;118(7):1848–51.
84. Hurley SF, McNeil JJ, Donnan GA, Forbes A, Salzberg M, Giles GG. Tobacco smoking and alcohol consumption as risk factors for glioma: a case-control study in Melbourne, Australia. J Epidemiol Community Health. 1996;50(4):442–6.
85. Holick CN, Giovannucci EL, Rosner B, Stampfer MJ, Michaud DS. Prospective study of cigarette smoking and adult glioma: dosage, duration, and latency. Neuro Oncol. 2007;9(3):326–34.

# Pathology and Molecular Pathology of Brain Cancer

# 16

Keith L. Ligon, Kim Wilkinson and Charles D. Stiles

## 16.1 Introduction

In the accompanying chapter, the clinical picture, risk factors, and epidemiology of several central nervous system (CNS) cancers were discussed with a focus on high-grade gliomas in adults. The salient point of these preceding sections is that high-grade gliomas are infrequent but lethal. Anaplastic astrocytomas or malignant gliomas (WHO grades III and IV) account for less than 1.5 % of all new cancers reported in the United States (US) each year [1]. However, these cancers can never be completely resected by surgery and they are notoriously resistant to radiotherapy and genotoxic drugs. For these reasons, astrocytic tumors are the third leading cause of cancer-related deaths among middle-aged men and the fourth leading cause of death for women between 15 and 34 years of age [2]. For children, the impact of astrocytomas is likewise severe. Primary cancers of the brain have recently surpassed leukemias as the number one cause of cancer-related deaths in children and astrocytomas are the most common type of brain cancer in children. Pediatric astrocytomas tend to be lower in grade and less aggressive than their adult counterparts. Unlike high-grade gliomas in adults, some pediatric astrocytomas can be cured by surgery and others respond well to radiation or chemotherapy. However, the clinical sequelae of surgery, radiation, and genotoxic drugs on growing children can be significant. In this chapter, we introduce readers to contemporary concepts in the biology and pathobiology of these tumors with special emphasis on the interface between brain development and cancer.


K.L. Ligon
Harvard Medical School, 450 Brookline Avenue, Boston, MA 02115, USA

K. Wilkinson
Dana-Farber Cancer Institute, 450 Brookline Avenue, Boston, MA 02215, USA

C.D. Stiles (✉)
Department of Neurobiology, Harvard Medical School, Boston, MA, USA
e-mail: Charles_stiles@dfci.harvard.edu

C.D. Stiles
Department of Cancer Biology, Dana-Farber Cancer Institute, Boston, MA, USA


## 16.2 Embryology and Embryonic Stem Cells

As introduced in Chap. 3, the skin and nervous system are derived from the ectoderm layer of the developing embryo. In the third week of development, the neural plate, the anlage of the entire nervous system, forms along the dorsal aspect of the embryo. This neural plate folds to form a central groove, eventually forming the neural tube, which later differentially proliferates in segmental order to give rise to the brain and spinal cord. In contrast, the membranes that eventually surround the brain and spinal cord (i.e., the meninges) arise from both the mesoderm layer surrounding the neural tube (dura),


M. Loda et al. (eds.), *Pathology and Epidemiology of Cancer*,
DOI 10.1007/978-3-319-35153-7_16

and from the neural crest cells which delaminate to give rise to the leptomeninges and other structures (pia and arachnoid).

Analogies have been drawn between the development of blood (i.e., hematopoiesis) and development of the nervous system (i.e., neuropoiesis) [3]. In both tissues, a variety of lineage committed neural and hematopoietic stem cells originate from undifferentiated embryonic stem cells (ESC) during formation of the embryo and the neural plate. These lineage-restricted stem cells have the capacity to self renew but also to produce progeny cells with an even more restricted developmental potential [3]. In the embryonic brain, the neural stem cells reside in the ventricular zone region lining the entire ventricle of the developing brain and give rise to all constituents of the newborn brain including neurons, astrocytes, and oligodendrocytes.

At one time, it was believed that analogies between formation of blood and formation of brain broke down at birth. Formation of new blood cells continues throughout life whereas the brain was believed to be an organic "fait accompli" at birth or shortly after birth. However, that distinction between blood and brain was put to rest with the discovery of replication-competent neural stem cells in the postnatal brain in the early 1990s [4]. It is now widely conceded that in rodents, and perhaps to a lesser degree in humans, new olfactory neurons are produced continually from neural progenitor cells in the subventricular zone of the postnatal brain [5, 6]. Also in both mouse and man, neuronal turnover with birth of new neurons is observed locally within the adult dentate gyrus (for reviews see [7–9]). Additionally, oligodendrocyte progenitor cells, a.k.a. NG2 positive glia, are broadly distributed in the adult brain where they constitute the most abundant mitotically active cell type [10].

## 16.3 Anatomy and Histology of the Brain

Not surprisingly given its function, the brain exhibits greater diversity in its anatomical structures than any other organ in the human body. The brain is generally divided into several structures of relevance for categorization of the tumors: the dura and meningeal coverings; the frontal, temporal, parietal, and occipital lobes; the cerebellum; the brainstem; and the spinal cord and peripheral nerves.

The dura and other meninges are the outermost surface of the brain. They are not the technical part of the brain but instead cover the brain and are derived from the mesodermal germ layer with only a minor component derived from the neural crest cells of the neuroectoderm. As such, they contain distinctly different cell types than those seen within the brain parenchyma. The dura mater is the outermost layer and is a strong protective connective tissue covering composed of numerous fibroblasts and blood vessels in a highly structured and dense arrangement. Intimately associated with these elements are the meningothelial cells which line the inner surface of the dura and are the presumed origin of a majority of meningioma tumor types. The next layers are the arachnoid and pia that together are commonly referred to as the leptomeninges (Fig. 16.1a). These thin membranes cover nearly the entire surface of the brain and spinal cord. Currently, it is not formally known as to what degree they give rise to meningeal or other tumors but may serve to produce some of the more rare types compared to those associated with the dura.

The brain parenchyma proper is segmented into lobes which each can be affected by all major primary brain tumor types. However, the frequency of each tumor type can vary significantly based on location and the relative abundance of particular cell lineages. Each lobe contains a similar overall architecture with an outermost gray matter layer that consists of a cortical ribbon containing a large number of neurons relative to glia. Below this layer is the white matter that consists of a vast number of glia including myelinating oligodendrocytes and astrocytes (Fig. 16.1b and c, respectively). Throughout the brain, the NG2-cells represent the largest mitotically active cell population within the brain, and are OLIG2-expressing oligodendroglial progenitors [11]. The white matter serves as a "superhighway" of axons but

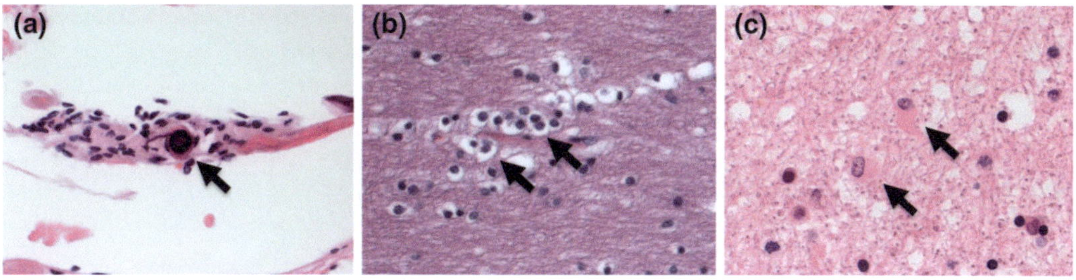


**Fig. 16.1** Histological features of normal cell types within the brain. **a** Arachnoidal collection of meningeal cells and psammomatous calcification similar to those seen in meningiomas, **b** oligodendrocytes within white matter with perinuclear halos analogous to those seen in oligodendrogliomas, **c** reactive astrocytes with eosinophilic cytoplasm similar to that in astrocytomas

very few neuronal bodies and is more abundant than the gray matter overall. The different incidence of cell types and different progenitor types within each region of the brain is thought to be the basis for differences where each brain tumor most commonly arises. Oligodendrogliomas arise most frequently in regions of the frontal lobes where white matter is most abundant. Astrocytomas or glioblastomas are felt to arise more frequently from deeper locations around the ventricles presumably closer to where stem cells of the brain reside. Such patterns have recently also been supported based on imaging studies of brain tumor growth patterns [12].

The brainstem and cerebellum are some of the most primitive and basic functional regions of the brain, and together reside in the area of the cranium termed the posterior fossa. While adult primary brain tumors rarely arise in either of these regions, these structures are one of the most frequent sites of primary brain tumors within children. This is possibly due to the normally late stem and progenitor cell development that occurs in these structures [13]. The spinal cord extends for most of the length of the spinal canal and contains similar anatomic structures and cell types of neurons, astrocytes, and oligodendrocytes as is seen in other regions of the brain. The spinal cord is an infrequent site for tumors in adults and children although several specialized types of tumors are known to occur there, some almost exclusively (e.g., myxopapillary ependymoma).

## 16.4 General Neuropathology Principles

### 16.4.1 Brain Tumor Classification

Brain tumors were first classified based on the resemblance of their microscopic cell morphology to normal mature cell types and structures within the brain (Fig. 16.1). The first widely employed classification system was established in this manner by Bailey and Cushing who were particularly struck by the similarities of brain tumor cells to normal differentiated cells and undifferentiated progenitor cells of the developing brain [14]. The names most commonly used today still derive from their conventions with the "blastoma" terminology a reminder of the original developmental perspective. This strategy for classification was a useful one for many years, and as the development of the brain and its tumors has become better understood it is even more clear that the tumors resemble earlier stages of developing progenitor or neural stem cells more than they have analogy to differentiated cell types [15].

Histomorphologic assessment remains the most widely used biomarker classification of brain tumors in clinical practice. This is one of the most important functions performed by pathologists in their care of patients with important distinctions made by the initial assignment of tumors as gliomas versus other tumor types. This assessment is based on microscopic examination of hematoxylin and eosin (H&E) staining which is the standard approach to the pathologic evaluation of most cancers. Tumors are first assessed for the class of the tumor that is commonly referred to as the diagnosis (Fig. 16.2). Some of the most common diagnoses or classes of primary brain tumors are gliomas (glioblastoma, astrocytoma) or meningiomas but the range of diagnostic categories is extremely broad and there are currently more than 100 distinct brain tumor types recognized by the WHO classification schema [16]. Many of these are rare tumors with distinct appearances and biology associated with them, particularly those that occur in children.

### 16.4.2 Brain Tumor Grading

Once a class of tumor lineage has been specified, then the associated grading criteria are applied to the tumor. As such the grade can only be accurately assigned once a tumor class or classes have been decided upon (Fig. 16.2). Grading criteria are most commonly based on the degree of cellularity, atypia, nuclear size relative to the cytoplasm, and degree of proliferative or mitotic activity. The higher the histologic grade, then the more aggressive the tumor is generally and with a less favorable prognosis implied with respect to recurrence risk or survival metrics. The WHO grading assignments are from Grade I to Grade IV with the higher the number the more aggressive the tumor appears. Tumors sometimes contain more than one grade type, and in these instances, the highest grade determines the overall grade of the tumor, even if most of the tumor is made up of low-grade cells. While some general categories of tumors are felt to progress with recurrence or time to a higher histologic grade as part of their natural progression, (e.g., Grade II diffuse astrocytoma progressing to a Grade III anaplastic astrocytoma) most tumors do not have more than one associated grade or show progression. The clinical course of each grade can be varied but generally WHO Grade I brain tumors are slow growing and often curable by surgical resection alone if completely excised, while Grade IV lesions grow rapidly and are fatal diseases if not treated.

### 16.4.3 Brain Tumor Staging

One of the more unique aspects of brain and spinal cord tumors is that they rarely spread or metastasize to other parts of the body. This is hypothesized to be due in part to the blood–brain barrier or other specialized aspects of the brain vasculature. However, circulating GBM and other

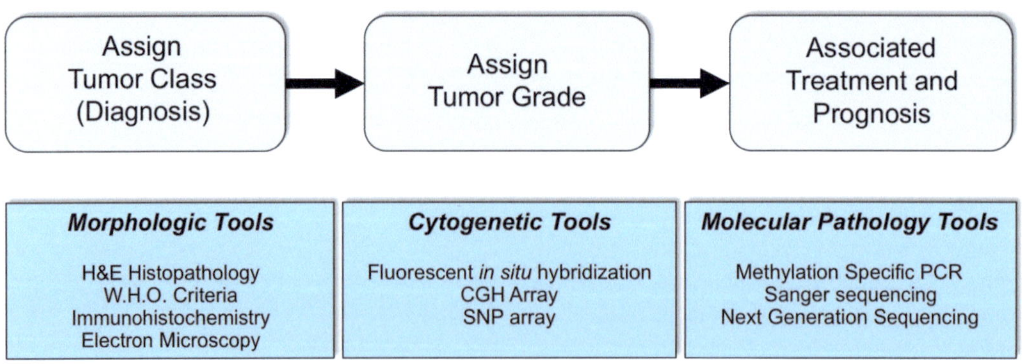


**Fig. 16.2** Diagnostic approach and tools used in neuropathology practice. *Gray bubbles* indicate the workflow of assigning brain tumor class and grade by morphologic assessment. Such assignments lead to information on treatment and prognosis for individual patients. In *blue* are the clinical tools commonly utilized in modern pathology practice to aid in classification and prognostic assignment of patients and their tumors

tumor cells have recently been described in peripheral blood, albeit at much lower levels than seen in other cancers. As a result of their lack of spread, there is no formal staging system that is used in clinical practice. However, CNS tumors are capable of spreading within the central nervous system and affecting other parts of the brain and spinal cord. This is most often felt to occur through sloughing of cells into the cerebrospinal fluid and "seeding" most typically occurs in the spinal region due to gravity and positional factors. This form of staging is usually assessed by examination of cerebrospinal fluid or on radiologic imaging. Another important aspect of evaluation of brain tumors is imaging taken after surgery to assess the extent of resection of a tumor. While the most common tumors of the CNS cannot be fully removed due to their diffuse infiltration of normal structures (e.g., diffuse astrocytoma, glioblastoma) the extent of resection of the tumor and amount left behind can affect the patient's outcome significantly in certain tumor types. Treatment planning is also commonly affected by whether a tumor is completely resected (gross total resection) or subtotally resected.

## 16.5 Pathology of Brain Tumors

Given the breadth of brain tumor types, we will only cover the most common tumors and grades encountered in practice for meningioma, oligodendrogliomas, astrocytomas, and glioblastomas, each described separately below.

### 16.5.1 Meningioma

Meningiomas are one of the most common intracranial tumors and represent approximately 30 % of brain tumors in the US and occur almost exclusively in adults [16]. Generally, such tumors are slow growing and benign and may be discovered incidentally on MRI or at autopsy. These tumors can arise in any location where there are meningeal coverings to the brain. Occasional tumors may arise in the choroid plexus or spinal nerve root location causing some diagnostic confusion if not carefully assessed. The most common sites are the convexity of the brain (falx, parasagittal regions) and skull base regions (olfactory region, sphenoid, suprasellar).

Meningiomas are grossly well circumscribed and tend to push aside the brain parenchyma without invading or destroying it. They are most commonly attached to the inner surface of the dura and sometimes erode the overlying bone of the skull. The tumors rarely spread within the intracranial compartment or to other sites in the body. They typically occur singly but when present as multiple lesions these can occur sporadically or in the setting of neurofibromatosis type 2 (NF2) syndrome.

Currently, the WHO recognizes 15 distinct histological subclasses of meningioma with the most common being WHO Grade I tumors such as meningothelial, fibrous, or transitional classes (Fig. 16.3a). Each have a strong morphologic resemblance to the normal arachnoidal meningothelial cells seen within the dura and leptomeninges (Fig. 16.1a). Other WHO Grade I subtypes include psammomatous, angiomatous, microcystic, secretory, lymphoplasmacyte-rich, and metaplastic meningiomas. Grade II tumors have increased proliferative rate, cellularity, atypia, or nuclear cytoplasmic ratio, or loss of a normal morphologic pattern. Several specific subclasses are also automatically categorized as Grade II due to an increased recurrence risk including chordoid and clear cell variants; tumors with further evidence of mitotic activity, necrosis, and near complete loss of meningiomatous differentiation. Specific variants, namely the anaplastic, papillary, and rhabdoid meningiomas are also assessed WHO Grade III due to their strong recurrence risk. The presence of brain invasion is considered an independent recurrence risk sign.

Meningiomas are often recognized by their H&E morphology, however, in cases where the diagnosis is in question this can be aided by use of immunohistochemistry (IHC). The most common positive identification marker expressed in meningiomas is epithelial membrane antigen (EMA) [17]. The expression of this marker, however, is often weak and diffuse and therefore

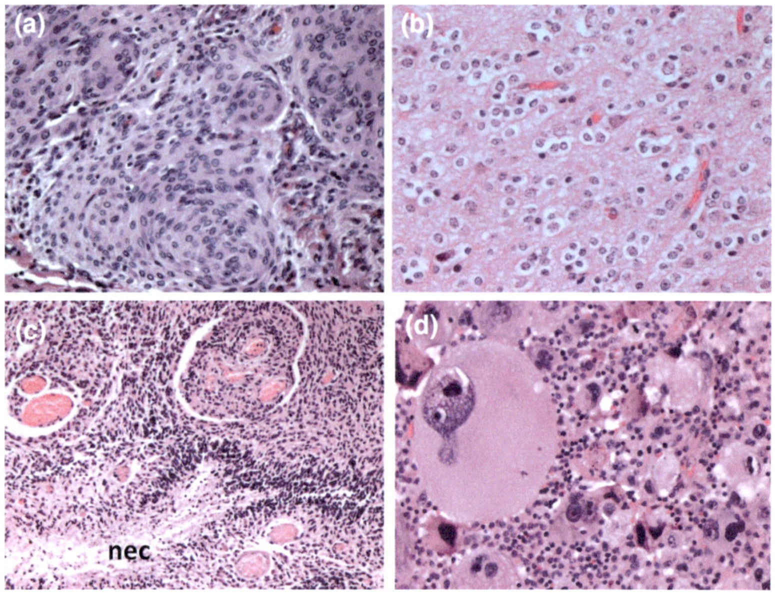


**Fig. 16.3** Histological features of common brain tumors **a** meningioma with typical whorls, **b** oligodendroglioma with diffuse infiltrating cells surrounded by perinuclear halos, **c** glioblastoma with pseudopalisading necrosis (nec) and dense atypical tumor cells, **d** glioblastoma with markedly atypical giant cells

not always reliable for identification. This is particularly true in that higher grade meningiomas lose the expression of this marker. Given their origin from non-neuroepithelial cell types, meningiomas lack expression of primary brain tumor markers such as the glial markers OLIG2 or GFAP which also may be helpful in their distinction from these entities. Electron microscopy is occasionally used to identify key features of the mesenchymal phenotype when other markers may have failed. Future studies to identify specific markers of the lineage for meningeal cells in addition to EMA are needed and would aid in the diagnostic evaluation of these tumors. Further study of the complex developmental origins of the normal meningeal cells and the different cell types in the brain's coverings will likely lead to such discoveries as they have in other tumors.

## 16.5.2 Oligodendroglioma

Oligodendrogliomas are rare diffusely infiltrating tumors of adults that have a very distinctive histological appearance on H&E staining and a similar appearance to normal differentiated oligodendrocytes of the brain. The tumor cells are highly uniform with rounded nuclei and a monomorphic or "clonal" appearance overall. Their most distinctive feature is the presence of clear spaces in formalin-fixed paraffin-embedded (FFPE) sections around each cell which is described as a perinuclear halo or the "fried egg" appearance (Fig. 16.3b). The nucleus in oligodendroglioma cells is very centrally located within the halo and is present in abundance throughout the tumor. This should be distinguished from the common appearance of halos seen around numerous cells in the normal reactive brain or other types of tumors where the nucleus is often not centrally located and the phenotype is seen only focally and not all throughout the tumor. Oligodendrogliomas are frequently associated with microcalcifications and the tumor cells can cluster around vessels and neurons in a visually striking "satellite" arrangement.

IHC, while not officially part of the current WHO diagnostic criteria for oligodendrogliomas, is extremely helpful in identifying these tumors. All oligodendrogiomas are positive for the oligodendroglial lineage marker OLIG2 as well as IDH1(R132H) (see molecular profile below) [18]. The tumors should lack expression of TP53 in most cases. Each of these markers typically has a homogeneous expression pattern with the marker being detected in essentially all of the tumor cells.

Grading of oligodendrogliomas follows WHO guidelines using cellularity, anaplasia, mitoses, vascular proliferation, and necrosis as the main criteria for assignment. Most oligodendroglioma WHO Grade II have few to no mitoses, no more than moderate cell density and generally very minimal atypia. Tumors are categorized as anaplastic oligodendroglioma WHO Grade III if brisk mitotic activity, necrosis, or vascular proliferation is present. The criteria for this are somewhat controversial and distinction from higher grade glioblastoma can sometimes be a challenge as not all of these features are generally present in high-grade tumors. Specific stipulations as to how many should be present are not yet established. No grading-specific markers have been identified for oligodendrogliomas but could be useful for future evaluation of these tumors.

### 16.5.3 Astrocytoma

Astrocytomas are diffuse gliomas that are common in children and young adults. They were originally named based on the tumor cell resemblance to non-tumorigenic reactive astrocytes in the brain. However, more recent studies have shown that astrocytomas invariably contain a mixture of both astrocytic and oligodendroglial lineage cell types with varying degrees of abundance of each. The WHO entity termed "Oligoastrocytoma" or mixed glioma, was created to further distinguish tumors which had more obvious oligodendroglial features [16]. However, oligoastrocytoma is now considered to lie within the spectrum of morphology seen in the category of astrocytomas, and in practice is not reliably able to be distinguished as a separate entity morphologically, biologically, or genetically. As a result, astrocytoma and oligoastrocytoma are expected to be merged into one category, defined by the presence of IDH1 mutations and absence of 1p/19q codeletion, in order to achieve a category with highly uniform genetics, biology, and clinical prognosis. These more rigorous requirements are the generally stated goal for all class distinctions in the next WHO consensus [19].

Overall, astrocytomas are easily recognized as distinct from oligodendrogliomas by their more pleomorphic and irregular nuclear cytology with mixtures of rounded and elongate cell shapes, the frequent presence of eosinophilic cytoplasm, and lack of significant perinuclear halos in tumor cells. The tumors are sometimes difficult to recognize in their lower grade forms and particularly in children [20]. The background neuropil is frequently intact due to their infiltrative nature although several variants and grades exist with varying alteration in the background.

Grading of astrocytomas is based on mitotic rate and the presence of vascular proliferation or necrosis. Diffuse astrocytomas WHO Grade II are the lowest form of this particular category and generally are slow growing with usually no mitoses identified in the tumor, a low cell density and pleomorphism. The presence of multiple mitoses designates a tumor in this class as WHO Grade III and is considered "high grade" for clinical treatment purposes. The presence of vascular proliferation or necrosis prior to treatment of the tumor promotes a tumor to WHO Grade IV and is then referred to as glioblastoma by current criteria.

### 16.5.4 Glioblastoma

The histology of the majority of glioblastomas (GBMs) is quite distinct from the lower grade astrocytomas (Grade II and Grade III) and this likely reflects their unique biology and general absence of mutations in IDH1/2 genes [21]. The typical case is de novo in origin with no prior precursor lesion in the patient (e.g., primary GBM). Secondary GBM can arise from a lower grade IDH1 mutant astrocytoma but it is estimated that fewer than 10 % of GBMs may arise from such progression. The histology of GBM is diverse. This was even indicated by their originally assigned name "glioblastoma multiforme." Although the diversity is a key feature, they do in fact share several histological and clinical features which make them morphologically distinguishable from oligodendrogliomas and other CNS tumors.

By WHO 2007 classification, GBMs all fall under the larger category of diffuse astrocytomas and are the highest most aggressive Grade IV within this category. While originally thought to represent a "pure" astrocytic tumor and potentially even derived from astrocytes, recent studies have shown that all astrocytomas are somewhat misnamed. Like their lower grade astrocytoma counterparts, GBMs are all of a mixed astrocytic and oligodendroglial lineage as defined by multiple markers for these cell types (e.g., GFAP, OLIG2). They are likely to originate from several normally dividing populations of stem or progenitor cell types within the normal brain [18]. Like other diffuse gliomas, GBMs exhibit striking diffuse infiltration and intermingling with the preexisting brain structures making them incurable by surgical resection.

GBMs by definition have marked anaplasia, mitotic activity, necrosis, and microvascular proliferation which are the essential features used to distinguish GBM from lower grade astrocytomas (Fig. 16.3c and d). These features automatically allow grading as WHO Grade IV and this is the maximum grade allowed by the WHO. However, currently WHO guidelines recognize three histological subclasses: classic GBM, gliosarcoma, and giant cell GBM. Additional subclasses have been described in the literature and include GBM with oligodendroglioma component, small cell GBM, and epithelioid/adenoid GBM but have not yet been accepted as distinct or able to be reliably distinguished as clinico-pathologic or molecular entities. Although these subclasses are commonly referred to in practice, as yet none are associated with significant prognostic value.

## 16.6 Molecular Pathology of Brain Tumors

### 16.6.1 Diagnostic Tools

There has been an exciting explosion in the number and type of assays available to pathologists for improved molecular and cytogenetic analysis of cancer in the past decade and many of these are now routinely used in brain tumor diagnostics to provide prognostic and treatment information [22]. While all the methods were first pioneered in the research setting, the advancement to clinical use requires careful consideration of factors such as accuracy, cost, and turnaround time, which, in the clinical realm can often be quite different than that required in the research laboratory setting. Assays to determine copy number, single nucleotide variants, and epigenetic assessments are all commonly used in the clinical setting and described below.

#### 16.6.1.1 RNA and Protein Expression

RNA is a labile biomarker which has seen rather limited adoption in the clinical realm. RNA in situ hybridization is utilized in a research setting but no brain tumor biomarkers of clinical relevance have been discovered. While RNA expression profiling has been useful in the research setting, it has not seen rapid adoption with modern techniques commonly used in the clinic and most clinical assessments are performed to detect protein expression via IHC methods (Fig. 16.4a and b) [23, 24].

#### 16.6.1.2 Cytogenetics

Cytogenetic techniques were some of the first methods employed to clinically distinguish gains and losses of DNA and specific regions of the genome or genes that were altered. Karyotypes from acutely cultured cells were used to determine associated gains or losses of chromosomes (e.g., gain of Chr 7 and loss of Chr 10 in GBM) in several brain tumors, however the application of this in practice was fairly limited compared to other cancers due to its fairly limited clinical value and is no longer commonly used in the diagnostic setting.

Fluorescence in situ hybridization (FISH), introduced in Chap. 2, is an extremely useful technique for hybridization and counting of copies of specific probes on tissue sections at targeted regions of the genome. This analysis has the advantage of being analyzed at a single cell level and, therefore, is extremely sensitive and specific. It is one of the most commonly used methods today to assess brain tumor copy aberrations, but given its targeted nature of analysis

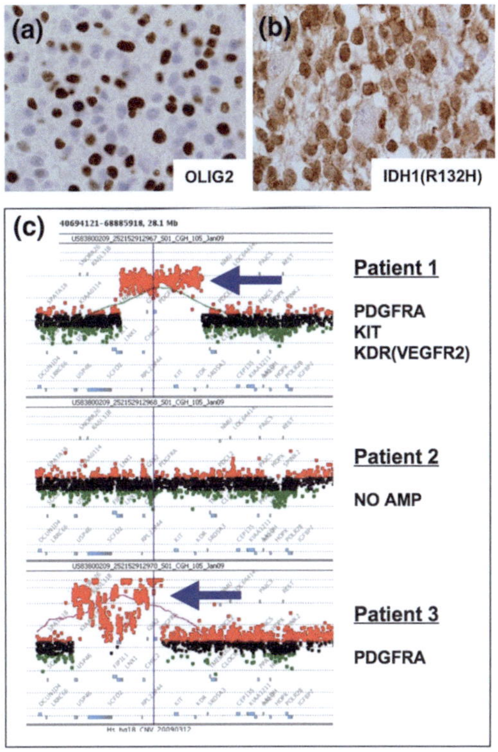


**Fig. 16.4** Immunohistochemical and molecular methods for assessing brain tumors. **a** Immunohistochemistry for OLIG2 is positive (*brown*) in a subpopulation of astrocytoma cells. **b** IDH1(R132H) mutation (*brown*) in an astrocytoma demonstrates strong uniform expression by IHC, **c** array CGH results showing PDGFRA locus in patients with glioblastoma. Patient 1 shows amplification involving PDGFRA, KIT, and KDR, patient 2 shows no aberration, and patient 3 shows amplification involving only the PDGFRA gene with a complex pattern. *Red probes* = gains, *green probes* = losses, *black* = neutral; *blue arrows* indicate amplification events

of only single regions of the genome, it is being replaced by more multiplexed or whole genome approaches.

Hybridization based copy number arrays (i.e., array CGH, SNP arrays) are highly sophisticated tools for multiplex assessment of the number of copies of genes or loci in the genome. These technologies now allow pathologists to simultaneously query up to one million different locations with good sensitivity and specificity in clinical FFPE samples (Fig. 16.4c) [25]. Next generation sequencing panels are also now used to assess copy number with good results for large (arm level) copy changes in cancers but has more limited ability to reliably detect smaller size and low level changes in copy number [26]. As example, there is often poor sensitivity for small single copy gains and deletions due to the cost of reaching broad coverage as is typically achieved in array based assays. New analytical methods and lowered sequencing costs are expected to improve this situation in the future.

#### 16.6.1.3 Sequencing

Detection of single nucleotide variants (e.g., point mutations) was originally performed using Sanger sequencing and other targeted methods and has been most commonly used to identify mutations in IDH genes [27]. Targeted sequencing panels based on PCR amplification of multiple genes of interest are also commonly used in current practice [28]. However, the advent of next generation multiplex sequencing methods has rapidly advanced the ability to detect multiple mutations in cancer simultaneously and has been rapidly adopted for diagnostics and prognostic use in brain tumors [26, 29].

Most sequencing is currently focused on specific known oncogenes and tumor suppressors either at "hotspots" or at all exonic sequences of these genes, in order to efficiently detect mutations of interest in FFPE samples. The use of the targeted exome sequencing approach is particularly useful for brain tumors where tumor suppressor genes are frequently altered and lack consistent "hotspots" for effective evaluation. Current focus on next generation sequencing methods are on reduced turnaround time to allow more rapid use in clinical practice attempting to balance increased gene coverage with time required for analysis. The next generation sequencing techniques also allow for assessment of gene insertions or deletions (indels) and gene fusion events, however, analytical tools for such events are currently not reliable or widely available for clinical practice. These are expected to evolve rapidly to enable such events to be detected with clinical standards of sensitivity and specificity in the near future.

#### 16.6.1.4 Epigenetics

Brain cancers were one of the first indications where assessment of epigenetic modifications of the genome were of clinical relevance, specifically with respect to methylation status of the methylguanine methyltransferase (MGMT) gene in gliomas [30, 31]. Methylation events are characteristically assessed clinically using methylation specific PCR strategy with bisulfite conversion of methylated cytosine to uracil. The converted DNA is then sequenced using Sanger or pyrosequencing methods to infer which residues in the target regions of interest were methylated. As for copy number and single nucleotide variant detection methods, the use of multiplexed methylation arrays capable of whole genome methylation assessments has recently been entering into clinical use. Given the association of global methylation patterns with cell lineage, methylation arrays are most commonly used currently for recognition of diagnostic patterns of methylation in specific tumor types [32, 33]. Also such methods can reliably report out copy number information at the same time, in a similar manner to dedicated copy arrays. Interestingly, use of methylation arrays has not replaced targeted approaches for MGMT promoter assessment, although with further improvements and experience this is expected to be possible in the future.

### 16.6.2 Cytogenomic Features of Brain Tumor Types

#### 16.6.2.1 Meningiomas

The most common molecular events in meningiomas are loss of the tumor suppressor gene NF2 on chromosome 22q, which occurs in approximately 50 % of sporadic and nearly 100 % of syndromic tumors. Atypical WHO Grade II meningiomas are associated with losses of 1p and 14q most commonly. The overall burden of copy number changes appears to increase with histologic grade and recurrence risk in atypical meningioma and following gross total resection could be predicted using array CGH or other multiplex copy number assays to determine the overall genomic copy number changes within the tumor [34].

Recent sequencing studies of meningioma have identified strong clinico-pathologic correlation of mutational genotypes with histologic subtypes of meningiomas likely due to cell of origin differences. Oncogenic mutations in AKT1 and SMO were typically associated with an anterior/medial skull base location [35]. Mutations in TRAF7, a pro-apoptotic E3 ubiquitin ligase, have also been described in approximately 25 % of meningiomas as a common recurrent event and were present in association with KLF4 mutations or less commonly AKT1 mutations [36]. Drugs targeting the NF2, SMO, and AKT1 pathways are currently being evaluated in clinical trials of meningiomas and may help to address the lack of effective medical therapies for this disease.

#### 16.6.2.2 Oligodendroglioma

These tumors were one of the first brain tumors to be recognized as associated with a specific diagnostic and prognostic molecular genetic signature. Essentially, all oligodendrogliomas exhibit codeletion of chromosomes 1p and 19q which are present in greater than 90 % of cases in carefully selected series [21]. The codeletion of these chromosomal arms was noted to result from a balanced translocation with subsequent loss of one rearranged allele leading to overall single copy loss of these regions [37]. More recently, these tumors were also noted to universally harbor mutations in either the IDH1 or IDH2 genes [21, 38]. These mutations alter the enzymatic activity of the IDH proteins leading to high levels of the onco-metabolite 2-hydroxyglutarate (2HG) which is normally expressed at only low levels in cells and may promote tumor growth. Furthermore, IDH mutations are strongly associated with a distinct methylation profile termed G-CIMP [39]. Such a profile may be induced by the IDH mutations or may represent the profile of the cell of origin for these tumors that has yet to be defined.

Other mutations in CIC, an HMG-box transcription factor with repressor activity, are identified in approximately 70 % of oligodendrogliomas and are unique to the disease. The presence of mutations in ATRX, TP53, and amplification of

EGFR, hallmarks of other gliomas, are extremely rare in oligodendroglioma and typically 1p/19q codeletion are mutually exclusive with each of these events. While not yet formally utilized in the grading of oligodendrogliomas, the presence of CDKN2A loss is the most prominent change identified.

#### 16.6.2.3 Astrocytoma

Astrocytomas are first defined by the presence of IDH1 mutations, most commonly the IDH1 (R132H) variant (more than 80 % of patients). At a very high rate IDH1 mutations in this disease co-occur with loss of function mutations in the tumor suppressor genes ATRX and TP53 [21]. The most common mutations in ATRX lead to loss of expression of the protein in tumor cells as detected by IHC staining for the wild-type protein and is a useful and cost effective assay in clinical practice. Conversely, the most common mutations in TP53 lead to increased expression of P53 in tumor cells by IHC, with more than 10 % of cells generally having high-level nuclear expression. However, the use of these IHC markers as surrogates for detection of IDH1/2, ATRX, or TP53 mutations in oligodendrogliomas or astrocytomas can miss approximately 10–20 % of those cases with variant mutations whose nature does not alter protein expression in the predicted manner. Multiplex sequencing methods are, therefore, highly useful for reliable assignment of prognosis and diagnosis in all gliomas.

#### 16.6.2.4 Glioblastoma

GBM is one of the best characterized cancers at the molecular level [40] and was one of the first cancers to be studied in large scale integrative genomic approaches of The Cancer Genome Atlas (TCGA) program established by the National Cancer Institute [41]. Collectively, these molecular studies have consistently identified three core signaling pathways which are disrupted in GBM: increased activation of receptor tyrosine kinase/RAS/PI3 K signaling, loss of function in P53 signaling, and reduced signaling of the RB pathway. Aberrant activation of RTK/RAS/PI3 K signaling is evident in 88 % of GBMs and most characteristically occurs due to amplification of the EGFR gene, along with rearrangements and overexpression of mutant EGFRvIII and extracellular domain mutants [42]. This pathway is also activated by amplification of other well known oncogenes including PDGRA, MET, AKT, or PIK3CA and/or aberrations that lead to loss of the PTEN tumor suppressor gene function. Alterations in the tumor suppressor TP53 are common in a subset of adult GBM including at least 42 % of adult tumors, either through direct mutations in TP53 or by amplification of MDM2 or MDM4 each seen in approximately 10 % of patients. Interestingly, with time studies have shown that TP53 mutations are more common in young adult astrocytomas and pediatric high-grade gliomas where they are seen in the majority of these tumors [43, 44]. The RB pathway is frequently targeted by different routes, including genomic losses combined with mutation of RB, as well as genomic losses of the CDKN2 family of tumor suppressor genes, or amplification of the negative regulators of the RB pathway, such as CDK4.

Molecular tools are most helpful in the setting of GBM to support the diagnosis and prognosis of patients. Patients with IDH1 mutation have an improved outcome relative to wild-type IDH patients, and likely marks those secondary GBM tumors that have evolved from lower grade astrocytomas. The most common method of detection is IHC using a mutation-specific antibody to IDH1(R132H), but research groups have recently utilized next generation sequencing to identify R132H as well as other rare variants [45]. Methylation of the promoter of the MGMT gene is also associated with increased progression free survival in adult GBM and partially overlaps the IDH1-mutated population. This finding is most commonly assessed using methylation specific PCR and sequencing. The presence of a high level of EGFR amplification (>10 copies) is a hallmark event (>40 % of GBM) that is diagnostically useful in identifying GBM. However, it is not effective as a prognostic biomarker if evaluated in GBM cohorts where the IDH1 mutant tumors have been excluded [46]. Younger adults and older children with "pediatric" type GBM harbor mutations in the histone gene H3F3A as a distinct and common event and generally lack the prototypical combined aberrations associated with patients over the age of 40 which includes combined EGFR amplification, CDKN2A homozygous deletion, Chr 10/PTEN loss, and Chr 13/RB1 loss. Given the complexity and breadth of such aberrations needed to be detected in GBM, the use of combined multiplex technologies for copy detection and single nucleotide variants by array technologies and next generation sequencing have demonstrated potential for efficient delivery of these results both in the clinical and clinical trial settings.

## 16.7 Neural Development and Stem Cells in Brain Tumors

### 16.7.1 Neural Stem Cells and Fate Choice: The Role of Transcription Factors

If one agrees that the multiple, different cell types of the adult brain and brain tumors are ultimately derived from uncommitted progenitor cells or subsets of partially restricted neural progenitors, then it becomes interesting to think about the molecular biology of the fate choice decision and how this might also drive tumor formation. How does a neural stem cell decide whether to become a neuron, an astrocyte or an oligodendrocyte?

Genetically accessible organisms such as Drosophila have highlighted the prominent role of transcription factors in the fate choice decision to form neurons. Transcription factors belonging to the basic helix-loop-helix (bHLH) family play an especially prominent role in neurogenesis [47]. As shown in Fig. 16.5, these neurogenic bHLH transcription factors can be parsed into two columns: the proneural transcription factors (so-called because if you knock them out, you get a fly that is missing some neurons) and the antineural transcription factors (so-called, because if you knock them out you get a fly with excessive neurons). By pushing against each other, these proneural and antineural transcription

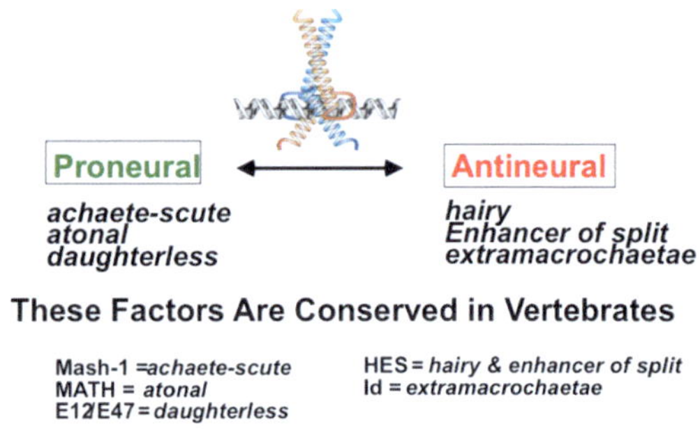


**Fig. 16.5** Neural Stem Cells and Fate Choice: The Role of Transcription Factors. In Drosophila the fate choice decision to form neurons from multipotent progenitors is controlled by an oppositional relationship between proneural and antineural transcription factors of the basic helix-loop-helix (bHLH) family. These neurogenic bHLH transcription factors are conserved in form and also in function within vertebrates [47]

factors generate a fly with the right number of neurons, at the right time, and in the right place.

The value of the fly as an engine of discovery for neural development is that these proneural and antineural bHLH transcription factors have been conserved both in structure and in function, all the way up the phylogenetic ladder to the vertebrate central nervous system. As examples, the mammalian homologue of the Drosophila proneural transcription factors achaete-scute or atonal function as proneural transcription factors in the central nervous system of developing mice. Targeted disruption of Mammalian Achaete-Scute Homologue (MASH-1) or Mammalian Atonal Homologue (MATH) generates mice that are missing a few neurons [47]. This conservation of form and function really comes as no surprise because a neuron in a fly and a neuron in an elephant is basically the same kind of cell. Invertebrate and vertebrate neurons are both configured to push an electrical signal down a long, skinny cytoplasmic tube (the axon) by sequentially opening and closing ion channels.

### 16.7.2 Transcription Factors for Glial Development

Forward genetic screens with invertebrate model organisms have proved generally uninformative about glia, one of the major components of the human brain. Glial cells missing (gcm) was found in 1995 by three independent groups that were screening for genes involved in axonal pathfinding [48–50]. Subsequent studies showed that gcm encodes a transcription factor that is expressed in nearly all of the developing glial cells in Drosophila and is both necessary and conditionally sufficient for their development [48, 49, 51]. Unfortunately, the conservation of form and function noted for neurogenic bHLH transcription factors is not seen with gcm. There are two gcm orthologs in mammals [52, 53]. However, in mammals, the major expression sites of both gcm genes are not within the CNS [54] and targeted disruption studies suggest that neither gene plays a role in brain development [55–57]. Thus it appears that the primordial gcm gene discovered in Drosophila has been adapted to other missions in vertebrate animals.

Against that backdrop one might ask, "are there bHLH transcription factors that specify the glial lineages?" The answer is "yes" and the names of the two key factors are "OLIG1″ and "OLIG2″. The bHLH transcription factors OLIG1 and OLIG2 are co-localized to within 40 kb of each other on human chromosome 21 within the Down syndrome critical region of that chromosome. Unlike classic proneural bHLH transcription factors (e.g., Mash, Math, Ngn1/2) that are never seen in mature neurons [58–60], initial in situ hybridization images showed the two genes to be expressed in white matter tracts of corpus callosum, optic nerve, and cerebellum (hence the name "OLIG" [61–63]. The two OLIG genes are expressed exclusively within the central nervous system.

The primary amino acid sequences of OLIG1 and OLIG2 are very similar within the DNA-targeting bHLH domain, suggesting that they might have some functions in common. However, outside the bHLH domain, OLIG1 and OLIG2 are quite different proteins suggesting some unique, non-overlapping biological functions as well and this suggestion is born out in reality. Targeted disruption studies show that OLIG1 plays very specific roles in the maturation of committed oligodendrocyte progenitor cells into mature myelinating glia [64, 65]. The biological functions of OLIG2 are broader in scope. OLIG2 is expressed exclusively within the CNS. In the developing CNS, OLIG2 is expressed in progenitor cells that give rise to neurons (somatic motor neurons and forebrain cholinergic neurons) and oligodendrocytes [65–70]. Furthermore, targeted disruption of OLIG2 results in a total loss of oligodendrocyte progenitors and motor neurons within the developing spinal cord (Fig. 16.6) [65, 70].

In the adult CNS, OLIG2 is expressed only in myelinating oligodendrocytes and in neural

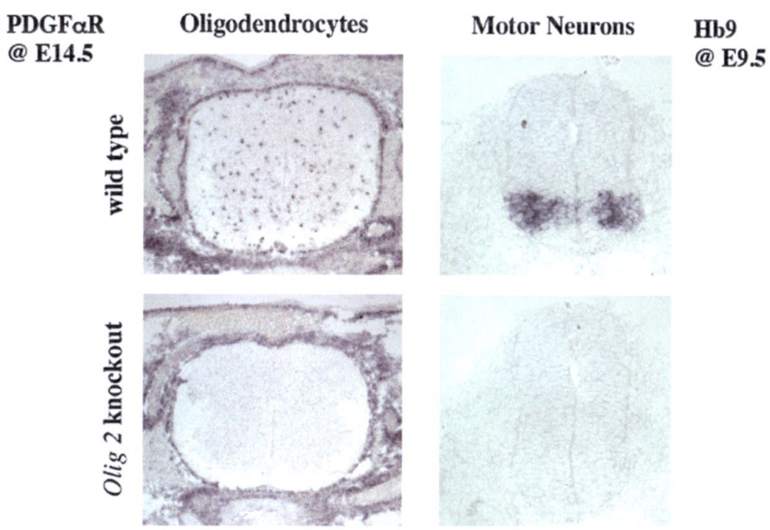


**Fig. 16.6** Developmental Functions of OLIG2. Targeted disruption of Olig2 during development ablates oligodendrocytes (marked by Pdgfra) and certain types of neurons, notably motor neurons (marked by Hb9). Images adapted from Lu et al. [104]. See text for details

progenitor cells. Focusing specifically on adult neural progenitors, OLIG2 is expressed in the transit amplifying type C progenitors of the subventricular zone (SVZ) [71]. In addition, OLIG2 is expressed in essentially 100 % of the NG2-positive glia and is required for formation of these cells [72]. NG2-positive glia are the most prevalent cycling progenitor cells of the adult CNS. They have been proposed to play critical roles in brain repair and regeneration and are the primary responding neural cell type to brain injury [9, 73]. Transit amplifying cells of the SVZ and NG2-positive glia have both been proposed as plausible "cells-of-origin" for human gliomas which makes sense given that cycling cells in other cancers are often more prone to transformation (reviewed in Stiles and Rowitch) [74].

### 16.7.3 OLIG2 and Glioma Stem Cells

During early stages of neural development, a key role of OLIG2 is to maintain the replication-competent state so as to expand the pool of available progenitors for motor neuron and oligodendrocyte development [75]. An emerging body of literature documents an "evil twin" of this pro-mitogenic function in development. In particular, immunohistochemical analysis shows that OLIG2 is expressed in 100 percent of human diffuse gliomas irrespective of grade [18]. Gliomas—especially high-grade adult gliomas—contain a heterogeneous mixture of cell types. However, a highly tumorigenic subpopulation of cells in both adult and pediatric astrocytomas is frequently marked by CD133 (a.k.a. Prominin-1)—a cell surface antigen that is also a known marker of multipotent stem cells in blood and other tissues including the brain [76–78]. Essentially 100 % of these tumorigenic, CD133-positive stem cells in fresh surgical isolates of human astrocytomas express OLIG2 [79].

Beyond simply marking glioma stem cells, OLIG2 function is actually required for tumor formation in "genetically relevant" murine models of adult glioma [79] and in freshly isolated human "gliomasphere" cultures [80]. A final link between OLIG2 and the stem-like cells in glioma comes from a very interesting experiment. Preservation of the stem-like state in glioma cells requires them to be propagated under conditions identical to those used to maintain normal neural stem cells. Specifically, they must be cultured under non-adherent conditions in serum-free medium supplemented with the growth factors EGF and FGF. If glioma stem cells are allowed to attach to the surface of the culture dish and cultured in the presence of serum (or BMP4) the expression of stem-like marker proteins is lost and the cells lose tumorigenic potential. Under normal circumstances these losses are irreversible [81–85].

Against this backdrop, Suva et al. asked whether there might be a transcription factor code for restoring the stem-like state and tumorigenic potential of serum-treated glioma cells. A functional analogy for such a code would be found in the Nobel-prize winning studies of Shinya Yamanaka who found a set of four transcription factors (SOX2, OCT4, KLF4, and MYC) could induce normal human fibroblasts to form pluripotent stem cells (iPSC). Suva et al. found a set of three transcription factors SOX2, POU3F2, and SALL2 were sufficient to restore most stem-like features of the phenotype but the resulting cells were still not tumorigenic. A fourth transcription factor, OLIG2, was required to restore the tumorigenic component of the phenotype [86].

## 16.8 Resistance to Therapy and the Cancer Stem Cell Hypothesis

Malignant gliomas in adults and also in children typically show an initial response to radiation and chemotherapy, however the tumors inevitably recur. Conventional wisdom attributes resistance to radiation or genotoxic drugs to the selection of preexisting (and/or therapy-induced) mutant cells within the tumor. In accord with this point of view, tumor cells, especially in high-grade gliomas, are genetically unstable. Moreover, conventional therapeutic modalities are highly mutagenic. However, there is an alternative explanation for the failures of treatment. This alternative explanation is known as the "cancer stem cell hypothesis" [87, 88].

Key features of the cancer stem cell hypothesis are that solid tumors arise from developmentally stalled progenitor cells (as is the case with leukemias). Accordingly, the tumor cell population is heterogeneous, being composed of a minor population of highly tumorigenic, self renewing "cancer stem cells" and a majority population of partially committed (and presumably less tumorigenic) progenitor cells. The central idea of the cancer stem cell hypothesis is that therapies for solid tumors usually fail because the minor population of "stem-like" cells within the tumor is intrinsically resistant to radiation and genotoxic drugs [89].

Biological support for the cancer stem cell hypothesis comes from the observation that normal stem cells are known to cycle slowly, a property that could account for resistance to radiotherapy. Moreover, normal stem cells have high levels of drug export transporter proteins, a feature that could account for resistance to chemotherapy. However, there are other observations that are inconsistent with the hypothesis. For example, tumors that recur following radiation and chemotherapy are often genetically quite different from the original lesions, whereas the cancer stem cell hypothesis predicts that recurrent tumors would be genetically similar to the primary cancer. Moreover, some tumors that are clearly of embryonic stem cell origin (e.g., testicular cancers) are quite sensitive to conventional chemotherapy [90].

Glioblastomas (adult and pediatric) have become one of the favored scientific vehicles for testing the cancer stem cell hypothesis. There are three reasons for this. First, the so-called "wall charts" of knowledge describing the stages of brain development ("neuropoiesis") are surpassed in detail only by those for blood development ("hematopoiesis"). As is the case with blood progenitors, multipotent neural progenitors can be isolated from developing embryos or from the postnatal brain and placed into culture where their differentiation can be manipulated and observed [4]. Second, as noted by Cushing, Bailey and other pioneers in neurosurgery in the early 1920s, high-grade adult gliomas contain a heterogeneous mixture of cell types that actually look like the product of a developmentally arrested neural stem cell. Hence the term "glioblastoma multiforme" (GBM) [91], as mentioned previously. Third, GBMs contain a subpopulation of cells with stem-like properties. These glioma stem cells can be manipulated like their normal counterparts to grow as neurospheres or to undergo at least partial differentiation into cells that express neuronal or glial marker proteins [76–78, 92, 93].

## 16.9 Conflict of Interest: An Oppositional Relationship Between OLIG2 and TP53

Why are gliomas so intrinsically resistant to radiation and chemotherapy? In healthy cells, the responses to genetic damage, cellular stress, or inappropriate mitogenic cues (e.g., oncogenes) are triggered by P53—the so-called "guardian of the genome." The TP53 gene product is a transcription factor and activation of P53 transcriptional functions results in programmed cell death, transient growth arrest, or permanent growth arrest (senescence), depending on cell type, strength of the inducing stimulus, and other factors. The pro-death, anti-growth functions of P53 are obviously not in the best interest of tumor cells. Accordingly, genetic ablation of TP53 is a frequent feature of adult solid tumors [94].

Malignant gliomas are notoriously insensitive to radiation and genotoxic drugs. Paradoxically, the TP53 gene is structurally intact in the majority of primary gliomas [41, 80]. Resistance to genotoxic modalities in P53-positive gliomas has been attributed to attenuation of P53 functions by other mutations within a P53 signaling axis that includes the Ink4a/ARF, MDM2, and ATM gene products [27, 41, 95, 96] (see Fig. 16.7). However, as noted earlier in this chapter high-grade gliomas are a heterogeneous mixture of cell types. Bao et al. observed that it is

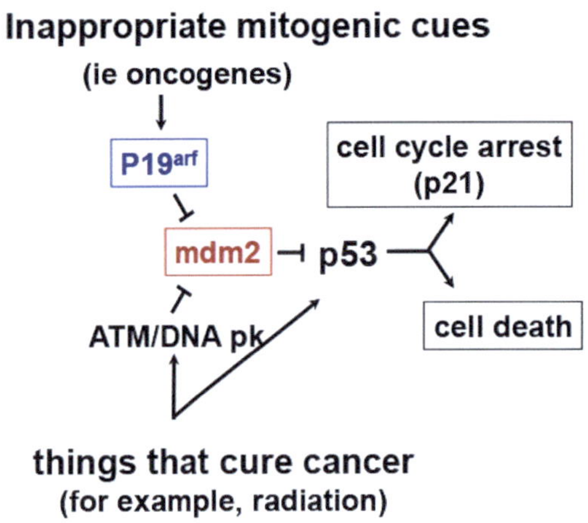


**Fig. 16.7** A P53 signaling axis protects from genotoxic insults and inappropriate mitogenic cues. See text for details

the subpopulation of stem-like cells within high-grade gliomas that have significantly enhanced resistance to radiation [97]. Since any mutations within the P53 signaling axis would be present in all tumor cells, the resistance of gliomas to genotoxic agents may reflect mechanisms specific to these stem-like tumor-initiating cells.

One such mechanism is suggested by the studies of Mehta et al. [80]. These investigators showed that OLIG2 confers a radioresistant phenotype to both normal and malignant neural progenitors by opposing P53-mediated responses to DNA gamma irradiation [80]. In the absence of OLIG2, even attenuated levels of p53 function (as seen for example in Ink4a/ARF null cells) are adequate to suppress growth of irradiated cells. The oppositional relationship between OLIG2 and P53 reflects, at least in part, a suppression of P53 transcriptional functions. In the presence of OLIG2, expression of a key P53 effector gene, $p21^{WAF1/CIP1}$ (hereafter called "p21") is much reduced [79, 80].

Why would normal neural progenitor cells express a transcription factor that opposes the guardian functions of P53 to genotoxic damage and inappropriate mitogenic cues? A compelling incentive is provided by other functions of the P53/p21 signaling axis in formation and maintenance of stemness. Studies on the production of induced pluripotent stem (iPS) cells from normal fibroblasts highlight an oppositional relationship between P53, p21 and the acquisition of stemness [98]. Ablation of TP53 greatly enhances the efficiency of iPS formation from normal fibroblast cells and p21 is an important component of this outcome [98–101]. The iPS work resonates with earlier studies showing that targeted disruption of either TP53 [102] or p21 [103] compromises the relative quiescence of neural progenitors and accelerates self-renewal. The work of Mehta et al. suggests that OLIG2-mediated vulnerability to DNA damage and malignant transformation might be the price paid by the adult CNS to maintain replication and sustain a reserve of neural progenitors for response to injury and for normal turnover of certain neuronal populations.

## 16.10 Conclusions and the Road Ahead

Brain cancers remain as some of the most challenging cancers to treat and many important questions about their biology remain to be discovered. Where do tumors of the brain come from? How do tumors arise in an organ that is isolated from the environment and, to a first approximation, mitotically inert? Why is the most lethal form of brain cancer, the high-grade glioma, GBM, so notoriously resistant to radiotherapy and chemotherapy? We have seen that the answers to these questions may be rooted in a better understanding of brain development. Developmental principles early on had a significant impact on improved classification of brain tumors from the time of their first study and such principles are likely to inform treatment in today's era. Analogies can be drawn between the process of blood development and brain development in that multiple cell types of these organs arise from smaller subsets of multipotent stem cells and partially committed (but still replication competent) progenitors derived from these stem cells. Stem cells have an intrinsic oppositional relationship with P53 and its downstream effectors which are mutated often in gliomas. High-grade gliomas contain a subpopulation of highly tumorigenic cells with stem-like properties. A broad body of data suggests that these stem-like, tumor-initiating cells have co-opted the molecular mechanisms used by normal neural progenitors to sustain replication competence and oppose the actions of P53.

Challenges for the road ahead will be to convert this knowledge into power. Transcription factors with bHLH domains regulate proliferation and differentiation of neural stem cells. One bHLH transcription factor known as OLIG2 is expressed in high-grade gliomas and is essential for the tumorigenic component of the glioma stem cell phenotype. For this reason, OLIG2 would seem to be an attractive target for drug development. However, transcription factors do not lend themselves readily to the development of small molecule antagonists because their interactions with DNA and with co-regulator proteins involve large and complex surface area contacts.

Finally, there have been a great many advances in recent years in understanding the molecular and genetic pathology of all kinds of brain tumors, but especially GBMs. This information has already led to identification of glioma patients with IDH mutations as a greatly improved prognostic group, as well as a vastly improved diagnostic classification of tumors such as oligodendroglioma by their histologic appearance and molecular alterations. It is now possible to match molecularly identified drug targets within individual patient tumors to the experimental treatments that may be effective in those exact patients. This form of precision medicine has already enabled improved categorization and understanding of brain tumors and is now able to serve as the foundation for improved treatments that are sorely needed in these diseases.

## References


1. Central Brain Tumor Registry Of The United States. Available From: http://Www.Cbtrus.Org.
2. Prados MD, Wilson CD. Neoplasms of the central nervous system. In: Holland Jf, Frei E, Bast Rc, Kufe Dw, Morton Dl, Weichselbaum Rr, Editors. Cancer Medicine. 1. 3 Ed. Philadelphia, London: Lea And Febiger;1993. p. 1080–119.
3. Anderson DJ. The neural crest cell lineage problem: neuropoiesis? Neuron. 1989;3(1):1–12.
4. Reynolds BA, Weiss S. Generation of neurons and astrocytes from isolated cells of the adult mammalian central nervous system. Science. 1992;255 (5052):1707–10.
5. Lois C, Alvarez-Buylla A. Long-distance neuronal migration in the adult mammalian brain. Science. 1994;264(5162):1145–8.
6. Doetsch F, Caille I, Lim Da, Garcia-Verdugo Jm, Alvarez-Buylla A. Subventricular zone astrocytes are neural stem cells in the adult mammalian brain. Cell. 1999;97(6):703–16.
7. Alvarez-Buylla A, Lim DA. For the long run: maintaining germinal niches in the adult brain. Neuron. 2004;41(5):683–6.
8. Zhao C, Deng W, Gage FH. Mechanisms and functional implications of adult neurogenesis. Cell. 2008;132(4):645–60.
9. Nishiyama A, Komitova M, Suzuki R, Zhu X. Polydendrocytes (Ng2 Cells): multifunctional cells with lineage plasticity. Nat Rev Neurosci. 2009;10 (1):9–22.
10. Polito A, Reynolds R. Ng2-expressing cells as oligodendrocyte progenitors in the normal and demyelinated adult central nervous system. J Anat. 2005;207(6):707–16.
11. Dawson MR, Polito A, Levine JM, Reynolds R. Ng2-expressing glial progenitor cells: an abundant and widespread population of cycling cells in the adult rat Cns. Mol Cell Neurosci. 2003;24(2):476–88.
12. Lim DA, Cha S, Mayo MC, Chen MH, Keles E, Vandenberg S, et al. Relationship of glioblastoma multiforme to neural stem cell regions predicts invasive and multifocal tumor phenotype. Neuro-Oncology. 2007;9(4):424–9.
13. Carpenter MB, Sutin J. Human neuroanatomy. 8th Ed. Baltimore: Williams & Wilkins;1983. Xiv, 872 pp.
14. Bailey P, Cushing HA Classification of the tumors of the glioma group on a histogenetic basis with a correlated study of prognosis.: Philadelphia, London [Etc.]: J.B. Lippincott Company;1926.
15. Sanai N, Alvarez-Buylla A, Berger MS. Neural stem cells and the origin of gliomas. N Engl J Med. 2005;353(8):811–22.
16. Dn Louis, Ohgaki H, Wiestler Od, Wk Cavenee, Pc Burger, Jouvet A, et al. The 2007 who classification of tumours of the central nervous system. Acta Neuropathol. 2007;114(2):97–109.
17. Schnitt SJ, Vogel H. Meningiomas. Diagnostic value of immunoperoxidase staining for epithelial membrane antigen. Am J Surg Pathol. 1986;10 (9):640–9.
18. Kl Ligon, Ja Alberta, At Kho, Weiss J, Mr Kwaan, Cl Nutt, et al. The oligodendroglial lineage marker Olig2 Is universally expressed in diffuse gliomas. J Neuropathol Exp Neurol. 2004;63(5):499–509.

19. Louis DN, Perry A, Burger P, Ellison DW, Reifenberger G, Von Deimling A, et al. International society of neuropathology — haarlem consensus guidelines for nervous system tumor classification and grading. Brain Pathol. 2014;24(5):429–35.
20. Ramkissoon LA, Horowitz PM, Craig JM, Ramkissoon SH, Rich BE, Schumacher SE, et al. Genomic analysis of diffuse pediatric low-grade gliomas identifies recurrent oncogenic truncating rearrangements in the transcription factor Mybl1. Proc Natl Acad Sci U S A. 2013;110(20):8188–93.
21. Cancer Genome Atlas Research N, Brat DJ, Verhaak RG, Aldape KD, Yung WK, Salama SR, et al. Comprehensive, integrative genomic analysis of diffuse lower-grade gliomas. N Engl J Med. 2015;372(26):2481–98.
22. Brandner S, Von Deimling A. Diagnostic, prognostic and predictive relevance of molecular markers in gliomas. Neuropathol Appl Neurobiol. 2015;41 (6):694–720.
23. Phillips HS, Kharbanda S, Chen R, Forrest WF, Soriano RH, Wu TD, et al. Molecular subclasses of high-grade glioma predict prognosis, delineate a pattern of disease progression, and resemble stages in neurogenesis. Cancer Cell. 2006;9(3):157–73.
24. Verhaak RG, Hoadley KA, Purdom E, Wang V, Qi Y, Wilkerson MD, et al. Integrated genomic analysis identifies clinically relevant subtypes of glioblastoma characterized by abnormalities in Pdgfra, Idh1, Egfr, And Nf1. Cancer Cell. 2010;17 (1):98–110.
25. Ramkissoon SH, Bi WI, Schumacher SE, Ramkissoon LA, Haidar S, Knoff D, et al. Clinical implementation of integrated whole-genome copy number and mutation profiling for glioblastoma. Neuro-Oncology. 2015;17(10):1344–55.
26. Cryan JB, Haidar S, Ramkissoon LA, Bi WI, Knoff DS, Schultz N, et al. Clinical multiplexed exome sequencing distinguishes adult oligodendroglial neoplasms from astrocytic and mixed lineage gliomas. Oncotarget. 2014;5(18):8083–92.
27. Parsons DW, Jones S, Zhang X, Lin JC, Leary RJ, Angenendt P, et al. An integrated genomic analysis of human glioblastoma multiforme. Science. 2008;321(5897):1807–12.
28. Chi AS, Batchelor TT, Dias-Santagata D, Borger D, Stiles CD, Batchelor DI, et al. Prospective, high-throughput molecular profiling of human gliomas. J Neurooncol. 2012;110(1):89–98.
29. Cheng DT, Mitchell TN, Zehir A, Shah RH, Benayed R, Syed A, et al. Memorial sloan kettering-integrated mutation profiling of actionable cancer targets (Msk-Impact): a hybridization capture-based next-generation sequencing clinical assay for solid tumor molecular oncology. J Mol Diagn Jmd. 2015;17(3):251–64.
30. Hegi ME, Diserens AC, Gorlia T, Hamou MF, De Tribolet N, Weller M, et al. Mgmt gene silencing and benefit from temozolomide in glioblastoma. N Engl J Med. 2005;352(10):997–1003.
31. Stupp R, WP Mason, Van Den Bent MJ, Weller M, Fisher B, Taphoorn MJ, et al. Radiotherapy plus concomitant and adjuvant temozolomide for glioblastoma. N Engl J Med. 2005;352(10):987–96.
32. Hovestadt V, Remke M, Kool M, Pietsch T, Pa Northcott, Fischer R, et al. Robust molecular subgrouping and copy-number profiling of medulloblastoma from small amounts of archival tumour material using high-density dna methylation arrays. Acta Neuropathol. 2013;125(6):913–6.
33. Wiestler B, Capper D, Sill M, Weller DT, Hovestadt V, Sturm D, et al. Integrated Dna methylation and copy-number profiling identify three clinically and biologically relevant groups of anaplastic glioma. Acta Neuropathol. 2014;128(4):561–71.
34. Aizer AA, Abedalthagafi M, Bi WL, Horvath MC, Arvold ND, Al-Mefty O, et al. A prognostic cytogenetic scoring system to guide the adjuvant management of patients with atypical meningioma. Neuro-Oncology. 2015.
35. Brastianos PK, Horowitz PM, Santagata S, Jones RT, Mckenna A, Getz G, et al. Genomic sequencing of meningiomas identifies oncogenic smo and Akt1 mutations. Nat Genet. 2013;45 (3):285–9.
36. Clark VE, Erson-Omay EZ, Serin A, Yin J, Cotney J, Ozduman K, et al. Genomic analysis of non-NF2 meningiomas reveals mutations in Traf7, Klf4, Akt1, and SMO. Science. 2013;339(6123):1077–80.
37. Rb Jenkins, Blair H, Kv Ballman, Giannini C, Rm Arusell, Law M, et al. A T(1;19)(Q10;P10) mediates the combined deletions of 1p and 19q and predicts a better prognosis of patients with oligodendroglioma. Cancer Res. 2006;66(20):9852–61.
38. Yan H, Parsons DW, Jin G, Mclendon R, Rasheed BA, Yuan W, et al. Idh1 And Idh2 mutations in gliomas. N Engl J Med. 2009;360 (8):765–73.
39. Noushmehr H, Weisenberger DJ, Diefes K, Phillips HS, Pujara K, Berman BP, et al. Identification of a Cpg Island methylator phenotype that defines a distinct subgroup of glioma. Cancer Cell. 2010;17(5):510–22.
40. Furnari FB, Fenton T, Bachoo RM, Mukasa A, Stommel JM, Stegh A, et al. Malignant astrocytic glioma: genetics, biology and paths to treatment. Genes Dev. 2007;21(21):2683–710.
41. Tcga. Comprehensive genomic characterization defines human glioblastoma genes and core pathways. Nature. 2008;455(7216):1061–8.
42. Maire CL, Ligon KL. Molecular pathologic diagnosis of epidermal growth factor receptor. Neuro-Oncology. 2014;16 Suppl 8:Viii1–6.
43. Zhang J, Wu G, Miller CP, Tatevossian RG, Dalton JD, Tang B, et al. Whole-genome sequencing identifies genetic alterations in pediatric low-grade gliomas. Nat Genet. 2013;45(6):602–12.
44. Schwartzentruber J, Korshunov A, Liu XY, Jones DT, Pfaff E, Jacob K, et al. Driver mutations in histone H3.3 and chromatin remodelling genes in

paediatric glioblastoma. Nature. 2012;482(7384):226–31.
45. Capper D, Weissert S, Balss J, Habel A, Meyer J, Jager D, et al. Characterization Of R132 h mutation-specific Idhl antibody binding in brain tumors. Brain Pathol. 2010;20(1):245–54.
46. Simmons MI, Lamborn KR, Takahashi M, Chen P, Israel MA, Berger MS, et al. Analysis of complex relationships between age, P53, epidermal growth factor receptor, and survival in glioblastoma patients. Cancer Res. 2001;61(3):1122–8.
47. Se Ross, Me Greenberg, Stiles Cd. Basic Helix-Loop-Helix factors in cortical development. Neuron. 2003;39(1):13–25.
48. Jones BW, Fetter RD, Tear G, Goodman CS. Glial cells missing: a genetic switch that controls glial versus neuronal fate. Cell. 1995;82(6):1013–23.
49. Hosoya T, Takizawa K, Nitta K, Hotta Y. Glial cells missing: a binary switch between neuronal and glial determination in drosophila. Cell. 1995;82(6):1025–36.
50. Kania A, Salzberg A, Bhat M, D'evelyn D, He Y, Kiss I, Et Al. P-element mutations affecting embryonic peripheral nervous system development in drosophila melanogaster. Genetics. 1995;139(4):1663–78.
51. Pfrieger FW, Ba Barres. What the fly's glia tell the fly's brain. Cell. 1995;83(5):671–4.
52. Akiyama Y, Hosoya T, Poole AM, Hotta Y. The Gcm-Motif: a novel dna-binding motif conserved in drosophila and mammals. Proc Natl Acad Sci U S A. 1996;93(25):14912–6.
53. Altshuller Y, Copeland NG, Gilbert DJ, Jenkins NA, Frohman MA. Gcml, a mammalian homolog of drosophila glial cells missing. Febs Lett. 1996;393(2–3):201–4.
54. Kim J, Jones BW, Zock C, Chen Z, Wang H, Goodman CS, et al. Isolation and characterization of mammalian homologs of the drosophila gene glial cells missing. Proc Natl Acad Sci U S A. 1998;95(21):12364–9.
55. Schreiber J, Riethmacher-Sonnenberg E, Riethmacher D, Tuerk EE, Enderich J, Bosl MR, et al. Placental failure in mice lacking the mammalian homolog of glial cells missing Gcma. Mol Cell Biol. 2000;20(7):2466–74.
56. Anson-Cartwright L, Dawson K, Holmyard D, Fisher SJ, Lazzarini RA, Cross JC. The glial cells missing-1 protein is essential for branching morphogenesis in the chorioallantoic placenta. Nat Genet. 2000;25(3):311–4.
57. Gunther T, Chen ZF, Kim J, Priemel M, Rueger JM, Amling M, et al. Genetic ablation of parathyroid glands reveals another source of parathyroid hormone. Nature. 2000;406(6792):199–203.
58. Lee JE. Neurod and neurogenesis. Dev Neurosci. 1997;19(1):27–32.
59. Parras CM, Galli R, Britz O, Soares S, Galichet C, Battiste J, et al. Mashl specifies neurons and oligodendrocytes in the postnatal brain. Embo J. 2004;23(22):4495–505.
60. Kageyama R, Ishibashi M, Takebayashi K, Tomita K. Bhlh transcription factors and mammalian neuronal differentiation. Int J Biochem Cell Biol. 1997;29(12):1389–99.
61. Lu QR, Yuk D, Alberta JA, Zhu Z, Pawlitzky I, Chan J, et al. Sonic hedgehog — regulated oligodendrocyte lineage genes encoding Bhlh proteins in the mammalian central nervous system. Neuron. 2000;25(2):317–29.
62. Takebayashi H, Yoshida S, Sugimori M, Kosako H, Kominami R, Nakafuku M, et al. Dynamic expression of basic helix-loop-helix olig family members: implication of Olig2 in neuron and oligodendrocyte differentiation and identification of a new member, Olig3. Mech Dev. 2000;99(1–2):143–8.
63. Zhou Q, Wang S, Anderson DJ. Identification of a novel family of oligodendrocyte lineage-specific basic helix-loop-helix transcription factors. Neuron. 2000;25(2):331–43.
64. Meijer Dh, Mf Kane, Mehta S, Liu H, Harrington E, Cm Taylor, et al. Separated at birth? the functional and molecular divergence of Oligl And Olig2. Nat Rev Neurosci. 2012;13(12):819–31.
65. Lu QR, Sun T, Zhu Z, Ma N, Garcia M, Stiles CD, et al. Common developmental requirement for olig function indicates a motor neuron/oligodendrocyte connection. Cell. 2002;109(1):75–86.
66. Furusho M, Ono K, Takebayashi H, Masahira N, Kagawa T, Ikeda K, et al. Involvement of the olig2 transcription factor in cholinergic neuron development of the basal forebrain. Dev Biol. 2006;293(2):348–57.
67. Hack MA, Saghatelyan A, De Chevigny A, Pfeifer A, Ashery-Padan R, Lledo PM, et al. Neuronal fate determinants of adult olfactory bulb neurogenesis. Nat Neurosci. 2005;8(7):865–72.
68. Novitch BG, Chen AI, Jessell TM. Coordinate regulation of motor neuron subtype identity and pan-neuronal properties by the Bhlh repressor Olig2. Neuron. 2001;31(5):773–89.
69. Takebayashi H, Nabeshima Y, Yoshida S, Chisaka O, Ikenaka K. The basic helix-loop-helix factor olig2 is essential for the development of motoneuron and oligodendrocyte lineages. Curr Biol. 2002;12(13):1157–63.
70. Zhou Q, Anderson DJ. The Bhlh transcription factors olig2 and oligl couple neuronal and glial subtype specification. Cell. 2002;109(1):61–73.
71. Jackson EI, Garcia-Verdugo JM, Gil-Perotin S, Roy M, Quinones-Hinojosa A, Vandenberg S, et al. Pdgfr alpha-positive B cells are neural stem cells in the adult Svz that form glioma-like growths in response to increased Pdgf signaling. Neuron. 2006;51(2):187–99.
72. Ligon KI, Kesari S, Kitada M, Sun T, Arnett HA, Alberta JA, et al. Development Of Ng2 neural progenitor cells requires olig gene function. Proc Natl Acad Sci U S A. 2006;103(20):7853–8.
73. Belachew S, Chittajallu R, Aguirre AA, Yuan X, Kirby M, Anderson S, et al. Postnatal Ng2

proteoglycan-expressing progenitor cells are intrinsically multipotent and generate functional neurons. J Cell Biol. 2003;161(1):169–86.
74. Stiles CD, Rowitch DH. Glioma stem cells: a midterm exam. Neuron. 2008;58(6):832–46.
75. Lee Sk, Lee B, Ruiz Ec, Pfaff Sl. Olig2 And Ngn2 function in opposition to modulate gene expression in motor neuron progenitor cells. Genes Dev. 2005;19(2):282–94.
76. Galli R, Binda E, Orfanelli U, Cipelletti B, Gritti A, De Vitis S, et al. Isolation and characterization of tumorigenic, stem-like neural precursors from human glioblastoma. Cancer Res. 2004;64 (19):7011–21.
77. Singh SK, Clarke ID, Terasaki M, Bonn VE, Hawkins C, Squire J, et al. Identification of a cancer stem cell in human brain tumors. Cancer Res. 2003;63(18):5821–8.
78. Yuan X, Curtin J, Xiong Y, Liu G, Waschsmann-Hogiu S, Dl Farkas, et al. Isolation of cancer stem cells from adult glioblastoma multiforme. Oncogene. 2004;23(58):9392–400.
79. Kl Ligon, Huillard E, Mehta S, Kesari S, Liu H, Ja Alberta, et al. Olig2-regulated lineage-restricted pathway controls replication competence in neural stem cells and malignant glioma. Neuron. 2007;53 (4):503–17.
80. Mehta S, Huillard E, Kesari S, Cl Maire, Golebiowski D, Harrington EP, et al. The central nervous system-restricted transcription factor olig2 opposes P53 responses to genotoxic damage in neural progenitors and malignant glioma. Cancer Cell. 2011;19(3):359–71.
81. Piccirillo SG, Reynolds BA, Zanetti N, Lamorte G, Binda E, Broggi G, et al. Bone morphogenetic proteins inhibit the tumorigenic potential of human brain tumour-initiating cells. Nature. 2006;444 (7120):761–5.
82. Lee J, Kotliarova S, Kotliarov Y, Li A, Su Q, Nm Donin, et al. Tumor stem cells derived from glioblastomas cultured in Bfgf and Egf more closely mirror the phenotype and genotype of primary tumors than do serum-cultured cell lines. Cancer Cell. 2006;9(5):391–403.
83. Natsume A, Ito M, Katsushima K, Ohka F, Hatanaka A, Shinjo K, et al. Chromatin regulator Prc2 is a key regulator of epigenetic plasticity in glioblastoma. Cancer Res. 2013;73(14):4559–70.
84. Janiszewska M, Suva Ml, Riggi N, Houtkooper RH, Auwerx J, Clement-Schatlo V, et al. Imp2 controls oxidative phosphorylation and is crucial for preserving glioblastoma cancer stem cells. Genes Dev. 2012;26(17):1926–44.
85. Wakimoto H, Kesari S, Farrell Cj, Curry Wt, Jr., Zaupa C, Aghi M, et al. Human glioblastoma-derived cancer stem cells: establishment of invasive glioma models and treatment with oncolytic herpes simplex virus vectors. Cancer Res. 2009;69(8):3472–81.
86. Ml Suva, Rheinbay E, Sm Gillespie, Ap Patel, Wakimoto H, Rabkin Sd, et al. Reconstructing and reprogramming the tumor-propagating potential of glioblastoma stem-like cells. Cell. 2014;157(3):580–94.
87. Reya T, Morrison Sj, Clarke Mf, Weissman Il. Stem cells, cancer, and cancer stem cells. Nature. 2001;414(6859):105–11.
88. Tan BT, Park CY, Ailles LE, Weissman IL. The cancer stem cell hypothesis: a work in progress. Lab Invest. 2006;86(12):1203–7.
89. Al-Hajj M, Becker Mw, Wicha M, Weissman I, Clarke Mf. Therapeutic implications of cancer stem cells. Curr Opin Genet Dev. 2004;14(1):43–7.
90. Polyak K, Hahn WC. Roots and stems: stem cells in cancer. Nat Med. 2006;12(3):296–300.
91. Mallory FB. The principles of pathologic histology. London: Uk; 1914.
92. HD Hemmati, Nakano I, Lazareff JA, Masterman-Smith M, Geschwind DH, Bronner-Fraser M, et al. Cancerous stem cells can arise from pediatric brain tumors. Proc Natl Acad Sci U S A. 2003;100(25):15178–83.
93. Ignatova TN, Kukekov VG, Laywell ED, Suslov ON, Vrionis FD, Steindler DA. human cortical glial tumors contain neural stem-like cells expressing astroglial and neuronal markers in vitro. Glia. 2002;39(3):193–206.
94. Olivier M, Hollstein M, Hainaut P. Tp53 mutations in human cancers: origins, consequences, and clinical use. Cold Spring Harb Perspect Biol. 2010;2(1): A001008.
95. Nakamura M, Watanabe T, Klangby U, Asker C, Wiman K, Yonekawa Y, et al. P14arf deletion and methylation in genetic pathways to glioblastomas. Brain Pathol. 2001;11(2):159–68.
96. Reifenberger G, Liu L, Ichimura K, Schmidt EE, Collins VP. Amplification and overexpression of the Mdm2 gene in a subset of human malignant gliomas without P53 mutations. Cancer Res. 1993;53 (12):2736–9.
97. Bao S, Wu Q, Re Mclendon, Hao Y, Shi Q, Ab Hjelmeland, et al. Glioma stem cells promote radioresistance by preferential activation of the dna damage response. Nature. 2006;444(7120):756–60.
98. Hong H, Takahashi K, Ichisaka T, Aoi T, Kanagawa O, Nakagawa M, et al. Suppression of induced pluripotent stem cell generation by the P53-P21 pathway. Nature. 2009;460(7259):1132–5.
99. Utikal J, Jm Polo, Stadtfeld M, Maherali N, Kulalert W, Rm Walsh, et al. Immortalization eliminates a roadblock during cellular reprogramming into Ips cells. Nature. 2009;460(7259):1145–8.
100. Kawamura T, Suzuki J, Wang YV, Menendez S, Morera LB, Raya A, et al. Linking the P53 tumour

suppressor pathway to somatic cell reprogramming. Nature. 2009;460(7259):1140–4.
101. Li H, Collado M, Villasante A, Strati K, Ortega S, Canamero M, et al. The Ink4/Arf locus is a barrier for Ips cell reprogramming. Nature. 2009;460(7259):1136–9.
102. Meletis K, Wirta V, Hede Sm, Nister M, Lundeberg J, Frisen J. P53 suppresses the self-renewal of adult neural stem cells. Development. 2006;133(2):363–9.
103. Kippin TE, Martens DJ, Van Der Kooy D. P21 loss compromises the relative quiescence of forebrain stem cell proliferation leading to exhaustion of their proliferation capacity. Genes Dev. 2005;19(6):756–67.
104. Lu QR, Sun T, Zhu Z, Ma N, Garcia M, Stiles CD, et al. Common developmental requirement for Olig function indicates a motor neuron/oligodendrocyte connection. Cell. 2002;109(1):75–86.

# Epidemiology of Renal Cell Carcinoma

17

Alejandro Sanchez, Christopher B. Allard and Kathryn M. Wilson

## 17.1 Overview

In the last two decades, there has been an expansion of knowledge surrounding the underlying genetics of RCC, which has led to significant drug discovery [1]. RCC incidence has increased in the United States (U.S.) and worldwide in recent decades. Patients are being identified at earlier stages due to the increasing use of radiologic imaging, and localized disease accounts for most of the increase in incidence [2]. Mortality rates increased along with incidence rates from the 1970s until the early 2000s in most countries; however, mortality rates have stabilized or decreased in the past 10 years [3]. In this section, we review the current clinical picture of the disease including presenting symptoms, diagnosis, staging, and treatment options. We then review the descriptive epidemiology of RCC, followed by a review of major risk factors for the disease. Upper tract urothelial carcinoma (pelvis/ureter), which comprises approximately 10 % of kidney cancer cases, will not be covered in this chapter.

## 17.2 Clinical Presentation

The kidney is located within the retroperitoneum and as a result many renal masses remain asymptomatic and nonpalpable until they are advanced. Traditionally, patients with RCC presented with local, systemic, and/or paraneoplastic symptoms. The classic triad of symptoms (aka the "too late triad") including flank pain, gross hematuria, and palpable mass is now a rare presentation of this disease. With the increasing use of diagnostic imaging, these tumors are more readily identified incidentally at early stages. However, despite the increase in detection at earlier stages, there has been an overall increase in RCC incidence and mortality in the U.S. over the last several decades [2].

Table 17.1 summarizes the clinical characteristics of patients presenting with RCC. Perirenal hemorrhage/hematoma is an important presentation, as patients with a spontaneous perirenal hemorrhage may have up to a 50 % risk of having an underlying renal tumor [4]. About 20 % of patients with RCC exhibit a paraneoplastic syndrome at presentation (Table 17.1). History and physical exam, including a family history of RCC, are critical components of the

A. Sanchez
Department of Urology, Massachusetts General Hospital, 55 Fruit St, Boston, MA 02114, USA

C.B. Allard
Department of Urology, Massachusetts General Hospital, 45 Francis St, Boston, MA 02114, USA

K.M. Wilson (✉)
Harvard T.H. Chan School of Public Health, 677 Huntington Ave., Kresge 920a, Boston, MA 02115, USA
e-mail: kwilson@hsph.harvard.edu


M. Loda et al. (eds.), *Pathology and Epidemiology of Cancer*,
DOI 10.1007/978-3-319-35153-7_17

**Table 17.1** Clinical symptoms in patients presenting with RCC

| RCC growth | Clinical manifestations |
|---|---|
| Local tumor growth | Hematuria, flank pain, abdominal mass, perirenal hematoma |
| Metastases | Persistent cough, bone pain, cervical lymphadenopathy, constitutional symptoms (weight loss/fever/malaise) |
| Obstruction of the inferior vena cava | Bilateral lower extremity edema, nonreducing or right-sided varicocele |
| Paraneoplastic syndromes [132] | Elevated ESR (56 %), hypertension (38 %), anemia (36 %), cachexia/weight loss (35 %), pyrexia (17 %), abnormal liver function (14 %), hypercalcemia (5 %), polycythemia (4 %), neuromyopathy (3 %), amyloidosis (2 %) |

*ESR* erythrocyte sedimentation rate

**Table 17.2** Extrarenal manifestations of patients with familial RCC syndromes

| Syndrome | Gene | Chromosome | Type of RCC | Extra-renal manifestations |
|---|---|---|---|---|
| Von Hippel–Lindau | VHL | 3p25–26 | Clear cell RCC | Hemangioblastomas of the central nervous system, retinal angiomas, pheochromocytoma, epididymal cyst adenomas, pancreatic cysts |
| Hereditary papillary RCC | c-Met | 7q31 | Type 1 papillary RCC | None |
| Familial leiomyomatosis and RCC | Fumarate hydratase | 1q42 | Typ2 2 papillary RCC | Cutaneous leiomyomas, uterine leiomyomas |
| Birt–Hogg–Dubé | BHD1 | 17p12q11 | Chromophobe RCC, oncocytoma, hybrid/oncocytic tumors, occasional clear cell RCC | Cutaneous fibrofolliculomas, lung cysts, spontaneous pneumothorax |

evaluation of patients with a renal mass. Physical exam may reveal extrarenal manifestations found in familial RCC syndromes (Table 17.2).

### 17.2.1 Diagnosis

Prior to computed tomography (CT), renal masses were diagnosed with intravenous pyelogram or renal arteriography. The majority of renal masses are now detected with routine imaging. Focused imaging with CT or magnetic resonance imaging (MRI) can be used for diagnosis and staging renal masses. It is important to consider that, depending on the size of the mass, up to 22 % of small renal masses (≤4 cm) treated surgically are found to be benign [5]. Imaging is generally used to clinically stage all patients (Table 17.3).

Renal mass biopsy has become an important tool for initial evaluation and management of incidentally found small renal masses. Renal mass biopsies are typically considered in patients with suspected infection, extrarenal metastases, lymphoma, and/or in poor operative candidates in preparation for minimally invasive treatments (radiofrequency ablation vs. cryotherapy) or active surveillance. Recent reports have demonstrated a false negative rate of less than 1 % for renal mass biopsies [6].

The most common histologic subtype of RCC is clear cell (70–80 %) followed by papillary RCC (10–15 %), chromophobe RCC (3–5 %), collecting duct carcinoma (<1 %), multilocular cystic ccRCC (uncommon), renal medullary carcinoma (rare), RCC associated with Xp11.2 translocations/TFE3 gene fusions (rare), mucinous tubular and

**Table 17.3** TNM staging of renal cell carcinoma [133]

| Stage | Tumor size | Localization | Description |
|---|---|---|---|
| T1 | Diameter ≤7 cm | Localized | Tumor confined to the kidney |
| T2 | Diameter >7 cm | Localized | Tumor confined to the kidney |
| T3 | Any size | Regional | Tumor extends into major veins or perinephric tissues but not into the ipsilateral adrenal gland and not beyond Gerota fascia |
| T4 | Any size | Metastatic | Tumor invades beyond the Gerota fascia |

spindle cell carcinoma (rare), and unclassified RCC (1–3 %) [7]. Additional details on the histology can be found in Chap. 18.

### 17.2.2 Screening

Due to the relatively low incidence of RCC in the general population, there is no indication for widespread implementation of screening. Screening patients with RCC for somatic mutations has been recommended in younger patients (45 years or younger) presenting with a family history of RCC, bilateral/multifocal renal masses, associated extrarenal clinical manifestations, and/or certain specific tumor histologies [8].

### 17.2.3 Treatment

The initial treatment provided to patients presenting with RCC depends on clinical stage and performance status at presentation. For localized kidney cancer (non-metastatic) the standard of care is radical or partial nephrectomy. Recent studies have focused on the long-term outcomes of patients receiving nephron-sparing surgery (NSS) due to the increased risk of chronic renal disease after radical nephrectomy. Alternatively, outcomes for partial nephrectomy for renal masses that are less than or equal to 7 cm but confined to the kidney (clinical stage T1) have been shown to have equivalent oncologic outcomes compared to radiofrequency ablation (RFA) [9]. Other option for management of small renal masses (≤4 cm) includes active surveillance where masses are followed with serial imaging to establish growth kinetics and intervention is performed if the mass demonstrates significant growth.

The role of cytoreductive nephrectomy or "debulking nephrectomy" in patients with metastatic disease evidence of metastatic disease has also been established [10]. Studies demonstrated an improvement in progression-free survival in patients who underwent cytoreductive nephrectomy for metastatic RCC. Ongoing clinical trials are evaluating the role of cytoreductive nephrectomy in the era of targeted drug therapy. Furthermore, clinical trials are ongoing to evaluate the role of neoadjuvant targeted therapy to improve the chances of performing nephron-sparing surgery in patients who would otherwise only be candidates for a radical nephrectomy [11].

### 17.2.4 Prognosis

Pathologic stage, tumor size, nuclear grade, histologic subtype, and molecular subtypes have the greatest utility for prognosis. Pathologic stage is the most important prognostic factor [12]. Five-year relative survival among RCC patients diagnosed between 1992 and 2007 in the U.S. is 73 % among white patients and 68 % among black patients [13]. Among white patients, 5-year relative survival is 93 % for those diagnosed with localized disease, 66 % for regionally spread disease, and 10 % for distant metastasis [13].

## 17.3 Descriptive Epidemiology

Globally, there were an estimated 337,860 new cases of kidney cancer and 143,406 deaths due to kidney cancer in 2012 [14]. U.S. and international cancer statistics generally combine RCC and cancer of the renal pelvis, with RCC comprising approximately 90 % of total kidney cancer.

There is considerable variation in incidence of kidney cancer between geographical regions (Fig. 17.1). Age-standardized incidence rates vary from 1.2 per 100,000 in Africa to 11.7 per 100,000 in North America [14]. Variation in mortality rates is lower, with age-standardized mortality rates ranging from 1.0 per 100,000 in Africa to 3.1 per 100,000 in Europe [14].

Rates of kidney cancer are higher among men than women worldwide, with rates approximate two times higher in men (Fig. 17.1). Among men and women, and in countries with lower and higher incidence, incidence rates begin to increase around age 30 and continue to rise until after age 70. There is some suggestion of a plateau or decrease in incidence rates by age 80 (Figs. 17.2 and 17.3).

Incidence of kidney cancer has been increasing in recent decades worldwide. Age-standardized incidence rates of kidney cancer over time are shown for several countries in

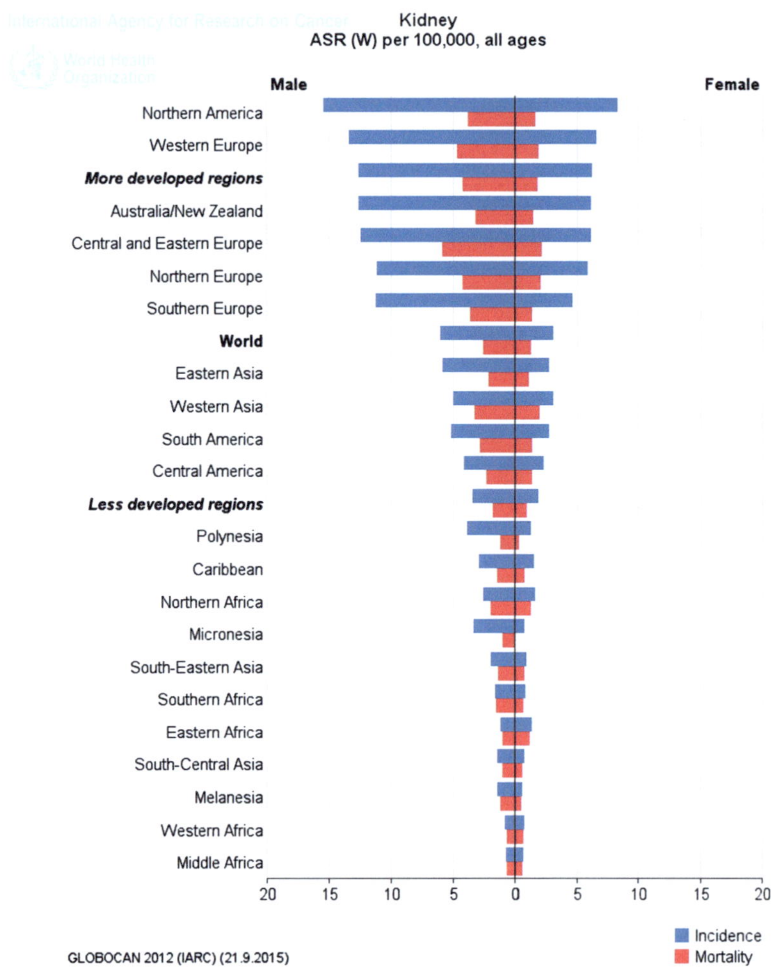


**Fig. 17.1** Age-standardized incidence and mortality rates of kidney cancer by region, 2012 [14]

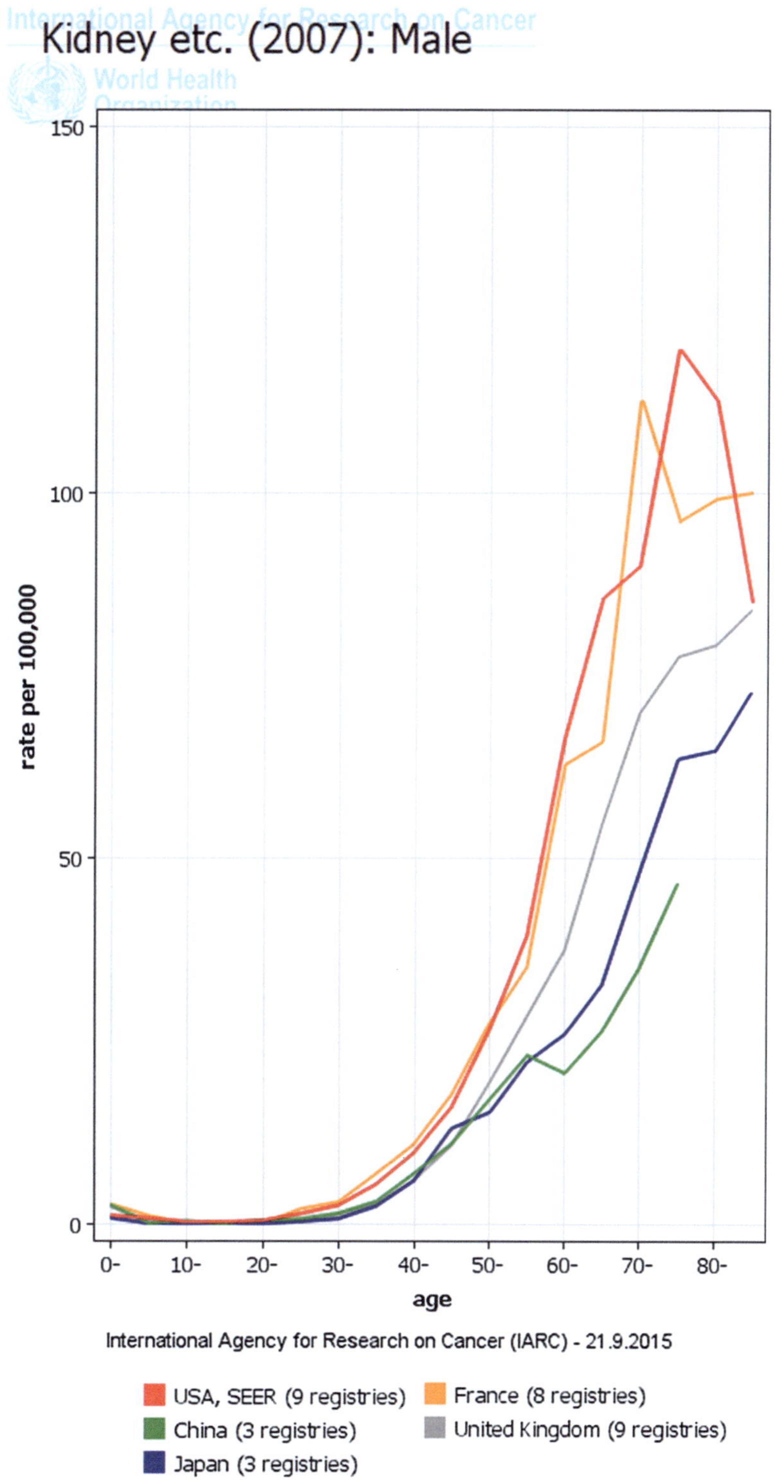


**Fig. 17.2** Kidney cancer incidence by age among men [14]

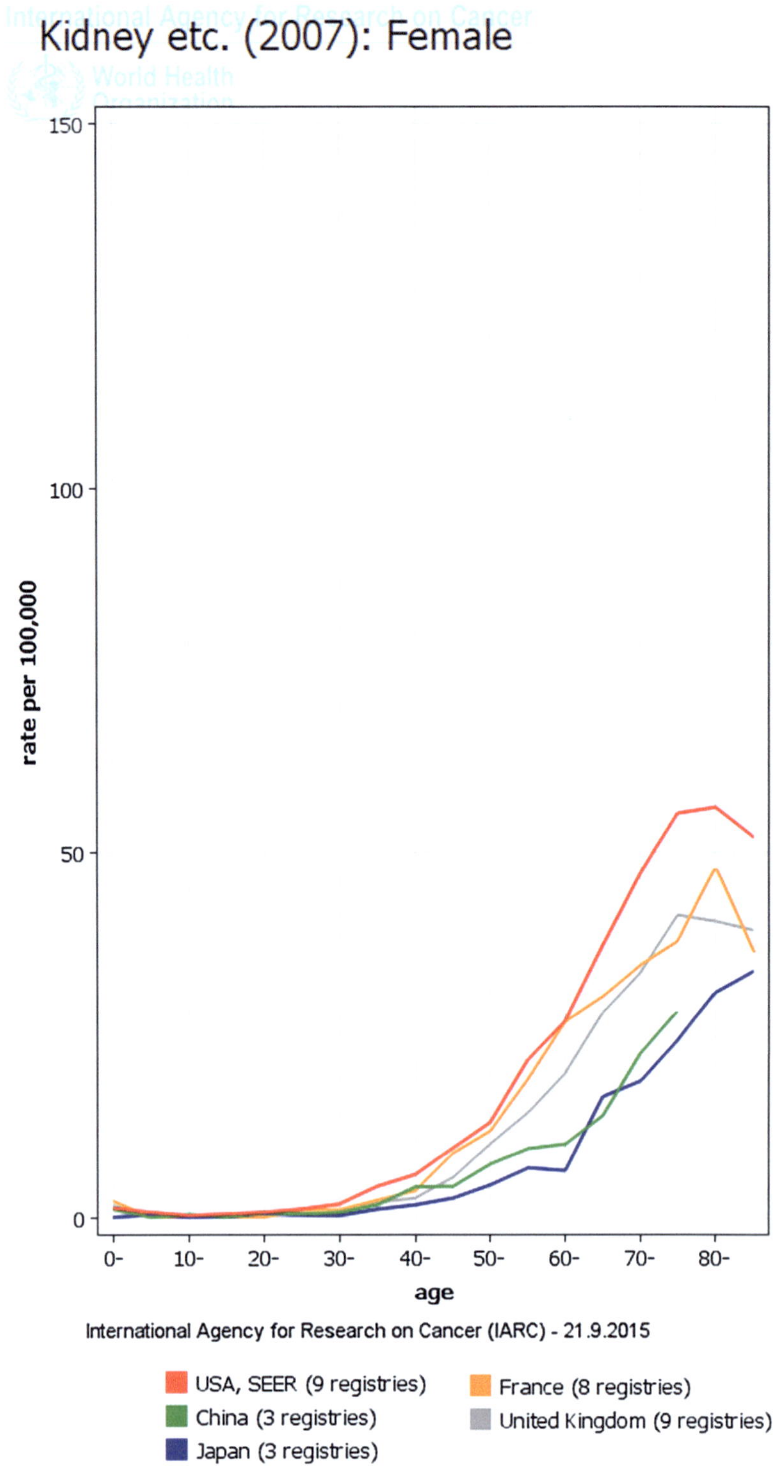


**Fig. 17.3** Kidney cancer incidence by age among women [14]

Fig. 17.4. In the U.S., the annual percent change in age-standardized incidence rates has been approximately 2 % per year throughout the period from 1975 to 2012 [15]. The increase in incidence is likely due in part to increased diagnosis related to use of imaging technologies including ultrasound and computer tomography (CT) scanning. Increased prevalence of risk factors for kidney cancer including diabetes, obesity, and hypertension may also play a role. The increased incidence of kidney cancer in the U.S. has been largely seen for localized cancers, suggesting that increased diagnosis due to increased use of imaging has played a key role in the increase in incidence; however, smaller increases in the diagnosis of more advanced cancers suggests that other factors have also played a role [2]. Mortality rates increased with incidence rates from the 1970s to the 2000s, but have plateaued in the past decade in the US and Europe, with suggestions of a decrease in some countries in Northern and Western Europe [3, 15, 16]. In the U.S., the annual percent change in age-standardized kidney cancer mortality rates was 1.3 % for 1980–1989 and 0.4 % for 1990–2012, and then decreased −0.9 % from 2003 to 2012 [15].

In the U.S., kidney cancer incidence and mortality rates are higher in blacks than in whites (Fig. 17.5). Incidence and mortality rates have also increased more since the 1970s for blacks than for whites, with an age-standardized annual percent change of 2.8 % for blacks and 2.1 % for whites from 1975 to 2012, and an increase in mortality rates of 0.8 % in blacks and 0.3 % in whites from 1969 to 2012 [15]. Median age at diagnosis is lower in blacks than whites [13, 17]. The reasons for these differences are not clear; however, the prevalence of obesity and hypertension are higher in blacks than in whites [18].

## 17.4 Risk Factors

Smoking and obesity-related traits including obesity, hypertension, and diabetes have been consistently identified as risk factors for RCC. Several reproductive factors among women also appear to be associated with RCC risk. There is also some evidence for associations with analgesic use, physical activity, alcohol intake, and other aspects of diet. The association between these factors and RCC incidence is discussed below; data on the associations with survival after diagnosis are also discussed when available.

### 17.4.1 Smoking

The International Agency for Research on Cancer (IARC) considers there to be "sufficient" evidence that cigarette smoking causes RCC [19]. A 2005 meta-analysis [20] of 19 case-control and 5 cohort studies found the relative risk of RCC for ever smokers versus never smokers was 1.38 (95 % CI 1.27–1.50), with a dose-dependent increase in risk for number of cigarettes per day (Table 17.4). Associations of cigarette smoking were stronger in cohort and population-based case-control studies compared to hospital-based case-control studies. The IARC review from 2004 found adjustment for hypertension or body mass index (BMI) did not appear to have a large impact on the smoking associations in studies that reported relative risks with and without adjustment for these other risk factors [20].

The relative risk for ever smoking was slightly higher in men than in women among the five cohort studies included in the 2005 meta-analysis (Men: Relative Risk (RR) 1.54, 95 % Confidence Interval (CI) 1.42–1.68; Women: RR 1.22, 95 % CI 1.09–1.36). Three cohort studies have been published since the meta-analysis. One in the Nurses' Health Study (women) and the Health Professionals Follow-up Study (men) found a significant trend for increased risk of RCC with increasing pack-years of smoking among men ($p = 0.003$) and a borderline significant trend among women ($p = 0.09$) [21]. Another found significant trends in both men and women ($p < 0.001$ for men, $p = 0.02$ for women) [22]. A third found significantly increased risk for 22.5 or more pack-years of smoking among men and women combined [23].

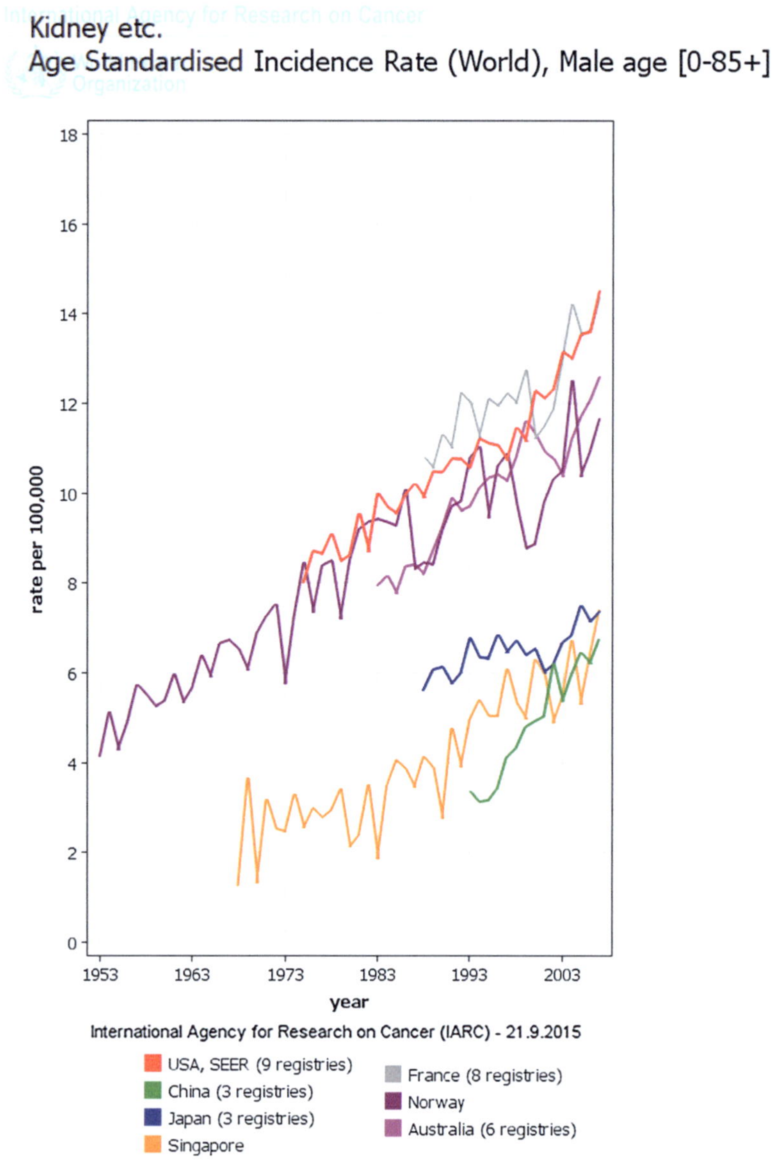


**Fig. 17.4** Incidence of kidney cancer over time in selected countries [14]

The 2005 meta-analysis found some suggestion that increased years since quitting smoking was associated with a lower risk of RCC; however, this was seen only among men, and there was significant heterogeneity between studies. Five [24–28] of six [29] studies included in the 2004 IARC review found significant negative trends with increasing number of years since quitting. The time required for the relative risk to return to that of never smokers varied across studies and ranged from 10 years to greater than 20. A study published after the IARC review found that years since quitting was associated with a linear decrease in risk of RCC, but that 30 years were required for risk to return to that of never smokers [30].

Two studies have examined smoking and risk of specific histological subtypes of RCC. One study based on data from two large case-control studies, one in the US with population-based controls and one in Europe with hospital-based controls, found no association overall between

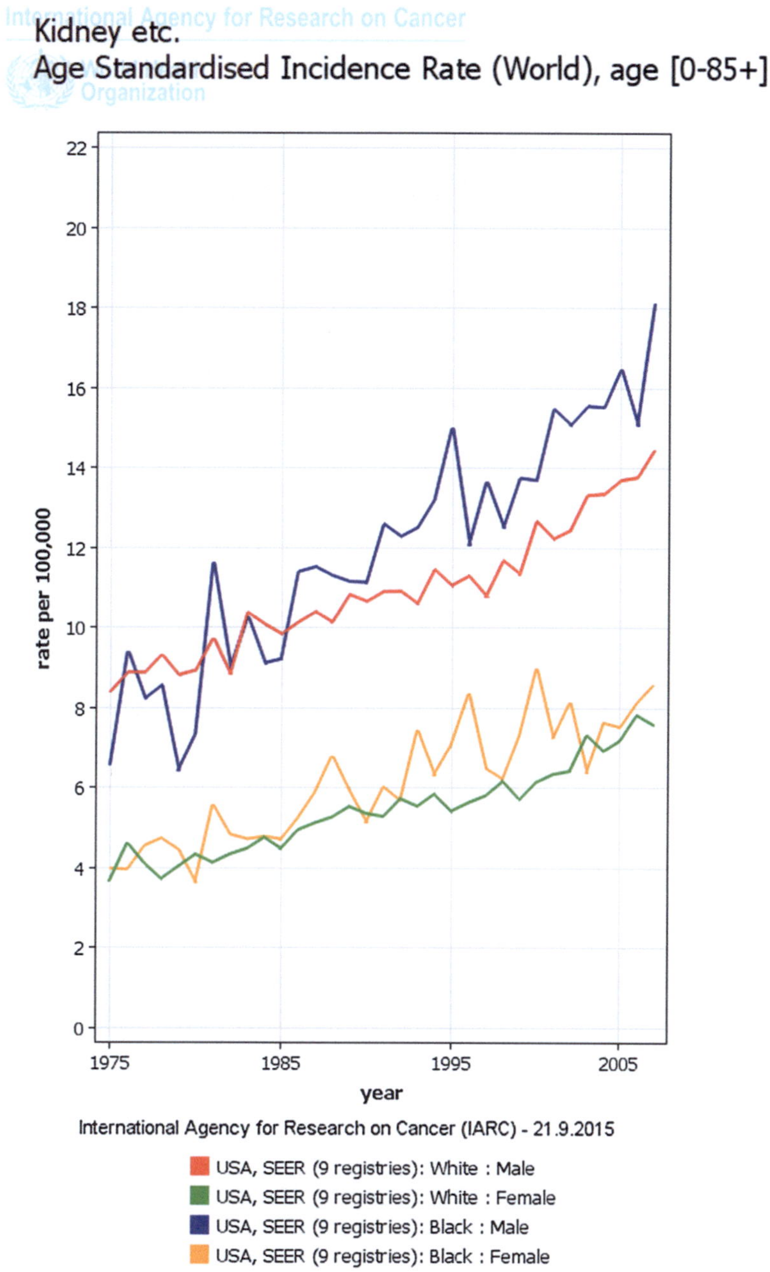


**Fig. 17.5** Kidney cancer incidence over time in the US by race and sex [14]

**Table 17.4** Relative risk of RCC by smoking history, from meta-analysis of 24 studies [20]

| Smoking category | Men | Women |
|---|---|---|
| Never smokers | 1.00 (reference) | 1.00 (reference) |
| Ever smoker, 1–9 cigs/day | 1.60 (1.21–2.12) | 0.98 (0.71–1.35) |
| Ever smoker, 10–19 cigs/day | 1.83 (1.30–2.57) | 1.38 (0.90–2.11) |
| Ever smoker, 20+ cigs/day | 2.03 (1.51–2.74) | 1.58 (1.14–2.20) |

smoking and any subtype of RCC; however, the US study found an increased risk of clear cell and papillary subtypes, but not of chromophobe [31]. Consistent with this, a study comparing 705 consecutive RCC cases with 111 cancer-free nephrectomy patients found that smoking was associated with clear cell and papillary, but not with chromophobe RCC [32].

An IARC review of involuntary smoking in 2004 found no evidence on environmental tobacco smoke (ETS) and RCC risk. At that time only one study had examined the issue and found nonsignificant increased risks for both men and women reporting more than 8 h per day of ETS exposure [29]. Two more recent population-based case-control studies found significantly increased risks of RCC with both home and occupational ETS exposure [33, 34].

Survival. A meta-analysis of 5 studies examining smoking and disease-specific mortality in RCC patients found that current smoking was significantly associated with poorer survival, with a hazard ratio of 1.5 (95 % CI 1.1–2.1) compared to never smokers [35]. Current smoking was also associated with increased overall mortality, poorer overall survival, poorer cancer-specific survival, and worse progression-free survival across 14 studies of smoking in RCC patients.

Current smokers tend to be diagnosed with more advanced (higher stage) disease, and it is not clear whether smoking is associated with poorer prognosis independent of stage at diagnosis. A study of 2242 surgically treated clear cell RCC patients from the Mayo Clinic found no association between current smoking and risk of RCC-specific death after adjustment for stage, with a hazard ratio of 1.03 (95 % CI 0.85–1.23) [36]. However, a clinical study of 1809 patients found that smoking was independently associated with survival among patients diagnosed with non-metastatic cancer, but not among those with metastasis at diagnosis. Among patients with non-metastatic disease, each pack-year of smoking was associated with a 1 % increased risk of cancer-specific death ($p = 0.008$), with adjustment for stage, grade, and other clinical prognostic factors [37]. As in the Mayo Clinic study, smoking was associated with higher stage and grade at diagnosis in this study population. The role of smoking in RCC survival warrants further investigation, perhaps with incorporation of tumor biomarkers that could shed light on whether smoking plays an independent role in cancer progression after diagnosis and treatment.

### 17.4.2 Hypertension and Renal Cell Carcinoma

Hypertension has been consistently associated with risk of RCC, independent of obesity, smoking, and diabetes. Because RCC may increase the risk of hypertension through tumor secretion of renin, renal artery stenosis, renal failure, or other means, the direction of observed associations in epidemiological studies has not always been clear. However, a sufficient number of studies have found that a long-term history of hypertension is associated with RCC risk and that there is a dose–response relationship between higher blood pressure and RCC risk to conclude that hypertension is a risk factor for the development of RCC [21, 22, 38–42]. In addition, a Swedish cohort study of men [38] with multiple measures of blood pressure over time found that increases over time were associated with increased risk and decreases were associated with decreased risk, suggesting that effectively controlling hypertension may reduce risk of RCC. A study of risk factors according to histological subtypes of RCC found no differences in the hypertension–RCC relationship between subtypes [31].

The relative risk of RCC for those with a history of hypertension diagnosis compared to those without is approximately 1.5–2.0 across studies [21, 22, 41, 42]. The relative risk associated with high systolic blood pressure (with definitions ranging from >130 to >160 mm Hg) compared to normal (typically <120 mm Hg) ranges from 1.5 to 2.5 across studies. For high diastolic blood pressure (>90 or >100 mm Hg, compared to <80 mm Hg), the relative risk ranges from 1.5 to 2.3 [38–40, 42]. These risks associated with hypertension are independent of obesity and smoking.

Some studies reported an increased risk of RCC with use of antihypertensive medications; however, the EPIC study, which prospectively measured blood pressure in nearly 300,000 people, found that medication use was only associated with RCC when hypertension was poorly controlled, suggesting that hypertension itself drives the observed associations with antihypertensive medications [40].

A population-based case-control study in Detroit and Chicago found that hypertension diagnosis was associated with RCC risk in both whites and blacks, and that risk increased with increasing time since diagnosis, reaching a 4.1-fold (95 % CI 2.3–7.4) for blacks and 2.6-fold (95 % CI 1.7–4.1) for whites higher risk after 25 years [43]. Another study based in the Kaiser Permanente Northern California health care network also found similar associations between hypertension and RCC risk across races [42]. This suggests that the increased incidence of RCC among African-Americans may be due, in part, to the increased prevalence of hypertension in this group. Interestingly, the angiotensin receptor inhibitors, a class of antihypertensives, has been shown to improve overall and progression-free survival in patients with metastatic RCC [44].

### 17.4.3 Obesity

A meta-analysis of cohort studies with 15,144 cases and 9,080,052 participants found increased risks of RCC with increased BMI [45]. The pooled relative risk for overweight (BMI 25–<30 kg/m$^2$) was 1.28 (95 % CI 1.24–1.33), and for obese (BMI ≥30 kg/m$^2$) was 1.77 (95 % CI 1.68–1.87) compared to BMI of 18.5–<25 kg/m$^2$). There was no evidence of heterogeneity across studies. Relative risks were somewhat stronger for women than for men (RR for obesity of 1.63, 95 % CI 1.50–1.77 for men; 1.95, 95 % CI 1.81–2.10 for women).

An analysis of two case-control studies of RCC examining risk factors by histological subtype found that BMI was associated with risk of clear cell (odds ratio (OR) 1.2, 95 % CI 1.1–1.3 per 5 kg/m$^2$ increase) and chromophobe (OR 1.2, 95 % CI 1.1–1.4) but not papillary RCC (OR 1.1, 95 % CI 1.0–1.2, *p*-value for difference from clear cell = 0.006) [31]. An Italian hospital-based case-control study also found a suggestion that higher BMI at age 30 was more strongly associated with clear cell than with non-clear cell histology (*p*-value for interaction = 0.08) [46].

Survival. Multiple clinical cohorts of patients treated for RCC, usually with surgery, have found that obesity is associated with improved survival [47, 48]. This has given rise to an "obesity paradox" which stipulates that while obese people are more likely to be diagnosed with RCC, they appear less likely to die of the disease.

A meta-analysis [47] of 15 studies of BMI and cancer-specific mortality found a pooled relative risk of 0.66 (95 % CI 0.53–0.81) for a 5 kg/m$^2$ increase in BMI, with evidence of significant heterogeneity across studies. The heterogeneity may be partially explained by geographical differences, with a stronger association in Asian compared to European and American studies, and to adjustment for presence of symptoms at diagnosis, with a stronger association in studies that adjusted for symptom presence. Sex does not appear to have been examined as a source of heterogeneity in the meta-analysis. However, a Japanese study of 435 patients surgically treated for RCC found that obesity was associated with better prognosis in men, but not in women [49]. A study of 2769 patients surgically treated for non-metastatic RCC in Korea found that higher BMI was associated with significantly improved cancer-specific survival in clear cell RCC, with significantly worse cancer-specific survival in chromophobe RCC, and was not associated with survival in papillary RCC [50]. The lack of association with papillary RCC is consistent with the observations for incidence as well.

It has been hypothesized that obese patients develop a biologically less aggressive disease. Supporting this, a study in a subset of 126 patients from a clinical cohort surgically treated at Memorial Sloan-Kettering Cancer Center who had data available from The Cancer Genome Atlas Project found significantly lower gene expression of fatty acid synthase (FASN) in obese patients [48]. FASN expression, in turn is associated with increased cancer-specific mortality in clear cell RCC.

However, the "obesity paradox" may also be explained by methodological problems with the study designs, rather than by any underlying biology. Reverse causation is a major concern, given that the evidence comes from clinical cohorts with measures of obesity at the time of diagnosis or treatment. At that point, there may have been weight loss due to undiagnosed disease, which likely correlates with disease severity. In addition, these clinical cohorts likely suffer from selection bias, as they tend to be based among surgically treated RCC patients, rather than among all patients diagnosed with RCC, regardless of treatment strategy. Finally, another form of selection bias is a methodological problem in studies of disease survival when the exposure of interest is also a risk factor for disease incidence [51]. Given these potential limitations, more study is needed to understand the role of obesity in RCC survival.

### 17.4.4 Height

A recent review and meta-analysis for the World Cancer Research Fund (WCRF)/American Institute for Cancer Research (AICR) Continuous Update Project found a significant positive association between adult attained height and kidney cancer risk. Across 10 studies included in the dose–response meta-analysis, with 9874 cases, a 5 cm increase in height was associated with a 10 % increased risk of kidney cancer (95 % CI 1.08–1.12) [52]. The association was similar for men and for women. In two studies of kidney cancer mortality, height was nonsignificantly inversely associated with mortality in one study [53] and nonsignificantly inversely associated with mortality in the other [54].

### 17.4.5 Diabetes and Diabetes Medications

Diabetes mellitus (DM) has been associated with an increased risk of various cancers [55]. There is particular interest in studying this association in renal cell carcinoma (RCC) due to advances in genetic sequencing that have allowed the identification of metabolic alterations that are key drivers of disease [1]. Furthermore, the most well-established risk factors for RCC are hypertension and obesity, which are components of the metabolic syndrome. The proposed mechanisms through which diabetes may exert an effect on RCC risk include insulin resistance, hyperinsulinemia, increased growth factors, and inflammation [56].

Studies evaluating the association of DM on incident and fatal RCC have yielded conflicting results. A recent meta-analysis of 18 case-control and cohort studies found an increased risk of RCC in patients with DM (RR 1.40, 95 % CI 1.16–1.69). The risk for women was somewhat higher than for men (RR for women 1.47, 95 % CI 1.18–1.83; RR for men 1.28, 95 % CI 1.10–1.48) [57]. Results were very similar in the 5 studies that adjusted for BMI, smoking, and alcohol intake (Only 3 of 18 studies adjusted for hypertension, and the effect of this adjustment was not examined in the meta-analysis). Among 8 cohort studies that evaluated the association between DM and RCC mortality, the pooled RR was 1.12 and not statistically significant (95 % CI 0.99–1.20) [57].

Studies varied in their ascertainment of diabetic status (physician confirmed vs. self-reported) and exclusion of type 1 diabetes. Most studies were not able to assess the association between severity of DM (i.e., HgA1c or diabetic complications) and incident or fatal RCC.

Survival. A common challenge faced in published series analyzing differences in outcomes between diabetic and nondiabetic cases is the potential for confounding introduced by a substantial imbalance in clinical demographic features between the two groups. Most studies have not found a difference in RCC histologic subtype, grade, or stage at presentation in patients with DM compared to those without diabetes [58–60]. However, two studies found that patients with DM presented with higher grade disease [61, 62].

Several studies suggest a positive association between DM at time of surgery and survival outcomes. The largest study comes from a multi-institution retrospective cohort of 2597 patients (14 % with DM) with localized RCC (pT1–2). Patients with DM had a worse recurrence free survival (RFS), cancer-specific survival (CSS), and overall survival (OS) [63]. However, this study was limited by short follow-up, with a median of 3 years. Studies with longer follow-up have reported conflicting results. A study from the Mayo clinic of 257 patients with diabetes and matched nondiabetic patients treated surgically for RCC found that those patients with DM had worse DSS and OS over a median 8.7 years of follow-up [64]. However, other studies have reported similar outcomes between patients with and without DM [60, 62]. Further studies with longer follow-up and competing risks analyses are needed to assess the effect of DM on survival outcomes in patients with DM.

### 17.4.6 Analgesic Use

Both acetaminophen (Tylenol) and nonaspirin NSAIDs (ibuprofen, naproxen) have been associated with risk of RCC. A 2013 meta-analysis of 11 case-control studies and 3 cohort studies found a pooled relative risk for acetaminophen use (any use or regular use, depending on the study, compared to no use) of 1.28 (95 % CI 1.15–1.44) [65]. There were no significant differences by study design, country, or outcome (5 of the 14 studies used combined "kidney cancer" as the outcome rather than RCC). Nine of the studies also examined high intake of acetaminophen and found a pooled RR of 1.68 (95 % CI 1.22–2.30). In five studies that assessed duration of use there was no association between longer duration and risk of kidney cancer. Results were similar among 10 studies that adjusted for (at least) BMI and smoking.

Use of nonaspirin NSAIDs was also associated with significantly increased risk, with a pooled RR of 1.25 (95 % CI 1.06–1.46) for any or regular use compared to no use across five studies, 3 case-control and 2 cohort. Two of these studies [66] looked at higher intakes, with a significantly increased risk, and one [67] looked at duration of use, with an increased risk for 10 more years of use. Among 3 studies that adjusted for (at least) BMI and smoking, results were somewhat stronger (pooled RR 1.38, 95 % CI 1.16–1.65). However, an additional large cohort study, the NIH-AARP cohort, with 1084 cases of RCC, was published after this meta-analysis and found no association between nonaspirin NSAID use and RCC [68].

The 2013 meta-analysis found no association between aspirin use and kidney cancer across 14 studies [65]. Since then, the NIH-AARP cohort also found no association between aspirin use and RCC risk [68].

The biological mechanisms for the observed associations for acetaminophen and nonaspirin NSAIDs are not clear. Acetaminophen is a metabolite of phenacetin, an analgesic banned in the US since 1983, which causes renal failure and cancers of the renal pelvis [69]. However, the association for acetaminophen was seen in studies focusing on RCC in the meta-analysis [65], and it has been shown to induce kidney tumors in mice [70, 71]. NSAIDs inhibit renal synthesis of prostaglandins, which can result in chronic subacute renal injuries [72–74]; this could theoretically lead to carcinogenesis.

### 17.4.7 Reproductive Factors and Hormones

Sex differences in renal cell carcinoma (RCC) incidence suggest the possible role of hormonal factors. As a result of this observed difference, hormonal therapy for advanced RCC was the subject of clinical trials in the 1970s and 1980s, preceding the advent of cytokines; however, response rates were low and there is currently no established role for hormonal agents in the management of RCC [75].

The androgen receptor has consistently been found to be expressed in RCC [76, 77]. Furthermore, androgen receptor mRNA expression has been associated with prognosis [78]. Recent studies have also linked androgen receptor function to RCC progression through its influence on HIF2α[alpha]/VEGF signaling [79].

Polymorphisms in the estrogen receptor have been associated with RCC [80]. Recently, studies have evaluated the role of estrogen as a possible inhibitor of carcinogenesis in RCC. Estrogen receptor-β[beta] may have a role in decreasing cell proliferation and inducing apoptosis [81]. Estrogen receptor-α[alpha] expression has also been implicated in RCC risk [82]. Animal and in vitro models have demonstrated a potential role of estrogen and progesterone in the development of RCC [83].

Parity has been associated with risk of RCC, with a 10–15 % higher risk of RCC per childbirth, and an increased risk for earlier age at first birth [84–88]. However, other studies have found no such associations [89]. Associations with oral contraceptive use and postmenopausal hormone use have been inconsistent [84–86, 90–92].

### 17.4.8 Physical Activity

A meta-analysis of physical activity and kidney cancer from 2013 based on 19 studies found a relative risk for "high" versus "low" physical activity of 0.88 (95 % CI 0.79–0.97) [93]. Among studies in the top tertile of the methodologic quality score the estimate was somewhat stronger (RR 0.78, 95 % CI 0.66–0.92). Estimates were virtually identical for cohort and case-control studies, and for studies with and without adjustment for obesity, diabetes, hypertension, and smoking. Estimates for total physical activity, occupational activity, and recreational physical activity were also similar. The meta-analysis included all kidney cancer as an outcome, but 9 of the 12 estimates from cohort studies of kidney cancer incidence were based only on renal cell carcinoma, while 3 included all "kidney cancer."

Survival. Data on the association between physical activity and survival among kidney cancer patients is lacking. However, one study among 703 patients with kidney cancer found a significant improvement in quality of life among those engaging in 150 min or more per week of moderate physical activity compared to those who were completely sedentary [94].

### 17.4.9 Diet

A number of dietary factors have been studied with respect to kidney cancer risk. Of these, only alcohol intake has been consistently linked to risk.

#### 17.4.9.1 Alcohol

A recent report from the AICR Continuous Update Project found "strong evidence" that alcohol intake up to 30 g per day, or about 2 drinks, decreases the risk of kidney cancer [95]. A dose–response meta-analysis of 7 studies found a relative risk of 0.92 (95 % CI 0.86–0.97) per 10 g of alcohol per day; this was based on 3525 RCC cases.

A pooled analysis of 12 prospective cohort studies (3 of which were also included in the AICR meta-analysis) [96] found that compared with nondrinking, ≥15 g alcohol consumption per day (equivalent to slightly more than one drink) was associated with a 28 % lower risk of RCC (95 % CI for RR 0.60–0.86). There were statistically significant inverse trends for both women and men. There were no significant differences by type of alcohol (beer, wine, or liquor).

The AICR review concluded that there was insufficient evidence regarding intakes of alcohol beyond 30 g per day. The pooled analysis found significant evidence of nonlinearity in the association, with a linear inverse association up to 30 g per day and a fairly flat relationship beyond that [96]. The European Prospective Investigation into Cancer and Nutrition (EPIC) [96], found significantly decreased risks for intakes up to 60 g per day, but no association for greater than 60 g.

#### 17.4.9.2 Other Beverages and Total Fluid Intake

It has been hypothesized that the inverse association with alcohol intake might be due to a diluting effect on carcinogens through higher total fluid intake. However, three prospective cohort studies found no association between total fluid intake and risk of RCC [97–99].

There is some suggestion that tea and coffee intake is associated with lower risk of RCC. A pooled analysis of 13 prospective cohort studies with 7–20 years of follow-up and a total of 1478 incident renal cell carcinoma cases found a pooled relative risk of 0.84 (95 % CI 0.67–1.05) for 3 or more cups of coffee per day compared to less than one cup per day [100]. The relative risk for one or more cups of tea per day compared to nondrinkers was 0.85 (95 % CI 0.71–1.02). The inverse association for coffee (3 or more cups/day vs. <1 cup/day) was statistically significant among women (RR 0.71, 95 % CI 0.53–0.97), but not among men (RR 1.00, 95 % CI 0.73–1.37), although this difference by sex was not statistically significant. In contrast, the association for tea was statistically significant among men (RR 0.72, 95 % CI 0.55–0.94), but not women (RR 0.98, 95 % CI 0.77–1.26). The differences by sex raise the possibility that the results are due to chance; however, it is also possible that dietary risk factors for RCC vary by sex, as has been observed for other risk factors.

#### 17.4.9.3 Other Foods and Nutrients

A pooled analysis of 13 prospective cohort studies found evidence that fruit and vegetable intake was associated with lower risk of RCC [101], with a relative risk of 0.68 (95 % CI 0.54–0.87) for 600 or more grams per day compared to less than 200 g/day. Inverse associations were seen for carotenoid intakes, with a significantly reduced risk with higher intake of beta-carotene. In the same set of 13 studies, intakes of meat, fat, and protein were not associated with risk [102]. However, a report from the National Institutes of Health (NIH)-AARP Diet and Health Study cohort found a significantly increased risk of RCC among those with higher intakes of meat cooking carcinogens (heterocyclic amines and polycyclic aromatic hydrocarbons) [103]. This association was observed specifically for papillary RCC, but not for clear cell RCC.

### 17.4.10 Other Medical Conditions

#### 17.4.10.1 Chronic Kidney Disease

Acquired cystic kidney disease, which occurs in end-stage renal disease (ESRD) with progressive development of cysts in a poorly functioning or nonfunctioning kidney, is strongly associated with the development of RCC [104]. Acquired cystic disease is seen in 7–22 % of patients with ESRD prior to dialysis, but the proportion increases to about 90 % after 10 years of dialysis. RCC arising in ESRD patients has distinct clinical and pathological features from RCC diagnosed in the general population.

The incidence of RCC in patients with ESRD has been reported to be up to 40–100-times higher than in the general population [105]. This increased risk continues even after renal transplantation. RCCs arising in native kidneys of patients with ESRD are less aggressive than renal tumors in the sporadic or non-ESRD setting [106]. Furthermore, patients with chronic renal insufficiency have been shown to have an increased risk of papillary RCC [107]. It is unclear if the underlying renal damage is itself a carcinogenic event or else the cause of normal cells transformation is related to other mechanisms such as the presence of circulating carcinogens or an immune system dysfunction [108].

#### 17.4.10.2 Previous Cancers and Cancer Treatment

Increased risk of kidney cancer as a second primary malignancy has been observed for people with Hodgkin's lymphoma, testicular cancer, cervical cancer [109], and breast cancer [110]. It is not clear whether the increased risk is due to shared genetic or environmental risk factors, increased screening or imaging studies, or treatment-related effects from the first cancer [109]. A recent study in Taiwan found that patients treated for thyroid cancer with radioactive iodine were at significantly increased risk of kidney cancers, along with other cancers, particularly among those exposed to higher cumulative radioactive iodine doses [111].

## 17.5 Genetics of Renal Cell Carcinoma

Individuals with a family history of RCC in a first degree relative have a 2-fold increased risk of developing RCC themselves [112]. Familial RCC syndromes have provided a great deal of understanding of the genetic predisposition to RCC. All known familial RCC syndromes have an autosomal dominant inheritance pattern. Genes relating to these syndromes are summarized below and presented in greater detail in Chap. 18. Most of the familial risk, however, is due to more common genetic variants with lower penetrance.

### 17.5.1 Von Hippel Lindau (VHL)

VHL is located on the short arm of chromosome 3 [113]. Under normal conditions, the VHL protein targets hypoxia-inducible factors (HIFs) HIF1α[alpha] and HIF2α[alpha] for ubiquitin-mediated degradation. The HIFs are overexpressed under conditions of hypoxia and act as transcription factors to regulate vascular endothelial growth factor (VEGF) and other growth factors [114].

Germline VHL mutations are responsible for the VHL syndrome, which is characterized by a high risk of clear cell RCC as well as pheochromocytoma, pancreatic islet cell tumors, central nervous system hemangioblastomas, retinal angiomas, endolymphatic sac tumors of the inner ear, and epididymal cystadenomas. RCC associated with the VHL syndrome tends to be bilateral and multifocal, and often develops at an early age. VHL inactivation is also responsible for the majority of sporadic (non-inherited) clear cell RCCs [115]. The VHL/VEGF pathway is the basis for modern targeted therapies for advanced RCC; the tyrosine kinase inhibitors inhibit the VEGF receptor, while bevacizumab—a monoclonal antibody—targets the VEGF ligand, and the mammalian target of rapamycin (mTOR) inhibitors inhibit the translation and stability of HIF. Prior to elucidation of these pathways, advanced RCCs were exclusively treated with nonspecific immunotherapies.

### 17.5.2 c-Met

The c-Met proto-oncogene, located on chromosome 7, is the cell surface receptor for hepatocyte growth factor (HGF). Germline Met derangements are responsible for hereditary papillary RCC (HPRC) [116]. This disorder is associated with bilateral, multifocal papillary (type 1) RCC [117]. Met mutations have also been implicated in some sporadic papillary RCCs [118].

### 17.5.3 Fumarate Hydratase

Fumarate hydratase (FH) is a tumor suppressor gene located on chromosome 1 which is responsible for the conversion of fumarate to malate in the Krebs/TCA cycle. Mutations of this tumor suppressor gene are responsible for the hereditary leiomyomatosis and RCC syndrome (HLRCC). Affected individuals are at high risk for papillary renal cell carcinoma (type 2), in addition to cutaneous and uterine leiomyomas [119]. RCC associated with HLRCC is of relatively poor prognosis.

### 17.5.4 FLCN

FLCN encodes folliculin and is located on chromosome 17. Germline mutations are responsible for the Birt Hogg Dubé (BHD) syndrome [120], resulting in an increased risk of chromophobe and oncocytic renal tumors [121]. Non-renal manifestations include cutaneous fibrofolliculomas, pulmonary cysts, pneumothorax, and possibly colonic polyps.

### 17.5.5 TSC1 and TSC2

Tuberous sclerosis complex (TSC) is caused by germline mutations of TSC1 or TSC2 on chromosomes 9 and 16 which produce hamartin and tuberin, respectively. This syndrome is associated with a wide range of clinical manifestations. Its renal manifestations include renal angiomyolipoma, cysts, and less commonly, RCC [122]. Clear cell, chromophobe, and oncocytic renal tumors have been observed.

PTEN. Cowden syndrome is characterized by autosomal dominant germline mutations in phosphatase and tensin homolog PTEN. Patients develop multiple hamartomas and are at increased risk of renal, breast, endometrial and thyroid cancers. Clear cell, chromophobe, and papillary neoplasms have been observed [8].

### 17.5.6 GWAS-Identified Susceptibility Genes

The rare germline mutations discussed above do not account for a large part of the familial risk of RCC [123]. Several genome-wide association studies (GWAS) have been done in RCC to identify single-nucleotide polymorphisms (SNPs) associated with more modest increases in disease risk [124–127]. Risk SNPs have been identified at chromosome locations 2p21, 2q22.3, 8q24.21, 11q13.3, 12p11.33 and 12q24.31. A recent meta-analysis of published GWAS also identified a susceptibility locus at 1q24.1, mapping to ALDH9A1 [128].

### 17.5.7 Other Gene Candidates

Germline mutations of the succinate dehydrogenase (SDH) gene results in the familial pheochromocytoma/paraganglioma syndrome and an increased risk of clear cell RCC [129]. Mutations in chromatin remodeling genes, including PBRM1, SETD2, and BAP1, are also associated with clear cell RCC [130, 131]. Genes encoding histone-modifying proteins have been implicated in somatic RCC mutations [114]. A recent study that profiled the molecular signatures of 446 renal tumors identified 19 individual genes that were mutated with various frequencies [130].

## 17.6 Summary

Insights into the molecular underpinnings of RCC have contributed significantly to the understanding of the underlying mechanisms involved in the pathogenesis of RCC. These studies have lead to the development of molecularly targeted treatments for advanced RCC. Several lifestyle factors including obesity, hypertension, smoking, and diabetes have been consistently associated with an increased risk of RCC. Work remains to be done to elucidate the underlying mechanisms by which these risk factors increase the incidence of RCC and how these factors affect survival outcomes in this patient population.

## References


1. Linehan WM, Ricketts CJ. The metabolic basis of kidney cancer. Semin Cancer Biol. 2013;23(1):46–55.
2. Gandaglia G, Ravi P, Abdollah F, Abd-El-Barr AE, Becker A, Popa I, et al. Contemporary incidence and mortality rates of kidney cancer in the United States. Can Urol Assoc J. 2014;8(7–8):247–52.
3. Znaor A, Lortet-Tieulent J, Laversanne M, Jemal A, Bray F. International variations and trends in renal cell carcinoma incidence and mortality. Eur Urol. 2015;67(3):519–30.
4. Zhang JQ, Fielding JR, Zou KH. Etiology of spontaneous perirenal hemorrhage: a meta-analysis. J Urol. 2002;167(4):1593–6.
5. Hilton S, Jones LP. Recent advances in imaging cancer of the kidney and urinary tract. Surg Oncol Clin N Am. 2014;23(4):863–910.
6. Lim A, O'Neil B, Heilbrun ME, Dechet C, Lowrance WT. The contemporary role of renal

mass biopsy in the management of small renal tumors. Front Oncol. 2012;2:106.
7. Lopez-Beltran A, Scarpelli M, Montironi R, Kirkali Z. 2004 WHO classification of the renal tumors of the adults. Eur Urol. 2006;49(5):798–805.
8. Shuch B, Vourganti S, Ricketts CJ, Middleton L, Peterson J, Merino MJ, et al. Defining early-onset kidney cancer: implications for germline and somatic mutation testing and clinical management. J Clin Oncol. 2014;32(5):431–7.
9. Thompson RH, Atwell T, Schmit G, Lohse CM, Kurup AN, Weisbrod A, et al. Comparison of partial nephrectomy and percutaneous ablation for cT1 renal masses. Eur Urol. 2015;67(2):252–9.
10. Margulis V, Shariat SF, Rapoport Y, Rink M, Sjoberg DD, Tannir NM, et al. Development of accurate models for individualized prediction of survival after cytoreductive nephrectomy for metastatic renal cell carcinoma. Eur Urol. 2013;63(5):947–52.
11. Karam JA, Devine CE, Urbauer DL, Lozano M, Maity T, Ahrar K, et al. Phase 2 trial of neoadjuvant axitinib in patients with locally advanced nonmetastatic clear cell renal cell carcinoma. Eur Urol. 2014;66(5):874–80.
12. Meskawi M, Sun M, Trinh QD, Bianchi M, Hansen J, Tian Z, et al. A review of integrated staging systems for renal cell carcinoma. Eur Urol. 2012;62(2):303–14.
13. Chow WH, Shuch B, Linehan WM, Devesa SS. Racial disparity in renal cell carcinoma patient survival according to demographic and clinical characteristics. Cancer. 2013;119(2):388–94.
14. Ferlay J, Soerjomataram I, Dikshit R, Eser S, Mathers C, Rebelo M, et al. Cancer incidence and mortality worldwide: sources, methods and major patterns in GLOBOCAN 2012. Int J Cancer. 2015;136(5):E359–86.
15. Surveillance E, End Results (SEER) Program. NCHS mortality—all COD, aggregated with state, total U.S. (1969–2012). 2015.
16. Levi F, Ferlay J, Galeone C, Lucchini F, Negri E, Boyle P, et al. The changing pattern of kidney cancer incidence and mortality in Europe. BJU Int. 2008;101(8):949–58.
17. Stafford HS, Saltzstein SL, Shimasaki S, Sanders C, Downs TM, Sadler GR. Racial/ethnic and gender disparities in renal cell carcinoma incidence and survival. J Urol. 2008;179(5):1704–8.
18. Chow WH, Devesa SS, Warren JL, Fraumeni JF Jr. Rising incidence of renal cell cancer in the United States. JAMA. 1999;281(17):1628–31.
19. Humans IWGotEoCRt. Tobacco smoke and involuntary smoking. IARC Monogr Eval Carcinog Risks Hum. 2004;83:1–1438.
20. Hunt JD, van der Hel OL, McMillan GP, Boffetta P, Brennan P. Renal cell carcinoma in relation to cigarette smoking: meta-analysis of 24 studies. Int J Cancer. 2005;114(1):101–8.
21. Flaherty KT, Fuchs CS, Colditz GA, Stampfer MJ, Speizer FE, Willett WC, et al. A prospective study of body mass index, hypertension, and smoking and the risk of renal cell carcinoma (United States). Cancer Causes Control. 2005;16(9):1099–106.
22. Setiawan VW, Stram DO, Nomura AM, Kolonel LN, Henderson BE. Risk factors for renal cell cancer: the multiethnic cohort. Am J Epidemiol. 2007;166(8):932–40.
23. Macleod LC, Hotaling JM, Wright JL, Davenport MT, Gore JL, Harper J, et al. Risk factors for renal cell carcinoma in the VITAL study. J Urol. 2013;190(5):1657–61.
24. McLaughlin JK, Mandel JS, Blot WJ, Schuman LM, Mehl ES, Fraumeni JF Jr. A population-based case-control study of renal cell carcinoma. J Natl Cancer Inst. 1984;72(2):275–84.
25. La Vecchia C, Negri E, D'Avanzo B, Franceschi S. Smoking and renal cell carcinoma. Cancer Res. 1990;50(17):5231–3.
26. McCredie M, Stewart JH. Risk factors for kidney cancer in New South Wales—I. Cigarette smoking. Eur J Cancer. 1992;28A(12):2050–4.
27. McLaughlin JK, Lindblad P, Mellemgaard A, McCredie M, Mandel JS, Schlehofer B, et al. International renal-cell cancer study. I. Tobacco use. Int J Cancer. 1995;60(2):194–8.
28. Yuan JM, Castelao JE, Gago-Dominguez M, Yu MC, Ross RK. Tobacco use in relation to renal cell carcinoma. Cancer Epidemiol Biomarkers Prev. 1998;7(5):429–33.
29. Kreiger N, Marrett LD, Dodds L, Hilditch S, Darlington GA. Risk factors for renal cell carcinoma: results of a population-based case-control study. Cancer Causes Control. 1993;4(2):101–10.
30. Parker AS, Cerhan JR, Janney CA, Lynch CF, Cantor KP. Smoking cessation and renal cell carcinoma. Ann Epidemiol. 2003;13(4):245–51.
31. Liao LM, Weinstein SJ, Pollak M, Li Z, Virtamo J, Albanes D, et al. Prediagnostic circulating adipokine concentrations and risk of renal cell carcinoma in male smokers. Carcinogenesis. 2013;34(1):109–12.
32. Patel NH, Attwood KM, Hanzly M, Creighton TT, Mehedint DC, Schwaab T, et al. Comparative analysis of smoking as a risk factor among renal cell carcinoma histological subtypes. J Urol. 2015;194(3):640–6.
33. Hu J, Ugnat AM, Canadian Cancer Registries Epidemiology Research G. Active and passive smoking and risk of renal cell carcinoma in Canada. Eur J Cancer. 2005;41(5):770–8.
34. Theis RP, Dolwick Grieb SM, Burr D, Siddiqui T, Asal NR. Smoking, environmental tobacco smoke, and risk of renal cell cancer: a population-based case-control study. BMC Cancer. 2008;8:387.
35. Xu Y, Qi Y, Zhang J, Lu Y, Song J, Dong B, et al. The impact of smoking on survival in renal cell carcinoma: a systematic review and meta-analysis. Tumour Biol. 2014;35(7):6633–40.

36. Parker A, Lohse C, Cheville J, Leibovich B, Igel T, Blute M. Evaluation of the association of current cigarette smoking and outcome for patients with clear cell renal cell carcinoma. Int J Urol. 2008;15 (4):304–8.
37. Kroeger N, Klatte T, Birkhauser FD, Rampersaud EN, Seligson DB, Zomorodian N, et al. Smoking negatively impacts renal cell carcinoma overall and cancer-specific survival. Cancer. 2012;118(7):1795–802.
38. Chow WH, Gridley G, Fraumeni JF Jr, Jarvholm B. Obesity, hypertension, and the risk of kidney cancer in men. N Engl J Med. 2000;343(18):1305–11.
39. Vatten LJ, Trichopoulos D, Holmen J, Nilsen TI. Blood pressure and renal cancer risk: the HUNT study in Norway. Br J Cancer. 2007;97(1):112–4.
40. Weikert S, Boeing H, Pischon T, Weikert C, Olsen A, Tjonneland A, et al. Blood pressure and risk of renal cell carcinoma in the European prospective investigation into cancer and nutrition. Am J Epidemiol. 2008;167(4):438–46.
41. Shen T, Shu XO, Xiang YB, Li HL, Cai H, Gao YT, et al. Association of hypertension and obesity with renal cell carcinoma risk: a report from the Shanghai Men's and Women's Health Studies. Cancer Causes Control. 2015;26(8):1173–80.
42. Hofmann JN, Corley DA, Zhao WK, Colt JS, Shuch B, Chow WH, et al. Chronic kidney disease and risk of renal cell carcinoma: differences by race. Epidemiology. 2015;26(1):59–67.
43. Colt JS, Schwartz K, Graubard BI, Davis F, Ruterbusch J, DiGaetano R, et al. Hypertension and risk of renal cell carcinoma among white and black Americans. Epidemiology. 2011;22(6):797–804.
44. Izzedine H, Derosa L, Le Teuff G, Albiges L, Escudier B. Hypertension and angiotensin system inhibitors: impact on outcome in sunitinib-treated patients for metastatic renal cell carcinoma. Ann Oncol. 2015;26(6):1128–33.
45. Wang F, Xu Y. Body mass index and risk of renal cell cancer: a dose-response meta-analysis of published cohort studies. Int J Cancer. 2014;135 (7):1673–86.
46. Dal Maso L, Zucchetto A, Tavani A, Montella M, Ramazzotti V, Talamini R, et al. Renal cell cancer and body size at different ages: an Italian multicenter case-control study. Am J Epidemiol. 2007;166 (5):582–91.
47. Choi Y, Park B, Jeong BC, Seo SI, Jeon SS, Choi HY, et al. Body mass index and survival in patients with renal cell carcinoma: a clinical-based cohort and meta-analysis. Int J Cancer. 2013;132 (3):625–34.
48. Hakimi AA, Furberg H, Zabor EC, Jacobsen A, Schultz N, Ciriello G, et al. An epidemiologic and genomic investigation into the obesity paradox in renal cell carcinoma. J Natl Cancer Inst. 2013;105 (24):1862–70.
49. Ohno Y, Nakashima J, Nakagami Y, Satake N, Gondo T, Ohori M, et al. Sex and the clinical value of body mass index in patients with clear cell renal cell carcinoma. Br J Cancer. 2013;109(7):1899–903.
50. Lee WK, Hong SK, Lee S, Kwak C, Oh JJ, Jeong CW, et al. Prognostic value of body mass index according to histologic subtype in nonmetastatic renal cell carcinoma: a large cohort analysis. Clin Genitourin Cancer. 2015;13(5):461–8.
51. Lajous M, Bijon A, Fagherazzi G, Boutron-Ruault MC, Balkau B, Clavel-Chapelon F, et al. Body mass index, diabetes, and mortality in French women: explaining away a "paradox". Epidemiology. 2014;25(1):10–4.
52. World Cancer Research Fund International/ American Institute for Cancer Research. Continuous update project report: diet, nutrition, physical activity and kidney cancer. 2015. Available at: wcrf.org/ kidney-cancer-2015.
53. Batty GD, Shipley MJ, Langenberg C, Marmot MG, Davey Smith G. Adult height in relation to mortality from 14 cancer sites in men in London (UK): evidence from the original Whitehall study. Ann Oncol. 2006;17(1):157–66.
54. Fujino Y, Japan Collaborative Cohort Study for Evaluation of C. Anthropometry, development history and mortality in the Japan Collaborative Cohort Study for Evaluation of Cancer (JACC). Asian Pac J Cancer Prev. 2007;8(Suppl):105–12.
55. Vigneri P, Frasca F, Sciacca L, Pandini G, Vigneri R. Diabetes and cancer. Endocr Relat Cancer. 2009;16(4):1103–23.
56. Zhang GM, Zhu Y, Ye DW. Metabolic syndrome and renal cell carcinoma. World J Surg Oncol. 2014;12:236.
57. Bao C, Yang X, Xu W, Luo H, Xu Z, Su C, et al. Diabetes mellitus and incidence and mortality of kidney cancer: a meta-analysis. J Diab Complications. 2013;27(4):357–64.
58. Lee S, Hong SK, Kwak C, Kim HH, Lee SE. Prognostic significance of diabetes mellitus in localized renal cell carcinoma. Jpn J Clin Oncol. 2012;42(4):318–24.
59. Hofner T, Zeier M, Hatiboglu G, Eisen C, Schonberg G, Hadaschik B, et al. The impact of type 2 diabetes on the outcome of localized renal cell carcinoma. World J Urol. 2014;32(6):1537–42.
60. Suer E, Ozturk E, Gulpinar O, Kayis A, Baltaci S. Effect of type 2 diabetes mellitus on prognosis of nonmetastatic renal cell cancer. Korean J Urol. 2013;54(8):499–503.
61. Otunctemur A, Ozbek E, Sahin S, Dursun M, Besiroglu H, Koklu I, et al. Diabetes mellitus as a risk factor for high grade renal cell carcinoma. Asian Pac J Cancer Prev. 2014;15(9):3993–6.
62. Antonelli A, Arrighi N, Corti S, Zanotelli T, Cozzoli A, Cosciani Cunico S, et al. Pre-existing type-2 diabetes is not an adverse prognostic factor in patients with renal cell carcinoma: a single-center retrospective study. Urol Oncol. 2013;31(7):1310–5.
63. Ha YS, Kim WT, Yun SJ, Lee SC, Kim WJ, Park YH, et al. Multi-institutional analysis of

localized renal cell carcinoma that demonstrates the impact of diabetic status on prognosis after nephrectomy. Ann Surg Oncol. 2013;20(11):3662–8.
64. Psutka SP, Stewart SB, Boorjian SA, Lohse CM, Tollefson MK, Cheville JC, et al. Diabetes mellitus is independently associated with an increased risk of mortality in patients with clear cell renal cell carcinoma. J Urol. 2014;192(6):1620–7.
65. Choueiri TK, Je Y, Cho E. Analgesic use and the risk of kidney cancer: a meta-analysis of epidemiologic studies. Int J Cancer. 2014;134(2):384–96.
66. Gago-Dominguez M, Yuan JM, Castelao JE, Ross RK, Yu MC. Regular use of analgesics is a risk factor for renal cell carcinoma. Br J Cancer. 1999;81(3):542–8.
67. Cho E, Curhan G, Hankinson SE, Kantoff P, Atkins MB, Stampfer M, et al. Prospective evaluation of analgesic use and risk of renal cell cancer. Arch Intern Med. 2011;171(16):1487–93.
68. Liu W, Park Y, Purdue MP, Giovannucci E, Cho E. A large cohort study of nonsteroidal anti-inflammatory drugs and renal cell carcinoma incidence in the National Institutes of Health-AARP Diet and Health Study. Cancer Causes Control. 2013;24(10):1865–73.
69. Human Drugs: Prescription and over-the-counter drug products containing phenacetin; withdrawal of approval of new drug application. Fed Reg. 1983;48:139–42.
70. Tsuda H, Sakata T, Masui T, Imaida K, Ito N. Modifying effects of butylated hydroxyanisole, ethoxyquin and acetaminophen on induction of neoplastic lesions in rat liver and kidney initiated by N-ethyl-N-hydroxyethylnitrosamine. Carcinogenesis. 1984;5(4):525–31.
71. Luijten M, Speksnijder EN, van Alphen N, Westerman A, Heisterkamp SH, van Benthem J, et al. Phenacetin acts as a weak genotoxic compound preferentially in the kidney of DNA repair deficient Xpa mice. Mutat Res. 2006;596(1–2):143–50.
72. Sandler DP, Burr FR, Weinberg CR. Nonsteroidal anti-inflammatory drugs and the risk for chronic renal disease. Ann Intern Med. 1991;115(3):165–72.
73. Segasothy M, Samad SA, Zulfigar A, Bennett WM. Chronic renal disease and papillary necrosis associated with the long-term use of nonsteroidal anti-inflammatory drugs as the sole or predominant analgesic. Am J Kidney Dis. 1994;24(1):17–24.
74. Ou YC, Yang CR, Cheng CL, Li JR, Raung SL, Hung YY, et al. Indomethacin causes renal epithelial cell injury involving Mcl-1 down-regulation. Biochem Biophys Res Commun. 2009;380(3):531–6.
75. Harris DT. Hormonal therapy and chemotherapy of renal-cell carcinoma. Semin Oncol. 1983;10(4):422–30.
76. Langner C, Ratschek M, Rehak P, Schips L, Zigeuner R. Steroid hormone receptor expression in renal cell carcinoma: an immunohistochemical analysis of 182 tumors. J Urol. 2004;171(2 Pt 1):611–4.
77. Zhu G, Liang L, Li L, Dang Q, Song W, Yeh S, et al. The expression and evaluation of androgen receptor in human renal cell carcinoma. Urology. 2014;83(2):510.e19–24.
78. Lucca I, Klatte T, Fajkovic H, de Martino M, Shariat SF. Gender differences in incidence and outcomes of urothelial and kidney cancer. Nat Rev Urol. 2015;12(10):585–92.
79. He D, Li L, Zhu G, Liang L, Guan Z, Chang L, et al. ASC-J9 suppresses renal cell carcinoma progression by targeting an androgen receptor-dependent HIF2alpha/VEGF signaling pathway. Cancer Res. 2014;74(16):4420–30.
80. Tanaka Y, Sasaki M, Kaneuchi M, Fujimoto S, Dahiya R. Estrogen receptor alpha polymorphisms and renal cell carcinoma—a possible risk. Mol Cell Endocrinol. 2003;202(1–2):109–16.
81. Yu CP, Ho JY, Huang YT, Cha TL, Sun GH, Yu DS, et al. Estrogen inhibits renal cell carcinoma cell progression through estrogen receptor-beta activation. PLoS ONE. 2013;8(2):e56667.
82. Jung YS, Lee SJ, Yoon MH, Ha NC, Park BJ. Estrogen receptor alpha is a novel target of the Von Hippel-Lindau protein and is responsible for the proliferation of VHL-deficient cells under hypoxic conditions. Cell Cycle (Georgetown, Tex). 2012;11(23):4462–73.
83. Liehr JG, Sirbasku DA, Jurka E, Randerath K, Randerath E. Inhibition of estrogen-induced renal carcinogenesis in male Syrian hamsters by tamoxifen without decrease in DNA adduct levels. Cancer Res. 1988;48(4):779–83.
84. Lee JE, Hankinson SE, Cho E. Reproductive factors and risk of renal cell cancer: the Nurses' Health Study. Am J Epidemiol. 2009;169(10):1243–50.
85. Zucchetto A, Talamini R, Dal Maso L, Negri E, Polesel J, Ramazzotti V, et al. Reproductive, menstrual, and other hormone-related factors and risk of renal cell cancer. Int J Cancer (Journal international du cancer). 2008;123(9):2213–6.
86. Purdue MP, Colt JS, Graubard B, Davis F, Ruterbusch JJ, Digaetano R, et al. A case-control study of reproductive factors and renal cell carcinoma among black and white women in the United States. Cancer Causes Control. 2011;22(11):1537–44.
87. Lambe M, Lindblad P, Wuu J, Remler R, Hsieh CC. Pregnancy and risk of renal cell cancer: a population-based study in Sweden. Br J Cancer. 2002;86(9):1425–9.
88. Kabat GC, Silvera SA, Miller AB, Rohan TE. A cohort study of reproductive and hormonal factors and renal cell cancer risk in women. Br J Cancer. 2007;96(5):845–9.
89. Setiawan VW, Kolonel LN, Henderson BE. Menstrual and reproductive factors and risk of renal cell cancer in the multiethnic cohort. Cancer Epidemiol Biomarkers Prev. 2009;18(1):337–40.
90. McLaughlin JK, Gao YT, Gao RN, Zheng W, Ji BT, Blot WJ, et al. Risk factors for renal-cell cancer in

Shanghai, China. Int J Cancer (Journal international du cancer). 1992;52(4):562–5.
91. Lindblad P, Mellemgaard A, Schlehofer B, Adami HO, McCredie M, McLaughlin JK, et al. International renal-cell cancer study. V. Reproductive factors, gynecologic operations and exogenous hormones. Int J Cancer (Journal international du cancer). 1995;61(2):192–8.
92. Nicodemus KK, Sweeney C, Folsom AR. Evaluation of dietary, medical and lifestyle risk factors for incident kidney cancer in postmenopausal women. Int J Cancer. 2004;108(1):115–21.
93. Behrens G, Leitzmann MF. The association between physical activity and renal cancer: systematic review and meta-analysis. Br J Cancer. 2013;108(4):798–811.
94. Trinh L, Plotnikoff RC, Rhodes RE, North S, Courneya KS. Associations between physical activity and quality of life in a population-based sample of kidney cancer survivors. Cancer Epidemiol Biomarkers Prev. 2011;20(5):859–68.
95. Research WCRFIAIfC. Continuous update project report: diet, nutrition, physical activity and kidney cancer. 2015.
96. Lee JE, Hunter DJ, Spiegelman D, Adami HO, Albanes D, Bernstein L, et al. Alcohol intake and renal cell cancer in a pooled analysis of 12 prospective studies. J Natl Cancer Inst. 2007;99 (10):801–10.
97. Deckers IA, van den Brandt PA, van Engeland M, Soetekouw PM, Baldewijns MM, Goldbohm RA, et al. Long-term dietary sodium, potassium and fluid intake; exploring potential novel risk factors for renal cell cancer in the Netherlands Cohort Study on diet and cancer. Br J Cancer. 2014;110(3):797–801.
98. Allen NE, Balkwill A, Beral V, Green J, Reeves G, Million Women Study C. Fluid intake and incidence of renal cell carcinoma in UK women. Br J Cancer. 2011;104(9):1487–92.
99. Lee JE, Giovannucci E, Smith-Warner SA, Spiegelman D, Willett WC, Curhan GC. Total fluid intake and use of individual beverages and risk of renal cell cancer in two large cohorts. Cancer Epidemiol Biomarkers Prev. 2006;15(6):1204–11.
100. Lee JE, Hunter DJ, Spiegelman D, Adami HO, Bernstein L, van den Brandt PA, et al. Intakes of coffee, tea, milk, soda and juice and renal cell cancer in a pooled analysis of 13 prospective studies. Int J Cancer. 2007;121(10):2246–53.
101. Lee JE, Mannisto S, Spiegelman D, Hunter DJ, Bernstein L, van den Brandt PA, et al. Intakes of fruit, vegetables, and carotenoids and renal cell cancer risk: a pooled analysis of 13 prospective studies. Cancer Epidemiol Biomarkers Prev. 2009;18(6):1730–9.
102. Lee JE, Spiegelman D, Hunter DJ, Albanes D, Bernstein L, van den Brandt PA, et al. Fat, protein, and meat consumption and renal cell cancer risk: a pooled analysis of 13 prospective studies. J Natl Cancer Inst. 2008;100(23):1695–706.
103. Daniel CR, Cross AJ, Graubard BI, Park Y, Ward MH, Rothman N, et al. Large prospective investigation of meat intake, related mutagens, and risk of renal cell carcinoma. Am J Clin Nutr. 2012;95(1):155–62.
104. Russo P. End stage and chronic kidney disease: associations with renal cancer. Front Oncol. 2012;2:28.
105. Denton MD, Magee CC, Ovuworie C, Mauiyyedi S, Pascual M, Colvin RB, et al. Prevalence of renal cell carcinoma in patients with ESRD pre-transplantation: a pathologic analysis. Kidney Int. 2002;61(6):2201–9.
106. Breda A, Lucarelli G, Rodriguez-Faba O, Guirado L, Facundo C, Bettocchi C, et al. Clinical and pathological outcomes of renal cell carcinoma (RCC) in native kidneys of patients with end-stage renal disease: a long-term comparative retrospective study with RCC diagnosed in the general population. World J Urol. 2015;33(1):1–7.
107. Woldu SL, Weinberg AC, RoyChoudhury A, Chase H, Kalloo SD, McKiernan JM, et al. Renal insufficiency is associated with an increased risk of papillary renal cell carcinoma histology. Int Urol Nephrol. 2014;46(11):2127–32.
108. Stengel B. Chronic kidney disease and cancer: a troubling connection. J Nephrol. 2010;23 (3):253–62.
109. Wood ME, Vogel V, Ng A, Foxhall L, Goodwin P, Travis LB. Second malignant neoplasms: assessment and strategies for risk reduction. J Clin Oncol. 2012;30(30):3734–45.
110. Ricceri F, Fasanelli F, Giraudo MT, Sieri S, Tumino R, Mattiello A, et al. Risk of second primary malignancies in women with breast cancer: results from the European prospective investigation into cancer and nutrition (EPIC). Int J Cancer. 2015;137(4):940–8.
111. Teng CJ, Hu YW, Chen SC, Yeh CM, Chiang HL, Chen TJ, et al. Use of radioactive iodine for thyroid cancer and risk of second primary malignancy: a nationwide population-based study. J Natl Cancer Inst. 2016;108(2).
112. Gago-Dominguez M, Yuan JM, Castelao JE, Ross RK, Yu MC. Family history and risk of renal cell carcinoma. Cancer Epidemiol Biomarkers Prev. 2001;10(9):1001–4.
113. Gnarra JR, Tory K, Weng Y, Schmidt L, Wei MH, Li H, et al. Mutations of the VHL tumour suppressor gene in renal carcinoma. Nat Genet. 1994;7 (1):85–90.
114. Jonasch E, Futreal PA, Davis IJ, Bailey ST, Kim WY, Brugarolas J, et al. State of the science: an update on renal cell carcinoma. Mol Cancer Res. 2012;10(7):859–80.
115. Moore LE, Nickerson ML, Brennan P, Toro JR, Jaeger E, Rinsky J, et al. Von Hippel-Lindau (VHL) inactivation in sporadic clear cell renal cancer: associations with germline VHL polymor-

phisms and etiologic risk factors. PLoS Genet. 2011;7(10):e1002312.
116. Schmidt L, Duh FM, Chen F, Kishida T, Glenn G, Choyke P, et al. Germline and somatic mutations in the tyrosine kinase domain of the MET proto-oncogene in papillary renal carcinomas. Nat Genet. 1997;16(1):68–73.
117. Lubensky IA, Schmidt L, Zhuang Z, Weirich G, Pack S, Zambrano N, et al. Hereditary and sporadic papillary renal carcinomas with c-met mutations share a distinct morphological phenotype. Am J Pathol. 1999;155(2):517–26.
118. Schmidt L, Junker K, Nakaigawa N, Kinjerski T, Weirich G, Miller M, et al. Novel mutations of the MET proto-oncogene in papillary renal carcinomas. Oncogene. 1999;18(14):2343–50.
119. Toro JR, Nickerson ML, Wei MH, Warren MB, Glenn GM, Turner ML, et al. Mutations in the fumarate hydratase gene cause hereditary leiomyomatosis and renal cell cancer in families in North America. Am J Hum Genet. 2003;73(1):95–106.
120. Birt AR, Hogg GR, Dube WJ. Hereditary multiple fibrofolliculomas with trichodiscomas and acrochordons. Arch Dermatol. 1977;113(12):1674–7.
121. Schmidt LS, Nickerson ML, Warren MB, Glenn GM, Toro JR, Merino MJ, et al. Germline BHD-mutation spectrum and phenotype analysis of a large cohort of families with Birt-Hogg-Dube syndrome. Am J Hum Genet. 2005;76(6):1023–33.
122. Crino P, Nathanson K, Henske E. The tuberous sclerosis complex. N Engl J Med. 2006;355 (13):1345–56.
123. Goldgar DE, Easton DF, Cannon-Albright LA, Skolnick MH. Systematic population-based assessment of cancer risk in first-degree relatives of cancer probands. J Natl Cancer Inst. 1994;86(21):1600–8.
124. Henrion M, Frampton M, Scelo G, Purdue M, Ye Y, Broderick P, et al. Common variation at 2q22.3 (ZEB2) influences the risk of renal cancer. Hum Mol Genet. 2013;22(4):825–31.
125. Wu X, Scelo G, Purdue MP, Rothman N, Johansson M, Ye Y, et al. A genome-wide association study identifies a novel susceptibility locus for renal cell carcinoma on 12p11.23. Hum Mol Genet. 2012;21(2):456–62.
126. Purdue MP, Johansson M, Zelenika D, Toro JR, Scelo G, Moore LE, et al. Genome-wide association study of renal cell carcinoma identifies two susceptibility loci on 2p21 and 11q13.3. Nat Genet. 2011;43(1):60–5.
127. Gudmundsson J, Sulem P, Gudbjartsson DF, Masson G, Petursdottir V, Hardarson S, et al. A common variant at 8q24.21 is associated with renal cell cancer. Nat Commun. 2013;4:2776.
128. Henrion MY, Purdue MP, Scelo G, Broderick P, Frampton M, Ritchie A, et al. Common variation at 1q24.1 (ALDH9A1) is a potential risk factor for renal cancer. PLoS ONE. 2015;10(3):e0122589.
129. Vanharanta S, Buchta M, McWhinney SR, Virta SK, Peczkowska M, Morrison CD, et al. Early-onset renal cell carcinoma as a novel extraparaganglial component of SDHB-associated heritable paraganglioma. Am J Hum Genet. 2004;74 (1):153–9.
130. Cancer Genome Atlas Research N. Comprehensive molecular characterization of clear cell renal cell carcinoma. Nature. 2013;499(7456):43–9.
131. Su D, Singer EA, Srinivasan R. Molecular pathways in renal cell carcinoma: recent advances in genetics and molecular biology. Curr Opin Oncol. 2015;27 (3):217–23.
132. Gold PJ, Fefer A, Thompson JA. Paraneoplastic manifestations of renal cell carcinoma. Semin Urol Oncol. 1996;14(4):216–22.
133. Edge SB, Compton CC. The American Joint Committee on Cancer: the 7th edition of the AJCC cancer staging manual and the future of TNM. Ann Surg Oncol. 2010;17(6):1471–4.

# Pathology and Molecular Pathology of Renal Cancer

# 18

Sue Chang and Massimo Loda

## 18.1 Introduction

We begin this chapter by reviewing the development of the kidney together with a description of the normal anatomy and histology. Disease states and benign tumors are discussed next, followed by an introduction to renal cell carcinoma and its main variants. Familial cancer syndromes associated with renal cell carcinoma are presented and finally, emerging histologic variants and their molecular underpinnings are also considered.

## 18.2 Embryology

From the fourth to fifth weeks of embryogenesis, the intermediate mesoderm layer forms three sequential kidney systems: the pronephros, mesonephros, and metanephros. The pronephros regresses by the end of the fourth week. Portions of the mesonephros remain as the Bowman's capsule around the eventual glomerulus and the ureteric bud. The metanephros is the permanent kidney system, and appears in the fifth week. The metanephros differentiates into the metanephric mesoderm and metanephric blastema. The ureteric bud and the metanephric blastema meet to develop into a functional permanent kidney by the twelfth week of embryogenesis. The ureteric bud develops into the collecting system, which consists of collecting ducts, major and minor calyces, and the ureter. The metanephric blastema eventually develops into the excretory system, which consists of glomeruli, proximal convoluted tubule, loop of Henle, and distal convoluted tubule. The renal tumors discussed in this chapter are all derivatives of the metanephric blastemal cells [1].

## 18.3 Anatomy and Histology

The kidney is encapsulated by a dense capsule of fibroblasts and myofibroblasts. Anatomically, it is separated into an outer cortex and an inner medulla. The cortex is composed of a renal corpuscles and a subsequent series of convoluted and straight tubules. The medulla contains the proximal and distal straight tubules and collecting ducts of the nephron, which collect the plasma ultrafiltrate into minor calices, major calices, and ultimately the renal pelvis.

The nephron is the basic functional unit of the kidney, and is composed of the renal corpuscle (glomerulus, Bowman's capsule), proximal

S. Chang (✉)
David Geffen School of Medicine, UCLA, 10833 Le Conte Ave., PO. Box 951732, CHS Los Angeles, CA 90095, USA
e-mail: suechang@mednet.ucla.edu

M. Loda
Department of Pathology, Dana Farber Cancer Institute & Brigham & Women's Hospital, Harvard Medical School, 450 Brookline Avenue., Boston, MA 02115, USA


M. Loda et al. (eds.), *Pathology and Epidemiology of Cancer*,
DOI 10.1007/978-3-319-35153-7_18

convoluted tubule, proximal straight tubule, descending thin limb, thick ascending limb, distal convoluted tubule, connecting tubule, and the collecting ducts. The nephron is accompanied by an equally complex vascular system of afferent arterioles, efferent arterioles, glomerular capillaries, and peritubular cortical venous plexi. Twenty-five percent of cardiac output at one time is located in the renal system, 90–95 % of which is in the renal cortex (Fig. 18.1).

## 18.4 Disease States and Benign Tumors

### 18.4.1 Autosomal Dominant Polycystic Kidney Disease (*PKD1* or *PKD2* Gene)

Autosomal dominant polycystic kidney disease (ADPKD) is an inherited developmental disorder

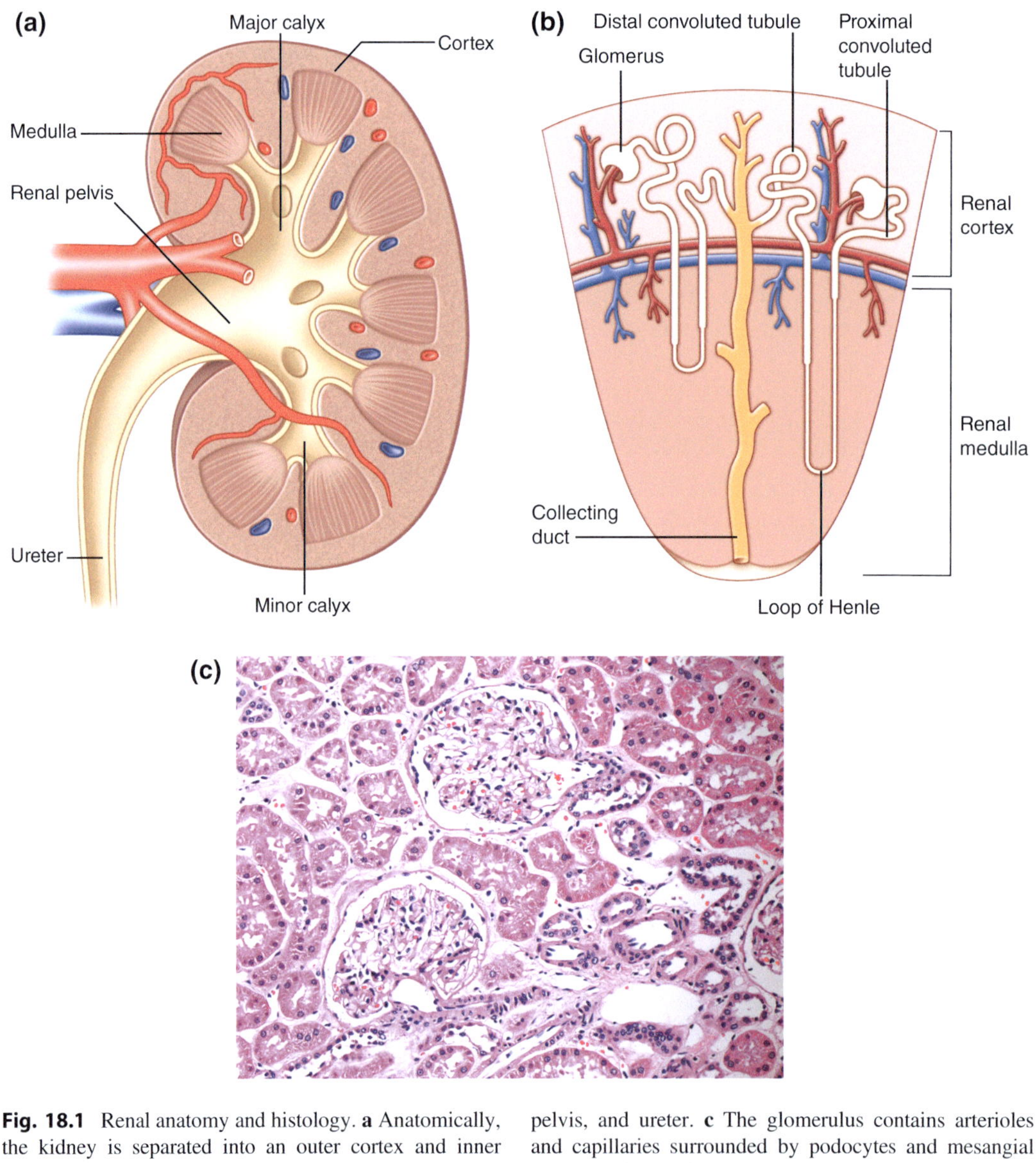


**Fig. 18.1** Renal anatomy and histology. **a** Anatomically, the kidney is separated into an outer cortex and inner medulla. The glomeruli and convoluted tubules are located in the cortex. **b** The plasma ultrafiltrate is concentrated through a series of tubes that empties into the collecting ducts and eventually into the calices, renal pelvis, and ureter. **c** The glomerulus contains arterioles and capillaries surrounded by podocytes and mesangial cells, and filters plasma. Adjacent to the glomeruli in this cross-section are proximal convoluted tubules and distal convoluted tubules

of the kidney that is characterized by the development of cysts. The *PKD1* gene is responsible for 85–90 % of cases, and the *PKD2* gene is responsible for the remaining 15–20 %. These genes encode for polycystin-1 and polycystin-2, respectively, which are used in the assembly of cilia in the renal tubules. The resultant cysts develop anywhere along the nephron and involve both renal cortex and medulla, and can also develop in the liver and pancreas. Cystic destruction of the nephrons and collection of necrotic fluid within the cysts causes renal failure, massively enlarged kidneys, hypertension, hematuria, and flank pain.

The cysts are lined with an attenuated columnar to cuboidal epithelium and thickened basement membrane, and involved by a mixed inflammatory infiltrate. Some studies have suggested an increased risk of renal cell carcinoma in those with ADPKD, but a definite causation has not been found.

### 18.4.2 Benign Cysts

Benign cysts can develop as part of the global effect of diseases such as von Hippel–Lindau disease or tuberous sclerosis. The cysts develop in the renal cortex, and are lined by hyperplastic epithelium, although there is no known association of these cortical cysts with malignancy.

### 18.4.3 Oncocytoma

Oncocytomas are benign neoplasms that grossly are well-circumscribed, brown in appearance, and may have a central scar. The oncocytoma cells have a dense nested architecture and individual cells are eosinophilic (oncocytic) due to their high mitochondrial content. The cells are typically uniform in appearance and positive for KIT by immunohistochemistry. Unlike many RCCs, oncocytoma cells are negative for cytokeratin 7. The main differential diagnosis is the eosinophilic variant of chromophobe RCC (Ch-RCC, discussed in more detail in Sect. 18.7.3), in which cytokeratin 7 is usually positive.

The mitochondrial accumulation in oncocytoma tumor cells has been hypothesized to be a compensatory mechanism for insufficient oxidative phosphorylation [2]. A number of molecular abnormalities have been detected in sporadic oncocytomas, chiefly loss of chromosomes Y and 1, and alterations of chromosome 11q13. The alterations of chromosome 11q13 involve the *CCDN1* gene, causing overexpression of cyclin D1 gene product. However, many oncocytomas have no detectable abnormalities by fluorescence in situ hybridization (FISH) or chromosomal karyotyping. When familial cases of oncocytoma have been reported, the patients have been found to have Birt–Hogg–Dubé syndrome (discussed in Sect. 18.7.3.2) [3, 4].

## 18.5 Renal Cell Carcinoma

Renal cell carcinoma (RCC) arises from the epithelium of the renal tubules, and represents more than 90 % of the renal malignancies in adults. Molecular and cytogenetic studies have elucidated specific genetic changes in the majority of histologic subtypes. Epithelial malignancies of the kidney are comprised of a number of histologically, immunophenotypically, and genetically distinct profiles, with resultant different clinical course and increasing numbers of targeted therapeutic agents.

Although the earliest models of kidney cancer utilized studies of inherited RCCs, hereditary cancer syndromes account for only an estimated 2–4 % of RCC cases. However, further comparison between RCCs in hereditary cancer syndromes and sporadic RCCs has shown common pathways and molecular alterations within each RCC type. Identification of the causal gene mutations has improved the diagnostic accuracy as well as underscored the importance of distinguishing between types of RCC.

### 18.5.1 Histologic Diagnosis and Immuno histochemistry

RCC variants have been categorized by morphology and histologic features, with immunohistochemical staining playing a somewhat limited role. Both benign renal tubules and RCCs in general show positive staining with Paired Box 8 (PAX8) antibodies. This feature is primarily useful in confirming the renal origin of metastatic RCC.

The histology and molecular alterations associated with the subtypes of clear cell RCC (CC-RCC), papillary RCC (P-RCC), and Ch-RCC will be discussed in particular in Sect. 18.7. However, it is important to note that a panel of cytokeratin 7, AMACR, carbonic anhydrase 9 (CA-9), and CD10 can be useful in delineating between CC-RCC, clear cell tubulopapillary RCC, and P-RCC. CC-RCCs are negative for cytokeratin 7, whereas P-RCC, tubulopapillary RCC, and Ch-RCC are positive for cytokeratin 7. CC-RCC is positive for CA-9, whereas most P-RCC and Ch-RCC are negative. While P-RCCs are positive for AMACR, 20 % of CC-RCCs are positive, and clear cell tubulopapillary RCCs are negative [5].

A separate panel of HNF1beta, CD10, and S100a1 is used to help differentiate between Ch-RCC and oncocytoma. Ch-RCC demonstrates a complete loss of HNF1beta protein expression, and is negative for S100a. Oncocytomas retain HNF1beta protein expression, S100a expression, and stain with Hale's colloidal iron. These staining differences are utilized to differentiate between the eosinophilic variant of Ch-RCC and oncocytomas [5].

### 18.5.2 Fuhrman Nuclear Grading

Fuhrman nuclear grading was the first important independent predictive factor of disease-free survival, and is still used in the grading of most RCCs [6]. This grading system was first published in 1982 and is based on size of nuclei, presence of nucleoli, and nuclear membrane contours. Fuhrman grade I nuclei are small (<10 microns in diameter), round, and with absent nucleoli. Fuhrman nuclear grade II includes nuclei 15 microns in diameter with slight nuclear irregularities and small nucleoli visible under high power magnification (40x). Grade III nuclei are larger (20 microns), with very irregular outlines and prominent nucleoli visible at low power magnification (10x). Grade IV nuclei are similar to Grade III nuclei, but with the addition of pleomorphic, bizarre, or multilobated nuclei with macronucleoli. The most severe nuclear grade seen in one high-powered field is reported [7].

Recent modifications to the grading schema have limited Fuhrman nuclear grading to clear cell and papillary types, emphasized the importance of nucleolar features in grading, and reserved nuclear grade IV for those with extreme nuclear pleomorphism, sarcomatoid differentiation, or rhabdoid differentiation [8].

### 18.5.3 Staging

The American Joint Committee on Cancer (AJCC) TNM staging for RCCs, last updated in 2010, stages RCCs based on primary tumor size, involvement of adjacent structures (adrenal gland, renal vein, Gerota's fascia, vena cava), involvement of lymph nodes, and distant metastasis for anatomic staging [9].

The pathologic stage has most consistently been shown to have prognostic utility in RCC. Pathologic staging after histologic typing has been shown to be even more effective in prognosticating outcomes. As discussed below, histologic types have been shown to behave differently. The size of tumor and extent of involvement leading to pathologic stage are powerful predictors of prognosis [10], with stage 1 RCC having a 5-year disease survival of 80–95 % [11].

## 18.6 The Warburg Effect and Molecular Alterations in Carcinoma

The Warburg effect is a theory of carcinomatosis first posited by Otto Warburg in the 1920s [12]. A fundamental aspect of cancer is the use of aerobic glycolysis for energy production. Tumors that switch to aerobic glycolysis instead of oxidative phosphorylation for energy production must fundamentally alter the expression of a number of genes involved in cell metabolism, oxygen sensing, and fatty acid synthesis [12]. As will be discussed in the remaining sections, a number of RCC types exemplify the Warburg effect of altering cell metabolism and respiration to suit tumor cell growth.

## 18.7 RCC Types and Associated Familial Cancer Syndromes

A large portion of the research into molecular underpinnings of renal tumors has been due to the study of familial cancer syndromes. While the syndromes represent a minority of the total renal carcinomas diagnosed, this initial research shed insight to the molecular drivers of sporadic renal cell carcinomas. It is estimated that 1–4 % of RCCs are due to hereditary cancer syndromes [13].

### 18.7.1 Clear Cell Carcinoma and Von Hippel–Lindau (VHL) Syndrome

#### 18.7.1.1 Histology of CC-RCC

CC-RCC accounts for approximately 75 % of the cancers in the kidney [14, 15]. Sporadic CC-RCC is the most common and most aggressive subtype of RCCs. CC-RCC is grossly bright golden-yellow due to a high lipid content in the lesional cells. The tumor cells are arranged in nests, sheets, and tubules. The cells have clear cytoplasm and single round nuclei. Intralesional hemorrhage is common, due to the highly vascular septations coursing between tumor nests. Higher grade lesions have eosinophilic granular cytoplasm and higher grade nuclei, and can have focal sarcomatoid features (Fig. 18.2).

#### 18.7.1.2 von Hippel–Lindau Syndrome

Von Hippel–Lindau (VHL) syndrome is caused by the inherited autosomal dominant germline mutation of the *VHL* tumor suppressor gene on chromosome 3p25. The syndrome has a high degree of penetrance, and an estimated incidence of 1/36,000–1/45,500. The inherited germline mutation silences one allele of the *VHL* gene, and a point mutation, deletion, or promoter hypermethylation of the remaining allele results in clinical manifestations of the disease. As a result, the onset of malignancies is earlier in those with von Hippel–Lindau syndrome than with sporadic tumors of the same kind, with an average onset in the third to fourth decade versus seventh decade of life. Tumors include cerebellar hemangioblastoma, retinal blastoma, pheochromocytoma, pancreatic cysts, inner ear cysts, and multiple renal cysts and CC-RCC [16]. The *VHL*

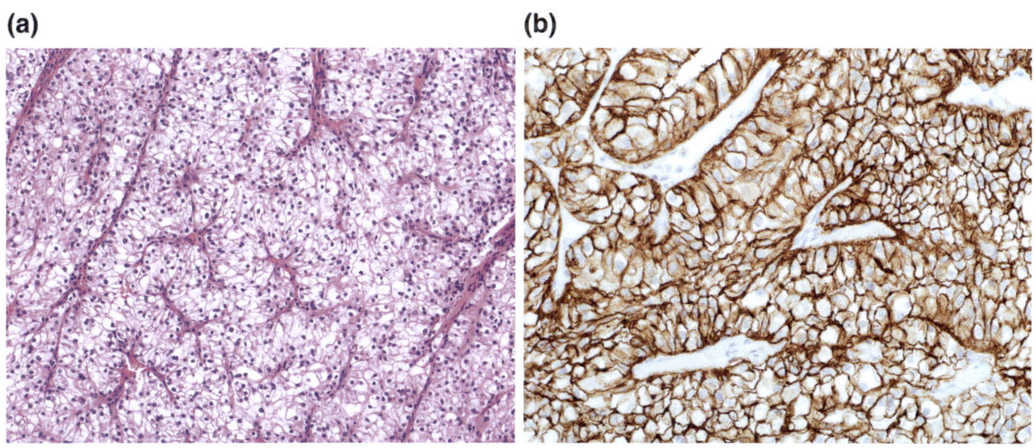


**Fig. 18.2** Clear cell renal cell carcinoma (CC-RCC) and CA-9 staining. **a** CC-RCC is composed of clear cells with nested architecture with intervening small capillaries. Fuhrman nuclear grading is determined by nuclear size, nucleoli, and nuclear contours. **b** Cytoplasmic and membranous accrual of CA-9, visualized by immunohistochemistry is a feature of CC-RCC, but can be seen in peri-necrotic tumoral cells in similarly low oxygen conditions

gene was detected in 1993, and a peripheral blood test for the germline mutation is now commonly used in detection within at-risk families.

The gene product pVHL is necessary for degradation of hypoxia-inducible factor HIF-1α[alpha], which itself controls downstream angiogenesis. When pVHL is faulty, HIF-1α[alpha] accumulates and leads to stabilization of its downstream targets such as vascular endothelial growth factor (*VEGF*), glucose transporter 1 (*GLUT1*), carbonic anhydrase 9 (CA-9), epidermal derived growth factor (*EGFR*), and platelet derived growth factors (*PDGFR*). These genes are involved in angiogenesis, cell migration, and cell metabolism in eventual tumor formation [17].

#### 18.7.1.3 mTOR Complex

The mammalian target of rapamycin protein (mTOR) is a protein that exists as a dimer and functions, in part, in regulation of cell growth and metabolism. The mTORC1 complex has been activated in a cited 60–85 % of CC-RCCs, and through unknown mechanisms also drives the progression of CC-RCCs. The mTOR pathway has been more clearly discussed in tumors of tuberous sclerosis complex and angiomyolipomas [17].

#### 18.7.1.4 Sporadic CC-RCC

The Cancer Genome Project recently sequenced 417 cases of CC-RCC. The underlying genetic changes included alterations in the genes controlling cellular oxygen sensing and maintenance of chromatin states. Whole-exome sequencing of the tumors from the Cancer Genome Project identified over 36,000 somatic mutations. There were 19 significantly mutated genes, most commonly *VHL*, *PBRMI*, *BAP1*, and *SETD2*. Loss of 14q, which leads to loss of HIF-1α[alpha], was associated with more aggressive disease. Most sporadic CC-RCCs contain alteration of the *VHL* gene, with up to 87 % of CC-RCCs showing *VHL* inactivation through sequence alterations or promoter methylation [18]. In addition to mutation in *VHL*, up to 41 % of sporadic CC-RCCs have mutation in *PBRM1* leading to altered chromatin biology. Copy number variations (CNVs) are common with CC-RCC. Alterations in *BAP1* and *SETD2* are correlated with poor survival.

The Cancer Genome Project reports that 7 % of CC-RCC showed an epigenetic silencing of *VHL* that was unassociated with somatic *VHL* mutation. An additional 289 other genes showed evidence of epigenetic silencing, and increased promoter hypermethylation correlated with higher clinical stage and grade. RNA expression in this study identified four subsets of mRNA classes, the m1 subtype of which was associated with chromatin remodeling and higher frequency of *PBRM1* mutations [19].

CC-RCCs that harbor no *VHL* mutations and have low CA-9 expression are clinically more aggressive [20], and may be a different disease altogether from its phenotypically similar CC-RCC with *VHL* loss.

#### 18.7.1.5 *VHL* Targets HIF1α and EPAS1 (also Called HIF2α)

Renal cell carcinomas share similar pathways relating to oxygen utilization and nutrient sensing. In conjunction with derangements in the mTOR pathway, which also controls HIF synthesis, alterations in VHF and HIF-1α[alpha] pathways lead to significant increased accrual of CA-9. This increased CA-9 has been found in CC-RCC tumor cells and in the hypoxia-driven necrosis found in other RCCs. Nonclear cell RCCs share common metabolic pathways, and some relate back to the HIF1α[alpha] pathway. The genes *VHL*, *MET*, *FLCN*, *FH*, *TSC1*, *TSC2*, *TFE3*, and *SDH* are used to some extent by the cell to sense oxygen or energy. This commonality has led to the hypothesis that RCCs are fundamentally a metabolic disease, following the Warburg effect [12].

### 18.7.2 Papillary Renal Cell Carcinoma (P-RCC) and Hereditary Papillary Type 1 RCC

#### 18.7.2.1 Histology of P-RCC

P-RCC is the second most common carcinoma of the renal parenchyma, and this subtype is further

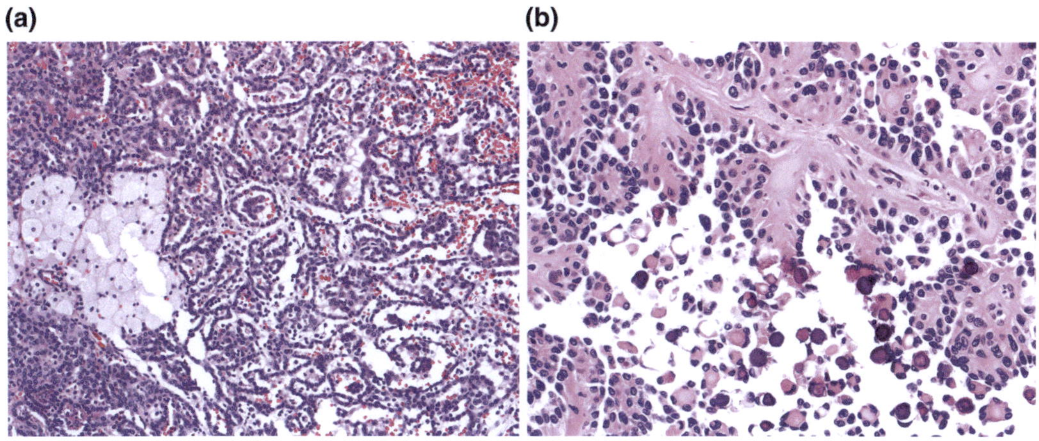

**Fig. 18.3** Type 1 papillary renal cell carcinoma (P-RCC) and type 2 P-RCC. Both demonstrate papillary and tubular architecture. **a** Type 1 contains lower grade nuclei with variable foamy cells. **b** Type 2 has eosinophilic cytoplasm and occasional psammoma bodies

divided into two types. Both types of P-RCC exhibit a mixture of papillary, tubular, and solid patterns. The papillae have central fibrovascular cores and frequently contain foamy macrophages. Type 1 P-RCC is predominantly composed of small cells with scant pale cytoplasm, while type 2 P-RCC demonstrates larger tumor cells with higher nuclear grade, and a more eosinophilic cytoplasm. P-RCC is positive for AMACR in a diffuse cytoplasmic pattern and for CD10 in a membranous pattern. Expression of CA-9 is localized to necrotic areas, which can be found in papillae tips [21].

While the categorization of P-RCC into type 1 and type 2 tumors is accepted, no consensus has been reached as to whether one type has a more favorable prognosis (Fig. 18.3) [22].

#### 18.7.2.2 Hereditary P-RCC

Two familial syndromes are associated with P-RCCs. Hereditary P-RCC is an inherited autosomal dominant activating mutation of the *MET* proto-oncogene, located on chromosome 7q31. A missense mutation in the tyrosine kinase domain of the gene leads to constitutive activation of MET protein and development of Type 1 P-RCC [23]. The mutation is highly penetrant; patients have a 90 % chance of developing P-RCC by their eighth decade of life [21].

The syndrome of hereditary leiomyomatosis and renal cell carcinoma (HLRCC) is caused by the autosomal dominant mutation in the fumarate hydratase (*FH*) gene on 1p42. HLRCC-associated renal carcinomas are more aggressive and invasive than other forms of RCC. The *FH* mutation leads to a remodeling of the Krebs cycle in a shift away from oxidative phosphorylation and towards aerobic glycolysis. Patients with HLRCC develop uterine leiomyomas, cutaneous leiomyomas, and aggressively invasive type II P-RCCs [13, 24].

#### 18.7.2.3 Sporadic P-RCC

The same *MET* mutation from Hereditary P-RCC is present in only 13 % of sporadic P-RCCs [25]. Many sporadic Type 1 P-RCCs show gains of 7 and 17, as well as loss of chromosome Y. Sporadic Type 2 P-RCCs have been shown to have loss of chromosomes 8, 11, and 18. The genetic driver behind the majority of sporadic P-RCC is currently unknown.

### 18.7.3 Chromophobe RCC (Ch-RCC) and Birt-Hogg-Dubé Syndrome

#### 18.7.3.1 Histology of Ch-RCC

Ch-RCC was first described in 1985 by Thoenes [26]. Ch-RCC originates from the intercalated cells of the distal nephron, compared with the

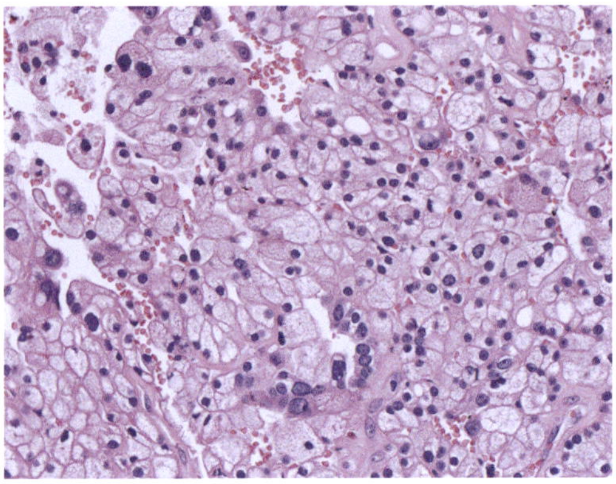

**Fig. 18.4** Chromophobe RCC. Ch-RCC is composed of polygonal variably sized cells with pale to clear cytoplasm, wrinkled hyperchromatic nuclei, perinuclear halos, and prominent thick cell membranes. Cells are arranged in a solid growth pattern and admixed with smaller cells with eosinophilic cytoplasm

more proximal tubule origin of CC-RCC. The tumor is characterized by polygonal variably sized cells with pale to clear cytoplasm, wrinkled hyperchromatic nuclei, perinuclear halos, and prominent thick cell membranes. Cells are arranged in a solid growth pattern and admixed with smaller cells with eosinophilic cytoplasm. Ch-RCC is thought to have a better prognosis than CC-RCC or P-RCC, even with metastatic disease. The eosinophilic variant of Ch-RCC is often in the differential when diagnosing an oncocytoma (see Sect. 18.4.3) (Fig. 18.4).

#### 18.7.3.2 Birt–Hogg–Dubé Syndrome

Birt–Hogg–Dubé syndrome is caused by an inherited autosomal dominant mutation of incomplete penetrance on the folliculin gene (*FLCN*), located on chromosome 17p11.2. The *FCLN* gene acts as a tumor suppressor in the AMPK/tuberous sclerosis complex/mTOR pathway. Up to 70 % of patients with Birt–Hogg–Dubé tumors have a mutation in the gene. The syndrome is characterized by cutaneous fibrofolliculomas, trichodiscomas, and acrochordons, as well as medullary thyroid carcinoma, colorectal neoplasms, lipomas, and bilateral renal tumors. Ch-RCC occurs in 33 % of Birt–Hogg–Dubé syndrome patients, but papillary, clear cell, and oncocytomas have been reported. Up to 50 % of renal tumors in the syndrome will have a hybrid chromophobe-oncocytoma histology [16]. Regardless of histologic subtype, renal tumors occur in 15–27 % of patients with the syndrome. Frequently, the cutaneous tumors appear by the third decade of life, three decades before the renal tumors are detected.

#### 18.7.3.3 Cowden Syndrome

Ch-RCC has been associated with the germline mutation in *PTEN* that is responsible for Cowden syndrome. Cowden syndrome is characterized by the development of multiple hamartomas, most commonly in the skin and mucous membranes. While breast carcinoma, follicular thyroid carcinoma, and endometrial carcinomas are the most common malignancies and considered part of the major criteria for diagnosis, RCCs are a minor criterion [27].

#### 18.7.3.4 Sporadic Ch-RCC

Ch-RCCs frequently have multiple chromosomal losses, leading to a hypodiploid DNA content. A recent study of 66 Ch-RCC specimens by the Cancer Genome Atlas found losses of whole chromosomes 1, 2, 6, 10, 13, and 17 in 86 % of tumors studied. Other reported losses include chromosomes 21 and Y. The molecular alterations involve mitochondrial DNA responsible for the Krebs cycle and the electron transport chain. The Cancer Genome Atlas also observed recurrent DNA rearrangement breakpoints within the Telomerase Reverse Transcriptase (*TERT*) promoter region in a subset of tumors, leading to increased *TERT* expression. *TERT* is known as having a role in telomerase maintenance and DNA repair, and is found in many cancers. Whole-exome sequencing of Ch-RCC has shown an overall lower median rate of somatic mutations compared to CC-RCC [28]. A single specific mutational event in sporadic Ch-RCC is unknown.

### 18.7.4 MITF/TFE Family of RCCs/ Translocation-Associated Carcinoma

The MITF/TFE family of RCCs includes mutations in *TFE3*, *TFEB*, and *MITF*. These genes are

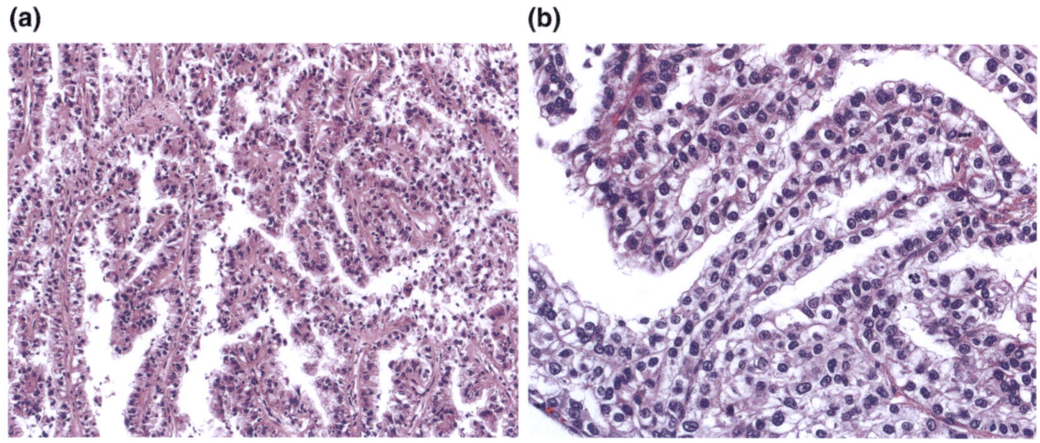

**Fig. 18.5** *TFE3* translocation RCC. **a** This tumor is characterized by papillary or alveolar nested architecture, high nuclear grade. **b** Abundant clear cytoplasm, and psammoma bodies

part of the microphthalmia transcription factor (*MITF*)/transcription factor E family of basic helix-loop-helix leucine zipper transcription factors. These carcinomas have prominent papillary or alveolar nested architecture, high nuclear grade, abundant clear cytoplasm, and psammoma bodies. While there is a bias toward younger patients, there is no gender bias. No hereditary syndrome has been associated with these translocation-associated RCCs (Fig. 18.5) [29].

#### 18.7.4.1 *TFE3/Xp11.2* Translocation RCCs

*TFE3* is involved in transforming growth factor β (TGF-β) signal transduction, and is located on chromosome Xp11.2. This translocation has multiple partners, including 1q21 (*PRCC*), 17q25 (*ASPL*), 1p34 (*PSF*), Xq12 (*NonO*), 17q23 (*CLTC*), and 17q25 (*RCC17*). Many new fusion partners are being reported, and some of these may alter mitotic checkpoint control. These tumors are uncommon but constitute 20–45 % of RCCs in children and young adults [25]. The TFE3 immunohistochemical nuclear stain is a surrogate marker for Xp11 translocation. These tumors also express melanocytic immunohistochemical markers such as HMB45 and Melan-A.

#### 18.7.4.2 Alpha-TFEB/t(6, 11) RCC

The *TFEB* gene is utilized in placental vascularization, and its fusion partner is the alpha gene of unknown function, located on 11q12. The resultant t(6;11) RCCs frequently have a biphasic appearance of both large and small epithelial cells. The tumors are also predominantly found in children. The fusion gene product results in dysregulated (increased) expression of the TFEB protein, detectable by immunohistochemistry. These tumors also express HMB45 and Melan-A [29].

#### 18.7.4.3 MITF-Related RCC

A germline missense mutation in *MITF* has been implicated in translocation-associated RCCs and melanoma, and has been found in families with increased risk of developing melanoma and RCC. *MITF* has been proposed to act as a melanoma oncogene, and also stimulates the transcription of hypoxia-inducible factor (HIF1A) [30].

## 18.8 Genetics of Other Hereditary Syndromes with RCC

### 18.8.1 Tuberous Sclerosis

Tuberous sclerosis complex is a familial hamartomatous syndrome caused by mutations in the *TSC1* gene (located on chromosome 9q34) or *TSC2* gene (located on 16p13). Affected family members develop facial angiofibromas, central nervous system (CNS) cortical tubers, CNS subependymal giant cell tumors, cardiac rhabdomyomas, angiomyolipomas, pulmonary lymphangiomas, and renal tumors. The renal tumors include CC-RCCS, type 2 P-RCCs, and Ch-RCCs. The gene products of *TSC1* (hamartin) and *TSC2* (tuberin) form a heterodimer that inhibits the mTOR pathway. Mutations in either gene cause activation of the mTOR pathway and affects translational control of downstream HIF1α.

## 18.9 Emerging Histologic Variants and Their Molecular Underpinnings

There are a number of rare subtypes whose underlying molecular drivers have yet to be determined in a definitive manner. These entities include succinate dehydrogenase mutation-associated RCC, tubulocystic RCC (TC-RCC), clear cell (tubulo)papillary RCC, acquired cystic disease-associated RCC (ACD-RCC), and medullary carcinoma.

### 18.9.1 Succinate Dehydrogenase Deficiency-Associated RCC and Pheochromocytoma/Paraganglioma Syndrome Type 4 (PGL4)

Patients with succinate dehydrogenase (SDH) deficiency have germline mutations in the *SDHA*, *SDHB*, *SDHC*, or *SDHD* genes, and are associated with the pheochromocytoma/paraganglioma syndrome type 4 (PGL4). SDH is composed of SDHA, SDHB, SDHC, and SDHD subunits. The subunits act as enzymes in the Krebs cycle and the electron transport chain. Mutations in the *SDH* gene lead to impaired cell oxidative phosphorylation, causing the cell to shift to aerobic glycolysis. The net effect is accumulation of HIF1α and its downstream effects, previously described in the section on CC-RCC.

SDH deficiency-associated RCC is found particularly in patients with *SDHB* and *SDHD* mutations. Affected individuals develop pheochromocytomas, paragangliomas, and gastrointestinal stromal tumors, and have a 14 % lifetime risk of renal carcinoma [24]. The RCC is characterized by compact nests of eosinophilic polygonal cells, with frequent vacuolated cytoplasm or cytoplasmic inclusions. A loss of SDHB protein expression by immunohistochemical staining is reported as a specific marker for the entity. A limited number of cases have been reported, and the entity is still under provisional status.

### 18.9.2 Tubulocystic Renal Cell Carcinoma (TC-RCC)

Tubulocystic renal cell carcinoma (TC-RCC) was originally described as a low-grade collecting duct carcinoma. They are typically situated in kidney cortex or at the corticomedullary junction, with well-circumscribed borders. The tumor is composed of dilated simple tubules lined with cuboidal cells that have cytologically high grade nucleoli, and may overlap in morphology with P-RCC or collecting duct carcinoma. Mitotic activity is low, lending to the low-grade appearance. Ultrastructural studies of the tubules demonstrates features of proximal convoluted tubules and distal tubules. The tumors have a strong male predominance (>7:1), and present as asymptomatic complex cysts [24].

Molecular studies of TC-RCC have been limited by the small number of cases. Some have been shown to have molecular clustering with P-RCC and similar gains of chromosomes 7 and 17, or loss of Y chromosome, which reiterates the possible relationship between the two types. Other studies have shown no commonality with CC-RCC or Ch-RCC.

### 18.9.3 Renal Medullary Carcinoma

Renal medullary carcinoma was first described in 1995 in patients with sickle trait or Hemoglobin SC disease. The carcinoma is aggressive, with an average 15 weeks of survival after first diagnosis. There is increased expression of DNA topoisomerase alpha [31], increased HIF1α[alpha] expression, and loss of INI1 labeling by immunohistochemistry [24]. Most of the genes involved in renal medullary carcinoma are important in the hypoxia-induced signaling pathways and altered glycolysis. A specific mutation or affected gene has not been consistently implicated.

In addition, provisional tumor entities such as thyroid-like follicular RCC and *ALK* translocation RCC have yet to be entirely accepted and have been reported sporadically in the literature [24].

## 18.10 Future Directions

The discovery of specific molecular alterations and pathways has resulted in the development of targeted therapeutic drugs. These therapies target either the circulating active VEGF (bevacizumab) or VEGF receptors on tumor cells (sorafenib, sunitinib, pazopanib, axitinib). Antibodies targeting VEGF have shown increased time to disease progression. Additionally, another class of agents (temsirolimus, everolimus) acts along the mTOR pathway by binding to the upstream prolyl isomerase FKBP12, thereby inhibiting mTOR activity. An increasing list of drugs targeting the VHL pathway are under clinical trial, with some showing increased survival and measurable antitumor activity in CC-RCC [32].

The typing of RCCs was born out of histologic differences and prognostic differences, and these differences have been confirmed by the genetic alterations in each type. The work done in exome sequencing of CC-RCC and Ch-RCC has shown a wide heterogeneity across tumors of the same histologic subtype and even within individual tumor masses. The heterogeneity of gene mutation profiles, RNA expression profiles, and protein expression profiles suggests that future studies will further stratify RCC based on genomics and proteomics.

## 18.11 Summary

Our understanding of RCCs has benefited from studies of familial syndromes and hereditary cancers. The changes found in these hereditary tumors have been similar, but not entirely identical, to the alterations found in sporadic tumors.

More recent sequencing of CC-RCCs and Ch-RCCs has confirmed the cell environment of renal carcinomas, showing that tumorogenesis involves mutations in oxygen sensing, cell metabolism, and maintenance of chromatin states. The specific genetic mutations and protein expression profiles have diagnostic and prognostic value, as well as being potential therapeutic targets. While these discoveries have created strides in the treatment of RCCs, the efficacy and long-term use of these molecular targeted therapies is unknown.

## References

1. Sadler TW. Urognital system. In: Langman's medical embryology, 13th ed. New York: Wolters Kluwer Health; 2015. p. 250–77.
2. Simonnet H, Demont J, Pfeiffer K, Guenaneche L, Bouvier R, Brandt U, Schagger H, Godinot C. Mitochondrial complex I is deficient in renal oncocytomas. Carcinogenesis. 2003;24:1461–66.
3. Sukov WR, et al. CCND1 rearrangements and cyclin D1 overexpression in renal oncocytomas: frequency, clinicopathologic features, and utility in differentiation from chromophobe renal cell carcinoma. Hum Pathol. 2009.
4. Fuzesi L, et al. Cytogenetic analysis of 11 renal oncocytomas: further evidence of structural rearrangements of 11q13 as a characteristic chromosomal anomaly. Cancer Genet Cytogenet. 1998;107(1):1–6.
5. Serrano MF, Humphrey PA. Adult renal neoplasms. In: Humphrey PA, Dehner LP, Pfeifer JD, editors. The Washington manual of surgical pathology. 2nd ed. New York: LWW; 2012.
6. Ficarra V1, Righetti R, Martignoni G, D'Amico A, Pilloni S, Rubilotta E, Malossini G, Mobilio G. Prognostic value of renal cell carcinoma nuclear grading: multivariate analysis of 333 cases. Urol Int. 2001;67(2):130–4.
7. Fuhrman SA, Lasky LC, Limas C. Prognostic significance of morphologic parameters in renal cell carcinoma. Am J Surg Pathol. 1982;6(7):655–63.

8. Delahunt B, Cheville JC, Martignoni G, Humphrey PA, Magi-Galluzzi C, McKenney J, Egevad L, Algaba F, Moch H, Grignon DJ, Montironi R, Srigley JR. Members of the ISUP Renal Tumor Panel. The International Society of Urological Pathology (ISUP) grading system for renal cell carcinoma and other prognostic parameters. Am J Surg Pathol. 2013;37(10):1490–504.
9. Edge S, Byrd DR, Compton CC, Greene FL, Trotti A, editors. AJCC cancer staging manual, 7th ed. New York: Springer; 2011.
10. Moch H, et al. Prognostic utility of the recently recommended histologic classification and revised TNM staging system of renal cell carcinoma. A Swiss experience with 588 tumors. Cancer. 2000;89:604–14.
11. Elmore JM, Kadesky KT, Koeneman KS, Sagalowsky AI. Reassessment of the 1997 TNM classification system for renal cell carcinoma. Cancer. 2003;98(11):2329–34.
12. Linehan WM. Genetic basis of kidney cancer: role of genomics for the development of disease-based therapeutics. Genome Res. 2012;22:2089–100.
13. Byler TK, Bratslavsky G. Hereditary renal cell carcinoma: genetics, clinical features, and surgical considerations. World J Urol. 2014;32:623–30.
14. American Cancer Society: Cancer facts and figures. 2014. http://www.cancer.org/acs/groups/cid/documents/webcontent/003107-pdf.pdf.
15. Patard JJ, Leray E, Rioux-Leclercq N, et al. Prognostic value of histologic subtypes in renal cell carcinoma: a multicenter experience. J Clin Oncol. 2005;23(12):2763–71.
16. Hagenkord JM, Gatalica Z, Jonasch E, Monzon FA. Clinical genomics of renal epithelial tumors. Cancer Genet. 2011;204:285–97.
17. Keefe SM, Nathanson KL, Rathmell WK. The molecular biology of renal cell carcinoma. Semin Oncol. 2013;40:421–8.
18. Moore LE, Nickerson ML, Brennan P, Toro JR, Jaeger E, Rinsky J, Han SS, Zaridze D, Matveev V, Janout V, Kollarova H, Bencko V, Navratilova M, Szeszenia-Dabrowska N, Mates D, Schmidt LS, Lenz P, Karami S, Linehan WM, Merino M, Chanock S, Boffetta P, Chow WH, Waldman FM, Rothman N. Von Hippel-Lindau (VHL) inactivation in sporadic clear cell renal cancer: associations with germline VHL polymorphisms and etiologic risk factors. PLoS Genet. 2011;7(10):e1002312. doi:10.1371/journal.pgen.1002312.
19. The Cancer Genome Atlas Research Network. Comprehensive molecular characterization of clear cell renal cell carcinoma. Nature. 2013;499:43–8.
20. Patard JJ, Fergelot P, Karakiewicz PI, Klatte T, Trinh QD, Rioux-Leclercq N, Said JW, Belldegrun AS, Pantuck AJ. Low CAIX expression and absence of VHL gene mutation are associated with tumor aggressiveness and poor survival of clear cell renal cell carcinoma. Int J Cancer. 2008;123(2):395–400.
21. Tickoo SK, Reuter VE. Papillary renal cell carcinoma. In: Amin MB, McKenney JK, Tickoo SK, Paner GP, Shen SS, Velazquez EF, Cubilla AL, Ro JY, Reuter VE, editors. Diagnostic pathology: genitourinary. Manitoba, Canada: Amirsys; 2010.
22. Delahunt B, Srigley JR, Montironi R, Egevad L. Advances in renal neoplasia: recommendations from the 2012 International Society of Urological Pathology consensus conference. Urology. 2012;2014(83):969–74.
23. Schmidt LS, Duh FM, Chen F, Kishida T, Glenn GM, Choyke P, Scherer SW, Shuang Z, Luibensky I, Dean M, Allikmets R, Chidambaram A, Bergerheim UR, Feltis JT, Casadevall C, Zamarron A, Bernues M, Richard S, Lips CJ, Walther MM, Tsui LC, Geil L, Orcutt ML, Stackhouse T, Lipan J, Slife L, Brauch H, Decker J, Niehans G, Hughson MD, Moch H, Storkel S, Lerman MI, Linehan WM, Zbar B. Germline and somatic mutations in the tyrosine kinase domain of the MET proto-oncogene in papillary renal carcinomas. Nat Genet. 1997;16:68–73.
24. Srigley JR, Delahunt B, Eble JN, Egevad L, Epstein JI, Grignon D, Hes O, Moch H, Montironi R, Tickoo SK, Zhou M, Argani P, The ISUP Renal Tumor Panel. The International Society of Urological Pathology (ISUP) Vancouver classification of renal neoplasia. Am J Surg Pathol. 2013;37(10):1469–89.
25. Linehan WM, Srinivasan R, Garvia JA. Non-clear cell renal cancer: disease-based management and opportunities for targeted therapeutic approaches. Semin Oncol. 2013;40(4):511–20.
26. Thoenes W, Storkel S, Rumpelt HJ. Human chromophobe renal cell carcinoma. Virchows Arch B Cell Pathol. 1985;48:207–11.
27. Eng C. Will the real Cowden syndrome please stand up: revised diagnostic criteria. Med Genet. 2000;37:828–30.
28. Davis CF, Ricketts CJ, Wang M, et al. The somatic genomic landscape of chromophobe renal cell carcinoma. Cancer Cell. 2014;26(3):319–30.
29. Argani P. MiT family translocation renal cell carcinoma. Semin Diagn Pathol. 2015;32(2):103–13.
30. Bertolotto C, et al. A SUMOylation-defective MITF germline mutation predisposes to melanoma and renal carcinoma. Nature. 2011;480(7375):94–8.
31. Davis CJ Jr, Mostofi FK, Sesterhenn IA. Renal medullary carcinoma. The seventh sickle cell nephropathy. Am J Surg Pathol. 1995;19:1–11.
32. Jonasch E, Gao J, Rathmell WK. State of the art review: renal cell carcinoma. BMJ. 2014;348:g4797.

# 19 Epidemiology of Lung Cancer

Dimitra Repana and James Spicer

## 19.1 Clinical Picture of the Disease

### 19.1.1 Introduction

Only a small portion of patients with lung cancer will present with localized disease amenable to radical treatment; the majority have metastatic disease and currently must be treated with palliative intent. Developments in imaging with the addition of PET (Positron Emission Tomography), and in techniques for sampling mediastinal lymph nodes have led to more accurate staging. Recognition of driver mutations and introduction of novel targeted treatments have changed the way lung cancer is treated and heralded a new paradigm for personalized treatment in cancer.

### 19.1.2 Symptoms

The median age of diagnosis for lung cancer is 70 for both men and women [1]. The most common symptoms are respiratory and include cough, dyspnoea, haemoptysis and chest pain. Further local symptoms can appear and are related to the extent of the tumour to local structures, such as hoarseness due to involvement of the recurrent laryngeal nerve, superior vena cava obstruction, dysphagia or involvement of the branchial plexus (Pancoast syndrome) [2]. According to the United States (U.S.) Surveillance, Epidemiology, and End Results (SEER) registry for years 2005–2011, 57 % of patients present with metastatic disease [1]. Liver, bones, adrenals and brain are the most common sites of metastases [2]. Lung cancer may also present with a variety of paraneoplastic syndromes such as hypercalcaemia, syndrome of inappropriate antidiuretic hormone secretion (SIADH), neurologic syndromes (most commonly associated with small cell lung cancer) such as Lambert–Eaton myasthenic syndrome, cerebellar ataxia, sensory neuropathy, limbic encephalitis, autonomic neuropathy and many others, haematologic disorders and hypercoagulability disorders, hypertrophic osteoarthropathy, dermatomyositis and polymyositis and Cushing's syndrome [3].

### 19.1.3 Diagnosis and Staging

For patients with suspected lung cancer, a computed tomography (CT) scan of the chest is the first step to evaluate the primary tumour and mediastinal lymph nodes. Staging is completed with a CT scan of the abdomen and further imaging of brain or skeleton if clinically

D. Repana
Guy's and St. Thomas' NHS Foundation Trust,
Great Maze Pond, London SE1 9RT, UK

J. Spicer (✉)
Guy's Hospital, King's College London,
Bermondsey Wing, London SE1 9RT, UK
e-mail: james.spicer@kcl.ac.uk


M. Loda et al. (eds.), *Pathology and Epidemiology of Cancer*,
DOI 10.1007/978-3-319-35153-7_19

indicated. PET and integrated PET/CT imaging have become the standard of care in the last decade and is more accurate for staging the mediastinum, but also for revealing otherwise not suspected metastatic disease which alters the therapeutic approach [4, 5].

The biopsy technique will be chosen according to the location of the tumour or metastatic spread and the safest and least-invasive procedure should be preferred. The primary tumour can be assessed either by bronchoscopy if located centrally or by transthoracic needle biopsy if located in the periphery of the lung. In certain cases, more invasive procedures like video-assisted thoracoscopic surgery (VATS) may be necessary [6].

Accurate mediastinal lymph node sampling is crucial for early stage tumours since this will determine the modality of further management. Endobronchial ultrasonography with transbronchial needle aspiration (EBUS/TBNA) is now the most commonly used technique, and has high sensitivity and specificity. Other techniques including transesophageal ultrasonography or mediastinoscopy may also be used if indicated [7].

Cytologic samples can reliably provide the diagnosis and can also be used for further molecular testing; however, due to the increasing number of necessary diagnostic molecular tests it is preferable to try to maximize the amount of tissue obtained when this is deemed feasible and safe [8, 9].

The 7th edition of the American Joint Committee on Cancer Staging Manual [10] is currently used for lung cancer staging. To inform treatment modality, lung cancer stage can be grouped as early, locally advanced or metastatic disease.

### 19.1.4 Pathology and Molecular Testing

Lung cancer is divided into two major categories: non-small cell lung cancer (NSCLC) and small cell lung cancer (SCLC) with different therapeutic approaches for each. As described in the pathology chapter, NSCLC is further characterized into major histological subtypes, by microscopy and immunohistochemistry, as adenocarcinoma, squamous and large cell carcinoma. At least 10 % of NSCLC cannot be classified and is referred as NSCLC not otherwise specified (NSCLC-NOS) [9]. Pathologic subtype has been found to be predictive for response to specific treatments. For example, pemetrexed chemotherapy showed better efficacy in adenocarcinoma versus squamous carcinoma [11], and the antiangiogenic antibody bevacizumab was associated with significant bleeding risk when used in squamous carcinoma [12] and is only indicated for adenocarcinomas.

Certain driver mutations are also associated with specific histological subtypes. These have led to the development of therapeutic agents directed towards these specific targets, and have introduced a new paradigm of tailored treatment for lung cancer directed by molecular testing.

The epidermal growth factor receptor (EGFR) is commonly overexpressed in lung cancers [13]. When EGFR inhibitors were first used in unselected population clinical characteristics such as female sex, non-smoker status, adenocarcinoma histology and East Asian ethnicity were predictive for response [14, 15]. Eventually, it was demonstrated that the predictive biomarker for response, in turn associated with these demographic factors, was an activating mutation in the EGFR gene which is found in 10–16 % of the western population [16, 17], and in up to 62 % of East Asians [18]. Mutations in EGFR are more common in women than in men (69.7 %), in those who have never smoked (66.7 %) and adenocarcinomas since very rarely are found in squamous carcinomas [17]. Deletions in chromosome 19 and an L858R substitution in exon 21 are the most common mutations, between them accounting for almost 90 % of the total [19].

Another molecular subset of NSCLC is characterized by a fusion protein resulting from a chromosomal rearrangement joining together part of the echinoderm microtubule-associated protein like 4 (EML4) gene to anaplastic lymphoma kinase (ALK), and the resulting fusion gene product acts as an oncogene [20]. Patients with ALK rearrangements are usually significantly younger than the average age for lung cancer patients, have a light or never smoking history, and adenocarcinoma histology. It is estimated that around 3–5 % of all adenocarcinomas will have an ALK rearrangement [21–23].

A translocation in ROS1 is another potent oncogenic driver found in 1 % of lung cancers [24]. Inhibitors of ALK have also shown efficacy in this subset of patients, who have clinical characteristics similar to the ALK rearranged subtype (younger age, light history of smoking and adenocarcinoma histology) [25].

KRAS mutations are found in 20–25 % of lung cancers and are usually associated with smoking history [26]. Their presence seems to be predictive of a worse outcome in the metastatic setting as shown in retrospective trials [26–28]. A variety of other genetic alterations in genes such as BRAF, HER2, PI3KCA, RET, MET, FGFR1 and DDR2 have been found in lung cancer and research is focusing on novel agents against these targets [29].

### 19.1.5 Treatment of Non-small Cell Lung Cancer and Prognosis

For early stage, NSCLC tumours (stage I and II), surgical resection is the standard of care. For patients with stage IB and II, cisplatin-based adjuvant chemotherapy following surgery can reduce the risk of recurrence by 5.4 % (hazard ratio (HR) 0.89, 95 % CI 0.82–0.96) at five years compared to surgery alone [30]. For patients who are not surgical candidates due to poor lung function or co-morbidities, radiotherapy is a reasonable alternative and new techniques like stereotactic body radiation therapy (SBRT) have shown excellent long-term control rates [31].

For locally advanced tumours (stage III), radical treatment with concurrent chemoradiotherapy is the treatment of choice, whilst there is a small selected subset of patients that may benefit from trimodality treatment that also includes surgery [32, 33].

In metastatic disease, a variety of treatments options guided by pathology and molecular testing such as chemotherapy and molecularly targeted agents are included in the armamentarium of the oncologists [34], with most recently the addition of novel immunotherapy with check point inhibitors [35]. A meta-analysis published in 2008 showed an absolute benefit for chemotherapy compared to best supportive care of 4.5–6 months in median survival, and an increase in 1-year survival from 20 to 29 % [36].

Patients with EGFR mutations have high response rates to first- and second-generation EGFR inhibitors (gefitinib, erlotinib, afatinib) and longer progression-free survival (PFS) compared to treatment with standard chemotherapy [37]. However, eventually all these patients develop resistance to treatment, and for approximately 50 % of them this will be the result of a de novo T790M mutation in exon 20. Other mechanisms of resistance have also been described, and therapies for use in this scenario have been developed and are in clinical trials [38].

Crizotinib, an ALK inhibitor, has shown superior outcomes in both the first and second line of treatment compared to standard chemotherapy [39, 40] in patients with an ALK rearrangement, but also has activity in patients with ROS1 rearrangement [25]. More potent second-generation ALK inhibitors such as ceritinib have also been approved and are entering clinical practice [41]. Ongoing trials are trying to establish the role of other targeted agents for lung cancer based on mutation status, and enrolment of patients in clinical trials is strongly encouraged.

### 19.1.6 Treatment of Small Cell Carcinoma of the Lung

SCLC is characterized by rapid growth and aggressive behaviour. Traditionally, SCLC has been staged as limited or extensive, distinguishing the subset of patients that would be amenable to a radical approach with concurrent chemoradiotherapy. Only one third of patients will present with localized disease in the thorax, and for those patients 5-year survival is around 20–25 %. The role for surgery in SCLC is controversial, but is generally offered to the small proportion of patients (no more than 5 %) presenting with very early stage tumours (stage I). For metastatic disease, despite high rates of response to platinum-based chemotherapy, resistance will inevitably occur and median overall survival is around 10 months [42–44]. Due to a high rate of brain metastases, prophylactic cranial irradiation (PCI) is used for those patients who respond to treatment, and has been shown to decrease the rate of symptomatic brain metastases with a modest increase in overall survival [45]. Thoracic consolidation radiotherapy following response to first-line chemotherapy has also shown improvement in survival at 2 years and is commonly used [46].

### 19.1.7 Prognosis

Prognosis in NSCLC is largely dependent on stage at presentation, and for stage I disease, overall survival at 5 years is around 50–70 %, for stage II 36–46 % and for stage III 9–24 % [47]. Prognosis for metastatic lung cancer is generally poor and has been reported between 4 and 13 % at 5 years [1, 48]. According to SEER data, of patients diagnosed with lung cancer in the U.S. between 2005–2011, only 17.4 % will be alive at 5 years [1]. For the period 2010–2011 in England and Wales, 1-year survival was 32.1 %, 5-year survival 9.4 % and 10 year survival 4.9 % [48].

Clinical parameters such as performance status [49] or continuation of smoking following diagnosis have a negative impact on survival [50, 51]. Patients with EGFR mutations have a better overall survival compared to patients without a mutation [52], and those with an exon 19 deletion seem to benefit more than those with the L858R mutation in exon 21 [53]. Similarly, patients with ALK rearrangements have superior overall survival compared to those without this driver mutation [40].

### 19.1.8 Screening

Screening programs using chest radiography and sputum cytology failed to show improvement in lung cancer mortality. The PLCO (Prostate, Lung, Colorectal and Ovarian) screening study randomized more than 150,000 participants aged 55–74 to annual chest radiograph for 4 years or standard care. After 13 years of follow up, no difference was observed in lung cancer mortality between the two groups [54]. Another major screening study was the Mayo Lung Project, which randomized more than 9000 male smokers to an intervention arm which consisted of chest radiography and sputum cytology every 4 months for 6 years, or to the standard arm where the same tests were performed annually. Before randomization, all participants had a baseline chest radiograph and sputum cytology. Similarly, no difference in lung cancer mortality was seen between the arms, even after extended follow-up [55].

In 2011, the National Lung Screening Trial Research Team reported the results of a randomized trial with more than 50,000 individuals aged 55–74. Eligible participants had a history of at least 30 pack-years, and were either current smokers or ex-smokers who had quit within the previous 15 years. Participants were randomized to low-dose CT of the chest or chest radiography once annually for 3 years. This was the first screening study that reported a relative reduction of 20 % in lung cancer mortality, and of 6.7 % in all-cause mortality, in favour of the low-dose CT scan arm [56]. Following these results, various organizations have issued lung cancer screening guidelines for high-risk individuals. The American Cancer Society recommends annual low-dose CT for current or ex-smokers (quit within last 15 years) who are in good health and have at least a 30 pack-year smoking history [57]. Over diagnosis and unnecessary invasive procedures causing patient distress are the major disadvantages of lung cancer screening, and the European Society for Medical Oncology (ESMO) recommends that screening should be offered only in high-volume centres with expertize in thoracic oncology [4]. Several randomized trials are currently ongoing to elucidate the role of low-dose CT scan screening [58].

## 19.2 Descriptive Epidemiology

### 19.2.1 Introduction

Lung cancer incidence varies significantly by geography, sex, race and socioeconomic status and important changes have been documented over time. The major trends will be discussed in this chapter.

### 19.2.2 Trends by Geography

In 2012, there were 1.8 million new cases of lung cancer worldwide, making it the most common cancer and accounting for 13 % of all cancer diagnoses. For the same year, lung cancer caused 1.2 million deaths accounting for 20 % of all cancer deaths. There is an up to 80-fold variation in incidence amongst countries with North America, Europe and East Asia having some of the highest rates, whilst some African and Asian countries having low incidence [59] (Fig. 19.1). Approximately, one third of all new lung cancer cases are diagnosed in China.

In the U.S., lung cancer ranks second in estimated new cancer cases in men and women, after prostate and breast cancer, respectively. For 2015, an estimated 115,610 men will be diagnosed, which accounts for 14 % of all cancer new cases, and 105,590 women, accounting for 13 % of all cancer new cases. Lung cancer remains the first cause of cancer-related mortality for both sexes, accounting for 28 % of all cancer deaths in males and 26 % in women [60].

Similar patterns of incidence and mortality are seen in Europe [61]. Lung cancer is the leading cause of cancer-related death for males (25 % of all cancer deaths) and is predicted to be the leading cause of cancer-related death for women as well in 2015 (14 % of all cancer deaths) [62].

In China, lung cancer is the most common cancer in men and the second most common cancer in women after breast cancer, with 416,333 new male cases and 189,613 new female cases in 2010. It is the first cause of cancer-related death in both sexes [63].

Africa has the lowest incidence of lung cancer worldwide [59], but the increased prevalence of smoking in adolescents in Africa is of concern [64].

### 19.2.3 Trends Over Time

The incidence of lung cancer has varied considerably in populations over time, and is determined by smoking habits over time. Lung cancer was a rare disease up until the end of the nineteenth century, with few case reports published in the literature until 1900. The incidence started to increase significantly in the U.S. and other countries in the second half of the nineteenth century and the first decade of the twentieth century. In the 1930s, the first case-control studies indicated a correlation with smoking, and in the next decades several reports in Europe and U.S. were published regarding the lethal hazards of tobacco. Scientists from the Nazi Germany had recognized the addictive nature of tobacco and the associated potential hazard of lung cancer and an antismoking campaign was launched [65]. In 1950, some of the milestone epidemiological studies were published, particularly from Sir Richard Doll, the British epidemiologist, who interviewed patients with lung cancer and suggested that lung cancer risk is associated with the amount of cigarettes smoked and the duration of smoking [66], and from Wynder and Graham [67] in the U.S. who reported similar observations. In 1954, Sir Doll confirmed the association between smoking habits and lung cancer risk within the British Doctor's Study, a prospective cohort of 40,000 doctors [68]. In 1962, the Royal College of Physicians issued a report on the association of smoking and lung

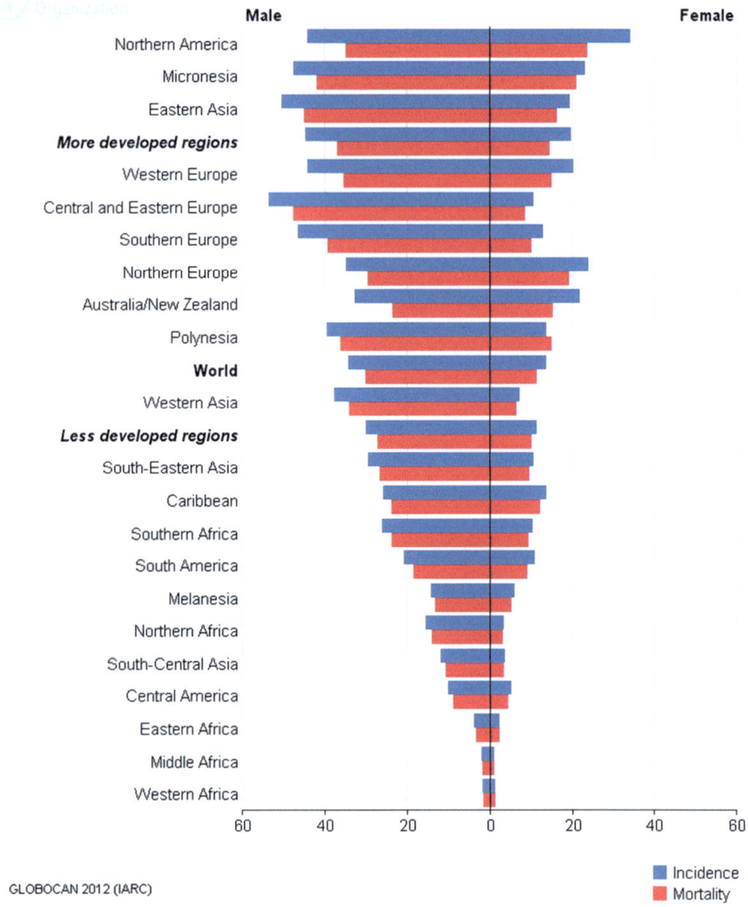


**Fig. 19.1** Lung cancer incidence and mortality worldwide from Ref. [59]

cancer [69], and in 1964 the famous U.S. General Surgeon's report was published and smoking was formally accepted as a definite cause for lung cancer [70]. A plethora of epidemiologic studies, animal experiments and pathologic studies have provided robust evidence that tobacco causes precancerous lesions in lung tissue [71].

Lung cancer incidence in the U.S. continued to increase for males until the mid-1980s and then gradually started to decline. For women, due to later adoption of smoking reaching a peak two decades later, incidence continued to increase until late 1990s, and then followed a similar declining pattern [1] (Fig. 19.2).

The frequency of specific subtypes of lung cancer has also changed during the years reflecting changes in smoking habits. Squamous carcinoma was the most common histologic type until the 1980s and then the incidence of squamous, small cell and large cell histology started to decline, whilst adenocarcinoma started to increase and now accounts for more than half of all cases [72] (Fig. 19.3). This increase was attributed to changes in smoking behaviour associated with the introduction of filtered cigarettes. These cause smokers to take longer and deeper inhalations in order to compensate for reduced inhaled nicotine concentrations, resulting in increased deposition of smoke in the peripheral lung where adenocarcinoma usually arises [73].

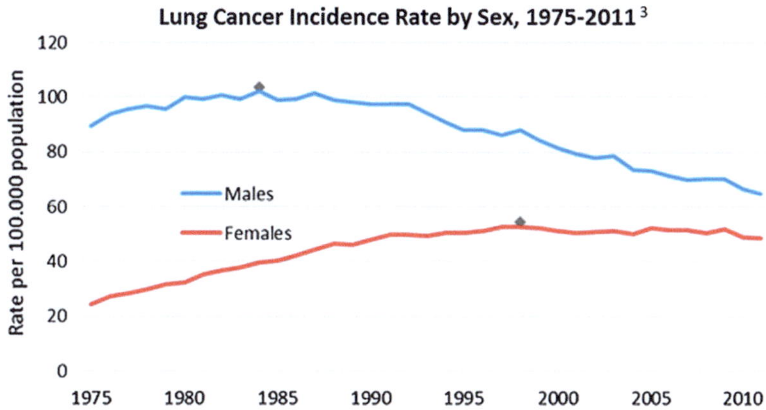


**Fig. 19.2** Lung cancer incidence rate by sex, 1975–2011. National Institutes of Health Ref. [1]

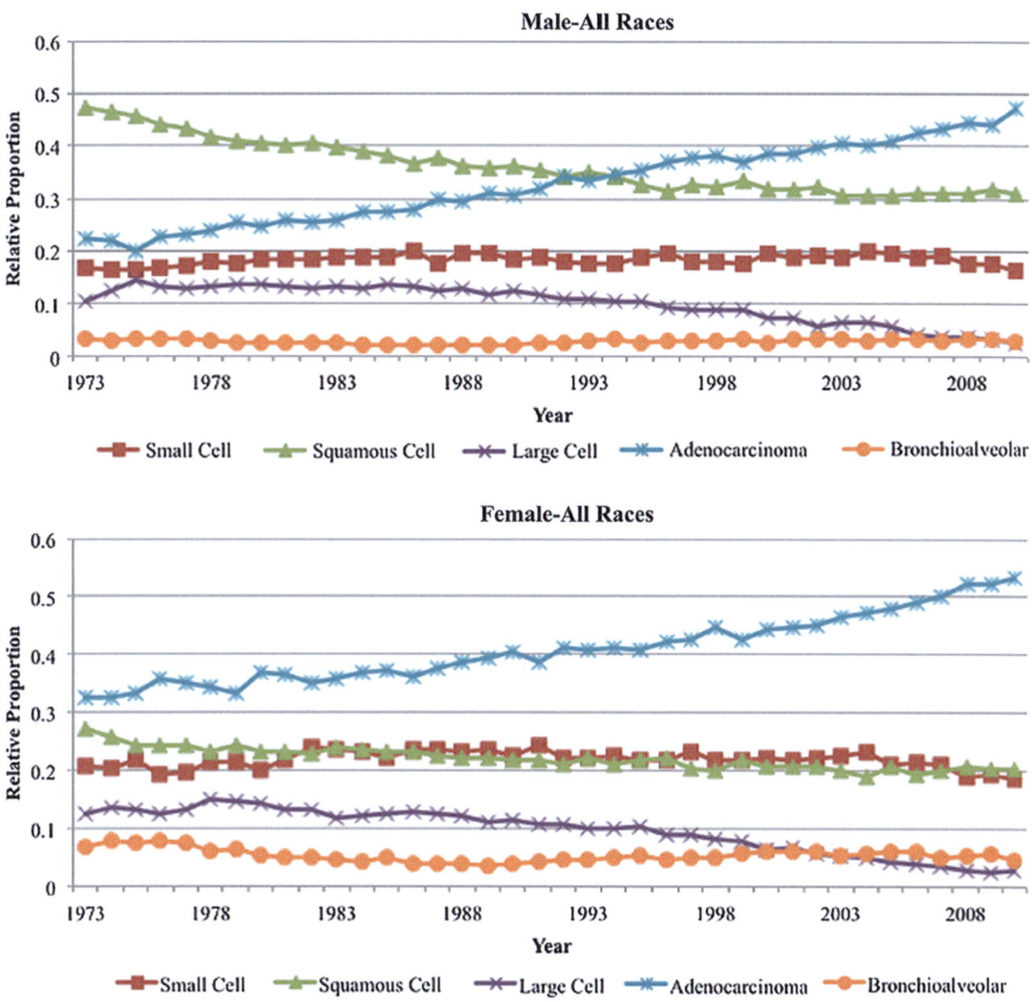


**Fig. 19.3** Histology over time in U.S. from Ref. [72]

### 19.2.4 Differences by Sex

Globally, deaths from lung cancer continue to be more common in men than women since smoking continues to be more common in men. The absolute difference has become smaller over the years since lung cancer incidence peaked earlier for men whilst continued to increase for women [74].

There are no significant differences between men and women in the effect of smoking for a given life time exposure [75]. One significant difference by sex is the incidence of lung cancer amongst never smokers [76]. There are significant geographical variations in the proportion of lung cancer cancers among women that have never smoked, varying from 83 % in South Asia to 15 % in the U.S. Among women who are never smokers, passive smoking exposure at home in adult life and especially prolonged exposure of ≥30 years increases the risk of lung cancer compared to women without home exposure (HR 1.61, CI 95 % 1.00–2.58) [77]. Another risk factor is indoor pollution produced by burning wood and coal for cooking and heating. A retrospective study in China showed that the use of smoky coal compared to smokeless coal increases the risk for lung cancer by more than 30 fold [78].

There is a strong, positive association between hormone replacement therapy (HRT) with oestrogen and progestin and lung cancer risk, as shown from the analysis of several randomized trials. HRT compared to placebo increases the risk of lung cancer and this increase is proportional to the duration of exposure to HRT, with a 50 % increased risk for ≥10 versus <10 years of use (HR 1.48, 95 % CI 1.03–2.12), as well as an association with advanced stage at diagnosis [79].

Although in recent years adenocarcinoma has become the most common histology for both sexes, proportionally more women will be diagnosed with adenocarcinoma compared to men, with the opposite for squamous carcinoma [80]. Women are also more likely to have an EGFR mutation compared to men [17].

In several studies, women have better prognosis than men irrespective of stage at diagnosis and treatment. Data from SEER of more than 18,000 elderly patients with stage I and stage II NSCLC diagnosed during 1991–1999 showed that irrespective of treatment and after adjusting for confounding factors, women had significantly superior outcomes, including lung cancer-specific survival [81]. A meta-analysis of 39 studies and a total of 86,800 patients found that women had better overall survival regardless of stage, histology and smoking status [82]. Women worldwide have a longer life expectancy than men but this alone cannot explain the differences, and it seems that several other biologic factors contribute to this result, warranting further research in this field [83].

### 19.2.5 Racial Differences

An early observation was that lung cancer incidence and mortality was higher in African-American men compared to white men [84]. The U.S. Multiethnic Cohort examined racial/ethnic differences in lung cancer risk in five different ethnic groups in the U.S.: African-Americans, Japanese-Americans, native Hawaiians, Latinos and white men and women. They found no differences in risk for those who smoked more than 30 cigarettes per day, but for those who smoked less than 30 cigarettes per day there was an increased risk of lung cancer for African-Americans and native Hawaiians even after adjusting for occupational risk factors, diet and socioeconomic status [85]. Possible explanations include different smoking styles with deeper and longer inhalations for African-Americans, but also biological differences of the effect of carcinogens.

Higher mortality from lung cancer in African-Americans has been directly associated with socioeconomic status and lack of access to health services, lack of information and differences in cultural beliefs towards lung cancer, as well as access to care [86]. EGFR mutations vary significantly among races and have been reported as occurring in 10–16 % of Whites, 50–60 % of East Asians [18, 87] and 22 % of Indians [88].

### 19.2.6 Trends in Mortality

Over time there has been a steady, but small improvement in lung cancer survival. In the U.S., mortality has shown a steady decline over time, with a 2.6 % annual reduction in mortality for men and a 1.3 % reduction for women between 2002 and 2011 [89]. In the United Kingdom (U. K.), there was an increase in the 5-year-survival rate for the period 1971–2011, from 5 to 8 % in men and from 4 to 12 % in women [90].

## 19.3 Risk Factors

### 19.3.1 Introduction

Lung cancer is one of the most characteristic examples of a relationship between exposure and disease in all of epidemiological science. The major risk factor of smoking is now well-established, and a latency period measured in decades explains the relationship of incidence and mortality with smoking habits around the world and over time.

### 19.3.2 Smoking

More than 1 billion people smoke worldwide and 80 % of them are in low- and middle-income countries. Smoking is responsible for 6 million deaths per year worldwide. In 2008, the World Health Organization introduced the 6 MPOWER measures to fight the tobacco epidemic:

- Monitor tobacco use and prevention policies
- Protect people from tobacco use
- Offer help to quit tobacco use
- Warn about the dangers of tobacco
- Enforce bans on tobacco advertising, promotion and sponsorship
- Raise taxes on tobacco [91].

#### 19.3.2.1 Cigarettes

Tobacco contains at least 98 known hazardous compounds that have been extensively studied [92], belonging to one of the following categories: polycyclic aromatic hydrocarbons, azaarenes, *N*-nitrosamines, aromatic amines, heterocyclic aromatic amines, aldehydes, miscellaneous organic compounds and inorganic compounds [93, 94].

Smokers have at least a 20-fold increased risk of developing lung cancer compared to lifelong non-smokers [95]. Lung cancer risk increases with the number of cigarettes smoked per day in an almost linear way [96], but duration of smoking has the strongest impact [97]. Small cell and squamous histology have the strongest association with smoking with essentially all cases associated with smoking whilst adenocarcinoma is the most common histology in non-smokers [98].

#### 19.3.2.2 Cigars and Pipes

Cigars contain tobacco that is wrapped in a tobacco leaf. Their association with lung cancer has been well documented and they contain many of the carcinogens that are found in cigarettes. Specific carcinogens like nicotine, *N*-nitrosamines, benzene, benzopyrene, carbon monoxide and nitrogen oxide are found in higher levels in cigars than cigarettes due to the curing and fermentation process of cigar manufacture [99]. Cigars are associated with a lower overall risk for lung cancer compared to cigarettes, but risk varies according to intensity, duration and depth of inhalation. Moreover, lung cancer mortality risks for non-smokers range from 1.59 to 7.64, as compared to non-smokers [100]. Data from the U.S. report that cigar use has doubled between years 2000 and 2011, possibly explained by their portrayal in the media as a symbol of success and luxury [101].

Pipe smoking is most prevalent among the elderly and is associated with at least a fivefold increased risk of lung cancer, independent of cigarette smoking. Again intensity, duration and depth of inhalation are important factors that determine risk [102].

#### 19.3.2.3 Marijuana

Marijuana contains carcinogens such as benzopyrene and benzanthracene in higher concentrations than cigarettes. It is usually smoked without a filter and with deeper and longer inhalations resulting in increased retention of tar in the lungs compared to cigarette smoking [103]. Endobronchial biopsies of individuals who smoke marijuana have shown histopathologic changes such as squamous metaplasia and cellular atypia which are known precursor lesions of lung cancer. However, epidemiological data linking marijuana smoking and lung cancer have been conflicting [104]. Several limitations have been recognized in most of these studies, including quality of data available for use of an illegal product, confounding factors such as cigarette smoking, confounding by age since marijuana is most prevalent in the young and short follow up. Notably, a Swedish cohort study of more than 49,000 men with a follow up of 40 years showed that lung cancer risk is doubled (HR 2.12, 95 % CI 1.08–4.14) for heavy marijuana users compared to non-users, even after adjusting for confounding factors such as tobacco and alcohol use, respiratory conditions and socioeconomic status. Heavy marijuana use was defined as more than 50 times during lifetime [105]. Further studies are warranted to answer questions regarding the risks of marijuana smoking, and there is a need for systematic collection of data regarding use. Public health initiatives to inform patients about potential hazards of marijuana smoking, including lung cancer, are required [106].

#### 19.3.2.4 Electronic Nicotine Delivery Systems

Electronic cigarettes were launched in China in 2003 and then patented internationally in 2007. They consist of a battery-operated heating device and a nicotine cartridge; heat converts nicotine to a vapour which is inhaled by the user. Although electronic cigarettes appear less harmful since they do not contain the variety of carcinogens that are found in traditional cigarettes, long-term data regarding their safety are not available and there are many concerns since they are produced by different companies without any independent quality check [107]. Controversy exists regarding their use as smoking cessation aid, with some proponents advocating their use as being less harmful than tobacco products, with a counter-argument that they may encourage continuation in smokers who would otherwise be willing to quit. Electronic cigarettes are not approved by the U.S. Food and Drug Administration (FDA) as a smoking cessation aid, and many prefer to encourage patients to use evidence-based smoking cessation methods and inform smokers about the lack of data in this field [108]. The American Association of Cancer Research (AACR) and the American Society of Clinical Oncology (ASCO) published a policy statement in March 2015 to address these concerns [109].

#### 19.3.2.5 Second-Hand Smoke

Second-hand smoke exposure was recognized much later than primary use as a lung cancer risk factor. The first two studies of passive smoking were published in 1981 and reported increased risk of lung cancer in non-smoking wives of heavy smokers [110, 111]. A meta-analysis of 55 studies calculated a relative risk for lung cancer of 1.27 for non-smoking women exposed to second-hand smoke from their spouses, compared to women not exposed to spousal environmental tobacco smoke [112]. It is estimated that 21,400 lung cancer deaths worldwide are attributable to passive exposure to tobacco products, which includes exposure at home, work and in public places [113].

#### 19.3.2.6 Smoking Cessation

Smoking cessation reduces risk of lung cancer at any age of quitting resulting in added years of life expectancy, and advice on this subject can been viewed as one of the most cost-effective of all healthcare interventions made by physicians [114–116]. The risk of lung cancer decreases with years of abstinence, although it never becomes identical to that for non-smokers even 40 years after smoking cessation [117]. Smoking reduction is also beneficial and results in reduction of lung cancer risk [118]. Smoking cessation is beneficial even after lung cancer diagnosis. Continuing smoking following treatment for early lung cancer is associated with an increased risk of recurrence, second primary tumours and all-cause mortality [51].

### 19.3.3 Radiation

It is well established that lung cancer risk is significantly increased in patients treated for Hodgkin's lymphoma, with a median relative risk that ranges from 2.2 to 7 [119]. Both radiotherapy and chemotherapy with alkylating agents associated with treatment of Hodgkin's increase the risk. The risk is related to dose of therapies, whilst combination of chemotherapy and radiotherapy acts synergistically in this context. In a large population-cohort study that followed more than 19,000 Hodgkin's lymphoma survivors, median time to diagnosis of lung cancer after treatment for lymphoma was 10 years (range 1–28 years). The risk is even greater for smokers, with the highest risk in those who are moderate to heavy smokers, and were treated with both radiotherapy and chemotherapy (relative risk 49.0) [120].

For women treated for breast cancer, adjuvant radiotherapy post-mastectomy almost doubles the risk of subsequent ipsilateral lung cancer. Smoking is again the most important concomitant risk factor, and women who are ever smokers have almost a 40-fold increased risk compared to non-smokers who received adjuvant radiotherapy [121]. Lung cancer typically develops at least ten years after radiotherapy treatment for breast cancer [122].

### 19.3.4 Occupational Exposure

According to the International Agency for Research in Cancer (IARC), lung cancer is the most common occupational cancer worldwide [123]. Several occupations are associated with an increased lung cancer risk, including work involved in aluminium production, coal gasification, coke production, hematite mining, iron and steel founding, painting and rubber production [124]. It was estimated that 102,000 lung cancer deaths in the year 2000 could be attributed to occupational carcinogens [125], and every tenth lung cancer death is related to work-related factors [126].

Asbestos, radon and silica are the most common lung cancer carcinogens, but there are several others chemicals and mixtures like bis-(chloromethyl) ether and chloromethyl methyl ether, coal tar pitch, soot, sulphur mustard, diesel exhausts and metals like arsenic, beryllium, cadmium, chromium, nickel and their compounds [124].

Asbestos is characterized by flame resistance and has been used widely in the building and manufacturing industries [127]. Exposure increases the risk of lung cancer by at least threefold, and smoking acts synergistically [128, 129]. The risk is proportional to the concentration and frequency of the exposure [130].

Radon is a radioactive gas produced during the decay of uranium and is associated with a threefold increased risk of lung cancer in uranium miners [128]. Air pollution from radon of geological origin, which can accumulate indoors, has also been linked with risk of lung cancer in certain geographical areas [131].

### 19.3.5 Environmental Exposure

Air pollution in urban areas is associated with an increased risk of lung cancer [132]. The European Study for Cohorts for Air Pollution Effect was a large prospective study which confirmed that exposure to particulate matter air pollution contributes to lung cancer risk [133]. Indoor pollution from burning coal or cooking oil fumes without appropriate ventilation, which is common in less-developed countries, has also been linked with an increased risk especially in non-smoking women [76].

Water contaminated with high concentrations of arsenic has been linked with various types of cancer including lung cancer [134]. A characteristic epidemiologic example has been described in the Antofagasta area of Chile, where water was supplied by rivers arising from springs in the Andes with a high concentration in arsenic. A change in the water supply in 1958 led to a large increase in arsenic concentration in drinking water. For the period 1958–1970, a rise in the number of deaths from lung and bladder cancer was observed and linked to the high levels of arsenic in drinking water. In 1971, water treatment plants were installed and arsenic levels were lowered. Despite this change, cancer deaths continued to increase until the late 1990s, but then started to decrease, indicating a long latency period [135]. Cancer deaths attributed to arsenic exposure have also been described in Taiwan, Bangladesh and West India, and the World Health Organization has published guidelines regarding drinking water quality requirements [136].

### 19.3.6 HIV

Individuals with HIV infection have a threefold increase of risk for lung cancer compared to the general population. In contrast with AIDS-defining malignancies, for which incidence has decreased since the introduction of highly active antiretroviral therapy (HAART), lung cancer risk has increased among HIV-infected individuals, and this is attributed to the increase in life expectancy and aging of this population. Smoking remains the most significant risk factor for individuals with HIV infection, who have a higher smoking prevalence than the general population [137]. Despite these high smoking rates, additional factors seem to contribute to lung carcinogenesis, possibly including chronic pulmonary inflammation associated with repeated infections [138].

Introduction of HAART has resulted in csurvival improvements for AIDS-defining malignancies, but this is not the case for lung cancer where outcomes remain poor and unchanged over time [139]. Various studies have reported worse prognosis in HIV-infected lung cancer patients compared to non-HIV infected (HR 1.28, 95 % CI 1.17–1.39) even after adjusting for stage and treatment, indicating that immunosuppression may contribute to the more aggressive behaviour of the disease [140, 141].

### 19.3.7 Benign Lung Disease

Inflammation has been recognized as one of the hallmarks of cancer [142] and previous lung disease predisposes to lung cancer. A pooled analysis of 17 studies from the International Lung Cancer Consortium looked at the relationship of lung cancer with emphysema, chronic bronchitis, pneumonia and tuberculosis. Relative risk of lung cancer after adjusting for smoking was 2.44 for emphysema, 1.47 for chronic bronchitis, 1.57 for pneumonia and 1.48 for tuberculosis, as compared to those without any of these benign lung diseases [143].

Chronic obstructive pulmonary disease (COPD) and its subtypes chronic bronchitis, emphysema and chronic obstructive asthma, is a common disease in smokers, although 20 % of patients with COPD do not have a history of smoking [144]. Among never smokers, both emphysema and the combination of emphysema and chronic bronchitis together are associated with increased lung cancer risk, but not chronic bronchitis alone [145].

Pulmonary fibrosis is associated with a poor prognosis (median survival 2–3 years), with the most common cause of death being respiratory failure arising from the disease itself [146]. Several studies have shown that patients with pulmonary fibrosis have a higher risk of developing lung cancer with an incidence that ranges from 4.8 to 48 % [147–150]. A retrospective study showed a cumulative incidence of lung cancer at 1, 5 and 10 years of 3.3, 15.4 and 54.7 %, respectively [151].

Alpha1-antitrypsin deficiency carriers (heterozygous) may not have severe symptoms and may often be undiagnosed; however, they appear also to have an increased risk of lung cancer compared to non-carriers, with an odds ratio of 1.7 (95 % CI 1.2–2.4) [152].

### 19.3.8 Genetic Factors

Individuals with a first-degree relative with lung cancer have a 1.5-fold increase of lung cancer after adjustment for confounding factors, with a higher risk when a sibling rather than a parent is affected, as shown in a pooled analysis including 24 case-control studies from the International Lung Cancer Consortium. As expected, family history is a stronger risk factors for smokers compared to non-smokers, with more than a threefold difference [153]. Diagnosis of lung cancer in more than one relative, or in a relative at a younger age, have also been associated with increased risk, although different age cut-offs have been used in the various studies [154].

For never smokers, family history of lung cancer again appears to increase the risk of lung cancer [155], and in turn is associated with an increased frequency of EGFR mutations [156].

### 19.3.9 Diet

In the early 1980s, two large randomized studies with beta-carotene supplementation were conducted to examine its potential as chemoprevention. The CARET study (Beta-Carotene and Retinol Efficacy Trial) recruited more than 18,000 American men and women who were current or recent ex-smokers, or men with asbestos exposure [157]. The ATBC [158] (Alpha-Tocopherol Beta-Carotene Cancer Prevention) trial in Finland studied more than 23,000 male current smokers. Both of these trials actually showed a detrimental effect, with increased lung cancer incidence in individuals randomized to beta-carotene compared to placebo (by 28 % in the CARET trial and 16 % in ATBC). Interestingly, in the Physicians' Health Study, which included more than 22,000 male physicians of whom half were never smokers, neither harm or benefit was found with beta-carotene supplementation [159]. There are several cohort, case-control and ecological studies that are looking into the protective effect of fruit consumption against lung cancer. Most of the studies report a protective effect for higher versus lower fruit intake; however, a significant limitation for all of these studies is that very few have adjusted for smoking status, highly likely to be a confounding variable in this context [160].

A meta-analysis of 20 prospective cohort studies showed that the risk of lung cancer is reduced by 3 % for every one serving per day of vegetables, 5 % for every one serving of fruit and 3 % for every one serving of fruit or vegetables. Risk was not reduced further after the threshold of two servings per day of fruit or vegetables [161]. Consumption of red meat has been linked with lung cancer and seems to have an adverse effect as shown in a meta-analysis of 34 studies, where it was estimated that there is almost a 35 % increase in lung cancer risk for the higher versus the lower consumption of red meat. Interestingly, all of the studies included were adjusted for smoking or involved non-smokers [162].

There is some evidence supporting physical activity as having a protective effect against lung cancer, including case-control and cohort studies, although there is significant heterogeneity in this data. Physical activity is a general term that is challenging to measure. Physical exercise may arise from occupational, household, transportation or recreational activities, and variability also arises from factors like frequency, intensity and duration [163]. In a prospective study of more than 500,000 men and women, Leitzmann et al. found a 23 % reduction in lung cancer risk for current smokers, and a 22 % reduction for ex-smokers, when comparing individuals with the highest and lowest levels of physical activity (≥5 times per week versus inactive). No difference was seen in never smokers. Physical activity was defined as any activity with duration ≥20 min that resulted in increased heart rate, respiratory rate or sweating [164]. A meta-analysis of 12 cohort studies and 6 case-control studies, including a total of almost 2.5 million participants, showed that any amount of physical activity compared to no activity resulted in a relative risk of 0.79 for lung cancer (95 % CI 0.68–0.86). Similarly to previously reported studies the benefit was seen in smokers and former smokers but not in never smokers [165].

## 19.3.10 Summary

Lung cancer accounts for millions of deaths every year worldwide. Despite recent advances in understanding the molecular mechanisms behind its pathogenesis, outcomes remain poor. Smoking is responsible for the majority of lung cancer cases and smoking cessation is the most cost-effective prevention measure.

## References


1. Howlader NNA, Krapcho M et al. SEER Cancer Statistics Review, 1975–2012, National Cancer Institute. Bethesda, MD. http://seer.cancer.gov/csr/1975_2012/, based on November 2014 SEER data submission, posted to the SEER web site, April 2015.
2. Hyde L, Hyde CI. Clinical manifestations of lung cancer. Chest. 1974;65(3):299–306.
3. Midthun DE. Overview of the risk factors, pathology and clinical manifestations of lung cancer. In: UpToDate, Post TW, editors. Waltham, MA: UpToDate. Assessed on 20 July 2015.
4. Vansteenkiste J, Crino L, Dooms C, Douillard JY, Faivre-Finn C, Lim E, et al. 2nd ESMO consensus conference on lung cancer: early-stage non-small-cell lung cancer consensus on diagnosis, treatment and follow-up. Ann oncol (Official Journal of the European Society for Medical Oncology/ESMO). 2014;25(8):1462–74.
5. Silvestri GA, Gonzalez AV, Jantz MA, Margolis ML, Gould MK, Tanoue LT, et al. Methods for staging non-small cell lung cancer: diagnosis and management of lung cancer, 3rd ed: American College of chest physicians evidence-based clinical practice guidelines. Chest. 2013;143(5 Suppl):e211S–50S.
6. Rivera MP, Mehta AC, Wahidi MM. Establishing the diagnosis of lung cancer: diagnosis and management of lung cancer, 3rd ed: American College of chest physicians evidence-based clinical practice guidelines. Chest. 2013;143(5 Suppl):e142S–65S.
7. De Leyn P, Dooms C, Kuzdzal J, Lardinois D, Passlick B, Rami-Porta R, et al. Revised ESTS guidelines for preoperative mediastinal lymph node staging for non-small-cell lung cancer. Eur J Cardiothorac Surg (Official Journal of the European Association for Cardio-thoracic Surgery). 2014;45(5):787–98.
8. Thunnissen E, Kerr KM, Herth FJ, Lantuejoul S, Papotti M, Rintoul RC, et al. The challenge of NSCLC diagnosis and predictive analysis on small samples. Practical approach of a working group. Lung Cancer. 2012;76(1):1–18.
9. Kerr KM, Bubendorf L, Edelman MJ, Marchetti A, Mok T, Novello S, et al. Second ESMO consensus conference on lung cancer: pathology and molecular biomarkers for non-small-cell lung cancer. Ann Oncol (Official Journal of the European Society for Medical Oncology/ESMO). 2014;25(9):1681–90.
10. Gauglhofer C, Paur J, Schrottmaier WC, Wingelhofer B, Huber D, Naegelen I, et al. Fibroblast growth factor receptor 4: a putative key driver for the aggressive phenotype of hepatocellular carcinoma. Carcinogenesis. 2014;35(10):2331–8.
11. Syrigos KN, Vansteenkiste J, Parikh P, von Pawel J, Manegold C, Martins RG, et al. Prognostic and predictive factors in a randomized phase III trial comparing cisplatin-pemetrexed versus cisplatin-gemcitabine in advanced non-small-cell lung cancer. Ann Oncol (Official Journal of the European Society for Medical Oncology/ESMO). 2010;21(3):556–61.
12. Johnson DH, Fehrenbacher L, Novotny WF, Herbst RS, Nemunaitis JJ, Jablons DM, et al. Randomized phase II trial comparing bevacizumab plus carboplatin and paclitaxel with carboplatin and paclitaxel alone in previously untreated locally advanced or metastatic non-small-cell lung cancer. J Clin Oncol (Official Journal of the American Society of Clinical Oncology). 2004;22(11):2184–91.
13. Salomon DS, Brandt R, Ciardiello F, Normanno N. Epidermal growth factor-related peptides and their receptors in human malignancies. Crit Rev Oncol Hematol. 1995;19(3):183–232.
14. Tsao MS, Sakurada A, Cutz JC, Zhu CQ, Kamel-Reid S, Squire J, et al. Erlotinib in lung cancer—molecular and clinical predictors of outcome. N Engl J Med. 2005;353(2):133–44.
15. Fukuoka M, Yano S, Giaccone G, Tamura T, Nakagawa K, Douillard JY, et al. Multi-institutional randomized phase II trial of gefitinib for previously treated patients with advanced non-small-cell lung cancer (The IDEAL 1 Trial) [corrected]. J Clin Oncol (Official Journal of the American Society of Clinical Oncology). 2003;21(12):2237–46.
16. Marchetti A, Martella C, Felicioni L, Barassi F, Salvatore S, Chella A, et al. EGFR mutations in non-small-cell lung cancer: analysis of a large series of cases and development of a rapid and sensitive method for diagnostic screening with potential implications on pharmacologic treatment. J Clin Oncol (Official Journal of the American Society of Clinical Oncology). 2005;23(4):857–65.
17. Rosell R, Moran T, Queralt C, Porta R, Cardenal F, Camps C, et al. Screening for epidermal growth factor receptor mutations in lung cancer. N Engl J Med. 2009;361(10):958–67.
18. Shi Y, Au JS, Thongprasert S, Srinivasan S, Tsai CM, Khoa MT, et al. A prospective, molecular epidemiology study of EGFR mutations in Asian patients with advanced non-small-cell lung cancer of

adenocarcinoma histology (PIONEER). J Thorac Oncol (Official Publication of the International Association for the Study of Lung Cancer). 2014;9 (2):154–62.
19. Sharma SV, Bell DW, Settleman J, Haber DA. Epidermal growth factor receptor mutations in lung cancer. Nat Rev Cancer. 2007;7(3):169–81.
20. Soda M, Choi YL, Enomoto M, Takada S, Yamashita Y, Ishikawa S, et al. Identification of the transforming EML4-ALK fusion gene in non-small-cell lung cancer. Nature. 2007;448 (7153):561–6.
21. Shaw AT, Yeap BY, Mino-Kenudson M, Digumarthy SR, Costa DB, Heist RS, et al. Clinical features and outcome of patients with non-small-cell lung cancer who harbor EML4-ALK. J Clin Oncol (Official Journal of the American Society of Clinical Oncology). 2009;27(26):4247–53.
22. Rodig SJ, Mino-Kenudson M, Dacic S, Yeap BY, Shaw A, Barletta JA, et al. Unique clinicopathologic features characterize ALK-rearranged lung adenocarcinoma in the western population. Clin Cancer Res (An Official Journal of the American Association for Cancer Research). 2009;15(16):5216–23.
23. Kwak EL, Bang YJ, Camidge DR, Shaw AT, Solomon B, Maki RG, et al. Anaplastic lymphoma kinase inhibition in non-small-cell lung cancer. N Engl J Med. 2010;363(18):1693–703.
24. Gainor JF, Shaw AT. Novel targets in non-small cell lung cancer: ROS1 and RET fusions. Oncologist. 2013;18(7):865–75.
25. Shaw AT, Ou SH, Bang YJ, Camidge DR, Solomon BJ, Salgia R, et al. Crizotinib in ROS1-rearranged non-small-cell lung cancer. N Engl J Med. 2014;371(21):1963–71.
26. Johnson ML, Sima CS, Chaft J, Paik PK, Pao W, Kris MG, et al. Association of KRAS and EGFR mutations with survival in patients with advanced lung adenocarcinomas. Cancer. 2013;119(2):356–62.
27. Brugger W, Triller N, Blasinska-Morawiec M, Curescu S, Sakalauskas R, Manikhas GM, et al. Prospective molecular marker analyses of EGFR and KRAS from a randomized, placebo-controlled study of erlotinib maintenance therapy in advanced non-small-cell lung cancer. J Clin Oncol (Official Journal of the American Society of Clinical Oncology). 2011;29(31):4113–20.
28. Mascaux C, Iannino N, Martin B, Paesmans M, Berghmans T, Dusart M, et al. The role of RAS oncogene in survival of patients with lung cancer: a systematic review of the literature with meta-analysis. Br J Cancer. 2005;92(1):131–9.
29. Califano R, Abidin A, Tariq NU, Economopoulou P, Metro G, Mountzios G. Beyond EGFR and ALK inhibition: unravelling and exploiting novel genetic alterations in advanced non small-cell lung cancer. Cancer Treat Rev. 2015;41(5):401–11.
30. Pignon JP, Tribodet H, Scagliotti GV, Douillard JY, Shepherd FA, Stephens RJ, et al. Lung adjuvant cisplatin evaluation: a pooled analysis by the LACE Collaborative Group. J Clin Oncol (Official Journal of the American Society of Clinical Oncology). 2008;26(21):3552–9.
31. Senthi S, Lagerwaard FJ, Haasbeek CJ, Slotman BJ, Senan S. Patterns of disease recurrence after stereotactic ablative radiotherapy for early stage non-small-cell lung cancer: a retrospective analysis. Lancet Oncol. 2012;13(8):802–9.
32. Bezjak A, Temin S, Franklin G, Giaccone G, Govindan R, Johnson ML, et al. Definitive and adjuvant radiotherapy in locally advanced non-small-cell lung cancer: American Society of clinical oncology clinical practice guideline endorsement of the American Society for radiation oncology evidence-based clinical practice guideline. J Clin Oncol (Official Journal of the American Society of Clinical Oncology). 2015;33(18):2100–5.
33. Eberhardt WE, De Ruysscher D, Weder W, Le Pechoux C, De Leyn P, Hoffmann H, et al. 2nd ESMO consensus conference in lung cancer: locally advanced stage III non-small-cell lung cancer. Ann Oncol (Official Journal of the European Society for Medical Oncology/ESMO). 2015;26(8):1573–88.
34. Besse B, Adjei A, Baas P, Meldgaard P, Nicolson M, Paz-Ares L, et al. 2nd ESMO consensus conference on lung cancer: non-small-cell lung cancer first-line/second and further lines of treatment in advanced disease. Ann Oncol (Official Journal of the European Society for Medical Oncology/ESMO). 2014;25(8):1475–84.
35. Nivolumab approved for lung cancer. Cancer Discovery. 2015;5(5):OF1.
36. Group NM-AC. Chemotherapy in addition to supportive care improves survival in advanced non-small-cell lung cancer: a systematic review and meta-analysis of individual patient data from 16 randomized controlled trials. J Clin Oncol (Official Journal of the American Society of Clinical Oncology). 2008;26(28):4617–25.
37. Lee CK, Brown C, Gralla RJ, Hirsh V, Thongprasert S, Tsai CM, et al. Impact of EGFR inhibitor in non-small cell lung cancer on progression-free and overall survival: a meta-analysis. J Natl Cancer Inst. 2013;105(9):595–605.
38. Stewart EL, Tan SZ, Liu G, Tsao MS. Known and putative mechanisms of resistance to EGFR targeted therapies in NSCLC patients with EGFR mutations-a review. Transl Lung Cancer Res. 2015;4(1):67–81.
39. Shaw AT, Kim DW, Nakagawa K, Seto T, Crino L, Ahn MJ, et al. Crizotinib versus chemotherapy in advanced ALK-positive lung cancer. N Engl J Med. 2013;368(25):2385–94.
40. Solomon BJ, Mok T, Kim DW, Wu YL, Nakagawa K, Mekhail T, et al. First-line crizotinib versus chemotherapy in ALK-positive lung cancer. N Engl J Med. 2014;371(23):2167–77.
41. Shaw AT, Kim DW, Mehra R, Tan DS, Felip E, Chow LQ, et al. Ceritinib in ALK-rearranged

non-small-cell lung cancer. N Engl J Med. 2014;370(13):1189–97.
42. Jett JR, Schild SE, Kesler KA, Kalemkerian GP. Treatment of small cell lung cancer: diagnosis and management of lung cancer, 3rd ed: American College of chest physicians evidence-based clinical practice guidelines. Chest. 2013;143(5 Suppl):e400S–19S.
43. Kalemkerian GP, Akerley W, Bogner P, Borghaei H, Chow LQ, Downey RJ, et al. Small cell lung cancer. J Natl Compr Cancer Netw JNCCN. 2013;11(1):78–98.
44. Fruh M, De Ruysscher D, Popat S, Crino L, Peters S, Felip E, et al. Small-cell lung cancer (SCLC): ESMO clinical practice guidelines for diagnosis, treatment and follow-up. Ann Oncol (Official Journal of the European Society for Medical Oncology/ESMO). 2013;24 Suppl 6:vi99–105.
45. Zhang W, Jiang W, Luan L, Wang L, Zheng X, Wang G. Prophylactic cranial irradiation for patients with small-cell lung cancer: a systematic review of the literature with meta-analysis. BMC Cancer. 2014;14:793.
46. Slotman BJ, van Tinteren H, Praag JO, Knegjens JL, El Sharouni SY, Hatton M, et al. Use of thoracic radiotherapy for extensive stage small-cell lung cancer: a phase 3 randomised controlled trial. Lancet. 2015;385(9962):36–42.
47. Groome PA, Bolejack V, Crowley JJ, Kennedy C, Krasnik M, Sobin LH, et al. The IASLC lung cancer staging project: validation of the proposals for revision of the T, N, and M descriptors and consequent stage groupings in the forthcoming (seventh) edition of the TNM classification of malignant tumours. J Thorac Oncol (Official Publication of the International Association for the Study of Lung Cancer). 2007;2(8):694–705.
48. Cancer Research UK, England and Wales Survival (2010–2011) Summary—April 2014 [Internet].
49. Gridelli C. Lung cancer: defining standard of care for performance status 2—don't rush. Nat Rev Clin Oncol. 2013;10(9):490–2.
50. Videtic GM, Stitt LW, Dar AR, Kocha WI, Tomiak AT, Truong PT, et al. Continued cigarette smoking by patients receiving concurrent chemoradiotherapy for limited-stage small-cell lung cancer is associated with decreased survival. J Clin Oncol (Official Journal of the American Society of Clinical Oncology). 2003;21(8):1544–9.
51. Parsons A, Daley A, Begh R, Aveyard P. Influence of smoking cessation after diagnosis of early stage lung cancer on prognosis: systematic review of observational studies with meta-analysis. BMJ. 2010;340:b5569.
52. Fukuoka M, Wu YL, Thongprasert S, Sunpawaravong P, Leong SS, Sriuranpong V, et al. Biomarker analyses and final overall survival results from a phase III, randomized, open-label, first-line study of gefitinib versus carboplatin/paclitaxel in clinically selected patients with advanced non-small-cell lung cancer in Asia (IPASS). J Clin Oncol (Official Journal of the American Society of Clinical Oncology). 2011;29(21):2866–74.
53. Lee CK, Wu YL, Ding PN, Lord SJ, Inoue A, Zhou C, et al. Impact of specific epidermal growth factor receptor (EGFR) mutations and clinical characteristics on outcomes after treatment with EGFR tyrosine kinase inhibitors versus chemotherapy in EGFR-mutant lung cancer: a meta-analysis. J Clin Oncol (Official Journal of the American Society of Clinical Oncology). 2015;33(17):1958–65.
54. Oken MM, Hocking WG, Kvale PA, Andriole GL, Buys SS, Church TR, et al. Screening by chest radiograph and lung cancer mortality: the prostate, lung, colorectal, and ovarian (PLCO) randomized trial. JAMA. 2011;306(17):1865–73.
55. Marcus PM, Bergstralh EJ, Fagerstrom RM, Williams DE, Fontana R, Taylor WF, et al. Lung cancer mortality in the Mayo Lung Project: impact of extended follow-up. J Natl Cancer Inst. 2000;92(16):1308–16.
56. National Lung Screening Trial Research Team, Aberle DR, Adams AM, Berg CD, Black WC, Clapp JD, et al. Reduced lung-cancer mortality with low-dose computed tomographic screening. N Engl J Med. 2011;365(5):395–409.
57. Smith RA, Manassaram-Baptiste D, Brooks D, Doroshenk M, Fedewa S, Saslow D, et al. Cancer screening in the United States, 2015: a review of current American Cancer Society guidelines and current issues in cancer screening. CA Cancer J Clin. 2015;65(1):30–54.
58. Field JK, van Klaveren R, Pedersen JH, Pastorino U, Paci E, Becker N, et al. European randomized lung cancer screening trials: post NLST. J Surg Oncol. 2013;108(5):280–6.
59. Ferlay JSI, Ervik M, et al. GLOBOCAN 2012 v1.1, cancer incidence and mortality worldwide: IARC CancerBase No. 11 [Internet]. Lyon, France: International Agency for Research on Cancer; 2014. Available from: http://globocan.iarc.fr. Accessed on 16 Jan 2015.
60. Siegel RL, Miller KD, Jemal A. Cancer statistics, 2015. CA Cancer J Clin. 2015;65(1):5–29.
61. Ferlay J, Steliarova-Foucher E, Lortet-Tieulent J, Rosso S, Coebergh JW, Comber H, et al. Cancer incidence and mortality patterns in Europe: estimates for 40 countries in 2012. Eur J Cancer. 2013;49(6):1374–403.
62. Malvezzi M, Bertuccio P, Rosso T, Rota M, Levi F, La Vecchia C, et al. European cancer mortality predictions for the year 2015: does lung cancer have the highest death rate in EU women? Ann Oncol (Official Journal of the European Society for Medical Oncology/ESMO). 2015;26(4):779–86.
63. Zhou C. Lung cancer molecular epidemiology in China: recent trends. Transl Lung Cancer Res. 2014;3(5):270–9.
64. Mashita RJ, Themane MJ, Monyeki KD, Kemper HC. Current smoking behaviour among rural

South African children: Ellisras Longitudinal Study. BMC Pediatr. 2011;11:58.
65. Proctor RN. The Nazi war on tobacco: ideology, evidence, and possible cancer consequences. Bull Hist Med. 1997;71(3):435–88.
66. Doll R, Hill AB. Smoking and carcinoma of the lung; preliminary report. Br Med J. 1950;2 (4682):739–48.
67. Wynder EL, Graham EA. Tobacco smoking as a possible etiologic factor in bronchiogenic carcinoma; a study of 684 proved cases. J Am Med Assoc. 1950;143(4):329–36.
68. Doll R, Hill AB. The mortality of doctors in relation to their smoking habits; a preliminary report. Br Med J. 1954;1(4877):1451–5.
69. Evans PA. Smoking and health. Summary of a report of the Royal College of Physicians of London on smoking in relation to cancer of the lung and other diseases. Cent Afr J Med. 1962;8:234–6.
70. U.S. Department of Health, Education, and Welfare. Smoking and health. Report of the Advisory Committee to the surgeon general of the Public Health Service. Washington: U.S. Department of Health, Education, and Welfare, Public Health Service, Center for Disease Control; 1964. PHS Publication No. 1103.
71. Proctor RN. The history of the discovery of the cigarette-lung cancer link: evidentiary traditions, corporate denial, global toll. Tob Control. 2012;21 (2):87–91.
72. Meza R, Meernik C, Jeon J, Cote ML. Lung cancer incidence trends by gender, race and histology in the United States, 1973–2010. PLoS ONE. 2015;10(3): e0121323.
73. Janssen-Heijnen ML, Coebergh JW, Klinkhamer PJ, Schipper RM, Splinter TA, Mooi WJ. Is there a common etiology for the rising incidence of and decreasing survival with adenocarcinoma of the lung? Epidemiology. 2001;12(2):256–8.
74. Alberg AJ, Wallace K, Silvestri GA, Brock MV. Invited commentary: the etiology of lung cancer in men compared with women. Am J Epidemiol. 2013;177(7):613–6.
75. De Matteis S, Consonni D, Pesatori AC, Bergen AW, Bertazzi PA, Caporaso NE, et al. Are women who smoke at higher risk for lung cancer than men who smoke? Am J Epidemiol. 2013;177 (7):601–12.
76. Sun S, Schiller JH, Gazdar AF. Lung cancer in never smokers—a different disease. Nat Rev Cancer. 2007;7(10):778–90.
77. Wang A, Kubo J, Luo J, Desai M, Hedlin H, Henderson M, et al. Active and passive smoking in relation to lung cancer incidence in the women's health initiative observational study prospective cohort. Ann Oncol (Official Journal of the European Society for Medical Oncology/ESMO). 2015;26 (1):221–30.
78. Barone-Adesi F, Chapman RS, Silverman DT, He X, Hu W, Vermeulen R, et al. Risk of lung cancer associated with domestic use of coal in Xuanwei, China: retrospective cohort study. BMJ. 2012;345:e5414.
79. Slatore CG, Chien JW, Au DH, Satia JA, White E. Lung cancer and hormone replacement therapy: association in the vitamins and lifestyle study. J Clin Oncol (Official Journal of the American Society of Clinical Oncology). 2010;28(9):1540–6.
80. Fu JB, Kau TY, Severson RK, Kalemkerian GP. Lung cancer in women: analysis of the national surveillance, epidemiology, and end results database. Chest. 2005;127(3):768–77.
81. Wisnivesky JP, Halm EA. Sex differences in lung cancer survival: do tumors behave differently in elderly women? J Clin Oncology (Official Journal of the American Society of Clinical Oncology). 2007;25(13):1705–12.
82. Nakamura H, Ando K, Shinmyo T, Morita K, Mochizuki A, Kurimoto N, et al. Female gender is an independent prognostic factor in non-small-cell lung cancer: a meta-analysis. Ann Thorac Cardiovasc Surg (Official Journal of the Association of Thoracic and Cardiovascular Surgeons of Asia). 2011;17(5):469–80.
83. Donington JS, Le QT, Wakelee HA. Lung cancer in women: exploring sex differences in susceptibility, biology, and therapeutic response. Clin Lung Cancer. 2006;8(1):22–9.
84. Walker B, Figgs LW, Zahm SH. Differences in cancer incidence, mortality, and survival between African Americans and whites. Environ Health Perspect. 1995;103 Suppl 8:275–81.
85. Haiman CA, Stram DO, Wilkens LR, Pike MC, Kolonel LN, Henderson BE, et al. Ethnic and racial differences in the smoking-related risk of lung cancer. N Engl J Med. 2006;354(4):333–42.
86. Coughlin SS, Matthews-Juarez P, Juarez PD, Melton CE, King M. Opportunities to address lung cancer disparities among African Americans. Cancer Med. 2014;3(6):1467–76.
87. Xia N, An J, Jiang QQ, Li M, Tan J, Hu CP. Analysis of EGFR, EML4-ALK, KRAS, and c-MET mutations in Chinese lung adenocarcinoma patients. Exp Lung Res. 2013;39(8):328–35.
88. Chougule A, Prabhash K, Noronha V, Joshi A, Thavamani A, Chandrani P, et al. Frequency of EGFR mutations in 907 lung adenocarcioma patients of Indian ethnicity. PLoS ONE. 2013;8 (10):e76164.
89. Kohler BA, Sherman RL, Howlader N, Jemal A, Ryerson AB, Henry KA, et al. Annual report to the nation on the status of cancer, 1975–2011, featuring incidence of breast cancer subtypes by race/ethnicity, poverty, and state. J Natl Cancer Inst. 2015;107(6):djv048.
90. Cancer Research UK Cancer Survival Group, London School of Hygiene and Tropical Medicine. Personal communication, 2014. Accessed June 2015.

91. WHO, Tobacco. Fact sheet N°339. Updated July 2015, 6 July 2015.
92. Talhout R, Schulz T, Florek E, van Benthem J, Wester P, Opperhuizen A. Hazardous compounds in tobacco smoke. Int J Environ Res Public Health. 2011;8(2):613–28.
93. Hecht SS. Tobacco smoke carcinogens and lung cancer. J Natl Cancer Inst. 1999;91(14):1194–210.
94. Humans IWGotEoCRt. Tobacco smoke and involuntary smoking. IARC monographs on the evaluation of carcinogenic risks to humans. vol. 83. World Health Organization, International Agency for Research on Cancer; 2004. pp. 1–1438.
95. Alberg AJ, Brock MV, Ford JG, Samet JM, Spivack SD. Epidemiology of lung cancer: diagnosis and management of lung cancer, 3rd ed: American College of chest physicians evidence-based clinical practice guidelines. Chest. 2013;143(5 Suppl):e1S–29S.
96. Law MR, Morris JK, Watt HC, Wald NJ. The dose-response relationship between cigarette consumption, biochemical markers and risk of lung cancer. Br J Cancer. 1997;75(11):1690–3.
97. Flanders WD, Lally CA, Zhu BP, Henley SJ, Thun MJ. Lung cancer mortality in relation to age, duration of smoking, and daily cigarette consumption: results from Cancer Prevention Study II. Cancer Res. 2003;63(19):6556–62.
98. Khuder SA. Effect of cigarette smoking on major histological types of lung cancer: a meta-analysis. Lung Cancer. 2001;31(2–3):139–48.
99. Iribarren C, Tekawa IS, Sidney S, Friedman GD. Effect of cigar smoking on the risk of cardiovascular disease, chronic obstructive pulmonary disease, and cancer in men. N Engl J Med. 1999;340(23):1773–80.
100. Chang CM, Corey CG, Rostron BL, Apelberg BJ. Systematic review of cigar smoking and all cause and smoking related mortality. BMC Public Health. 2015;15:390.
101. Centers for Disease Control and Prevention (CDC). Consumption of cigarettes and combustible tobacco–United States, 2000–2011. MMWR Morb Mortal Wkly Rep. 2012;61:565–9.
102. Henley SJ, Thun MJ, Chao A, Calle EE. Association between exclusive pipe smoking and mortality from cancer and other diseases. J Natl Cancer Inst. 2004;96(11):853–61.
103. Tashkin DP. Effects of marijuana smoking on the lung. Ann Am Thorac Soc. 2013;10(3):239–47.
104. Mehra R, Moore BA, Crothers K, Tetrault J, Fiellin DA. The association between marijuana smoking and lung cancer: a systematic review. Arch Int Med. 2006;166(13):1359–67.
105. Callaghan RC, Allebeck P, Sidorchuk A. Marijuana use and risk of lung cancer: a 40-year cohort study. Cancer Causes Control CCC. 2013;24(10):1811–20.
106. Biehl JR, Burnham EL. Cannabis smoking in 2015: a concern for lung health? Chest. 2015.
107. Glynn TJ. E-cigarettes and the future of tobacco control. CA Cancer J Clin. 2014;64(3):164–8.
108. Cummings KM, Dresler CM, Field JK, Fox J, Gritz ER, Hanna NH, et al. E-cigarettes and cancer patients. J Thorac Oncol (Official Publication of the International Association for the Study of Lung Cancer). 2014;9(4):438–41.
109. Brandon TH, Goniewicz ML, Hanna NH, Hatsukami DK, Herbst RS, Hobin JA, et al. Electronic nicotine delivery systems: a policy statement from the American Association for cancer research and the American Society of clinical oncology. J Clin Oncol (Official Journal of the American Society of Clinical Oncology). 2015;33(8):952–63.
110. Hirayama T. Non-smoking wives of heavy smokers have a higher risk of lung cancer: a study from Japan. Br Med J (Clin Res Ed). 1981;282 (6259):183–5.
111. Trichopoulos D, Kalandidi A, Sparros L, MacMahon B. Lung cancer and passive smoking. Int J Cancer (Journal International du Cancer). 1981;27(1):1–4.
112. Taylor R, Najafi F, Dobson A. Meta-analysis of studies of passive smoking and lung cancer: effects of study type and continent. Int J Epidemiol. 2007;36(5):1048–59.
113. Oberg M, Jaakkola MS, Woodward A, Peruga A, Pruss-Ustun A. Worldwide burden of disease from exposure to second-hand smoke: a retrospective analysis of data from 192 countries. Lancet. 2011;377(9760):139–46.
114. Jha P, Ramasundarahettige C, Landsman V, Rostron B, Thun M, Anderson RN, et al. 21st-century hazards of smoking and benefits of cessation in the United States. N Engl J Med. 2013;368(4):341–50.
115. Jha P. The 21st century benefits of smoking cessation in Europe. Eur J Epidemiol. 2013;28 (8):617–9.
116. Fellows JL, Bush T, McAfee T, Dickerson J. Cost effectiveness of the Oregon quitline “free patch initiative.” Tob Control. 2007;16 Suppl 1:i47–52.
117. Samet JM. The 1990 report of the surgeon general: the health benefits of smoking cessation. Am Rev Respir Dis. 1990;142(5):993–4.
118. Godtfredsen NS, Prescott E, Osler M. Effect of smoking reduction on lung cancer risk. JAMA. 2005;294(12):1505–10.
119. Lorigan P, Radford J, Howell A, Thatcher N. Lung cancer after treatment for Hodgkin’s lymphoma: a systematic review. Lancet Oncol. 2005;6(10):773–9.
120. Travis LB, Gospodarowicz M, Curtis RE, Clarke EA, Andersson M, Glimelius B, et al. Lung cancer following chemotherapy and radiotherapy for Hodgkin’s disease. J Natl Cancer Inst. 2002;94 (3):182–92.
121. Kaufman EL, Jacobson JS, Hershman DL, Desai M, Neugut AI. Effect of breast cancer radiotherapy and cigarette smoking on risk of second primary lung cancer. J Clin Oncol (Official Journal of the

American Society of Clinical Oncology). 2008;26 (3):392–8.
122. Darby SC, McGale P, Taylor CW, Peto R. Long-term mortality from heart disease and lung cancer after radiotherapy for early breast cancer: prospective cohort study of about 300,000 women in US SEER cancer registries. Lancet Oncol. 2005;6 (8):557–65.
123. Espina C, Straif K, Friis S, Kogevinas M, Saracci R, Vainio H, et al. European code against cancer 4th edition: environment, occupation and cancer. Cancer Epidemiol. 2015.
124. Field RW, Withers BL. Occupational and environmental causes of lung cancer. Clin Chest Med. 2012;33(4):681–703.
125. Driscoll T, Nelson DI, Steenland K, Leigh J, Concha-Barrientos M, Fingerhut M, et al. The global burden of disease due to occupational carcinogens. Am J Ind Med. 2005;48(6):419–31.
126. WHO: Environmental and occupational cancers. Fact sheet N°350. March 2011.
127. Niklinski J, Niklinska W, Chyczewska E, Laudanski J, Naumnik W, Chyczewski L, et al. The epidemiology of asbestos-related diseases. Lung Cancer. 2004;45 Suppl 1:S7-15.
128. Hubaux R, Becker-Santos DD, Enfield KS, Lam S, Lam WL, Martinez VD. Arsenic, asbestos and radon: emerging players in lung tumorigenesis. Environ Health (A Global Access Science Source). 2012;11:89.
129. Ngamwong Y, Tangamornsuksan W, Lohitnavy O, Chaiyakunapruk N, Scholfield CN, Reisfeld B, et al. Additive synergism between asbestos and smoking in lung cancer risk: a systematic review and meta-analysis. PLoS ONE. 2015;10(8):e0135798.
130. Villeneuve PJ, Parent ME, Harris SA, Johnson KC, Canadian Cancer Registries Epidemiology Research Group. Occupational exposure to asbestos and lung cancer in men: evidence from a population-based case-control study in eight Canadian provinces. BMC Cancer. 2012;12:595.
131. Darby S, Hill D, Auvinen A, Barros-Dios JM, Baysson H, Bochicchio F, et al. Radon in homes and risk of lung cancer: collaborative analysis of individual data from 13 European case-control studies. BMJ. 2005;330(7485):223.
132. Dockery DW, Pope CA 3rd, Xu X, Spengler JD, Ware JH, Fay ME, et al. An association between air pollution and mortality in six U.S. cities. N Engl J Med. 1993;329(24):1753–9.
133. Raaschou-Nielsen O, Andersen ZJ, Beelen R, Samoli E, Stafoggia M, Weinmayr G, et al. Air pollution and lung cancer incidence in 17 European cohorts: prospective analyses from the European study of cohorts for air pollution effects (ESCAPE). Lancet Oncol. 2013;14(9):813–22.
134. Celik I, Gallicchio L, Boyd K, Lam TK, Matanoski G, Tao X, et al. Arsenic in drinking water and lung cancer: a systematic review. Environ Res. 2008;108(1):48–55.
135. Marshall G, Ferreccio C, Yuan Y, Bates MN, Steinmaus C, Selvin S, et al. Fifty-year study of lung and bladder cancer mortality in Chile related to arsenic in drinking water. J Natl Cancer Inst. 2007;99(12):920–8.
136. WHO: Arsenic in Drinking Water. WHO/SDE/WSH/03.04/75/rev. 1 2012.
137. Shcherba M, Shuter J, Haigentz M Jr. Current questions in HIV-associated lung cancer. Curr Opin Oncol. 2013;25(5):511–7.
138. Hessol NA, Martinez-Maza O, Levine AM, Morris A, Margolick JB, Cohen MH, et al. Lung cancer incidence and survival among HIV-infected and uninfected women and men. Aids. 2015;29 (10):1183–93.
139. Hleyhel M, Belot A, Bouvier AM, Tattevin P, Pacanowski J, Genet P, et al. Trends in survival after cancer diagnosis among HIV-infected individuals between 1992 and 2009. Results from the FHDH-ANRS C04 cohort. Int J Cancer (Journal International du Cancer). 2015.
140. Coghill AE, Shiels MS, Suneja G, Engels EA. Elevated cancer-specific mortality among HIV-infected patients in the United States. J Clin Oncol (Official Journal of the American Society of Clinical Oncology). 2015;33(21):2376–83.
141. Sigel K, Crothers K, Dubrow R, Krauskopf K, Jao J, Sigel C, et al. Prognosis in HIV-infected patients with non-small cell lung cancer. Br J Cancer. 2013;109(7):1974–80.
142. Hanahan D, Weinberg RA. Hallmarks of cancer: the next generation. Cell. 2011;144(5):646–74.
143. Brenner DR, Boffetta P, Duell EJ, Bickeboller H, Rosenberger A, McCormack V, et al. Previous lung diseases and lung cancer risk: a pooled analysis from the International Lung Cancer Consortium. Am J Epidemiol. 2012;176(7):573–85.
144. Lamprecht B, McBurnie MA, Vollmer WM, Gudmundsson G, Welte T, Nizankowska-Mogilnicka E, et al. COPD in never smokers: results from the population-based burden of obstructive lung disease study. Chest. 2011;139(4):752–63.
145. Turner MC, Chen Y, Krewski D, Calle EE, Thun MJ. Chronic obstructive pulmonary disease is associated with lung cancer mortality in a prospective study of never smokers. Am J Respir Crit Care Med. 2007;176(3):285–90.
146. Kim HJ, Perlman D, Tomic R. Natural history of idiopathic pulmonary fibrosis. Respir Med. 2015;109(6):661–70.
147. Wells C, Mannino DM. Pulmonary fibrosis and lung cancer in the United States: analysis of the multiple cause of death mortality data, 1979 through 1991. South Med J. 1996;89(5):505–10.
148. Hubbard R, Venn A, Lewis S, Britton J. Lung cancer and cryptogenic fibrosing alveolitis. A population-based cohort study. Am J Respir Crit Care Med. 2000;161(1):5–8.
149. Le Jeune I, Gribbin J, West J, Smith C, Cullinan P, Hubbard R. The incidence of cancer in patients with

idiopathic pulmonary fibrosis and sarcoidosis in the UK. Respir Med. 2007;101(12):2534–40.
150. Matsushita H, Tanaka S, Saiki Y, Hara M, Nakata K, Tanimura S, et al. Lung cancer associated with usual interstitial pneumonia. Pathol Int. 1995;45(12):925–32.
151. Ozawa Y, Suda T, Naito T, Enomoto N, Hashimoto D, Fujisawa T, et al. Cumulative incidence of and predictive factors for lung cancer in IPF. Respirology. 2009;14(5):723–8.
152. Yang P, Sun Z, Krowka MJ, Aubry MC, Bamlet WR, Wampfler JA, et al. Alpha1-antitrypsin deficiency carriers, tobacco smoke, chronic obstructive pulmonary disease, and lung cancer risk. Arch Intern Med. 2008;168(10):1097–103.
153. Cote ML, Liu M, Bonassi S, Neri M, Schwartz AG, Christiani DC, et al. Increased risk of lung cancer in individuals with a family history of the disease: a pooled analysis from the International Lung Cancer Consortium. Eur J Cancer. 2012;48(13):1957–68.
154. Matakidou A, Eisen T, Houlston RS. Systematic review of the relationship between family history and lung cancer risk. Br J Cancer. 2005;93(7):825–33.
155. Lin H, Huang YS, Yan HH, Yang XN, Zhong WZ, Ye HW, et al. A family history of cancer and lung cancer risk in never-smokers: a clinic-based case-control study. Lung Cancer. 2015;89(2):94–8.
156. Gaughan EM, Cryer SK, Yeap BY, Jackman DM, Costa DB. Family history of lung cancer in never smokers with non-small-cell lung cancer and its association with tumors harboring EGFR mutations. Lung Cancer. 2013;79(3):193–7.
157. Omenn GS, Goodman G, Thornquist M, Grizzle J, Rosenstock L, Barnhart S, et al. The beta-carotene and retinol efficacy trial (CARET) for chemoprevention of lung cancer in high risk populations: smokers and asbestos-exposed workers. Cancer Res. 1994;54(7 Suppl):2038s–43s.
158. The effect of vitamin E and beta carotene on the incidence of lung cancer and other cancers in male smokers. The alpha-tocopherol, beta carotene cancer prevention study group. N Engl J Med. 1994;330(15):1029–35.
159. Hennekens CH, Buring JE, Manson JE, Stampfer M, Rosner B, Cook NR, et al. Lack of effect of long-term supplementation with beta carotene on the incidence of malignant neoplasms and cardiovascular disease. N Engl J Med. 1996;334(18):1145–9.
160. World Cancer Research Fund/American Institute for Cancer Research. Food, nutrition, physical activity, and the prevention of cancer: a global perspective. Washington DC: AICR; 2007.
161. Wang Y, Li F, Wang Z, Qiu T, Shen Y, Wang M. Fruit and vegetable consumption and risk of lung cancer: a dose-response meta-analysis of prospective cohort studies. Lung Cancer. 2015;88(2):124–30.
162. Yang WS, Wong MY, Vogtmann E, Tang RQ, Xie L, Yang YS, et al. Meat consumption and risk of lung cancer: evidence from observational studies. Ann Oncol (Official Journal of the European Society for Medical Oncology/ESMO). 2012;23(12):3163–70.
163. Leitzmann M, Powers H, Anderson AS, Scoccianti C, Berrino F, Boutron-Ruault MC, et al. European code against cancer 4th edition: physical activity and cancer. Cancer Epidemiol. 2015.
164. Leitzmann MF, Koebnick C, Abnet CC, Freedman ND, Park Y, Hollenbeck A, et al. Prospective study of physical activity and lung cancer by histologic type in current, former, and never smokers. Am J Epidemiol. 2009;169(5):542–53.
165. Zhong S, Ma T, Chen L, Chen W, Lv M, Zhang X, et al. Physical activity and risk of lung cancer: a meta-analysis. Clin J Sport Med (Official Journal of the Canadian Academy of Sport Medicine). 2015.

# Pathology and Molecular Pathology of Lung Cancer

# 20

Lucian R. Chirieac and Lester Kobzik

## 20.1 Introduction

The new molecular markers and assays reviewed in this chapter are the first wave of an exciting new phase in lung cancer pathology and treatment. The optimal interpretation and clinical application of these remarkable advances requires proficiency in lung anatomy and histopathology. Hence, we begin with a brief overview of lung structure, followed by a discussion of histological types of lung cancer, and the principles of grading and staging. We then focus on the molecular abnormalities of lung cancer, targeted therapies and the molecular biomarkers that help identify patients likely to benefit from these targeted therapies, the basic molecular biology principles behind these therapies, selected molecular diagnostic techniques, and the pathological features correlated with molecular abnormalities in lung cancer. Lastly, we discuss predictive biomarkers and their corresponding drugs that are currently under investigation in various phases of clinical trials. The investigation and analysis of lung cancer for particular abnormalities expands the expertise of the pulmonary oncologic pathologist, who in addition to conventional pathologic analysis of surgical lung specimens will determine predictive biomarkers for lung cancer-targeted therapies [1–4].

L.R. Chirieac (✉) · L. Kobzik
Harvard Medical School, Brigham and Women's Hospital, 75 Francis St., Boston, MA 02115, USA
e-mail: l.chirieac@partners.org

## 20.2 Embryology and Development of the Lung

Human lung development begins with a budding longitudinal groove from the ventral side of the primitive foregut around 26 days after fertilization. This bud progressively bifurcates and grows on either side of the foregut as the embryonic lungs, with successive branching giving rise to the lobar, segmental, and other divisions of the airways. The developing lungs grow into the coelomic cavity, the mesothelial lining of which forms from the mesoderm early in gestation. After the development of the diaphragm and a pleuropericardial membrane, the lungs become confined to the pleural cavities.

The phases of lung development and the major events are assigned to: (1) an embryonic stage that lasts until 6 weeks' gestation; (2) a pseudoglandular stage in which the primitive air spaces are widely separated by abundant mesenchyme and feature a vacuolated, glycogen-rich columnar or cuboidal epithelium; after division of the conductive airways down to the terminal bronchiole is complete ($\sim$16 weeks) this phase ends; (3) a canalicular stage which includes appearance of air spaces with an attenuated lining epithelium. Although no true alveoli are yet


M. Loda et al. (eds.), *Pathology and Epidemiology of Cancer*,
DOI 10.1007/978-3-319-35153-7_20

present, respiration is possible near the end of the canalicular stage (~28 weeks). The development of true alveoli is first readily apparent at ~36 weeks' gestation, but it is worth noting that alveoli develop mainly, but not completely, in the first year of life. This process, along with remodeling of airways, continues slowly through childhood and possibly into adolescence.

We finish by noting that the relationship of the biology of lung development to tumorigenesis has long sparked interest. For example, the familiar tumor marker TTF-1 (Nkx2.1) has critical functions as a transcription factor in lung budding and development. Especially relevant to the molecular pathology focus of this chapter are recent reviews that explore both specific molecular pathways [5] and potential roles for normal/cancer stem cells [6].

## 20.3 Anatomy of the Lung

The main bronchi divide to five lobar bronchi, and subsequent divisions supply the 19 lung segments [7, 8]. It is certainly worthwhile to be familiar with their terminology/nomenclature so as to correlate pathology with input from radiologists and surgeons. However, for the pathologist, this is perhaps the perfect situation in which to apply Einstein's advice to "never memorize something that you can look up."

After multiple airway divisions (following asymmetrical dichotomy—two smaller but unequal branches), eventually respiratory bronchioles alveolar ducts and spaces are formed where gas exchange occurs. One instance of somewhat confusing anatomic terminology merits discussion. The lung parenchyma can be formally divided into primary and secondary lobules. Primary lobules in the original sense correspond to one acinus, the unit supplied by one terminal bronchiole. The secondary lobule is a visible portion of lung surrounded by fibrous septa that comprises several primary lobules. However, in practice the septation of secondary lobules varies and the ability to precisely define these structures is limited. The term lobule is commonly used to refer to either true primary or secondary lobules, and must be interpreted in the context of the discussion.

The vasculature of the lungs features both branches of the pulmonary artery that bifurcate along with the airways (the 'broncho-vascular bundle') and pulmonary veins that run in the interlobular septa at the periphery of the lungs but join the artery and airway more proximally near the hilum. Other components include the bronchial artery circulation derived from the aorta that supplies oxygenated blood to the larger airways, and the lymphatics. The latter include the valved channels that begin at the bronchiolar level and the pleural plexus that both ultimately drain into the hilar and mediastinal lymph nodes. The importance of the mediastinal lymph nodes to modern lung cancer staging is paramount. The precise mapping of the inferior, aortic, and superior groups of nodes is well-summarized and illustrated elsewhere [9, 10].

## 20.4 Histology of the Lung

The microscopic structure of the lung is beautifully complex, and we cannot do it justice in this cursory review. Especially, germane to lung cancer biology are: (1) pseudostratified ciliated columnar epithelium of the airways; (2) the abundant mucus-secreting goblet cells within airways and in the submucosal glands present more proximally in the larger airways; (3) the close proximity of rich lymphatic vasculature to the airway epithelium, facilitating invasion by malignant tumors. Newer microanatomy research in mice identifies specialized bronchoalveolar stem cells at the junction of the airways and alveoli. One hypothesis is that these cells may also act as cancer stem cells, but their role in human lung cancer development remains to be characterized [7].

## 20.5 Classification of Lung Cancer

Lung cancer is generally divided into two major categories: non-small cell lung cancer (NSCLC) and small cell lung cancer (SCLC). The recent tumor classification system (Table 20.1) issued in 2011 [9] addresses the inadequacies of the 2004 classification and provides the foundation for tumor diagnosis and patient therapy and a critical basis for epidemiologic, molecular, and clinical studies [11]. The most important changes to the 2004 revised classification system were:

- eliminate bronchioloalveolar carcinoma,
- define the term adenocarcinoma in situ (AIS),
- define the term minimally invasive adenocarcinoma (MIA),
- revive the term "lepidic,"
- promote comprehensive histologic subtyping,

**Table 20.1** WHO classification of lung tumors

| Histologic type and subtypes | | | |
|---|---|---|---|
| EPITHELIAL TUMORS | | | |
| Adenocarcinoma | | | |
| | Lepidic adenocarcinoma | | |
| | Acinar adenocarcinoma | | |
| | Papillary adenocarcinoma | | |
| | Micropapillary adenocarcinoma | | |
| | Solid adenocarcinoma | | |
| | Invasive mucinous adenocarcinoma | | |
| | | Mixed invasive mucinous and | |
| | | Non-Mucinous adenocarcinoma | |
| | Colloid adenocarcinoma | | |
| | Fetal adenocarcinoma | | |
| | Enteric adenocarcinoma | | |
| | Minimally invasive adenocarcinoma | | |
| | | Non-Mucinous | |
| | | Mucinous | |
| | Preinvasive lesions | | |
| | | Atypical adenomatous hyperplasia | |
| | | Adenocarcinoma *in situ* | |
| | | | Non-mucinous |
| | | | Mucinous |
| Squamous cell carcinoma | | | |
| | Keratinizing squamous cell carcinoma | | |
| | Non-keratinizing squamous cell carcinoma | | |
| | Basaloid squamous cell carcinoma | | |
| | Pre-invasive lesion | | |
| | | Squamous cell carcinoma *in situ* | |
| Neuroendocrine Tumors | | | |
| Small cell carcinoma | | | |
| | Combined small cell carcinoma | | |
| Large cell neuroendocrine carcinoma | | | |
| | Combined large cell neuroendocrine carcinoma | | |
| Carcinoid tumors | | | |
| | Typical carcinoid tumor | | |
| | Atypical carcinoid tumor | | |
| Pre-invasive lesion | | | |
| | Diffuse idiopathic pulmonary neuroendocrine | | |
| | Cell hyperplasia | | |
| Large cell carcinoma | | | |
| Adenosquamous carcinoma | | | |
| Sarcomatoid carcinomas | | | |
| | Pleomorphic carcinoma | | |
| | Spindle cell carcinoma | | |
| | Giant cell carcinoma | | |
| | Carcinosarcoma | | |
| | Pulmonary blastoma | | |
| Other and Unclassified carcinomas | | | |
| | Lymphoepithelioma-like carcinoma | | |
| | NUT carcinoma | | |
| Salivary gland-type tumors | | | |
| | Mucoepidermoid carcinoma | | |

**Table 20.1** (continued)

| |
|---|
| Adenoid cystic carcinoma |
| Epithelial-myoepithelial carcinoma |
| Pleomorphic adenoma |
| Papillomas |
| Squamous cell papilloma |
| Exophytic |
| Inverted |
| Glandular papilloma |
| Mixed squamous and glandular papilloma |
| Adenomas |
| Sclerosing pneumocytoma |
| Alveolar adenoma |
| Papillary adenoma |
| Mucinous cystadenoma |
| Mucous gland adenoma |
| Mesenchymal Tumors |
| Pulmonary hamartoma |
| Chondroma |
| PEComatous tumors |
| Lymphangioleiomyomatosis |
| PEComa, benign |
| Clear cell tumor |
| PEComa, malignant |
| Congenital Peribronchial myofibroblastic tumor |
| Diffuse pulmonary lymphangiomatosis |
| Inflammatory myofibroblastic tumor |
| Epithelioid hemangioendothelioma |
| Pleuropulmonary blastoma |
| Synovial sarcoma |
| Pulmonary artery intimal sarcoma |
| Pulmonary myxoid sarcoma with EWSR1-CREB 1 translocation |
| Myoepithelial tumors |
| Myoepithelioma |
| Myoepithelial carcinoma |
| Lymphohistiocytic tumors |
| Extranodal marginal zone lymphomas of mucosa associated |
| Lymphoid tissue (MALT lymphoma) |
| Diffuse large-cell lymphoma |
| Lymphomatoid granulomatosis |
| Intravascular large B-cell lymphoma |
| Pulmonary Langerhans cell histiocytosis |
| Erdheim-Chester disease |
| Tumors of ectopic origin |
| Germ cell tumors |
| Teratoma, mature |
| Teratoma, immature |
| Intrapulmonary thymoma |
| Melanoma |
| Meningioma, NOS |
| Metastatic tumors |

- emphasize and introduce the term micropapillary carcinoma,
- detach the term mucinous adenocarcinoma, and
- discourage use of the term NSCLC and subclassify the tumors in as much detail as possible.

The histological features of each are described separately below.

## 20.5.1 Non-small Cell Carcinoma

### 20.5.1.1 Adenocarcinoma

#### Clinical Features

Adenocarcinoma is the most frequent cell type of lung cancer, accounting for over 50 % of cancers in most recent series. To date, most validated and investigational predictive biomarkers have been identified in adenocarcinoma as compared to other cell types and a new subtype classification of adenocarcinoma has been proposed by the International Association for the Study of Lung Cancer, American Thoracic Society and the European Respiratory Society that takes into account the molecular pathology of these tumors [9]. The current classification of lung adenocarcinoma by the World Health Organization recognizes several distinct morphologic subtypes of adenocarcinoma: papillary (Fig. 20.1), micropapillary (Fig. 20.2), acinar (Fig. 20.3), solid (Fig. 20.4), and lepidic (Fig. 20.5) [11]. The majority of lung adenocarcinomas exhibit combinations of morphologic patterns [12–14]. While the biologic basis for the histologic subtypes remains an area of active investigation [14], there is evidence that some subtypes may be associated with specific molecular alterations [14–19] or a better outcome [20–22].

## Pathologic Features

### Gross Findings

Grossly, adenocarcinoma typically has an irregularly lobulated configuration, with a gray-white cut appearance. As they are predominantly peripheral parenchymal masses, adenocarcinomas, in contrast to squamous cell carcinoma of the lung, are rarely associated with large airways. Anthracotic pigment is commonly entrapped in the tumor mass. Gross necrosis is uncommon except in larger masses. They may be found in association with fibrosis and pleural puckering. The penetration of the pleura may require additional studies such as elastic stains are important in tumor staging (see below).

### Microscopic Findings

Adenocarcinoma in situ (AIS): AIS, formerly BAC, is an important subtype of pulmonary adenocarcinoma (Fig. 20.6). This cancer has received increasing attention in recent years owing to its increasing incidence and rate of sensitivity to epidermal growth factor–tyrosine kinase inhibitors [24]. AIS is a primary lung tumor with a peripheral location, well-differentiated cytology, lepidic growth pattern, and a tendency for both aerogenous and lymphatic spread. The key feature is preservation of the underlying architecture of the lung with no invasion.

Minimally invasive adenocarcinoma (MIA): MIA was introduced to define patients with a near 100 % 5-year disease-free survival. It is defined as a lepidic predominant tumor measuring 3 cm or less that has an invasive component

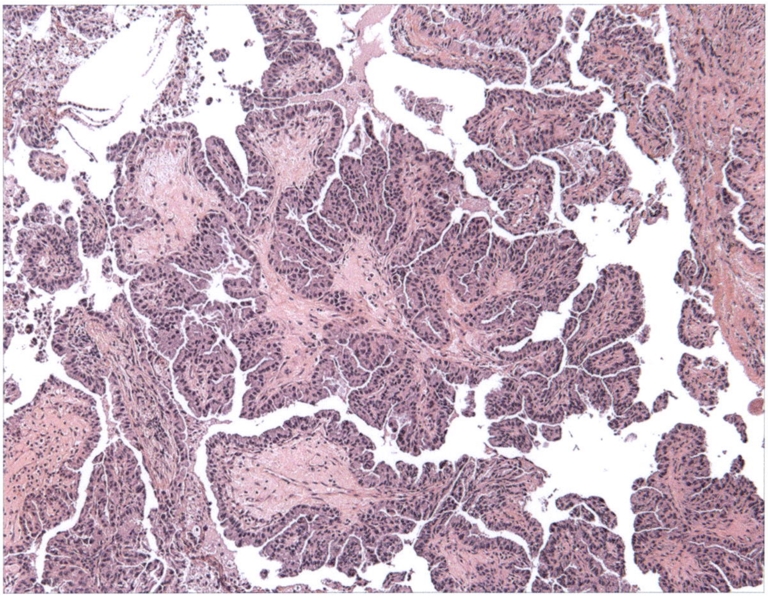

**Fig. 20.1** Papillary adenocarcinoma is characterized by finger-like projections of tumor cells connected by a stromal core abundant in vascular structures, with tumor cells protruding to the outside of these core structures. Due to the spatial 3D arrangements, some of these papillae give the false impression of floating into the tumor spaces

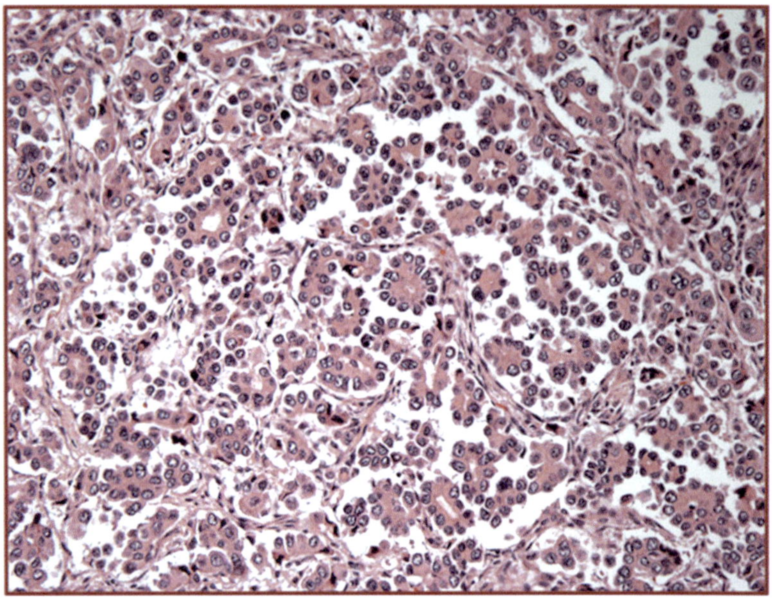

**Fig. 20.2** Micropapillary adenocarcinoma is characterized by a piling up and clustering of tumor cells in the alveoli. These clusters miss the vascular cores, in contrast to papillary adenocarcinoma

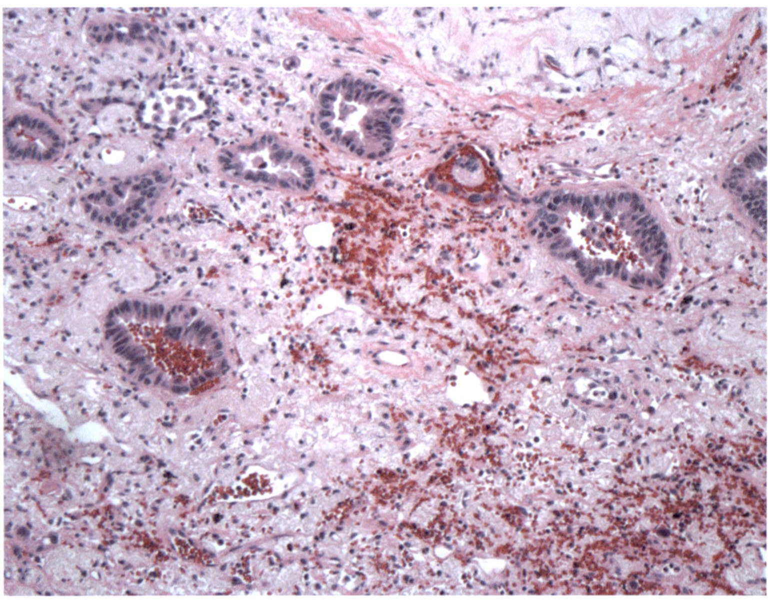

**Fig. 20.3** Acinar adenocarcinoma is the classical pattern of adenocarcinoma with tumor cells arranged in tubular architectures also called acinar

of 5 mm or less [25]. MIA is characterized by a combination of ground glass opacity (GGO) and a central solid opacity, with the solid component measuring 5 mm or less. Nonmucinous MIA is more common than mucinous MIA and most often appears as a GGO. Mucinous MIA appears radiologically as a solid or part-solid nodule.

Invasive adenocarcinoma: Changes were inserted in the classification of invasive adenocarcinomas. The current classification of lung adenocarcinoma by the World Health Organization recognizes several distinct morphologic subtypes of adenocarcinoma: papillary, micropapillary, acinar, solid, and lepidic [11].

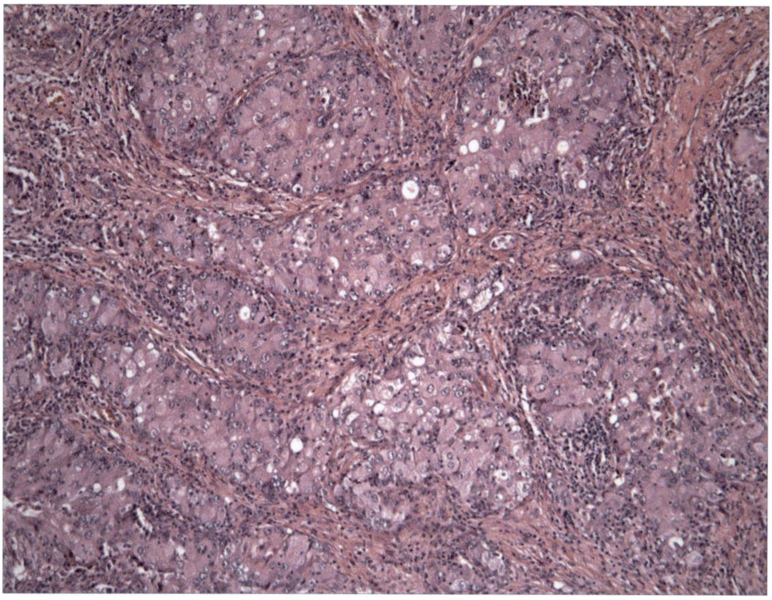

**Fig. 20.4** Solid lung adenocarcinoma characterized by irregularly shaped islands of tumor cells sometimes separated by stroma (desmoplastic reaction). This architectural pattern has no particular shapes, and is considered one of the worst behaving and aggressive type of lung cancer

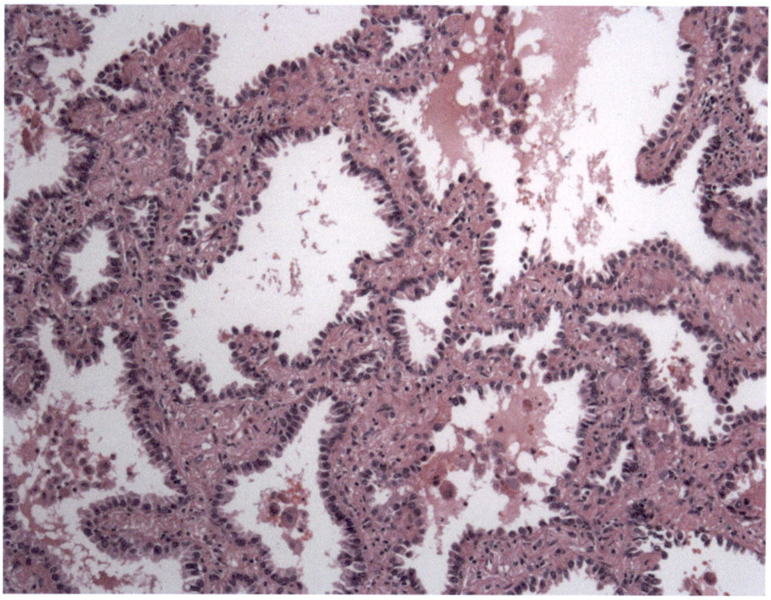

**Fig. 20.5** The lepidic pattern characterized by the tumor cells missing any invasion into the stroma. In this architectural pattern, the tumor cells spread along the alveolar walls, replacing the normal pneumocytes type II. This pattern is most commonly associated with some of the other previously mentioned patterns

The majority of lung adenocarcinomas exhibit combinations of morphologic patterns [12–14]. While the biologic basis for the histologic subtypes remains an area of active investigation [14], there is evidence that some subtypes may be associated with specific molecular alterations [14–18] or a better outcome [20–23]. Overtly invasive adenocarcinomas are classified according to the predominant subtype after the use of comprehensive histologic subtyping to estimate the percentages of the various components in a semiquantitative fashion in 5–10 % increments. The adenocarcinoma patterns are: lepidic, acinar, papillary, micropapillary, and solid. The invasive adenocarcinoma variants are mucinous adenocarcinoma (Fig. 20.7), colloid, fetal, and enteric morphologies. The term lepidic predominant adenocarcinoma consists of mixed

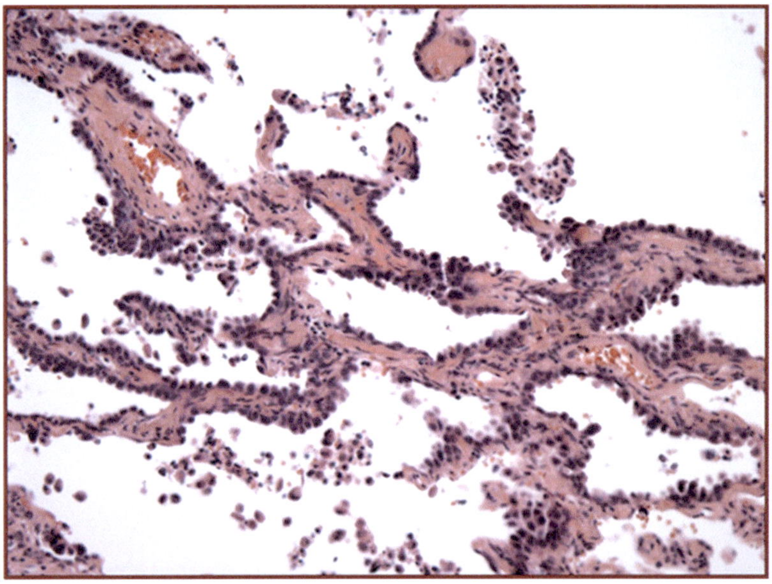

**Fig. 20.6** Adenoma in situ (AIS) This tumor is composed only by the noninvasive lepidic pattern. No invasion is identified into stroma, lymphatics, or pleura. Patients with this type of in situ carcinoma are considered to be cured by simple surgical excision

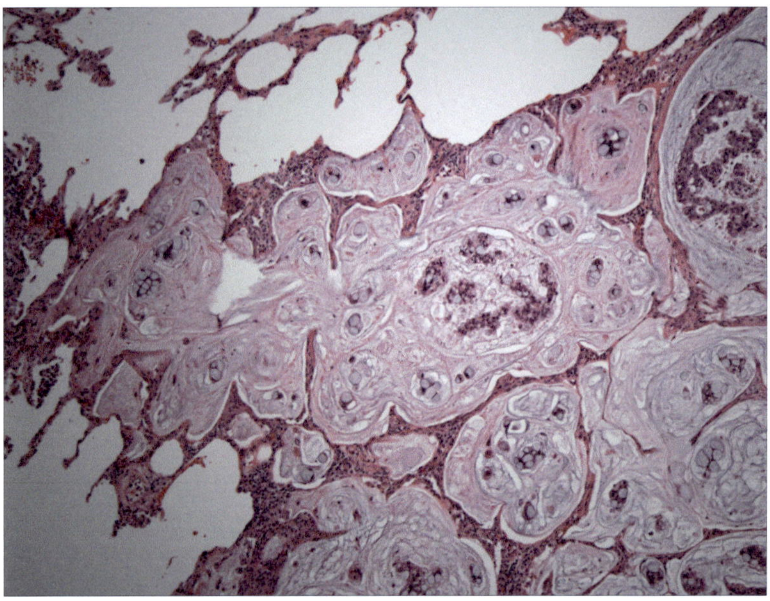

**Fig. 20.7** Mucinous adenocarcinoma composed of tumor cells secreting large amounts of mucus in the alveolar spaces and stroma

subtype tumors containing a predominant lepidic growth pattern of type II pneumocytes and/or Clara cells (formerly known as nonmucinous BAC) that have an invasive component >5 mm. A micropapillary predominant subtype is added because it has been recognized as a poor prognostic category. Signet ring (Fig. 20.8) and clear cell carcinoma (Fig. 20.9) subtypes are now recorded as cytologic features whenever present with a comment about the percentage identified.

High-power examination reveals that the tumor cells are typically polygonal with large vesicular nuclei, prominent nucleoli, and moderately abundant cytoplasm. Unlike SCC, the cytoplasmic borders are often poorly defined or indistinct. In addition, a variety of cell types have

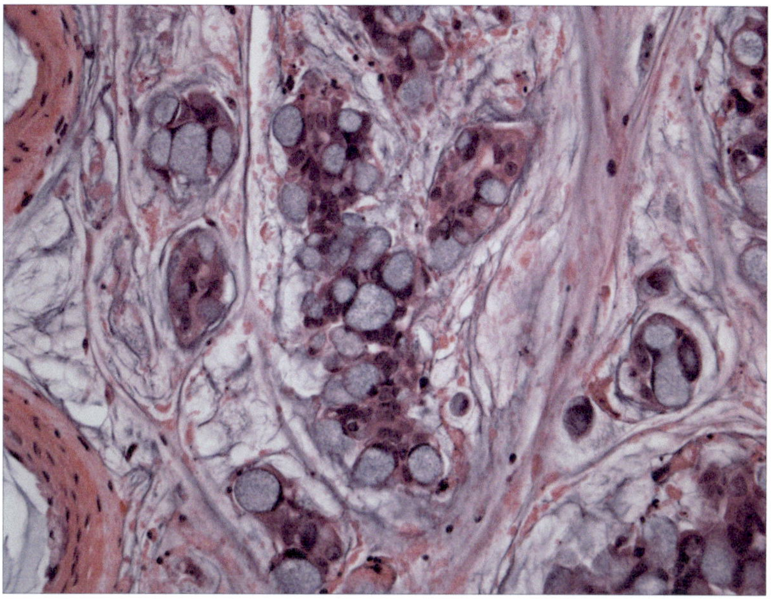

**Fig. 20.8** Signet ring cell carcinoma are cells with a peculiar cytology: the mucin is collected in the cytoplasm, pushing the nucleus to the periphery of the cell, giving the cell a peculiar appearance resembling an archaic ring with an attached seal structure

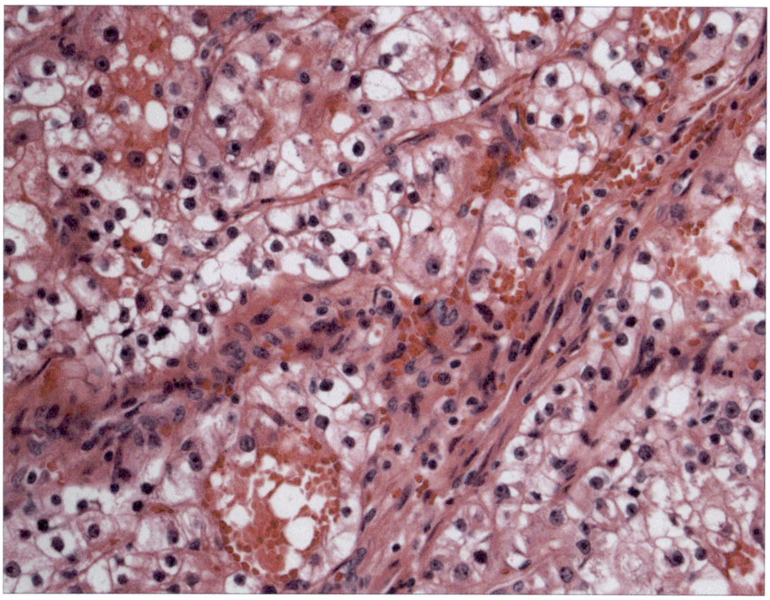

**Fig. 20.9** Some of the carcinomas have a glycogenized (clear) cytoplasm. This particular type of cytology (clear cell cytology) has no particular prognostic or biological significance

been described, such as clear cell, mucinous, fetal, sarcomatoid, and signet-ring cell. Moreover, it is not uncommon for adenocarcinoma to be present in association with other types of lung carcinoma (e.g., combined with SCLC, or large-cell neuroendocrine carcinoma, or SCC, or sarcomatoid carcinoma).

#### 20.5.1.2 Squamous Cell Carcinoma

**Clinical Features**

SCCs represent approximately 30 % of all NSCLC. Their incidence has been decreasing compared to adenocarcinomas, possibly due to changes in smoking habits. It is strongly linked to a history of cigarette smoking.

## Pathologic Features

### Gross Findings

Most SCCs arise centrally from the segmental or subsegmental bronchi. However, the incidence of SCC of the peripheral lung is increasing. Grossly, tumor masses are usually gray white to yellow tan. SCC is the most common type to give rise to a thick-walled irregular cavity with central necrosis. The texture can be firm or gritty and may be surrounded by areas of obstructive consolidation. Some of the proximal tumors have an exophytic, papillary, and endobronchial growth pattern. Because of their frequent central location (in the mainstem, lobar, or segmental bronchi), diagnosis by cytological examination of sputum, bronchoalveolar lavage (BAL), bronchial brushing and washing, or endoscopic biopsies can be performed with generally satisfactory results. Also, because of their central location, direct extension of the primary tumor mass into the adjacent hilar lymph nodes is common.

### Microscopic Findings

SCC is a malignant epithelial tumor with keratinization and/or intercellular bridges (Fig. 20.10). SCC is graded as well differentiated if prominent keratinization, intercellular bridges, or pearl formation is present. They are moderately differentiated if these features are easily seen but not extensive. Poorly differentiated SCCs have only focal morphologic features of squamous differentiation. Keratinization may take the form of squamous pearls or individual cells with markedly eosinophilic dense cytoplasm. The presence of intracellular mucin in a few cells does not exclude tumors from this category. In situ SCC may be seen in the adjacent airway mucosa.

SCCs can present as histological variants which include: papillary, clear cell, small cell, and basaloid patterns. Rarely, these patterns are seen throughout the tumor, but more commonly, they are focal. Even though invasive growth is not identified, papillary SCC can be diagnosed if there is sufficient cytological atypia. However, small biopsy specimens that show a very well-differentiated papillary squamous epithelium should be interpreted with caution since separation of a papillary squamous carcinoma from a papilloma can be difficult. Furthermore, the squamous epithelium of a squamous papilloma may extend into the bronchial glands and should not be confused with invasion.

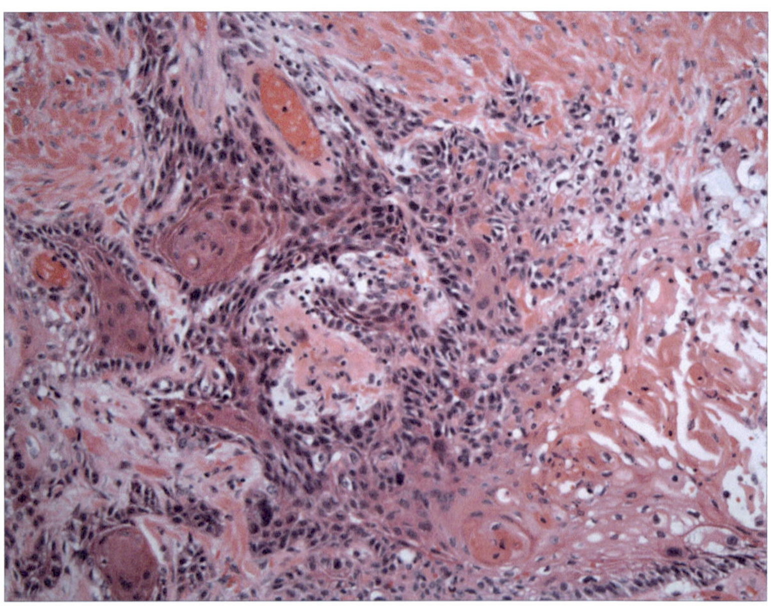

**Fig. 20.10** Squamous cell carcinoma is a malignant epithelial tumor with keratinization and/or intercellular bridges

#### 20.5.1.3 Large-Cell Carcinoma

Large-cell carcinoma (LCC) is an undifferentiated malignant epithelial tumor accounting for approximately 10 % of all lung cancers in many series. It is strongly associated with cigarette smoking. The lesion tends to occur in the periphery and grows rapidly. It usually presents at a late stage, resulting in a poor outcome.

**Pathologic Features**

Gross Findings

About half of the cases display a relationship to a large airway. Grossly, LCCs are frequently greater than 5 cm in size and have a white to gray or fish-flesh cut appearance. Gross tumor necrosis is commonly appreciable.

Microscopic Findings

LCC is a diagnosis of exclusion made after ruling out the presence of a component of SCC, adenocarcinoma, or SCLC. Because of this, the diagnosis is best made on resection specimens, and not on small biopsy specimens. In general, the cells typically have large nuclei, prominent nucleoli, and a moderate amount of cytoplasm. They are typically arranged in sheets or large nests, frequently revealing foci of necrosis.

Morphologically, they have lobular, trabecular, or palisading growth patterns surrounding typically centrally located comedo-type necrosis. These tumors consist of relatively small monomorphic cuboidal to fusiform cells with moderately hyperchromatic nuclei, finely granular chromatin, absent or only focal nucleoli, scant cytoplasm, and a high mitotic rate. Neither intercellular bridges nor individual cell keratinization are present. A high percentage of cases have associated carcinoma in situ. Immunohistochemical stains for adenocarcinoma, SCC, and neuroendocrine markers are negative. About half of the tumors with this histological pattern are pure basaloid carcinomas. Immunohistochemistry (IHC) reveals that these neoplasms commonly express cytokeratin, but do not express TTF-1 or p63. Ultrastructural analysis of LCCs usually shows evidence of squamous, glandular, or neuroendocrine differentiation, suggesting that these are, in fact, very poorly differentiated NSCLCs.

### 20.5.2 Small Cell Carcinoma

SCLC is defined as a neuroendocrine tumor with more than 10 mitoses per 2 $mm^2$ and small cell cytologic features. Cells have an oval or vaguely spindled shape and have scant cytoplasm. Nuclei are hyperchromatic and have absent or very small nucleoli (Fig. 20.11). Crush artifact may be prominent on small biopsies, but this is not pathognomonic for the diagnosis of SCLC. In larger core biopsies or resected specimens, the cells may appear slightly larger than in a transbronchial biopsy and may have distinct cytoplasm. Numerous prominent nucleoli and large cells should not be seen.

## 20.6 Lung Cancer Grading

Histological grading (well, moderately, and poorly differentiated adenocarcinomas) has prognostic significance and recently published analyses have been validated in clinical practice using histological and cytologic criteria [26].

## 20.7 Lung Cancer Staging

Stage is the most important prognostic and predictive factor for patients with lung neoplasms, and pathologists are expected to provide accurate staging information in lung-resection specimens. Separate staging systems have been proposed for patients with NSCLC and SCLC. The seventh edition of the published guidelines of the American Joint Commission on Cancer (AJCC) compiles the most recent clinical and pathologic staging of patients with lung cancer and other neoplasms (Table 20.1). The designation "T" refers to a primary tumor that has not been previously treated. The symbol "p" refers to the pathologic classification of the tumor/node/metastasis (TNM), as opposed to the clinical classification, and is based on gross and

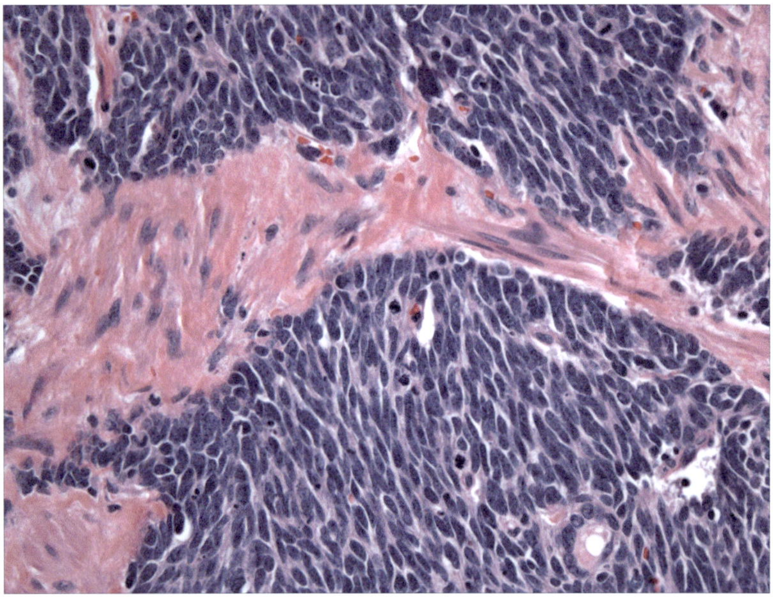

**Fig. 20.11** Small cell carcinomas are extremely aggressive tumors with a neuroendocrine differentiation. Cells have an oval or vaguely spindle shape with very scant cytoplasm, nuclear molding with absent or very small nucleoli

microscopic examination. "pT" entails a resection of the primary tumor or biopsy adequate to evaluate the highest pT category, pN entails removal of nodes adequate to validate lymph node metastasis, and pM implies microscopic examination of distant lesions. Overall survival for lung cancer is 16 %; however, survival is stage-dependent. Overall survival rates for patients with stages I–IV NSCLC are 60–80, 25–50, 10–40, and 4:5 %, respectively.

Patients with Stage I usually undergo surgical resection with lobectomy, segmentectomy, or wedge resection. Usually for patients with stage II disease (and for those with Stage III disease diagnosed upon final pathology following resection), treatment should include anatomic surgical resection followed by adjuvant chemotherapy. Patients with resectable disease (Stages I and II) who have medical contraindications for surgery are candidates for curative radiation therapy. Patients with stage IIIA NSCLC receive neoadjuvant chemotherapy followed by surgical resection. Patients with Stage IIIB disease are treated with radiation therapy in combination with chemotherapy; those with stage IV disease receive this regimen, predominantly as palliative therapy.

## 20.8 Molecular Alterations and Precision Oncology in Lung Cancer

NSCLC is the second most common cancer diagnosed in the United States and the leading cause of cancer-related mortality, with an estimated 221,200 new cases and 158,040 deaths anticipated in 2015 [27]. Lung cancer was the leading cause of cancer death among men in 2012 [28]. Among women, lung cancer was the leading cause of cancer death in more developed countries, and the second leading cause of cancer death in less developed countries [28]. Globally, the overall lifetime risk of lung cancer is about 1 in 13 for men and 1 in 16 for women. The risk is significantly higher for smokers and lower for nonsmokers [29–31]. However, lung cancer rates in Chinese women (20.4 cases per 100,000 women) were higher than rates among women in some European countries despite a lower prevalence of smoking. This is thought to reflect indoor air pollution from unventilated coal-fueled stoves and cooking fumes [32]. Other known risk factors for lung cancer include exposure to occupational and environmental carcinogens such as asbestos, arsenic, radon, and polycyclic aromatic hydrocarbons [33]. Recently, outdoor pollution has also been determined to cause lung cancer [34, 35]. More than one-half of the lung cancer deaths attributable to ambient fine particles were projected to have been in East Asian countries [36].

In the past years, given the development of new targeted therapies, tremendous efforts have been directed towards identifying potentially druggable molecular alterations, especially against known activating mutations. Although numerous mutations have been described in lung adenocarcinoma [37], the mutation status remains unknown in more than 50 % of cases [38]. So far, we can identify at the present moment therapeutic targets in only 20 % of lung cancers.

Molecular profiling has become the standard of care for advanced (metastatic) lung cancer. For nonsquamous NSCLC, which accounts for more than half of all lung cancer cases, routine testing for epidermal growth factor receptor (EGFR) mutations and anaplastic lymphoma kinase (ALK) rearrangements is recommended. In cases with identified EGFR (approximately 15 % of NSCLC) or ALK alterations (approximately 5 % of NSCLC), molecularly targeted therapy with EGFR-or ALK-targeting drugs is now the preferred initial approach to treatment [39].

### 20.8.1 Targeted Therapies in Lung Cancers with Epidermal Growth Factor Receptor (EGFR) Abnormalities

#### 20.8.1.1 EGFR

Recognized mechanisms of EGFR gain of function in NSCLC include somatic activating mutations in the exons encoding the tyrosine kinase domain and EGFR gene amplification [40–42]. The EGFR mutation status is best determined by gene sequencing abnormalities of EGFR status may also be observed with gene copy number determined by fluorescence in situ hybridization (FISH) or chromogenic in situ hybridization (CISH), and protein expression determined by IHC with mutation-specific antibodies. Several mutations have been recently described in the tyrosine kinase domain of EGFR [40, 43]. EGFR is expressed in 50 % of NSCLCs, and its expression is correlated with poor prognosis [44]. These two factors make EGFR and its family members prime candidates for the development of targeted therapeutics [45]. EGFR kinase domain mutations target four exons [18–21], which encode part of the tyrosine kinase domain (the entire kinase domain is encoded by exons 18–24) and are clustered around the ATP-binding pocket of the enzyme [46–50].

EGFR gene amplification is detected in some EGFR-mutation-positive patients as well [51]. A subset of lung adenocarcinomas has activation of growth factor receptor (EGFR) by mutations and/or amplification but the interaction between them is complex and unclear. Some EGFR-amplified lung adenocarcinomas have distinct genetic alterations, unique clinicopathologic features, and worsened prognosis [51, 52]. Furthermore, EGFR amplification and EGFR mutations are heterogeneously distributed within any given tumor. These are novel and important findings with implications for the efficacy of treatment with tyrosine kinase inhibitors in patients with EGFR-mutant lung adenocarcinoma [51].

Recent discoveries described EGFR mutation-specific antibodies that could help in the rapid screening of lung cancers with EGFR mutations (Fig. 20.12) [52].

Mutations in the tyrosine kinase domain of the EGFR have prognostic significance since patients with EGFR-mutant NSCLC have prolonged disease-free survival compared with those with wild-type disease, regardless of the treatment received [2, 53, 54]. Although EGFR mutations are predictive of response to EGFR tyrosine kinase inhibitor (TKI) therapy, they do not appear to be predictive of a differential effect on survival [54].

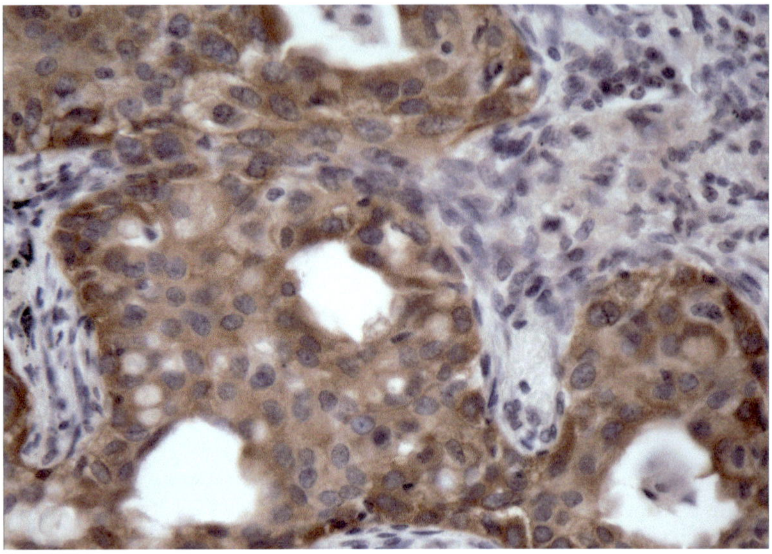

**Fig. 20.12** Using exon 19 deletion-specific antibody recently described one is able to directly visualize the location of tumor cells with EGFR exon 19 deletion mutations and show heterogeneity in receptor overexpression among different tumor cells (Immunohistochemistry, exon 19 deletion-specific antibody, 200× magnification)

#### 20.8.1.2 Targeted Agents Against Lung Cancer with EGFR Mutations

EGFR-mutant NSCLC generally refers to cases with sensitizing mutations in the EGFR kinase domain (exon 19 deletions or exon 21 L858R substitutions). These activating mutations result in constitutive activity of the EGFR kinase domain, generating survival and proliferative signals through the PI3 K-Akt-mTOR and Ras-Raf-MEK pathways. In these cases, EGFR inhibitors such as erlotinib, gefitinib, and afatinib in the first-line setting yield response rates in excess of 75 %, and overall survival exceeding 2 years [39]. In contrast, in NSCLC without actionable molecular alterations treated with conventional chemotherapy, response rates are approximately 30 %, median overall survival is 12 months. The first two TKI agents approved for use in lung cancer that target lung cancer with EGFR mutations were gefitinib (2002) and erlotinib (2003). EGFR mutation is a specific target for therapy by TKIs and is a validated biomarker of treatment response [42]. The clinical utility of this biomarker is supported by prospective clinical trials that have demonstrated a progression-free survival benefit of TKI as first-line therapy in EGFR-mutant patients [55]. Based on current data, predictive biomarker tests for EGFR should involve mutational analysis. Molecular profiling has become the standard of care for advanced (metastatic) lung cancer and routine testing for EGFR is recommended [56, 57]. In cases with identified EGFR alterations, molecularly targeted therapy with EGFR-targeting drugs is now the preferred initial approach to treatment.

Resistance to TKI therapy is associated with KRAS mutation and specific acquired EGFR mutations such as T790 M [58, 59]. These molecular events, as well as other genetic alterations in cMet (amplification), ERBB3 (overexpression), and epiregulin (autocrine loop activation), account for approximately 50 % of cases of TKI-resistance [50, 60–62].

### 20.8.2 Genotype-Phenotype Correlations

In patients with lung adenocarcinoma treated with erlotinib and gefitinib, favorable responses were associated with adenocarcinoma with lepidic patterns [18]. This finding led to trials of gefitinib and erlotinib in patients that showed that 17–22 % of patients had a response to gefitinib [63, 64]. The weak correlation between EGFR mutation status and adenocarcinoma subtypes [65, 66] led to the adoption of reflex genetic

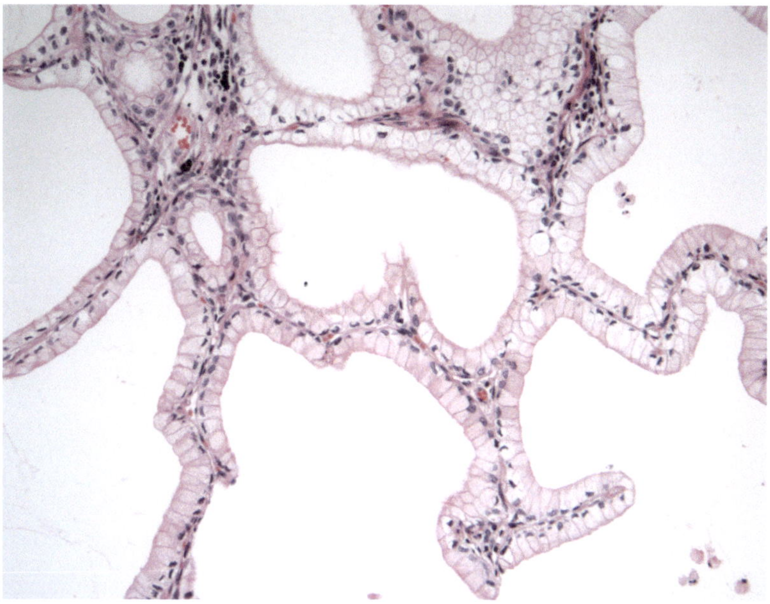

**Fig. 20.13** Mucinous adenocarcinoma is exclusively TTF1 negative, EGFR mutation negative but may have Ras mutation (H&E, 100× magnification)

testing of all lung cancers and investigation for treatable molecular targets [67]. Genetic abnormalities can be seen in different histology types although with various frequency. One characteristic correlation is that mucinous adenocarcinoma (Fig. 20.13) may be exclusively TTF1 negative, EGFR mutation negative but may have Ras mutation, and expresses CDX2 possibly because of their presumed derivation from bronchiolar mucinous goblet cells [15, 68]. However, more recently, molecular genetic analyses of lung adenocarcinoma have recently become the standard of care for treatment selection [57].

### 20.8.3 Targeted Therapies with Angiogenesis Inhibitors in Nonsquamous NSCLC

Recent studies show that NSCLC with histology types other than SCC appear to be more associated with response to treatment with bevacizumab. Bevacizumab (Avastin®) is a monoclonal antibody with high affinity for VEGF. Despite the potential benefit of bevacizumab for some patients with previously untreated advanced NSCLC [69, 70], the appropriate clinical setting for the use of this antiangiogenic agent is stringent, due to safety issues raised in patients with SCC, which requires an accurate diagnosis on the pretreatment biopsy specimens. The clinical activity of bevacizumab in inoperable locally advanced, metastatic, or recurrent NSCLC was first shown in chemotherapy-naive patients [71]. Patients with nonsquamous NSCLC histology are the only patients who benefit from treatment with bevacizumab in combination with chemotherapy [69].

Bevacizumab is currently contraindicated in patients with SCC on the basis of the results of a recently published phase II trial [71] that 31 % of patients with SCC histology developed a life-threatening or fatal hemoptysis associated with bevacizumab, although it is still not clear whether histology alone is the reason for increased bleeding risk. Excluding patients with SCC appeared to markedly limit the risk of life-threatening bleeding complications associated with bevacizumab.

### 20.8.4 Targeted Therapies in Lung Cancers with Anaplastic Lymphoma Kinase (ALK) Abnormalities

On August 26, 2011, the Food and Drug Administration (FDA) approved crizotinib for the treatment of patients with locally advanced or metastatic NSCLC that is ALK-positive by FISH (Fig. 20.14). Anaplastic large-cell lymphoma kinase gene (ALK) was originally identified through cloning of the t [2, 5] (p23;35) translocation found in a subset of anaplastic large-cell lymphomas (ALCLs), a tumor of T-cell lineage [72, 73]. ALK encodes a tyrosine kinase receptor that is normally expressed only in select neuronal cell types. In ALK-rearranged ALCLs, the intracytoplasmic portion of ALK is fused to the N-terminal portion of nucleophosmin (NPM) resulting in a chimeric protein with constitutive kinase activity. Several other balanced translocations involving ALK have been discovered in ALCLs; however, the various resulting chimeric proteins all retain the ALK kinase domain [74]. The importance of the kinase activity is exemplified by ALK-rearranged ALCL cell lines which are dependent upon ALK enzymatic activity for growth and survival.

Recently, ALK rearrangements were identified in rare NSCLC cell lines and in isolated primary adenocarcinomas from Japanese and Chinese populations [75, 76]. The majority of the ALK rearrangements within NSCLCs result from an interstitial deletion and inversion in chromosome 2p and result in the EML4-ALK fusion gene product [75, 76]. Murine tumors, human cell lines, and a recent published clinical trial have shown that lung cancers expressing EML-ALK are sensitive to inhibitors of ALK kinase activity [77–79]. Thus, it is critical to efficiently and accurately identify those lung adenocarcinomas that harbor ALK rearrangements in routine practice in order to guide the appropriate clinical therapy.

None of the ALK-rearranged adenocarcinomas showed coexistent mutations in EGFR. Published studies show that ALK-rearranged adenocarcinomas are more likely to present in younger patients with a history of never-smoking, and at higher stage relative to those without ALK rearrangements (ALK germline) [80]. The majority of ALK-rearranged adenocarcinomas had a distinct histology represented by solid tumor growth and frequent signet-ring cells with abundant intracellular mucin (Fig. 20.15) [80].

The developing evidence-based guideline recommendations of the College of American

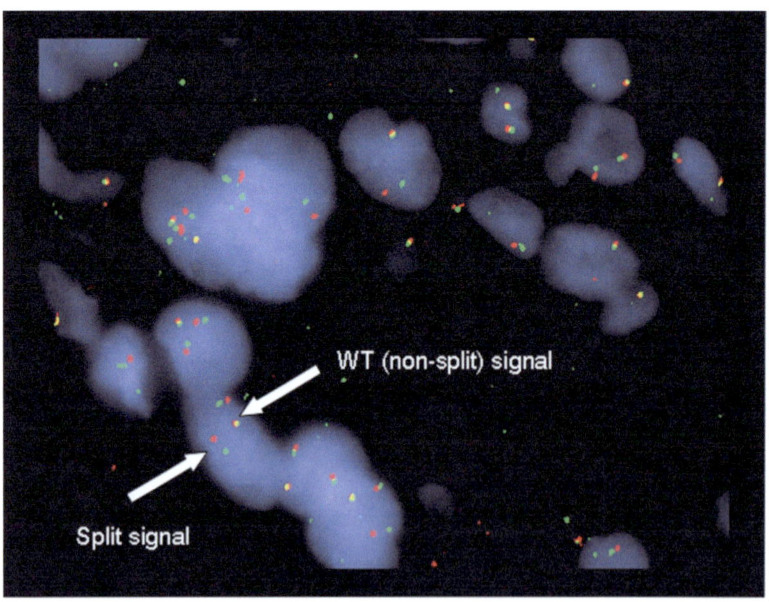


**Fig. 20.14** Identification of lung cancers with chromosomal translocations involving ALK requires fluorescence in situ hybridization on formalin-fixed, paraffin-embedded tumor tissues using a break-apart probe to the ALK gene (FISH, 1000× magnification)

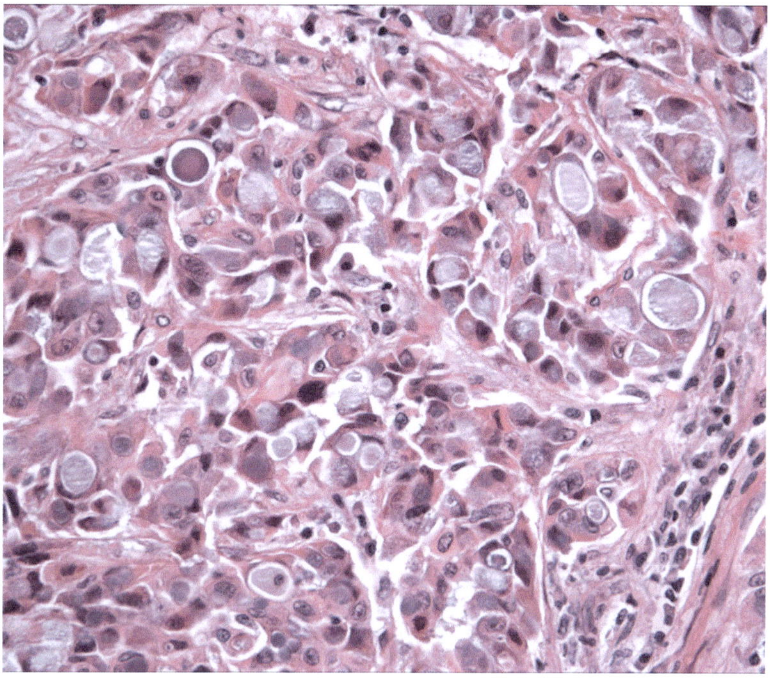

**Fig. 20.15** The majority of ALK-rearranged adenocarcinomas had a distinct histology represented by solid tumor growth and frequent signet-ring cells with abundant intracellular mucin (H&E, 100× magnification)

Pathologists (CAP), the International Association for the Study of Lung Cancer (IASLC), and the Association for Molecular Pathology (AMP) for the molecular testing of lung cancers will likely conclude that ALK rearrangements be typically assessed by molecular cytogenetic techniques such as FISH and the currently commercially available ALK monoclonal antibodies help in screening lung cancers for ALK rearrangements. Therefore, IHC is useful in identifying cases with ALK rearrangements at the present time [80–83].

### 20.8.5 Other Molecular Abnormalities that Show Promise for Targeted Therapies in Lung Cancer

#### 20.8.5.1 ROS1

Due to multiplex genomic analysis, such as that conducted by the Lung Cancer Mutation Consortium, and the emergence of clinically available Next Generation sequencing, understanding of the molecular underpinnings and vulnerabilities of lung cancer has evolved well beyond EGFR and ALK [84]. ROS1 rearrangements occur in 1–2 % of NSCLC [85]. ROS1 has a high degree of homology with ALK (approximately 50 % within the kinase domain and 75 % within the ATP-binding site), and the majority of cases respond to the first-generation ALK inhibitor crizotinib; however, certain other ALK inhibitors such as alectinib do not appear to have activity against ROS1-positive cases. ROS1 is a distinct receptor with a kinase domain that is phylogenetically related to the anaplastic lymphoma kinase/lymphocyte-specific protein tyrosine kinase (ALK/LTK) and insulin receptor (INSR) RTK families, suggesting that tyrosine kinase inhibitors for these receptors could have cross-activity against ROS1 [86].

#### 20.8.5.2 Her-2

Unlike the other members of the HER family, HER-2 is not strictly a receptor tyrosine kinase because no high-affinity endogenous ligand has been identified. HER-2 acts as a signaling network coordinator and amplifier when it heterodimerizes with other HER family members. HER-2 mutations (in contrast to amplification, which is the pathogenic event in breast and gastroesophageal cancer) occur in 2–4 % of NSCLC. Dual EGFR/HER2 inhibitors such as afatinib and lapatinib, as well as the anti-HER2 antibody trastuzumab, have activity against these cases [87, 88]. They are in-frame insertions in exon 20 and have targeted the corresponding TK domain region, as in EGFR-insertion mutations. These mutations occur in the same subpopulation as those with EGFR mutations (adenocarcinoma, never-smoker, East Asian, and women) [66]. Although HER-2 mutations occur in only 2 % of patients, HER-2 is frequently overexpressed to some degree in NSCLC and appears to be associated with drug resistance, increased metastatic potential, increased production of vascular endothelial growth factor (VEGF), and poor prognosis [89]. HER-2–mediated resistance to DNA-damaging agents requires the activation of Akt, which phosphorylates murine double minute 2 (MDM2) and therefore enhances MDM2-mediated ubiquitination and degradation of p53. Blocking the Akt pathway mediated by HER-2 increases the cytotoxic effect of DNA-damaging drugs in tumor cells with wild-type p53. Furthermore, recent studies have shown that the G/G genotype of the MDM2 polymorphism is associated with worse overall survival among early-stage NSCLC patients, particularly those with squamous cell histology [90].

Trastuzumab (Herceptin) is a chimerized monoclonal antibody against HER-2. Combinations of trastuzumab and chemotherapy are well tolerated, with response rates of 21–40 % [91]. One trial showed that patients whose tumors highly overexpressed HER-2 (3+) by IHC or evidence of amplification by FISH showed a good response. It appears that highly overexpressing HER-2 cases of NSCLC (3+ by IHC), although relatively infrequent (3–9 %), may show benefit with treatment with trastuzumab.

#### 20.8.5.3 MET Proto-oncogene

MET can be activated by mutations, autocrine/paracrine growth, overexpression by gene amplification, or decreased degradation [92]. Germline and somatic MET gene mutations have been reported in hereditary and sporadic papillary renal cell cancers [93]. MET gene mutations and amplifications have been reported in other cancers to be predictors of response to therapy [94]. Expression of MET and phospho-MET has been studied in lung cancer, and recently it was shown that 40 % of lung cancer tissues overexpressed MET [95–97]. Recent studies have shown that survival in NSCLC patients with ≥5 copies/cell is worse than those with less than 5 copies/cell and that MET gene amplification leads to EGFR tyrosine kinase resistance in EGFR-mutant patients [62]. Anti-HGF antibodies, anti-MET antibodies, and small-molecule MET TKI inhibitors are all in various stages of development, and predictive biomarkers for MET inhibitors will be important to elucidate for future trials and treatment decisions [92].

#### 20.8.5.4 Other Targeted Molecular Therapies

There has been tremendous research and investment in the development of small molecules that target key proteins in cell signaling pathways that are aberrantly altered in disease, particularly in carcinogenesis. For instance, receptor tyrosine kinases (RTKs) serve as potential therapeutic targets in several solid tumors, including lung cancer.

Gene fusions involving the rearranged during transfection (RET) gene occur in approximately 1 % of NSCLC. Several commercially available multitargeted kinase inhibitors have RET activity (e.g., vandetanib, sorafenib, sunitinib, cabozantinib); cabozantinib has led to radiographic responses.

BRAF mutations occur in 1–3 % of NSCLC. Of these, approximately 50 % are V600 and respond to BRAF inhibitors such as vermurafenib and dabrafenib, both currently approved for V600 BRAF-mutant melanoma.

The RTK c-kit is highly expressed in SCLC (although it is not mutated), and this has led to clinical trials with the specific c-KIT inhibitor (STI-571, Glivec, Novartis), alone and in combination therapy. However, these trials have failed to show a meaningful benefit from the imatinib treatment [98]. Antibodies against the angiogenic factor VEGF and small molecules against VEGF receptors, such as SU5416, which is an inhibitor of Flk-1 receptor, are being tested in NSCLC and other tumor types. More recently, modification of gene expression using siRNAs has the promise of being the most powerful tool yet.

## 20.9 Conclusion

Surgical excision remains the only therapeutic modality that can cure selected lung cancer patients. Pathologists play an important role in the surgical management of patients with lung cancer, from preoperative diagnosis and staging, to intraoperative evaluation and postoperative assessment of tumor genetic alterations. With the design of new targeted therapies, pathologists are required to identify the "targeted" population or the subset of patients that benefits the most from these novel therapies.

The clinical application of molecular diagnostic techniques has allowed a more precise and rapid assessment of lung cancer and will triage the patient to "personalized" therapies that will have the highest rates of eradicating the tumor. The convergence of genetics, informatics, and imaging, along with other novel technologies, like circulating free DNA (cfDNA) is rapidly expanding the area of "precision medicine" by refining the classification of lung cancer, often with important prognostic and treatment implications [99–101]. Among these new technologies, genetics and next-generation DNA sequencing methods are having the greatest effect. The prospect of sequencing whole genomes at lesser costs reshapes our approaches to genetic testing [102]. The clinical implications will be greatest when the results of genetic testing are actionable to inform about prognosis and prediction to therapeutic modalities [103].

Our knowledge about lung cancer changed radically in the past decade, and progress mainly depends on identifying new predictive biomarkers. We need to better understand both the tumor and the host biology that underlies tumor sensitivity and resistance in order to provide a rationale for specific targeted therapy.

## References


1. Cappuzzo F, Hirsch FR, Rossi E, Bartolini S, Ceresoli GL, Bemis L, et al. Epidermal growth factor receptor gene and protein and gefitinib sensitivity in non-small-cell lung cancer. J Natl Cancer Inst. 2005;97(9):643–55 (PubMed PMID: 15870435).
2. Eberhard DA, Johnson BE, Amler LC, Goddard AD, Heldens SL, Herbst RS, et al. Mutations in the epidermal growth factor receptor and in KRAS are predictive and prognostic indicators in patients with non-small-cell lung cancer treated with chemotherapy alone and in combination with erlotinib. J Clin Oncol. 2005;23(25):5900–9 (PubMed PMID: 16043828).
3. Ettinger DS, Wood DE, Akerley W, Bazhenova LA, Borghaei H, Camidge DR, et al. Non-small cell lung cancer, version 6.2015. J Nat Compr Cancer Netw JNCCN. 2015;13(5):515–24 (PubMed PMID: 25964637).
4. Imielinski M, Berger AH, Hammerman PS, Hernandez B, Pugh TJ, Hodis E, et al. Mapping the hallmarks of lung adenocarcinoma with massively parallel sequencing. Cell. 2012;150(6):1107–20 (PubMed PMID: 22980975. Pubmed Central PMCID: 3557932).
5. Powers S, Mu D. Genetic similarities between organogenesis and tumorigenesis of the lung. Cell Cycle. 2008;7(2):200–4 (PubMed PMID: 18256532).
6. Lundin A, Driscoll B. Lung cancer stem cells: progress and prospects. Cancer Lett. 2013;338 (1):89–93 (PubMed PMID: 22906416. Pubmed Central PMCID: 3686996).
7. Leeman KT, Fillmore CM, Kim CF. Lung stem and progenitor cells in tissue homeostasis and disease. Current topics in developmental biology. 2014;107: 207–33 (PubMed PMID: 24439808. Pubmed Central PMCID: 4038302).
8. Corrin B, Nicholson, A. The structure of the normal lungs. In: Corrin B, Nicholson A, editors. The pathology of the lungs. 3rd ed. London: Churchill Livingstone; 2011.
9. Travis WD, Brambilla E, Noguchi M, Nicholson AG, Geisinger KR, Yatabe Y, et al. International Association for the Study of Lung Cancer/American Thoracic Society/European Respiratory Society International multidisciplinary classification of lung adenocarcinoma. J Thorac Oncol. Official publication of the International Association for the Study of Lung Cancer. 2011 Feb;6(2):244–85 (PubMed PMID: 21252716. Pubmed Central PMCID: 4513953).
10. Edge SB, Byrd DR, Compton CC, et al., editors. AJCC cancer staging handbook. 7th ed. New York: Springer; 2009.
11. Travis W, Brambilla A, Burke A, Marx A, Nicholson A. Pathology and genetics of tumors of the lung, pleura, thymus and heart. World Health

Organization Classification of Tumors. Lyon, France: IARC Press; 2015.
12. Travis W. Pathology and genetics of tumours of the lung, pleura, thymus and heart. Lyon: IARC Press; 2004.
13. Barletta JA, Perner S, Iafrate AJ, Yeap BY, Weir BA, Johnson LA, et al. Clinical significance of TTF-1 protein expression and TTF-1 gene amplification in lung adenocarcinoma. J Cell Mol Med. 2008 (PubMed PMID: 19040416. Epub 2008/12/02. Eng).
14. Motoi N, Szoke J, Riely GJ, Seshan VE, Kris MG, Rusch VW, et al. Lung adenocarcinoma: modification of the 2004 WHO mixed subtype to include the major histologic subtype suggests correlations between papillary and micropapillary adenocarcinoma subtypes, EGFR mutations and gene expression analysis. Am J Surg Pathol. 2008;32(6):810–27.
15. Finberg KE, Sequist LV, Joshi VA, Muzikansky A, Miller JM, Han M, et al. Mucinous differentiation correlates with absence of EGFR mutation and presence of KRAS mutation in lung adenocarcinomas with bronchioloalveolar features. J Mol Diagn. 2007;9(3):320–6 (PubMed PMID: 17591931).
16. Haneda H, Sasaki H, Shimizu S, Endo K, Suzuki E, Yukiue H, et al. Epidermal growth factor receptor gene mutation defines distinct subsets among small adenocarcinomas of the lung. Lung Cancer. 2006;52 (1):47–52 (PubMed PMID: 16503085. Epub 2006/03/01. eng).
17. Hsieh RK, Lim KH, Kuo HT, Tzen CY, Huang MJ. Female sex and bronchioloalveolar pathologic subtype predict EGFR mutations in non-small cell lung cancer. Chest. 2005;128 (1):317–21 (PubMed PMID: 16002952. Epub 2005/07/09. eng).
18. Miller VA, Kris MG, Shah N, Patel J, Azzoli C, Gomez J, et al. Bronchioloalveolar pathologic subtype and smoking history predict sensitivity to gefitinib in advanced non-small-cell lung cancer. J Clin Oncol. 2004;22(6):1103–9 (PubMed PMID: 15020612. Epub 2004/03/17. eng).
19. Matsumoto S, Iwakawa R, Kohno T, Suzuki K, Matsuno Y, Yamamoto S, et al. Frequent EGFR mutations in noninvasive bronchioloalveolar carcinoma. Int J Cancer. 2006;118(10):2498–504 (PubMed PMID: 16353158. Epub 2005/12/15. eng).
20. Jackman DM, Chirieac LR, Janne PA. Bronchioloalveolar carcinoma: a review of the epidemiology, pathology, and treatment. Seminars in respiratory and critical care medicine. 2005;26 (3):342–52 (PubMed PMID: 16052436).
21. Noguchi M, Morikawa A, Kawasaki M, Matsuno Y, Yamada T, Hirohashi S, et al. Small adenocarcinoma of the lung. Histologic characteristics and prognosis. Cancer. 1995;75(12):2844–52 (PubMed PMID: 7773933. Epub 1995/06/15. eng).
22. Sakao Y, Miyamoto H, Sakuraba M, Oh T, Shiomi K, Sonobe S, et al. Prognostic significance of a histologic subtype in small adenocarcinoma of the lung: the impact of nonbronchioloalveolar carcinoma components. Ann Thorac Surg. 2007;83 (1):209–14 (PubMed PMID: 17184664. Epub 2006/12/23. eng).
23. Warth A, Muley T, Meister M, Stenzinger A, Thomas M, Schirmacher P, et al. The novel histologic International Association for the Study of Lung Cancer/American Thoracic Society/European Respiratory Society classification system of lung adenocarcinoma is a stage-independent predictor of survival. J Clin Oncol. 2012;30(13):1438–46 (PubMed PMID: 22393100).
24. Jackman DM, Chirieac LR, Jänne, Pasi A. Bronchioloalveolar carcinoma: A Review of the Epidemiology, Pathology, and Treatment. Seminars in respiratory and critical care medicine. 2005; VO—26(03):342–52.
25. Terasaki H, Niki T, Matsuno Y, Yamada T, Maeshima A, Asamura H, et al. Lung adenocarcinoma with mixed bronchioloalveolar and invasive components: clinicopathological features, subclassification by extent of invasive foci, and immunohistochemical characterization. Am J Surg Pathol. 2003;27(7):937–51 (PubMed PMID: 12826886).
26. Barletta JA, Yeap BY, Chirieac LR. Prognostic significance of grading in lung adenocarcinoma. Cancer. 2010;116(3):659–69 (PubMed PMID: 20014400. Pubmed Central PMCID: 2846761).
27. Siegel RL, Miller KD, Jemal A. Cancer statistics, 2015. CA Cancer J Clin. 2015;65(1):5–29 (PubMed PMID: 25559415).
28. Torre LA, Bray F, Siegel RL, Ferlay J, Lortet-Tieulent J, Jemal A. Global cancer statistics, 2012. CA: A Cancer J Clinic. 2015;65(2):87–108.
29. Sato M, Shames DS, Gazdar AF, Minna JD. A translational view of the molecular pathogenesis of lung cancer. J. Thorac. Oncol. Official publication of the International Association for the Study of Lung Cancer. 2007;2(4):327–43 (PubMed PMID: 17409807).
30. Hampton T. Lung cancer trials probe effects of maintenance therapy, targeted agents. JAMA. 2009;302(6):616, 8 (PubMed PMID: 19671897. Epub 2009/08/13. eng).
31. Carter BD, Abnet CC, Feskanich D, Freedman ND, Hartge P, Lewis CE, et al. Smoking and mortality—beyond established causes. N Engl J Med. 2015;372 (7):631–40 (PubMed PMID: 25671255).
32. Secretan B, Straif K, Baan R, Grosse Y, El Ghissassi F, Bouvard V, et al. A review of human carcinogens—Part E: tobacco, areca nut, alcohol, coal smoke, and salted fish. Lancet Oncol. 10 (11):1033–4.
33. Schottenfeld D, Fraumeni JF. Cancer epidemiology and prevention. 3rd ed. Oxford; New York: Oxford University Press; 2006. xviii, 1392 p.
34. Hamra GB, Guha N, Cohen A, Laden F, Raaschou-Nielsen O, Samet JM, et al. Outdoor particulate matter exposure and lung cancer: a systematic review and meta-analysis. Environ

Health Perspect. 2014;122(9):906–11 (PubMed PMID: 24911630. Pubmed Central PMCID: 4154221).

35. Turner MC, Krewski D, Pope CA 3rd, Chen Y, Gapstur SM, Thun MJ. Long-term ambient fine particulate matter air pollution and lung cancer in a large cohort of never-smokers. Am J Respir Critical care Med. 2011;184(12):1374–81 (PubMed PMID: 21980033).
36. Straif K, Cohen A, Samet J. IARC Scientific Publication No. 161: Air Pollution and Cancer 2015. Available from: http://www.iarc.fr/en/publications/books/sp161/Air%20Pollution%20and%20Cancer.epub.
37. Ding L, Getz G, Wheeler DA, Mardis ER, McLellan MD, Cibulskis K, et al. Somatic mutations affect key pathways in lung adenocarcinoma. Nature. 2008;455(7216):1069–75 (PubMed PMID: 18948947. Pubmed Central PMCID: 2694412. Epub 2008/10/25. eng).
38. Herbst RS, Heymach JV, Lippman SM. Lung cancer. N Engl J Med. 2008;359(13):1367–80 (PubMed PMID: 18815398. Epub 2008/09/26. eng).
39. Gerber DE, Oxnard GR, Govindan R. ALCHEMIST: bringing genomic discovery and targeted therapies to early-stage lung cancer. Clin Pharmacol Ther. 2015;97(5):447–50 (PubMed PMID: 25677079).
40. Lynch TJ, Bell DW, Sordella R, Gurubhagavatula S, Okimoto RA, Brannigan BW, et al. Activating mutations in the epidermal growth factor receptor underlying responsiveness of non-small-cell lung cancer to gefitinib. N Engl J Med. 2004;350 (21):2129–39 (PubMed PMID: 15118073).
41. Testa JR, Siegfried JM. Chromosome abnormalities in human non-small cell lung cancer. Cancer Res. 1992;52(9 Suppl):2702s–6s (PubMed PMID: 1314134).
42. Riely GJ, Yu HA. EGFR: The paradigm of an oncogene-driven lung cancer. Clin Cancer Res. 2015;21(10):2221–6 (PubMed PMID: 25979928. Pubmed Central PMCID: 4435716).
43. Paez JG, Janne PA, Lee JC, Tracy S, Greulich H, Gabriel S, et al. EGFR mutations in lung cancer: correlation with clinical response to gefitinib therapy. Science. 2004;304(5676):1497–500 (PubMed PMID: 15118125).
44. Tsao MS, Sakurada A, Cutz JC, Zhu CQ, Kamel-Reid S, Squire J, et al. Erlotinib in lung cancer—molecular and clinical predictors of outcome. N Engl J Med. 2005;353(2):133–44 (PubMed PMID: 16014883).
45. Sharma SV, Bell DW, Settleman J, Haber DA. Epidermal growth factor receptor mutations in lung cancer. Nat Rev Cancer. 2007;7(3):169–81 (PubMed PMID: 17318210).
46. Gow CH, Shih JY, Chang YL, Yu CJ. Acquired gefitinib-resistant mutation of EGFR in a chemonaive lung adenocarcinoma harboring gefitinib-sensitive mutation L858R. PLoS Med. 2005;2(9):e269 (PubMed PMID: 16173832).
47. Inukai M, Toyooka S, Ito S, Asano H, Ichihara S, Soh J, et al. Presence of epidermal growth factor receptor gene T790 M mutation as a minor clone in non-small cell lung cancer. Cancer Res. 2006;66 (16):7854–8 (PubMed PMID: 16912157).
48. Kobayashi S, Boggon TJ, Dayaram T, Janne PA, Kocher O, Meyerson M, et al. EGFR mutation and resistance of non-small-cell lung cancer to gefitinib. N Engl J Med. 2005;352(8):786–92 (PubMed PMID: 15728811).
49. Kwak EL, Sordella R, Bell DW, Godin-Heymann N, Okimoto RA, Brannigan BW, et al. Irreversible inhibitors of the EGF receptor may circumvent acquired resistance to gefitinib. Proc Natl Acad Sci U S A. 2005;102(21):7665–70 (PubMed PMID: 15897464).
50. Pao W, Miller VA, Politi KA, Riely GJ, Somwar R, Zakowski MF, et al. Acquired resistance of lung adenocarcinomas to gefitinib or erlotinib is associated with a second mutation in the EGFR kinase domain. PLoS Med. 2005;2(3):e73 (PubMed PMID: 15737014).
51. Sholl LM, Yeap BY, Iafrate AJ, Holmes-Tisch AJ, Chou YP, Wu MT, et al. Lung adenocarcinoma with EGFR amplification has distinct clinicopathologic and molecular features in never-smokers. Cancer Res. 2009;69(21):8341–8 (PubMed PMID: 19826035. Pubmed Central PMCID: 2783286. Epub 2009/10/15. eng).
52. Yu J, Kane S, Wu J, Benedettini E, Li D, Reeves C, et al. Mutation-specific antibodies for the detection of EGFR mutations in non-small-cell lung cancer. Clin Cancer Res. 2009;15(9):3023–8 (PubMed PMID: 19366827. Epub 2009/04/16. eng).
53. Sasaki H, Shimizu S, Endo K, Takada M, Kawahara M, Tanaka H, et al. EGFR and erbB2 mutation status in Japanese lung cancer patients. Int J Cancer. 2006;118(1):180–4 (PubMed PMID: 16003726).
54. Mok TS, Wu YL, Thongprasert S, Yang CH, Chu DT, Saijo N, et al. Gefitinib or carboplatin-paclitaxel in pulmonary adenocarcinoma. N Engl J Med. 2009;361(10):947–57 (PubMed PMID: 19692680. Epub 2009/08/21. eng).
55. Mok TS, Wu YL, Yu CJ, Zhou C, Chen YM, Zhang L, et al. Randomized, placebo-controlled, phase II study of sequential erlotinib and chemotherapy as first-line treatment for advanced non-small-cell lung cancer. J Clin Oncol. 2009;27(30):5080–7 (PubMed PMID: 19738125. Epub 2009/09/10. eng).
56. Sequist LV, Heist RS, Shaw AT, Fidias P, Rosovsky R, Temel JS, et al. Implementing multiplexed genotyping of non-small-cell lung cancers into routine clinical practice. Ann. Oncol. Off J Eur Soc Med Oncol/ESMO. 2011;22(12):2616–24 (PubMed PMID: 22071650. Pubmed Central PMCID: 3493130).
57. Sholl LM, Aisner DL, Varella-Garcia M, Berry LD, Dias-Santagata D, Wistuba, II, et al. Multi-institutional oncogenic driver mutation analysis in lung adenocarcinoma: the lung cancer

mutation consortium experience. J Thorac Oncol. Official publication of the International Association for the Study of Lung Cancer. 2015;10(5):768–77 (PubMed PMID: 25738220. Pubmed Central PMCID: 4410843).
58. Zhu CQ, da Cunha Santos G, Ding K, Sakurada A, Cutz JC, Liu N, et al. Role of KRAS and EGFR as biomarkers of response to erlotinib in National Cancer Institute of Canada Clinical Trials Group Study BR.21. J Clin Oncol. 2008;26(26):4268–75 (PubMed PMID: 18626007. Epub 2008/07/16. eng).
59. Sequist LV, Waltman BA, Dias-Santagata D, Digumarthy S, Turke AB, Fidias P, et al. Genotypic and histological evolution of lung cancers acquiring resistance to EGFR inhibitors. Sci Transl Med. 2011;3(75):75ra26 (PubMed PMID: 21430269. Pubmed Central PMCID: 3132801).
60. Tang Z, Du R, Jiang S, Wu C, Barkauskas DS, Richey J, et al. Dual MET-EGFR combinatorial inhibition against T790 M-EGFR-mediated erlotinib-resistant lung cancer. Br J Cancer. 2008;99(6):911–22 (PubMed PMID: 19238632. Pubmed Central PMCID: 2538758. Epub 2009/02/25. eng).
61. Cappuzzo F, Janne PA, Skokan M, Finocchiaro G, Rossi E, Ligorio C, et al. MET increased gene copy number and primary resistance to gefitinib therapy in non-small-cell lung cancer patients. Ann Oncol Off J Eur Soc Med Oncol/ESMO. 2009;20(2):298–304 (PubMed PMID: 18836087. Pubmed Central PMCID: 2733067. Epub 2008/10/07. eng).
62. Engelman JA, Zejnullahu K, Mitsudomi T, Song Y, Hyland C, Park JO, et al. MET amplification leads to gefitinib resistance in lung cancer by activating ERBB3 signaling. Science. 2007;316(5827):1039–43 (PubMed PMID: 17463250. Epub 2007/04/28. eng).
63. West HL, Franklin WA, McCoy J, Gumerlock PH, Vance R, Lau DH, et al. Gefitinib therapy in advanced bronchioloalveolar carcinoma: Southwest Oncology Group Study S0126. J Clin Oncol. 2006;24(12):1807–13 (PubMed PMID: 16622257. Epub 2006/04/20. eng).
64. Miller VA, Riely GJ, Zakowski MF, Li AR, Patel JD, Heelan RT, et al. Molecular characteristics of bronchioloalveolar carcinoma and adenocarcinoma, bronchioloalveolar carcinoma subtype, predict response to erlotinib. J Clin Oncol. 2008;26(9):1472–8 (PubMed PMID: 18349398. Epub 2008/03/20. eng).
65. Marchetti A, Martella C, Felicioni L, Barassi F, Salvatore S, Chella A, et al. EGFR mutations in non-small-cell lung cancer: analysis of a large series of cases and development of a rapid and sensitive method for diagnostic screening with potential implications on pharmacologic treatment. J Clin Oncol. 2005;23(4):857–65 (PubMed PMID: 15681531. Epub 2005/02/01. eng).
66. Shigematsu H, Lin L, Takahashi T, Nomura M, Suzuki M, Wistuba II, et al. Clinical and biological features associated with epidermal growth factor receptor gene mutations in lung cancers. J Natl Cancer Inst. 2005;97(5):339–46 (PubMed PMID: 15741570).
67. Lindeman NI, Cagle PT, Beasley MB, Chitale DA, Dacic S, Giaccone G, et al. Molecular testing guideline for selection of lung cancer patients for EGFR and ALK tyrosine kinase inhibitors: guideline from the College of American Pathologists, International Association for the Study of Lung Cancer, and Association for Molecular Pathology. J Thorac Oncol. Official publication of the International Association for the Study of Lung Cancer. 2013;8(7):823–59 (PubMed PMID: 23552377. Pubmed Central PMCID: 4159960).
68. Kadota K, Yeh YC, D'Angelo SP, Moreira AL, Kuk D, Sima CS, et al. Associations between mutations and histologic patterns of mucin in lung adenocarcinoma: invasive mucinous pattern and extracellular mucin are associated with KRAS mutation. Am J Surg Pathol. 2014;38(8):1118–27 (PubMed PMID: 25029118).
69. Sandler A, Gray R, Perry MC, Brahmer J, Schiller JH, Dowlati A, et al. Paclitaxel-carboplatin alone or with bevacizumab for non-small-cell lung cancer. N Engl J Med. 2006;355(24):2542–50 (PubMed PMID: 17167137. Epub 2006/12/15. eng).
70. West H, Harpole D, Travis W. Histologic considerations for individualized systemic therapy approaches for the management of non-small cell lung cancer. Chest. 2009;136(4):1112–8 (PubMed PMID: 19809052. Epub 2009/10/08. eng).
71. Johnson DH, Fehrenbacher L, Novotny WF, Herbst RS, Nemunaitis JJ, Jablons DM, et al. Randomized phase II trial comparing bevacizumab plus carboplatin and paclitaxel with carboplatin and paclitaxel alone in previously untreated locally advanced or metastatic non-small-cell lung cancer. J Clin Oncol. 2004;22(11):2184–91 (PubMed PMID: 15169807. Epub 2004/06/01. eng).
72. Lamant L, Meggetto F, al Saati T, Brugieres L, de Paillerets BB, Dastugue N, et al. High incidence of the t(2;5)(p23;q35) translocation in anaplastic large cell lymphoma and its lack of detection in Hodgkin's disease. Comparison of cytogenetic analysis, reverse transcriptase-polymerase chain reaction, and P-80 immunostaining. Blood. 1996;87(1):284–91 (PubMed PMID: 8547653).
73. Morris SW, Kirstein MN, Valentine MB, Dittmer KG, Shapiro DN, Saltman DL, et al. Fusion of a kinase gene, ALK, to a nucleolar protein gene, NPM, in non-Hodgkin's lymphoma. Science. 1994;263(5151):1281–4 (PubMed PMID: 8122112).
74. Swerdlow SH, International Agency for Research on Cancer., World Health Organization. WHO classification of tumours of haematopoietic and lymphoid tissues. 4th ed. Lyon, France: International Agency for Research on Cancer; 2008. 439 p.
75. Soda M, Choi YL, Enomoto M, Takada S, Yamashita Y, Ishikawa S, et al. Identification of the transforming EML4-ALK fusion gene in

non-small-cell lung cancer. Nature. 2007;448(7153):561–6 (PubMed PMID: 17625570).
76. Rikova K, Guo A, Zeng Q, Possemato A, Yu J, Haack H, et al. Global survey of phosphotyrosine signaling identifies oncogenic kinases in lung cancer. Cell. 2007;131(6):1190–203 (PubMed PMID: 18083107).
77. Soda M, Takada S, Takeuchi K, Choi YL, Enomoto M, Ueno T, et al. A mouse model for EML4-ALK-positive lung cancer. Proc Natl Acad Sci U S A. 2008;105(50):19893–7 (PubMed PMID: 19064915).
78. Koivunen JP, Mermel C, Zejnullahu K, Murphy C, Lifshits E, Holmes AJ, et al. EML4-ALK fusion gene and efficacy of an ALK kinase inhibitor in lung cancer. Clin Cancer Res. 2008;14(13):4275–83 (PubMed PMID: 18594010).
79. Kwak EL, Bang YJ, Camidge DR, Shaw AT, Solomon B, Maki RG, et al. Anaplastic lymphoma kinase inhibition in non-small-cell lung cancer. N Engl J Med. 2010;363(18):1693–703 (PubMed PMID: 20979469. Pubmed Central PMCID: 3014291. Epub 2010/10/29. eng).
80. Rodig SJ, Mino-Kenudson M, Dacic S, Yeap BY, Shaw A, Barletta JA, et al. Unique clinicopathologic features characterize ALK-rearranged lung adenocarcinoma in the western population. Clin Cancer Res. 2009;15(16):5216–23 (PubMed PMID: 19671850. Epub 2009/08/13. eng).
81. Mino-Kenudson M, Chirieac LR, Law K, Hornick JL, Lindeman N, Mark EJ, et al. A Novel, Highly sensitive antibody allows for the routine detection of ALK-rearranged lung adenocarcinomas by Standard Immunohistochemistry. Clin Cancer Res. 2010;16(5):1561–71 (PubMed PMID: 20179225. Pubmed Central PMCID: 2831135. Epub 2010/02/25. eng).
82. Sasaki T, Rodig SJ, Chirieac LR, Janne PA. The biology and treatment of EML4-ALK non-small cell lung cancer. Eur J Cancer. 2010;46(10):1773–80 (PubMed PMID: 20418096. Pubmed Central PMCID: 2888755).
83. Shaw AT, Yeap BY, Mino-Kenudson M, Digumarthy SR, Costa DB, Heist RS, et al. Clinical features and outcome of patients with non-small-cell lung cancer who harbor EML4-ALK. J Clin Oncol. 2009;27(26):4247–53 (PubMed PMID: 19667264. Pubmed Central PMCID: 2744268).
84. Kris MG, Johnson BE, Berry LD, Kwiatkowski DJ, Iafrate AJ, Wistuba, II, et al. Using multiplexed assays of oncogenic drivers in lung cancers to select targeted drugs. JAMA. 2014;311(19):1998–2006 (PubMed PMID: 24846037. Pubmed Central PMCID: 4163053).
85. Shaw AT, Ou SH, Bang YJ, Camidge DR, Solomon BJ, Salgia R, et al. Crizotinib in ROS1-rearranged non-small-cell lung cancer. N Engl J Med. 2014;371(21):1963–71 (PubMed PMID: 25264305. Pubmed Central PMCID: 4264527).
86. Robinson DR, Wu YM, Lin SF. The protein tyrosine kinase family of the human genome. Oncogene. 2000;19(49):5548–57 (PubMed PMID: 11114734).
87. Shigematsu H, Takahashi T, Nomura M, Majmudar K, Suzuki M, Lee H, et al. Somatic mutations of the HER2 kinase domain in lung adenocarcinomas. Cancer Res. 2005;65(5):1642–6 (PubMed PMID: 15753357).
88. Stephens P, Hunter C, Bignell G, Edkins S, Davies H, Teague J, et al. Lung cancer: intragenic ERBB2 kinase mutations in tumours. Nature. 2004;431(7008):525–6 (PubMed PMID: 15457249).
89. Giaccone G. Clinical impact of novel treatment strategies. Oncogene. 2002;21(45):6970–81 (PubMed PMID: 12362278).
90. Heist RS, Zhou W, Chirieac LR, Cogan-Drew T, Liu G, Su L, et al. MDM2 polymorphism, survival, and histology in early-stage non-small-cell lung cancer. J Clin Oncol. 2007;25(16):2243–7.
91. Azzoli CG, Krug LM, Miller VA, Kris MG, Mass R. Trastuzumab in the treatment of non-small cell lung cancer. Semin Oncol. 2002;29(1 Suppl 4):59–65 (PubMed PMID: 11894015).
92. Sadiq AA, Geynisman DM, Salgia R. Inhibition of MET receptor tyrosine kinase and its ligand hepatocyte growth factor. J Thorac Oncol. 2011;6(11 Suppl 4):S1810–1 (PubMed PMID: 22005540. Epub 2011/10/26. eng).
93. Lindor NM, Dechet CB, Greene MH, Jenkins RB, Zincke MT, Weaver AL, et al. Papillary renal cell carcinoma: analysis of germline mutations in the MET proto-oncogene in a clinic-based population. Genet Test. 2001;5(2):101–6 (PubMed PMID: 11551094. Epub 2001/09/12. eng).
94. Comoglio PM, Giordano S, Trusolino L. Drug development of MET inhibitors: targeting oncogene addiction and expedience. Nat Rev Drug Discov. 2008;7(6):504–16 (PubMed PMID: 18511928. Epub 2008/05/31. eng).
95. Ma PC, Tretiakova MS, MacKinnon AC, Ramnath N, Johnson C, Dietrich S, et al. Expression and mutational analysis of MET in human solid cancers. Genes Chromosom Cancer. 2008;47(12):1025–37 (PubMed PMID: 18709663. Pubmed Central PMCID: 2583960. Epub 2008/08/19. eng).
96. Benedettini E, Sholl LM, Peyton M, Reilly J, Ware C, Davis L, et al. Met activation in non-small cell lung cancer is associated with de novo resistance to EGFR inhibitors and the development of brain metastasis. Am J Pathol. 2010;177(1):415–23 (PubMed PMID: 20489150. Pubmed Central PMCID: 2893683. Epub 2010/05/22. eng).
97. Yoshida T, Okamoto I, Okamoto W, Hatashita E, Yamada Y, Kuwata K, et al. Effects of Src inhibitors on cell growth and epidermal growth factor receptor and MET signaling in gefitinib-resistant non-small cell lung cancer cells with acquired MET amplification. Cancer Sci. 2009;101(1):167–72 (PubMed PMID: 19804422. Epub 2009/10/07. eng).

98. Spigel DR, Hainsworth JD, Simons L, Meng C, Burris HA, 3rd, Yardley DA, et al. Irinotecan, carboplatin, and imatinib in untreated extensive-stage small-cell lung cancer: a phase II trial of the Minnie Pearl Cancer Research Network. J Thorac Oncol. 2007;2(9):854–61 (PubMed PMID: 17805064. Epub 2007/09/07. eng).
99. Jameson JL, Longo DL. Precision medicine—personalized, problematic, and promising. N Engl J Med. 2015;372(23):2229–34 (PubMed PMID: 26014593).
100. Toward Precision Medicine: Building a Knowledge Network for Biomedical Research and a New Taxonomy of Disease. The National Academies Collection: Reports funded by National Institutes of Health. Washington (DC) 2011.
101. Morgensztern D, Politi K, Herbst RS. EGFR mutations in non-small-cell lung cancer: find, divide, and conquer. JAMA Oncol. 2015;1(2):146–8 (PubMed PMID: 26181013).
102. Hayden EC. Technology: The $1,000 genome. Nature. 2014;507(7492):294–5 (PubMed PMID: 24646979).
103. Lander ES. Cutting the Gordian helix—regulating genomic testing in the era of precision medicine. N Engl J Med. 2015;372(13):1185–6 (PubMed PMID: 25689017).

# 21 Epidemiology of Colorectal Cancer

NaNa Keum and Edward L. Giovannucci

## 21.1 Clinical Picture of the Disease: Symptoms, Diagnosis, Treatment, and Prognosis

Colorectal cancer (CRC) refers to malignant tumors that develop from cells in the large intestine. In the early stages, CRC usually produces no symptoms. In the later stages, however, CRC may cause symptoms, including changes in bowel habits (e.g., diarrhea, constipation, and narrow stools), persistent abdominal discomfort (e.g., cramps, gas, and pain), bloody stools, fatigue, and unintended weight loss. These symptoms manifest in varying degrees depending on the location, size, and stage of CRC. To diagnose CRC, several tests are available including fecal occult blood test, flexible sigmoidoscopy, colonoscopy, computed tomography CT colonography, and barium enema. Surgical removal, chemotherapy, and radiation therapy are the three major treatment options that may be used alone or in combination. The prognosis largely depends on the stage and grade of CRC at diagnosis. An estimated five-year survival rate in the United States (U.S.) between 2004 and 2010 ranges from 12 % for the most advanced CRC (i.e., stage IV) to 90 % for the least advanced CRC (i.e., stage I) [1]. Early detection of asymptomatic CRC through screening enhances CRC survival.

## 21.2 Descriptive Epidemiology

CRC is the third most commonly diagnosed cancer and the fourth leading cause of cancer death worldwide [2]. Globally, over 1.3 million new cases (9.7 % of all cancer diagnosis excluding non-melanoma skin cancer) were diagnosed and approximately 690,000 deaths (8.5 % of all cancer deaths excluding non-melanoma skin cancer) were attributed to this malignancy in 2012 [2]. Before the 1900s, CRC was relatively uncommon, but incidence rates have risen dramatically in parallel with economic development and adoption of the sedentary lifestyle and Western diet. Over the last few decades, incidence rates invariably increased in economically transitioning countries, but stabilized in the majority of developed countries and decreased in the U.S. [3].

Geographically, the higher rates of CRC in economically developed countries drive a wide variation in age-adjusted incidence rates across the countries worldwide. In 2012, there was an estimated 11-fold difference in age-standardized incidence rates between the highest observed in South Korea (45.0/100,000 person-years) and the lowest observed in Western Africa (4.1/100,000 person-years) (Fig. 21.1) [4]. The variation was approximately 2.5-fold between more developed

N.N. Keum · E.L. Giovannucci (✉)
Harvard T.H. Chan School of Public Health, 655 Huntington Avenue, Boston, MA 02115, USA
e-mail: egiovann@hsph.harvard.edu


M. Loda et al. (eds.), *Pathology and Epidemiology of Cancer*,
DOI 10.1007/978-3-319-35153-7_21

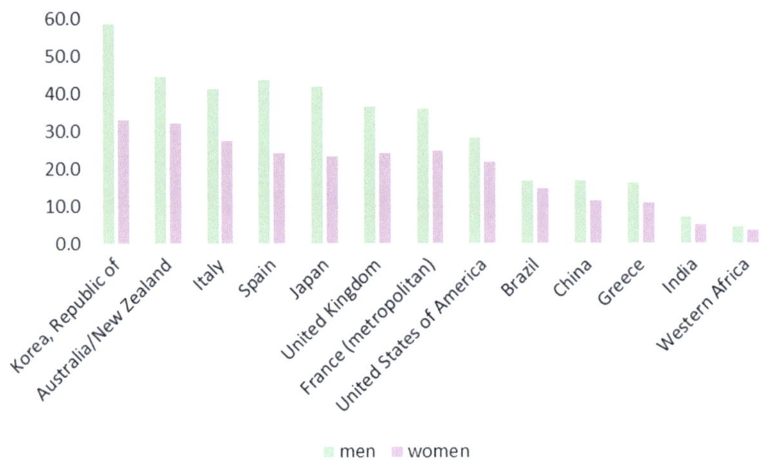


**Fig. 21.1** Age-standardized incidence rates per 100,000 person-years. Rates were standardized to the World Standard Population. Data based on [4]

countries (29.2/100,000 person-years, in all regions of Europe plus Northern America, Australia/New Zealand and Japan) and less developed countries (11.7/100,000 person-years, in all regions of Africa, Asia [excluding Japan], Latin America and the Caribbean, Melanesia, Micronesia and Polynesia) [4].

CRC is rare under the age of 40, but incidence rates increase with age thereafter. With the rates rising rapidly after age 50 years, approximately 90 % of CRC cases are diagnosed in people age 50 years or older [5]. For an average individual in a typical Western population (e.g., U.S.), the lifetime risk of developing CRC is approximately 5 % [6].

Globally, the age-standardized incidence rate is about 1.4 times higher in men (20.6/100,000 person-years) than in women (14.3/100,000 person-years) [7]. Yet, women on average live longer than men, with the median age at diagnosis higher (73 years) than that in men (69 years) [8]. There are only slightly more CRC deaths in men (373,631/year) than in women (320,250/year) [7].

In the U.S., the highest age-standardized incidence rate has been observed in African Americans (men: 62.3/100,000 person-years, women: 47.5/100,000 person-years) among all racial and ethnic groups, followed by White, Hispanic, Asian/Pacific Islander, and American Indian/Alaska Native people [9].

## 21.3 Risk Factors

The risk of CRC is predominantly attributed to unhealthy lifestyle factors associated with Westernization. Migrant studies observed that CRC incidence and mortality rates among immigrants who moved from low-risk to high-risk countries converged to the rates of the host countries [10, 11], indicating the importance of environmental influences in colorectal carcinogenesis. In the U.S., as much as 50 % of CRC cases were estimated to be attributable to an unhealthy diet, physical inactivity, and excess body weight [12]. Additionally including a lower intake of folic acid from supplements and long-term smoking, more than two-thirds of CRC cases can be attributable to lifestyle factors [13]. Considering that these lifestyle factors are modifiable, such a large population attributable risk percent implies a considerable number of CRC cases preventable through lifestyle modifications.

In contrast, some factors that predispose people to CRC are non-modifiable. While explaining a smaller proportion of CRC cases, non-modifiable risk factors including a positive family history and inherited genetic factors are responsible for differences in the susceptibility of individuals to CRC (Table 21.1) [14–16], and contribute particularly to early-onset CRC.

**Table 21.1** Non-modifiable risk factors for colorectal cancer

| Groups | Lifetime risk of colorectal cancer (%) | Contribution to colorectal cancer cases (%) |
|---|---|---|
| Average population in the U.S.[a] | 5 | 75 |
| High-risk populations | | |
| Familial adenomatous polyposis | 100 | 1 |
| Hereditary non-polyposis colorectal cancer syndrome | 50 | 5 |
| Inflammatory bowel disease | 5–50 | 1 |
| Family history of colorectal cancer | 10–15 | 15–20 |
| Past history of adenomas | 10–20 | Variable |

*Note* Reference for table [14–16]
[a]Asymptomatic men and women age 50 years or older with no special risk factors

### 21.3.1 Non-modifiable Risk Factors

#### 21.3.1.1 Hereditary Syndromes

Most CRC occurs sporadically, (i.e., without known hereditary causes). Less than 10 % of CRC occur due to inherited gene mutations [17], and these cancers tend to manifest at younger ages (under age 50). The two most common subtypes of familial syndromes that predispose people to CRC are hereditary non-polyposis colorectal cancer (HNPCC) syndrome and familial adenomatous polyposis (FAP).

As the most common familial syndrome, HNPCC syndrome (also known as Lynch syndrome) is caused by mutations in DNA mismatch repair genes that are inherited in an autosomal dominant fashion [17]. HNPCC syndrome increases CRC risk by elevating malignant potential of adenomas (described below); affected individuals do not have an unusual number of adenomas but the adenomas they develop are more likely to progress into CRC. Approximately half of individuals with HNPCC syndrome develop CRC and the mean age at diagnosis is 44 years [18].

FAP is an autosomal dominant disease caused by an inherited mutation in the adenomatous polyposis coli (APC) gene, a tumor suppressor. APC mutations are hallmarks of early-stage colorectal carcinogenesis. Unlike HNPCC syndrome, individuals with FAP develop hundreds to thousands of adenomas in their teens or early adulthood. While fewer than 10 % of adenomas progress into cancer [19], some adenomas that occur with FAP inevitably undergo malignant transformation due to the presence of such a high number of adenomas. Thus, affected individuals have an almost 100 % chance of getting CRC by the age of 40 if not treated via colectomy. Yet, FAP is a rare disease (about 1 in 10,000 people) [20] and accounts for less than 1 % of CRC cases [21].

#### 21.3.1.2 Personal History of Inflammatory Bowel Diseases (Particularly, Ulcerative Colitis)

Ulcerative colitis and Crohn's disease, the two most common forms of inflammatory bowel disease, cause chronic inflammation in all or a part of the gastrointestinal tract. Chronic inflammation in the colon triggers compensatory proliferation to regenerate damaged tissue, which increases opportunities for mutations to occur [22]. Accumulation of genetic alterations can lead to dysplasia and subsequently to cancer. Thus, the pathogenesis of CRC in these patients does not evolve through adenoma, the classical precursor, but rather through dysplasia induced by chronic inflammation. Although CRC risk varies by duration and anatomic extent of the disease, affected individuals, particularly those with ulcerative colitis, can have up to a 50 % chance of developing CRC [23] if not treated, and the mean age at diagnosis is between age 40 and 50 years [24]. Thus, along with HNPCC syndrome and FAP, ulcerative colitis defines the top three subgroups at high risk of developing CRC.

#### 21.3.1.3 Family History of Colorectal Cancer

The lifetime risk of CRC is roughly doubled for individuals with a family history of CRC in a first-degree relative (parents, siblings, or children). The risk is even higher if the first-degree relative is diagnosed young or if more than one first-degree relative is affected [25, 26]. The increased risk may be attributable to inherited genes, shared environmental factors, or combinations of the two. A family history of CRC is found in approximately 15–20 % of CRC patients.

#### 21.3.1.4 Adenomas (Adenomatous Polyps)

A colorectal polyp is an abnormal growth of tissue rising from the inner lining of the large intestine into the lumen. When classified histologically, the most common types include hyperplastic and adenomatous polyps [27]. Hyperplastic polyps are generally non-cancerous. Of note, once considered as a subgroup of hyperplastic polyps, serrated polyps are now recognized as a distinctive type with malignant potential (i.e., precancerous).

Adenomatous polyps (adenomas), originating from the mucus-secreting epithelial cells of the colorectum, are benign by themselves but harbor malignant potential. While fewer than 10 % of adenomas progress to cancers, more than 95 % of sporadic CRC develop from adenomas [19, 28]. Thus, adenomas represent the classical precursor lesions of CRC, with the carcinogenic pathway termed as adenoma-carcinoma sequence. Defined as advanced adenoma, an adenoma with a large size (≥1 cm in diameter), a villous component, or high-grade dysplasia has a particularly high propensity to develop into cancer [29].

Adenomas are common in Western countries. About one-third of asymptomatic average risk individuals of age 50–75 years harbor adenomas [30]. A significant proportion of people with initial adenomectomy develops recurrent adenomas within three years [31]. Although varying by characteristics of adenomas and subsequent surveillance by colonoscopy, the lifetime risk of CRC among individuals with a history of adenomas is estimated to be 10–20 %.

### 21.3.2 Modifiable Risk/Protective Factors

#### 21.3.2.1 Risk Factors Associated with Westernization (Obesity, Sedentary Lifestyle, Western Dietary Pattern)

Invariably high incidence rates of CRC in Westernized countries have been incontrovertibly attributed to environmental factors associated with Westernization. A number of risk factors have been identified with relatively high consistency, although the underlying mechanisms are not fully understood. Initial hypotheses focused on the direct effects of high-fat and low-fiber intakes on the colorectal lumen, suggesting that their influences on fecal contents may drive colorectal carcinogenesis [32, 33]. However, accumulating evidence lends support to an alternative hypothesis, which implicates insulin and insulin-like growth factor 1 (IGF-1) [32]. The observation that risk factors for hyperinsulinemia are also risk factors for CRC led to the hypothesis that hyperinsulinemia and a corresponding increase in free IGF-1 may be the underlying mediators linking lifestyle risk factors associated with Westernization to CRC (Fig. 21.2) [32]. Insulin and IGF-1 increase proliferation and decrease apoptosis of colorectal epithelial cells, thereby promoting colorectal carcinogenesis [33, 34]. This insulin/IGF hypothesis is supported by a nested case-control study conducted among non-diabetic people, which observed a lower risk of CRC among those with lower plasma levels of both c-peptide (a marker of insulin secretion) and IGF-1/IGF-1 binding protein ratio [35].

##### Obesity

Evidence from prospective cohort studies consistently indicates that excess adiposity may elevate the risk of CRC, with the association

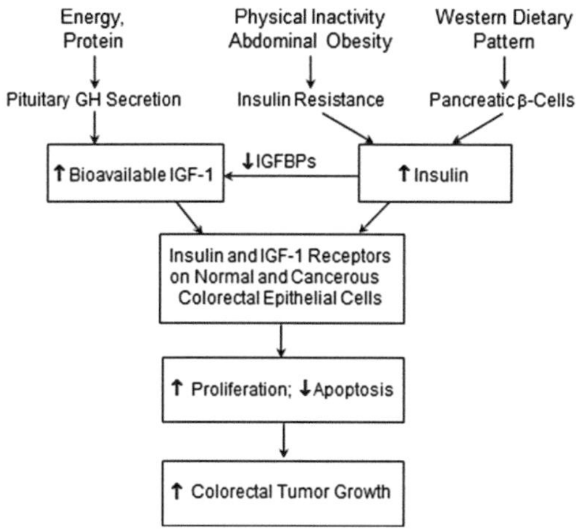


**Fig. 21.2** Proposed mechanism presenting insulin and IGF-1 as the key mediators linking Western lifestyle and dietary factors to colorectal cancer risk [33, 78]

stronger for colon cancer (CC) than for rectal cancer (RC) and stronger in men than in women. The association has been investigated based on anthropometric measures including body mass index (BMI), waist circumference (WC) or waist-to-hip ratio (WHR), and adult weight change. While correlated with each other, these measures capture different aspects of adiposity, with each representing a degree of overall body fatness, abdominal fatness, and time-integrated fat accumulation, respectively. Generally, overweight or obese men experience a 1.5- to 2-fold increased risk of CC compared with those in the normal or low range of BMI. The adverse effect of excess adiposity is more consistently seen with WC or WHR across epidemiologic studies, suggesting particular relevance of abdominal adiposity to CC risk. For instance, in the European Prospective Investigation Into Cancer and Nutrition (EPIC) study, a 55 % increased risk of CC was observed for men with BMI $\geq$29.4 kg/m$^2$ compared to men with BMI <23.6 kg/m$^2$, but no significant association was found in women [36]. In contrast, in an analysis comparing extreme quintiles of WHR, an approximately 50 % increased risk of CC was observed for both men and women [36]. Annual weight gain (kg/year) during adulthood (from age 20 to 50) was positively associated with CC, especially if it resulted in an increase in WC [37].

Substantial evidence points to abdominal fat, specifically visceral adipose tissue (VAT), as a principal etiologic factor for CC pathogenesis. First, an association between excess adiposity and CC risk is stronger in men than in women [38], and men have a tendency towards abdominal distribution of fat [39, 40]. Second, a positive association of WC or WHR with CC risk remained significant even after adjustment for BMI, whereas that of BMI with CC risk became non-significant after adjustment for WC or WHR [36, 41]. Third, in studies that used VAT and subcutaneous adipose tissue (SAT) in the abdominal area as measured by CT, VAT but not SAT was associated with adenomas, particularly with advanced colorectal adenomas [42–44]. Finally, in a study that simultaneously included BMI, WC, and VAT in the same regression model, only VAT was a statistically significant predictor of adenomas [45]. Taken altogether, VAT may be the underlying mediator of excess adiposity and colorectal neoplasia and thus, the presence of an association between WC and CC independent of BMI may be attributable to the better capability of WC to capture VAT than BMI.

Biological mechanisms explaining the specific contribution of VAT to CC risk are well-aligned with the insulin/IGF hypothesis. Compared to SAT, VAT secrets more pro-inflammatory cytokines (e.g., interleukin-6, tumor necrosis factor-$\alpha$) and less insulin-sensitizing adiponectin [46–48]. Furthermore, VAT more readily effuses free fatty acids into the circulation and, in response, the liver and muscle preferentially uptake fatty acids over glucose, becoming less responsive to insulin [48–50]. Hence, VAT is more strongly associated with insulin resistance and consequently, more relevant to the etiology of CC. In view of this, VAT may be linked to differential incident rates of colorectal neoplasia by sex and race. The amount of VAT is higher in men than in women on average [40, 51]. Asians, who have a tendency towards lower muscle mass and greater abdominal adiposity, have more VAT than Caucasians for a given BMI [52]. Consistently, higher incidence rates of colorectal neoplasia are observed in men than women and in Asians in the normal range of BMI than Caucasians with equivalent BMIs [53].

#### Sedentary Lifestyle

With Westernization, people become highly sedentary in diverse domains of life including occupation and recreation. Distinct from physical inactivity describing the absence of structured and purposive exercise, sedentary behavior is characterized by prolonged sitting.

Sedentary lifestyle is emerging as an independent risk factor for CRC by initiating qualitatively unique cellular and molecular responses in the skeletal muscle that may be relevant to colorectal carcinogenesis [54, 55]. To date, the largest prospective study was conducted in the NIH-AARP Diet and Health Study cohort in which sedentary lifestyle was captured by time spent watching TV. Independent of physical activity, those with ≥63 h/week of TV viewing time had an approximately 1.5-fold increased risk of CC compared with those with <21 h/week of TV viewing time [56].

#### Western Dietary Pattern

People eat foods not in isolation but in combination. Thus, the analysis of dietary patterns that captures the combined effects of nutrients and foods as they are consumed in a population has some advantages over the single nutrient or food approach in epidemiologic studies [57]. With regard to CRC, one of the most commonly identified dietary patterns is an unhealthy Western diet characterized by high consumption of red/processed meats and refined grains [58]. Although studies varied in the inclusion of high-fat dairy products and French fries in defining the Western dietary pattern, a deleterious effect of the Western dietary pattern on CRC risk has been consistently reported [58]. In a recent meta-analysis based on 11 observational studies, the risk of CC, but not RC, increased with higher accordance with the Western dietary pattern [59].

High consumption of red/processed meat that predominately characterizes the Western dietary pattern may contribute to carcinogenesis not only through its adverse effects on circulating insulin concentration [60], but also through heme-iron [61] and carcinogens (N-nitroso compounds [62]; heterocyclic amines and polycyclic aromatic hydrocarbons that are produced when meats are cooked at high temperatures [63]).

### 21.3.2.2 Other Risk Factors

#### Smoking

A wide range of carcinogens in cigarette smoke can readily reach the colorectal mucosa through the circulatory system or direct ingestion, inducing genetic mutations or epigenetic alterations [64]. Long-term heavy smoking, chiefly initiated at an early age, is a strong risk factor for CRC, particularly for RC. According to a recent dose–response meta-analysis, CRC risk decreased by 4 % for each 10-year delay in smoking initiation, but increased by 20 % for each 40-year increase in smoking duration [65]. Of note, a positive association between smoking and CRC risk was not supported by earlier studies conducted soon after the smoking epidemic (around the 1920s for men and the late 1940s for women), but was demonstrated in later studies that allowed for 35–40 years of an induction period between the booming of smoking and CRC diagnosis [64, 66–68]. Thus, smoking appears to act predominantly at an early stage of carcinogenesis.

The mutagenic effect of tobacco smoke has been generally considered irreversible, as most studies have shown that past smokers carry a persistently higher risk than non-smokers [66]. However, recent epidemiologic studies on duration of smoking cessation and CRC risk have suggested reversibility of the effect. For instance, in a large cohort study conducted in the U.S., CRC risk among past smokers reversed to lifelong non-smoker level after at least 31 years of cessation [69]. A more recent study accounting for molecular heterogeneity of CRC observed that the benefit of smoking cessation was limited to CpG island methylator pathway (CIMP)-high CRC, a distinct molecular phenotype of CRC characterized by selected cancer-related genes hypermethylated at CpG islands (promoter regions of genes rich in a linear sequence of a cytosine nucleotide and a guanine nucleotide that are linked via phosphate) [70, 71]. Thus, smoking may also act at a relatively late stage of carcinogenesis related to DNA methylation pathways leading to CIMP-high CRC.

#### Heavy Alcohol Intake

Heavy alcohol intake is a fairly consistent risk factor for CRC. In a pooled analysis of eight large prospective studies that included 4687 CRC cases among nearly 500,000 participants, individuals who consumed ≥45 g/day of alcohol (i.e., ethanol content in 3 drinks/day in the U.S.) had a 42 % elevated risk of CRC relative to non-drinkers, irrespective of types of alcoholic drinks [72]. In general, the association is stronger among men than among women, probably due to a higher range of alcohol intake in men, although a true gender difference cannot be excluded. Consistent with the well-known biological interaction between alcohol and folate (i.e., alcohol antagonizes the effect of folate by impairing the intestinal absorption of folate or by suppressing intracellular folate metabolism) [73], alcohol intake was noticeably harmful when folate status was poor, especially among men [74].

The adverse effect of alcohol on CRC risk may be chiefly mediated by the first metabolite of ethanol [75, 76], acetaldehyde, which is classified as a carcinogen. Absorbed alcohol is rapidly distributed to the body through the bloodstream, reaching colonocytes. Intestinal microflora with alcohol dehydrogenase (ADH) activity oxidizes the supplied ethanol to acetaldehyde, thereby serving as a major determinant of acetaldehyde concentration in colonocytes [75–77]. As colonocytes lack the ability to detoxify acetaldehyde, the accumulated acetaldehyde in these cells may induce colorectal carcinogenesis by causing DNA damage, promoting excessive growth of the colonic mucosa [75, 76], or antagonizing the protective effect of folate particularly by directly destructing intracellular folate [74]. The central contribution of acetaldehyde to colorectal carcinogenesis is also supported by the findings that the positive association between alcohol intake and CRC risk is pronounced among people carrying genetic polymorphisms leading to acetaldehyde accumulations (e.g., ADH1C alleles encoding ADH enzymes with high activity [i.e., quick conversion of ethanol to acetaldehyde] or ALDH2 alleles encoding ALDH enzymes with low activity [i.e., slow elimination of acetaldehyde]) [75].

### 21.3.2.3 Protective Factors

#### Physical Activity

Physical activity is a well-accepted protective factor, particularly against CC. The investigation of physical activity in relation to CRC has been hindered to some extent due to difficulties measuring activity accurately as well as a narrow range of physical activity in many study populations. These factors have likely generated heterogeneity in the magnitude and statistical significance of epidemiologic findings. Nevertheless an inverse association with CRC was consistently observed across sex, study designs, countries, domains of physical activity (e.g., occupational, recreational, household), and stage of carcinogenesis (i.e., adenomas, cancer) [79, 80].

Physical activity has diverse dimensions including intensity, duration, and frequency. The total Metabolic Equivalent Task (MET)-hours/week, a composite score incorporating all of the aforementioned dimensions of total weekly physical activity, is the most comprehensive measure of physical activity. In large cohort studies of men (the Health Professionals Follow-up Study) and women (the Nurses' Health Study) that used validated questionnaires to assess leisure-time physical activity, a higher MET-hours/week was associated with a substantially reduced risk of CC after accounting for BMI and other important risk factors [81, 82]. Given that walking was the most common type of leisure-time physical activity in these cohorts, people may have accumulated most of their MET-hours/week through walking. Thus, these findings suggest that activities of moderate intensity such as brisk walking may be sufficient to reduce CC risk. Yet several lines of evidence indicate that a greater intensity of physical activity may provide additional benefit [83, 84].

The beneficial effect of physical activity on CC risk may be mediated through the insulin/IGF pathway as well as through other mechanisms. Physical activity lowers circulating insulin levels directly by increasing insulin sensitivity [85] and indirectly by decreasing abdominal fatness (particularly VAT). Evidence supporting the direct effect not mediated through adiposity comes from some, but not all biomarker studies, which showed a significant inverse relationship, particularly among men, between physical activity and circulating C-peptide, insulin, or leptin concentrations adjusting for measures of adiposity [86, 87]. With regard to the indirect effect, VAT, which is more lipolytic than SAT, is preferentially lost in response to physical activity [48, 88–90]. Controlled trials have demonstrated that moderate-intensity physical activities reduce VAT [88, 89], with vigorous activities inducing greater VAT loss [91], and a significant reduction in VAT occurs even in the absence of weight loss [88, 89]. Other proposed mechanisms whereby physical activity lowers CC risk include reducing chronic inflammation, improving immune function, and decreasing colonic exposure to carcinogens by increasing colonic motility [92, 93].

### Fiber

A possible beneficial role of dietary fiber intake in the prevention of CRC was first hypothesized by Burkitt in the late 1960s upon observing a low incidence of CRC in southern Africa, where dietary fiber intake was high [94]. Several mechanistic explanations have been proposed to support the potential association. Insoluble fibers may reduce CRC risk by increasing stool bulk, shortening stool transit time, and decreasing exposure of the colorectal mucosa to potential carcinogens [95]. Soluble fibers, which can be fermented by the anaerobic intestinal microbiota into short-chain fatty acids (e.g., acetate, propionate, and butyrate), may protect against CRC by reducing colorectal pH or through butyrate, which inhibits proliferation and induces apoptosis of CRC cells [95, 96].

Observational studies overall suggest an inverse association, with a recent dose–response meta-analysis of prospective studies estimating an approximately 10 % reduced risk of CRC for a 10 g/day increased intake of total fiber, particularly cereal fiber intake [97]. However, some cohort studies that adjusted for physical activity and dietary intake of folate and calcium did not observe an association [98–101]. No appreciable protective association was observed in most randomized trials of fiber supplements and risk of adenoma [102]. Yet, at present, we cannot entirely dismiss a potential protective role of fiber intake against CRC, because its efficacy may be limited to specific types (insoluble, soluble) or food sources (fruits, vegetables, and cereal fibers) and may vary depending on intestinal microflora profiles [96, 102]. Additionally, in modern diets in which refined food is a dominant component, the amount and quality of fiber is much lower than in traditional diets.

## Micronutrients: Folate, Calcium, Vitamin D

### Folate (Vitamin $B_9$)

For decades, folate has been proposed to protect against CRC due to its critical roles in maintaining the integrity of DNA synthesis and regulating DNA methylation that controls proto-oncogene expression. In contrast, in recent years, a role in facilitating the progression of preneoplastic lesions has been suggested [103–106].

Particularly required to observe a benefit of folate on CRC risk is a long induction period, which is evident from both molecular mechanistic perspectives and epidemiologic studies. For example, in colonocytes, total folate concentration significantly decreases from the normal tissue without adenoma, to the normal tissue adjacent to adenoma, and to the adenoma itself [107]. The lower folate level in the normal appearing, non-neoplastic tissue corresponds to an increase in uracil misincorporation into DNA and DNA hypomethylation [107], which suggests the presence of a field defect surrounding adenomas. As the progression from precursor lesions to cancer diagnosis generally takes at least 10 years [108], macroscopically undetectable abnormalities due to folate deficiency would occur at least a decade before CRC is apparent. In a pooled analysis of 13 prospective studies with follow-up periods ranging from 7 to 20 years, total folate intake of ≥560 mcg/day relative to <240 mcg/day was significantly associated with a 13 % lower risk of CC [109]. In a cohort study that directly tested several potential induction periods between folate intake and CRC risk, a significant inverse association emerged only after a lag period of at least 12–16 years [110]. In view of this finding, randomized placebo-controlled trials (RCTs) of folic acid supplementations, if having a short follow-up period, are unlikely to observe a benefit on CRC. On the contrary, for the outcome of initial adenomas, a benefit was observed after 3 years of intervention in a recent RCT in China [111].

Regarding the potential adverse effect of excess folic acid on lesions at late stages of colorectal carcinogenesis, a recent meta-analysis of 13 RCTs provided rebutting evidence that folic acid supplementations at considerably high doses (500–40,000 mcg/day) did not substantially increase the risk of CRC during the first 5 years of intervention compared to placebos (Relative Risk (RR) 1.07, 95 % CI 0.83–1.37) [112]. In RCTs, most of the cancers diagnosed in the first 5 years are likely to originate from covert advanced adenomas or latent cancers. Thus, if folic acid supplementation were to promote the progression of such existing neoplasms, a significantly increased risk of CRC should have been observed in the supplement arms, which was not the case. Even in the subgroup analysis restricted to three trials conducted among participants with a prior history of adenomas and thus at high risk for CRC, no evidence of a positive association was apparent (RR 0.76, 95 % CI 0.32–1.80) [112].

### Calcium

There is strong mechanistic evidence that calcium may decrease the risk of colorectal neoplasia. Experimental studies in animals and humans suggest that calcium may bind secondary bile acids or ionized fatty acids in the colorectal lumen, diminishing their carcinogenic effects on the colorectal mucosa [113, 114]. Alternatively, evidence from in vivo and in vitro human colonic epithelial cells suggests that calcium may reduce cell proliferation and promote cell differentiation by modulating cell signaling pathways [115, 116]. Calcium supplementation of 2000 mg/day was demonstrated to induce favorable changes on gene expression in the APC/β-catenin pathway in the normal-appearing mucosa of colorectal adenoma patients [117]. Perturbation of this pathway is a common early event in colorectal carcinogenesis.

RCTs also provide strong evidence supporting the chemopreventive potential of calcium against adenomas. In a meta-analysis of RCTs, those assigned to take 1200–2000 mg of calcium supplements without co-administered vitamin D over 3–4 years had an approximately 20 % reduced risk of adenoma recurrence compared to those assigned to the placebo group [118]. In the Calcium Polyp Prevention Study, the largest RCT (930 participants) included in the meta-analysis, 1200 mg/day supplementation of calcium carbonate over 4 years was most protective against histologically advanced neoplasms, suggesting its benefit may extend to cancer outcome as well [119].

Contrary to expectation, a meta-analysis of eight RCTs on CRC outcome showed no benefit of calcium supplementation over 4 years [120]. These inconsistent results by adenoma versus cancer outcome may be potentially explained by the presence of a long induction period between adequate calcium intake and CRC diagnosis. Given that the prevention or removal of an adenoma leads to the prevention of CRC and that the progression from adenoma to CRC spans at least 10 years, a longer follow-up would be required for the observed benefit of calcium against adenoma outcome [118] to extend to cancer outcome. Consistent with this hypothesis, a pooled analysis of 10 cohort studies [121] whose follow-up periods ranged between 6 and 16 years showed a significant inverse association between total calcium intake and CRC [121].

While the potential of calcium as a chemopreventive agent against CRC appears promising, the optimal dose remains to be identified. More specifically, a pooled analysis indicated no additional benefit at a total calcium intake beyond 1000 mg/day [121], whereas a recent dose–response meta-analysis of prospective observational studies suggested that CRC risk may continue to decrease beyond 1000 mg/day [122].

### Vitamin D Status

Observational studies provide consistent evidence indicating an etiologic role of a poor vitamin D status in CRC development and, particularly, in CRC progression [123]. The potential anti-carcinogenic effect of vitamin D was first proposed by Garland and Garland in 1980, upon noticing that CC mortality rates in the U.S. were generally higher in the northern regions that have lower exposure to solar ultraviolet B (UV-B) radiation, which is required to synthesize vitamin D in the skin [124]. The hypothesis is further corroborated by molecular studies showing that vitamin D, beyond its conventional role in calcium and phosphorus homeostasis, influences cellular signaling pathways leading to an inhibition of proliferation and angiogenesis and an activation of apoptosis [125].

In epidemiologic studies, the hypothesis has been investigated using a variety of surrogate measures of vitamin D status such as UV-B radiation exposure, dietary and supplementary vitamin D intake, and predicted or measured levels of circulating 25(OH)vitamin D (25(OH) D), which is considered the best biochemical indicator of vitamin D status. Regardless of the measures used, results have been generally consistent across studies [123, 126]. For example, in the largest available study on circulating 25(OH) D and CRC endpoints, the previously mentioned EPIC study, prediagnostic 25(OH)D concentration in the highest quintile was associated with an approximately 40 % reduced risk of developing CRC [127] and a 31 % improved survival of patients with CRC [128] compared with the level in the lowest quintile.

The optimal levels of 25(OH)D to reduce CRC incidence and mortality are not yet established, but available studies suggest the range of 30–40 ng/mL [123]. This range is consistent with the clinical practice guidelines on vitamin D deficiency by the Endocrine Society, which recommends a minimum level of 30 ng/mL but 40–60 ng/mL to guarantee sufficiency [129].

The potential of vitamin D supplements as a chemopreventive agent against CRC endpoints (e.g., incidence and mortality) is best assessed with an RCT. Yet, published RCTs on vitamin D are inadequate to evaluate its chemopreventive potential due to the low doses of vitamin D supplements tested, the short durations of follow-up that might not account for a potentially long induction period between an adequate vitamin D status and CRC occurrence, sufficient vitamin D status of participants at baseline who may not gain benefit from additional supplements, and/or the interaction with other testing agents such as hormone replacement therapy that may mask the benefit from vitamin D supplements. There are ongoing RCTs on vitamin D that will provide important information in the future.

## NSAIDs Including Aspirin

There is definitive evidence from both cohort studies and RCTs that long-term, regular use of non-steroidal anti-inflammatory drugs (NSAIDs), particularly aspirin, lowers the risk of CRC [130–132]. One of the largest RCTs of aspirin, the Women's Health Study (WHS), demonstrated that even a low dose of aspirin (100 mg) on alternate days could confer protection against CRC [130]. Of note, the benefit was evident only after a decade of follow-up. In the WHS, at the end of the 10-year active intervention, CRC incidence was not different across the intervention and placebo groups [133]. However, upon extended post-trial follow-up leading to a median cumulative follow-up of 18 years, a 10 % reduced risk of CRC was observed in the aspirin group [130]. The long latency indicates that aspirin may operate at the early stages of colorectal carcinogenesis but may be no longer effective at a stage in adenoma progression. Indeed, aspirin reduced the risk of colorectal adenoma [134].

NSAIDs, including aspirin, may affect colorectal carcinogenesis by inhibiting the prostaglandin-endoperoxide synthase 2 (PTGS2 or COX2) pathway, a pro-inflammatory pathway whose downstream signaling promotes cell proliferation, inhibits apoptosis, and stimulates angiogenesis [135, 136]. Indeed, in a large cohort of men and women, the benefit of regular aspirin use was limited to a molecular subtype of CRC that overexpresses COX-2 [137]. Thus, strong evidence that NSAIDs as well as inflammatory bowel disease (particularly ulcerative colitis) are implicated in colorectal carcinogenesis presents chronic inflammation as an important pathogenic pathway of CRC. Although aspirin is proven to protect against CRC, long-term use in average risk populations is not generally recommended currently due to the potential adverse effects, mainly bleeding episodes in the upper gastrointestinal tract particularly.

### Estrogens and/or Progestogen (Natural or Bioidentical Progesterone + Synthetic Progestin)

While the observation of higher CRC incidence rates among men than women raised a possibility that estrogen and/or progesterone may confer protection against CRC, evidence remains inconclusive.

The Women's Health Initiative (WHI) hormone therapy trial, the first large RCT on postmenopausal hormone treatment, provides some evidence supporting a protective role of exogenous female sex hormones when estrogen and progestin are administered together. During the intervention phase, postmenopausal women assigned to estrogen plus progestin had a 38 % reduced risk of CRC, whereas those assigned to estrogen-alone did not receive any benefit [138]. However, over an extended post-intervention follow-up (a median cumulative follow-up of 13 years), the magnitude of protection conferred by the estrogen plus progestin treatment diminished with borderline statistical significance and the estrogen-alone treatment continued to show no benefit [138]. In a recent meta-analysis that pooled the results across case-control studies, cohort studies and RCTs, the combined estrogen-progestogen therapy was associated with a reduced risk of CRC regardless of the recency of the therapy but, for estrogen-alone therapy, only current use was associated with a reduced risk [139].

Evidence regarding endogenous estrogen among postmenopausal women is even more inconsistent. The WHI Observational Study was the first to examine the relationship between circulating endogenous estradiol levels and incident CRC and, contrary to the expectation, found an adverse association [140]. A subsequent study also suggested an increased risk of CRC associated with higher circulating estrone levels [141]. Yet, in another nested case-control study, the ratio of estradiol to testosterone (a marker of estradiol production by aromatase) was inversely associated with CRC risk even after adjusting for BMI and C-peptide in postmenopausal women not on hormone therapy [142]. Interestingly, when stratified by BMI, the inverse association was limited to women with a normal weight.

The apparent heterogeneous findings between exogenous and endogenous hormones may be biologically plausible. First, oral administration of exogenous estrogens in a large bolus was shown to decrease hepatic production of insulin and IGF-1, thereby mitigating their cancer-promoting effects [140, 143]. Second, considering that circulating progesterone levels are low among postmenopausal women, a possible protective role of exogenous progestin in colorectal carcinogenesis cannot be discounted [144].

### Screening

CRC incidence rates in the U.S. have been declining over the past two decades and about half the declines have been ascribed to increased screening practices [145].

Screening confers benefits on both incidence and mortality: the detection and removal of precancerous adenomas lead to the secondary prevention of CRC, and an early detection of asymptomatic CRC improves prognosis. In a landmark randomized trial that evaluated the effect of sigmoidoscopy compared to the usual care group, incidence rates of CRC remained lower in the screening group after an initial temporary rise in the rates due to an increased detection of prevalent cancers [146]. Of note, since CC develops most frequently in the sigmoid colon (around 25 %) and cecum (around 20 %) [147], colonoscopy that examines the entire colon and rectum is preferred over sigmoidoscopy that screens only the sigmoid colon and rectum. However, evidence suggests that colonoscopy may not be as effective in detecting proximal lesions as detecting distal lesions, possibly due to inherent biologic differences or greater difficulty in visualizing the proximal colon, especially the cecum [148].

Screening for CRC is strongly encouraged by major societies of medical professionals including the American Cancer Society and the U.S. Preventive Services Task Force. According to the American Cancer Society, individuals at average risk for CRC are recommended to undergo one of the following screening test beginning at age 50: colonoscopy every 10 years, sigmoidoscopy every 5 years, CT colonography every 5 years, or barium enema every 5 years, with the last three requiring colonoscopy when the test results are positive. Individuals at higher risk for CRC (e.g., a personal history of CRC, adenomas, or inflammatory bowel disease and a family history of CRC, adenomas, and hereditary syndromes) are recommended to begin screening before age 50 and/or be screened more frequently [149].

## 21.4 Summary

CRC is a major cause of cancer incidence and mortality, and its burden is expected to rise with increasing Westernization of countries. Now there are a number of well-established risk factors for CRC, most of which are modifiable. Moreover, CRC screening is associated with reduced CRC incidence and mortality. As such, much of the burden of this common cancer could be eliminated with changes in lifestyle and screening patterns.

## References


1. What are the survival rates for colorectal cancer by stage? [internet]. Updated 31 Dec 2014; cited 17 Jan 2015. Available from http://www.cancer.org/cancer/colonandrectumcancer/detailedguide/colorectal-cancer-survival-rates.
2. Estimated incidence, mortality and 5-year prevalence: both sexes [internet]. Updated 2015; cited 17 Jan 2015. Available from http://globocan.iarc.fr/Pages/fact_sheets_population.aspx.
3. Center MM, Jemal A, Ward E. International trends in colorectal cancer incidence rates. Cancer Epidemiol Biomarkers Prev. 2009;18:1688–94.
4. Incidence/mortality rates: cancers by population [internet]. Updated 2015; cited 17 Jan 2015. Available from http://globocan.iarc.fr/Pages/summary_table_pop_sel.aspx.
5. Siegel R, Desantis C, Jemal A. Colorectal cancer statistics, 2014. CA Cancer J Clin. 2014;64:104–17.
6. SEER stat fact sheets: colon and rectum cancer [internet]. Updated 2014; cited 17 Jan 2015. Available from http://seer.cancer.gov/statfacts/html/colorect.html.
7. Incidence/mortality: age-specific table [internet]. Updated 2015; cited 17 Jan 2015. Available from http://globocan.iarc.fr/Pages/age-specific_table_sel.aspx.
8. American Cancer Society. Colorectal cancer facts & figures 2014–2016. Atlanta: American Cancer Society; 2014.
9. Colorectal cancer rates by race and ethnicity [internet]. Updated 27 Aug 2014; cited 17 Jan 2015. Available from http://www.cdc.gov/cancer/colorectal/statistics/race.htm.
10. McMichael AJ, Giles GG. Cancer in migrants to Australia: extending the descriptive epidemiological data. Cancer Res. 1988;48:751–6.
11. Dunn JE. Cancer epidemiology in populations of the United States—with emphasis on Hawaii and California—and Japan. Cancer Res. 1975;35:3240–5.
12. Estimating Preventability [internet]. Updated 2015; cited 2015 17 Jan. Available from http://www.aicr.org/research/research_science_policy_report.html.
13. Platz EA, Willett WC, Colditz GA, Rimm EB, Spiegelman D, Giovannucci E. Proportion of colon cancer risk that might be preventable in a cohort of middle-aged US men. Cancer Causes Control. 2000;11:579–88.
14. Winawer SJ, Schottenfeld D, Flehinger BJ. Colorectal cancer screening. J Natl Cancer Inst. 1991;83:243–53.
15. Winawer S, Fletcher R, Rex D, Bond J, Burt R, Ferrucci J, Ganiats T, Levin T, Woolf S, Johnson D, Kirk L, Litin S, et al. Colorectal cancer screening and surveillance: clinical guidelines and rationale-update based on new evidence. Gastroenterology. 2003;124:544–60.
16. Winawer SJ, Fletcher RH, Miller L, Godlee F, Stolar MH, Mulrow CD, Woolf SH, Glick SN, Ganiats TG, Bond JH, Rosen L, Zapka JG, et al. Colorectal cancer screening: clinical guidelines and rationale. Gastroenterology. 1997;112:594–642.
17. Lynch syndrome [internet]. Updated 2015; 17 Jan cited 2015. Available from http://www.cancer.ca/en/cancer-information/cancer-101/what-is-a-risk-factor/genetic-risk/lynch-syndrome/?region=on.

18. National Cancer Institute. What you need to know about cancer of the colon and rectum. Washington: NIH publication; 2006.
19. Risio M. The natural history of adenomas. Best Pract Res Clin Gastroenterol. 2010;24:271–80.
20. Familial Adenomatous Polyposis (FAP) [internet]. Updated 2015; cited 17 Jan 2015. Available from http://www.cancer.ca/en/cancer-information/cancer-101/what-is-a-risk-factor/genetic-risk/fap/?region=on.
21. Risk factors for colorectal cancer [internet]. [updated 5 Feb 2015; cited 17 Feb 2015. Available from: http://www.cancer.org/cancer/colonandrectumcancer/moreinformation/colonandrectumcancerearlydetection/colorectal-cancer-early-detection-risk-factors-for-crc.
22. Coussens LM, Werb Z. Inflammation and cancer. Nature. 2002;420:860–7.
23. Triantafillidis JK, Nasioulas G, Kosmidis PA. Colorectal cancer and inflammatory bowel disease: epidemiology, risk factors, mechanisms of carcinogenesis and prevention strategies. Anticancer Res. 2009;29:2727–37.
24. Mattar MC, Lough D, Pishvaian MJ, Charabaty A. Current management of inflammatory bowel disease and colorectal cancer. Gastrointest Cancer Res. 2011;4:53–61.
25. Butterworth AS, Higgins JP, Pharoah P. Relative and absolute risk of colorectal cancer for individuals with a family history: a meta-analysis. Eur J Cancer. 2006;42:216–27.
26. Johns LE, Houlston RS. A systematic review and meta-analysis of familial colorectal cancer risk. Am J Gastroenterol. 2001;96:2992–3003.
27. Shussman N, Wexner SD. Colorectal polyps and polyposis syndromes. Gastroenterol Rep. 2014;2:1–15.
28. Neugut AI, Jacobson JS, De Vivo I. Epidemiology of colorectal adenomatous polyps. Cancer Epidemiol Biomarkers Prev. 1993;2:159–76.
29. Muto T, Bussey HJ, Morson BC. The evolution of cancer of the colon and rectum. Cancer. 1975;36:2251–70.
30. Heitman SJ, Ronksley PE, Hilsden RJ, Manns BJ, Rostom A, Hemmelgarn BR. Prevalence of adenomas and colorectal cancer in average risk individuals: a systematic review and meta-analysis. Clin Gastroenterol Hepatol. 2009;7:1272–8.
31. Neugut AI, Jacobson JS, Ahsan H, Santos J, Garbowski GC, Forde KA, Treat MR, Waye J. Incidence and recurrence rates of colorectal adenomas: a prospective study. Gastroenterology. 1995;108:402–8.
32. Giovannucci E. Insulin and colon cancer. Cancer Causes Control. 1995;6:164–79.
33. Giovannucci E. Insulin, insulin-like growth factors and colon cancer: a review of the evidence. J Nutr. 2001;131:3109S–20S.
34. Sandhu MS, Dunger DB, Giovannucci EL. Insulin, insulin-like growth factor-I (IGF-I), IGF binding proteins, their biologic interactions, and colorectal cancer. J Natl Cancer Inst. 2002;94:972–80.
35. Wei EK, Ma J, Pollak MN, Rifai N, Fuchs CS, Hankinson SE, Giovannucci E. A prospective study of C-peptide, insulin-like growth factor-I, insulin-like growth factor binding protein-1, and the risk of colorectal cancer in women. Cancer Epidemiol Biomarkers Prev. 2005;14:850–5.
36. Pischon T, Lahmann PH, Boeing H, Friedenreich C, Norat T, Tjonneland A, Halkjaer J, Overvad K, Clavel-Chapelon F, Boutron-Ruault MC, Guernec G, Bergmann MM, et al. Body size and risk of colon and rectal cancer in the European Prospective Investigation Into Cancer and Nutrition (EPIC). J Natl Cancer Inst. 2006;98:920–31.
37. Aleksandrova K, Pischon T, Buijsse B, May AM, Peeters PH, Bueno-de-Mesquita HB, Jenab M, Fedirko V, Dahm CC, Siersema PD, Freisling H, Ferrari P, et al. Adult weight change and risk of colorectal cancer in the European Prospective Investigation into Cancer and Nutrition. Eur J Cancer. 2013;49:3526–36.
38. Keum N, Greenwood DC, Lee DH, Kim R, Aune D, Ju W, Hu FB, Giovannucci EL. Adult weight gain and adiposity-related cancers: a dose-response meta-analysis of prospective observational studies. J Natl Cancer Inst 2015;107.
39. Despres JP. Abdominal obesity: the most prevalent cause of the metabolic syndrome and related cardiometabolic risk. Eur Heart J Suppl. 2006;8: B4–12.
40. Geer EB, Shen W. Gender differences in insulin resistance, body composition, and energy balance. Gend Med. 2009;6(Suppl 1):60–75.
41. MacInnis RJ, English DR, Hopper JL, Haydon AM, Gertig DM, Giles GG. Body size and composition and colon cancer risk in men. Cancer Epidemiol Biomarkers Prev. 2004;13:553–9.
42. Nagata N, Sakamoto K, Arai T, Niikura R, Shimbo T, Shinozaki M, Aoki T, Kishida Y, Sekine K, Tanaka S, Okubo H, Watanabe K, et al. Visceral abdominal fat measured by computed tomography is associated with an increased risk of colorectal adenoma. Int J Cancer. 2014;135 (10):2273–81.
43. Kang HW, Kim D, Kim HJ, Kim CH, Kim YS, Park MJ, Kim JS, Cho SH, Sung MW, Jung HC, Lee HS, Song IS. Visceral obesity and insulin resistance as risk factors for colorectal adenoma: a cross-sectional, case-control study. Am J Gastroenterol. 2010;105:178–87.
44. Keum N, Lee DH, Kim R, Greenwood DC, Giovannucci EL. Visceral adiposity and colorectal adenomas: dose-response meta-analysis of observational studies. Ann Oncol. 2015;26:1101–9.
45. Yamaji Y, Mitsushima T, Koike K. Pulse-wave velocity, the ankle-brachial index, and the visceral fat area are highly associated with colorectal adenoma. Dig Liver Dis. 2014;46(10):943–9.

46. Riondino S, Roselli M, Palmirotta R, Della-Morte D, Ferroni P, Guadagni F. Obesity and colorectal cancer: role of adipokines in tumor initiation and progression. World J Gastroenterol. 2014;20:5177–90.
47. Donohoe CL, Doyle SL, Reynolds JV. Visceral adiposity, insulin resistance and cancer risk. Diabetol Metab Syndr. 2011;3:12.
48. Ibrahim MM. Subcutaneous and visceral adipose tissue: structural and functional differences. Obesity Rev. 2010;11:11–8.
49. Wajchenberg BL. Subcutaneous and visceral adipose tissue: their relation to the metabolic syndrome. Endocrine Rev. 2000;21:697–738.
50. Kaaks R, Lukanova A, Kurzer MS. Obesity, endogenous hormones, and endometrial cancer risk: a synthetic review. Cancer Epidemiol Biomarkers Prev. 2002;11:1531–43.
51. Janssen I, Heymsfield SB, Allison DB, Kotler DP, Ross R. Body mass index and waist circumference independently contribute to the prediction of non-abdominal, abdominal subcutaneous, and visceral fat. Am J Clin Nutr. 2002;75:683–8.
52. Hu FB. Globalization of diabetes: the role of diet, lifestyle, and genes. Diabetes Care. 2011;34:1249–57.
53. Ning Y, Wang L, Giovannucci EL. A quantitative analysis of body mass index and colorectal cancer: findings from 56 observational studies. Obesity Rev. 2010;11:19–30.
54. Schmid D, Leitzmann MF. Television viewing and time spent sedentary in relation to cancer risk: a meta-analysis. J Natl Cancer Inst 2014;106.
55. Hamilton MT, Hamilton DG, Zderic TW. Role of low energy expenditure and sitting in obesity, metabolic syndrome, type 2 diabetes, and cardiovascular disease. Diabetes. 2007;56:2655–67.
56. Howard RA, Freedman DM, Park Y, Hollenbeck A, Schatzkin A, Leitzmann MF. Physical activity, sedentary behavior, and the risk of colon and rectal cancer in the NIH-AARP diet and health study. Cancer Causes Control. 2008;19:939–53.
57. Hu FB. Dietary pattern analysis: a new direction in nutritional epidemiology. Curr Opin Lipidol. 2002;13:3–9.
58. Fung TT, Brown LS. Dietary patterns and the risk of colorectal cancer. Curr Nutr Rep. 2013;2:48–55.
59. Magalhaes B, Peleteiro B, Lunet N. Dietary patterns and colorectal cancer: systematic review and meta-analysis. Eur J Cancer Prev. 2012;21:15–23.
60. Ley SH, Sun Q, Willett WC, Eliassen AH, Wu KN, Pan A, Grodstein F, Hu FB. Associations between red meat intake and biomarkers of inflammation and glucose metabolism in women. Am J Clin Nutr. 2014;99:352–60.
61. Knobel Y, Weise A, Glei M, Sendt W, Claussen U, Pool-Zobel BL. Ferric iron is genotoxic in non-transformed and preneoplastic human colon cells. Food Chem Toxicol. 2007;45:804–11.
62. Dubrow R, Darefsky AS, Park Y, Mayne ST, Moore SC, Kilfoy B, Cross AJ, Sinha R, Hollenbeck AR, Schatzkin A, Ward MH. Dietary components related to N-nitroso compound formation: a prospective study of adult glioma. Cancer Epidemiol Biomarkers Prev. 2010;19:1709–22.
63. Butler LM, Sinha R, Millikan RC, Martin CF, Newman B, Gammon MD, Ammerman AS, Sandler RS. Heterocyclic amines, meat intake, and association with colon cancer in a population-based study. Am J Epidemiol. 2003;157:434–45.
64. Giovannucci E, Martinez ME. Tobacco, colorectal cancer, and adenomas: a review of the evidence. J Natl Cancer Inst. 1996;88:1717–30.
65. Liang PS, Chen TY, Giovannucci E. Cigarette smoking and colorectal cancer incidence and mortality: systematic review and meta-analysis. Int J Cancer. 2009;124:2406–15.
66. Giovannucci E. An updated review of the epidemiological evidence that cigarette smoking increases risk of colorectal cancer. Cancer Epidemiol Biomarkers Prev. 2001;10:725–31.
67. Giovannucci E, Colditz GA, Stampfer MJ, Hunter D, Rosner BA, Willett WC, Speizer FE. A prospective study of cigarette smoking and risk of colorectal adenoma and colorectal cancer in U.S. women. J Natl Cancer Inst 1994;86:192–9.
68. Giovannucci E, Rimm EB, Stampfer MJ, Colditz GA, Ascherio A, Kearney J, Willett WC. A prospective study of cigarette smoking and risk of colorectal adenoma and colorectal cancer in U.S. men. J Natl Cancer Inst 1994;86:183–91.
69. Hannan LM, Jacobs EJ, Thun MJ. The association between cigarette smoking and risk of colorectal cancer in a large prospective cohort from the United States. Cancer Epidemiol Biomarkers Prev. 2009;18:3362–7.
70. Nishihara R, Morikawa T, Kuchiba A, Lochhead P, Yamauchi M, Liao X, Imamura Y, Nosho K, Shima K, Kawachi I, Qian ZR, Fuchs CS, et al. A prospective study of duration of smoking cessation and colorectal cancer risk by epigenetics-related tumor classification. Am J Epidemiol. 2013;178:84–100.
71. Curtin K, Slattery ML, Samowitz WS. CpG island methylation in colorectal cancer: past, present and future. Pathology Res Int. 2011;2011:902674.
72. Cho E, Smith-Warner SA, Ritz J, van den Brandt PA, Colditz GA, Folsom AR, Freudenheim JL, Giovannucci E, Goldbohm RA, Graham S, Holmberg L, Kim DH, et al. Alcohol intake and colorectal cancer: a pooled analysis of 8 cohort studies. Ann Intern Med. 2004;140:603–13.
73. Hillman RS, Steinberg SE. The effects of alcohol on folate metabolism. Annu Rev Med. 1982;33:345–54.
74. Giovannucci E. Alcohol, one-carbon metabolism, and colorectal cancer: recent insights from molecular studies. J Nutr. 2004;134:2475S–81S.

75. Seitz HK, Becker P. Alcohol metabolism and cancer risk. Alcohol Res Health. 2007;30(38–41):4–7.
76. Jokelainen K, Nosova T, Koivisto T, Vakevainen S, Jousimies-Somer H, Heine R, Salaspuro M. Inhibition of bacteriocolonic pathway for ethanol oxidation by ciprofloxacin in rats. Life Sci. 1997;61:1755–62.
77. Salaspuro M. Acetaldehyde as a common denominator and cumulative carcinogen in digestive tract cancers. Scand J Gastroenterol. 2009;44:912–25.
78. Giovannucci E. Nutrition, insulin, insulin-like growth factors and cancer. Horm Metab Res. 2003;35:694–704.
79. Boyle T, Keegel T, Bull F, Heyworth J, Fritschi L. Physical activity and risks of proximal and distal colon cancers: a systematic review and meta-analysis. J Natl Cancer Inst. 2012;104:1548–61.
80. Wolin KY, Yan Y, Colditz GA. Physical activity and risk of colon adenoma: a meta-analysis. Br J Cancer. 2011;104:882–5.
81. Giovannucci E, Ascherio A, Rimm EB, Colditz GA, Stampfer MJ, Willett WC. Physical activity, obesity, and risk for colon cancer and adenoma in men. Ann Intern Med. 1995;122:327–34.
82. Martinez ME, Giovannucci E, Spiegelman D, Hunter DJ, Willett WC, Colditz GA. Leisure-time physical activity, body size, and colon cancer in women. Nurses' Health Study Research Group. J Natl Cancer Inst. 1997;89:948–55.
83. Giovannucci E, Colditz GA, Stampfer MJ, Willett WC. Physical activity, obesity, and risk of colorectal adenoma in women (United States). Cancer Causes Control. 1996;7:253–63.
84. Wolin KY, Lee IM, Colditz GA, Glynn RJ, Fuchs C, Giovannucci E. Leisure-time physical activity patterns and risk of colon cancer in women. Int J Cancer. 2007;121:2776–81.
85. Turcotte LP, Fisher JS. Skeletal muscle insulin resistance: roles of fatty acid metabolism and exercise. Physical Ther. 2008;88:1279–96.
86. Regensteiner JG, Mayer EJ, Shetterly SM, Eckel RH, Haskell WL, Marshall JA, Baxter J, Hamman RF. Relationship between habitual physical activity and insulin levels among nondiabetic men and women—San-Luis Valley Diabetes Study. Diabetes Care. 1991;14:1066–74.
87. Fung TT, Hu FB, Yu J, Chu NF, Spiegelman D, Tofler GH, Willett WC, Rimm EB. Leisure-time physical activity, television watching, and plasma biomarkers of obesity and cardiovascular disease risk. Am J Epidemiol. 2000;152:1171–8.
88. Ross R, Dagnone D, Jones PJ, Smith H, Paddags A, Hudson R, Janssen I. Reduction in obesity and related comorbid conditions after diet-induced weight loss or exercise-induced weight loss in men. A randomized, controlled trial. Ann Intern Med. 2000;133:92–103.
89. Ross R, Janssen I, Dawson J, Kungl AM, Kuk JL, Wong SL, Nguyen-Duy TB, Lee S, Kilpatrick K, Hudson R. Exercise-induced reduction in obesity and insulin resistance in women: a randomized controlled trial. Obesity Res. 2004;12:789–98.
90. Thomas EL, Brynes AE, McCarthy J, Goldstone AP, Hajnal JV, Saeed N, Frost G, Bell JD. Preferential loss of visceral fat following aerobic exercise, measured by magnetic resonance imaging. Lipids. 2000;35:769–76.
91. Irving BA, Davis CK, Brock DW, Weltman JY, Swift D, Barrett EJ, Gaesser GA, Weltman A. Effect of exercise training intensity on abdominal visceral fat and body composition. Med Sci Sports Exerc. 2008;40:1863–72.
92. Harriss DJ, Cable NT, George K, Reilly T, Renehan AG, Haboubi N. Physical activity before and after diagnosis of colorectal cancer—disease risk, clinical outcomes, response pathways and biomarkers. Sports Med. 2007;37:947–60.
93. Slattery ML. Physical activity and colorectal cancer. Sports Med. 2004;34:239–52.
94. Burkitt DP. Related disease—related cause? Lancet. 1969;2:1229–31.
95. Kritchevsky D. Epidemiology of fibre, resistant starch and colorectal cancer. Eur J Cancer Prev. 1995;4:345–52.
96. Sebastian C, Mostoslavsky R. Untangling the fiber yarn: butyrate feeds warburg to suppress colorectal cancer. Cancer Discov. 2014;4:1368–70.
97. Aune D, Chan DS, Lau R, Vieira R, Greenwood DC, Kampman E, Norat T. Dietary fibre, whole grains, and risk of colorectal cancer: systematic review and dose-response meta-analysis of prospective studies. BMJ. 2011;343:d6617.
98. Michels KB, Fuchs CS, Giovannucci E, Colditz GA, Hunter DJ, Stampfer MJ, Willett WC. Fiber intake and incidence of colorectal cancer among 76,947 women and 47,279 men. Cancer Epidemiol Biomarkers Prev. 2005;14:842–9.
99. Schatzkin A, Mouw T, Park Y, Subar AF, Kipnis V, Hollenbeck A, Leitzmann MF, Thompson FE. Dietary fiber and whole-grain consumption in relation to colorectal cancer in the NIH-AARP Diet and Health Study. Am J Clin Nutr. 2007;85:1353–60.
100. Otani T, Iwasaki M, Ishihara J, Sasazuki S, Inoue M, Tsugane S, Japan Public Health Center-Based Prospective Study G. Dietary fiber intake and subsequent risk of colorectal cancer: the Japan Public Health Center-based prospective study. Int J Cancer. 2006;119:1475–80.
101. Wakai K, Date C, Fukui M, Tamakoshi K, Watanabe Y, Hayakawa N, Kojima M, Kawado M, Suzuki K, Hashimoto S, Tokudome S, Ozasa K, et al. Dietary fiber and risk of colorectal cancer in the Japan collaborative cohort study. Cancer Epidemiol Biomarkers Prev. 2007;16:668–75.
102. Baron JA. Dietary fiber and colorectal cancer: an ongoing saga. JAMA. 2005;294:2904–6.
103. Ulrich CM. Folate and cancer prevention: a closer look at a complex picture. Am J Clin Nutr. 2007;86:271–3.

104. Kim YI. Folate: a magic bullet or a double edged sword for colorectal cancer prevention? Gut. 2006;55:1387–9.
105. Mason JB, Dickstein A, Jacques PF, Haggarty P, Selhub J, Dallal G, Rosenberg IH. A temporal association between folic acid fortification and an increase in colorectal cancer rates may be illuminating important biological principles: a hypothesis. Cancer Epidemiol Biomarkers Prev. 2007;16:1325–9.
106. Osterhues A, Holzgreve W, Michels KB. Shall we put the world on folate? Lancet. 2009;374:959–61.
107. McGlynn AP, Wasson GR, O'Reilly SL, McNulty H, Downes CS, Chang CK, Hoey L, Molloy AM, Ward M, Strain JJ, McKerr G, Weir DG, et al. Low colonocyte folate is associated with uracil misincorporation and global DNA hypomethylation in human colorectum. J Nutr;143:27–33.
108. Winawer SJ. Natural history of colorectal cancer. Am J Med 1999;106:3S-6S (discussion 50S-1S).
109. Kim DH, Smith-Warner SA, Spiegelman D, Yaun SS, Colditz GA, Freudenheim JL, Giovannucci E, Goldbohm RA, Graham S, Harnack L, Jacobs EJ, Leitzmann M, et al. Pooled analyses of 13 prospective cohort studies on folate intake and colon cancer. Cancer Causes Control;21:1919–30.
110. Lee JE, Willett WC, Fuchs CS, Smith-Warner SA, Wu K, Ma J, Giovannucci E. Folate intake and risk of colorectal cancer and adenoma: modification by time. Am J Clin Nutr. 2011;93:817–25.
111. Gao QY, Chen HM, Chen YX, Wang YC, Wang ZH, Tang JT, Ge ZZ, Chen XY, Sheng JQ, Fang DC, Yu CG, Zheng P, et al. Folic acid prevents the initial occurrence of sporadic colorectal adenoma in Chinese older than 50 years of age: a randomized clinical trial. Cancer Prev Res. 2013;6:744–52.
112. Vollset SE, Clarke R, Lewington S, Ebbing M, Halsey J, Lonn E, Armitage J, Manson JE, Hankey GJ, Spence JD, Galan P, Bonaa KH, et al. Effects of folic acid supplementation on overall and site-specific cancer incidence during the randomised trials: meta-analyses of data on 50 000 individuals. Lancet. 2013;381:1029–36.
113. Wargovich MJ, Eng VW, Newmark HL. Calcium inhibits the damaging and compensatory proliferative effects of fatty acids on mouse colon epithelium. Cancer Lett. 1984;23:253–8.
114. Newmark HL, Wargovich MJ, Bruce WR. Colon cancer and dietary fat, phosphate, and calcium: a hypothesis. J Natl Cancer Inst. 1984;72:1323–5.
115. Lamprecht SA, Lipkin M. Cellular mechanisms of calcium and vitamin D in the inhibition of colorectal carcinogenesis. Ann N Y Acad Sci. 2001;952:73–87.
116. Buset M, Lipkin M, Winawer S, Swaroop S, Friedman E. Inhibition of human colonic epithelial cell proliferation in vivo and in vitro by calcium. Cancer Res. 1986;46:5426–30.
117. Ahearn TU, Shaukat A, Flanders WD, Rutherford RE, Bostick RM. A randomized clinical trial of the effects of supplemental calcium and vitamin D3 on the APC/beta-catenin pathway in the normal mucosa of colorectal adenoma patients. Cancer Prev Res (Phila). 2012;5:1247–56.
118. Shaukat A, Scouras N, Schunemann HJ. Role of supplemental calcium in the recurrence of colorectal adenomas: a metaanalysis of randomized controlled trials. Am J Gastroenterol. 2005;100:390–4.
119. Wallace K, Baron JA, Cole BF, Sandler RS, Karagas MR, Beach MA, Haile RW, Burke CA, Pearson LH, Mandel JS, Rothstein R, Snover DC. Effect of calcium supplementation on the risk of large bowel polyps. J Natl Cancer Inst. 2004;96:921–5.
120. Bristow SM, Bolland MJ, Maclennan GS, Avenell A, Grey A, Gamble GD, Reid IR. Calcium supplements and cancer risk: a meta-analysis of randomised controlled trials. Br J Nutr. 2013;110:1384–93.
121. Cho E, Smith-Warner SA, Spiegelman D, Beeson WL, van den Brandt PA, Colditz GA, Folsom AR, Fraser GE, Freudenheim JL, Giovannucci E, Goldbohm RA, Graham S, et al. Dairy foods, calcium, and colorectal cancer: a pooled analysis of 10 cohort studies. J Natl Cancer Inst. 2004;96:1015–22.
122. Keum N, Aune D, Greenwood DC, Ju W, Giovannucci EL. Calcium intake and colorectal cancer risk: dose-response meta-analysis of prospective observational studies. Int J Cancer. 2014;135:1940–8.
123. Giovannucci E. Epidemiology of vitamin D and colorectal cancer. Anticancer Agents Med Chem. 2013;13:11–9.
124. Garland CF, Garland FC. Do sunlight and vitamin D reduce the likelihood of colon cancer? Int J Epidemiol. 1980;9:227–31.
125. Feldman D, Krishnan AV, Swami S, Giovannucci E, Feldman BJ. The role of vitamin D in reducing cancer risk and progression. Nat Rev Cancer. 2014;14:342–57.
126. Giovannucci E. Vitamin D and cancer incidence in the Harvard cohorts. Ann Epidemiol. 2009;19:84–8.
127. Jenab M, Bueno-de-Mesquita HB, Ferrari P, van Duijnhoven FJ, Norat T, Pischon T, Jansen EH, Slimani N, Byrnes G, Rinaldi S, Tjonneland A, Olsen A, et al. Association between pre-diagnostic circulating vitamin D concentration and risk of colorectal cancer in European populations: a nested case-control study. BMJ. 2010;340:b5500.
128. Fedirko V, Riboli E, Tjonneland A, Ferrari P, Olsen A, Bueno-de-Mesquita HB, van Duijnhoven FJ, Norat T, Jansen EH, Dahm CC, Overvad K, Boutron-Ruault MC, et al. Prediagnostic 25-hydroxyvitamin D, VDR and CASR polymorphisms, and survival in patients with colorectal cancer in western European ppulations. Cancer Epidemiol Biomarkers Prev. 2012;21:582–93.
129. Holick MF, Binkley NC, Bischoff-Ferrari HA, Gordon CM, Hanley DA, Heaney RP, Murad MH, Weaver CM, Endocrine S. Evaluation, treatment, and prevention of vitamin D deficiency: an

Endocrine Society clinical practice guideline. J Endocrinol Metab. 2011;96:1911–30.
130. Cook NR, Lee IM, Zhang SMM, Moorthy MV, Buring JE. Alternate-day, low-dose aspirin and cancer risk: long-term observational follow-up of a randomized trial. Ann Intern Med. 2013;159:77.
131. Rothwell PM, Wilson M, Elwin CE, Norrving B, Algra A, Warlow CP, Meade TW. Long-term effect of aspirin on colorectal cancer incidence and mortality: 20-year follow-up of five randomised trials. Lancet. 2010;376:1741–50.
132. Chan AT, Giovannucci EL, Meyerhardt JA, Schernhammer ES, Curhan GC, Fuchs CS. Long-term use of aspirin and nonsteroidal anti-inflammatory drugs and risk of colorectal cancer. JAMA. 2005;294:914–23.
133. Cook NR, Lee IM, Gaziano JM, Gordon D, Ridker PM, Manson JE, Hennekens CH, Buring JE. Low-dose aspirin in the primary prevention of cancer the women's health study: a randomized controlled trial. JAMA. 2005;294:47–55.
134. Cole BF, Logan RF, Halabi S, Benamouzig R, Sandler RS, Grainge MJ, Chaussade S, Baron JA. Aspirin for the chemoprevention of colorectal adenomas: meta-analysis of the randomized trials. J Natl Cancer Inst. 2009;101:256–66.
135. Sostres C, Gargallo CJ, Lanas A. Aspirin, cyclooxygenase inhibition and colorectal cancer. World J Gastrointest Pharmacol Ther. 2014;5:40–9.
136. Rajakariar R, Yaqoob MM, Gilroy DW. COX-2 in inflammation and resolution. Mol Interv. 2006;6:199–207.
137. Chan AT, Ogino S, Fuchs CS. Aspirin and the risk of colorectal cancer in relation to the expression of COX-2. N Engl J Med. 2007;356:2131–42.
138. Manson JE, Chlebowski RT, Stefanick ML, Aragaki AK, Rossouw JE, Prentice RL, Anderson G, Howard BV, Thomson CA, LaCroix AZ, Wactawski-Wende J, Jackson RD, et al. Menopausal hormone therapy and health outcomes during the intervention and extended poststopping phases of the Women's Health Initiative randomized trials. JAMA. 2013;310:1353–68.
139. Lin KJ, Cheung WY, Lai JY, Giovannucci EL. The effect of estrogen vs. combined estrogen-progestogen therapy on the risk of colorectal cancer. Int J Cancer. 2012;130:419–30.
140. Gunter MJ, Hoover DR, Yu H, Wassertheil-Smoller S, Rohan TE, Manson JE, Howard BV, Wylie-Rosett J, Anderson GL, Ho GY, Kaplan RC, Li J, et al. Insulin, insulin-like growth factor-I, endogenous estradiol, and risk of colorectal cancer in postmenopausal women. Cancer Res. 2008;68:329–37.
141. Clendenen TV, Koenig KL, Shore RE, Levitz M, Arslan AA, Zeleniuch-Jacquotte A. Postmenopausal levels of endogenous sex hormones and risk of colorectal cancer. Cancer Epidemiol Biomarkers Prev. 2009;18:275–81.
142. Lin JH, Zhang SMM, Rexrode KM, Manson JE, Chan AT, Wu KN, Tworoger SS, Hankinson SE, Fuchs C, Gaziano JM, Buring JE, Giovannucci E. Association between sex hormones and colorectal cancer risk in men and women. Clin Gastroenterol Hepatol. 2013;11:419–24.
143. Campagnoli C, Biglia N, Altare F, Lanza MG, Lesca L, Cantamessa C, Peris C, Fiorucci GC, Sismondi P. Differential effects of oral conjugated estrogens and transdermal estradiol on insulin-like growth factor 1, growth hormone and sex hormone binding globulin serum levels. Gynecol Endocrinol. 1993;7:251–8.
144. Barzi A, Lenz AM, Labonte MJ, Lenz HJ. Molecular pathways: estrogen pathway in colorectal cancer. Clin Cancer Res. 2013;19:5842–8.
145. Edwards BK, Ward E, Kohler BA, Eheman C, Zauber AG, Anderson RN, Jemal A, Schymura MJ, Lansdorp-Vogelaar I, Seeff LC, van Ballegooijen M, Goede SL, et al. Annual report to the nation on the status of cancer, 1975–2006, featuring colorectal cancer trends and impact of interventions (risk factors, screening, and treatment) to reduce future rates. Cancer. 2010;116:544–73.
146. Schoen RE, Pinsky PF, Weissfeld JL, Yokochi LA, Church T, Laiyemo AO, Bresalier R, Andriole GL, Buys SS, Crawford ED, Fouad MN, Isaacs C, et al. Colorectal-cancer incidence and mortality with screening flexible sigmoidoscopy. N Engl J Med;366:2345–57.
147. Giovannucci E, Wu K. Cancers of the colon and rectum. In: Schottenfeld D, Fraumeni JF Jr, editors. Cancer epidemiology and prevention. 3rd ed. New York: Oxford University Press;2006. p. 809–29.
148. Brenner H, Stock C, Hoffmeister M. Effect of screening sigmoidoscopy and screening colonoscopy on colorectal cancer incidence and mortality: systematic review and meta-analysis of randomised controlled trials and observational studies. BMJ. 2014;348:g2467.
149. American Cancer Society recommendations for colorectal cancer early detection [internet]. Updated 5 Feb 2015; cited 17 Feb 2015. Available from http://www.cancer.org/cancer/colonandrectumcancer/moreinformation/colonandrectumcancerearlydetection/colorectal-cancer-early-detection-acs-recommendations.

# 22 Pathology and Molecular Pathology of Colorectal Cancer

Emily Jean Poulin, Jeanne Shen, Jessica J. Gierut and Kevin M. Haigis

## 22.1 Introduction

The preceding chapter discussed the epidemiology of colorectal cancer (CRC), as well as the utility of established and emerging molecular markers for the diagnosis, prognosis, and prediction of response to treatment of CRC. This chapter begins with a basic introduction to the embryology, anatomy, and histology of the colon and rectum, followed by an overview of the various premalignant colorectal lesions and their molecular pathologic correlates. We will then cover the histopathology of CRC, focusing on those routinely assessed features with established prognostic or predictive value. The remainder of the chapter will be devoted to the genetics and molecular pathology of the disease. For most of the twentieth and early twenty-first centuries, CRC was diagnosed, staged, and treated as a single pathologic entity, but recent advances in our understanding of its molecular pathogenesis suggest that "colorectal cancer" might better be thought of as a heterogeneous collection of primary epithelial malignancies, each arising via distinct (although sometimes partially overlapping) pathways of tumorigenesis, many of which are associated with specific precursor lesions and clinicopathologic phenotypes.

## 22.2 Embryology and Development of the Colon and Rectum

The primary function of the colon is to absorb water and electrolytes, while that of the rectum is to provide a repository for stool. The large surface area of the colon and rectum also provides an interface for interactions between the innate and adaptive immune systems and the "outside world" represented by the luminal contents. The colon and rectum are derived from all three embryonic germ layers: the endoderm gives rise to the epithelial lining, the mesoderm gives rise to the muscle layers and connective tissue, and the ectoderm gives rise to the enteric nervous system. During the first month of embryogenesis, the three germ layers are formed, and, by the fourth week, the trilaminar germ disc has formed an elongated tube consisting of the foregut, midgut, and hindgut [1, 2]. These segments are defined by

E.J. Poulin
Beth Israel Deaconess Medical Center, 3 Blackfan Circle, CLS417, Boston, MA 02115, USA

J. Shen
Southwestern Medical Center, University of Texas, 6201 Harry Hines Blvd, RM UH 04.245, Dallas, TX 75390, USA

J.J. Gierut
Beth Israel Deaconess Medical Center, 59 Bishop Allen Dr., Apt. 3, Cambridge, MA 02139, USA

K.M. Haigis (✉)
Harvard Medical School, Cancer Research Institute, Beth Israel Deaconess Medical Center, 3 Blackfan Circle, Boston, MA 02115, USA
e-mail: khaigis@bidmc.harvard.edu


M. Loda et al. (eds.), *Pathology and Epidemiology of Cancer*,
DOI 10.1007/978-3-319-35153-7_22

their blood supply, which comes from the celiac, superior mesenteric, and inferior mesenteric arteries, respectively. The proximal part of the colon is therefore derived from the midgut, whereas the remainder of the colon and the rectum are derived from the hindgut. The formation and proliferation of the crypts that line the luminal surface of the colon and rectum appear to involve Wnt and Bone Morphogenetic Protein (BMP) signaling pathways, while crypt epithelial differentiation involves the Notch pathway [3].

The pathways that cooperate to promote colonic development are now being exploited to generate experimentally tractable in vitro systems. Intestine-like tissue can be generated from human embryonic stem (ES) and induced pluripotent stem (iPS) cells following differentiation into definitive endoderm and treatment with Wnt3A and FGF4 [4, 5]. This method results in hindgut differentiation and morphogenesis into gut spheroids, which resemble the embryonic gut. When placed in a specialized 3D culture system, spheroids mature and form organoids, structures composed of crypt and villus-like units that contain all of the differentiated intestinal cell types. Organoids can also be generated from multiple primary tissue types, including isolated crypts from human and mouse intestinal tissue, as well as from purified intestinal stem cells [6, 7].

## 22.3 Anatomy of the Colon and Rectum

The average lengths of the colon and rectum in the normal adult are 80–110 cm and 10–15 cm, respectively [8, 9]. The colon begins at the ileocecal valve and, proximally to distally, is comprised of the following segments: the cecum, ascending, transverse, descending, and sigmoid colon (Fig. 22.1). The transitions from the ascending to transverse, and from the transverse

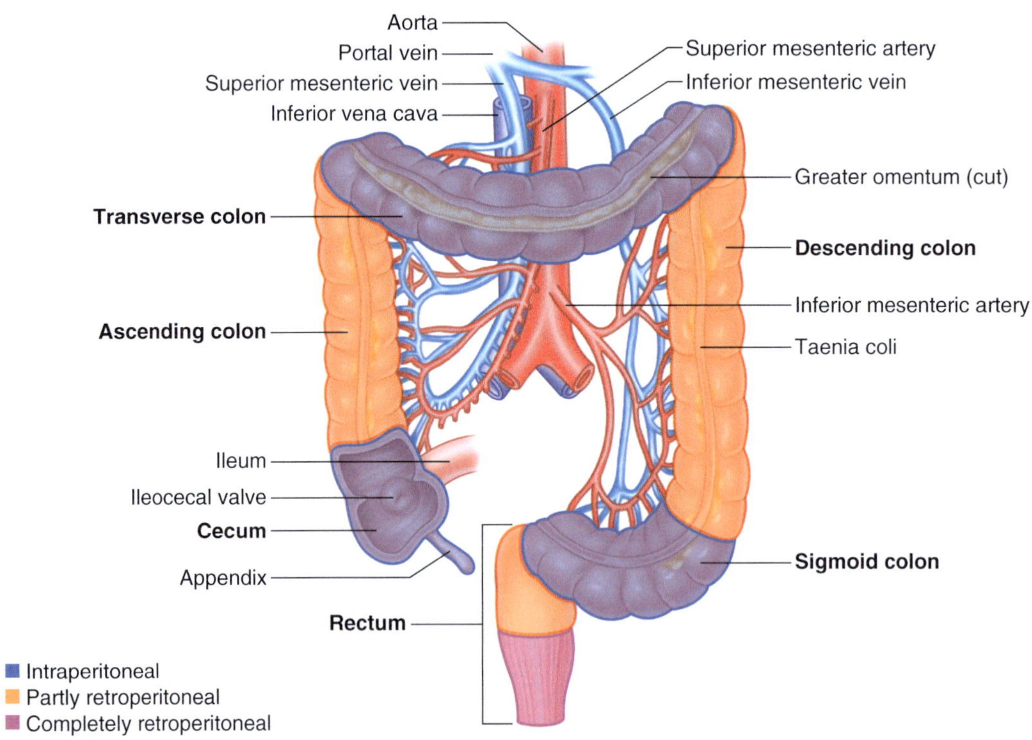


**Fig. 22.1** Anatomy of the colon and rectum with segments labeled. Peritonealized and non-peritonealized segments are shaded different colors

to descending colon, occur at the hepatic and splenic flexures, respectively. The appendix arises from the cecum, below the level of the ileocecal valve. Three bands of muscle (the taenia coli) run longitudinally along the entire outer surface of the colon, converging to form a circumferential muscular coat at the junction of the sigmoid colon and rectum. The rectum extends distally to the dentate line, which marks the anatomic junction between the rectum and anus. The cecum, ascending, and proximal two-thirds of the transverse colon are collectively referred to as the right colon, while the distal one-third of the transverse colon through the sigmoid are referred to as the left colon, reflecting their respective embryologic derivations from the midgut and hindgut.

The transverse colon is suspended from a sheet of peritonealized fat, the lesser omentum, which originates along the greater curvature of the stomach. Hanging from the surface of the transverse colon is a second larger sheet of peritonealized fat, the greater omentum. When staging and assessing the adequacy of resection of CRC (particularly the status of the radial resection margin), it is important to know which anatomic segments are involved by tumor and whether a peritoneal lining covers these segments. The cecum is entirely peritonealized, while the ascending colon, hepatic and splenic flexures, and descending colon are only covered by peritoneum along their anterolateral surfaces. The transverse and sigmoid colon are almost completely surrounded by peritoneum, except at their mesenteric attachments. The upper two-thirds of the rectum is covered laterally and/or anteriorly by peritoneum, whereas the lower third is below the peritoneal reflection and therefore has no peritoneal covering.

The blood supply to the cecum, ascending, and proximal transverse colon comes from branches of the superior mesenteric artery, while branches of the inferior mesenteric artery supply the remainder of the colon and the proximal rectum (Fig. 22.1). The distal rectum is supplied by the internal iliac and pudendal arteries [8]. Venous drainage of the colon and rectum is to the portal vein via the superior and inferior mesenteric veins. Lymph nodes are located within the pericolorectal connective tissue, as well as along the branches of the mesenteric vessels. From here, lymphatic drainage converges on major nodal groups around the roots of the superior and inferior mesenteric arteries, ultimately emptying into the thoracic duct. The lymphatic drainage of the distal rectum is to the hypogastric, obturator, and internal iliac nodes via the paraaortic nodes [8].

## 22.4 Histology of the Benign Colon and Rectum

Figure 22.2 illustrates the layers of the colorectum, which, from inner (luminal) to outer surface, are the (1) mucosa, composed of the epithelium, lamina propria, and muscularis mucosae, (2) submucosa, (3) muscularis propria, with inner circular and outer longitudinal muscle layers, (4) pericolorectal adipose tissue or subserosa (subserosa being present in the peritonealized segments of colon and rectum, and (5) serosa, the outermost peritonealized wrapping of the intestine (where present). The luminal surface of the large intestine is lined by straight, nonbranching crypts oriented parallel to one another and perpendicular to the lumen, like test tubes in a rack. The epithelial cells lining these crypts are of two main types (Fig. 22.3a): (1) absorptive enterocytes, which are tall, columnar cells with basally oriented, ovoid to elongated nuclei, pale eosinophilic cytoplasmic mucin, and an apical microvillus brush border, and (2) mucus secreting goblet cells, so named because they resemble drinking goblets in shape, which are interspersed between the absorptive enterocytes and contain small, basally oriented nuclei and cytoplasm distended by pale basophilic acid mucin, which stains positive by Alcian blue at pH 2.5 (in contrast to the neutral mucin present in absorptive enterocytes, which is negative with Alcian blue and stains magenta on a periodic acid Schiff stain). The mucus secreted by goblet cells forms a protective layer against luminal pathogens.

Another cell type normally present within the epithelium, and found at the crypt base, is the

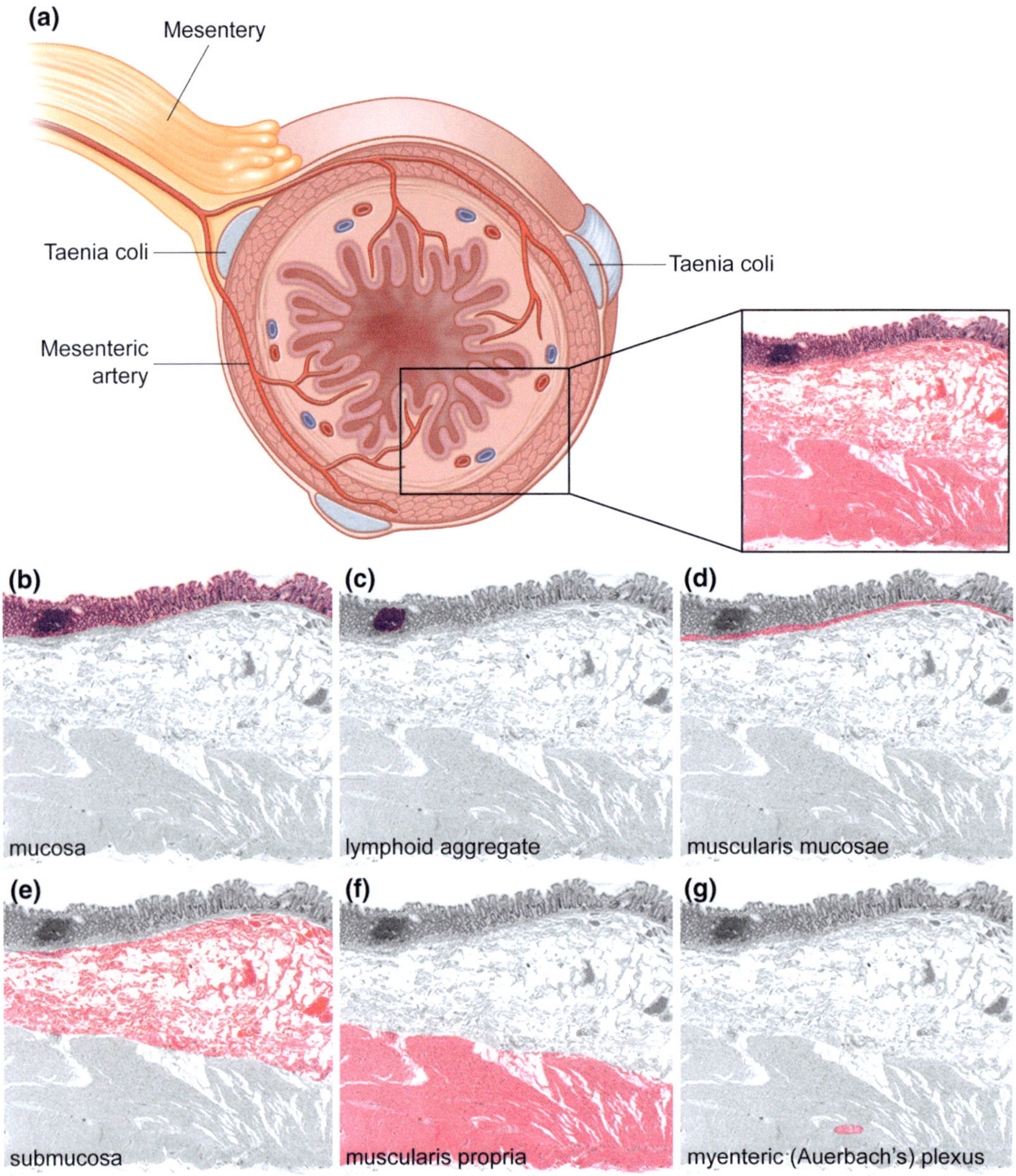


**Fig. 22.2** Histology of the colon. **a** Cross section of intestinal wall illustrating the different histologic layers. Tissue is stained with hematoxylin and eosin (H&E). **b**–**g** Histologic layers of the colon. In each panel, a single histologic portion is highlighted in color

endocrine cell, of which there are two morphologic types: (1) the enterochromaffin cell (Fig. 22.3b), which secretes serotonin and is easily identified on hematoxylin and eosin (H&E) stained sections as having a pyramidal or spindled shape, a luminally oriented nucleus, and basally oriented, reddish-orange granules, and (2) the non-enterochromaffin endocrine cell, which is not easily identified on H&E stained sections, and has a basally located nucleus and inconspicuous, apically oriented granules. Endocrine cells secrete various peptides that include somatostatin, cholecystokinin, motilin, secretin, vasoactive intestinal polypeptide (VIP), substance P, glucagon, neurotensin, and gastric

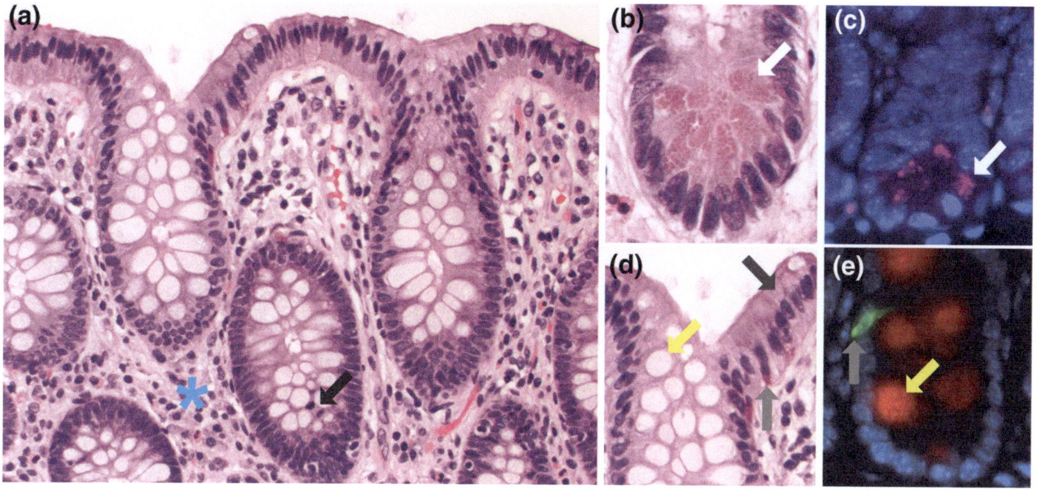


**Fig. 22.3** Cell types of the normal colorectum. **a** H&E stain of normal colon. A mitotic figure is highlighted by a *black arrow*. The *blue asterisk* is in the lamina propria. **b** High magnification view of the *bottom* of the colonic crypt. A Paneth cell is highlighted with a *white arrow*. **c** Immunofluorescence for lysozyme (*purple*), a marker of Paneth cells. The Paneth cell is highlighted with a *white arrow*. **d** High magnification view of the *top* of the colonic crypt. A goblet cell is highlighted with a *yellow arrow*. Enterocytes (*black arrow*) and enteroendocrine cells (*gray arrow*) are also highlighted. **e** Immunofluorescence for mucin 2 (*red*), a marker of goblet cells and serotonin (*green*), a marker for some enteroendocrine cells. The goblet cell is highlighted with a *yellow arrow*. The enteroendocrine cell is highlighted with a *gray arrow*

inhibitory polypeptide, among others, with differences in the distributions of these cell types depending on location in the colon and rectum [8].

Also present within the bases of the crypts are Paneth cells (Fig. 22.3c), characterized by luminally oriented eosinophilic granules that contain antimicrobial proteins (defensins). Paneth cells play an important role in the innate and adaptive immune responses against luminal bacteria, and express two main types of innate pattern recognition receptors: the transmembrane toll-like receptors (TLRs) and the cytosolic nucleotide-binding oligomerization domain-2 (NOD2) receptors, both of which recognize highly conserved pathogen-associated molecular patterns (PAMPs) [10]. Paneth cells are normally found in the right colon, but may be seen in the left colon in the setting of chronic inflammation and injury. In humans, defects in Paneth cells have been linked to diseases such as neonatal necrotizing enterocolitis and inflammatory bowel disease [11, 12].

The last differentiated cell type characteristically lining the colorectum is the microfold (or M) cell, which is often found above mucosal lymphoid aggregates and functions to transport luminal antigens into the underlying lymphoid aggregate for recognition by the immune system [13, 14].

Colonic stem cells, morphologically distinctive cells in the crypt base (often called crypt base columnar cells or CBCs), can be identified by expression of LGR5/GPR49, a G protein-coupled receptor that binds the Wnt agonist R-spondin and potentiates Wnt signaling [15–18]. Each stem cell divides to yield a daughter stem cell and a committed progenitor cell, sometimes referred to as a transit amplifying cell (TAC) due to its rapid proliferation rate. These TACs can be identified by their incorporation of bromodeoxyuridine (BrdU) or tritiated thymidine, or by Ki-67 immunostaining. They are located along the sides of the crypts, immediately above the Paneth cells. As these cells proliferate and mature, they migrate upward along the crypt wall, eventually reaching the surface, where they differentiate into absorptive enterocytes, goblet cells, endocrine cells, and M cells. Some of these cells also migrate downward toward the crypt base to become Paneth cells.

Each colorectal crypt is surrounded by a basement membrane, to which the epithelial cells are anchored. The lamina propria is a supportive connective tissue stroma in between the individual crypts, which contains capillaries, small nerves, lymphatics (mainly within the deepest part of the lamina propria), and an inflammatory infiltrate containing plasma cells, histiocytes, lymphocytes, eosinophils, mast cells, and the rare neutrophil. The muscularis mucosae forms the deepest extent of the mucosa. Mucosal lymphoid aggregates are normally present within the lamina propria, sometimes extending through the muscularis mucosae and into the submucosa.

The submucosa contains loose connective tissue, blood vessels, lymphatics, nerves, and the ganglion cells of the submucosal (Meissner's and Henle's) plexuses. Beneath this is the muscularis propria, which contains an inner circular layer and outer longitudinal layer, between which is located the myenteric (Auerbach's) plexus. The taenia coli are actually localized thickenings of this outer longitudinal muscle layer. External to the muscularis propria is pericolorectal connective tissue containing blood vessels, nerves, and a variable amount of fat. In segments of the colorectum that are covered by peritoneum, this layer is referred to as the subserosa, and the outermost layer, the serosa, is characterized by a single layer of mesothelial cells that forms the smooth outer surface of the intestine.

## 22.5 Premalignant Lesions of the Colon and Rectum

The earliest potential precursor to colorectal neoplasia is the aberrant crypt focus (ACF), a discrete focus of one or more crypts distinguishable from the surrounding normal crypts by a larger than normal crypt diameter and increased numbers of lining epithelial cells [19, 20]. ACF are invisible to the unaided eye, but may be seen under a dissecting microscope after methylene blue staining (Fig. 22.4), or by chromoendoscopy based on their unique pit patterns [21]. Multiple types of ACF exist, including non-dysplastic hyperplastic ACF and dysplastic ACF, although it is controversial whether all ACF progress to adenoma [22–24].

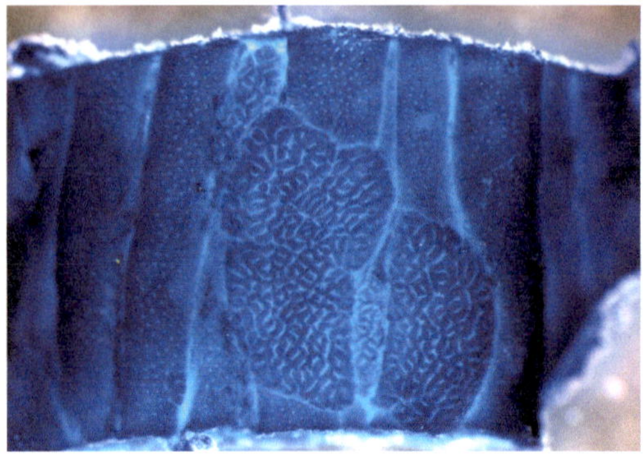

**Fig. 22.4** Wholemount methylene blue staining of an aberrant crypt focus from a mouse. From Velho and Haigis [25]

The universally accepted precursor to carcinoma is dysplasia, which is formally defined as unequivocally neoplastic epithelium that is confined to the basement membrane [26]. This is in contrast to carcinoma, which represents spread of neoplastic cells beyond the basement membrane into the surrounding lamina propria (intramucosal adenocarcinoma) or beyond. Dysplasia may be grossly or endoscopically visible, or only seen on microscopic evaluation. Visible dysplasia may take on many different appearances, from polypoid lesions to ill-defined carpet-like patches, and is classified according to the Paris classification [27]. Microscopically, dysplasia is classified into several morphologic subtypes, including conventional (intestinal type), serrated, and villous/hypermucinous or gastric type [26]. Dysplasia may arise sporadically, or as a consequence of chronic inflammation, such as that seen in inflammatory bowel disease (IBD). The standardized classification used for microscopic evaluation of dysplasia (both sporadic and IBD-associated) in the United States is the Dysplasia Morphology Study Group [28], whereas the Vienna classification is used in Asia and most of Europe [29].

### 22.5.1 Adenomas

The most common type of dysplastic lesion is the adenoma, which often takes the gross appearance of a sessile or pedunculated polyp. However, non-polypoid flat or depressed adenomas can also occur. An estimated 12 % of individuals undergoing screening colonoscopy in the United States will develop adenomas by age 50, with the prevalence increasing up to 50 % beyond age 50 [30]. Adenomas are the classical precursor in the chromosomal instability (CIN) pathway of colorectal carcinogenesis described by Vogelstein and Fearon (see section on Molecular Pathology for details), and it is well established that endoscopic removal of these lesions reduces the risk of subsequent colorectal carcinoma [31, 32].

The morphologic subtype of dysplasia seen in adenomas is often referred to as "conventional" or intestinal type dysplasia, which may be low or high grade (Fig. 22.5). In low-grade dysplasia, the tubular architecture of the crypts is generally maintained, but the epithelial cells have enlarged, pencillate (pencil-shaped), hyperchromatic nuclei arranged in a crowded or stratified pattern. Mitotic figures and apoptotic bodies may be present. High-grade dysplasia is characterized by the presence of architectural changes such as glandular crowding and cribriform or back-to-back glands. The epithelial cells in high-grade

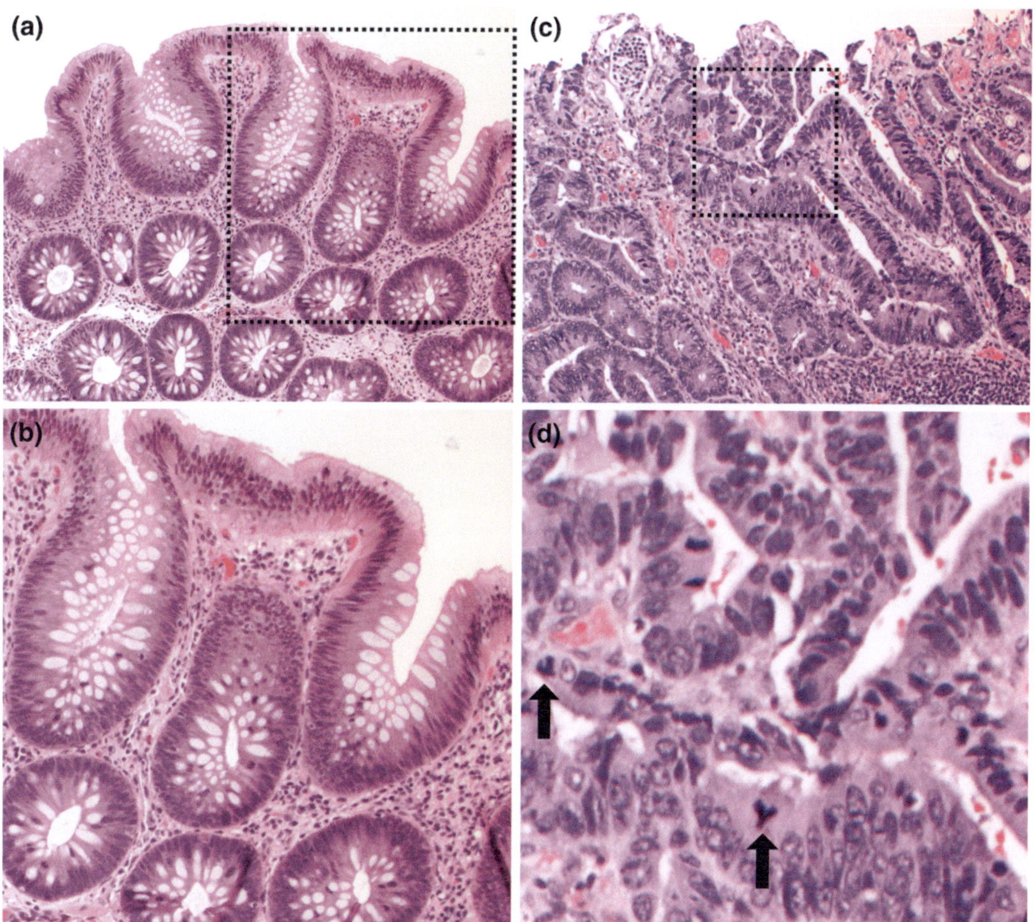


**Fig. 22.5** Colonic adenoma. **a** Low magnification of low grade dysplasia. **b** High magnification of the region outlined in *black* in panel A. **c** Low magnification of a high grade dysplasia. **d** High magnification of the region outlined in *black* in panel C

dysplasia also appear more atypical, with more nuclear enlargement, pleomorphism, prominent nucleoli, atypical mitotic figures, and loss of nuclear polarity. Dysplasia usually extends from the crypt base all the way to the surface, but may be limited to the crypt base in early cases ("crypt dysplasia"). Although the classical adenoma-carcinoma sequence is thought of as progressing sequentially from low-grade to high-grade dysplasia and then to carcinoma, some carcinomas have been observed to arise directly from low-grade dysplasia without passing through an intermediate high-grade dysplastic stage [33].

In addition to grade, adenomas are also frequently subclassified into three types based on the amount of the adenoma that exhibits a villous architecture (villi = finger or leaf-like projections). Tubular adenomas (the most common type) are 0–25 % villous, tubulovillous adenomas are 26–74 % villous, and villous adenomas are >75 % villous. The villous component appears to increase with increasing size of the adenoma [34].

Although adenomas are relatively common, most do not progress to carcinoma. Those with an increased risk of progression are referred to as "advanced adenomas," and are defined as those that are larger than 1 cm in size, contain a villous component of >25 % (i.e., tubulovillous or villous adenomas), and/or have high-grade dysplasia [35]. For adenomas that are >1 cm in size, the estimated 10-year risk of progression to carcinoma is 10–15 % [36]. The exact rate of progression from an adenoma to carcinoma is difficult to assess because most adenomas are removed at diagnosis, but it likely varies according to the degree of dysplasia and molecular background in which the adenoma arose, with one often-cited study demonstrating a 15-year interval from the detection of an adenoma to the development of carcinoma [37].

## 22.5.2 Serrated Polyps

Serrated polyps comprise a heterogeneous group of neoplastic lesions that differ with regard to their morphology, predominant location within the colon (right- versus left-sided), and molecular pathology. A common feature of serrated polyps is a serrated luminal contour, thought to result from increased cellular senescence, leading to an accumulation or "piling up" of crypt epithelial cells and a serrated (sawtooth) crypt shape [38]. According to the standardized nomenclature and classification of serrated polyps proposed by the World Health Organization (WHO) Classification of Tumours of the Digestive System [39], the three main types of serrated polyps are the: (1) hyperplastic polyp, (2) sessile serrated polyp/adenoma (SSP/A), and (3) traditional serrated adenoma (TSA).

### 22.5.2.1 Hyperplastic Polyps

Hyperplastic polyps are, by far, the most common type of serrated polyp, comprising at least 75 % of all serrated polyps, and are present in up to 35 % of screening colonoscopy patients aged 50 or older [40]. They are sessile, often small (<5 mm) polyps that are found throughout the colorectum, but occur most often on the left side. Three morphologic subtypes of hyperplastic polyp are recognized: microvesicular (MVHP), goblet cell rich (GCHP), and mucin poor (MPHP). The most common subtype, the MVHP (Fig. 22.6), is characterized by crypts that are narrow at the base, wide at the surface, and lined by enterocytes containing fine (microvesicular) mucin droplets, interspersed with variable numbers of goblet cells. The nuclei of the enterocytes are small and arranged in a single row, without stratification or hyperchromasia. On cross-sectional analysis, the crypts of the MVHP exhibit a characteristic stellate shape. The GCHP is much less common than the MVHP, and is characterized by mildly dilated crypts lined predominantly by goblet cells, with minimal to no luminal serration. The least common subtype, the MPHP, looks like a mucin-depleted MVHP, and may, in fact, represent a MVHP with reactive changes. Although all hyperplastic polyps are considered non-dysplastic, some are precursors to other serrated polyp types (SSP/A and TSA) with established premalignant potential (see below).

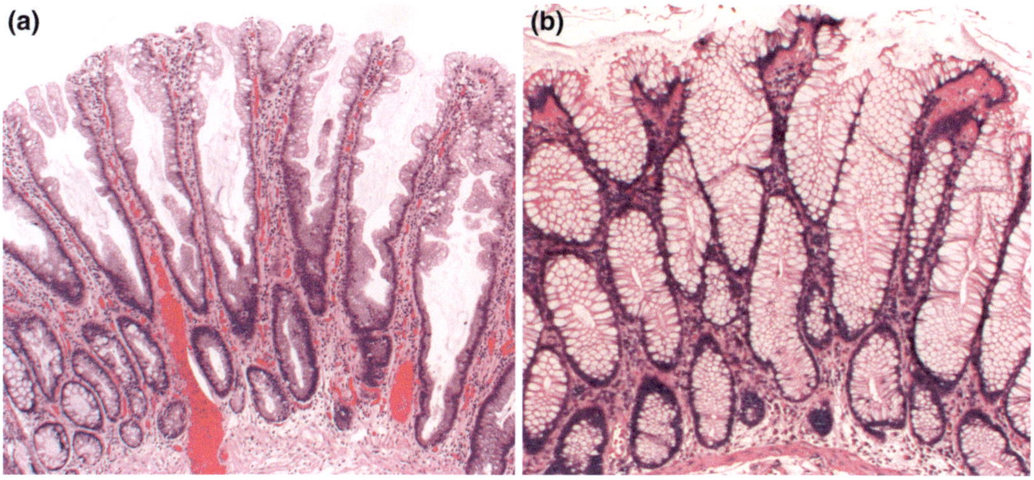


**Fig. 22.6** Hyperplastic polyps. **a** Microvesicular. **b** Goblet cell rich

#### 22.5.2.2 Sessile Serrated Polyps/Adenomas

SSP/As represent about 9–40 % of all serrated polyps and have also been referred to as "sessile serrated adenoma (SSA)" or "sessile serrated polyp (SSP)" [39, 41]. Grossly, they are sessile, usually >5 mm in size, often surfaced by a yellow mucin cap, and more commonly found in the right colon. Histologically, they resemble hyperplastic polyps, but are additionally characterized by abnormal maturation, with mature goblet cells and serration found within the crypt base, and abnormal crypt architecture, with basal dilation of crypts and a horizontal growth pattern, resulting in "boot" and "anchor"-shaped crypts (Fig. 22.7). They may contain dystrophic goblet cells with nuclei oriented toward the crypt lumen rather than toward the base. When evaluating studies of SSPs, one should be aware that there is high interobserver variability in the distinction between MVHP and SSP [42, 43]. According to a recent expert consensus panel, the presence of one or more crypts exhibiting the characteristic features of an SSP is sufficient to make the diagnosis [44].

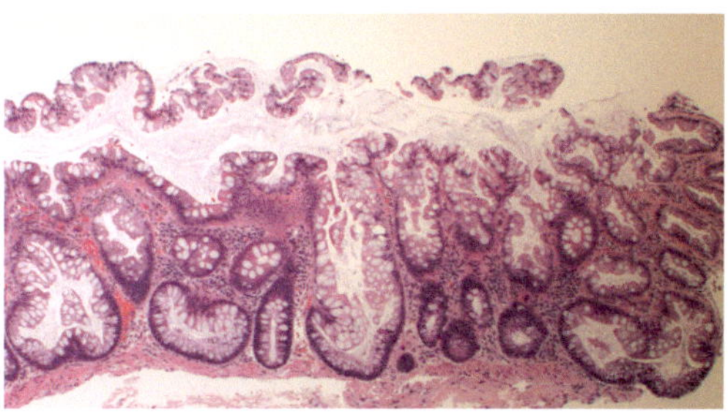

**Fig. 22.7** Sessile serrated polyp

Although SSPs are considered non-dysplastic lesions, they may also give rise to dysplasia and carcinoma, and are considered precursor lesions to the serrated pathway of colorectal carcinogenesis [45]. Given that these polyps are more common in the right colon, and are more easily missed endoscopically or during pathologic evaluation, it is not surprising that screening colonoscopy has been much less effective at preventing right-sided (relative to left-sided) colorectal cancers [46, 47]. It has been theorized that the MVHP may be a precursor of SSP, since these two types of polyps share not only a morphologic resemblance, but also activating mutations in BRAF, a serine–threonine kinase involved in cell proliferation and anti-apoptotic pathways [48]. The development of dysplasia in SSPs has been linked to methylation of the MLH1 promoter, which results in microsatellite instability (MSI, see section on Molecular Pathology for details) [49]. It is thought that progression to invasive carcinoma occurs more rapidly in SSPs with dysplasia than in conventional adenomas [39, 50]. Currently, the exact risk and time interval of progression from an SSP to carcinoma is unknown.

#### 22.5.2.3 Traditional Serrated Adenomas

The TSA is the least common type of serrated polyp, comprising <1 % of all serrated polyps, and is characterized by villiform architecture, tall, columnar epithelial cells with very eosinophilic cytoplasm, and ectopic crypt foci (small, horizontally oriented, bud-like crypts arising in the sides of villi that do not exhibit the usual anchorage to the muscularis mucosae) (Fig. 22.8). The nuclei of TSAs are ovoid or pencillate, with open chromatin and one or more small nucleoli, and do not exhibit the dense hyperchromasia or prominent nuclear stratification present in conventional adenomas. Unlike hyperplastic polyps and SSPs, TSAs are considered dysplastic lesions. Molecularly, TSAs appear to be a heterogeneous group. Based on their left-sided predominance and the presence of shared KRAS mutations, a subset of TSAs may arise from GCHPs [51, 52]. However, hyperplastic polyps and SSPs have also been proposed to be precursors to TSAs (particularly those found in the right colon), based on the presence of shared BRAF mutations [53]. Most carcinomas arising from TSAs are either microsatellite stable (MSS) or have a low degree of microsatellite instability (MSI-L) [54], and are associated with KRAS mutations and methylation of the MGMT gene [55].

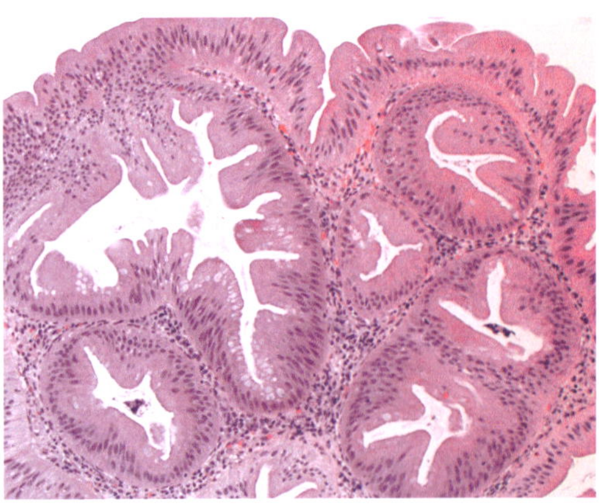

**Fig. 22.8** Traditional serrated adenoma

#### 22.5.2.4 Mixed and Unclassifiable Polyps

Although they do not comprise a formal category within the current WHO classification, serrated polyps that exhibit features consistent with more than one type of serrated polyp (mixed polyps, e.g., mixed SSP/TSA or unclassifiable polyps), or serrated polyp-like areas admixed with conventional adenomatous areas (mixed adenoma/TSA), do occur and are difficult to classify. These may reflect a yet-to-be-clarified biologic relationship between these polyp types.

### 22.5.3 Other Premalignant Lesions

Dysplasia of the conventional (intestinal) type may develop in two special polyp types, juvenile polyps and Peutz-Jeghers polyps, which are characteristic of their respective polyposis syndromes and will be described in the section on the genetics of hereditary colorectal cancer (see Sects. 22.7.5 and 22.7.6). Patients with IBD also have a well-recognized increased risk of colorectal carcinoma [56]. Dysplasia occurring in the setting of IBD has a special endoscopic-pathologic classification, which is beyond the scope of this chapter. For further information, the reader is referred to a number of excellent reviews on the subject [26, 57, 58].

## 22.6 Histopathology of Colorectal Cancer

Grossly and endoscopically, colorectal carcinoma may exhibit many different appearances, which are classified using the same Paris classification as dysplasia. Most carcinomas are sessile, centrally ulcerated lesions with irregular raised borders (Fig. 22.9). As they grow along the intestinal wall, they may involve the full circumference of the lumen, resulting in luminal narrowing and a classic "napkin ring" appearance. The cut surface

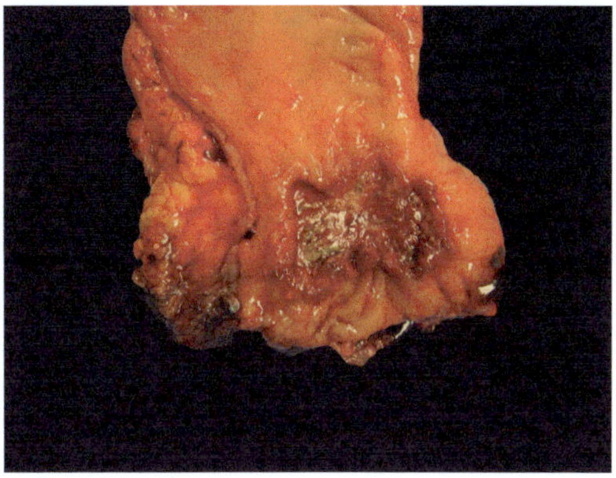

**Fig. 22.9** Gross appearance of colorectal cancer. *Image courtesy* of Dr. Jeffrey C. Perumean, University of Texas, Southwestern Medical Center, Dallas, TX

of the tumor is firm, with a solid white to yellow-tan color. Mucinous cancers may have a villous surface covered by abundant mucin, with mucin oozing from pools in the cut surface of the tumor. Invasion beyond the muscularis propria is characterized by irregular areas of induration extending from the main mass into the surrounding pericolorectal fat, and peritoneal involvement is characterized by induration and puckering of the serosa overlying the involved segment of bowel.

Many histopathologic features have been assessed for colorectal carcinoma prognostication and prediction of response to therapy. Among the features with independent prognostic and/or predictive value in multivariate analyses that are frequently reported during pathologic evaluation are the histologic grade, stage, morphologic subtype, presence of lymphovascular and extramural venous invasion, peri- and intratumoral lymphocytic response, tumor budding, surgical margin status, and certain molecular markers.

### 22.6.1 Histologic Grading

The histologic grade of a colorectal tumor reflects the degree of differentiation (based on percentage gland formation) of the tumor and correlates inversely with prognosis. Traditionally, a three-tiered system has been used, where tumors with >95 % gland formation are well differentiated, those with 50–95 % gland formation are moderately differentiated, and those with <50 % gland formation are poorly differentiated. Due to high intra- and interobserver variability using this system, the American Joint Committee on Cancer (AJCC) has adopted a two-tiered system for evaluation of resected CRC specimens. Under this two-tiered system, well- and moderately differentiated tumors (which have been shown to have similar outcomes in most multivariate analyses) fall into the low grade category, whereas poorly differentiated tumors are considered high grade [9] (Fig. 22.10).

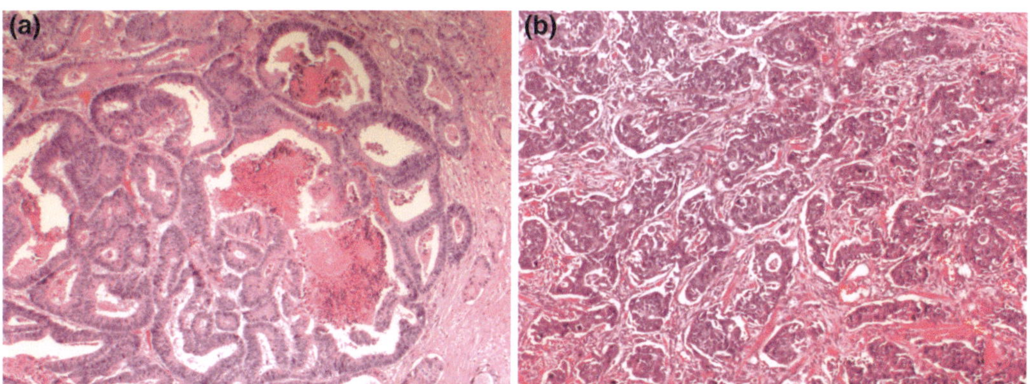


**Fig. 22.10** Conventional adenocarcinoma. **a** Low grade. **b** High grade

### 22.6.2 Histologic Spread and Staging

The histologic stage is currently the most powerful predictor of CRC prognosis. The older Dukes' staging system graded tumors from A to D. Using this staging system, stage A tumors are limited to the mucosa, stage B tumors reach into (B1) or through (B2) the muscularis propria, stage C tumors extend into (C1) or through (C2) the muscularis propria with positive lymph node involvement, and stage D tumors exhibit distant metastases [59]. The most widely used staging system today (TNM) is based on the depth of invasion into the bowel wall (T stage), as well as the presence or absence of metastasis to regional lymph nodes (N stage) and distant sites (M stage). For more detailed TNM staging information, as well as guidelines for the processing and evaluation of resected colorectal cancer specimens, the reader is referred to the most recent edition of the AJCC staging manual, as well as the College of American Pathologists' protocol for the examination of these specimens [9, 60]. Although carcinoma is defined as any spread of neoplastic cells beyond the basement membrane, in the colon and rectum the risk of local and distant metastasis is close to zero for carcinoma that is restricted to the lamina propria or muscularis mucosae (intramucosal carcinoma) [9]. Therefore, intramucosal carcinoma is assigned the same T stage as high-grade dysplasia, and both are considered "in situ" lesions treatable by endoscopic excision, whereas the term "invasive carcinoma" is used to refer only to carcinoma invasive into or beyond the submucosa, which usually warrants surgical resection of the involved segment of colon or rectum.

Intraabdominal/pelvic spread of cancer occurs after perforation of the serosa. The first site of distant metastasis via a hematogenous route is the liver, via venous invasion. Extramural venous invasion portends the highest risk, and is an independent predictor of cancer recurrence and decreased survival [61]. The use of an elastic stain to highlight the elastic lamina, or a smooth muscle stain to highlight the muscular vessel wall, may be helpful in identifying vascular invasion in difficult cases. Lymphatic spread of a tumor occurs first to the pericolonic/perirectal lymph nodes, then to the mesenteric lymph nodes, and finally to the lungs and systemic circulation via the thoracic duct. Lymphovascular invasion (LVI) is identified by tumor emboli within endothelial-lined spaces (Fig. 22.11), sometimes surrounded by a layer of endothelial cells ("endothelial wrapping"). Endothelial markers, such as CD31 and CD34, may be used to stain endothelial cells, and the lymphatic-specific endothelial marker D2–40 may be used to highlight lymphatic invasion. Tumors may also spread via perineural invasion, which is characterized by tumor within the perineural sheath surrounding nerves.

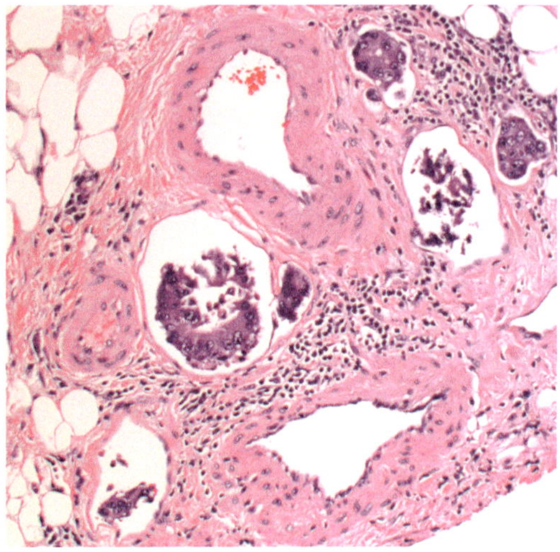

**Fig. 22.11** Lymphovascular invasion by colorectal cancer

### 22.6.3 Histologic Subtypes

Many histologic subtypes of CRC exist, and an exhaustive discussion of all of these subtypes is beyond the scope of this chapter. Therefore, the following discussion pertains only to the most common subtypes, as well as those with established molecular pathologic relevance. For information on other subtypes, the reader is referred to the WHO Classification of Tumours of the Digestive System [62].

#### 22.6.3.1 Conventional or "Usual" Type Adenocarcinoma

Conventional type adenocarcinoma is the most common histologic subtype of CRC. These account for 75–80 % of colorectal carcinomas, and are characterized by irregularly shaped, haphazardly arranged glands lined by tall columnar epithelial cells with variable amounts of intra- and extracellular mucin (Fig. 22.10). The glands may exhibit cribriform architecture and central inspissated mucus mixed with necrotic cellular debris ("dirty necrosis"). Poorly differentiated tumors may exhibit a predominantly solid growth pattern. In addition to the tall columnar enterocytes, tumors may contain variable numbers of goblet cells, Paneth cells, endocrine cells, and even benign squamous cells, melanocytes, and trophoblasts. Invasion of glands into the submucosa is typically accompanied by a stromal desmoplastic response characterized by a cellular, myxoid-appearing stroma containing plump myofibroblasts, collagen, and a variable inflammatory infiltrate, which surrounds the neoplastic glands in a streaming pattern.

#### 22.6.3.2 Mucinous Adenocarcinoma

Mucinous, also known as colloid, adenocarcinomas represent 8–10 % of CRCs and are defined by WHO criteria as adenocarcinomas where >50 % of the tumor volume is occupied by extracellular mucin (tumors where the quantity of mucin is <50 % are diagnosed as adenocarcinomas with mucinous differentiation or mucinous features). These tumors are graded the same way as conventional carcinomas, where the low-grade end of the spectrum is represented by mucin pools lined by well-differentiated, tall, columnar epithelium, and the high-grade end is represented by poorly differentiated and/or signet ring cells floating within mucin pools (Fig. 22.12). With the exception of those that have a high degree of microsatellite instability (MSI-H), mucinous carcinomas tend to present at more advanced stages, and have also been associated with a worse prognosis, independent of stage and other prognostic factors [63, 64]. The molecular basis for the mucinous phenotype has been linked to the expression of the intestinal epithelial transcription factor, MATH1, which activates MUC2 expression [65]. In addition, the transforming growth factor beta (TGFβ) pathway has been associated with the mucinous phenotype. SMAD4 mutation is associated with mucinous CRC in humans [66] and mice [67], and loss of Tgfbr2 and Pten results in development of mucinous adenocarcinoma in genetically engineered mice [68].

#### 22.6.3.3 Signet Ring Cell Adenocarcinoma

Signet ring cell adenocarcinoma comprises up to 1 % of all CRCs, and is defined by the presence

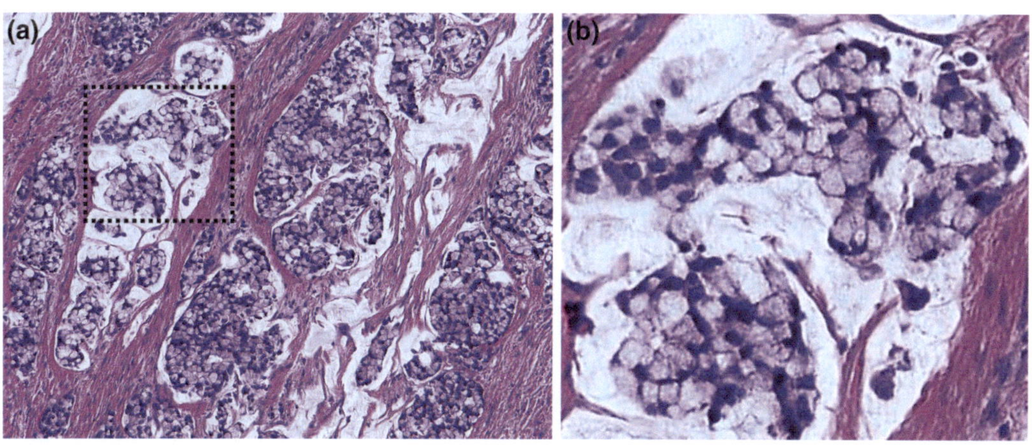


**Fig. 22.12** Mucinous adenocarcinoma with signet ring cells. **a** Low magnification. **b** High magnification of region outlined in panel A

of at least 50 % signet ring cells within the tumor. Grossly, it may appear to be a well-defined mass or, if diffusely infiltrative, an ill-defined area of mural thickening or stricturing. Microscopically, it is characterized by single infiltrating cells and clusters of cells, often with a large cytoplasmic vacuole containing mucin or other cytoplasmic contents, which gives the cell a round-to-ovoid shape and indents the nucleus, forming a profile that resembles a signet ring (Fig. 22.12). When an adenocarcinoma contains >50 % extracellular mucin and signet ring cells, it should be classified as a signet ring cell carcinoma. By definition, signet ring cell carcinoma is considered high grade, and, except when MSI-H, has been associated with a poor prognosis, with a 5-year survival rate of <10 % [69, 70].

#### 22.6.3.4 Medullary Carcinoma

Also referred to as lymphoepithelioma-like carcinoma and large cell minimally differentiated carcinoma, medullary carcinoma is the subtype most classically associated with MSI-H, and may occur in the setting of both hereditary (Lynch syndrome, usually with MSH2 mutations) and sporadic colorectal cancer (MLH1 promoter methylation) [71]. Medullary carcinomas are associated with right-sided location, female gender (usually older females), lower incidence of lymph node metastasis, and typically have a good prognosis [72–74]. Morphologically, these are poorly differentiated or undifferentiated tumors composed of solid sheets, nests, or trabeculae of polygonal tumor cells with vesicular chromatin, prominent nucleoli, and abundant cytoplasm with a syncytial appearance. These are associated with a prominent peritumoral and intratumoral lymphocytic infiltrate [72] (Fig. 22.13). These tumors are well circumscribed, with a broad-front pushing (as opposed to an infiltrative) growth pattern. Unlike typical colorectal carcinomas, which are $CK7^-$, $CK20^+$, and $CDX\text{-}2^+$ by immunohistochemistry, medullary carcinomas tend to be CK7 variably positive, $CK20^-$, and $CDX\text{-}2^-$ or only weakly positive [71], and are positive for the mesothelial marker Calretinin [74].

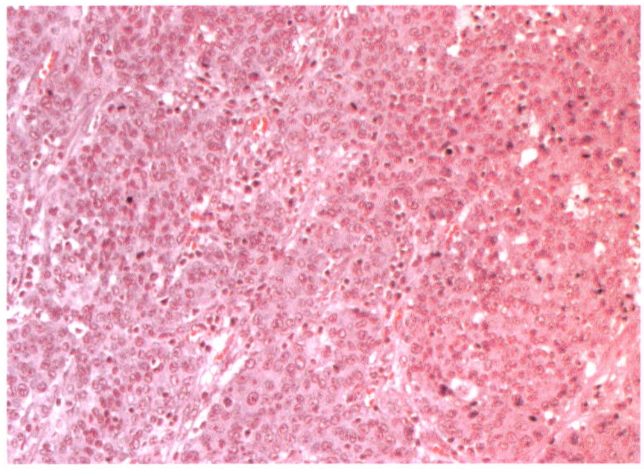

**Fig. 22.13** Medullary carcinoma

#### 22.6.3.5 MSI-H Adenocarcinoma

Approximately 10–15 % of colorectal cancers are MSI-H [75], and, although these tumors do not comprise a single distinct morphologic subtype, they nonetheless have been shown to share some common histopathologic features. MSI-H CRC may be any one of the previously discussed morphologic subtypes (conventional, mucinous, medullary, and signet ring cell), are frequently right-sided, polypoid, exophytic masses that lack the “dirty necrosis” typical of most colorectal carcinomas, and have in common the presence of tumor-infiltrating lymphocytes (TILs), the most sensitive and specific morphologic marker, to date, of MSI-H status [76–78]. TILs are $CD3^+$ $CD8^+$ cytotoxic T cells that infiltrate between epithelial cells within the tumor. According to the current (2016 edition) CAP criteria, a colorectal carcinoma is positive for TILs if it contains at least three TILs per high-power field (HPF). Other features that have been associated with MSI-H status include a peritumoral lymphocytic infiltrate, which surrounds, but does not infiltrate, malignant glands and cell clusters, and may form nodular aggregates at the deep periphery of the tumor, resembling the lymphoid aggregates seen in Crohn’s disease (peritumoral “Crohn’s-like” infiltrate). Poor tumor differentiation and a mucinous phenotype have also been associated with MSI-H tumors [79, 80].

As an aside, tumors with a CpG island methylation phenotype ($CIMP^+$, see Molecular

Pathology section for details) may be either MSI-H or not. Those that are MSI-H share the same features as other MSI-H tumors. Those that are not MSI-H are more often right-sided and poorly differentiated but, unlike MSI-H tumors, do exhibit "dirty necrosis" and tend to lack TILs or a peritumoral lymphocytic response [81].

As a group, patients with MSI-H tumors have better overall survival compared to non-MSI-H tumors, when controlled for stage as well as multiple other prognostic factors [82–85]. This may in part reflect a stronger antitumor immune response, as evidenced by the prominent peri- and intratumoral lymphocytic infiltrates seen in these tumors. Because the commonly used chemotherapeutic agents 5-FU and oxaliplatin require a functional DNA mismatch repair (MMR) system in order to promote tumor cell apoptosis, MSI-H tumors are less responsive to these agents, but have been shown to be sensitive to the topoisomerase inhibitor irinotecan, which induces double-stranded DNA breaks. In fact, current National Comprehensive Cancer Network guidelines state that adjuvant 5-FU therapy should not be offered to stage II colorectal cancer patients with MSI-H cancers, as it is not only of little additional benefit (since these patients have a very good prognosis to begin with), but may also cause harm [86].

#### 22.6.3.6 Serrated Adenocarcinoma

"Serrated adenocarcinoma" is not a distinct subtype in the current WHO classification, and has been used to refer to both adenocarcinomas with a serrated morphology and adenocarcinomas arising via the serrated pathway of carcinogenesis. Morphologically, these tumors often exhibit a serrated luminal contour to their glands and frequently produce mucin. Epithelial cells within these tumors are characterized by eosinophilic or clear cytoplasm, and vesicular, ovoid nuclei with a peripheral rim of condensed chromatin (Fig. 22.14). Serrated carcinomas usually lack the "dirty necrosis" of conventional colorectal carcinoma [87, 88]. Clinicopathologically and molecularly, these comprise at least two distinct subgroups: those that are MSI-H right-sided cancers arising in SSPs, and those that are MSS/MSI-L left-sided cancers arising in TSAs [87].

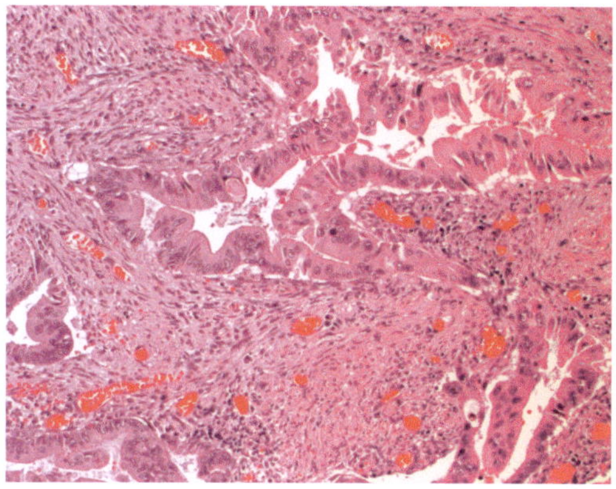

**Fig. 22.14** Serrated adenocarcinoma

### 22.6.4 Other Prognostic Histopathologic Markers

Tumor budding, which is thought to represent dedifferentiation to a more stem cell-like phenotype (often referred to as epithelial–mesenchymal transition), has been independently associated with a worse prognosis [89]. Although this is currently a controversial area due to the lack of standardized criteria or grading systems, the generally accepted definition of tumor budding is the presence of single epithelial tumor cells or clusters of <5 cells lacking well-formed glandular architecture, which are present at the deep invasive front of a tumor and associated with a characteristic desmoplastic reaction (Fig. 22.15) [90]. There is currently no standardized threshold for the degree of budding that is prognostically significant, with one study defining high-grade tumor budding as the presence of >1 bud in 5 (20x) HPF, or >10 tumor buds in 10 HPF [91]. Regardless, the presence of tumor budding has become increasingly incorporated as a routine component in pathology reports.

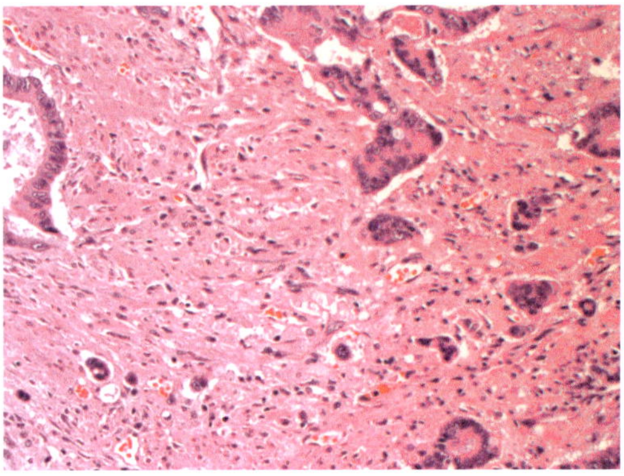

**Fig. 22.15** Tumor budding

The antitumor immune response, particularly the lymphocyte-mediated component, has gained prominence in recent years due to its association with MSI-H colorectal cancer and an improved prognosis in multivariate analyses [92, 93]. Components of the lymphocytic antitumor immune response include TILs, peritumoral lymphocytes and lymphoid aggregates, and reactive regional lymph nodes. It has been hypothesized that, in MSI-H tumors, the T-cell mediated response is directed against frameshift-induced neopeptides [94]. Although several grading systems for the degree of antitumor lymphocytic response have been proposed, there is currently no standardized, clinically validated grading system, although one will likely emerge in the future.

The concept of cancer-initiating stem cells, which are characterized by the ability to self-renew and differentiate into multi-lineage progenitors, has gained prominence over the last two decades due to their observed resistance to therapy and purported role as the cells that initiate and drive tumorigenesis and recurrence [95–97]. Although several potential stem cell markers have been identified in colorectal cancer (CD133, LGR5, CD166, CD24, MSI-1, BMI, Aldehyde dehydrogenase, and DCM kinase-like II) [98], none of these currently have established value as prognostic markers or therapeutic targets, and this is an area requiring further investigation.

## 22.7 Genetics of Hereditary Colorectal Cancer

A number of hereditary syndromes are associated with an increased risk of CRC and will be discussed in the following section. Two other syndromes, Cowden syndrome and Cronkhite–Canada syndrome, are associated with the development of colorectal polyps; for a discussion of the former, the reader is referred to the chapters on ovarian/uterine and breast cancer. The latter syndrome does not have an established association with CRC, and will not be discussed further.

### 22.7.1 Familial Adenomatous Polyposis (FAP) and Variants

One of the best-known familial CRC syndromes is familial adenomatous polyposis (FAP), an autosomal dominant condition characterized by germline mutations in the Adenomatous Polyposis Coli (APC) gene on chromosome 5q. Over 800 disease-associated mutations in APC have been identified, >90 % of which are nonsense or frameshift mutations leading to a truncated protein product [99]. Many of these mutations have specific phenotypic correlations. Patients develop hundreds (and usually thousands) of colorectal adenomas within the first two decades of life that eventually progress to CRC (on average, by age 45) unless the entire colorectum is prophylactically resected. Additionally, patients with FAP may develop a number of extracolorectal manifestations, including desmoid tumors, small intestinal adenomas, dysplastic gastric fundic gland polyps, jaw osteomas, congenital hypertrophy of the retinal pigment epithelium (CHRPE), epidermal cysts, multiple endocrine neoplasia (MEN), pancreatic adenocarcinoma, hepatoblastoma, extrahepatic cholangiocarcinoma, and papillary thyroid carcinoma [100]. Gardner's syndrome is an FAP variant with extraintestinal manifestations, particularly desmoid tumors and osteomas. One of the two variants of Turcot's syndrome is another variant of FAP, where patients also present with medulloblastoma. Attenuated FAP (AFAP) is a milder variant of FAP characterized by the presence of <100 adenomas, a milder phenotype, and later clinical presentation.

### 22.7.2 MUTYH-Associated Polyposis (MAP)

MUTYH-associated polyposis (MAP) is an autosomal recessive colorectal polyposis syndrome characterized by biallelic germline mutations in the MUTYH gene on chromosome 1p34.1, which encodes a glycosylase involved in DNA repair after oxidative damage [101]. Defective MUTYH function causes an increased rate of G > T transversions in tumor suppressors and oncogenes, with the most common target being APC. Patients have a variable number of polyps (including adenomas and serrated polyps) and an increased risk of CRC. The most common phenotype resembles that seen in AFAP patients, with most patients having <100 polyps, some having >100, and a few having >1000 polyps. The average age at presentation is 45 years [102, 103], and 60 % of patients have CRC at initial presentation [104]. MAP may account for up to 42 % of patients with adenomatous polyposis who do not have APC mutations [105].

### 22.7.3 Lynch Syndrome

Lynch syndrome is an autosomal dominant cancer syndrome caused by germline mutations in one of four DNA MMR genes (MLH1, PMS2, MSH2, and MSH6) (discussed in more detail in Sect. 22.13). In addition, EPCAM mutations have been implicated in Lynch syndrome, as they lead to transcriptional silencing of MSH2 via promoter methylation [106]. Rarely, heritable epigenetic silencing of MLH1 or MSH2 may occur [107–109]. Lynch syndrome accounts for 10–25 % of familial colorectal cancer and 3 % of all colorectal carcinomas [110]. The vast majority of patients (~90 %) have mutations in MLH1 or MSH2 [111–113], with PMS2 and MSH6 mutations being much rarer. Mutations in another MMR gene, MSH3, have also been documented in the setting of MLH1 deficiency, but to date, no germline mutations in MSH3 have been found [114, 115]. In total, over 1000 different mutations in the MMR genes have been identified [116].

The term hereditary nonpolyposis colorectal cancer (HNPCC) is often used synonymously with Lynch syndrome, which is not entirely accurate on two counts. First, HNPCC is a clinical designation for patients with familial colorectal cancer predisposition meeting the Amsterdam criteria [117], and actually encompasses two patient groups: those with Lynch syndrome and those of currently unknown genetic cause, referred to as Familial Colorectal Cancer Type X [118]. Second, Lynch syndrome patients do also develop colonic polyps, albeit not as many as in the adenomatous polyposis syndromes (usually <15 polyps).

The lifetime risk of CRC with Lynch syndrome ranges from 10 to 53 %, depending on the MMR gene that is mutated [119], with the average age at development of CRC being 40–50 years. In approximately 20 % of patients, multiple CRCs may develop, either in a synchronous or metachronous fashion. Most Lynch syndrome patients are born with a mutation in one allele, with the second allele lost by somatic mutation, loss of heterozygosity, or epigenetic silencing. Rare patients are born with biallelic mutations and are characterized by pediatric CRC, as well as other pediatric hematologic and brain malignancies [120–122].

In addition to the increased risk of CRC, Lynch syndrome patients are also at increased risk for endometrial carcinoma (15–44 %), ovarian carcinoma, gastric and small intestinal adenocarcinoma, carcinomas of the renal pelvis and ureter, hepatocellular carcinoma, cholangiocarcinoma, sebaceous tumors, and brain tumors, among others [123–126]. Lynch syndrome patients with primary brain tumors (often glioblastoma) comprise one of the two subtypes of Turcot syndrome (the other being FAP patients with medulloblastoma). Muir–Torre syndrome is a variant of Lynch syndrome characterized by the presence of sebaceous tumors and keratoacanthomas.

#### 22.7.3.1 Routine Testing for Lynch Syndrome in Colorectal Cancer

Although there are well-established clinical criteria for identifying patients at risk for Lynch syndrome (Amsterdam II criteria [127] and the Revised Bethesda Criteria [128]), they will miss a significant number of patients [129–131]. Furthermore, although Lynch-associated CRCs often exhibit morphologic features of MSI-H tumors, none of these features are helpful in distinguishing Lynch-associated from sporadic MSI-H tumors. Therefore, the primary screening method for the detection of Lynch syndrome in CRC patients is immunohistochemistry for protein expression of components of the MMR complex. This method is >90 % sensitive and nearly 100 % specific for identifying MMR deficiency when one or more of the four MMR proteins (MLH1, PMS2, MSH2, and MSH6) are lost [132, 133]. Immunohistochemistry may rarely miss some mutations that lead to loss of function without loss of immunoreactivity (for example, when the mutation occurs away from the epitope detected by the antibody).

As MLH1/PMS2 and MSH2/MSH6 function as dimer pairs, loss of immunostaining of the dominant partner, MLH1 or MSH2, results in concurrent loss of immunostaining of the nondominant partner, PMS2 or MSH6. Conversely, when PMS2 or MSH6 are lost, there is still some retained immunostaining for MLH1 and MSH2, because the latter two proteins may form dimers with other partners, such as MLH3 [134, 135]. In most cases, the background stromal and inflammatory cells or adjacent normal crypts retain MMR protein expression, serving as positive controls when determining whether there is MMR protein loss in tumor cells (Fig. 22.16). This occurs because the allele carrying the wild-type MMR gene is only inactivated in tumor cells, allowing detection of the wild-type protein in surrounding normal tissue where the unaffected allele is still intact. An exception occurs when a patient has biallelic germline mutations, which result in loss of immunostaining in normal tissues as well.

MLH1 deficiency may be due to either Lynch syndrome or MLH1 promoter methylation, so loss of MLH1 expression by immunohistochemistry may be followed up by MLH1 promoter methylation testing by PCR, or germline testing for MLH1 mutations. Because the BRAF V600E mutation is often associated with MLH1 methylation in sporadic colorectal cancers, but almost never present in Lynch-associated colorectal cancers, testing for this mutation may be helpful in distinguishing sporadic from Lynch-associated tumors with MLH1 deficiency [136]. When MMR immunohistochemistry is inconclusive or is positive for all four MMR proteins in a patient with high clinical suspicion for Lynch syndrome, MSI testing by PCR or germline testing for MMR mutations may be useful.

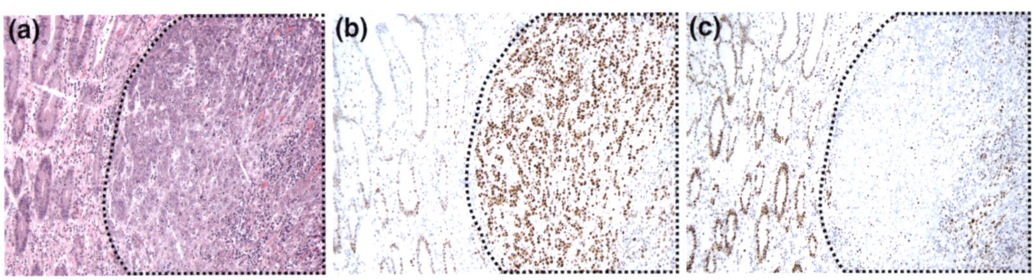


**Fig. 22.16** Mismatch repair immunohistochemistry. **a** H&E staining of the border between normal colon and CRC. **b** Immunohistochemistry for MLH1. Both tumor cells and normal cells stain positively. **c** Immunohistochemistry for MSH2. Normal cells are positive, but tumor cells are negative. Positively staining cells within the tumor mass are non-epithelial cells. In all panels, the cancer is outlined in *black*

Current guidelines recommend testing all newly diagnosed colorectal cancers for Lynch syndrome by MMR immunohistochemistry, MSI PCR, or both of these methods [137, 138].

### 22.7.4 Serrated Polyposis Syndrome (SPS)

Serrated polyposis syndrome (SPS) is a hereditary CRC syndrome of unknown molecular etiology, characterized by the presence of increased numbers of serrated polyps (either hyperplastic polyps or SSPs) and an estimated lifetime risk of developing colorectal cancer of 50 % (97). Two subtypes have been identified: Type 1 is characterized by mostly hyperplastic polyps and a low risk of CRC, while Type 2 is characterized by mostly SSPs, with or without hyperplastic polyps, and a higher risk of CRC. Approximately 33–53 % of colorectal cancers in SPS patients are thought to arise via the CIMP pathway [139–141]. Conventional adenomas may also be present in >80 % of SPS patients.

### 22.7.5 Peutz-Jeghers Syndrome

Peutz-Jeghers syndrome (PJS) is an autosomal dominant syndrome characterized by mucocutaneous pigmentation and GI tract polyps with a unique arborizing smooth muscle network. These Peutz-Jeghers polyps usually present within the first two decades of life (Fig. 22.17). Patients have an increased risk of carcinoma of the colorectum, small intestine, stomach, pancreas, esophagus, ovary, uterus, lung, and breast [142]. The syndrome is associated with germline mutations in LKB1 (also called STK11), a tumor suppressor gene encoding a serine–threonine kinase that modulates cellular energy pathways and maintains cell polarity [143]. About 15–30 % of Peutz–Jeghers syndrome patients develop dysplastic polyps and carcinoma [144, 145]. Rare occurrences of solitary sporadic Peutz–Jeghers polyps have been reported. The clinical significance of these lesions remains unclear, but removal is recommended, as these polyps may harbor dysplasia [146, 147].

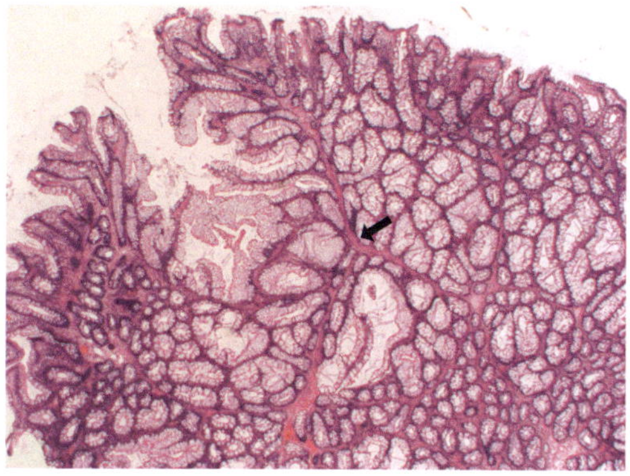

**Fig. 22.17** Peutz-Jeghers polyp

### 22.7.6 Juvenile Polyposis Syndrome (JPS)

Juvenile polyposis syndrome (JPS) is an autosomal dominant GI polyposis syndrome characterized by the presence of multiple juvenile polyps within the colorectum and proximal GI tract, including the stomach. JPS is most commonly associated with mutations in the TGFβ signaling pathway, including germline mutations in SMAD4 (15 %) and Bone Morphogenetic Protein Receptor Type 1A (BMPR1A; 25 %), which occur in 40–60 % of JPS [148, 149]. A small percentage of JPS patients (5 %) have mutations in PTEN, a negative regulator of AKT signaling [150, 151]. Juvenile polyps are a morphologically distinct type of non-neoplastic mucosal polyp characterized by disorganized, cystically dilated glands filled with mucus, surrounded by an edematous, inflamed stroma (Fig. 22.18). Although juvenile polyps may occur sporadically in children, and as solitary sporadic polyps in adults, there is no increased risk of carcinoma unless the polyps occur in the syndromic setting (for detailed clinical criteria for the diagnosis of JPS, please refer to [152]). In syndromic patients, there is an increased risk of colorectal, gastric, and pancreatic carcinoma, with an estimated 17–22 % risk of colorectal cancer by age 35 years, which increases to 68 %

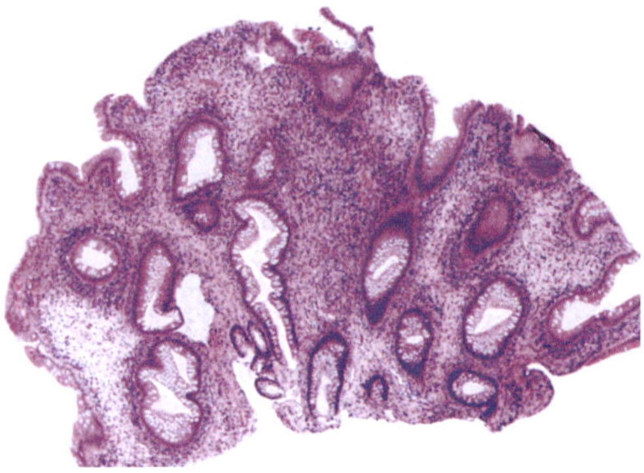

**Fig. 22.18** Juvenile polyp

by age 60 [142]. Carcinogenesis occurs by the adenoma-carcinoma pathway in juvenile polyps that develop conventional dysplasia.

### 22.7.7 Mixed Polyposis Syndrome

Mixed polyposis syndrome is an autosomal dominant syndrome characterized by the presence of "mixed" polyps with adenomatous, juvenile, and hyperplastic features, and an increased risk of CRC [153]. It is currently unclear whether this represents a variant of juvenile polyposis syndrome or serrated polyposis syndrome, but associations with BMPR1A and GREM-1 mutations have been described [154–156].

## 22.8 The Molecular Pathology of Colorectal Cancer

CRC presents as sporadic, inherited, or familial disease. Sporadic disease, in which there is no family history, accounts for approximately 70 % of all CRC cases. This pattern of disease derives from stochastic somatic mutations that are driven by both dietary and environmental risk factors. Inherited disease (described above) accounts for only 5 % of CRC cases and typically presents with well-characterized predisposing mutations. These cases are subdivided based on whether the patients display colonic polyps, as in familial adenomatous polyposis (FAP) or MUTYH-associated polyposis (MAP), or those without polyps (Lynch syndrome). The third, and least understood, pattern is familial CRC, which accounts for 25 % of cases. Affected patients have a family history of CRC, similar to patients with inherited disease, but the genetic etiology has not been defined. The molecular pathogenesis of CRC has been extensively studied and will be described in detail in this chapter. CRC can develop via one of two major molecular pathways of tumorigenesis: CIN or MSI. These CRC subclasses have different mutational spectra and are therefore likely to require distinct therapeutic strategies as targeted therapies become available.

## 22.9 Initiation of CRC Tumorigenesis

Specific genetic mutations are thought to drive the transformation from normal colonic epithelium to invasive cancer. In 1990, Fearon and Vogelstein described a multistep genetic model of colorectal carcinogenesis in which each accumulated genetic alteration contributed to the progression from normal epithelium to adenoma to invasive cancer (Fig. 22.19) [157]. Although not the complete story, this model has become the basis for much of our understanding of how CRC tumorigenesis progresses. It was quickly observed that cancer cells contained a high incidence of mutations compared to normal cells, which contained very few mutations. This observation led to the idea that one of the initial steps in carcinogenesis occurs when cancer cells acquire mutations in genes that maintain genomic instability, which facilitates an increased mutation rate and thereby, a "mutator phenotype" [158]. In contrast, a second theory suggests that tumor initiation depends less upon increased mutational rate than on positive clonal selection driven by an advantageous mutation [159]. In this case, a normal mutation rate is sufficient for tumor initiation if positive selection is accounted for [159, 160]. Recent work suggests that since normal somatic cells acquire mutations at the same rate [161], the risk of cancer developing in an organ is associated with the number of lifetime stem cell divisions that occur within a given

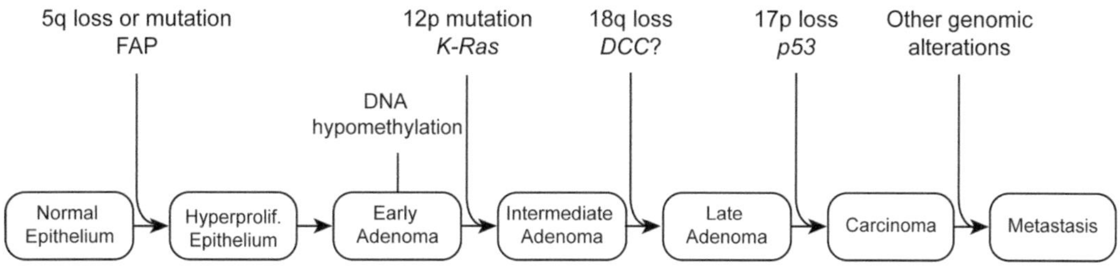


**Fig. 22.19** Original genetic model of CRC development as proposed by Fearon and Vogelstein in 1990. Adapted from Fearon and Vogelstein [163]

tissue [162]. Based on this theory, mutations accumulate randomly by chance during DNA replication. Since the colonic epithelium is continuously turning over, the risk of mutation in this tissue is high compared to less proliferative tissues. It is estimated that the time between adenoma initiation and carcinoma development in the colon is about 17 years, while the time between carcinoma and metastasis is only 1.8 years [161]. These estimates suggest that for adenoma to progress, it has to acquire more mutations and undergo more clonal expansions than a carcinoma that metastasizes. Based on this model, it is likely that most, if not all, of the mutations required for metastasis are present in the parent carcinoma.

## 22.10 Mechanisms Leading to Chromosomal Instability (CIN)

CRCs can be divided into classes based on the general state of the genome. The majority (~70 %) of sporadic CRCs contain a high frequency of gain and/or loss of whole chromosomes or chromosomal segments. This phenotype is known as chromosomal instability (CIN) and is characterized by chromosomal translocations, amplifications, deletions, and insertions [164, 165]. One of the underlying causes of such aneuploidy is improper segregation of chromosomes. The mitotic spindle checkpoint maintains fidelity of chromosome segregation by delaying anaphase onset until chromosomes are properly aligned on the mitotic spindle [166]. A second mechanism of aneuploidy is driven by centrosomes, which secure cytoplasmic microtubules to the mitotic spindle during cell division. Abnormal centrosome number results in mulitpolar spindles, ultimately causing chromosomal missegregation [166]. Genomic integrity is also maintained by a second checkpoint, the DNA damage checkpoint, which protects cells against postreplicative DNA damage and genotoxic stress. Failed DNA damage checkpoints allow damaged DNA to progress through mitosis, which can result in chromosomal structural alterations [164]. Loss of heterozygosity (LOH) is one of the primary outcomes of chromosomal instability, and CRC tumors may display up to 40 % of parental allelic loss [164, 167]. The most frequent CRC-associated chromosomal losses occur at 1p, 4q, 5q, 8p, 14q, 15q, 17p and q, 18p and q, 20p, and 22q [168]. Of note, loss of 18p and q (containing SMAD4) or 17p and q (containing TP53), the most common chromosomal deletions, occur in up to 66 and 56 % of CRC, respectively [168]. The exact mechanisms that underlie LOH are not well understood, but several pathways have been implicated, including mitotic nondisjunction, homologous recombination, and chromosomal structural changes originating from recombinations and deletions following DNA double strand break repair [164, 169].

## 22.11 Genetic Alterations in CIN-Type CRC

A complex genetic landscape of CRC has emerged in recent years as the sequences of whole cancer genomes have become available. Integrated analysis of single nucleotide mutations, amplifications, and deletions indicates that each CRC carries approximately 80 mutations. The most frequently mutated genes in CIN-type CRC include APC, KRAS, PIK3CA, TP53, SMAD4 and FBXW7, resulting in the activation of oncogenic pathways and suppression of tumor suppressor pathways [168, 170, 171].

### 22.11.1 Wnt Signaling

Perhaps the earliest and most critical genetic alteration in CRC is activation of the Wnt signaling pathway through disruption of Adenomatous Polyposis Coli (APC) [172]. A single germline mutation in APC is responsible for the development of FAP (discussed in 22.17.1) and somatic APC mutations occur in approximately 81 % of non-hypermutated tumors and 51 % of hypermutated tumors [168]. APC is a critical modulator of the Wnt signaling pathway (Fig. 22.20). APC binds the axin–conductin scaffold and, along with the other members of the "β-catenin destruction complex" (glycogen synthase kinase 3β (GSK-3β) and casein kinase 1α/ε (CK1α/ε)), mediates β-catenin phosphorylation, ubiquitylation, and proteasomal degradation, preventing activation of Wnt signaling [173]. APC loss disrupts complex formation and relieves β-catenin degradation, resulting in accumulation of cytoplasmic β-catenin even in the absence of Wnt ligands. Translocation of β-catenin into the nucleus releases repression of the T-cell factor and lymphoid enhancing factor (TCF/LEF) transcription factors, driving transcription of multiple Wnt target genes, some of which have been implicated in tumor growth and invasion [174]. APC inactivation can also be achieved by epigenetic silencing of the APC promoter by hypermethylation [175]. APC is a multi-domain scaffolding protein that participates in various cellular functions independent of its role in Wnt signaling, and therefore, APC mutations affect more than just the Wnt pathway. In particular, largely in vitro studies have implicated APC in aspects of cell polarity [176], microtubule and cytoskeletal dynamics [177, 178], mitotic spindle dynamics [176–178], and apoptosis [179]. Other components of the canonical Wnt signaling pathway are altered in CRC, including β-catenin (CTNNB1) itself. Stabilizing gain-of-function mutations in CTNNB1 occur in about 5–7 % of CRC and disrupt key regulatory phosphorylation and ubiquitylation domains, resulting in the inability of β-catenin to be phosphorylated and degraded [151, 168]. Additionally, reported inactivating mutations or deletions in SOX9, AXIN2, DKK protein family members, TCF7L2, FBXW7, ARID1A, and FAM123B, all negative regulatory mechanisms, and increased expression of FZD10, which encodes the Wnt co-receptor, represent alternative mechanisms for increased Wnt signaling [168].

### 22.11.2 RAS

The RAS subfamily of small, monomeric GTPases includes K-RAS4A, K-RAS4B, N-RAS and H-RAS. RAS proteins are binary molecular switches that oscillate between active (GTP-bound) and inactive (GDP-bound) states. Gain-of-function point mutations lock RAS proteins into the active state, resulting in sustained activation of multiple downstream signaling pathways [180], the most well-known of which is the Raf-MEK-ERK pathway (Fig. 22.21) [181]. Canonical activating mutations in K-RAS occur in about 40 % of CRC, while N-RAS mutations occur in only 3–5 %. Mutations in H-RAS have not yet been identified in CRC [168, 182, 183]. The genetic basis for mutation distribution among RAS isoforms is not understood, but functional studies in genetically engineered mouse models support the idea that K-RAS and N-RAS have different roles in CRC tumor biology. In particular, expression of mutationally activated K-Ras promoted progression of colonic adenocarcinoma in an APC-deficient model, whereas N-Ras did not

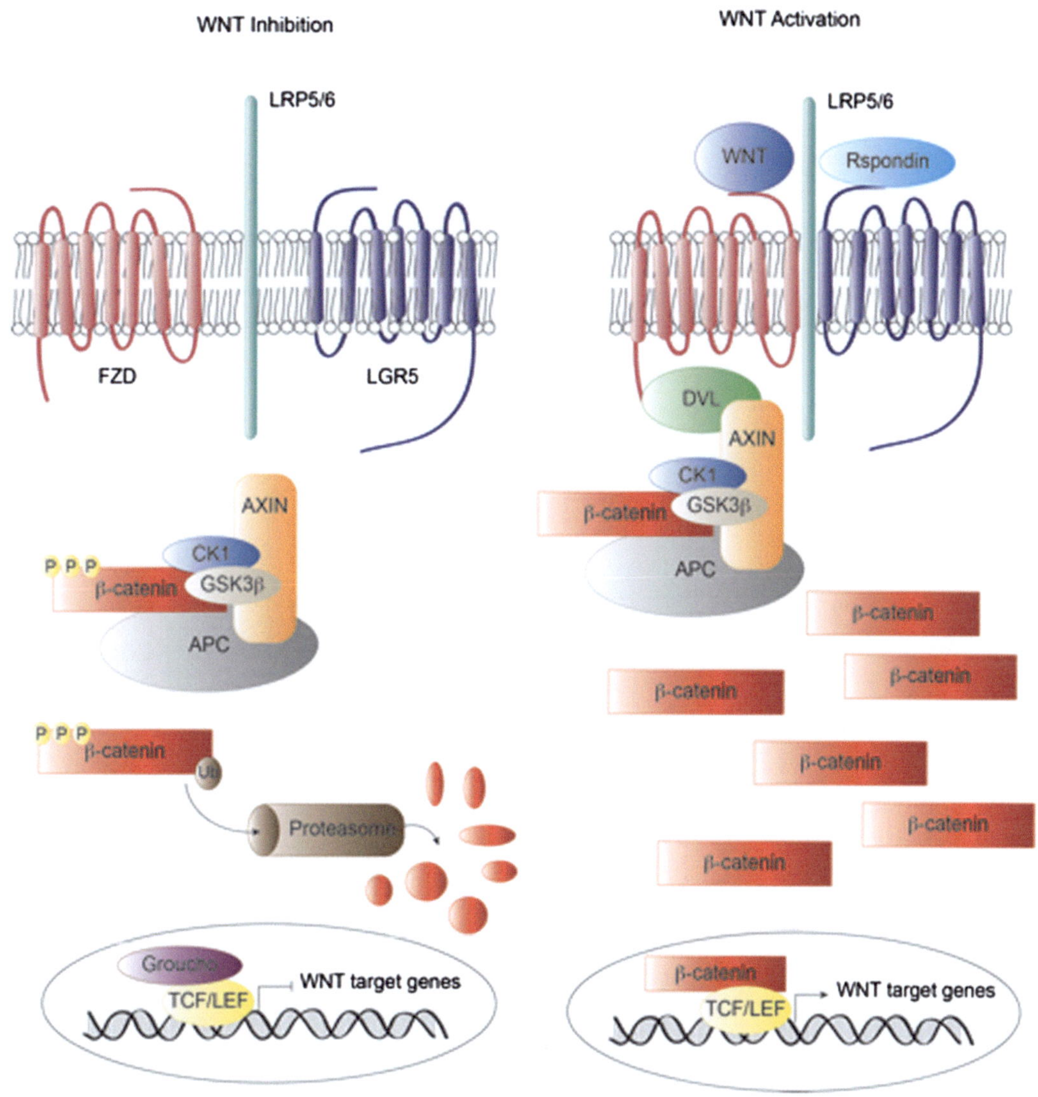


**Fig. 22.20** The WNT signaling pathway. In the absence of WNT ligands, the β-catenin destruction complex (AXIN, CK1, GSK3β, and APC) mediates phosphorylation of β-catenin, stimulating its ubiquitylation and degradation by the proteasome. Groucho inhibits TCF/LEF transcription factors, preventing transcription of WNT target genes. WNT ligands bind the co-receptors Frizzled (FZD) and LRP5/6, while the WNT agonist R-Spondin binds LGR receptors. When WNT ligands are present, the β-catenin destruction complex is inhibited and β-catenin accumulates. Translocation of β-catenin into the nucleus relieves Groucho inhibition of TCF/LEF and WNT target gene transcription is activated

[184]. Instead, mutant N-Ras may promote CRC driven by inflammation [185]. The majority (75 %) of reported K-RAS mutations are in codon 12, while mutations at codons 13(10 %), 146 (4 %), and 61(2 %) occur less often [186, 187]. The RAS effectors BRAF and PIK3CA are implicated in CRC pathogenesis and will be discussed below.

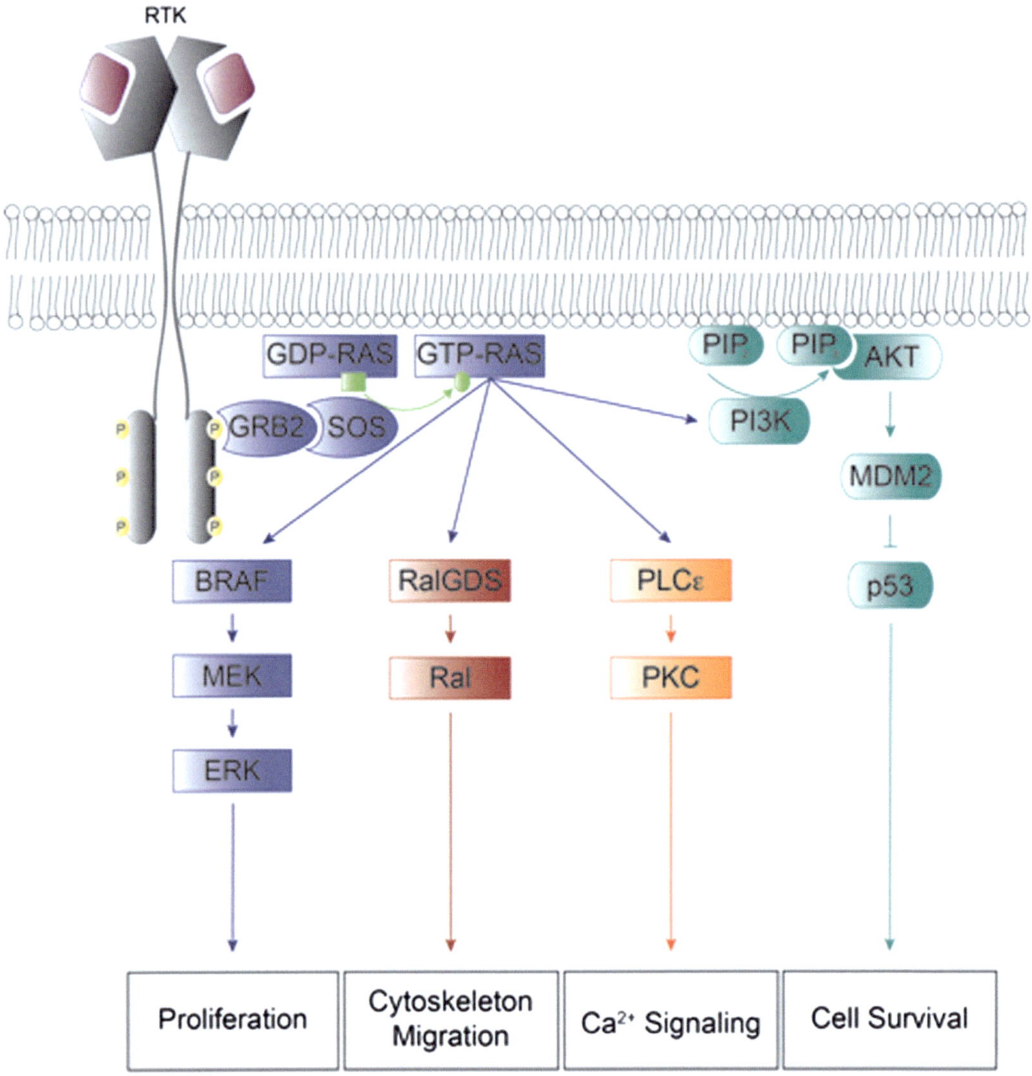


**Fig. 22.21** RAS signaling pathways. Ligand-activated receptor tyrosine kinases (RTKs) recruit the adaptor protein, GRB2, which activates the RAS guanine nucleotide exchange factor SOS. SOS mediates nucleotide exchange, activating RAS and inducing multiple RAS effector pathways. RAS-induced cellular responses are mediated by kinase cascades (BRAF-MEK-ERK), cytoskeletal dynamics (Ral), calcium signaling (protein kinase C, PKC), and the PI3K/AKT survival pathway. Mutationally activated RAS signals in the absence of upstream inputs from RTKs

### 22.11.3 PI3K

The phosphatidylinositol 3-kinase (PI3K) pathway is critical for cell survival and cell growth. PI3K is a lipid kinase that catalyzes the phosphorylation of phosphatidylinositol 4,5-bisphosphate ($PIP_2$) to generate phosphatidylinositol 3,4,5-triphosphate ($PIP_3$). AKT binds $PIP_3$ via its Pleckstrin Homology domain and is recruited to the membrane where it is activated [188]. PI3K is composed of a catalytic subunit (p110) and a regulatory subunit (p85), which binds SH2 domains on receptor tyrosine kinases, as well as activated RAS. Activating mutations in PIK3CA, the gene that encodes the p110α catalytic subunit of PI3K, occurs almost exclusively in established carcinomas in 15–18 % of CRC cases (COSMIC Database) [168]. In CRC, the three most common mutations include H1047R, and E545 K and E542 K mutations [189]. These missense mutations cause constitutive lipid kinase activity, resulting in cell growth promotion and invasion in cancer cells [190]. Aside from mutational activation of PI3KCA, the PI3K pathway can be activated by loss or mutation of the tumor suppressor, PTEN, a negative regulator of AKT. Approximately 4–10 % of sporadic CRCs exhibit somatic PTEN mutations, which most likely cause activation of the AKT survival pathway [168, 191]. PI3K is also activated downstream of ligand-activated receptor tyrosine kinases (RTKs). CRC-associated alterations in certain RTKs, including Epidermal Growth Factor Receptor (EGFR) and its family members (ERBB2/4), KIT, and MET (COSMIC Database), result in dysregulation of PI3K activity. Additionally, PI3K is a canonical RAS effector and is activated through direct interaction of the p110 subunit with RAS (Fig. 22.21) [192]. Interestingly, although they signal through the same pathway, RAS and PI3K mutations are not mutually exclusive in CRC [193, 194].

### 22.11.4 p53

TP53 is one of the most commonly mutated genes in human cancers and is mutated in about 60 % of CRC, making it the second most commonly mutated gene in CRC [168]. p53 is a transcription factor that is activated in response to cellular stress by repressing cell cycle progression and inducing apoptosis. Under normal conditions, p53 undergoes ubiquitin-mediated proteasomal degradation, but upon onset of cellular stress, it is stabilized and activates transcription of genes involved in cell cycle checkpoints, cell cycle arrest, and apoptosis [195]. The majority ($\sim$80 %) of TP53 mutations are missense mutations (GC to AT transitions) that occur most often in codons 175, 245, 248, 273, and 282. These mutations disrupt the ability of p53 to bind target DNA sequences, effectively inactivating a host of cellular protection mechanisms. Homozygous inactivation of p53 occurs by LOH, and in fact, loss of chromosome 17p and q (which includes the TP53 gene), is observed in 56 % of CRC [151, 168, 183, 196, 197]. p53 loss occurs in 4–26 % of adenomas, 50 % of adenomas with invasive foci, and in 50–75 % of CRCs, suggesting that inactivation of the p53-dependent cell cycle arrest and apoptotic pathways might facilitate tumor progression [198].

### 22.11.5 Allelic Loss of 18q

Allelic loss of chromosome 18q is one of the most commonly lost chromosomal regions in CRC and is observed in 65–70 % of primary colorectal tumors, particularly in advanced stages [157, 168]. 18q contains the coding region for SMAD2 and SMAD4, downstream mediators of the transmembrane transforming growth factor-β (TGFβ) receptors, encoded by TGFBRI and TGFBR2. Upon activation of the TGF β pathway, SMAD proteins translocate into the nucleus and activate target gene transcription. Loss of 18q results in loss of downstream TGFβ signaling, which represses cell proliferation, induces apoptosis, and regulates aspects of epithelial-to-mesenchymal transition (EMT), cell motility, and adhesion [199, 200]. In addition to chromosomal loss, inactivating mutations in SMAD2 and SMAD4 are reported in 6 and 10 % of CIN-type CRC, respectively [168].

### 22.11.6 Fbxw7

The FBXW7 gene encodes the FBW7 tumor suppressor F-box protein, a component of SCF (Skp1, Cullin-1, F-box protein) E3 ligase complexes. FBW7 recognizes a specific phosphodegron domain (Cdc4 phosphodegron, CPD) on its substrates and mediates ubiquitylation of mulitple oncogenic targets, including c-myc, the Notch1 intracellular domain (NICD), cyclin E, and c-Jun [201]. FBXW7 mutations occur in about 11 % of CRC [168] and typically are missense mutations in key arginine residues that interact with substrate CPD phosphate residues, therefore disrupting substrate interaction and subsequent ubiquitylation. Given the general oncogenic properties of its major targets, loss of FBW7 function affects cell cycle dynamics, as well as cell growth, proliferation and differentiation [201].

## 22.12 Order of Mutations

Fearon and Vogelstein's seminal work on the genetics of CRC identified four genomic alterations that were associated with the transformation of normal colonic epithelium to cancer (Fig. 22.19) [157]. Although their original model is often construed as a linear accumulation of mutations that results in stepwise tumor progression, Fearon and Vogelstein argue that "the total accumulation of changes, rather than their order with respect to one another, is responsible for determining the tumor's biologic properties" [157]. This misunderstanding likely stems from the absence of non-neoplastic endpoints, such as aberrant crypt foci, from the original study. Today, we have a wealth of information gleaned from studying CRC tumorigenesis in genetically engineered mouse models, in which we can examine the pathogenesis of both precursor lesions and advanced CRC. Using this information, which includes non-neoplastic endpoints, application of a nonlinear model of the stepwise transition from normal colon to malignant cancer demonstrates that tumor progression is independent of the order in which mutations are acquired (Fig. 22.22).

## 22.13 Microsatellite Instability (MSI)

In 1993, the Perucho and Thibodeau groups independently described the phenomenon of microsatellite instability (MIN or MSI) in CRC [202, 203]. Further work demonstrated that CIN and MSI represent separate mechanisms of

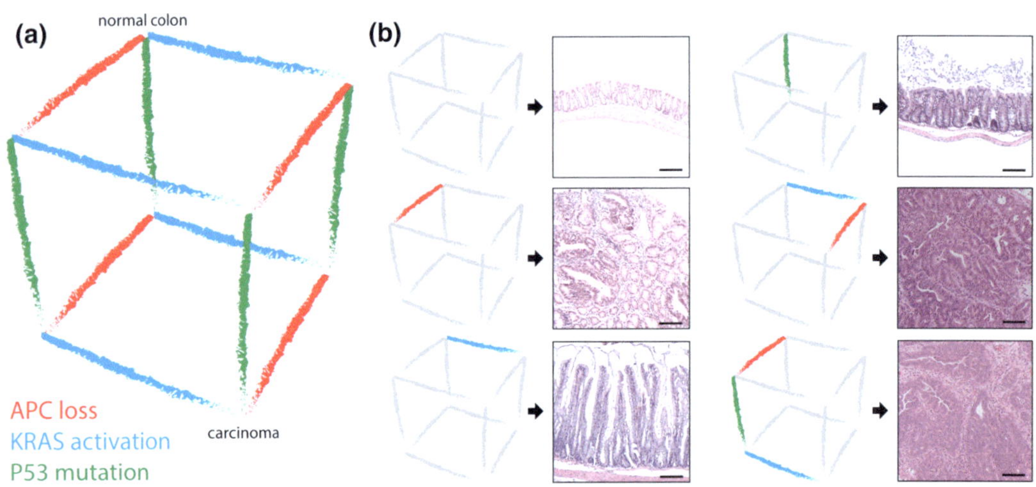


**Fig. 22.22** A nonlinear model for CRC development. **a** Normal colonic epithelium can progress to carcinoma in situ through three genetics events (APC loss, KRAS activation, and p53 mutation), regardless of the order in which these events occur. This concept is represented schematically with a cube, where movement from one corner to another takes three steps, regardless of which route is taken. **b** Histologic states produced by defined genetic mutations, as revealed by genetically engineered mouse models. The exact genetic change associated with each histologic state is highlighted by the cube diagram. Mutation of KRAS or p53 produces a non-neoplastic endpoint. Loss of APC, the first step in the original Fearon and Vogelstein model (Fig. 22.19), is required for neoplasia. Mutation of APC and KRAS creates an advanced/dysplastic adenoma. Mutation of APC, KRAS, and p53 creates frank cancer

colorectal tumorigenesis with distinct mutational and phenotypic characteristics. Microsatellites are small DNA segments (100–200 bp in length) that consist of stretches of nucleotide repeats, most often composed of adenine $(A)_n$ mononucleotide repeats or cytosine–adenine $(CA)_n$ dinucleotide repeats. These repetitive sequences are prone to DNA polymerase strand slippage during replication, which results in DNA mismatches or extrahelical loops. This causes the DNA strands to reanneal incorrectly within the microsatellite and can result in protein truncations if not repaired. The microsatellite-associated instability of any given allele is related to the length of the nucleotide repeats and therefore, genes that contain longer stretches of nucleotide repeats are considered less stable [135, 204, 205]. The DNA MMR pathway maintains the fidelity of DNA replication by correcting mismatches, insertions, and deletions following replication. Mutation or epigenetic silencing of MMR pathway components renders this pathway defective and leads to established MSI. In particular, in contrast to CIN-type tumors, MSI tumors largely display a normal diploid karytype and do not manifest large chromosomal aberrations [164].

### 22.13.1 Hereditary MSI CRC

MSI is detected in about 15 % of all colorectal cancers; 3 % are of which are associated with Lynch syndrome (discussed in Sect. 22.7.3). Identification of the causative mutations of Lynch syndrome revealed the role of MMR pathway components in MSI development. MLH1 (Mut L homologue) and MSH2 (Mut S Homologue) are the most commonly mutated genes in Lynch syndrome and result in MMR failure. Additional Lynch syndrome-associated inactivating mutations occur in the MMR genes MSH6 and PMS2 (Postmeiotic segregation-2) (Fig. 22.23) [135, 151, 204]. Loss of MMR pathway genes results in failure of the complex to recognize DNA replication errors or prevents repair, rendering the MMR pathway inactive. Only 65 % of Lynch syndrome cases contain mutations in MMR genes, while the other 35 % are considered to be MMR-proficient and characterized as familial colorectal cancer type X (FCCTX) [118, 206]. Mutations in TGFBR2 have been identified in families with FCCTX, however, additional disease-associated genes and mechanisms of tumorigenesis are unknown [207].

### 22.13.2 Sporadic MSI CRC

The majority of MSI CRC is sporadic (about 12 % of all CRC) and is largely associated with epigenetic silencing of MLH1 [135, 168, 204]. Following MLH1 silencing and MMR inactivation, a number of genes develop subsequent mutations due to MSI. Of the hypermutated tumors characterized by The Cancer Genome Atlas (TCGA), mutations were reported in ACVR2A (63 %), TGFBR2 (51 %), BRAF (46 %), MSH3 (40 %), MSH6 (40 %), and TCF7L2 (31 %) [168]. The TGFβ pathway is particularly affected by MSI, with two receptor types being the most highly mutated genes in this tumor subset. MSI-associated ACVR2A and TGFBR2 mutations result in truncated inactive proteins, which prevent normal growth suppressive TGFβ signaling [208]. One of the underlying causes of MSI is the CpG island methylator phenotype, or CIMP. CIMP is exhibited by cytosine hypermethylation within promotor CG-rich islands, resulting in epigenetic gene silencing of tumor suppressor genes [209]. In particular, MLH1 silencing, a readout of CIMP, was observed in 83 % of MSI-H tumors in the TCGA and in hypermutated tumors, was associated with an increased frameshift mutation rate compared with hypermutated tumors lacking MLH1 silencing [168]. BRAF mutations are also associated with CIMP, the most common mutation being a T:A transversion, resulting in a V600E mutation [210–212]. BRAF is a downstream RAS effector and activating V600E mutations result in constitutive activation of downstream growth-promoting pathways, similar to the effect of point mutations in RAS. Interestingly, BRAF-mutant cell lines can induce proliferation independent of RAS, and BRAF mutations are typically mutually exclusive with KRAS mutations [211, 213].

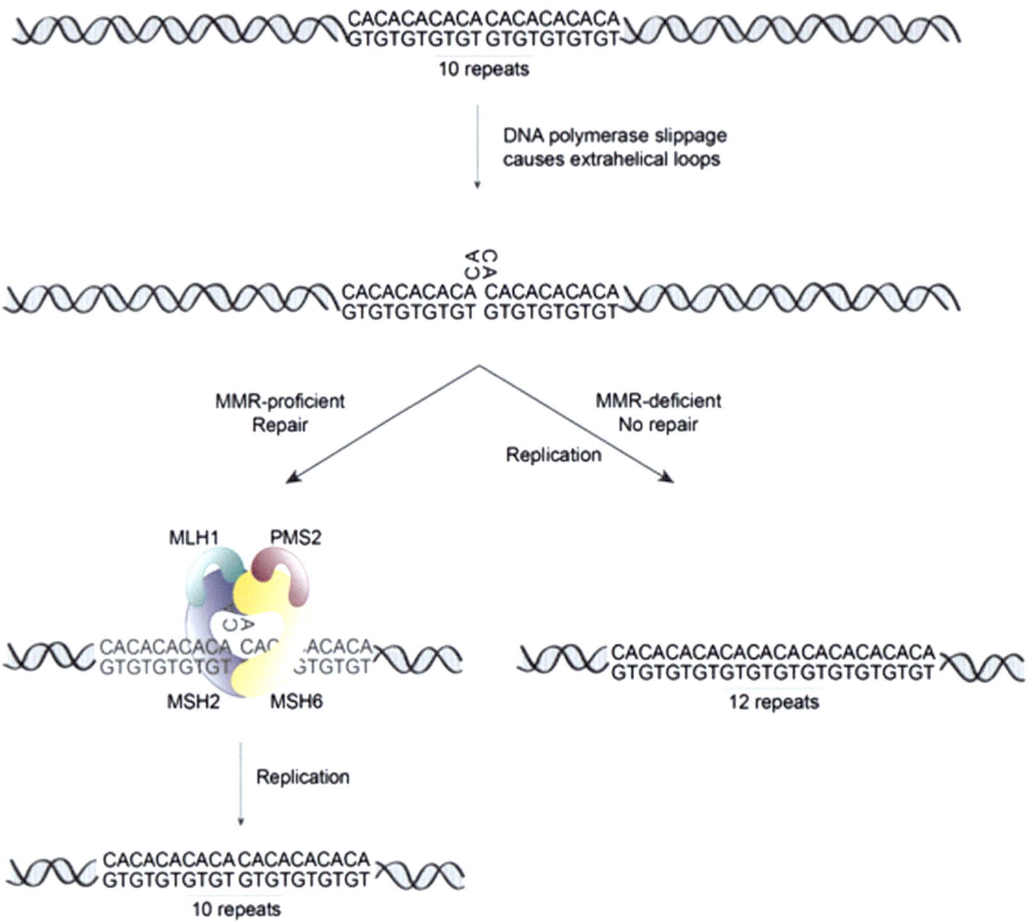


**Fig. 22.23** The DNA mismatch repair (MMR) pathway and microsatellite instability (MSI). DNA polymerase slippage can occur at repetitive DNA sequences called microsatellites. The MMR complex is composed of two heterodimers (MSH2-MLH1 and MSH6-PMS2) that recognize mismatches and extrahelical loops and mediate nucleotide excision and repair by exonuclease I, DNA polymerase, and DNA ligase. Loss of MMR components results in failure of repair and alteration of microsatellite length

## 22.14 Alternative Pathways to CRC

### 22.14.1 MicroRNAs (MiRNAs)

MicroRNAs (miRNAs) are short (20–25 nucleotides) noncoding RNAs that posttranscriptionally regulate protein expression by preventing mRNA translation. A number of miRNAs are altered in CRC, suggesting that they may have a functional role in tumor pathogenesis [214]. Depending on their targets, miRNAs can be considered either tumor suppressors or oncogenes. For example, intestinal-specific deletion of Let-7 in mice results in development of adenocarcinomas associated with an increased stem cell phenotype, confirming that Let-7 acts as a tumor suppressor and promotes differentiation [215]. This study identified a number of oncogenic Let-7 targets, including Hmga2, Mycn and E2f2, upregulation of which were also observed in human CRC. In contrast, LIN28A and LIN28B, which negatively regulate Let-7 levels, are upregulated in human CRC and positively correlated with tumor progression and worse survival. Ectopic intestinal overexpression of LIN28a or LIN28B caused small intestinal adenocarcinoma development that correlated with decreased

Let-7 levels [216]. As more CRC-associated miRNAs are identified, it is clear that miRNAs have the potential to affect multiple tumorigenesis-related pathways and the relative expression of particular miRNAs has provided useful biomarker and prognostic information [217].

### 22.14.2 Colitis-Associated CRC

One of the most significant environmental risk factors for CRC is chronic intestinal inflammation. Patients with chronic IBD have a 20 % risk of developing colitis-associated CRC (CAC), which has a >50 % mortality rate [218]. The pathogenesis of CAC reveals key insights into the tumor-promoting effects of immune system components. Although CAC involves many of the same oncogenic pathways as sporadic CRC, tumorigenesis is likely initiated by the effects of chronic inflammation. In particular, the enhanced presence of reactive oxygen species (ROS) causes DNA oxidative damage, contributing to increased DNA mutagenesis. In particular, mutations in TP53 are common early alterations. Epigenetic silencing is also associated with gene silencing in CAC, and genomic instability can be amplified by methylation of MMR gene promoters [218–220]. CAC tumor progression is enhanced by inflammatory pathways that promote cell proliferation and survival, including NF-κ[kappa]B, STAT3, and cytokine signaling. Mouse models of inflammation using the epithelial irritant, dextran sodium sulfate (DSS), combined with the carcinogen, azoxymethane (AOM), have provided significant insight into the inflammatory mediators that support CAC progression [221]. NF-κ[kappa]B is a multi-subunit transcription factor that is activated by cytokine signaling and promotes cell survival and cytokine production [218]. In a mouse model of CAC, NF-κ[kappa]B inactivation in intestinal epithelial cells or in myeloid cells resulted in diminished tumorigenesis, suggesting that NF-κ[kappa]B signaling is an important intersection between inflammation and CAC [222]. The proinflammatory, NF-κ[kappa]B-regulated cytokine, interleukin-6 (IL-6), stimulates downstream pathways that include PI3K and STAT3. In an AOM/DSS mouse model of CAC, Il-6 deficiency resulted in decreased tumorigenesis, suggesting that it acts as a tumor promoter. In addition, Il-6 production by immune cells within the intestinal epithelium promoted proliferation and survival of intestinal epithelial cells in a STAT3-dependent manner [223]. The STAT3 transcription factor is a critical downstream mediator of IL-6 pro-inflammatory signals. In the same AOM/DSS mouse model, Stat3 knockout in intestinal epithelial cells also decreased tumor size and number [223]. STAT3 can also activate anti-apoptotic and proliferative pathways, which contribute to tumor promotion [218]. Finally, the expression of tumor necrosis family (TNF) cytokines has been associated with pro-inflammatory and angiogenic properties in CAC [218]. In particular, TNF-α, which is increased in human IBD, was significantly increased in the colon of mice treated with AOM/DSS, and knockout of the TNF-α receptor resulted in decreased tumor burden and immune infiltrate [218, 224].

## 22.15 Summary

This chapter covered the normal anatomy, and embryological development of the colon and rectum, followed by the pathology, pathogenesis, and molecular drivers of CRC tumorigenesis. CRC is a complex disease that develops via multiple routes, each with its own molecular profile of genetic and epigenetic alterations. Classic early studies by Fearon and Vogelstein identified the most commonly altered pathways in CRC, validation of which has been upheld by the recent TCGA studies. The vast heterogeneity of CRC presents a therapeutic challenge, but our understanding of the disease is constantly evolving, and advances from emerging areas, such as the miRNA field, may provide additional insight into the pathogenesis of CRC and inform therapeutic strategies.

## References


1. Wells JM, Melton DA. Vertebrate endoderm development. Annu Rev Cell Dev Biol. 1999;15:393–410 Epub 1999/12/28.
2. de Santa Barbara P, van den Brink GR, Roberts DJ. Development and differentiation of the intestinal epithelium. Cell Mol Life Sci CMLS. 2003;60(7):1322–32. Epub 2003/08/29.
3. Noah TK, Donahue B, Shroyer NF. Intestinal development and differentiation. Exp Cell Res. 2011;317(19):2702–10 Epub 2011/10/08.
4. D'Amour KA, Agulnick AD, Eliazer S, Kelly OG, Kroon E, Baetge EE. Efficient differentiation of human embryonic stem cells to definitive endoderm. Nat Biotechnol. 2005;23(12):1534–41 Epub 2005/11/01.
5. Spence JR, Mayhew CN, Rankin SA, Kuhar MF, Vallance JE, Tolle K, et al. Directed differentiation of human pluripotent stem cells into intestinal tissue in vitro. Nature. 2011;470(7332):105–9 Epub 2010/12/15.
6. Sato T, Vries RG, Snippert HJ, van de Wetering M, Barker N, Stange DE, et al. Single Lgr5 stem cells build crypt-villus structures in vitro without a mesenchymal niche. Nature. 2009;459(7244):262–5 Epub 2009/03/31.
7. Sato T, Stange DE, Ferrante M, Vries RG, Van Es JH, Van den Brink S, et al. Long-term expansion of epithelial organoids from human colon, adenoma, adenocarcinoma, and Barrett's epithelium. Gastroenterology. 2011;141(5):1762–72 Epub 2011/09/06.
8. Adler D, Farraye F, Crawford J. Gastrointestinal tract endoscopic and tissue processing techniques and normal histology. In: Odze RD, Goldblum JR, editors. Odze and Goldblum surgical pathology of the GI tract, liver, biliary tract, and pancreas. Third edition. ed: Saunders, Elsevier Inc.; 2015. p. 19, 1612.
9. Edge SB, American Joint Committee on Cancer. AJCC cancer staging manual. 7th ed. New York: Springer;2010. xiv, 648 pp.
10. Ostaff MJ, Stange EF, Wehkamp J. Antimicrobial peptides and gut microbiota in homeostasis and pathology. EMBO Mol Med. 2013;5(10):1465–83 Epub 2013/09/17.
11. Underwood MA. Paneth cells and necrotizing enterocolitis. Gut Microbes. 2012;3(6):562–5 Epub 2012/08/17.
12. Adolph TE, Tomczak MF, Niederreiter L, Ko HJ, Bock J, Martinez-Naves E, et al. Paneth cells as a site of origin for intestinal inflammation. Nature. 2013;503(7475):272–6 Epub 2013/10/04.
13. Neutra MR, Kraehenbuhl JP. The role of transepithelial transport by M cells in microbial invasion and host defense. J Cell Sci Suppl. 1993;17:209–15 Epub 1993/01/01.
14. Mabbott NA, Donaldson DS, Ohno H, Williams IR, Mahajan A. Microfold (M) cells: important immunosurveillance posts in the intestinal epithelium. Mucosal Immunol. 2013;6(4):666–77 Epub 2013/05/23.
15. Barker N, van Es JH, Kuipers J, Kujala P, van den Born M, Cozijnsen M, et al. Identification of stem cells in small intestine and colon by marker gene Lgr5. Nature. 2007;449(7165):1003–7 Epub 2007/10/16.
16. de Lau W, Barker N, Low TY, Koo BK, Li VS, Teunissen H, et al. Lgr5 homologues associate with Wnt receptors and mediate R-spondin signalling. Nature. 2011;476(7360):293–7 Epub 2011/07/06.
17. Carmon KS, Gong X, Lin Q, Thomas A, Liu Q. R-spondins function as ligands of the orphan receptors LGR4 and LGR5 to regulate Wnt/beta-catenin signaling. Proc Natl Acad Sci USA. 2011;108(28):11452–7 Epub 2011/06/23.
18. Glinka A, Dolde C, Kirsch N, Huang YL, Kazanskaya O, Ingelfinger D, et al. LGR4 and LGR5 are R-spondin receptors mediating Wnt/beta-catenin and Wnt/PCP signalling. EMBO Rep. 2011;12 (10):1055–61 Epub 2011/09/13.
19. Bird RP. Observation and quantification of aberrant crypts in the murine colon treated with a colon carcinogen: preliminary findings. Cancer Lett. 1987;37(2):147–51 Epub 1987/10/30.
20. Lopez-Ceron M, Pellise M. Biology and diagnosis of aberrant crypt foci. Colorectal Dis Official J Assoc Coloproctol Great Br Irel. 2012;14(4):e157–64 Epub 2011/12/21.
21. Takayama T, Katsuki S, Takahashi Y, Ohi M, Nojiri S, Sakamaki S, et al. Aberrant crypt foci of the colon as precursors of adenoma and cancer. New England J Med. 1998;339(18):1277–84 Epub 1998/10/29.
22. Pretlow TP, Pretlow TG. Mutant KRAS in aberrant crypt foci (ACF): initiation of colorectal cancer? Biochim Biophys Acta. 2005;1756(2):83–96 Epub 2005/10/13.
23. Takayama T, Ohi M, Hayashi T, Miyanishi K, Nobuoka A, Nakajima T, et al. Analysis of K-ras, APC, and beta-catenin in aberrant crypt foci in sporadic adenoma, cancer, and familial adenomatous polyposis. Gastroenterology. 2001;121(3):599–611 Epub 2001/08/28.
24. Otori K, Sugiyama K, Hasebe T, Fukushima S, Esumi H. Emergence of adenomatous aberrant crypt foci (ACF) from hyperplastic ACF with concomitant increase in cell proliferation. Cancer Res. 1995;55 (21):4743–6 Epub 1995/11/01.
25. Velho S, Haigis KM. Regulation of homeostasis and oncogenesis in the intestinal epithelium by Ras. Exp Cell Res. 2011;317(19):2732–9.

26. Patil D, Greenson J, RD O. Inflammatory disorders of the large intestine. In: Odze RD, Goldblum JR, editors. Odze and Goldblum Surgical pathology of the GI tract, liver, biliary tract, and pancreas. Third edition. ed: Saunders, Elsevier Inc. 2015. p. 460.
27. Update on the paris classification of superficial neoplastic lesions in the digestive tract. Endoscopy. 2005;37(6):570–8. Epub 2005/06/04.
28. Riddell RH, Goldman H, Ransohoff DF, Appelman HD, Fenoglio CM, Haggitt RC, et al. Dysplasia in inflammatory bowel disease: standardized classification with provisional clinical applications. Hum Pathol. 1983;14(11):931–68 Epub 1983/11/01.
29. Schlemper RJ, Riddell RH, Kato Y, Borchard F, Cooper HS, Dawsey SM, et al. The Vienna classification of gastrointestinal epithelial neoplasia. Gut. 2000;47(2):251–5.
30. Imperiale TF, Wagner DR, Lin CY, Larkin GN, Rogge JD, Ransohoff DF. Results of screening colonoscopy among persons 40 to 49 years of age. N Engl J Med. 2002;346(23):1781–5 Epub 2002/06/07.
31. Winawer SJ, Zauber AG, Ho MN, O'Brien MJ, Gottlieb LS, Sternberg SS, et al. Prevention of colorectal cancer by colonoscopic polypectomy. The National Polyp Study Workgroup. N Engl J Med. 1993;329(27):1977–81 Epub 1993/12/30.
32. Zauber AG, Winawer SJ, O'Brien MJ, Lansdorp-Vogelaar I, van Ballegooijen M, Hankey BF, et al. Colonoscopic polypectomy and long-term prevention of colorectal-cancer deaths. N Engl J Med. 2012;366(8):687–96 Epub 2012/02/24.
33. Levi GS, Harpaz N. Intestinal low-grade tubuloglandular adenocarcinoma in inflammatory bowel disease. Am J Surg Pathol. 2006;30(8):1022–9 Epub 2006/07/25.
34. Hornick J, Odze R. Polyps of the large intestine. In: Odze RD, Goldblum JR, editors. Odze and Goldblum surgical pathology of the GI tract, liver, biliary tract, and pancreas. Third edition. ed: Saunders, Elsevier Inc; 2015. p. xix, 1612 pages.
35. Lieberman D. Colorectal cancer screening: practice guidelines. Dig Dis. 2012;30(Suppl 2):34–8 Epub 2012/12/12.
36. Stryker SJ, Wolff BG, Culp CE, Libbe SD, Ilstrup DM, MacCarty RL. Natural history of untreated colonic polyps. Gastroenterology. 1987;93(5):1009–13 Epub 1987/11/01.
37. Roncucci L, Modica S, Pedroni M, Tamassia MG, Ghidoni M, Losi L, et al. Aberrant crypt foci in patients with colorectal cancer. Br J Cancer. 1998;77 (12):2343–8 Epub 1998/07/02.
38. Rosty C, Hewett DG, Brown IS, Leggett BA, Whitehall VL. Serrated polyps of the large intestine: current understanding of diagnosis, pathogenesis, and clinical management. J Gastroenterol. 2013;48 (3):287–302 Epub 2012/12/05.
39. Snover DC. Update on the serrated pathway to colorectal carcinoma. Hum Pathol. 2011;42(1):1–10 Epub 2010/09/28.
40. Huang CS, O'Brien MJ, Yang S, Farraye FA. Hyperplastic polyps, serrated adenomas, and the serrated polyp neoplasia pathway. Am J Gastroenterol. 2004;99(11):2242–55 Epub 2004/11/24.
41. Spring KJ, Zhao ZZ, Karamatic R, Walsh MD, Whitehall VL, Pike T, et al. High prevalence of sessile serrated adenomas with BRAF mutations: a prospective study of patients undergoing colonoscopy. Gastroenterology. 2006;131(5):1400–7 Epub 2006/11/15.
42. Glatz K, Pritt B, Glatz D, Hartmann A, O'Brien MJ, Blaszyk H. A multinational, internet-based assessment of observer variability in the diagnosis of serrated colorectal polyps. Am J Clin Pathol. 2007;127(6):938–45 Epub 2007/05/19.
43. Farris AB, Misdraji J, Srivastava A, Muzikansky A, Deshpande V, Lauwers GY, et al. Sessile serrated adenoma: challenging discrimination from other serrated colonic polyps. Am J Surg Pathol. 2008;32(1):30–5 Epub 2007/12/29.
44. Rex DK, Ahnen DJ, Baron JA, Batts KP, Burke CA, Burt RW, et al. Serrated lesions of the colorectum: review and recommendations from an expert panel. Am J Gastroenterol. 2012;107(9):1315–29; quiz 4, 30. Epub 2012/06/20.
45. Bettington M, Walker N, Clouston A, Brown I, Leggett B, Whitehall V. The serrated pathway to colorectal carcinoma: current concepts and challenges. Histopathology. 2013;62(3):367–86 Epub 2013/01/24.
46. Singh H, Nugent Z, Demers AA, Kliewer EV, Mahmud SM, Bernstein CN. The reduction in colorectal cancer mortality after colonoscopy varies by site of the cancer. Gastroenterology. 2010;139 (4):1128–37 Epub 2010/07/06.
47. Brenner H, Chang-Claude J, Seiler CM, Rickert A, Hoffmeister M. Protection from colorectal cancer after colonoscopy: a population-based, case-control study. Ann Intern Med. 2011;154(1):22–30 Epub 2011/01/05.
48. Chan TL, Zhao W, Leung SY, Yuen ST. BRAF and KRAS mutations in colorectal hyperplastic polyps and serrated adenomas. Cancer Res. 2003;63 (16):4878–81 Epub 2003/08/28.
49. Kim KM, Lee EJ, Ha S, Kang SY, Jang KT, Park CK, et al. Molecular features of colorectal hyperplastic polyps and sessile serrated adenoma/polyps from Korea. Am J Surg Pathol. 2011;35(9):1274–86 Epub 2011/08/13.
50. Goldstein NS. Serrated pathway and APC (conventional)-type colorectal polyps: molecular-morphologic correlations, genetic pathways, and implications for classification. Am J Clin Pathol. 2006;125(1):146–53 Epub 2006/02/18.
51. Higashidani Y, Tamura S, Morita T, Tadokoro T, Yokoyama Y, Miyazaki J, et al. Analysis of K-ras codon 12 mutation in flat and nodular variants of serrated adenoma in the colon. Dis Colon Rectum. 2003;46(3):327–32 Epub 2003/03/11.
52. Carr NJ, Mahajan H, Tan KL, Hawkins NJ, Ward RL. Serrated and non-serrated polyps of the

colorectum: their prevalence in an unselected case series and correlation of BRAF mutation analysis with the diagnosis of sessile serrated adenoma. J Clin Pathol. 2009;62(6):516–8 Epub 2009/01/08.
53. Kim KM, Lee EJ, Kim YH, Chang DK, Odze RD. KRAS mutations in traditional serrated adenomas from Korea herald an aggressive phenotype. Am J Surg Pathol. 2010;34(5):667–75 Epub 2010/03/23.
54. East JE, Saunders BP, Jass JR. Sporadic and syndromic hyperplastic polyps and serrated adenomas of the colon: classification, molecular genetics, natural history, and clinical management. Gastroenterology clinics of North America. 2008;37(1):25–46, v. Epub 2008/03/04.
55. Kim JC, Choi JS, Roh SA, Cho DH, Kim TW, Kim YS. Promoter methylation of specific genes is associated with the phenotype and progression of colorectal adenocarcinomas. Ann Surg Oncol. 2010;17(7):1767–76 Epub 2010/01/16.
56. Dyson JK, Rutter MD. Colorectal cancer in inflammatory bowel disease: what is the real magnitude of the risk? World J Gastroenterol WJG. 2012;18 (29):3839–48 Epub 2012/08/10.
57. Harpaz N, Ward SC, Mescoli C, Itzkowitz SH, Polydorides AD. Precancerous lesions in inflammatory bowel disease. Best Pract Res Clin Gastroenterol. 2013;27(2):257–67 Epub 2013/07/03.
58. Rutter MD, Riddell RH. Colorectal dysplasia in inflammatory bowel disease: a clinicopathologic perspective. Clin Gastroenterol Hepatol Official Clin Pract J Am Gastroenterological Assoc. 2014;12 (3):359–67 Epub 2013/06/13.
59. Berlin JW, Gore RM, Yaghmai V, Newmark GM, Miller FH. Staging of colorectal cancer. Semin Roentgenol. 2000;35(4):370–84 Epub 2000/11/04.
60. Washington MK, Berlin J, Branton P, Burgart LJ, Carter DK, Fitzgibbons PL, et al. Protocol for the examination of specimens from patients with primary carcinoma of the colon and rectum. Arch Pathol Lab Med. 2009;133(10):1539–51 Epub 2009/10/02.
61. Messenger DE, Driman DK, Kirsch R. Developments in the assessment of venous invasion in colorectal cancer: implications for future practice and patient outcome. Hum Pathol. 2012;43(7):965–73 Epub 2012/03/13.
62. Bosman FT, World Health Organization., International Agency for Research on Cancer. WHO classification of tumours of the digestive system. 4th ed. Lyon: International Agency for Research on Cancer;2010. 417 pp.
63. Verhulst J, Ferdinande L, Demetter P, Ceelen W. Mucinous subtype as prognostic factor in colorectal cancer: a systematic review and meta-analysis. J Clin Pathol. 2012;65(5):381–8 Epub 2012/01/20.
64. Hyngstrom JR, Hu CY, Xing Y, You YN, Feig BW, Skibber JM, et al. Clinicopathology and outcomes for mucinous and signet ring colorectal adenocarcinoma: analysis from the National Cancer Data Base. Ann Surg Oncol. 2012;19(9):2814–21 Epub 2012/04/06.
65. Park ET, Oh HK, Gum JR Jr, Crawley SC, Kakar S, Engel J, et al. HATH1 expression in mucinous cancers of the colorectum and related lesions. Clin Cancer Res Official J Am Assoc Cancer Res. 2006;12(18):5403–10 Epub 2006/09/27.
66. Fleming NI, Jorissen RN, Mouradov D, Christie M, Sakthianandeswaren A, Palmieri M, et al. SMAD2, SMAD3 and SMAD4 mutations in colorectal cancer. Cancer Res. 2013;73(2):725–35 Epub 2012/11/10.
67. Takaku K, Oshima M, Miyoshi H, Matsui M, Seldin MF, Taketo MM. Intestinal tumorigenesis in compound mutant mice of both Dpc4 (Smad4) and Apc genes. Cell. 1998;92(5):645–56 Epub 1998/03/20.
68. Yu M, Trobridge P, Wang Y, Kanngurn S, Morris SM, Knoblaugh S, et al. Inactivation of TGF-beta signaling and loss of PTEN cooperate to induce colon cancer in vivo. Oncogene. 2014;33(12):1538–47 Epub 2013/04/23.
69. Messerini L, Palomba A, Zampi G. Primary signet-ring cell carcinoma of the colon and rectum. Dis Colon Rectum. 1995;38(11):1189–92 Epub 1995/11/01.
70. Psathakis D, Schiedeck TH, Krug F, Oevermann E, Kujath P, Bruch HP. Ordinary colorectal adenocarcinomavs. primary colorectal signet-ring cell carcinoma: study matched for age, gender, grade, and stage. Dis Colon Rectum. 1999;42(12):1618–25 Epub 1999/12/29.
71. Hinoi T, Tani M, Lucas PC, Caca K, Dunn RL, Macri E, et al. Loss of CDX2 expression and microsatellite instability are prominent features of large cell minimally differentiated carcinomas of the colon. Am J Pathol. 2001;159(6):2239–48 Epub 2001/12/26.
72. Jessurun J, Romero-Guadarrama M, Manivel JC. Medullary adenocarcinoma of the colon: clinicopathologic study of 11 cases. Hum Pathol. 1999;30 (7):843–8 Epub 1999/07/22.
73. Wick MR, Vitsky JL, Ritter JH, Swanson PE, Mills SE. Sporadic medullary carcinoma of the colon: a clinicopathologic comparison with nonhereditary poorly differentiated enteric-type adenocarcinoma and neuroendocrine colorectal carcinoma. Am J Clin Pathol. 2005;123(1):56–65 Epub 2005/03/15.
74. Winn B, Tavares R, Matoso A, Noble L, Fanion J, Waldman SA, et al. Expression of the intestinal biomarkers Guanylyl cyclase C and CDX2 in poorly differentiated colorectal carcinomas. Hum Pathol. 2010;41(1):123–8 Epub 2009/10/06.
75. Sinicrope FA, Sargent DJ. Molecular pathways: microsatellite instability in colorectal cancer: prognostic, predictive, and therapeutic implications. Clin Cancer Res Official J Am Assoc Cancer Res. 2012;18(6):1506–12 Epub 2012/02/04.

76. Alexander J, Watanabe T, Wu TT, Rashid A, Li S, Hamilton SR. Histopathological identification of colon cancer with microsatellite instability. Am J Pathol. 2001;158(2):527–35 Epub 2001/02/13.
77. Smyrk TC, Watson P, Kaul K, Lynch HT. Tumor-infiltrating lymphocytes are a marker for microsatellite instability in colorectal carcinoma. Cancer. 2001;91(12):2417–22 Epub 2001/06/20.
78. Bellizzi AM, Frankel WL. Colorectal cancer due to deficiency in DNA mismatch repair function: a review. Adv Anat Pathol. 2009;16(6):405–17 Epub 2009/10/24.
79. Jass JR. HNPCC and sporadic MSI-H colorectal cancer: a review of the morphological similarities and differences. Fam Cancer. 2004;3(2):93–100 Epub 2004/09/02.
80. Greenson JK, Huang SC, Herron C, Moreno V, Bonner JD, Tomsho LP, et al. Pathologic predictors of microsatellite instability in colorectal cancer. Am J Surg Pathol. 2009;33(1):126–33 Epub 2008/10/03.
81. Chirieac LR, Shen L, Catalano PJ, Issa JP, Hamilton SR. Phenotype of microsatellite-stable colorectal carcinomas with CpG island methylation. Am J Surg Pathol. 2005;29(4):429–36 Epub 2005/03/16.
82. Halling KC, French AJ, McDonnell SK, Burgart LJ, Schaid DJ, Peterson BJ, et al. Microsatellite instability and 8p allelic imbalance in stage B2 and C colorectal cancers. J Natl Cancer Inst. 1999;91 (15):1295–303 Epub 1999/08/05.
83. Gryfe R, Kim H, Hsieh ET, Aronson MD, Holowaty EJ, Bull SB, et al. Tumor microsatellite instability and clinical outcome in young patients with colorectal cancer. N Engl J Med. 2000;342 (2):69–77 Epub 2000/01/13.
84. Lin CC, Lai YL, Lin TC, Chen WS, Jiang JK, Yang SH, et al. Clinicopathologic features and prognostic analysis of MSI-high colon cancer. Int J Colorectal Dis. 2012;27(3):277–86 Epub 2011/11/15.
85. Li D, Peng X, Yan D, Tang H, Huang F, Yang Y, et al. Msi-1 is a predictor of survival and a novel therapeutic target in colon cancer. Ann Surg Oncol. 2011;18(7):2074–83 Epub 2011/03/29.
86. NCCN Guidelines Version 2.2016: Colon Cancer. 11/24/15. National Comprehensive Cancer Network, Inc.2015. Printed on 24 May 2016
87. Makinen MJ. Colorectal serrated adenocarcinoma. Histopathology. 2007;50(1):131–50 Epub 2007/01/06.
88. Garcia-Solano J, Perez-Guillermo M, Conesa-Zamora P, Acosta-Ortega J, Trujillo-Santos J, Cerezuela-Fuentes P, et al. Clinicopathologic study of 85 colorectal serrated adenocarcinomas: further insights into the full recognition of a new subset of colorectal carcinoma. Hum Pathol. 2010;41(10):1359–68 Epub 2010/07/03.
89. Mitrovic B, Schaeffer DF, Riddell RH, Kirsch R. Tumor budding in colorectal carcinoma: time to take notice. Mod Pathol Official J United States Can Acad Pathol Inc. 2012;25(10):1315–25. Epub 2012/07/14.
90. Prall F. Tumour budding in colorectal carcinoma. Histopathology. 2007;50(1):151–62 Epub 2007/01/06.
91. Horcic M, Koelzer VH, Karamitopoulou E, Terracciano L, Puppa G, Zlobec I, et al. Tumor budding score based on 10 high-power fields is a promising basis for a standardized prognostic scoring system in stage II colorectal cancer. Hum Pathol. 2013;44 (5):697–705 Epub 2012/11/20.
92. Nosho K, Baba Y, Tanaka N, Shima K, Hayashi M, Meyerhardt JA, et al. Tumour-infiltrating T-cell subsets, molecular changes in colorectal cancer, and prognosis: cohort study and literature review. J Pathol. 2010;222(4):350–66 Epub 2010/10/12.
93. Mlecnik B, Tosolini M, Kirilovsky A, Berger A, Bindea G, Meatchi T, et al. Histopathologic-based prognostic factors of colorectal cancers are associated with the state of the local immune reaction. J Clin Oncol Official J Am Soc Clin Oncol. 2011;29 (6):610–8 Epub 2011/01/20.
94. Bauer K, Nelius N, Reuschenbach M, Koch M, Weitz J, Steinert G, et al. T cell responses against microsatellite instability-induced frameshift peptides and influence of regulatory T cells in colorectal cancer. Cancer Immunol Immunother CII. 2013;62 (1):27–37 Epub 2012/06/26.
95. Lin SP, Lee YT, Yang SH, Miller SA, Chiou SH, Hung MC, et al. Colon cancer stem cells resist antiangiogenesis therapy-induced apoptosis. Cancer Lett. 2013;328(2):226–34 Epub 2012/09/29.
96. Iovino F, Meraviglia S, Spina M, Orlando V, Saladino V, Dieli F, et al. Immunotherapy targeting colon cancer stem cells. Immunotherapy. 2011;3 (1):97–106 Epub 2010/12/23.
97. Sanders MA, Majumdar AP. Colon cancer stem cells: implications in carcinogenesis. Front Biosci (Landmark Ed). 2011;16:1651–62 Epub 2011/01/05.
98. Abdul Khalek FJ, Gallicano GI, Mishra L. Colon cancer stem cells. Gastrointestinal cancer research: GCR. 2010(Suppl 1):S16–23. Epub 2011/04/08.
99. Jain D, Riddell R. Small and large bowel polyps and tumors. In: Riddell R, Jain D, editors. Lewin, Weinstein and Riddell's Gastrointestinal Pathology and its Clinical Implications. Second ed: Lippincott Williams & Wilkins; 2014.
100. O'Sullivan MJ, McCarthy TV, Doyle CT. Familial adenomatous polyposis: from bedside to benchside. Am J Clin Pathol. 1998;109(5):521–6 Epub 1998/05/12.
101. Al-Tassan N, Chmiel NH, Maynard J, Fleming N, Livingston AL, Williams GT, et al. Inherited variants of MYH associated with somatic G:C — > T: A mutations in colorectal tumors. Nat Genet. 2002;30(2):227–32 Epub 2002/01/31.
102. Nielsen M, de Miranda NF, van Puijenbroek M, Jordanova ES, Middeldorp A, van Wezel T, et al. Colorectal carcinomas in MUTYH-associated

polyposis display histopathological similarities to microsatellite unstable carcinomas. BMC Cancer. 2009;9:184 Epub 2009/06/17.
103. Nielsen M, Joerink-van de Beld MC, Jones N, Vogt S, Tops CM, Vasen HF, et al. Analysis of MUTYH genotypes and colorectal phenotypes in patients With MUTYH-associated polyposis. Gastroenterology. 2009;136(2):471–6 Epub 2008/11/27.
104. Lubbe SJ, Di Bernardo MC, Chandler IP, Houlston RS. Clinical implications of the colorectal cancer risk associated with MUTYH mutation. J Clin Oncol Official J Am Soc Clin Oncol. 2009;27(24):3975–80 Epub 2009/07/22.
105. Lindor NM. Hereditary colorectal cancer: MYH-associated polyposis and other newly identified disorders. Best Pract Res Clin Gastroenterol. 2009;23(1):75–87 Epub 2009/03/05.
106. Huth C, Kloor M, Voigt AY, Bozukova G, Evers C, Gaspar H, et al. The molecular basis of EPCAM expression loss in Lynch syndrome-associated tumors. Mod Pathol Official J US Canad Acad Pathol Inc. 2012;25(6):911–6. Epub 2012/03/06.
107. Chan TL, Yuen ST, Kong CK, Chan YW, Chan AS, Ng WF, et al. Heritable germline epimutation of MSH2 in a family with hereditary nonpolyposis colorectal cancer. Nat Genet. 2006;38(10):1178–83 Epub 2006/09/05.
108. Banno K, Kisu I, Yanokura M, Tsuji K, Masuda K, Ueki A, et al. Epimutation and cancer: a new carcinogenic mechanism of Lynch syndrome (Review). Int J Oncol. 2012;41(3):793–7 Epub 2012/06/28.
109. Lagerstedt KK, Staaf J, Jonsson G, Hansson E, Lonnroth C, Kressner U, et al. Tumor genome wide DNA alterations assessed by array CGH in patients with poor and excellent survival following operation for colorectal cancer. Cancer Inform. 2007;3:341–55 Epub 2007/01/01.
110. Lynch HT, Lynch PM, Lanspa SJ, Snyder CL, Lynch JF, Boland CR. Review of the Lynch syndrome: history, molecular genetics, screening, differential diagnosis, and medicolegal ramifications. Clin Genet. 2009;76(1):1–18 Epub 2009/08/08.
111. Peltomaki P, Vasen H. Mutations associated with HNPCC predisposition — Update of ICG-HNPCC/INSiGHT mutation database. Dis Markers. 2004;20(4–5):269–76 Epub 2004/11/06.
112. Woods MO, Williams P, Careen A, Edwards L, Bartlett S, McLaughlin JR, et al. A new variant database for mismatch repair genes associated with Lynch syndrome. Hum Mutat. 2007;28(7):669–73 Epub 2007/03/10.
113. Bonadona V, Bonaiti B, Olschwang S, Grandjouan S, Huiart L, Longy M, et al. Cancer risks associated with germline mutations in MLH1, MSH2, and MSH6 genes in Lynch syndrome. JAMA: J Am Med Assoc. 2011;305(22):2304–10. Epub 2011/06/07.
114. Plaschke J, Kruger S, Jeske B, Theissig F, Kreuz FR, Pistorius S, et al. Loss of MSH3 protein expression is frequent in MLH1-deficient colorectal cancer and is associated with disease progression. Cancer Res. 2004;64(3):864–70 Epub 2004/02/12.
115. Laghi L, Bianchi P, Delconte G, Celesti G, Di Caro G, Pedroni M, et al. MSH3 protein expression and nodal status in MLH1-deficient colorectal cancers. Clin Cancer Res Official J Am Assoc Cancer Res. 2012;18(11):3142–53 Epub 2012/04/13.
116. Borras E, Pineda M, Blanco I, Jewett EM, Wang F, Teule A, et al. MLH1 founder mutations with moderate penetrance in Spanish Lynch syndrome families. Cancer Res. 2010;70(19):7379–91 Epub 2010/09/23.
117. Vasen HF, Wijnen J. Clinical implications of genetic testing of hereditary nonpolyposis colorectal cancer. Cytogenet Cell Genet. 1999;86(2):136–9 Epub 1999/11/05.
118. Lindor NM, Rabe K, Petersen GM, Haile R, Casey G, Baron J, et al. Lower cancer incidence in Amsterdam-I criteria families without mismatch repair deficiency: familial colorectal cancer type X. JAMA J Am Med Assoc. 2005;293(16):1979–85. Epub 2005/04/28.
119. Vasen HF. Clinical description of the Lynch syndrome [hereditary nonpolyposis colorectal cancer (HNPCC)]. Fam Cancer. 2005;4(3):219–25 Epub 2005/09/02.
120. Felton KE, Gilchrist DM, Andrew SE. Constitutive deficiency in DNA mismatch repair. Clin Genet. 2007;71(6):483–98 Epub 2007/06/02.
121. Wimmer K, Kratz CP. Constitutional mismatch repair-deficiency syndrome. Haematologica. 2010;95(5):699–701 Epub 2010/05/06.
122. Jasperson KW, Samowitz WS, Burt RW. Constitutional mismatch repair-deficiency syndrome presenting as colonic adenomatous polyposis: clues from the skin. Clin Genet. 2011;80(4):394–7 Epub 2010/11/03.
123. Chen S, Wang W, Lee S, Nafa K, Lee J, Romans K, et al. Prediction of germline mutations and cancer risk in the Lynch syndrome. JAMA J Am Med Assoc. 2006;296(12):1479–87. Epub 2006/09/28.
124. Watson P, Vasen HF, Mecklin JP, Bernstein I, Aarnio M, Jarvinen HJ, et al. The risk of extra-colonic, extra-endometrial cancer in the Lynch syndrome. Int J Cancer J Int du Cancer. 2008;123 (2):444–9 Epub 2008/04/10.
125. Senter L, Clendenning M, Sotamaa K, Hampel H, Green J, Potter JD, et al. The clinical phenotype of Lynch syndrome due to germ-line PMS2 mutations. Gastroenterology. 2008;135(2):419–28 Epub 2008/07/08.
126. Baglietto L, Lindor NM, Dowty JG, White DM, Wagner A, Gomez Garcia EB, et al. Risks of lynch syndrome cancers for MSH6 mutation carriers.

J Natl Cancer Inst. 2010;102(3):193–201 Epub 2009/12/24.

127. Vasen HF, Watson P, Mecklin JP, Lynch HT. New clinical criteria for hereditary nonpolyposis colorectal cancer (HNPCC, Lynch syndrome) proposed by the international collaborative group on HNPCC. Gastroenterology. 1999;116(6):1453–6 Epub 1999/05/29.
128. Umar A, Boland CR, Terdiman JP, Syngal S, de la Chapelle A, Ruschoff J, et al. Revised Bethesda guidelines for hereditary nonpolyposis colorectal cancer (Lynch syndrome) and microsatellite instability. J Natl Cancer Inst. 2004;96(4):261–8 Epub 2004/02/19.
129. Kievit W, de Bruin JH, Adang EM, Ligtenberg MJ, Nagengast FM, van Krieken JH, et al. Current clinical selection strategies for identification of hereditary non-polyposis colorectal cancer families are inadequate: a meta-analysis. Clin Genet. 2004;65 (4):308–16 Epub 2004/03/18.
130. Hampel H, Frankel WL, Martin E, Arnold M, Khanduja K, Kuebler P, et al. Screening for the Lynch syndrome (hereditary nonpolyposis colorectal cancer). N Engl J Med. 2005;352(18):1851–60 Epub 2005/05/06.
131. Moreira L, Balaguer F, Lindor N, de la Chapelle A, Hampel H, Aaltonen LA, et al. Identification of Lynch syndrome among patients with colorectal cancer. JAMA J Am Med Assoc. 2012;308 (15):1555–65. Epub 2012/10/18.
132. Bartley AN, Luthra R, Saraiya DS, Urbauer DL, Broaddus RR. Identification of cancer patients with Lynch syndrome: clinically significant discordances and problems in tissue-based mismatch repair testing. Cancer Prev Res (Phila). 2012;5(2):320–7 Epub 2011/11/17.
133. Lindor NM, Burgart LJ, Leontovich O, Goldberg RM, Cunningham JM, Sargent DJ, et al. Immunohistochemistry versus microsatellite instability testing in phenotyping colorectal tumors. J Clin Oncol Official J Am Soc Clin Oncol. 2002;20(4):1043–8 Epub 2002/02/15.
134. Jiricny J. MutLalpha: at the cutting edge of mismatch repair. Cell. 2006;126(2):239–41 Epub 2006/07/29.
135. Boland CR, Goel A. Microsatellite instability in colorectal cancer. Gastroenterology. 2010;138 (6):2073–87 e3. Epub 2010/04/28.
136. McGivern A, Wynter CV, Whitehall VL, Kambara T, Spring KJ, Walsh MD, et al. Promoter hypermethylation frequency and BRAF mutations distinguish hereditary non-polyposis colon cancer from sporadic MSI-H colon cancer. Fam Cancer. 2004;3(2):101–7 Epub 2004/09/02.
137. Funkhouser WK Jr, Lubin IM, Monzon FA, Zehnbauer BA, Evans JP, Ogino S, et al. Relevance, pathogenesis, and testing algorithm for mismatch repair-defective colorectal carcinomas: a report of the association for molecular pathology. J Mol Diagn JMD. 2012;14(2):91–103 Epub 2012/01/21.
138. Bellcross CA, Bedrosian SR, Daniels E, Duquette D, Hampel H, Jasperson K, et al. Implementing screening for Lynch syndrome among patients with newly diagnosed colorectal cancer: summary of a public health/clinical collaborative meeting. Genet Med Official J Am Coll Med Genet. 2012;14 (1):152–62 Epub 2012/01/13.
139. Buchanan D, Young J. A perspective on bi-allelic MUTYH mutations in patients with hyperplastic polyposis syndrome. Gastroenterology. 2009;136 (7):2407–8 Epub 2009/05/02.
140. Boparai KS, Dekker E, Polak MM, Musler AR, van Eeden S, van Noesel CJ. A serrated colorectal cancer pathway predominates over the classic WNT pathway in patients with hyperplastic polyposis syndrome. Am J Pathol. 2011;178(6):2700–7 Epub 2011/06/07.
141. Rosty C, Buchanan DD, Walsh MD, Pearson SA, Pavluk E, Walters RJ, et al. Phenotype and polyp landscape in serrated polyposis syndrome: a series of 100 patients from genetics clinics. Am J Surg Pathol. 2012;36(6):876–82 Epub 2012/04/19.
142. Schreibman IR, Baker M, Amos C, McGarrity TJ. The hamartomatous polyposis syndromes: a clinical and molecular review. Am J Gastroenterol. 2005;100(2):476–90 Epub 2005/01/26.
143. Hezel AF, Bardeesy N. LKB1; linking cell structure and tumor suppression. Oncogene. 2008;27 (55):6908–19 Epub 2008/11/26.
144. Perzin KH, Bridge MF. Adenomatous and carcinomatous changes in hamartomatous polyps of the small intestine (Peutz-Jeghers syndrome): report of a case and review of the literature. Cancer. 1982;49 (5):971–83 Epub 1982/03/01.
145. Narita T, Eto T, Ito T. Peutz-Jeghers syndrome with adenomas and adenocarcinomas in colonic polyps. Am J Surg Pathol. 1987;11(1):76–81 Epub 1987/01/01.
146. Burkart AL, Sheridan T, Lewin M, Fenton H, Ali NJ, Montgomery E. Do sporadic Peutz-Jeghers polyps exist? Experience of a large teaching hospital. Am J Surg Pathol. 2007;31(8):1209–14 Epub 2007/08/02.
147. Suzuki S, Hirasaki S, Ikeda F, Yumoto E, Yamane H, Matsubara M. Three cases of Solitary Peutz-Jeghers-type hamartomatous polyp in the duodenum. World J Gastroenterol WJG. 2008;14 (6):944–7 Epub 2008/02/02.
148. Calva D, Howe J. Juvenile Polyposis. In: Riegert-Johnson DL, Boardman LA, Hefferon T, Roberts M, editors. Cancer Syndromes. Bethesda (MD);2009.
149. Brosens LA, Langeveld D, van Hattem WA, Giardiello FM, Offerhaus GJ. Juvenile polyposis syndrome. World J Gastroenterol WJG. 2011;17 (44):4839–44 Epub 2011/12/16.

150. van Hattem WA, Brosens LA, de Leng WW, Morsink FH, Lens S, Carvalho R, et al. Large genomic deletions of SMAD4, BMPR1A and PTEN in juvenile polyposis. Gut. 2008;57(5):623–7 Epub 2008/01/08.
151. Fearon ER. Molecular genetics of colorectal cancer. Annu Rev Pathol. 2011;6:479–507 Epub 2010/11/26.
152. Giardiello FM, Hamilton SR, Kern SE, Offerhaus GJ, Green PA, Celano P, et al. Colorectal neoplasia in juvenile polyposis or juvenile polyps. Arch Dis Child. 1991;66(8):971–5 Epub 1991/08/01.
153. Rozen P, Samuel Z, Brazowski E. A prospective study of the clinical, genetic, screening, and pathologic features of a family with hereditary mixed polyposis syndrome. Am J Gastroenterol. 2003;98 (10):2317–20 Epub 2003/10/24.
154. Cheah PY, Wong YH, Chau YP, Loi C, Lim KH, Lim JF, et al. Germline bone morphogenesis protein receptor 1A mutation causes colorectal tumorigenesis in hereditary mixed polyposis syndrome. Am J Gastroenterol. 2009;104(12):3027–33 Epub 2009/09/24.
155. O'Riordan JM, O'Donoghue D, Green A, Keegan D, Hawkes LA, Payne SJ, et al. Hereditary mixed polyposis syndrome due to a BMPR1A mutation. Colorectal Dis Official J Assoc Coloproctol Great Br Irel. 2010;12(6):570–3 Epub 2009/05/15.
156. Jaeger E, Leedham S, Lewis A, Segditsas S, Becker M, Cuadrado PR, et al. Hereditary mixed polyposis syndrome is caused by a 40-kb upstream duplication that leads to increased and ectopic expression of the BMP antagonist GREM1. Nat Genet. 2012;44(6):699–703 Epub 2012/05/09.
157. Fearon ER, Vogelstein B. A genetic model for colorectal tumorigenesis. Cell. 1990;61(5):759–67.
158. Loeb LA. A mutator phenotype in cancer. Cancer Res. 2001;61(8):3230–9 Epub 2001/04/20.
159. Tomlinson IP, Novelli MR, Bodmer WF. The mutation rate and cancer. Proc Natl Acad Sci USA. 1996;93(25):14800–3 Epub 1996/12/10.
160. Tomlinson I, Bodmer W. Selection, the mutation rate and cancer: ensuring that the tail does not wag the dog. Nat Med. 1999;5(1):11–2 Epub 1999/01/12.
161. Jones S, Chen WD, Parmigiani G, Diehl F, Beerenwinkel N, Antal T, et al. Comparative lesion sequencing provides insights into tumor evolution. Proc Natl Acad Sci USA. 2008;105(11):4283–8 Epub 2008/03/14.
162. Tomasetti C, Vogelstein B. Cancer etiology. Variation in cancer risk among tissues can be explained by the number of stem cell divisions. Science. 2015;347(6217):78–81 Epub 2015/01/03.
163. Fearon ER, Vogelstein B. A genetic model for colorectal tumorigenesis. Cell. 1990;61(5):759–67.
164. Lengauer C, Kinzler KW, Vogelstein B. Genetic instabilities in human cancers. Nature. 1998;396 (6712):643–9 Epub 1999/01/01.
165. Rajagopalan H, Lengauer C. Aneuploidy and cancer. Nature. 2004;432(7015):338–41 Epub 2004/11/19.
166. Kops GJ, Weaver BA, Cleveland DW. On the road to cancer: aneuploidy and the mitotic checkpoint. Nat Rev Cancer. 2005;5(10):773–85 Epub 2005/10/01.
167. Vogelstein B, Fearon ER, Kern SE, Hamilton SR, Preisinger AC, Nakamura Y, et al. Allelotype of colorectal carcinomas. Science. 1989;244 (4901):207–11 Epub 1989/04/14.
168. Comprehensive molecular characterization of human colon and rectal cancer. Nature. 2012;487 (7407):330–7. Epub 2012/07/20.
169. Thiagalingam S, Laken S, Willson JK, Markowitz SD, Kinzler KW, Vogelstein B, et al. Mechanisms underlying losses of heterozygosity in human colorectal cancers. Proc Natl Acad Sci USA. 2001;98(5):2698–702 Epub 2001/02/28.
170. Wood LD, Parsons DW, Jones S, Lin J, Sjoblom T, Leary RJ, et al. The genomic landscapes of human breast and colorectal cancers. Science. 2007;318 (5853):1108–13 Epub 2007/10/13.
171. Seshagiri S, Stawiski EW, Durinck S, Modrusan Z, Storm EE, Conboy CB, et al. Recurrent R-spondin fusions in colon cancer. Nature. 2012;488 (7413):660–4 Epub 2012/08/17.
172. Powell SM, Zilz N, Beazer-Barclay Y, Bryan TM, Hamilton SR, Thibodeau SN, et al. APC mutations occur early during colorectal tumorigenesis. Nature. 1992;359(6392):235–7 Epub 1992/09/17.
173. Nusse R. Wnt signaling. Cold spring harbor perspectives in biology. 2012;4(5). Epub 2012/05/03.
174. Clevers H, Nusse R. Wnt/beta-catenin signaling and disease. Cell. 2012;149(6):1192–205 Epub 2012/06/12.
175. Esteller M, Sparks A, Toyota M, Sanchez-Cespedes M, Capella G, Peinado MA, et al. Analysis of adenomatous polyposis coli promoter hypermethylation in human cancer. Cancer Res. 2000;60 (16):4366–71 Epub 2000/09/02.
176. Bellis J, Duluc I, Romagnolo B, Perret C, Faux MC, Dujardin D, et al. The tumor suppressor Apc controls planar cell polarities central to gut homeostasis. J Cell Biol. 2012;198(3):331–41 Epub 2012/08/02.
177. Dikovskaya D, Li Z, Newton IP, Davidson I, Hutchins JR, Kalab P, et al. Microtubule assembly by the Apc protein is regulated by importin-beta — RanGTP. J Cell Sci. 2010;123(Pt 5):736–46 Epub 2010/02/11.
178. Green RA, Kaplan KB. Chromosome instability in colorectal tumor cells is associated with defects in microtubule plus-end attachments caused by a dominant mutation in APC. J Cell Biol. 2003;163 (5):949–61 Epub 2003/12/10.
179. Morin PJ, Vogelstein B, Kinzler KW. Apoptosis and APC in colorectal tumorigenesis. Proc Natl Acad Sci USA. 1996;93(15):7950–4 Epub 1996/07/23.

180. Downward J. Targeting RAS signalling pathways in cancer therapy. Nat Rev Cancer. 2003;3(1):11–22 Epub 2003/01/02.
181. Campbell SL, Khosravi-Far R, Rossman KL, Clark GJ, Der CJ. Increasing complexity of Ras signaling. Oncogene. 1998;17(11 Reviews):1395–413. Epub 1998/10/21.
182. Bos JL, Fearon ER, Hamilton SR, Verlaan-de Vries M, van Boom JH, van der Eb AJ, et al. Prevalence of ras gene mutations in human colorectal cancers. Nature. 1987;327(6120):293–7 Epub 1987/05/03.
183. Vogelstein B, Fearon ER, Hamilton SR, Kern SE, Preisinger AC, Leppert M, et al. Genetic alterations during colorectal-tumor development. N Engl J Med. 1988;319(9):525–32 Epub 1988/09/01.
184. Haigis KM, Kendall KR, Wang Y, Cheung A, Haigis MC, Glickman JN, et al. Differential effects of oncogenic K-Ras and N-Ras on proliferation, differentiation and tumor progression in the colon. Nat Genet. 2008;40(5):600–8 Epub 2008/04/01.
185. Wang Y, Velho S, Vakiani E, Peng S, Bass AJ, Chu GC, et al. Mutant N-RAS protects colorectal cancer cells from stress-induced apoptosis and contributes to cancer development and progression. Cancer Discov. 2013;3(3):294–307 Epub 2013/01/01.
186. Zenker M, Lehmann K, Schulz AL, Barth H, Hansmann D, Koenig R, et al. Expansion of the genotypic and phenotypic spectrum in patients with KRAS germline mutations. J Med Genet. 2007;44 (2):131–5 Epub 2006/10/24.
187. Edkins S, O'Meara S, Parker A, Stevens C, Reis M, Jones S, et al. Recurrent KRAS codon 146 mutations in human colorectal cancer. Cancer Biol Ther. 2006;5(8).928–32 Epub 2006/09/14.
188. Cantley LC. The phosphoinositide 3-kinase pathway. Science. 2002;296(5573):1655–7 Epub 2002/06/01.
189. Samuels Y, Velculescu VE. Oncogenic mutations of PIK3CA in human cancers. Cell Cycle. 2004;3 (10):1221–4 Epub 2004/10/07.
190. Samuels Y, Diaz LA Jr, Schmidt-Kittler O, Cummins JM, Delong L, Cheong I, et al. Mutant PIK3CA promotes cell growth and invasion of human cancer cells. Cancer Cell. 2005;7(6):561–73 Epub 2005/06/14.
191. Chalhoub N, Baker SJ. PTEN and the PI3-kinase pathway in cancer. Annu Rev Pathol. 2009;4:127–50 Epub 2008/09/05.
192. Rodriguez-Viciana P, Warne PH, Dhand R, Vanhaesebroeck B, Gout I, Fry MJ, et al. Phosphatidylinositol-3-OH kinase as a direct target of Ras. Nature. 1994;370(6490):527–32 Epub 1994/08/18.
193. Velho S, Oliveira C, Ferreira A, Ferreira AC, Suriano G, Schwartz S Jr, et al. The prevalence of PIK3CA mutations in gastric and colon cancer. Eur J Cancer. 2005;41(11):1649–54 Epub 2005/07/05.
194. De Roock W, Claes B, Bernasconi D, De Schutter J, Biesmans B, Fountzilas G, et al. Effects of KRAS, BRAF, NRAS, and PIK3CA mutations on the efficacy of cetuximab plus chemotherapy in chemotherapy-refractory metastatic colorectal cancer: a retrospective consortium analysis. Lancet Oncol. 2010;11(8):753–62 Epub 2010/07/14.
195. Vogelstein B, Kinzler KW. Cancer genes and the pathways they control. Nat Med. 2004;10(8):789–99 Epub 2004/08/03.
196. Beroud C, Soussi T. The UMD-p53 database: new mutations and analysis tools. Hum Mutat. 2003;21 (3):176–81 Epub 2003/03/06.
197. Russo A, Bazan V, Iacopetta B, Kerr D, Soussi T, Gebbia N. The TP53 colorectal cancer international collaborative study on the prognostic and predictive significance of p53 mutation: influence of tumor site, type of mutation, and adjuvant treatment. J Clin Oncol Official J Am Soc Clin Oncol. 2005;23 (30):7518–28 Epub 2005/09/21.
198. Leslie A, Carey FA, Pratt NR, Steele RJ. The colorectal adenoma-carcinoma sequence. Br J Surg. 2002;89(7):845–60 Epub 2002/06/26.
199. Grady WM, Markowitz SD. Genetic and epigenetic alterations in colon cancer. Annu Rev Genomics Hum Genet. 2002;3:101–28 Epub 2002/07/27.
200. Lamouille S, Xu J, Derynck R. Molecular mechanisms of epithelial-mesenchymal transition. Nat Rev Mol Cell Biol. 2014;15(3):178–96 Epub 2014/02/22.
201. Davis RJ, Welcker M, Clurman BE. Tumor suppression by the Fbw7 ubiquitin ligase: mechanisms and opportunities. Cancer Cell. 2014;26(4):455–64 Epub 2014/10/15.
202. Ionov Y, Peinado MA, Malkhosyan S, Shibata D, Perucho M. Ubiquitous somatic mutations in simple repeated sequences reveal a new mechanism for colonic carcinogenesis. Nature. 1993;363 (6429):558–61 Epub 1993/06/10.
203. Thibodeau SN, Bren G, Schaid D. Microsatellite instability in cancer of the proximal colon. Science. 1993;260(5109):816–9 Epub 1993/05/07.
204. Vilar E, Gruber SB. Microsatellite instability in colorectal cancer-the stable evidence. Nat Rev Clin Oncol. 2010;7(3):153–62 Epub 2010/02/10.
205. Dieringer D, Schlotterer C. Two distinct modes of microsatellite mutation processes: evidence from the complete genomic sequences of nine species. Genome Res. 2003;13(10):2242–51 Epub 2003/10/04.
206. Jass JR. Hereditary Non-Polyposis Colorectal Cancer: the rise and fall of a confusing term. World J Gastroenterol WJG. 2006;12(31):4943–50 Epub 2006/08/29.
207. Lu SL, Kawabata M, Imamura T, Akiyama Y, Nomizu T, Miyazono K, et al. HNPCC associated with germline mutation in the TGF-beta type II receptor gene. Nat Genet. 1998;19(1):17–8 Epub 1998/05/20.

208. de Miranda NF, van Dinther M, van den Akker BE, van Wezel T, ten Dijke P, Morreau H. Transforming Growth Factor beta Signaling in Colorectal Cancer Cells With Microsatellite Instability Despite Biallelic Mutations in TGFBR2. Gastroenterology. 2015;148(7):1427–37 e8. Epub 2015/03/05.
209. Toyota M, Ahuja N, Ohe-Toyota M, Herman JG, Baylin SB, Issa JP. CpG island methylator phenotype in colorectal cancer. Proc Natl Acad Sci USA. 1999;96(15):8681–6 Epub 1999/07/21.
210. Weisenberger DJ, Siegmund KD, Campan M, Young J, Long TI, Faasse MA, et al. CpG island methylator phenotype underlies sporadic microsatellite instability and is tightly associated with BRAF mutation in colorectal cancer. Nat Genet. 2006;38 (7):787–93 Epub 2006/06/29.
211. Rajagopalan H, Bardelli A, Lengauer C, Kinzler KW, Vogelstein B, Velculescu VE. Tumorigenesis: RAF/RAS oncogenes and mismatch-repair status. Nature. 2002;418(6901):934. Epub 2002/08/29.
212. Oliveira C, Pinto M, Duval A, Brennetot C, Domingo E, Espin E, et al. BRAF mutations characterize colon but not gastric cancer with mismatch repair deficiency. Oncogene. 2003;22 (57):9192–6 Epub 2003/12/12.
213. Davies H, Bignell GR, Cox C, Stephens P, Edkins S, Clegg S, et al. Mutations of the BRAF gene in human cancer. Nature. 2002;417(6892):949–54 Epub 2002/06/18.
214. Schetter AJ, Okayama H, Harris CC. The role of microRNAs in colorectal cancer. Cancer J. 2012;18 (3):244–52 Epub 2012/06/01.
215. Madison BB, Jeganathan AN, Mizuno R, Winslow MM, Castells A, Cuatrecasas M, et al. Let-7 Represses Carcinogenesis and a Stem Cell Phenotype in the Intestine via Regulation of Hmga2. PLoS genetics. 2015;11(8):e1005408. Epub 2015/08/06.
216. Tu HC, Schwitalla S, Qian Z, LaPier GS, Yermalovich A, Ku YC, et al. LIN28 cooperates with WNT signaling to drive invasive intestinal and colorectal adenocarcinoma in mice and humans. Genes Dev. 2015;29(10):1074–86 Epub 2015/05/10.
217. Schee K, Fodstad O, Flatmark K. MicroRNAs as biomarkers in colorectal cancer. Am J Pathol. 2010;177(4):1592–9 Epub 2010/09/11.
218. Terzic J, Grivennikov S, Karin E, Karin M. Inflammation and colon cancer. Gastroenterology. 2010;138(6):2101–14 e5. Epub 2010/04/28.
219. Karatzas PS, Gazouli M, Safioleas M, Mantzaris GJ. DNA methylation changes in inflammatory bowel disease. Ann Gastroenterol Q Publ Hellenic Soc Gastroenterol. 2014;27(2):125–32 Epub 2014/04/16.
220. Cooke MS, Evans MD, Dizdaroglu M, Lunec J. Oxidative DNA damage: mechanisms, mutation, and disease. FASEB J Official Publ Fed Am Soc Exp Biol. 2003;17(10):1195–214 Epub 2003/07/02.
221. De Robertis M, Massi E, Poeta ML, Carotti S, Morini S, Cecchetelli L, et al. The AOM/DSS murine model for the study of colon carcinogenesis: from pathways to diagnosis and therapy studies. J Carcinog. 2011;10:9 Epub 2011/04/13.
222. Greten FR, Eckmann L, Greten TF, Park JM, Li ZW, Egan LJ, et al. IKKbeta links inflammation and tumorigenesis in a mouse model of colitis-associated cancer. Cell. 2004;118(3):285–96 Epub 2004/08/06.
223. Grivennikov S, Karin E, Terzic J, Mucida D, Yu GY, Vallabhapurapu S, et al. IL-6 and Stat3 are required for survival of intestinal epithelial cells and development of colitis-associated cancer. Cancer Cell. 2009;15(2):103–13 Epub 2009/02/03.
224. Popivanova BK, Kitamura K, Wu Y, Kondo T, Kagaya T, Kaneko S, et al. Blocking TNF-alpha in mice reduces colorectal carcinogenesis associated with chronic colitis. J Clin Investig. 2008;118 (2):560–70 Epub 2008/01/26.

# Epidemiology of Hepatocellular Carcinoma

# 23

Francesca Giunchi, Francesco Vasuri and Michelangelo Fiorentino

## 23.1 Clinical Picture

### 23.1.1 Symptoms

Apart from the specific symptoms of the underlying cirrhosis or chronic hepatitis that accompany most hepatocellular carcinomas (HCC), these tumors are generally asymptomatic at earlier stages. Abdominal pain, weight loss, jaundice, ascites, and other general symptoms may appear in non-cirrhotic patients in advanced cases, but they are often masked by the symptoms of hepatic decompensation in cirrhotic patients [1, 2]. In the most advanced cases, signs of visceral hemorrhage and hepatic or portal vein thrombosis can appear. Patients with symptomatic HCC are usually at advanced stage, and not eligible for surgical therapy [2].

### 23.1.2 Diagnosis and Staging

In cirrhotic patients, imaging techniques and serum Alpha-fetoprotein (AFP) are used for the diagnosis of HCC, albeit the diagnosis of early lesions (less than 1–2 cm in diameter) still remains a challenge, and diagnostic biopsy is frequently requested. The radiological and contrastographic characteristics of early and advanced HCC go beyond the purposes of the present chapter, and they will not be discussed.

According to the American Joint Committee on Cancer (AJCC, 7th edition, 2010) [3], the stage of HCC is related to tumor dimensions and microvascular invasion.

- T1: solitary tumors without microvascular invasion;
- T2: tumors with microvascular invasion or multiple lesions less than cm 5 in diameter;
- T3a: multiple tumors more than cm 5 in diameter;
- T3b: invasion of a major branch of the hepatic vein or portal vein;
- T4: direct invasion of adjacent organs (rather than gallbladder) or perforation of the peritoneum.

## 23.2 Descriptive Epidemiology

HCC is the most common primary liver cancer, and represents one of the most frequent malignancies in the world; each year 782,000 men and women are diagnosed with the disease. It is the fifth most common malignancy in men and the eighth most common in women worldwide [1, 4]. Age-standardized incidence and mortality rates of HCC are much high in men than women (Fig. 23.1a, b).

F. Giunchi · F. Vasuri · M. Fiorentino (✉)
S.Orsola-Malpighi Teaching Hospital, Addarii Institute of Oncology, 40138 Bologna, Italy
e-mail: michelangelo.fiorentino@aosp.bo.it


M. Loda et al. (eds.), *Pathology and Epidemiology of Cancer*,
DOI 10.1007/978-3-319-35153-7_23

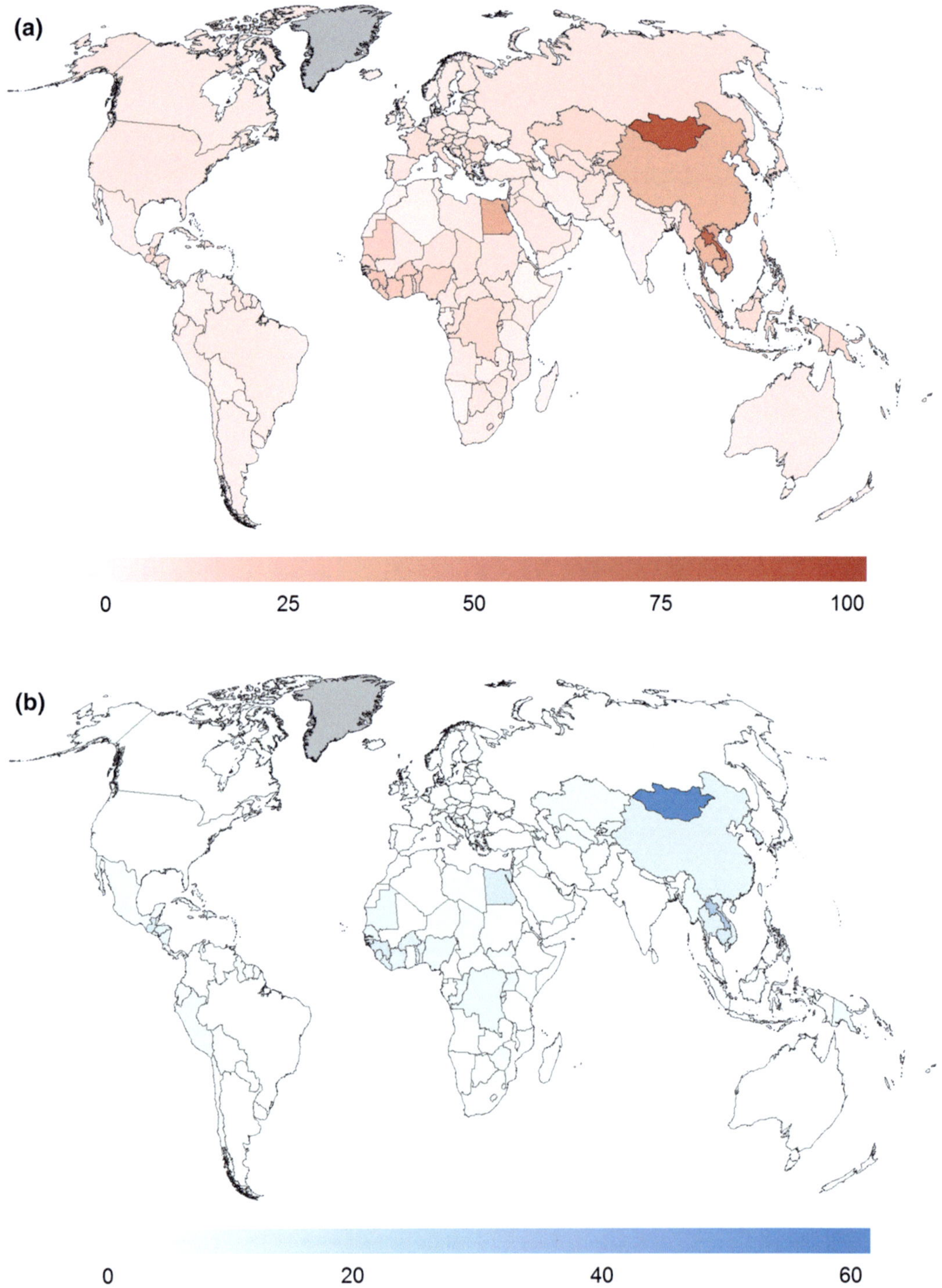


**Fig. 23.1** **a** A comparison of age-standardized mortality rates (per 100,000) of HCC among men, across different countries. Data are taken from the Globocan 2012 database. **b** A comparison of age-standardized mortality rates (per 100,000) of HCC among women, across different countries. Data are taken from the Globocan 2012 database

### 23.2.1 Trends by Geography

The age-standardized incidence rates of HCC show tremendous geographic differences, with a relatively low incidence in Europe and North America (2–7 per 100,000 cases per year), and a higher incidence in East Asia and Africa (>20 per 100,000 cases per year) (Fig. 23.1a, b) [1]. More than half ($N = 395{,}000$) of the cases of HCC occur in China, where it is the second most common cancer, primarily due to the endemic presence of the Hepatitis B Virus (HBV) (described under risk factors below) [4]. In Japan, the high incidence is due to the high prevalence of Hepatitis C virus (HCV) among the population [5].

More than 80 % of all cases occur in developing countries. In Africa HCC is the 4th most common cancer, with a particularly high incidence in the sub-Saharan area [4, 5]. The age of onset of HCC also varies considerably. In North America, the mean age at the diagnosis of HCC is 60 years, in Africa 35 years, and in Taiwan around 50 years [1]. In low-risk populations, the highest incidence of HCC is generally recorded among individuals of 75 years or older. HCC is rarely seen during the first 4 decades of life, except in population where infection with HBV is endemic.

### 23.2.2 Prognosis and Trends in Mortality

HCC is a highly lethal cancer, and thus the mortality rates of HCC shows the same geographic distribution of incidence [6]: over the last decades HCC mortality has decreased in Southern European countries, but increased in former low mortality areas of central and Northern Europe. Trends in HCC mortality in the United States (U.S.) and Australia were similar to those observed in central and Northern Europe, with an increased mortality rate in individuals 45–64 years. These mortality changes across Europe could be related to the decrease of HBV infection rates after the adoption of vaccination programs and the reduction of alcohol intake and tobacco smoking in Southern Europe. On the other hand, the increasing prevalence of HCV infection and alcohol consumption explains the opposite trend in central and Northern Europe. The increase of diabetes and the consequent overweight and obesity in several populations may have had a role in the recent adverse outcomes in HCC particularly in North America [7].

## 23.3 Pathogenesis and Risk Factors

HCC is a multistep process involving various risk factors, leading to chronic liver damage and genetic alterations. The most important single risk factor for HCC is cirrhosis, which can occur as a consequence of HBV and HCV infections as well as hemochromatosis and α1-antitrypsin deficit showed an independent risk of developing HCC [1]. Less studied independent risk factors include the exposure to aflatoxin and hormones, and recently also tobacco smoking has been shown to correlate with HCC [8].

### 23.3.1 Cirrhosis

Cirrhosis is a major health problem worldwide, and apart from the clinical problem of the end-stage disease, 80–90 % of HCC both in Asia and in Western Countries are estimated to be associated with this condition [9, 10]. The risk of HCC among individuals with cirrhosis ranges from 30 % at 5 years in Asian patients with HCV-driven cirrhosis, 20 % in patients with hemochromatosis, to 10 % for cirrhosis associated with alcohol and primary biliary disease [10]. In Western countries, alcohol consumption is a leading cause of cirrhosis, and its association with HCC is reinforced by other causes of chronic liver damage [11]. The oncogenic risk of cirrhosis largely derives from the hyperproliferative status of the regenerative hepatocytic nodules, but other molecular mechanisms have been proposed. For example, in cirrhotic liver tissue the activated stellate cells are able to activate both mitogen-activated protein kinase and the phoshatidylinositol-3 kinase pathways in response to circulating and paracrine pro-inflammatory cytokines (platelet-derived growth factor and transforming growth factor-ß, among others). Notably, the protein kinase and the phoshatidylinositol-3 kinase pathways represent the signal cascade typically seen in hepatocarcinogenesis [12].

In Western countries, with lower prevalence of HBV infection, 15–20 % of HCC arises without cirrhosis [13, 14]. HCC without cirrhosis is more likely occurring among men, has a bimodal distribution of age (peaking at 2nd and 7th decades) and a diagnosis typically at a symptomatic advanced stage, compared to patients with cirrhosis. Clinically, these HCC differ as well, including presenting as single large masses and lower serum α-fetoprotein level [13, 15]. Genomically, these cancers have a lower rate of p53 mutation and a higher prevalence of β[-beta]-catenin mutation, p14 inactivation and global gene methylation [13].

Despite the higher stage at the diagnosis, patients with non-cirrhotic HCC have a better overall survival and disease-free survival than cirrhotic patients, due to the underlying liver conditions [14].

### 23.3.2 HBV

HBV is a Hepadnavirus infecting 350,000,000 people worldwide [5], representing the primary cause of liver cirrhosis and HCC in Asia, Africa, and South America, although incidence varies considerably across countries [4, 16]. In endemic areas such as China, HBV infection causes approximately 80 % of HCC cases [6]. Moreover, HBV is associated with virtually all HCC cases in children. In Western Countries where the mass vaccination against HBV has taken place since the 1980s, the incidence of HBV-related hepatitis and HCC has lowered [17].

Eight different HBV genotypes have been identified (A–H), with distinct geographical and ethnic distribution. In Asia end-stage liver disease, cirrhosis, and HCC are much more associated with the genotype C than genotype B; in Western Europe and North America genotype D has a greater incidence in severe liver disease and HCC than genotype A. Furthermore, recent data associated genotype B with an earlier onset of HCC and with a non-cirrhotic background [6]. The lifetime risk of developing HCC in patients with chronic HBV liver disease is 10–25 % [18], with a 20–30 years lag time from HBV infection to diagnosis of HCC [19].

Most HBV-positive patients develop HCC in a cirrhotic background, although there is evidence of a direct carcinogenic effect of HBV in non-cirrhotic patients: 20 % of all HBV-related HCC developed in liver without cirrhosis and, in some cases, even without chronic hepatitis [20]. Among the proposed mechanisms of this direct viral carcinogenicity are the integration of the double-stranded DNA of HBV in the genome of the infected hepatocytes and the binding of the product of the viral HBx gene to product of the p53 gene causing its inactivation [21, 22]. HBx does not bind the DNA but makes protein–protein interaction, triggering the transcriptional activation of several viral and cellular promoters and enhancers. HBx also deregulates the expression of oncogenes (e.g., c-Myc and c-Jun), cytokines and transcription factors, and modulates cytoplasmic signal transduction pathways, and is involved in a number of oncogenic effects [20, 21, 23–25]. It was recently shown that HBX may also have an important role in modulating the epigenetic control of viral and cellular genes, including a number of tumor suppressor genes [18].

HBV-related HCC commonly exhibits a higher rate of chromosomal abnormalities than liver tumors linked to other risk factor, and it has been suggested that HBV might generate genomic instability either through viral DNA integration or the activity of its proteins [18].

### 23.3.3 HCV

Hepatitis C virus (HCV) is a single-stranded RNA virus of the Flaviviridae family, with at least six known genotypes characterized by high genetic variability. The highest prevalence of HCV infection worldwide is recorded in the

Middle East and Africa, the lowest in Europe and America [26]. In North America, Japan, and some European countries like Italy and Spain, HCV is the main risk factor of cirrhosis and HCC [18]. The incidence of HCC in patients infected with HCV was estimated up to 4 %, with a mean time from HCV infection to the diagnosis of HCC of 30 years [10, 27].

HCV-associated HCC arises almost invariably in livers with cirrhosis, since there do not appear to be direct carcinogenic roles in HCC [5] and the pathogenesis is more likely to be related to the HCV-driven chronic hepatitis and fibrosis. HCV leads to chronic inflammation, immune-mediate hepatocytes death, tissue damage, fibrosis activation by hepatic stellate cells, and replicative senescence due to telomerase shortening. In addition, factors such as oxidative stress, steatosis, and insulin resistance accelerate the evolution to cirrhosis and HCC [28] There is no evidence that the HCV integrates into the host genome, although HCV proteins interact with many host-cell proteins at a cytoplasmic level, thereby influencing cell signaling, transcription, proliferation, apoptosis, and translational regulation [29]. In vitro studies show a possible oncogenic role for four HCV proteins: core protein, NS3, NS5b, and NS5a [21, 23, 29, 30]. In particular, HCV core protein and NS3 are able to modulate the expression of p21 and p53, with alterations in the apoptosis and cell proliferation [23]. Another proposed mechanism for HCV carcinogenesis is an increased cellular oxidative stress by compromising the endoplasmic reticulum [23].

### 23.3.4 Hemochromatosis

Hemochromatosis is a relatively frequent metabolic disorder leading to iron accumulation and cirrhosis, and it is associated with an increased risk of developing HCC, compared to the general population [2]. HCC accounts for as many as 45 % of disease-related deaths in patients afflicted by hemochromatosis; the estimated risk of HCC in hemochromatosis is 8–10 % in most studies, although lower risk (as low as 1.7 %) has been recorded [31]. Although the effect of hemochromatosis is largely attributed to its effect on cirrhosis, some studies have found that intracellular iron overload can have a role in HCC pathogenesis [32]. HCC in hemochromatosis is commonly attributed to the iron overload in the hepatocytes that leads to direct oxidative damage by the creation of free radicals: this damage associated with an activation of stellate cells leads to liver fibrosis, cirrhosis, and then HCC [33]. Very recently, a "two-hit" hypothesis has been proposed from the observation of iron-free dysplastic nodules and HCC: according to this hypothesis the tumor initiation might be represented by the iron overload leading to cirrhosis and a second step, tumor promotion, might be represented by the proliferation of iron-free malignant hepatocytes in the effort to avoid the citotoxic effect of iron [28].

### 23.3.5 Aflatoxin B1

Aflatoxin B1 is a toxin produced by fungi of the Aspergillus family, and can be present in contaminated foods including wheat, nuts, and corn. It is found in moist climates, including parts of Asia and sub-Saharan Africa. Among the four known Aflatoxins (B1, B2, G1 and G2), Aflatoxin B1 is the most potent hepatocarcinogen in animal models [7].

High levels of Aflatoxin B1 are linked to a high risk of developing HCC, in particular, in association with chronic hepatitis of other etiologies, HBV above all [33, 34]. Prospective studies in Chinese populations showed that the urinary excretion of Aflatoxin metabolites was associated with a fourfold increase of risk of developing HCC. HBV-positive subjects who also excreted Aflatoxin metabolites had as much as a 60-folder increased risk of HCC [6]. Changes in food policy and economic development have led to a decline in the Aflatoxin exposure in some endemic regions, such as Taiwan and selected China areas with a consequent reduction in HCC mortality [7].

Aflatoxin is metabolized in AFB1–5,9-oxide by cytochromes, and this active form of the toxin binds the DNA, causing mutations in the onco-suppressor gene TP53, mutations which are highly represented in the HCC arisen in patients exposed to the toxin [4, 19, 35].

### 23.3.6 Other Risk Factors

*Alcohol*: Heavy alcohol intake (>50–70 g/day, less in women) is likely to be independently associated with an increased risk of HCC. While alcoholic cirrhosis is associated with the risk of developing HCC, the direct hepatocarcinogenic role of alcohol is unknown, especially in the non-cirrhotic livers [28]. There is evidence of a synergistic effect between heavy alcohol intake and HCV or HBV infection [6]. Ethanol showed no direct carcinogenic effects in mice and rats, but acetaldehyde, an ethanol metabolite, has demonstrated carcinogenic properties through DNA binding. Production of acetaldehyde depends on CYP2E1, a cytochrome induced by ethanol consumption, which converts the ethanol into acetaldehyde but also in reactive oxygen species. Chronic oxidative stress induced by ethanol intake and cytokine production, in the context of chronic inflammation, could lead to the accumulation of reactive species of oxygen, lipid peroxidation, and the creation of DNA adducts [28]. Other proposed mechanisms of alcohol-mediated carcinogenesis are represented by downregulation of retinoic acid levels and modulation of DNA methylation [28].

*Metabolic disorders*: Diabetes, obesity, and lipid disorders can lead to nonalcoholic steatohepatitis (NASH) and both diabetes and obesity are associated with a two to threefold increased risk of HCC. Epidemiological studies have demonstrated an association between diabetes and an increased risk of HCC (as well as HCC-related mortality), compared to non-diabetic cirrhotic patients. Changes in hepatic activity related to metabolic alterations or impaired liver function in diabetics are a possible explanation of this association [12]. In particular, diabetes increases the risk of nonalcoholic fatty liver disease that can evolve through NASH to fibrosis and then to HCC [7].

The pathogenesis of NASH and diabetes is related to various factor that caused the activation of insulin-mediated proliferative pathways, release of specific cytokines, intracellular lipid species, oxidative stress, and mitochondrial damage, with mutations of p53 and other genes [2, 36]. Moreover, hyperinsulinemia and insulin resistance cause an upregulation of the insulin-like growth factor-1 system, which stimulates cellular proliferation and inhibits apoptosis within the liver [7].

*Hormones*: The use of contraceptive steroids has been reported to play a possible role in liver carcinogenesis [37]. Albeit demonstrated in mouse models [38], the relationship between sex hormones and HCC in patients without other causes of liver damage is still much less clear than hepatocellular adenoma, and the long-term public-health implication of any modest excess liver cancer risk among current oral contraceptive users is likely to be small [7].

Cigarette smoking: Several studies analyzed the relationship between cigarette smoking and HCC, with discordant results. Among the studies describing positive associations, most found that the effects were limited to individuals with HBV and HCV infection: meta-analyses that evaluated the interaction among HBV, HCV, cigarette smoking, and HCC revealed a more than additive interaction between HBV and smoking and a more than multiplicative interaction between HCV and smoking [6, 7].

The molecular epidemiology of HCC is still largely unexplored and the available data today is derived mainly from selected geographical areas. Further studies involving larger and varied well-annotated populations with stronger functional evidences could provide potential germline biomarkers for HCC risk assessment.

## 23.4 Summary

Epidemiological studies have established several of the major causes of HCC, including HBV, HCV, alcohol, obesity, and exposure to aflatoxin. Virtually all of these risk factors are avoidable or modifiable, and thus a substantial proportion of the burden of HCC could be reduced. The prevalence of these diverse risk factors varies across countries and accounts for the observed geographic variability of this cancer. The development of liver cirrhosis is likely to represent the key mechanism of action in all these risk factors, at least in most HCC cases.

## References

1. Odze RD, Goldblum JR editors. Surgical pathology of the GI tract, liver, biliary tract, and pancreas. 2nd ed. Saunders, Elsevier; 2009.
2. Burt AD, Portmann BC, Ferrell LD editors. MacSween's pathology of the liver. 6th ed. Elsevier, Churchill Livingstone; 2012.
3. Edge S, Byrd D.R, Compton CC, Fritz AG, Greene FL, Trotti A. American Joint Committee on Cancer (AJCC) cancer staging manual. 7th ed. Berlin: Springer; 2010.
4. Nordenstedt H, White DL, El-Serag HB. The changing pattern of epidemiology in hepatocellular carcinoma. Dig Liver Dis. 2010;42:S206–14.
5. IARC/WHO classification of tumours of the digestive system. 4th ed. Lyon; 2010.
6. El-Serag HB. Epidemiology of viral hepatitis and hepatocellular carcinoma. Gastroenterology. 2012;142:1264–73.
7. Bosetti C, Turati F, La Vecchia C. Hepatocellular carcinoma epidemiology. Best Pract Res Clin Gastroenterol. 2014;28:753–70.
8. IARC Working Group on the Evaluation of Carcinogenic. Risks to humans. Tobacco smoke and involuntary smoking. IARC Monogr Eval Carcinog Risks Hum. 2004;83:1–1438.
9. Colombo M, de Franchis R, Del Ninno E, et al. Hepatocellular carcinoma in Italian patients with cirrhosis. N Engl J Med. 1991;325:675–80.
10. Fattovich G, Stroffolini T, Zagni I, et al. Hepatocellular carcinoma in cirrhosis: incidence and risk factors. Gastroenterology. 2004;127(5 Suppl 1):S35–50.
11. Franceschi S, Montella M, Polesel J, et al. Hepatitis viruses, alcohol, and tobacco in the etiology of hepatocellular carcinoma in Italy. Cancer Epidemiol Biomarkers Prev. 2006;15:683–9.
12. Kim DY, Han KH. Epidemiology and surveillance of hepatocellular carcinoma. Liver Cancer. 2012;1:2–14.
13. Trevisani F, Frigerio M, Santi V, et al. Hepatocellular carcinoma in non-cirrhotic liver: a reappraisal. Dig Liver Dis. 2010;42:341–7.
14. Gaddikeri S, McNeeley MF, Wang CL, et al. Hepatocellular carcinoma in the noncirrhotic liver. Am J Roentgenol. 2014;203:W34–47.
15. Giannini EG, Marenco S, Bruzzone L, et al. Hepatocellular carcinoma in patients without cirrhosis in Italy. Dig Liver Dis. 2013;45:164–9.
16. Raza SA, Clifford GM, Franceschi S. Worldwide variation in the relative importance of hepatitis B and hepatitis C viruses in hepatocellular carcinoma: a systematic review. Br J Cancer. 2007;96:1127–34.
17. Chang MH, Chen CJ, Lai MS, et al. Universal hepatitis B vaccination in Taiwan and the incidence of hepatocellular carcinoma in children. Taiwan Childhood Hepatoma Study Group. N Engl J Med. 1997;336:1855–9.
18. Seeger C, Mason WS. Hepatitis B virus biology. Microbiol Mol Biol Rev. 2000;64:51–68.
19. Lee WM. Hepatitis B virus infection. N Engl J Med. 1997;337:1733–45.
20. Pollicino T, Saitta C, Raimondo G. Hepatocellular carcinoma: the point of view of the hepatitis B virus. Carcinogenesis. 2011;32:1122–32.
21. Feitelson MA, Sun B, Satiroglu Tufan NL, et al. Genetic mechanisms of hepatocarcinogenesis. Oncogene. 2002;21:2593–604.
22. Huo TI, Wang XW, Forgues M, et al. Hepatitis B virus X mutants derived from human hepatocellular carcinoma retain the ability to abrogate p53-induced apoptosis. Oncogene. 2001;20:3620–8.
23. Block TM, Mehta AS, Fimmel CJ, et al. Molecular viral oncology of hepatocellular carcinoma. Oncogene. 2003;22:5093–107.
24. Hu Z, Zhang Z, Doo E, et al. Hepatitis B virus X protein is both a substrate and a potential inhibitor of the proteasome complex. J Virol. 1999;73:7231–40.
25. Rahmani Z, Huh KW, Lasher R, et al. Hepatitis B virus X protein colocalizes to mitochondria with a human voltage-dependent anion channel, HVDAC3, and alters its transmembrane potential. J Virol. 2000;74:2840–6.
26. World Hepatitis Alliance. Viral hepatitis: global policy. London: World Hepatitis Alliance, 2014.
27. Tong MJ, el-Farra NS, Reikes AR, Co RL. Clinical outcomes after transfusion-associated hepatitis C. N Engl J Med. 1995;332:1463–6.
28. Nault JC. Pathogenesis of hepatocellular carcinoma according to aetiology. Best Pract Res Clin Gastroenterol. 2014;28:937–47.
29. Herzer K, Gerken G, Hofmann TG. Hepatitis C-associated liver carcinogenesis: role of PML nuclear bodies. World J Gastroenterol. 2014;20:12367–71.
30. Liang TJ, Heller T. Pathogenesis of hepatitis C-associated hepatocellular carcinoma. Gastroenterology. 2004;127(5 Suppl 1):S62–71.
31. Kew MC. Hepatic iron overload and hepatocellular carcinoma. Liver Cancer. 2014;3:31–40.
32. Harrison SA, Bacon BR. Relation of hemochromatosis with hepatocellular carcinoma: epidemiology, natural history, pathophysiology, screening,

treatment, and prevention. Med Clin North Am. 2005;89:391–409.
33. Sun Z, Lu P, Gail MH, et al. Increased risk of hepatocellular carcinoma in male hepatitis B surface antigen carriers with chronic hepatitis who have detectable urinary aflatoxin metabolite M1. Hepatology. 1999;30:379–83.
34. Ross RK, Yuan JM, Yu MC, et al. Urinary aflatoxin biomarkers and risk of hepatocellular carcinoma. Lancet. 1992;339(8799):943–6.
35. Montesano R, Hainaut P, Wild CP. Hepatocellular carcinoma: from gene to public health. J Natl Cancer Inst. 1997;89:1844–51.
36. Dongiovanni P, Romeo S, Valenti L. Hepatocellular carcinoma in nonalcoholic fatty liver: role of environmental and genetic factors. World J Gastroenterol. 2014;20:12945–55.
37. Kalra M, Mayes J, Assefa S, et al. Role of sex steroid receptors in pathobiology of hepatocellular carcinoma. World J Gastroenterol. 2008;14:5945–61.
38. Shimomura M, Higashi S, Mizumoto R. 32P-postlabeling analysis of DNA adducts in rats during estrogen-induced hepatocarcinogenesis and effect of tamoxifen on DNA adduct level. Jpn J Cancer Res. 1992;83:438–44.

# The Pathology and Molecular Pathology of Hepatocellular Carcinoma

# 24

Francesco Vasuri, Francesca Giunchi and Michelangelo Fiorentino

## 24.1 Introduction

In this chapter, we begin by introducing the embryology and anatomy of the normal human liver. Next, we discuss the histology and molecular pathology of precursor lesions including regenerative and dysplastic liver nodules. We subsequently concentrate on the macro- and microscopic features of hepatocellular carcinoma and conclude with a discussion of the molecular pathways involved in hepatocarcinogenesis.

## 24.2 Liver Embryology

Human liver develops at the end of the third gestational week of the human embryo. The liver bud ("hepatic diverticulum") derives from the endodermal tissue of the distant foregut, next to the duodenal bud. The hepatic bud is originally a hollow structure, then it merges with the mesenchymal tissue derived from the septum transversum and the coelomic cavity, acquiring the shape of a solid organ at the 4–5th gestational week. Mature liver stroma, capsule, hematopoietic elements, and Kupffer cell derive from this mesenchymal component. At this time the development of the sinusoidal system occurs: from the liver bud, buds of endothelial sinusoidal cells develop radially throughout the mesenchymal component. After the 4th week, a portion of the liver bud which has not merged in the septum transversum gives rise to the extrahepatic biliary tree, including the cystic duct and the gallbladder. The synthesis of α[alpha]-fetoprotein by hepatoblasts starts very early, at the 4–5th week, while glycogen storage begins at the 8th gestational week [1]. Apart from α[alpha]-fetoprotein, bipotential hepatoblasts, precursors of both hepatocyte and cholangiocytes, express HNF4α[alpha], keratin 8 (K8), K18, K19, K14, E-cadherin, EpCAM, CD133, DLK1, and synthesize Albumin transcripts [2]. The bile canaliculi appear as spaces lined by hepatoblasts in the 5–6th week, when the epithelial cells in contact with the mesenchyma surrounding the portal vein generate the ductal plate, that in turn gives rise to the intrahepatic biliary tree [3]. According to some authors, the ductal plate harbors the fetal hepatic stem cells, within the bipotential hepatoblasts: these stem cells are negative for both Albumin transcripts and α[alpha]-fetoprotein, and express stemness markers such as CD133, CD90 CD34, and c-Kit, in addition to Claudin-3, NCAM1, and EpCAM [4]. Fetal hepatic stem cells are able to generate hepatoblasts in vitro [5].

F. Vasuri · F. Giunchi · M. Fiorentino (✉)
S.Orsola-Malpighi Teaching Hospital, Addarii Institute of Oncology, Bologna 40138, Italy
e-mail: michelangelo.fiorentino@aosp.bo.it


M. Loda et al. (eds.), *Pathology and Epidemiology of Cancer*,
DOI 10.1007/978-3-319-35153-7_24

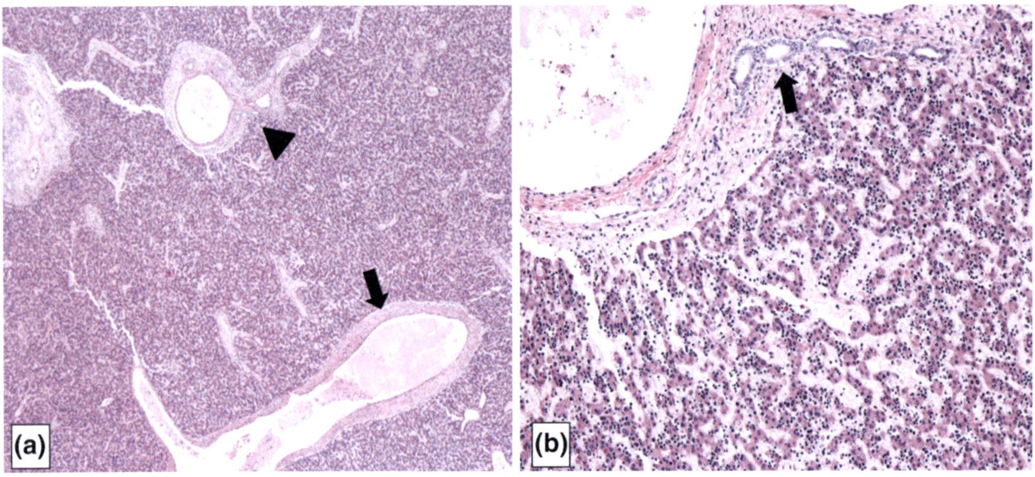


**Fig. 24.1** Fetal liver (3° trimester). **a** Periportal space (*arrowhead*) and hepatic vein (*arrow*). **b** Intrahepatic hemopoiesis and mature biliary ducts (*arrow*) haematoxylin–eosin stain

During the 6th gestational week, hematopoiesis becomes visible and by the 12th week the liver becomes the main hematopoietic organ of the fetus. Within the third trimester the hepatic liver hemopoietic function stops (Fig. 24.1).

From 5th to 9th gestational week a rapid growing of the organ is seen, with the liver reaching up to 10 % of the overall embryo weight. At the third trimester this growth slackens, and the liver reaches the 5 % of the final newborn's weight [3]. The development of intrahepatic biliary tree proceeds centrifugally from the porta hepatis to the peripheral branches of portal tracts. At birth, the biliary system is still not fully developed, and Keratin 7-positive structures may not be visible in peripheral portal structures until the first month after birth [1].

Many molecules and several molecular pathways have been identified to play a role in liver development. The Nodal signaling pathway—included in the TGFβ[beta] family—is involved in the progression of the definitive endoderm, via Alk4, Alk7, or ActRII A and B, driving the commitment of the endodermic and mesodermic layers [2]. The Nodal signaling is finely regulated by several molecules, such as fibroblast growth factor (FGF) or bone morphogenetic protein (BMP) [2]. The foregut formation in animal models seems to be related to the low expression of β[beta]-catenin in the anterior side of endoderm during gastrulation phase: the low β[beta]-catenin activity in this region causes an upregulation of the Wnt antagonists of the secreted frizzled-related protein (SFRP) family, as well as DKK-1, key molecules in foregut development [2].

The bipotential differentiation od hepatoblasts in hepatocyte or biliocyte is regulated by TGFβ[beta], Wnt, Notch, FGF, and BMP, among others. In particular, an increased TGFβ[beta] activity around the portal tracts together with an increased activity of the Notch and FGF signaling are linked to biliocyte differentiation and proliferation, while the HGF and TNFα[alpha] signaling promote hepatocyte differentiation [6, 7].

Furthermore, mouse models revealed other mechanisms involved in the following phases of hepatocytic differentiation, including Wnt signaling, FGF, BMP, and FoxA2 [2, 8–10], while c-Jun, the hepatoma-derived growth factor (HDGF), the nuclear factor kB (NFkB), the c-Met pathway, and HNF4α[alpha] were all described during the phases of liver growing [11, 12].

## 24.3 Anatomy and Histology of the Liver

In adults, the liver accounts for approximately the 2 % of the total body weight, with a mean weight of 1400–1600 g [3]. The liver is located in the right upper quadrant of the abdomen, completely covered by the ribs in nonpathological conditions. The canonical anatomic division of the liver includes a right and a left lobe, respectively, lateral and medial to the falciform ligament, as well as the smaller quadrate and caudate lobes. This topographical division has been progressively replaced by the more recent division in eight functional segments: each segment depends on specific branches of the hepatic artery, the portal vein, and the biliary tree. Thus, this division follows the anatomy of the vascular and biliary branching and not the gross anatomy of the organ. Therefore, it is important for surgical reasons, especially when small portions of the liver parenchyma must be resected. According to the segmental division, the liver is not divided in left and right lobes (with the falciform ligament as mark), but in left and right hemiliver, according to the first branching of the portal vein at hilum. The left hemiliver incorporates segments II, III, and IV, the right hemiliver the V, VI, VII, and VIII; caudate lobe is segment I. According to some authors, the quadrate lobe should be regarded as a separate segment [1, 13].

Histologically, the liver lobule was described as the fundamental unit of the liver since seventeenth century [1]: the canonical lobule is conceived as an hexagon with portal tracts in each angle—each made by a terminal branch of the liver artery, a terminal branch of the portal vein, and a bile duct—and a central vein. The modern acinar concept was introduced by Rappaport in the middle-twentieth century: the "Rappaport acinus" is seen as a triangle, with two portal tracts in two angles and the hepatic vein (called "terminal vein") in the third (Fig. 24.2). This approach is more functional, since it is based on the blood flow (and the hepatocyte oxygenation), which goes from the portal tracts to the terminal vein dividing the hepatocytes layers into three zones, from "zone 1" (closer to the portal tract and more oxygenated) to "zone 3" (closer to the vein and less oxygenated) [14]. Although many other systems have been proposed for the classification of the liver functional unit, the "Rappaport acinus" still represents the most utilized since it provides a strong link

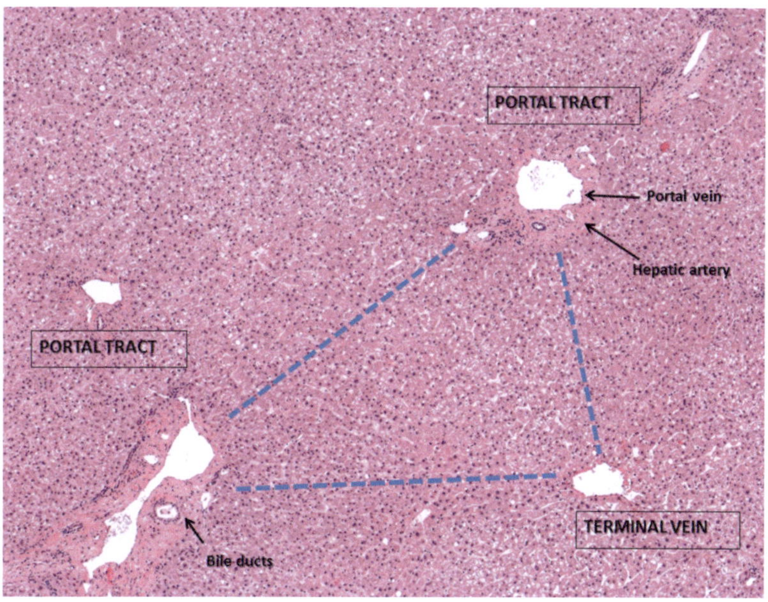


**Fig. 24.2** The "Rappaport acinus" in a liver with normal architecture is represented as a triangle, with two portal tracts in two angles and the terminal vein in the third, according to the blood flow. Haematoxylin–eosin stain

between liver microanatomy and liver pathology (e.g., fibrosis progression, toxic and ischemic damage, etc.) [1].

In adults, hepatocytes are large polygonal cells, with wide eosinophilic cytoplasms, disposed in one-cell thick layers. Each heaptocyte has vascular and biliary "poles" that are properly shaped to create the sinusoids and the terminal bile ductules when the cells are juxta posed in layers. The sinusoids form a vascular network between these hepatocyte plates, surrounding them with a very specialized endothelium, with sieve-like cytoplasm and without a continuous basal membrane, that leads the blood flow toward the terminal veins. The biliary spaces created between two hepatocytes bring the bile toward the portal tracts where it flows into the bile ductules [3]. All these microanatomical features warrant the interrelations among hepatocytes portal blood flow and bile flow.

## 24.4 Histopathology of Precursor Lesions: Regenerative Nodules and Dysplastic Nodules

Liver cirrhosis represents the main clinical condition leading to hepatocellular carcinoma (HCC) in developed countries (Fig. 24.3). Foci of cells with atypical features can be present in cirrhotic regenerative nodules (RNs). Although these hepatocytes may be enlarged with prominent nuclei, their nucleus-to-cytoplasm (NC) ratio is generally preserved with only a little reduction of the amount of cytoplasm present. These features can be ascribed to hepatocyte degenerative and/or regenerative changes rather than to true cell dysplasia [15]. Therefore, the diagnosis of dysplasia is recommended only when these changes involve an entire liver nodule or at least a large focus within a nodule [16].

The distinction between a large RN and a low-grade dysplastic nodule (LGDN) is not achievable at the clinical or histopathological level, since the only significant difference is that an LGDN is clonal while an RN is not. Because

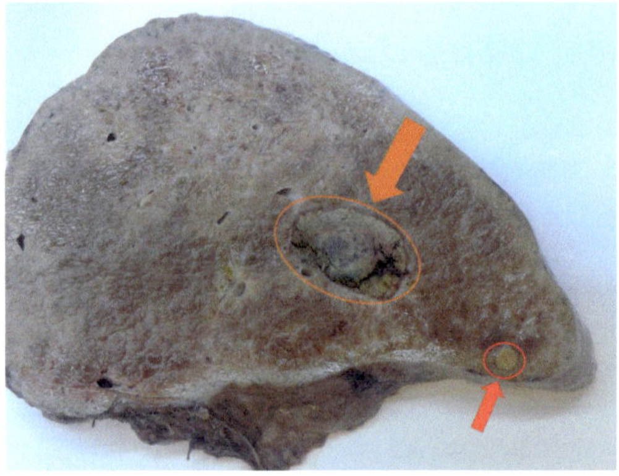

**Fig. 24.3** Gross features of a cirrhotic liver: the parenchyma is replaced by regenerative nodules, with evidence of a major HCC nodule (*orange arrow*); a distant neoplastic thrombosis is visible as well (*red arrow*)

clonality assays are expensive and time-consuming, and since an LGDN is considered a lesion with very low risk of progression to HCC, the distinction between an RN and an LGDN is considered of little clinical relevance. These lesions are by definition bigger than 1 cm in size (alternatively, they must be clearly identifiable in the cirrhotic parenchyma regardless of size), and should lack the histological and architectural characteristics of a high-grade dysplastic nodule (HGDN) [17, 18]. LGDNs still display hepatocyte plates no more than two-cell thick, without trabeculae or acinar structures and surrounded by a preserved reticulin network. The portal structures are readily recognizable within the nodules.

HGDNs are macroscopically similar to RNs and LGDNs, but they are characterized by dysplastic cytology. Histologically, these nodules are often hypercellular [19]. HGDNs may focally loose the reticulin network, with hepatocellular plates comprised of three or more cells in thickness. Unpaired arteries can be seen associated with an increased sinusoidal endothelialization [20]. These subtle features may not always be visible a small biopsy, making diagnosis challenging when diagnostic specimen are limited. The most important differential diagnosis of a HGDN is with early HCC (eHCC). This will be discussed below.

### 24.4.1 Molecular Pathology of Precursor Lesions

The process of tumor initiation and progression in cirrhosis is well established. Progressive accumulation of genetic and epigenetic abnormalities marks the temporal steps of hepatocarcinogenesis [1]. In the early stages of hepatocarcinogenesis the epigenetic mechanisms of gene expression upregulation prevail, involving particularly the transforming growth factor-α[alpha] (TGF α[alpha]) and the insulin-like growth factor-2 (IGF-2). Chronic hepatitis (both viral and nonviral) and cirrhosis are considered predisposing situations, since the combined action of inflammatory cytokines, viral transactivation, and regenerative response can determine changes in telomerase length, gene methylation, microsatellite instability, and fraction of alleles that were observed both in hepatocytes during chronic inflammation and cirrhosis, and in dysplastic nodules as well as HCC [21]. In these inflammatory situations there is an increased expression of DNA methyltransferases (DNMTs), that catalyze the methylation and demethylation of CpG groups, and both DNMT1 and DNMT3a are upregulated in HCC [21].

Loss of heterozygosity (LOH) and microsatellite instability (MSI) are detectable in preneoplastic liver lesions, with the most frequent deletion occurring at 8p, both in HGDNs and HCC. Similarly, RNs, dysplastic nodules, and adjacent HCC may share an MSI phenotype at identical loci and similar allelic deletions or gene mutations. Oncogene activation (v-akt homolog 2), loss of tumor suppressor genes (WT1, RB1), and DNA repair genes are deregulated in HGDN. Similarly, the expression at the protein level of growth factors (EGFR, IGFBP3) cytokines, adhesion proteins, signal transduction proteins, transcription factors, and housekeeping genes is altered in precursor and neoplastic hepatocellular lesions [1, 21–23]. The AKT-mTORC1 signaling pathway is activated in cirrhotic liver with and without HCC. As a result, increased expression of phospho-S6 (an AKT effector) was observed, while PTEN staining was negative in 24 % of HCC cases [23].

Allelic deletions are rare in chronic hepatitis and cirrhosis, but their number steadily increases in dysplastic nodules and in HCC and that suggests that HCCs often arise as clonal outgrowths of cirrhotic (dysplastic) nodules. Although several of the these structural alterations in preneoplastic cell populations differ from those found in adjacent HCCs, suggesting that most of the cells harboring the early genomic aberrations do not evolve into the malignant phenotype [21].

Hepatocyte dysplasia, especially the "large-cell" HGDN, is often accompanied by alterations that can also be found in HCC, such as telomerase shortening, microsatellite instability, inactivation of cell cycle checkpoints CDKN2A and CDKN1A, resulting in increased proliferation index [22]. However, the most common genetic alterations in HCC (i.e., β[beta]-catenin or TP53 mutations) have never been described in preneoplastic liver lesions [24].

## 24.5 Hepatocellular Carcinoma

### 24.5.1 Gross Appearance

The macroscopic features of HCC largely depend on the status of the surrounding liver. In cirrhotic livers, HCC is seen as a single lesion or as multiple lesions, and is encapsulated, resembling larger cirrhotic nodules (Fig. 24.3). The neoplastic nodules are soft, with a grayish-white to greenish color, according to the amount of bile content. The macroscopic diagnosis of HCC in a cirrhotic background is not always easy, especially in eHCC < 2 cm in size. HCC in noncirrhotic livers is usually a single large mass, with irregular, poorly defined borders, sometimes with satellite lesions and grossly visible vascular invasion. Vascular invasion is a common feature of any HCC at the time of diagnosis, involving the portal and the hepatic veins, or the vena cava [15].

### 24.5.2 Histopathology of "Early HCC"

Small nodules (≤2 cm) within a cirrhotic background with the architectural and cytologic atypical of a HGDN, but affecting 100 % of the cells, is classified as (eHCC) [19]. eHCC is by definition grade 1 according to the Edmondson and Steiner grading system (see below). At a histological level, hypercellularity, loss of reticulin staining, and the presence of hepatocellular trabeculae more than two-cell thick are helpful diagnostic criteria if present throughout. In addition, no portal tracts are visible within the lesion. Nearly half of eHCCs show fatty changes (macrovescicular steatosis) [24]. An important diagnostic feature is the occurrence of stromal invasion, defined as the presence of HCC cells in the stroma of the surrounding portal tracts (or in fibrous septa) [25]. Although stromal invasion is the most specific histological feature for the diagnosis of eHCC, it is not always recognizable on needle biopsies, and even when present, it is difficult to distinguish from non-neoplastic hepatocytes entrapped within the fibrosis.

### 24.5.3 Histopathology of Advanced HCC

Unlike many other solid tumors the neoplastic cells in HCC are quite similar to the non-neoplastic hepatocytes, i.e., they are polygonal cells with a variable amount of eosinophilic granular cytoplasm and large polymorphic vesicular nuclei, with open chromatin and one or more prominent nucleoli. Many cell features seen in not-neoplastic hepatocytes are common also in HCC, such as Mallory bodies, hyaline bodies, glycogen inclusions, fatty changes, bile production, and nuclear inclusions [15].

HCCs are routinely graded according to the original Edmondson and Steiner system of 1954 [26]. Like the grading systems of other cancers, Edmondson and Steiner's is based on the dedifferentiation of the neoplastic hepatocytes, i.e., their progressive loss of resemblance with the normal hepatocyte (Fig. 24.4). More practically, the grade of HCC is based on the nuclear-to-cytoplasmic ration, nuclear hyperchromasia, and cell polymorphism. Grade 1 HCCs are characterized by neoplastic hepatocytes with preserved eosinophilic cytoplasms, mild (or no) atypia, while in grade 2 and 3 HCCs a progression of the above-mentioned features is observed. Grade 4 HCCs include anaplastic, small-cell and giant-cell tumors.

The architectural appearance of HCC is variable, with different, often intermingled patterns. The trabecular, acinar, and solid growth patterns are recognized. The most common pattern, particularly in well-to-moderately differentiated HCC, is the trabecular pattern, characterized by cords of neoplastic cells surrounded by vascular sinusoids [15]. These cords resemble the normal hepatic trabeculae, but are generally more than two neoplastic cell thick, with a high N/C ratio and nuclear overcrowding. With tumor progression, the neoplastic trabeculae can become distorted, creating a pseudoglandular pattern. In the acinar growth pattern, the neoplastic hepatocytes form luminal structures resembling bile canaliculi. Interestingly, bile production is detectable in

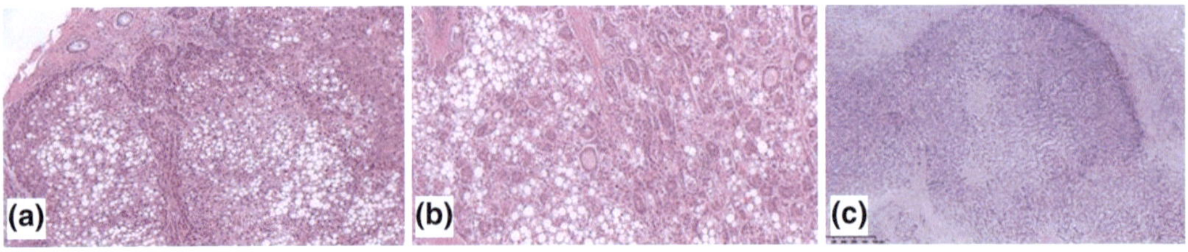


**Fig. 24.4** Examples of HCCs with progressive Edmonson and Steiner grade. G1 HCCs (**a**) are characterized by hepatocytes with low atypia and low nuclear-to-cytoplasmic ratio; early HCC often show steatosis. G2 (**b**) and G3 (**c**) HCCs show a progressive increase in atypia, nuclear-to-cytoplasmic ratio, and polymorphism. Haematoxylin–eosin stain

nearly half of the cases simply by light microscopy [1]. These acini are not true glands, hence the term pseudoglandular, and may contain bile. Their biliary derivation can be demonstrated by electron microscopy or by the immunoreaction for CEA. The acini may contain PAS-positive proteinaceous fluid as well. The solid pattern is the least frequent, and occurs in poorly differentiated HCC, generally in association with at least one of the other architectural patterns. In the solid pattern the tumor cells are densely aggregated, without visible trabeculae or sinusoids. Immunostaining for CD34, however, often reveals compressed sinusoids within the lesion [22]. Other histological variants of HCC include: fibrolamellar HCC, scirrhous HCC, sarcomatoid HCC, clear-cell HCC, giant-cell HCC, HCC with a predominant inflammatory infiltrate, HCC with predominant hematopoiesis [1].

### 24.5.4 Diagnostic Markers

Due to the lack of strong diagnostic criteria, the differential diagnosis between HGDN and eHCC is challenging, and often requires immunohistochemistry (IHC). The American Association for the Study of Liver Diseases (AASLD) and the European Association for the Study of the Liver/European Organization for Research and Treatment of Cancer EASL/EORTC guidelines recommend the use of the three markers: Glutamine Synthase (GS), Glypican-3 and heat shock protein-70 (HSP70) [17, 18]. The combination of these three markers is considered sensitive and specific for the diagnosis of HCC in small liver lesions, while their diagnostic value in biopsy samples is less valuable (Fig. 24.5a–l). Although additional markers, such as EZH2 (Fig. 24.5m–o) [27] and Agrin [28] have been proposed, the differential diagnosis between a HGDN and eHCC in needle biopsies still represents a major challenge for the pathologist. A new promising class of biomarkers is represented by the cleavage products of the endoplasmatic reticulum protein, in particular calreticulin, PDIA3, PDI, and GRP78 [23]. A role as potential biomarker for HCC was postulated for the ubiquitin-conjugating enzyme E2C (UBE2C) due to its overexpression in human HCC at significantly higher levels compared to normal hepatocytes. In addition, high-UBE2C levels were associated with significantly lower disease-free survival rates [23].

### 24.5.5 Molecular Pathology of HCC

Although the list of prognostic and predictive HCC gene signatures is unceasingly growing in the literature, little is known about the molecular mechanisms leading to liver carcinogenesis.

The main molecular abnormalities involved in HCC can be summarized as follows:

- Deregulation of the cell cycle through somatic mutations/LOH of the *p53* gene, silencing of CDKN2a and RB1 or overexpression of CCND1.
- Increased angiogenesis caused by overexpression or amplification of vascular endothelial growth factor (VEGF), platelet-derived growth factor (PDGF) or angiopoietin 2 (ANGPT2).
- Activation of survival signals, such as the NFkB pathway, with relative inhibition of apoptosis.
- Reactivation of telomerase reverse transcriptase (TERT) [1].

These biological processes are involved in the majority of HCC with heterogeneity of molecular alterations. Mutations, DNA methylations, and alterations of gene expression have been described during the multiple steps of liver carcinogenesis and represent possible targets of biological therapies (Fig. 24.6) [29].

### 24.5.6 Pathway Activation Involving Cell Differentiation

Different signaling pathways involved in proliferation, differentiation, inflammation, and angiogenesis are frequently deregulated in HCC, resulting in tumor complexity and heterogeneity

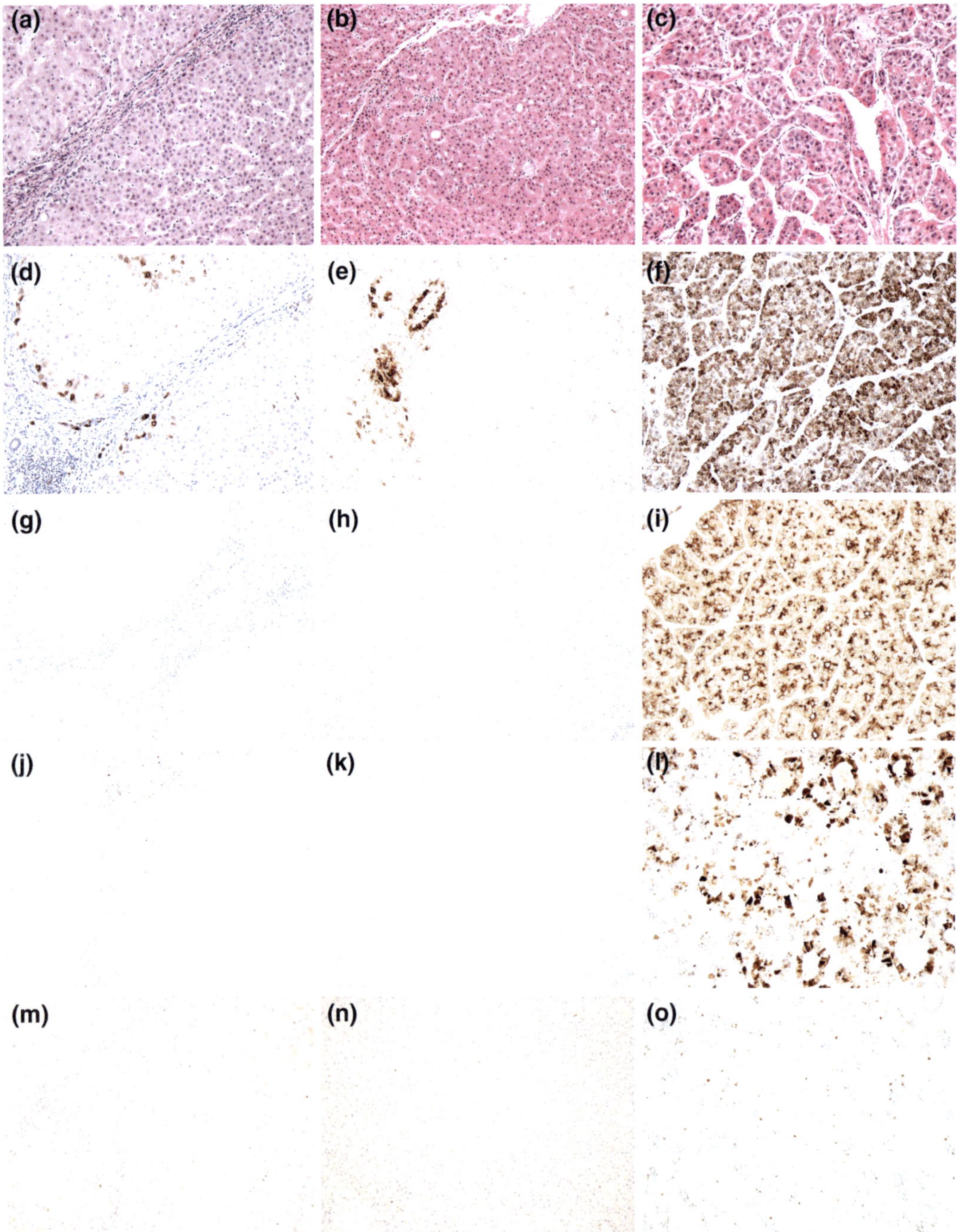


**Fig. 24.5** Histological appearance at haematoxylin–eosin of a (**a**) Regenerative Nodule (RN), a (**b**) High-Grade Displastic Nodule (HGDN), and a (**c**) Hepatocellular Carcinoma (HCC), and expected immunohistochemistry pattern. Glutamine Synthase shows positive reaction only in periportale, perivascular, and perifibrotic areas in RN (**d**) and HGDN (**e**), while it is diffusely positive in HCC (**f**). Glypican-3 is completely negative in RN (**g**) and HGDN (**h**) and diffusely positive in HCC (**i**). Heat shock protein-70 is completely negative in RN (**j**) and HGDN (**k**) and variably positive in HCC **l**. EZH2 is completely negative in RN (**m**) and HGDN (**n**) and shows variable nuclear positivity in HCC (**o**)

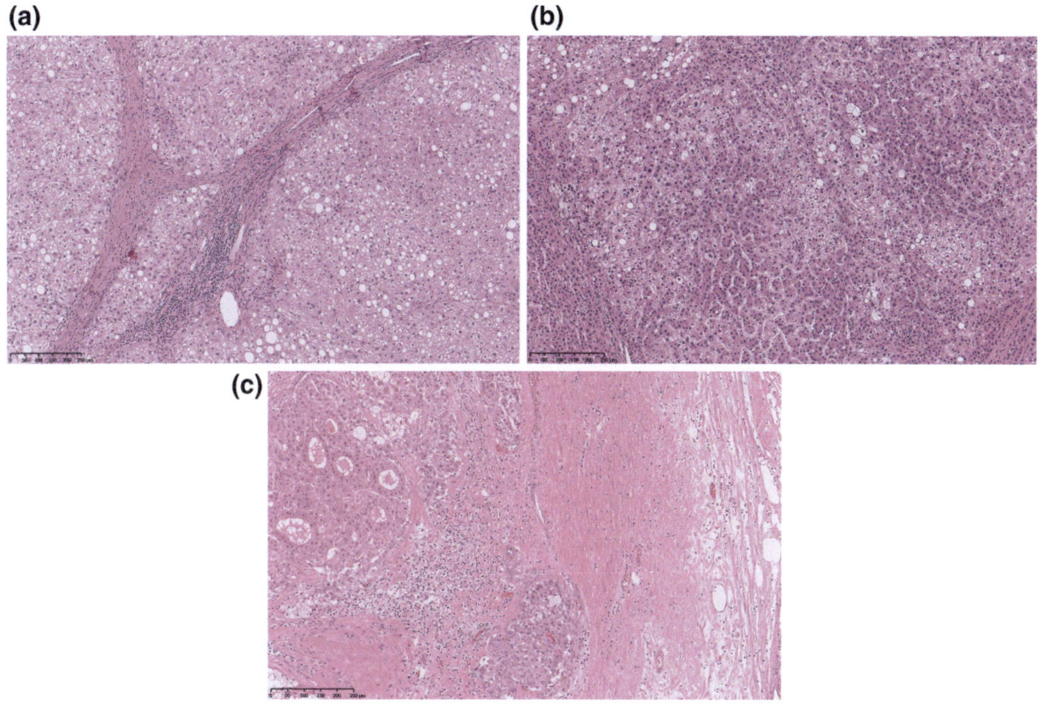


**Fig. 24.6** Flowchart of the major molecular mechanisms involved in HCC cancerogenesis and progression, with special emphasis on pathways activation (*green*), inflammatory pathways (*red*), and neoangiogenesis (*blue*). See text for details

[29]. An early event in liver carcinogenesis is the mutation of β[beta]-catenin, an essential component of Wnt signaling pathway. The disruption of Wnt/β[beta]-catenin signaling results from both genetic and epigenetic changes and is observed in about one-third of HCC [30]. Activation of the Wnt pathway implies an overload of β[beta]-catenin, that translocates in the nucleus and finally activates c-Myc, c-Jun, cyclin D1, fibronectin, and matrix metalloproteases (MMPs). Mutations in the genes Axin 1 and Axin 2, that negatively regulate the Wnt pathway, have been found in HCC [23]. Furthermore, β[beta]-catenin mutations are common in HCC associated with HCV and HBV infection, aflatoxin contamination, and alcoholic liver cirrhosis [23].

### 24.5.7 Pathway Activation Involving Cell Proliferation

The epidermal growth factor (EGF) pathway is one of the most thoroughly studied signaling cascades in HCC. Single nucleotide polymorphisms (SNPs) of the EGF gene are significantly associated with risk of developing HCC. EGF was ranked among the most frequently upregulated genes in a gene signature that was able to identify HCC with high risk of developing a late recurrence. In a genotoxic model of HCC the application of the EGFR tyrosine kinase inhibitor Gefitinib was able to significantly decrease HCC incidence [31].

The Insulin-like Growth Factor (IGF) pathway is activated by the binding of two small proteins, IGF1 and IGF2, secreted by the liver in an autocrine and paracrine manner, to the IGF1 receptor (IGF1R). This binding, in turn, promotes tumor invasion, anchorage-independent cell growth and inhibition of apoptosis. Constitutive activation of this pathway is described in approximately 20 % of HCCs, predominantly via IGF2 or IGFR1 overexpression. IGFR2 inhibits cell growth by enabling the inactivation of IGF2: 10–20 % of HCCs demonstrate inactivation mutations and deletions of IGFR2 associated with LOH, underpinning the tumor suppressor role of IGFR2 [32].

Hepatocyte growth factor (HGF), the sole ligand for the Met receptor, is a strong mitogen for hepatocytes. HGF is involved in cell proliferation, migration, and angiogenesis, and its binding to c-Met mediates the phosphorylation of GRB2, GAB and the recruitment of son of sevenless (SOS). SOS in turn induces the activation of the RAS/RAF/MEK/ERK pathway [31]. The HGF/Met signaling pathway regulates multiple cellular functions, and seems to be a key promoter of metastasis formation and tumor angiogenesis. Experimental studies suggested that HGF/Met and other distinct but functionally homologous signaling pathways, could share the same effector genes. For instance Met, EGF, and TGF-α[alpha] induce similar intracellular response. Some Met target genes may be cross-regulated by the TGF-α[alpha]/EGFR pathway or other tyrosine kinases and, at the same time, activation of the EGF receptor may lead to direct transactivation of Met in transformed cells [33]. High levels of the Met receptor are described in 20–40 % of advanced HCC, are associated with vascular invasion and metastases, and can be therefore eligible for specific therapies targeting Met [31, 33].

Apart from EGFR, IGFR, or Met activation, extracellular signals can be transduced through either the phosphoinositide 3 kinase (PI3K) or the mitogen-activated protein kinase (MAPK) signaling pathways. Effectors belonging to both signaling cascades (AKT or KRAS) are well-known oncogenes in human cancers [1, 23]. Genes of the RAS family (HRAS, NRAS, and KRAS) present single point mutations in HCC that may be induced by various chemical agents. RAS interacts with a downstream serine/threonine kinase Raf-1 leading to its activation and downstream signaling, which includes activation of MAPK kinase MEK1 and MEK2, thus regulating proliferation and apoptosis. Increased expression of oncogenes such as CRAS or NRAS and c-MYC has been related to invasiveness and metastasis in HCC [32]. Proteins related to HCV, HBV, and HEV may modulate MAPK signaling by targeting multiple steps along the pathway. These proteins are involved in cell survival, differentiation, adhesion, and proliferation. In HCC the expression of the Spread protein, an inhibitor of RAS/Raf-1/MAPK-ERK pathway, is deregulated and the forced expression of Spread causes reduction in tumor cell proliferation. These findings suggest that Spread might represent a potential therapeutic target for HCC [1, 23].

The PI3K/AKT/mTOR pathway is activated by growth factor receptors such as IGFR or EGFR in approximately 30–40 % of HCCs and it is significantly associated with eHCC recurrence [32]. It has been shown that AKT expression is able to predict HCC recurrence after surgery in a large cohort of Japanese patients [31]. mTOR activation is also associated with tumor recurrence in a small subset of HCCs and might be targeted by mTOR inhibitors [31, 32]. Another pathway involved in the control of proliferation in HCC both in experimental models and in humans is the Hedgehog signaling pathway (HH), activated by the overexpression of the Sonic or the Indian ligands and/or the inactivation of the HH inhibitory protein [31].

The tumor suppressor retinoblastoma protein (pRB1) controls cell cycle progression via repression of the E2F transcription factor family of proteins. pRB phosphorylation leads to G1/S transition of the cell cycle through the activity of several cyclin-dependent kinases (CDKs). Alterations of the CDK inhibitors p16, p21, and p27 are described in approximately 90 % of HCCs. p16 is predominantly inactivated in the early stages of hepatocarcinogenesis and in late disease progression where it is frequently hypermethylated, while p21 expression is associated with *p53* gene mutations [1, 23]. In a single series almost all aggressive HCC cases showed methylation of the CDKN2A promoter and inactivation of both p53 and RB pathways [32].

SALL4 (human homologue of the Drosophila spalt homeotic gene) plays a critical role in hepatic cell lineage commitment and characterizes a subtype of HCC with progenitor-like features and generally poor prognosis. SALL4 is a key factor for the maintenance of pluripotency and self-renewal capability by embryonic stem cells, through interaction with the transcription factors Oct4, Sox2, and Nanog. Moreover, SALL4 interacts with the PTEN/PI3K/AKT through the nucleosome remodeling deacetylase (NuRD) and histone deacetylase (HDAC) complex. In this mechanism, SALL4 inhibits PTEN, reducing pAKT levels and blocking PI3K survival signaling in HCC. These tumors show high HDAC activity and chemosensitivity to HDAC inhibitors [32].

#### 24.5.7.1 Inflammatory Pathways

Transforming Growth Factor β[beta] (TGF-β[beta]) plays an important role in the control of cell proliferation, adhesion, migration, differentiation, as well as in the control of the cellular microenvironment in both cancer cells and stem cells. TGF-β[beta] may exhibit both tumor suppressive and oncogenic properties in early (inhibiting proliferation and inducing apoptosis) and late (tumor progression) stages of HCC development. Patients with a late TGF-β[beta] signature show shorter survival times and increased tumor recurrence rates compared to early TGF-β[beta] signatures. This may suggest a direct association between TGF-β[beta] and tumor metastasis mediators [34]. In addition, TGF-β[beta] signaling contributes to patterns of HCC differentiation and development through the activation of interleukin-6 (IL6) in hepatic stem cells. Hepatocytes are important immunogenic cells and represent a major target for the immune system as revealed by the huge cytokine production and inflammation in response to JAK/STAT (Janus kinase/signal transducer and activator of transcription) activation in the liver. IL6 recruits JAK1 and JAK2 that phosphorylate STAT1 and STAT3, inducing their homodimerization or heterodimerization, allowing their activation in the nucleus as transcription factors. Moreover, small in-frame deletions of the IL6 gene at its binding site permanently activate the IL6/JAK/STAT pathway, independently of the ligand in tumor hepatocytes [32]. Pathway activation downstream IL6 has been associated with poor survival and de novo tumor formation in surgically treated HCC. The NFkB cascade has been strongly linked to inflammation and liver oncogenesis [31].

#### 24.5.7.2 Pathways Involved in Neoangiogenesis

Vascular endothelial growth factor (VEGF), and (FGFs) are the major drivers of angiogenesis and frequently cross talk in human HCC. VEGF is often overexpressed in HCC, as well as its receptors VEGFR-1 and VEGFR-2. Moreover, high VEGF serum levels have been consistently associated with a more aggressive tumor course. Besides these pathways, angiopoietin (Ang) has been involved in normal and aberrant vascular formation through its interaction with the receptor Tie-2 in HCC [31].

#### 24.5.7.3 Epigenetic Changes

DNA hypomethylation, CpG island hypermethylation, and histone modification represent the epigenetic markers of malignant transformation [23]. In HCC, there is no clear understanding of the methylome and epidrivers. Epigenetic regulation of gene expression in HCC seems to act indirectly by selecting a population of hepatocytes containing structurally and functionally aberrant genes. Panels of hypermethylated promoter targets (APC, GSTP1, RASSF1A, and SFRP1) have been found to be different in plasma from HCC patients compared to nonneoplastic controls. Hypermethylation of CDKL2, CDKN2A, HIST1H3G, STEAP4 gene, and ZNF154 was variably detected in the serum free DNA 37–63 % HCC [35]. A gene-oncology analysis of hypermethylated genes revealed an enrichment of genes that are either involved in metabolic processes or are commonly altered in cancer [36]. Frequent hypermethylation and subsequent loss of expression have also been demonstrated in tumor suppressor genes, such as p16, p14, p15, SOCS1, RIZ1, E-cadherin, gluthatione S-transferase (GSTP1), Rb1, SPINT2, and 14–3–3 sigma. E-cadherin is epigenetically inactivated in 13–50 % of HCC, and it was associated with β[beta]-catenin mutations in tumor with satellite nodules [23, 32, 37].

Genes such as IGF, PI3K, TGF-β[beta], and WNT are deregulated by DNA methylation in HCC [38]. In the above-mentioned IL6/JAK/STAT pathway epigenetics changes may also occur, such as the silencing of suppressor cytokine signaling 1 (SOCS1) by promoter methylation, occurring in 61 % of HCC. Some HCC cells lines exhibit SOCS1 inactivation, resulting in the sensitization to inhibition of the JAK/STAT pathway by a JAK2 inhibitor, and the restoring of SOCS1 in these cells line leads to growth arrest [32]. JAK/STAT pathway activation induced by SOCS3 and NORE1A methylation is reported in HCC, leading to a negative control of JAK/STAT and Ras/MAPK pathways, respectively [32].

Highly conserved Polycomb-group proteins promote gene repression through modification of the chromatin structure and form multiple Polycomb Repressive Complexes (PRC) with an intrinsic histone methyltransferase activity. The latter results in maintenance of the core histone methylation. PRC2 has been related to stem cell and cancer biology, and recently it was observed to promote hypermethylation also in HCC [36]. Aberrant CpG island hypermethylation in gene promoter regions is associated with inactivation and loss of tumor suppressor gene function, which appears to be a key molecular mechanism of HCC. Indeed, a growing number of genes that undergo CpG island hypermethylation were linked to hepatocarcinogenesis [39]. Some reports demonstrated specific methylations in different CpG sites between HBV- or HCV-associated HCC [36]. Methylation profiling may significantly contribute to a comprehensive molecular classification of human HCC [36].

#### 24.5.7.4 "Driver" Genetic Alterations

Unlike other gastrointestinal tumors there is no single "driver" mutation or genetic alteration in human HCC. Cancer progression is rather influenced by a series of concomitant "bystander" alterations that are directly related to cancer cells or involve the surrounding non-neoplastic liver tissue. One of the most frequently reported genetic abnormalities in advanced HCC is the gain of the 8q22–24 region that includes the Myc locus. Amplification of Myc is reported in 40–60 % of HCCs [33]. Copy number variation in this region is one of the earliest genomic events associated with HCC development. In animal models inactivation of Myc in invasive HCC leads to tumor regression with proliferation arrest, differentiation, and apoptosis of tumor cells [33]. Inactivation of the tumor suppressor p53 is detected in 10–61 % of HCCs, occurring via deletion, missense, or nonsense mutations, and is frequently associated with LOH of the second allele [33]. Mutations of p53 play a critical role particularly in HCC associated with HBV infection, Wilson's disease, and hemochromatosis [35]. The p14arf protein, encoded by the same locus of p16, is able to antagonize MDM2-mediated ubiquitination and degradation of p53, leading to cell cycle arrest. Genetic and epigenetic alterations of p14arf have been described in 5 and 40 % of HCC, respectively. [32] Another major genetic alteration typical of HCC involves the Wnt/β[beta]-catenin pathway. β[beta]-catenin mutations present in HCC can be associated with the rare mutation of the glycoprotein 130 (gp130), a gene frequently mutated in liver inflammatory adenomas. These HCC probably result from malignant transformation of a hepatocellular adenoma. Activating mutations of β[beta]-catenin have been frequently reported in HCC, generally in association with a chromosomal stable phenotype [35, 36]. Although alterations of the ErbB family of class 1 tyrosine kinase receptors are well recognized drivers of tumorigenesis, only a small percentage of HCCs harbor exceptional activating mutations of ErbB2 and EGFR. By contrast, missense mutations of c-MET, inducing a constitutive activation in the absence of ligand, have been described in 30 % of HCCs. These tumors could be targeted by small molecule inhibitors of c-MET. As described earlier in this chapter, a small subset of HCCs harbor driver genetic alterations involving the PI3K/AKT/mTOR pathway. PTEN inactivating mutations, deletions, or insertions associated with LOH are found in 5–8 % of HCCs and lead to AKT/mTOR pathway activation. Similarly, activating mutations of PIK3CA, an oncogene that encodes for the alpha subunit of p110 in the PI3K pathway, are described in 5 % of HCCs: this small subset of tumors can be potentially treated with specific mTOR inhibitors such as sirolimus [32].

#### 24.5.7.5 Alteration of Micro-RNA Expression

Micro-RNAs (miRNAs) can undergo aberrant regulation during carcinogenesis and can act as oncogenes or tumor suppressor genes. These alterations may result from distorted epigenetic regulations of miRNA expression, abnormalities in processing genes and proteins, and the location of miRNA at cancer-associated genomic regions. Abnormal miRNA expression is reported in HCC with both tumor suppressor (miR-122, miR-26, miR-223) and oncogenic (miR-130b, miR-221, miR-222) activity [30]. The most studied miRNA in liver is miR-122, which is involved in cell stress response, hepatocarcinogenesis and inhibition of HCV replication. In animal models, the downregulation of miR-122 is associated with hepatocarcinogenesis and represents a candidate biomarker for human liver cancer. From a mechanistic point of view some reports revealed that miR-122a may modulate cyclin G1 expression [23]. In a cell line model the inhibition of miR-21 increased the expression of the tumor suppressor PTEN, with decreased cell proliferation, migration, and invasion. Conversely, enhanced miR-21 expression resulted in increased cell proliferation, migration, and invasion. Furthermore, the modulation of this miRNA altered the adhesion kinase phosphorylalation and the expression of metalloproteases 2 and 9, downstream mediators of PTEN [23]. MiR-181 promotes the acquisition of stem cell features in HCC cells, by targeting the mRNA encoding the caudal type of homeobox transcription factor 2 (CDX2) and GATA6 (that are both hepatic transcriptional regulators of differentiation). At the same time miR-181 inhibits the mRNA that encodes NLK, an inhibitor of Wnt/β[beta]-catenin signaling. All these alterations help maintain the HCC "stemness" and contribute to metastasis and drug resistance [30].

## 24.6 Summary

The pathology of HCC has changed in the last 20 years at least in developed countries. The vast majority of HCC nodules are currently diagnosed in the context of the cirrhotic liver and the median tumor size is generally <5 cm. We are now familiar with the diagnosis of precancerous liver nodules with different degrees of atypia that were not recognized previously. Despite many efforts in the search of prognostic and predictive biomarkers of HCC, none of them are currently utilized in the clinical practice. The main challenges in the next few years will be to substantiate the promising experimental data on early markers of liver carcinogenesis and on new targets for biological therapies. To achieve such goals the availability of well epidemiologically annotated series of HCC patients will represent a necessary requirement.

## References

1. Burt A D, Portmann B C, Ferrell L D, editors. MacSween's pathology of the liver. 6th ed. Elsevier, Churchill Livingstone; 2012.
2. Gordillo M, Evans T, Gouon-Evans V. Orchestrating liver development. Development. 2015;142(12): 2094–108.
3. Mills, Stacey E. Histology for pathologists, 3rd ed. Lippincott Williams & Wilkins Editors; 2007.
4. Nava S, Westgren M, Jaksch M, Tibell A, Broome U, Ericzon B-G, Sumitran-Holgersson S. Characterization of cells in the developing human liver. Differentiation. 2005;73:249–60.
5. Schmelzer E, Zhang L, Bruce A, Wauthier E, Ludlow J, Yao HL, Moss N, Melhem A, McClelland R,

Turner W, et al. Human hepatic stem cells from fetal and postnatal donors. J Exp Med. 2007;204:1973–87.
6. Antoniou A, Raynaud P, Cordi S, Zong Y, Tronche F, Stanger BZ, Jacquemin P, Pierreux CE, Clotman F, Lemaigre FP. Intrahepatic bile ducts develop according to a new mode of tubulogenesis regulated by the transcription factor SOX9. Gastroenterology. 2009;136:2325–33.
7. McDaniell R, Warthen DM, Sanchez-Lara PA, Pai A, Krantz ID, Piccoli DA, Spinner NB. NOTCH2 mutations cause Alagille syndrome, a heterogeneous disorder of the notch signaling pathway. Am J Hum Genet. 2006;79:169–73.
8. Li Y, Rankin SA, Sinner D, Kenny AP, Krieg PA, Zorn AM. Sfrp5 coordinates foregut specification and morphogenesis by antagonizing both canonical and noncanonical Wnt11 signaling. Genes Dev. 2008;22(21):3050–63.
9. Keng VW, Yagi H, Ikawa M, Nagano T, Myint Z, Yamada K, Tanaka T, Sato A, Muramatsu I, Okabe M, Sato M, Noguchi T. Homeobox gene Hex is essential for onset of mouse embryonic liver development and differentiation of the monocyte lineage. Biochem Biophys Res Commun. 2000;276(3):1155–61.
10. Zaret KS. Regulatory phases of early liver development: paradigms of organogenesis. Nat Rev Genet. 2002;3(7):499–512.
11. Lemaigre FP. Mechanisms of liver development: concepts for understanding liver disorders and design of novel therapies. Gastroenterology. 2009;137(1):62–79.
12. Tanimizu N, Miyajima A. Molecular mechanism of liver development and regeneration. Int Rev Cytol. 2007;259:1–48.
13. Kodama Y, Hijikata M, Kageyama R, Shimotohno K, Chiba T. The role of notch signaling in the development of intrahepatic bile ducts. Gastroenterology. 2004;127(6):1775–86.
14. Rappaport AM, Borowy ZJ, Lougheed WM, Lotto WN. Subdivision of hexagonal liver lobules into a structural and functional unit; role in hepatic physiology and pathology. Anat Rec. 1954;119(1):11–33.
15. IARC/WHO classification of tumours of the digestive system. 4th ed. Lyon; 2010.
16. International Working Party. Terminology of nodular lesions of the liver. Hepatology. 1995;22:983–93.
17. Bruix J, Sherman M. Management of hepatocellular carcinoma: an update. Hepatology. 2011;53:1020–2.
18. European Association for the Study of the Liver. EASL-EORTC clinical practice guidelines: management of hepatocellular carcinoma. J Hepatol. 2012;56:908–43.
19. Kim H, Park YN. Role of biopsy sampling for diagnosis of early and progressed hepatocellular carcinoma. Best Pract Res Clin Gastroenterol. 2014;28:813–29.
20. Park YN, Yang CP, Fernandez GJ, Cubukcu O, Thung SN, Theise ND. Neoangiogenesis and sinusoidal "capillarization" in dysplastic nodules of the liver. Am J Surg Pathol. 1998;22:656–62.
21. Thorgeirsson SS, Grisham JW. Molecular pathogenesis of human hepatocellular carcinoma. Nat Genet. 2002;31:339–46.
22. Odze R D, Goldblum J R editors. Surgical pathology of the Gi tract, liver, biliary tract, and pancreas 2nd ed. Saunders, Elsevier; 2009.
23. Aravalli RN, Steer CJ, Cressman EN. Molecular mechanisms of hepatocellular carcinoma. Hepatology. 2008;48(6):2047–63.
24. Kutami R, Nakashima Y, Nakashima O, Shiota K, Kojiro M. Pathomorphologic study on the mechanism of fatty change in small hepatocellular carcinoma of humans. J Hepatol. 2000;33:282–9.
25. Park YN, Kojiro M, Di Tommaso L, et al. Ductular reaction is helpful in defining early stromal invasion, small hepatocellular carcinomas, and dysplastic nodules. Cancer. 2007;109:915–23.
26. Edmondson HA, Steiner PE. Primary carcinoma of the liver: a study of 100 cases among 48,900 necropsies. Cancer. 1954;7:462–503.
27. Cai MY, Tong ZT, Zheng F, et al. EZH2 protein: a promising immunomarker for the detection of hepatocellular carcinomas in liver needle biopsies. Gut. 2011;60:967–76.
28. Tatrai P, Somoracz A, Batmunkh E, et al. Agrin and CD34 immunohistochemistry for the discrimination of benign versus malignant hepatocellular lesions. Am J Surg Pathol. 2009;33:874–85.
29. Woo HG, Park ES, Thorgeirsson SS, Kim YJ. Exploring genomic profiles of hepatocellular carcinoma. Mol Carcinog. 2011; 50(4):235–43.
30. Oishi N, Yamashita T, Kaneko S. Molecular biology of liver cancer stem cells. Liver Cancer. 2014;3(2):71–84.
31. Zender L, Villanueva A, Tovar V, et al. Cancer gene discovery in hepatocellular carcinoma. J Hepatol. 2010;52(6):921–9.
32. Nault JC, Zucman-Rossi J. Genetics of hepatobiliary carcinogenesis. Semin Liver Dis. 2011;31(2):173–87.
33. Kaposi-Novak P, Libbrecht L, Woo HG, et al. Central role of c-Myc during malignant conversion in human hepatocarcinogenesis. Cancer Res. 2009;69(7):2775–82.
34. Coulouarn C, Factor VM, Thorgeirsson SS. Transforming growth factor-beta gene expression signature in mouse hepatocytes predicts clinical outcome in human cancer. Hepatology. 2008;47(6):2059–67.
35. Hoshida Y, Villanueva A, Kobayashi M, et al. Gene expression in fixed tissues and outcome in hepatocellular carcinoma. N Engl J Med. 2008;359(19):1995–2004.

36. Neumann O, Kesselmeier M, Geffers R, et al. Methylome analysis and integrative profiling of human HCCs identify novel protumorigenic factors. Hepatology. 2012;56(5):1817–27.
37. Revill K, Wang T, Lachenmayer A, et al. Genome-wide methylation analysis and epigenetic unmasking identify tumor suppressor genes in hepatocellular carcinoma. Gastroenterology. *201; 145:1424–35, e1–25.
38. Villanueva A, Hoshida Y, Battiston C, et al. Combining clinical, pathology, and gene expression data to predict recurrence of hepatocellular carcinoma. Gastroenterology. 2011;140(1501–12):e2.
39. Lee S, Lee HJ, Kim JH, et al. Aberrant CpG island hypermethylation along multistep hepatocarcinogenesis. Am J Pathol. 2003;163:1371–8.

# Epidemiology of Pancreatic Cancer 25

Dominique S. Michaud

## 25.1 Descriptive Data

As with many cancers, incidence rates for pancreatic cancer increase exponentially with age and a diagnosis of pancreatic cancer before the age of 50 years is rare. By the age of 85 years, incidence rates reach 100 or more cases per 100,000 (Figs. 25.1 and 25.2). Age-adjusted incidence rates are slightly higher for men than women (14 per 100,000 for men vs. 11 per 100,000 for women) and differences by sex are consistent across racial groups [1]. In the United States (US), African-American men have the highest incidence rates for pancreatic cancer (15.4 per 100,000) and Asians have the lowest rates (9.7 per 100,000 for both sexes) (Figs. 25.1 and 25.2) [1]. Pancreatic cancer is the most rapidly fatal cancer.

In the US, the 5-year survival rate is 7 % [2]. Mortality rates are very similar to incidence rates across all sexes and racial groups (age-adjusted rates 12.5 per 100,000 for men and 9.5 per 100,000 for women) [1].

In the US in 2015, an estimated 48,960 people will be diagnosed with pancreatic cancer and 40,560 will die of this disease [2]. Overall rates (incidence and mortality) for pancreatic cancer have been relatively constant over the past 3–4 decades, although a small increase in rates have been observed in the past decade [1]. Globally, pancreatic cancer incidence rates are higher in more developed countries particularly high rates are found in Eastern Europe, with the highest rates reported in Armenia, Hungary, Slovakia, and the Czech Republic (Globocan, 2015). An estimated 330,391 individuals die of pancreatic cancer annually in the world, accounting for 4 % of all cancer deaths. The majority of new cases are diagnosed in Asia (42 %), followed by Europe (30 %) and North America (14 %).

## 25.2 Risk Factors

### 25.2.1 Introduction

Our understanding of what causes pancreatic cancer improved dramatically over the past two decades as many prospective cohort studies achieved critical numbers of pancreatic cancer cases that allowed for more detailed analyses of suspected risk factors. In addition, a number of pooled and meta-analyses have been conducted to solidify strengths of associations. There now exists a large body of evidence for risk factors that are considered established risk factors of pancreatic cancer; these include family history, AB blood type, chronic pancreatitis, tobacco smoke, long-standing type 2 diabetes, high alcohol consumption, and obesity.

D.S. Michaud (✉)
Tufts University School of Medicine, 136 Harrison Avenue, Boston, MA 02111, USA
e-mail: Dominique.Michaud@tufts.edu


M. Loda et al. (eds.), *Pathology and Epidemiology of Cancer*,
DOI 10.1007/978-3-319-35153-7_25

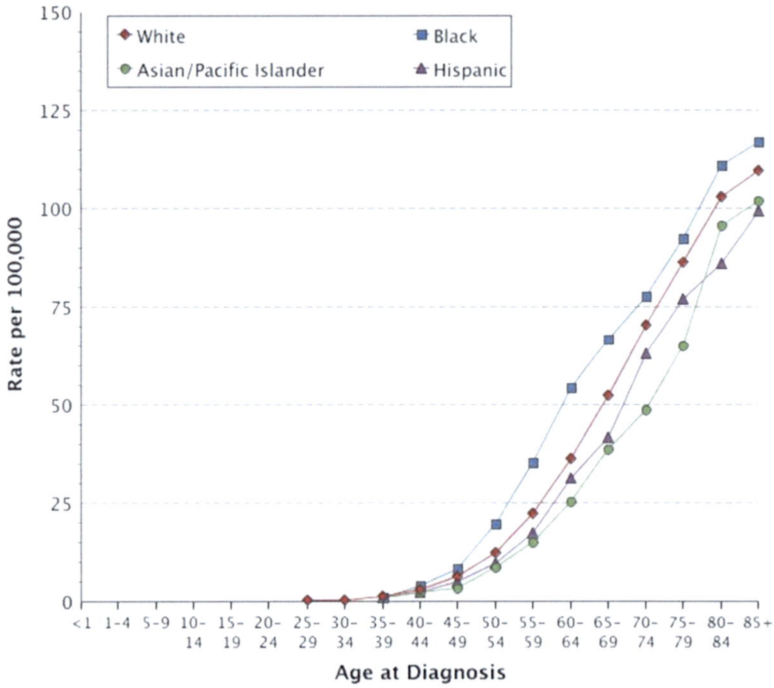


**Fig. 25.1** Age-specific incidence rates of pancreatic cancer among men in the U.S., 2008–2012, by race/ethnicity. Data are taken from the U.S. SEER Registry

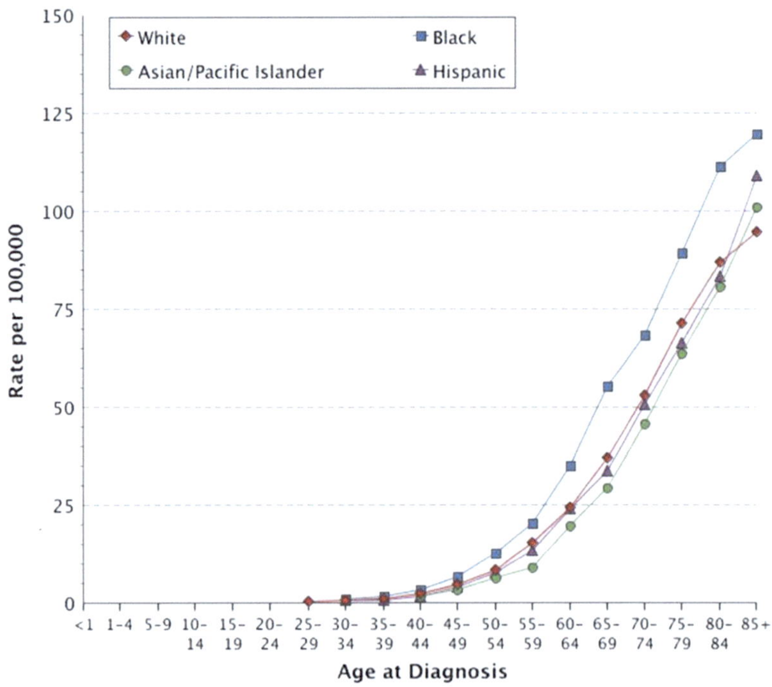


**Fig. 25.2** Age-specific incidence rates of pancreatic cancer among women in the U.S., 2008–2012, by race/ethnicity. Data are taken from the U.S. SEER Registry

### 25.2.2 Tobacco Smoke

In a large pooled analysis from the International Pancreatic Cancer Case-Control Consortium (Panc4), 12 case-control studies were combined to include 6507 pancreatic cancer cases and 12,890 controls [3]. In this pooled analysis, the odds ratio (OR) for current smokers was 2.2, compared to never smokers, with a 95 % confidence interval (CI) of 1.7–2.8, which is consistent with results from prospective cohort studies where relative risks (RRs) have ranged between 1.7 and 2.5 [4–7]. Risk of pancreatic cancer increases with higher cigarette dose such that current smokers smoking 30 or more cigarettes per day, compared to never smokers, have a 2–3 fold higher risk [3, 4]. Former smokers have a much lower risk of pancreatic cancer; however, this estimate is largely dependent on time since smoking cessation among former smokers, as risk of pancreatic cancer nears that of never smokers 15–20 years after smoking cessation [3, 4]. Other cohort studies have reported shorter time periods after cessation (less than 10 years) for rates among former smokers to be similar to never smokers [6, 7].

Smoking cigars has also been associated with elevated pancreatic cancer risk for cigar smokers; the pooled OR from cigar-only smokers in a large case-control consortium (Panc4) was 1.62 (95 % CI 1.15–2.29), compared to never tobacco users [8]. In contrast, pipe smokers do not appear to have an increase risk in pancreatic cancer [8].

Measuring the association between smokeless tobacco (e.g., chewing tobacco/snuff) and pancreatic cancer poses a particular challenge, as confounding by smoking needs to be ruled out; consequently, the ideal populations to examine these associations are never smokers. Types of smokeless tobacco differ in the US and Europe and may have different impacts on risk. Three recent meta-analyses were performed on smokeless tobacco and cancer, but results for pancreatic cancer have been inconsistent [9–11]. A twofold increase in risk of pancreatic cancer among never smokers was reported in a retrospective cohort study of snuff users (RR 2.0, 95 % CI 1.2–3.3) [12], but another large cohort study did not report an increase in risk of pancreatic cancer among never smokers (only among smokers) [13]. At this time, the data on smokeless tobacco and pancreatic cancer are inconclusive.

For passive smoking, early exposure to cigarette smoke appears to increase risk of pancreatic cancer among never smokers, while environmental smoke from the workplace or home in adulthood do not appear to be associated with pancreatic cancer [14, 15]; other studies are needed to confirm these findings.

Examining mutation profiles in tumors may confirm associations and provide insight into mechanistic data to inform on causality. Mutations in the K-ras gene are common and arise early in the development of pancreatic cancer. To date, several studies have examined K-ras mutations in association with risk factors of pancreatic cancer, although they have often been limited by small number of cases with available tissue samples. In one study, smoking was associated with K-ras positive mutations in codon 12, but was not associated with K-ras negative tumors [16]. A positive association between smoking and K-ras mutations in pancreatic cancer was observed in two other studies [17, 18], but no associations were noted in three studies [19–21]. Recently, large mutational analysis (entailing sequencing of protein coding exons for >20,000 genes) compared pancreatic tumor tissue from smokers and nonsmokers; in this study, only one case (out of 114) did not have K-ras mutations [22]. Mutations in driver genes (KRAS, TP53, SMAD4, and CDKN2A/p16) were not more common in smokers, but smokers were observed to have more nonsynonymous mutations ($p = 0.04$) and more overall mutations ($p = 0.06$) than nonsmokers [22]. These findings suggest that smoking impacts pancreatic tissue in a way that is different from the lung tissue, as smoking has been strongly associated with K-ras mutations in lung tumor tissue [23, 24]. Lack of difference in mutations of driver genes suggest that smoking is not acting in the initiation events of pancreatic cancer. These findings are consistent with the observational data, given that smoking is not as strongly associated with pancreatic cancer as it is with lung cancer, and that a substantial risk reduction occurs after 10–15 years of smoking cessation. It is possible that smoking acts late in tumor development and through a number of non-specific biological pathways.

### 25.2.3 Diabetes Type 2 and Hyperglycemia

It is well established that type 2 diabetes can develop as a consequence of pancreatic tumors and is often detected before the cancer diagnosis, but there has been substantial controversy over the role of diabetes as a cause of this cancer. There is considerable evidence supporting the role of glucose in the development of pancreatic cancer, both from epidemiologic and animal studies [25]. Prospective studies examining pre-diagnostic blood glucose levels have observed that individuals with elevated blood glucose a decade or more prior to diagnosis have a higher risk of pancreatic cancer [26]. The same observation has been made for long-standing type 2 diabetes; risk for pancreatic cancer remains elevated over 20 years post-diagnosis of diabetes type 2, and thus likely plays a causal role in carcinogenesis [27–29]. While a very high risk of pancreatic cancer is observed when diabetes type 2 is diagnosed within 1 year of the tumor diagnosis (RR 5.38, 95 % 3.49–8.30), a 50 % increase in risk is observed in individuals with a diabetes diagnosis 10 or more years before cancer diagnosis (RR 1.51, 95 % CI 1.16–1.96 [27]; RR 1.47, 95 % CI 0.94–2.31 [30]; RR 1.30, 95 % CI 1.03–1.63, compared to those without a diabetes diagnosis [28]) suggests that diabetes is both a consequence and cause of pancreatic cancer. Long-standing diabetes (>4 years) has also been associated with poorer survival after pancreatic cancer diagnosis [31].

In four prospective cohort studies, higher risk of pancreatic cancer was reported among participants with elevated blood glucose at baseline, compared to those with normal blood glucose [26, 32–34]. Two of these studies examined the temporal association between blood glucose and pancreatic cancer diagnosis [26, 34]. One study, including only nondiabetics, reported a statistically significant association for glucose levels above 198 mg/dL, compared to levels below 118.8, after excluding cases diagnosed in the first 5 years of follow-up (RR 1.97, 95 % CI 1.08–3.57) [34]. In the second study conducted among male smokers (ATBC study), risk remained elevated for those who had high blood glucose (>107 mg/dL), compared to <93 mg/dL, 10 or more years prior to diagnosis (RR 2.16, 95 % CI 1.05–4.42) [26].

In a large European cohort study, elevated glycated hemoglobin (HbA1c), a marker of prolonged elevated average glucose, was positively associated with pancreatic cancer risk (OR 2.42, 95 % CI 1.33–4.39, for highest [≥6.5 %, 48 mmol/mol] versus lowest [≤5.4 %, 36 mmol/mol] category) [35]. Risk with elevated HbA1C was slightly higher for those who were diagnosed 5 years or more after start of follow-up, suggesting that changes in glucose levels were not impacted by onset of disease. In another study pooling five prospective cohorts, HbA1c was also associated with risk of pancreatic cancer (OR = 1.79, 95 % CI = 1.17–2.72, comparing top to bottom quintile) and did not vary by time to diagnosis [36].

Insulin levels and insulin resistance have also been associated with higher pancreatic cancer risk in two studies [26, 37], whereas C-peptide levels (markers of insulin production) have not been consistently associated with risk of pancreatic [35, 37].

### 25.2.4 Obesity

Obesity has only recently been widely accepted as a risk factor for pancreatic cancer as many earlier case-control studies had not observed any associations for obesity. Over the past 15 years, prospective cohorts have published consistent positive associations for obesity and pancreatic cancer. To date, four meta-analyses and two pooled analyses of prospective cohort studies have been conducted examining body mass index (BMI) and pancreatic cancer risk. In the most recent and largest meta-analysis, which included 23 prospective studies and 9,504 pancreatic cancer cases, a 10 % increase in risk was reported for each 5-unit increase in BMI (RR 1.10, 95 % CI 1.07–1.14) [38]. A similar magnitude between obesity and pancreatic cancer was reported in other summary analyses (e.g., pooled study with 2170 cases of pancreatic cancer and 2209 matched controls: RR 1.13, 95 % CI 1.11–1.14 for a 5-unit increase in BMI [39]). The association between BMI and pancreatic cancer is not linear, however, and the increase in risk is more dramatic among very obese patients (RR 1.55, 95 % CI 1.16–2.07, comparing BMI >35 to BMI <25 kg/m$^2$) [39]. Associations with obesity and pancreatic cancer are similar in men and women and by geographical region, but appear to be stronger among never smokers as the increase in risk is already apparent among those who were overweight [38].

Abdominal obesity (measured using waist circumference or waist-to-hip ratio) is also strongly associated with pancreatic cancer [38, 40]. The summary RR for a 10-cm increase in waist circumference was 1.11 (95 % CI 1.05–1.18) and for a 0.1-unit increment in waist-to-hip ratio was 1.19 (95 % CI 1.09–1.31) in the largest meta-analysis [38].

Obesity has also been associated with poorer survival from pancreatic cancer. The risk of death is roughly 50 % higher among patients with BMI >35 kg/m$^2$ prior to diagnosis compared to those with normal BMI (<25 kg/m$^2$) [41].

### 25.2.5 Chronic Pancreatitis

As with diabetes, pancreatitis is often diagnosed close to cancer diagnosis, making it difficult to disentangle cause and effect. The most convincing evidence for a causal role for chronic pancreatitis comes from studies on patients with hereditary pancreatitis. Symptoms of pancreatitis in patients with hereditary pancreatitis occur at a mean age of 10 years [42], and in these patients, rates of pancreatic cancer are not only much higher than those in the general population, but cancer develops at younger ages. In a national series study of 200 patients with hereditary pancreatitis (conducted in France), the average age at cancer onset was 55 years, and the standardized incidence ratio for this group, compared to the general population, was 87 (95 % CI, 42–113) [43]. Other hereditary pancreatitis case series, including two multi-site studies, reported slightly lower, but of similar magnitude, SIRs (53 [44] and 67 [45]). Smoking increases the risk of pancreatic cancer among those with hereditary pancreatitis and is associated with substantially earlier age at onset (50 years for ever smokers vs. 70 years for never smokers) [46]. As smoking prevalence was higher in the study conducted in France [43] (51 vs. 40 % in Lowenfels) [46], it may explain the higher relative risks noted in that study.

The incredibly high rates of pancreatic cancer observed among cases with hereditary pancreatitis, with known genetic mutations that lead to inflammation of the pancreas, provide support for a causal association between chronic pancreatitis (nonhereditary) and pancreatic cancer. The difficulty in studying chronic pancreatitis and pancreatic cancer risk includes small numbers, reverse causality, and confounding; studies that have examined the temporal relationship between chronic pancreatitis and diagnosis of pancreatic cancer find decreasing rates of cancer as the lag period increases [47]. As with type II diabetes, however, long-standing chronic pancreatitis is associated with a significant increase in risk; a 5.8-fold increase in risk of pancreatic cancer was estimated from six studies that excluded pancreatic cancer cases diagnosed within 2 years from chronic pancreatitis diagnosis (95 % CI 2.1–15.9) [47]. With a ten-year lag, the largest cohort study based on registry data reported a relative risk of 2.2, and although it was not statistically significant (95 % CI 0.9–4.4), only eight cases were available for this group [48].

Genetically modified pancreatic mouse models (developed to mimic human pancreatic adenocarcinoma) also support a role for chronic pancreatitis; induction of pancreatitis causes dramatic acceleration of pancreatic carcinoma in these mouse models [49]. These mouse models have provided many new insights into the mechanisms underlying pancreatic carcinogenesis and overwhelming evidence supports a role for inflammation [49].

### 25.2.6 Family History of Pancreatic Cancer and Genetic Susceptibility

Family history of pancreatic cancer explains a small fraction of pancreatic cases (<5 %); individuals with a parent, sibling, or child with pancreatic cancer have a moderately higher risk than those without family history (RR 1.76, 95 % CI 1.19–2.61) [50]. Family history of prostate cancer also appears to increase the risk of pancreatic cancer (RR 1.45, 95 % CI 1.12–1.89), but no associations were noted for those with family history of ovarian, breast, or colorectal cancers in one large study [50], despite the known elevated risk among those with BRCA2 mutations [51]. Germline mutations in BRCA2, CDKN2A, STK11, PRSS1, SPINK1, PRSS2, CTRC, and DNA mismatch repair have been associated with pancreatic cancer risk [52].

Genome-wide association studies (GWAS) have uncovered new areas of genetic susceptibility for pancreatic cancer. A region in the ABO gene (rs687289 at 9q34.2) was identified as being strongly associated with risk of pancreatic cancer in three GWAS analyses (PanScan III: ABO, OR 1.27, 95 % CI 1.20–1.35, P $1.6 \times 10^{-16}$) [53–55]. These findings are consistent with studies that examined the role of blood groups directly, including suggestions for these associations as early as 1960 [56]. In a recent analysis, individuals with blood group A (HR 1.32, 95 % CI 1.02–1.72), AB (HR 1.51, 95 % CI 1.02–2.23), or B (HR 1.72, 95 % CI 1.25–2.38) had a higher risk of pancreatic cancer than those with blood group O [57]. Using these results, it was estimated that 17 % of the pancreatic cancer cases were attributable to inheriting a non-O blood group (blood group A, B, or AB) [57]. A number of other studies have confirmed the associations between blood type and pancreatic cancer risk [58–61].

In addition to ABO, three regions have been associated with pancreatic cancer risk using GWAS analyses (PanScan I and II) [53] and confirmed in PanScan III: rs9543325 at 13q22.1 (KLF5/KLF12, OR 1.23, 95 % CI 1.18–1.30), rs10919791 at 1q32.1 (NR5A2, OR 0.79, 95 % CI 0.75–0.85) and rs31490 at 5p15.33 (CLPTM1L, OR 1.20, 95 % CI 1.14–1.27) [55]. Recently, five new regions were identified in PanScan III: rs2736098 at 5p15.33 (a second signal in TERT), rs6971499 at 7q32.3 (LINC-PINT), rs7190458 at 16q23.1 (BCAR1/CTRB1/CTRB2), rs9581943 at 13q12.2 (PDX1), rs16986825 at 22q12.1 (ZNRF3) [55]. Several of the new loci identified in PanScan III are located in genes that have been implicated in pancreas development, pancreatic beta-cell function and predisposition to diabetes. While these findings provide insight into the pathways that are of importance for pancreatic carcinogenesis, the small effects observed for these susceptibility regions cannot (alone or combined) contribute predictive ability in determining who is at higher risk of pancreatic cancer [62].

### 25.2.7 *Helicobacter pylori* Infection

Positive associations between *Helicobacter pylori* and pancreatic cancer have been observed in some observational studies. The first report of an association between *H. pylori* and pancreatic cancer risk came from a case-control study with a small control group (n = 27); a twofold increase in risk was observed (OR 2.1, 95 % CI 1.09–4.05) [63]. In the Alpha-Tocopherol, Beta-Carotene Cancer Prevention study (ATBC), a prospective cohort study of male smokers, men with *H. pylori* antibodies or CagA-positive strains had about a twofold elevated risk of pancreatic cancer, compared to men who were seronegative for those antibodies (OR 1.87, 95 % CI 1.05–3.34; OR 2.01, 95 % CI 1.09–3.70, respectively) [64]. Findings from subsequent studies with large case numbers have been weaker and not statistically significant [60, 65–67]. Results from recent meta-analyses are inconsistent and largely influenced by authors' decisions regarding which studies to include/exclude [68, 69]. Overall, the evidence for an association for *H. pylori* and pancreatic cancer, when examined critically, is weak.

### 25.2.8 Peptic Ulcers

A number of epidemiological studies have examined the relationship between peptic ulcers and risk of pancreatic cancer. Results from cohort studies with large number of pancreatic cancer cases and detailed information on type of peptic ulcers (i.e., gastric vs. duodenal) observed positive associations with gastric ulcers, but not duodenal ulcers [70, 71]. While *H. pylori* infections are associated with both types of peptic ulcers, gastric ulcers are associated with low acid production while duodenal ulcers are associated with hyperacidity; consequently, nitrosamine levels are higher in individuals with gastric ulcers and may explain the association with pancreatic cancer risk. Low acidity, however, also allows for the colonization of other bacteria, which may provide an opportunity for oral bacteria to move into the stomach and the gut.

### 25.2.9 Periodontal Disease

Periodontal disease is an inflammatory disease of the gums; with advanced disease, inflammation in the gums can lead to gum recession, soft tissue damage, bone loss and tooth loss (severe periodontitis) [72]. As with many chronic diseases, periodontal disease has multiple risk factors, including smoking and diabetes, and several bacteria have been linked to the severity and progression of periodontitis [72]. *Porphyromonas gingivalis*, a periodontal pathogen, has been extensively studied due to its unique ability to evade the immune response [73]. Positive associations between periodontitis and pancreatic cancer risk have been reported in three separate cohort studies [74–76]. In the National Health and Nutrition Examination Survey (NHANES) I Epidemiologic Follow-up Study, individuals with periodontitis at baseline had a higher risk of fatal pancreatic cancer compared to those with healthy periodontium (RR 1.77, 95 % CI 0.85–1.85), after controlling for age and sex [74]. A strong positive association between periodontal disease and pancreatic cancer was reported in a prospective cohort study of male health professionals [75]. Participants self-reported tooth loss and periodontal disease at baseline and were subsequently followed for 16 years. During that period, 216 cases of pancreatic cancer were newly diagnosed. After adjusting for age, smoking, diabetes, body mass index, and a number of other dietary factors, men with bone loss from periodontal disease had a 64 % higher risk of pancreatic cancer compared to those reporting no bone loss from periodontal disease. Among never smokers, a twofold increase in pancreatic cancer risk was observed (RR 2.09, 95 % CI 1.18–3.71), ruling out the possibility that the overall association was confounded by smoking. Furthermore, the association was stronger among dentists (RR 1.91, 95 % CI 1.31–2.78) who more accurately report history of periodontal disease [77]. In a recent analysis of the NHANES III data, a fourfold increase in risk of pancreatic cancer was observed among those with severe periodontitis, although the association was not statistically significant due to small numbers of cases (RR 4.56, 95 % CI 0.93–22.3) [76].

The association between antibodies to periodontal pathogens and risk of pancreatic cancer has been examined in a large European cohort (EPIC) [78] using blood samples stored on over 385,000 men and women at baseline (i.e., prior to disease). Using a nested case-control study design of 405 pancreatic cancer cases and 410 controls, a greater than twofold increase in risk of pancreatic cancer was observed among those with high levels of antibodies to a pathogenic strain of *P. gingivalis* (OR 2.38, 95 % CI 1.16–4.90, comparing >200 ng/ml vs. <200 ng/ml) after adjusting for known risk factors [78]. In the NHANES III cohort study, elevated antibodies to *P. gingivalis* (>69.1 EU, compared to less than 69.1) were associated with a threefold increase risk of orodigestive cancer mortality (RR 3.03, 95 % CI 0.99–9.31). Removing subjects with clinically apparent periodontal disease only decreased the association slightly (RR 2.25, 95 % CI 1.23–4.14). A separate examination of *P. gingivalis* with pancreatic cancer mortality could not be conducted in that study due to insufficient case numbers [76].

## 25.2.10 Allergies and Immune Response

A number of studies have examined the potential role of allergies in pancreatic cancer. A meta-analysis conducted in 2005 reported a 18 % reduction in risk of pancreatic cancer (OR 0.82, 95 % CI 0.68–0.99) [79]. In a multicenter case-control study, a 36 % decrease in risk was noted for individuals with a history of allergies (OR 0.64, 95 % CI 0.50–0.82) [80]. A more recent pooled analysis of 10 case-control studies, including 3567 cases and 9145 controls, harmonized the allergy variables to examine "any allergies" and reported a borderline statistically significant inverse association (OR 0.79, 95 % CI 0.62–1.00), controlling for known risk factors [81]. Statistically significant inverse associations were also found for hay fever and allergies to animals in the pooled analysis, while associations for asthma, eczema and allergies to food were weaker [81]. In a case-control study, 30–50 % reduced risk of pancreatic cancer were observed among subjects who had positive skin prick results for hay fever allergens, dust/mold, or animal allergens [82]; moreover, age of allergy onset did not impact associations, suggesting cancer was not likely causing allergy onset.

A 45 % lower risk of pancreatic cancer was reported in individuals with high levels of 22 oral antibodies, when compared with those individuals with overall lower levels of those antibodies, in a nested case-control study (OR 0.55, 95 % CI 0.36–0.83) [78]. Given that both allergies and high antibodies to bacteria are associated with a Th2 immune response, these findings support the hypothesis that the adaptive immune response may play an important role in pancreatic carcinogenesis.

## 25.2.11 Diet

A large number of studies have examined various aspects of diet in relation to pancreatic cancer risk, but overall, findings have been largely inconsistent or null. This section will focus on results from cohort studies, as case-control studies on diet are prone to bias, and have often reported findings that have not been reproduced in cohort studies.

### 25.2.11.1 Alcohol

Several large pooled analyses on alcohol intake and pancreatic cancer have been conducted which have provided strong data on dose-response relationships [83–85]. Results from a pooled analysis, two pooling data from prospective cohort studies [83, 84] and one from case-control studies [85], consistently indicate that alcohol is only a risk factor for pancreatic cancer at higher intake levels. In the largest pooling study, including over 5500 cases, a statistically significant elevated risk was observed for those consuming 6 or more drinks/day, compared to those consuming 0 to <1 drinks/day (OR 1.46, 95 % CI 1.16–1.83) [85]; no interaction was noted for the most common risk factors, including age, sex, race, and smoking status. In a pooling second study, a statistically significant 22 % increase in risk was observed among those consuming >30 g of alcohol/day (about three drinks/day), compared to nondrinkers [84]; however, numbers were too small to examine associations at higher intake levels. In a third pooling study, alcohol intake at >60 g/day (about 6 drinks/day) was associated with a nonsignificant increase in risk (OR 1.38, 95 % CI 0.86–2.23), compared to those who drank some alcohol (>0 to <5 g/day) [83]. The pooled OR for total alcohol for >30 g/day was similar to the other pooled analysis by Genkinger et al. (OR 1.23, 95 % CI 0.97–1.57) [84]. Finally, a large prospective cohort study (NIH-AARP), including 1,149 cases of pancreatic cancer, reported a 45 % increased risk among those consuming >3 drinks (95 % CI 1.17–1.80) and a 55 % increase in risk among those consuming 6 or more drinks of alcohol/day compared to >0 to <1 drinks/day (RR 1.55, 95 % CI 1.13–2.13, P-trend = 0.004) [86].

Based on these findings, no increase in risk is observed for moderate alcohol drinkers (i.e., 1–2 drinks/day), and an elevated risk is only observed among heavy drinkers, with increasing risk at higher levels. Two of the large studies also reported that liquor consumption was more strongly associated with risk than other alcoholic beverages [83, 86], while one study found wine to be more strongly associated with risk [85]. It is unclear if these are linked to behavioral patterns, or are markers for heavier alcoholic consumption.

#### 25.2.11.2 Folate Intake

A pooled analysis of 14 prospective cohort studies with 2915 pancreatic cancer cases reported no association with folate intake (RR 1.06, 95 % CI 0.80–1.16, comparing highest vs. lowest quintile of total folate intake, i.e., dietary and supplement use) [87]. Similarly, no association was observed for dietary folate without supplements, or for supplemental use (RR 0.94, 95 % CI 0.73–1.22, for highest vs. lowest tertile of folic acid supplement use). Several meta-analyses have reported inverse associations between dietary folate intake and pancreatic cancer [88–90]; however, publication bias appears to be a major issue in these meta-analyses, as demonstrated in a forest plot of a recent meta-analysis [90] and suggested by the number of publications from cohort studies (6 published studies vs. 14 cohorts included in the pooling project). Furthermore, these meta-analyses included both case-control and cohort studies; summary estimates from cohort studies were weaker than summary estimates for case-control data [89, 90].

Studies using blood to measure folate levels have been inconsistent [91–93]. In the EPIC study, a U-shaped relationship was observed, with modest elevated risk at both ends of the spectrum of plasma folate levels [92]. An inverse association was observed in the ATBC study [93], while no statistically significant associations were noted in a nested case-control analysis of four prospective cohorts [91].

Taken together, these data do not support a dose-response between folate intake and pancreatic cancer and higher intake of folate may not result in lower risk of pancreatic cancer, as previously observed.

#### 25.2.11.3 Fruits and Vegetables

Numerous case-control studies have reported inverse associations between fruit and vegetable intake and pancreatic cancer, but as with many other cancers, these findings have not been replicated in cohort studies. The largest pooled analysis of 14 cohort studies reported no association for fruit, vegetable, or total fruit and vegetable consumption and pancreatic cancer [94]. In this pooling project, no inverse associations were noted for individual fruit or vegetable consumption. Similarly, no associations were reported for fruit and vegetable consumption in a large European cohort study (EPIC) that was not included in the pooling study, with 555 pancreatic cancer cases [95]. Inverse associations for subgroups of vegetables have been observed in a couple of cohort studies, but the associations were observed for different subgroups and thus likely to be chance findings. It is important to note that meta-analyses on fruit and vegetable consumption including case-control studies are likely to provide misleading conclusions, given that case-control studies have consistently reported inverse findings, which are likely the result of selection and reporting biases. Moreover, a recent meta-analysis of cohort studies reporting statistically significant inverse associations for fruits and vegetables only included three cohort studies (out of >12 published studies) and is highly misrepresentative [96].

#### 25.2.11.4 Meat Consumption

Meat consumption has been linked to pancreatic cancer in several cohort studies [97–102], although the type of meat associated with risk has not been entirely consistent across these studies, and three cohort studies reported no associations [103–105]. The inconsistency in findings for meat intake may be due to differences in cooking practices that can influence the level of carcinogens consumed. Cooking meats at high temperatures, or on an open fire, can result in production of compounds that are known carcinogens, such as heterocyclic amines (e.g., DiMelQx and MelQx) and polycyclic aromatic hydrocarbons (e.g., benzo(a)pyrene, BaP). Details on cooking practices for meats (e.g., doneness preferences and cooking method) have been incorporated in some cohort questionnaires to address this issue. Results from two large cohort studies with detailed cooking data reported significant positive associations with elevated levels of DiMelQx, MelQx and mutagenic activity, but not with BaP levels [97, 98]. In one study, consuming a high amount of meat cook at high temperatures was associated with a 52 % increase in risk of pancreatic cancer in men (top vs. bottom quintile of meat cooked at high temperature, RR 1.52, 95 % CI 1.12–2.06), although no association was observed in women [98]. In the second cohort, consuming very well-done red meat was associated with a 60 % increase in risk (RR 1.60, 95 % CI 1.01–2.54), compared to those consuming meat that is medium or rare [97]. Although only two cohort studies have examined cooking doneness, their findings are consistent with previous case-control studies [106, 107]. Moreover, the fact that most cohort studies did not report a positive association with processed meats (only one study observed a statistically significant association for processed meats [99]), but many reported positive associations with red meat, supports the possibility that cooking doneness plays a role in etiology.

#### 25.2.11.5 Vitamin D

Strong correlations exist between incidence of pancreatic cancer and UVB irradiance in ecological studies, and incidence rates for pancreatic cancer are higher at higher latitudes in both hemispheres [108]. The first cohort studies on dietary vitamin D intake and pancreatic cancer risk reported inverse associations with higher intake [109, 110], but findings from subsequent studies using circulating levels of vitamin D have been highly conflicting. To date, three large cohort studies have published their findings individually on blood 25-hydroxyvitamin D [25-(OH)D] [111–113]. A strong inverse association for higher levels of plasma 25(OH)D was observed in a study combining five prospective cohort studies (>75 nmol/L vs. <50 nmol/L: OR 0.71, 95 % CI 0.52–0.97) [112]. In contrast, two other studies have found positive associations with higher 25(OH)D levels; in the ATBC study, a threefold increase in risk was noted for those with >65.6 nmol/L compared with <32 nmol/L, and in the PLCO study, a fourfold increase in risk was observed in the highest quartile of 25 (OH)D levels, as compared to the lowest, among those living in low sun exposure residential areas. One study reported a different association of 25(OH)D levels based on the vitamin D binding protein (DBP) [114], which could perhaps explain the discrepancy in the findings from measuring blood 25(OH) D levels. More research is necessary to resolve these conflicting findings, but supplementation with vitamin D at this time is not recommended for pancreatic cancer prevention.

### 25.2.12 Physical Activity

Despite the positive associations observed with obesity and elevated blood glucose, physical activity has not consistently been associated with pancreatic cancer risk. One of the first prospective studies to examine this association reported an inverse association for moderate levels of activity (RR 0.45, 95 % CI 0.29–0.70 for the highest vs. lowest categories of activity), but not for total physical activity. Among overweight individuals, total physical activity was associated with lower risk of pancreatic cancer (RR 0.59, 95 % CI 0.37–0.94 for the top vs. bottom tertiles of total physical activity); no associations were observed for individuals with normal weight [115]. Other subsequent prospective analyses, including a number of very large cohorts, have not replicated the associations with physical activity [116–122].

### 25.2.13 Pesticides

Chemicals used in pesticides and herbicides have (historically) contained compounds that are carcinogenic in animal models, but establishing a link to human cancer is difficult as exposures are typically low in the general population and not easily measured. Several population case-control studies have measured self-reported history of pesticide use and observed higher risk of pancreatic cancer with history of exposure to organochlorines, such as dichlorodiphenyltricholorethane (DDT) and ethylan [123], and herbicides and fungicides [124]. In case-control study with measures of serum organochorines, a fourfold increase in risk of pancreatic cancer was observed for those in the top tertile of polychlorinated byphenyls (PCBs) levels (OR 4.2, 95 % CI 1.8–9.4), and a significant dose-response was noted with increasing PCB levels [125]. After controlling for PCBs, serum DDE was not associated with risk in that study [125]. In another study on serum organochlorines and pancreatic cancer, p,p′-DDT and p,p′-DDE levels did not vary by case-control status, but both were associated with more frequent K-ras mutations among cases [126]. To address potential bias associated with case-control studies, a large cohort of agricultural workers (Agricultural Health Study cohort [AHS]) was undertaken to measure pesticide exposures and cancer risk. Results from the AHS support the possible link between herbicides and pancreatic cancer, with a higher risk observed among workers with the highest lifetime use of EPTC (RR 2.5, 95 % CI 1.1–5.4) and pendimethalin (RR 3.0, 95 % CI 1.3–7.2), both herbicides, compared to never users [127]. However, no associations were observed for DDT or other organochlorines in that study. A recent analysis in the AHS reported a positive association with a new herbicide, acetochlor, and pancreatic cancer [128]. Findings from the AHS will need to be confirmed in other studies as the number of cases were relatively small (<100).

## 25.3 Summary

Table 25.1 summarizes the risk factors for pancreatic cancer with the current level of evidence available for each factor. Many of the established risk factors, including smoking, obesity, and type II diabetes, are modifiable and although incidence rates have not changed much over time, the rise in obesity and diabetes could translate to higher incident rates over the next decade or two. Other risk factors for which the level of evidence is not as strong suggest that the immune response and bacterial infections are likely to play a role in the etiology of this cancer; however, more research is needed to clarify these associations.

**Table 25.1** Risk factors for pancreatic cancer

| Established | Suggestive evidence | No association |
|---|---|---|
| Current cigarette smoke | Allergies (protective) | Folate intake |
| Long-standing diabetes type II | Periodontal disease | Fruit & vegetable intake |
| Chronic pancreatitis | Gastric ulcers | |
| Obesity | *H. pylori* infection | |
| Heavy alcohol intake | Well-done red meat | |
| ABO blood type | | Limited evidence |
| Family hstory | | Physical activity |
| | | Vitamin D |
| | | Pesticide exposure |

## References


1. Surveillance, Epidemiology, and End Results (SEER) Program (www.seer.cancer.gov) SEER*Stat Database: Incidence—SEER 9 Regs Research Data, Nov 2013 Sub (1973-2011) <Katrina/Rita Population Adjustment>—Linked To County Attributes—Total U.S., 1969-2012 Counties, National Cancer Institute, DCCPS, Surveillance Research Program, Surveillance Systems Branch, released April 2014. National Cancer Institute, DCCPS, Surveillance Research Program;2014.
2. ACS. Cancer Facts & Figures 2015. Atlanta: American Cancer Society, Inc.;2015.
3. Bosetti C, Lucenteforte E, Silverman DT, Petersen G, Bracci PM, Ji BT, et al. Cigarette smoking and pancreatic cancer: an analysis from the

International Pancreatic Cancer Case-Control Consortium (Panc4). Annals Oncol (Off J Eur Soc Med Oncol ESMO). 2012;23:1880–8.
4. Lynch SM, Vrieling A, Lubin JH, Kraft P, Mendelsohn JB, Hartge P, et al. Cigarette smoking and pancreatic cancer: a pooled analysis from the pancreatic cancer cohort consortium. Am J Epidemiol. 2009;170:403–13.
5. Heinen MM, Verhage BA, Goldbohm RA, van den Brandt PA. Active and passive smoking and the risk of pancreatic cancer in the Netherlands Cohort Study. Cancer Epidemiol Biomarkers Prev. 2010;19:1612–22 (A publication of the American Association for Cancer Research, cosponsored by the American Society of Preventive Oncology).
6. Vrieling A, Bueno-de-Mesquita HB, Boshuizen HC, Michaud DS, Severinsen MT, Overvad K, et al. Cigarette smoking, environmental tobacco smoke exposure and pancreatic cancer risk in the European Prospective Investigation into Cancer and Nutrition. Int J Cancer (Journal international du cancer). 2010;126:2394–403.
7. Fuchs CS, Colditz GA, Stampfer MJ, Giovannucci EL, Hunter DJ, Rimm EB, et al. A prospective study of cigarette smoking and the risk of pancreatic cancer. Arch Intern Med. 1996;156:2255–60.
8. Bertuccio P, La Vecchia C, Silverman DT, Petersen GM, Bracci PM, Negri E, et al. Cigar and pipe smoking, smokeless tobacco use and pancreatic cancer: an analysis from the International Pancreatic Cancer Case-Control Consortium (PanC4). Annals Oncol (Off J Eur Soc Med Oncol ESMO). 2011;22:1420–6.
9. Lee PN, Hamling J. Systematic review of the relation between smokeless tobacco and cancer in Europe and North America. BMC Med. 2009;7:36.
10. Boffetta P, Hecht S, Gray N, Gupta P, Straif K. Smokeless tobacco and cancer. Lancet Oncol. 2008;9:667–75.
11. Sponsiello-Wang Z, Weitkunat R, Lee PN. Systematic review of the relation between smokeless tobacco and cancer of the pancreas in Europe and North America. BMC Cancer. 2008;8:356.
12. Luo J, Ye W, Zendehdel K, Adami J, Adami HO, Boffetta P, et al. Oral use of Swedish moist snuff (snus) and risk for cancer of the mouth, lung, and pancreas in male construction workers: a retrospective cohort study. Lancet. 2007;369:2015–20.
13. Boffetta P, Aagnes B, Weiderpass E, Andersen A. Smokeless tobacco use and risk of cancer of the pancreas and other organs. Int J Cancer (Journal International du Cancer). 2005;114:992–5.
14. Bao Y, Giovannucci E, Fuchs CS, Michaud DS. Passive smoking and pancreatic cancer in women: a prospective cohort study. Cancer Epidemiol Biomarkers Prev. 2009;18:2292–6 (A publication of the American Association for Cancer Research, cosponsored by the American Society of Preventive Oncology).
15. Lin Y, Yagyu K, Ueda J, Kurosawa M, Tamakoshi A, Kikuchi S, et al. Active and passive smoking and risk of death from pancreatic cancer: findings from the Japan Collaborative Cohort Study. Pancreatology (Off J Int Assoc Pancreatol). 2013;13:279–84.
16. Fryzek JP, Garabrant DH, Schenk M, Kinnard M, Greenson JK, Sarkar FH. The association between selected risk factors for pancreatic cancer and the expression of p53 and K-ras codon 12 mutations. Int J Gastrointestinal Cancer. 2006;37:139–45.
17. Hruban RH, van Mansfeld AD, Offerhaus GJ, van Weering DH, Allison DC, Goodman SN, et al. K-ras oncogene activation in adenocarcinoma of the human pancreas. A study of 82 carcinomas using a combination of mutant-enriched polymerase chain reaction analysis and allele-specific oligonucleotide hybridization. Am J Pathol. 1993;143:545–54.
18. Jiao L, Zhu J, Hassan MM, Evans DB, Abbruzzese JL, Li D. K-ras mutation and p16 and preproenkephalin promoter hypermethylation in plasma DNA of pancreatic cancer patients: in relation to cigarette smoking. Pancreas. 2007;34:55–62.
19. Slebos RJ, Hoppin JA, Tolbert PE, Holly EA, Brock JW, Zhang RH, et al. K-ras and p53 in pancreatic cancer: association with medical history, histopathology, and environmental exposures in a population-based study. Cancer Epidemiol Biomarkers Prev. 2000;9:1223–32 (A publication of the American Association for Cancer Research, cosponsored by the American Society of Preventive Oncology).
20. Crous-Bou M, Porta M, Lopez T, Jariod M, Malats N, Alguacil J, et al. Lifetime history of tobacco consumption and K-ras mutations in exocrine pancreatic cancer. Pancreas. 2007;35:135–41.
21. Malats N, Porta M, Corominas JM, Pinol JL, Rifa J, Real FX. Ki-ras mutations in exocrine pancreatic cancer: association with clinico-pathological characteristics and with tobacco and alcohol consumption. PANK-ras I Project Investigators. Int J Cancer (Journal International du Cancer). 1997;70:661–7.
22. Blackford A, Parmigiani G, Kensler TW, Wolfgang C, Jones S, Zhang X, et al. Genetic mutations associated with cigarette smoking in pancreatic cancer. Cancer Res. 2009;69:3681–8.
23. Slebos RJ, Hruban RH, Dalesio O, Mooi WJ, Offerhaus GJ, Rodenhuis S. Relationship between K-ras oncogene activation and smoking in adenocarcinoma of the human lung. J Natl Cancer Inst. 1991;83:1024–7.
24. Porta M, Crous-Bou M, Wark PA, Vineis P, Real FX, Malats N, et al. Cigarette smoking and K-ras mutations in pancreas, lung and colorectal adenocarcinomas: etiopathogenic similarities, differences and paradoxes. Mutat Res. 2009;682:83–93.
25. Giovannucci E, Michaud D. The role of obesity and related metabolic disturbances in cancers of the

colon, prostate, and pancreas. Gastroenterology. 2007;132:2208–25.
26. Stolzenberg-Solomon RZ, Graubard BI, Chari S, Limburg P, Taylor PR, Virtamo J, et al. Insulin, glucose, insulin resistance, and pancreatic cancer in male smokers. JAMA. 2005;294:2872–8.
27. Huxley R, Ansary-Moghaddam A. Berrington de Gonzalez A, Barzi F, Woodward M. Type-II diabetes and pancreatic cancer: a meta-analysis of 36 studies. Br J Cancer. 2005;92:2076–83.
28. Bosetti C, Rosato V, Li D, Silverman D, Petersen GM, Bracci PM, et al. Diabetes, antidiabetic medications, and pancreatic cancer risk: an analysis from the International Pancreatic Cancer Case-Control Consortium. Annals Oncol (Off J Eur Soc Med Oncol ESMO). 2014;25:2065–72.
29. Ben Q, Cai Q, Li Z, Yuan Y, Ning X, Deng S, et al. The relationship between new-onset diabetes mellitus and pancreatic cancer risk: a case-control study. Eur J Cancer. 2011;47:248–54.
30. Ben Q, Xu M, Ning X, Liu J, Hong S, Huang W, et al. Diabetes mellitus and risk of pancreatic cancer: A meta-analysis of cohort studies. Eur J Cancer. 2011;47:1928–37.
31. Yuan C, Rubinson DA, Qian ZR, Wu C, Kraft P, Bao Y, et al. Survival among patients with pancreatic cancer and long-standing or recent-onset diabetes mellitus. J Clin Oncol (Off J Am Soc Clin Oncol). 2015;33:29–35.
32. Batty GD, Shipley MJ, Marmot M, Smith GD. Diabetes status and post-load plasma glucose concentration in relation to site-specific cancer mortality: findings from the original Whitehall study. Cancer Causes Control (CCC). 2004;15:873–81.
33. Jee SH, Ohrr H, Sull JW, Yun JE, Ji M, Samet JM. Fasting serum glucose level and cancer risk in Korean men and women. JAMA. 2005;293: 194–202.
34. Gapstur SM, Gann PH, Lowe W, Liu K, Colangelo L, Dyer A. Abnormal glucose metabolism and pancreatic cancer mortality. JAMA. 2000;283:2552–8.
35. Grote VA, Rohrmann S, Nieters A, Dossus L, Tjonneland A, Halkjaer J, et al. Diabetes mellitus, glycated haemoglobin and C-peptide levels in relation to pancreatic cancer risk: a study within the European Prospective Investigation into Cancer and Nutrition (EPIC) cohort. Diabetologia. 2011;54:3037–46.
36. Wolpin BM, Bao Y, Qian ZR, Wu C, Kraft P, Ogino S, et al. Hyperglycemia, insulin resistance, impaired pancreatic beta-cell function, and risk of pancreatic cancer. J Natl Cancer Inst. 2013;105:1027–35.
37. Michaud DS, Wolpin B, Giovannucci E, Liu S, Cochrane B, Manson JE, et al. Prediagnostic plasma C-peptide and pancreatic cancer risk in men and women. Cancer Epidemiol Biomarkers Prev. 2007;16:2101–9 (A publication of the American Association for Cancer Research, cosponsored by the American Society of Preventive Oncology).
38. Aune D, Greenwood DC, Chan DS, Vieira R, Vieira AR, Navarro Rosenblatt DA, et al. Body mass index, abdominal fatness and pancreatic cancer risk: a systematic review and non-linear dose-response meta-analysis of prospective studies. Annals Oncol (Off J Eur Soc Med Oncol ESMO). 2012;23:843–52.
39. Arslan AA, Helzlsouer KJ, Kooperberg C, Shu XO, Steplowski E, Bueno-de-Mesquita HB, et al. Anthropometric measures, body mass index, and pancreatic cancer: a pooled analysis from the Pancreatic Cancer Cohort Consortium (PanScan). Arch Intern Med. 2010;170:791–802.
40. Genkinger JM, Spiegelman D, Anderson KE, Bernstein L, van den Brandt PA, Calle EE, et al. A pooled analysis of 14 cohort studies of anthropometric factors and pancreatic cancer risk. Int J Cancer (Journal International du Cancer). 2011;129:1708–17.
41. Yuan C, Bao Y, Wu C, Kraft P, Ogino S, Ng K, et al. Prediagnostic body mass index and pancreatic cancer survival. J Clin oncol (Off J Am Soc Clin Oncol). 2013;31:4229–34.
42. Rebours V, Boutron-Ruault MC, Schnee M, Ferec C, Le Marechal C, Hentic O, et al. The natural history of hereditary pancreatitis: a national series. Gut. 2009;58:97–103.
43. Rebours V, Boutron-Ruault MC, Schnee M, Ferec C, Maire F, Hammel P, et al. Risk of pancreatic adenocarcinoma in patients with hereditary pancreatitis: a national exhaustive series. Am J Gastroenterology. 2008;103:111–9.
44. Lowenfels AB, Maisonneuve P, DiMagno EP, Elitsur Y, Gates LK Jr, Perrault J, et al. Hereditary pancreatitis and the risk of pancreatic cancer. International Hereditary Pancreatitis Study Group. J Natl Cancer Inst. 1997;89:442–6.
45. Howes N, Lerch MM, Greenhalf W, Stocken DD, Ellis I, Simon P, et al. Clinical and genetic characteristics of hereditary pancreatitis in Europe. Clin Gastroenterology Hepatol (Off Clin Practice J Am Gastroenterological Assoc). 2004;2:252–61.
46. Lowenfels AB, Maisonneuve P, Whitcomb DC, Lerch MM, DiMagno EP. Cigarette smoking as a risk factor for pancreatic cancer in patients with hereditary pancreatitis. JAMA. 2001;286:169–70.
47. Raimondi S, Lowenfels AB, Morselli-Labate AM, Maisonneuve P, Pezzilli R. Pancreatic cancer in chronic pancreatitis; aetiology, incidence, and early detection. Best Pract Res Clin Gastroenterol. 2010;24:349–58.
48. Karlson BM, Ekbom A, Josefsson S, McLaughlin JK, Fraumeni JF Jr, Nyren O. The risk of pancreatic cancer following pancreatitis: an association due to confounding? Gastroenterology. 1997;113:587–92.

49. Murtaugh LC. Pathogenesis of pancreatic cancer: lessons from animal models. Toxicol Pathol. 2014;42:217–28.
50. Jacobs EJ, Chanock SJ, Fuchs CS, Lacroix A, McWilliams RR, Steplowski E, et al. Family history of cancer and risk of pancreatic cancer: a pooled analysis from the Pancreatic Cancer Cohort Consortium (PanScan). Int J Cancer (Journal International du Cancer). 2010.
51. Naderi A, Couch FJ. BRCA2 and pancreatic cancer. Int J Gastrointestinal cancer. 2002;31:99–106.
52. Landi S. Genetic predisposition and environmental risk factors to pancreatic cancer: A review of the literature. Mutat Res. 2009;681:299–307.
53. Amundadottir L, Kraft P, Stolzenberg-Solomon RZ, Fuchs CS, Petersen GM, Arslan AA, et al. Genome-wide association study identifies variants in the ABO locus associated with susceptibility to pancreatic cancer. Nat Genet. 2009;41:986–90.
54. Petersen GM, Amundadottir L, Fuchs CS, Kraft P, Stolzenberg-Solomon RZ, Jacobs KB, et al. A genome-wide association study identifies pancreatic cancer susceptibility loci on chromosomes 13q22.1, 1q32.1 and 5p15.33. Nat Genet. 2010;42:224–8.
55. Wolpin BM, Rizzato C, Kraft P, Kooperberg C, Petersen GM, Wang Z, et al. Genome-wide association study identifies multiple susceptibility loci for pancreatic cancer. Nat Genet. 2014;46:994–1000.
56. Aird I, Lee DR, Roberts JA. ABO blood groups and cancer of oesophagus, cancerof pancreas, and pituitary adenoma. Br Med J. 1960;1:1163–6.
57. Wolpin BM, Chan AT, Hartge P, Chanock SJ, Kraft P, Hunter DJ, et al. ABO blood group and the risk of pancreatic cancer. J Natl Cancer Inst. 2009;101:424–31.
58. Wolpin BM, Kraft P, Gross M, Helzlsouer K, Bueno-de-Mesquita HB, Steplowski E, et al. Pancreatic cancer risk and ABO blood group alleles: results from the pancreatic cancer cohort consortium. Cancer Res. 2010;70:1015–23.
59. Wolpin BM, Kraft P, Xu M, Steplowski E, Olsson ML, Arslan AA, et al. Variant ABO blood group alleles, secretor status, and risk of pancreatic cancer: results from the pancreatic cancer cohort consortium. Cancer Epidemiol Biomarkers Prev. 2010;19:3140–9 (A publication of the American Association for Cancer Research, cosponsored by the American Society of Preventive Oncology).
60. Risch HA, Yu H, Lu L, Kidd MS. ABO blood group, Helicobacter pylori seropositivity, and risk of pancreatic cancer: a case-control study. J Natl Cancer Inst. 2010;102:502–5.
61. Sun W, Wen CP, Lin J, Wen C, Pu X, Huang M, et al. ABO blood types and cancer risk-A cohort study of 339,432 subjects in Taiwan. Cancer Epidemiol. 2015;39:150–6.
62. Klein AP, Lindstrom S, Mendelsohn JB, Steplowski E, Arslan AA, Bueno-de-Mesquita HB, et al. An absolute risk model to identify individuals at elevated risk for pancreatic cancer in the general population. PLoS ONE. 2013;8:e72311.
63. Raderer M, Wrba F, Kornek G, Maca T, Koller DY, Weinlaender G, et al. Association between Helicobacter pylori infection and pancreatic cancer. Oncology. 1998;55:16–9.
64. Stolzenberg-Solomon RZ, Blaser MJ, Limburg PJ, Perez-Perez G, Taylor PR, Virtamo J, et al. Helicobacter pylori seropositivity as a risk factor for pancreatic cancer. J Natl Cancer Inst. 2001;93: 937–41.
65. de Martel C, Llosa AE, Friedman GD, Vogelman JH, Orentreich N, Stolzenberg-Solomon RZ, et al. *Helicobacter pylori* infection and development of pancreatic cancer. Cancer Epidemiol Biomarkers Prev. 2008;17:1188–94 (A publication of the American Association for Cancer Research, cosponsored by the American Society of Preventive Oncology).
66. Lindkvist B, Johansen D, Borgstrom A, Manjer J. A prospective study of *Helicobacter pylori* in relation to the risk for pancreatic cancer. BMC Cancer. 2008;8:321.
67. Yu G, Murphy G, Michel A, Weinstein SJ, Mannisto S, Albanes D, et al. Seropositivity to Helicobacter pylori and risk of pancreatic cancer. Cancer Epidemiol Biomarkers Prev. 2013;22:2416–9 (A publication of the American Association for Cancer Research, cosponsored by the American Society of Preventive Oncology).
68. Wang Y, Zhang FC, Wang YJ. Helicobacter pylori and pancreatic cancer risk: a meta- analysis based on 2,049 cases and 2,861 controls. Asian Pacific J Cancer Prev (APJCP). 2014;15:4449–54.
69. Xiao M, Wang Y, Gao Y. Association between Helicobacter pylori infection and pancreatic cancer development: a meta-analysis. PLoS ONE. 2013;8: e75559.
70. Bao Y, Spiegelman D, Li R, Giovannucci E, Fuchs CS, Michaud DS. History of peptic ulcer disease and pancreatic cancer risk in men. Gastroenterology. 2010;138:541–9.
71. Luo J, Nordenvall C, Nyren O, Adami HO, Permert J, Ye W. The risk of pancreatic cancer in patients with gastric or duodenal ulcer disease. Int J Cancer (Journal International du Cancer). 2007;120:368–72.
72. Papapanou PN. Periodontal diseases: epidemiology. Ann Periodontol. 1996;1:1–36.
73. Tribble GD, Kerr JE, Wang BY. Genetic diversity in the oral pathogen Porphyromonas gingivalis: molecular mechanisms and biological consequences. Future Microbiol. 2013;8:607–20.
74. Hujoel PP, Drangsholt M, Spiekerman C, Weiss NS. An exploration of the periodontitis-cancer association. Ann Epidemiol. 2003;13:312–6.
75. Michaud DS, Joshipura K, Giovannucci E, Fuchs CS. A prospective study of periodontal disease and pancreatic cancer in US male health professionals. J Natl Cancer Inst. 2007;99:171–5.

76. Ahn J, Segers S, Hayes RB. Periodontal disease, Porphyromonas gingivalis serum antibody levels and orodigestive cancer mortality. Carcinogenesis. 2012;33:1055–8.
77. Joshipura KJ, Douglass CW, Garcia RI, Valachovic R, Willett WC. Validity of a self-reported periodontal disease measure. J Public Health Dent. 1996;56:205–12.
78. Michaud DS, Izard J, Wilhelm-Benartzi CS, You DH, Grote VA, Tjonneland A, et al. Plasma antibodies to oral bacteria and risk of pancreatic cancer in a large European prospective cohort study. Gut. 2013;62:1764–70.
79. Gandini S, Lowenfels AB, Jaffee EM, Armstrong TD, Maisonneuve P. Allergies and the risk of pancreatic cancer: a meta-analysis with review of epidemiology and biological mechanisms. Cancer Epidemiol Biomarkers Prev. 2005;14:1908–16 (A publication of the American Association for Cancer Research, cosponsored by the American Society of Preventive Oncology).
80. Maisonneuve P, Lowenfels AB, Bueno-de-Mesquita HB, Ghadirian P, Baghurst PA, Zatonski WA, et al. Past medical history and pancreatic cancer risk: results from a multicenter case-control study. Ann Epidemiol. 2010;20:92–8.
81. Olson SH, Hsu M, Satagopan JM, Maisonneuve P, Silverman DT, Lucenteforte E, et al. Allergies and risk of pancreatic cancer: a pooled analysis from the Pancreatic Cancer Case-Control Consortium. Am J Epidemiol. 2013;178:691–700.
82. Cotterchio M, Lowcock E, Hudson TJ, Greenwood C, Gallinger S. Association between allergies and risk of pancreatic cancer. Cancer Epidemiol Biomarkers Prev. 2014;23:469–80 (A publication of the American Association for Cancer Research, cosponsored by the American Society of Preventive Oncology).
83. Michaud DS, Vrieling A, Jiao L, Mendelsohn JB, Steplowski E, Lynch SM, et al. Alcohol intake and pancreatic cancer: a pooled analysis from the pancreatic cancer cohort consortium (PanScan). Cancer causes & control: CCC. 2010;21:1213–25.
84. Genkinger JM, Spiegelman D, Anderson KE, Bergkvist L, Bernstein L, van den Brandt PA, et al. Alcohol intake and pancreatic cancer risk: a pooled analysis of fourteen cohort studies. Cancer Epidemiol Biomarkers Prev. 2009;18:765–76 (A publication of the American Association for Cancer Research, cosponsored by the American Society of Preventive Oncology).
85. Lucenteforte E, La Vecchia C, Silverman D, Petersen GM, Bracci PM, Ji BT, et al. Alcohol consumption and pancreatic cancer: a pooled analysis in the International Pancreatic Cancer Case-Control Consortium (PanC4). Annals Oncol (Off J Eur Soc Med Oncol ESMO). 2012;23:374–82.
86. Jiao L, Silverman DT, Schairer C, Thiebaut AC, Hollenbeck AR, Leitzmann MF, et al. Alcohol Use and Risk of Pancreatic Cancer: The NIH-AARP Diet and Health Study. Am J Epidemiol. 2009.
87. Bao Y, Michaud DS, Spiegelman D, Albanes D, Anderson KE, Bernstein L, et al. Folate intake and risk of pancreatic cancer: pooled analysis of prospective cohort studies. J Natl Cancer Inst. 2011;103:1840–50.
88. Larsson SC, Giovannucci E, Wolk A. Folate intake, MTHFR polymorphisms, and risk of esophageal, gastric, and pancreatic cancer: a meta-analysis. Gastroenterology. 2006;131:1271–83.
89. Tio M, Andrici J, Cox MR, Eslick GD. Folate intake and the risk of upper gastrointestinal cancers: a systematic review and meta-analysis. J Gastroenterol Hepatol. 2014;29:250–8.
90. Lin HL, An QZ, Wang QZ, Liu CX. Folate intake and pancreatic cancer risk: an overall and dose-response meta-analysis. Public health. 2013;127:607–13.
91. Schernhammer E, Wolpin B, Rifai N, Cochrane B, Manson JA, Ma J, et al. Plasma folate, vitamin B6, vitamin B12, and homocysteine and pancreatic cancer risk in four large cohorts. Cancer Res. 2007;67:5553–60.
92. Chuang SC, Stolzenberg-Solomon R, Ueland PM, Vollset SE, Midttun O, Olsen A, et al. A U-shaped relationship between plasma folate and pancreatic cancer risk in the European Prospective Investigation into Cancer and Nutrition. Eur J Cancer. 2011;47:1808–16.
93. Stolzenberg-Solomon RZ, Albanes D, Nieto FJ, Hartman TJ, Tangrea JA, Rautalahti M, et al. Pancreatic cancer risk and nutrition-related methyl-group availability indicators in male smokers. J Natl Cancer Inst. 1999;91:535–41.
94. Koushik A, Spiegelman D, Albanes D, Anderson KE, Bernstein L, van den Brandt PA, et al. Intake of fruits and vegetables and risk of pancreatic cancer in a pooled analysis of 14 cohort studies. Am J Epidemiol. 2012;176:373–86.
95. Vrieling A, Verhage BA, van Duijnhoven FJ, Jenab M, Overvad K, Tjonneland A, et al. Fruit and vegetable consumption and pancreatic cancer risk in the European Prospective Investigation into Cancer and Nutrition. Int J Cancer (Journal International du Cancer). 2009;124:1926–34.
96. Alsamarrai A, Das SL, Windsor JA, Petrov MS. Factors that affect risk for pancreatic disease in the general population: a systematic review and meta-analysis of prospective cohort studies. Clin Gastroenterol Hepatol (Off Clin Practice J Am Gastroenterol Assoc). 2014;12:1635–44 (e5; quiz e103).
97. Anderson KE, Mongin SJ, Sinha R, Stolzenberg-Solomon R, Gross MD, Ziegler RG, et al. Pancreatic cancer risk: associations with meat-derived carcinogen intake in the prostate, lung, colorectal, and ovarian cancer screening trial (PLCO) cohort. Mol Carcinog. 2012;51:128–37.

98. Stolzenberg-Solomon RZ, Cross AJ, Silverman DT, Schairer C, Thompson FE, Kipnis V, et al. Meat and meat-mutagen intake and pancreatic cancer risk in the NIH-AARP cohort. Cancer Epidemiol Biomarkers Prev. 2007;16:2664–75 (A publication of the American Association for Cancer Research, cosponsored by the American Society of Preventive Oncology).
99. Nothlings U, Wilkens LR, Murphy SP, Hankin JH, Henderson BE, Kolonel LN. Meat and fat intake as risk factors for pancreatic cancer: the multiethnic cohort study. J Natl Cancer Inst. 2005;97:1458–65.
100. Larsson SC, Hakanson N, Permert J, Wolk A. Meat, fish, poultry and egg consumption in relation to risk of pancreatic cancer: a prospective study. Int J Cancer (Journal International du Cancer). 2006;118:2866–70.
101. Zheng W, McLaughlin JK, Gridley G, Bjelke E, Schuman LM, Silverman DT, et al. A cohort study of smoking, alcohol consumption, and dietary factors for pancreatic cancer (United States). Cancer Causes Control (CCC). 1993;4:477–82.
102. Rohrmann S, Linseisen J, Nothlings U, Overvad K, Egeberg R, Tjonneland A, et al. Meat and fish consumption and risk of pancreatic cancer: results from the European Prospective Investigation into Cancer and Nutrition. Int J Cancer (Journal International du Cancer). 2013;132:617–24.
103. Michaud DS, Giovannucci E, Willett WC, Colditz GA, Fuchs CS. Dietary meat, dairy products, fat, and cholesterol and pancreatic cancer risk in a prospective study. Am J Epidemiol. 2003;157:1115–25.
104. Heinen MM, Verhage BA, Goldbohm RA, van den Brandt PA. Meat and fat intake and pancreatic cancer risk in the Netherlands Cohort Study. Int J Cancer (Journal International du Cancer). 2009;125:1118–26.
105. Stolzenberg-Solomon RZ, Pietinen P, Taylor PR, Virtamo J, Albanes D. Prospective study of diet and pancreatic cancer in male smokers. Am J Epidemiol. 2002;155:783–92.
106. Anderson KE, Sinha R, Kulldorff M, Gross M, Lang NP, Barber C, et al. Meat intake and cooking techniques: associations with pancreatic cancer. Mutat Res. 2002;506–507:225–31.
107. Anderson KE, Kadlubar FF, Kulldorff M, Harnack L, Gross M, Lang NP, et al. Dietary intake of heterocyclic amines and benzo(a)pyrene: associations with pancreatic cancer. Cancer Epidemiol Biomarkers Prev. 2005;14:2261–5 (A publication of the American Association for Cancer Research, cosponsored by the American Society of Preventive Oncology).
108. Mohr SB, Garland CF, Gorham ED, Grant WB, Garland FC. Ultraviolet B irradiance and vitamin D status are inversely associated with incidence rates of pancreatic cancer worldwide. Pancreas. 2010;39: 669–74.
109. Skinner HG, Michaud DS, Giovannucci E, Willett WC, Colditz GA, Fuchs CS. Vitamin D intake and the risk for pancreatic cancer in two cohort studies. Cancer Epidemiol Biomarkers Prev. 2006;15:1688–95 (A publication of the American Association for Cancer Research, cosponsored by the American Society of Preventive Oncology).
110. Giovannucci E, Liu Y, Rimm EB, Hollis BW, Fuchs CS, Stampfer MJ, et al. Prospective study of predictors of vitamin D status and cancer incidence and mortality in men. J Natl Cancer Inst. 2006;98:451–9.
111. Stolzenberg-Solomon RZ, Vieth R, Azad A, Pietinen P, Taylor PR, Virtamo J, et al. A prospective nested case-control study of vitamin D status and pancreatic cancer risk in male smokers. Cancer Res. 2006;66:10213–9.
112. Wolpin BM, Ng K, Bao Y, Kraft P, Stampfer MJ, Michaud DS, et al. Plasma 25-hydroxyvitamin D and risk of pancreatic cancer. Cancer Epidemiol Biomarkers Prev. 2012;21:82–91 (A publication of the American Association for Cancer Research, cosponsored by the American Society of Preventive Oncology).
113. Stolzenberg-Solomon RZ, Hayes RB, Horst RL, Anderson KE, Hollis BW, Silverman DT. Serum vitamin D and risk of pancreatic cancer in the prostate, lung, colorectal, and ovarian screening trial. Cancer Res. 2009;69:1439–47.
114. Weinstein SJ, Stolzenberg-Solomon RZ, Kopp W, Rager H, Virtamo J, Albanes D. Impact of circulating vitamin D binding protein levels on the association between 25-hydroxyvitamin D and pancreatic cancer risk: a nested case-control study. Cancer Res. 2012;72:1190–8.
115. Michaud DS, Giovannucci E, Willett WC, Colditz GA, Stampfer MJ, Fuchs CS. Physical activity, obesity, height and the risk of pancreatic cancer. JAMA. 2001;286:921–9.
116. Stolzenberg-Solomon RZ, Adams K, Leitzmann M, Schairer C, Michaud DS, Hollenbeck A, et al. Adiposity, physical activity, and pancreatic cancer in the National Institutes of Health-AARP Diet and Health Cohort. Am J Epidemiol. 2008;167:586–97.
117. Luo J, Iwasaki M, Inoue M, Sasazuki S, Otani T, Ye W, et al. Body mass index, physical activity and the risk of pancreatic cancer in relation to smoking status and history of diabetes: a large-scale population-based cohort study in Japan—the JPHC study. Cancer Causes Control (CCC). 2007;18: 603–12.
118. Lin Y, Kikuchi S, Tamakoshi A, Yagyu K, Obata Y, Inaba Y, et al. Obesity, physical activity and the risk of pancreatic cancer in a large Japanese cohort. Int J Cancer (Journal International du Cancer). 2007;120:2665–71.
119. Berrington de Gonzalez A, Spencer EA, Bueno-de-Mesquita HB, Roddam A, Stolzenberg-Solomon R, Halkjaer J, et al. Anthropometry,

physical activity, and the risk of pancreatic cancer in the European prospective investigation into cancer and nutrition. Cancer Epidemiol Biomarkers Prev. 2006;15:879–85 (A publication of the American Association for Cancer Research, cosponsored by the American Society of Preventive Oncology).

120. Nothlings U, Wilkens LR, Murphy SP, Hankin JH, Henderson BE, Kolonel LN. Body mass index and physical activity as risk factors for pancreatic cancer: the Multiethnic Cohort Study. Cancer causes & control: CCC. 2007;18:165–75.

121. Patel AV, Rodriguez C, Bernstein L, Chao A, Thun MJ, Calle EE. Obesity, recreational physical activity, and risk of pancreatic cancer in a large U.S. Cohort. Cancer Epidemiol Biomarkers Prev. 2005;14:459–66 (A publication of the American Association for Cancer Research, cosponsored by the American Society of Preventive Oncology).

122. Sinner PJ, Schmitz KH, Anderson KE, Folsom AR. Lack of association of physical activity and obesity with incident pancreatic cancer in elderly women. Cancer Epidemiol Biomarkers Prev. 2005;14:1571–3 (A publication of the American Association for Cancer Research, cosponsored by the American Society of Preventive Oncology).

123. Fryzek JP, Garabrant DH, Harlow SD, Severson RK, Gillespie BW, Schenk M, et al. A case-control study of self-reported exposures to pesticides and pancreas cancer in southeastern Michigan. Int J Cancer (Journal International du Cancer). 1997;72:62–7.

124. Ji BT, Silverman DT, Stewart PA, Blair A, Swanson GM, Baris D, et al. Occupational exposure to pesticides and pancreatic cancer. Am J Ind Med. 2001;39:92–9.

125. Hoppin JA, Tolbert PE, Holly EA, Brock JW, Korrick SA, Altshul LM, et al. Pancreatic cancer and serum organochlorine levels. Cancer Epidemiol Biomarkers Prev. 2000;9:199–205 (A publication of the American Association for Cancer Research, cosponsored by the American Society of Preventive Oncology).

126. Porta M, Malats N, Jariod M, Grimalt JO, Rifa J, Carrato A, et al. Serum concentrations of organochlorine compounds and K-ras mutations in exocrine pancreatic cancer. PANKRAS II Study Group. Lancet. 1999;354:2125–9.

127. Andreotti G, Freeman LE, Hou L, Coble J, Rusiecki J, Hoppin JA, et al. Agricultural pesticide use and pancreatic cancer risk in the Agricultural Health Study Cohort. Int J Cancer (Journal International du Cancer). 2009;124:2495–500.

128. Lerro CC, Koutros S, Andreotti G, Hines CJ, Blair A, Lubin J, et al. Use of acetochlor and cancer incidence in the Agricultural Health Study. Int J Cancer (Journal International du Cancer). 2015.

# 26 Pathology and Molecular Pathology of Pancreatic Cancer

Wai Chin Foo and Huamin Wang

## 26.1 Clinical Picture of Disease

### 26.1.1 Signs and Symptoms

Patients with pancreatic cancer most commonly present with pain, jaundice, and unexplained weight loss [1]. The pain that accompanies pancreatic cancer is typically epigastric and radiates to the back. However, the presentation varies depending on the location of the tumor; the majority (60–70 %) of which are localized to the head of the pancreas [2]. Other nonspecific symptoms include clay-colored stools, nausea, and migratory thrombophlebitis (also called Trousseau's syndrome) [3]. While depression is not uncommon in patients with pancreatic cancer, interestingly, there have been several studies where the diagnosis of depression precedes the diagnosis of carcinoma [4, 5]. Unfortunately, the presenting signs and symptoms are nonspecific; as such, the diagnosis is not made until late in the course of the disease after the cancer has already spread to other organs.

W.C. Foo · H. Wang (✉)
MD Anderson Cancer Center, University of Texas, 1515 Holcombe Blvd., Unit 085, Houston, TX 77030, USA
e-mail: hmwang@mdanderson.org

### 26.1.2 Diagnosis

The gold standard for the diagnosis of pancreatic cancer requires histologic confirmation. When based only on symptoms that are considered highly specific and sensitive for the disease, Di Magno et al. found a significant lack of specificity. The majority of patients had other diagnoses, including nonpancreatic cancers, pancreatitis, and nonpancreatic disorders, e.g., irritable bowel syndrome [6]. However, histologic confirmation is not always required as the diagnostic evaluation of a patient with suspected disease includes serologic evaluation of carcinoembryonic antigen (CEA) [7] and carbohydrate antigen 19–9 (CA19–9) [8, 9] and abdominal imaging.

Technological improvements in abdominal imaging have significantly increased the sensitivity. Multidetector computed tomography (MDCT) [7, 10–12], magnetic resonance imaging (MRI), and endoscopic ultrasound (EUS) [10, 13] can be used to visualize the tumor and its relationship to vessels. Both CT and MRI have similar sensitivities and specificities. In addition, positron emission tomography (PET) imaging also has a role in the evaluation [14] as staging is important for therapeutic considerations. When the imaging findings are not typical, EUS-guided or percutaneous aspirations or biopsies can be obtained for histologic confirmation. EUS-guided fine needle aspiration (FNA) is the best modality for obtaining a diagnosis with a sensitivity and specificity of 87 and 98 %, respectively [12].


M. Loda et al. (eds.), *Pathology and Epidemiology of Cancer*,
DOI 10.1007/978-3-319-35153-7_26

### 26.1.3 Treatment and Prognosis

In general, surgical resection offers the only potential cure for pancreatic cancer. However, the feasibility of surgery depends on accurate clinical staging. The American Joint Committee on Cancer (AJCC), together with the TNM classification, is the most widely used system for staging.

Patients with stage I and II disease are optimally treated with surgical resection followed by adjuvant therapy, using either chemotherapy (gemcitabine or 5-fluorouracil often in combination with other agents, e.g., irinotecan, cardocetaxel, paclitaxel, and capecitabine), radiation therapy, or combined approaches (chemoradiotherapy) [15]. Even with adjuvant therapy, the -5-year survival rates for patients are low, between 25 and 30 % for those with node-negative disease, and is lower, at approximately 10 %, for those with node-positive disease [16, 17]. Patients with stage III disease are treated with chemotherapy or chemoradiotherapy [15]. A subset of these patients may proceed to surgical resection depending on the extent of disease involvement with the superior mesenteric artery, the celiac axis, and the superior mesenteric and portal veins [18]. However, the vast majority of these patients develop metastatic disease. Patients with stage IV disease are either offered systemic chemotherapy or supportive therapy [19]. The type of resection, i.e., pancreaticoduodenectomy [20, 21], distal subtotal pancreatectomy [22], or total pancreatectomy [23], is dependent on the location of the tumor.

Despite offering the only potential cure, only 15–20 % of cases are likely to be resectable at presentation. Approximately 40 % have distant metastases, and another 30–40 % have locally advanced, unresectable disease. Furthermore, prognosis is poor, even for those undergoing complete resection (R0). The median survival rate ranges from 17 to 27 months, with a 5-year survival rate of approximately 20 % [20]. Moreover, the surgical procedures necessary to resect the disease are associated with significant morbidity (40–60 %) and mortality (2–3 %) [20, 24].

## 26.2 Pathology

### 26.2.1 Introduction

This section discusses the embryological development of the pancreas as well as the normal anatomy and histology. The remainder of the section focuses on pancreatic ductal adenocarcinoma (PDA) as it is synonymous with the term "pancreatic cancer." It is the cause for the dismal prognosis of pancreatic cancer in the United States [25] and has been the focus of intense research efforts to unravel its molecular pathogenesis. Precursors to PDA, including pancreatic intraepithelial neoplasia (PanIN), intraductal papillary mucinous neoplasms (IPMNs), and mucinous cystic neoplasms (MCNs), are also discussed.

### 26.2.2 Embryology and Development of the Pancreas

The pancreas arises from the fusion of two endoderm buds, the smaller ventral bud and the large dorsal bud, off the primitive duodenum. The ventral bud develops into the uncinate process and the inferior portion of the head, and the dorsal bud develops into the superior portion of the head and the remainder of the pancreas. The main pancreatic duct (also called the duct of Wirsung) forms from the fusion of the two buds. Occasionally, the proximal portion of the dorsal duct remains as the accessory pancreatic duct (also called the duct of Santorini).

### 26.2.3 Anatomy of the Pancreas

In adults, the pancreas measures approximately 15 cm and weighs 80 g. It is situated in the retroperitoneum and can be divided into the head, the body, and the tail. The head, which includes the uncinate process, lies within the curve of the duodenum and extends to the superior mesenteric vein. The uncinate process represents the inferior portion of the head and is wedged between the superior mesenteric vessels anteriorly and the aorta posteriorly. The body of the pancreas extends from the superior mesenteric vein to the aorta; the junction between the head and the body is arbitrarily designated as the neck. The tail extends from the aorta and tapers in the splenorenal ligament to the splenic hilum. Peritoneum covers the anterior and inferior aspects of the head and only covers the anterior aspect of the body and tail [26].

The arterial supply to the pancreas is provided by branches from the common hepatic and splenic arteries, which themselves are branches of the celiac artery; as well as the inferior pancreaticoduodenal artery, which is a branch from the superior mesenteric artery. The superior pancreaticoduodenal artery, which comes off the common hepatic artery, and its inferior counterpart form the pancreaticoduodenal arcade, which supplies the head and the uncinate process. The splenic artery supplies the body and tail of the pancreas. Venous drainage is via the superior and inferior pancreaticoduodenal veins that empty into the portal vein and superior mesenteric vein, respectively; and the inferior mesenteric vein that empties into the splenic vein. Lymphatic vessels in the pancreatic head drain into the pancreaticoduodenal lymph nodes and those in the hepatoduodenal ligament. The body and tail drain into the nodes along the middle colic, hepatic, and splenic arteries. The draining terminates into the celiac, superior mesenteric, para-aortic, and aortocaval lymph nodes. The pancreas is richly innervated. The sympathetic fibers come from T6 to T10 via the thoracic splanchnic nerves and celiac plexus. The parasympathetic fibers come from the posterior vagal trunk via its celiac branch [27].

Given that surgical resections provide the best hope for long-term survival [28], knowledge of the anatomy is crucial in the evaluation of the resection specimens.

### 26.2.4 Histology of the Pancreas

The pancreas is composed of two separate glandular components, the exocrine and endocrine pancreas, and a ductal system (see Fig. 26.1). The exocrine pancreas represents 80–85 % of the parenchyma and is composed of columnar to pyramidal epithelial cells forming an acinus. The endocrine pancreas is composed of the islets of Langerhans, which in turn consist of four major cell types (alpha, beta, delta, and pancreatic polypeptide cells) and two minor cell types (D1 and enterochromaffin cells). Together, these small round cells are responsible for the secretion of insulin, glucagon, somatostatin, pancreatic polypeptide, vasoactive polypeptide, and serotonin. The ductal system is composed of cuboidal to columnar cells, beginning at the small ducts that drain each acinus to the main pancreatic duct that joins the common bile duct before emptying into the duodenum [26]. The mesenchymal portion consists of connective tissue that separates the parenchyma into lobules and connective tissue, vasculature, and nerves that travel alongside the ductal system. Adipose tissue can also be seen in the connective tissue.

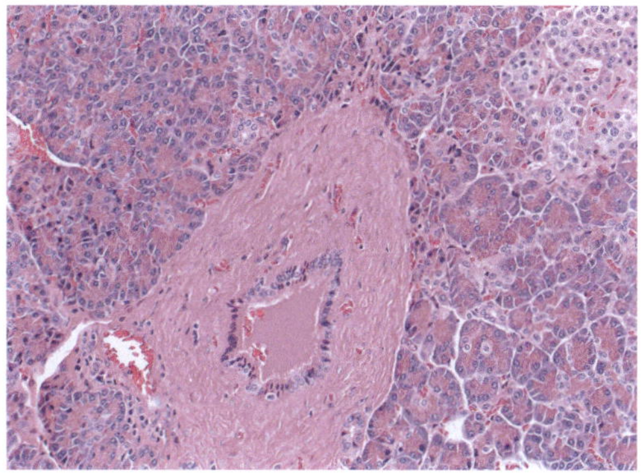

**Fig. 26.1** Normal pancreas

### 26.2.5 Pancreatic Cancer and Precursor Lesions

Despite representing only a small portion of the epithelial component, the vast majority (greater than 90 %) of pancreatic neoplasms have a ductal origin. PDA comprise 80–90 % of this group [29]. There are several similar entities that can develop into or are at least associated with PDA. These entities are PanIN, intraductal papillary mucinous neoplasia (IPMN), and mucinous cystic neoplasia (MCN). One of these entities, PanIN, was initially identified as a series of increasingly atypical proliferative changes in the epithelium of the pancreatic ducts. These lesions have long been recognized, but have only been reported previously with descriptive terminology [30–32].

#### 26.2.5.1 Pancreatic Intraepithelial Neoplasia

PanIN is a relatively common finding in resected specimens and is seen in association with PDA and other tumor types [33]. It represents a spectrum of proliferative lesions in the ductal system, all of which contain cytoplasmic mucin. The lesions are separated into three grades (PanIN-1, PanIN-2, and PanIN-3), depending on the extent of cytological atypia and architectural complexity (see Fig. 26.2). PanIN-1 lesions (previously called mucinous metaplasia or mucous cell hypertrophy) are characterized by tall columnar mucinous cells with nuclei lacking any atypia or loss of polarity. These lesions are further divided into PanIN-1A and PanIN-1B, based on the presence or absence of papillary and micropapillary formations and nuclear stratification. PanIN-1A lacks both the architectural and cytological features. PanIN-2 lesions (previously called atypical hyperplasia or moderate dysplasia) are characterized by prominent nuclear stratification, loss of polarity, and mild nuclear atypia. Finally, PanIN-3 lesions (previously called carcinoma in situ or severe dysplasia) show marked nuclear atypia, significant loss of polarity, frequent mitotic figures, and papillary tufts "floating" in the lumen [34]. Recently, a two-tier classification was recommended by an international consensus meeting to replace the current

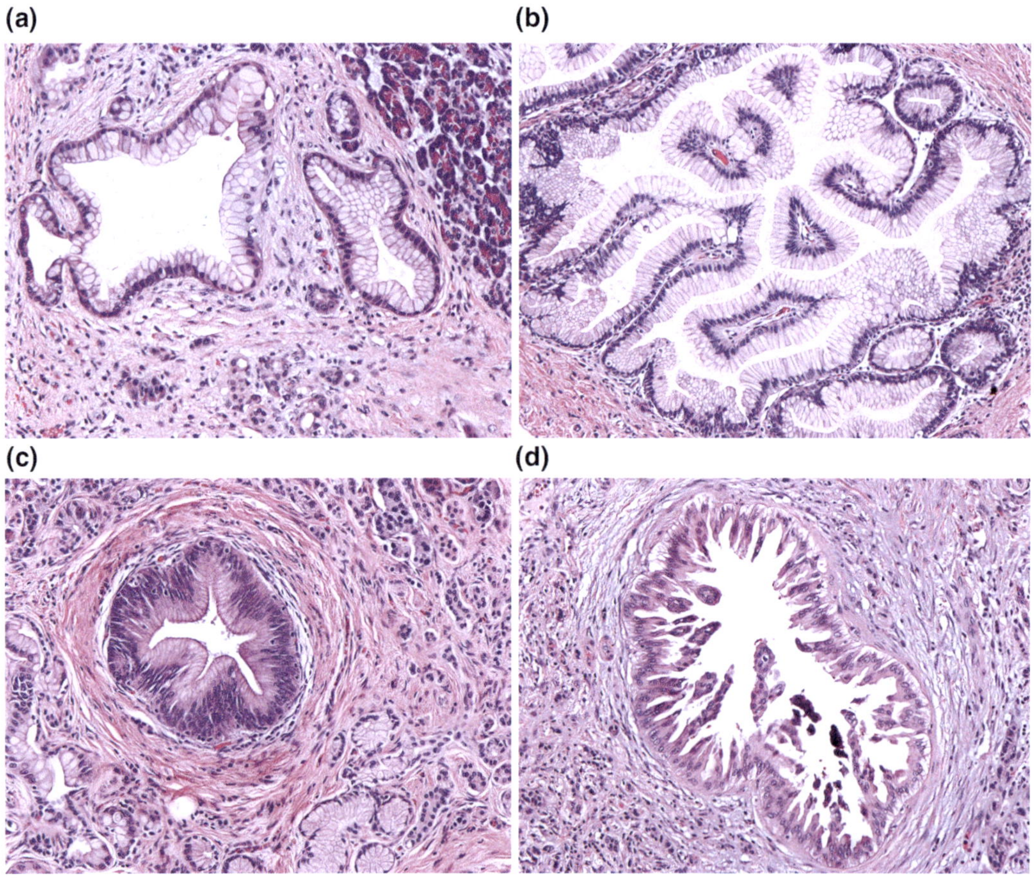


**Fig. 26.2** Pancreatic intraepithelial neoplasia (PanIN). **a** PanIN-1A; **b** PanIN-1B; **c** PanIN-2; **d** PanIN-3

three-tier classification used for PanIN as well as IPMN and MCN. In the revision, PanIN-1 and PanIN-2 are categorized as low grade. PanIN-3 is categorized as high grade PanIN. This change in classification brings the terminology in line with the clinical management of these lesions [35].

Lower grade lesions, i.e., PanIN-1, are often incidental findings in neoplastic and nonneoplastic pancreas resections. In one study, PanIN-1A lesions were identified in 43 % of older adults with nonneoplastic resections [36]. PainIN-2 and PanIN-3 are significantly more common in pancreases with PDA [33, 34]. Recent retrospective data suggest that there is a very low risk of cancer progression in PanIN-1 and PanIN-2 lesions [37]. However, Brat et al. [31] have reported rare cases where PanIN-2 lesions were identified before the development of PDA. The natural history of these lesions is exceedingly difficult to observe given the lack of reproducible radiographic findings and serologic markers.

#### 26.2.5.2 Intraductal Papillary Mucinous Neoplasms

IPMNs are intraductal papillary proliferations of mucin-producing cells. Intraluminal mucin is often evident, and accumulation of the mucin often leads to cystic dilation of the ducts, which may be localized to the immediate vicinity or may involve the entire ductal system. They may manifest as multilocular cystic masses or abundant papillary nodules in either the main duct or the branch duct and are classified accordingly [38–42].

IPMNs share many cytological features with PanIN and are characterized by mucinous cells with varying degrees of atypia. However, most IPMNs are larger and involve cystically dilated ducts that are at least 1 cm in diameter [34, 43, 44]. The atypia is stratified into three grades (i.e., low, intermediate, and high) on the basis of the most severely dysplastic area and parallel the spectrum of changes seen in PanIN-1 through PanIN-3. IPMNs with low grade dysplasia (also called intraductal papillary mucinous adenomas) are characterized by mucinous cells that lack any nuclear atypia or loss of polarity (see Fig. 26.3). Intermediate grade dysplasia is characterized by more prominent nuclear stratification, loss of

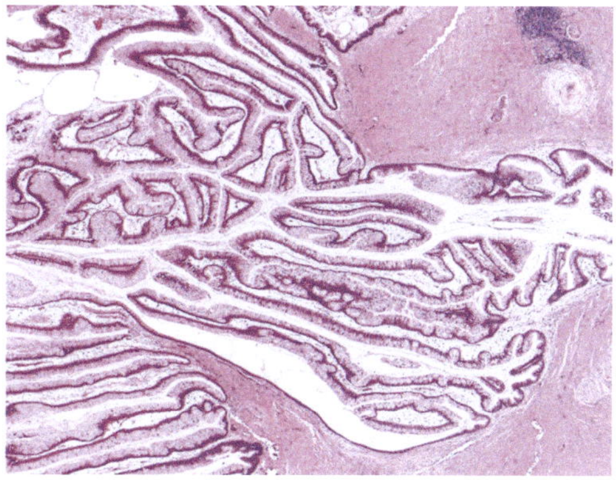

**Fig. 26.3** Intraductal papillary mucinous neoplasm (IPMN) with low grade dysplasia

polarity, and mild nuclear atypia. IPMNs with high grade dysplasia (also called intraductal papillary mucinous carcinoma in situ) have significant loss of polarity and marked nuclear atypia (see Fig. 26.4). Similar to the changes in PanIN terminology, both IPMNs with low grade dysplasia and intermediate grade dysplasia are now categorized as IPMN, low grade, in the revised classification and IPMN with high grade dysplasia as IPMN, high grade [35]. This two-tier classification scheme more accurately parallels the 2012 consensus guidelines of the International Association of Pancreatology for the management of IPMNs and MCNs, which recommend low and intermediate grade dysplasia as being amenable to observation and high grade dysplasia as requiring clinical attention and

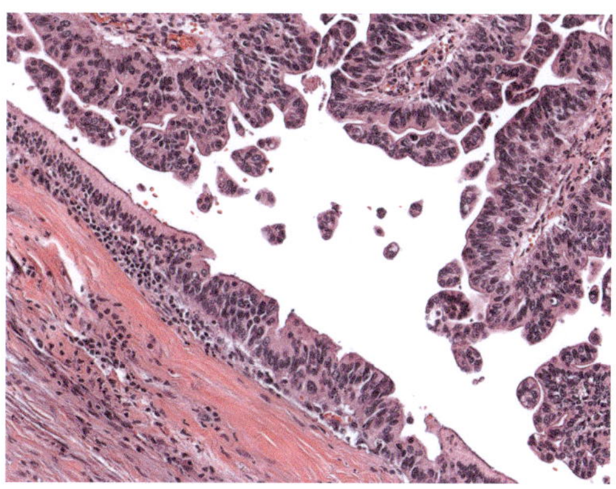

**Fig. 26.4** IPMN with high grade dysplasia

intervention [45]. In addition to the grading of dysplasia, three papillary patterns have been described [38, 46–48]. The most common pattern is the gastric type (50 %), which is characterized by cells that resemble gastric foveolar epithelium with basally oriented nuclei and abundant mucinous cytoplasm. The papillae in these lesions are often less exuberant and can be flat [49–52]. The second most frequently recognized pattern is the intestinal type (35 %), characterized by cells that resemble adenomas of the colorectum [53] with villiform papillae; like the adenomas in the colorectum, there is often nuclear pseudostratification and intracellular mucin. The least common pattern is the pancreaticobiliary type (15 %) that is characterized by cuboidal cells often with more complex papillae formation [54], including numerous branched papillae, micropapillae, and cribriform areas. The cells are typically not pseudostratified, but show marked variation in size, irregular contours, prominent nucleoli, and loss of nuclear polarity. While an overlap of these patterns can be seen, the intestinal and pancreaticobiliary patterns are not commonly seen in a single tumor.

IPMNs express keratins, such as CK7, CK8, CK18, and CK19. The degree of CK20 immunoreactivity depends on the histologic subtype [42, 55]. Expression of glycoproteins, MUC1 and MUC2, also varies with the histologic subtype [46, 47, 53, 56]. Expression of MUC1 is more common in pancreaticobiliary-type IPMNs, whereas MUC2 is more commonly expressed in intestinal-type IPMNs. Gastric-type IPMNs do not express MUC1 or MUC2. Moreover, MUC2 expression in intestinal-type IPMNs is paralleled by other markers of intestinal differentiation, such as CK20 and CDX-2. MUC6 is preferentially expressed in pancreaticobiliary-type IPMNs suggesting pyloric differentiation [57]. Almost all IPMNs, regardless of their histologic subtype, express CEA, CA19–9, and MUC5AC [58].

#### 26.2.5.3 Mucinous Cystic Neoplasms

MCNs are typically single, multilocular cystic masses with a thick, fibrotic capsule. The cysts are often large, greater than 10 cm. Unlike PDAs and IPMNs, these tumors do not communicate with the ductal system. The septa between cysts are often thin, but some can appear trabeculated and thickened. Papillations are not uncommon. Similar to IPMNs, the cysts contain mucinous material. Degenerative changes with hemorrhage may occur and may resemble a pseudocyst [59–62].

MCNs are characterized by a distinctive, subepithelial stroma that resembles ovarian stroma (see Fig. 26.5), composed of densely packed spindle cells with uniform, wavy nuclei, sparse cytoplasm, and scattered clusters of plump, epithelioid cells suggestive of luteinization [63, 64]. In

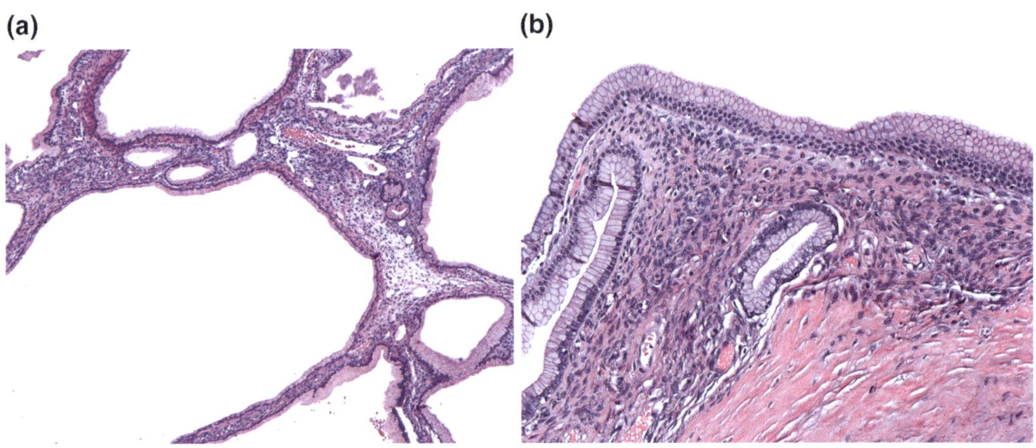


**Fig. 26.5** Mucinous cystic neoplasm (MCN) with low grade dysplasia. **a** Low-power. **b** High power of ovarian-type stroma

addition, the spindle cells demonstrate immunoreactivity for estrogen and progesterone receptors, inhibin, and melan-A [62, 63, 65]. The epithelium, when not denuded, is characterized by cuboidal to columnar cells with abundant apical mucin and basally located nuclei. These cells lie flat or form papillae, and occasionally, goblet cells, neuroendocrine cells, and Paneth cells can be found interspersed [66].

Similar to IPMNs, the degree of dysplasia in the epithelial component is graded as low, intermediate, or high based on the area with the most severe dysplasia. MCN with low grade dysplasia (also called mucinous cystadenoma) is characterized by minimal cytological atypia and minimal architectural complexity (see Fig. 26.5). MCN with intermediate grade dysplasia has moderate cytological atypia, including loss of nuclear polarity, and mild architectural complexity. MCN with high grade dysplasia (also called mucinous cystadenocarcinoma in situ) is characterized by marked cytological atypia, including irregular, hyperchromatic nuclei, mitotic figures and significant architectural complexity [61, 62] (see Fig. 26.6). The proposed change in classification for PanINs and IPMNs is also recommended for MCNs. MCNs with low grade dysplasia and intermediate grade dysplasia are categorized as MCN, low grade; MCN with high grade dysplasia is categorized as MCN, high grade, in the 2014, revised classification system [35].

Invasive adenocarcinoma, resembling conventional PDA, can develop in association with MCN, especially those with high grade dysplasia (see Fig. 26.7). Other histological variants, e.g., undifferentiated carcinoma with osteoclast-like giant cells, have also been reported [67]. MCN with an associated invasive carcinoma are also called invasive mucinous cystadenocarcinoma. Similar to IPMNs, the invasive component in MCNs can also be focal and often necessitates in toto submission of the tumor. Overall, the rate of malignancy in MCNs ranges from 10 to 28 % [68], and it has been suggested that they may be less aggressive than their conventional counterparts, especially, when there is only minimal invasion [69].

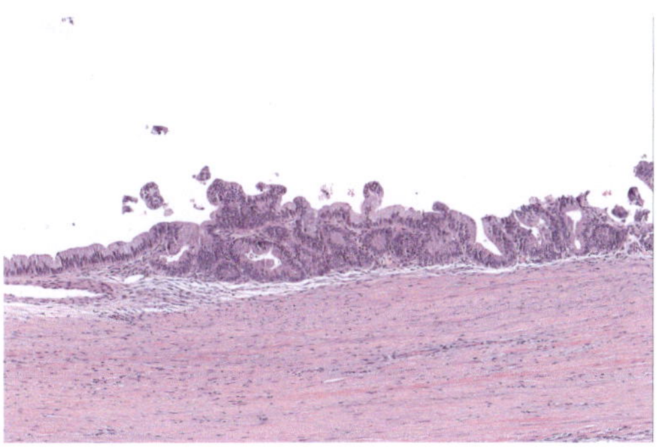

**Fig. 26.6** MCN with high grade dysplasia

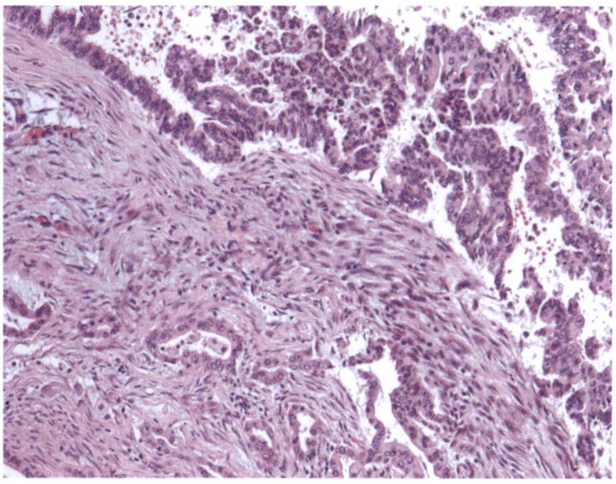

**Fig. 26.7** MCN associated with invasive carcinoma

MCNs show similar immunoreactivity to keratin and glycoprotein markers, e.g., CEA and CA19–9 [62, 63, 65, 70], as IPMNs. The similarities are seen in the expression of MUC5AC, MUC2, and CDX-2 with the latter only seen in the goblet cells [71].

#### 26.2.5.4 Conventional Ductal Adenocarcinoma

Most PDAs are firm, discrete tan-white masses with ill-defined, infiltrative borders. Frank necrosis and hemorrhage is infrequent. In an uninjured pancreas, the appearance of the tumor is easily distinguished from the normal pancreas, however, it can be difficult to visualize the mass-forming tumor in a background of fibrosis, secondary to either chronic pancreatitis or neoadjuvant therapy [26].

Conventional PDA represents one of two general groups of pancreatic cancer, the second group being histological variants, which are most likely of ductal origin. They are characterized by a proliferation of variably sized glands haphazardly set in a desmoplastic stroma [72]. The neoplastic glands are composed of cuboidal to columnar cells with variable amounts of cytoplasm and mucin. The nuclei vary in size and shape, and there is frequently a loss of polarity. Rarely, the nuclei may retain their normal basal orientation (see Fig. 26.8). In addition to the conventional (or tubular) pattern, PDAs can also demonstrate other morphological patterns, including a foamy gland pattern resembling low grade PanIN [73]; a large duct (or microcystic) pattern commonly seen with duodenal infiltration [74]; a vacuolated pattern resembling adipocytes or signet ring cells [75]; a solid, nested pattern resembling squamous cell carcinoma or hepatoid carcinoma [76]; a micropapillary pattern; and a lobular carcinoma-like pattern with neoplastic cells in cords or single files (see Fig. 26.9). With rare exceptions, these patterns are not considered to have clinical significance. PDAs with a micropapillary pattern are considered to be particularly aggressive. Regardless of the morphological patterns, the neoplastic glands destroy the normal lobular architecture. Lymphovascular and perineural invasion are almost invariably seen and infiltration into peripancreatic fat is also common [77]. Interestingly, colonization of the normal epithelium and the basement membrane

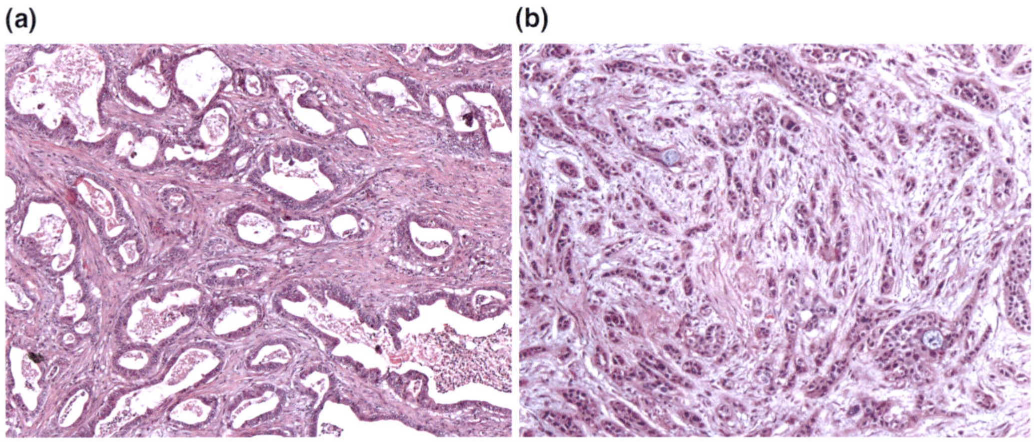


**Fig. 26.8** Conventional pancreatic ductal adenocarcinoma (PDA). **a** Well differentiated; **b** poorly differentiated

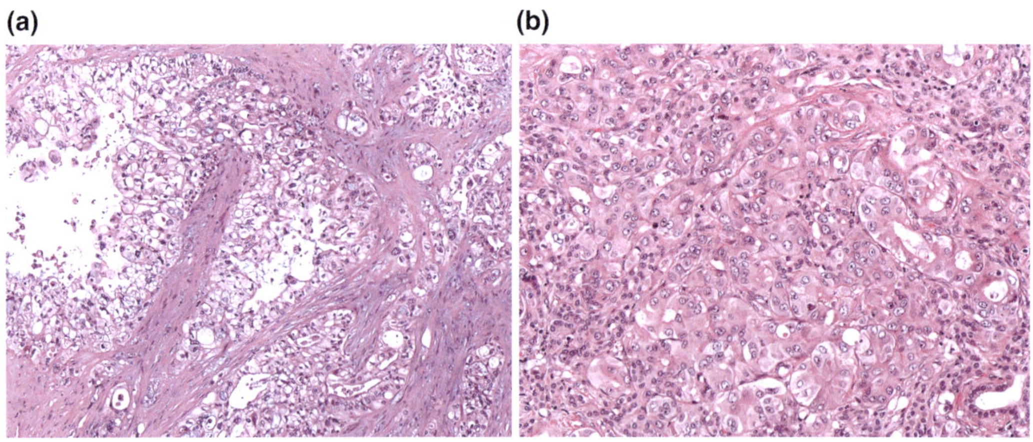


**Fig. 26.9** Other morphologic patterns of PDA. **a** Vacuolated pattern. **b** Solid, nested pattern

can be seen with invasion into the common bile duct, duodenum, and native pancreatic ducts, mimicking the appearance of a precursor lesion.

The TNM grading system is based on the extent of gland formation and is advocated by the College of American Pathologists. Well-differentiated (grade 1) adenocarcinomas are characterized by greater than 95 % gland formation. Adenocarcinomas with 50–95 % gland formation are considered moderately differentiated (grade 2). Poorly differentiated (grade 3) adenocarcinomas have less than 50 % gland formation. Undifferentiated carcinomas, which are discussed below, are considered grade 4. The Klöppel grading system, which is endorsed by the World Health Organization (WHO), assesses mucin production, mitoses, and nuclear atypia in addition to gland formation [78], however, it is cumbersome in practice. Adsay and colleagues [79] have proposed a third system based on the predominant and secondary patterns of infiltration, but it is not widely used. In a comparison between the TNM and Klöppel grading systems, no significant differences in predictive value were found [80]. Regardless of the grading system used, the histologic grade has been shown to be prognostically significant, with higher grades, i.e., grades 3 and 4, being more unfavorable [79, 80].

PDAs stain for CK7, CK8, CK18, CK19, and epithelial membrane antigen (EMA) [81]. CK20 is positive in 25 % of cases with typical immunoreactivity being focal and weak [82] except in mucinous noncystic carcinoma in which immunoreactivity is diffuse and strong [56]. In addition, conventional PDAs usually express the cell surface-associated mucin proteins, MUC1, MUC3, MUC4, and MUC5AC. However, MUC2 expression is only seen in those with intestinal differentiation and MUC6 is expressed in those with pyloric gland differentiation [56, 83, 84]. Similar to IPMNs and MCNs, immunohistochemical stains to glycoproteins, e.g., CA19–9, CEA, TAG-72, and CA-125, are often positive in conventional PDAs. Of these, the latter three proteins are expressed only to a limited degree in low grade PanIN lesions and are not typically expressed in the normal pancreas [85].

## Histological Variants of Adenocarcinoma

There are several histological variants that are most likely of ductal origin, as evident by an associated component of conventional PDA. These variants include mucinous noncystic carcinomas, squamous cell, and adenosquamous carcinomas, undifferentiated carcinomas, medullary carcinomas, and hepatoid carcinomas. With a few exceptions, the natural history and molecular alterations of these variants are not well understood due to their infrequency.

### Mucinous Noncystic Carcinoma

Mucinous noncystic carcinoma (also called colloid carcinoma) is a unique variant, both clinically and histologically. These lesions are characterized by pools of mucin, within which are flat strips of small clusters of neoplastic cells that are either attached to the edge of the mucin pool or floating within it (see Fig. 26.10). Signet ring cells are not infrequently seen in these pools [86]. They are typically associated with IPMNs or MCNs [87]. Relative to conventional PDA, mucinous noncystic carcinomas have a more favorable, protracted clinical course [86].

### Squamous Cell Carcinoma and Adenosquamous Carcinoma

Squamous cell and adenosquamous cell carcinomas represent a small minority of pancreatic cancers [88, 89]. They resemble squamous cell carcinomas from other sites, with variable

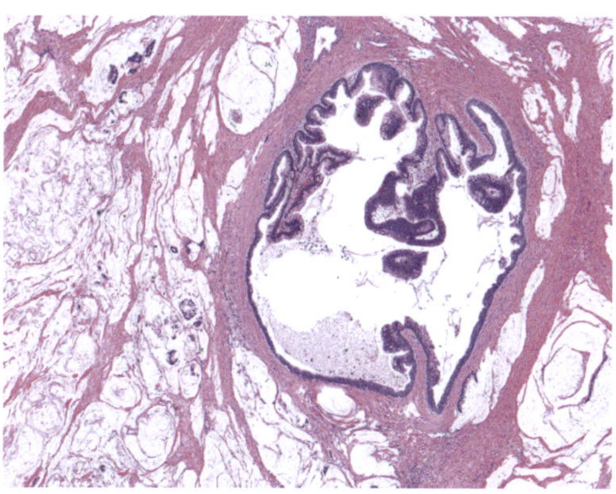

**Fig. 26.10** IPMN with mucinous noncystic carcinoma

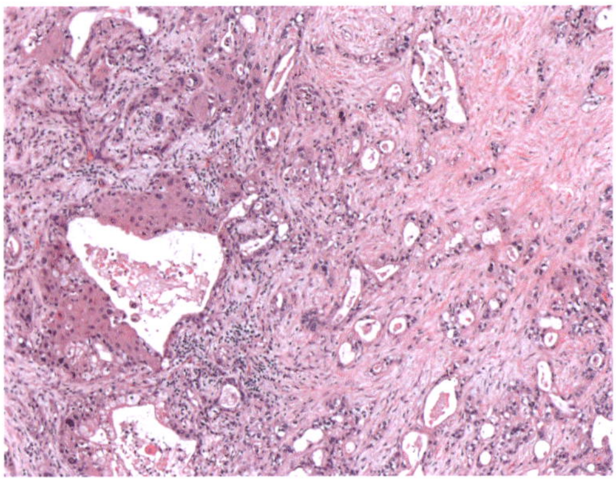

**Fig. 26.11** Adenosquamous carcinoma of the pancreas

keratinization. The diagnosis of adenosquamous carcinoma only requires an arbitrary 30 % squamous component [90] (see Fig. 26.11). As such, pure squamous cell carcinomas are exceedingly uncommon (see Fig. 26.12). Both squamous cell and adenosquamous cell carcinomas have a similar clinical course to conventional PDA.

### Undifferentiated Carcinoma

Undifferentiated carcinomas encompass a wide range of histological appearances, and include sarcomatoid (or spindle cell) carcinomas, anaplastic carcinomas, carcinosarcomas, and undifferentiated carcinomas with osteoclast-like giant cells [91]. Undifferentiated carcinomas are thought to be ductal in origin as a conventional PDA component or preinvasive component, e.g., PanIN or MCN [92–94], can be identified. The epithelioid component is frequently markedly atypical and dyshesive. The sarcomatoid component may consist of only spindle cells or large anaplastic giant cells (see Fig. 26.13). In addition, there may also be heterologous differentiation, e.g., skeletal muscle, cartilage, or bone formation [95]. Undifferentiated carcinomas with osteoclast-like giant cells, as the name indicates, contain a variable number of osteoclast-like giant cells scattered throughout the neoplasm [96]. The giant cells are immunoreactive for CD68, supporting a histiocytic origin [97] (see Fig. 26.14). Undifferentiated carcinomas are aggressive with most patients dying of disease within 2 years of initial diagnosis [98].

### Medullary Carcinoma

Medullary carcinomas are uncommon tumors that are characterized by a syncytial growth pattern composed of poorly differentiated neoplastic cells. These neoplasms are typically associated with a prominent inflammatory infiltrate containing neutrophils and lymphocytes. Similar to ones at other sites, medullary carcinomas can be associated with microsatellite instability [99–101] and are associated with a good prognosis [101, 102].

### Hepatoid Carcinoma

Hepatoid carcinomas are exceedingly rare and are characterized by polygonal cells with granular and eosinophilic cytoplasm, centrally located nuclei, and prominent nucleoli. The neoplastic cells, which have hepatocellular differentiation, are arranged in solid sheets, nests, or trabeculae [103–105].

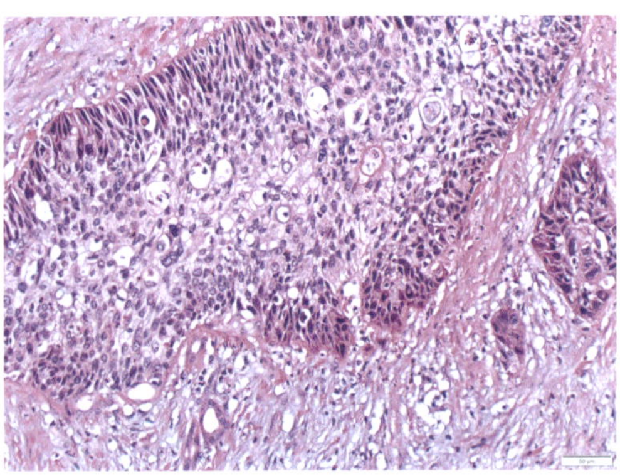

**Fig. 26.12** Squamous cell carcinoma of the pancreas

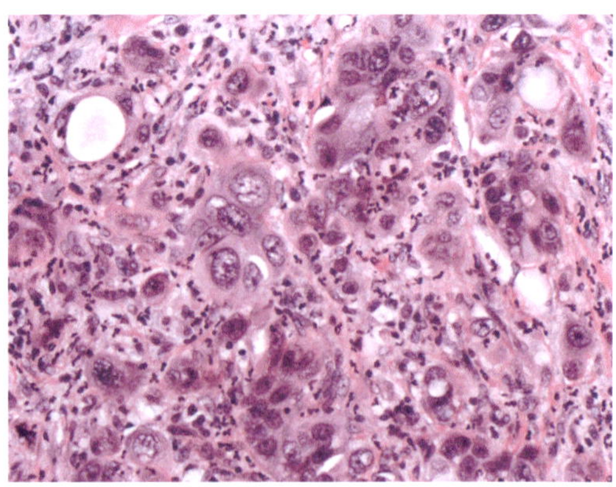

**Fig. 26.13** Undifferentiated carcinoma of the pancreas

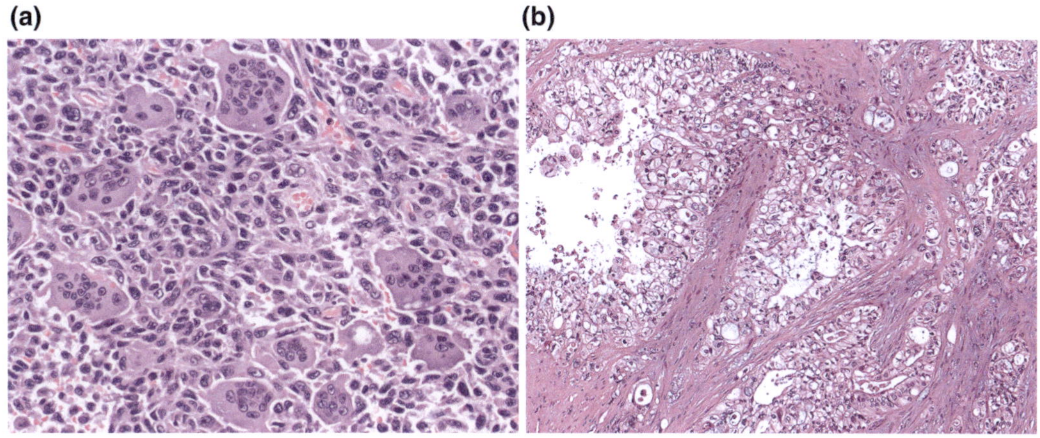

**Fig. 26.14** Undifferentiated carcinoma with osteoclast-like giant cells. **a** H&E; **b** CD68 immunohistochemistry highlighting giant cells

## 26.2.6 Molecular Pathogenesis

Cancer is fundamentally caused by accumulation of mutations in oncogenes and tumor suppressor genes. These mutations can be acquired somatically, or patients can inherit them. PDA is among the best tumors characterized molecularly. Numerous mutations in oncogenes and tumor suppressor genes, both inherited and acquired, have been identified and provide an unprecedented insight into the molecular pathogenesis of pancreatic cancer.

### 26.2.6.1 Somatic Alterations

The molecular alterations in sporadic pancreatic cancer have been extensively investigated in the last few decades and have given us insights into PDA tumorigenesis. Whole exome sequencing has revealed an average of 48 somatic mutations with four genes being altered in greater than 50 % of cases [106]. These four genes are KRAS, CDKN2A (also called CDKN, p16, and INK4A), TP53 and SMAD4 (also called DPC4) [107–109] (see Table 26.1).

KRAS is the most frequently altered oncogene in PDAs. The gene, which is located on chromosome 12, encodes for a small GTPase that is vital to several cell signaling pathways, including MAPK, ERK, and AKT pathways. The point mutations involve three target codons almost exclusively (12, 13, and 61). Mutations involving codon 12 are detected in greater than 90 % of PDAs [110]. KRAS mutations are also identified in histological variants, including undifferentiated carcinomas with osteoclast-like giant cells [111] and mucinous noncystic carcinomas [86], as well as both IPMNs and MCNs. In IPMNs, the frequency of KRAS mutations ranges from 30 to 80 % with increasing frequency in those with high grade dysplasia and invasive carcinoma [112–119]. A similar trend in frequency of KRAS mutation is observed in MCNs [115, 120, 121]. Notably, KRAS mutations have been identified in more than 90 % of

**Table 26.1** Most common sporadic alterations in pancreatic ductal adenocarcinoma

| Chromosome | Gene | Prevalence (%) | Mechanism |
|---|---|---|---|
| 12 | KRAS | 95 | Missense mutation |
| 9 | CDK2NA | 95 | LOH, homozygous deletion, promoter methylation |
| 17 | TP53 | 75 | LOH |
| 18 | SMAD4 | 55 | LOH, homozygous deletion |

the earliest PanIN lesions (PanIN-1), implicating KRAS mutations as a key initiating event in pancreatic neoplasia [122].

CDKN2A is the most frequently altered tumor suppressor gene in PDAs. The gene, which is located on chromosome 9p, encodes for the protein, p16, which is involved in cell cycle regulation. Inactivation of the gene is observed in 95 % of pancreatic cancers [106] and is thought to promote unrestricted growth. In addition to DNA alteration, CDKN2A inactivation can be accomplished through epigenetic silencing via aberrant methylation [123]. Similar to KRAS mutations, CDKN2A mutations occur early in the progression from PanIN to pancreatic cancer.

TP53 is another frequently altered tumor suppressor gene and is inactivated in 75 % of PDA [106]. The gene, which is located on chromosome 17p, encodes the protein p53, which plays an important role in cellular stress responses by activating DNA repair, inducing growth arrest, and triggering apoptosis. TP53 mutations are also found in MCNs and IPMNs [124]. In contrast to KRAS and CDKN2A, TP53 mutations occur later in the PanIN-to-invasive carcinoma sequence, being observed in high grade PanIN lesions (i.e., PanIN-3) [125].

SMAD4 is a tumor suppressor gene on chromosome 18q [126]. The protein, SMAD4, is involved in the transforming growth factor beta (TGF-β[beta]) signaling pathway. SMAD4 mutations are associated with a poor prognosis and widely metastatic disease [17, 127] and can be detected using immunohistochemistry. Loss of SMAD4 expression by immunohistochemistry is found in 55 % of PDAs [128] (see Fig. 26.15). Similar to TP53, mutations in SMAD4 are observed in PanIN-3 lesions, implicating its late role in the genetic progression to cancer [125].

In addition to somatic mutations, large chromosomal gains and losses, and complex karyotypic abnormalities occur frequently in PDA. Some changes target known driver genes, whereas others identify regions with specific loci that have unknown roles [129]. The chromosomal abnormalities may be related to early telomere shortening, which is reported to occur in approximately 90 % of low grade PanIN lesions [130]. Together, progression from PanIN to PDA and the sequential acquired alterations, telomere shortening, and chromosomal instability can be mapped (see Fig. 26.16).

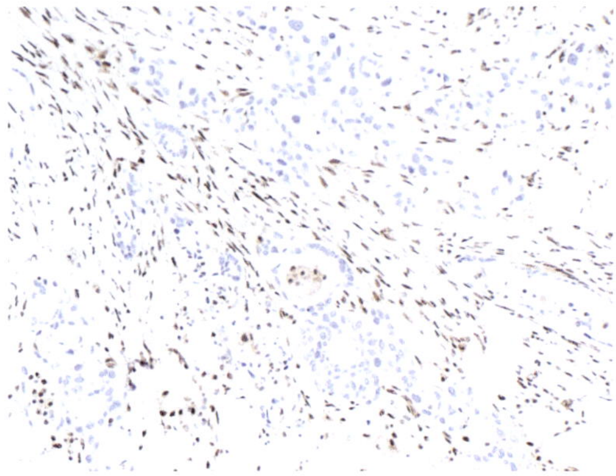

**Fig. 26.15** Loss of SMAD4 by immunohistochemistry in conventional PDA

Despite a common progression to pancreatic cancer, two somatic mutations unique to MCNs and IPMNs have been identified. The affected genes include GNAS and RNF43 [119, 124]. The latter, RNF43, encodes a protein with intrinsic E3 ubiquitin ligase activity. Mutations in RNF43 are found in both MCNs and IPMNs [124]. GNAS, which encodes for a stimulatory G-protein subunit, is unique in that GNAS mutations are only seen in IPMNs [119].

#### 26.2.6.2 Germline Alterations and Mutations

Approximately 10 % of PDAs have a familial basis, which is defined as at least a pair of first-degree relatives diagnosed with pancreatic cancer. Case-control and cohort studies have shown that those with a family history of pancreatic cancer have a 1.9 to 13-fold increased risk [131–134]. The genetic basis for the majority of familial pancreatic cancer is unknown [135]. Several germline genetic syndromes also lead to an increased risk for PDA (see Table 26.2).

Germline BRCA2 mutations are associated with a significantly elevated lifetime risk of breast, ovarian, prostate, and pancreatic cancer [136–139]. PALB2 (also called FANCN), which encodes for a protein that partners with the

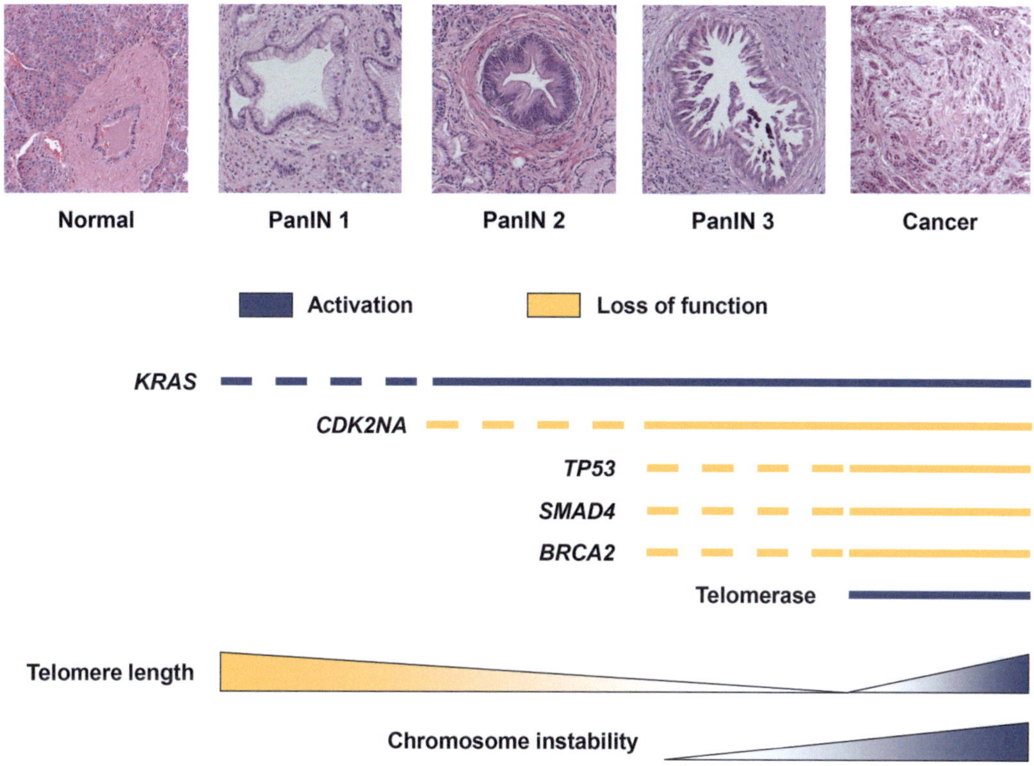


**Fig. 26.16** Sequential molecular alterations and increasing chromosomal instability

**Table 26.2** Germline alterations in pancreatic cancer and cancer risk

| Gene | Syndrome | Estimated lifetime cancer risk |
|---|---|---|
| BRCA1, BRCA2 | Familial breast cancer | BRCA1[a], 2.16 % (by age 80)<br>BRCA2, 3.36 % (by age 80) |
| PALB2 | Familial breast cancer | Elevated |
| CDK2NA | Familial atypical multiple mole melanoma (FAMMM) syndrome | 17 % (by age 75) |
| STK11 | Peutz-Jeghers syndrome (PJS) | 11–32 % |
| PRSS1 | Hereditary pancreatitis | 30–40 % (by age 70) |
| ATM | Ataxia-telangiectasia | Unknown |
| MSH2, MSH6, MLH1, PMS2 | Lynch syndrome | 1.45–5.88 % (by age 70) |

[a]Studies are conflicted; minimal or no increased cancer risk with BRCA1 mutations

BRCA2 protein, is also implicated in familial PDA. Germline mutations in this gene account for 1–3 % of familial PDA [140–142]. Both BRCA2 and PALB2 encode proteins crucial to the Fanconi anemia pathway, and germline mutations in other genes involved in the pathway, including FANCC and FANCG, have been reported in young patients diagnosed with PDA [143–145]. However, FANCA mutations have not been implicated [146]. In addition to conventional PDA, IPMNs have also been reported in patients with inherited BRCA2 mutations and a family history of PDA [147].

Unlike germline mutations in BRCA2 and PALB2, it is unclear if inherited mutations in BRCA1 are also at higher risk. Studies show conflicting results, with one large study conducted by the Breast Cancer Linkage Consortium observing a 2.26-fold increased risk [148]. Others have not shown an increased prevalence of BRCA1 mutations [149, 150].

Germline mutations in CDKN2A causes familial atypical multiple mole melanoma (FAMMM) syndrome, which results in a 38-fold increased risk of PDA and melanoma [151–156].

Peutz-Jeghers syndrome (PJS) results in the development of gastrointestinal hamartomas and pigmented macules on the lips, and buccal mucosa. It is associated with an increased risk of PDA. Germline mutations in STK11 (also called LKB1) explain 80 % of PJS cases [157]. PDAs in these patients also show somatic loss of the wild-type STK11 allele [158]. In addition, rare cases of IPMNs have been reported in patients with this syndrome [159].

Germline mutations in PRSS1 and SPINK1 cause hereditary pancreatitis [160, 161], which is characterized by repeated episodes of acute pancreatitis in young patients. As a result of the continuous and relapsing inflammation and repair, these patients have a markedly increased risk (58-fold) for PDA [162].

Germline mutations in ATM, which encodes a protein in cell cycle regulation and DNA damage response, have been identified in approximately 2 % of familial pancreatic cancer [163]. The germline mutation is heterozygous; biallelic germline mutations in the gene result in ataxia-telangiectasia, which is characterized by cerebellar ataxia and sensitivity to ionizing radiation.

Lynch syndrome (or hereditary nonpolyposis colorectal cancer, HNPCC) is caused by germline mutations in one of the DNA mismatch repair genes, MLH1, PMS2, MSH2, or MSH6, resulting in microsatellite instability. The syndrome is associated with an increased risk of carcinomas in the colorectum and other sites [164] and an 8.6-fold increased risk of PDA [165]. Rare cases of IPMN have also been reported in these patients [166].

## 26.3 Targeted Molecular Treatments

At this time, there are no targeted therapies for pancreatic cancer. Given the prevalence of KRAS mutations and its critical role, it is an enticing target, however, all attempts at targeting the protein have failed in the clinic [167]. More recent strategies have focused on targeting critical downstream targets in Ras-mediated pathways, including MAPK, ERK, and AKT [168, 169]. Perhaps the most intriguing discoveries have been in therapies targeting rare genetic alterations in pancreatic cancer. PDAs with biallelic inactivations in either BRCA2 [136, 170] or PALB2 [171, 172] have been shown to be exquisitely sensitive to DNA cross-linking agents, e.g., mitomycin C or poly (ADP-ribose) polymerase (PARP) inhibitors. PDAs with inactivating ATM mutations may also be similarly affected [173].

The future for the development of new treatments is promising. In the not-too-distant future, all patients with pancreatic cancer will have their tumors histologically analyzed and sequenced, with the information gleaned from both to be used for individualizing their care.

## References

1. Porta M, Fabregat X, Malats N, Guarner L, Carrato A, de Miguel A, et al. Exocrine pancreatic cancer: symptoms at presentation and their relation to tumour site and stage. Clin Transl Oncol: Official Publ Fed Spanish Oncol Soc Natl Cancer Insti Mexico. 2005;7(5):189–97.
2. Modolell I, Guarner L, Malagelada JR. Vagaries of clinical presentation of pancreatic and biliary tract cancer. Ann Oncol: Official J Eur Soc Med Oncol (ESMO). 1999;10(Suppl 4):82–4.
3. Okusaka T, Okada S, Ueno H, Ikeda M, Shimada K, Yamamoto J, et al. Abdominal pain in patients with resectable pancreatic cancer with reference to clinicopathologic findings. Pancreas. 2001;22(3):279–84.
4. Mayr M, Schmid RM. Pancreatic cancer and depression: myth and truth. BMC Cancer. 2010;10:569.

5. Makrilia N, Indeck B, Syrigos K, Saif MW. Depression and pancreatic cancer: a poorly understood link. J Pancreas (JOP). 2009;10(1):69–76.
6. DiMagno EP, Malagelada JR, Taylor WF, Go VL. A prospective comparison of current diagnostic tests for pancreatic cancer. N Engl J Med. 1977;297 (14):737–42.
7. Pasanen PA, Eskelinen M, Partanen K, Pikkarainen P, Penttila I, Alhava E. A prospective study of the value of imaging, serum markers and their combination in the diagnosis of pancreatic carcinoma in symptomatic patients. Anticancer Res. 1992;12(6B):2309–14.
8. van den Bosch RP, van Eijck CH, Mulder PG, Jeekel J. Serum CA19–9 determination in the management of pancreatic cancer. Hepatogastroenterology. 1996;43(9):710–3.
9. Pleskow DK, Berger HJ, Gyves J, Allen E, McLean A, Podolsky DK. Evaluation of a serologic marker, CA19–9, in the diagnosis of pancreatic cancer. Ann Intern Med. 1989;110(9):704–9.
10. Brand R. The diagnosis of pancreatic cancer. Cancer J. 2001;7(4):287–97.
11. Bluemke DA, Cameron JL, Hruban RH, Pitt HA, Siegelman SS, Soyer P, et al. Potentially resectable pancreatic adenocarcinoma: spiral CT assessment with surgical and pathologic correlation. Radiology. 1995;197(2):381–5.
12. Pasanen PA, Partanen KP, Pikkarainen PH, Alhava EM, Janatuinen EK, Pirinen AE. A comparison of ultrasound, computed tomography and endoscopic retrograde cholangiopancreatography in the differential diagnosis of benign and malignant jaundice and cholestasis. Eur J Surg Acta Chir. 1993;159(1):23–9.
13. Brambs HJ, Claussen CD. Pancreatic and ampullary carcinoma. Ultrasound, computed tomography, magnetic resonance imaging and angiography. Endoscopy. 1993;25(1):58–68.
14. Schellenberg D, Quon A, Minn AY, Graves EE, Kunz P, Ford JM, et al. 18 Fluorodeoxyglucose PET is prognostic of progression-free and overall survival in locally advanced pancreas cancer treated with stereotactic radiotherapy. Int J Radiat Oncol Biol Phys. 2010;77(5):1420–5.
15. Evans DB, Varadhachary GR, Crane CH, Sun CC, Lee JE, Pisters PW, et al. Preoperative gemcitabine-based chemoradiation for patients with resectable adenocarcinoma of the pancreatic head. J Clin Oncol: Official J Am Soc Clin Oncol. 2008;26 (21):3496–502.
16. Gnerlich JL, Luka SR, Deshpande AD, Dubray BJ, Weir JS, Carpenter DH, et al. Microscopic margins and patterns of treatment failure in resected pancreatic adenocarcinoma. Arch Surg. 2012;147(8):753–60.
17. Iacobuzio-Donahue CA, Fu B, Yachida S, Luo M, Abe H, Henderson CM, et al. DPC4 gene status of the primary carcinoma correlates with patterns of failure in patients with pancreatic cancer. J Clin Oncol: Official J Am Soc Clin Oncol. 2009;27 (11):1806–13.
18. Wolfgang CL, Herman JM, Laheru DA, Klein AP, Erdek MA, Fishman EK, et al. Recent progress in pancreatic cancer. CA Cancer J Clin. 2013;63 (5):318–48.
19. Tempero MA, Arnoletti JP, Behrman SW, Ben-Josef E, Benson AB, 3rd, Casper ES, et al. Pancreatic Adenocarcinoma, version 2.2012: featured updates to the NCCN Guidelines. J Natl Compr Cancer Netw: JNCCN. 2012;10(6):703–13.
20. Winter JM, Cameron JL, Campbell KA, Arnold MA, Chang DC, Coleman J, et al. 1423 pancreaticoduodenectomies for pancreatic cancer: a single-institution experience. J Gastrointest Surg Official J Soc Surg Aliment Tract. 2006;10(9):1199–210; discussion 210–1.
21. Schmidt CM, Powell ES, Yiannoutsos CT, Howard TJ, Wiebke EA, Wiesenauer CA, et al. Pancreaticoduodenectomy: a 20-year experience in 516 patients. Arch Surg. 2004;139(7):718–25; discussion 25–7.
22. Dalton RR, Sarr MG, van Heerden JA, Colby TV. Carcinoma of the body and tail of the pancreas: is curative resection justified? Surgery. 1992;111 (5):489–94.
23. Schmidt CM, Glant J, Winter JM, Kennard J, Dixon J, Zhao Q, et al. Total pancreatectomy (R0 resection) improves survival over subtotal pancreatectomy in isolated neck margin positive pancreatic adenocarcinoma. Surgery. 2007;142(4):572–8; discussion 8–80.
24. Sohn TA, Yeo CJ, Cameron JL, Koniaris L, Kaushal S, Abrams RA, et al. Resected adenocarcinoma of the pancreas-616 patients: results, outcomes, and prognostic indicators. J Gastrointest Surg Official J Soc Surg Alimentary Tract. 2000;4 (6):567–79.
25. Howlader NNA, Krapcho M, Garshell J, Miller D, Altekruse SF, Kosary CL, Yu M, Ruhl J, Tatalovich Z, Mariotto A, Lewis DR, Chen HS, Feuer EJ, Cronin KA (eds). SEER cancer statistics review. 2014.
26. Odze RD, Goldblum JR. Odze and Goldblum surgical pathology of the GI tract, liver, biliary tract, and pancreas. 3rd ed: Saunders; 2014. 1632 p.
27. Pandol SJ. The exocrine pancreas. San Rafael, CA: Morgan and Claypool Life Sciences; 2010 (cited 2015 03/31/2015). Available from: http://www.ncbi.nlm.nih.gov/books/NBK54134/.
28. Ferrone CR, Brennan MF, Gonen M, Coit DG, Fong Y, Chung S, et al. Pancreatic adenocarcinoma: the actual 5-year survivors. J Gastrointest Surg Official J Soc Surg Aliment Tract. 2008;12(4):701–6.
29. Fletcher CDM. Diagnostic histopathology of tumors, 4th ed. Philadelphia, PA: Churchill Livingstone; 2013. 2296 p.
30. Brockie E, Anand A, Albores-Saavedra J. Progression of atypical ductal hyperplasia/carcinoma in situ

of the pancreas to invasive adenocarcinoma. Ann Diagn Pathol. 1998;2(5):286–92.
31. Brat DJ, Lillemoe KD, Yeo CJ, Warfield PB, Hruban RH. Progression of pancreatic intraductal neoplasias to infiltrating adenocarcinoma of the pancreas. Am J Surg Pathol. 1998;22(2):163–9.
32. Kozuka S, Sassa R, Taki T, Masamoto K, Nagasawa S, Saga S, et al. Relation of pancreatic duct hyperplasia to carcinoma. Cancer. 1979;43(4):1418–28.
33. Stelow EB, Adams RB, Moskaluk CA. The prevalence of pancreatic intraepithelial neoplasia in pancreata with uncommon types of primary neoplasms. Am J Surg Pathol. 2006;30(1):36–41.
34. Hruban RH, Takaori K, Klimstra DS, Adsay NV, Albores-Saavedra J, Biankin AV, et al. An illustrated consensus on the classification of pancreatic intraepithelial neoplasia and intraductal papillary mucinous neoplasms. Am J Surg Pathol. 2004;28(8):977–87.
35. Basturk O, Hong SM, Wood LD, Adsay NV, Albores-Saavedra J, Biankin AV, et al. A revised classification system and recommendations from the baltimore consensus meeting for neoplastic precursor lesions in the pancreas. Am J Surg Pathol. 2015;39(12):1730–41.
36. Andea A, Sarkar F, Adsay VN. Clinicopathological correlates of pancreatic intraepithelial neoplasia: a comparative analysis of 82 cases with and 152 cases without pancreatic ductal adenocarcinoma. Mod Pathol Official J US Can Acad Pathol Inc. 2003;16(10):996–1006.
37. Konstantinidis IT, Vinuela EF, Tang LH, Klimstra DS, D'Angelica MI, Dematteo RP, et al. Incidentally discovered pancreatic intraepithelial neoplasia: what is its clinical significance? Ann Surg Oncol. 2013;20(11):3643–7.
38. Adsay NV, Conlon KC, Zee SY, Brennan MF, Klimstra DS. Intraductal papillary-mucinous neoplasms of the pancreas: an analysis of in situ and invasive carcinomas in 28 patients. Cancer. 2002;94(1):62–77.
39. Azar C, Van de Stadt J, Rickaert F, Deviere M, Baize M, Kloppel G, et al. Intraductal papillary mucinous tumours of the pancreas. Clinical and therapeutic issues in 32 patients. Gut. 1996;39(3):457–64.
40. Kloppel G. Clinicopathologic view of intraductal papillary-mucinous tumor of the pancreas. Hepatogastroenterology. 1998;45(24):1981–5.
41. Loftus EV Jr, Olivares-Pakzad BA, Batts KP, Adkins MC, Stephens DH, Sarr MG, et al. Intraductal papillary-mucinous tumors of the pancreas: clinicopathologic features, outcome, and nomenclature. Members of the Pancreas Clinic, and Pancreatic Surgeons of Mayo Clinic. Gastroenterology. 1996;110(6):1909–18.
42. Paal E, Thompson LD, Przygodzki RM, Bratthauer GL, Heffess CS. A clinicopathologic and immunohistochemical study of 22 intraductal papillary mucinous neoplasms of the pancreas, with a review of the literature. Mod Pathol Official J US Can Acad Pathol Inc, Inc. 1999;12(5):518–28.
43. Longnecker DS, Adsay NV, Fernandez-del Castillo C, Hruban RH, Kasugai T, Klimstra DS, et al. Histopathological diagnosis of pancreatic intraepithelial neoplasia and intraductal papillary-mucinous neoplasms: interobserver agreement. Pancreas. 2005;31(4):344–9.
44. Hruban RH, Adsay NV, Albores-Saavedra J, Compton C, Garrett ES, Goodman SN, et al. Pancreatic intraepithelial neoplasia: a new nomenclature and classification system for pancreatic duct lesions. Am J Surg Pathol. 2001;25(5):579–86.
45. Tanaka M, Fernandez-del Castillo C, Adsay V, Chari S, Falconi M, Jang JY, et al. International consensus guidelines 2012 for the management of IPMN and MCN of the pancreas. Pancreatology. 2012;12(3):183–97.
46. Furukawa T, Kloppel G, Volkan Adsay N, Albores-Saavedra J, Fukushima N, Horii A, et al. Classification of types of intraductal papillary-mucinous neoplasm of the pancreas: a consensus study. Virchows Arch Int J Pathol. 2005;447(5):794–9.
47. Adsay NV, Merati K, Andea A, Sarkar F, Hruban RH, Wilentz RE, et al. The dichotomy in the preinvasive neoplasia to invasive carcinoma sequence in the pancreas: differential expression of MUC1 and MUC2 supports the existence of two separate pathways of carcinogenesis. Mod Pathol Official J US Can Acad Pathol Inc. 2002;15(10):1087–95.
48. Adsay NV, Longnecker DS, Klimstra DS. Pancreatic tumors with cystic dilatation of the ducts: intraductal papillary mucinous neoplasms and intraductal oncocytic papillary neoplasms. Semin Diagn Pathol. 2000;17(1):16–30.
49. Nakayama Y, Inoue H, Hamada Y, Takeshita M, Iwasaki H, Maeshiro K, et al. Intraductal tubular adenoma of the pancreas, pyloric gland type: a clinicopathologic and immunohistochemical study of 6 cases. Am J Surg Pathol. 2005;29(5):607–16.
50. Albores-Saavedra J, Sheahan K, O'Riain C, Shukla D. Intraductal tubular adenoma, pyloric type, of the pancreas: additional observations on a new type of pancreatic neoplasm. Am J Surg Pathol. 2004;28(2):233–8.
51. Kato N, Akiyama S, Motoyama T. Pyloric gland-type tubular adenoma superimposed on intraductal papillary mucinous tumor of the pancreas. Pyloric gland adenoma of the pancreas. Virchows Arch Int J Pathol. 2002;440(2):205–8.
52. Bakotic BW, Robinson MJ, Sturm PD, Hruban RH, Offerhaus GJ, Albores-Saavedra J. Pyloric gland adenoma of the main pancreatic duct. Am J Surg Pathol. 1999;23(2):227–31.
53. Adsay NV, Merati K, Basturk O, Iacobuzio-Donahue C, Levi E, Cheng JD, et al. Pathologically and biologically distinct types of

epithelium in intraductal papillary mucinous neoplasms: delineation of an "intestinal" pathway of carcinogenesis in the pancreas. Am J Surg Pathol. 2004;28(7):839–48.
54. Ishida M, Egawa S, Aoki T, Sakata N, Mikami Y, Motoi F, et al. Characteristic clinicopathological features of the types of intraductal papillary-mucinous neoplasms of the pancreas. Pancreas. 2007;35(4):348–52.
55. Terada T, Ohta T, Kitamura Y, Ashida K, Matsunaga Y. Cell proliferative activity in intraductal papillary-mucinous neoplasms and invasive ductal adenocarcinomas of the pancreas: an immunohistochemical study. Arch Pathol Lab Med. 1998;122 (1):42–6.
56. Luttges J, Zamboni G, Longnecker D, Kloppel G. The immunohistochemical mucin expression pattern distinguishes different types of intraductal papillary mucinous neoplasms of the pancreas and determines their relationship to mucinous noncystic carcinoma and ductal adenocarcinoma. Am J Surg Pathol. 2001;25(7):942–8.
57. Basturk O, Khayyata S, Klimstra DS, Hruban RH, Zamboni G, Coban I, et al. Preferential expression of MUC6 in oncocytic and pancreatobiliary types of intraductal papillary neoplasms highlights a pyloropancreatic pathway, distinct from the intestinal pathway, in pancreatic carcinogenesis. Am J Surg Pathol. 2010;34(3):364–70.
58. Nagai E, Ueki T, Chijiiwa K, Tanaka M, Tsuneyoshi M. Intraductal papillary mucinous neoplasms of the pancreas associated with so-called "mucinous ductal ectasia". Histochemical and immunohistochemical analysis of 29 cases. Am J Surg Pathol. 1995;19(5):576–89.
59. Thompson LD, Becker RC, Przygodzki RM, Adair CF, Heffess CS. Mucinous cystic neoplasm (mucinous cystadenocarcinoma of low-grade malignant potential) of the pancreas: a clinicopathologic study of 130 cases. Am J Surg Pathol. 1999;23(1):1–16.
60. Compagno J, Oertel JE. Mucinous cystic neoplasms of the pancreas with overt and latent malignancy (cystadenocarcinoma and cystadenoma). A clinicopathologic study of 41 cases. Am J Clin Pathol. 1978;69(6):573–80.
61. Wilentz RE, Albores-Saavedra J, Hruban RH. Mucinous cystic neoplasms of the pancreas. Semin Diagn Pathol. 2000;17(1):31–42.
62. Zamboni G, Scarpa A, Bogina G, Iacono C, Bassi C, Talamini G, et al. Mucinous cystic tumors of the pancreas: clinicopathological features, prognosis, and relationship to other mucinous cystic tumors. Am J Surg Pathol. 1999;23(4):410–22.
63. Izumo A, Yamaguchi K, Eguchi T, Nishiyama K, Yamamoto H, Yonemasu H, et al. Mucinous cystic tumor of the pancreas: immunohistochemical assessment of "ovarian-type stroma". Oncol Rep. 2003;10 (3):515–25.
64. Hara T, Kawashima H, Ishigooka M, Kashiyama M, Takanashi S, Yamazaki S, et al. Mucinous cystic tumors of the pancreas. Surg Today. 2002;32 (11):965–9.
65. Ridder GJ, Maschek H, Flemming P, Nashan B, Klempnauer J. Ovarian-like stroma in an invasive mucinous cystadenocarcinoma of the pancreas positive for inhibin. A hint concerning its possible histogenesis. Virchows Arch Int J Pathol. 1998;432 (5):451–4.
66. Albores-Saavedra J, Gould EW, Angeles-Angeles A, Henson DE. Cystic tumors of the pancreas. Pathol Ann. 1990;25(Pt 2):19–50.
67. Hirano H, Morita K, Tachibana S, Okimura A, Fujisawa T, Ouchi S, et al. Undifferentiated carcinoma with osteoclast-like giant cells arising in a mucinous cystic neoplasm of the pancreas. Pathol Int. 2008;58(6):383–9.
68. Goh BK, Tan YM, Chung YF, Chow PK, Cheow PC, Wong WK, et al. A review of mucinous cystic neoplasms of the pancreas defined by ovarian-type stroma: clinicopathological features of 344 patients. World J Surg. 2006;30(12):2236–45.
69. Lewis GH, Wang H, Bellizzi AM, Klein AP, Askin FB, Schwartz LE, et al. Prognosis of minimally invasive carcinoma arising in mucinous cystic neoplasms of the pancreas. Am J Surg Pathol. 2013;37(4):601–5.
70. Ohta T, Nagakawa T, Fukushima W, Mori K, Kayahara M, Akiyama T, et al. Immunohistochemical study of carcinoembryonic antigen in mucinous cystic neoplasm of the pancreas. Eur Surg Res (Europaische Chirurgische Forschung Recherches Chirurgicales Europeennes). 1992;24(1):37–44.
71. Luttges J, Feyerabend B, Buchelt T, Pacena M, Kloppel G. The mucin profile of noninvasive and invasive mucinous cystic neoplasms of the pancreas. Am J Surg Pathol. 2002;26(4):466–71.
72. Chen J, Baithun SI. Morphological study of 391 cases of exocrine pancreatic tumours with special reference to the classification of exocrine pancreatic carcinoma. J Pathol. 1985;146(1):17–29.
73. Adsay V, Logani S, Sarkar F, Crissman J, Vaitkevicius V. Foamy gland pattern of pancreatic ductal adenocarcinoma: a deceptively benign-appearing variant. Am J Surg Pathol. 2000;24(4):493–504.
74. Bagci P, Andea AA, Basturk O, Jang KT, Erbarut I, Adsay V. Large duct type invasive adenocarcinoma of the pancreas with microcystic and papillary patterns: a potential microscopic mimic of non-invasive ductal neoplasia. Mod Pathol Official J US Can Acad Pathol Inc. 2012;25(3):439–48.
75. Dursun N, Feng J, Basturk O, Bandyopadhyay S, Cheng JD, Adsay VN. Vacuolated cell pattern of pancreatobiliary adenocarcinoma: a clinicopathological analysis of 24 cases of a poorly recognized distinctive morphologic variant important in the differential diagnosis. Virchows Arch Int J Pathol. 2010;457(6):643–9.

76. Zerbi A, De Nardi P, Braga M, Radice F, Sironi M, Di Carlo V. An oncocytic carcinoma of the pancreas with pulmonary and subcutaneous metastases. Pancreas. 1993;8(1):116–9.
77. Bandyopadhyay S, Basturk O, Coban I, Thirabanjasak D, Liang H, Altinel D, et al. Isolated solitary ducts (naked ducts) in adipose tissue: a specific but underappreciated finding of pancreatic adenocarcinoma and one of the potential reasons of understaging and high recurrence rate. Am J Surg Pathol. 2009;33(3):425–9.
78. Luttges J, Schemm S, Vogel I, Hedderich J, Kremer B, Kloppel G. The grade of pancreatic ductal carcinoma is an independent prognostic factor and is superior to the immunohistochemical assessment of proliferation. J Pathol. 2000;191(2):154–61.
79. Adsay NV, Basturk O, Bonnett M, Kilinc N, Andea AA, Feng J, et al. A proposal for a new and more practical grading scheme for pancreatic ductal adenocarcinoma. Am J Surg Pathol. 2005;29 (6):724–33.
80. Giulianotti PC, Boggi U, Fornaciari G, Bruno J, Rossi G, Giardino D, et al. Prognostic value of histological grading in ductal adenocarcinoma of the pancreas. Kloppel vs. TNM grading. Int J Pancreatol Official J Int Assoc Pancreatol. 1995;17(3):279–89.
81. Duval JV, Savas L, Banner BF. Expression of cytokeratins 7 and 20 in carcinomas of the extrahepatic biliary tract, pancreas, and gallbladder. Arch Pathol Lab Med. 2000;124(8):1196–200.
82. Lee MJ, Lee HS, Kim WH, Choi Y, Yang M. Expression of mucins and cytokeratins in primary carcinomas of the digestive system. Mod Pathol Official J US Can Acad Pathol Inc. 2003;16(5):403–10.
83. Andrianifahanana M, Moniaux N, Schmied BM, Ringel J, Friess H, Hollingsworth MA, et al. Mucin (MUC) gene expression in human pancreatic adenocarcinoma and chronic pancreatitis: a potential role of MUC4 as a tumor marker of diagnostic significance. Clin Cancer Res Official J Am Assoc Cancer Res. 2001;7(12):4033–40.
84. Monges GM, Mathoulin-Portier MP, Acres RB, Houvenaeghel GF, Giovannini MF, Seitz JF, et al. Differential MUC 1 expression in normal and neoplastic human pancreatic tissue. An immunohistochemical study of 60 samples. Am J Clin Pathol. 1999;112(5):635–40.
85. Klimstra DHMMAea. Ductal proliferative lesions associated with infiltrating ductal adenocarcinoma of the pancreas. Int J Pancreatol Official J Int Assoc Pancreatol. 1994(16):224–5.
86. Adsay NV, Pierson C, Sarkar F, Abrams J, Weaver D, Conlon KC, et al. Colloid (mucinous noncystic) carcinoma of the pancreas. Am J Surg Pathol. 2001;25(1):26–42.
87. Seidel G, Zahurak M, Iacobuzio-Donahue C, Sohn TA, Adsay NV, Yeo CJ, et al. Almost all infiltrating colloid carcinomas of the pancreas and periampullary region arise from in situ papillary neoplasms: a study of 39 cases. Am J Surg Pathol. 2002;26(1):56–63.
88. Kardon DE, Thompson LD, Przygodzki RM, Heffess CS. Adenosquamous carcinoma of the pancreas: a clinicopathologic series of 25 cases. Mod Pathol Official J US Can Acad Pathol Inc. 2001;14(5):443–51.
89. Yamaguchi K, Enjoji M. Adenosquamous carcinoma of the pancreas: a clinicopathologic study. J Surg Oncol. 1991;47(2):109–16.
90. Hruban RH, Pitman MB, Klimstra D. Tumors of the pancreas: American registry of pathology in collaboration with the Armed Forces Institute of Pathology; 2007.
91. Paal E, Thompson LD, Frommelt RA, Przygodzki RM, Heffess CS. A clinicopathologic and immunohistochemical study of 35 anaplastic carcinomas of the pancreas with a review of the literature. Ann Diagn Pathol. 2001;5(3):129–40.
92. van der Zwan JM, Trama A, Otter R, Larranaga N, Tavilla A, Marcos-Gragera R, et al. Rare neuroendocrine tumours: results of the surveillance of rare cancers in Europe project. Eur J Cancer. 2013;49 (11):2565–78.
93. Watanabe M, Miura H, Inoue H, Uzuki M, Noda Y, Fujita N, et al. Mixed osteoclastic/pleomorphic-type giant cell tumor of the pancreas with ductal adenocarcinoma: histochemical and immunohistochemical study with review of the literature. Pancreas. 1997;15(2):201–8.
94. Posen JA. Giant cell tumor of the pancreas of the osteoclastic type associated with a mucous secreting cystadenocarcinoma. Hum Pathol. 1981;12 (10):944–7.
95. Alguacil-Garcia A, Weiland LH. The histologic spectrum, prognosis, and histogenesis of the sarcomatoid carcinoma of the pancreas. Cancer. 1977;39 (3):1181–9.
96. Dhall D, Klimstra DS. The cellular composition of osteoclastlike giant cell-containing tumors of the pancreatobiliary tree. Am J Surg Pathol. 2008;32 (2):335–7; author response 7.
97. Dworak O, Wittekind C, Koerfgen HP, Gall FP. Osteoclastic giant cell tumor of the pancreas. An immunohistological study and review of the literature. Pathol Res Pract. 1993;189(2):228–31; discussion 32–4.
98. Han SS, Jang JY, Kim SW, Kim WH, Lee KU, Park YH. Analysis of long-term survivors after surgical resection for pancreatic cancer. Pancreas. 2006;32(3):271–5.
99. Banville N, Geraghty R, Fox E, Leahy DT, Green A, Keegan D, et al. Medullary carcinoma of the pancreas in a man with hereditary nonpolyposis colorectal cancer due to a mutation of the MSH2 mismatch repair gene. Hum Pathol. 2006;37 (11):1498–502.
100. Goggins M, Offerhaus GJ, Hilgers W, Griffin CA, Shekher M, Tang D, et al. Pancreatic adenocarcinomas with DNA replication errors (RER+) are

associated with wild-type K-ras and characteristic histopathology. Poor differentiation, a syncytial growth pattern, and pushing borders suggest RER +. Am J Pathol. 1998;152(6):1501–7.
101. Wilentz RE, Goggins M, Redston M, Marcus VA, Adsay NV, Sohn TA, et al. Genetic, immunohistochemical, and clinical features of medullary carcinoma of the pancreas: a newly described and characterized entity. Am J Pathol. 2000;156 (5):1641–51.
102. Nakata B, Wang YQ, Yashiro M, Nishioka N, Tanaka H, Ohira M, et al. Prognostic value of microsatellite instability in resectable pancreatic cancer. Clin Cancer Res Official J Am Assoc Cancer Res. 2002;8(8):2536–40.
103. Marchegiani G, Gareer H, Parisi A, Capelli P, Bassi C, Salvia R. Pancreatic hepatoid carcinoma: a review of the literature. Dig Surg. 2013;30(4–6):425–33.
104. Paner GP, Thompson KS, Reyes CV. Hepatoid carcinoma of the pancreas. Cancer. 2000;88 (7):1582–9.
105. Hruban RH, Molina JM, Reddy MN, Boitnott JK. A neoplasm with pancreatic and hepatocellular differentiation presenting with subcutaneous fat necrosis. Am J Clin Pathol. 1987;88(5):639–45.
106. Jones S, Zhang X, Parsons DW, Lin JC, Leary RJ, Angenendt P, et al. Core signaling pathways in human pancreatic cancers revealed by global genomic analyses. Science. 2008;321(5897):1801–6.
107. Shi C, Daniels JA, Hruban RH. Molecular characterization of pancreatic neoplasms. Adv Anat Pathol. 2008;15(4):185–95.
108. Hruban RH, Adsay NV. Molecular classification of neoplasms of the pancreas. Hum Pathol. 2009;40 (5):612–23.
109. Hong SM, Park JY, Hruban RH, Goggins M. Molecular signatures of pancreatic cancer. Arch Pathol Lab Med. 2011;135(6):716–27.
110. Hruban RH, Iacobuzio-Donahue C, Wilentz RE, Goggins M, Kern SE. Molecular pathology of pancreatic cancer. Cancer J. 2001;7(4):251–8.
111. Westra WH, Sturm P, Drillenburg P, Choti MA, Klimstra DS, Albores-Saavedra J, et al. K-ras oncogene mutations in osteoclast-like giant cell tumors of the pancreas and liver: genetic evidence to support origin from the duct epithelium. Am J Surg Pathol. 1998;22(10):1247–54.
112. Satoh K, Shimosegawa T, Moriizumi S, Koizumi M, Toyota T. K-ras mutation and p53 protein accumulation in intraductal mucin-hypersecreting neoplasms of the pancreas. Pancreas. 1996;12(4):362–8.
113. Z'Graggen K, Rivera JA, Compton CC, Pins M, Werner J, Fernandez-del Castillo C, et al. Prevalence of activating K-ras mutations in the evolutionary stages of neoplasia in intraductal papillary mucinous tumors of the pancreas. Ann Surg. 1997;226 (4):491–8; discussion 8–500.
114. Lubezky N, Ben-Haim M, Marmor S, Brazowsky E, Rechavi G, Klausner JM, et al. High-throughput mutation profiling in intraductal papillary mucinous neoplasm (IPMN). J Gastrointest Surg Official J Soc Surg Aliment Tract. 2011;15(3):503–11.
115. Yoshizawa K, Nagai H, Sakurai S, Hironaka M, Morinaga S, Saitoh K, et al. Clonality and K-ras mutation analyses of epithelia in intraductal papillary mucinous tumor and mucinous cystic tumor of the pancreas. Virchows Arch Int J Pathol. 2002;441 (5):437–43.
116. Schonleben F, Qiu W, Bruckman KC, Ciau NT, Li X, Lauerman MH, et al. BRAF and KRAS gene mutations in intraductal papillary mucinous neoplasm/carcinoma (IPMN/IPMC) of the pancreas. Cancer Lett. 2007;249(2):242–8.
117. Schonleben F, Allendorf JD, Qiu W, Li X, Ho DJ, Ciau NT, et al. Mutational analyses of multiple oncogenic pathways in intraductal papillary mucinous neoplasms of the pancreas. Pancreas. 2008;36 (2):168–72.
118. Chadwick B, Willmore-Payne C, Tripp S, Layfield LJ, Hirschowitz S, Holden J. Histologic, immunohistochemical, and molecular classification of 52 IPMNs of the pancreas. Appl Immunohistochem Mol Morphol AIMM/Official Publ Soc Appl Immunohistochem. 2009;17(1):31–9.
119. Wu J, Matthaei H, Maitra A, Dal Molin M, Wood LD, Eshleman JR, et al. Recurrent GNAS mutations define an unexpected pathway for pancreatic cyst development. Sci Transl Med. 2011;3 (92):92ra66.
120. Jimenez RE, Warshaw AL, Z'Graggen K, Hartwig W, Taylor DZ, Compton CC, et al. Sequential accumulation of K-ras mutations and p53 overexpression in the progression of pancreatic mucinous cystic neoplasms to malignancy. Annals of surgery. 1999;230(4):501–9; discussion 9–11.
121. Kim SG, Wu TT, Lee JH, Yun YK, Issa JP, Hamilton SR, et al. Comparison of epigenetic and genetic alterations in mucinous cystic neoplasm and serous microcystic adenoma of pancreas. Mod Pathol Official J US Can Acad Pathol Inc. 2003;16 (11):1086–94.
122. Kanda M, Matthaei H, Wu J, Hong SM, Yu J, Borges M, et al. Presence of somatic mutations in most early-stage pancreatic intraepithelial neoplasia. Gastroenterology. 2012;142(4):730–3 e9.
123. Omura N, Li CP, Li A, Hong SM, Walter K, Jimeno A, et al. Genome-wide profiling of methylated promoters in pancreatic adenocarcinoma. Cancer Biol Ther. 2008;7(7):1146–56.
124. Wu J, Jiao Y, Dal Molin M, Maitra A, de Wilde RF, Wood LD, et al. Whole-exome sequencing of neoplastic cysts of the pancreas reveals recurrent mutations in components of ubiquitin-dependent pathways. Proc Natl Acad Sci USA. 2011;108 (52):21188–93.
125. Hruban RH, Maitra A, Kern SE, Goggins M. Precursors to pancreatic cancer. Gastroenterol Clin North Am. 2007;36(4):831–49, vi.

126. Hahn SA, Schutte M, Hoque AT, Moskaluk CA, da Costa LT, Rozenblum E, et al. DPC4, a candidate tumor suppressor gene at human chromosome 18q21.1. Science. 1996;271(5247):350–3.
127. Blackford A, Serrano OK, Wolfgang CL, Parmigiani G, Jones S, Zhang X, et al. SMAD4 gene mutations are associated with poor prognosis in pancreatic cancer. Clin Cancer Res Official J Am Assoc Cancer Res. 2009;15(14):4674–9.
128. Wilentz RE, Su GH, Dai JL, Sparks AB, Argani P, Sohn TA, et al. Immunohistochemical labeling for dpc4 mirrors genetic status in pancreatic adenocarcinomas: a new marker of DPC4 inactivation. Am J Pathol. 2000;156(1):37–43.
129. Dewald GW, Smyrk TC, Thorland EC, McWilliams RR, Van Dyke DL, Keefe JG, et al. Fluorescence in situ hybridization to visualize genetic abnormalities in interphase cells of acinar cell carcinoma, ductal adenocarcinoma, and islet cell carcinoma of the pancreas. Mayo Clin Proc. 2009;84(9):801–10.
130. van Heek NT, Meeker AK, Kern SE, Yeo CJ, Lillemoe KD, Cameron JL, et al. Telomere shortening is nearly universal in pancreatic intraepithelial neoplasia. Am J Pathol. 2002;161(5):1541–7.
131. Jacobs EJ, Chanock SJ, Fuchs CS, Lacroix A, McWilliams RR, Steplowski E, et al. Family history of cancer and risk of pancreatic cancer: a pooled analysis from the Pancreatic Cancer Cohort Consortium (PanScan). Int J Cancer (Journal International du Cancer). 2010;127(6):1421–8.
132. Schenk M, Schwartz AG, O'Neal E, Kinnard M, Greenson JK, Fryzek JP, et al. Familial risk of pancreatic cancer. J Natl Cancer Inst. 2001;93(8):640–4.
133. Ghadirian P, Boyle P, Simard A, Baillargeon J, Maisonneuve P, Perret C. Reported family aggregation of pancreatic cancer within a population-based case-control study in the Francophone community in Montreal, Canada. Int J Pancreatol Official J Int Assoc Pancreatol. 1991;10(3–4):183–96.
134. Falk RT, Pickle LW, Fontham ET, Correa P, Fraumeni JF Jr. Life-style risk factors for pancreatic cancer in Louisiana: a case-control study. Am J Epidemiol. 1988;128(2):324–36.
135. Hruban RH, Canto MI, Goggins M, Schulick R, Klein AP. Update on familial pancreatic cancer. Adv Surg. 2010;44:293–311.
136. Lowery MA, Kelsen DP, Stadler ZK, Yu KH, Janjigian YY, Ludwig E, et al. An emerging entity: pancreatic adenocarcinoma associated with a known BRCA mutation: clinical descriptors, treatment implications, and future directions. Oncologist. 2011;16(10):1397–402.
137. Hahn SA, Greenhalf B, Ellis I, Sina-Frey M, Rieder H, Korte B, et al. BRCA2 germline mutations in familial pancreatic carcinoma. J Natl Cancer Inst. 2003;95(3):214–21.
138. Goggins M, Schutte M, Lu J, Moskaluk CA, Weinstein CL, Petersen GM, et al. Germline BRCA2 gene mutations in patients with apparently sporadic pancreatic carcinomas. Cancer Res. 1996;56(23):5360–4.
139. Consortium BCL. Cancer risks in BRCA2 mutation carriers. J Natl Cancer Inst. 1999;91(15):1310–6.
140. Slater EP, Langer P, Niemczyk E, Strauch K, Butler J, Habbe N, et al. PALB2 mutations in European familial pancreatic cancer families. Clin Genet. 2010;78(5):490–4.
141. Tischkowitz MD, Sabbaghian N, Hamel N, Borgida A, Rosner C, Taherian N, et al. Analysis of the gene coding for the BRCA2-interacting protein PALB2 in familial and sporadic pancreatic cancer. Gastroenterology. 2009;137(3):1183–6.
142. Jones S, Hruban RH, Kamiyama M, Borges M, Zhang X, Parsons DW, et al. Exomic sequencing identifies PALB2 as a pancreatic cancer susceptibility gene. Science. 2009;324(5924):217.
143. van der Heijden MS, Brody JR, Gallmeier E, Cunningham SC, Dezentje DA, Shen D, et al. Functional defects in the fanconi anemia pathway in pancreatic cancer cells. Am J Pathol. 2004;165(2):651–7.
144. Rogers CD, van der Heijden MS, Brune K, Yeo CJ, Hruban RH, Kern SE, et al. The genetics of FANCC and FANCG in familial pancreatic cancer. Cancer Biol Ther. 2004;3(2):167–9.
145. van der Heijden MS, Yeo CJ, Hruban RH, Kern SE. Fanconi anemia gene mutations in young-onset pancreatic cancer. Cancer Res. 2003;63(10):2585–8.
146. Rogers CD, Couch FJ, Brune K, Martin ST, Philips J, Murphy KM, et al. Genetics of the FANCA gene in familial pancreatic cancer. J Med Genet. 2004;41(12):e126.
147. Lubezky N, Ben-Haim M, Lahat G, Marmor S, Solar I, Brazowski E, et al. Intraductal papillary mucinous neoplasm of the pancreas: associated cancers, family history, genetic predisposition? Surgery. 2012;151(1):70–5.
148. Thompson D, Easton DF. Cancer Incidence in BRCA1 mutation carriers. J Natl Cancer Inst. 2002;94(18):1358–65.
149. Ferrone CR, Levine DA, Tang LH, Allen PJ, Jarnagin W, Brennan MF, et al. BRCA germline mutations in Jewish patients with pancreatic adenocarcinoma. J Clin Oncol: Official J Am Soc Clin Oncol. 2009;27(3):433–8.
150. Axilbund JE, Argani P, Kamiyama M, Palmisano E, Raben M, Borges M, et al. Absence of germline BRCA1 mutations in familial pancreatic cancer patients. Cancer Biol Ther. 2009;8(2):131–5.
151. Bartsch DK, Langer P, Habbe N, Matthai E, Chaloupka B, Sina M, et al. Clinical and genetic analysis of 18 pancreatic carcinoma/melanoma-prone families. Clin Genet. 2010;77(4):333–41.
152. de Snoo FA, Bishop DT, Bergman W, van Leeuwen I, van der Drift C, van Nieuwpoort FA, et al. Increased risk of cancer other than melanoma in CDKN2A founder mutation (p16-Leiden)-

positive melanoma families. Clin Cancer Res Official J Am Assoc Cancer Res. 2008;14(21):7151–7.
153. Vasen HF, Gruis NA, Frants RR, van Der Velden PA, Hille ET, Bergman W. Risk of developing pancreatic cancer in families with familial atypical multiple mole melanoma associated with a specific 19 deletion of p16 (p16-Leiden). Int J Cancer (Journal International du Cancer). 2000;87 (6):809–11.
154. Whelan AJ, Bartsch D, Goodfellow PJ. Brief report: a familial syndrome of pancreatic cancer and melanoma with a mutation in the CDKN2 tumor-suppressor gene. N Engl J Med. 1995;333 (15):975–7.
155. Goldstein AM, Fraser MC, Struewing JP, Hussussian CJ, Ranade K, Zametkin DP, et al. Increased risk of pancreatic cancer in melanoma-prone kindreds with p16INK4 mutations. N Engl J Med. 1995;333(15):970–4.
156. Goldstein AM, Struewing JP, Fraser MC, Smith MW, Tucker MA. Prospective risk of cancer in CDKN2A germline mutation carriers. J Med Genet. 2004;41(6):421–4.
157. Giardiello FM, Welsh SB, Hamilton SR, Offerhaus GJ, Gittelsohn AM, Booker SV, et al. Increased risk of cancer in the Peutz-Jeghers syndrome. N Engl J Med. 1987;316(24):1511–4.
158. Su GH, Hruban RH, Bansal RK, Bova GS, Tang DJ, Shekher MC, et al. Germline and somatic mutations of the STK11/LKB1 Peutz-Jeghers gene in pancreatic and biliary cancers. Am J Pathol. 1999;154 (6):1835–40.
159. Sato N, Rosty C, Jansen M, Fukushima N, Ueki T, Yeo CJ, et al. STK11/LKB1 Peutz-Jeghers gene inactivation in intraductal papillary-mucinous neoplasms of the pancreas. Am J Pathol. 2001;159 (6):2017–22.
160. Witt H, Luck W, Hennies HC, Classen M, Kage A, Lass U, et al. Mutations in the gene encoding the serine protease inhibitor, Kazal type 1 are associated with chronic pancreatitis. Nat Genet. 2000;25 (2):213–6.
161. Whitcomb DC, Gorry MC, Preston RA, Furey W, Sossenheimer MJ, Ulrich CD, et al. Hereditary pancreatitis is caused by a mutation in the cationic trypsinogen gene. Nat Genet. 1996;14(2):141–5.
162. Lowenfels AB, Maisonneuve P, DiMagno EP, Elitsur Y, Gates LK Jr, Perrault J, et al. Hereditary pancreatitis and the risk of pancreatic cancer. International Hereditary Pancreatitis Study Group. J Natl Cancer Inst. 1997;89(6):442–6.
163. Roberts NJ, Jiao Y, Yu J, Kopelovich L, Petersen GM, Bondy ML, et al. ATM mutations in patients with hereditary pancreatic cancer. Cancer Discovery. 2012;2(1):41–6.
164. Rustgi AK. The genetics of hereditary colon cancer. Genes Dev. 2007;21(20):2525–38.
165. Kastrinos F, Mukherjee B, Tayob N, Wang F, Sparr J, Raymond VM, et al. Risk of pancreatic cancer in families with Lynch syndrome. JAMA. 2009;302(16):1790–5.
166. Sparr JA, Bandipalliam P, Redston MS, Syngal S. Intraductal papillary mucinous neoplasm of the pancreas with loss of mismatch repair in a patient with Lynch syndrome. Am J Surg Pathol. 2009;33 (2):309–12.
167. Berndt N, Hamilton AD, Sebti SM. Targeting protein prenylation for cancer therapy. Nat Rev Cancer. 2011;11(11):775–91.
168. Yap JL, Worlikar S, MacKerell AD Jr, Shapiro P, Fletcher S. Small-molecule inhibitors of the ERK signaling pathway: towards novel anticancer therapeutics. ChemMedChem. 2011;6(1):38–48.
169. Collisson EA, Trejo CL, Silva JM, Gu S, Korkola JE, Heiser LM, et al. A central role for RAF→MEK→ERK signaling in the genesis of pancreatic ductal adenocarcinoma. Cancer Discovery. 2012;2(8):685–93.
170. van der Heijden MS, Brody JR, Dezentje DA, Gallmeier E, Cunningham SC, Swartz MJ, et al. In vivo therapeutic responses contingent on Fanconi anemia/BRCA2 status of the tumor. Clin Cancer Res Official J Am Assoc Cancer Res. 2005;11(20):7508–15.
171. Canto MI, Harinck F, Hruban RH, Offerhaus GJ, Poley JW, Kamel I, et al. International cancer of the pancreas screening (CAPS) consortium summit on the management of patients with increased risk for familial pancreatic cancer. Gut. 2013;62(3):339–47.
172. Villarroel MC, Rajeshkumar NV, Garrido-Laguna I, De Jesus-Acosta A, Jones S, Maitra A, et al. Personalizing cancer treatment in the age of global genomic analyses: PALB2 gene mutations and the response to DNA damaging agents in pancreatic cancer. Mol Cancer Ther. 2011;10(1):3–8.
173. Dedes KJ, Wilkerson PM, Wetterskog D, Weigelt B, Ashworth A, Reis-Filho JS. Synthetic lethality of PARP inhibition in cancers lacking BRCA1 and BRCA2 mutations. Cell Cycle. 2011;10(8):1192–9.

# Epidemiology of Bladder Cancer

# 27

Sarah M. Rudman and Danielle Crawley

## 27.1 Clinical Picture of the Disease

### 27.1.1 Introduction

Bladder cancer is the ninth commonest cancer worldwide, and in 2012, 430,000 cases were diagnosed. It has an incidence of 9.5 per 100,000 in the developed world [1]. Over 90 % of cases are Transitional Cell Carcinomas (TCC), previously known as Urothelial cancers.

The disease can be divided into: Non-Muscle-Invasive Bladder Cancer (NMIBC), Muscle-Invasive Bladder Cancer (MIBC), and Metastatic Bladder Cancer (MBC). All have their own diagnostic, treatment and prognostic pathways, which will be described in this section.

### 27.1.2 Symptoms

The most common presenting symptom among bladder cancer patients is painless haematuria. This can be either micro- or macroscopic. While the vast majority of patients presenting in this way will have a benign pathology, approximately 10 % will be diagnosed with a bladder cancer [2, 3]. Patients can also present with irritative symptoms, such as urgency and dysuria, as well as with abdominal pain. Less commonly they present with systemic symptoms as a result of metastatic disease.

### 27.1.3 Diagnosis

The diagnostic process begins with a thorough history, physical examination, basic laboratory tests, urinalysis and urine cytology. The gold standard diagnostic test is a cystoscopy, to allow direct visualization of the bladder, before proceeding to a Transurethral Resection of Bladder Tumor (TURBT) to obtain a final histological diagnosis. All biopsy specimens should contain muscle to enable the disease to be classified definitively as either NMIBC or MIBC.

If the histology confirms MIBC then further imaging with CT or MRI is required. If the clinical suspicion of an invasive tumor is high, then these should be performed pre-cystoscopy to avoid artifact produced by the procedure. Imaging of the upper urinary tracts with either CT urogram, retrograde or iv pyelogram is also recommended to exclude a synchronous upper tract tumor (which can occur in up to 2.5 % of cases) [4]. Further investigations including CT chest and bone scan may be indicated to complete staging.

S.M. Rudman (✉)
Guy's and St. Thomas' NHS Foundation Trust, Bermondsey Wing Guy's Hospital, 3rd Floor, Great Maze Pond, London SE1 9RT, UK
e-mail: sarah.rudman@gstt.nhs.uk

D. Crawley
Cancer Epidemiology Group, Research Oncology, Bermondsey Wing Guy's Hospital, King's College London, 3rd Floor, Great Maze Pond, London SE1 9RT, UK


M. Loda et al. (eds.), *Pathology and Epidemiology of Cancer*,
DOI 10.1007/978-3-319-35153-7_27

### 27.1.4 Staging

Bladder cancer is staged using the tumor-node-metastasis (TNM) classification. See Table 27.1 [5]. For more details see the bladder pathology chapter.

**Table 27.1** TNM classification bladder cancer [5]

| | |
|---|---|
| Primary tumor (T) | |
| TX | Primary tumor cannot be assessed |
| T0 | No evidence of primary tumor |
| Ta | Noninvasive papillary carcinoma |
| Tis | Carcinoma in situ: "flat tumor" |
| T1 | Tumor invades subepithelial connective tissue |
| T2 | Tumor invades muscularis propria |
| pT2a | Tumor invades superficial muscularis propria (inner half) |
| pT2b | Tumor invades deep muscularis propria (outer half) |
| T3 | Tumor invades perivesical tissue |
| pT3a | Microscopically |
| pT3b | Macroscopically (extravesical mass) |
| T4 | Tumor invades any of the following: prostatic stroma, seminal vesicles, uterus, vagina, pelvic wall, abdominal wall |
| T4a | Tumor invades prostatic stroma, uterus, vagina |
| T4b | Tumor invades pelvic wall, abdominal wall |
| **Regional lymph nodes (N)** | |
| NX | Lymph nodes cannot be assessed |
| N0 | No lymph node metastasis |
| N1 | Single regional lymph node metastasis in the true pelvis (hypogastric, obturator, external iliac, or presacral lymph node) |
| N2 | Multiple regional lymph node metastasis in the true pelvis (hypogastric, obturator, external iliac, or presacral lymph node metastasis) |
| N3 | Lymph node metastasis to the common iliac lymph nodes |
| **Distant metastasis (M)** | |
| MO | No distant metastasis |
| M1 | Distant metastasis |

### 27.1.5 Treatment

#### 27.1.5.1 Non-Muscle-Invasive Bladder Cancer (NMIBC)

**Transurethral Resection of Bladder Tumor (TURBT)**

The gold standard management of NMIBC is a TURBT, aiming to remove all visible lesions. Detailed guidance on best practice is provided by many bodies including the European Association of Urology (EAU) and American Urological Association (AUA) [6, 7]. Many studies have evaluated the benefits of using Photodynamic Diagnosis (PDD) during which either 5-aminolevulinic acid (ALA) or hexaminolevulinic acid (HAL) are instilled into the bladder before cystoscopy is performed using ultraviolet light. Results are mixed with some studies finding no improvement in recurrence rates [8, 9] and others reporting improvements [10–12]. Currently the use of PDD is not routinely included in treatment algorithms but is used in many centers [6].

**Risk Stratification/Surveillance**

Patients with Ta and T1 tumors can be divided into low-, intermediate-, and high-risk groups using the EORTC scoring system [13]. The factors included are number of tumors, size of tumors, prior recurrence rate, T category in TNM staging, presence of concurrent carcinoma in situ and tumor grade.

Stratification into low, intermediate, and high risks can be used to identify which patients may benefit most from maintenance treatments. Low-risk tumors may only require one immediate post-operative instillation of chemotherapy, whereas those with intermediate risk are recommended to receive a year of maintenance treatment. This can be with either intravesical Bacillus Calmette–Guerin (BCG) or chemotherapy. Patients with high-risk tumors are recommended to continue BCG instillations for up to 3 years.

This risk stratification also helps determine which surveillance schedule is recommended, although no clear consensus exists about the precise follow-up schedule. The EAU guidelines suggest patients with low-risk tumors undergo cystoscopy at 3 months, 9 months and then annually for 5 years. Patients with high-risk tumors should undergo both cystoscopy and urine cytology at 3 months, then 3 monthly for 2 years, 6 monthly for 5 years, and annual cystoscopy thereafter. Patients with intermediate-risk tumors should have a surveillance schedule in between that for low- and high-risk patients [6].

Evidence for the use of both immunotherapy and chemotherapy exists in NMIBC.

#### Adjuvant Intravesical Bacillus Calmette–Guerin (BCG) Immunotherapy

Intravesical instillations using BCG have been shown to reduce recurrence rates compared to TURBT alone [14] and compared to intravesical chemotherapy [15, 16]. No clear guidance for dosing, timing, and durations are published, although there is consensus that it should be given on a maintenance schedule and not as a single post-operative instillation, unlike chemotherapy which can in some cases be given as a single treatment [16, 17]. The AUA advocates the use of the SWOG regimen, a 6-week induction course of BCG followed by a 3-week maintenance course at 3, 6, 12, 18, 24, 30, and 36 months [7].

#### Adjuvant Intravesical Chemotherapy

Adjuvant treatment with intravesical chemotherapy using mitomycin C (MMC), epirubicin, thiopeta, and doxorubicin are all used. However, as a result of limited availability of the other drugs, MMC is the most widely used. Sylvester et al. performed a meta-analysis of a single immediate post-operative instillation, including 7 randomized trials using MMC, thiopeta, and epirubicin, with recurrence data on 1476 patients and showed a significant decrease in the risk of recurrence with an odds ratio (OR) 0.61 ($p < 0.0001$) [18]. The benefits of further instillations remain unclear. However, Sylvester et al. also performed a meta-analysis examining this and concluded that they may be beneficial in higher risk patients [19]. However, no clear guidance on the exact dosing, timing, and duration of ongoing instillations has been published.

#### Radical Cystectomy

In some cases of NMIBC radical cystectomy is indicated. These include all patients who have or develop BCG refractory disease. It should also be considered as a potential treatment in those with multiple, larger tumors, concurrent CIS, disease in the prostatic urethra, and micropapillary histology.

### 27.1.5.2 Muscle-Invasive Bladder Cancer

#### Radical Cystectomy + Lymph Node Dissection

Radical cystectomy with lymph node dissection is the gold standard treatment for MIBC if the patient is fit for surgery. Lymph node dissection can be either standard (regional bilateral node dissection), extended (to aortic bifurcation), or super-extended (to inferior mesenteric artery). Extended lymph node dissection has been shown in a number of retrospective studies to be beneficial [20, 21], but currently no clear recommendation for its use exists, pending the results of randomized controlled trials [22].

#### Neoadjuvant/Adjuvant Chemotherapy

Classically the 5-year survival rate, following radical cystectomy, is approximately 50 % [23] and so the benefits of adjuvant treatment have been investigated. Neoadjuvant chemotherapy, consisting of a cisplatin based regimen, has been shown to result in a 5 % improvement in overall survival in a number of meta-analyses for stage II and III disease [24–26]. The benefits of adjuvant chemotherapy, however, are not as well established, although a recent meta-analysis of nine randomized controlled trials revealed an overall survival benefit with a pooled hazard ratio of 0.77 (95 % confidence interval (CI) 0.59–0.99; $p = 0.049$). It is only currently recommended in high-risk patients who have not received neoadjuvant treatment [4].

#### Multi-modality Bladder Preservation

This approach is appropriate if patients are not fit for radical surgery or express a personal preference for a bladder conserving approach. Treatment is a combination of TURBT, chemotherapy, and radiotherapy, with a tri-modal combination being the preferred treatment option. The addition of cisplatin based chemotherapy improves outcomes with both radical and adjuvant radiotherapy following TURBT [27]. Five-year survivals between 50 and 60 % have been demonstrated, making this a feasible approach for some patients [28, 29]. These patients must undergo vigorous follow-up and bladder surveillance, as by retaining their bladder they remain at risk of both recurrence and new bladder tumors.

#### 27.1.5.3 Advanced/Metastatic

**First Line Treatment**

Currently chemotherapy combinations, containing cisplatin, form the basis of first line treatment in advanced or metastatic disease. Both Gemcitabine and Cisplatin (GC) and Methotrexate, Vinblastine, Adriamycin, and Cisplatin (MVAC) have shown prolonged median survival rates of 13.8 months and 14.8 months [30, 31]. GC is less toxic than MVAC and is therefore usually the regimen of choice [30]. It is estimated that up to 50 % patients with advanced or metastatic disease are not fit to receive cisplatin therapy, due to a variety of factors including poor renal function and performance status [32]. It is common practice to replace cisplatin with carboplatin in this situation and to use it in combination with gemcitabine [33].

**Second Line Treatment**

Patients progressing following first line chemotherapy have limited further treatment options. Vinflunine, a third generation vinca alkaloid, remains the only licensed treatment in the second line setting for MBC. However, despite phase III trial evidence [34] and the lack of other treatment options, vinflunine has not been widely adopted into clinical practice.

An alternate strategy if relapse or progression occurs greater than 6 months after first line chemotherapy is to rechallenge the patient with cisplatin based combination chemotherapy or enrolment in a clinical trial.

**New Treatments**

In recent years, a large number of cytotoxic and targeted therapies have been investigated in phase II and III clinical trials, although these studies have yet to result in any new licensed treatments. At present, however, immunotherapy is undergoing a renaissance in solid cancer treatment. In advanced bladder cancer, checkpoint inhibitors targeting programmed cell death (PD-1) and its primary ligand PD-L1 have demonstrated the most promising results in the last 30 years. Powles et al. reported results of an extended phase I trial of Atezolizumab (MPDL3280A) a PD-L-1 inhibitor which demonstrated significant activity in advanced bladder cancer. They showed it is the most active against tumors with PD-L1 positive tumor infiltrating immune cells with objective response rates 43 % (13/30 patients CI 26–63 %), but still demonstrated objective response rates of around 11 % (4/35 CI 4–26 %) in PD-L1 negative tumors [35]. Similarly promising results were reported by Plimack et al. using Pembrolizumab, an anti-PD1 antibody [36]. This led to the US FDA granting this class of agent 'breakthrough drug' status in 2014 and later stage clinical trials are currently ongoing.

**Palliative Treatment**

Both surgery and radiotherapy are used as palliative treatments to treat localized symptoms such as frank hematuria or pain. Skeletal related events can also be minimized for patients with bone metastases using intravenous bisphosphonates [37]. In the absence of multiple treatment options for patients with advanced or metastatic disease, best supportive care remains of great importance [38].

## 27.2 Descriptive Epidemiology

### 27.2.1 Introduction

Globally, 430,000 men and women are diagnosed with bladder cancer and 165,000 individuals die each year [1]. The epidemiology of bladder cancer shows geographical, sex and age variations in its incidence, as well as being an example of a cancer where environmental and occupational risk factors are important in its etiology. In this section, the descriptive epidemiology will be discussed followed by the known risk factors for bladder cancer.

### 27.2.2 Geography

The incidence of bladder cancer varies greatly between the developed and developing world. The age-standardized incidence is 9.5 per 100,000 in more developed countries compared to 3.3 per 100,000 in less developed countries. The highest incidence is seen in Europe, North America, North Africa, and the Middle East. Some of these geographical differences in incidence can be explained by different registration practices for low-grade NMIBC which may significantly increase the recorded incidence. However, many studies have demonstrated that a migrant's risk of developing bladder cancer approximates to that of their host country, suggesting different environmental factors are also important [39].

### 27.2.3 Sex

Bladder cancer incidence across the world is higher in males than females, with global rates of 9.0 per 100,000 versus 2.2 per 100,000. The sex difference in incidence tends to mirror that seen in lung cancer on a country-by-country basis, suggesting that the difference in smoking habits between the sexes is largely responsible for the differences rather than any hormonal or genetic influences [1].

### 27.2.4 Trends Over Time

The number of bladder cancer cases have risen over recent decades, with increases as large as 50 % in North America recorded between 1985 and 2005 [40]. These differences may reflect change in practice, with increasing registrations of lower grade tumors, rather than a true increase in numbers. Mortality rates in localized disease were static in the United States (U.S.) between 1973 and 2009, whereas they increased in metastatic disease by an estimated annual percentage increase of 1 % [41].

### 27.2.5 Age

Bladder cancer is a disease of aging, with the incidence gradually increasing with age in their 30s and 40s, before a sharp rise in both sexes after the age of 50. Most patients will not die from bladder cancer but will experience multiple recurrences. As a result, after prostate cancer older men with bladder cancer have the highest prevalence rates [1].

## 27.3 Risk Factors

### 27.3.1 Smoking

The most important risk factor for bladder cancer in the Western world is smoking and this association has been widely studied. There are over 60 carcinogens known to be in cigarettes, although the ones which are individually responsible for the increased risk of bladder cancer are not fully established [42]. The tobacco constituent 4-aminobiphenyl (4-ABP) however, is a well-established risk factor for bladder cancer. It is known to cause chromosomal instability in human cells [43]. It has also been shown that smokers of blond tobacco are at lower risk of bladder cancer than smokers of black tobacco, which is richer in 4-ABP [44].

One of the largest studies of the association between smoking and bladder cancer was the National Institutes of Health-AARP Diet and Health Study Cohort which included 280,000 men and 186,000 women who were followed from 1995 to 2006. The hazard ratio (95 % CI) for bladder cancer in current smokers compared to non-smokers was 4.1 (95 % CI 3.7–4.5) and remained elevated in ex-smokers (HR 2.2, 95 % CI 2.0–2.4). The authors estimated the population attributable risk of cigarette smoking to be 50 % in men and 52 % in women [45]. A meta-analysis including 11 case-control studies looking at ex-smokers and risk of bladder cancer showed a dose relationship with the number of cigarettes smoked per day and risk, although this reached a plateau at 15–20 cigarettes/day. This may be because of inaccurate recall when higher numbers of cigarettes are smoked per day or could be reflecting saturation in the biological mechanisms of carcinogenesis. They also reported a 30 % decrease in risk following cessation at 1–4 years and a 60 % reduction in risk at 25 years. But even at 25 years the risk did not return to that of never smokers [46].

The relationship between cigar and pipe smoking is not as well defined. One study looking particularly at cigar smoking failed to show any significant increased risk in all cigar smokers (Relative risk (RR) 1.0, 95 % CI 0.4–2.3), but did show an increased risk in a subset whom inhaled smoke (RR 3.6, 95 % CI 1.3–9.9) [47]. Therefore, both the method and type of smoking appear to be important in calculating the risk of developing bladder cancer.

Environmental tobacco smoke, also known as secondhand smoking, has been shown to increase risk of bladder cancer in lifelong non-smoking females, but not in non-smoking males. In a case-control study by Jiang et al. approximately twofold increased risks were seen among women living with a spouse or domestic partner who smoked for greater or equal to 10 years or who had a co-worker who smoked in an indoor environment for greater or equal to 10 years [48]. This may, in part be explained by higher quantities of the tobacco constituent 4-ABP in secondhand smoke [49].

### 27.3.2 Occupational Exposures

The association between occupational exposures and increased risk of bladder cancer was first described as early as 1895. The results of the first large occupational epidemiological study was published in 1954 [50] and examined the exposure of workman in the dye industry to aniline, benzidine, alpha-naphthylamine, and beta-naphthylamine. Occupations using these chemicals such as painters, mechanics, textile and metal workers have been shown to be at increased risk of bladder cancer. A RR as high as 40-fold has been reported for certain occupations but the intervention of health and safety executives, and the resulting reduction in exposure has contributed to a decrease in the incidence of occupational related bladder cancer in the developed world [51]. For a full list of known carcinogens please see Table 27.2.

**Table 27.2** List of agents known to cause urinary bladder cancer

| Carcinogenic agents with sufficient evidence in humans | Agents with limited evidence in humans |
|---|---|
| Aluminum production | 4-Chloro-ortho-toluidine |
| 4-Aminobiphenyl | Coal-tar pitch |
| Arsenic and inorganic arsenic compounds | Coffee |
| Auramine production | Dry cleaning |
| Benzidine | Engine exhaust, diesel |
| Chlornaphazine | Hairdressers and barbers, occupational exposure |
| Cyclophosphamide | Pioglitazone |
| Magenta production | Printing processes |
| 2-Naphthylamine | Soot |
| Painting | Textile manufacturing |
| Rubber production industry | Tetrachloroethylene |
| Schistosoma haematobium | |
| Tobacco smoking | |
| ortho-Toluidine | |
| X-radiation, gamma-radiation | |

From the IARC Monographs on the evaluation of carcinogenic risks to humans. Accessed June 2015 at http://monographs.iarc.fr/ENG/Classification/Table4.pdf; used with permission

#### 27.3.2.1 Polyaromatic Hydrocarbons (PAHs) and Diesel Engine Exhaust

PAHs arise from substances such as coal tar and diesel exhaust. Certain occupations including those working in aluminum and coal production as well as transport workers may have exposure to PAHs. There is evidence that 'heavy exposure,' variously defined but working with PAHs for at least 10 years or more, is associated with a moderately increased risk of bladder cancer [52, 53].

#### 27.3.2.2 Hair Dyes

It was established in the 1970s that some hair dyes contained carcinogens including aromatic amines and high molecular weight complexes among other chemicals. Some older epidemiological studies suggested an increased risk of bladder cancer in hairdressers with occupational exposure to hair dyes. Manufacture of hair dyes has now been modernized and there is little or no evidence that modern hair dyes increase the risk of bladder cancer [54]. A meta-analysis investigating personal hair dye use and bladder cancer concluded only a very marginal increased risk (HR 1.15, 95 % CI 1.05–1.27) [55] in ever users of hair dye compared to never users.

### 27.3.3 Infections

#### 27.3.3.1 Schistosomiasis

Schistosomiasis, also referred to as Bilharzia, is a parasitic infection caused by parasitic worms of the Schistosoma type. In endemic areas, snails play an important role in the life cycle initially, harboring sporocysts, before releasing infectious cercariae, which then contaminate the water supply. These become the source of infection to human populations. Clinically, Schistosomiasis can affect the liver, gastrointestinal and urinary tracts. In the urinary tract it causes hematuria, infections, strictures, and ultimately bladder cancer. In contrast to Western Europe where TCC of the bladder accounts for >90 % cases, in Africa, Squamous Cell Carcinoma (SCC) accounts for as many as 75 % of bladder cancer cases. The incidence of bladder cancer in Africa and the Middle East is higher reflecting this association. In Egypt, bladder cancer has been the most common cancer in men and is second only to breast cancer in women [56]. However, due to concerted public health efforts which have dramatically reduced the prevalence of Schistosomiasis from ~50 % in rural areas to 3.5 % in 2002, the epidemiology of bladder cancer has shifted in Egypt to reflect a picture much more like that seen in Western Europe [56].

#### 27.3.3.2 Human Papillomavirus (HPV)

HPV has been studied extensively in relation to risk of bladder cancer. Several meta-analysis have shown a small increased risk, the most recent by Li et al. included 19 cases control studies and reported an odds ratio of 2.84 (95 % CI 1.39–5.80) associated with HPV infection [57]. However, in the same study the prevalence of HPV in bladder cancer is estimated at only around 17 % and so HPV is unlikely to play a major etiological role in most cases of bladder cancer.

#### 27.3.3.3 Chronic Cystitis

Chronic cystitis, specifically in those with spinal cord injury and/or long-term indwelling catheters, has been associated with an increased risk of bladder cancer. Groah et al. conducted a retrospective cohort study of people with spinal cord injuries and found that those who had long-term indwelling catheters had an increased RR of 4.9 (95 % CI 1.3–13.8) compared to those managed without a long-term indwelling catheter [58]. The exact mechanisms underlying these associations are not known but several plausible explanations exist. First, chronic cystitis can lead to bacterial super infection, which results in the production of carcinogenic nitrosamines. Second, the continued presence of inflammatory cells can promote malignancy via changes in the cytokine environment and lastly inflammation can lead to genetic polymorphisms which may also increase the risk of bladder cancer [59].

### 27.3.4 Iatrogenic

#### 27.3.4.1 Cyclophosphamide

Cyclophosphamide is an alkylating agent used in the treatment of some cancers, particularly lymphomas, as well as in some autoimmune diseases. It has a number of significant side effects, one being haemorrhagic cystitis and a resulting increased risk of bladder cancer [60, 61]. This risk is clearly dose dependent and is caused by a urinary metabolite of cyclophosphamide; acrolein. The drug mesna binds to acrolein creating an inert thioether which is then safely excreted and reduces the risk of both haemorrhagic cystitis and bladder cancer [62].

#### 27.3.4.2 Analgesics

Phenacetin was a commonly used analgesic until its withdrawal in the 1980s when it was confirmed as a carcinogen and shown to increase the risk of urothelial cancers [63]. Phenacetin use has now been replaced by the use of paracetamol and non-steroidal anti-inflammatory drugs (NSAIDS) which have not been shown to increase bladder cancer. In fact, conversely, some studies have reported a protective effect of NSAIDS on risk of bladder cancer [64].

#### 27.3.4.3 Barbiturates

The use of barbiturates as anti-convulsants has in a small number of studies been associated with a decreased risk of bladder cancer [65]. It is hypothesized that the enzyme inducing effect of these drugs leads to increased metabolism of carcinogens in cigarette smoke and it is via this mechanism that the reduced risk is observed [66].

#### 27.3.4.4 Pioglitazone

Pioglitazone is a second generation thiazolidinedione, used in the treatment of diabetes. Its use has been shown to be associated with a small increased risk of bladder cancer which increases with cumulative dose and length of exposure [67, 68]. In 2011, it was withdrawn in France and Germany, but the European Medicines Agency and US Food and Drug Administration have withheld judgment on these findings, advising that it should not be used in those with history of bladder cancer. They await further data before a making a final judgment about its ongoing use [69].

### 27.3.5 Diet

#### 27.3.5.1 Fluid Intake

The Harvard Health Professionals Follow-up Study investigated fluid intake and risk of bladder cancer and showed that total daily fluid intake was inversely associated with the risk of bladder cancer; the multivariate RR was 0.51 (95 % CI 0.32–0.80) for the highest quintile of total daily fluid intake (>2531 mL per day) as compared with the lowest quintile (<1290 mL per day) [70]. The hypothesis being that increased fluid intake flushes out the bladder and reduces the time in which carcinogens are in contact with the bladder urothelium. However, these findings have not been reproduced in other studies [71, 72].

#### 27.3.5.2 Coffee/Tea/Sweeteners

Coffee consumption and risk of bladder cancer has been extensively studied, although results have been mixed. There is debate that the observed positive association may be largely due to potential confounding caused by smoking [73]. However, studies restricted to non-smokers show a small excess risk remains, although this was only observed in those drinking very high levels of coffee (over ten cups/day) [74]. Tea drinking has also been examined but little evidence of a positive or inverse association has been found [75]. Artificial sweeteners have also been linked to bladder cancer, as they have been shown to cause bladder tumors in rats; however the consensus is that no increased risk has been observed in humans [76].

#### 27.3.5.3 Chlorination

Chlorination is the process by which water is decontaminated and the relationship between the consumption of chlorinated water and bladder cancer has been studied. A meta-analysis including six case-control and two cohort studies concluded that a small significant increased risk was seen in men but not in women, with risk increasing over long durations of exposure [77]. The hypothesis is that chlorination produces trihalomethanes as a by-product and that these compounds can have adverse health consequences. Although the risks are small, as the vast majority of the world's population are exposed this could potentially be a clinically important association.

#### 27.3.5.4 Arsenic

In some areas, including Taiwan and Chile, very high levels of arsenic are found in the natural water supply. As a result of this geographical localization most of the epidemiological studies examining the relationship between arsenic exposure and bladder cancer have focused in these areas and many have found an increased risk. Some studies have shown relatives risks as high as 6 for men and 13 for women associated with arsenic in drinking water up to levels as high as 580 μg/L compared to normal levels of exposure [78]. These risks have been shown to reduce following measures to decrease the level of arsenic exposure in these populations [79]. The underlying mechanism is not fully understood but arsenic is known to interfere with DNA repair, leading indirectly to DNA damage and carcinogenesis [78].

### 27.3.6 Additional Risk Factors

Several studies have investigated a range of epidemiological factors, including alcohol [80], physical activity [81], obesity [81] and hormonal factors [82]. In general, the results of prospective studies suggest no association between these factors and the risk of bladder

## 27.4 Summary

Epidemiologically bladder cancer shows geographic, sex, and age-specific variations in incidence. A number of lifestyle and environmental factors have been identified, the most important of which is smoking. Bladder cancer is a clinically significant disease, causing considerable morbidity and mortality worldwide. Treatment options though varied and effective in early stage disease are limited for advanced and metastatic disease. The emergence of new immunotherapies, however, provides hope of improved outcomes for these patients.

## References

1. Torre LA, Bray F, Siegel RL, Ferlay J, Lortet-Tieulent J, Jemal A. Global cancer statistics, 2012. CA: A Cancer J Clin. 2015;65(2):87–108.
2. Khadra MH, Pickard RS, Charlton M, Powell PH, Neal DE. A prospective analysis of 1,930 patients with hematuria to evaluate current diagnostic practice. J Urol. 2000;163(2):524–7.
3. Mariani AJ, Mariani MC, Macchioni C, Stams UK, Hariharan A, Moriera A. The significance of adult hematuria: 1,000 hematuria evaluations including a risk-benefit and cost-effectiveness analysis. J Urol. 1989;141(2):350–5.
4. Bellmunt J, Orsola A, Leow JJ, Wiegel T, De Santis M, Horwich A. Bladder cancer: ESMO practice guidelines for diagnosis, treatment and follow-up. Ann Oncol Official J Eur Soc Med Oncol/ESMO. 2014;25 Suppl 3:iii40–8.
5. Cheng L, Montironi R, Davidson DD, Lopez-Beltran A. Staging and reporting of urothelial carcinoma of the urinary bladder. Mod Pathol Official J US Can Acad Pathol, Inc. 2009;22 Suppl 2:S70–95.
6. Babjuk M, Burger M, Zigeuner R, Shariat SF, van Rhijn BW, Comperat E, et al. EAU guidelines on non-muscle-invasive urothelial carcinoma of the bladder: update 2013. Eur Urol. 2013;64(4):639–53.
7. Association AU. Guideline for the management of nonmuscle invasive bladder cancer: stages Ta, T1 and Tis: An Update 2014.
8. Horstmann M, Banek S, Gakis G, Todenhofer T, Aufderklamm S, Hennenlotter J, et al. Prospective evaluation of fluorescence-guided cystoscopy to detect bladder cancer in a high-risk population: results from the UroScreen-study. SpringerPlus. 2014;3:24.
9. O'Brien T, Ray E, Chatterton K, Khan MS, Chandra A, Thomas K. Prospective randomized trial of hexylaminolevulinate photodynamic-assisted transurethral resection of bladder tumour (TURBT) plus single-shot intravesical mitomycin C vs conventional white-light TURBT plus mitomycin C in newly presenting non-muscle-invasive bladder cancer. BJU Int. 2013;112(8):1096–104.
10. Burger M, Grossman HB, Droller M, Schmidbauer J, Hermann G, Dragoescu O, et al. Photodynamic diagnosis of non-muscle-invasive bladder cancer with

hexaminolevulinate cystoscopy: a meta-analysis of detection and recurrence based on raw data. Eur Urol. 2013;64(5):846–54.
11. Grossman HB, Stenzl A, Fradet Y, Mynderse LA, Kriegmair M, Witjes JA, et al. Long-term decrease in bladder cancer recurrence with hexaminolevulinate enabled fluorescence cystoscopy. J Urol. 2012;188 (1):58–62.
12. Mowatt G, N'Dow J, Vale L, Nabi G, Boachie C, Cook JA, et al. Photodynamic diagnosis of bladder cancer compared with white light cystoscopy: systematic review and meta-analysis. Int J Technol Assess Health Care. 2011;27(1):3–10.
13. Sylvester RJ, van der Meijden AP, Oosterlinck W, Witjes JA, Bouffioux C, Denis L, et al. Predicting recurrence and progression in individual patients with stage Ta T1 bladder cancer using EORTC risk tables: a combined analysis of 2596 patients from seven EORTC trials. Eur Urol. 2006;49(3):466–5; (discussion 75–7).
14. Shelley MD, Kynaston H, Court J, Wilt TJ, Coles B, Burgon K, et al. A systematic review of intravesical bacillus Calmette-Guerin plus transurethral resection vs transurethral resection alone in Ta and T1 bladder cancer. BJU Int. 2001;88(3):209–16.
15. Shelley MD, Wilt TJ, Court J, Coles B, Kynaston H, Mason MD. Intravesical bacillus Calmette-Guerin is superior to mitomycin C in reducing tumour recurrence in high-risk superficial bladder cancer: a meta-analysis of randomized trials. BJU Int. 2004;93(4):485–90.
16. Bohle A, Jocham D, Bock PR. Intravesical bacillus Calmette-Guerin versus mitomycin C for superficial bladder cancer: a formal meta-analysis of comparative studies on recurrence and toxicity. J Urol. 2003;169 (1):90–5.
17. Sylvester RJ, van der MA, Lamm DL. Intravesical bacillus Calmette-Guerin reduces the risk of progression in patients with superficial bladder cancer: a meta-analysis of the published results of randomized clinical trials. J Urol. 2002;168(5):1964–70.
18. Sylvester RJ, Oosterlinck W, van der Meijden AP. A single immediate postoperative instillation of chemotherapy decreases the risk of recurrence in patients with stage Ta T1 bladder cancer: a meta-analysis of published results of randomized clinical trials. J Urol. 2004;171(6 Pt 1):2186–90, (quiz 435).
19. Sylvester RJ, Oosterlinck W, Witjes JA. The schedule and duration of intravesical chemotherapy in patients with non-muscle-invasive bladder cancer: a systematic review of the published results of randomized clinical trials. Eur Urol. 2008;53(4):709–19.
20. Abol-Enein H, Tilki D, Mosbah A, El-Baz M, Shokeir A, Nabeeh A, et al. Does the extent of lymphadenectomy in radical cystectomy for bladder cancer influence disease-free survival? A prospective single-center study. Eur Urol. 2011;60(3):572–7.
21. Eapen R, Liew MS, Tafreshi A, Papa N, Lawrentschuk N, Azad A, et al. Lymphadenectomy with radical cystectomy at an Australian tertiary referral institution: time trends and impact on oncological outcomes. ANZ J Surg. 2014.
22. Tilki D, Brausi M, Colombo R, Evans CP, Fradet Y, Fritsche HM, et al. Lymphadenectomy for bladder cancer at the time of radical cystectomy. Eur Urol. 2013;64(2):266–76.
23. May M, Helke C, Nitzke T, Vogler H, Hoschke B. Survival rates after radical cystectomy according to tumor stage of bladder carcinoma at first presentation. Urol Int. 2004;72(2):103–11.
24. Collaboration ABCM-a. Neoadjuvant chemotherapy in invasive bladder cancer: a systematic review and meta-analysis. Lancet. 2003;361(9373):1927–34.
25. Winquist E, Kirchner TS, Segal R, Chin J, Lukka H. Neoadjuvant chemotherapy for transitional cell carcinoma of the bladder: a systematic review and meta-analysis. J Urol. 2004;171(2 Pt 1):561–9.
26. Collaboration ABCM. Neoadjuvant chemotherapy in invasive bladder cancer: update of a systematic review and meta-analysis of individual patient data advanced bladder cancer (ABC) meta-analysis collaboration. Eur Urol. 2005;48(2):202–5; (discussion 5–6).
27. Coppin CM, Gospodarowicz MK, James K, Tannock IF, Zee B, Carson J, et al. Improved local control of invasive bladder cancer by concurrent cisplatin and preoperative or definitive radiation. The National Cancer Institute of Canada Clinical Trials Group. J Clin Oncol Official J Am Soc Clin Oncol. 1996;14(11):2901–7.
28. Rodel C, Grabenbauer GG, Kuhn R, Papadopoulos T, Dunst J, Meyer M, et al. Combined-modality treatment and selective organ preservation in invasive bladder cancer: long-term results. J Clin Oncol Official J Am Soc Clin Oncol. 2002;20(14):3061–71.
29. Shipley WU, Zietman AL, Kaufman DS, Coen JJ, Sandler HM. Selective bladder preservation by trimodality therapy for patients with muscularis propria-invasive bladder cancer and who are cystectomy candidates — the Massachusetts General Hospital and Radiation Therapy Oncology Group experiences. Semin Radiat Oncol. 2005;15(1):36–41.
30. von der Maase H, Hansen SW, Roberts JT, Dogliotti L, Oliver T, Moore MJ, et al. Gemcitabine and cisplatin versus methotrexate, vinblastine, doxorubicin, and cisplatin in advanced or metastatic bladder cancer: results of a large, randomized, multinational, multicenter, phase III study. J Clin Oncol Official J Am Soc Clin Oncol. 2000;18 (17):3068–77.
31. Sternberg CN, Yagoda A, Scher HI, Watson RC, Geller N, Herr HW, et al. Methotrexate, vinblastine, doxorubicin, and cisplatin for advanced transitional cell carcinoma of the urothelium. Efficacy and patterns of response and relapse. Cancer. 1989;64 (12):2448–58.
32. Dash A, Galsky MD, Vickers AJ, Serio AM, Koppie TM, Dalbagni G, et al. Impact of renal impairment on eligibility for adjuvant cisplatin-based

chemotherapy in patients with urothelial carcinoma of the bladder. Cancer. 2006;107(3):506–13.
33. De Santis M, Bellmunt J, Mead G, Kerst JM, Leahy M, Maroto P, et al. Randomized phase II/III trial assessing gemcitabine/carboplatin and methotrexate/carboplatin/vinblastine in patients with advanced urothelial cancer who are unfit for cisplatin-based chemotherapy: EORTC study 30986. J Clin Oncol Official J Am Soc Clin Oncol. 2012;30(2):191–9.
34. Bellmunt J, Theodore C, Demkov T, Komyakov B, Sengelov L, Daugaard G, et al. Phase III trial of vinflunine plus best supportive care compared with best supportive care alone after a platinum-containing regimen in patients with advanced transitional cell carcinoma of the urothelial tract. J Clin Oncol: Official J Am Soc Clin Oncol. 2009;27(27):4454–61.
35. Powles T, Eder JP, Fine GD, Braiteh FS, Loriot Y, Cruz C, et al. MPDL3280A (anti-PD-L1) treatment leads to clinical activity in metastatic bladder cancer. Nature. 2014;515(7528):558–62.
36. Plimack E, Gupta S, Bellmunt J, Berger R, Montgomery B, Gonzalez E.J., Pulini J, Dolled-Filhart M, Emancipator K, Pathiraja K, Gause C, Perini R, Cheng J, O'Donnell PH. A phase 1b study of pembrolizumab (pembro; mk-3475) in patients with advanced urothelial tract cancer. Ann Oncol 2014;25(suppl 4).
37. Witjes JA, Comperat E, Cowan NC, De Santis M, Gakis G, Lebret T, et al. EAU guidelines on muscle-invasive and metastatic bladder cancer: summary of the 2013 guidelines. Eur Urol. 2014;65 (4):778–92.
38. Rabow MW, Benner C, Shepard N, Meng MV. Concurrent urologic and palliative care after cystectomy for treatment of muscle-invasive bladder cancer. Urol Oncol. 2015;33(6):267.e23–9.
39. Hemminki K, Li X. Cancer risks in Nordic immigrants and their offspring in Sweden. Eur J Cancer (Oxford, England: 1990). 2002;38(18):2428–34.
40. Jemal A, Murray T, Ward E, Samuels A, Tiwari RC, Ghafoor A, et al. Cancer statistics, 2005. CA: A Cancer J Clin. 2005;55(1):10–30.
41. Abdollah F, Gandaglia G, Thuret R, Schmitges J, Tian Z, Jeldres C, et al. Incidence, survival and mortality rates of stage-specific bladder cancer in United States: a trend analysis. Cancer Epidemiol. 2013;37(3):219–25.
42. Hecht SS. Cigarette smoking: cancer risks, carcinogens, and mechanisms. Langenbeck's archives of surgery/Deutsche Gesellschaft fur Chirurgie. 2006;391(6):603–13.
43. Saletta F, Matullo G, Manuguerra M, Arena S, Bardelli A, Vineis P. Exposure to the tobacco smoke constituent 4-aminobiphenyl induces chromosomal instability in human cancer cells. Cancer Res. 2007;67(15):7088–94.
44. Malaveille C, Vineis P, Esteve J, Ohshima H, Brun G, Hautefeuille A, et al. Levels of mutagens in the urine of smokers of black and blond tobacco correlate with their risk of bladder cancer. Carcinogenesis. 1989;10 (3):577–86.
45. Freedman ND, Silverman DT, Hollenbeck AR, Schatzkin A, Abnet CC. Association between smoking and risk of bladder cancer among men and women. JAMA. 2011;306(7):737–45.
46. Brennan P, Bogillot O, Cordier S, Greiser E, Schill W, Vineis P, et al. Cigarette smoking and bladder cancer in men: a pooled analysis of 11 case-control studies. Int J Cancer (Journal international du cancer). 2000;86(2):289–94.
47. Shapiro JA, Jacobs EJ, Thun MJ. Cigar smoking in men and risk of death from tobacco-related cancers. J Natl Cancer Inst. 2000;92(4):333–7.
48. Jiang X, Yuan JM, Skipper PL, Tannenbaum SR, Yu MC. Environmental tobacco smoke and bladder cancer risk in never smokers of Los Angeles County. Cancer Res. 2007;67(15):7540–5.
49. Van Hemelrijck MJ, Michaud DS, Connolly GN, Kabir Z. Secondhand smoking, 4-aminobiphenyl, and bladder cancer: two meta-analyses. Cancer epidemiology, biomarkers & prevention: a publication of the American Association for Cancer Research, cosponsored by the American Society of Preventive Oncology. 2009;18(4):1312–20.
50. Case RA, Hosker ME, Mc DD, Pearson JT. Tumours of the urinary bladder in workmen engaged in the manufacture and use of certain dyestuff intermediates in the British chemical industry. I. The role of aniline, benzidine, alpha-naphthylamine, and beta-naphthylamine. Br J Ind Med. 1954;11(2):75–104.
51. Brown T, Slack R, Rushton L. Occupational cancer in Britain. Urinary tract cancers: bladder and kidney. Br J Cancer. 2012;107(Suppl 1):S76–84.
52. Boffetta P, Silverman DT. A meta-analysis of bladder cancer and diesel exhaust exposure. Epidemiology (Cambridge, Mass). 2001;12(1):125–30.
53. Zeegers MP, Swaen GM, Kant I, Goldbohm RA, van den Brandt PA. Occupational risk factors for male bladder cancer: results from a population based case cohort study in the Netherlands. Occup Environ Med. 2001;58(9):590–6.
54. Czene K, Tiikkaja S, Hemminki K. Cancer risks in hairdressers: assessment of carcinogenicity of hair dyes and gels. Int J Cancer (Journal international du cancer). 2003;105(1):108–12.
55. Takkouche B, Etminan M, Montes-Martinez A. Personal use of hair dyes and risk of cancer: a meta-analysis. JAMA. 2005;293(20):2516–25.
56. Salem S, Mitchell RE, El-Alim El-Dorey A, Smith JA, Barocas DA. Successful control of schistosomiasis and the changing epidemiology of bladder cancer in Egypt. BJU Int. 2011;107(2):206–11.
57. Li N, Yang L, Zhang Y, Zhao P, Zheng T, Dai M. Human papillomavirus infection and bladder cancer risk: a meta-analysis. J Infect Dis. 2011;204(2):217–23.
58. Groah SL, Weitzenkamp DA, Lammertse DP, Whiteneck GG, Lezotte DC, Hamman RF. Excess risk of bladder cancer in spinal cord injury: evidence for an association between indwelling catheter use and bladder cancer. Arch Phys Med Rehabil. 2002;83 (3):346–51.

59. Leibovici D, Grossman HB, Dinney CP, Millikan RE, Lerner S, Wang Y, et al. Polymorphisms in inflammation genes and bladder cancer: from initiation to recurrence, progression, and survival. J Clin Oncol Official J Am Soc Clin Oncol. 2005;23(24):5746–56.
60. Talar-Williams C, Hijazi YM, Walther MM, Linehan WM, Hallahan CW, Lubensky I, et al. Cyclophosphamide-induced cystitis and bladder cancer in patients with Wegener granulomatosis. Ann Intern Med. 1996;124(5):477–84.
61. Travis LB, Curtis RE, Glimelius B, Holowaty EJ, Van Leeuwen FE, Lynch CF, et al. Bladder and kidney cancer following cyclophosphamide therapy for non-Hodgkin's lymphoma. J Natl Cancer Inst. 1995;87(7):524–30.
62. Korkmaz A, Topal T, Oter S. Pathophysiological aspects of cyclophosphamide and ifosfamide induced hemorrhagic cystitis; implication of reactive oxygen and nitrogen species as well as PARP activation. Cell Biol Toxicol. 2007;23(5):303–12.
63. Piper JM, Tonascia J, Matanoski GM. Heavy phenacetin use and bladder cancer in women aged 20 to 49 years. N Engl J Med. 1985;313(5):292–5.
64. Fortuny J, Kogevinas M, Zens MS, Schned A, Andrew AS, Heaney J, et al. Analgesic and anti-inflammatory drug use and risk of bladder cancer: a population based case control study. BMC Urol. 2007;7:13.
65. Olsen JH, Boice JD Jr, Jensen JP, Fraumeni JF Jr. Cancer among epileptic patients exposed to anticonvulsant drugs. J Natl Cancer Inst. 1989;81(10):803–8.
66. Habel LA, Bull SA, Friedman GD. Barbiturates, smoking, and bladder cancer risk. Cancer epidemiology, biomarkers & prevention: a publication of the American Association for Cancer Research, cosponsored by the American Society of Preventive Oncology. 1998;7(11):1049–50.
67. Bosetti C, Rosato V, Buniato D, Zambon A, La Vecchia C, Corrao G. Cancer risk for patients using thiazolidinediones for type 2 diabetes: a meta-analysis. Oncologist. 2013;18(2):148–56.
68. Colmers IN, Bowker SL, Majumdar SR, Johnson JA. Use of thiazolidinediones and the risk of bladder cancer among people with type 2 diabetes: a meta-analysis. CMAJ: Can Med Assoc J (journal de l'Association medicale canadienne). 2012;184(12):E675–83.
69. Agency EM. European Medicines Agency recommends new contra-indications and warnings for pioglitazone to reduce small increased risk of bladder cancer. Press Release 2011. http://www.ema.europa.eu/ema/index.jsp?curl=pages/news_and_events/news/2011/07/news_detail_001311.jsp&mid=WC0b01ac058004d5c1.
70. Michaud DS, Spiegelman D, Clinton SK, Rimm EB, Curhan GC, Willett WC, et al. Fluid intake and the risk of bladder cancer in men. N Engl J Med. 1999;340(18):1390–7.
71. Jensen OM, Wahrendorf J, Knudsen JB, Sorensen BL. The Copenhagen case-control study of bladder cancer. II. Effect of coffee and other beverages. Int J Cancer (Journal international du cancer). 1986;37(5):651–7.
72. Geoffroy-Perez B, Cordier S. Fluid consumption and the risk of bladder cancer: results of a multicenter case-control study. Int J Cancer (Journal international du cancer). 2001;93(6):880–7.
73. Pelucchi C, La Vecchia C. Alcohol, coffee, and bladder cancer risk: a review of epidemiological studies. Eur J Cancer Prev Official J Eur Cancer Prev Organ (ECP). 2009;18(1):62–8.
74. Sala M, Cordier S, Chang-Claude J, Donato F, Escolar-Pujolar A, Fernandez F, et al. Coffee consumption and bladder cancer in nonsmokers: a pooled analysis of case-control studies in European countries. Cancer Causes & Control: CCC. 2000;11(10):925–31.
75. Zeegers MP, Tan FE, Goldbohm RA, van den Brandt PA. Are coffee and tea consumption associated with urinary tract cancer risk? A systematic review and meta-analysis. Int J Epidemiol. 2001;30(2):353–62.
76. Elcock M, Morgan RW. Update on artificial sweeteners and bladder cancer. Regul Toxicol Pharmacol: RTP. 1993;17(1):35–43.
77. Villanueva CM, Fernandez F, Malats N, Grimalt JO, Kogevinas M. Meta-analysis of studies on individual consumption of chlorinated drinking water and bladder cancer. J Epidemiol Community Health. 2003;57(3):166–73.
78. Marshall G, Ferreccio C, Yuan Y, Bates MN, Steinmaus C, Selvin S, et al. Fifty-year study of lung and bladder cancer mortality in Chile related to arsenic in drinking water. J Natl Cancer Inst. 2007;99(12):920–8.
79. Tsai SM, Wang TN, Ko YC. Cancer mortality trends in a blackfoot disease endemic community of Taiwan following water source replacement. J Toxicol Environ Health Part A. 1998;55(6):389–404.
80. Pelucchi C, Galeone C, Tramacere I, Bagnardi V, Negri E, Islami F, et al. Alcohol drinking and bladder cancer risk: a meta-analysis. Ann Oncol Official J Eur Soc Med Oncol/ESMO. 2012;23(6):1586–93.
81. Noguchi JL, Liss MA, Parsons JK. Obesity, Physical Activity and Bladder Cancer. Current Urol Rep. 2015;16(10):74.
82. Lucca I, Klatte T, Fajkovic H, de Martino M, Shariat SF. Gender differences in incidence and outcomes of urothelial and kidney cancer. Nat Rev Urol. 2015;12(10):585–92.

# 28 Bladder Cancer

Elizabeth L. Kehr and Justine A. Barletta

## 28.1 Introduction

In this chapter we will focus on the histopathology and molecular alterations of bladder cancer. As background, we begin with an introduction to bladder embryology, normal anatomy, and normal histology of the bladder. Neoplastic processes of the bladder will subsequently be outlined with an emphasis on papillary urothelial carcinoma. Immunohistochemistry of normal and neoplastic urothelium will be discussed, and the chapter will conclude with an appraisal of our current understanding of the molecular alterations of urothelial carcinoma.

## 28.2 Embryology and Development of the Urinary Bladder

The urinary bladder forms early in gestation. The allantois, the connection between the embryo and the yolk sac, plays a key role in bladder embryogenesis. As the embryo develops from a flat disc to a folded structure, the allantois is partially incorporated into the body of the embryo where it forms the endodermal-lined cloaca. During the fourth to seventh week of development, the cloaca divides into the urogenital sinus anteriorly and the anal canal posteriorly. The urogenital sinus is further divided into three components: the bladder, the membranous urethra and prostate, and the phallus. The majority of the bladder epithelium is derived from the cloacal endoderm. However, the trigone of the bladder develops from the dilatation, fusion and incorporation of the mesonephric ducts into the urogenital sinus. The posterior wall, bladder dome, and portions of the lateral walls arise from mesenchyme surrounding the urogenital sinus when the allantois divides into the urogenital sinus and the rectum. The anterior wall and portions of the lateral walls develop in conjunction with the closure of the infraumbilical abdominal wall. The point of connection between the allantois and the bladder, called the urachus, is at the anterior superior aspect of the bladder. The allantois involutes to become the median umbilical ligament in most adults [1, 2]. The ureteric buds form as caudal sprouts off the mesonephric ducts. The portion of the ureteric buds in contact with the metanephric blastema undergo successive rounds of branching morphogenesis to create the upper tract collecting system. The portion of the ureteric buds lying outside the metanephros differentiate into ureters. The upper and lower tract join when the ureters undergo transposition, moving their primary insertion sites from the mesonephric ducts to the urogenital sinus [3].

E.L. Kehr · J.A. Barletta (✉)
Brigham and Women's Hospital, Harvard Medical School, 75 Francis Street, Boston, MA 02115, USA
e-mail: jbarletta@partners.org


M. Loda et al. (eds.), *Pathology and Epidemiology of Cancer*,
DOI 10.1007/978-3-319-35153-7_28

## 28.3 Normal Anatomy of the Urinary Bladder

The bladder is a hollow, distensible organ that is located in the pelvis and acts as a receptacle for urine that is excreted by the kidneys and enters the bladder via the ureters. Gross photographs of cystectomy specimens are shown in Fig. 28.1. In both sexes, the bladder is bound anteriorly by the pubic symphysis and laterally by the internal obturator and levator ani muscles. The superior aspect, also referred to as the "dome," is lined by peritoneum in both sexes. This is the site of the embryologic urachus, remnants of which remain in many people. In children they can act as a source of infection, and in adults they can give rise to rare urachal adenocarcinomas. In men, the inferior aspect abuts the prostate, whereas in women, it abuts the muscles of the pelvic floor. The posterior bladder abuts the rectum and seminal vesicles in men and the cervix and vagina in women. The ureters enter the bladder through the internal ureteric orifices at the posterolateral walls to form the apices of the trigone, the triangle formed by the ureters and the internal urethra. The urethra is a sphincter-bound tube that drains the bladder.

The arterial blood supply of the bladder stems from the internal iliac arteries in the following manner: the inferior vesicle arteries arise directly from the internal iliac arteries and the superior vesicle arteries branch from the umbilical arteries. Smaller vessels branch from the uterine arteries in women and from obturator and internal pudendal arteries in both sexes. The blood from the bladder drains through the vesical venous plexus into the internal iliac veins.

Innervation of the bladder is predominantly autonomic [1, 4, 5], with both a parasympathetic and sympathetic nerve supply. The main bladder muscle (otherwise known as the detrusor muscle or muscularis propria) is dominated by parasympathetic activity stimulating motor activity and inhibiting the internal urethral sphincter, while sympathetic fibers are inhibitory to the detrusor and motor to the sphincter. Sympathetic nerves that innervate the bladder neck are also important in males, as they prevent the reflux of semen into the bladder during ejaculation. While the internal urethral sphincter (smooth muscle) is under autonomic nerve control, the external urethral sphincter (skeletal muscle) is under somatic nerve control by the pudendal nerve, and detects bladder distention via muscle tension and stretch receptors. Parasympathetic pre-ganglionic nerves are located in the sacral spinal cord and run via ventral roots to the parasympathetic ganglion next to the pelvic organs. In the bladder, the ganglia are located in the detrusor muscle and in the vesicle venous plexus. In contrast, the pre-ganglionic sympathetic

(a) (b) (c)

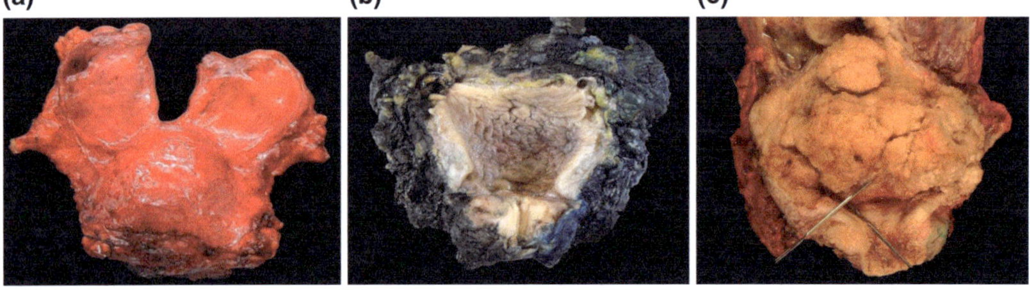

**Fig. 28.1** Gross photographs of cystectomy specimens. **a** Cystectomy specimen prior to opening of the bladder. The shiny surface represents serosa while the more ragged surfaces represent surgical margins. **b** A bivalved cystectomy specimen (coronally cut) reveals normal bladder mucosa. The blue and black surfaces were inked at the time of gross evaluation of the specimen. **c** A bivalved cystectomy specimen (coronally cut) reveals a large exophytic tumor involving the entire posterior wall of the bladder. The probes are passing through the ureteral orifices

nerves located in the thoraco-lumbar spine (T1-L2) connect to post-ganglionic fibers in the sympathetic trunk ganglion. The post-ganglionic sympathetic nerves run with the hypogastric nerve into the pelvis.

## 28.4 Normal Histology of the Urinary Bladder

Histologically, the bladder is comprised of four layers: urothelium, lamina propria, muscularis propria, and adventitia or serosa [5]. A photomicrograph of normal bladder histology is shown in Fig. 28.2. The urothelium is a specialized epithelium that is also referred to as "transitional cell epithelium." The normal urothelium averages about 5 layers in thickness, but varies between 2 and 7 layers depending on the degree of bladder distention. The layer closest to the lamina propria is the basal layer and is characterized by cuboidal cells with small round nuclei and relatively scant cytoplasm. The intermediate cell layers make up the majority of the thickness of the urothelium. These cells are polyhedral to columnar in shape with slightly more cytoplasm than the basal layer. The cells of the basal and intermediate layers have sparse desmosomes which facilitates their ability to flatten and slide over one another during bladder distension [1]. At the luminal surface of the urothelium is an umbrella cell layer comprised of large cells which often cover two or more intermediate cells. While umbrella cells can have large, irregular nuclei, the nuclear to cytoplasmic ratio of umbrella cells is low (averaging 1:4–5), which acts as a clue to their benignity [1]. Normal urothelium has an orderly, stratified appearance. A thin basement membrane separates the urothelium from the underlying lamina propria, which is a layer of connective tissue between the urothelium and muscularis propria. It is comprised of connective tissue containing a network of vessels, lymphatics, nerve endings and elastic fibers. Small vessels closely approximate the urothelial mucosa such that denudation or other mucosal disturbances often cause bleeding. Also within the lamina propria, there is a vestigial muscularis mucosae, analogous to the homonymous layer in the intestine, that is characterized by small discontinuous fascicles of smooth muscle. Adipose tissue may also be present within the lamina propria, a fact important to note for correct staging of urothelial tumors [1]. The thickness of the lamina propria changes with anatomic location and is notably narrow at the bladder neck. The muscularis propria is comprised of rounded bundles of smooth muscle with loosely anastomosing, ill-defined internal and external longitudinal layers and a prominent middle circular layer of muscle. It is distinguished from the muscularis mucosae by the bundling arrangement and the large caliber of the smooth muscle fascicles. In males, the muscularis propria at the bladder neck is continuous with the fibromuscular tissue of the prostate. In the region of the bladder neck, the muscularis propria gradually decreases in size and extends nearly to the mucosal surface. Accurate interpretation of the anatomic layers of the bladder

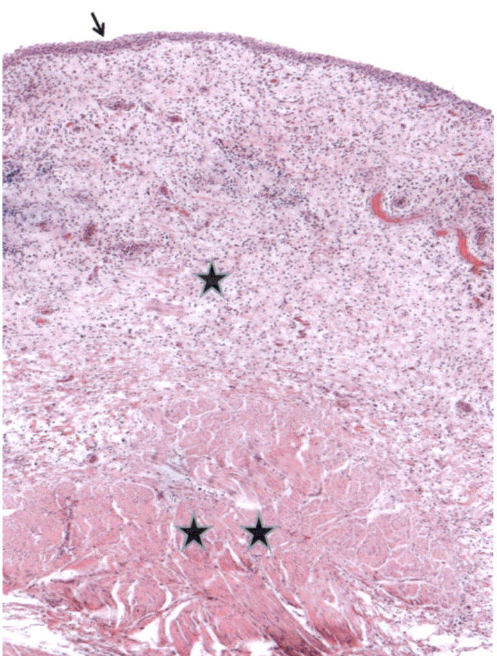

**Fig. 28.2** Normal bladder histology. The arrow indicates the urothelium. *One star* indicates lamina propria and *two stars* represent muscularis propria

wall is vital to correct staging of a patient with a primary bladder malignancy [1, 5].

## 28.5 Overview of Bladder Neoplasms

The urothelium gives rise to over 90 % of bladder tumors. Urothelial-derived bladder neoplasms can broadly be divided in two ways: papillary versus flat processes, and, for carcinoma, non-muscle invasive bladder cancer (NMIBC) and muscle invasive bladder cancer (MIBC). Papillary lesions include benign and malignant neoplastic processes, are generally easily identified at cystoscopy and histologically are characterized by urothelium lining lamina propria with fibrovascular cores. Carcinoma in situ (CIS) is the main flat neoplastic process. It can appear as a flat red lesion on cystoscopy or may be cystoscopically inapparent. NMIBC is defined as carcinoma that does not invade the muscularis propria (invasion of muscularis mucosae is allowed) and represents 75–85 % of bladder cancer [6]. NMIBC includes papillary urothelial carcinoma confined to the basement membrane and thus without invasion into lamina propria (70 % of NMIBC), papillary urothelial carcinoma with invasion into the lamina propria but without invasion of the muscularis propria (20 % of NMIBC), and CIS (10 % of NMIBC) [6]. NMIBC is often a multifocal disease and has a propensity for multiple recurrences. Overall, the prognosis of NMIBC is good, with a five-year survival rate approaching 90 % if treated with surgical resection and intravesical immunotherapy [7]. About 15 % of NMIBC progresses to MIBC, at which point the outcome is similar to that of tumors that presented initially as MIBC [8, 9]. The majority (85 %) of MIBC is diagnosed as muscle invasive at presentation [7]. MIBC predominantly arises from CIS, though roughly 15 % arises from papillary carcinoma [7]. MIBC is an aggressive disease, and at least 50 % of patients die from metastases within 2 years of diagnosis [7, 8].

## 28.6 Benign and Low Risk Papillary Lesions of the Bladder

The histologic subclassification of papillary lesions takes into account both the architecture and the cytologic features of the urothelium [5]. The pathologic diagnosis of the papillary lesion links histologic features with predictions of clinical behavior. Photomicrographs of benign and low risk papillary lesions of the bladder are shown in Fig. 28.3.

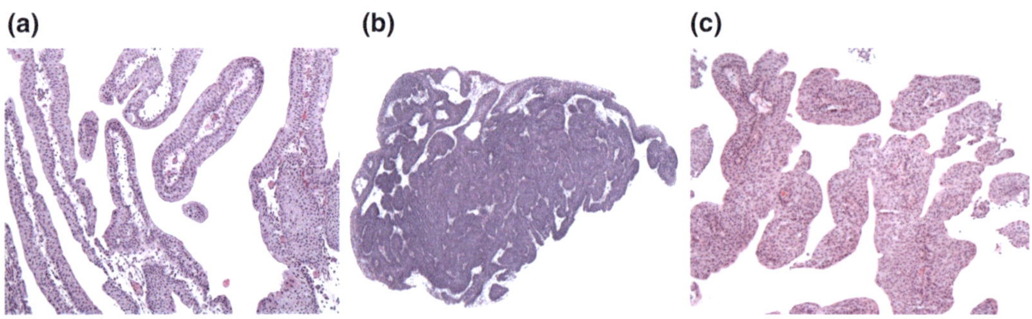


**Fig. 28.3** Benign and low risk papillary lesions of the bladder. **a** A papilloma demonstrating thin papillae with fibrovascular cores lined by urothelium of normal thickness and lacking cytologic atypia. **b** An inverted papilloma demonstrating endophytic growth and a characteristic complex architecture. **c** A papillary urothelial neoplasm of low malignant potential. The urothelium is slightly thicker than that of the papilloma, however, this tumor lacks significant cytologic atypia and mitoses are absent

### 28.6.1 Papillary Hyperplasia

Papillary hyperplasia (also termed urothelial proliferation of uncertain malignant potential) is a rare pathologic diagnosis. In one retrospective review an incidence of 0.4 % was reported [10]. Papillary hyperplasia lacks the complex architecture of a true papillary neoplasm. Instead, the urothelium is thicker than normal urothelium (i.e., there is an increase in the number of urothelial cell layers) and has an undulating, folded appearance. The hyperplastic urothelium is cytologically benign with a similar appearance to the adjacent normal urothelium [10]. Loss of heterozygosity of chromosome 9, one of the most common molecular changes in urothelial carcinoma, is seen in approximately half of cases of papillary hyperplasia [11]. Based on this finding, papillary hyperplasia has been considered a non-obligate precursor to papillary urothelial carcinoma.

### 28.6.2 Papilloma

Papillomas of the bladder are benign lesions with two distinct patterns including an exclusively exophytic variant ("papilloma") and an exclusively endophytic or inverted variant ("inverted papilloma"). Papillomas can occur in two settings: on their own (i.e., diagnosed in isolation) or in the context of a concurrent or prior urothelial neoplasm of higher malignant potential. The following discussion is limited to papillomas that are diagnosed in isolation. Papillomas are rare lesions that typically occur in younger patients. Histologically, they are composed of thin, finger-like papillae with central fibrovascular cores lined by urothelium of normal thickness lacking cytologic atypia or mitotic activity. Clinicopathologic studies of papillomas describe a benign clinical course. Roughly 10 % of papillomas recur, and only rare cases demonstrate malignant tumors on follow-up [12, 13]. Accurate histologic categorization of papillomas using strict diagnostic criteria is vital to correctly predict risk of recurrence and progression.

Inverted papillomas are uncommon urothelial neoplasms with a distinctive cystoscopic appearance and histomorphology [1]. Cystoscopically, inverted papillomas are solitary lesions with a raised or polypoid shape and a smooth surface. They are sharply demarcated and rarely exceed 3 cm in size, although cases as large as 8 cm have been described [1]. Histologically, inverted papillomas are endophytic lesions comprised of anastomosing nests of urothelium. Although the architecture appears complex, the urothelium remains well polarized (i.e,. normal stratification of urothelium is present) and while the cells may have a somewhat spindled appearance, they lack significant atypia or mitotic activity. Often the overlying urothelium is normal. In some cases the surrounding stroma may be fibrotic; however, inverted papillomas lack the desmoplastic stromal response (i.e., a loose fibroblastic proliferation similar in histologic appearance to an early scar) that is incited by invasive carcinoma. Inverted papillomas are benign lesions, and tumor recurrences are rare [14].

### 28.6.3 Papillary Urothelial Neoplasm of Low Malignant Potential

The category of papillary urothelial neoplasm of low malignant potential (PUNLMP) was created to describe papillary tumors that have abnormally thick urothelium but lack significant cytologic atypia [15]. Histologically, PUNLMPS have delicate fibrovascular stalks that are lined by a normal to slightly thickened urothelium (> 7 cells layers) with only slight cytologic atypia. PUNLMPs demonstrate evenly spaced urothelial cells with preserved polarity and an intact umbrella cell layer. The cells may be grooved, but lack prominent nucleoli or mitotic figures [1, 5]. PUNLMP is a fairly recently accepted diagnosis in urologic pathology [5]. Most tumors that are now diagnosed as PUNLMPs were previously diagnosed as low grade urothelial carcinomas. While the diagnosis of PUNLMP has been somewhat controversial due to inconsistent findings demonstrating a difference in clinical outcome parameters from low grade papillary urothelial carcinoma, the diagnosis has allowed pathologists to avoid labeling some patients with very low grade histologic tumors with cancer, while still rendering a diagnosis that prompts appropriate clinical follow-up. Overall, PUNLMPs are considered to have lower rates of disease recurrence and progression compared to low grade urothelial carcinoma. For example, in a study of 1,515 cases of NIBC, PUNLMPs represented 14 % of cases, had a recurrence rate of approximately 18 %, progressed in 2 % of cases, and had a mortality rate of zero. In contrast, low grade papillary urothelial carcinoma had a recurrence rate of 35 %, progressed in approximately 7 % of cases, and had a mortality rate of 2 % [16]. Given the recurrence potential of PUNLMPs, these patients are followed with regular bladder cystoscopies.

## 28.7 Urothelial Carcinoma

### 28.7.1 Papillary Urothelial Carcinoma

Papillary urothelial carcinoma accounts for 90 % of NMIBC [6]. These tumors are generally readily apparent cystoscopically. Papillary urothelial carcinoma is often multifocal. Histologically these tumors again have a papillary architecture, however, in contrast to papillomas and PUNLMPs, the urothelium lining the fibrovascular cores is generally markedly thickened (well over 7 cell layers thick) and the papillae often have a complex, branching architecture. Additionally, there is a variable degree of cytologic atypia and mitotic activity. Photomicrographs of papillary urothelial carcinoma are demonstrated in Fig. 28.4. The two main outcome parameters of clinical significance for NMIBC are disease recurrence and disease progression (with progression defined as deeper invasion on subsequent biopsy or resection). While clinical parameters such as number of foci of tumor, size of tumor, and prior recurrence rate are the most important factors influencing disease recurrence, pathologic findings including grade of tumor, presence of lamina propria invasion, and presence of CIS are the drivers of rate of disease progression for NMIBC [6, 17].

### 28.7.2 Histologic Grading

Papillary urothelial carcinoma is classified as high grade or low grade according to the 2004 World Health Organization/International Society of Urologic Pathology (WHO/ISUP) grading

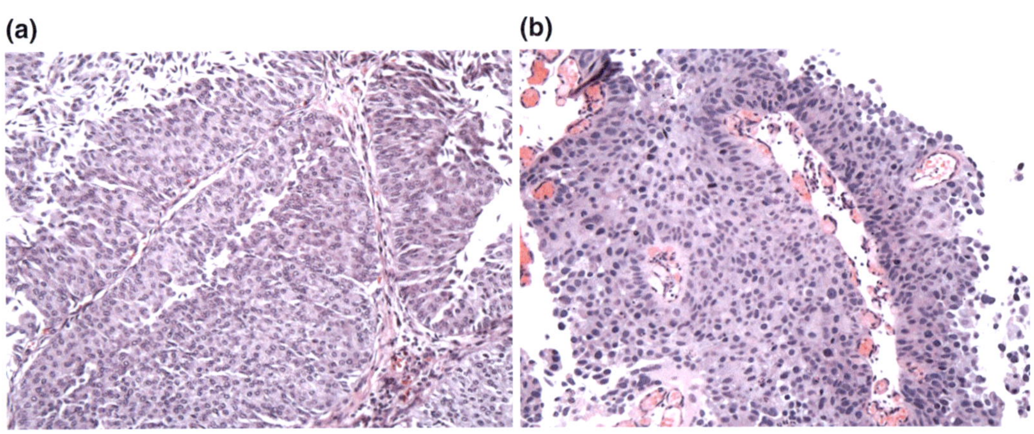


**Fig. 28.4** Papillary urothelial carcinoma. **a** Low grade papillary urothelial carcinoma. While the urothelium is clearly well over 7 cell layers, the polarity of the urothelium is maintained, the cells are uniform, the cytologic atypia is mild, and only scattered mitoses are present in the lower half of the urothelium. **b** High grade papillary urothelial carcinoma. The layers of urothelium are disorganized, the nuclei are markedly enlarged and demonstrate dark, smudgy chromatin. Mitoses and apoptotic cells are readily identified throughout all cell layers

system [15, 16]. This grading system is most useful for NMIBC since virtually all MIBC is high grade. The current system represents a revision of the 1973 WHO classification of urothelial tumors to a system identical to that proposed by ISUP. The 2004 WHO/ISUP system brought both important categorical and threshold changes to the 1973 WHO system. While the 1973 WHO system categorized urothelial carcinoma into three grades (grades 1–3), the 2004 WHO/ISUP system categorizes carcinomas as low or high grade and additionally added the diagnosis of PUNLMP (discussed above). Tumors that were previously grade 1 according to the 1973 WHO grading system are virtually all low grade under the 2004 WHO/ISUP system (rarely tumors that were previously considered grade 1 urothelial carcinomas are now diagnosed as PUNLMPs). Tumors that were previously grade 3 according to the 1973 WHO grading system are all high grade under the 2004 WHO/ISUP system. Finally, tumors that were previously grade 2 according to the 1973 WHO grading system now may be categorized as low or high grade under the 2004 WHO/ISUP grading system depending on the degree of architectural and cytologic atypia and the proliferative activity of the tumor [5]. The current system, reflecting a scheme proposed by Malmstrom et al. [18], describes a spectrum of cytologic and architectural features for high and low grade papillary urothelial carcinomas and PUNLMPs. Architectural features that are assessed include the complexity of the papillae and the overall organization of the cells (i.e., are the layers of urothelial cells polarized or is there a lack of organization of cell layers, resulting in a jumbled appearance). Cytologic features include nuclear size, shape, characterization of the chromatin, presence of nucleoli, and degree of variability of nuclear size and shape (i.e., nuclear pleomorphism). Proliferative indices include the number and location of mitoses and the presence of apoptotic cells or karyorrhectic debris. Tumors are graded based on the highest grade present within the tumor, even if it represents a small region of the tumor [5].

### 28.7.3 Low Grade Papillary Urothelial Carcinoma

The histomorphology of low grade papillary urothelial carcinoma is characterized by variably complex papillary structures lined by thickened urothelium. An overall orderly architecture is maintained. Tumor cells are uniform and show mild nuclear atypia with slight nuclear enlargement, vesicular chromatin, and small nucleoli. Mitotic figures and apoptotic cells can be seen but are infrequent and confined to the lower half of the urothelium [5]. Low grade papillary urothelial carcinoma represents approximately 40 % of cases of NMIBC [16]. The recurrence rate for low grade papillary urothelial carcinoma is approximately 35–50 % [15, 16]. In a study by Samaratunga et al., the rate of progression for low grade papillary tumors lacking lamina propria invasion was approximately 11.5 % in a 90-month follow-up period, with 10 % progressing to tumors with lamina propria invasion and 1.5 % progressing to MIBC [17]. The mortality rate for low grade papillary urothelial carcinoma is very low, at approximately 2.0 % [16].

### 28.7.4 High Grade Papillary Urothelial Carcinoma

High grade papillary urothelial carcinoma is characterized by moderate to marked architectural and cytologic atypia. The cells are disorganized with uneven spacing and a lack of stratification of cell layers. Frequently the cells are more discohesive than seen with low grade tumors. Cytologically, the nuclei demonstrate moderate to marked nucleomegaly, hyperchromasia, chromatin clumping, and often large nucleoli. Nuclear pleomorphism may also be marked. Mitotic activity and karyorrhectic debris is easily appreciated, and mitoses are seen at all levels of the urothelium [5]. High grade papillary urothelial carcinoma represents roughly 40 % of cases of NMIBC [16]. The recurrence rate for high grade papillary urothelial carcinoma is approximately 35–60 % [15, 16], and the rate of progression for high grade papillary tumors lacking lamina propria invasion is approximately 45 % in a 90-month follow-up period, with approximately one third progressing to tumors with lamina propria invasion and roughly 15 % progressing to MIBC [17]. The mortality rate for non-muscle invasive high grade papillary urothelial carcinoma is approximately 20 % [16].

### 28.7.5 Carcinoma In Situ

CIS is defined as a flat (generally 7 cell layers or less) neoplastic proliferation of urothelial cells without breach of the basement membrane that demonstrates severe cytologic atypia [5, 19]. It is frequently multifocal, can be coincident with high grade papillary urothelial carcinomas elsewhere in the bladder, and only rarely is seen in association with low grade papillary tumors [1]. The classic cystoscopic appearance of CIS is an area of erythematous mucosa, but it can also be cystoscopically inapparent. Histologically, CIS is characterized by nuclei that are roughly 5 times the size of a normal lymphocyte, with nuclear membrane irregularities, and often large, prominent nucleoli. Frequent mitotic figures including atypical forms can be seen. An additional finding supportive of CIS is neovascularization of the lamina propria directly beneath the neoplastic proliferation, a finding that explains the erythema seen on cystoscopy [5]. As a result of the discohesive nature of CIS, the affected area can be largely denuded with single neoplastic cells left clinging to the basement membrane. The primary histologic differential diagnosis is reactive atypia, a concerning differential given the vastly different clinical implications. Photomicrographs of CIS and reactive atypia are demonstrated in Fig. 28.5. Features of reactive atypia include moderate nucleomegaly with fine chromatin texture and a single central nucleolus. Overall, reactive proliferations tend to maintain their polarity, but often have increased mitotic activity —which can be equivalent to the level seen with CIS. CIS frequently harbors TP53 mutations, a frequent mutation in MIBC [20]. Approximately half of cases of CIS treated with resection or fulguration alone progress to MIBC within four years [21]. First-line therapy for CIS is intravesical Bacillus Calmette-Guerin (BCG), with radical cystectomy performed in the setting of BCG failure and recurrent CIS [15, 21].

### 28.7.6 Invasive Urothelial Carcinoma

The definition of invasive carcinoma is a malignant lesion that breaches the basement membrane

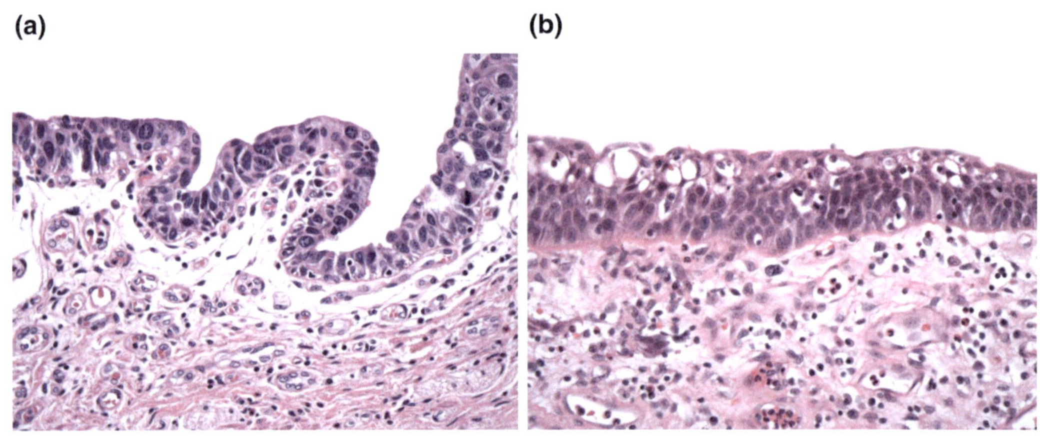


**Fig. 28.5** Carcinoma in situ (CIS) versus reactive change. **a** CIS demonstrating cells with large nuclei with course chromatin and scattered mitoses. **b** Reactive atypia. While the nuclei are large, the chromatin is delicate with variably present single, small nucleoli

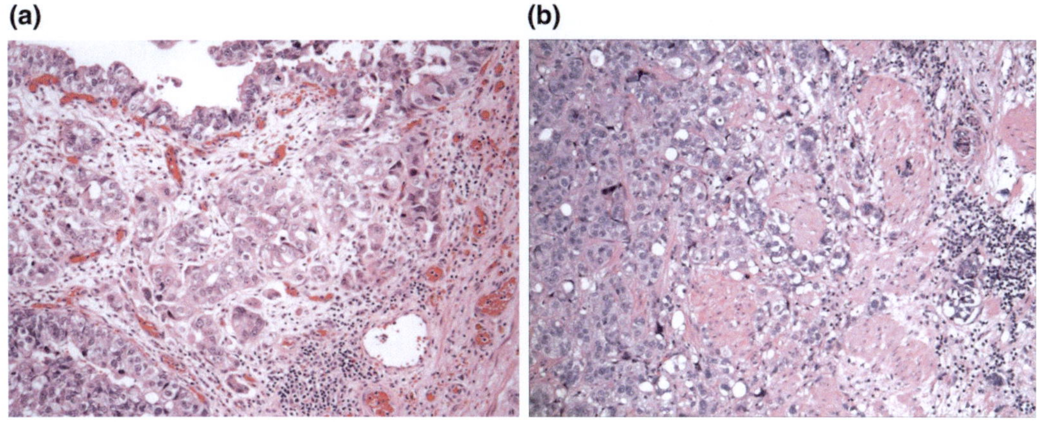


**Fig. 28.6** Urothelial carcinoma with invasion. **a** Tumor invading the lamina propria. Retraction artifact highlights small irregular nests and single cells present within the lamina propria. **b** Tumor invading the muscularis propria. Nests of tumor infiltrate the large caliber muscle bundle

of a mucosal surface. Invasive urothelial carcinomas present cystoscopically as polypoid, sessile, ulcerated, or infiltrative lesions [5]. The histology is variable with two distinct patterns. Most lamina propria invasive tumors are papillary, well differentiated, and have minimal invasion. Most muscle invasive lesions are nonpapillary, high grade, and have extensive invasion [4]. Photomicrographs of lamina propria invasion and muscularis propria invasion are demonstrated in Fig. 28.6.

In the case of early lamina propria invasion, which generally occurs at the base of the lesion but can also be present in the papillary stalk, invasion may be difficult to distinguish from tangential sectioning. Early invasion is characterized by small irregular nests or single cells infiltrating into the lamina propria. Histologic clues of lamina propria invasion include retraction around the infiltrating cells, paradoxical differentiation (e.g., invasive nests of cells with abundant eosinophilic cytoplasm), and a stromal reaction such as desmoplasia or inflammation [4]. The presence of lamina propria invasion is clinical significant. For example, while the difference in recurrence rate is not significantly different between noninvasive high grade papillary urothelial carcinoma and high grade papillary urothelial carcinoma with lamina propria invasion, the risk of progression to MIBC is significantly higher [6]. Current staging does not take into account the extent of lamina propria invasion, however, there are studies demonstrating different rates of disease progression to MIBC based on the extent of lamina propria invasion [22, 23]. For this reason, it is possible that in the future the extent of lamina propria invasion will become a staging parameter. Papillary tumors both with and without lamina propria invasion are usually managed with transurethral resection (i.e., local resection) with or without intravesical therapy.

Most MIBC demonstrates nests or sheets of high grade urothelial carcinoma unequivocally surrounding or obliterating bundles of muscularis propria. The cytology of infiltrating urothelial carcinoma is variable, but is most often characterized by polygonal tumor cells with moderate eosinophilic to amphophilic cytoplasm, large hyperchromatic nuclei, and numerous mitoses [4]. As previously indicated, MIBC is an aggressive disease with 50 % of patients dying of metastatic disease within two years of diagnosis [7, 8]. For MIBC, the stage (see below) is the most prognostically significant factor, however, other findings such as the presence of lymphovascular invasion and the margin status have also been shown to be prognostically significant and are therefore documented in the pathology report [24, 25]. MIBC is traditionally treated by cystectomy, however, in the last 10 years there has been a shift toward treatment with neoadjuvant chemotherapy. Grossman et al. [26] demonstrated that compared with radical cystectomy alone, the use of neoadjuvant chemotherapy (methotrexate, vinblastine, doxorubicin, and cisplatin) followed by radical cystectomy is associated with improved survival in patients with MIBC (median survival 77 compared to 46 months). Approximately 40 % of patients have a complete pathologic response, with no tumor found upon histologic evaluation of the cystectomy specimen. Patients with a complete response do significantly better than those with residual disease (85 % are alive at 5 years) [26].

### 28.7.7 Pathologic Staging

Pathologic staging of urothelial tumors links the pathologic findings to predictions of outcome and therefore is one of the principle variables in treatment decisions [5]. The American Joint Committee on Cancer/International Union Against Cancer (AJCC/UICC) TNM (tumor, node, metastasis) staging system is the most widely used staging system [24]. The T stage of urothelial carcinoma is based on the deepest extent of invasion of carcinoma into the bladder wall. T0 refers to a cystectomy specimen without evidence of residual carcinoma, in which case the entire lesion was presumably removed prior to cystectomy. Noninvasive disease (i.e., disease confined to the basement membrane) is designated as Ta if the lesion is papillary and Tis if the lesion is CIS. T1 tumors invade the lamina propria, but not the muscularis propria. T2 tumors invade the muscularis propria and are further subclassified as T2a, tumors involving the inner half of the muscularis propria, and T2b, tumors involving the outer half of the muscularis propria. T3 tumors invade the perivesical adipose tissue. T3 is also further subdivided with T3a defined as microscopic invasion of the perivesical adipose tissue and T3b defined as gross invasion of the perivesical adipose tissue. T4 tumors invade adjacent organs. N and M reflect lymph node and distant metastases, respectively. For tumors that were treated with neoadjuvant therapy a "y" will precede the TNM stage.

## 28.8 Histologic Variants of Urothelial Carcinomas and Other Malignancies Involving the Bladder

### 28.8.1 Histologic Variants

Urothelial carcinoma can express a wide range of morphologies. In the bladder, the term variant is used to denote microscopic forms of urothelial carcinoma that vary from the typical urothelial carcinoma that was described above [4, 27]. Approximately 25 % of bladder urothelial carcinomas have mixed histology with tumors showing areas of typical urothelial carcinoma and areas of variant morphology. Variant morphology is important to recognize since it can impact histologic interpretation, therapeutic approach and prognosis. The presence of any amount of variant morphology is reported. In general, variant expression tracts with high grade urothelial carcinoma [28].

While a complete discussion of all described histologic variants is outside the scope of this chapter, a few subtypes are described. Photomicrographs of some examples of histologic variants are shown in Fig. 28.7. Urothelial carcinomas with squamous and glandular differentiation represent the most common variants and are present in up to 60 % and 10 % of urothelial carcinomas, respectively. Squamous differentiation is defined as cells with keratin production and intercellular bridges. Tumors with glandular differentiation usually demonstrate small tubules composed of cuboidal cells, though occasionally urothelial carcinoma with glandular differentiation demonstrates enteric-like differentiation with tumor cells with cigar-shaped nuclei and associated necrosis. Rarely urothelial carcinomas with glandular differentiation have areas with signet ring cells or mucinous differentiation. While some studies have shown worse outcomes in these variants, this finding is not significant once the results are

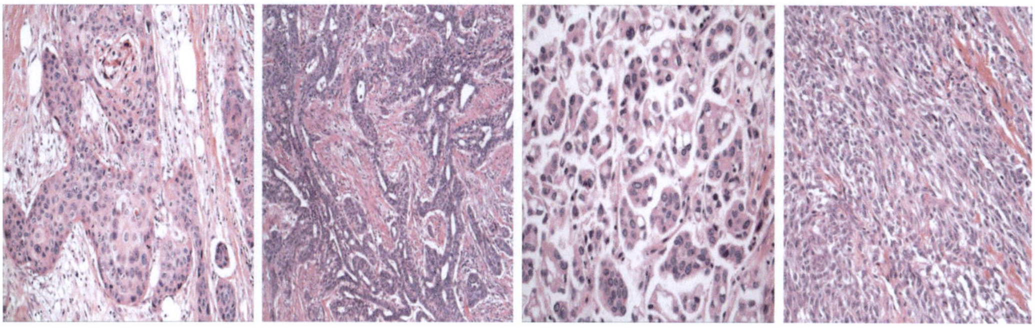


**Fig. 28.7** Urothelial carcinomas demonstrating variant morphology. **a** Squamous differentiation with keratinization. **b** Glandular differentiation with tubules lined by cuboidal cells. **c** Micropapillary differentiation. Small rings and nests of cells are present within lacunar spaces. **d** Sarcomatoid differentiation. The tumor has a spindled architecture

adjusted for tumor stage [28, 29]. The micropapillary variant of urothelial carcinoma comprises between 1 and 6 % of urothelial carcinomas, but is worth mentioning due to the worse disease-specific survival compared with patients with pure urothelial carcinoma [28, 29]. The micropapillary variant exhibits two distinct morphologic features. Surface tumors have slender fine papillary and filiform processes, while the invasive portion has tiny nests/rings of tumors cells that are contained within lacunae. Most of these tumors present as MIBC, and lymphovascular invasion and metastatic disease are both frequently associated with micropapillary urothelial carcinoma. Metastatic foci tend to retain a micropapillary architecture. The nested variant of urothelial carcinoma is described as "deceptively benign" with the nests of tumor closely resembling von Brunn nests (a benign, reactive process characterized by a proliferation of nests of urothelial cells beneath the surface urothelium). Awareness of this subtype can prevent the misclassification of the nested variant as a benign lesion. Finally, urothelial carcinoma with sarcomatoid differentiation represents < 1 % of bladder cancers. Studies show that this variant has a worse outcome than high grade urothelial carcinomas of the usual type [30]. Urothelial carcinoma with sarcomatoid differentiation represents a high grade urothelial carcinoma in which areas of the tumor demonstrate spindled growth characteristic of sarcomas. Some of these tumors even demonstrate heterologous differentiation with elements such as heterologous osteosarcoma present within the tumor [30].

### 28.8.2 Other Malignancies

Whereas 90 % of bladder cancer cases are urothelial carcinomas, the remaining 10 % represent a heterogeneous group of neoplasms [29]. Neoplasms considered separate from urothelial carcinoma include (but are not limited to) other epithelial neoplasms such as pure squamous cell carcinoma or pure adenocarcinoma, small cell carcinoma, secondary malignancies (i.e., tumors originating from other organs secondarily involving the bladder), and mesenchymal neoplasms [4]. Though a complete discussion of these tumor types is beyond the scope of this chapter, a few of the most common tumor types in adults are discussed below.

Pure squamous cell carcinoma of the bladder (i.e., a tumor that lacks a component of typical urothelial carcinoma) occurs in two forms: squamous cell carcinoma related to *Schistosoma hematobium* infection and non-schistosomal squamous cell carcinomas. Schistosomal squamous cell carcinoma (SCC) is common in Egypt and other countries harboring the trematode *S. hematobium*. Adult worms that reside in the veins draining pelvic organs release terminal spine eggs that penetrate the bladder and are excreted in the urine. Schistosome-related bladder cancer is a prevalent disease in Egypt, accounting for approximately one third of the total cancer incidence [31]. The 5 year survival rate is about 50 %. Non-schistosomal squamous cell carcinomas represent around 5 % of bladder cancers in western countries [32]. There is a strong association with long standing bladder irritation and the development of squamous cell carcinomas [32].

Pure adenocarcinomas of the bladder (i.e., a tumor that lacks a component of typical urothelial carcinoma) are rare, representing around 2.5 % of malignant bladder cancers. By definition, adenocarcinomas are comprised entirely of glandular elements and include both primary adenocarcinomas of the bladder and urachal carcinomas. Exclusion of secondary malignancies is always important when evaluating an adenocarcinoma involving the bladder. Most cases of primary bladder adenocarcinomas occur in the setting of intestinal metaplasia. Urachal carcinomas arise from urachal remnants and thus involve the muscular wall of the bladder dome. Classically, the histology of urachal adenocarcinomas is mucinous type, although signet ring, enteric and mixed types are also described [4].

Small cell carcinoma represents between 1 and 2 % of cases of bladder cancer. Recognition of small cell carcinoma is prognostically significant and affects management decisions such as chemotherapy regimen. Small cell carcinoma is defined as a malignant neuroendocrine neoplasm that mimics its pulmonary counterpart [4]. These carcinomas are comprised of invasive nests and sheets of uniform cells with scant cytoplasm, nuclear molding, finely stippled chromatin, and a very high proliferative rate. Small cell carcinoma is an aggressive disease, with over 90 % of patients presenting with muscle invasive disease and two-thirds developing systemic metastases. The reported 5 year survival ranges between 8 and 40 %, depending on treatment and disease stage [29].

Finally, secondary malignancies represent roughly 2 % of malignant tumors of the bladder and involve the bladder either by direct extension (accounting for roughly 70 % of cases) or as metastatic disease [33]. Tumors that most frequently involve the bladder by direct extension include prostate cancer, colorectal cancer, and Müllerian malignancies, while gastric cancer, malignant melanoma, lung cancer, and breast cancer are the most frequent malignancies to involve the bladder as metastatic disease [33]. Some of these tumors are clearly not bladder primaries based on the histomorphology, however, other cases may require immunohistochemical stains to determine site of primary (see below discussion of immunohistochemistry). Additionally, information regarding a history of malignancy as well as information regarding clinical and radiologic findings is essential and should always be provided to the pathologist. Photomicrographs of a case of malignant melanoma metastatic to the bladder mimicking a high grade papillary urothelial carcinoma is demonstrated in Fig. 28.8.

## 28.9 Immunohistochemistry

Urothelium is a stratified epithelium that demonstrates similarities with stratified squamous epithelium including overlapping immunohistochemical profiles. For example, high molecular weight keratins, such as CK5/6 and 34BE12, and p63 are frequently expressed in carcinomas of urothelial and squamous origin. Urothelial carcinoma is one of the few tumors that frequently co-express the cytokeratins CK7 and CK20, which can be helpful in excluding squamous cell carcinoma as well as carcinomas from other sites. GATA3 is a nuclear transcription factor expressed in 70–90 % of urothelial carcinomas. However, it is far from specific and should be interpreted in the context of a specific differential diagnosis. Non-urothelial tumors that are positive for GATA3 include squamous cell carcinoma of various sites (with percentage of cases positive varying by site), ductal and lobular breast carcinoma, pancreatic adenocarcinoma, paraganglioma, yolk sac tumor, and trophoblastic tumors. Of note, GATA3 is not expressed in prostate cancer; thus, GATA3 is often used to aid in the

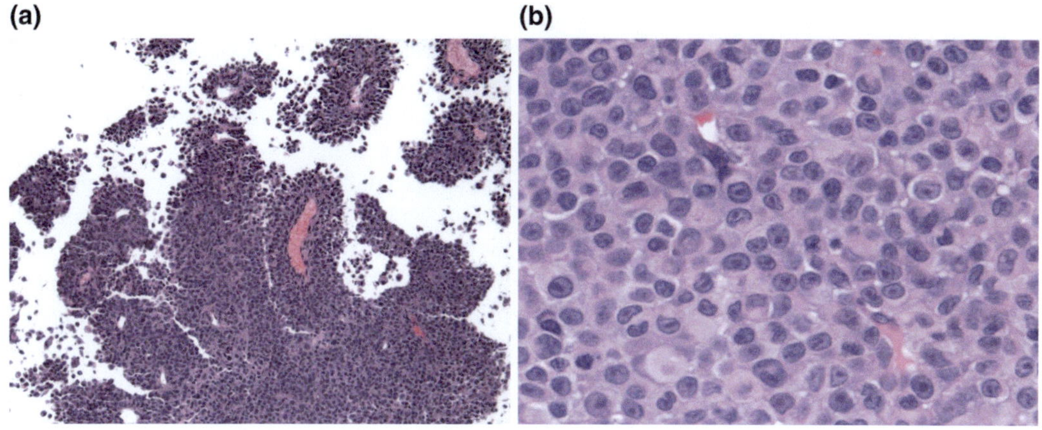

**Fig. 28.8** Malignant melanoma metastatic to the bladder mimicking a high grade papillary urothelial carcinoma. **a** At low power the tumor appears to have a papillary architecture. **b** At higher power the characteristic cherry-red nucleoli of melanoma are present. This case required immunohistochemical stains to confirm the diagnosis

differentiation of a poorly differentiated urothelial carcinoma and a poorly differentiated prostatic adenocarcinoma. Uroplakin III, a transmembrane protein expressed by urothelial lining cells is a more specific marker, but has a low sensitivity and tends to be expressed in lower grade tumors, which often do not require immunohistochemistry. When uroplakin III is positive, it should be membranous and plaque-like. Thrombomodulin is also expressed in 49–91 % of urothelial tumors but has a low specificity and is expressed in many other tumor types [34].

Immunohistochemistry may also be utilized to differentiate CIS from benign, reactive urothelium (see above discussion on CIS). Benign urothelium has the following pattern of antibody expression: CK20 positivity is limited to the umbrella cell layer; CD44 is limited to basal layers and p53 staining is weak and patchy. In contrast, CIS is characterized by aberrant expression of CK20 throughout all cell layers, CD44 is absent, and p53 is strong and diffusely positive in the neoplastic cells. Photomicrographs of CIS and p53 immunohistochemical staining are demonstrated in Fig. 28.9. Classically, urothelium with reactive atypia has CK20 expression limited to the umbrella cells, whereas CD44 expression is increased in all cell layers. P53 remains weak and patchy in reactive conditions [34]. While some cases show a definitive profile for benign/reactive

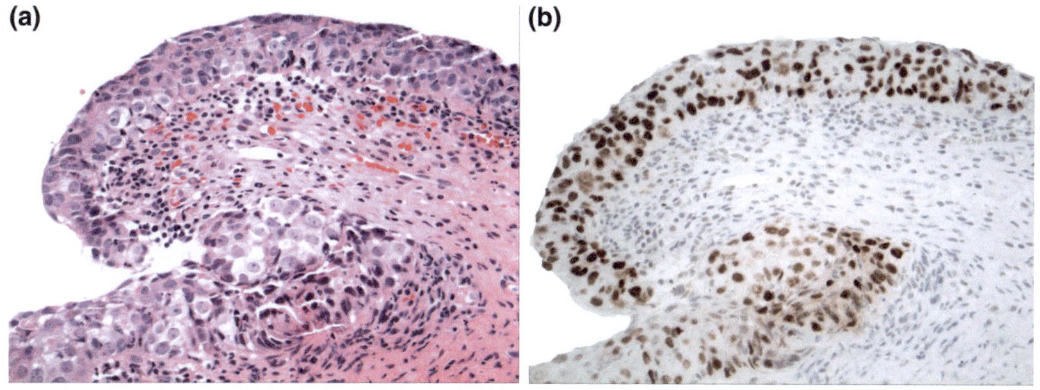

**Fig. 28.9** Carcinoma in situ (CIS) with a p53 immunohistochemical stain. **a** H&E photomicrograph of CIS. **b** p53 immunohistochemical stain of the same area of CIS demonstrating diffuse, strong positivity

or CIS, the staining pattern can be difficult to interpret, and thus the results of immunohistochemical staining must always be considered in conjunction with the histologic impression. Finally, at the present time, there is no immunohistochemical marker used in routine clinical practice to prognosticate or select therapy for typical urothelial carcinomas [34].

## 28.10 Molecular Genetics of Bladder Cancer

The molecular underpinnings of urothelial bladder cancer is proving to be a complex interplay of multiple genetic changes with different pathways involved in different subtypes of urothelial carcinoma.

### 28.10.1 Familial and Genetic Risk

Unlike other malignancies such as breast and colon, there are no highly penetrant single gene mutations that lead to a familial cancer syndrome of bladder cancer [35]. Instead, studies point towards mild increased risk in probands with family members with bladder cancer. Specifically, twin studies show that monozygotic twins are three times as likely to have bladder cancer compared to dizygotic twins, suggesting that there is a component of genetic risk [36]. In addition, population-based studies show odds ratios between 1.3 and 1.45 for the development of bladder cancer in first degree relatives of bladder cancer patients, compared to people with no family history of bladder cancer [37]. Overall, urothelial carcinoma is a complex disease that evolves in the context of multiple low-penetrant, low-risk genetic changes.

Single nucleotide polymorphisms (SNPs) are DNA sequence variations commonly occurring in the population in which a single nucleotide differs between members of a species or between chromosome pairs of an individual. Genome wide association studies (GWAS) scan the genome of a population to find SNPs statistically associated with specific diseases. Since 2009, several GWAS studies have identified sequence variances that confer susceptibility to urothelial carcinoma. The studies varied in ethnicity of populations studied, the numbers of cases and controls, and the SNP loci studied. Multiple major susceptibility loci for bladder cancer have been identified using this method. For example, genetic variation in 8q24.21, a non-genic location 30 KB upstream from the gene that encodes c-MYC has been reported [38]. Relative to non-carriers, heterozygous and homozygous carriers of the risk allele have an odd ratios of 1.22 and 1.49, respectively, for the development of urothelial carcinoma. c-MYC, a nuclear phosphoprotein involved in transcriptional regulation, has been implicated in many cancers, most notably Burkitt lymphoma. SNPs proximal to c-MYC (on 8q24), have also been associated with prostate cancer, colorectal cancer and breast cancer. Clustering of independent cancer-associated variants in this region suggests a common mechanism of susceptibility. A second described SNP for urothlial carcinoma resides at 3q28. Kiemeney and colleagues reported that the SNP on chromosome 3q28 is associated with an odds ratio of 1.19 for the development of urothelial carcinoma [38]. This locus is near TP63, a gene that encodes the p63 protein which is a cell cycle regulator. A third sequence variant that confers susceptibility to bladder cancer with an odds ratio of 1.15 is located at 8q24.311, an allele within exon 1 of PSCA gene. PSCA, a protein initially described in prostate cancer, is also overexpressed in bladder cancer [39]. Additional SNPs have been described in loci involving the TACC3-FGFR3 gene (chromosome 4p16.3); TERT-CLPTM1L (chromosome 5p15.33); cyclin E1 (chromosome 19q12); UDP-glucuronosyltransferase 1A gene (chromosome 2q37.1) and APOBEC3A chromosome 22q13.1 [38–41].

### 28.10.2 Molecular Genetics of Non-muscle Invasive Bladder Cancer

There are marked molecular differences between NMIBC and MIBC. Primary molecular alterations in NMIBC lead to constitutive activation of the MAP kinase and the PIK3 pathways by way of mutations in fibroblast growth factor receptor (FGFR), PIK3CA and HRAS [9]. Fibroblast growth factor receptors are a family of 4 tyrosine kinase receptors comprised of an extracellular domain, transmembrane domain, and a cytoplasmic domain. The most common FGFR mutations in urothelial carcinoma are missense mutations in FGFR3 that result in amino acid substitutions that occur in the extracellular and/or cytoplasmic domains and lead to ligand independent activation. Also described are mutations in the kinase domain leading to enhanced kinase activity. FGFR, once activated, acts as adaptor protein which initiates downstream activation of the MAP kinase pathway [42]. FGFR3 mutations are associated with low stage, low grade urothelial carcinomas. FGFR3 is mutated in 80 % of NMIBC and in less than 20 % of MIBC, FGFR mutations have not been described in CIS [7, 42, 43].

The phosphoinositide 3-kinases (PI3Ks) are key effector lipids that respond to cell stimulation by initiating cell growth, cell cycle entry, cell migration and cell survival. The PI3K family is composed of three members that differ in terms of substrate specificity, activation mechanisms, and expression patterns. Class 1A PI3K has been shown to be directly involved in carcinogenesis. The catalytic subunit of 1A PI3K is encoded by the gene PIK3CA, which has been found to be mutated in multiple tumor types including NMIBC [44]. PI3K, once activated, interacts with downstream enzymes that ultimately phosphorylate AKT. Phosporylated AKT targets a host of enzymes that affect cell growth, cell cycle entry and cell survival [45]. The prevalence of PIK3CA mutations decreases with increasing stage and grade [44]. PIK3CA mutations occur in 20 % of NMIBC and have a very low prevalence in MIBC [44]. They tend to occur in a subset of NMIBCs harboring FGFR3 mutations.

An activating HRAS mutation is the third genetic alteration frequently seen in NMIBC, and was the first human oncogene identified in urothelial carcinoma [9, 46]. The RAS superfamily of G proteins includes more than 100 members in humans, three of which (KRAS, NRAS, and HRAS) are commonly mutated in human cancers. All RAS proteins cycle between a GTP-bound active state and a GDP-bound inactive state. In the active state RAS binds and further stimulates effector proteins including MAP kinases. The most common HRAS mutations involve codons 12, 13, and 61 [7]. Transgenic mouse models with activated HRAS induce early onset urothelial proliferation leading to urothelial hyperplasia and low grade noninvasive papillary tumors [7]. Long term follow-up data indicate that the tumors remain low grade and do not invade. HRAS mutations occur in an estimated 10 % of bladder cancers [7, 46]. FGFR3 mutations and HRAS mutations are mutually exclusive, likely because they both signal through a common downstream pathway [9, 42]. While the changes described here are associated with NMIBC, approximately 15 % of NMIBC will progress to muscle invasive cancer. This progression is associated with the accumulation of abnormalities in cell cycle regulation, specifically the acquisition of mutations in TP53 and RB genes [9].

### 28.10.3 Molecular Genetics of Muscle Invasive Bladder Cancer

The molecular changes of MIBC include alterations in tumor suppressor genes involved in cell cycle control, such as TP53, RB, and PTEN [9, 47]. The tumor suppressor gene TP53 is located on chromosome 17p and encodes the protein p53. P53 inhibits cell cycle progression at the G1-S transition, partially mediated through transcriptional activation of p21. Most MIBCs exhibit loss of a single 17p allele with inactivation of the second gene [8]. A mouse model used the uroplakin promoter to drive expression of the SV40 large T antigen which leads to inactivation of p53 and RB [48]. Complete loss of p53 was associated with CIS. Progression to muscle invasion, however, required additional events such as loss of Rb or PTEN activity [49]. Many studies have shown that p53 pathway abnormalities contribute to poor clinical outcomes in bladder cancer [47].

The retinoblastoma gene (RB) is located on chromosome 13q14 and encodes the Rb protein, which regulates the G1-S transition via sequestration of the transcription factor E2F [8]. P53 and Rb both regulate the G1-S transition; however, they work through independent but interconnected pathways [50]. A 1998 study by Cote et al. showed increased recurrence and decreased survival rates in bladder cancer patients with abnormalities in RB. Furthermore, mutations in both RB and TP53 led to negative effects on recurrence and survival when compared to mutation of either one alone. TP53 and RB mutations seem to be early events in bladder cancer and set the stage for an accumulation of additional molecular abnormalities in MIBC.

Activation of the PTEN/PI3/AKT/mTOR pathway is another candidate driver of the muscle invasive phenotype [47]. Specifically, a mouse model with elimination of p53 and PTEN activity developed early CIS that uniformly progressed to muscle invasive tumors [47, 51]. Of note, a different group studied a mouse model with PTEN loss in a wild type p53 background without development of bladder cancer. This suggests that p53 loss may be an obligate cofactor for PTEN mutation dependent bladder cancer [47, 52].

## 28.10.4 Chromosomal Alterations

Chromosomal alterations are common in urothelial carcinomas and are characterized by aneuploidy, deletions, and amplifications that affect nearly all chromosomes [7, 53]. An early event in urothelial carcinogenesis appears to be loss of heterogeneity (LOH) of chromosome 9. Chromosome 9 deletions are found in normal appearing urothelium adjacent to tumor. Moreover, 9q abnormalities are more prevalent in low grade noninvasive papillary tumors; however, they are also found in MIBC and do not predict behavior. Finally, chromosome 9 deletions are seen both as the sole abnormality and concomitantly with complex changes in more aggressive tumors. These observations indicate that chromosome 9 abnormalities are an early and important change [53]. It is postulated that key tumor suppressors on chromosome 9 are lost and set the stage for future genetic changes. Several tumor suppressor genes are located on chromosome 9, including the tuberous sclerosis 1 complex (TSC1) at 9q34 [54]. TSC1 is well validated as a complex that negatively regulates the mTOR branch of the PI3K pathway. Other tumor suppressors on chromosome 9 include cyclin-dependent kinase inhibitor 2A (CDKN2A), cyclin-dependent kinase inhibitor 2B (CDKN2B), patched1 (PTCH1), and deleted in bladder cancer 1 (DBC1). Of note, no chromosome 9 genes show significant mutation frequency in exome sequencing studies [54]. It may be that epigenetic changes link chromosome 9 LOH and urothelial carcinoma. Loss of chromosome 8p is also described in 25–30 % of urothelial carcinomas and is primarily associated with high grade and late stage tumors. Finally, LOH of 15q occurs in 40 % of urothelial carcinomas and is another area of interest [7].

## 28.10.5 Epigenetic Regulation

Histone regulation, an epigenetic process that impacts chromatin remodeling, is commonly altered in bladder cancer. A 2011 study using whole exome genomic DNA sequencing found that tumors from 60 % of subjects with urothelial carcinoma harbored non-silent mutations in chromatin remodeling genes. Specifically, they found mutations in the following genes: UTX, a histone demethylase gene; CREBBP and EP300, two histone acetyltransferase genes; and the chromatin remodeling gene ARID1A [55]. The comprehensive molecular evaluation of urothelial bladder cancer confirmed that UTX, EP300, and ARID1A are significantly mutated in MIBC [56]. In addition, a novel mutation in MLL2, a gene encoding histone H3 lysine 4 (H3K4) methyltransferase was identified [56]. Overall, this study showed that 75 % of MIBC harbored an inactivating mutation in one or more chromatin regulatory genes. Further integrated network analysis revealed that mutations in genes known to influence epigenetic regulation affected the activity levels of transcription factors implicated in carcinogenesis. Currently, there are drugs in development that target chromatin modifications, thereby opening new possibilities for the treatment of bladder cancer [56].

### 28.10.6 Gene Expression Profiling

Molecular subtypes of urothelial carcinoma have been described using cluster analysis of gene expression. The cancer genome atlas research network comprehensively evaluated 131 chemotherapy-naive MIBCs. Using cluster analysis of RNA-sequencing data, they identified 4 distinct groups. Noteworthy patterns included tumors with a papillary morphology and frequent FGFR3 mutations (cluster 1) and tumors with high expression of estrogen receptor (ER) beta and HER2 (ERBB2) levels (clusters I and II). In addition, a third cluster showed basal/squamous-like morphology and immunohistochemistry, similar to that seen in basal-like breast cancers and basaloid squamous cancers of the head and neck [55]. A different group also published an mRNA cluster analysis of 73 cases of MIBC [57]. Modeled after molecular subtypes in breast cancer, this study described three groups including luminal, basal and p53-like subtypes. The luminal group was immunohistochemically characterized by CK20 positive cells and CK5/6 and CD44 negativity. The luminal molecular changes included FGFR3, ER and peroxisome proliferator activator receptor (PPAR) expression and clinically showed a poor response to chemotherapy. The basal subtype was enriched in sarcomatoid and squamous histology, CK5/6, and CD44 expression and was negative for CK20. Molecular features of the basal subtype included abnormalities in p63, MYC, STAT-3, NFK-beta, and EGFR pathways. This basal subtype was shown to present at a later clinical stage compared to luminal subtypes and demonstrated a good response to chemotherapy. The third group, called "p53-like" had a gene expression signature that predicted chemotherapy resistance, however, all groups had a similar proportion of TP53 mutations, suggesting that the p53-like gene expression signature did not correlate with TP53 mutation status [57]. Molecular subtyping has the potential to predict response to chemotherapy and identify candidates for targeted therapies.

## 28.11 Summary

In summary, we have set out to describe the histopathology together with the molecular alterations of bladder urothelial carcinoma. Hopefully, in the years to come, our understanding of the molecular basis of urothelial carcinoma will further expand treatment options for patients with bladder cancer.

## References


1. Murphy WM, Grignon DJ, Perlman EJ, Armed Forces Institute of Pathology (U.S.), Universities Associated for Research and Education in Pathology. Tumors of the kidney, bladder, and related urinary structures. Washington, D.C.: Armed Forces Institute of Pathology;2004. xiii, p. 394.
2. Sadler TW, Langman J. Langman's medical embryology. 10th ed. Philadelphia: Lippincott Williams & Wilkins;2006. xiii, p. 371.
3. Viana R, Batourina E, Huang H, Dressler GR, Kobayashi A, Behringer RR, et al. The development of the bladder trigone, the center of the anti-reflux mechanism. Development. 2007;134(20):3763–9.
4. Eble JN, World Health Organization., International Agency for Research on Cancer. Pathology and genetics of tumours of the urinary system and male genital organs. Lyon Oxford: IARC Press; Oxford University Press (distributor);2004. p. 359.
5. Epstein JI, Amin MB, Reuter VE, Epstein JI. Biopsy interpretation of the bladder. 2nd ed. Philadelphia: Wolters Kluwer/Lippincott Williams & Wilkins Health;2010. vii, p. 355.

6. van Rhijn BW, Burger M, Lotan Y, Solsona E, Stief CG, Sylvester RJ, et al. Recurrence and progression of disease in non-muscle-invasive bladder cancer: from epidemiology to treatment strategy. Eur Urol. 2009;56(3):430–42.
7. Wu XR. Urothelial tumorigenesis: a tale of divergent pathways. Nat Rev Cancer. 2005;5(9):713–25.
8. Mitra AP, Cote RJ. Molecular pathogenesis and diagnostics of bladder cancer. Annu Rev Pathol. 2009;4:251–85.
9. Netto GJ. Clinical applications of recent molecular advances in urologic malignancies: no longer chasing a "mirage"? Adv Anat Pathol. 2013;20(3):175–203.
10. Taylor DC, Bhagavan BS, Larsen MP, Cox JA, Epstein JI. Papillary urothelial hyperplasia. A precursor to papillary neoplasms. Am J Surg Pathol. 1996;20(12):1481–8.
11. Chow NH, Cairns P, Eisenberger CF, Schoenberg MP, Taylor DC, Epstein JI, et al. Papillary urothelial hyperplasia is a clonal precursor to papillary transitional cell bladder cancer. Int J Cancer. 2000;89(6):514–8.
12. Magi-Galluzzi C, Epstein JI. Urothelial papilloma of the bladder: a review of 34 de novo cases. Am J Surg Pathol. 2004;28(12):1615–20.
13. McKenney JK, Amin MB, Young RH. Urothelial (transitional cell) papilloma of the urinary bladder: a clinicopathologic study of 26 cases. Mod Pathol. 2003;16(7):623–9.
14. Eiber M, van Oers JM, Zwarthoff EC, van der Kwast TH, Ulrich O, Helpap B, et al. Low frequency of molecular changes and tumor recurrence in inverted papillomas of the urinary tract. Am J Surg Pathol. 2007;31(6):938–46.
15. Clark PE, Agarwal N, Biagioli MC, Eisenberger MA, Greenberg RE, Herr HW, et al. Bladder cancer. J Natl Compr Canc Netw. 2013;11(4):446–75.
16. Pan CC, Chang YH, Chen KK, Yu HJ, Sun CH, Ho DM. Prognostic significance of the 2004 WHO/ISUP classification for prediction of recurrence, progression, and cancer-specific mortality of non-muscle-invasive urothelial tumors of the urinary bladder: a clinicopathologic study of 1,515 cases. Am J Clin Pathol. 2010;133(5):788–95.
17. Samaratunga H, Makarov DV, Epstein JI. Comparison of WHO/ISUP and WHO classification of noninvasive papillary urothelial neoplasms for risk of progression. Urology. 2002;60(2):315–9.
18. Malmstrom PU, Busch C, Norlen BJ. Recurrence, progression and survival in bladder cancer. A retrospective analysis of 232 patients with greater than or equal to 5-year follow-up. Scand J Urol Nephrol. 1987;21(3):185–95.
19. Knowles MA. Molecular subtypes of bladder cancer: Jekyll and Hyde or chalk and cheese? Carcinogenesis. 2006;27(3):361–73.
20. Hopman AH, Kamps MA, Speel EJ, Schapers RF, Sauter G, Ramaekers FC. Identification of chromosome 9 alterations and p53 accumulation in isolated carcinoma in situ of the urinary bladder versus carcinoma in situ associated with carcinoma. Am J Pathol. 2002;161(4):1119–25.
21. Casey RG, Catto JW, Cheng L, Cookson MS, Herr H, Shariat S, et al. Diagnosis and management of urothelial carcinoma in situ of the lower urinary tract: A systematic review. Eur Urol. 2015;67(5):876–88.
22. Chang WC, Chang YH, Pan CC. Prognostic significance in substaging ofT1 urinary bladder urothelial carcinoma on transurethral resection. Am J Surg Pathol. 2012;36(3):454–61.
23. Cheng L, Neumann RM, Weaver AL, Spotts BE, Bostwick DG. Predicting cancer progression in patients with stage T1 bladder carcinoma. J Clin Oncol. 1999;17(10):3182–7.
24. Cheng L, Montironi R, Davidson DD, Lopez-Beltran A. Staging and reporting of urothelial carcinoma of the urinary bladder. Mod Pathol. 2009;22(Suppl 2):S70–95.
25. Xylinas E, Cha EK, Sun M, Rink M, Trinh QD, Novara G, et al. Risk stratification of pT1–3N0 patients after radical cystectomy for adjuvant chemotherapy counselling. Br J Cancer. 2012;107(11):1826–32.
26. Grossman HB, Natale RB, Tangen CM, Speights VO, Vogelzang NJ, Trump DL, et al. Neoadjuvant chemotherapy plus cystectomy compared with cystectomy alone for locally advanced bladder cancer. N Engl J Med. 2003;349(9):859–66.
27. Xylinas E, Rink M, Robinson BD, Lotan Y, Babjuk M, Brisuda A, et al. Impact of histological variants on oncological outcomes of patients with urothelial carcinoma of the bladder treated with radical cystectomy. Eur J Cancer. 2013;49(8):1889–97.
28. Amin MB. Histological variants of urothelial carcinoma: diagnostic, therapeutic and prognostic implications. Mod Pathol. 2009;22(Suppl 2):S96–118.
29. Pons F, Orsola A, Morote J, Bellmunt J. Variant forms of bladder cancer: basic considerations on treatment approaches. Curr Oncol Rep. 2011;13(3):216–21.
30. Wright JL, Black PC, Brown GA, Porter MP, Kamat AM, Dinney CP, et al. Differences in survival among patients with sarcomatoid carcinoma, carcinosarcoma and urothelial carcinoma of the bladder. J Urol. 2007;178(6):2302–6 (discussion 7).
31. Abd El-Aal AA, Bayoumy IR, Basyoni MM, Abd El-Aal AA, Emran AM, Abd El-Tawab MS, et al. Genomic instability in complicated and uncomplicated Egyptian schistosomiasis haematobium patients. Mol Cytogenet. 2015;8(1):1.
32. Kassouf W, Spiess PE, Siefker-Radtke A, Swanson D, Grossman HB, Kamat AM, et al. Outcome and patterns of recurrence of nonbilharzial pure squamous cell carcinoma of the bladder: a contemporary review of The University of Texas M D Anderson Cancer Center experience. Cancer. 2007;110(4):764–9.
33. Bates AW, Baithun SI. Secondary neoplasms of the bladder are histological mimics of nontransitional cell primary tumours: clinicopathological and histological

features of 282 cases. Histopathology. 2000;36 (1):32–40.

34. Amin MB, Trpkov K, Lopez-Beltran A, Grignon D. Members of the IIiDUPG. Best practices recommendations in the application of immunohistochemistry in the bladder lesions: report from the International Society of Urologic Pathology consensus conference. Am J Surg Pathol. 2014;38(8):e20–34.
35. Aben KK, Baglietto L, Baffoe-Bonnie A, Coebergh JW, Bailey-Wilson JE, Trink B, et al. Segregation analysis of urothelial cell carcinoma. Eur J Cancer. 2006;42(10):1428–33.
36. Lichtenstein P, Holm NV, Verkasalo PK, Iliadou A, Kaprio J, Koskenvuo M, et al. Environmental and heritable factors in the causation of cancer—analyses of cohorts of twins from Sweden, Denmark, and Finland. N Engl J Med. 2000;343(2):78–85.
37. Murta-Nascimento C, Silverman DT, Kogevinas M, Garcia-Closas M, Rothman N, Tardon A, et al. Risk of bladder cancer associated with family history of cancer: do low-penetrance polymorphisms account for the increase in risk? Cancer Epidemiol Biomarkers Prev. 2007;16(8):1595–600.
38. Kiemeney LA, Thorlacius S, Sulem P, Geller F, Aben KK, Stacey SN, et al. Sequence variant on 8q24 confers susceptibility to urinary bladder cancer. Nat Genet. 2008;40(11):1307–12.
39. Wu X, Ye Y, Kiemeney LA, Sulem P, Rafnar T, Matullo G, et al. Genetic variation in the prostate stem cell antigen gene PSCA confers susceptibility to urinary bladder cancer. Nat Genet. 2009;41(9):991–5.
40. Rafnar T, Sulem P, Stacey SN, Geller F, Gudmundsson J, Sigurdsson A, et al. Sequence variants at the TERT-CLPTM1L locus associate with many cancer types. Nat Genet. 2009;41(2):221–7.
41. Rothman N, Garcia-Closas M, Chatterjee N, Malats N, Wu X, Figueroa JD, et al. A multi-stage genome-wide association study of bladder cancer identifies multiple susceptibility loci. Nat Genet. 2010;42(11):978–84.
42. Dienstmann R, Rodon J, Prat A, Perez-Garcia J, Adamo B, Felip E, et al. Genomic aberrations in the FGFR pathway: opportunities for targeted therapies in solid tumors. Ann Oncol. 2014;25(3):552–63.
43. Castillo-Martin M, Domingo-Domenech J, Karni-Schmidt O, Matos T, Cordon-Cardo C. Molecular pathways of urothelial development and bladder tumorigenesis. Urol Oncol. 2010;28(4):401–8.
44. Lopez-Knowles E, Hernandez S, Malats N, Kogevinas M, Lloreta J, Carrato A, et al. PIK3CA mutations are an early genetic alteration associated with FGFR3 mutations in superficial papillary bladder tumors. Cancer Res. 2006;66(15):7401–4.
45. Cantley LC. The phosphoinositide 3-kinase pathway. Science. 2002;296(5573):1655–7.
46. Oxford G, Theodorescu D. The role of Ras superfamily proteins in bladder cancer progression. J Urol. 2003;170(5):1987–93.
47. McConkey DJ, Lee S, Choi W, Tran M, Majewski T, Lee S, et al. Molecular genetics of bladder cancer: Emerging mechanisms of tumor initiation and progression. Urol Oncol. 2010;28(4):429–40.
48. Zhang ZT, Pak J, Huang HY, Shapiro E, Sun TT, Pellicer A, et al. Role of Ha-ras activation in superficial papillary pathway of urothelial tumor formation. Oncogene. 2001;20(16):1973–80.
49. Cheng J, Huang H, Pak J, Shapiro E, Sun TT, Cordon-Cardo C, et al. Allelic loss of p53 gene is associated with genesis and maintenance, but not invasion, of mouse carcinoma in situ of the bladder. Cancer Res. 2003;63(1):179–85.
50. Cote RJ, Dunn MD, Chatterjee SJ, Stein JP, Shi SR, Tran QC, et al. Elevated and absent pRb expression is associated with bladder cancer progression and has cooperative effects with p53. Cancer Res. 1998;58 (6):1090–4.
51. Puzio-Kuter AM, Castillo-Martin M, Kinkade CW, Wang X, Shen TH, Matos T, et al. Inactivation of p53 and Pten promotes invasive bladder cancer. Genes Dev. 2009;23(6):675–80.
52. Yoo LI, Liu DW, Le Vu S, Bronson RT, Wu H, Yuan J. Pten deficiency activates distinct downstream signaling pathways in a tissue-specific manner. Cancer Res. 2006;66(4):1929–39.
53. Hoglund M. The bladder cancer genome; chromosomal changes as prognostic makers, opportunities, and obstacles. Urol Oncol. 2012;30(4):533–40.
54. Knowles MA, Hurst CD. Molecular biology of bladder cancer: new insights into pathogenesis and clinical diversity. Nat Rev Cancer. 2015;15(1):25–41.
55. Gui Y, Guo G, Huang Y, Hu X, Tang A, Gao S, et al. Frequent mutations of chromatin remodeling genes in transitional cell carcinoma of the bladder. Nat Genet. 2011;43(9):875–8.
56. Research Cancer Genome Atlas. N. Comprehensive molecular characterization of urothelial bladder carcinoma. Nature. 2014;507(7492):315–22.
57. Choi W, Porten S, Kim S, Willis D, Plimack ER, Hoffman-Censits J, et al. Identification of distinct basal and luminal subtypes of muscle-invasive bladder cancer with different sensitivities to frontline chemotherapy. Cancer Cell. 2014;25(2):152–65.

# 29 Epidemiology of Hematologic Malignancies

Julie L. Batista, Brenda M. Birmann and Mara Meyer Epstein

## 29.1 Introduction

Hematologic neoplasms comprise a broad category of malignancies originating from cells of the bone marrow and lymphatic system. Given the breadth of hematologic malignancy subtypes, it is not possible to adequately cover all myeloid and lymphoid malignancies within the scope of this chapter, and therefore we have chosen to focus specifically on three major subdivisions of lymphoid malignancies that are grouped according to biological and epidemiologic commonalities: Hodgkin lymphoma (HL), non-Hodgkin lymphoma (NHL), and multiple myeloma (MM). HL is distinct from NHL due to its different biological behavior, spread and response to treatment, as will be described below. MM, the most common type of plasma cell neoplasm, can be considered a histologic subtype of B-cell NHL, however, the distinct clinical presentation and descriptive epidemiology of the disease also warrant separate discussion. Since these subgroups of hematologic cancer still comprise a large spectrum of diseases and disorders, each with a distinct morphologic and molecular phenotype, we have described in detail only a selection of the malignancies that fall within these categories. For more detailed description of the epidemiology of myeloid cancers, the reader should refer to several recent publications [1–7].

After decades of inconsistency in the approach to classifying hematologic tumors, in 2001 the World Health Organization (WHO) published a novel, modern consensus approach developed by expert hematopathologists [8]. The consensus classification scheme, now considered the gold standard, considers not only cell morphology but also immunophenotype, clinical presentation and epidemiology, and thus incorporates insights from emerging molecular technological study of those tumors. Diagnoses are typically coded according to the International Classification of Diseases-Oncology, Third Edition (ICD-O-3), which is also issued by the WHO and periodically revised to accommodate updates to the tumor classification scheme as new tumor entities come to attention (http://codes.iarc.fr/abouticdo.php) [9, 10]. Although the WHO classification applies to all lymphoid and myeloid tumors, it has been particularly valuable for clarifying and unifying the classification of the diverse tumors categorized as NHL. A hierarchical nested classification of lymphoid neoplasms was published in 2007 and updated in 2010 by the International Lymphoma Epidemiology (InterLymph) Consortium Pathology Working Group specifically to facilitate

J.L. Batista · B.M. Birmann
Channing Division of Network Medicine, Department of Medicine, Brigham and Women's Hospital and Harvard Medical School, 181 Longwood Avenue, Boston, MA 02115, USA

M.M. Epstein (✉)
Department of Medicine, University of Massachusetts Medical School, 365 Plantation St, Biotech 1, Suite 100, Worcester, MA 01605, USA
e-mail: mara.epstein@umassmed.edu


M. Loda et al. (eds.), *Pathology and Epidemiology of Cancer*,
DOI 10.1007/978-3-319-35153-7_29

uniformity in the investigation of lymphoma subtypes in epidemiologic research [11, 12].

## 29.2 Description and Clinical Presentation of Hematologic Malignancies

### 29.2.1 Disease Description and Subtypes

#### 29.2.1.1 Hodgkin Lymphoma

HL is an uncommon B-cell malignancy that is classified into two main types: classical HL, accounting for the vast majority of HL cases diagnosed in Western countries, and nodular lymphocyte-predominant HL (NLPHL) accounting for approximately 5 % of cases [13, 14]. Classical HL is characterized by the presence of large, often multinucleated Hodgkin "Reed-Sternberg" (RS) cells surrounded by reactive infiltrating cells, especially T-cells [15–17]. Classical HL can be further divided into 4 histological subtypes: nodular sclerosis HL (the most common, approximately 60 % of cases), mixed cellularity HL (approximately 30 % of cases), lymphocyte-rich HL, and lymphocyte-depleted HL [18]. In contrast, NLPHL tumors are characterized by lymphocyte-predominant cells, also known as popcorn cells, in the absence of RS cells. Of note, the RS cells, which are now known to be abnormal germinal center B-lymphocytes, typically comprise only 1–2 % of HL tumor tissue [19]. The comparative abundance of reactive immune cells in the tumor microenvironment suggests a role for host immune deregulation/chronic inflammation in the pathogenesis of these tumors and implies that immune-modulating exposures are likely to influence risk.

#### 29.2.1.2 Non-Hodgkin Lymphoma

The term NHL collectively refers to more than 60 discrete histologic subtypes with heterogeneous biologic, clinical, etiologic, and epidemiologic features. NHL usually presents in the lymph nodes or other components of the lymphatic system; a minority of subtypes can present outside the lymphatic system, i.e. "extranodally," including on the skin, and in the central nervous system. The various NHL subtypes are characterized by the overproduction of different types of lymphocytes at varying stages of maturation, including B-cells, T-cells, and natural killer cells. The most common subtypes of NHL are those that originate in B-cells (90 % of all NHL), including the aggressive diffuse large B-cell lymphoma (DLBCL) and the more indolent follicular lymphoma (FL), which make up about 30 and 20 % of all NHL diagnoses in the United States (U.S.), respectively [20]. Other histologic types of B-cell NHL, such as marginal zone, mantle cell, Burkitt, Burkitt-like, Waldenström macroglobulinemia (WM) or lymphoplasmacytic lymphoma (LPL), comprise a comparatively small proportion of cases in the U.S. [21, 22]. Even less common in the U.S. are most histologic types of T-cell NHL, such as cutaneous T-cell lymphoma, hairy cell leukemia, mycosis fungoides, or adult T-cell leukemia/lymphoma (ATL), although some of those tumors comprise a larger proportion of NHLs in other populations [22–25].

#### 29.2.1.3 Multiple Myeloma

MM is a plasma cell neoplasm characterized by the monoclonal proliferation of plasma cells, which are mature B-cells that originate in the bone marrow and secrete antibodies in response to exposure to foreign antigens. Although in the literature MM has been considered a histologic subtype of B-cell NHL, MM has a distinct clinical presentation and descriptive epidemiology. MM arises from the expansion of a plasma cell clone secreting a single monoclonal immunoglobulin, called M-protein [26]. More than half of MMs are characterized by malignant cells producing IgG, while about 20 % of myelomas produce IgA [27]. However, up to 3 % of cases exhibit no detectable M-protein; these cases are known as non-secretory myeloma, and follow a similar clinical trajectory to that of secretory cases. MM is believed to be preceded by a condition known as monoclonal gammopathy of unknown significance (MGUS), which is identified by detection of lower M-protein and plasma cell concentrations in the absence of end-organ damage. MGUS can persist undetected for many years and may never come to clinical attention in a large proportion of patients [28]. In 70 % of cases, patients with MM will present with multiple lytic lesions or fractures of the bone [26]; if only one lesion is detected, in the absence of other symptoms and clinical features, the disease is known as solitary plasmacytoma of bone.

## 29.2.2 Symptoms

### 29.2.2.1 Hodgkin Lymphoma

Enlargement of one or more lymph nodes, usually painless and most often in the neck, armpit, or groin, is the most common symptom of HL. Non-specific ("B") symptoms of potential prognostic importance include persistent fever, drenching night sweats, and unexplained weight loss. However, systemic symptoms are present in only approximately one third of patients at diagnosis. Other non-specific symptoms are fatigue and rarely, severe itching and lymph node pain after consuming alcohol [13].

### 29.2.2.2 Non-Hodgkin Lymphoma

The most common symptom of NHL is one or more persistent, painless, and swollen lymph nodes often appearing on the neck, armpit, or groin. Other, more non-specific symptoms include night sweats, unexplained weight loss, abdominal pain or swelling, fatigue, fever, or shortness of breath. NHL cells in bone marrow may lead to altered blood cell and/or platelet counts. Some NHL subtypes are largely asymptomatic; for example, chronic lymphocytic leukemia (CLL) is often diagnosed following repeated observations of elevated white blood cell counts in the absence of other symptoms, with confirmation primarily by flow cytometry to detect a clone of B-cells expressing the indicative markers [29].

### 29.2.2.3 Multiple Myeloma

Patients generally experience non-specific symptoms prior to MM diagnosis. These vary somewhat by stage of disease but often include some combination of weakness, bone pain or fracture, anemia, recurrent infection, unexplained weight loss, and symptoms of renal failure. These non-specific symptoms correspond to the spectrum of clinical signs that inform the formal diagnosis of MM as described below.

## 29.2.3 Diagnosis

### 29.2.3.1 Hodgkin Lymphoma

The diagnosis of HL begins with a medical exam and blood test to detect infection or other more common reasons for enlarged lymph nodes or fever. On ruling out those alternative explanations, a definitive diagnosis is achieved through biopsy of the affected lymph node and histopathologic confirmation of the presence of RS cells.

### 29.2.3.2 Non-Hodgkin Lymphoma

NHL can be definitively diagnosed through a biopsy of the presenting swollen lymph nodes. Blood tests and flow cytometry, bone marrow biopsy, tumor genetic testing and imaging tests, including X-ray, computerized tomography (CT), magnetic resonance imaging (MRI) or positron emission tomography (PET), may also be administered to aid in diagnosis and determine the extent of disease. Through examination of biopsy specimens and the supporting clinical and laboratory data, a hematopathologist can determine the cell-type of origin (B-cell, T-cell, NK cell) and histologic subtype, which is crucial to guide treatment decisions.

Recent advances in molecular technology have identified molecular subtypes of DLBCL that correspond to differing cell of origin (germinal center B-cell versus post-germinal center B-cell) and to differing tumor aggressiveness and clinical prognosis [30–34]. Similarly, molecular features of typically indolent lymphomas such as CLL can help to identify more aggressive forms [35, 36]. Such molecular diagnostics seem likely to further elucidate tumor classification, pathogenesis and clinical prognosis and guide treatment decisions.

#### 29.2.3.3 Multiple Myeloma

Diagnostic criteria for MM have been established by the International Myeloma Working Group (IMWG) [37–39]. Traditionally, MM has been diagnosed by the presence of serum monoclonal protein ≥30 g/L, a bone marrow biopsy showing ≥10 % clonal bone marrow plasma cells, and evidence of end-organ damage including hypercalcemia, renal insufficiency, anemia, and lytic bone lesions ("CRAB" features) [37–40]. An updated version of the IMWG criteria in 2014 removed the monoclonal protein requirement, as 3 % of MM cases have non-secretory disease without a measurable M-protein on serum or urine immunofixation [38, 39]. The new guidelines also require one of the following biomarkers of malignancy in the absence of CRAB features: clonal bone marrow plasma cell percentage ≥60 %, serum free light chain ratio ≥100, or more than one focal lesion on an MRI study. MM cases are staged using a modification of the Durie-Salmon staging system called the International Staging System (ISS), which incorporates serum $B_2$-microglobulin and albumin levels and was recently updated to also include chromosomal abnormalities and lactate dehydrogenase levels [41].

As noted earlier, MM is preceded in nearly all cases by a premalignant condition known as MGUS, which is marked by serum M-protein levels <30 g/l and <10 % clonal bone marrow plasma cells in the absence of end-organ damage. MGUS may be present in at least 3 % of all adults in the U.S. older than 50 years [42, 43], with variability by race [44–47] and higher incidence among older adults and those with a family history; however, the condition is mostly diagnosed incidentally. Individuals with MGUS progress to myeloma at an estimated rate of 1 % per year [48, 49].

Smoldering, or asymptomatic, MM is an intermediate premalignant condition marked by rising serum M-protein levels ≥30 g/1, and/or ≥10 % clonal bone marrow plasma cells in the absence of end-organ damage [39]. Patients with smoldering MM have an estimated 10 % annual risk of progression to MM, and a median time to progression of 4.8 years [50]. Although smoldering MM is generally not treated, results from a recent randomized trial suggest early therapy may improve overall survival for patients with smoldering MM [51].

Solitary plasmacytoma, a variant of plasma cell myeloma, is characterized by the presence of a single plasma cell tumor on the interior surface of the bone. In contrast to MM, malignant plasma cells are often not disseminated throughout the bone marrow, and patients do not experience abnormal blood test results or CRAB features. Plasmacytomas are diagnosed through biopsy, with more extensive skeletal imaging tests showing no other bone or tissue lesions [9]. Patients may present with bone pain or fracture. Plasmacytomas comprise approximately 3–5 % of all plasma cell malignancies, and patients have a 10 % risk of progression to MM within 3 years [39].

Plasma cell leukemia (PCL) is another variant of plasma cell malignancy, accounting for less than 5 % of myeloma cases [26]. A PCL diagnosis is characterized by $>2\times10^9$ clonal plasma cells in the peripheral blood, or >20 % circulating plasma cells [52, 53]. Although rare, PCL tends to be aggressive, and patients generally have a poor prognosis in comparison to other plasma cell disorders, with less than 10 % of patients surviving 5 years following diagnosis [54]. PCL presents as two distinct clinical and biological conditions: primary PCL (60 % of cases), or secondary PCL, which develops in patients previously diagnosed with MM (40 % of cases) [52, 54, 55].

## 29.3 Descriptive Epidemiology of Hematologic Malignancies

### 29.3.1 Incidence (Age, Sex, Race/Ethnicity)

#### 29.3.1.1 Hodgkin Lymphoma

Hodgkin lymphoma is a relatively rare cancer with an estimated 9050 new diagnoses in the U.S. in 2015 (Fig. 29.1) [56]. The age-adjusted incidence

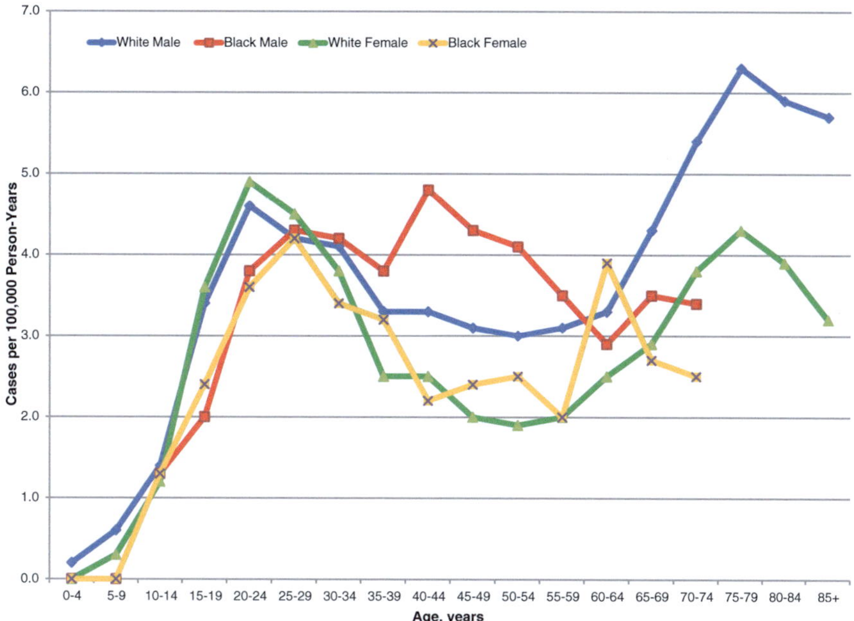


**Fig. 29.1** Age-Specific Incidence of Hodgkin Lymphoma in the United States (SEER areas) by Sex and Race, 2007–2012. *Source* Howlader et al. [60] Rates for black males and black females aged 75 and older were not calculated due to small case numbers in those intervals

rate in the U.S. from 2007–2011 was 2.7 per 100,000 overall, with a slightly higher incidence among men (3.1 per 100,000) compared to women (2.4 per 100,000) [57]. Incidence rates vary by race/ethnicity, with a higher incidence observed among non-Hispanic Caucasians (3.2 per 100,000) compared to African Americans (2.7 per 100,000), Asian/Pacific Islanders (1.3 per 100,000) and Hispanics (2.4 per 100,000).

Worldwide, approximately 66,000 new diagnoses of HL are made each year, with an overall age-adjusted incidence rate of 0.9 per 100,000 [58]. Incidence is higher among more developed (2.1 per 100,000) versus less developed (0.6 per 100,000) regions, with the highest rates in Europe and the Americas, and the lowest rates in Africa, Asia and the Pacific (Fig. 29.2).

HL incidence by age displays a characteristic bimodal shape, with two peak incidence rates at young and older adulthood. In the U.S., HL is the eighth most commonly diagnosed cancer among children ages 0–14 years, and is the leading diagnosis among adolescents ages 15–19 [59]. According to Surveillance, Epidemiology, and End Results (SEER) estimates from 2007–2011, age specific incidence rates peak at 4.5 per 100,000 for age 20–24 years, fall to 2.4 per 100,000 for age 45–54 years, and increase again to 4.5–4.7 per 100,000 for age 70–84 years [60].

#### 29.3.1.2 Non-Hodgkin Lymphoma

Overall, NHL is the sixth most commonly diagnosed cancer in the U.S. among both men and women. There were an estimated 71,850 new cases of NHL in the U.S. in 2015, accounting for more than 4 % of all cancer diagnoses (Fig. 29.3) [56, 60]. CLL is a leukemic condition that shares major pathological features with small lymphocytic lymphoma (SLL) and is classified with B-cell NHL. CLL accounted for an additional 14,620 NHL diagnoses in the U.S. in 2015 [60]. In addition, acute lymphoblastic leukemia (ALL) is a precursor lymphoid neoplasm of either pre-B (85 % of cases) or pre-T-cells (15 % of cases). In contrast to most other NHL subtypes, 75 % of ALL cases occur in children less than six years old, with more than 6,000 new cases diagnosed in the U.S. annually [26, 60]. The incidence rate for NHL is 19.7 per 100,000 men and women; however incidence rates for the

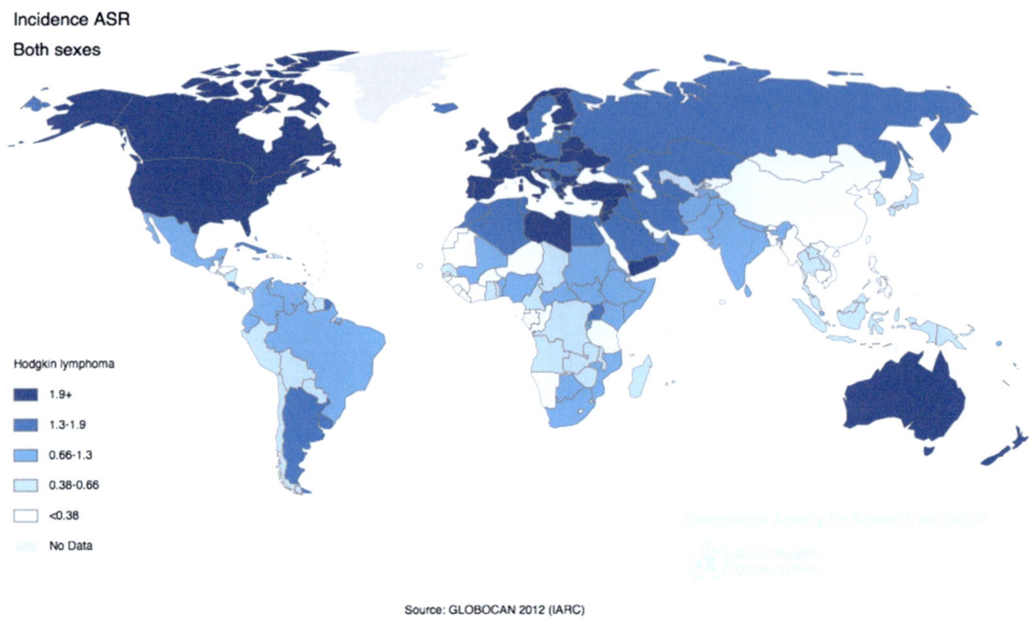


**Fig. 29.2** Age-Standardized Incidence Rates for Hodgkin Lymphoma by Country for Men and Women Combined, 2012. *Source* World Health Organization, International Agency for Cancer Research, GLOBOCAN 2012

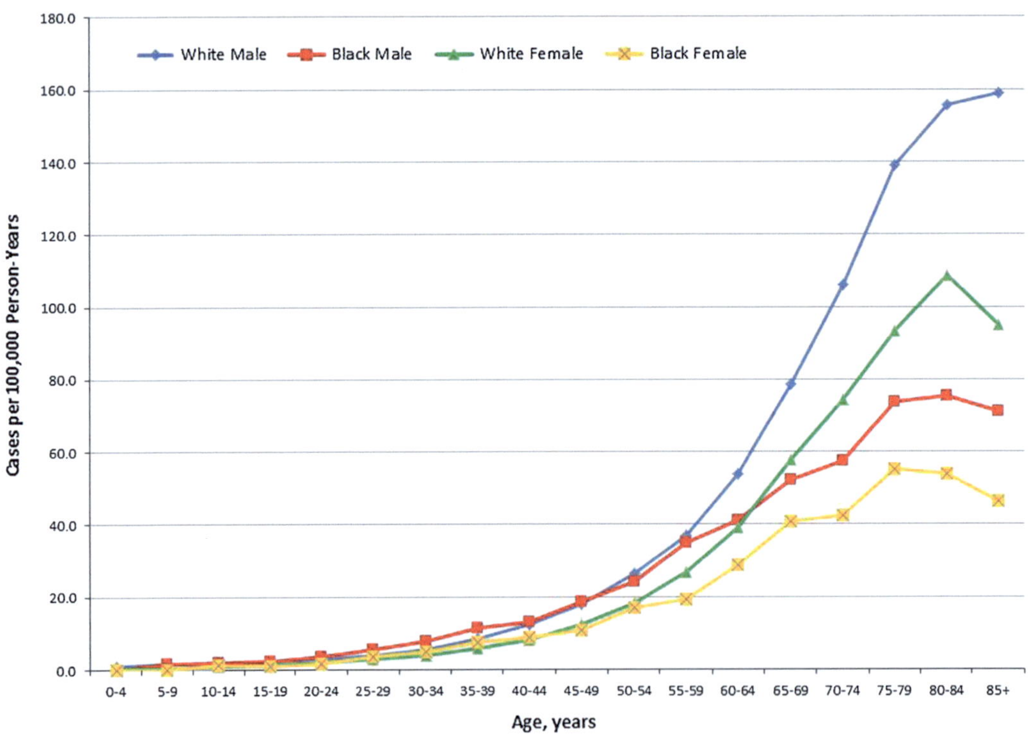


**Fig. 29.3** Age-Specific Incidence of Non-Hodgkin Lymphoma in the United States (SEER areas) by Sex and Race, 2007–2012. *Source* Howlader et al. [60]

various subtypes of NHL differ dramatically. Incidence rates of NHL in the U.S. and many other countries rose steadily during the second half of the twentieth century; the rate increases seem to have slowed overall in recent years, but the rising trend continues for the two most common subtypes, DLBCL and FL [61, 62].

Worldwide, 385,741 cases of NHL were reported by IARC in 2012, with an age-standardized incidence rate of 5.0 per 100,000 people [58]. The more developed regions of the world experience a NHL incidence rate more than twice that of less developed regions, with the highest rates in the U.S., Western Europe, and Australia, and the lowest rates in Asia and Africa. However, the incidence of different NHL subtypes varies by geographic region; for example, DLBCL and FL are more common in North America and Europe than in Asia, whereas T-cell lymphomas are more common in Asia than in Western countries (Fig. 29.4) [61].

#### 29.3.1.3 Multiple Myeloma

An estimated 26,850 men and women in the U.S. were diagnosed with MM in 2015, accounting for more than 1.5 % of new cancer diagnoses, and more than 15 % of hematologic cancer diagnoses (Fig. 29.5) [60]. Based on data from the U.S. SEER Registry from 2008 to 2012, the overall incidence rate for MM is 6.3 cases per 100,000 men and women; however, men exhibit higher incidence rates than women (7.9 vs. 5.1 cases per 100,000), and African Americans have higher incidence rates (15.1 per 100,000 men and 11.2 per 100,000 women) compared to whites (7.5 per 100,000 men and 4.5 per 100,000 women) [60]. Studies suggest that people of African or African American descent also have twice the prevalence of precursor MGUS compared to white and Hispanic populations [46, 63, 64].

Most cases of MM are diagnosed in adults age 65–74 years, with a median age at diagnosis of

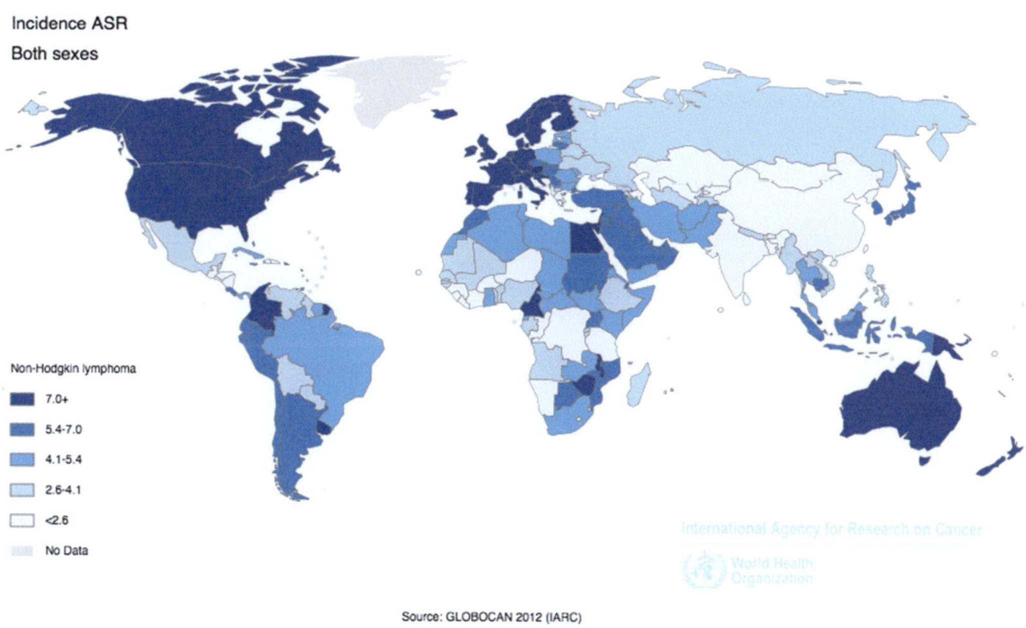


**Fig. 29.4** Age-Standardized Incidence Rates for Non-Hodgkin Lymphoma by Country for Men and Women Combined, 2012. *Source* World Health Organization, International Agency for Cancer Research, GLOBOCAN 2012

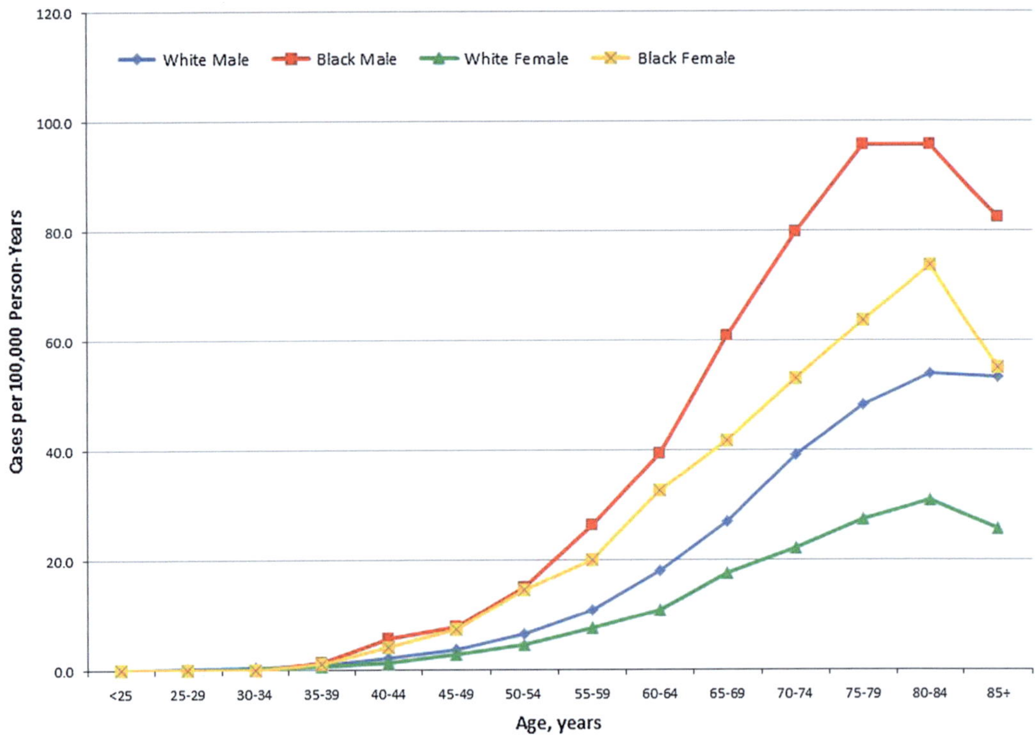


**Fig. 29.5** Age-Specific Incidence of Multiple Myeloma in the United States (SEER areas) by Sex and Race, 2007–2012. *Source* Howlader et al. [60]

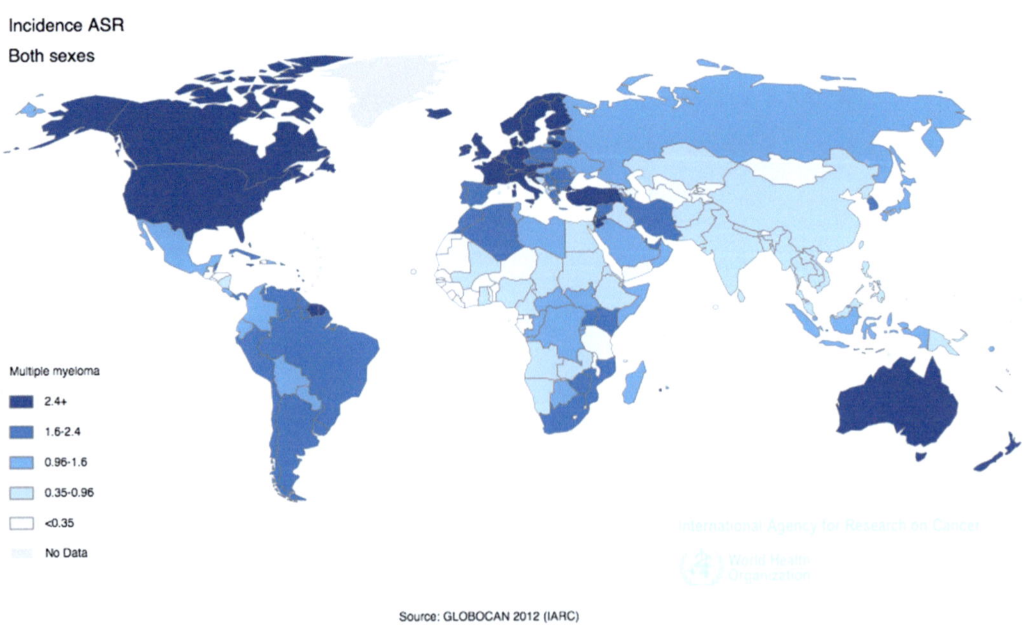


**Fig. 29.6** Age-Standardized Incidence Rates for Multiple Myeloma by Country for Men and Women Combined, 2012. *Source* World Health Organization, International Agency for Cancer Research, GLOBOCAN 2012

69 years [60]. Only about 3 % of cases are diagnosed in people younger than 40 years old [65, 66]. Older age at diagnosis is associated with poorer survival, as well as more advanced stage at diagnosis [67]. Conversely, younger patients often present at an earlier stage, and have less frequent adverse prognostic factors, such as low hemoglobin and high C-reactive protein [66]. As the population ages, more adults are expected to be diagnosed with MM, and the elderly will likely comprise a larger proportion of patients [68, 69].

### 29.3.2 Mortality

#### 29.3.2.1 Hodgkin Lymphoma

An estimated 1150 HL deaths occurred in the U.S. in 2015, with death rates steadily declining over the past four decades [56]. The 5-year relative survival rate for patients diagnosed between 2004 and 2010 was 85 %, with a 94 % survival for patients diagnosed at less than 45 years of age versus 53 % for patients diagnosed at 65 years of age or older [57]. Five-year survival rates were generally similar between men (84 %) and women (86 %), and between Caucasian (86 %) and African American (82 %) patients. Worldwide, approximately 25,500 people die annually due to HL [58]. Age-standardized mortality rates are similar in more developed and less developed regions of the world (0.3 per 100,000).

Among long-term survivors of HL, second primary malignancies, largely due to complications of primary HL therapy, are a major cause of morbidity and mortality [70]. Among the highest relative risks are second hematologic malignancies, particularly leukemia and NHL [71, 72]. HL survivors also have an excess risk of certain solid cancers compared to the general population, including breast and colorectal cancers [73].

#### 29.3.2.2 Non-Hodgkin Lymphoma

NHL is the ninth most common cause of cancer death in men, and the eighth most common cause of cancer death in women in the U.S., with approximately 20,000 deaths expected in 2015, accounting for 3–4 % of all cancer deaths [56, 60].

About 200,000 people worldwide die from NHL annually, with an age-standardized mortality rate of 2.5 per 100,000 men and women. Overall, mortality rates are somewhat more uniform between countries, in contrast to incidence rates, although slightly higher mortality rates are observed in the Middle East and Africa, and lower mortality rates are observed in Asia and Europe [58]. For all NHL subtypes combined, 70 % of people will survive at least 5 years following diagnosis [60]. However, as observed for incidence rates, mortality rates vary considerably across histologic types and tumor molecular subtypes [22].

#### 29.3.2.3 Multiple Myeloma

Approximately 11,000 deaths due to MM occur in the U.S. annually, accounting for more than 2 % of cancer deaths, with a median age at death from MM of 75 years [60]. Following decades of unchanged survival rates, the development of new MM treatments in the early 2000s, including immunomodulatory drugs and protease inhibitors, has led to marked increases in survival among MM patients of all age groups [74–77]. Between 1995 and 2001, 5-year relative survival rates for MM were only 32 %, with slightly higher rates for men [60]. According to more recent analyses from the U.S. SEER Registry, 47 % (2005–2011) to 53 % (2008–2010) of patients diagnosed with MM now survive at least 5 years, an increase of more than 14 % over a ten-year period, although disparities between ethnic groups persist [60, 78, 79]. A study of 1038 patients from the Mayo Clinic observed improved survival for patients diagnosed between 2006 and 2010 compared to those diagnosed just a few years earlier (2001–2005), with the largest improvements among patients aged older than 65 years [80].

Worldwide, over 114,250 incident cases of MM were diagnosed in 2012, and over 80,000 people died from the disease, accounting for less than 1 % of all new cancer cases and cancer deaths (Fig. 29.6). The highest incidence and mortality rates occur in more developed countries compared to less developed countries [58]. The U.S. and Europe have similar MM incidence

(3.6 and 2.6 per 100,000, respectively) and mortality (1.9 and 1.4 per 100,000) rates, more than twice the rates observed in Africa and Asia [58]. Some of the geographical differences in incidence and mortality rates may be due to detection bias, resulting from the challenges of diagnosing MM in low resource settings [81].

## 29.4 Risk Factors for Hematologic Malignancies

### 29.4.1 Risk Factors for Hodgkin Lymphoma

#### 29.4.1.1 Introduction

The most consistent risk factors for HL are family history of HL and delayed exposure to the Epstein-Barr virus (EBV). More modest associations with increased risk of HL have been reported for childhood social environment, smoking, and higher BMI, whereas aspirin has been associated with reduced risk of HL. Each of these HL risk factors is hypothesized to act via immune-mediating mechanisms, such as by influencing the inflammatory milieu or by delaying the maturation of the immune responses that protect against the oncogenic effects of infectious agents like EBV. The following section will focus on risk factors for classical HL since NLPHL has a different pathogenesis and natural history [14].

#### 29.4.1.2 Family History and Genetic Susceptibility

Risk of HL in monozygotic twins with one twin affected is approximately 100-fold higher than the general population [82]. A more modest 4-fold increased risk of HL is seen among first-degree relatives of HL patients [83]. Furthermore, risk of HL is higher among relatives with an affected sibling compared to an affected parent [84]. Family history may increase risk of HL through shared childhood exposures, such as EBV infection, common inherited genetic susceptibility, or a combination of these factors.

Evidence from family as well as population-based studies has indicated a consistent association between human leukocyte antigen (HLA) type and HL risk [85]. Genome-wide association studies (GWAS) have replicated an association between genetic variation in HLA-related genes and HL risk, while also identifying a limited number of novel susceptibility loci [86–88].

#### 29.4.1.3 Infections

As noted above, exposure (and timing of exposure) to common childhood infections may influence risk of HL [89–94]. Hypotheses focus in particular on EBV infection, a ubiquitous herpesvirus and well-studied infectious risk factor for HL. Delayed first exposure to EBV can lead to infectious mononucleosis, which is characterized by EBV antibody profiles that indicate poor host immune control of EBV, similar to persons with severe or chronic EBV [95, 96]. It is plausible that such persons have diminished immune protection against the known oncogenic effects of EBV [97, 98]. Several cohort studies performed in the 1970s indicated a consistent 3-fold increased risk of HL in young adults with serologically confirmed infectious mononucleosis [99]. Subsequent studies have also shown an association between history of infectious mononucleosis and HL risk [100], as well as an approximately 3 to 4-fold higher risk for developing EBV-positive HL [101, 102]. Of interest, one serologic study suggested that chronic or more severe EBV is a risk factor for HL risk independent of history of infectious mononucleosis [97], confirming other studies of EBV serology and HL risk [98]. Furthermore, the EBV genome and expression of EBV-encoded latent gene products have been detected in approximately one third to more than two thirds of HL tumors depending on histologic subtype [103–108]. Whether EBV contributes to the etiology or pathogenesis of EBV-negative HL cases is unclear.

Human immunodeficiency virus (HIV)-infected individuals are at increased risk of developing HL as well as certain subtypes of NHL, including DLBCL and Burkitt lymphoma [109, 110]. Numerous other viral infections have been explored in relation to HL risk, though no consistent associations have been reported [111–113].

#### 29.4.1.4 Lifestyle Factors

Childhood environment. Many studies have shown a link between HL risk and factors related to childhood environment, which are hypothesized to reflect the timing of exposure to common infectious agents. The timing of such exposures is believed to influence underlying immune competency, especially with regard to the maturation of the immune responses that protect against the oncogenic effect of infections; those immune responses are not well developed at birth but are understood to be induced and matured through exposure to common infections during childhood [114, 115]. Several studies have linked increased sibship size and later birth order, each of which would increase the chances for earlier exposure to common infections, with reduced risk of HL in young adulthood [89, 91–94]. Nursery school or daycare attendance for at least one year, which could also increase the likelihood of (earlier) exposure to infections, was also associated with a reduced risk of HL among young adults in a large case-control study in Massachusetts and Connecticut [90]. Taken together, these studies suggest that decreased or delayed exposure to common childhood infection may increase risk of HL, possibly due to their underlying influence on host immune system maturation.

Aspirin. Regular use of aspirin, which has anti-inflammatory properties, was inversely associated with risk of HL in a large case-control study, whereas acetaminophen use was positively associated with HL risk [116]. A nationwide study of more than 6 million Danish residents also reported a modest inverse association between long-term aspirin use and HL risk, and a positive association for selective cyclooxygenase-2 inhibitors and other non-steroidal anti-inflammatory drugs with HL risk [117].

Smoking. A number of studies have linked smoking with increased risk of HL [90, 118–121]. In addition, a pooled analysis conducted by the InterLymph Consortium, including 3335 HL cases, reported a modest 10 % increased risk of HL among ever versus never smokers, particularly for mixed cellularity HL (Odds Ratio [OR] 1.60, 95 % Confidence Interval [CI] 1.29–1.99) and EBV-positive HL (OR 1.81, 95 % CI 1.27–2.56) among current smokers [122].

Obesity. Studies of body mass index (BMI) and HL have shown mixed results. In a meta-analysis of 7 prospective studies, the association with HL risk was null for BMI values considered to indicate overweight, but positive for BMI values associated with obesity (summary relative risk [RR] 1.41, 95 % CI 1.14–1.75) [123], a condition characterized by heightened inflammation and deregulation of several immune mediators and growth factors. A large cohort study of 1.3 million women in the United Kingdom, including 267 HL cases, reported that higher BMI was associated with an increased risk of HL [124]. In a case-control study, BMI was positively associated with HL among women aged 19–44 years, but inversely associated with HL risk at older ages [125]. In another case-control study, increasing BMI was associated with higher risk of HL among women younger than 35 years, but lower risk of HL among women aged 35 years or older; no association was found among men [126].

Diet. Associations between dietary factors and HL risk have been reported in few studies. A large case-control study of dietary patterns found that a diet high in desserts/sweets was associated with younger adult and EBV-negative, younger adult HL risk, while a diet high in meat was associated with older adult and EBV-negative, older adult HL risk [127]. The same case-control study reported a positive association of dietary intake of saturated fat, and an inverse association of monounsaturated fat, with younger adult HL risk [128]. A "Western" style dietary pattern and diets high in saturated fat have been associated with pro-inflammatory factors [129, 130], thereby suggesting a potential immune-modulating mechanism of diet on HL risk. Alcohol intake has been associated with reduced risk of HL in four studies [121, 131–133], while a fifth study found no association [120]. Potential mechanisms for alcohol and reduced risk of HL include improved immune function and the antioxidant properties of certain alcoholic beverages [134]. Additional dietary studies of HL risk have shown largely null or modest associations [135–139].

#### 29.4.1.5 Occupational and Environmental Exposures

Very little evidence exists linking occupational or environmental chemical exposures to HL risk. Weak positive associations have been noted for wood-related exposures and woodworking occupations, though many studies found no association [reviewed in 15]. Studies involving workplace exposure to chemicals have also been inconclusive [reviewed in 140].

### 29.4.2 Risk Factors for Non-Hodgkin Lymphoma

#### 29.4.2.1 Introduction

The most established risk factor for NHL is severe immune compromise, such as that associated with HIV infection or post-transplant medication use. However, most cases of NHL occur in persons with no overt immune compromise, suggesting that other immune-modulating exposures influence their development. NHL is a group of etiologically and epidemiologically distinct disease subtypes; as such, reported associations between some risk factors and NHL appear to vary by subtype. Other risk factors appear to be consistently associated with risk of most NHL subtypes, including family history of hematologic cancer, autoimmune and atopic diseases, certain lifestyle factors, and sun exposure [141–143]. The incidence of most NHL subtypes increases with age; however, certain subtypes, such as endemic Burkitt lymphoma and ALL, are more prevalent among children and adolescents. Overall, men experience higher incidence rates of NHL compared to women, but some histologic types do not demonstrate gender differences in incidence rates. The following section will summarize the literature on NHL risk factors, focusing on associations with the most common subtypes of the disease to reflect the significant etiologic complexity of NHL [141].

#### 29.4.2.2 Family History and Genetic Susceptibility

Individuals with a family history of a first-degree relative diagnosed with HL and NHL, leukemia, or MM are at an increased risk for being diagnosed with NHL; in particular, a family history of NHL or HL is associated with an increased risk of a B-cell or T-cell NHL diagnosis [141]. A family history of MM is associated with an increased risk of mycosis fungoides and Sézary syndrome, which are T-cell lymphomas that originate in the skin [24]. Individual studies of candidate genes and also pooled genetic studies by the InterLymph Consortium have reported genetic susceptibility from common variants in genes related to the host immune response, cell cycle control, DNA repair and other plausible pathways [144–150]; more recently, numerous novel susceptibility loci in immunoregulatory, human leukocyte antigen-related, and other genes have been reported for several major histologic types of NHL by the ongoing, large international GWAS of NHL [151–154].

#### 29.4.2.3 Immunosuppression, Autoimmune Disease, and Infections

Immunosuppression has long been associated with an increased risk of NHL. NHL is among the most common cancers diagnosed in recipients of solid organ transplant and is an AIDS-defining malignancy in individuals infected with HIV [155–158]. Furthermore, individuals diagnosed with certain autoimmune diseases, including rheumatoid arthritis, systemic lupus erythematosus, and Sjögren's syndrome, are at an increased risk for NHL [159, 160]. Allergies, hay fever, or atopic disease have been associated with a reduced risk of several NHL subtypes including FL, CLL/SLL, peripheral T-cell lymphomas, mantle cell lymphoma, LPL/WM and sporadic Burkitt lymphoma (i.e. non-African Burkitt lymphoma) [161–166]. In contrast, eczema may increase the risk of T-cell lymphomas and sporadic Burkitt lymphoma.

Given that some of the evidence for the allergy/atopy association with NHL derives from case-control studies, prospective data are needed to assess the influence of subclinical disease (i.e., reverse causality) on the published observations.

Several subtypes are associated with infection by a specific agent. For example, human T-cell lymphotropic virus-type 1 (HTLV-1) is associated with and a necessary cause of ATL, although incidence varies greatly by geographic area among HTLV-1-endemic populations [167, 168]. Also, infection with Helicobacter pylori is associated with an increased risk of MALT lymphoma [169], and infection with EBV is associated with 97 % of endemic Burkitt lymphoma tumors in equatorial Africa [170, 171]. Sporadic Burkitt lymphoma does not exhibit such a strong causal relationship with EBV, and is often diagnosed at older ages [172, 173]. EBV has also been associated with other subtypes of NHL, although generally not with most NHL [174]. Infection with Hepatitis C (HCV) has also been associated with an increased risk of B-cell NHL, including DLBCL, CLL/SLL, LPL/WM, marginal zone lymphoma and sporadic Burkitt lymphoma in populations with higher HCV prevalence [141, 163, 165, 166, 175–177].

#### 29.4.2.4 Lifestyle Factors

Obesity/Physical Activity. Several studies have reported an association between obesity and other anthropometric factors and NHL risk [178]; however, published associations across specific subtypes are inconsistent, and thus reported associations with all NHL types combined are challenging to interpret. No association was reported between obesity and all NHL types combined in a pooled analysis of 18 case-control studies; however, an 80 % increased risk of DLBCL was associated with severe obesity (BMI ≥ 40 kg/m$^2$) [179]. A meta-analysis of prospective studies reported a 7 % increased risk of all NHL combined, and a 14 % increased risk of overall NHL mortality associated with a 5 kg/m$^2$ increase in BMI. When restricted to studies reporting NHL subtype, increased BMI was only significantly associated with an increased risk of DLBCL [123]. Higher young adult BMI has also been associated with an increased risk of DLBCL and FL [161, 177]. In summary, a relationship between obesity and NHL risk appears strongest among patients diagnosed with DLBCL, the most commonly diagnosed NHL subtype; it is less clear whether obesity is associated with risk of (or mortality from) other histologic types.

Studies have not consistently reported an association between physical activity and NHL risk, overall or by subtype [180, 181]. A recent meta-analysis of 23 studies observed a borderline significant 9 % decreased risk of all NHL subtypes combined when comparing high versus low physical activity levels, but found no association with risk of DLBCL or FL subtype [182].

Sun Exposure and Vitamin D. Several epidemiologic studies have reported a positive association between high levels of ambient ultraviolet (UV) exposure and NHL risk, with variation by subtype and race/ethnicity [183]; however, other studies have not observed such an association. In contrast, a large international pooled study by the InterLymph Consortium recently reported an inverse association of recreational sun exposure with risk of NHL that did not demonstrate significant variability by histologic type [141]. A strong relationship between dietary or serum Vitamin D and NHL risk is not apparent in the literature [142, 143, 184, 185].

Diet. Analyses of diet and NHL risk have been inconsistent and vary by subtype. Studies of dietary pattern suggest a positive association between a Western-style diet high in fats and meat and risk of NHL, in particular an increased risk of FL [186, 187], while a Mediterranean diet was not associated with NHL risk [188]. These findings are consistent with other reports of an increased risk of NHL in persons with higher consumption of total and trans fats, and of red meats [189]. Dietary phytocompounds were not associated with B-cell NHL in a cohort of women [135]. Relatively consistent reports suggest an inverse association of vegetables [190, 191] and of marine fatty acids [192] and risk of some NHL subtypes.

Smoking. Smoking is associated with some NHL subtypes, and within those subtypes, has shown variation by anatomical site. For example, smoking is associated with DLBCL that presents in the CNS [193]. Cigarette smoking is associated with an increased risk of FL, mycosis fungoides and Sézary syndrome [162, 165], peripheral T-cell lymphoma, and LPL/WM, and an inverse association with the rare subtype hairy cell leukemia [24, 141, 161, 194].

Alcohol. A pooled analysis of 9 international case-control studies reported a protective association between people who currently drink alcohol and risk of NHL, compared to nondrinkers, with weaker associations for former drinkers [195]. The authors did not observe a dose-response relationship between the amount of alcohol consumed and NHL risk. An updated pooled analysis of case-control studies reported protective associations between ever-drinkers of alcohol and risk of overall NHL, DLBCL, and FL, when compared to nondrinkers, although no association was observed for risk of MZL or CLL/SLL [141]. However, a prospective analysis found an elevated risk of overall B-cell NHL and of FL in women who were former alcohol drinkers, suggesting the timing of alcohol consumption, as well as factors associated with quitting drinking, may be relevant to lymphomagenesis [196].

#### 29.4.2.5 Occupational and Environmental Exposures

A large European study including 866 cases of NHL observed an increased risk of disease among car repair workers and butchers, possibly reflecting exposure to solvents or zoonotic viruses [197]. A pooled analysis of four case-control studies including more than 3700 cases of NHL observed an increased risk associated with occupational exposure to trichloroethylene (TCE) [198]. Similarly, a meta-analysis of cohort and case-control studies also observed an increased risk of NHL with occupational TCE exposure [199].

Teachers may have a reduced risk of being diagnosed with multiple subtypes of NHL, including marginal zone and FL, while painters and farm workers may be at an increased risk [141, 200]. Positive associations have been observed between farm workers and DLBCL particularly among women, CLL/SLL, hairy cell leukemia, mantle cell lymphoma, and mycosis fungoides and Sézary syndrome. An inverse association was observed between farm work and peripheral T-cell lymphoma in a pooled analysis of 15 case-control studies [24, 162, 164, 166, 177, 194]. Other occupations with consistent positive associations reported with NHL subtypes, including DLBCL, across studies include seamstress or embroiderer and hairdresser [162, 166, 177]. Reported associations between NHL risk and exposure to lead or cadmium have been inconsistent [201, 202].

### 29.4.3 Risk Factors for Multiple Myeloma

#### 29.4.3.1 Introduction

Few modifiable risk factors for MM have been identified. Since this malignancy is often diagnosed at an advanced stage, and survival rates remain modest despite recent improvements in treatment, identification of risk factors and means of prevention are paramount to reducing the burden of MM. In addition to the growing literature linking genetic markers to MM risk and survival, characterization of the role of cytokine and growth factor signaling in the tumor microenvironment is becoming increasingly important to understanding myeloma pathogenesis. In particular, insights from the well-characterized pathogenic roles for some cytokine and growth factor pathways may inform etiologic hypotheses. The following section summarizes the epidemiologic evidence elucidating risk factors for MM.

#### 29.4.3.2 Family History and Genetic Susceptibility

Individuals with a first-degree relative diagnosed with MGUS or MM have a 2–3 fold increased risk of developing either disease themselves [48, 203]. A family history of B-cell lymphoproliferative diseases is also associated with an increased risk of MM [204, 205]. Studies of familial MM have also observed an increased incidence of both hematologic and solid tumors in relatives of patients with MM, as well as an early age at myeloma diagnosis in successive generations (i.e. "anticipation"), further suggesting a role for genetic heritability [206].

Two GWAS identified 7 single nucleotide polymorphisms (SNP) significantly associated with MM risk — 6 SNPs were positively associated and 1 SNP was inversely associated with myeloma risk [207–210]. The first GWAS of MM survival identified variants at 16p13 near the gene FOPNL associated with survival [211]. Certain genetic mutations and chromosomal aberrations have been consistently associated with poor survival among patients with MM, including t(4;14) and 17p13 deletions [75, 212, 213]. Evidence is inconclusive for t(14;16) and chromosome 1q21 amplification [75].

Variability in several genes has been associated with an increased risk of MM; however, the 11 strongest associations reported between specific SNPs and MM risk from candidate gene studies were not validated in a recent analysis by the International Multiple Myeloma rESEarch (IMMEnSE) consortium [214]. Variability in genes that encode molecules on the insulin-like growth factor (IGF)-1 and interleukin (IL)-6 signaling pathways, including the gene encoding insulin receptor substrate-1 (IRS-1), and the gene encoding IL-6 (IL6) and the IL-6 receptor (IL6R), have been associated with risk of MM in U.S. studies [215, 216] supporting earlier investigations linking immune system modulation and serum levels of cytokines to MM development [217, 218]. Numerous prognostic gene expression profiling signatures for MM have been reported; currently, the IMWG is working to unify the prognostic signatures through prognostic modeling [75].

Cytokine and growth factor signaling in the tumor microenvironment may also be important to myeloma development, and this is a developing area of interest for etiologic study. Prediagnostic serologic levels of IGF-binding protein (BP)-1 were positively associated with MM risk within 3 years of diagnosis, and sIL-6R with MM risk within 6 years of diagnosis in a pooled analysis of 8 cohort studies [219].

#### 29.4.3.3 Lifestyle Factors

Obesity/Physical Activity. When considering the associations between potentially modifiable lifestyle factors and MM, the most consistently reported association has been that between obesity and MM. A pooled analysis of 20 prospective studies including 1388 MM deaths observed a positive association of mortality from myeloma with both young-adult and cohort-entry BMI. A separate positive association was observed between MM mortality and high waist circumference. Women who reported both young-adult BMI > 25 and cohort-entry BMI > 30 had a nearly 2-fold increased risk of dying from MM compared to women with normal BMI at both time points; this association was not evident in men [220]. Obesity and overweight BMI have also been associated with an increased risk of incident myeloma [181, 221]. Of note, these findings for BMI are consistent with the aforementioned data from serologic studies of other hormonal pathways that are altered in obese individuals and important in MM pathogenesis. Data on physical activity and sedentary behavior is more limited, but suggest an inverse association with physical activity and a positive association with increased sitting time [181, 221, 222].

The findings for energy balance-related risk factors appear to be corroborated by biomarker studies, including the observation that circulating adipokines, which are cytokines secreted by adipose tissue, may play a role in myelomagenesis, and that adiponectin levels in particular are inversely associated with myeloma risk in several studies [223–225].

Aspirin. Aspirin use has been associated with a significant 37–39 % reduced risk of MM in a prospective analysis, among adults taking ≥5 adult strength tablets per week, and those with ≥11 years of continuous use [226]. Although another large prospective study did not observe an association of aspirin use with MM risk [227], the aforementioned findings are plausible given the known role of NF-κB-mediated pathways in MM pathogenesis and the down-regulatory effect of aspirin on NF-κB and several of its downstream targets [228–230]. However, a hospital-based case-control study did not observe an association between aspirin use and MM risk, although regular acetaminophen use was associated with a nearly 3-fold increased risk of myeloma in that population [231].

Alcohol and Smoking. A pooled analysis of 6 case-control studies including 1567 MM cases suggests an inverse association between self-reported alcohol use and myeloma risk [232]. A separate pooled analysis of 9 international case-control studies including 2670 cases found no evidence of an association between cigarette smoking and risk of MM [233].

#### 29.4.3.4 Occupational and Environmental Exposures

Reports of positive associations between MM risk and a number of environmental and occupational exposures have long been suggested in the literature. Studies have focused on agricultural workers, employees of the petroleum and leather industries, cosmetologists, and firefighters, with the most consistent associations reported following exposure to pesticides [234–242]. Studies have also reported a suggested increased risk of MM in radiation workers [243–245] and among hairdressers and those exposed to hair dye [246]. However, the associations between occupational and environmental factors and MM risk have been modest and inconsistent across studies, possibly due to inconsistent or biased methods for assessing exposure, inconsistent ranges of exposure across populations, or perhaps suggesting these factors may not be strongly related to MM.

## 29.5 Summary

Together, hematologic malignancies account for about 10 % of all cancers diagnosed in the U.S. each year [56]. However, hematologic malignancies are in fact a diverse group of diseases, with distinct combinations of cell of origin, clinical presentation, descriptive epidemiology, and known risk factors, as detailed above. The emerging knowledge of molecular subtypes within specific histologic types of these tumors further adds complexity to the study of their etiology and prevention. Several of these cancers share common aspects in their pathogenesis, for example a role for inflammation and immune activation, even if specific signaling pathway alterations may vary across types. As a result, a given immune-modulating risk factor, such as autoimmune disease, infection, obesity or immunosuppression may demonstrate an association with more than one type, but not necessarily with all types, of hematologic cancer.

As a group, hematologic malignancies appear to share some epidemiologic risk factors, such as a family history of hematologic cancer. The knowledge of increased risk among first-degree relatives of hematologic cancer or MGUS patients further underscores the urgency of identifying modifiable risk factors and prevention strategies for these malignancies. However, to date many putative risk factors do not show a universal association with all subtypes. For example, while certain histologic subtypes of both HL and NHL have been linked to infection with EBV, others show no association with infectious agents. Some tumors have known and increasingly well-characterized pre-malignant conditions that are believed to precede all malignant diagnoses (i.e., MGUS prior to MM or WM), but we know very little regarding risk factors for progression from pre-malignancy. Similarly, we do not know whether putative risk factors vary between more and less aggressive molecular subtypes that have been identified and found to hold clinical relevance for some histologic types of hematologic cancer (such as DLBCL, CLL, FL and others). Studies that incorporate the collection and molecular interrogation of diagnostic tumor tissue, and that evaluate risk factors separately by molecular type, will likely offer valuable insights regarding opportunities to prevent and diminish the severity of those tumors. Further, much of the current understanding of risk factors for hematologic malignancies derives from case-control studies that are vulnerable to survival, recall and other biases such as reverse causality. Thus, clarification of purported etiologic associations and exploration of questions such as the most relevant timing of exposure for influencing disease risk are urgently needed from prospective cohort studies.

## References


1. Ma X. Epidemiology of myelodysplastic syndromes. Am J Med. 2012;125(7 Suppl):S2–5.
2. Estey E, Dohner H. Acute myeloid leukaemia. Lancet. 2006;368(9550):1894–907.
3. Dohner H, Weisdorf DJ, Bloomfield CD. Acute myeloid leukemia. N Engl J Med. 2015;373 (12):1136–52.
4. Rubnitz JE, Gibson B, Smith FO. Acute myeloid leukemia. Pediatr Clin N Am. 2008;55(1):21–51, ix.
5. Hoglund M, Sandin F, Simonsson B. Epidemiology of chronic myeloid leukaemia: an update. Ann Hematol. 2015;94(Suppl 2):S241–7.
6. Linet MS, Devesa SS, Morgan GJ. The Leukemias. In: Schottenfeld D, Fraumeni Jr. JF, editors. Cancer Epidemiology and Prevention, 3rd ed. New York. Oxford University Press; 2006.
7. Visser O, Trama A, Maynadie M, Stiller C, Marcos-Gragera R, De Angelis R, et al. Incidence, survival and prevalence of myeloid malignancies in Europe. Eur J Cancer. 2012;48(17):3257–66.
8. Jaffe ES, Harris NL, Stein H, Vardiman JW (eds). Pathology and genetics of tumours of haematopoietic and lymphoid tissues. World Health Organization Classification of Tumours. Lyon, France: IARC Press 2001.
9. Swerdlow SH, Campo E, Harris NL, Jaffe ES, Pileri SA, Stein H, et al. WHO classification of tumours of haematopoietic and lymphoid tissues. Lyon, France: IARC Press; 2008.
10. Swerdlow SH, Campo E, Pileri SA, Harris NL, Stein H, Siebert R, et al. The 2016 revision to the World Health Organization classification of lymphoid neoplasms. Blood. 2016;127(20):2375–90.
11. Morton LM, Turner JJ, Cerhan JR, Linet MS, Treseler PA, Clarke CA, et al. Proposed classification of lymphoid neoplasms for epidemiologic research from the pathology working group of the international lymphoma epidemiology consortium (InterLymph). Blood. 2007;110(2):695–708.
12. Turner JJ, Morton LM, Linet MS, Clarke CA, Kadin ME, Vajdic CM, et al. InterLymph hierarchical classification of lymphoid neoplasms for epidemiologic research based on the WHO classification (2008): update and future directions. Blood. 2010;116(20):e90–8.
13. Gobbi PG, Ferreri AJ, Ponzoni M, Levis A. Hodgkin lymphoma. Crit Rev Oncol Hematol. 2013;85(2):216–37.
14. Hoppe RT, Advani RH, Ai WZ, Ambinder RF, Aoun P, Bello CM, et al. Hodgkin lymphoma, version 2.2015. J Nat Compr Cancer Netw JNCCN. 2015;13(5):554–86.
15. Mueller NE, Grufferman S. Hodgkin lymphoma. In: Schottenfeld D, Fraumeni Jr. JF, editors. Cancer epidemiology and prevention, 3rd ed. New York. Oxford University Press;2006.
16. Marshall NA, Christie LE, Munro LR, Culligan DJ, Johnston PW, Barker RN, et al. Immunosuppressive regulatory T cells are abundant in the reactive lymphocytes of Hodgkin lymphoma. Blood. 2004;103(5):1755–62.
17. Poppema S. Immunobiology and pathophysiology of Hodgkin lymphomas. Hematol Educ Program Am Soc Hematol Am Soc Hematol Educ Program. 2005:231–8.
18. Kuppers R. The biology of Hodgkin's lymphoma. Nat Rev Cancer. 2009;9(1):15–27.
19. Kuppers R, Klein U, Hansmann ML, Rajewsky K. Cellular origin of human B-cell lymphomas. N Engl J Med. 1999;341(20):1520–9.
20. A clinical evaluation of the International Lymphoma Study Group classification of non-Hodgkin's lymphoma. The Non-Hodgkin's lymphoma classification project. Blood. 1997;89(11):3909–18.
21. Muller AM, Ihorst G, Mertelsmann R, Engelhardt M. Epidemiology of non-Hodgkin's lymphoma (NHL): trends, geographic distribution, and etiology. Ann Hematol. 2005;84(1):1–12.
22. Al-Hamadani M, Habermann TM, Cerhan JR, Macon WR, Maurer MJ, Go RS. Non-Hodgkin lymphoma subtype distribution, geodemographic patterns, and survival in the US: A longitudinal analysis of the national cancer data base from 1998 to 2011. Am J Hematol. 2015;90(9):790–5.
23. Anderson JR, Armitage JO, Weisenburger DD. Epidemiology of the non-Hodgkin's lymphomas: distributions of the major subtypes differ by geographic locations. Non-Hodgkin's Lymphoma Classification Project. Ann Oncol Official J Eur Soc Med Oncol ESMO. 1998;9(7):717–20.
24. Aschebrook-Kilfoy B, Cocco P, La Vecchia C, Chang ET, Vajdic CM, Kadin ME, et al. Medical history, lifestyle, family history, and occupational risk factors for mycosis fungoides and Sezary syndrome: the InterLymph Non-Hodgkin Lymphoma Subtypes Project. J Nat Cancer Inst Monogr. 2014;2014(48):98–105.
25. Huh J. Epidemiologic overview of malignant lymphoma. Korean J Hematol. 2012;47(2):92–104.
26. Swerdlow SH, Campo E, Harris NL, Jaffe ES, Pileri SA, Stein H, Thiele J, Vardiman JW, editors. WHO Classification of tumours of haematopoietic and lymphoid tissues. 4th ed. International Agency for Research on Cancer: Lyon; 2008.

27. Kyle RA, Gertz MA, Witzig TE, Lust JA, Lacy MQ, Dispenzieri A, et al. Review of 1027 patients with newly diagnosed multiple myeloma. Mayo Clin Proc. 2003;78(1):21–33.
28. Landgren O, Kyle RA, Pfeiffer RM, Katzmann JA, Caporaso NE, Hayes RB, et al. Monoclonal gammopathy of undetermined significance (MGUS) consistently precedes multiple myeloma: a prospective study. Blood. 2009;113(22):5412–7.
29. Muller-Hermelink HK, Montserrat E, Catovsky D, Campo E, Harris NL, Stein H. Chronic lymphocytic leukemia/small lymphocytic lymphoma. In: Swerdlow SH, Campo E, Harris NL, Jaffe ES, Pileri SA, Stein H, Thiele H, Vardiman JW, editors. WHO classification of tumors of haematopoietic and lymphoid tissue. Lyon: International Agency for Research on Cancer; 2008. p. 180–93.
30. Choi WW, Weisenburger DD, Greiner TC, Piris MA, Banham AH, Delabie J, et al. A new immunostain algorithm classifies diffuse large B-cell lymphoma into molecular subtypes with high accuracy. Clin Cancer Res Official J Am Assoc Cancer Res. 2009;15(17):5494–502.
31. Meyer PN, Fu K, Greiner TC, Smith LM, Delabie J, Gascoyne RD, et al. Immunohistochemical methods for predicting cell of origin and survival in patients with diffuse large B-cell lymphoma treated with rituximab. J Clin Oncol Official J Am Soc Clin Oncol. 2011;29(2):200–7.
32. Hans CP, Weisenburger DD, Greiner TC, Gascoyne RD, Delabie J, Ott G, et al. Confirmation of the molecular classification of diffuse large B-cell lymphoma by immunohistochemistry using a tissue microarray. Blood. 2004;103(1):275–82.
33. Lenz G. Insights into the molecular pathogenesis of activated B-cell-like diffuse large B-Cell lymphoma and its therapeutic implications. Cancers (Basel). 2015;7(2):811–22.
34. Lenz G, Wright GW, Emre NC, Kohlhammer H, Dave SS, Davis RE, et al. Molecular subtypes of diffuse large B-cell lymphoma arise by distinct genetic pathways. Proc Natl Acad Sci U S A. 2008;105(36):13520–5.
35. Sakata S, Tsuyama N, Takeuchi K. Pathology of indolent B-cell neoplasms other than follicular lymphoma. J Clin Exp Hematop. 2014;54(1):11–22.
36. Kridel R, Sehn LH, Gascoyne RD. Pathogenesis of follicular lymphoma. J Clin Invest. 2012;122(10):3424–31.
37. Kyle RA, Durie BG, Rajkumar SV, Landgren O, Blade J, Merlini G, et al. Monoclonal gammopathy of undetermined significance (MGUS) and smoldering (asymptomatic) multiple myeloma: IMWG consensus perspectives risk factors for progression and guidelines for monitoring and management. Leukemia. 2010;24(6):1121–7.
38. Rajkumar SV, Dimopoulos MA, Palumbo A, Blade J, Merlini G, Mateos MV, et al. International myeloma working group updated criteria for the diagnosis of multiple myeloma. Lancet Oncol. 2014;15(12):e538–48.
39. International Myeloma Working Group. Criteria for the classification of monoclonal gammopathies, multiple myeloma and related disorders: a report of the International myeloma working group. Br J Haematol. 2003;121(5):749–57.
40. Kyle RA, Rajkumar SV. Treatment of multiple myeloma: a comprehensive review. Clin Lymphoma Myeloma. 2009;9(4):278–88.
41. Palumbo A, Avet-Loiseau H, Oliva S, Lokhorst HM, Goldschmidt H, Rosinol L, et al. Revised international staging system for multiple myeloma: a report from international myeloma working group. J Clin Oncol Official J Am Soc Clin Oncol. 2015;33(26):2863–9.
42. Weiss BM, Abadie J, Verma P, Howard RS, Kuehl WM. A monoclonal gammopathy precedes multiple myeloma in most patients. Blood. 2009;113(22):5418–22.
43. Landgren O, Kyle RA, Hoppin JA, Beane Freeman LE, Cerhan JR, Katzmann JA, et al. Pesticide exposure and risk of monoclonal gammopathy of undetermined significance in the agricultural health study. Blood. 2009;113(25):6386–91.
44. Landgren O, Weiss BM. Patterns of monoclonal gammopathy of undetermined significance and multiple myeloma in various ethnic/racial groups: support for genetic factors in pathogenesis. Leukemia. 2009;23(10):1691–7.
45. Watanaboonyongcharoen P, Nakorn TN, Rojnuckarin P, Lawasut P, Intragumtornchai T. Prevalence of monoclonal gammopathy of undetermined significance in Thailand. Int J Hematol. 2012;95(2):176–81.
46. Landgren O, Katzmann JA, Hsing AW, Pfeiffer RM, Kyle RA, Yeboah ED, et al. Prevalence of monoclonal gammopathy of undetermined significance among men in Ghana. Mayo Clin Proc. 2007;82(12):1468–73.
47. Wu SP, Minter A, Costello R, Zingone A, Lee CK, Au WY, et al. MGUS prevalence in an ethnically Chinese population in Hong Kong. Blood. 2013;121(12):2363–4.
48. Vachon CM, Kyle RA, Therneau TM, Foreman BJ, Larson DR, Colby CL, et al. Increased risk of monoclonal gammopathy in first-degree relatives of patients with multiple myeloma or monoclonal gammopathy of undetermined significance. Blood. 2009;114(4):785–90.
49. Landgren O, Kristinsson SY, Goldin LR, Caporaso NE, Blimark C, Mellqvist UH, et al. Risk of plasma cell and lymphoproliferative disorders among 14621 first-degree relatives of 4458 patients with monoclonal gammopathy of undetermined significance in Sweden. Blood. 2009;114(4):791–5.
50. Kyle RA, Remstein ED, Therneau TM, Dispenzieri A, Kurtin PJ, Hodnefield JM, et al. Clinical course and prognosis of smoldering (asymptomatic)

multiple myeloma. N Engl J Med. 2007;356(25):2582–90.
51. Mateos MV, Hernandez MT, Giraldo P, de la Rubia J, de Arriba F, Lopez Corral L, et al. Lenalidomide plus dexamethasone for high-risk smoldering multiple myeloma. N Engl J Med. 2013;369(5):438–47.
52. van de Donk NW, Lokhorst HM, Anderson KC, Richardson PG. How I treat plasma cell leukemia. Blood. 2012;120(12):2376–89.
53. Kyle RA, Maldonado JE, Bayrd ED. Plasma cell leukemia. Report on 17 cases. Arch Intern Med. 1974;133(5):813–8.
54. Fernandez de Larrea C, Kyle RA, Durie BG, Ludwig H, Usmani S, Vesole DH, et al. Plasma cell leukemia: consensus statement on diagnostic requirements, response criteria and treatment recommendations by the international myeloma working group. Leukemia. 2013;27(4):780–91.
55. Noel P, Kyle RA. Plasma cell leukemia: an evaluation of response to therapy. Am J Med. 1987;83(6):1062–8.
56. American Cancer Society. Cancer facts and figures 2015. Atlanta, GA: American Cancer Society; 2015.
57. Howlader N, Noone AM, Krapcho M, Garshell J, Miller D, Altekruse SF, Kosary CL, Yu M, Ruhl J, Tatalovich Z, Mariotto A, Lewis DR, Chen HS, Feuer EJ, Cronin KA (eds). SEER Cancer Statistics Review, 1975–2011, National Cancer Institute. Bethesda, MD. http://seer.cancer.gov/csr/1975_2011/, based on November 2013 SEER data submission, posted to the SEER web site, April 2014.
58. Ferlay J, Soerjomataram I, Ervik M, Dikshit R, Eser S, Mathers C, Rebelo M, Parkin DM, Forman D, Bray, F. GLOBOCAN 2012 v1.0, Cancer Incidence and Mortality Worldwide: IARC CancerBase No. 11 [Internet]. Lyon, France: International Agency for Research on Cancer; 2013. Available from: http://globocan.iarc.fr. Accessed on 20 July 2015.
59. American Cancer Society. Cancer facts and figures 2014. Atlanta: American Cancer Society; 2014.
60. Howlader N, Noone AM, Krapcho M, Garshell J, Miller D, Altekruse SF, Kosary CL, Yu M, Ruhl J, Tatalovich Z,Mariotto A, Lewis DR, Chen HS, Feuer EJ, Cronin KA (eds). SEER cancer statistics review, 1975–2012, National Cancer Institute. Bethesda, MD, http://seer.cancer.gov/csr/1975_2012/, based on November 2014 SEER data submission, posted to the SEER web site, April 2015..
61. Boffetta PI. Epidemiology of adult non-Hodgkin lymphoma. Ann Oncol Official J Eur Soc Med Oncol ESMO. 2011;22(suppl 4):iv27–iv31.
62. Howlader N, Noone AM, Krapcho M, Garshell J, Neyman N, Altekruse SF, Kosary CL, Yu M, Ruhl J, Tatalovich Z, Cho H, Mariotto A, Lewis DR, Chen HS, Feuer EJ, Cronin KA (eds). SEER cancer statistics review, 1975–2010, National Cancer Institute. Bethesda, MD, http://seer.cancer.gov/csr/1975_2010/, based on November 2012 SEER data submission, posted to the SEER web site, April 2013.
63. Landgren O, Gridley G, Turesson I, Caporaso NE, Goldin LR, Baris D, et al. Risk of monoclonal gammopathy of undetermined significance (MGUS) and subsequent multiple myeloma among African American and white veterans in the United States. Blood. 2006;107(3):904–6.
64. Landgren O, Graubard BI, Katzmann JA, Kyle RA, Ahmadizadeh I, Clark R, et al. Racial disparities in the prevalence of monoclonal gammopathies: a population-based study of 12,482 persons from the national health and nutritional examination survey. Leukemia. 2014;28(7):1537–42.
65. Corso A, Klersy C, Lazzarino M, Bernasconi C. Multiple myeloma in younger patients: the role of age as prognostic factor. Ann Hematol. 1998;76(2):67–72.
66. Ludwig H, Durie BG, Bolejack V, Turesson I, Kyle RA, Blade J, et al. Myeloma in patients younger than age 50 years presents with more favorable features and shows better survival: an analysis of 10 549 patients from the International Myeloma Working Group. Blood. 2008;111(8):4039–47.
67. Ludwig H, Bolejack V, Crowley J, Blade J, Miguel JS, Kyle RA, et al. Survival and years of life lost in different age cohorts of patients with multiple myeloma. J Clin Oncol Official J Am Soc Clin Oncol. 2010;28(9):1599–605.
68. Turesson I, Velez R, Kristinsson SY, Landgren O. Patterns of multiple myeloma during the past 5 decades: stable incidence rates for all age groups in the population but rapidly changing age distribution in the clinic. Mayo Clin Proc. 2010;85(3):225–30.
69. Tuchman SA, Shapiro GR, Ershler WB, Badros A, Cohen HJ, Dispenzieri A, et al. Multiple myeloma in the very old: an IASIA conference report. J Nat Cancer Inst. 2014;106(5).
70. Baxi SS, Matasar MJ. State-of-the-art issues in Hodgkin's lymphoma survivorship. Curr Oncol Rep. 2010;12(6):366–73.
71. van Leeuwen FE, Klokman WJ, Hagenbeek A, Noyon R, van den Belt-Dusebout AW, van Kerkhoff EH, et al. Second cancer risk following Hodgkin's disease: a 20-year follow-up study. J Clin Oncol Official J Am Soc Clin Oncol. 1994;12(2):312–25.
72. Kaldor JM, Day NE, Clarke EA, Van Leeuwen FE, Henry-Amar M, Fiorentino MV, et al. Leukemia following Hodgkin's disease. N Engl J Med. 1990;322(1):7–13.
73. Hodgson DC, Gilbert ES, Dores GM, Schonfeld SJ, Lynch CF, Storm H, et al. Long-term solid cancer risk among 5-year survivors of Hodgkin's lymphoma. J Clin Oncol Official J Am Soc Clin Oncol. 2007;25(12):1489–97.

74. Kumar SK, Rajkumar SV, Dispenzieri A, Lacy MQ, Hayman SR, Buadi FK, et al. Improved survival in multiple myeloma and the impact of novel therapies. Blood. 2008;111(5):2516–20.
75. Chng WJ, Dispenzieri A, Chim CS, Fonseca R, Goldschmidt H, Lentzsch S, et al. IMWG consensus on risk stratification in multiple myeloma. Leukemia. 2014;28(2):269–77.
76. Warren JL, Harlan LC, Stevens J, Little RF, Abel GA. Multiple myeloma treatment transformed: a population-based study of changes in initial management approaches in the United States. J Clin Oncol Official J Am Soc Clin Oncol. 2013;31(16):1984–9.
77. Turesson I, Velez R, Kristinsson SY, Landgren O. Patterns of improved survival in patients with multiple myeloma in the twenty-first century: a population-based study. J Clin Oncol Official J Am Soc Clin Oncol. 2010;28(5):830–4.
78. Ailawadhi S, Aldoss IT, Yang D, Razavi P, Cozen W, Sher T, et al. Outcome disparities in multiple myeloma: a SEER-based comparative analysis of ethnic subgroups. Br J Haematol. 2012;158(1):91–8.
79. Pulte D, Jansen L, Castro FA, Emrich K, Katalinic A, Holleczek B, et al. Trends in survival of multiple myeloma patients in Germany and the United States in the first decade of the 21st century. Br J Haematol. 2015.
80. Kumar SK, Dispenzieri A, Lacy MQ, Gertz MA, Buadi FK, Pandey S, et al. Continued improvement in survival in multiple myeloma: changes in early mortality and outcomes in older patients. Leukemia. 2014;28(5):1122–8.
81. De Roos AJ, Baris D, Weiss NS, Herrington LJ. Multiple Myeloma. In: Schottenfeld D, Fraumeni Jr. JF, editors. Cancer epidemiology and prevention, 3rd ed. New York. Oxford University Press;2006.
82. Mack TM, Cozen W, Shibata DK, Weiss LM, Nathwani BN, Hernandez AM, et al. Concordance for Hodgkin's disease in identical twins suggesting genetic susceptibility to the young-adult form of the disease. N Engl J Med. 1995;332(7):413–8.
83. Goldin LR, Bjorkholm M, Kristinsson SY, Turesson I, Landgren O. Highly increased familial risks for specific lymphoma subtypes. Br J Haematol. 2009;146(1):91–4.
84. Goldin LR, Pfeiffer RM, Gridley G, Gail MH, Li X, Mellemkjaer L, et al. Familial aggregation of Hodgkin lymphoma and related tumors. Cancer. 2004;100(9):1902–8.
85. Diepstra A, Niens M, te Meerman GJ, Poppema S, van den Berg A. Genetic susceptibility to Hodgkin's lymphoma associated with the human leukocyte antigen region. Eur J Haematol Suppl. 2005;66:34–41.
86. Enciso-Mora V, Broderick P, Ma Y, Jarrett RF, Hjalgrim H, Hemminki K, et al. A genome-wide association study of Hodgkin's lymphoma identifies new susceptibility loci at 2p16.1 (REL), 8q24.21 and 10p14 (GATA3). Nat Genet. 2010;42(12):1126–30.
87. Urayama KY, Jarrett RF, Hjalgrim H, Diepstra A, Kamatani Y, Chabrier A, et al. Genome-wide association study of classical Hodgkin lymphoma and Epstein-Barr virus status-defined subgroups. J Natl Cancer Inst. 2012;104(3):240–53.
88. Frampton M, da Silva Filho MI, Broderick P, Thomsen H, Forsti A, Vijayakrishnan J, et al. Variation at 3p24.1 and 6q23.3 influences the risk of Hodgkin's lymphoma. Nat Commun. 2013;4:2549.
89. Westergaard T, Melbye M, Pedersen JB, Frisch M, Olsen JH, Andersen PK. Birth order, sibship size and risk of Hodgkin's disease in children and young adults: a population-based study of 31 million person-years. Int J Cancer J Int du Cancer. 1997;72(6):977–81.
90. Chang ET, Zheng T, Weir EG, Borowitz M, Mann RB, Spiegelman D, et al. Childhood social environment and Hodgkin's lymphoma: new findings from a population-based case-control study. Cancer Epidemiol Biomarkers Prev Publ Am Assoc Cancer Res Cosponsored Am Soc Prev Oncol. 2004;13(8):1361–70.
91. Chang ET, Montgomery SM, Richiardi L, Ehlin A, Ekbom A, Lambe M. Number of siblings and risk of Hodgkin's lymphoma. Cancer Epidemiol Biomarkers Prev Publ Am Assoc Cancer Res Cosponsored Am Soc Prev Oncol. 2004;13(7):1236–43.
92. Gutensohn N, Cole P. Childhood social environment and Hodgkin's disease. N Engl J Med. 1981;304(3):135–40.
93. Mack TM, Norman JE, Jr., Rappaport E, Cozen W. Childhood Determination of Hodgkin lymphoma among U.S. servicemen. Cancer Epidemiol Biomarkers Prev Publ Am Assoc Cancer Res Cosponsored Am Soc Prev Oncol. 2015.
94. Glaser SL, Clarke CA, Nugent RA, Stearns CB, Dorfman RF. Social class and risk of Hodgkin's disease in young-adult women in 1988–94. Int J Cancer J Int du Cancer. 2002;98(1):110–7.
95. Lennette ET, Rymo L, Yadav M, Masucci G, Merk K, Timar L, et al. Disease-related differences in antibody patterns against EBV-encoded nuclear antigens EBNA 1, EBNA 2 and EBNA 6. Eur J Cancer. 1993;29A(11):1584–9.
96. Henle W, Henle G, Andersson J, Ernberg I, Klein G, Horwitz CA, et al. Antibody responses to Epstein-Barr virus-determined nuclear antigen (EBNA)-1 and EBNA-2 in acute and chronic Epstein-Barr virus infection. Proc Natl Acad Sci U S A. 1987;84(2):570–4.
97. Mueller NE, Lennette ET, Dupnik K, Birmann BM. Antibody titers against EBNA1 and EBNA2 in relation to Hodgkin lymphoma and history of infectious mononucleosis. Int J Cancer J Int du Cancer. 2012;130(12):2886–91.
98. Mueller N, Evans A, Harris NL, Comstock GW, Jellum E, Magnus K, et al. Hodgkin's disease and

Epstein-Barr virus. Altered antibody pattern before diagnosis. N Engl J Med. 1989;320(11):689–95.
99. Gutensohn N, Cole P. Epidemiology of Hodgkin's disease. Semin Oncol. 1980;7(2):92–102.
100. Jarrett RF. Viruses and Hodgkin's lymphoma. Ann Oncol Official J Eur Soc Med Oncol ESMO. 2002;13(Suppl 1):23–9.
101. Hjalgrim H, Askling J, Rostgaard K, Hamilton-Dutoit S, Frisch M, Zhang JS, et al. Characteristics of Hodgkin's lymphoma after infectious mononucleosis. N Engl J Med. 2003;349(14): 1324–32.
102. Hjalgrim H, Smedby KE, Rostgaard K, Molin D, Hamilton-Dutoit S, Chang ET, et al. Infectious mononucleosis, childhood social environment, and risk of Hodgkin lymphoma. Cancer Res. 2007;67 (5):2382–8.
103. Elgui de Oliveira D, Bacchi MM, Abreu ES, Niero-Melo L, Bacchi CE. Hodgkin disease in adult and juvenile groups from two different geographic regions in Brazil: characterization of clinicopathologic aspects and relationship with Epstein-Barr virus infection. Am J Clin Pathol. 2002;118(1): 25–30.
104. Glaser SL, Lin RJ, Stewart SL, Ambinder RF, Jarrett RF, Brousset P, et al. Epstein-Barr virus-associated Hodgkin's disease: epidemiologic characteristics in international data. Int J Cancer J Int du Cancer. 1997;70(4):375–82.
105. Herbst H, Dallenbach F, Hummel M, Niedobitek G, Pileri S, Muller-Lantzsch N, et al. Epstein-Barr virus latent membrane protein expression in Hodgkin and Reed-Sternberg cells. Proc Natl Acad Sci U S A. 1991;88(11):4766–70.
106. Herbst H, Niedobitek G, Kneba M, Hummel M, Finn T, Anagnostopoulos I, et al. High incidence of Epstein-Barr virus genomes in Hodgkin's disease. Am J Pathol. 1990;137(1):13–8.
107. Huh J, Park C, Juhng S, Kim CE, Poppema S, Kim C. A pathologic study of Hodgkin's disease in Korea and its association with Epstein-Barr virus infection. Cancer. 1996;77(5):949–55.
108. Pallesen G, Hamilton-Dutoit SJ, Rowe M, Young LS. Expression of Epstein-Barr virus latent gene products in tumour cells of Hodgkin's disease. Lancet. 1991;337(8737):320–2.
109. Diebold J, Raphael M, Prevot S, Audouin J. Lymphomas associated with HIV infection. Cancer Surv. 1997;30:263–93.
110. Riedel DJ, Rositch AF, Redfield RR, Blattner WA. HIV-associated lymphoma sub-type distribution, immunophenotypes and survival in an urban clinic population. Leuk lymphoma. 2015:1–7.
111. Benharroch D, Shemer-Avni Y, Levy A, Myint YY, Ariad S, Rager B, et al. New candidate virus in association with Hodgkin's disease. Leuk Lymphoma. 2003;44(4):605–10.
112. Maggio E, Benharroch D, Gopas J, Dittmer U, Hansmann ML, Kuppers R. Absence of measles virus genome and transcripts in Hodgkin-Reed/Sternberg cells of a cohort of Hodgkin lymphoma patients. Int J Cancer J Int du Cancer. 2007;121(2):448–53.
113. Wilson KS, Freeland JM, Gallagher A, Cosby SL, Earle JA, Alexander FE, et al. Measles virus and classical Hodgkin lymphoma: no evidence for a direct association. Int J Cancer J Int du Cancer. 2007;121(2):442–7.
114. Kramer U, Heinrich J, Wjst M, Wichmann HE. Age of entry to day nursery and allergy in later childhood. Lancet. 1999;353(9151):450–4.
115. Ball TM, Castro-Rodriguez JA, Griffith KA, Holberg CJ, Martinez FD, Wright AL. Siblings, day-care attendance, and the risk of asthma and wheezing during childhood. N Engl J Med. 2000;343(8):538–43.
116. Chang ET, Zheng T, Weir EG, Borowitz M, Mann RB, Spiegelman D, et al. Aspirin and the risk of Hodgkin's lymphoma in a population-based case-control study. J Natl Cancer Inst. 2004;96 (4):305–15.
117. Chang ET, Froslev T, Sorensen HT, Pedersen L. A nationwide study of aspirin, other non-steroidal anti-inflammatory drugs, and Hodgkin lymphoma risk in Denmark. Br J Cancer. 2011;105(11): 1776–82.
118. Briggs NC, Hall HI, Brann EA, Moriarty CJ, Levine RS. Cigarette smoking and risk of Hodgkin's disease: a population-based case-control study. Am J Epidemiol. 2002;156(11):1011–20.
119. Kroll ME, Murphy F, Pirie K, Reeves GK, Green J, Beral V, et al. Alcohol drinking, tobacco smoking and subtypes of haematological malignancy in the UK Million Women Study. Br J Cancer. 2012;107 (5):879–87.
120. Lim U, Morton LM, Subar AF, Baris D, Stolzenberg-Solomon R, Leitzmann M, et al. Alcohol, smoking, and body size in relation to incident Hodgkin's and non-Hodgkin's lymphoma risk. Am J Epidemiol. 2007;166(6):697–708.
121. Nieters A, Deeg E, Becker N. Tobacco and alcohol consumption and risk of lymphoma: results of a population-based case-control study in Germany. Int J Cancer J Int du Cancer. 2006;118(2):422–30.
122. Kamper-Jorgensen M, Rostgaard K, Glaser SL, Zahm SH, Cozen W, Smedby KE, et al. Cigarette smoking and risk of Hodgkin lymphoma and its subtypes: a pooled analysis from the international lymphoma epidemiology consortium (InterLymph). Ann Oncol Official J Eur Soc Med Oncol ESMO. 2013;24(9):2245–55.
123. Larsson SC, Wolk A. Body mass index and risk of non-Hodgkin's and Hodgkin's lymphoma: a meta-analysis of prospective studies. Eur J Cancer. 2011;47(16):2422–30.
124. Murphy F, Kroll ME, Pirie K, Reeves G, Green J, Beral V. Body size in relation to incidence of subtypes of haematological malignancy in the

prospective million women study. Br J Cancer. 2013;108(11):2390–8.
125. Keegan TH, Glaser SL, Clarke CA, Dorfman RF, Mann RB, DiGiuseppe JA, et al. Body size, physical activity, and risk of Hodgkin's lymphoma in women. Cancer Epidemiol Biomarkers Prev Publ Am Assoc Cancer Res Cosponsored Am Soc Prev Oncol. 2006;15(6):1095–101.
126. Li Q, Chang ET, Bassig BA, Dai M, Qin Q, Gao Y, et al. Body size and risk of Hodgkin's lymphoma by age and gender: a population-based case-control study in Connecticut and Massachusetts. Cancer Causes Control CCC. 2013;24(2):287–95.
127. Epstein MM, Chang ET, Zhang Y, Fung TT, Batista JL, Ambinder RF, et al. Dietary pattern and risk of hodgkin lymphoma in a population-based case-control study. Am J Epidemiol. 2015;182(5):405–16.
128. Gao Y, Li Q, Bassig BA, Chang ET, Dai M, Qin Q, et al. Subtype of dietary fat in relation to risk of Hodgkin lymphoma: a population-based case-control study in Connecticut and Massachusetts. Cancer Causes Control CCC. 2013;24(3):485–94.
129. Lopez-Garcia E, Schulze MB, Fung TT, Meigs JB, Rifai N, Manson JE, et al. Major dietary patterns are related to plasma concentrations of markers of inflammation and endothelial dysfunction. Am J Clin Nutr. 2004;80(4):1029–35.
130. Nettleton JA, Steffen LM, Mayer-Davis EJ, Jenny NS, Jiang R, Herrington DM, et al. Dietary patterns are associated with biochemical markers of inflammation and endothelial activation in the multi-ethnic study of atherosclerosis (MESA). Am J Clin Nutr. 2006;83(6):1369–79.
131. Besson H, Brennan P, Becker N, De Sanjose S, Nieters A, Font R, et al. Tobacco smoking, alcohol drinking and Hodgkin's lymphoma: a European multi-centre case-control study (EPILYMPH). Br J Cancer. 2006;95(3):378–84.
132. Gorini G, Stagnaro E, Fontana V, Miligi L, Ramazzotti V, Amadori D, et al. Alcohol consumption and risk of Hodgkin's lymphoma and multiple myeloma: a multicentre case-control study. Ann Oncol Official J Eur Soc Med Oncol ESMO. 2007;18(1):143–8.
133. Monnereau A, Orsi L, Troussard X, Berthou C, Fenaux P, Soubeyran P, et al. Cigarette smoking, alcohol drinking, and risk of lymphoid neoplasms: results of a French case-control study. Cancer Causes Control CCC. 2008;19(10):1147–60.
134. Romeo J, Warnberg J, Nova E, Diaz LE, Gomez-Martinez S, Marcos A. Moderate alcohol consumption and the immune system: a review. Br J Nutr. 2007;98(Suppl 1):S111–5.
135. Chang ET, Canchola AJ, Clarke CA, Lu Y, West DW, Bernstein L, et al. Dietary phytocompounds and risk of lymphoid malignancies in the California teachers study cohort. Cancer Causes Control CCC. 2011;22(2):237–49.
136. Kasperzyk JL, Chang ET, Birmann BM, Kraft P, Zheng T, Mueller NE. Nutrients and genetic variation involved in one-carbon metabolism and Hodgkin lymphoma risk: a population-based case-control study. Am J Epidemiol. 2011;174(7):816–27.
137. La Vecchia C, Chatenoud L, Negri E, Franceschi S. Session: whole cereal grains, fibre and human cancer wholegrain cereals and cancer in Italy. Proc Nutr Soc. 2003;62(1):45–9.
138. Negri E, La Vecchia C, Franceschi S, D'Avanzo B, Parazzini F. Vegetable and fruit consumption and cancer risk. Int J Cancer J Int du Cancer. 1991;48(3):350–4.
139. Tavani A, Pregnolato A, Negri E, Franceschi S, Serraino D, Carbone A, et al. Diet and risk of lymphoid neoplasms and soft tissue sarcomas. Nutr Cancer. 1997;27(3):256–60.
140. McCunney RJ. Hodgkin's disease, work, and the environment. A review. J Occup Environ Med Am Coll Occup Environ Med. 1999;41(1):36–46.
141. Morton LM, Slager SL, Cerhan JR, Wang SS, Vajdic CM, Skibola CF, et al. Etiologic heterogeneity among non-Hodgkin lymphoma subtypes: the InterLymph Non-Hodgkin lymphoma subtypes project. J Nat Cancer Inst Monogr. 2014;2014(48):130–44.
142. Bertrand KA, Chang ET, Abel GA, Zhang SM, Spiegelman D, Qureshi AA, et al. Sunlight exposure, vitamin D, and risk of non-Hodgkin lymphoma in the Nurses' Health Study. Cancer Causes Control CCC. 2011;22(12):1731–41.
143. Chang ET, Canchola AJ, Cockburn M, Lu Y, Wang SS, Bernstein L, et al. Adulthood residential ultraviolet radiation, sun sensitivity, dietary vitamin D, and risk of lymphoid malignancies in the California Teachers Study. Blood. 2011;118(6):1591–9.
144. Skibola CF, Bracci PM, Nieters A, Brooks-Wilson A, de Sanjose S, Hughes AM, et al. Tumor necrosis factor (TNF) and lymphotoxin-alpha (LTA) polymorphisms and risk of non-Hodgkin lymphoma in the InterLymph Consortium. Am J Epidemiol. 2010;171(3):267–76.
145. Lan Q, Wang SS, Menashe I, Armstrong B, Zhang Y, Hartge P, et al. Genetic variation in Th1/Th2 pathway genes and risk of non-Hodgkin lymphoma: a pooled analysis of three population-based case-control studies. Br J Haematol. 2011;153(3):341–50.
146. Lan Q, Zheng T, Shen M, Zhang Y, Wang SS, Zahm SH, et al. Genetic polymorphisms in the oxidative stress pathway and susceptibility to non-Hodgkin lymphoma. Hum Genet. 2007;121(2):161–8.

147. Morton LM, Purdue MP, Zheng T, Wang SS, Armstrong B, Zhang Y, et al. Risk of non-Hodgkin lymphoma associated with germline variation in genes that regulate the cell cycle, apoptosis, and lymphocyte development. Cancer Epidemiol Biomarkers Prev Publ Am Assoc Cancer Res Cosponsored Am Soc Prev Oncol. 2009;18(4):1259–70.
148. Rothman N, Skibola CF, Wang SS, Morgan G, Lan Q, Smith MT, et al. Genetic variation in TNF and IL10 and risk of non-Hodgkin lymphoma: a report from the InterLymph Consortium. Lancet Oncol. 2006;7(1):27–38.
149. Shen M, Menashe I, Morton LM, Zhang Y, Armstrong B, Wang SS, et al. Polymorphisms in DNA repair genes and risk of non-Hodgkin lymphoma in a pooled analysis of three studies. Br J Haematol. 2010;151(3):239–44.
150. Wang SS, Purdue MP, Cerhan JR, Zheng T, Menashe I, Armstrong BK, et al. Common gene variants in the tumor necrosis factor (TNF) and TNF receptor superfamilies and NF-kB transcription factors and non-Hodgkin lymphoma risk. PLoS ONE. 2009;4(4):e5360.
151. Berndt SI, Skibola CF, Joseph V, Camp NJ, Nieters A, Wang Z, et al. Genome-wide association study identifies multiple risk loci for chronic lymphocytic leukemia. Nat Genet. 2013;45(8):868–76.
152. Cerhan JR, Berndt SI, Vijai J, Ghesquieres H, McKay J, Wang SS, et al. Genome-wide association study identifies multiple susceptibility loci for diffuse large B cell lymphoma. Nat Genet. 2014;46(11):1233–8.
153. Skibola CF, Berndt SI, Vijai J, Conde L, Wang Z, Yeager M, et al. Genome-wide association study identifies five susceptibility loci for follicular lymphoma outside the HLA region. Am J Hum Genet. 2014;95(4):462–71.
154. Vijai J, Wang Z, Berndt SI, Skibola CF, Slager SL, de Sanjose S, et al. A genome-wide association study of marginal zone lymphoma shows association to the HLA region. Nat Commun. 2015;6:5751.
155. Clifford GM, Polesel J, Rickenbach M, Dal Maso L, Keiser O, Kofler A, et al. Cancer risk in the Swiss HIV cohort study: associations with immunodeficiency, smoking, and highly active antiretroviral therapy. J Natl Cancer Inst. 2005;97(6):425–32.
156. Silverberg MJ, Chao C, Leyden WA, Xu L, Horberg MA, Klein D, et al. HIV infection, immunodeficiency, viral replication, and the risk of cancer. Cancer Epidemiol Biomarkers Prev Publ Am Assoc Cancer Res Cosponsored Am Soc Prev Oncol. 2011;20(12):2551–9.
157. Eltom MA, Jemal A, Mbulaiteye SM, Devesa SS, Biggar RJ. Trends in Kaposi's sarcoma and non-Hodgkin's lymphoma incidence in the United States from 1973 through 1998. J Natl Cancer Inst. 2002;94(16):1204–10.
158. Opelz G, Henderson R. Incidence of non-Hodgkin lymphoma in kidney and heart transplant recipients. Lancet. 1993;342(8886–8887):1514–6.
159. Ekstrom Smedby K, Vajdic CM, Falster M, Engels EA, Martinez-Maza O, Turner J, et al. Autoimmune disorders and risk of non-Hodgkin lymphoma subtypes: a pooled analysis within the InterLymph Consortium. Blood. 2008;111(8):4029–38.
160. Fallah M, Liu X, Ji J, Forsti A, Sundquist K, Hemminki K. Autoimmune diseases associated with non-Hodgkin lymphoma: a nationwide cohort study. Ann Oncol Official J Eur Soc Med Oncol ESMO. 2014;25(10):2025–30.
161. Linet MS, Vajdic CM, Morton LM, de Roos AJ, Skibola CF, Boffetta P, et al. Medical history, lifestyle, family history, and occupational risk factors for follicular lymphoma: the InterLymph Non-Hodgkin Lymphoma Subtypes Project. J Nat Cancer Inst Monogr. 2014;2014(48):26–40.
162. Wang SS, Flowers CR, Kadin ME, Chang ET, Hughes AM, Ansell SM, et al. Medical history, lifestyle, family history, and occupational risk factors for peripheral T-cell lymphomas: the InterLymph Non-Hodgkin Lymphoma Subtypes Project. J Nat Cancer Inst Monogr. 2014;2014(48):66–75.
163. Mbulaiteye SM, Morton LM, Sampson JN, Chang ET, Costas L, de Sanjose S, et al. Medical history, lifestyle, family history, and occupational risk factors for sporadic Burkitt lymphoma/leukemia: the Interlymph Non-Hodgkin lymphoma subtypes project. J Nat Cancer Inst Monogr. 2014;2014(48):106–14.
164. Smedby KE, Sampson JN, Turner JJ, Slager SL, Maynadie M, Roman E, et al. Medical history, lifestyle, family history, and occupational risk factors for mantle cell lymphoma: the InterLymph Non-Hodgkin Lymphoma subtypes project. J Nat Cancer Inst Monogr. 2014;2014(48):76–86.
165. Vajdic CM, Landgren O, McMaster ML, Slager SL, Brooks-Wilson A, Smith A, et al. Medical history, lifestyle, family history, and occupational risk factors for lymphoplasmacytic lymphoma/Waldenstrom's macroglobulinemia: the InterLymph Non-Hodgkin Lymphoma Subtypes Project. J Nat Cancer Inst Monogr. 2014;2014(48):87–97.
166. Slager SL, Benavente Y, Blair A, Vermeulen R, Cerhan JR, Costantini AS, et al. Medical history, lifestyle, family history, and occupational risk factors for chronic lymphocytic leukemia/small lymphocytic lymphoma: the InterLymph Non-Hodgkin Lymphoma subtypes project. J Nat Cancer Inst Monogr. 2014;2014(48):41–51.
167. International Agency for Research on Cancer (IARC). Human T-cell lymphotropic viruses. In: IARC. Monographs on the evaluation of carcinogenic risks to humans. Vol 67. Lyon: IARC, 1996:261–390.

168. Mueller N, Birmann B, Parsonnet J, Schiffman M, Stuver S. Infectious agents. In: Schottenfeld D, Fraumeni Jr. JF, editors. Cancer epidemiology and prevention, 3rd ed. New York. Oxford University Press; 2006.
169. Zucca E, Bertoni F, Vannata B, Cavalli F. Emerging role of infectious etiologies in the pathogenesis of marginal zone B-cell lymphomas. Clin Cancer Res Official J Am Assoc Cancer Res. 2014;20(20): 5207–16.
170. Piriou E, Asito AS, Sumba PO, Fiore N, Middeldorp JM, Moormann AM, et al. Early age at time of primary Epstein-Barr virus infection results in poorly controlled viral infection in infants from Western Kenya: clues to the etiology of endemic Burkitt lymphoma. J Infect Dis. 2012;205(6): 906–13.
171. de-The G. Epstein-Barr virus and Burkitt's lymphoma worldwide: the causal relationship revisited. IARC scientific publications. 1985(60):165–76.
172. Bornkamm GW. Epstein-Barr virus and the pathogenesis of Burkitt's lymphoma: more questions than answers. Int J Cancer J Int du Cancer. 2009;124 (8):1745–55.
173. Satou A, Asano N, Nakazawa A, Osumi T, Tsurusawa M, Ishiguro A, et al. Epstein-Barr virus (EBV)-positive sporadic burkitt lymphoma: an age-related lymphoproliferative disorder? Am J Surg Pathol. 2015;39(2):227–35.
174. Teras LR, Rollison DE, Pawlita M, Michel A, Brozy J, de Sanjose S, et al. Epstein-Barr virus and risk of non-Hodgkin lymphoma in the cancer prevention study-II and a meta-analysis of serologic studies. Int J Cancer J Int du Cancer. 2015;136 (1):108–16.
175. Peveling-Oberhag J, Arcaini L, Hansmann ML, Zeuzem S. Hepatitis C-associated B-cell non-Hodgkin lymphomas. Epidemiology, molecular signature and clinical management. J Hepatol. 2013;59(1):169–77.
176. Bracci PM, Benavente Y, Turner JJ, Paltiel O, Slager SL, Vajdic CM, et al. Medical history, lifestyle, family history, and occupational risk factors for marginal zone lymphoma: the InterLymph Non-Hodgkin Lymphoma subtypes project. J Nat Cancer Inst Monogr. 2014;2014(48):52–65.
177. Cerhan JR, Kricker A, Paltiel O, Flowers CR, Wang SS, Monnereau A, et al. Medical history, lifestyle, family history, and occupational risk factors for diffuse large B-cell lymphoma: the InterLymph Non-Hodgkin lymphoma subtypes project. J Nat Cancer Inst Monogr. 2014;2014(48): 15–25.
178. Kabat GC, Kim MY, Jean Wactawski W, Bea JW, Edlefsen KL, Adams-Campbell LL, et al. Anthropometric factors, physical activity, and risk of non-Hodgkin's lymphoma in the Women's Health Initiative. Cancer Epidemiol. 2012;36(1):52–9.
179. Willett EV, Morton LM, Hartge P, Becker N, Bernstein L, Boffetta P, et al. Non-Hodgkin lymphoma and obesity: a pooled analysis from the InterLymph Consortium. Int J Cancer J Int du Cancer. 2008;122(9):2062–70.
180. van Veldhoven CM, Khan AE, Teucher B, Rohrmann S, Raaschou-Nielsen O, Tjonneland A, et al. Physical activity and lymphoid neoplasms in the European prospective investigation into cancer and nutrition (EPIC). Eur J Cancer. 2011;47(5): 748–60.
181. Teras LR, Gapstur SM, Diver WR, Birmann BM, Patel AV. Recreational physical activity, leisure sitting time and risk of non-Hodgkin lymphoid neoplasms in the american cancer society cancer prevention study II cohort. Int J Cancer J International du Cancer. 2012;131(8):1912–20.
182. Jochem C, Leitzmann MF, Keimling M, Schmid D, Behrens G. Physical activity in relation to risk of hematologic cancers: a systematic review and meta-analysis. Cancer Epidemiol Biomarkers Prev Publ Am Assoc Cancer Res Cosponsored Am Soc Prev Oncol. 2014;23(5):833–46.
183. Cahoon EK, Pfeiffer RM, Wheeler DC, Arhancet J, Lin SW, Alexander BH, et al. Relationship between ambient ultraviolet radiation and non-Hodgkin lymphoma subtypes: a U.S. population-based study of racial and ethnic groups. Int J Cancer J Int du Cancer. 2015;136(5):E432–41.
184. Luczynska A, Kaaks R, Rohrmann S, Becker S, Linseisen J, Buijsse B, et al. Plasma 25-hydroxyvitamin D concentration and lymphoma risk: results of the European prospective investigation into cancer and nutrition. Am J Clin Nutr. 2013;98(3):827–38.
185. Erber E, Maskarinec G, Lim U, Kolonel LN. Dietary vitamin D and risk of non-Hodgkin lymphoma: the multiethnic cohort. Br J Nutr. 2010;103(4):581–4.
186. Ollberding NJ, Aschebrook-Kilfoy B, Caces DB, Smith SM, Weisenburger DD, Chiu BC. Dietary patterns and the risk of non-Hodgkin lymphoma. Pub Health Nutr. 2014;17(7):1531–7.
187. Erber E, Maskarinec G, Gill JK, Park SY, Kolonel LN. Dietary patterns and the risk of non-Hodgkin lymphoma: the Multiethnic Cohort. Leuk Lymphoma. 2009;50(8):1269–75.
188. Campagna M, Cocco P, Zucca M, Angelucci E, Gabbas A, Latte GC, et al. Risk of lymphoma subtypes and dietary habits in a Mediterranean area. Cancer Epidemiol. 2015.
189. Zhang S, Hunter DJ, Rosner BA, Colditz GA, Fuchs CS, Speizer FE, et al. Dietary fat and protein in relation to risk of non-Hodgkin's lymphoma among women. J Natl Cancer Inst. 1999;91 (20):1751–8.
190. Zhang SM, Hunter DJ, Rosner BA, Giovannucci EL, Colditz GA, Speizer FE, et al. Intakes of fruits, vegetables, and related nutrients and the risk of non-Hodgkin's lymphoma among women. Cancer Epidemiol Biomarkers Prev Publ Am Assoc Cancer Res Cosponsored Am Soc Prev Oncol. 2000;9 (5):477–85.

191. Chen GC, Lv DB, Pang Z, Liu QF. Fruits and vegetables consumption and risk of non-Hodgkin's lymphoma: a meta-analysis of observational studies. Int J Cancer J Int du Cancer. 2013;133(1):190–200.
192. Charbonneau B, O'Connor HM, Wang AH, Liebow M, Thompson CA, Fredericksen ZS, et al. Trans fatty acid intake is associated with increased risk and n3 fatty acid intake with reduced risk of non-hodgkin lymphoma. J Nutr. 2013;143(5):672–81.
193. Morton LM, Hartge P, Holford TR, Holly EA, Chiu BC, Vineis P, et al. Cigarette smoking and risk of non-Hodgkin lymphoma: a pooled analysis from the international lymphoma epidemiology consortium (interlymph). Cancer Epidemiol Biomarkers Prev Publ Am Assoc Cancer Res Cosponsored Am Soc Prev Oncol. 2005;14(4):925–33.
194. Monnereau A, Slager SL, Hughes AM, Smith A, Glimelius B, Habermann TM, et al. Medical history, lifestyle, and occupational risk factors for hairy cell leukemia: the InterLymph Non-Hodgkin Lymphoma subtypes project. J Nat Cancer Inst Monogr. 2014;2014(48):115–24.
195. Morton LM, Zheng T, Holford TR, Holly EA, Chiu BC, Costantini AS, et al. Alcohol consumption and risk of non-Hodgkin lymphoma: a pooled analysis. Lancet Oncol. 2005;6(7):469–76.
196. Chang ET, Clarke CA, Canchola AJ, Lu Y, Wang SS, Ursin G, et al. Alcohol consumption over time and risk of lymphoid malignancies in the California Teachers Study cohort. Am J Epidemiol. 2010;172(12):1373–83.
197. Neasham D, Sifi A, Nielsen KR, Overvad K, Raaschou-Nielsen O, Tjonneland A, et al. Occupation and risk of lymphoma: a multicentre prospective cohort study (EPIC). Occup Environ Med. 2011;68(1):77–81.
198. Cocco P, Vermeulen R, Flore V, Nonne T, Campagna M, Purdue M, et al. Occupational exposure to trichloroethylene and risk of non-Hodgkin lymphoma and its major subtypes: a pooled InterLymph [correction of IinterLlymph] analysis. Occup Environ Med. 2013;70(11):795–802.
199. Karami S, Bassig B, Stewart PA, Lee KM, Rothman N, Moore LE, et al. Occupational trichloroethylene exposure and risk of lymphatic and haematopoietic cancers: a meta-analysis. Occup Environ Med. 2013;70(8):591–9.
200. Schenk M, Purdue MP, Colt JS, Hartge P, Blair A, Stewart P, et al. Occupation/industry and risk of non-Hodgkin's lymphoma in the United States. Occup Environ Med. 2009;66(1):23–31.
201. Kelly RS, Lundh T, Porta M, Bergdahl IA, Palli D, Johansson AS, et al. Blood erythrocyte concentrations of cadmium and lead and the risk of B-cell non-Hodgkin's lymphoma and multiple myeloma: a nested case-control study. PLoS ONE. 2013;8(11):e81892.
202. Ramis R, Diggle P, Boldo E, Garcia-Perez J, Fernandez-Navarro P, Lopez-Abente G. Analysis of matched geographical areas to study potential links between environmental exposure to oil refineries and non-Hodgkin lymphoma mortality in Spain. Int J Health Geogr. 2012;11:4.
203. Greenberg AJ, Rajkumar SV, Vachon CM. Familial monoclonal gammopathy of undetermined significance and multiple myeloma: epidemiology, risk factors, and biological characteristics. Blood. 2012;119(23):5359–66.
204. Alexander DD, Mink PJ, Adami HO, Cole P, Mandel JS, Oken MM, et al. Multiple myeloma: a review of the epidemiologic literature. Int J Cancer J Int du Cancer. 2007;120(Suppl 12):40–61.
205. Ogmundsdottir HM, Haraldsdottirm V, Johannesson GM, Olafsdottir G, Bjarnadottir K, Sigvaldason H, et al. Familiality of benign and malignant paraproteinemias. A population-based cancer-registry study of multiple myeloma families. Haematologica. 2005;90(1):66–71.
206. Lynch HT, Watson P, Tarantolo S, Wiernik PH, Quinn-Laquer B, Isgur Bergsagel K, et al. Phenotypic heterogeneity in multiple myeloma families. J Clin Oncol Official J Am Soc Clin Oncol. 2005;23(4):685–93.
207. Broderick P, Chubb D, Johnson DC, Weinhold N, Forsti A, Lloyd A, et al. Common variation at 3p22.1 and 7p15.3 influences multiple myeloma risk. Nat Genet. 2012;44(1):58–61.
208. Chubb D, Weinhold N, Broderick P, Chen B, Johnson DC, Forsti A, et al. Common variation at 3q26.2, 6p21.33, 17p11.2 and 22q13.1 influences multiple myeloma risk. Nat Genet. 2013;45(10):1221–5.
209. Morgan GJ, Johnson DC, Weinhold N, Goldschmidt H, Landgren O, Lynch HT, et al. Inherited genetic susceptibility to multiple myeloma. Leukemia. 2014;28(3):518–24.
210. Weinhold N, Johnson DC, Chubb D, Chen B, Forsti A, Hosking FJ, et al. The CCND1 c.870G > A polymorphism is a risk factor for t(11;14)(q13;q32) multiple myeloma. Nat Genet. 2013;45(5):522–5.
211. Ziv E, Dean E, Hu D, Martino A, Serie D, Curtin K, et al. Genome-wide association study identifies variants at 16p13 associated with survival in multiple myeloma patients. Nat Commun. 2015;6:7539.
212. Avet-Loiseau H, Durie BG, Cavo M, Attal M, Gutierrez N, Haessler J, et al. Combining fluorescent in situ hybridization data with ISS staging improves risk assessment in myeloma: an international myeloma working group collaborative project. Leukemia. 2013;27(3):711–7.
213. Neben K, Jauch A, Bertsch U, Heiss C, Hielscher T, Seckinger A, et al. Combining information regarding chromosomal aberrations t(4;14) and del(17p13) with the International Staging System classification allows stratification of myeloma patients undergoing autologous stem cell transplantation. Haematologica. 2010;95(7):1150–7.
214. Martino A, Campa D, Jurczyszyn A, Martinez-Lopez J, Moreno MJ, Varkonyi J, et al. Genetic

variants and multiple myeloma risk: IMMEnSE validation of the best reported associations — an extensive replication of the associations from the candidate gene era. Cancer Epidemiol Biomarkers Prev Publ Am Assoc Cancer Res Cosponsored Am Soc Prev Oncol. 2014;23(4):670–4.
215. Birmann BM, Tamimi RM, Giovannucci E, Rosner B, Hunter DJ, Kraft P, et al. Insulin-like growth factor-1-and interleukin-6-related gene variation and risk of multiple myeloma. Cancer Epidemiol Biomarkers Prev Publ Am Assoc Cancer Res Cosponsored Am Soc Prev Oncol. 2009;18 (1):282–8.
216. Cozen W, Gebregziabher M, Conti DV, Van Den Berg DJ, Coetzee GA, Wang SS, et al. Interleukin-6-related genotypes, body mass index, and risk of multiple myeloma and plasmacytoma. Cancer Epidemiol Biomarkers Prev Publ Am Assoc Cancer Res Cosponsored Am Soc Prev Oncol. 2006;15(11):2285–91.
217. Lauta VM. A review of the cytokine network in multiple myeloma: diagnostic, prognostic, and therapeutic implications. Cancer. 2003;97(10):2440–52.
218. Papadaki H, Kyriakou D, Foudoulakis A, Markidou F, Alexandrakis M, Eliopoulos GD. Serum levels of soluble IL-6 receptor in multiple myeloma as indicator of disease activity. Acta Haematol. 1997;97(4):191–5.
219. Birmann BM, Neuhouser ML, Rosner B, Albanes D, Buring JE, Giles GG, et al. Prediagnosis biomarkers of insulin-like growth factor-1, insulin, and interleukin-6 dysregulation and multiple myeloma risk in the Multiple Myeloma Cohort Consortium. Blood. 2012;120(25):4929–37.
220. Teras LR, Kitahara CM, Birmann BM, Hartge PA, Wang SS, Robien K, et al. Body size and multiple myeloma mortality: a pooled analysis of 20 prospective studies. Br J Haematol. 2014;166(5):667–76.
221. Birmann BM, Giovannucci E, Rosner B, Anderson KC, Colditz GA. Body mass index, physical activity, and risk of multiple myeloma. Cancer Epidemiol Biomarkers Prev Publ Am Assoc Cancer Res Cosponsored Am Soc Prev Oncol. 2007;16 (7):1474–8.
222. Khan MM, Mori M, Sakauchi F, Matsuo K, Ozasa K, Tamakoshi A, et al. Risk factors for multiple myeloma: evidence from the Japan Collaborative Cohort (JACC) study. Asian Pac J Cancer Prev. 2006;7(4):575–81.
223. Hofmann JN, Liao LM, Pollak MN, Wang Y, Pfeiffer RM, Baris D, et al. A prospective study of circulating adipokine levels and risk of multiple myeloma. Blood. 2012;120(22):4418–20.
224. Dalamaga M, Karmaniolas K, Panagiotou A, Hsi A, Chamberland J, Dimas C, et al. Low circulating adiponectin and resistin, but not leptin, levels are associated with multiple myeloma risk: a case-control study. Cancer Causes Control CCC. 2009;20(2):193–9.
225. Reseland JE, Reppe S, Olstad OK, Hjorth-Hansen H, Brenne AT, Syversen U, et al. Abnormal adipokine levels and leptin-induced changes in gene expression profiles in multiple myeloma. Eur J Haematol. 2009;83(5):460–70.
226. Birmann BM, Giovannucci EL, Rosner BA, Colditz GA. Regular aspirin use and risk of multiple myeloma: a prospective analysis in the Health Professionals Follow-up Study and Nurses' Health Study. Cancer Prev Res. 2013.
227. Teras LR, Gapstur SM, Patel AV, Thun MJ, Diver WR, Zhai Y, et al. Aspirin and other nonsteroidal anti-inflammatory drugs and risk of non-hodgkin lymphoma. Cancer Epidemiol Biomarkers Prev. 2013;22:422–8.
228. Naugler WE, Karin M. NF-kappaB and cancer-identifying targets and mechanisms. Curr Opin Genet Dev. 2008;18(1):19–26.
229. Demchenko YN, Glebov OK, Zingone A, Keats JJ, Bergsagel PL, Kuehl WM. Classical and/or alternative NF-kappaB pathway activation in multiple myeloma. Blood. 2010;115(17):3541–52.
230. Klein B. Positioning NK-kappaB in multiple myeloma. Blood. 2010;115(17):3422–4.
231. Moysich KB, Bonner MR, Beehler GP, Marshall JR, Menezes RJ, Baker JA, et al. Regular analgesic use and risk of multiple myeloma. Leuk Res. 2007;31 (4):547–51.
232. Andreotti G, Birmann B, De Roos AJ, Spinelli J, Cozen W, Camp NJ, et al. A pooled analysis of alcohol consumption and risk of multiple myeloma in the international multiple myeloma consortium. Cancer Epidemiol Biomarkers Prev Publ Am Assoc Cancer Res Cosponsored Am Soc Prev Oncol. 2013;22(9):1620–7.
233. Andreotti G, Birmann BM, Cozen W, De Roos AJ, Chiu BC, Costas L, et al. A pooled analysis of cigarette smoking and risk of multiple myeloma from the international multiple myeloma consortium. Cancer Epidemiol Biomarkers Prev Publ Am Assoc Cancer Res Cosponsored Am Soc Prev Oncol. 2015;24(3):631–4.
234. LeMasters GK, Genaidy AM, Succop P, Deddens J, Sobeih T, Barriera-Viruet H, et al. Cancer risk among firefighters: a review and meta-analysis of 32 studies. J Occup Environ Med Am Coll Occup Environ Med. 2006;48(11):1189–202.
235. Pukkala E, Martinsen JI, Weiderpass E, Kjaerheim K, Lynge E, Tryggvadottir L, et al. Cancer incidence among firefighters: 45 years of follow-up in five Nordic countries. Occup Environ Med. 2014;71(6):398–404.
236. Perrotta C, Kleefeld S, Staines A, Tewari P, De Roos AJ, Baris D, et al. Multiple myeloma and occupation: a pooled analysis by the international multiple myeloma consortium. Cancer Epidemiol. 2013;37(3):300–5.
237. Perrotta C, Staines A, Codd M, Kleefeld S, Crowley D, T'Mannetje A, et al. Multiple myeloma and

lifetime occupation: results from the EPILYMPH study. J Occup Med Toxicol. 2012;7(1):25.
238. Baris D, Silverman DT, Brown LM, Swanson GM, Hayes RB, Schwartz AG, et al. Occupation, pesticide exposure and risk of multiple myeloma. Scand J Work Environ Health. 2004;30(3):215–22.
239. Ruder AM, Hein MJ, Hopf NB, Waters MA. Mortality among 24,865 workers exposed to polychlorinated biphenyls (PCBs) in three electrical capacitor manufacturing plants: a ten-year update. Int J Hyg Environ Health. 2014;217(2–3):176–87.
240. Stenehjem JS, Kjaerheim K, Bratveit M, Samuelsen SO, Barone-Adesi F, Rothman N, et al. Benzene exposure and risk of lymphohaematopoietic cancers in 25 000 offshore oil industry workers. Br J Cancer. 2015;112(9):1603–12.
241. Wong O, Raabe GK. Multiple myeloma and benzene exposure in a multinational cohort of more than 250,000 petroleum workers. Regul Toxicol Pharmacol RTP. 1997;26(2):188–99.
242. Tsai RJ, Luckhaupt SE, Schumacher P, Cress RD, Deapen DM, Calvert GM. Risk of cancer among firefighters in California, 1988–2007. Am J Ind Med. 2015;58(7):715–29.
243. Cardis E, Vrijheid M, Blettner M, Gilbert E, Hakama M, Hill C, et al. The 15-country collaborative study of cancer risk among radiation workers in the nuclear industry: estimates of radiation-related cancer risks. Radiat Res. 2007;167(4):396–416.
244. Muirhead CR, O'Hagan JA, Haylock RG, Phillipson MA, Willcock T, Berridge GL, et al. Mortality and cancer incidence following occupational radiation exposure: third analysis of the national registry for radiation workers. Br J Cancer. 2009;100(1):206–12.
245. Schubauer-Berigan MK, Daniels RD, Bertke SJ, Tseng CY, Richardson DB. Cancer Mortality through 2005 among a pooled cohort of U.S. Nuclear workers exposed to external ionizing radiation. Radiat Res. 2015;183(6):620–31.
246. Takkouche B, Regueira-Mendez C, Montes-Martinez A. Risk of cancer among hairdressers and related workers: a meta-analysis. Int J Epidemiol. 2009;38(6):1512–31.

# Pathology and Molecular Pathology of Hematologic Malignancies

30

Dinesh Rao and Jonathan Said

## 30.1 Introduction

Hematologic malignancies are a diverse group of diseases, but can generally be characterized by clinical presentation into leukemias and lymphomas. Leukemia involves the peripheral blood, while lymphomas involve solid tissues or lymph nodes. Leukemias are most often comprised of immature hematopoietic elements while lymphomas are composed of B-cells, T-cells, or natural killer (NK) cells of varying degrees of maturity. Since lymphomas may become leukemic and involve the peripheral blood, the distinction between leukemia and lymphoma is not rigid, and the diseases are best considered according to their pathologic features, pathogenesis, and molecular changes. Most hematologic malignancies are clonal neoplasms and have specific somatic genetic and molecular characteristics which in turn may influence therapeutic response and prognosis. The molecular basis of each tumor type is becoming quintessential to the diagnosis, in addition to determining therapy and prognosis. Leukemias may also be characterized as acute and immediately life threatening, or chronic, indolent, and not necessarily requiring initial therapy. Acute leukemia includes those derived from lymphoid and myeloid progenitors, which constitute the traditional two branches in the hierarchy of hematopoietic development. Chronic leukemia can also be derived from lymphoid or myeloid-derived cells, but generally these show a range of maturation, as opposed to the monotonous composition of progenitor cells in acute leukemia. Hence, the hematologic malignancies comprise a diverse group of diseases, ranging in cell of origin and clinical presentation.

A consensus classification system has been developed by an expert group of hematopathologists that combines clinical, pathologic, and molecular features of the disease to classify the hematologic malignancies [1]. This system, and others like it, divides hematologic malignancies, where possible, into lymphoid and myeloid-derived diseases, and these broad categories of disease are discussed in this chapter.

### 30.1.1 Development, Anatomy, and Histology of the Lymphatic System

Successive waves of hematopoiesis occur during embryonic development, and definitive adult hematopoiesis begins in the period around birth. Anatomically, the bone marrow is the site of

J. Said (✉)
David Geffen School of Medicine, UCLA Medical Center, 10833 Le Conte Avenue,13-222 CHS, Los Angeles, CA 90095, USA
e-mail: jsaid@mednet.ucla.edu

D. Rao
David Geffen School of Medicine, UCLA Medical Center, 10833 Le Conte Avenue,13-222 CHS, Los Angeles, CA 90095, USA


M. Loda et al. (eds.), *Pathology and Epidemiology of Cancer*,
DOI 10.1007/978-3-319-35153-7_30

primary hematopoiesis in the adult. The bone marrow cellular population includes hematopoietic stem cells, which through successive steps of differentiation, give rise to the mature blood cells, briefly introduced in Chap. 2. The major lineages in the peripheral blood include lymphoid cells (B-and T-cells), myeloid cells (granulocytic and monocytic cells) as well as anucleate red blood cells and platelets. In addition to the true pluripotent hematopoietic stem cells, there are progenitor cells within the different lineages (i.e., lymphoid, myeloid, etc.) that have stem-like properties, and are capable of producing cells of one or more of the mature blood cell lineages [2]. The development of these cells has been the focus of many years of research, and developmental and lineage-specific cells can be recognized morphologically in the bone marrow (Fig. 30.1). Mature cells exit the bone marrow and persist in the systemic circulation for varying periods of time. The longest lived cells are generally the lymphoid cells, which additionally colonize secondary lymphoid tissues, such as the spleen and lymph nodes. In these secondary lymphoid tissues, B-and T-lymphocytes are organized into distinctive structures that can be recognized morphologically and immunophenotypically. The normal lymph node is encapsulated, and can be divided into cortical and medullary areas. The cortex contains the primary follicles, while the medulla contains a mixture of blood vessels, lymphocytes and other cells. Primary follicles in the lymphoid tissues consist of a mixture of B-cells that are positive for the cell surface antigen CD20 and T-cells that are positive for CD3 (Fig. 30.2). It is important to note that the antigens expressed by lymphoid cells are much more complex than this simple dichotomy, and the subtypes of cells and their morphologic and functional characteristics, as well as their immunologic role, are infinitely more diverse. Moreover, these cells also colonize extranodal sites including the skin and mucosal surfaces. Together these lymphoid tissues perform the function of immune surveillance and the lymphoid cells drain into lymphatic channels which form a low-pressure, slow secondary circulatory system with an eventual return to the systemic circulation via the thoracic duct. The cells described here form the majority of the cells of origin in hematologic malignancies.

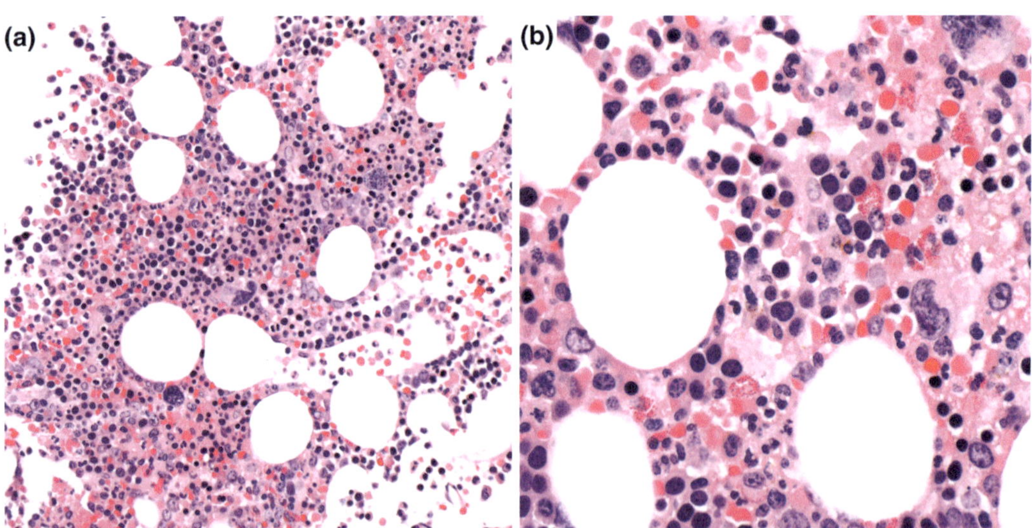


**Fig. 30.1** Normal bone marrow. Normal elements of the bone marrow include megakaryocytes, erythroid cells and myeloid cells. Megakaryocytes are large cells with multilobated nuclei. Erythroid cells show dark round nuclei that condense and are extruded as the cells mature. Myeloid cells show indented and segmented nuclei, with variable granularity. Hematoxylin and eosin stain, original magnification ×400 (**a**) and ×1000 (**b**)

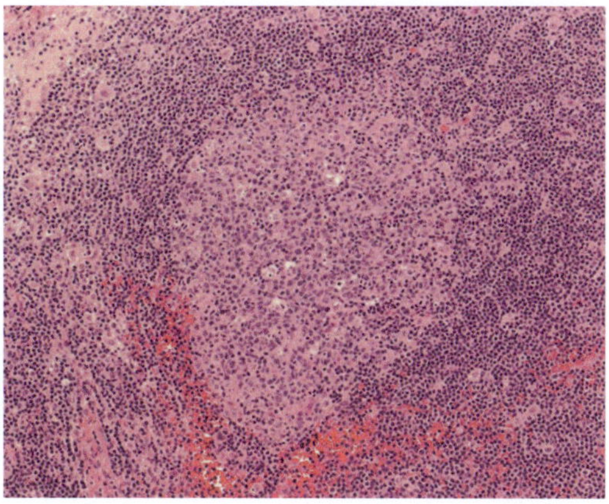

**Fig. 30.2** Normal lymph node. This section shows the quintessential lymph node structure, a follicle with a germinal center. The germinal center is the central region of a follicle, with a mixture of small and large cells, mitotic figures and tingible body macrophages. Surrounding the germinal center is a mantle zone, a compact region of small lymphocytes. Between follicles are interfollicular zones, which can be appreciated at the edges of the photomicrograph. Hematoxylin and eosin stain, original magnification ×400

### 30.1.2 Lymphoid Activation and the Basis of Pathogenetic Translocations

To understand the cell of origin in lymphoma, it is necessary to consider lymphoid activation. B-cells begin their lives in the bone marrow as precursors, and as they mature they move out into the periphery where they populate the lymph node cortex. When antigenic stimulation leads to activation of the immune system, the follicles in the lymph nodes develop germinal centers. In these germinal centers, B-cells undergo somatic hypermutation and class-switch recombination, and some eventually develop into plasma cells capable of producing specific immunoglobulins. Both processes constitute tightly controlled DNA mutation of the immunoglobulin loci, catalyzed by specific enzymes including activation-induced deaminase (AID) [3]. During, this maturation process and controlled DNA mutation, most of the B-cells acquire mutations that do not increase their affinity for antigen, and undergo programmed cell death. A subset survives which is selected for an increased affinity of their immunoglobulin for specific antigens. Occasionally, however, these mechanisms of control and selection fail, and a B-cell with a pathological DNA mutation survives the germinal center reaction with a potential for additional genetic events and neoplastic transformation. Based on the stage of differentiation of the cell that acquires the mutation, different subtypes of lymphoma can develop (Fig. 30.3).

Many genetic alterations involve breaks in the immunoglobulin heavy chain (*IGH*) gene locus on chromosome 14 which can result in a translocation with an oncogene such as *MYC* [4]. This results in a balanced translocation and abnormal expression of MYC protein. *MYC* is a powerful oncogene which drives cells into the cell cycle. In cells with t(14;18), *MYC* is juxtaposed between the *IGH* locus, which is highly active in B-cells, resulting in a massive overexpression of this oncogene which in turn drives cell growth and proliferation. Similarly, translocations involving the *IGH* gene and *BCL2* are typical of follicular lymphoma (FL) [5, 6]. In this case, the *IGH* locus is juxtaposed against *BCL2* whose overexpression prevents cell death of lymphoma cells. Abnormalities in BCL6 can

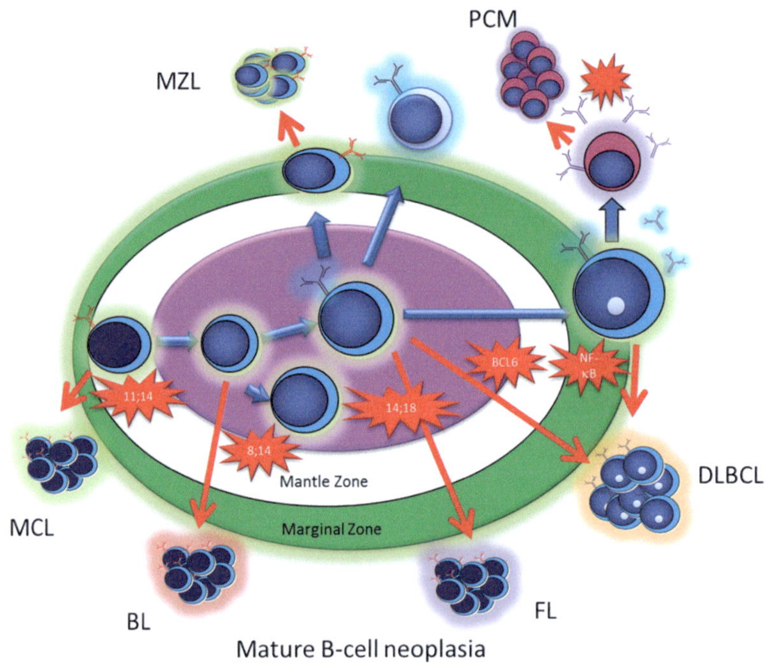


**Fig. 30.3** A schematic of antigen-driven B-cell maturation and pathologic translocations that can lead to the different subtypes of B-cell lymphoma. Abbreviations: MCL, Mantle cell lymphoma; BL, Burkitt Lymphoma; FL, Follicular lymphoma; DLBCL, Diffuse large B-cell lymphoma; PCM, Plasma cell myeloma; MZL, Marginal Zone lymphoma

arise from point mutations, and can involve the same *IGH* locus. BCL6 is a transcription factor and when over expressed can block B-cell differentiation resulting in long-lived proliferative cells [7]. This brief discussion illustrates neoplastic progression from increased survival or a block in apoptosis with BCL2 translocations, increased growth from oncogenes that drive cells into cell cycle like *MYC*, and a potential block in differentiation from genes such as *BCL6* which govern the germinal center cell reaction. These three concepts, survival, growth, and differentiation, are thought to involve all lymphoid tumors to various degrees.

### 30.1.3 Epstein-Barr Virus and Lymphomagenesis

The infectious agent most often associated with lymphomas is the Epstein-Barr virus (EBV). Over 90 % of the population is infected with EBV by the time they reach adulthood [8], and EBV persists for the lifetime of the host in the latent form by embedding itself in a very small percentage of B-cells, where it is thought to be held in check by T-cell-mediated immunity. When there is impairment of the immune system for example due to iatrogenic drugs, transplantation, or infections such as human immunodeficiency virus (HIV), the EBV-infected B-cells proliferate and are subject to additional genetic alterations which can lead to lymphoma. In an EBV-positive lymphoma not only is the B-cell clonal, but the EBV within that B-cell is clonal as well. In patients over the age of 60, lymphomatous proliferations known as EBV-positive diffuse large B-cell lymphoma (DLBCL) of the elderly may occur, presumably related to senescence of the immune system [9]. Immune dysregulation is clearly a key factor, which contributes to lymphomagenesis in many circumstances. Immune senescence, geography, congenital, or acquired immune defects may all contribute to the neoplastic process.

## 30.2 Neoplastic Lymphoid Proliferations

Given the breadth of hematologic malignancy subtypes, this discussion will be restricted to mature lymphoid neoplasms, which tend to involve the lymph nodes and other tissues, but the malignant cells may also circulate in the peripheral blood as leukemias. These neoplasms can also be classified as Hodgkin lymphoma (HL) and non-Hodgkin lymphoma (NHL), the latter a large group derived from mature B-cells, T-cells, or NK cells. B-cell NHL is far more common than T-cell NHL, and comprises approximately 80 % of lymphomas in Europe and the United States [10]. Although heterogeneous in their etiology and pathogenesis, many are derived from the germinal center reaction as mentioned above, and certain types are associated with infectious agents. In the following discussion, we will briefly introduce HL and T-cell NHL, and summarize the relevant immunological considerations and molecular pathogenesis of the most common B-cell NHL subtypes (also summarized in Tables 30.1 and 30.2) [11–22].

### 30.2.1 Hodgkin Lymphoma and T-Cell Lymphoma

Together, HL and T-cell NHL constitute a minority of cases of lymphoma in the United States and Europe, but their incidence differs in other parts of the world [10]. Briefly, HL was the first defined type of lymphoma in the early nineteenth century, and it was recognized as such due to a unique and stereotyped presentation and anatomic pattern of involvement. Many years later, histopathologists identified the malignant cell, the "Reed-Sternberg" (RS) cell that has a characteristic morphologic appearance (Fig. 30.4). Histologically, HL contains a minority of these neoplastic cells with an accompanying infiltrate composed of a diverse group of immune cells including histiocytes, plasma cells, small lymphocytes, and eosinophils. Molecular studies of

**Table 30.1** A partial list of b-cell lymphoma with immunohistochemical staining patterns

| Lymphoma type | CD20 | CD5 | CD10 | CD23 | BCL1 | BCL2 | BCL6 |
|---|---|---|---|---|---|---|---|
| Mantle cell lymphoma | + | + | – | – | + | – | – |
| CLL/SLL | + | + | – | + | – | + | – |
| Marginal zone lymphoma | + | – | – | – | – | + | – |
| Burkitt lymphoma | + | – | + | – | – | – | + |
| Follicular lymphoma | + | – | + | ± | – | + | + |
| DLBCL, GC type | + | – | + | – | – | ± | ± |
| DLBCL, non GC type | + | – | – | – | – | ± | ± |

*CLL/SLL* chronic lymphocytic leukemia/small lymphocytic lymphoma; *DLBCL* Diffuse Large B-cell lymphoma; *GC* germinal center B-cell

**Table 30.2** Molecular genetic abnormalities seen in lymphoid malignancies

| Lymphoma | Molecular alterations |
|---|---|
| Small lymphocytic lymphoma/CLL | Del 13q14.3 most common followed by trisomy 12 and deletions of 11q22–23, 17p13 and 6q21 [11]. Possible targets in the 13q14.3 region are two microRNA genes, miR-16–1 and miR-15a, ATM in the 11q22–23 region and *TP53* in the 17p13 region |
| Follicular lymphoma | Translocation t(14;18)(q32;q21) and *BCL2* gene rearrangements are the hallmark of follicular lymphoma [12]. Abnormalities of 3q27 and/or *BCL6* rearrangement are found in 5–15 % of cases Other genetic alterations include loss of 1p, 6q, 10q and 17p, and gains of chromosomes 1, 6p, 7, 8, 12q, X among others |
| Plasma cell myeloma | Multiple numerical and structural abnormalities are found, including trisomys, whole or partial chromosome deletions and translocations [13]. Complex cytogenetic abnormalities are common. The most frequent chromosome translocations involve the heavy chain locus *(IGH@)* on chromosome 14q32. Major recurrent oncogenes involved in 14q32 translocations include cyclin D1 (11q13) |
| Mantle cell lymphoma | The t(11;14)(q13;q32) between *IGH* and cyclin D1 (*CCND1*) is the hallmark of mantle cell lymphoma [14], and is found in about 95 % of cases. MCL also carries a high number of non-random secondary chromosomal abnormalities including loss of *TP53* and trisomy 12 |
| Diffuse large B-cell lymphoma | Most commonly there are abnormalities, including translocations, involving the 3q27 and the *BCL6* gene [15]. Rearrangements of the *MYC* gene occur in about 10 %, and are usually associated with a complex pattern of additional molecular alterations. Other mutations in DLBCL include frequent mutations in *MYD88*, *CD79A/B*, *CARD11*, A20 loss, and *TP53* [16] |
| Burkitt lymphoma | The molecular hallmark of BL is a translocation involving the *MYC* gene at band q24, from chromosome 8 to the Ig heavy chain region on chromosome 14 [t(8;14)] or less commonly at light chain loci on 2p12 [t(2;8)] or 22q11[t(8;22)] [17] |
| Marginal zone B-cell lymphoma (MZL) | Approximately 30 % of splenic MZL show 7q deletion<br>Splenic marginal zone lymphomas may have mutations involving *NOTCH2*. Chromosomal translocations associated with extranodal marginal zone lymphomas (lymphomas of mucosal associated lymphoid tissues or MALT) include t(11;18), t(1;14), t(14;18) and t(3;14), resulting in the production of a chimeric protein (API2-MALT1) [18]. Trisomy's including chromosome 3 and 18 also occur |
| Hairy cell leukemia (HCL) | *BRAF*V600E mutations are the most important genetic abnormalities which characterize HCL [19] |
| Peripheral T-cell lymphoma (PTCL) | PTCL tend to have multiple abnormalities and complex karyotypes [20]. Recurrent chromosomal gains have been observed in chromosomes 7q, 8q, 17q and 22q, and recurrent losses in chromosomes 4q, 5q,and 12q among others. PTCL involving T-follicular helper cells, such as angioimmunoblastic T-cell lymphoma (AITL), have frequent mutations involving *RHOA*, *TET2*, and *DNMT3A* |
| Anaplastic large cell lymphoma (ALCL) | The hallmark of ALK positive ALCL is fusion of the *ALK* gene with various translocation partners. The most frequent genetic alteration is a translocation, t(2;5)(p23;q35), between the *ALK* gene on chromosome 2 and the nucleophosmin (*NPM*) gene on chromosome 5 [21]. Variant translocations involving *ALK* and other partner genes occur less frequently. In ALK negative ALCL the DUSP 22 translocation characterizes an important subgroup with improved outcome, while those with TP63 do poorly [22] |

singly isolated RS cells revealed that these were cells of B-cell origin that had undergone crippling mutations of their immunoglobulin genes and activation of NFκ[kappa]B. Moreover, recent studies have identified that amplifications in 9p24 led to overexpression of the genes encoding the immunoregulatory proteins PD-L1 and PD-L2, and this is now the basis of a novel therapeutic

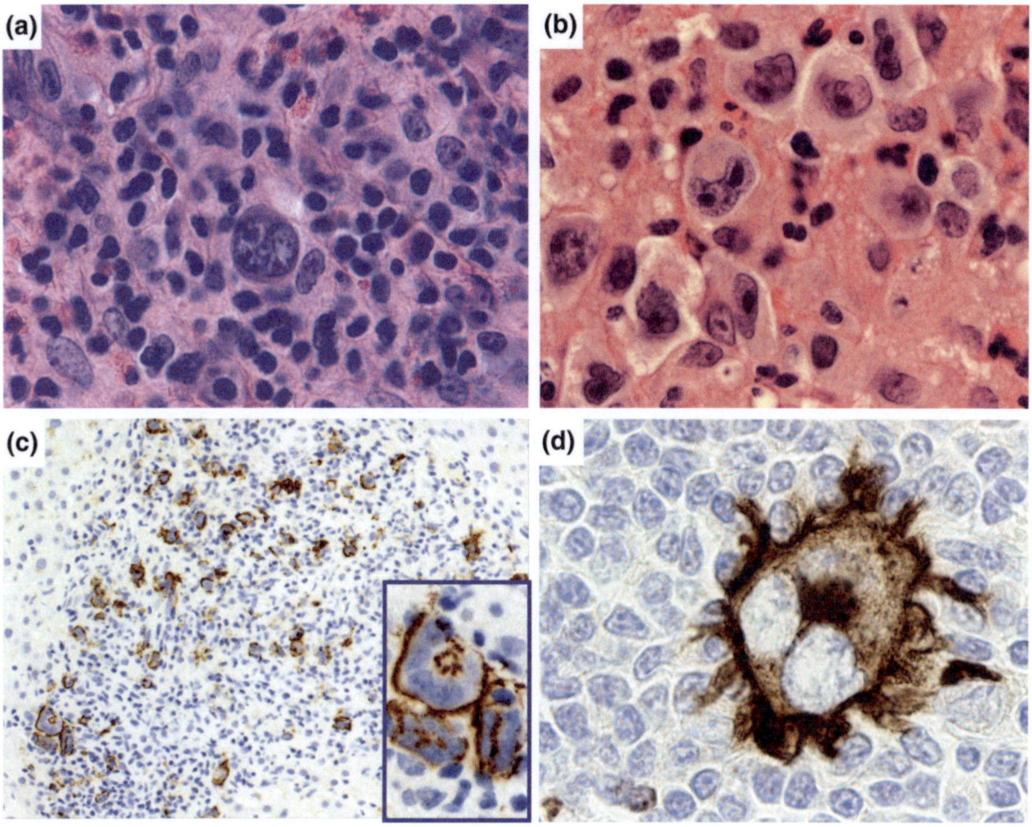


**Fig. 30.4** Classical Hodgkin lymphoma. **a**, **b** Histologically, classical Hodgkin lymphoma is composed of a mixture of cells, including small lymphocytes, plasma cells, histiocytes, and eosinophils, as well as the large, multilobated Reed-Sternberg (RS) cell. Hematoxylin counterstain, original magnification ×1000. **c** The RS cell classically stains for the activation marker CD30 (immunohistochemistry with anti-CD30 antibody, hematoxylin counterstain, original magnification, ×200 and inset, ×1000). **d** The RS cell also stains for CD15, a myeloid related antigen (immunohistochemistry with anti-CD15 antibody, hematoxylin counterstain, original magnification, ×1000)

approach [23]. HL can be divided into two major subgroups including nodular lymphocyte predominant HL, where tumor cells retain immunophenotypic features of germinal center B-cells, and classical HL, which fail to express these gene products. Classical HL is further subdivided into nodular sclerosis classical HL, mixed cellularity classical HL, lymphocyte-rich classical HL and lymphocyte depleted classical HL, definitions of which are beyond the scope of this book chapter. The reader can refer to a recent review of this disease for further information [24].

Another major lymphoma subtype is T-cell NHL. T-cell lymphomas constitute a diverse array of lymphoma types that encompass different cells of origin, many different mutations, and show a range of histologic features from very bland to highly pleomorphic. Recent progress in the field has focused on more specific definitions of subtypes of T-cell lymphoma that may have a specific histogenesis. For example, Angioimmunoblastic T-cell lymphoma (AITL) is thought to derive from a follicular T-helper cell, which shows expression of specific markers including

PD1 and CXCL13. The proliferation of the T-cells leads to recruitment of additional immune cells and vascular proliferation, generating the classic appearance of this disease (Fig. 30.5). In addition, there are scattered B-immunoblasts within the infiltrate that are generally positive for EBV, and the latter may represent the underlying etiological agent responsible for this disease. Other forms of T-cell lymphoma include anaplastic large cell lymphoma and peripheral T-cell lymphoma, along with a number of others. In general, T-cell lymphoma is thought to be more common in Asian populations, where an association with EBV has been noted [10].

### 30.2.2 Non-Hodgkin B-Cell Lymphoma

#### 30.2.2.1 Chronic Lymphocytic Leukemia/Small Lymphocytic Lymphoma

Chronic lymphocytic leukemia/small lymphocytic lymphoma (CLL/SLL) is a neoplastic proliferation comprised of small regular lymphoid cells similar to those found in normal tissues but present in increased numbers and effacing normal architecture in the bone marrow, peripheral blood, lymph nodes, spleen, and other organs

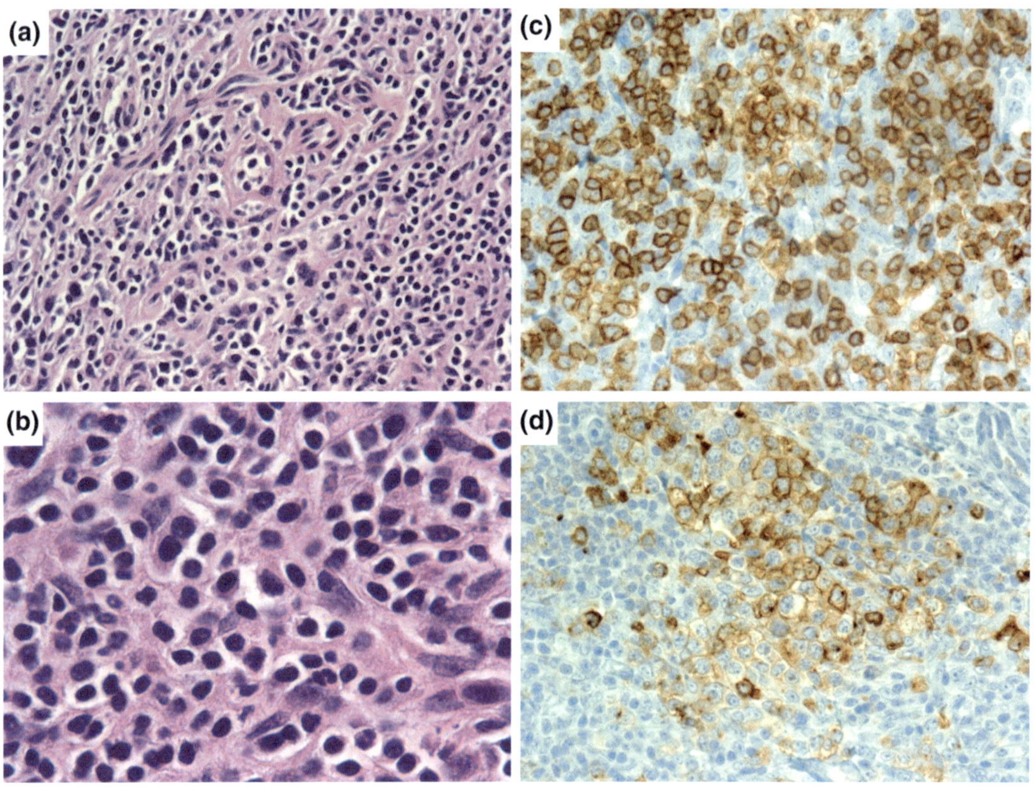


**Fig. 30.5** Angioimmunoblastic T-cell lymphoma. **a**, **b** AITL consists of a mixture of cells, including malignant T-cells and a proliferation of high endothelial venules and follicular dendritic cells. The T cells generally show "water-clear cytoplasm" as can be appreciated in **b**. Hematoxylin and eosin stain, original magnification, ×200 (**a**) and ×1000 (**b**). The T cells stain with antibodies to CD3e (**c**) and to CD10 (**d**), a follicular marker, pointing to a T-follicular helper cell origin for these lymphomas. Hematoxylin counterstain, original magnification ×400

(Fig. 30.6). CLL is defined as greater than $5 \times 10^9$ L circulating clonal B-cells with appropriate morphology and immunophenotype. When the disease presents in lymph nodes it is called SLL. CLL has a relatively high genetic predisposition, with a family history of CLL in up to 10 % of cases. Immunoglobulin genes are rearranged and with somatic hypermutation in 50–60 % of cases [25], the remainder has unmutated immunoglobulin genes. Most cases have cytogenetic abnormalities, particularly del (13q) in about 50 % of cases, trisomy 12, as well as deletions of 11q, 17p13, and 6q21. For many years the candidate tumor suppressor on chromosome 13q was unknown, as the deleted region did not show many protein coding genes. However, the recognition of non-coding RNA and microRNA in the last decade led to the identification of a candidate tumor suppressor microRNA gene on chromosome 13q [26]. Recent work in mouse models has suggested that this microRNA, miR-15a/16 may be pathogenetic in causing low-grade B-lymphocytosis and eventually CLL/SLL [27]. In addition to the pathogenetic implications, many of these genetic findings have prognostic implications. CLL patients with mutated IGH loci have a better prognosis, whereas those with deletions of 17p in the region of the TP53 tumor suppressor gene generally have a worse prognosis.

#### 30.2.2.2 Burkitt Lymphoma

Burkitt lymphoma (BL) is an example of lymphomagenesis with specific molecular alterations, typically but not necessarily associated with EBV. It is the most common non-Hodgkin lymphoma in children and adolescents. Pathologically, it is a monomorphic proliferation of medium sized transformed germinal center related B-cells with round nuclei, clumped chromatin, basophilic cytoplasm, squared-off cell borders, cytoplasmic vacuoles, medium-sized paracentral nucleoli, and a starry sky pattern caused by the presence of numerous phagocytic histiocytes (Fig. 30.7). Translocations involving *MYC* are characteristic but not specific for BL. Although greater than 90 % of African BL cases are positive for EBV, only about 30 % of cases in Europe and the United States contain EBV [28].

BL has a characteristic chromosome abnormality of translocation at chromosome 8q24 involving *MYC* usually with chromosome 14q32 involving IGH [4, 29–31]. Variant translocations occur with the lambda light chain gene (IGL) at chromosome 22q11 or the kappa gene (IGK) at chromosome 2p12 in up to 16 % of cases [32–34]. The *MYC* break-apart probe is generally used in interphase fluorescence in situ hybridization (FISH) to detect the translocations, since it is not dependent on the specific

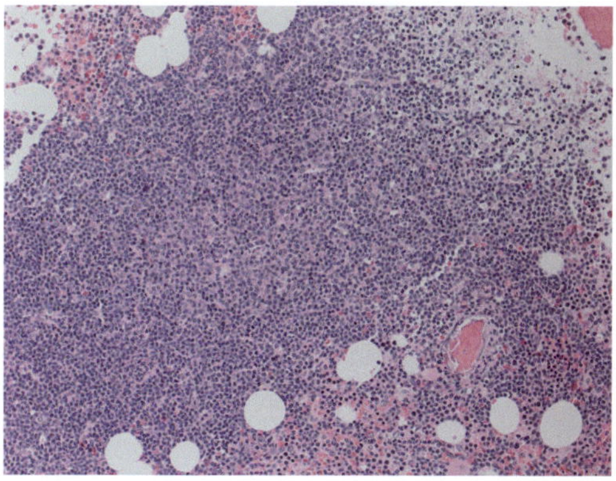

**Fig. 30.6** Bone marrow involved by chronic lymphocytic leukemia. There is an aggregate of small neoplastic lymphoid cells with a central proliferation center. Hematoxylin and Eosin stain, original magnification ×400

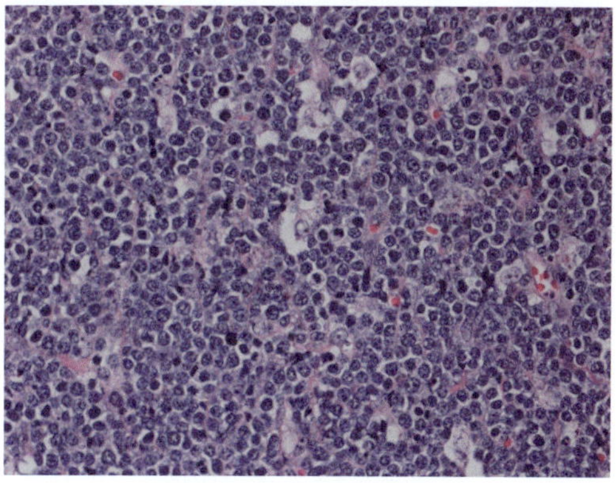

**Fig. 30.7** Burkitt lymphoma showing a diffuse proliferation of cohesive intermediate sized cells with distinct nucleoli and frequent mitoses. There are scattered phagocytic histiocytes lend a starry sky appearance to the infiltrate. Hematoxylin and Eosin stain, original magnification ×400

translocation partner. The breakpoints may vary in different types of BL. For example, in endemic BL the *MYC* breakpoint is generally far 5′ centromeric of *MYC*, while in HIV-related BL the breakpoint tends to occur between exon and intron 1. Nonetheless, it is thought that all of these translocations result in a massive, unregulated overexpression of MYC.

#### MYC and Its Role in Burkitt Lymphoma

Recent studies have suggested that MYC is a global amplifier of all active promoters and enhancers in the genome, rather than a conventional transcription factor [35]. In this setting, MYC is able to drive proliferation by increasing glucose utilization and increasing protein synthesis. However, this interpretation of the data has been questioned by a second, more recent study, which reiterates the capacity of MYC to target specific genes and drive tumorigenesis [36]. Nonetheless, it is apparent that neither EBV infection nor *MYC* translocation are sufficient to initiate and maintain neoplastic proliferations in BL, and t(8;14) has even been detected in normal individuals [37]. There is considerable new data regarding additional genetic aberrations which contribute to the pathogenesis of BL. For example, co-activation of MYC-PI3K selects for stabilizing mutations in cyclin D3 (CCND3), which is a key regulator of the cell cycle in germinal center B-cells [38], and cyclin D3 is commonly overexpressed in BL. Abrogating PI3K signaling or cyclin D3 leads to BL cell death [39], highlighting the importance of this pathway and providing a potential for use of therapeutic agents that target the PI3K signal transduction cascade. Thus, the relationship between c-MYC and PI3K signaling has far-reaching effects in BL diagnosis, prognosis, and treatment and bears further study.

### 30.2.2.3 Lymphoma of Mucosal Associated Lymphoid Tissues

Mucosal associated lymphoid tissue (MALT) lymphomas typically involve extranodal tissues including the stomach, orbit, skin, thyroid, and other sites. It is comprised of small B-lymphocytes most commonly resembling those in the marginal zone of the lymphoid follicle, often called centrocyte-like cells. The cells are small with mildly irregular nuclei and increased often clear cytoplasm. They frequently show plasma cell differentiation. Where, they involve epithelial sites the neoplastic cells typically infiltrate the adjacent epithelium forming so-called lymphoepithelial lesions (Fig. 30.8).

MALT lymphomas serve as a model of lymphomagenesis related to infection, antigenic stimulation, and genetic dysregulation. The clearest example occurs in the case of gastric MALT lymphomas which are often preceded by gastritis due to infection with the bacterial organism *Helicobacter pylori*. This causes an inflammatory gastritis which includes formation of reactive germinal centers, and over time genetic abnormalities can result in an uncontrolled clonal proliferation of B-cells whose antigen receptors recognize *H. pylori*. This process can be reversed and the lymphoma eradicated by using antibiotics effective against *H. pylori* [40]. There are several genetic abnormalities which characterize MALT lymphomas including the translocation t(11;18), t(14;18)(q32; q21) not involving the *BCL2* gene, and t(3;14) which results in the production of a chimeric protein known as API-MALT1 [41]. Many of the

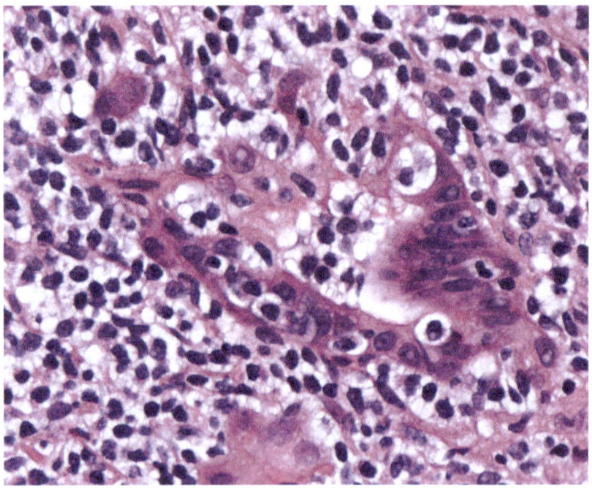

**Fig. 30.8** MALT lymphoma of the stomach. The neoplastic cells are small with mildly irregular nuclei and clear cytoplasm. They are infiltrating a gastric gland forming a lymphoepithelial lesion. Hematoxylin and Eosin stain, original magnification ×400

translocations are thought to result in the constitutive activation of the NFκ[kappa]B pathway, which results in increased cell growth and inhibition of apoptosis. In MALT lymphoma there are different genetic abnormalities associated with various sites of disease and geographic variability, and many of these abnormalities have prognostic implications including the likelihood of responding the antibiotic therapy. The latter therapy is one of the few examples where removal of an inciting agent can result in the regression of a malignancy, and highlights the close relationship between normal immune system function and development of B-cell neoplasia.

#### 30.2.2.4 Diffuse Large B-Cell Lymphoma

DLBCL is a common histologic subtype of lymphoma and comprises 30 % of adult non-Hodgkin lymphomas in the west and an even higher percentage in developing countries [10]. DLBCL is an aggressive form of lymphoma and despite improved treatment regimens about 40 % of patients fail to respond or relapse following chemotherapy. DLBCL is generally characterized by clonal rearrangements of *IGH*, *IGK*, and *IGL* genes, with many cases also showing somatic hypermutation, indicative of a post-germinal center cell origin. The most common cytogenetic findings are abnormalities in chromosome 3q27 involving *BCL6* in 30 % of cases, abnormalities of *BCL2* with translocation t (14;18) in 20 %, and *MYC* rearrangement in 10 % of cases [42, 43]. DLBCL can occur at nodal or extranodal sites. There is an ongoing effort to tailor therapy to individual patients with subtypes of DLBCL, and prognostic markers are becoming increasingly important.

DLBCL is characterized by diffuse proliferation by sheets of large cells of B-cell phenotype (Fig. 30.9). In many cases the cells resemble the large germinal center cells or centroblasts present in normal germinal centers. Centroblasts are intermediate to large with round or oval nuclei, vesicular chromatin, and 2–4 nucleoli often at the nuclear membrane. There is moderate amphophilic to basophilic cytoplasm. Based on gene expression profiling DLBCL can be characterized in two main groups, germinal center B-cell (GCB) with signature of germinal center B-cells (50 % cases), and activated B-cell (ABC), the latter demonstrating a gene expression profile similar to that induced by in vitro activation of peripheral blood B-cells, and this group has been associated with an adverse prognosis. Cases of DLBCL with ABC genotype exhibit constitutive NFκ[kappa]B signaling, and it appears that activation of this pathway is fundamental to the growth of this neoplasm [44]. Proteasome inhibitors such as Bortezomib have shown some activity against these types of lymphomas given the dependence of NFκ[kappa]B signaling on the proteasome. This approach is also being used to treat multiple myeloma, where constitutive activation of NFκ[kappa]B can also be seen.

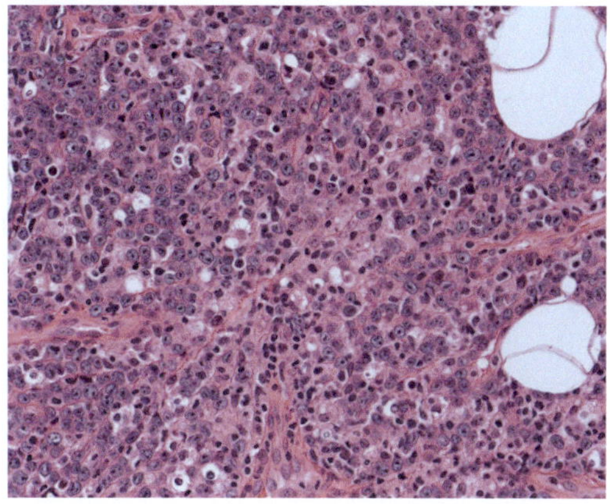

**Fig. 30.9** Diffuse large B-cell lymphoma showing sheets of large centroblastic appearing cells with prominent nucleoli infiltrating soft tissue. Hematoxylin and Eosin stain, original magnification ×400

#### 30.2.2.5 Mantle Cell Lymphoma

Mantle cell lymphoma (MCL) is a mature B-cell neoplasm derived from the mantle zone of the normal germinal follicle. It consists of monomorphic small to medium-sized atypical lymphocytes arising from naïve mantle zone B-cells, with a generally aggressive clinical course and a median survival of 3–5 years.

Mantle cell lymphoma makes up about 5 % of adult non-Hodgkin lymphomas, and is more common in males. The critical cytogenetic event in MCL is the t(11;14)(q13;q32) between the *IGH* and *CCND1* genes [6]. This results in dysregulated cyclin D1 expression and progression through the G1-S cell cycle checkpoint (Fig. 30.10). This is a relatively specific marker of B-cell neoplasia, as Cyclin D1 expression is generally restricted to non-B-lymphocytes. While this cell cycle alteration alone is insufficient to cause MCL, it sets the stage for other genetic events that contribute to a clonal proliferation of neoplastic lymphoid cells and mantle cell lymphoma. Because of the aggressive nature of the disease (most relapse within 2 years of treatment), there are a number of clinical trials with a more biological approach to treatment. These include drugs that affect the microenvironment of the tumor (thalidomide, lenalidomide) small molecule inhibitors like pan-BCL2 inhibitors, and proteasome inhibitors. Most recently, inhibition of Bruton's tyrosine kinase (BTK) with novel small molecule drugs has also shown promise in this difficult-to-treat lymphoma.

#### 30.2.2.6 Follicular Lymphoma

FL accounts for about 20 % of all lymphomas, and has its highest incidence in the United States and Western Europe [45]. It is rare in individuals under 20, and usually involves adults in the 6th decade. Patients usually present with generalized lymphadenopathy, and the disease often involves the bone marrow. Morphologically, FL frequently has a follicular growth pattern resembling the germinal centers from which the malignant cells are derived (Fig. 30.11). It is thought that there is a block in further differentiation of the lymphoma cells, resulting in a uniform proliferation of centrocytes and centroblasts that fail to differentiate into plasma cells and memory B-cells. The immunoglobulin heavy and light chain genes are rearranged in FL, and the variable region genes may show ongoing somatic hypermutation. FL is characterized in 90 % of cases by the t(14;18) translocation involving the *IGH* gene and *BCL2* gene on chromosome 18 [6]. Since the *IGH-BCL2* fusions may be found in healthy individuals, other genetic events are required to develop lymphoma [46]. Abnormalities of *BCL6* and 3q27 are found in up to 15 % of FL, as well as other genetic alterations. New therapeutic approaches targeting the BCL2 pathway are being explored, as small molecule inhibitors of BCL2 family proteins show some promise in clinical trials. Similarly, high-throughput studies are showing the importance of non-B-cells in the

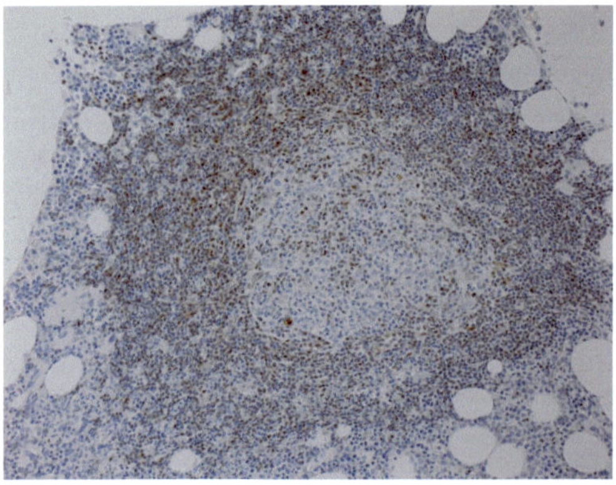

**Fig. 30.10** Mantle cell lymphoma involving the bone marrow. The cells are present in the mantle zone of a germinal center and show nuclear expression of cyclin D1 (*brown*). Hematoxylin counterstain, original magnification ×400

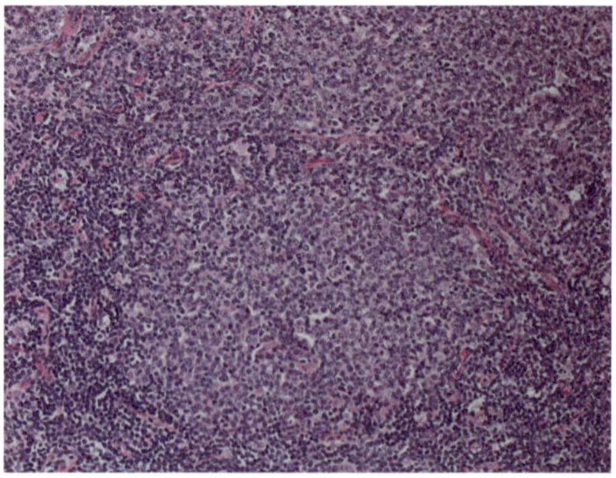

**Fig. 30.11** Follicular lymphoma showing a neoplastic nodule resembling a follicle but comprised of a uniform population of neoplastic centrocytes and centroblasts. Hematoxylin and Eosin stain, original magnification ×400

tumors, as these seem to create a distinct microenvironment that seems to influence prognosis in distinct ways [47].

## 30.3 Myeloid Neoplasms

Bone marrow stem and progenitor populations are thought to be the cell of origin in neoplasms of the myeloid lineage [48, 49]. It is useful to think of the myeloid neoplasms in terms of two properties of these progenitor cells, namely, their proliferative capacity and their ability to differentiate. The three major subtypes of myeloid neoplasms we will consider in this discussion are acute myelogenous leukemia (AML), myeloproliferative neoplasms (MPN), and myelodysplastic syndromes (MDS). In AML, myeloid differentiation is arrested and the progenitor cells (such as the myeloblast or promyelocyte) have a proliferative advantage. The clonal proliferation of blastic cells expands to fill the bone marrow and spills into the peripheral blood. Typically, patients are anemic and thrombocytopenic, and depending on whether the blasts exit the bone marrow or not, the peripheral white blood cell count may be increased or decreased. In contrast, myeloproliferative disorders have acquired a proliferative advantage in the stem cells or the committed progenitors, but without a block in differentiation, resulting in increased hematopoiesis with higher than normal numbers of mature cells in the bone marrow or peripheral blood. The myelodysplastic syndromes are characterized by decreased differentiation or decreased overall hematopoiesis. This results when the stem cells are absent or abnormal and cannot differentiate into progenitor cells for neutrophils and erythrocytes, but do not necessarily have a major proliferative advantage. The clinical presentation is usually as a bone marrow failure syndrome, characterized by anemia, thrombocytopenia, and low white cell count.

### 30.3.1 Acute Myeloid Leukemia

In AML, the normal heterogeneous population of maturing hematopoietic progenitor cells within the bone marrow is replaced by a uniform population of blast cells. Blasts are primitive cells characterized morphologically by high nuclear-to-cytoplasmic ratio and immature nuclei with prominent nucleoli. They may contain primary granules which can contain abnormal rod-like structures called Auer rods (Fig. 30.12). In contrast to leukemic blasts, normal hematopoietic stem cells are rare and only weakly proliferative, giving rise to committed progenitor cells which are much more proliferative and committed along different lineages. Some

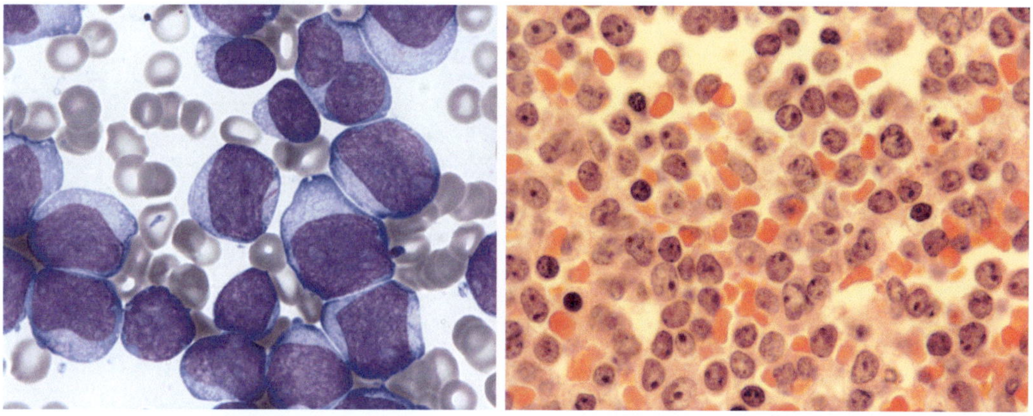

**Fig. 30.12** Acute myelogenous leukemia. *Left hand panel* The peripheral blood shows numerous blasts with prominent nucleoli and immature chromatin. The cytoplasm includes the presence of a few sparse granules and a central cell with an Auer rod. Giemsa stain, original magnification ×1000. The *right hand panel* shows a tissue section with sheets of blast cells. Hematoxylin and eosin stain, original magnification, ×400

studies implicate these committed progenitor cell as the cell of origin for much AML [50].

Hematopoietic development has been studied mostly in mice, and carefully performed studies have revealed that different transcription factors can govern the fate of committed progenitor cells. For example, B-cell differentiation requires the transcription factor PAX5 and its experimental deletion in transgenic mice leads to an accumulation of immature B-lineage cells [51]. Similarly, there are transcription factors that govern the proliferation and differentiation of precursors in the myeloid lineage and that are associated with specific diseases. Genetic mutations of these factors usually result in a loss-of-function which may block differentiation. Mutations can arise from chromosomal translocations, and lead to the formation of aberrant fusion proteins, a fairly common event in AML. For example, in a subtype of AML known as acute promyelocytic leukemia (APL) (Fig. 30.13), there is a balanced translocation between chromosome 15 and chromosome 17 involving the *PML* gene and the retinoic acid receptor gene (*RARA*) [52, 53]. The resulting fusion oncogene, *PML-RAR-α[alpha]*, is an oncogene which produces a protein that does not normally exist in nature, and is expressed in the neoplastic myeloid precursor cells. In addition to translocations, other genetic abnormalities common in leukemia include an increase in copy numbers, single-base pair substitutions, deletions, and insertions among others [54, 55].

Once the genetic alteration has been identified, it becomes possible to develop therapies that specifically target the aberrant cells. In the case of APL, the *PML-RARα[alpha]* fusion oncogene produces a protein that disrupts the normal role of the retinoic acid receptor, which in turn causes mis-expression of various target genes that mediate differentiation in the myeloid lineage. Remarkably, by utilizing supra-physiologic levels of all-trans retinoic acid (ATRA), the mutant transcriptional complex can be de-repressed, allowing terminal differentiation of the APL cells [56]. Presumably, ATRA binds to the fusion protein inside the cells, and blocks its function. In this way, the description and characterization of the molecular lesions involved in a particular disease have allowed for the development of a targeted therapy.

In addition to this fairly well-characterized fusion gene product, there are also several other fusion proteins that are commonly encountered in AML, many of which target genes that are important in the differentiation or proliferation of myeloid progenitor cells. In addition, recent high-throughput genetic studies have revealed deletions and point mutations that appear to be

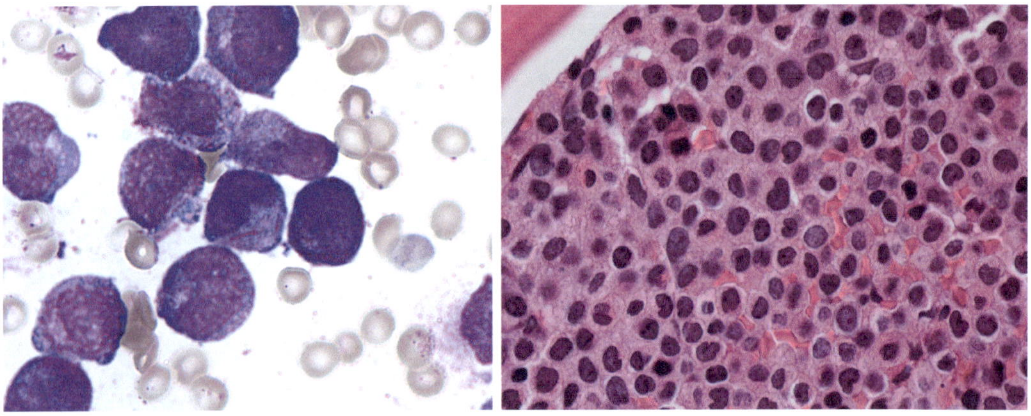

**Fig. 30.13** Acute promyelocytic leukemia. The *left hand panel* shows peripheral blood with blasts that demonstrate hypergranular cytoplasm and Auer rods. Giemsa stain, original magnification ×1000. The *right hand panel* shows a tissue section with sheets of immature cells, with indented nuclei and abundant amounts of cytoplasm. Hematoxylin and eosin stain, original magnification, ×400

important in the pathogenesis of this disease. Many of these genes are involved in the transcriptional regulation of myeloid development, as are the genes involved in some of the most common translocations (Table 30.3) [55]. Evidence from experimental models of leukemogenesis points out that many of these genes are not sufficient to cause the full-blown cancer phenotype, but can cause leukemia in conjunction with other genes (Table 30.4) [57]. Hence, it will be important in the future to consider the combinations of mutations and/or translocations that occur in a given patient, in order to guide therapy [58]. Indeed, targeted inhibition of particular mutations have led to preliminary results that are impressive and may in the future lead to more effective therapies for acute myeloid leukemia.

### 30.3.2 Myeloproliferative Neoplasms

The MPNs have been an area of rapid progress in molecular pathogenesis and therapy. The prime example in this category is chronic myelogenous leukemia (CML), a chronic form of leukemia in which understanding of genetic abnormalities have led to striking new therapies and improved outcome. In this MPN, a mutation occurs in a hematopoietic stem or progenitor cell, and there is a strong proliferative advantage in the neoplastic clone. Unlike AML, differentiation continues to progress and the myeloid lineage produces mature forms (Fig. 30.14). CML is characterized by a translocation between chromosome 9 and chromosome 22 resulting in the so-called Philadelphia chromosome, the t(9;22) balanced translocation [59, 60]. This produces a fusion gene encoding a constitutively active tyrosine kinase, *BCR-ABL*, which leads to increased cell proliferation [61]. The normal function of the ABL tyrosine kinase is to promote cell proliferation, but the kinase activity is tightly regulated by various signals, including extracellular growth factors. Cells that have this mutant fusion protein are capable of growth factor-independent proliferation and survival. To counter this autonomous cell proliferation, specific tyrosine kinase inhibitors were developed and tested. Gleevec (Imatinib) emerged as a highly successful single-agent targeted therapeutic, blocking signaling downstream of the constitutively activated kinase [62]. Remarkably, Imatinib can also target translocations involving the platelet derived growth factor receptor (PDGFR-α[alpha]) and PDGFRβ[beta], which can cause a different myeloproliferative neoplasm (Table 30.3).

Similar pathogenetic mechanisms are operant in other MPN, including polycythemia vera (PV), in which there is excess production of red blood cells. Erythropoietin (EPO) normally binds to erythroid precursors where it helps drive proliferation and increased survival of red cell precursors. Signaling through the EPO receptor is mediated mainly by JAK2, which is a member of a group of tyrosine kinases, collectively referred to as the Janus kinases [63]. In patients with PV, there is an activating mutation within *JAK2* which causes unregulated red cell production [64]. Unlike CML, these mutations that confer constitutive tyrosine kinase activity are point mutations rather than chromosomal translocations. Interestingly, *JAK2* mutations have also been described in other MPNs, and various mutants have been identified, all conferring the same constitutive tyrosine kinase activity. Clinical trials are ongoing to develop an effective JAK2 inhibitor for use in PV and other myeloproliferative neoplasms. More recently, point mutations in additional genes important for myeloid cell development and/or signaling have also been described, for example, *MPL* and *CALR* [65, 66]. Hence, aberrant and increased signaling, predominantly through the normally tightly regulated tyrosine kinases, is a major mechanism in the pathogenesis of myeloproliferative disorders.

**Table 30.3** List of disease defining molecular pathologic abnormalities in myeloid diseases

| Malignancy | Type | Characteristic mutation | Pathogenetic Mechanism | Targeted therapy |
|---|---|---|---|---|
| Chronic myelogenous leukemia | Translocation | t(9;22) *BCR-ABL* | Constitutively activated tyrosine kinase | Imantinib, other Tyrosine kinase inhibitors |
| Myeloproliferative neoplasms | Point mutations | *JAK2* V617F, exon 12<br>*MPL* W515<br>*CALR* | Constitutively activated tyrosine kinase | JAK2 inhibitor ruxolitinib, others under investigation |
| Myeloid neoplasms with eosinophilia | Translocation | Rearrangement of *PDGFRA*, *PDGFRB*, *FGFR1* | Constitutively activated tyrosine kinase | Imantinib, other tyrosine kinase inhibitors |
| Juvenile myelomonocytic leukemia | Point mutation | *NF1* mutations | Loss of function of a tumor suppressor gene | Under investigation |
| Myelodysplastic syndrome with isolated del (5q) | Deletion | Del (5q31–33) *RPS14*, miR-145/146a | Loss of multiple tumor suppressor genes | Under investigation |
| Acute myeloid leukemia | Translocations | t(8;21); *RUNX1-RUNX1T1* | Transcriptional dysregulation of core-binding factor activity | Under investigation |
| | | Inv (16); *CBFB-MYH11* | Transcriptional dysregulation of core-binding factor activity | Under investigation |
| | | t(15;17); *PML-RARA* | Arrest of differentiation | All-trans retinoic acid (ATRA) |
| | | t(9;11); *MLLT3-MLL* | Dysregulated histone methylase activity | Under investigation |
| | Internal tandem duplication | *FLT3* | Constitutive tyrosine kinase activity | Under investigation; FLT3 inhibitors |
| | Point mutations | *NPM1* | Loss-of-function of a multifunctional nucleolar protein | Under investigation |
| | | *CEBPA* | Transcription factor; biallelic mutation defines clinicopathologic entity | Under investigation |

Note the preponderance of tyrosine kinase mutations in chronic myeloproliferative disorders and transcriptional dysregulation in acute myeloid leukemia. A more complete description of point mutations in AML is provided in Table 30.4

### 30.3.3 Myelodysplastic Syndromes

The MDS are a group of myeloid neoplasms that are characterized by failure of bone marrow cells to produce differentiated, functional cells of the peripheral blood. Nonetheless, the early clonally mutated progenitor cells that cause MDS have enough of a proliferative advantage to effectively crowd out normal progenitor cells. Basically, deranged differentiation leads to apoptosis of bone marrow progenitors at some downstream point in their differentiation, resulting in ineffective hematopoiesis. These are predominantly indolent diseases, and develop mainly in elderly individuals who present with low numbers of circulating red blood cells, platelets, and/or white

**Table 30.4** Sequence level mutations in acute myeloid leukemia

| Mutated gene | Function | Frequency (%) | Clinical comments |
|---|---|---|---|
| *NPM1* | Multifunctional nucleolar proteins | 25–35 | Associated with translocations; overall favorable outcome |
| *CEBPA* | Transcription factor | 6–10 | Favorable outcome; associated with cytogenetically normal AML |
| *RUNX1* | Transcription factor | 5–15 | Associated with secondary AML arising from MDS |
| *FLT3-ITD* | Cell Surface receptor, tyrosine kinase | 20 | Unfavorable outcome, frequent in cytogenetically normal AML |
| *KIT* | Cell surface receptor, tyrosine kinase | <5 | Tyrosine kinase inhibitors in development, associated with poor prognosis in certain cytogenetic subtypes |
| *NRAS* | Small GTPase | 15 | May be predictive of response to cytarabine |
| *DNMT3A* | DNA methyltransferase | 18–22 | Early event in leukemogenesis; Associated with clonal hematopoiesis in elderly persons [57] |
| *ASXL1* | Putative polycomb group protein | 5–17 | Early event in leukemogenesis; associated with AML evolving from MDS and seen in clonal hematopoiesis |
| *IDH1* and *IDH2* | Isocitrate dehydrogenases | 7–14; 8–19 | Common in cytogenetically normal AML; inhibitors in clinical development; BCL2 inhibition may be effective in these patients |
| *TET2* | DNA methyltransferase | 7–25 | Early event in leukemogenesis, associated with clonal hematopoiesis in elderly persons [57] |
| *KMT2A*-PTD | Also known as MLL (epigenetic regulator); partial tandem duplication | 5 | Associated with cytogenetically normal AML |
| *TP53* | Tumor suppressor protein | 8 | Associated with very poor outcome |

Many of these are of uncertain clinical significance and have been revealed by high-throughput techniques. Please see Ref. [58] for an excellent review of acute myeloid leukemia

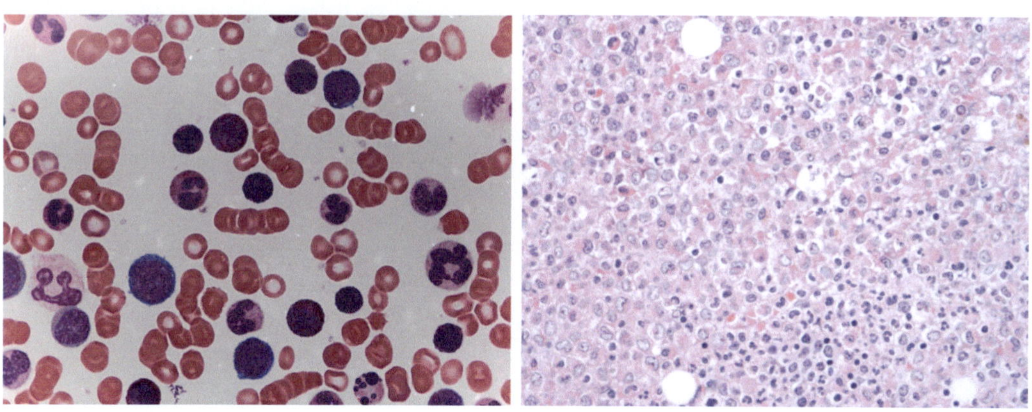

**Fig. 30.14** The *left hand panel* shows peripheral blood with chronic myelogenous leukemia. There are increased neutrophils including hypersegmented forms and basophils. Blast cells are not increased. Giemsa stain, original magnification ×1000. The *right hand panel* shows a tissue section with sheets of myeloid cells, showing a range of maturation. There are cells with round nuclei, as well as band forms and segmented forms. Hematoxylin and eosin stain, original magnification, ×200

blood cells. Because the progenitor cells have a clonal advantage, there is a risk of developing AML in certain subtypes of MDS, presumably as the progenitor cells accumulate additional mutations.

MDS are a heterogeneous group of conditions, many of which have specific genetic associations. Thus far, chromosomal abnormalities, most often deletions, have been associated with MDS. One of the most common deletions involves loss of the short arm of chromosome 5 known as 5q-syndrome or MDS with isolated del (5q). Features of myelodysplasia include hypogranular and bilobed neutrophils (pseudo Pelger-Huët cells) and hypolobated small megakaryocytes. Several genes located on 5q seem to be important in regulating the differentiation and proliferation of myeloid progenitors. Candidate genes that are under investigation include protein coding genes such as *RPS14* and novel genetic elements such as the microRNA miR-145 and miR-146a [67, 68]. In addition, some studies point toward activation of the NFκ[kappa]B as being a possible pathogenetic mechanism in MDS. These mechanisms require further study, but promise to bring forward new therapeutics in this disease.

Recent high-throughput sequencing studies of MDS have identified recurrent point mutations in approximately 20 genes in many cases of MDS. These genes include many different functional categories of genes, but a panel of these genes may serve as an important adjunct to diagnosis. Some of the genes that have been found include those coding for transcription factors (*TP53* or *ETV6*), epigenetic regulators involved in the methylation (*DNMT3A*) or hydroxymethylation (*TET2*, *IDH1*, *IDH2*) of cytosines, or in covalent modifications of histones (*EZH2*, *UTX*, *ASXL1*) (reviewed in [69]). These mutations would all be predicted to cause abnormal changes in gene expression via various direct and indirect mechanisms. In addition, a recent set of studies have found that genes that encode components of the mRNA splicing machinery (i.e., a complex of RNA and protein that mediates the processing of messenger RNA) [70]. In a large proportion of MDS cases, mutations predicted to lead to gain-or-loss-of-function have been identified in genes, such as *SF3B1*, *U2AF1*, and *SRSF2*. Downstream, these mutations would be predicted to cause changes in the overall splicing of mRNA, and therefore aberrant patterns of gene expression. Hence, there are many pathways affected in this diverse disease, but many of the identified mutations lead to gene expression dysregulation and abnormal myeloid development (summarized in Table 30.3). It should be noted that there is a great deal of overlap between the mutations found in these different types of myeloid disorders, highlighting relationships in the cellular ontogeny. From a clinical standpoint, fully recognizing, and utilizing the potential of these markers for diagnosis, prognosis, and treatment of myeloid malignancies has begun but requires additional work [71].

## 30.4 Summary

In this brief discussion, we have highlighted the cell of origin, pathologic features, and most significant molecular changes in the pathogenesis of this diverse group of malignancies. In addition, specific molecular alterations have been detected, showing promise for specific therapeutic interventions either affecting the malignant cells or tumor microenvironment. These new developments are bound to change our understanding of these diseases, leading to reclassification and refinement of therapeutics in both myeloid and lymphoid malignancies.

**Acknowledgment** We are grateful to Dr. Scott Rodig of the Brigham and Women's Hospital, Harvard Medical School for review of lecture notes in preparation for this chapter.

## References

1. Vardiman JW, et al. The 2008 revision of the World Health Organization (WHO) classification of myeloid neoplasms and acute leukemia: rationale and important changes. Blood. 2009;114(5):937–51.
2. Akashi K, et al. A clonogenic common myeloid progenitor that gives rise to all myeloid lineages. Nature. 2000;404(6774):193–7.
3. Muramatsu M, et al. Class switch recombination and hypermutation require activation-induced cytidine

deaminase (AID), a potential RNA editing enzyme. Cell. 2000;102(5):553–63.
4. Adams JM, et al. Cellular myc oncogene is altered by chromosome translocation to an immunoglobulin locus in murine plasmacytomas and is rearranged similarly in human Burkitt lymphomas. Proc Natl Acad Sci USA. 1983;80(7):1982–6.
5. Bakhshi A, et al. Cloning the chromosomal breakpoint of t(14;18) human lymphomas: clustering around JH on chromosome 14 and near a transcriptional unit on 18. Cell. 1985;41(3):899–906.
6. Tsujimoto Y, et al. Cloning of the chromosome breakpoint of neoplastic B cells with the t(14;18) chromosome translocation. Science. 1984;226 (4678):1097–9.
7. Kerckaert JP, et al. LAZ3, a novel zinc-finger encoding gene, is disrupted by recurring chromosome 3q27 translocations in human lymphomas. Nat Genet. 1993;5(1):66–70.
8. Thompson MP, Kurzrock R. Epstein-Barr virus and cancer. Clin Cancer Res. 2004;10(3):803–21.
9. Shimoyama Y, et al. Senile Epstein-Barr virus-associated B-cell lymphoproliferative disorders: a mini review. J Clin Exp Hematop. 2006;46(1):1–4.
10. Muller AM, et al. Epidemiology of non-Hodgkin's lymphoma (NHL): trends, geographic distribution, and etiology. Ann Hematol. 2005;84(1):1–12.
11. Dohner H, et al. Genomic aberrations and survival in chronic lymphocytic leukemia. N Engl J Med. 2000;343(26):1910–6.
12. Horsman DE, et al. Comparison of cytogenetic analysis, southern analysis, and polymerase chain reaction for the detection of t(14; 18) in follicular lymphoma. Am J Clin Pathol. 1995;103(4):472–8.
13. Rack K, Vidrequin S, Dargent, JL. Genomic profiling of myeloma: the best approach, a comparison of cytogenetics, FISH and array-CGH of 112 myeloma cases. J Clin Pathol. 2015.
14. Li JY, et al. Detection of translocation t(11;14)(q13; q32) in mantle cell lymphoma by fluorescence in situ hybridization. Am J Pathol. 1999;154(5):1449–52.
15. Kendrick S, et al. Diffuse large B-cell lymphoma cell-of-origin classification using the Lymph2Cx assay in the context of BCL2 and MYC expression status. Leuk Lymphoma. 2015;1–4.
16. Shaffer 3rd AL, Young RM, Staudt LM. Pathogenesis of human B cell lymphomas. Ann Rev Immunol. 2012;30:565–610.
17. Said J, Lones M, Yea S. Burkitt lymphoma and MYC: what else is new? Adv Anat Pathol. 2014;21 (3):160–5.
18. Baens M, et al. Structure of the MLT gene and molecular characterization of the genomic breakpoint junctions in the t(11;18)(q21;q21) of marginal zone B-cell lymphomas of MALT type. Genes Chromosom Cancer. 2000;29(4):281–91.
19. Turakhia S, et al. Immunohistochemistry for BRAF V600E in the differential diagnosis of hairy cell leukemia vs other splenic B-cell lymphomas. Am J Clin Pathol. 2015;144(1):87–93.
20. Iqbal J, et al. Genomic signatures in T-cell lymphoma: how can these improve precision in diagnosis and inform prognosis? Blood Rev. 2015.
21. Lamant L, et al. High incidence of the t(2;5)(p23;q35) translocation in anaplastic large cell lymphoma and its lack of detection in Hodgkin's disease. Comparison of cytogenetic analysis, reverse transcriptase-polymerase chain reaction, and P-80 immunostaining. Blood. 1996;87(1):284–91.
22. Parrilla-Castellar ER, et al. ALK-negative anaplastic large cell lymphoma is a genetically heterogeneous disease with widely disparate clinical outcomes. Blood. 2014;124(9):1473–80.
23. Ansell SM, et al. PD-1 blockade with nivolumab in relapsed or refractory Hodgkin's lymphoma. N Engl J Med. 2015;372(4):311–9.
24. Armitage JO. Early-stage Hodgkin's lymphoma. N Engl J Med. 2010;363(7):653–62.
25. Hamblin TJ, et al. Unmutated Ig V(H) genes are associated with a more aggressive form of chronic lymphocytic leukemia. Blood. 1999;94(6):1848–54.
26. Calin GA, et al. Frequent deletions and down-regulation of micro-RNA genes miR15 and miR16 at 13q14 in chronic lymphocytic leukemia. Proc Natl Acad Sci USA. 2002;99(24):15524–9.
27. Klein U, et al. The DLEU2/miR-15a/16–1 cluster controls B cell proliferation and its deletion leads to chronic lymphocytic leukemia. Cancer Cell. 2010;17 (1):28–40.
28. Magrath I. Epidemiology: clues to the pathogenesis of Burkitt lymphoma. Br J Haematol. 2012;156 (6):744–56.
29. Dalla-Favera R, et al. Human c-myc onc gene is located on the region of chromosome 8 that is translocated in Burkitt lymphoma cells. Proc Natl Acad Sci USA. 1982;79(24):7824–7.
30. Battey J, et al. The human c-myc oncogene: structural consequences of translocation into the IgH locus in Burkitt lymphoma. Cell. 1983;34(3):779–87.
31. Rabbitts TH, Hamlyn PH, Baer R. Altered nucleotide sequences of a translocated c-myc gene in Burkitt lymphoma. Nature. 1983;306(5945):760–5.
32. Onciu M, et al. Secondary chromosomal abnormalities predict outcome in pediatric and adult high-stage Burkitt lymphoma. Cancer. 2006;107(5):1084–92.
33. Poirel HA, et al. Specific cytogenetic abnormalities are associated with a significantly inferior outcome in children and adolescents with mature B-cell non-Hodgkin's lymphoma: results of the FAB/LMB 96 international study. Leukemia. 2009;23(2):323–31.
34. Boerma EG, et al. Translocations involving 8q24 in Burkitt lymphoma and other malignant lymphomas: a historical review of cytogenetics in the light of todays knowledge. Leukemia. 2009;23(2):225–34.
35. Lin CY, et al. Transcriptional amplification in tumor cells with elevated c-Myc. Cell. 2012;151(1):56–67.
36. Sabo A, et al. Selective transcriptional regulation by Myc in cellular growth control and lymphomagenesis. Nature. 2014;511(7510):488–92.

37. Janz S, Potter M, Rabkin CS. Lymphoma-and leukemia-associated chromosomal translocations in healthy individuals. Genes Chromosom Cancer. 2003;36(3):211–23.
38. Peled JU, et al. Requirement for cyclin D3 in germinal center formation and function. Cell Res. 2010;20(6):631–46.
39. Schmitz R, et al. Burkitt lymphoma pathogenesis and therapeutic targets from structural and functional genomics. Nature. 2012;490(7418):116–20.
40. Wotherspoon AC, et al. Regression of primary low-grade B-cell gastric lymphoma of mucosa-associated lymphoid tissue type after eradication of *Helicobacter pylori*. Lancet. 1993;342(8871):575–7.
41. Dierlamm J, et al. Genetic abnormalities in marginal zone B-cell lymphoma. Hematol Oncol. 2000;18 (1):1–13.
42. Hummel M, et al. A biologic definition of Burkitt's lymphoma from transcriptional and genomic profiling. N Engl J Med. 2006;354(23):2419–30.
43. Bastard C, et al. LAZ3 rearrangements in non-Hodgkin's lymphoma: correlation with histology, immunophenotype, karyotype, and clinical outcome in 217 patients. Blood. 1994;83(9):2423–7.
44. Alizadeh AA, et al. Distinct types of diffuse large B-cell lymphoma identified by gene expression profiling. Nature. 2000;403(6769):503–11.
45. Biagi JJ, Seymour, JF. Insights into the molecular pathogenesis of follicular lymphoma arising from analysis of geographic variation. Blood. 2002;99 (12):4265–75.
46. Limpens J, et al. Lymphoma-associated translocation t(14;18) in blood B cells of normal individuals. Blood. 1995;85(9):2528–36.
47. Dave SS, et al. Prediction of survival in follicular lymphoma based on molecular features of tumor-infiltrating immune cells. N Engl J Med. 2004;351(21):2159–69.
48. Dick JE. Stem cell concepts renew cancer research. Blood. 2008;112(13):4793–807.
49. Lapidot T, et al. A cell initiating human acute myeloid leukaemia after transplantation into SCID mice. Nature. 1994;367(6464):645–8.
50. Goardon N, et al. Coexistence of LMPP-like and GMP-like leukemia stem cells in acute myeloid leukemia. Cancer Cell. 2011;19(1):138–52.
51. Nutt SL, et al. Commitment to the B-lymphoid lineage depends on the transcription factor Pax5. Nature. 1999;401(6753):556–62.
52. Fu S, et al. PML/RARalpha, a fusion protein in acute promyelocytic leukemia, prevents growth factor withdrawal-induced apoptosis in TF-1 cells. Clin Cancer Res. 1995;1(6):583–90.
53. Rowley JD, Golomb HM, Dougherty C. 15/17 translocation, a consistent chromosomal change in acute promyelocytic leukaemia. Lancet. 1977;1 (8010):549–50.
54. Ley TJ, et al. DNA sequencing of a cytogenetically normal acute myeloid leukaemia genome. Nature. 2008;456(7218):66–72.
55. Welch JS, et al. The origin and evolution of mutations in acute myeloid leukemia. Cell. 2012;150(2):264–78.
56. Huang ME, et al. Use of all-trans retinoic acid in the treatment of acute promyelocytic leukemia. Blood. 1988;72(2):567–72.
57. Genovese G, et al. Clonal hematopoiesis and blood-cancer risk inferred from blood DNA sequence. N Engl J Med. 2014;371(26):2477–87.
58. Dohner H, Weisdorf DJ, Bloomfield CD. Acute myeloid leukemia. N Engl J Med. 2015;373 (12):1136–52.
59. Rowley JD. Letter: a new consistent chromosomal abnormality in chronic myelogenous leukaemia identified by quinacrine fluorescence and Giemsa staining. Nature. 1973;243(5405):290–3.
60. Nowell PC, Hungerford DA. Chromosome studies on normal and leukemic human leukocytes. J Natl Cancer Inst. 1960;25:85–109.
61. Ben-Neriah Y, et al. The chronic myelogenous leukemia-specific P210 protein is the product of the bcr/abl hybrid gene. Science. 1986;233(4760):212–4.
62. Druker BJ, et al. Efficacy and safety of a specific inhibitor of the BCR-ABL tyrosine kinase in chronic myeloid leukemia. N Engl J Med. 2001;344 (14):1031–7.
63. Witthuhn BA, et al. JAK2 associates with the erythropoietin receptor and is tyrosine phosphorylated and activated following stimulation with erythropoietin. Cell. 1993;74(2):227–36.
64. Kralovics R, et al. A gain-of-function mutation of JAK2 in myeloproliferative disorders. N Engl J Med. 2005;352(17):1779–90.
65. Pikman Y, et al. MPLW515L is a novel somatic activating mutation in myelofibrosis with myeloid metaplasia. PLoS Med. 2006;3(7):e270.
66. Klampfl T, et al. Somatic mutations of calreticulin in myeloproliferative neoplasms. N Engl J Med. 2013;369(25):2379–90.
67. Starczynowski DT, et al. Identification of miR-145 and miR-146a as mediators of the 5q-syndrome phenotype. Nat Med. 2010;16(1):49–58.
68. ellagatti A, et al. Haploinsufficiency of RPS14 in 5q-syndrome is associated with deregulation of ribosomal-and translation-related genes. Br J Haematol. 2008;142(1):57–64.
69. Ades L, Itzykson R, Fenaux P. Myelodysplastic syndromes. Lancet. 2014;383(9936):2239–52.
70. Yoshida K, et al. Frequent pathway mutations of splicing machinery in myelodysplasia. Nature. 2011;478(7367):64–9.
71. Pastore F, et al. Combined molecular and clinical prognostic index for relapse and survival in cytogenetically normal acute myeloid leukemia. J Clin Oncol. 2014;32(15):1586–94.

# 31 Epidemiology of Melanoma

Sameer Gupta and Hensin Tsao

## 31.1 Clinical Picture of the Disease

### 31.1.1 Introduction

Melanocytes specialize in the synthesis of melanin, a pigmented polymer that protects against environmental genotoxicity by absorbing UV light and serving as a free-radical scavenger [1]. Melanocytes are found in highest numbers in the epidermis and hair follicles, with a cutaneous density ranging from 550 to 1200 cells/mm$^2$ [2], but are also found in the inner ear, eye, brain, and heart, among other locations [3–7]. Melanomas arise in any anatomic location containing melanocytes, but cutaneous melanoma is the most common site of origin and is the focus of this chapter. Ocular and mucosal melanomas are, however, briefly discussed at the end of this section.

### 31.1.2 Signs and Symptoms

Cutaneous melanoma lesions may itch, bleed, and be tender [8]. However, early lesions are often asymptomatic but have distinctive physical features codified by the ABCDE acronym [9]:

S. Gupta · H. Tsao (✉)
Harvard Medical School, 50 Staniford Street, Boston, MA 02114, USA
e-mail: htsao@mgh.harvard.edu

- A is for Asymmetry: Melanomas typically grow at variable rates throughout the lesion, resulting in asymmetry.
- B is for Border: The aforementioned variable growth rate results in irregular borders.
- C is for Color: Melanomas typically contain hues of tan, brown, black, red, and white while benign lesions typically have a uniform color.
- D is for Diameter: Malignant melanomas will typically have a diameter of at least 6 mm.
- E is for Evolution and Elevation: Melanomas can change over time, particularly in regard to symptoms, pigmentation, shape, and size.

In addition, melanoma lesions will often have more distorted and atypical features that distinguish it from surrounding banal nevi—a clue for malignancy that has been termed "the ugly duckling sign."

### 31.1.3 Diagnosis

Early diagnosis is critical in the management of melanoma (see Fig. 31.1). A skin exam with close attention to the features described in the ABCDE criteria is important. The ABCDE criteria have proven to be clinically useful with high inter-rater reliability. However, a reasonable number of melanomas will not fulfill any of the ABCDE criteria (small amelanotic nodular melanomas for instance), and many irregular lesions that fulfill ABCDE ultimately prove not to be melanoma but rather dysplastic nevi. When a lesion concerns melanoma, the gold standard


M. Loda et al. (eds.), *Pathology and Epidemiology of Cancer*,
DOI 10.1007/978-3-319-35153-7_31

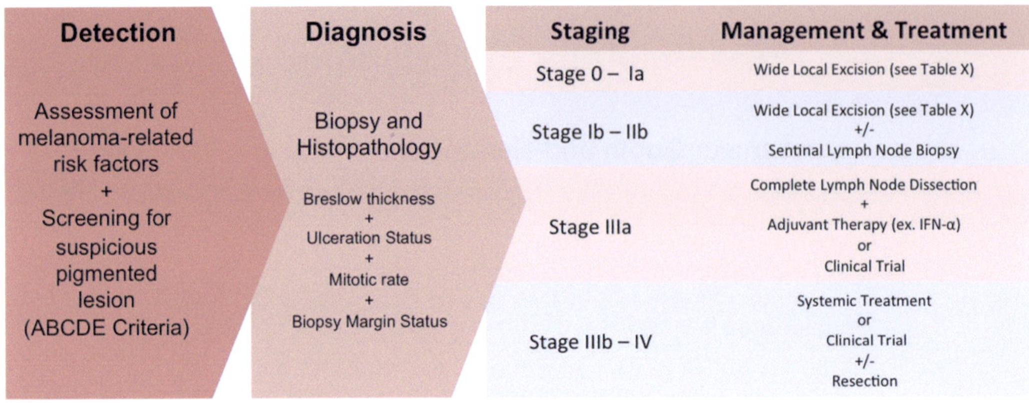


**Fig. 31.1** Clinical flowchart for melanoma management. This diagram is simplified from the 2014 National Comprehensive Cancer Network Clinical Practice Guidelines [24]

method of diagnosis is biopsy and histopathological examination (see the Pathology Chapter).

Light-based visual technologies were developed in the 1990s to help with the visualization of the skin. Dermoscopy (also called dermatoscopy) uses a hand-held, magnifier (typically 10×) with polarized light as a non-invasive method to better examine pigmented lesions. Contact dermoscopy (or epiluminescence) uses a thin layer of oil or alcohol to reduce light reflectance while newer, non-contact dermatoscopes use polarizing light filters to achieve a similar effect. With the removal of this obscuring visual "noise," it becomes possible to appreciate skin structures that would otherwise be unappreciated by the eye alone [10]. Pattern recognition across a number of variables and a general overall impression of a lesion, rather than a strict set of guidelines for required characteristics, is the most widely used method and has proven to be useful in the detection of melanoma.

Several novel techniques are being developed to help in the early diagnosis of melanoma, including digitized dermoscopy, laser-based imaging methods, ultrasound techniques, and lesion electrical conductance measurement [9]. One preliminary molecular method involves extracting mRNA from superficial epidermal cells removed by applying an adhesive tape over the lesion. mRNA is isolated from these cells to perform gene expression profiling to identify specific molecular alterations [11].

Wachsman and colleagues used this tape-stripping method to compare gene expression levels between melanoma and benign nevi, identifying 17 genes that might be used to detect malignancy. They evaluated this 17-gene classifier on a test dataset of 39 melanomas and 89 nevi, resulting in 100 % sensitivity and 88 % specificity. The majority of the genes are involved with cell death and cellular development [11].

### 31.1.4 Clinical Subtypes of Melanoma

Melanoma is a heterogeneous cancer, manifesting in a range of clinical subtypes. This heterogeneity has been classically captured by four subtypes characterized by morphological aspects of the radial growth phase and body site of the primary tumor: superficial spreading melanoma (SSM), lentigo maligna melanoma (LMM), nodular melanoma (NM), and acral lentiginous melanoma (ALM). The clinical characteristics and epidemiology of these four subtypes are described below.

SSM is the most common subtype, responsible for 50–80 % of all melanomas [12, 13]. This subtype often arises from a preexisting collection of benign melanocytes called nevi, slowly evolving into a malignancy over months to years. SSM typically have irregular borders and pigmentation, containing hues of brown, black, red, and white [14]. NM is the second most common subtype of melanoma, accounting for 20–30 % of all melanomas [12, 13]. These lesions are found most commonly on the trunk and may not be associated with preexisting nevi. In contrast to SSM, NM rapidly develops over weeks to months. NM tends to be raised on palpation and can appear dark blue-to-black, but may also appear amelanotic. Notably, analysis from SEER data from 1978 to 2007 found that SSM accounted for 66 % of incident melanomas and 46 % of fatal melanomas, NM accounted for 14 % of all melanomas and 37 % of melanoma deaths. Thus, despite accounting for fewer melanomas in total, NM accounts for a similar proportion of mortality as SSM. These findings are concordant with the observation that tumor thickness is an important prognostic factor in melanoma [13].

LMM is the third most common melanoma, accounting for 10–15 % of melanomas [12, 15]. LMM typically occurs in the elderly on chronically sun-damaged skin such as the head, neck, and forearms. Clinically, LMM presents as a flat, brown tumor and may typically appear with irregular borders. LMM differs from NM and SSM in that it typically does not have a nevus precursor, occurs more frequently in older individuals, and has a longer prolonged period of intra-epidermal growth as compared with SSM [14]. In regards to trends over time, SEER analysis revealed that LM is the most prevalent in situ subtype (79–83 %) and that incidence of LMM is increasing at a higher rate than other subtypes, especially in older men [15].

Finally, ALM is the least common subtype of melanomas, but drastically differs in prevalence among ethnicities [16]. This subtype accounts for the majority of melanomas among individuals with darker pigment, representing the majority of melanomas in African Americans and Asian populations [17–20]. In Caucasian populations, ALM only represents approximately 5 % of melanomas [16]. ALM is most commonly found on hairless areas of the body, including the soles, palms, and subungual regions. ALM typically appears as a flat lesion that develops irregular borders but can present with variable pigmentation and nail plate destruction. Subungual ALM typically arises from the nail matrix on the great toe or thumb and develops into a longitudinal black band. Notably, ALM appears to have a greater frequency of focused gene amplifications as compared to other subtypes, such as SSM [21].

### 31.1.5 Staging

When melanoma is confirmed by histopathology, staging of disease is important for treatment decisions and prognosis estimation. The American Joint Committee on Cancer (AJCC) has developed a tumor-node-metastasis (TNM) based clinical and histologic scheme that incorporates important prognostic indicators such as metastases, tumor thickness, ulceration, and mitotic activity [22]. The details of this staging scheme will be discussed in greater length in Sect. 31.4. Clinically, macroscopic metastasis can often be detected by physical exam, looking for palpable lymph or dermal/subcutaneous nodules around the area of the primary lesion. If a nodule concerning macroscopic metastasis is appreciated, fine needle aspiration or biopsy of the nodule is indicated. Subclinical, microscopic metastases can only be identified by lymph node biopsy.

In aggregated, localized melanoma (Stage I or II) is associated with 5-year survival rates of >90 % and 80 % respectively [23]; however, many features such as ulceration and lymphovascular invasion may reduce the survival rate. In contrast, survival decreases to 30–50 % with localized lymph node metastases (Stage III) disease. Distant metastatic disease (Stage IV) has historically been linked to a median survival of 8–9 months and a 15 % 3-year survival rate [22]. However, this survival data, reported in the most recent AJCC guidelines in 2009, predates much of the recent exciting advances in melanoma treatment.

### 31.1.6 Treatment

#### 31.1.6.1 Surgery

Wide surgical excision is the main treatment for primary cutaneous melanoma, with the goal of directly resecting all malignant cells from the primary site. The National Comprehensive Cancer Network (NCCN) Clinical Practice Recommendations vary regarding the extent of clean surgical margins required for melanoma based on tumor depth: melanoma in situ and primary melanomas of depth ≤1.0, 1.01–2, 2.01–4, and >4 mm are recommended to be excised with clean margins of 0.5–1, 1, 1–2, 2, and 2 cm, respectively [24]. In the presence of regional metastasis, complete lymph node dissection is recommended of the surrounding regional lymph node basin. Interferon-α [alpha] 2b (both pegylated and non-pegylated) and ipilimumab have been approved by the US Food and Drug Administration (FDA) for adjuvant treatment in the setting of Stage III disease.

#### 31.1.6.2 Systemic Therapy

Cytotoxic chemotherapy was the mainstay of treatment for metastatic melanoma for over three decades, but treatment options are fast evolving with the development of novel agents propelled by the revolution in our molecular understanding of melanoma and immunology (see Fig. 31.2).

#### 31.1.6.3 Chemotherapy

Alkylating agents, platinum analogs, and microtubular toxins have all been used to treat melanoma [25]. However, dacarbazine, an alkylating agent, is the only FDA-approved chemotherapeutic agent for metastatic melanoma, with a demonstrated 10–20 % response rate that has limited durability [25, 26]. In addition to monotherapy, combinations of multiple chemotherapeutic agents have also been tested. These combined regimens have yielded very modest response rates that are typically less than 20 % in first and second line settings [24].

#### 31.1.6.4 Immune Therapy

Several clinical observations suggest that melanoma is an immunogenic tumor. Melanoma has been transmitted from immunocompetent organ donors to immunosuppressed organ recipients [27–30]. Spontaneous regression of melanoma lesions has been observed, and approximately 3 % of all melanomas present with metastatic disease without a known primary lesion, which may be due to regression of the primary lesion from immunosurveillance [31, 32]. In addition, the presence of tumor-infiltrating lymphocytes in lesions may be associated with longer survival [32, 33].

Before 2011, immunotherapy was limited to cytokine-based treatment with IFN-α[alpha] in the adjuvant setting and high-dose interleukin-2 (IL-2) in Stage IV disease. Interferon's mechanism of action as an adjuvant therapy remains unclear as it is an immunomodulatory and

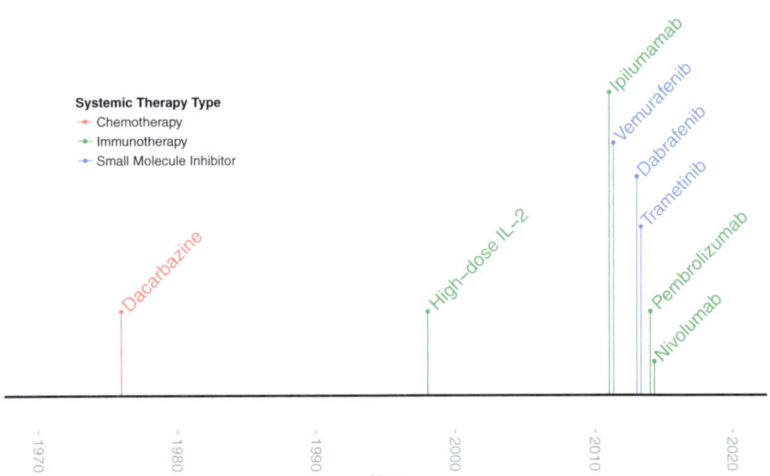


**Fig. 31.2** Timeline of FDA-approval of melanoma drugs

anti-angiogenic cytokine that can promote melanocyte apoptosis [36–39]. Moreover, its use remains controversial given that it has been demonstrated to improve disease free survival but not overall survival, while having a severe side-effect profile including neuropsychiatric, constitutional, and hepatic toxicity [34, 35]. Studies have not demonstrated consistent survival benefit [36, 37]. IL-2, a T cell growth factor, was subsequently developed and approved to treat patients with metastatic melanoma, with clinical trials demonstrating a modest but durable 16 % response rate [38, 39].

Since 2011, monoclonal antibody targeting of immune activating checkpoint blockade has emerged as a field-changing addition to our therapeutic armamentarium. Immunosurveillance begins with T cell recognition of malignancy-generated, intracellular peptides that are presented in the context of MHC-I molecules on the surface of tumor and other antigen presenting cells. This interaction is facilitated by receptor-specific, T cell binding of B7 on antigen presenting cells. CD28 and CTLA-4 receptors on the T cell surface compete to bind to B7, either resulting in a T cell stimulatory or inhibitory signal, respectively. Thus, CTLA-4 serves as a "brake," regulating the amplitude of early T cell activation [40, 41]. In 2011 and 2015, the FDA approved the first monoclonal antibody to CTLA-4, ipilimumab, for use in Stage IV and Stage III melanoma, respectively. Ipilimumab removes this brake and results in a more immune-activated state. Another relevant immune checkpoint is the inhibitory receptor programmed cell death receptor-1 (PD-1) on activated T cells. This inhibitory receptor, when bound to tumor associated PD-L1, downregulates the immune response [40]. The FDA approved two monoclonal antibodies targeting PD-1, pembrolizumab and nivolumab, in 2014.

#### 31.1.6.5 Targeted Therapy

Therapies that target a specific mutated protein implicated in melanomagenesis have become an important addition to our therapeutic arsenal. In 2002, it was discovered that 50 % of all melanomas carry a mutated protein kinase B-raf (BRAF), which constitutively activates the pro-proliferative mitogen-activated protein kinase (MAPK) pathway [42]. 75 % of BRAF mutations are the result of a valine to glutamate substitution at the 600th amino acid (V600E)[43]. Given the proportion of tumors bearing this mutation and the mutation's oncogenic potential, researchers began to search for an inhibitor. Sorafenib, an early BRAF inhibitor, proved to be a poor treatment for melanoma given an intolerable side-effect profile due to inhibition of wild type BRAF and other off-target effects [44]. In contrast, V600E-specific BRAF inhibitors—including vemurafenib and dabrafenib—have been developed with stunning results. In the phase III study, vemurafenib yielded a 48 % response rate and an overall survival advantage among metastatic melanoma, as compared with dacarbazine [45].

Despite these triumphs, the greatest obstacle for targeted therapy has been the rapid development of drug resistance after months of treatment. Approximately 50 % of patients treated with BRAF inhibitors demonstrate progression of disease after several months [46, 47]. Genomic profiling of tumors have helped identify pathways to resistance, including mutations in the downstream kinase, mitogen/extracellular signal-regulated kinase (MEK), that results in recovery of MAPK [48]. Other pathways to resistance that have been identified include mutational activation of NRAS, emergence of alternative BRAF splice products resistant to treatment, and BRAF amplification [49].

Combination therapy with BRAF inhibitors and an inhibitor of MEK has been tested with promising results. For instance, the combination of dabrafenib with trametinib, a MEK inhibitor, yields higher response rate and longer progression-free survival as compared to the BRAF inhibitor alone [50]. Studies exploring different combinations are currently underway and will hopefully yield higher, durable response rates.

### 31.1.7 Prognosis

Beyond the prognostic information that staging provides, several studies have searched for molecular signatures that may be predictive of clinic course. Whole-genome expression profiling has been conducted in both local and aggressive melanomas to find genes that might predict melanoma progression. An early study identified 254 genes that were associated with progression [51]. Another study using RNA profiling of primary tumor samples demonstrated that high-grade tumors had increased expression of proliferation and DNA damage signaling genes, while low-grade melanomas had increased expression of immune genes [49, 52]. Another study found a nine gene-signature predicted overall survival and distant metastasis-free patient survival that was independent of AJCC staging. Many of these genes coded for stromal-related proteins, suggesting an important interaction between the malignancy and its surrounding micro-environment in predicting prognosis [53].

### 31.1.8 Other Melanoma Types

While the focus of the chapter is on cutaneous melanoma, we will briefly discuss two other forms of melanoma, ocular and mucosal, to present a comprehensive overview of melanoma.

#### 31.1.8.1 Ocular Melanoma

Ocular melanoma is the most common intraocular cancer and the second most common melanoma after cutaneous melanoma, accounting for 3.7 % of all melanoma cases [54]. In the United States, incidence of ocular melanoma is six per a million [54]. Ocular melanoma rates are 8–10 times higher in Whites as compared to Blacks, but it should be noted that the difference is less pronounced when compared to cutaneous melanoma, in which 16 times higher rates among Caucasians are noted [54]. These melanomas arise from melanocytes positioned in the conjunctival membrane and the uveal tract of the eye [55]. However, the uvea is the most common site of origin of primary ocular melanomas, representing over 80 % of all ocular melanomas [54].

While melanoma can arise anywhere along the uveal tract, choroidal melanoma accounts for the majority of uveal melanomas, representing over 80 % of cases [54]. The clinical presentation of uveal melanoma depends primarily on the size and location, ranging from asymptomatic to complete vision loss in the affected eye. The most frequently reported symptoms include blurred vision, visual field defects, and irritation [55]. Diagnosis is frequently made by ophthalmic examination, ranging from slit lamp biomicroscopy and indirect ophthalmoscopy to ancillary testing such as ultrasonography [55]. Management of tumors varies between careful observation to orbital exenteration, depending on the size and location. Most individuals will obtain radiotherapy or enucleation. Regarding prognosis, less than 4 % of patients with uveal melanoma have detectable metastatic disease, but over half will develop metastatic disease due to the lack of effective systemic treatments [56]. Size of tumors is one of the most important prognostic factors for uveal melanoma. Risk of metastasis appears to increase with each millimeter increase in tumor thickness [57].

Similar to cutaneous melanoma, much has been learned about the molecular etiology of ocular melanoma. Among uveal melanomas, 80 % carry activating mutations in either GNAQ or GNA11. These genes encode for the GTP-binding protein α [alpha]—subunits that link G-protein-coupled receptor signaling to the MAPK pathway, resulting in constitutive activation of the MAPK pathway [58, 59]. In addition, inactivating mutations in BAP1 (BRCA1 associated protein-1), a deubquitinating enzyme, has been linked to a subgroup of uveal melanomas that are most likely to metastasize [60].

#### 31.1.8.2 Mucosal Melanoma

Mucosal melanomas are relatively rare, representing only 1 % of all melanomas [54]. In the United States, there are approximately 2.2 mucosal melanoma cases per million per year [54]. Primary mucosal melanomas may arise from melanocytes lining the respiratory, gastrointestinal, or urogenital tracts [61]. Mucosal melanomas in the respiratory tract are most frequently found in the nasal cavity and paranasal sinuses, and the most common symptoms include nasal obstruction and epistaxis, with patients first exhibiting epistaxis before exhibiting obstructive symptoms [62]. In contrast, mucosal melanomas arising from the gastrointestinal tract occur most frequently in the anorectal and oropharyngeal regions, and the most common symptoms include bleeding, pain, and discomfort [63]. Finally, mucosal melanomas of the urogenital tract are more common in females, and occur most frequently on the vulva and vaginal canal, with common presenting symptoms of vaginal bleeding and discharge [54]. In general, mucosal melanomas are typically discovered after work-up of presenting symptoms and are more advanced in nature than cutaneous melanomas given their hidden locations.

Mucosal melanomas differ from cutaneous and ocular melanomas in several notable ways. Unlike other subtypes of melanoma, there is a female predominance as compared to men (2.8 vs. 1.5 per million per year), which is related to the higher rates of genital tract melanomas found in women [54]. In addition, sun radiation is an important risk factor for cutaneous and uveal melanomas (see Sect. 31.3 for discussion), but is not thought to contribute risk for mucosal melanoma, given the shielded nature of the sites. Prognostically, primary mucosal melanomas are much more aggressive, with one study finding that the 5-year survival rates for cutaneous, ocular, and mucosal melanomas were 80.8, 74.6, and 25 %, respectively [61]. Finally, the molecular etiology of mucosal melanoma appears to be different than either cutaneous and ocular melanomas. Mucosal melanomas often harbor KIT (receptor tyrosine kinase) amplifications or mutations. KIT-inhibitors, such as imatinib and sunitinib, have been trialed with some limited response [64, 65]. Given the rarity of mucosal melanoma and that only a subset carry KIT aberrations, no large clinical trials have been possible [62]. Like cutaneous melanoma, surgery is the primary treatment with potential roles for chemotherapy and immunotherapy [62].

## 31.2 Descriptive Epidemiology

### 31.2.1 Geographic Trends

According to World Health Organization, there were over 230,000 new cases of melanoma and over 55,000 melanoma-related deaths worldwide in 2012 [66]. Notably, the burden is not constant across populations; geographic heterogeneity in melanoma incidence is a striking feature of this malignancy. Australia and New Zealand report the highest reported incidence rates with approximately 50 cases per 100,000 individuals per year [67–69]. In the United States, rates are approximately 20 cases per 100,000 per year,

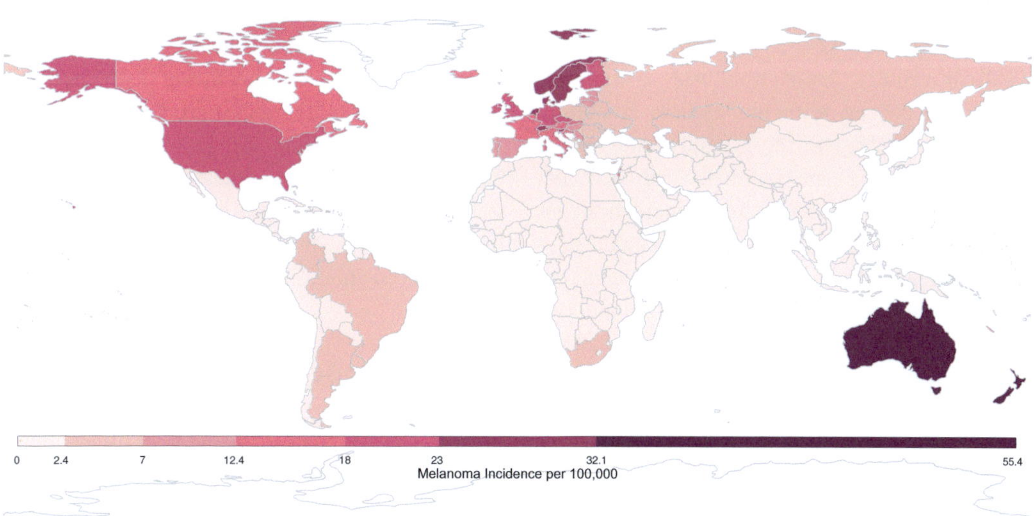


**Fig. 31.3** Global crude melanoma incidence rates. Data presented here was downloaded from the International Agency for Research on Cancer [66]

while Qatar cataloged only 0.3 cases per 100,000 per year (see Fig. 31.3) [66].

Generally, individuals living at lower latitudes experience higher rates of melanoma than those in higher latitudes. For instance, in Australia, residents of Queensland (capital city Brisbane, latitude 27°S) have higher rates of melanoma (65:100,000 per year) than residents of Victoria (capital city Melbourne, latitude 38°S; 36:100,000 per year) [70]. Indeed, similar latitude gradients have been observed in other regions, including the United States, New Zealand, and Scandinavia [69–72]. These observations may be explained, in part, by higher rates of ultraviolet (UV) exposure at lower latitudes, but as discussed in Sect. 31.3, the relationship between UV exposure and melanoma risk is not a simple dose response. Notably, a latitude gradient is not observed in all areas; for instance, in Europe, southern populations experience lower rates of melanoma than populations in the north, which is thought to be related to the protective effects of darker pigmentation of populations in this southern region [70, 73].

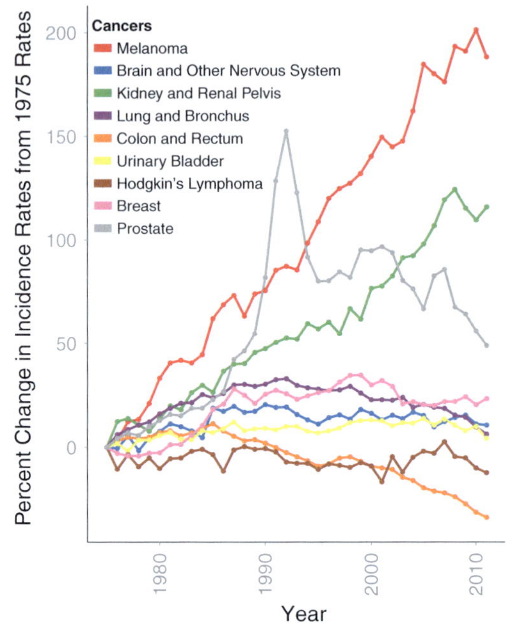


**Fig. 31.4** Percent change of age-adjusted incidence from 1975 rates among 9 cancers. Data downloaded from SEER database [81]. Notably, melanoma has the highest percent increase in incidence

### 31.2.2 Trends Over Time

Melanoma is an ancient disease: skeletons in Peru, dating over 2000 years old, have findings consistent with metastatic melanoma bone lesions [74]. However, at the beginning of the twentieth century, melanoma was a relatively uncommon cancer, with the earliest epidemiological report dated in the United States to the 1930s, estimating an incidence of 1.0 per 100,000 [75, 76]. Melanoma's incidence has been steadily increasing in the fair-skinned population since the mid-1960 s and has been projected to increase for, at least, another two decades [77–79]. The annual incidence increase varies across populations, ranging from 3 to 7 %, but indicates a doubling of rates approximately every 10–20 years [80]. In the United States, melanoma incidence rates are among the fastest growing cancers based on Surveillance Epidemiology and End Results (SEER) data (Fig. 31.4) [81]. This trend is thought to be driven by changes in risk-modifying behavior, including clothing, sun-seeking behavior, and skin awareness/ screening efforts (see Sect. 31.3) [82, 83].

Mortality rates appear to have peaked in the late 1980s to early 1990s in Australia, the United States, and parts of Europe (See rates in Fig. 31.5) [84]. Some have suggested that this dissociation between increasing incidence and stabilizing mortality rates may be the result of "over-diagnosis" [85, 86]. However, the distribution of melanoma thickness has remained stable and an alternative hypothesis is that the improvement in intervention may contribute to the observed mortality trend [87, 88].

### 31.2.3 Trends by Sex

In the United States, melanoma incidence is higher in females than in males in the age group less than 44 and is the most common cancer in females in the age group between 15 and 24 (See Fig. 31.6) [84, 89]. Indeed, melanoma incidence

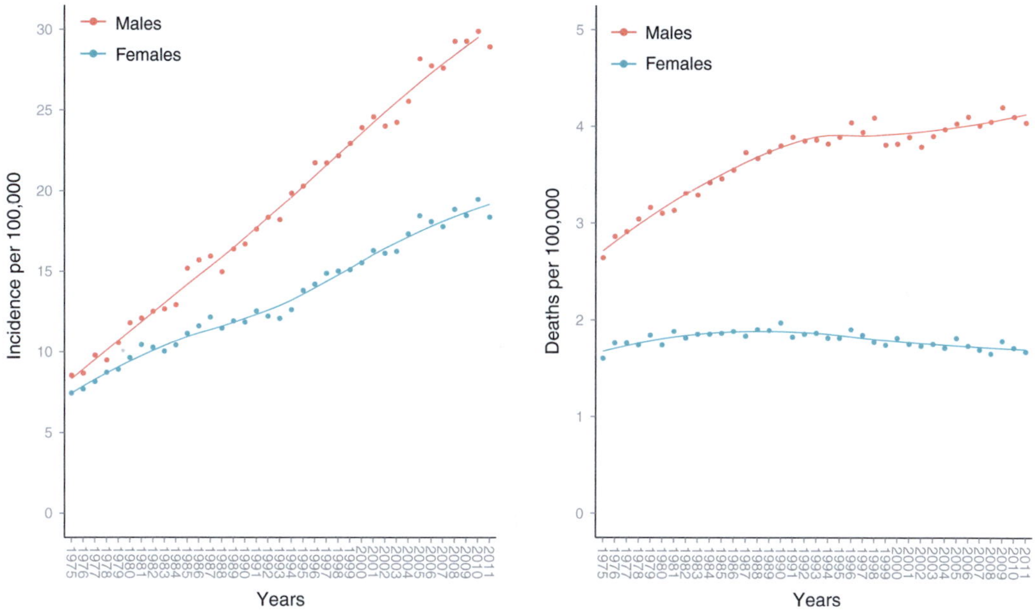


**Fig. 31.5** Age-adjusted incidence and mortality rates in the United States. Data collected and analyzed by SEER [81]. Age-adjusted to the 2000 US standard population

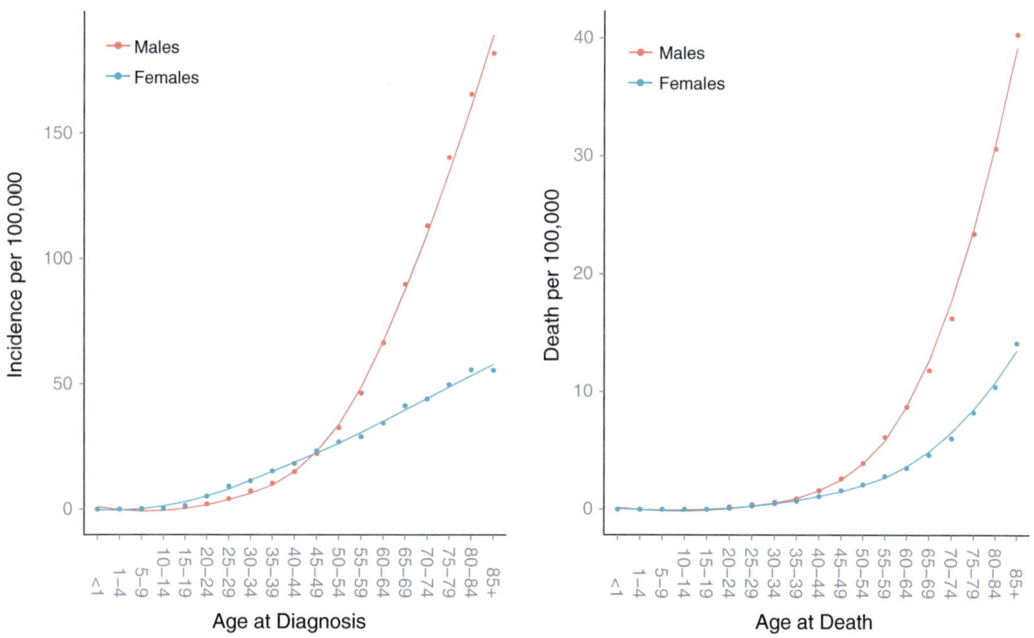


**Fig. 31.6** Age-specific incidence and death rates in the United States between 2007 and 2011. Data collected and analyzed by SEER [81]

appears to be increasing at alarming rates in young women. An analysis of the SEER data suggests that the age-adjusted rate of melanoma more than doubled from 8.1 to 17.4 per 100,000 between 1975 and 2008 among a relatively young age group (20–49 years), while incidence among men in this age group changed more modestly from 8.3 to 12.5 per 100,000 during the same time interval [89, 90]. These observations are most likely related to indoor tanning use among the young female demographic. However, the burden of deaths remain higher among young males versus young females in the United States, even after adjusting for tumor thickness and other prognostic factors [91]. Young men accounted for approximately 40 % of all cases but 60 % of all melanoma deaths among young adults. Indeed, males uniformly appear to do worse at all tumor thickness, histological subtypes, and anatomic sites [92]. A recent study observed that melanomas excised from men were more likely to have greater missense mutation burden than tumors from women, even after adjusting for important confounders such as age, stage, and site of primary tumor [93]. This observation supports the possibility of a biological mechanism underpinning the epidemiological sex discrepancy in melanoma outcome, which remains to be elucidated. One possibility is that sex differences in immunity may lead to more effective cancer surveillance and tumor clearance in women [93].

### 31.2.4 Patterns by Anatomic Site

The odds of developing melanoma by a specific anatomic site vary by sex. In general, women have an increased proportion of lesions on their legs, while males have an increased proportion of cancers on their trunk [94–96]. These differences are thought to be related, in part, to behaviors that are sex-specific, such as clothing, hairstyles, sun-seeking behavior, among others. Indeed, epidemiological trends suggest a non-biological component. As previously noted, melanoma incidence has been increasing at an alarming rate among young women. The increase in incidence has been most pronounced on the trunk region—a site hypothesized to be less shielded to UV exposure due to increasing popularity for full-body, indoor sun tanning [90].

When adjusting for anatomic surface area and thus yielding a proportion of melanoma per a unit of area, both sexes have highest propensity on the face [97]. The head and neck region accounts for 9 % of total body surface area, yet accounts for approximately 12–20 % of all cutaneous melanomas [98]. This increased propensity is likely related to the fact that the head and neck region has one of the highest densities of melanocytes and also is exposed to the highest level of UV radiation [98, 99]. Interestingly, sex-related differences on the head and neck location have also been noted: in a study of patients in France, head and neck melanomas in males were found in peripheral areas of the head and neck (scalp, forehead, temples, ears and neck), while these melanomas were more likely to be found in central facial regions (nose and cheeks) in women [98]. These gender differences likely suggest a photo-protective effect of hair.

The distribution of melanomas by anatomic site appears to significantly differ by age. Two studies in different populations both found that area-adjusted incidence of melanoma were much higher on the trunk among young adults than on the head and neck. The area-adjusted incidence on the head and neck appeared to increase rapidly in older individuals [69, 100]. This distribution is notable as it potentially suggests two distinct pathways for melanoma development that will be discussed more in Sect. 31.2.6.

There is some evidence that melanoma prognosis may differ by anatomic site. As noted in Sect. 31.1, melanoma prognosis and staging is calculated using thickness of tumor, ulceration, and nodal and distant metastases. However, individuals with melanoma of the scalp and neck died of melanoma almost at twice the rate of melanomas found on the extremities, even after controlling for age, tumor thickness, gender, and ulceration [101].

### 31.2.5 Patterns of Somatic Mutations

Mutations in genes in the pro-proliferative MAPK pathway play a significant role in melanomagenesis. In particular, mutations in BRAF and NRAS genes have been identified in a majority of melanomas. Constitutively activating BRAF mutations and NRAS mutations have been found in approximately 40 % and 18 % in cutaneous melanomas, respectively [102]. Approximately 75 % of all BRAF mutations appear to be the result of a T1799A transversion in exon 15 of the gene, causing a valine to glutamate amino acid substitution in the protein. It is important to note that 80 % of melanocytic precancerous nevi contain BRAF mutations, suggesting that these mutations may be necessary but not sufficient to drive malignant transformation [103]. The most common NRAS mutations occur in codon 61 and result in a glutamine to arginine or lysine transition that results in NRAS activation [102].

Notably, mutations in BRAF and NRAS are thought to be mutually exclusive and do not appear to co-exist in the same tumors, suggesting that double-hits in both genes may not be well tolerated. BRAF-mutated melanomas have been associated with younger age at diagnosis, localization to trunk site, and absence of chronic sun damage. In contrast, NRAS-mutated melanomas have been associated with thicker tumors and occurrence on the extremities [102].

### 31.2.6 Implications of Epidemiology Trends

The observation that melanomas in non-chronically exposed areas, such as the trunk, peak in incidence at earlier ages than chronically sun-exposed areas, such as the head and neck, led to the hypothesis of "two divergent pathways" in melanomagenesis [104]. This model hypothesizes that individuals with low propensity for melanocyte proliferation—i.e., those persons with low density of benign melanocytic tumors called nevi—need chronic sun exposure leading to melanomas on anatomic sites with habitual exposure such as on the head and neck. On the other hand, some patients have a high propensity for melanocyte proliferation—i.e., those individuals with high density of nevi—develop melanomas on sites with intermittent sun exposure such as on the trunk [70, 104]. Molecular studies that subsequently demonstrated higher proportion of BRAF mutations in truncal melanomas and higher proportions of NRAS mutations in more chronically exposed regions reified the notion that different processes likely contribute to melanomagenesis in these respective populations. However, differences in risk factors and causes of these pathways are not well understood.

## 31.3 Risk Factors

The "divergent pathway" hypothesis incorporates our knowledge about two risk factors, sun exposure and nevi (one is environmental and the other, in part, genetic). It is clear that melanoma develops from a complex interaction between environmental and host-specific factors, which are explored further in this section. Traditional host factors, such as pigmentation, nevi, and highly penetrant risk genes, are discussed first, followed by non-hereditary risk factors, including ultraviolet (UV) radiation—the most important environmental risk factor for melanoma—as well as evidence for risk modulation by commonly used drugs.

### 31.3.1 Host Factors

#### 31.3.1.1 Pigmentation

Fair-skinned individuals have higher rates of melanoma. Photoprotection of the skin is based primarily on the amount and type of melanin produced in the skin. The melanocortin 1 receptor (MC1R) is an important regulator of skin pigmentation, and is also the most studied germline melanoma-susceptibility locus. MC1R codes for a seven-transmembrane G-protein coupled receptor that, when bound to α[alpha]-melanocyte-stimulating hormone (MSH), increases cAMP levels within the cell. MSH binding ultimately regulates the type of pigment produced, switching from the production of the potentially mutagenic red/ yellow pheomelanin pigment to the photo-protective brown/ black eumelanin pigment [105].

MC1R is highly polymorphic, with over 75 allelic variants, and contributes to variation in skin phenotypes in humans [106]. Wild type MC1R is predominantly expressed in Africa, where high levels of eumelanin are important for photoprotection. It is thought that photoprotection could not have evolved to protect against skin cancer, given that these cancers only rarely cause death to individuals during ages of reproductive age. The evolutionary pressure that led to the development of photoprotection is debated, but the folate deficiency hypothesis suggests that folate, an essential nutrient required for nucleotide and hence DNA biosynthesis, is degraded by UV light, and dark pigmentation evolved to protect against this degradation [107].

Several MC1R variants that result in amino acid substitutions typically decrease receptor function, increasing the amount of pheomelanin in the cell. These increased levels result in red hair, pale skin, poor tanning ability, and freckles. MC1R variants likely contribute to melanoma risk by two mechanisms: modulating pigmentary contribution to UV protection as well as generation of reactive oxygen species [49]. Indeed, individuals with darker phenotypes still carry increased risk of melanoma with MC1R mutations [108]. Interestingly, some MC1R alleles appear to be double risk of melanoma for each allele carried by a patient, with an additive effect observed when carrying multiple alleles. In one meta-analysis, five MC1R alleles (D84E, R142H, R151C, I155T, R160 W) were statistically significantly associated with developing melanoma and the phenotype of red hair color and fair skin [109]. The meta-analysis also identified two variants, R163Q (rs885479) and D294H (rs1805009), that increased melanoma risk, but were not associated with red hair color, supporting the existence of a secondary pathway leading to melanoma risk other than the UV-protective effect of pigmentation [109]. The allele frequency distribution of R163Q is depicted in Fig. 31.7 to show that this allele is found globally and not restricted to populations typically expressing the red hair phenotype (data from [110]). Notably, other pigmentation-related genes have been associated with melanoma susceptibility, including SLC45A2, OCA2, TYR, and ASIP. Variants in these five genes have been reported to explain up to 50 % of the observed difference in melanoma risk contributed by the skin pigmentation phenotype [111].

#### 31.3.1.2 Nevi

Another melanoma risk factor, which is pigment related, is nevus density. Nevi are benign clonal proliferations of melanocytic cells [112]. Higher number of nevi increases risk of melanoma, with some estimates suggesting that each additional common nevus increases risk by 1.02 [113–115]. Twins studies support the hypothesis that the propensity for nevi development is a genetic trait, as monozygotic twins had higher correlations of nevus counts than dizygotic twins [116]. More recently, genome-wide association studies (GWAS) identified two genetic variants, 9p21 and 22q13, associated with nevus development and melanoma [117].

#### 31.3.1.3 Sex

Sex-specific differences in melanoma incidence and mortality are well established: women accounted for 42 % of the 73,870 new melanoma cases but only for 33 % of the 9710

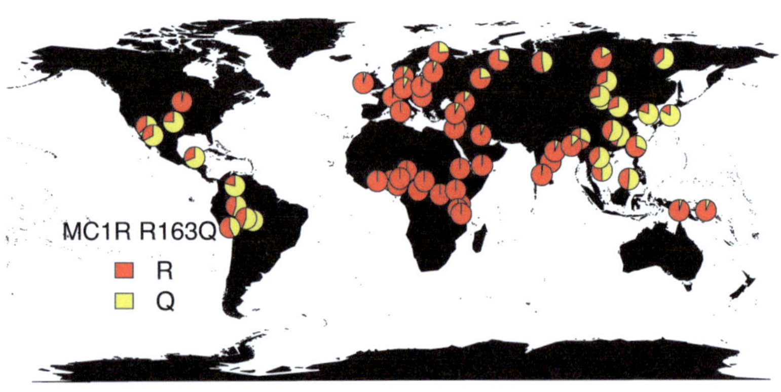


**Fig. 31.7** Global distribution of MC1R R163Q Allele frequency. Data was extracted from the Allele frequency database [110]

melanoma-related deaths recorded in the United States in 2015 [118]. Population-based studies in other countries reflect a similar trend [119–123]. For instance, a population-based study of 11,000 melanoma patients in Germany found that female melanoma patients were 38 % more likely to survive and 42 % less likely to progress than males [123]. Among patients who did experience disease progression, women kept a 20 % survival advantage [123].

Differences in clothing, sun-seeking behavior, UV exposure prevention, and skin screening have been hypothesized as non-biological sources of sex disparity in melanoma outcome [123–126]. However, women, stage-for-stage, have a 30 % relative advantage over men in both melanoma-specific survival and risk of metastases [124]. Furthermore, sex remains an independent predictor of outcome from other factors that would be strongly correlated to the hypothesized non-biological confounders [120, 121, 124–128]. For instance, a large study in the Netherlands found that men were almost twice as likely to die from melanoma than women, even after adjusting for time period of diagnosis, age, tumor thickness, histologic subtype, body site, and metastases status [120].

Thus, mounting evidence suggests that innate biological differences may contribute to sex-specific differences in melanoma outcome. Differences in immunosurveillance, hormonal influence, vitamin D metabolism, genome surveillance for mutations, and epigenetic mechanisms have all been proposed as possible biological drivers of sex disparity in melanoma [124, 129]. However, even the most widely studied of these hypotheses, the role of hormones, remains inconclusive [124, 130–134].

#### 31.3.1.4 Other

In addition to pigmentation-related genes and genes implicated in familial cases of melanoma, other melanoma-susceptibility genes involved in immunity, DNA repair, metabolism, and vitamin D receptor polymorphisms have been reported [49, 135]. It must be noted that the majority of these studies report modest associations and are in general underpowered to identify genes. However, cheaper sequencing will allow for larger population samples with deeper sequencing coverage [104].

#### 31.3.1.5 Genetic Risk Factors

Approximately 10 % of melanoma cases are thought to be familial, related to highly penetrant risk alleles. A well-established high-risk locus for melanoma susceptibility is the cyclin-dependent kinase (CDK) inhibitor 2A gene (CDKN2A), with 40 % of familial melanoma patients carrying a CDKN2A mutation. This gene encodes two tumor suppressor proteins, p16 and p14ARF, that regulate progression from $G_1$ of the cell cycle. p16 binds cyclin-dependent kinase 4 (CDK4), preventing this kinase from phosphorylating the retinoblastoma (Rb) tumor suppressor and consequently preventing E2F restriction at $G_1$ from being released. p14ARF increases the expression of p53. Thus, mutations in CDKN2A can inhibit the function of two tumor suppressor pathways that increase malignancy susceptibility [105]. Patients with mutations in CDKN2A have approximately 70 % increased lifetime risk of developing melanoma [49].

Another high-risk melanoma-susceptibility gene is CDK4, which codes for the target of p16 in the Rb tumor suppression pathway described above. However, fewer than 15 families containing germline mutations in CDK4 are known to exist worldwide, suggesting that germline mutations in CDKN2A are much more common. Other rare hereditary conditions may also increase risk for melanoma susceptibility, including the well-studied condition xeroderma pigmentosum resulting from inactivating mutations in one of eight nucleotide excision repair pathway genes.

The BRCA1 associated protein 1 (BAP1), a deubiquitinating enzyme, has recently been added to the list of germline high-risk melanoma-susceptibility genes. Multiple groups have found associations of germline BAP1 mutations in families predisposed to ocular or cutaneous melanomas [60, 136–139]. BAP1 has been thought to be a tumor suppressor gene, and germline loss-of-function mutations have been linked to increased risk for several malignancies, including melanoma, associated with the tumor predisposition syndrome [137]. More recent investigations highlight the complex role BAP1 plays in cancer biology. Despite the tumor suppressor role and consequences of germline mutations just described, higher BAP1 expression appears to have a pro-survival effect and potential growth-sustaining role in cutaneous melanoma [138]. Table 31.1 delineates genes implicated in melanoma risk [135, 140–142].

### 31.3.2 Environmental Risk Factor

UV radiation is probably the most widely recognized environmental risk factor. Sunlight is composed of the UV spectrum (200–400 nm wavelength), visible light spectrum (400–700 nm wavelength), and the infrared spectrum (>700 nm). Within the UV spectrum, there are three subcategories: UVA (400–320 nm), UVB (320–290 nm), and UVC (290–200 nm). UVB is the predominant culprit in causing skin damage, but UVA also contributes. About 95 % of UVA and 10 % of UVB radiation make it through the atmosphere and strike the earth surface [143]. Tanning devices generate both UVB and UVA radiation, but the ratio of emission type has changed over time, shifting away from UVB to more UVA to reduce risk of burning [144].

When UV radiation strikes the skin, photons damage DNA both directly, by being absorbed by nucleic acids, as well as indirectly by generating secondary genotoxic side-products. UVB most frequently facilitates a chemical reaction in two adjacent pyrimidines, resulting in a covalent bond between the C5 or C4 of one pyrimidine to the C6 position of the other. When these photoproducts are not repaired correctly, a C-to-T transition mutation results. In contrast, UVA radiation leads to G-to-T transversions via reactive oxygen species. UV radiation has been demonstrated to suppress the human immune system—an important check against malignancies—and hence may not only induce DNA damage, but also limit bodily defenses by dialing down immunosurveillance [143].

The role of UV radiation and melanoma risk is more complicated than greater exposure simply translating into greater risk. Some melanomas develop in regions with minimal or no sun exposure—such as in mucosal, acral, and uveal surfaces—suggesting that UV-induced carcinogenesis may not always be required. Indeed, sequencing studies cataloging somatic mutations among melanoma tumors found fewer signature UV mutations in mucosal, acral, and uveal melanomas than in melanomas with more chronically exposed UV radiation as expected [145].

The pattern of sun exposure may shape melanoma risk. While there are no standard definitions of sun exposure in the literature, intermittent sun exposure generally refers to sporadic exposure, resulting from recreational activities only in individuals whose work keeps them inside, shielded from the sun. In contrast, chronic sun exposure refers to extended, daily exposure that typically occurs in individuals with outdoor work or those who spend a considerable time in the outdoors. One systematic review of 25 studies found that intermittent exposure increased risk (Odds Ratio (OR) 1.57, 95 % Confidence Interval (CI) 1.29–1.91) as compared to other exposure patterns, but surprisingly chronic exposure reduced risk (OR 0.73, 95 % CI 0.60–0.89) as compared to other

**Table 31.1** Genes implicated in melanoma susceptibility [135, 140–142]

| Penetrance | Genes |
|---|---|
| High | CDKN2A, CDK4, BAP1, TERT, POT1, ACD, TERF2IP |
| Moderate | MC1R, MITF |
| Low | ASIP, TYR, TYRP1, OCA2, SLC45A2, MY07A, NID1, KIT, |
| | KITLG, IRF, HERC2, PAX3, EDNRB, ADTB3A, CHS1, MLANA, |
| | ATRN, SOX10, HPS, MRGN1, MYO5A, SLC24A4, PLA2G6, |
| | CYP1B1, CDKAL1, AGR3, TMEM38B, OBFC1, CCND1, OCA2 |

exposure patterns [146]. A more recent systematic analysis found the magnitude of risk associated with chronic sun exposure increased as populations were positioned closer to the equator [147]. Given the heterogeneity of results and the lack of experimental evidence, it is difficult to make conclusive statements about risk and exposure pattern. Risk associated with exposure pattern may be modified by anatomic site and genotype, as suggested by the "divergent pathway hypothesis" [104].

It is interesting to note that signature UV mutations are less frequently observed in mutated oncogenes in melanomas, as compared to other skin cancers. A whole-exome sequencing study of melanoma tumor samples found that 46 % of mutations in 21 oncogenes were due to signature UV mutations such as C-to-T and G-to-T mutations. When mutations in BRAF and NRAS were excluded, percentage of driver mutations increased to 67 % [148]. These results imply that activating mutations are constrained to specific codons and particular changes, as empirically observed with the majority of driver BRAF mutations constrained to the V600E region. In contrast, tumor suppressor genes, such as CDKN2A, have loss-of-function mutations that may be caused by signature UV mutations [149].

### 31.3.3 Medications and Coffee

There is some epidemiological evidence that common medications are associated with melanoma risk. TNF-α [alpha] inhibitors and sildenafil citrate are associated with an increased risk, while non-steroidal anti-inflammatory (NSAID) use is associated with reduced risk. TNF-α [alpha] inhibitors are used for a range of inflammatory conditions, including inflammatory bowel disease (IBD) and rheumatoid arthritis. One study focused on IBD treatment and demonstrated an increased risk with TNF-α [alpha] inhibitor use (OR 1.88, 95 % CI 1.08–3.29) as compared to no use. No risk was observed among individuals with relatively short term use, defined as under 120 days, while long-term users were at significantly increased risk (OR 3.93, 95 % CI 1.82–8.5) when compared to those with short term use [150]. Similarly, a study of 25,848 men in the Health Professionals Follow-up Study found that ever use of sildenafil increased melanoma risk [hazard ratio (HR) 1.92, 95 % CI 1.14–3.22] as compared to never use [151]. Sildenafil citrate is a phosphodiesterase (PDE) 5A inhibitor that is used for treatment of erectile dysfunction. Speculatively, PDE5A levels appear to be downregulated by BRAF activation, and low PDE5A levels associated with BRAF activation or sildenafil use may promote melanoma progression [151].

In contrast, coffee intake and NSAID drug use have also been linked to modest protection against skin cancer. In a prospective cohort study including 447,357 Whites, high coffee intake (greater than four cups per day) was associated with modest reduced risk of melanoma (HR 0.80, 95 % CI 0.68–0.93) as compared to no coffee intake [152]. Experimental evidence suggests plausible biological mechanisms for this observed association. Bioactive compounds in coffee may prevent UVB-induced carcinogenesis. For instance, cyclooxygenase (COX2) is known to be over-expressed in response to UVB exposure and melanoma cells and thought to have a functional role in melanomagenesis [153, 154]. 5-*O*-caffeyolquinic acid and its metabolites have been shown to suppress COX2 levels in epidermal cells in murine models [155]. Given the potential link between COX2 and melanomagenesis, the association between NSAIDs and melanoma risk has been the focus of several studies. In a case-control study in Denmark, NSAID exposures (defined as >2 prescriptions) among 3242 melanoma cases versus 32,400 age-and-sex matched controls was associated with an incidence rate ratio of 0.87 (95 % CI 0.8–0.95) [156]. These results seem to be supported by a meta-analysis of six case-control studies in the literature, but a similar meta-analysis of six cohort studies did not provide evidence for an association [157].

## 31.4 Summary

This chapter started by reflecting on the amazing pace of discovery in the field of melanoma research. Only a short nine years existed between the discovery of the driver V600E mutation in BRAF and the development of the first anti-BRAF targeted-drug. Similarly, technology has begun to provide molecular insight potentially underpinning trends observed in melanoma epidemiology. For instance, the perplexing female survival advantage in melanoma outcome has been documented since the 1960s [158]. The development of cost-effective sequencing has allowed us to link the long-observed, sex discrepancy in melanoma outcome to a difference in mutation burden in tumors.

The field of melanoma epidemiology is fast-moving, but one lesson remains clear: the complexity of melanoma epidemiology mirrors the heterogeneity of the disease. Better understanding of the molecular epidemiology of the disease will hopefully translate into better public health efforts and patient care focused on subtypes of melanoma. The rising melanoma incidence rates underscore that future efforts to control these trends will require efforts in prevention, early detection, and more effective treatments with an ever-expanding arsenal of therapeutics.

## References


1. Costin G-E, Hearing VJ. Human skin pigmentation: melanocytes modulate skin color in response to stress. FASEB J. 2007;21(4):976–94.
2. Szabo G. The regional anatomy of the human integument with special reference to the distribution of hair follicles, sweat glands and melanocytes. Philos Trans R Soc Lond B Biol Sci. 1967;252 (779):447–85.
3. Brito FC, Kos L. Timeline and distribution of melanocyte precursors in the mouse heart. Pigment Cell Melanoma Res. 2008;21(4):464–70.
4. Goldgeier MH, Klein LE, Klein-Angerer S, Moellmann G, Nordlund JJ. The distribution of melanocytes in the leptomeninges of the human brain. J Invest Dermatol. 1984;82(3):235–8.
5. Tachibana M. Sound needs sound melanocytes to be heard. Pigment Cell Res Spons Eur Soc Pigment Cell Res Int Pigment Cell Soc. 1999;12(6):344–54.
6. Yajima I, Larue L. The location of heart melanocytes is specified and the level of pigmentation in the heart may correlate with coat color. Pigment Cell Melanoma Res. 2008;21(4):471–6.
7. Holbrook KA, Underwood RA, Vogel AM, Gown AM, Kimball H. The appearance, density and distribution of melanocytes in human embryonic and fetal skin revealed by the anti-melanoma monoclonal antibody, HMB-45. Anat Embryol (Berl). 1989;180(5):443–55.
8. Schwartz JL, Wang TS, Hamilton TA, Lowe L, Sondak VK, Johnson TM. Thin primary cutaneous melanomas: associated detection patterns, lesion characteristics, and patient characteristics. Cancer. 2002;95(7):1562–8.
9. Rigel DS, Russak J, Friedman R. The evolution of melanoma diagnosis: 25 years beyond the ABCDs. CA Cancer J Clin. 2010;60(5):301–16.
10. Noor O, Nanda A, Rao BK. A dermoscopy survey to assess who is using it and why it is or is not being used. Int J Dermatol. 2009;48(9):951–2.
11. Wachsman W, Morhenn V, Palmer T, Walls L, Hata T, Zalla J, et al. Noninvasive genomic detection of melanoma. Br J Dermatol. 2011;164 (4):797–806.
12. Au A, Ariyan S. Melanoma of the head and neck. J Craniofac Surg. 2011;22(2):421–9.
13. Shaikh WR, Xiong M, Weinstock MA. The contribution of nodular subtype to melanoma mortality in the United States, 1978 to 2007. Arch Dermatol. 2012;148(1):30–6.
14. Duncan LM. The classification of cutaneous melanoma. Hematol Oncol Clin North Am. 2009;23 (3):501–13.
15. Swetter SM, Boldrick JC, Jung SY, Egbert BM, Harvell JD. Increasing incidence of lentigo maligna melanoma subtypes: northern California and national trends 1990–2000. J Invest Dermatol. 2005;125(4):685–91.
16. Bradford PT, Goldstein AM, McMaster ML, Tucker MA. Acral lentiginous melanoma: incidence and survival patterns in the United States, 1986–2005. Arch Dermatol. 2009;145(4):427–34.
17. Shoo BA, Kashani-Sabet M. Melanoma arising in African-, Asian-, Latino- and Native-American populations. Semin Cutan Med Surg. 2009;28 (2):96–102.
18. Stubblefield J, Kelly B. Melanoma in non-Caucasian populations. Surg Clin North Am. 2014;94(5):1115–26.
19. Kabigting FD, Nelson FP, Kauffman CL, Popoveniuc G, Dasanu CA, Alexandrescu DT. Malignant melanoma in African-Americans. Dermatol Online J. 2009;15(2):3.
20. Bellew S, Del Rosso JQ, Kim GK. Skin cancer in Asians. J Clin Aesthetic Dermatol. 2009;2(10):34–6.
21. Bastian BC, Kashani-Sabet M, Hamm H, Godfrey T, Moore DH, Bröcker EB, et al. Gene amplifications characterize acral melanoma and permit the detection of occult tumor cells in the surrounding skin. Cancer Res. 2000;60(7):1968–73.
22. Balch CM, Gershenwald JE, Soong S, Thompson JF, Atkins MB, Byrd DR, et al. Final version of 2009

AJCC melanoma staging and classification. J Clin Oncol. 2009;27(36):6199–206.
23. Garbe C, Eigentler TK, Keilholz U, Hauschild A, Kirkwood JM. Systematic review of medical treatment in melanoma: current status and future prospects. Oncologist. 2011;16(1):5–24.
24. Coit DG, Thompson JA, Andtbacka R, Anker CJ, Bichakjian CK, Carson WE, et al. Melanoma, version 4.2014. J Natl Compr Cancer Netw JNCCN. 2014;12(5):621–9.
25. Bhatia S, Tykodi SS, Thompson JA. Treatment of metastatic melanoma: an overview. Oncol Williston Park N. 2009;23(6):488–96.
26. Tarhini AA, Agarwala SS. Cutaneous melanoma: available therapy for metastatic disease. Dermatol Ther. 2006;19(1):19–25.
27. Elder GJ, Hersey P, Branley P. Remission of transplanted melanoma–clinical course and tumour cell characterisation. Clin Transplant. 1997;11 (6):565–8.
28. MacKie RM, Reid R, Junor B. Fatal melanoma transferred in a donated kidney 16 years after melanoma surgery. N Engl J Med. 2003;348(6):567–8.
29. Penn I. Malignant melanoma in organ allograft recipients. Transplantation. 1996;61(2):274–8.
30. Suranyi MG, Hogan PG, Falk MC, Axelsen RA, Rigby R, Hawley C, et al. Advanced donor-origin melanoma in a renal transplant recipient: immunotherapy, cure, and retransplantation. Transplantation. 1998;66(5):655–61.
31. Katz KA, Jonasch E, Hodi FS, Soiffer R, Kwitkiwski K, Sober AJ, et al. Melanoma of unknown primary: experience at Massachusetts General Hospital and Dana-Farber Cancer Institute. Melanoma Res. 2005;15(1):77–82.
32. McCardle TW, Messina JL, Sondak VK. Completely regressed cutaneous melanocytic lesion revisited. Semin Oncol. 2009;36(6):498–503.
33. Oble DA, Loewe R, Yu P, Mihm MC. Focus on TILs: prognostic significance of tumor infiltrating lymphocytes in human melanoma. Cancer Immun. 2009;9:3.
34. McArthur GA. Adjuvant interferon in melanoma: is duration of therapy important? J Clin Oncol. 2014;32(3):171–3.
35. Petrella T, Verma S, Spithoff K, Quirt I, McCready D. Adjuvant interferon therapy for patients at high risk for recurrent melanoma: an updated systematic review and practice guideline. Clin Oncol. 2012;24(6):413–23.
36. Kirkwood JM, Manola J, Ibrahim J, Sondak V, Ernstoff MS, Rao U, et al. A pooled analysis of eastern cooperative oncology group and intergroup trials of adjuvant high-dose interferon for melanoma. Clin Cancer Res Off J Am Assoc Cancer Res. 2004;10(5):1670–7.
37. Eggermont AMM, Suciu S, Testori A, Santinami M, Kruit WHJ, Marsden J, et al. Long-term results of the randomized phase III trial EORTC 18991 of adjuvant therapy with pegylated interferon alfa-2b versus observation in resected stage III melanoma. J Clin Oncol Off J Am Soc Clin Oncol. 2012;30 (31):3810–8.
38. Rosenberg SA, Lotze MT, Muul LM, Leitman S, Chang AE, Ettinghausen SE, et al. Observations on the systemic administration of autologous lymphokine-activated killer cells and recombinant interleukin-2 to patients with metastatic cancer. N Engl J Med. 1985;313(23):1485–92.
39. Atkins MB, Lotze MT, Dutcher JP, Fisher RI, Weiss G, Margolin K, et al. High-dose recombinant interleukin 2 therapy for patients with metastatic melanoma: analysis of 270 patients treated between 1985 and 1993. J Clin Oncol Off J Am Soc Clin Oncol. 1999;17(7):2105–16.
40. Pardoll DM. The blockade of immune checkpoints in cancer immunotherapy. Nat Rev Cancer. 2012;12 (4):252–64.
41. Hölzel M, Bovier A, Tüting T. Plasticity of tumour and immune cells: a source of heterogeneity and a cause for therapy resistance? Nat Rev Cancer. 2013;13(5):365–76.
42. Davies H, Bignell GR, Cox C, Stephens P, Edkins S, Clegg S, et al. Mutations of the BRAF gene in human cancer. Nature. 2002;417(6892):949–54.
43. Long GV, Menzies AM, Nagrial AM, Haydu LE, Hamilton AL, Mann GJ, et al. Prognostic and clinicopathologic associations of oncogenic BRAF in metastatic melanoma. J Clin Oncol Off J Am Soc Clin Oncol. 2011;29(10):1239–46.
44. Tronnier M, Mitteldorf C. Treating advanced melanoma: current insights and opportunities. Cancer Manag Res. 2014;6:349–56.
45. Chapman PB, Hauschild A, Robert C, Haanen JB, Ascierto P, Larkin J, et al. Improved survival with vemurafenib in melanoma with BRAF V600E mutation. N Engl J Med. 2011;364(26):2507–16.
46. Hauschild A, Grob J-J, Demidov LV, Jouary T, Gutzmer R, Millward M, et al. Dabrafenib in BRAF-mutated metastatic melanoma: a multicentre, open-label, phase 3 randomised controlled trial. Lancet. 2012;380(9839):358–65.
47. Sosman JA, Kim KB, Schuchter L, Gonzalez R, Pavlick AC, Weber JS, et al. Survival in BRAF V600-mutant advanced melanoma treated with vemurafenib. N Engl J Med. 2012;366(8):707–14.
48. Wagle N, Emery C, Berger MF, Davis MJ, Sawyer A, Pochanard P, et al. Dissecting therapeutic resistance to RAF inhibition in melanoma by tumor genomic profiling. J Clin Oncol Off J Am Soc Clin Oncol. 2011;29(22):3085–96.
49. Hawryluk EB, Tsao H. Melanoma: clinical features and genomic insights. Cold Spring Harb Perspect Med. 2014;4(9):a015388.
50. Flaherty KT, Infante JR, Daud A, Gonzalez R, Kefford RF, Sosman J, et al. Combined BRAF and MEK inhibition in melanoma with BRAF V600 mutations. N Engl J Med. 2012;367(18):1694–703.
51. Winnepenninckx V, Lazar V, Michiels S, Dessen P, Stas M, Alonso SR, et al. Gene expression profiling

of primary cutaneous melanoma and clinical outcome. J Natl Cancer Inst. 2006;98(7):472–82.
52. Harbst K, Staaf J, Lauss M, Karlsson A, Måsbäck A, Johansson I, et al. Molecular profiling reveals low- and high-grade forms of primary melanoma. Clin Cancer Res. 2012;18(15):4026–36.
53. Brunner G, Reitz M, Heinecke A, Lippold A, Berking C, Suter L, et al. A nine-gene signature predicting clinical outcome in cutaneous melanoma. J Cancer Res Clin Oncol. 2013;139(2):249–58.
54. McLaughlin CC, Wu X-C, Jemal A, Martin HJ, Roche LM, Chen VW. Incidence of noncutaneous melanomas in the U.S. Cancer. 2005;103(5):1000–7.
55. Jovanovic P, Mihajlovic M, Djordjevic-Jocic J, Vlajkovic S, Cekic S, Stefanovic V. Ocular melanoma: an overview of the current status. Int J Clin Exp Pathol. 2013;6(7):1230–44.
56. Finger PT, Kurli M, Reddy S, Tena LB, Pavlick AC. Whole body PET/CT for initial staging of choroidal melanoma. Br J Ophthalmol. 2005;89(10):1270–4.
57. Shields CL, Furuta M, Thangappan A, Nagori S, Mashayekhi A, Lally DR, et al. Metastasis of uveal melanoma millimeter-by-millimeter in 8033 consecutive eyes. Arch Ophthalmol. 2009;127(8):989–98.
58. Van Raamsdonk CD, Griewank KG, Crosby MB, Garrido MC, Vemula S, Wiesner T, et al. Mutations in GNA11 in uveal melanoma. N Engl J Med. 2010;363(23):2191–9.
59. Van Raamsdonk CD, Bezrookove V, Green G, Bauer J, Gaugler L, O'Brien JM, et al. Frequent somatic mutations of GNAQ in uveal melanoma and blue naevi. Nature. 2009;457(7229):599–602.
60. Harbour JW, Onken MD, Roberson EDO, Duan S, Cao L, Worley LA, et al. Frequent mutation of BAP1 in metastasizing uveal melanomas. Science. 2010;330(6009):1410–3.
61. Chang AE, Karnell LH, Menck HR. The National Cancer Data Base report on cutaneous and noncutaneous melanoma: a summary of 84,836 cases from the past decade. The American College of Surgeons Commission on Cancer and the American Cancer Society. Cancer. 1998;83(8):1664–78.
62. Mihajlovic M, Vlajkovic S, Jovanovic P, Stefanovic V. Primary mucosal melanomas: a comprehensive review. Int J Clin Exp Pathol. 2012;5 (8):739–53.
63. Cheung MC, Perez EA, Molina MA, Jin X, Gutierrez JC, Franceschi D, et al. Defining the role of surgery for primary gastrointestinal tract melanoma. J Gastrointest Surg Off J Soc Surg Aliment Tract. 2008;12(4):731–8.
64. Hodi FS, Friedlander P, Corless CL, Heinrich MC, Mac Rae S, Kruse A, et al. Major response to imatinib mesylate in KIT-mutated melanoma. J Clin Oncol Off J Am Soc Clin Oncol. 2008;26(12):2046–51.
65. Minor DR, Kashani-Sabet M, Garrido M, O'Day SJ, Hamid O, Bastian BC. Sunitinib therapy for melanoma patients with KIT mutations. Clin Cancer Res Off J Am Assoc Cancer Res. 2012;18(5):1457–63.
66. Ferlay J, Soerjomataram I, Ervik M, Dikshit R, Eser S, Mathers C, et al. GLOBOCAN 2012 v1.0, cancer incidence and mortality worldwide: IARC CancerBase No. 11 [Internet]. Lyon, France: International Agency for Research on Cancer; 2013. [Internet]. [cited 2014 Nov 3]. Available from: http://globocan.iarc.fr.
67. AIHW. Cancer in Australia: an overview 2012 [Internet]. 2012. Available from: http://www.aihw.gov.au/cancer/cancer-in-australia-overview-2012/.
68. Coory M, Baade P, Aitken J, Smithers M, McLeod GRC, Ring I. Trends for in situ and invasive melanoma in Queensland, Australia, 1982–2002. Cancer Causes Control CCC. 2006;17(1):21–7.
69. Bulliard JL, Cox B. Cutaneous malignant melanoma in New Zealand: trends by anatomical site, 1969–1993. Int J Epidemiol. 2000;29(3):416–23.
70. David Whiteman AG. Epidemiology of Malignant Melanoma. 2010;13–26.
71. Crombie IK. Variation of melanoma incidence with latitude in North America and Europe. Br J Cancer. 1979;40(5):774–81.
72. Magnus K. Incidence of malignant melanoma of the skin in Norway, 1955–1970. Variations in time and space and solar radiation. Cancer. 1973;32(5):1275–86.
73. Armstrong BK. Epidemiology of malignant melanoma: intermittent or total accumulated exposure to the sun? J Dermatol Surg Oncol. 1988;14(8):835–49.
74. Rebecca VW, Sondak VK, Smalley KSM. A brief history of melanoma: from mummies to mutations. Melanoma Res. 2012;22(2):114–22.
75. Rigel DS. Epidemiology of melanoma. Semin Cutan Med Surg. 2010;29(4):204–9.
76. Sober AJ, Haluska FG. American cancer society atlas of clinical oncology: skin cancer (Book with CD-ROM) with CDROM. Hamilton, Ont: B.C. Decker; 2001 350p.
77. de Vries E, van de Poll-Franse LV, Louwman WJ, de Gruijl FR, Coebergh JWW. Predictions of skin cancer incidence in the Netherlands up to 2015. Br J Dermatol. 2005;152(3):481–8.
78. Diffey BL. The future incidence of cutaneous melanoma within the UK. Br J Dermatol. 2004;151(4):868–72.
79. Erdmann F, Lortet-Tieulent J, Schüz J, Zeeb H, Greinert R, Breitbart EW, et al. International trends in the incidence of malignant melanoma 1953–2008—are recent generations at higher or lower risk? Int J Cancer. 2013;132(2):385–400.
80. Garbe C, Leiter U. Melanoma epidemiology and trends. Clin Dermatol. 2009;27(1):3–9.
81. Howlander N, Noone A, Krapcho M, Garshell J, Miller D, Altekruse S, et al. SEER cancer statistics review, 1975–2011, National Cancer Institute. [Internet]. Available from: http://seer.cancer.gov/csr/1975_2011/, based on November 2013 SEER data submission.

82. Bentham G, Aase A. Incidence of malignant melanoma of the skin in Norway, 1955–1989: associations with solar ultraviolet radiation, income and holidays abroad. Int J Epidemiol. 1996;25 (6):1132–8.
83. de Vries E, Coebergh JW. Cutaneous malignant melanoma in Europe. Eur J Cancer Oxf Engl 1990. 2004;40(16):2355–66.
84. Nikolaou V, Stratigos AJ. Emerging trends in the epidemiology of melanoma. Br J Dermatol. 2014;170(1):11–9.
85. Erickson C, Driscoll MS. Melanoma epidemic: facts and controversies. Clin Dermatol. 2010;28(3):281–6.
86. Welch HG, Black WC. Overdiagnosis in cancer. J Natl Cancer Inst. 2010;102(9):605–13.
87. Criscione VD, Weinstock MA. Melanoma thickness trends in the United States, 1988–2006. J Invest Dermatol. 2010;130(3):793–7.
88. Linos E, Swetter SM, Cockburn MG, Colditz GA, Clarke CA. Increasing burden of melanoma in the United States. J Invest Dermatol. 2009;129 (7):1666–74.
89. Purdue MP, Freeman LEB, Anderson WF, Tucker MA. Recent trends in incidence of cutaneous melanoma among US Caucasian young adults. J Invest Dermatol. 2008;128(12):2905–8.
90. Bradford PT, Anderson WF, Purdue MP, Goldstein AM, Tucker MA. Rising melanoma incidence rates of the trunk among younger women in the United States. Cancer Epidemiol Biomark Prev Publ Am Assoc Cancer Res Cosponsored Am Soc Prev Oncol. 2010;19(9):2401–6.
91. Gamba CS, Clarke CA, Keegan THM, Tao L, Swetter SM. Melanoma survival disadvantage in young, non-hispanic white males compared with females. JAMA Dermatol. 2013;149(8):912–20.
92. Fisher DE, Geller AC. Disproportionate burden of melanoma mortality in young us men: the possible role of biology and behavior. JAMA Dermatol. 2013;149(8):903–4.
93. Gupta S, Artomov M, Goggins W, Daly M, Tsao H. Gender disparity and mutation burden in metastatic melanoma. J Natl Cancer Inst. 2015;107(11).
94. Elwood JM, Lee JA. Recent data on the epidemiology of malignant melanoma. Semin Oncol. 1975;2 (2):149–54.
95. Houghton A, Flannery J, Viola MV. Malignant melanoma in Connecticut and Denmark. Int J Cancer J Int Cancer. 1980;25(1):95–104.
96. Clark LN, Shin DB, Troxel AB, Khan S, Sober AJ, Ming ME. Association between the anatomic distribution of melanoma and sex. J Am Acad Dermatol. 2007;56(5):768–73.
97. Green A, MacLennan R, Youl P, Martin N. Site distribution of cutaneous melanoma in Queensland. Int J Cancer J Int Cancer. 1993;53(2):232–6.
98. Lesage C, Barbe C, Le Clainche A, Lesage F-X, Bernard P, Grange F. Sex-related location of head and neck melanoma strongly argues for a major role of sun exposure in cars and photoprotection by hair. J Invest Dermatol. 2013;133(5):1205–11.
99. Whiteman DC, Pavan WJ, Bastian BC. The melanomas: a synthesis of epidemiological, clinical, histopathological, genetic, and biological aspects, supporting distinct subtypes, causal pathways, and cells of origin. Pigment Cell Melanoma Res. 2011;24(5):879–97.
100. Elwood JM, Gallagher RP. Body site distribution of cutaneous malignant melanoma in relationship to patterns of sun exposure. Int J Cancer. 1998;78 (3):276–80.
101. Lachiewicz AM, Berwick M, Wiggins CL, Thomas NE. Survival differences between patients with scalp or neck melanoma and those with melanoma of other sites in the surveillance, epidemiology, and end results (SEER) program. Arch Dermatol. 2008;144(4):515–21.
102. Lee J-H, Choi J-W, Kim Y-S. Frequencies of BRAF and NRAS mutations are different in histological types and sites of origin of cutaneous melanoma: a meta-analysis. Br J Dermatol. 2011;164(4):776–84.
103. Tsao H, Chin L, Garraway LA, Fisher DE. Melanoma: from mutations to medicine. Genes Dev. 2012;26(11):1131–55.
104. Whiteman DC, Watt P, Purdie DM, Hughes MC, Hayward NK, Green AC. Melanocytic nevi, solar keratoses, and divergent pathways to cutaneous melanoma. J Natl Cancer Inst. 2003;95(11):806–12.
105. Ward KA, Lazovich D, Hordinsky MK. Germline melanoma susceptibility and prognostic genes: a review of the literature. J Am Acad Dermatol. 2012;67(5):1055–67.
106. García-Borrón JC, Sánchez-Laorden BL, Jiménez--Cervantes C. Melanocortin-1 receptor structure and functional regulation. Pigment Cell Res Spons Eur Soc Pigment Cell Res Int Pigment Cell Soc. 2005;18 (6):393–410.
107. Jablonski NG, Chaplin G. Human skin pigmentation as an adaptation to UV radiation. Proc Natl Acad Sci USA. 2010;107 Suppl 2:8962–8.
108. Palmer JS, Duffy DL, Box NF, Aitken JF, O'Gorman LE, Green AC, et al. Melanocortin-1 receptor polymorphisms and risk of melanoma: is the association explained solely by pigmentation phenotype? Am J Hum Genet. 2000;66(1):176–86.
109. Raimondi S, Sera F, Gandini S, Iodice S, Caini S, Maisonneuve P, et al. MC1R variants, melanoma and red hair color phenotype: a meta-analysis. Int J Cancer. 2008;122(12):2753–60.
110. Rajeevan H, Soundararajan U, Kidd JR, Pakstis AJ, Kidd KK. ALFRED: an allele frequency resource for research and teaching. Nucleic Acids Res. 2012;40(Database issue):D1010–5.
111. Duffy DL, Zhao ZZ, Sturm RA, Hayward NK, Martin NG, Montgomery GW. Multiple pigmentation gene polymorphisms account for a substantial proportion of risk of cutaneous malignant melanoma. J Invest Dermatol. 2009;130(2):520–8.

112. Magana-Garcia M, Ackerman AB. What are nevus cells? Am J Dermatopathol. 1990;12(1):93–102.
113. Garbe C, Büttner P, Weiss J, Soyer HP, Stocker U, Krüger S, et al. Risk factors for developing cutaneous melanoma and criteria for identifying persons at risk: multicenter case-control study of the central malignant melanoma registry of the German dermatological society. J Invest Dermatol. 1994;102 (5):695–9.
114. Holly EA, Kelly JW, Shpall SN, Chiu SH. Number of melanocytic nevi as a major risk factor for malignant melanoma. J Am Acad Dermatol. 1987;17(3):459–68.
115. Olsen CM, Carroll HJ, Whiteman DC. Estimating the attributable fraction for cancer: a meta-analysis of nevi and melanoma. Cancer Prev Res Phila Pa. 2010;3(2):233–45.
116. Wachsmuth RC, Gaut RM, Barrett JH, Saunders CL, Randerson-Moor JA, Eldridge A, et al. Heritability and gene-environment interactions for melanocytic nevus density examined in a U.K. adolescent twin study. J Invest Dermatol. 2001;117(2):348–52.
117. Falchi M, Bataille V, Hayward NK, Duffy DL, Bishop JAN, Pastinen T, et al. Genome-wide association study identifies variants at 9p21 and 22q13 associated with development of cutaneous nevi. Nat Genet. 2009;41(8):915–9.
118. Siegel R, Ma J, Zou Z, Jemal A. Cancer statistics, 2014. CA Cancer J Clin. 2014;64(1):9–29.
119. Kölmel KF, Kulle B, Lippold A, Seebacher C. Survival probabilities and hazard functions of malignant melanoma in Germany 1972–1996, an analysis of 10433 patients. Evolution of gender differences and malignancy. Eur J Cancer Oxf Engl 1990. 2002;38(10):1388–94.
120. de Vries E, Nijsten TEC, Visser O, Bastiaannet E, van Hattem S, Janssen-Heijnen ML, et al. Superior survival of females among 10 538 Dutch melanoma patients is independent of Breslow thickness, histologic type and tumor site. Ann Oncol. 2008;19 (3):583–9.
121. Lasithiotakis K, Leiter U, Meier F, Eigentler T, Metzler G, Moehrle M, et al. Age and gender are significant independent predictors of survival in primary cutaneous melanoma. Cancer. 2008;112 (8):1795–804.
122. Micheli A, Ciampichini R, Oberaigner W, Ciccolallo L, de Vries E, Izarzugaza I, et al. The advantage of women in cancer survival: an analysis of EUROCARE-4 data. Eur J Cancer Oxf Engl 1990. 2009;45(6):1017–27.
123. Joosse A, de Vries E, Eckel R, Nijsten T, Eggermont AMM, Hölzel D, et al. Gender differences in melanoma survival: female patients have a decreased risk of metastasis. J Invest Dermatol. 2011;131(3):719–26.
124. Joosse A, Collette S, Suciu S, Nijsten T, Lejeune F, Kleeberg UR, et al. Superior outcome of women with stage I/II cutaneous melanoma: pooled analysis of four European organisation for research and treatment of cancer phase III trials. J Clin Oncol. 2012;30(18):2240–7.
125. Metelitsa AI, Dover DC, Smylie M, de Gara CJ, Lauzon GJ. A population-based study of cutaneous melanoma in Alberta, Canada (1993–2002). J Am Acad Dermatol. 2010;62(2):227–32.
126. Rueth NM, Groth SS, Tuttle TM, Virnig BA, Al-Refaie WB, Habermann EB. Conditional survival after surgical treatment of melanoma: an analysis of the surveillance, epidemiology, and end results database. Ann Surg Oncol. 2010;17(6):1662–8.
127. Xing Y, Bronstein Y, Ross MI, Askew RL, Lee JE, Gershenwald JE, et al. Contemporary diagnostic imaging modalities for the staging and surveillance of melanoma patients: a meta-analysis. J Natl Cancer Inst. 2011;103(2):129–42.
128. Kemeny MM, Busch E, Stewart AK, Menck HR. Superior survival of young women with malignant melanoma. Am J Surg. 1998;175(6):437–44; discussion 444–5.
129. Dorak MT, Karpuzoglu E. Gender differences in cancer susceptibility: an inadequately addressed issue. Front Genet [Internet]. 2012 Nov 28 [cited 2014 Sep 24];3. Available from: http://www.ncbi.nlm.nih.gov/pmc/articles/PMC3508426/.
130. Gandini S, Iodice S, Koomen E, Di Pietro A, Sera F, Caini S. Hormonal and reproductive factors in relation to melanoma in women: current review and meta-analysis. Eur J Cancer Oxf Engl 1990. 2011;47(17):2607–17.
131. Tang JY, Spaunhurst KM, Chlebowski RT, Wactawski-Wende J, Keiser E, Thomas F, et al. Menopausal hormone therapy and risks of melanoma and nonmelanoma skin cancers: women's health initiative randomized trials. J Natl Cancer Inst. 2011;103(19):1469–75.
132. Koomen ER, Joosse A, Herings RMC, Casparie MK, Guchelaar HJ, Nijsten T. Does use of estrogens decrease the Breslow thickness of melanoma of the skin? Oral contraceptives and hormonal replacement therapy. Melanoma Res. 2009;19(5):327–32.
133. Spanogle JP, Clarke CA, Aroner S, Swetter SM. Risk of second primary malignancies following cutaneous melanoma diagnosis: a population-based study. J Am Acad Dermatol. 2010;62(5):757–67.
134. Li W-Q, Qureshi AA, Ma J, Goldstein AM, Giovannucci EL, Stampfer MJ, et al. Personal history of prostate cancer and increased risk of incident melanoma in the United States. J Clin Oncol Off J Am Soc Clin Oncol. 2013;31(35):4394–9.
135. Ward KA, Lazovich D, Hordinsky MK. Germline melanoma susceptibility and prognostic genes: a review of the literature. J Am Acad Dermatol. 2012;67(5):1055–67.
136. Abdel-Rahman MH, Pilarski R, Cebulla CM, Massengill JB, Christopher BN, Boru G, et al. Germline BAP1 mutation predisposes to uveal melanoma, lung adenocarcinoma, meningioma, and other cancers. J Med Genet. 2011;48(12):856–9.

137. Wiesner T, Obenauf AC, Murali R, Fried I, Griewank KG, Ulz P, et al. Germline mutations in BAP1 predispose to melanocytic tumors. Nat Genet. 2011;43(10):1018–21.
138. Kumar R, Taylor M, Miao B, Ji Z, Njauw JC-N, Jönsson G, et al. BAP1 Has a survival role in cutaneous melanoma. J Invest Dermatol [Internet]. 2015 Jan 22 [cited 2015 Feb 1]; Available from: http://www.nature.com/jid/journal/vaop/ncurrent/abs/jid2014528a.html.
139. Njauw C-NJ, Kim I, Piris A, Gabree M, Taylor M, Lane AM, et al. Germline BAP1 inactivation is preferentially associated with metastatic ocular melanoma and cutaneous-ocular melanoma families. PLoS ONE. 2012;7(4):e35295.
140. Robles-Espinoza CD, Harland M, Ramsay AJ, Aoude LG, Quesada V, Ding Z, et al. POT1 loss-of-function variants predispose to familial melanoma. Nat Genet. 2014;46(5):478–81.
141. Aoude LG, Pritchard AL, Robles-Espinoza CD, Wadt K, Harland M, Choi J, et al. Nonsense mutations in the shelterin complex genes ACD and TERF2IP in familial melanoma. J Natl Cancer Inst. 2015;107(2).
142. Law MH, Bishop DT, Lee JE, Brossard M, Martin NG, Moses EK, et al. Genome-wide meta-analysis identifies five new susceptibility loci for cutaneous malignant melanoma. Nat Genet. 2015;47(9):987–95.
143. Garibyan L, Fisher DE. How sunlight causes melanoma. Curr Oncol Rep. 2010;12(5):319–26.
144. Erdei E, Torres SM. A new understanding in the epidemiology of melanoma. Expert Rev Anticancer Ther. 2010;10(11):1811–23.
145. Krauthammer M, Kong Y, Ha BH, Evans P, Bacchiocchi A, McCusker JP, et al. Exome sequencing identifies recurrent somatic RAC1 mutations in melanoma. Nat Genet. 2012;44(9):1006–14.
146. Nelemans PJ, Rampen FH, Ruiter DJ, Verbeek AL. An addition to the controversy on sunlight exposure and melanoma risk: a meta-analytical approach. J Clin Epidemiol. 1995;48(11):1331–42.
147. Gandini S, Sera F, Cattaruzza MS, Pasquini P, Picconi O, Boyle P, et al. Meta-analysis of risk factors for cutaneous melanoma: II. Sun exposure. Eur J Cancer Oxf Engl 1990. 2005;41(1):45–60.
148. Hodis E, Watson IR, Kryukov GV, Arold ST, Imielinski M, Theurillat J-P, et al. A landscape of driver mutations in melanoma. Cell. 2012;150 (2):251–63.
149. Hocker T, Tsao H. Ultraviolet radiation and melanoma: a systematic review and analysis of reported sequence variants. Hum Mutat. 2007;28 (6):578–88.
150. Long MD, Martin CF, Pipkin CA, Herfarth HH, Sandler RS, Kappelman MD. Risk of melanoma and nonmelanoma skin cancer among patients with inflammatory bowel disease. Gastroenterology. 2012;143(2):390–9. e1.
151. Li W, Qureshi AA, Robinson KC, Han J. Sildenafil use and increased risk of incident melanoma in us men: a prospective cohort study. JAMA Intern Med. 2014;174(6):964–70.
152. Loftfield E, Freedman ND, Graubard BI, Hollenbeck AR, Shebl FM, Mayne ST, et al. Coffee drinking and cutaneous melanoma risk in the NIH-AARP diet and health study. J Natl Cancer Inst. 2015;107(2):dju421.
153. Buckman SY, Gresham A, Hale P, Hruza G, Anast J, Masferrer J, et al. COX-2 expression is induced by UVB exposure in human skin: implications for the development of skin cancer. Carcinogenesis. 1998;19(5):723–9.
154. Denkert C, Köbel M, Berger S, Siegert A, Leclere A, Trefzer U, et al. Expression of cyclooxygenase 2 in human malignant melanoma. Cancer Res. 2001;61(1):303–8.
155. Kang NJ, Lee KW, Shin BJ, Jung SK, Hwang MK, Bode AM, et al. Caffeic acid, a phenolic phytochemical in coffee, directly inhibits Fyn kinase activity and UVB-induced COX-2 expression. Carcinogenesis. 2009;30(2):321–30.
156. Johannesdottir SA, Chang ET, Mehnert F, Schmidt M, Olesen AB, Sørensen HT. Nonsteroidal anti-inflammatory drugs and the risk of skin cancer. Cancer. 2012;118(19):4768–76.
157. Li S, Liu Y, Zeng Z, Peng Q, Li R, Xie L, et al. Association between non-steroidal anti-inflammatory drug use and melanoma risk: a meta-analysis of 13 studies. Cancer Causes Control CCC. 2013;24(8):1505–16.
158. Clark WH, From L, Bernardino EA, Mihm MC. The histogenesis and biologic behavior of primary human malignant melanomas of the skin. Cancer Res. 1969;29(3):705–27.

# Pathology and Molecular Pathology of Melanoma

# 32

Amrita Goyal and Lyn McDivitt Duncan

## 32.1 Introduction

Melanoma is a cancer of melanocytes that most often arises in the skin but may also develop in the eye and mucosal sites. If caught early, melanoma can be successfully treated by surgery, however, those patients diagnosed with advanced stage disease have a poor prognosis. In this chapter, we begin by discussing the embryological development of the skin with an attention to melanocytic migration, followed by a description of normal cutaneous anatomy and histology. The neoplastic processes of cutaneous melanocytes will then be outlined, and we will devote the remainder of the discussion to malignant disease. Finally, we will focus on the molecular pathogenesis of melanoma. This expanding and evolving melanoma research arena has furnished us with promising diagnostic and therapeutic applications. This chapter will focus mostly on cutaneous melanoma but will also touch on more rare forms of melanoma including ocular melanoma, and genitourinary, gastrointestinal, and upper respiratory mucosal melanoma.

Melanoma has significant societal impact because it affects patients at a relatively young age and when found to be metastatic there is no reliably effective treatment. The histopathological findings in the primary tumor are the principle factors that determine treatment, guiding surgical plans and in some cases adjuvant therapy. Melanoma staging is dependent upon histological features observed in routinely processed formalin-fixed and paraffin-embedded tissue sections. Promising recent discoveries in the pathogenesis of melanoma and identification of distinct molecular phenotypes may lead to the development of molecularly based staging schemas. Indeed, current molecular tests, in particular the identification of a BRAFV600E mutation, determine targeted therapy in patients with advanced stage disease. Current research studies promise a more molecularly integrated approach to melanoma treatment in the future.

## 32.2 Basic Anatomy of Skin

During embryogenesis melanocytes are derived from the neural crest. Coordinated cell–cell signaling cascades lead to selective activation of transcription factors that direct key morphological events in fetal skin development. By the end of the first trimester, morphogenesis of the skin is

A. Goyal
University of Minnesota, 516 Delaware Street. S.E., Phillips-Wangensteen Building, Suite 4-240, Minneapolis, MN 55455, USA

L.M. Duncan (✉)
WRN825 Dermatopathology Unit, Massachusetts General Hospital, 55 Fruit Street, Boston, MA 02114, USA
e-mail: duncan@helix.mgh.harvard.edu


M. Loda et al. (eds.), *Pathology and Epidemiology of Cancer*,
DOI 10.1007/978-3-319-35153-7_32

well underway and during the second trimester differentiation of the epidermis and adnexal structures occurs. Early during embryogenesis two specialized cell types migrate to the epidermis: melanocytes and Langerhans cells. Melanocytes are derived from the neural crest that forms along the dorsal neural tube. The embryogenic mechanisms by which melanocytes populate the epidermis and hair follicle epithelium are not clearly understood. However, it is likely that melanoblasts (melanocyte precursors) participate in an intermediate mesenchymal stage wherein they exist in intradermal, intraepidermal, and intrafollicular compartments [1]. Examinations in human fetal skin reveal well-defined melanoblast progression from intradermal (6–8 weeks gestation) to intraepidermal (12–15 weeks gestation) to intrafollicular (18–20 weeks gestation) localization [1]. Melanoblasts have been thought to arise from the dorsal neural tube and migrate dorsolaterally along the ectodermatome and ventrally through the developing dermis to their final destinations in the skin. Of note, new evidence supports the hypothesis that Schwann cell precursors also may serve as the cells of origin for a major fraction of skin melanocytes [2, 3].

## 32.3 Basic Histology of Skin

### 32.3.1 Epidermis

The epidermis is a stratified squamous epithelium composed of layers of keratinocytes. The external or surface stratum corneum has underlying stratum granulosum, stratum spinosum and, at the base of the epidermis, the stratum basale. Intermingled within the pavement-like arrangement of keratinocytes in the epidermis are several specialized nonkeratinocytic cells including: (1) Merkel cells, CK20 + neuroendocrine cells found in the stratum basale and associated with nerve endings from the dermis, (2) Langerhans cells, CD1a+ and Langerin + dendritic cells found in the stratum spinosum that function in antigen presentation, and (3) Melanocytes, S100+ MITF+ cells found in the stratum basale that function to transfer melanin pigment to keratinocytes.

### 32.3.2 Dermis

Directly underlying the epidermis is the dermis, separated from the epidermis by a basement membrane zone. The dermis is composed of a superficial papillary dermis and deep reticular dermis. The papillary dermis includes the dermal papillae which are intercalated with the epidermal rete ridges. The papillary dermis is composed of delicate, pale eosinophilic collagen fibers and contains free nerve endings and Meissner's corpuscles. The superficial venular plexus separates the papillary dermis from the underlying reticular dermis. The reticular dermis is composed of course, deeply eosinophilic collagen fibers and contains the deep vascular plexus, adnexal structures, nerve trunks, Pacinian corpuscles, and glomus bodies. These structures vary in their density and distribution with different cutaneous sites. For example, apocrine units are more common in the axilla, eccrine units are common on the palms and soles, and hair follicles are common on the skin of the head and neck.

### 32.3.3 Subcutis

The epidermis and dermis are considered the outermost or superficial layers of the skin. Beneath these two layers is the subcutis. The subcutis is composed of subcutaneous fat lobules that are separated by fibrous septa that extend from the deepest aspect of the reticular dermis. The subcutaneous fat contains few adnexal structures, including anagen hair bulbs and medium-sized arterioles and veins. Occasionally, lymph nodes may also be present in the subcutaneous fat.

## 32.4 Benign Neoplastic Pathology/Morphology

### 32.4.1 Histology of Melanocytic Hyperplasia

There are several clinical lesions that may be manifested histopathologically as a non-nested epidermal melanocytic hyperplasia. The most common of these is lentigo, which is subclassified by some as solar lentigo and lentigo simplex. These lesions are characterized by the presence of a lentiginous proliferation of melanocytes and increased pigmentation of basal layer keratinocytes. No melanocytic clustering or nesting is observed in lentigo. The term "lentiginous" is used to describe a pattern of individual melanocytes in increased numbers at the dermoepidermal junction. In addition to being present in lentigo, a lentiginous melanocytic proliferation is also observed in other processes including dysplastic nevus, lentigo maligna (Fig. 32.1a), acral lentiginous melanoma, and mucosal melanoma. The cytological characteristics of the melanocytes and associated patterns of growth differ for each of these diagnoses.

### 32.4.2 Histopathology of Benign Nevi

The histopathology of benign melanocytic tumors is quite variable. These nested proliferations of melanocytes may be confined to the epidermis (junctional nevus), present in both the epidermis and dermis (compound nevus) or limited to the dermis (dermal nevus). The most common nevi are composed of nests of benign appearing melanocytes with smooth nuclear contours, inconspicuous nucleoli, and variable amounts of amphophilic to eosinophilic cytoplasm. The pattern of melanocytes in the epidermis of benign nevi may be lentiginous (a proliferation of individual cells along the dermoepidermal junction) or nested (aggregates of three or more nevomelanocytes). In benign junctional and compound nevi the intraepidermal melanocytic proliferation is usually localized to the base of the epidermis (Fig. 32.1b). However, in some cases the melanocytes may be present in the upper half of the epidermis; this growth pattern is termed "pagetoid" when individual melanocytes are seen in the upper levels of the epidermis because it mimics Paget's disease of the breast. While a pagetoid growth pattern maybe seen in specific rare subsets of benign nevi, it is a characteristic feature of melanoma in situ (Fig. 32.1c). In compound and

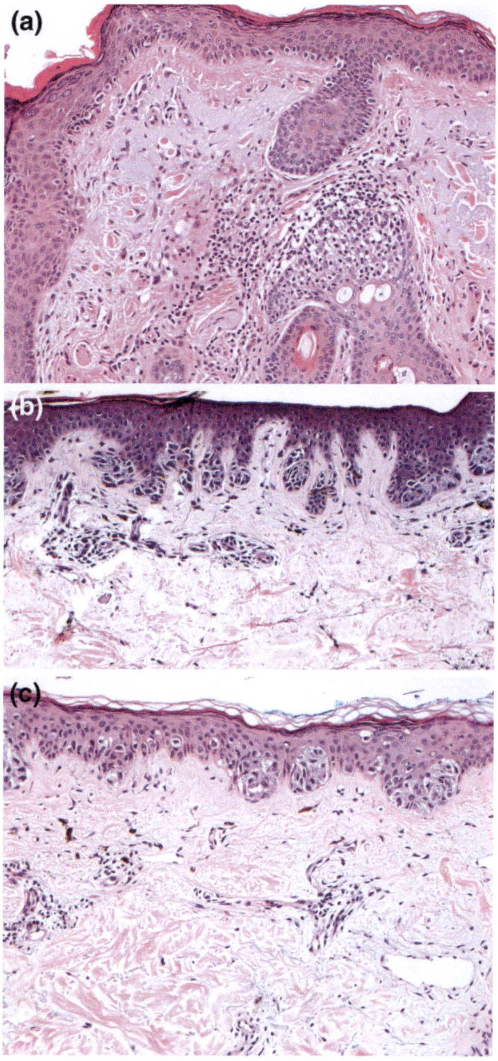


**Fig. 32.1** Intraepidermal melanocytic proliferations. **a** Atypical lentiginous melanocytic hyperplasia in lentigo maligna, **b** nested intraepidermal melanocytic proliferation in a compound dysplastic nevus with slight cytological atypia, **c** nested and pagetoid intraepidermal growth pattern in melanoma in situ

dermal nevi, the melanocytes in the superficial dermis have round to oval nuclei with small nucleoli and occasionally delicately pigmented cytoplasm, these are termed as type A nevomelanocytes. With increasing depth in the dermis the tumor cells have more round nuclei, less cytoplasm, and inconspicuous nucleoli, termed as type B nevomelanocytes. At the deepest aspect of the nevus, the type C nevomelanocytes have small round nuclei with minimal cytoplasm and may mimic lymphocytes or fibroblasts. This transition of cytological appearances from superficial cells with larger nuclei, open chromatin, more cytoplasm, to smaller cells with minimal cytoplasm, is termed as "maturation" and is a feature of benign nevi that is helpful in distinguishing benign melanocytic tumors from melanoma (Fig. 32.2a).

In addition to the common benign nevi, there are several histologically distinct types of benign melanocytic nevi; these include nevi with

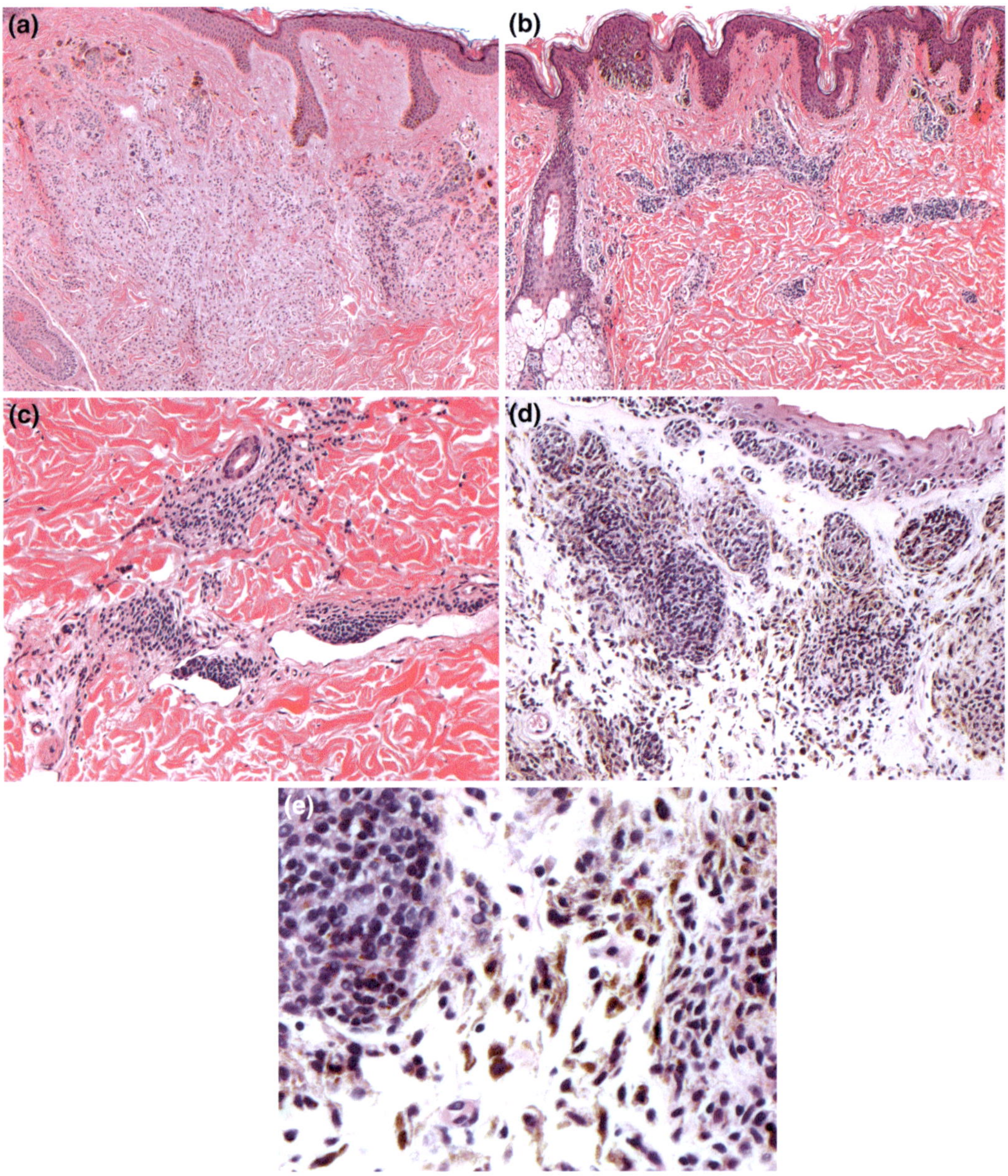


**Fig. 32.2** Benign intradermal melanocytic proliferations. **a** Melanocytic maturation in the dermal component of congenital nevus, **b** perivascular growth pattern in congenital nevus, **c** subendothelial growth pattern in congenital nevus, **d** combined dermal and blue nevus of the conjunctiva, **e** epithelioid dermal nevus cells admixed with pigmented dendritic blue nevus cells in combined conjunctival nevus

features of congenital onset, dermal dendritic melanocytic proliferations (blue nevi and variants), nevi with cytological and architectural disorder (dysplastic nevi), and spindled and epithelioid cell nevi (Spitz nevi). While in most cases these melanocytic nevus variants may be readily identified as benign, in some cases these tumors may display histopathological features that mimic findings characteristic of melanoma.

Melanocytic nevi with features of congenital onset have variable dermal patterns of growth that are distinctive. Congenital nevi are usually compound or dermal and characteristically involve the reticular dermis. The dermal growth patterns of congenital melanocytic nevi may be strikingly perivascular (Fig. 32.2b), diffusely infiltrate throughout the reticular dermis to form a plaque, extend around adnexal structures, infiltrate into the arrector pili smooth muscle, and occasionally display a subendothelial growth pattern mimicking vascular invasion (Fig. 32.2c).

Most benign melanocytic nevi are composed of melanocytes with round or oval nuclei and variable amounts of cytoplasm, termed "epithelioid" because of the resemblance to epithelial cells. However, there also is a distinctive category of nevi that are composed predominantly of melanocytes with small round to oval nuclei and delicately elongated dendritic cytoplasm. The cytoplasm of these melanocytes is most apparent when pigmented (Fig. 32.2d). The presence of cells with pigmented cytoplasm in the dermis gives a blue hue to the skin due to the Tyndall effect of light scatter. These tumors are termed "blue nevi" because of this distinctive clinical appearance. Variants of blue nevi include the nevus of Ito which typically occurs on the shoulder or upper arm, the nevus of Ota which typically occurs on skin innervated by the ophthalmic and maxillary branches of the trigeminal nerve and the mongolian spot which typically occurs in lumbosacral region. Blue nevi may also occur in combination with other types of benign nevi, these combined blue nevi are most commonly found in the skin of the eyelid [4] (Fig. 32.2e).

### 32.4.3 Histology of Dysplastic Nevi

Dysplastic melanocytic nevi were first identified as a distinctive type of atypical nevus in the setting of familial melanoma [5, 6]. The description of dysplastic nevi as sporadic nevi in patients without familial melanoma has stimulated considerable controversy. A significant body of science has established dysplastic nevi as an intermediate between common benign nevi and malignant melanoma. These tumors with cytological and architectural disorder are clinically and histologically indistinguishable from dysplastic nevi that occur in patients with familial melanoma. While controversy regarding the criteria for diagnosis continues, the most straightforward approach to the histopathological diagnosis of dysplastic melanocytic nevi resulted from a consensus conference on this topic more than two decades ago [7]. Using these guidelines, the histological diagnosis of a dysplastic nevus is based upon cytological features of the melanocytes, architectural growth patterns, and the host response to the tumor. Adherence to these criteria allows for consistent diagnosis of dysplastic nevi. Both major criteria are required and at least two of the minor criteria are required. The two major criteria are: a lentiginous and nested cytologically atypical intraepidermal melanocytic proliferation, and a nested intraepidermal component that extends three rete ridges beyond the dermal component (termed a "shoulder" with the body representing the dermal melanocytic tumor). This second criterion is not required in purely epidermal (junctional) dysplastic nevi. The minor criteria include:

- a prominent superficial vascular plexus,
- superficial papillary dermal fibrosis in either an eosinophilic concentric or lamellar pattern,
- a lymphocytic inflammatory infiltrate about the superficial vascular plexus, and
- nest expansion and fusion at the dermoepidermal junction, termed "bridging."

Cytological atypia is required for the diagnosis of a dysplastic nevus; however, there is a range of degrees of cytological atypia. While some will use architectural disorder, including limited pagetoid spread, to contribute to grading; most dysplastic nevi are graded based on the degree of melanocytic atypia. These tumors show nuclear variability along a continuum with gradual nuclear enlargement, increased chromatin density or clumping, nucleolar enlargement, and pleomorphism. With increasing degrees of atypia, the size of the nuclei increase; a good reference for size is the nucleus of the mid-layer keratinocyte. Because melanocytic atypia represents a spectrum, precise criteria are not possible. In general, mildly atypical cells have cytoplasmic retraction with a perinuclear halo and inconspicuous cytoplasm, the nuclei are oval, sickle shaped, angulated or rhomboidal, often with dense nuclear chromatin and do not exceed the size of a mid-layer keratinocyte nucleus. Moderately atypical nevomelanocytes have amphophilic or delicately pigmented cytoplasm, with enlarged rhomboidal, angulated, or oval nuclei, usually larger than a mid-layer keratinocyte nucleus, there is nuclear pleomorphism and nucleoli may be visible (Fig. 32.3). Severely atypical nevomelanocytes have features of melanoma cells, the cytoplasm may be amphophilic, eosinophilic or inconspicuous, nuclei are larger than mid-layer keratinocyte nuclei and have irregular folding, undulating nuclear contours, chromatin clumping or dense chromatin, nucleoli maybe be prominent and eosinophilic, and there is marked nuclear pleomorphism [8]. A limited pagetoid growth pattern with rare individual cells in the lower two-thirds of the epidermis may be seen in dysplastic nevi. Extensive pagetoid involvement of the upper layers of the epidermis is considered by many to be diagnostic of melanoma in situ, while others allow for a certain degree of pagetoid spread in severely atypical dysplastic nevi [9].

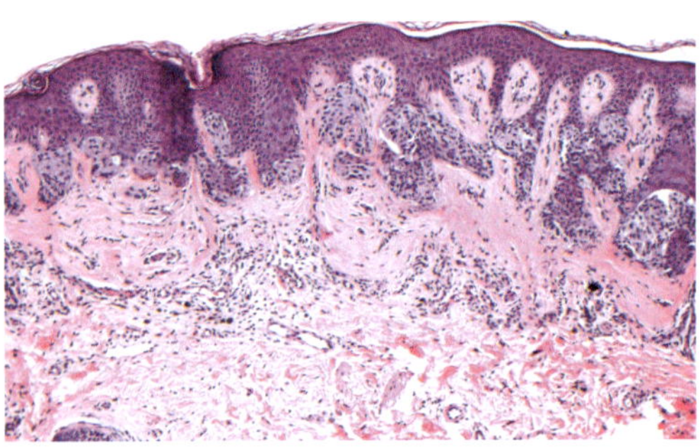

**Fig. 32.3** Compound dysplastic nevus with moderate cytological atypia. There is bridging of intrapepidermal nests, subepithelial fibrosis, lymphocytic infiltrate and moderate cytological atypia of the intraepidermal melanocytes

### 32.4.4 Histology of Spindled and Epithelioid Cell (Spitz) Nevi

The diagnosis, predicted prognosis, and optimal therapy continue to be controversial for atypical melanocytic tumors with features of the spindled and epithelioid cell melanocytic tumors described by Spitz [10]. While most cases may be identified histopathologically as either a benign Spitz nevus or a melanoma, a subset of melanocytic tumors are histologically ambiguous sharing features of Spitz nevus and melanoma. These cases are important to identify as either atypical Spitz tumors (with a high risk of local metastasis but minimal risk of spread beyond the sentinel lymph node basin) or spitzoid melanoma (with prognosis likely similar to conventional melanoma). There has been debate regarding the terminology of these atypical melanocytic tumors [11, 12], nevertheless, the histopathological criteria for distinguishing atypical Spitz tumor, spitzoid melanoma, and conventional melanoma are more well-defined now than in past decade [13]. Extensive research on this topic has led to evolution of the standards of care for patients with atypical Spitz tumor and spitzoid melanoma [14–16]. Nevertheless, the histological criteria for the distinction of atypical Spitz tumor from melanoma are not consistently defined [17–19]. Some will use the term "atypical Spitz tumor" for all Spitz nevi with any atypical histological feature, and "melanoma" for melanomas with spitzoid features (including spitzoid melanomas) [13]. Consensus meetings have led to some unifying concepts [17, 18]:

- histologically conventional benign Spitz nevi are not associated with metastasis (Fig. 32.4) [20, 21],
- atypical spindled and epithelioid melanocytic proliferations resembling Spitz nevi but with

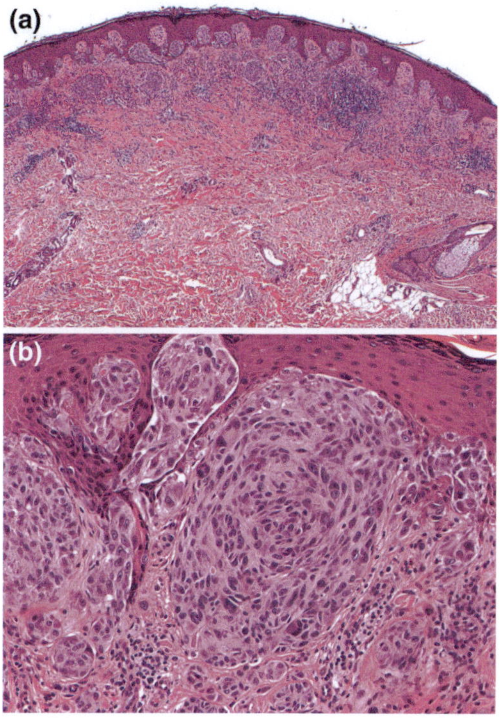


**Fig. 32.4** Compound spindled and epithelioid cell nevus (Spitz). **a** A wedge-shaped intraepidermal and dermal melanocytic proliferation extends into the reticular dermis, **b** Kamino bodies (intraepidermal eosinophilic deposits) are present among the epithelioid melanocytes

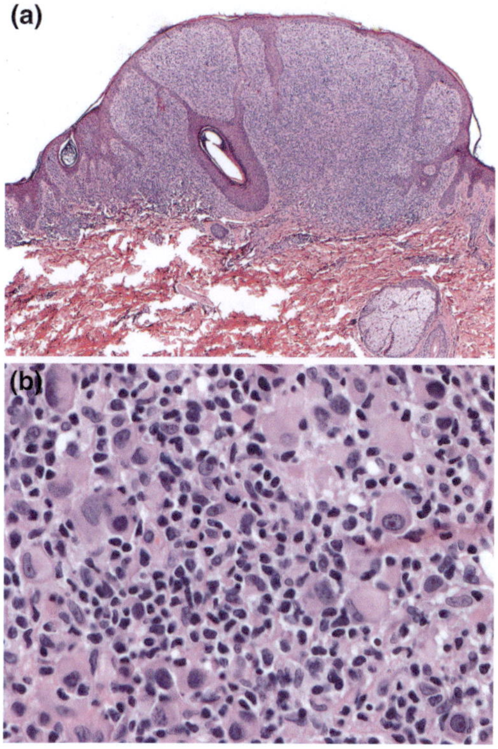


**Fig. 32.5** Spitzoid melanoma. **a** A predominantly epithelioid intraepidermal and dermal melanocytic proliferation, **b** cytologically pleomorphic epithelioid melanocytes, lymphocytes, and mitosis

cytological and/or architectural atypia (e.g., atypical Spitz tumors) are frequently associated with sentinel lymph node metastases [12, 18, 22–33], and

- spitzoid melanomas represent atypical spindled and epithelioid melanocytic tumors that share many cytological and architectural features with melanoma and are considered to be melanoma, possibly with a slightly better prognosis than conventional melanoma [34, 35].

Clear distinctions between these categories remain difficult and controversial because atypical Spitz tumor and spitzoid melanoma may exist on a continuum with Spitz nevi on one end and melanoma at the other. Histopathologically, atypical Spitz tumor has an overall resemblance to spindled and epithelioid cell nevus (Spitz nevus) but also displays a few atypical features. Spitzoid melanoma differs by showing more atypical features, with atypical findings sufficient for a diagnosis of melanoma (Fig. 32.5). The atypical histopathological findings include abnormalities of cellular organization, proliferation, and cytological atypia. Disorders of organization are manifested as: diameter > 10 mm, disordered intraepidermal growth pattern without supra-nest clefts, prominent pagetoid growth, ulceration, absence of Kamino bodies, confluent growth pattern, high cellular density, asymmetry at scanning magnification, poor circumscription, lack of maturation, lack of zonation (horizontal cytological consistency), and extension into the subcutaneous fat [13, 36]. Proliferation criteria include: more than two mitoses in the tumor or mitoses at the deep advancing margin of the tumor [37] and Ki-67 staining of more than 20 % of tumor cells [38]. Cytological atypia is manifested as: high nuclear-to-cytoplasmic ratio, "dusty" granular cytoplasmic pigmentation, nuclear chromatin clumping, nuclear membrane thickening, and irregularly shaped or enlarged nucleoli. In additional to the histopathological features, clinical characteristics including the age of the patient, and the color, size, and overall clinical appearance of the tumor also influence the final diagnosis.

### 32.4.5 Histology of Other Benign Melanoma Mimics

As noted above in the description for dysplastic nevi and spitzoid neoplasms, the diagnosis of melanoma can be challenging in a subset of melanocytic tumors that are benign but share histopathological features with melanoma. One of the histopathological features that may be observed in both benign and malignant melanocytic tumors is pagetoid intraepidermal spread, a proliferation of individual melanocytes present in the upper levels of the epidermis. There are several types of benign melanocytic proliferations that also display pagetoid spread and can be distinguished from melanoma. Pagetoid growth pattern is seen in the following benign melanocytic tumors: congenital nevus in children less than 5 years of age, spindled and epithelioid cell (Spitz) nevus, pigmented spindled cell nevus, recurrent nevus, excoriated nevus, acral nevus (also known as MANIAC = melanocytic acral nevus with intraepidermal ascent of cells [39]), and occasionally dysplastic nevi. In most of these cases the histopathological attributes including the pattern of melanocytic nesting and absence of marked cytological atypia will aid in arriving at the correct diagnosis.

Another category of benign melanoma mimics includes combined nevi. These complex melanocytic tumors share features of two or more types of melanocytic nevi. The most diagnostically challenging of these are the combined nevi with deep pigmented melanocytes including the deep penetrating/plexiform spindled cell nevi (Fig. 32.6) and the clonal/inverted type A nevi [4, 40, 41]. Despite some histopathological similarities to melanoma these nevi are entirely benign and can be identified based on histopathological and IHC analysis.

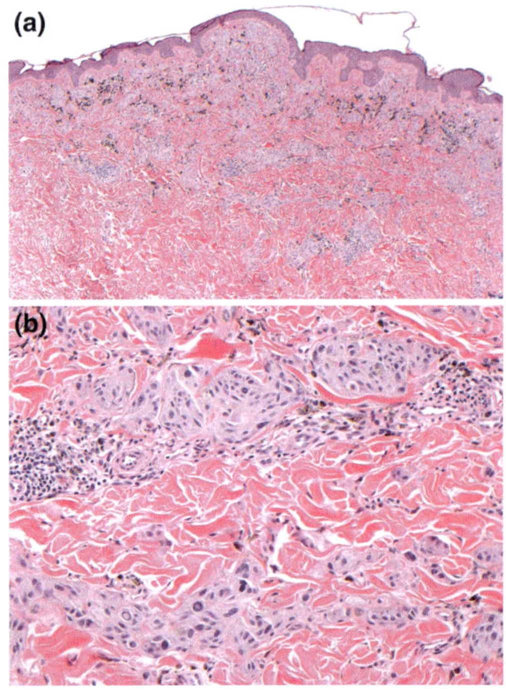


**Fig. 32.6** Deep penetrating/plexiform spindled cell nevus. **a** A wedge-shaped dermal proliferation of lightly pigmented melanocytes, extending along neurovascular bundles with admixed pigment-laden macrophages. **b** The delicately pigmented tumor cells are admixed with coarsely pigmented macrophages and surround vascular and neural structures, extending deep into the dermis with plexiform growth pattern

## 32.5 Histopathology of Melanoma

### 32.5.1 Melanoma Subtypes

The initial classification of primary cutaneous melanoma occurred nearly 50 years ago based upon detailed observations and descriptions of the clinical and histopathological findings [42, 43]. There have been some additions to the classification scheme, however, the original descriptions form the foundation for the currently recognized subtypes of cutaneous melanoma [42, 44–47]. The original descriptions of cutaneous melanoma included the clinical appearance, age, gender, anatomic sites, and degree of sun exposure. The authors also described in detail the histological features including patterns of tumor cell growth in the epidermis and dermis, cytologic features, epidermal changes including atrophy and ulceration, the presence of solar elastosis, the anatomic level of invasion, the maximal tumor thickness as measured perpendicular to the epidermal surface from the top of the granular cell layer, vascular invasion, mitotic activity, and the pattern and density of lymphocytic host response. These clinical and pathological variables formed the framework for cutaneous melanoma classification into three major subtypes: invasive melanoma with adjacent intraepidermal component of superficial spreading type (superficial spreading melanoma, SSM), invasive melanoma with adjacent intraepidermal component of Hutchinson's melanotic freckle type (lentigo maligna melanoma, LMM), and invasive melanoma without adjacent intraepidermal component (nodular melanoma, NMM). The adjacent intraepidermal component was defined as extending at least three rete ridges within the epidermis beyond the edge of the dermal component. The extension of melanoma in the epidermis and superficial dermis is termed "radial growth phase," when the dermal invasive component become more prominent or mitotically active it is termed "vertical growth phase" (Table 32.1). In the years since this original description, additional cutaneous melanoma subtypes, including acral lentiginous, mucosal lentiginous, nevoid, desmoplastic, spindled, and others have been described. Each has distinctive clinical and histopathological features (Table 32.2).

**Table 32.1** Definitions of radial and vertical growth phase in primary cutaneous melanoma

| |
|---|
| *Characteristics of radial growth phase (RGP) melanoma* |
| • Single cell dermal invasion<br>• Small invasive nests (dermal nests smaller than intraepidermal nests)<br>• No dermal tumor cell mitoses<br>• Inflammatory infiltrate present<br>• Papillary dermis involved (Clark level II)<br>• No expansile nodule (s) |
| *Characteristics of vertical growth phase (VGP) melanoma* |
| • Expansile nodule, nests in dermis larger than epidermis<br>• Dermal mitoses<br>• Stromal changes (desmoplastic)<br>• Present in papillary dermis, +/– reticular dermis, +/– fat (Clark level III, IV or V) |

#### 32.5.1.1 Superficial Spreading Melanoma

The clinical appearance of superficial spreading melanoma is variable and includes a broad range of colors including tan, brown, gray, black,

**Table 32.2** Distinctive histopathological findings, frequency, and growth patterns in primary cutaneous melanoma

| | % of cases | Distinctive findings | RGP | VGP |
|---|---|---|---|---|
| Superficial Spreading | >70 | Pagetoid RGP, epithelioid VGP | + | −/+ |
| Lentigo Maligna | 3 | Sun-damaged skin, lentiginous RGP extension down hair follicles, epithelioid or spindled VGP | + | −/+ |
| Acral Lentiginous | 2 | Acral sites, lentiginous RGP extension down eccrine units, epithelioid VGP | + | −/+ |
| Nodular | 20 | No radial growth phase, epithelioid VGP | − | + |
| Desmoplastic | <1 | Often associated with lentigo maligna, scar-like VGP | +/- | + |
| Nevoid | <1 | Symmetric nevic like dermal component, epithelioid VGP | −/+ | + |
| Spindled | <1 | Often associated with lentigo maligna, spindled VGP | +/- | + |
| Mucosal lentiginous | <1 | Oral, genitourinary, and gastrointestinal mucosa | + | −/+ |

violaceous, pink, and occasionally blue or white. The edge of the tumor with adjacent normal skin is usually sharply marginated with a few irregular peninsula-like protrusions. The surface may be slightly elevated or have a palpable papule or a nodule that extends several millimeters above the skin surface.

The histopathological features of superficial spreading melanoma include an intraepidermal component with a pagetoid and nested growth patterns at all levels of the epidermis (Fig. 32.7). The intraepidermal tumor cells have a prominent epithelioid cytology with abundant cytoplasm that is eosinophilic, amphophilic or has a "dusty"

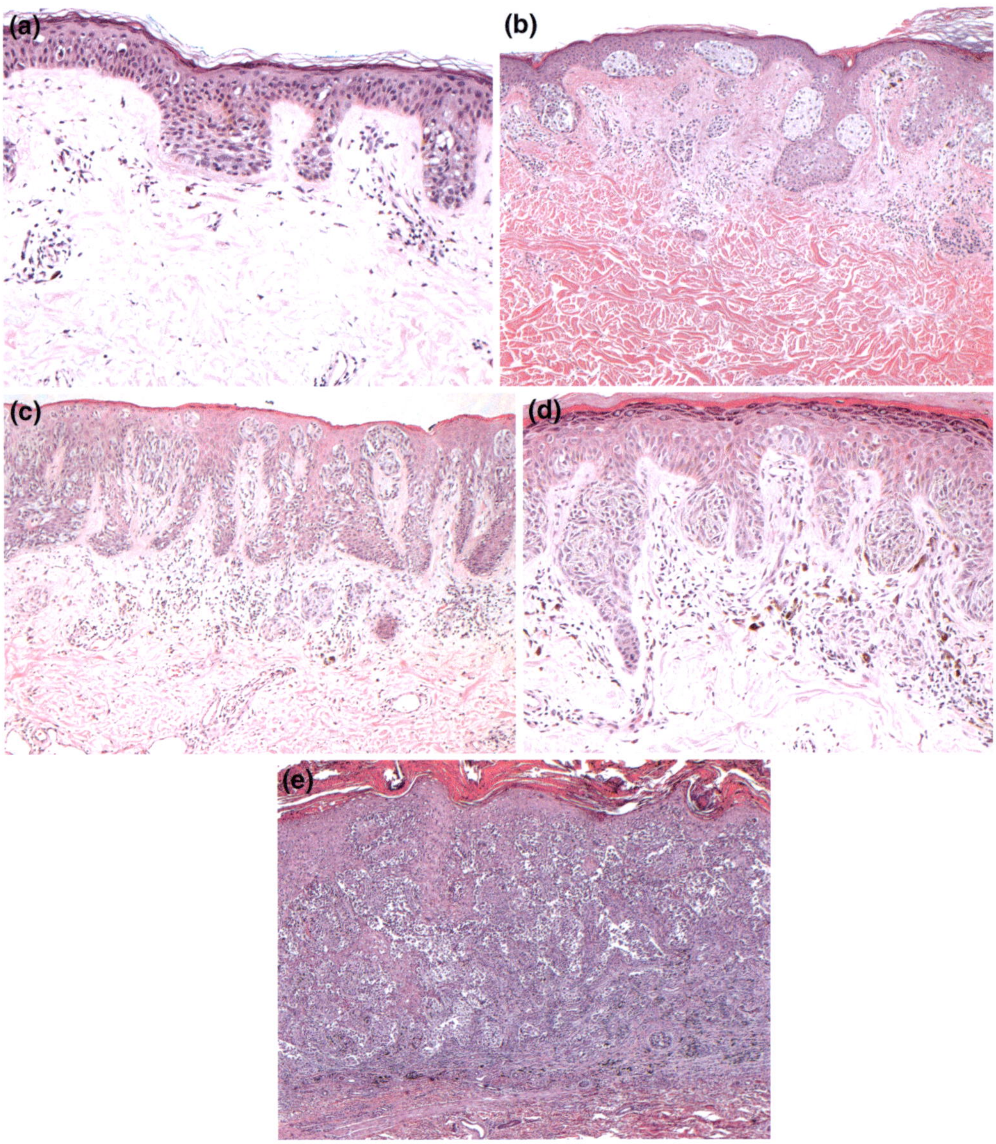


**Fig. 32.7** Superficial spreading melanoma. **a** In situ melanoma with pagetoid intraepidermal growth pattern, **b** in situ nested and pagetoid intraepidermal proliferation with superficial dermal nevus, **c** pagetoid and nested intraepidermal melanoma with invasion of the superficial dermis, **d** the cytology of the intradermal tumor cells is similar to that of the intraepidermal component, **e** intraepidermal and dermal melanoma with multifocal dermal invasion

distribution of fine cytoplasmic melanin granules. The nuclei may be large, have irregular nuclear contours, and contain one or more prominent nucleoli. These intraepidermal tumor cells are relatively uniform and appear cytologically similar to one another. In invasive dermal melanoma, the tumor cells have cytological features similar to the intraepidermal tumor cells. The dermal melanoma component may also be composed of numerous variably sized nests that may be associated with expansile nodule formation. In cases with abundant dermal tumor the cytological heterogeneity becomes more apparent, often with striking variation in cell morphology from one tumor nest to the next.

#### 32.5.1.2 Nodular Melanoma

The clinical appearance of nodular melanoma is usually a relatively uniform brown, black or blue-black elevated lesion. Nodular melanoma may be a smoothly surfaced cutaneous nodule, an elevated plaque with irregular outlines, or a polypoidal ulcerated exophytic tumor. In contrast to superficial spreading and lentigo maligna types of melanoma, there is no surrounding flat pigmented lesion associated with the tumor.

The histopathology of nodular melanoma is that of a predominantly dermal tumor. When an intraepidermal component is present it directly overlies the invasive melanoma (Fig. 32.8). Occasionally, the epidermal component is so minimal as to suggest the possibility that the tumor represents a dermal metastasis. The vertical growth phase of nodular melanoma is composed of small nests and aggregates of tumor cells that together form the overall tumor nodule.

#### 32.5.1.3 Lentigo Maligna Melanoma

The clinical appearance of lentigo maligna melanoma is that of a relatively large, mostly flat, pigmented lesion with a variegated coloration that includes tan, brown, and black and may have flecks of black or brown. The outline of the tumor is irregular and may merge with the surrounding skin tones. Wood's light examination may be helpful in identifying the edge of the melanocytic proliferation. Although lentigo maligna melanoma is predominantly flat, foci of invasion may be detected as a slightly raised papule that may be best detected by side lighting [48].

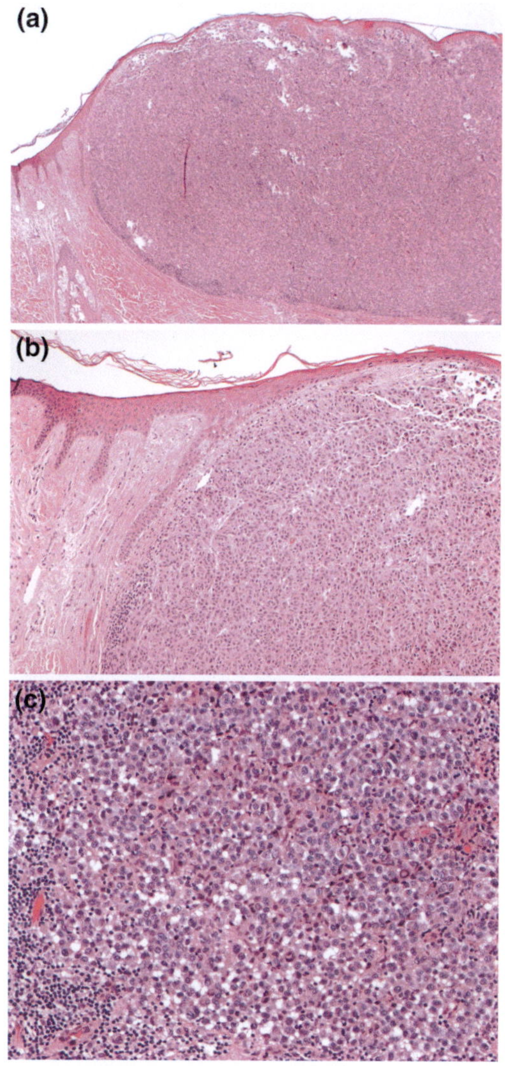


**Fig. 32.8** Nodular melanoma. **a** Epithelioid tumor cells form an expansile dermal nodule with only focal involvement of the overlying epidermis, this is pure vertical growth phase tumor, **b** there is no radial growth phase melanocytic proliferation, **c** the epithelioid tumor cells have pleomorphic nuclei with prominent nucleoli and a lymphocytic infiltrate

Lentigo maligna melanoma arises in the setting of lentigo maligna. There is a spectrum of atypical melanocytic proliferations that occur in

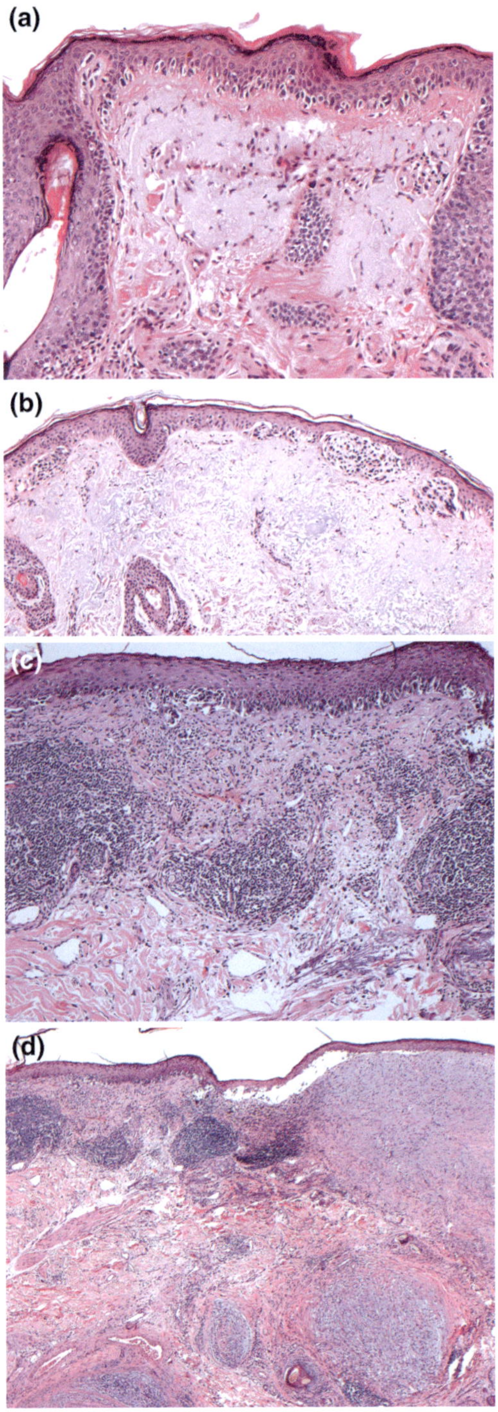


**Fig. 32.9** Lentigo Maligna Melanoma. **a** In situ melanoma composed of a confluent lentiginous intraepidermal proliferation of cytologically atypical melanocytes associated with epidermal atrophy and solar elastosis, **b** in situ melanoma with a lentiginous and nested intraepidermal proliferation of cytologically atypical melanocytes, **c** intraepidermal and dermal melanoma with a confluent lentiginous and nested intraepidermal component and underlying dermal invasion with associated lymphocytic infiltrate, **d** invasive desmoplastic neurotropic melanoma arising the setting of lentigo maligna melanoma

sun-damaged skin of the elderly, usually on the face, scalp, and neck. These proliferations arise in a background of epidermal atrophy with marked solar elastosis. Histopathologically, the melanocytic proliferations range from subtle atypical melanocytic hyperplasia, to lentigo maligna, to lentigo maligna melanoma in situ, to invasive lentigo maligna melanoma (Fig. 32.9). The histopathology of lentigo maligna melanoma is characterized by an intraepidermal predominantly individual cell melanocytic proliferation localized to the basal layers of the epidermis. The cytology of the lentiginous proliferation is strikingly atypical; tumor cells have large densely chromatic nuclei and are occasionally multinucleated. When the proliferation is confluent the basal keratinocytes appear to be replaced by a continuous line of severely atypical melanocytes. The lentiginous proliferation extends down hair follicle epithelium, maintaining close approximation to the basal layer. Intraepidermal nests and pagetoid spread may be observed, however, these features are subtle when present, in contrast to superficial spreading melanoma, which is characterized by a nested and pagetoid intraepidermal component. While some observers include lentigo maligna and lentigo maligna melanoma in situ in one diagnostic category, others separate these two based on histopathological findings [49, 50]. Criteria for lentigo maligna melanoma in situ include a severely atypical lentiginous melanocytic proliferation with two or more of the following:

(1) intraepidermal melanocytic nests, (2) pagetoid growth pattern, (3) confluence of melanocytes along the dermal epidermal junction [51]. When tumor cells extend into the dermis, the diagnosis is lentigo maligna melanoma. The dermal invasive component of lentigo maligna melanoma may display spindled cells and tumor cell pigmentation, have a scar-like desmoplastic appearance or may show features similar to those observed in the dermal component of superficial spreading and nodular melanoma. Desmoplasia and neurotropism are more commonly found in lentigo maligna melanoma than in other melanoma types.

#### 32.5.1.4 Melanoma: Rare Subtypes

In addition to superficial spreading, lentigo maligna and nodular melanoma, there are several rare, clinically and histopathologically distinctive subtypes. These include acral lentiginous melanoma (Fig. 32.10), mucosal lentiginous melanoma, nevoid melanoma, desmoplastic melanoma (mixed and pure), and spindled melanoma. Acral lentiginous and mucosal lentiginous melanomas arise in anatomically distinctive sites: hands and feet in the case of acral melanoma, and oral, genital, and gastrointestinal mucosa in the case of mucosal melanoma. Some of these rare forms of melanoma may fit into an existing subtype, for example, nevoid melanoma is considered by some to be a variant of nodular melanoma [36, 52–54] (Fig. 32.11). There may be overlap between nevoid melanoma with Spitz-like tumors and tumors associated with BAP-1 loss; further research is needed to fully understand this subset of tumors [55, 56]. Desmoplastic melanomas are usually lentigo maligna type [57–62] (Fig. 32.12). It is now known that even further subcategorization is helpful when determining the treatment for desmoplastic melanoma. Patients with pure desmoplastic melanoma have a lower risk of metastases than those with mixed desmoplastic and conventional melanoma [59]. Other rare forms of melanoma include the pigmented epithelioid melanocytoma [63, 64] and Spitzoid melanoma [20, 21, 37, 65–69]. The treatment plan for these rare tumors is based on reports of similar tumors rather than the guidelines from the American Joint Commission on Cancer (AJCC) or other schemes.

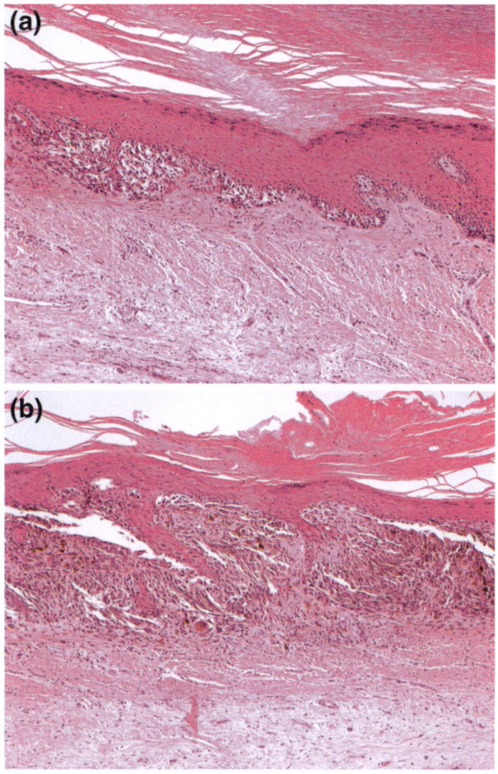


**Fig. 32.10** Acral Lentiginous Melanoma. **a** In situ melanoma in a lentiginous and nested pattern, **b** the intraepidermal tumor cells are cytologically atypical with high nuclear-to-cytoplasmic ratios and nuclear enlargement

### 32.5.2 Cutaneous Melanoma Histopathological Prognostic Factors

The histopathological features of cutaneous melanoma serve as the foundation for staging. Factors included in the AJCC staging system include primary tumor thickness, mitogenicity, and presence or absence of ulceration (Fig. 32.13). The histological identification of microscopic metastases as microscopic satellites, in-transit metastases, and sentinel lymph node metastases also factor into the AJCC stage. Other

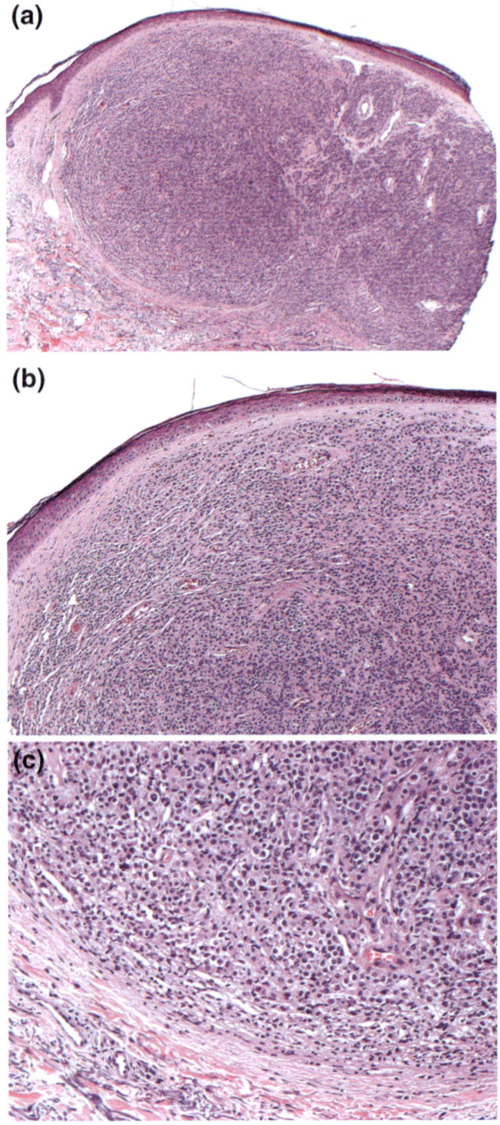


**Fig. 32.11** Nevoid Melanoma. **a** This dermal proliferation of melanocytes is not associated with an overlying epidermal melanocytic proliferation, **b** the tumor cells have small to medium-sized nuclei without significant cytological atypia, resembling dermal nevus cells, **c** there is no maturation of tumor cells with increasing dermal depth, the tumor cells at the base have similar features to those at the superficial aspect of the tumor

histopathological prognostic factors that may guide patient therapy include Clark level of invasion, tumor infiltrating lymphocytes, lymphovascular invasion, regression, and neurotropism (Fig. 32.13).

#### 32.5.2.1 Primary Tumor Thickness (Breslow)

The maximal thickness of primary cutaneous melanoma, as measured under the microscope using an intraocular ruler, is one of the most powerful predictors of melanoma survival [47]. This measurement is a critical component of the pathology report and must be determined consistently. A calibration table is used to ensure accurate measurement across different brands and models of microscopes. This melanoma measurement is taken perpendicular to the epidermis from the top of the epidermal granular cell layer overlying the thickest part of the tumor to the deepest invasive melanoma cell. When ulceration is present, the measurement is taken from the topmost viable tumor cell at the base of the ulcer to the deepest point of tumor cell invasion. The deepest measured melanoma cell must be free and clear of adnexal structures: tumor cells that extend along perineural and periadnexal (adventitial) dermis and perivascular or intravascular extension are not included in the primary tumor thickness measurement. When the primary tumor has a polypoidal architecture, the Breslow thickness is obtained by measuring across the largest diameter of the lesion perpendicular to the skin surface [70].

#### 32.5.2.2 Primary Tumor Mitogenicity

It has long been known that increased proliferative activity of invasive melanoma is associated with poor prognosis [71, 72]. Mitogenicity has been found to be of greatest prognostic power in patients with thin melanomas less than 1 mm thick [73–75]. Additionally, mitogenicity may be a powerful prognostic factor in patients with negative sentinel lymph nodes [76]. The presence of one mitosis in the dermal component of melanoma with thickness < 1 mm leads to upstaging from T1a to T1b. This upstaging is associated with more therapeutic intervention, usually sentinel lymph node removal and adjuvant therapy. Because the studies that revealed mitoses as clinically significant were retrospective studies that employed the "hot spot" technique of evaluating tumor mitogenicity, the AJCC Melanoma Staging Committee recommends that mitotic

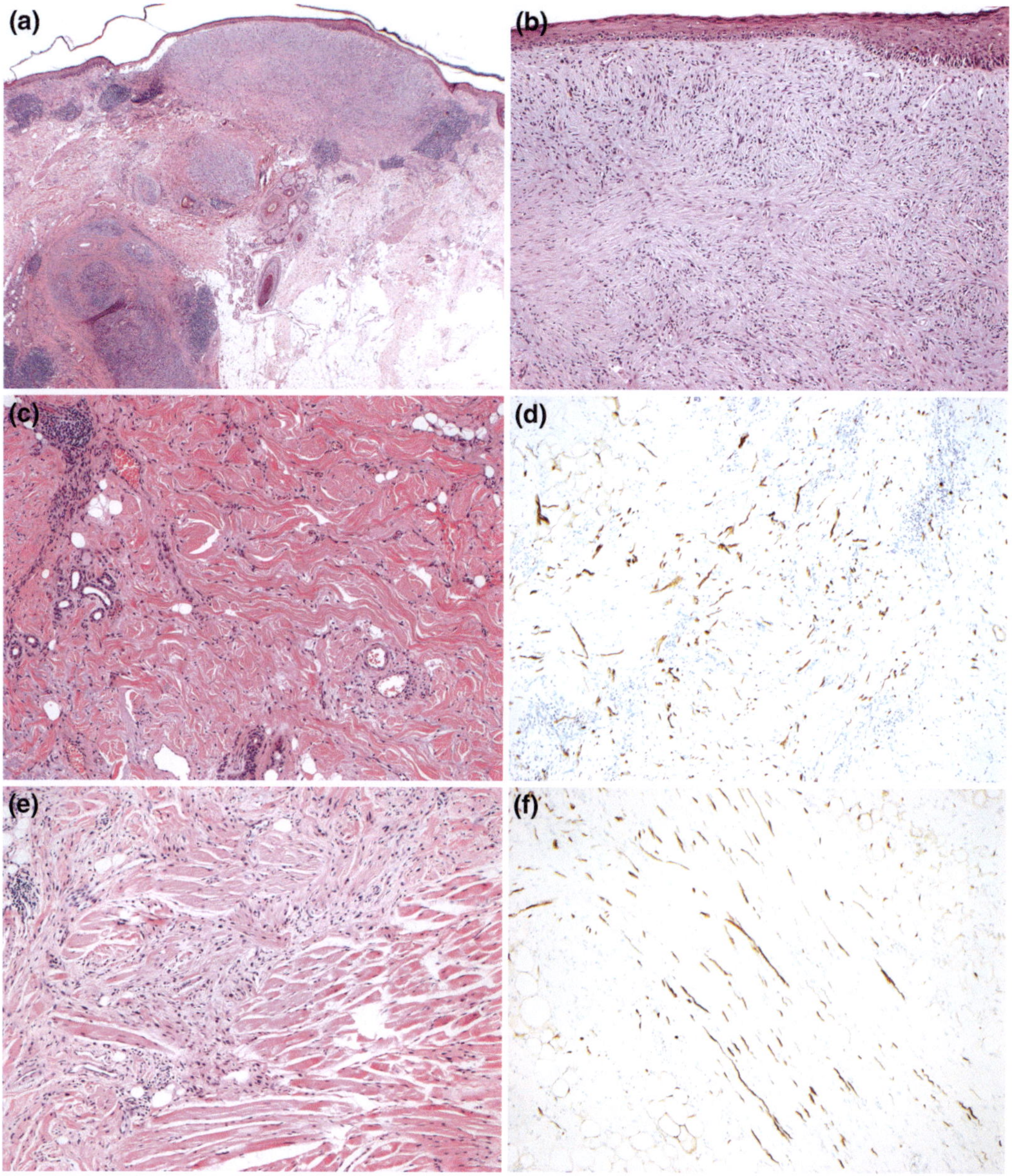


**Fig. 32.12** Desmoplastic melanoma. **a** There is an extensive infiltrative spindled and desmoplastic tumor arising in a background of lentigo maligna, the presence of lymphoid aggregates in the deep dermis and subcutis are a clue to the presence of neurotropism, **b** the superficial dermal tumor cells in this desmoplastic focus have spindled and dendritic cytology resembling scar, C. In some regions the tumor cell density is low, **d** an S100 stain in the region shown in panel C displays numerous desmoplastic melanoma cells, **e** the desmoplastic tumor invaded skeletal muscle, **f** an S100 stain highlights the dendritic melanoma cells infiltrating fat

count be determined by this approach and reported as the number of mitoses per square millimeter of the primary tumor [77]. Determining mitogenicity is accomplished by examination of routine hematoxylin and eosin (H&E)-stained tissue sections. It is not necessary to do exhaustive tissue sectioning. After review of the invasive tumor the area with the most mitotic figures

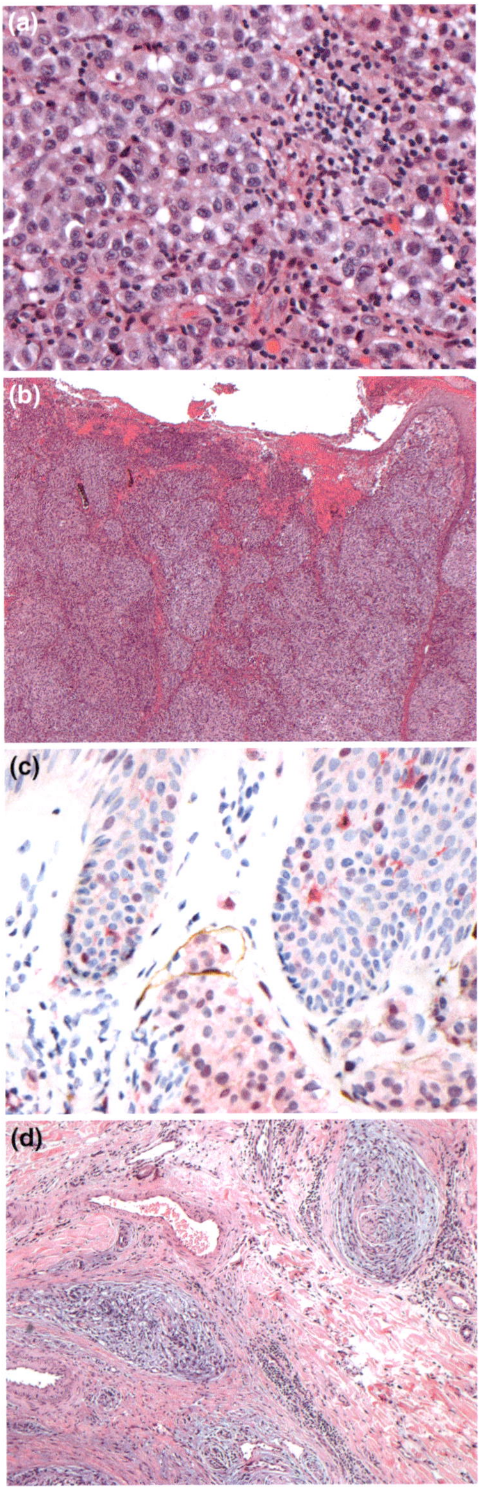


**Fig. 32.13** Melanoma prognostic factors. **a** mitoses, **b** ulceration, **c** lymphovascular invasion (*brown* D240, *pink* S100 staining), **d** neurotropism

("hot spot") is identified, and the count starts using a high power 40× objective (Fig. 32.13a). After determining the number of mitoses in the first high power field the count is extended to adjacent fields until an area of 1 $mm^2$ is assessed. To ensure accuracy across observers, individual microscopes are calibrated to determine the number of high power fields corresponding to 1 $mm^2$. Mitogenicity is reported as n/$mm^2$. If only one mitosis is found in the entire invasive component, the report states 1/$mm^2$. If no dermal mitoses are identified the count is reported as 0/$mm^2$. When the invasive component of the tumor measures less than 1 $mm^2$ the count is performed on the entire dermal tumor and reported as n/$mm^2$. The AJCC Melanoma Staging Committee strongly discourages the use of "<1/$mm^2$" in reporting melanoma. It is important to distinguish between the reporting function, which should always be in a whole number/$mm^2$ and the staging language which is described as a range, e.g., greater than or equal to 1/$mm^2$.

#### 32.5.2.3 Primary Tumor Ulceration

In the past decade ulceration has been identified as an important adverse prognostic factor in primary cutaneous melanoma. Unlike mitogenicity which is most powerful in thin tumors, ulceration serves as more powerful discriminator in tumors >1 mm thickness, perhaps in part because ulceration is very rare in tumors <1 mm. It may be difficult to distinguish thick-scale crust from an ulcer clinically, thus, in primary cutaneous melanoma staging ulceration is defined histopathologically. Tumor ulceration is defined as full-thickness interruption of the epidermis by tumor without prior history of mechanical trauma or surgery at the site. The epidermal disruption is associated with fibrin, inflammatory cells, and granulation tissue (Fig. 32.13b).

#### 32.5.2.4 Tumor Infiltrating Lymphocytes in Primary Cutaneous Melanoma

The presence of lymphocytes infiltrating the vertical growth phase of malignant melanoma has been associated with a better prognosis [72,

78, 79]. The pattern of tumor infiltrating lymphocytes is graded as "brisk" when the lymphocytes diffusely infiltrate throughout the vertical growth phase tumor or form a continuous inflammatory front along the entire advancing tumor front, "non-brisk" when there is focal or multifocal infiltration of the vertical growth phase, and "absent" when there are no lymphocytes infiltrating the tumoral compartment. Notably, if a dense inflammatory infiltrate is present in the specimen adjacent to, but not infiltrating the melanoma, this is also termed "absent." The presence of "brisk" tumor infiltrating lymphocytes is associated with a better prognosis, however, some of these patients may still develop metastasis and disease progression. Several additional studies have further characterized these as T cells [80]. Although tumor infiltrating lymphocyte grade is not currently part of cutaneous melanoma staging, advances in immunotherapy and further understanding of the role of lymphocytic subsets may lead to changes in future staging algorithms [81, 82].

#### 32.5.2.5 Lymphovascular Invasion in Primary Cutaneous Melanoma

The identification of lymphovascular invasion in a primary melanoma correlates with an increased risk of metastasis [83–86] (Fig. 32.13c). Lymphovascular invasion is observed as the presence of tumor cells within the lumen of a lymphatic vessel. In some cases tumor cells may be observed surrounding vascular structures, a phenomenon known as "extravascular migratory metastasis" or "angiotropism" [87, 88]. Lymphovascular invasion has been demonstrated as prognostic factor in early stage melanoma [89].

#### 32.5.2.6 Microscopic Satellites in Cutaneous Melanoma

The identification of microscopic melanoma metastases may occur in the tissue section with the primary tumor (microscopic satellite), in the skin within 5 cm of the primary tumor (satellite metastasis) and in the skin or soft tissue in a region between 5 cm from the cutaneous site and the regional lymph node basin (in-transit metastasis). Microscopic satellites identified in the primary tumor tissue section are defined as a discontinuous group or nest of melanoma cells, greater than 0.05 mm in diameter, that are located at least 0.3 mm from the dermal invasive tumor mass and separated from it by normal dermis or panniculus not affected by fibrosis or inflammation [90]. Described as a primary tumor prognostic factor, microscopic satellites are associated with poor prognosis [91, 92]. The AJCC staging system includes microscopic satellites, satellite metastases, and in-transit metastases along with intralymphatic metastases [93]. Primary tumor microsatellites are an independent predictor of reduced disease-free survival in patients with positive sentinel lymph nodes [94]. The terms "intralymphatic regional metastases" and "in-transit metastases/satellites" are now used to describe these patterns of local spread [77].

#### 32.5.2.7 Neural Involvement in Primary Cutaneous Melanoma

Perineural invasion and neurotropism may be observed in melanoma. Neurotropism is most commonly associated with spindled or desmoplastic melanoma as an extension of tumor cells around and within cutaneous nerves (Fig. 32.13 d). This neurotropic pattern of growth is often accompanied by a lymphocytic infiltrate and is most common on melanomas of the head and neck.

#### 32.5.2.8 Anatomic Level of Invasion (Clark)

The level of invasion in relationship to the anatomical boundaries of the papillary and reticular dermis and subcutaneous fat were originally defined by Clark in 1969 [42]. Increasing levels of invasion correlate with poor outcome. Clark level I tumors are limited to the epidermis, level II tumors display individual cell and small melanoma cell nests in the papillary dermis, level III tumors have expansile nodules in the papillary dermis that push upon but do not invade the reticular dermis, level IV tumors have tumor cells in the reticular dermis, and level V tumors invade the subcutaneous fat. While this metric has largely been replaced by mitogenicity in the AJCC staging schema, Clark level remains important in cases where mitoses are difficult to assess due to poor processing or when tumor thickness cannot be accurately measured due to poor tissue orientation. In these rare cases, the identification of melanoma in the reticular dermis (Clark level IV) or subcutaneous fat (Clark level V) can assist in tumor staging.

#### 32.5.2.9 Regression in Primary Cutaneous Melanoma

The phenomenon of regression was described by Clark as an important prognostic indicator and it has been associated with increased likelihood of metastasis, even in thin melanomas [72, 95]. Regression in primary cutaneous melanoma has the most prognostic significance when it is consistently defined. Regression occurs in the radial growth phase of the tumor and is characterized by complete absence of melanoma cells in the epidermis and dermis, flanked by viable melanoma cells. The regressed focus often displays epidermal atrophy with underlying fibroplasia, vascular prominence, chronic inflammation, and melanophages. The use of less stringent criteria for regression may contribute to the poor concordance between studies of tumor regression's prognostic significance.

### 32.5.3 Melanoma Metastases and Sentinel Lymph Nodes

Sentinel lymph node mapping was described over two decades ago as a minimally invasive means of identifying microscopic metastases in regional lymph nodes [96]. This technique allows for more precise staging of patients with clinically localized cutaneous melanoma. The identification of even a single melanoma cell in a sentinel lymph node leads to upstaging from AJCC stage II to stage III. The anatomy of cutaneous lymphatic drainage allows for identification of the first or sentinel lymph node to drain a specific cutaneous site. The use of isosulfan blue dye injected at the primary cutaneous tumor site allows for identification of the draining blue lymph node in the regional lymph node basin. This sentinel node is the most likely lymph node to contain metastatic melanoma. The addition of lymphoscintigraphy with technetium-99 m–(99 mTc) labeled sulfur colloid, followed by intraoperative identification of sentinel lymph nodes using a handheld scanner with a gamma-sensor probe to detect 99 mTc has refined the technique for identifying early regional lymph node metastases. After the hottest lymph node is identified and the 99 mTc counts quantified, additional lymph nodes with >10 % of the counts of this hottest lymph node are removed and also considered sentinel [97]. On average, two to three sentinel lymph nodes are removed [76, 98], if melanoma is not detected in the sentinel lymph nodes, the remaining lymph nodes in the basin are unlikely to contain melanoma [99]. Patients diagnosed with sentinel lymph node metastases are usually offered complete lymphadenectomy and adjuvant therapy, although recently there has been a move away from offering completion lymphadenectomy to select patients with small sentinel lymph node deposits.

The detection of melanoma metastases in sentinel lymph nodes is best performed by careful examination of H&E-stained tissue sections, along with levels into the lymph nodes and IHC (Fig. 32.14), as occasionally tumor is most apparent on IHC sections. The diagnostic criteria for lymph node metastasis rely upon the identification of cytological features as well as the location and pattern of intranodal tumor spread: (1) the presence of individual cells or nests of epithelioid or spindled cells foreign to the lymph node, (2) cytological atypia including large pleomorphic nuclei with prominent nucleoli, variable cytoplasm occasionally with dusty

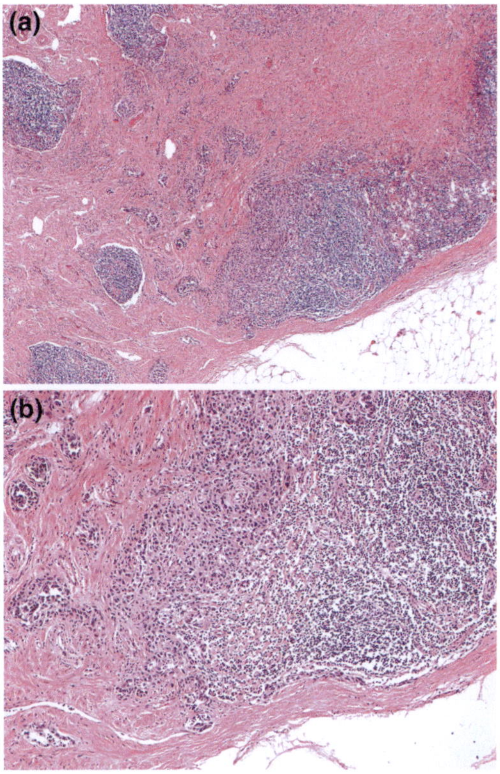

**Fig. 32.14** Metastatic melanoma in lymph node parenchyma. **a** The tumor is present as nests embedded within fibrous tissue and associated with lymph node parenchyma, **b** The tumor cell nuclei are large and pleomorphic in comparison with the lymphocytes

cytoplasmic melanin granules, and (3) positive staining for one or more melanocytic markers (e.g., S–100, MART-1, Melan-A, HMB-45, MITF). Sentinel lymph node metastases are identified in 15–20 % of patients who undergo this procedure [100]. The differential diagnosis of metastasis in this setting includes benign melanocytic rests (nodal nevus) usually observed in the lymph node capsule or fibrous trabeculae. The reported frequency of these benign melanocytic deposits ranges from a few percent to more than 20 % [101]. The diagnostic criteria for benign melanocytic nevi in lymph nodes are: (1) individual cells in a linear array or nests of epithelioid or spindle cells foreign to the lymph node, (2) round or oval uniform nuclei without cytological atypia, (3) positive staining for one or more melanocytic marker (S-100, MART-1, MELAN-A, MITF; nodal nevi are usually negative for HMB-45), (4) identification of the cells on H&E-stained sections, and (5) cells are usually present in the fibrous capsule or trabeculae (Fig. 32.15). Overall these criteria allow for distinction of nodal melanocytic nevi from melanoma in most cases.

### 32.5.4 Melanoma at Other Sites

#### 32.5.4.1 Ocular Melanoma

Ocular melanoma is a rare form of melanoma representing approximately 3 % of melanomas, compared to >90 % of melanomas arising in the skin. Primary ocular melanoma usually arises in the uveal tract (85 %), but may also arise in the conjunctiva (5 %) or other ocular sites (10 %) [102]. Uveal melanoma is distinct from cutaneous melanoma in that treatment is largely based on clinical diagnosis and cytogenetic analysis. The most common form of ocular melanoma is that of the uveal tract, an intraocular structure composed of the iris, ciliary body, and choroid. The iris and ciliary body are located anterior to the retina, and contiguous with the choroid. The choroid, responsible for nourishing the retina and found between the epithelium and sclera, is composed of blood vessels, nerve fibers, and pigmented cells within a connective tissue matrix. Although uveal melanomas usually arise in the choroid (90 %), they can arise in any part of the uveal tract including the iris (4 %) and the ciliary body (6 %) [102, 103].

Clinical diagnosis of ocular melanoma based on ophthalmic examination, slit lamp examination, indirect ophthalmoscopy, and other ancillary testing has an accuracy of more than 99 % [104]. Ancillary diagnostic tests include ultrasonography, fluorescein angiography, indocyanine green angiography, optical coherence tomography, fundus autofluorescence, and fine needle aspiration (FNA). FNA allows for cytological evaluation and provides tumor cells for genetic analysis.

Histopathologic analysis of uveal melanoma is often based on cytological features observed on FNA and includes assessment of cell size/shape,

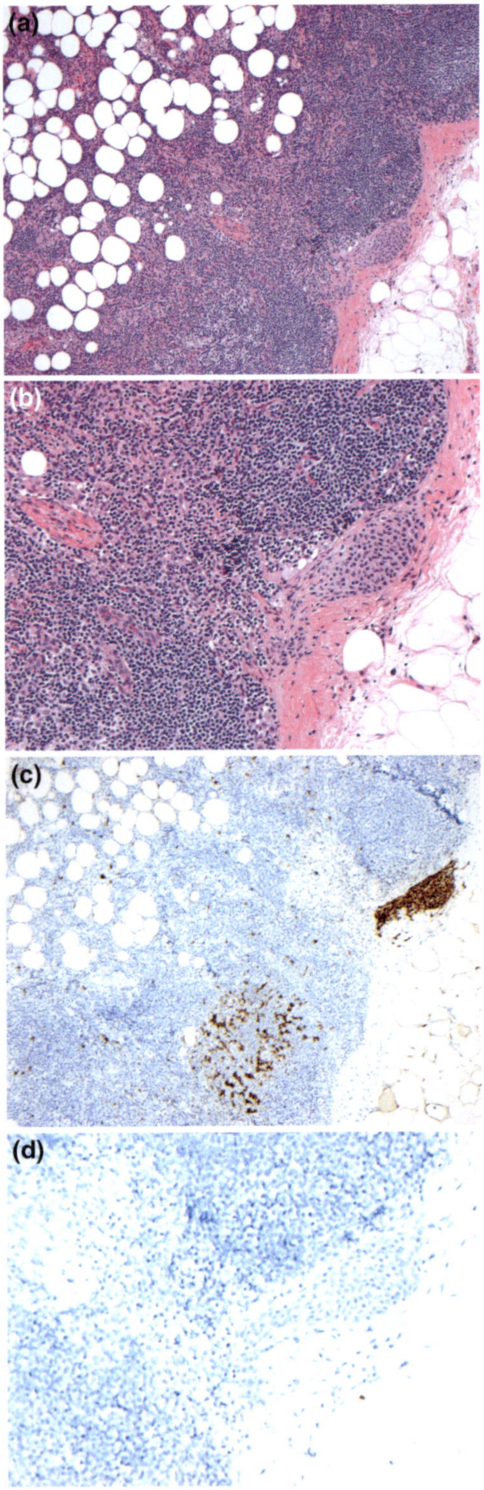


**Fig. 32.15** Benign nevus rests in lymph node capsule. **b** Cytologically banal nest of melanocytes are present in the fibrous capsule surrounding the lymphnode, **b** the nevic nuclei are small round to oval and uniform with little pleomorphism, **c** the nodal nevus and dendritic cells stain positively for S100, D. The nodal nevus does not stain for HMB-45 (same field as panel **c**)

cytoplasmic characteristics, nuclear and nucleolar features, degree of loss of cohesion, and proportions of cell types. In the past orbital enucleation allowed for more detailed histopathological analysis of ocular melanomas. Three histopathological subtypes of uveal melanoma have been described: spindle, mixed, and epithelioid (in order of worsening prognosis) [105]. Tumors composed predominantly of spindled cells comprise approximately 9 % of uveal melanomas, are slow growing, tightly cohesive and associated with a good prognosis. Approximately 5 % of uveal melanomas are composed of >50 % epithelioid cells with prominent eosinophilic nucleoli and ample cytoplasm are usually mitotically active, with dyscohesion and are associated with a poor prognosis. Mixed tumors, with 10–50 % epithelioid cells comprise the majority of uveal melanomas. Vasculogenic mimicry including the presence of closed vascular loops back to back on Periodic Acid–Schiff (PAS) stain is associated with increased mortality, and is often found in association with other negative prognostic indicators including epithelioid cell type and increased numbers of mitotic figures [106, 107].

Overall, important histopathologic features of ocular melanoma include mitotic count per 40 high power fields, mean diameter of the largest 10 nucleoli, presence of vasculogenic mimicry patterns including loops or other complex patterns, tumor infiltrating lymphocytes (more than 100 lymphocytes per 20 high power fields), and tumor infiltrating macrophages. In contrast to cutaneous melanoma, the presence of tumor infiltrating lymphocytes is a poor prognostic indicator in uveal melanoma. Basal tumor diameter, tumor height, presence of scleral invasion, and ciliary body involvement are important factors. Commonly seen characteristics include rupture of

Bruch's membrane (87.7 %), invasion of the retina (49.1 %), tumor cells in the vitreous (25.2 %), vortex vein invasion (8.9 %), invasion of tumor vessels by tumor cells (13.8 %), invasion into emissary canals (55.0 %), scleral invasion (55.7 %), and extrascleral extension 8.2 % [105]. Current AJCC staging guidelines are based primarily on tumor size and degree of extraocular spread.

#### 32.5.4.2 Mucosal Melanoma

Melanoma rarely arises in mucosal sites, in particular the genitourinary, oral, and sinonasal mucosa and gastrointestinal mucosa [108–114]. These tumors have a very poor prognosis. The in situ phase of mucosal melanomas is usually similar to that seen in acral lentiginous melanoma: an often subtle lentiginous intraepithelial melanocytic proliferation. Some mucosal melanomas have more extensive intraepidermal tumor with pagetoid spread and nesting similar to superficial spreading melanoma. Adequate biopsy and local control are challenges in mucosal sites. IHC stains may be helpful in assessing the specimen margins [108]. The invasive component of mucosal melanoma is usually composed of nests and expansile nodules of pleomorphic tumor cells. The tumor cells may have epithelioid or spindled morphology, and occasionally plasmacytoid, rhabdoid, or neural differentiation is seen. Numerous pigment-laden macrophages may be present, and the tumor cells may have variable degrees of pigmentation. There are numerous mitotic figures and extensive tumor necrosis may be seen.

### 32.5.5 The Use of Immunohistochemistry in Melanoma and Melanocytic Neoplasia

#### 32.5.5.1 Overview

IHC is a useful tool in the approach to melanocytic tumor diagnosis and staging. It is important to note that there are no reliable markers that distinguish benign from malignant melanocytic tumors. Most of the antibodies considered to detect "melanoma markers" actually detect pigment-related proteins that are present in normal melanocytes, benign nevi, and melanomas. These include Melan-A, MART-1, HMB-45, tyrosinase, and MITF (Fig. 32.16). S100 is the most sensitive IHC marker, present in 100 % of nevi and >99 % of melanomas whether primary or metastatic. The other markers show heterogeneous staining of 75–90 % of melanocytic tumors. These markers may be helpful in detection of small metastases in sentinel lymph nodes and also in discriminating metastatic melanoma from metastases of other tumor types. It is important to use a panel of markers when evaluating melanocytic proliferations because none of these markers routinely stain 100 % of the tumor cells. IHC stains to detect proliferation such as Ki-67 may identify a brisk proliferative activity in melanoma, an unusual finding in benign nevi. Markers that highlight the vasculature may aid in the detection of lymphovascular invasion. Finally, a few rare markers may be helpful in determining the presence or absence of specific mutations including BRAF V600E and BAP-1 or potential deletions such as with p16.

#### 32.5.5.2 Cutaneous Melanocytic Tumors

There are several settings wherein the use of immunohistochemistry (IHC) in primary cutaneous melanocytic tumors is a helpful addition to routine H&E staining: (1) evaluation of intraepidermal melanocytic proliferations in sun-damaged skin, (2) discrimination of spindled and desmoplastic dermal melanocytic proliferations, (3) evaluation of proliferation in dermal melanocytic proliferations, (4) detection of lymphovascular invasion, (5) evaluation of dermal epithelioid cell proliferations, and (6) detection of therapeutic targets.

In sun-damaged skin solar-induced cytological atypia of keratinocytes and melanocytes may obscure an intraepidermal melanocytic proliferation. In this setting, HMB-45 and MITF are useful markers for highlighting the extent of the intraepidermal melanocytic proliferation. A cytokeratin stain may also be helpful in confirming the presence of atypical keratinocytes and allow

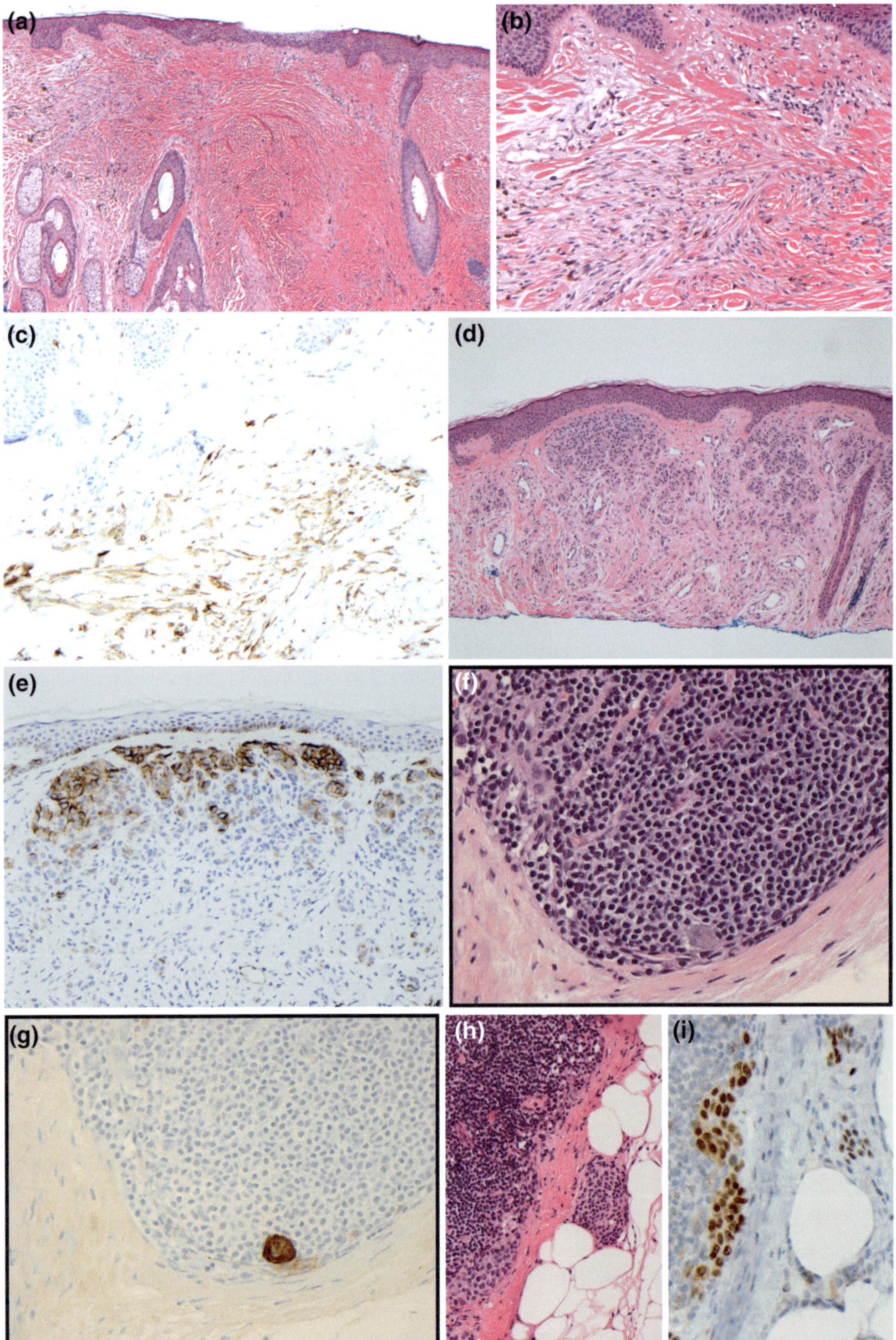
(a)
(b)
(c)
(d)
(e)
(f)
(g)
(h)
(i)

◀ **Fig. 32.16** Immunohistochemical staining in melanocytic tumors. **a** Blue nevus, a pigmented dendritic proliferation is present in the dermis, **b** blue nevus, the dermal tumor also has spindled cells, the differential diagnosis includes scar and desmoplastic melanoma, **c** blue nevus, the tumor stains positively for HMB-45 in a pattern supporting the diagnosis of blue nevus, **d** dermal nevus, **e** dermal nevus, HMB-45 stains only the superficial dermal tumor cells, **f** metastatic melanoma in a sentinel lymph node, **g** a single metastatic melanoma cell is identified on S100 stain, **h** nodal nevus and metastatic melanoma, **i** an MITF stain highlights the banal nuclear cytology of the intracapsular nodal nevus, in contrast to the pleomorphic large nuclei of the melanoma present in the subcapsular sinus

for the detection of nonstaining nests corresponding to the MITF or HMB-45 positive melanocytes. Notably, MART-1 and Melan-A stains are sensitive cytoplasmic stains that highlight melanocytic dendrites and in some cases may detect keratinocytic pigmentation. These stains may over estimate the density of an intraepidermal melanocytic proliferation in sun-damaged skin.

The differential diagnosis of dermal spindled and dendritic cell proliferations may include blue nevus, desmoplastic melanoma, and scar. IHC can be very helpful in this setting, the melanocytes of blue nevus will stain positively for S100 and HMB-45, whereas desmoplastic melanoma is usually positive for S100 but without staining for HMB-45, scar will not have large atypical S100 positive cells, but may have scattered S100 positive dermal dendrocytes, scar does not stain for HMB-45. Some authors have reported SOX-10 as a useful stain in discriminating desmoplastic melanoma from scar, however recent studies have shown SOX-10 staining in scar. The distinction of desmoplastic melanoma from neurofibroma may also be challenging, both have an S100+, HMB-45-immunophenotype. In the mixed form of desmoplastic melanoma, other melanocytic markers including MART-1 and Melan-A may help to highlight the nondesmoplastic portion of the tumor.

The presence of lymphovascular invasion in primary cutaneous melanoma is associated with a poor prognosis [89]. It is often difficult to distinguish vascular invasion from a peritumoral stromal retraction. Dual IHC for D240 and S100 allows for the identification of melanoma cells within lymphovascular spaces [115] (Fig. 32.13 C). CD31 may also be helpful in the evaluation of tumor in close apposition to vascular spaces as described in extravascular migratory metastases [87].

Occasionally cutaneous melanocytic proliferations are composed of dermal proliferations of pleomorphic epithelioid cells that may raise a broad differential diagnosis. In these cases there may be histopathological overlap between Spitz nevus, atypical Spitz tumor, nevoid melanoma, spitzoid melanoma, and nodular melanoma. There is no reliable IHC stain to routinely distinguish these cases, however in some cases the immunophenotype may contribute to the histopathological evaluation. HMB-45 shows a distinctive staining pattern in benign melanocytic nevi characterized by staining of the superficial dermal melanocytic proliferation and diminished staining with increasing tumor cell depth in the dermis. No tumor cell staining for HMB-45 is seen at the deep dermal aspect of the nevus. Melanoma on the other hand displays a patchy positive staining pattern for HMB-45 rather than the gradual gradient of staining observed in nevi. A subset of epithelioid cell melanocytic tumors that resemble Spitz nevi or nevoid melanomas have been described in patients with cutaneous/ocular melanoma, atypical melanocytic proliferations, and other internal neoplasms (COMMON syndrome), these tumors display loss of staining for BAP-1 [55]. Loss of p16 has also been described as a prognostic factor in melanoma, and some observers use the absence of p16 staining to triage cases for fluorescence in situ hybridization (FISH) analysis [116–118]. This is based on the result that in FISH analysis homozygous deletion of chromosome 9p21 is associated with a poor prognosis. In these cases p16 loss is seen immunohistochemically [119, 120]. Loss of p16, however, is not diagnostic of homozygous 9p21 deletion; in situ hybridization is helpful to confirm the copy number status of 9p21.

An evaluation of tumor cell proliferation may also be helpful in discriminating benign from malignant dermal melanocytic proliferations. Staining for Ki-67 when present in more than 20 % of the tumor cells supports a diagnosis of melanoma. This stain is also useful in identifying hot spots of proliferation in dermal melanocytic tumors. In cases with a prominent inflammatory infiltrate, it may be difficult to determine if Ki-67 is highlighting melanocytes or lymphocytes. Dual staining for a melanocytic marker, such as MART-1 with Ki-67 may be particularly helpful in this setting. Importantly, melanoma may have a low proliferation rate, therefore, the absence of significant Ki-67 positivity does not exclude the diagnosis of melanoma. Another stain that holds promise is anti-phosphohistone H3 (PHH3). By highlighting mitotic figures at any stage of mitosis this is a highly sensitive detection method. Additional studies that identify appropriate quantitative thresholds for reporting the results of anti-PHH3 are needed before it will be used in routine reporting of primary melanoma.

Finally, IHC staining may aid in identifying therapeutic targets. The detection of BRAFV600E in cutaneous melanoma or melanoma metastases may help to guide therapy [121]. Future studies may lead to evaluation of immune markers such as PD-1 and PD-L1, however, the clinical utility of these stains is yet to be demonstrated in a robust clinical study. Staining for c-Kit does not predict mutational status or sensitivity to tyrosine kinase inhibitors.

#### 32.5.5.3 Ocular Melanoma

Similar to cutaneous melanomas, melanocytic markers are the most sensitive IHC markers for uveal melanoma. These tumors stain uniformly positively for S100 and HMB-45, most uveal melanomas also stain for MART-1, MITF, melan-A, and tyrosinase [122]. Of note, melan-A and tyrosinase also stain normal uveal melanocytes in a variable manner, but with less intensity [122]. Given the importance of identifying mitotic figures in assessing ocular melanoma prognosis, staining for PHH3 may help assess mitotic count [123]. Some markers have been ascribed prognostic significance, including cyclin D1, which is associated with more aggressive course and histologically unfavorable disease. Cyclin D1 expression is present in 1–30 % of cases and is an independent risk factor for metastasis [124]. It is also possible to assess loss of expression of the BAP1 gene using IHC probes for its protein product; loss of BAP1 expression is correlated with poorer survival in uveal melanoma [56, 125].

#### 32.5.5.4 Melanoma Metastases and Sentinel Lymph Nodes

IHC is most helpful in the detection of microscopic metastases in sentinel lymph nodes and in the differentiation of metastatic melanoma from nonmelanoma metastases. S100 is the most sensitive melanoma marker, present in more than 99 % of melanomas; however, S100 is also present in some carcinomas and sarcomas and thus lacks specificity. Other markers of melanocytic differentiation, including HMB-45, MART-1, Melan-A, and MITF, are less sensitive (staining 75–90 % of melanomas) but are more specific than S100. A panel of stains is the most effective way of using these tests to assist in arriving at the correct diagnosis.

The histopathological analysis of sentinel lymph nodes includes the evaluation of H&E-stained tissue sections and IHC from multiple levels of each lymph node [101, 126, 127]. While there is marked variability in analytical platforms for the detection of melanoma in sentinel lymph nodes, some common practices exist: (1) submit the lymph node tissue entirely, (2) perform level sections deep into the block (beyond the initial set of tissue sections), and (3) use IHC. S100 and either MART-1 or Melan-A or most commonly employed; other less frequently used markers include HMB-45 and MITF. Additionally, some laboratories apply a cocktail of reagents including Melan-A/MART1 and HMB-45/Melan-A/Tyrosinase. Protocols that do not include levels into the block or IHC are associated with a false negative rate of approximately 15 % [101, 128–130]. Intraoperative frozen sections analysis is not a sensitive means of detecting melanoma in lymph nodes and is not recommended [131].

In some cases the presence of intranodal melanocytic rests may present a diagnostic challenge. To date most reports indicate that HMB-45 does not stain nodal nevi, similar to the absence of staining for HMB-45 deep dermal nevomelanocytes of a cutaneous nevus. Others have reported an absence of Ki-67 staining in nodal nevi, leading to the adoption of a double MART1/Ki-67 stain by some laboratories [132]. MITF and SOX10 are also helpful in the evaluation of nodal melanocytic tumors, these nuclear stains allow for the evaluation of nuclear size and pleomorphism.

Occasionally, melanoma metastases must be distinguished from metastatic carcinoma or sarcoma. Metastatic melanoma nearly always shows at least focal staining for S100 and in more than 75 % of cases will also demonstrate staining for other melanocytic markers including MART-1, Melan-A, HMB-45, MITF, SOX10, and tyrosinase. It is important to use a panel of markers because S100 may be present in some carcinomas and sarcomas and none of the melanocytic markers is 100 % specific for melanoma (Table 32.3).

**Table 32.3** Immunohistochemical markers in melanocytic neoplasia

| Antibody (protein) | Positive staining | Notable negatives | Comment |
|---|---|---|---|
| S100 | Diffusely positive in all nevi and primary melanoma, >99 % of melanoma metastases<br>Also positive in neural cells, dendritic cells, Langerhans cells, some carcinomas and sarcomas | <1 % of melanoma metastases | Highly sensitive, but not specific for melanocytic neoplasms, nuclear and cytoplasmic |
| HMB-45 (gp100, PMEL17) | Superficial aspect of cutaneous nevi, blue nevi, cutaneous melanoma, >75 % of metastatic melanoma<br>Perivascular epithelioid cell tumor (PEComa), Clear cell sarcoma, renal cell carcinoma | Deep dermal nevi, nevus rests in lymph nodes, desmoplastic melanoma<br>Neurofibroma | Patchy staining in cutaneous melanomas in contrast to gradient of diminished dermal staining in cutaneous nevi, except blue nevi which are diffusely positive<br>Nuclear and cytoplasmic |
| Melan-A (MART-1) | Diffusely positive in all nevi and primary melanoma, >85 % of melanoma metastases, nodal nevi<br>PEComa, clear cell sarcoma<br>A103 (Melan-A) stains adrenal cortical tumors and sex cord–gonadal tumors, M2–7C10 does not | Desmoplastic melanoma<br>Neurofibroma | Cytoplasmic stain usually highlights dendritic processes of intraepidermal melanocytes |
| MITF | Diffusely positive in nevi and primary melanoma, >90 % of melanoma metastases, nodal nevi<br>Histiocytes, follicular dendritic cells, mast cells, Schwann cells, clear cell sarcoma | | Nuclear stain |

(continued)

**Table 32.3** (continued)

| Antibody (protein) | Positive staining | Notable negatives | Comment |
|---|---|---|---|
| Tyrosinase | Superficial aspect of cutaneous nevi, cutaneous melanoma, >75 % of metastatic melanoma<br>Clear cell sarcoma | Deep aspect of cutaneous nevi, desmoplastic melanoma | Similar to HMB-45 |
| SOX-10 | Diffusely positive in all nevi and primary melanoma, >90 % of melanoma metastases<br>Also positive in neural cells, dermal dendritic cells | Follicular dendritic cells | Nuclear stain |
| MIB-1 (Ki-67 | <1 % in benign nevi and nodal nevi<br>>20 % in some melanomas<br>Lymphocytes<br>Basal keratinocytes | | Marker of cell proliferation |
| BRAFV600E | Positive in a subset of benign nevi, primary and metastatic melanomas | Negative in a subset of benign nevi, primary and metastatic melanomas | Reliable correlation with BRAFV600E mutation sequencing results |
| c-KIT | Present in a subset of mucosal melanomas | Absent in most melanomas | Poor correlation with c-KIT mutation |
| BAP-1 | Present in most nevi and melanomas, and subset of ocular melanoma<br>Present in lymphocytes | Loss in a subset of epithelioid dermal tumors with spitzoid or nevoid melanoma-like features, loss in some ocular melanoma | BAP-1 loss in cutaneous tumors is not associated with the poor survival that is seen in ocular melanomas with BAP-1 loss |
| P16 | Most cutaneous nevi including Spitz nevi, some melanomas | Loss in a subset of melanomas | Loss of p16 may prompt FISH analysis to look for homozygous 9p21 deletion |

## 32.6 Melanoma Staging

Melanoma staging as set forth by the AJCC is a tumor, node, metastasis (TNM) based scheme that segregates patients into prognostic categories [77, 133]. To date the therapeutic response data has little to contribute to melanoma staging, given that historically the best therapies were associated with complete responses in fewer than 25 % of patients. In the future, integration of newly identified immune and molecular features may lead to a staging scheme that segregates tumors by predicted outcome and potential response to specific therapy.

### 32.6.1 Cutaneous Melanoma

Most patients diagnosed with primary cutaneous melanoma do not have clinical evidence of metastases at the time of initial diagnosis. Staging in these patients is based upon the histopathological characteristics of the primary tumor and when appropriate evaluation of sentinel lymph nodes (Table 32.4). Primary tumor thickness, mitogenicity, and ulceration are the principle factors in cutaneous melanoma staging. The tumor (T) stage thickness cutoffs are 1.0, 2.0, and 4.0 mm. The mitotic count, expressed as number of dermal mitoses/mm$^2$ (n/mm$^2$), represents a strong and independent prognostic factor

**Table 32.4** Primary cutaneous melanoma staging (TNM, AJCC)

| | |
|---|---|
| T1a | <1.0 mm thickness, without ulceration, mitoses <1/mm$^2$ |
| T1b | <1.0 mm thickness, with ulceration and/or mitoses > or = to 1/mm$^2$ |
| T2a | 1.01–2.0 mm thickness, without ulceration |
| T2b | 1.01–2.0 mm thickness, with ulceration |
| T3a | 2.01–4.0 mm thickness, without ulceration |
| T3b | 2.01–4.0 mm thickness, with ulceration |
| T4a | >4.0 mm thickness, without ulceration |
| T4b | >4.0 mm thickness, with ulceration |
| N0 | No regional lymph node metastases detected |
| N1a | 1 regional lymph node metastasis, micrometastasis[a] |
| N1b | 1 regional lymph node metastasis, macrometastasis[b] |
| N2a | 2-3 regional lymph node metastases, micrometastasis[a] |
| N2b | 2-3 regional lymph node metastases, macrometastasis[b] |
| N2c | 2-3 regional lymph node metastases, in-transit met(s)/satellite(s) without metastatic nodes |
| N3 | >4 regional lymph node metastases, or matted nodes, or in-transit met(s)/satellite(s) with metastatic lymph node(s) |
| M0 | No detectable evidence of distant metastasis |
| M1a | Metastases to skin, subcutaneous or distant lymph nodes |
| M1b | Metastases to lung |
| M1c | Metastases to all other visceral sites or distant metastases to any site combined with an elevated serum LDH |

[a]Micrometastases are diagnosed after sentinel lymph node biopsy and completion lymphadenectomy (if performed)
[b]Macrometastases are defined as clinically detectable nodal metastases confirmed by therapeutic lymphadenectomy or when nodal metastasis exhibits gross extracapsular extension
Balch et al. [77]. Reprinted with permission. 

in this patient subset, particularly those with thin <1.0 mm tumors [74, 75, 134]. The presence of ulceration in the primary tumor has been described as a highly significant poor prognostic indicator as early as 1953 [135, 136]. Ulceration upgrades the T classification from "Ta" to "Tb." For T1 melanomas ulceration is used along with the presence or absence of dermal mitoses. The anatomic level of invasion as described by Wallace Clark has been discontinued as a primary determinant of T staging because level of invasion loses significance when mitotic count and ulceration are included in the analysis [77]. Currently, the presence of dermal tumor cell mitosis and/or ulceration are used as the principle criteria for considering a melanoma as T1b. The Clark level of invasion is only used when the mitotic count cannot be reliably determined due to poor preparation of the histology slides, e.g., when tissue sections are cut too thick or overstained and mitotic figures cannot be distinguished, or in poorly oriented tissues where thickness cannot be measured accurately. In the histopathological and staging analysis, patients with intralymphatic metastases (microscopic satellites, satellite metastases or in-transit metastases) without nodal metastases are classified as N2c (stage IIIB or IIIC), while those with combined intralymphatic metastases and nodal metastases are N3 (stage IIIC) and have a lower survival rate (Table 32.5).

The identification of clinically occult nodal metastases is the most important independent predictor of prognosis in patients with stage I and II melanoma [137]. In general, patients with clinically localized AJCC stage Ib or greater primary tumors are offered sentinel lymph node sampling. Exceptions may include patients with

**Table 32.5** Anatomic stage groupings for cutaneous melanoma (AJCC 7th edition)

| | Clinical Staging[a] | | | | Pathologic Staging[b] | | |
|---|---|---|---|---|---|---|---|
| | T | N | M | | T | N | M |
| 0 | Tis | N0 | M0 | 0 | Tis | N0 | M0 |
| IA | T1a | N0 | M0 | IA | T1a | N0 | M0 |
| IB | T1b<br>T2a | N0<br>N0 | M0<br>M0 | IB | T1b<br>T2a | N0<br>N0 | M0<br>M0 |
| IIA | T2b<br>T3a | N0<br>N0 | M0<br>M0 | IIA | T2b<br>T3a | N0<br>N0 | M0<br>M0 |
| IIB | T3b<br>T4a | N0<br>N0 | M0<br>M0 | IIB | T3b<br>T4a | N0<br>N0 | M0<br>M0 |
| IIC | T4b | N0 | M0 | IIC | T4b | N0 | M0 |
| III | Any T | N > N0 | M0 | IIIA | T1–4a<br>T1–4a | N1a<br>N2a | M0<br>M0 |
| | | | | IIIB | T1–4b<br>T1–4b<br>T1–4a<br>T1–4a<br>T1–4a | N1a<br>N2a<br>N1b<br>N2b<br>N2c | M0<br>M0<br>M0<br>M0<br>M0 |
| | | | | IIIC | T1–4b<br>T1–4b<br>T1–4b<br>Any T | N1b<br>N2b<br>N2c<br>N3 | M0<br>M0<br>M0<br>M0 |
| IV | Any T | Any N | M1 | IV | Any T | Any N | M1 |

[a]Clinical staging includes microstaging of the primary melanoma and clinical/radiologic evaluation for metastases. By convention, it should be used after complete excision of the primary melanoma with clinical assessment for regional and distant metastases

[b]Pathologic staging includes microstaging of the primary melanoma and pathologic information about the regional lymph nodes after partial (i.e., sentinel node biopsy) or complete lymphadenectomy. Pathologic stage 0 or stage IA patients are the exception; they do not require pathologic evaluation of their lymph nodes



very thin tumors and only one mitosis. In the current AJCC guidelines, the diagnosis of metastatic melanoma may be made if one or more melanoma cell is identified on any tissue section, whether stained with H&E or with IHC [77]. Patients with sentinel lymph node metastases are generally offered completion lymphadenectomy and adjuvant therapy.

The aims of sentinel lymph node biopsy include staging and regional disease control. Complications occur in 4–10 % of patients compared to the complication rate of 23–37 % associated with complete lymph node dissection [138]. The false negative rate is estimated to be less than 5 % [139]. The recurrence rate for patients with positive sentinel nodes and subsequent completion lymphadenectomy is less than 10 %, in contrast to reported recurrence rates of 20–50 % in patients with lymphadenectomy for clinically palpable tumor [138, 140]. Nevertheless, a clear survival benefit of sentinel lymph node mapping has not been demonstrated in a statistically robust randomized trial [141]. On the other hand, the staging and prognostic value of sentinel lymph node status is not disputed. Tumor burden, regardless of the method of measurement, has been demonstrated to correlate with risk of positive lymph nodes in the remainder of the basin, and also with overall survival [142–144]. Most patients with regional lymph node metastases (AJCC Stage III) have micrometastases detected histopathologically, only 20 % of patients are diagnosed with clinically detectable macrometastases. The 5-year survival for patients with histopathological and clinical detection of regional lymph node metastasis is 67 and 43 %, respectively. Patients with microscopic metastases have widely varied prognosis; in these patients multivariate analysis reveals that the number of tumor-containing lymph nodes, total metastasis size, primary tumor thickness, ulceration, tissue site, and patient age are independent predictors of survival. On the other hand, for patients with nodal macrometastases independent predictors of survival are number of tumor-containing nodes, primary tumor ulceration, and patient age.

Elevation of serum lactate dehydrogenase (LDH) has been identified as an independent and highly significant predictor of survival in patients with stage IV melanoma. LDH and the site of distant metastasis are used to define the M categories. M1a includes patients with metastases to the skin, subcutaneous tissues, nonregional lymph nodes and with a normal serum LDH. M1b includes patients with metastases to the lung and normal LDH levels. M1c is used for patients with elevated serum LDH and/or visceral metastases (other than the lung) [77]. While the number of metastases has been documented as an important prognostic factor, this was not included in the current schema due to significant variability in the use of tests to detect distant metastases.

### 32.6.2 Ocular Melanoma

Uveal melanoma differs from cutaneous melanoma; there is no in situ component or basement membrane zone invasion. Lymphatics are not present in the eye, metastases occur hematogenously usually to liver, lung, and bone. The mortality rate at 15 years is approximately 50 %, and over 93 % of patients who die have liver metastases [145]. The location of uveal melanoma is of prognostic significance. Tumors of the iris have the best prognosis, followed by choroidal melanoma, with ciliary body melanomas having the worst prognosis. The AJCC Staging of uveal melanoma takes into account tumor diameter, thickness, degree of ciliary or extraocular involvement, and presence of nodal or distant metastases (Fig. 32.17; Table 32.6).

### 32.6.3 Mucosal Melanoma

Melanoma arising in mucosa of the head and neck, genitourinary tract, and gastrointestinal tract is rare. The AJCC describes a staging scheme for mucosal melanoma of the head and neck that is based on the extent of tumor invasion, regional and distant metastases. The primary tumor T stage is defined as T3: mucosal disease, T4a: moderately advanced disease involving the deep soft tissue, cartilage, bone or overlying skin, and T4b: very advance disease

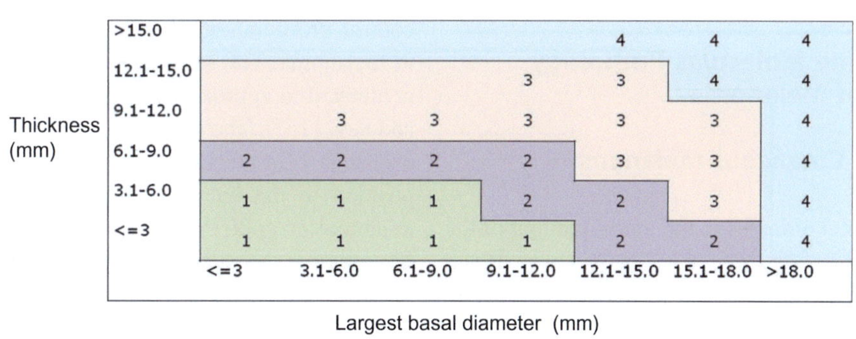


**Fig. 32.17** AJCC T stage of ciliary body and choroid uveal melanoma based on largest basal diameter and thickness in millimeters. *X*-axis is largest basal diameter, and *Y*-axis is thickness in millimeters (mm). Size categorization into categories 1–4 is then used in TNM staging and prognostic groups (see Table 32.6)

**Table 32.6** Uveal melanoma anatomic stage and prognostic groups

| | | | |
|---|---|---|---|
| Stage I | T1a | N0 | M0 |
| Stage IIA | T1b-d<br>T2a | N0 | M0 |
| Stage IIB | T2b<br>T3a | N0 | M0 |
| Stage IIIA | T2c-d<br>T3b-c<br>T4a | N0 | M0 |
| Stage IIIB | T3d<br>T4d-e | N0 | M0 |
| Stage IIIC | T4d-e | N0 | M0 |
| Stage IV | Any T<br>Any T | N1<br>Any N | M0<br>M1a-c |

T stage as seen in Fig. 32.17. a-d are varying degrees of ciliary and extraocular involvement

involving brain, dura, skull base, lower cranial nerves (IX, X,XI,XII), masticator space, carotid artery, prevertebral space, or mediastinal structures. Regional lymph nodes are defined as NX: regional lymph nodes cannot be assessed, N0: no regional node metastasis, and N1: regional lymph node metastasis present. Distant metastases are defined as M0: No distant metastasis, and M1: distant metastasis present. Patients with mucosal melanoma of the head and neck without extension to underlying or overlying structures and without lymph node or distant metastases are Stage III; patients with locally infiltrative disease and/or metastases have Stage IV melanoma.

## 32.7 The Molecular Pathology of Melanoma

### 32.7.1 Cutaneous Melanoma

Recent breakthroughs in the understanding of the molecular pathogenesis of melanoma have driven unprecedented advances in melanoma treatment. Melanoma is derived from melanocytes, pigment-synthesizing neural crest derived cells. Melanocytes in the skin function to synthesize melanin and transfer mature melanosomes to keratinocytes via extensive cytoplasmic dendritic processes. The pigments produced by melanocytes have been classified as pheomelanin (red/blonde) and eumelanin (brown/black). Pheomelanin has been associated with an increase in reactive oxygen species in the skin, whereas eumelanin may provide UV protection [146]. UV light-induced DNA damage is followed by stabilization of p53 and activation of the proopiomelanocortin (*POMC*) gene [147, 148]. Posttranslational processing of POMC into small peptides, including melanocyte-stimulating hormone (MSH), leads to stimulation of the melanocortin receptor 1 (MC1R) in melanocytes. Activated MC1R leads to melanocytic cyclic adenosine monophosphate induction and stimulation of microphthalmia-associated transcription factor (MITF).

MITF, known as the master regulator of melanocyte development, is responsible for activation of pigment producing enzymes (including HMB-45 and MART-1) and melanosome packaging and secretion [149]. Genomic amplification of *MITF* activates melanoma oncogenes [150]. MITF also directly regulates the antiapoptotic genes BCL2 and BCL2A1 and the cell cycle regulator CDK2 [151]. MITF is directly phosphorylated by mitogen-activated protein kinase (MAPK) and is thus linked to BRAF and NRAS [152].

Multiple molecular events occur in the development of melanoma from melanocytes, including mutation and amplification of oncogenes. Several such oncogenes have been described to lead to activation of signaling pathways that control proliferation, survival, and angiogenesis in melanoma. The serine/threonine kinase *BRAF* is mutated in approximately 50 % of melanomas and nevi [153]. The most common mutation is an exchange of glutamic acid for valine at the 600 position in the kinase domain (V600E). Other mutations at V600 (V600K) have been described. Mutations at this site account for more than 90 % of *BRAF* mutations in melanoma and are associated with constitutive activation of the MAPK pathway [154, 155]. *NRAS* mutation is observed in 15–25 % of patients with melanoma and congenital nevi and is usually mutually exclusive to BRAF mutation [154]. NRAS is regulated by the tumor suppressor gene neurofibromin 1 (NF1) and mutations of NF1 may also be associated with activation of the MAPK and phosphoinositol-3-kinase (PI3K) pathways [156, 157]. The tyrosine kinase receptor c-Kit is most commonly mutated in acral lentiginous and mucosal lentiginous melanoma, although in a minority of cases [158–162]. c-Kit mutations are usually seen in the transmembrane domains coded in exons 11 and 13 and lead to constitutive activation of the receptor. Signal transducing G-protein-coupled receptors are mutated in most patients with uveal melanoma. Activating mutations in GNAQ and GNA11 lead to protein kinase C activation, MEK phosphorylation, and triggering of the MAPK pathway [163, 164]. GNAQ mutations are also common in blue nevi [163]. Mutations in phosphatase and tensin homologue (*PTEN*) are found in 25 % of melanoma patients, usually along with mutant BRAF [165]. PTEN loss has been associated with PI3 K signaling leading to reduced effectiveness of BRAF inhibitor therapy [166]. HRAS mutations have been described most commonly in Spitz nevi, but in fewer than 20 % of these tumors. BRCA1-Associated Protein1 (BAP-1) is a tumor suppressor that has been found to be mutated in atypical epithelioid Spitz tumors/nevoid melanoma and uveal melanoma [55, 167–170].

Dysregulation of the cell cycle is common in melanoma. The most common abnormalities are in $p16^{INK4A}$ expression or function. More than 50 % of melanomas have deletion or mutation of CDNK2A which codes for $p16^{INK4A}$ and loss of p16 occurs in approximately 15 % of cases [171, 172]. $p16^{INK4A}$ is a cyclin-dependent kinase (CDK) inhibitor, and dysregulation leads to *CDK4* overexpression and cell cycle abnormalities.

Abnormalities in the above-noted genes lead to dysfunction in cell signaling pathways. The MAPK signaling pathway plays a critical role in the pathobiology of melanoma [154]. Canonical activation of the MAPK pathway begins with cell surface receptor–ligand engagement which leads to RAS activation, and binding of activated RAS to BRAF triggers dimerization and activation of the RAF serine–threonine kinase domain [173]. RAF phosphorylates MAPK kinase (MEK) and activated MEK leads to activation of extracellular signal–regulated kinase (ERK). ERK activation results in cell cycle progression and survival. In the setting of BRAF mutation the CDK inhibitor $p27^{KIP1}$ is downregulated and cyclin D1 is constitutively activated leading to increased cell proliferation [174]. MAPK signaling also leads to reduced apoptosis in melanoma. The PI3 K pathway results in Akt activation which also promotes cell proliferation, increased cell survival and angiogenesis. PI3 K pathway activation may occur via NRAS activation, NF1 mutation, AKT amplification, or loss of PTEN function [175]. The Wnt pathway activator nuclear β[beta]-catenin is detected in nearly 30 % of melanomas [176, 177]. Targeted therapy has shown great promise in patients with metastatic melanoma based on the molecular abnormalities described above. Similarly, resistance mechanisms have been elucidated through identification of cross talk between signaling pathways.

Epigenetics is defined as the processes leading to changes in gene expression other than those caused by alterations in the DNA sequence. The epigenome determines which genes are expressed or kept silent through altering chromatin structure by covalent modification of DNA bases or histone proteins, or regulating mRNA translation through noncoding RNA. Cancer pathogenesis is directed, in part, through dysregulated DNA methylation, DNA demethlyation/hydroxymethylation, histone modification, and noncoding RNAs. It is hypothesized that the epigenome may thus provide a link between hardwired genetics and factors such as environment and aging.

DNA methylation occurs on cytosine residues preceding guanine on the ipsilateral strand forming CpG pair. Regions enriched in CpG repeats are termed CpG islands and tend to cluster near gene promoters [178]. For the most part promoter methylation is associated with gene silencing [179]. Hypermethylation of *CDKN2a* occurs in melanoma and is associated with increased proliferation and reduced patient survival [180]. In contrast to DNA methylation, DNA demethylation is poorly understood. The first step appears to be oxidation of 5-methylcytosine (5-mC) to 5-hydroxymethylcytosine (5-hmC) via the Ten

Eleven Translocase (TET) family dioxygenase enzymes [181]. The activity and expression of 5-hmC and TET are tightly regulated during embryonic stem cell differentiation. TET may be considered a guardian of the epigenome due to its role in maintaining DNA methylation fidelity through enabling DNA demethylation repair [182]. Loss of TET function and 5-hmC with hypermethylation of tumor suppressor genes may have a role in melanocytic tumor progression. Indeed, loss of 5-hmC is observed immunohistochemically with melanoma tumor progression; 5-hmC is expressed in high levels in benign nevi [183]. Histone modifications may activate or silence transcription by controlling the access of DNA to the transcriptional machinery [184]. Histone hypoacetylation downregulates proapoptotic proteins including the Bcl-2 family (Bim, Bax and Bak) as well as tumor suppressor genes that negatively regulate the PI3 K signaling pathway [185, 186]. Aberrant histone methylation may lead to loss of p16 in melanoma via increased expression of EZH2 [187]. Noncoding RNAs comprise more than 90 % of the genome [188, 189]. Those noncoding RNAs less than 200 nucleotides in length are termed microRNAs (miRNAs) and those between 200 and 200 kilobases in size are termed long noncoding RNAs (lncRNAs) [190]. Numerous miRNAs have been described in melanoma with wide ranging effects including tumor suppression, pro-oncogenic, and prometastatic. miR-200c is reported to be downregulated in primary melanomas and metastases compared to nevi [191]. miR-200c overexpression downregulates Bmi-1 and inhibits melanoma metastasis in vitro. miR-1908, miR-199a-5p, and miR-199a-3p target apolipoprotein E which suppresses invasion and metastasis [192]. Patients with elevated levels of these miRNAs in their primary tumors have shorter metastasis-free survival. Similar to miRNAs, lncRNAs may promote melanomagenesis via tumor suppression, pro-oncogenic and prometastatic functions. HOTAIR is an lncRNA that is overexpressed in metastatic melanoma compared to primary tumors. The lncRNA may facilitate changes to chromatin structure through scaffolding interactions with histone-modifying enzymes [193]. Overall, as the melanoma epigenome continues to unfold this knowledge is likely to translate into significant therapeutic breakthroughs in the treatment of melanoma.

In addition to providing targets for development of novel therapies, cytogenetic analysis may aid in the discrimination of benign from malignant melanocytic proliferations. The gold standard for diagnosis is histological evaluation of H&E-stained tissue sections, and this analysis yields the correct diagnosis in most cases. There exists, however, a subset of melanocytic proliferations with diagnostically challenging histopathological features for which even experts cannot arrive at a consensus diagnosis. Advances in molecular and cytogenetic analysis have gained popularity in this realm.

Comparative genomic hybridization (CGH) is a method that evaluates copy number changes in the genome. CGH may be performed on formalin-fixed, paraffin-embedded tissues, however, the amount of tumor tissue needed may exceed that available for small melanocytic tumors. In cases where sufficient tumor DNA is isolated, most melanomas display chromosomal aberrations, including copy number gains of 1q, 6p, 7, 8q, 17q, and 20q and losses of chromosomes 6q, 8p, 9p, and 10q [194, 195]. Nevi generally do not have copy number variations detected by CGH with the exception of spindled and epithelioid cell nevi (Spitz) which may have gains at 11p or loss of chromosome 3 [168, 196, 197]. Although generally the above distinctions exist for many cases, even lethal melanomas occasionally will not have aberrations detected by CGH. Cytogenetic results should be evaluated as one of many factors including the findings on routine histology, the clinical features, and occasionally IHC results.

Fluorescence in situ hybridization (FISH) uses oligonucleotide probes to detect specific chromosomal targets. This method can also be performed on formalin-fixed, paraffin-embedded tissues and has the advantage over CGH of requiring less tumor tissue. In contrast to the genome-wide, unbiased approach of aCGH, FISH analysis is limited to only 4–6 preselected targets. FISH probes may target the centromeric region of a specific chromosome, allowing for copy number analysis, or may target a specific sequence, such as CDKN21 (p16) on 9p21 [198]. Recent advances have led to the development of FISH tests that identify genomic abnormalities in melanocytic tumors [119, 120, 199]. While single deletions of 9p21 may be seen in benign nevi, homozygous deletions are rare. The use of FISH to evaluate tumors for homozygous deletion of 9p21 has significantly increased the sensitivity of this test for melanoma. Additionally, the use of markers to detect multiple distinct chromosomes allows for the detection of tetraploidy, reducing the risk of false positives that occur when only one or two chromosomes are targeted. FISH test targets include RREB1 on 6p25, MYC on 8q24, p16 on 9p21, CCND1 on 11q13, MYB on 6q23 and Cen9, targeting the chromosome 9 centromere. Copy number gains in MYC and CCND1 correlate with poor prognosis, and homozygous deletion of p16 strongly supports the diagnosis of melanoma [120, 199]. Limitations of the technology are the interobserver variations in analyzing the nuclei for FISH signal, variations in cutoff values considered to represent a positive test, and technical variability. FISH tests that target chromosome 3 may be useful in evaluating uveal melanoma and also in the evaluation of nevoid melanomas and spitzoid melanocytic tumors in patients with cutaneous/ocular melanoma, atypical melanocytic proliferations, and other internal neoplasms (COMMON syndrome) [55, 168–170, 200]. FISH analysis of EWSR1 translocation 22q21 is also helpful in the diagnosis of clear cell sarcoma [201].

### 32.7.2 Ocular Melanoma

Improved understanding of the genetic mutations, chromosomal aberrations, and alterations in gene expression profiles have resulted in a dramatically improved ability to prognosticate ocular, particularly uveal, melanomas. Although somatic mutations in *BRAF* and *NRAS* are often found in cutaneous melanoma, the mutational profile of uveal melanoma differs significantly. There are three common gene mutations identified in uveal melanoma: GNAQ [164], GNA11 [163], and BAP1 [200]. Mutation in GNAQ, a gene coding for a Gαq stimulatory subunit, is an early event in tumorigenesis and is present in 46 % of uveal melanomas. Mutations occur almost exclusively in codon 209 of the protein's Ras-like domain, resulting in constitutive activation and transformation the GNAQ gene into a dominant oncogene [164]. GNA11, a paralogue of GNAQ, is mutated in 32 % of primary uveal melanomas and 57 % of uveal melanoma metastases. Of note, mutations of GNAQ and GNA11 are mutually exclusive [163]. Finally, inactivating somatic mutations of BAP1 (BRCA1-associated protein 1; located on chromosome 3p21.1) are believed to be a key event in development of metastatic potential in uveal melanoma. Although GNAQ mutations occur early in melanomagenesis and do not correlate with prognosis, BAP1 mutations are strongly associated with poor outcome. Studies have shown significantly lower expression levels in tumors that metastasize as compared to those that do not [200].

Karyotypic analysis has revealed recurrent abnormalities on chromosomes 3, 6, 8, and 1 [202]. Abnormalities in these chromosomes, particularly chromosome 3, have been found to be predictive of metastatic potential and patient outcome [202–204]. Monosomy 3 is present in 50 % of uveal melanomas overall [202, 205]. Loss of chromosome 3 is seen in 70 % of uveal melanomas that metastasize, but only 20 % in those that do not [204]; it hence correlated with metastatic disease and poor survival [202, 203, 206]. Of note, monosomy 3, associated with poor prognosis and increased metastatic potential, and gain of 6p, which correlates with improved prognosis, are mutually exclusive. Because both occur early in tumorigenesis, this suggests an early bifurcation in the path toward melanoma and metastasis [207].

Gene expression profiling is a powerful tool, and has allowed separation of uveal melanomas into two classes: class 1, low grade tumors with low risk of metastasis, and class 2 tumors, high grade tumors with high risk of metastasis. These tumor types have markedly different clinical profiles: type 1 tumors have a 92-month survival rate of 95 %, while class 2 have a survival rate of 31 % [208, 209]. Of note, 85 % of class 2 tumors have BAP1 mutation; BAP1 mutation may herald conversion to class 2 [200]. In uveal melanoma, gene expression profiling has been shown to be superior to both chromosomal analysis and immunohistochemical analysis in predicting patient outcomes [210, 211], and has the added advantage that it can be performed on small amounts of tissue acquired via fine needle aspiration [212].

## 32.8 Summary

Melanoma has significant societal impact because it affects patients at a relatively young age and when found to be metastatic there is no reliably effective treatment. The histopathological findings in the primary tumor are the principle factors that determine treatment, guiding surgical plans, and in some cases adjuvant therapy. Melanoma staging is dependent upon histological features observed in routinely processed formalin-fixed and paraffin-embedded tissue sections. Promising recent discoveries in the pathogenesis of melanoma and identification of distinct molecular phenotypes may lead to the development of molecularly based staging schemas. Current research studies promise a more molecularly integrated approach to melanoma treatment in the future.

## References

1. Gleason BC, Crum CP, Murphy GF. Expression patterns of MITF during human cutaneous embryogenesis: evidence for bulge epithelial expression and persistence of dermal melanoblasts. J Cutan Pathol. 2008;35:615–22.
2. Adameyko I, Lallemend F, Aquino JB, Pereira JA, Topilko P, Muller T, et al. Schwann cell precursors from nerve innervation are a cellular origin of melanocytes in skin. Cell. 2009;139:366–79.
3. Kawakami A, Fisher DE. Key discoveries in melanocyte development. J Invest Dermatol. 2011;131:E2–4.
4. Baran JL, Duncan LM. Combined melanocytic nevi: histologic variants and melanoma mimics. Am J Surg Pathol. 2011;35:1540–8.
5. Lynch HT, Frichot BC, Lynch JF. Familial atypical multiple mole-melanoma syndrome. J Med Genet. 1978;15:352–6.
6. Clark WH Jr, Reimer RR, Greene M, Ainsworth AM, Mastrangelo MJ. Origin of familial malignant melanomas from heritable melanocytic lesions. 'The B-K mole syndrome'. Arch Dermatol. 1978;114:732–8.
7. Clemente C, Cochran AJ, Elder DE, Levene A, MacKie RM, Mihm MC, et al. Histopathologic diagnosis of dysplastic nevi: concordance among pathologists convened by the World Health Organization Melanoma Programme. Hum Pathol. 1991;22:313–9.
8. Duncan LM, Berwick M, Bruijn JA, Byers HR, Mihm MC Jr, Barnhill RL. Histopathologic recognition and grading of dysplastic melanocytic nevi: an interobserver agreement study. J Invest Dermatol. 1993;100:318S–21S.
9. Frangos J, Duncan L, Piris A, Nazarian R, Mihm M, Hoang M, et al. Increased diagnosis of thin superficial spreading melanomas: a 20 year study. J Am Acad Dermatol. 2012;67:387–94.
10. Spitz S. Melanomas of childhood. Am J Pathol. 1948;24:591–609.
11. Barnhill RL, Barnhill MA, Berwick M, Mihm MC Jr. The histologic spectrum of pigmented spindle cell nevus: a review of 120 cases with emphasis on atypical variants. Human Pathol. 1991;22:52–8.
12. Smith KJ, Barrett TL, Skelton HG 3rd, Lupton GP, Graham JH. Spindle cell and epithelioid cell nevi with atypia and metastasis (malignant Spitz nevus). Am J Surg Pathol. 1989;13:931–9.
13. Barnhill RL. The spitzoid lesion: the importance of atypical variants and risk assessment. Am J Dermatopathol. 2006;28:75–83.
14. Hung T, Piris A, Lobo A, Mihm MC Jr, Sober AJ, Tsao H, et al. Sentinel lymph node metastasis is not predictive of poor outcome in patients with problematic spitzoid melanocytic tumors. Hum Pathol. 2013;44:87–94.
15. Luo S, Sepehr A, Tsao H. Spitz nevi and other Spitzoid lesions part II. Natural history and management. J Am Acad Dermatol. 2011;65:1087–92.
16. Luo S, Sepehr A, Tsao H. Spitz nevi and other Spitzoid lesions part I. Background and diagnoses. J Am Acad Dermatol. 2011;65:1073–84.
17. Barnhill RL, Argenyi ZB, From L, Glass LF, Maize JC, Mihm MC Jr, et al. Atypical Spitz nevi/tumors: lack of consensus for diagnosis,

discrimination from melanoma, and prediction of outcome. Hum Pathol. 1999;30:513–20.
18. Cerroni L, Barnhill R, Elder D, Gottlieb G, Heenan P, Kutzner H, et al. Melanocytic tumors of uncertain malignant potential: results of a tutorial held at the XXIX symposium of the International Society of Dermatopathology in Graz, October 2008. Am J Surg Pathol. 2010;34:314–26.
19. Wechsler J, Bastuji-Garin S, Spatz A, Bailly C, Cribier B, Andrac-Meyer L, et al. Reliability of the histopathologic diagnosis of malignant melanoma in childhood. Arch Dermatol. 2002;138:625–8.
20. Barnhill RL. The Spitzoid lesion: rethinking Spitz tumors, atypical variants, 'Spitzoid melanoma' and risk assessment. Mod Pathol. 2006;19(Suppl 2):S21–33.
21. Zedek DC, McCalmont TH. Spitz nevi, atypical spitzoid neoplasms, and spitzoid melanoma. Clin Lab Med. 2011;31:311–20.
22. Ali L, Helm T, Cheney R, Conroy J, Sait S, Guitart J, et al. Correlating array comparative genomic hybridization findings with histology and outcome in spitzoid melanocytic neoplasms. Int J Clin Exp Pathol. 2010;3:593–9.
23. Busam KJ, Murali R, Pulitzer M, McCarthy SW, Thompson JF, Shaw HM, et al. Atypical spitzoid melanocytic tumors with positive sentinel lymph nodes in children and teenagers, and comparison with histologically unambiguous and lethal melanomas. Am J Surg Pathol. 2009;33:1386–95.
24. Gamblin TC, Edington H, Kirkwood JM, Rao UN. Sentinel lymph node biopsy for atypical melanocytic lesions with spitzoid features. Ann Surg Oncol. 2006;13:1664–70.
25. Lohmann CM, Coit DG, Brady MS, Berwick M, Busam KJ. Sentinel lymph node biopsy in patients with diagnostically controversial spitzoid melanocytic tumors. Am J Surg Pathol. 2002;26:47–55.
26. Ludgate MW, Fullen DR, Lee J, Lowe L, Bradford C, Geiger J, et al. The atypical Spitz tumor of uncertain biologic potential: a series of 67 patients from a single institution. Cancer. 2009;115:631–41.
27. Murali R, Sharma RN, Thompson JF, Stretch JR, Lee CS, McCarthy SW, et al. Sentinel lymph node biopsy in histologically ambiguous melanocytic tumors with spitzoid features (so-called atypical spitzoid tumors). Ann Surg Oncol. 2008;15:302–9.
28. Raskin L, Ludgate M, Iyer R, Ackley T, Bradford C, Johnson T, et al. Copy number variations and clinical outcome in atypical spitz tumors. Am J Surg Pathol. 2011;35:243–52.
29. Sepehr A, Chao E, Trefrey B, Blackford A, Duncan LM, Flotte TJ, et al. Long-term outcome of Spitz-type melanocytic tumors. Arch Dermatol. 2011;147:1173–9.
30. Su LD, Fullen DR, Sondak VK, Johnson TM, Lowe L. Sentinel lymph node biopsy for patients with problematic spitzoid melanocytic lesions: a report on 18 patients. Cancer. 2003;97:499–507.
31. Tom WL, Hsu JW, Eichenfield LF, Friedlander SF. Pediatric "STUMP" lesions: evaluation and management of difficult atypical Spitzoid lesions in children. J Am Acad Dermatol. 2011;64:559–72.
32. Urso C, Borgognoni L, Saieva C, Ferrara G, Tinacci G, Begliomini B, et al. Sentinel lymph node biopsy in patients with "atypical Spitz tumors". A report on 12 cases. Hum Pathol. 2006;37:816–23.
33. Berk DR, Johnson DL, Uzieblo A, Kiernan M, Swetter SM. Sentinel lymph node biopsy for cutaneous melanoma: The Stanford experience, 1997–2004. Arch Dermatol. 2005;141:1016–22.
34. Kamino H. Spitzoid melanoma. Clin Dermatol. 2009;27:545–55.
35. Paradela S, Fonseca E, Pita-Fernandez S, Kantrow SM, Diwan AH, Herzog C, et al. Prognostic factors for melanoma in children and adolescents: a clinicopathologic, single-center study of 137 Patients. Cancer. 2008;116:4334–44.
36. Barnhill RL, Gupta K. Unusual variants of malignant melanoma. Clin Dermatol. 2009;27:564–87.
37. Crotty KA, Scolyer RA, Li L, Palmer AA, Wang L, McCarthy SW. Spitz naevus versus Spitzoid melanoma: when and how can they be distinguished? [see comment]. Pathol (Phila). 2002;34:6–12.
38. Kapur P, Selim MA, Roy LC, Yegappan M, Weinberg AG, Hoang MP. Spitz nevi and atypical Spitz nevi/tumors: a histologic and immunohistochemical analysis. Mod Pathol. 2005;18:197–204.
39. McCalmont TH, Brinsko R, LeBoit PE. Melanocytic acral nevi with intraepidermal ascent of cells (MANIACs): a reappraisal of melanocytic lesions from acral sites. J Cutan Pathol. 1991;18:378.
40. Winnepenninckx V, van den Oord JJ. Immunophenotype and possible origin of nevi with phenotypical heterogeneity. Arch Dermatol Res. 2004;296:49–53.
41. Ball NJ, Golitz LE. Melanocytic nevi with focal atypical epithelioid cell components: a review of seventy-three cases. J Am Acad Dermatol. 1994;30:724–9.
42. Clark WH Jr, From L, Bernardino EA, Mihm MC. The histogenesis and biologic behavior of primary human malignant melanomas of the skin. Cancer Res. 1969;29:705–27.
43. McGovern VJ. The classification of melanoma and its relationship with prognosis. Pathol (Phila). 1970;2:85–98.
44. McGovern VJ, Mihm MC Jr, Bailly C, Booth JC, Clark WH Jr, Cochran AJ, et al. The classification of malignant melanoma and its histologic reporting. Cancer. 1973;32:1446–57.
45. Mihm MC Jr, Clark WH Jr, From L. The clinical diagnosis, classification and histogenetic concepts of the early stages of cutaneous malignant melanomas. N Engl J Med. 1971;284:1078–82.
46. Trapl J, Palecek L, Ebel J, Kucera M. Tentative new classification of melanoma of the skin. Acta Derm Venereol. 1966;46:443–6.

47. Breslow A. Thickness, cross-sectional area and depth of invasion in the prognosis of cutaneous melanoma. Ann Surg. 1970;172:902–8.
48. Wolff K, Johnson R, Suurmond D, editors. Fitzpatrick's color atlas and synopsis of clinical dermatology. 5th ed. New York: McGraw-Hill;2005.
49. Clark WH Jr, Mihm MC Jr. Lentigo maligna and lentigo-maligna melanoma. Am J Pathol. 1969;55:39–67.
50. Flotte TJ, Mihm MC Jr. Lentigo maligna and malignant melanoma in situ, lentigo maligna type. Hum Pathol. 1999;30:533–6.
51. Tannous ZS, Lerner LH, Duncan LM, Mihm MC Jr, Flotte TJ. Progression to invasive melanoma from malignant melanoma in situ, lentigo maligna type. Hum Pathol. 2000;31:705–8.
52. Magro CM, Crowson AN, Mihm MC. Unusual variants of malignant melanoma. Mod Pathol. 2006;19(Suppl 2):S41–70.
53. Wong TY, Suster S, Duncan LM, Mihm MC Jr. Nevoid melanoma: a clinicopathological study of seven cases of malignant melanoma mimicking spindle and epithelioid cell nevus and verrucous dermal nevus. Hum Pathol. 1995;26:171–9.
54. Zembowicz A, McCusker M, Chiarelli C, Dei Tos AP, Granter SR, Calonje E, et al. Morphological analysis of nevoid melanoma: a study of 20 cases with a review of the literature. Am J Dermatopathol. 2001;23:167–75.
55. Njauw CN, Kim I, Piris A, Gabree M, Taylor M, Lane AM, et al. Germline BAP1 inactivation is preferentially associated with metastatic ocular melanoma and cutaneous-ocular melanoma families. PLoS ONE. 2012;7:e35295.
56. Shah AA, Bourne TD, Murali R. BAP1 protein loss by immunohistochemistry: a potentially useful tool for prognostic prediction in patients with uveal melanoma. Pathol (Phila). 2013;45:651–6.
57. Tsao H, Sober AJ, Barnhill RL. Desmoplastic neurotropic melanoma. Semin Cutan Med Surg. 1997;16:131–6.
58. Broer PN, Walker ME, Goldberg C, Buonocore S, Braddock DT, Lazova R, et al. Desmoplastic melanoma: a 12-year experience with sentinel lymph node biopsy. Eur J Surg Oncol. 2013;39:681–5.
59. Busam KJ. Cutaneous desmoplastic melanoma. Adv Anat Pathol. 2005;12:92–102.
60. Carlson JA, Dickersin GR, Sober AJ, Barnhill RL. Desmoplastic neurotropic melanoma. A clinicopathologic analysis of 28 cases. Cancer. 1995;75: 478–94.
61. Labrecque PG, Hu CH, Winkelmann RK. On the nature of desmoplastic melanoma. Cancer. 1976;38:1205–13.
62. Posther KE, Selim MA, Mosca PJ, Stanley WE, Johnson JL, Tyler DS, et al. Histopathologic characteristics, recurrence patterns, and survival of 129 patients with desmoplastic melanoma. Ann Surg Oncol. 2006;13:728–39.
63. Scolyer RA, Thompson JF, Warnke K, McCarthy SW. Pigmented epithelioid melanocytoma. Am J Surg Pathol. 2004;28:1114–5 (author reply 5–6).
64. Zembowicz A, Carney JA, Mihm MC. Pigmented epithelioid melanocytoma: a low-grade melanocytic tumor with metastatic potential indistinguishable from animal-type melanoma and epithelioid blue nevus. Am J Surg Pathol. 2004;28(5):31–40.
65. Gill M, Cohen J, Renwick N, Mones JM, Silvers DN, Celebi JT. Genetic similarities between Spitz nevus and Spitzoid melanoma in children. Cancer. 2004;101:2636–40.
66. King MS, Porchia SJ, Hiatt KM. Differentiating spitzoid melanomas from Spitz nevi through CD99 expression. J Cutan Pathol. 2007;34:576–80.
67. Massi D, Cesinaro AM, Tomasini C, Paglierani M, Bettelli S, Dal Maso L, et al. Atypical Spitzoid melanocytic tumors: a morphological, mutational, and FISH analysis. J Am Acad Dermatol. 2011;64:919–35.
68. Mooi WJ, Krausz T. Spitz nevus versus spitzoid melanoma: diagnostic difficulties, conceptual controversies. Adv Anat Pathol. 2006;13:147–56.
69. Lee DA, Cohen JA, Twaddell WS, Palacios G, Gill M, Levit E, et al. Are all melanomas the same? Spitzoid melanoma is a distinct subtype of melanoma. Cancer. 2006;106:907–13.
70. Piris A, Mihm MC Jr, Duncan LM. AJCC melanoma staging update: impact on dermatopathology practice and patient management. J Cutan Pathol. 2011;38:394–400.
71. Elder D. Tumor progression, early diagnosis and prognosis of melanoma. Acta Oncol. 1999;38:535–47.
72. Clark WH Jr, Elder DE, Guerry D IV, Braitman LE, Trock BJ, Schultz D, et al. Model predicting survival in stage I melanoma based on tumor progression. J Natl Cancer Inst. 1989;81:1893–904.
73. Thompson JF, Soong SJ, Balch CM, Gershenwald JE, Ding S, Coit DG, et al. Prognostic significance of mitotic rate in localized primary cutaneous melanoma: an analysis of patients in the multi-institutional American Joint Committee on Cancer Melanoma Staging Database. J Clin Oncol. 2011;29:2199–205.
74. Kesmodel SB, Karakousis GC, Botbyl JD, Canter RJ, Lewis RT, Wahl PM, et al. Mitotic rate as a predictor of sentinel lymph node positivity in patients with thin melanomas. Ann Surg Oncol. 2005;12:449–58.
75. Azzola MF, Shaw HM, Thompson JF, Soong SJ, Scolyer RA, Watson GF, et al. Tumor mitotic rate is a more powerful prognostic indicator than ulceration in patients with primary cutaneous melanoma: an analysis of 3661 patients from a single center. Cancer. 2003;97:1488–98.
76. Luo S, Lobo A, Tanabe K, Muzikansky A, Durazzo T, Sober A, et al. Clinical significance of

microscopic melanoma metastases in nonhottest sentinel lymph nodes. JAMA Surg. 2015;150:465–72.
77. Balch CM, Gershenwald JE, Soong SJ, et al. Final version of 2009 AJCC melanoma staging and classification. J Clin Oncol. 2009;27:6199–206.
78. Busam KJ, Antonescu CR, Marghoob AA, Nehal KS, Sachs DL, Shia J, et al. Histologic classification of tumor-infiltrating lymphocytes in primary cutaneous malignant melanoma. A study of interobserver agreement. Am J Clin Pathol. 2001;115:856–60.
79. Clemente CG, Mihm MC Jr, Bufalino R, Zurrida S, Collini P, Cascinelli N. Prognostic value of tumor infiltrating lymphocytes in the vertical growth phase of primary cutaneous melanoma. Cancer. 1996;77:1303–10.
80. Jacobs J, Nierkens S, Figdor C, de Vries I, Adema G. Regulatory T cells in melanoma: the final hurdle towards effective immunotherapy? Lancet Oncol. 2012;13:e32–42.
81. Topalian SL, Sznol M, McDermott DF, Kluger HM, Carvajal RD, Sharfman WH, et al. Survival, durable tumor remission, and long-term safety in patients with advanced melanoma receiving nivolumab. J Clin Oncol. 2014;32:1020–30.
82. Herbst RS, Soria JC, Kowanetz M, Fine GD, Hamid O, Gordon MS, et al. Predictive correlates of response to the anti-PD-L1 antibody MPDL3280A in cancer patients. Nature. 2014;515:563–7.
83. Petersson F, Diwan AH, Ivan D, Gershenwald JE, Johnson MM, Harrell R, et al. Immunohistochemical detection of lymphovascular invasion with D2–40 in melanoma correlates with sentinel lymph node status, metastasis and survival. J Cutan Pathol. 2009;36:1157–63.
84. Schmidt CR, Panageas KS, Coit DG, Patel A, Brady MS. An increased number of sentinel lymph nodes is associated with advanced Breslow depth and lymphovascular invasion in patients with primary melanoma. Ann Surg Oncol. 2009;16:948–52.
85. Doeden K, Ma Z, Narasimhan B, Swetter SM, Detmar M, Dadras SS. Lymphatic invasion in cutaneous melanoma is associated with sentinel lymph node metastasis. J Cutan Pathol. 2009;36:772–80.
86. Barnhill RL, Fine JA, Roush GC, Berwick M. Predicting five-year outcome for patients with cutaneous melanoma in a population-based study. Cancer. 1996;78:427–32.
87. Lugassy C, Barnhill RL. Angiotropic melanoma and extravascular migratory metastasis: a review. Adv Anat Pathol. 2007;14:195–201.
88. Barnhill RL, Lugassy C. Angiotropic malignant melanoma and extravascular migratory metastasis: description of 36 cases with emphasis on a new mechanism of tumour spread. Pathol (Phila). 2004;36:485–90.
89. Xu X, Chen L, Guerry D, Dawson P, Hwang W, Vanbelle P, et al. Lymphatic invasion is independently prognostic of metastasis in primary cutaneous melanoma. Clin Cancer Res. 2012;18:229–37.
90. Gershenwald JE, Soong SJ, Balch CM, American Joint Committee on Cancer Melanoma Staging C. TNM staging system for cutaneous melanoma... and beyond. Ann Surg Oncol. 2010;17:1475–7.
91. Harrist TJ, Rigel DS, Day CL Jr, Sober AJ, Lew RA, Rhodes AR, et al. "Microscopic satellites" are more highly associated with regional lymph node metastases than is primary melanoma thickness. Cancer. 1984;53:2183–7.
92. Day CL Jr, Harrist TJ, Gorstein F, Sober AJ, Lew RA, Friedman RJ, et al. Malignant melanoma. Prognostic significance of "microscopic satellites" in the reticular dermis and subcutaneous fat. Ann Surg. 1981;194:108–12.
93. Balch CM, Buzaid AC, Soong SJ, Atkins MB, Cascinelli N, Coit DG, et al. Final version of the American Joint Committee on Cancer staging system for cutaneous melanoma. J Clin Oncol. 2001;19:3635–48.
94. Murali R, Desilva C, Thompson JF, Scolyer RA. Factors predicting recurrence and survival in sentinel lymph node-positive melanoma patients. Ann Surg. 2011;253:1155–64.
95. Ronan SG, Eng AM, Briele HA, Shioura NN. Das Gupta TK. Thin malignant melanomas with regression and metastases. Arch Dermatol. 1987;123:1326–30.
96. Morton D, Wen D-R, Wong J, Economou J, Cagle L, Storm K, et al. Technical details of intraoperative lymphatic mapping for early stage melanoma. Arch Surg. 1992;127:392–9.
97. Liu Y, Truini C, Ariyan S. A randomized study comparing the effectiveness of methylene blue dye with lymphazurin blue dye in sentinel lymph node biopsy for the treatment of cutaneous melanoma. Ann Surg Oncol. 2008;15:2412–7.
98. Kroon HM, Lowe L, Wong S, Fullen D, Su L, Cimmino V, et al. What is a sentinel node? Re-evaluating the 10 % rule for sentinel lymph node biopsy in melanoma. J Surg Oncol. 2007;95:623–8.
99. Reintgen D, Cruse CW, Wells K, Berman C, Fenske N, Glass F, et al. The orderly progression of melanoma nodal metastases. Ann Surg. 1994;220:759–67.
100. Thompson JF. The value of sentinel node biopsy in patients with primary cutaneous melanoma. Dermatol Surg. 2008;34:550–4.
101. Yu LL, Flotte TJ, Tanabe KK, Gadd MA, Cosimi AB, Sober AJ, et al. Detection of microscopic melanoma metastases in sentinel lymph nodes. Cancer. 1999;86:617–27.
102. Shields CL, Kels JC, Shields JA. Melanoma of the eye: revealing hidden secrets, one at a time. Clin Dermatol. 2015;33:183–96.

103. Laver NV, McLaughlin ME, Duker JS. Ocular melanoma. Arch Pathol Lab Med. 2010;134:1778–84.
104. Shields CL, Kaliki S, Furuta M, Fulco E, Alarcon C, Shields JA. American Joint Committee on Cancer classification of posterior uveal melanoma (tumor size category) predicts prognosis in 7731 patients. Ophthalmology. 2013;120:2066–71.
105. Albert DM, Dierner-West M, Robinson NL, Grossniklaus HE, Green WR. Histopathologic characteristics of uveal melanomas in eyes enucleated from the collaborative ocular melanoma study COMS Report No. 6. Am J Ophthalmol. 1998;125:745–66.
106. Folberg R, Rummelt V, Parys-Van Ginderdeuren R, Hwang T, Woolson RF, Pe'er J, et al. The prognostic value of tumor blood vessel morphology in primary uveal melanoma. Ophthalmology. 1993;100:1389–98.
107. Lin AY, Maniotis AJ, Valyi-Nagy K, Majumdar D, Setty S, Kadkol S, et al. Distinguishing fibrovascular septa from vasculogenic mimicry patterns. Arch Pathol Lab Med. 2005;129:884–92.
108. Mochel MC, Duncan LM, Piris A, Kraft S. Primary mucosal melanoma of the sinonasal tract: a clinicopathological and immunohistochemical study of thirty-two cases. Head Neck Pathol. 2014;2:236–43.
109. Gru AA, Becker N, Dehner LP, Pfeifer JD. Mucosal melanoma: correlation of clinicopathologic, prognostic, and molecular features. Melanoma Res. 2014;24:360–70.
110. Tajudeen BA, Vorasubin N, Sanaiha Y, Palma-Diaz MF, Suh JD, Wang MB. Sinonasal mucosal melanoma: 20-year experience at a tertiary referral center. Int Forum Allergy Rhinol. 2014;4:592–7.
111. Lazarev S, Gupta V, Hu K, Harrison LB, Bakst R. Mucosal melanoma of the head and neck: a systematic review of the literature. Int J Radiat Oncol Biol Phys. 2014;90:1108–18.
112. Lourenco SV, Fernandes JD, Hsieh R, Coutinho-Camillo CM, Bologna S, Sangueza M, et al. Head and neck mucosal melanoma: a review. Am J Dermatopathol. 2014;36:578–87.
113. Sun CZ, Li QL, Hu ZD, Jiang YE, Song M, Yang AK. Treatment and prognosis in sinonasal mucosal melanoma: a retrospective analysis of 65 patients from a single cancer center. Head Neck. 2014;36:675–81.
114. Tacastacas JD, Bray J, Cohen YK, Arbesman J, Kim J, Koon HB, et al. Update on primary mucosal melanoma. J Am Acad Dermatol. 2014;71:366–75.
115. Abraham RM, Karakousis G, Acs G, Ziober AF, Cerroni L, Mihm MC Jr, et al. Lymphatic invasion predicts aggressive behavior in melanocytic tumors of uncertain malignant potential (MELTUMP). Am J Surg Pathol. 2013;37:669–75.
116. Flores JF, Walker GJ, Glendening JM, Haluska FG, Castresana JS, Rubio MP, et al. Loss of the p16INK4a and p15INK4b genes, as well as neighboring 9p21 markers, in sporadic melanoma. Cancer Res. 1996;56:5023–32.
117. Hilliard NJ, Krahl D, Sellheyer K. p16 expression differentiates between desmoplastic Spitz nevus and desmoplastic melanoma. J Cutan Pathol. 2009;36:753–9.
118. Al Dhaybi R, Agoumi M, Gagne I, McCuaig C, Powell J, Kokta V. p16 expression: a marker of differentiation between childhood malignant melanomas and Spitz nevi. J Am Acad Dermatol. 2011;65:357–63.
119. Gammon B, Beilfuss B, Guitart J, Gerami P. Enhanced detection of spitzoid melanomas using fluorescence in situ hybridization with 9p21 as an adjunctive probe. Am J Surg Pathol. 2012;36:81–8.
120. Gerami P, Scolyer RA, Xu X, Elder DE, Abraham RM, Fullen D, et al. Risk assessment for atypical spitzoid melanocytic neoplasms using FISH to identify chromosomal copy number aberrations. Am J Surg Pathol. 2013;37:676–84.
121. Long GV, Wilmott JS, Capper D, Preusser M, Zhang YE, Thompson JF, et al. Immunohistochemistry is highly sensitive and specific for the detection of V600E BRAF mutation in melanoma. Am J Surg Pathol. 2013;37:61–5.
122. Fernandes BF, Odashiro N, Saraiva VS, Logan P, Antecka E, Burnier MN. Immunohistochemical expression of Melan-A and tyrosinase in uveal melanoma. J Carcinogenesis. 2007;6:1–6.
123. Angi M, Damato B, Kalirai H, Dodson A, Taktak A, Coupland SE. Immunohistochemical assessment of mitotic count in uveal melanoma. Acta Ophthalmol (Oxf). 2011;89:e155–60.
124. Coupland SE, Anastassiou G, Stang A, Schilling H, Anagnostopoulos I, Bornfeld N, et al. The prognostic value of cyclin D1, p53, and MDM2 protein expression in uveal melanoma. J Pathol. 2000;191:120–6.
125. Koopmans AE, Verdijk RM, Brouwer RWW, van den Bosch TPP, van den Berg MMP, Vaarwater J, et al. Clinical significance of immunohistochemistry for detection of BAP1 mutations in uveal melanoma. Mod Pathol. 2014;27:1321–30.
126. Scolyer RA, Murali R, McCarthy SW, Thompson JF. Pathologic examination of sentinel lymph nodes from melanoma patients. Semin Diagn Pathol. 2008;25:100–11.
127. Dekker J, Duncan L. Lack of standards for the detection of melanoma in sentinel lymph nodes: a survey and recommendations. Arch Pathol Lab Med. 2013;137:1603–9.
128. Shidham VB, Qi DY, Acker S, Kampalath B, Chang CC, George V, et al. Evaluation of micrometastases in sentinel lymph nodes of cutaneous melanoma: higher diagnostic accuracy with Melan-A and MART-1 compared with S-100 protein and HMB-45. Am J Surg Pathol. 2001;25:1039–46.
129. Clary BM, Brady MS, Lewis JJ, Coit DG. Sentinel lymph node biopsy in the management of patients with primary cutaneous melanoma: review of a large

single-institutional experience with an emphasis on recurrence. Ann Surg. 2001;233:250–8.
130. Abrahamsen HN, Hamilton-Dutoit SJ, Larsen J, Steiniche T. Sentinel lymph nodes in malignant melanoma: extended histopathologic evaluation improves diagnostic precision. Cancer. 2004;100:1683–91.
131. Scolyer RA, Thompson JF, McCarthy SW, Gershenwald JE, Ross MI, Cochran AJ. Intraoperative frozen-section evaluation can reduce accuracy of pathologic assessment of sentinel nodes in melanoma patients. J Am Coll Surg. 2005;201:821–3.
132. Biddle DA, Evans HL, Kemp BL, El-Naggar AK, Harvell JD, White WL, et al. Intraparenchymal nevus cell aggregates in lymph nodes: a possible diagnostic pitfall with malignant melanoma and carcinoma. Am J Surg Pathol. 2003;27:673–81.
133. Balch CM, Soong SJ, Atkins MB, Buzaid AC, Cascinelli N, Coit DG, et al. An evidence-based staging system for cutaneous melanoma. CA Cancer J Clin. 2004;54:131–49.
134. Karakousis GC, Gimotty PA, Botbyl JD, Kesmodel SB, Elder DE, Elenitsas R, et al. Predictors of regional nodal disease in patients with thin melanomas. Ann Surg Oncol. 2006;13:533–41.
135. Allen AC, Spitz S. malignant melanoma A clinicopathological analysis of the criteria for diagnosis and prognosis. Cancer. 1953;6:1–45.
136. Balch CM, Murad Tm, Soong S-J, Ingalls AL, Halpern NB, Maddox WA. A multifactorial analysis of melanoma: prognostic histopathological features comparing Clark's and Breslow's staging methods. Ann Surg. 1978;188:732–42.
137. Gershenwald J, Thompson W, Mansfield PF, et al. Multi-institutional melanoma lymphatic mapping experience: the prognostic value of sentinel lymph node status in 612 stage I or II melanoma patients. J Clin Oncol. 1999;27:976–83.
138. Gershenwald JE, Ross MI. Sentinel-lymph-node biopsy for cutaneous melanoma. N Engl J Med. 2011;364:1738–45.
139. Veenstra HJ, Wouters MJ, Kroon BB, Olmos RA, Nieweg OE. Less false-negative sentinel node procedures in melanoma patients with experience and proper collaboration. J Surg Oncol. 2011;104:454–7.
140. Gershenwald JE, Colome MI, Lee JE, Mansfield PF, Tseng C, Lee JJ, et al. Patterns of recurrence following a negative sentinel lymph node biopsy in 243 patients with stage I or II melanoma. J Clin Oncol. 1998;16:2253–60.
141. Kanzler MH. Sentinel node biopsy and standard of care for melanoma: a re-evaluation of the evidence. J Am Acad Dermatol. 2010;62:880–4.
142. Gershenwald JE, Andtbacka RH, Prieto VG, Johnson MM, Diwan AH, Lee JE, et al. Microscopic tumor burden in sentinel lymph nodes predicts synchronous nonsentinel lymph node involvement in patients with melanoma. J Clin Oncol. 2008;26:4296–303.
143. Cadili A, Scolyer RA, Brown PT, Dabbs K, Thompson JF. Total sentinel lymph node tumor size predicts nonsentinel node metastasis and survival in patients with melanoma. Ann Surg Oncol. 2010;17:3015–20.
144. Cadili A, McKinnon G, Wright F, Hanna W, Macintosh E, Abhari Z, et al. Validation of a scoring system to predict non-sentinel lymph node metastasis in melanoma. J Surg Oncol. 2010;101:191–4.
145. Schoenfield L. Uveal melanoma: A pathologist's perspective and review of translational developments. Adv Anat Pathol. 2014;21:138–43.
146. Mitra D, Luo X, Morgan A, Wang J, Hoang MP, Lo J, et al. An ultraviolet-radiation-independent pathway to melanoma carcinogenesis in the red hair/fair skin background. Nature. 2012;491:449–53.
147. D'Orazio JA, Nobuhisa T, Cui R, Arya M, Spry M, Wakamatsu K, et al. Topical drug rescue strategy and skin protection based on the role of Mc1r in UV-induced tanning. Nature. 2006;443:340–4.
148. Cui R, Widlund HR, Feige E, Lin JY, Wilensky DL, Igras VE, et al. Central role of p53 in the suntan response and pathologic hyperpigmentation. Cell. 2007;128:853–64.
149. Levy C, Khaled M, Fisher DE. MITF: master regulator of melanocyte development and melanoma oncogene. Trends Mol Med. 2006;12:406–14.
150. Garraway LA, Widlund HR, Rubin MA, Getz G, Berger AJ, Ramaswamy S, et al. Integrative genomic analyses identify MITF as a lineage survival oncogene amplified in malignant melanoma. Nature. 2005;436:117–22.
151. Haq R, Yokoyama S, Hawryluk EB, Jonsson GB, Frederick DT, McHenry K, et al. BCL2A1 is a lineage-specific antiapoptotic melanoma oncogene that confers resistance to BRAF inhibition. Proc Natl Acad Sci USA. 2013;110:4321–6.
152. Hemesath TJ, Price ER, Takemoto C, Badalian T, Fisher DE. MAP kinase links the transcription factor Microphthalmia to c-Kit signalling in melanocytes. Nature. 1998;391:298–301.
153. Davies H, Bignell GR, Cox C, Stephens P, Edkins S, Clegg S, et al. Mutations of the BRAF gene in human cancer. Nature. 2002;417:949–54.
154. Omholt K, Platz A, Kanter L, Ringborg U, Hansson J. NRAS and BRAF mutations arise early during melanoma pathogenesis and are preserved throughout tumor progression. Clin Cancer Res. 2003;9:6483–8.
155. Solus JF, Kraft S. Ras, Raf, and MAP kinase in melanoma. Adv Anat Pathol. 2013;20:217–26.
156. Tsao H, Goel V, Wu H, Yang G, Haluska FG. Genetic interaction between NRAS and BRAF mutations and PTEN/MMAC1 inactivation in melanoma. J Invest Dermatol. 2004;122:337–41.
157. Maertens O, Johnson B, Hollstein P, Frederick DT, Cooper ZA, Messiaen L, et al. Elucidating distinct roles for NF1 in melanomagenesis. Cancer Discov. 2013;3:338–49.

158. Halaban R, Fan B, Ahn J, Funasaka Y, Gitay-Goren H, Neufeld G. Growth factors, receptor kinases, and protein tyrosine phosphatases in normal and malignant melanocytes. J Immunotherapy. 1992;12:154–61.
159. Curtin JA, Busam K, Pinkel D, Bastian BC. Somatic activation of KIT in distinct subtypes of melanoma. J Clin Oncol. 2006;24:4340–6.
160. Hodi FS, Friedlander P, Corless CL, Heinrich MC, Mac Rae S, Kruse A, et al. Major response to imatinib mesylate in KIT-mutated melanoma. J Clin Oncol. 2008;26:2046–51.
161. Carvajal RD, Antonescu CR, Wolchok JD, Chapman PB, Roman RA, Teitcher J, et al. KIT as a therapeutic target in metastatic melanoma. JAMA. 2011;305:2327–34.
162. Hodi FS, Corless CL, Giobbie-Hurder A, Fletcher JA, Zhu M, Marino-Enriquez A, et al. Imatinib for melanomas harboring mutationally activated or amplified KIT arising on mucosal, acral, and chronically sun-damaged skin. J Clin Oncol. 2013;31:3182–90.
163. Van Raamsdonk CD, Griewank KG, Crosby MB, Garrido MC, Vemula S, Wiesner T, et al. Mutations in GNA11 in uveal melanoma. N Engl J Med. 2010;363:2191–9.
164. Van Raamsdonk CD, Bezrookove V, Green G, Bauer J, Gaugler L, O'Brien JM, et al. Frequent somatic mutations of GNAQ in uveal melanoma and blue naevi. Nature. 2009;457:599–602.
165. Hodis E, Watson IR, Kryukov GV, Arold ST, Imielinski M, Theurillat JP, et al. A landscape of driver mutations in melanoma. Cell. 2012;150:251–63.
166. Shi H, Hugo W, Kong X, Hong A, Koya RC, Moriceau G, et al. Acquired resistance and clonal evolution in melanoma during BRAF inhibitor therapy. Cancer Discov. 2014;4:80–93.
167. Murali R, Wilmott JS, Jakrot V, Al-Ahmadie HA, Wiesner T, McCarthy SW, et al. BAP1 expression in cutaneous melanoma: a pilot study. Pathol (Phila). 2013;45:606–9.
168. Wiesner T, Murali R, Fried I, Cerroni L, Busam K, Kutzner H, et al. A distinct subset of atypical Spitz tumors is characterized by BRAF mutation and loss of BAP1 expression. Am J Surg Pathol. 2012;36:818–30.
169. Busam KJ, Wanna M, Wiesner T. Multiple epithelioid Spitz nevi or tumors with loss of BAP1 expression: a clue to a hereditary tumor syndrome. JAMA Dermatol. 2013;149:335–9.
170. Murali R, Wiesner T, Scolyer RA. Tumours associated with BAP1 mutations. Pathol (Phila). 2013;45:116–26.
171. Tsao H, Benoit E, Sober A, Thiele C, Haluska FG. Novel mutations in the p16/CDKN2A binding region of the cyclin-dependent kinase-4 gene. Cancer Res. 1998;58:109–13.
172. Muthusamy V, Hobbs C, Nogueira C, Cordon-Cardo C, McKee PH, Chin L, et al. Amplification of CDK4 and MDM2 in malignant melanoma. Genes Chromosom Cancer. 2006;45: 447–54.
173. Sullivan RJ, Flaherty K. MAP kinase signaling and inhibition in melanoma. Oncogene. 2013;32:2373–9.
174. Bhatt KV, Hu R, Spofford LS, Aplin AE. Mutant B-RAF signaling and cyclin D1 regulate Cks1/S-phase kinase-associated protein 2-mediated degradation of p27Kip1 in human melanoma cells. Oncogene. 2007;26:1056–66.
175. Tsao H, Zhang X, Fowlkes K, Haluska FG. Relative reciprocity of NRAS and PTEN/MMAC1 alterations in cutaneous melanoma cell lines. Cancer Res. 2000;60:1800–4.
176. Damsky WE, Curley DP, Santhanakrishnan M, Rosenbaum LE, Platt JT, Gould Rothberg BE, et al. beta-catenin signaling controls metastasis in Braf-activated Pten-deficient melanomas. Cancer Cell. 2011;20:741–54.
177. Rimm DL, Caca K, Hu G, Harrison FB, Fearon ER. Frequent nuclear/cytoplasmic localization of beta-catenin without exon 3 mutations in malignant melanoma. Am J Pathol. 1999;154:325–9.
178. Bird AP. CpG-rich islands and the function of DNA methylation. Nature. 1986;321:209–13.
179. Maunakea AK, Nagarajan RP, Bilenky M, Ballinger TJ, D'Souza C, Fouse SD, et al. Conserved role of intragenic DNA methylation in regulating alternative promoters. Nature. 2010;466: 253–7.
180. Straume O, Smeds J, Kumar R, Hemminki K, Akslen LA. Significant impact of promoter hypermethylation and the 540 C > T polymorphism of CDKN2A in cutaneous melanoma of the vertical growth phase. Am J Pathol. 2002;161:229–37.
181. Ito S, D'Alessio AC, Taranova OV, Hong K, Sowers LC, Zhang Y. Role of Tet proteins in 5mC to 5hmC conversion, ES-cell self-renewal and inner cell mass specification. Nature. 2010;466:1129–33.
182. Lee JJ, Murphy GF, Lian CG. Melanoma epigenetics: novel mechanisms, markers, and medicines. Lab Invest. 2014;94:822–38.
183. Lian CG, Xu Y, Ceol C, Wu F, Larson A, Dresser K, et al. Loss of 5-hydroxymethylcytosine is an epigenetic hallmark of melanoma. Cell. 2012;150:1135–46.
184. Shi Y, Lan F, Matson C, Mulligan P, Whetstine JR, Cole PA, et al. Histone demethylation mediated by the nuclear amine oxidase homolog LSD1. Cell. 2004;119:941–53.
185. Zhang XD, Gillespie SK, Borrow JM, Hersey P. The histone deacetylase inhibitor suberic bishydroxamate regulates the expression of multiple apoptotic mediators and induces mitochondria-dependent apoptosis of melanoma cells. Mol Cancer Ther. 2004;3:425–35.
186. Ye Y, Jin L, Wilmott JS, Hu WL, Yosufi B, Thorne RF, et al. PI(4,5)P2 5-phosphatase A regulates PI3 K/Akt signalling and has a tumour

suppressive role in human melanoma. Nat Commun. 2013;4:1508.
187. Bachmann IM, Halvorsen OJ, Collett K, Stefansson IM, Straume O, Haukaas SA, et al. EZH2 expression is associated with high proliferation rate and aggressive tumor subgroups in cutaneous melanoma and cancers of the endometrium, prostate, and breast. J Clin Oncol. 2006;24:268–73.
188. Qi P, Du X. The long non-coding RNAs, a new cancer diagnostic and therapeutic gold mine. Mod Pathol. 2013;26:155–65.
189. Costa FF. Non-coding RNAs: new players in eukaryotic biology. Gene. 2005;357:83–94.
190. Bartel DP. MicroRNAs: target recognition and regulatory functions. Cell. 2009;136:215–33.
191. Liu S, Tetzlaff MT, Cui R, Xu X. miR-200c inhibits melanoma progression and drug resistance through down-regulation of BMI-1. Am J Pathol. 2012;181:1823–35.
192. Pencheva N, Tran H, Buss C, Huh D, Drobnjak M, Busam K, et al. Convergent multi-miRNA targeting of ApoE drives LRP1/LRP8-dependent melanoma metastasis and angiogenesis. Cell. 2012;151:1068–82.
193. Tang L, Zhang W, Su B, Yu B. Long noncoding RNA HOTAIR is associated with motility, invasion, and metastatic potential of metastatic melanoma. Biomed Res Int. 2013;2013:251098.
194. Bastian BC, Olshen AB, LeBoit PE, Pinkel D. Classifying melanocytic tumors based on DNA copy number changes. Am J Pathol. 2003;163:1765–70.
195. Bauer J, Bastian BC. Distinguishing melanocytic nevi from melanoma by DNA copy number changes: comparative genomic hybridization as a research and diagnostic tool. Dermatol Ther. 2006;19:40–9.
196. Bastian BC, Wesselmann U, Pinkel D, Leboit PE. Molecular cytogenetic analysis of Spitz nevi shows clear differences to melanoma. J Invest Dermatol. 1999;113:1065–9.
197. Bastian BC, LeBoit PE, Pinkel D. Mutations and copy number increase of HRAS in Spitz nevi with distinctive histopathological features. Am J Pathol. 2000;157:967–72.
198. Gerami P, Wass A, Mafee M, Fang Y, Pulitzer MP, Busam KJ. Fluorescence in situ hybridization for distinguishing nevoid melanomas from mitotically active nevi. Am J Surg Pathol. 2009;33:1783–8.
199. Gerami P, Cooper C, Bajaj S, Wagner A, Fullen D, Busam K, et al. Outcomes of atypical spitz tumors with chromosomal copy number aberrations and conventional melanomas in children. Am J Surg Pathol. 2013;37:1387–94.
200. Harbour JW, Onken MD, Roberson ED, Duan S, Cao L, Worley LA, et al. Frequent mutation of BAP1 in metastasizing uveal melanomas. Science. 2010;330:1410–3.
201. Hantschke M, Mentzel T, Rutten A, Palmedo G, Calonje E, Lazar AJ, et al. Cutaneous clear cell sarcoma: a clinicopathologic, immunohistochemical, and molecular analysis of 12 cases emphasizing its distinction from dermal melanoma. Am J Surg Pathol. 2010;34:216–22.
202. Prescher G, Bornfeld N, Becher R. Nonrandom chromosomal abnormalities in primary uveal melanoma. J Natl Cancer Inst. 1990;82:1765–9.
203. Sisley K, Rennie IG, Parsons MA, Jacques R, Hammond DW, Bell SM, et al. Abnormalities of chromosomes 3 and 8 in posterior uveal melanoma correlate with prognosis. Genes Chromosom Cancer. 1997;19:22–8.
204. Aalto Y, Eriksson L, Seregard S, Larsson O, Knuutila S. Concomitant loss of chromosome 3 and whole arm losses and gains of chromosome 1, 6, or 8 in metastasizing primary uveal melanoma. Invest Ophthalmol Vis Sci. 2001;42:313–7.
205. Scholes AG, Damato BE, Nunn J, Hiscott P, Grierson I, Field JK. Monosomy 3 in uveal melanoma: correlation with clinical and histologic predictors of survival. Invest Ophthalmol Vis Sci. 2003;44:1008–11.
206. White VA, Chambers JD, Courtright PD, Chang WY, Horsman DE. Correlation of cytogenetic abnormalities with the outcome of patients with uveal melanoma. Cancer. 1998;83:354–9.
207. Hoglund M, Gisselsson D, Hansen GB, White VA, Sall T, Mitelman F, et al. Dissecting karyotypic patterns in malignant melanomas: temporal clustering of losses and gains in melanoma karyotypic evolution. Int J Cancer. 2004;108:57–65.
208. Tschentscher F, Husing J, Holter T, Kruse E, Dresen IG, Jockel KH, et al. Tumor classification based on gene expression profiling shows that uveal melanomas with and without monosomy 3 represent two distinct entities. Cancer Res. 2003;63:2578–84.
209. Onken MD, Worley LA, Ehlers JP, Harbour JW. Gene expression profiling in uveal melanoma reveals two molecular classes and predicts metastatic death. Cancer Res. 2004;64:7205–9.
210. Onken MD, Worley LA, Char DH, Augsburger JJ, Correa ZM, Nudleman E, et al. Collaborative Ocular Oncology Group report number 1: prospective validation of a multi-gene prognostic assay in uveal melanoma. Ophthalmology. 2012;119:1596–603.
211. Worley LA, Onken MD, Person E, Robirds D, Branson J, Char DH, et al. Transcriptomic versus chromosomal prognostic markers and clinical outcome in uveal melanoma. Clin Cancer Res. 2007;13:1466–71.
212. Onken MD, Worley LA, Davila RM, Char DH, Harbour JW. Prognostic testing in uveal melanoma by transcriptomic profiling of fine needle biopsy specimens. J Mol Diagn. 2006;8:567–73.

# Index

Note: Page numbers followed by *f* and *t* refer to figures and tables, respectively





M. Loda et al. (eds.), *Pathology and Epidemiology of Cancer*,
DOI 10.1007/978-3-319-35153-7

**N**

## O

MIX
Papier aus verantwortungsvollen Quellen
Paper from responsible sources
FSC® C105338

If you have any concerns about our products,
you can contact us on
**ProductSafety@springernature.com**

In case Publisher is established outside the EU,
the EU authorized representative is:
**Springer Nature Customer Service Center GmbH**
**Europaplatz 3, 69115 Heidelberg, Germany**

Printed by Libri Plureos GmbH
in Hamburg, Germany